Birkhäuser

Operator Theory: Advances and Applications

Volume 296

Founded in 1979 by Israel Gohberg

Valentin A. Zagrebnov • Hagen Neidhardt •
Takashi Ichinose

Trotter-Kato Product Formulæ

Birkhäuser

Valentin A. Zagrebnov
Université d'Aix-Marseille - CNRS
Institute de Mathématiques de Marseille
Marseille Cedex 13, France

Hagen Neidhardt
Berlin, Germany

Takashi Ichinose
Department of Mathematics
Kanazawa University
Kanazawa, Japan

ISSN 0255-0156 ISSN 2296-4878 (electronic)
Operator Theory: Advances and Applications
ISBN 978-3-031-56719-3 ISBN 978-3-031-56720-9 (eBook)
https://doi.org/10.1007/978-3-031-56720-9

Mathematics Subject Classification: 47D03, 47D06, 47A60, 47B44, 47A12, 35K90, 46A04, 46E50, 45D05

This book is published under the imprint Birkhäuser, www.birkhauser-science.com by the registered company Springer Nature Switzerland AG
The registered company address is: Gewerbestrasse 11, 6330 Cham, Switzerland

To
Galina, Hille, Michiko

Introduction: What Is This Book About? (De quoi parle ce livre?)

1. History and Motivation

Since 1875 it is owing to Sophus Lie we know that for any pair of (noncommutative) finite square matrices $A, B \in \mathcal{M}(\mathbb{R}^d)$ on the d-dimensional Euclidean space $\mathbb{R}^d$, one has for $n \in \mathbb{N}$ the estimates

$$\left\| \left(e^{-tA/n} e^{-tB/n} \right)^n - e^{-t(A+B)} \right\| \leq O(1/n), \tag{1}$$

$$\left\| \left(e^{-tA/2n} e^{-tB/n} e^{-tA/2n} \right)^n - e^{-t(A+B)} \right\| \leq O(1/n^2) . \tag{2}$$

for $n \to \infty$. Here $\| \cdot \|$ is *any* norm on the matrix space $\mathcal{M}(\mathbb{R}^d)$. Note that the powers $\{(e^{-tA/n} e^{-tB/n})^n\}_{n\geq 1}$ are called the *products* (or the product *approximants*), and the powers $\{(e^{-tA/2n} e^{-tB/n} e^{-tA/2n})^n\}_{n\geq 1}$ are called the *symmetric* (or *symmetrised*) *products* (or *approximants*) for Lie product formulæ (1) and (2). Extension of the Lie-type product formulæ to infinite-dimensional spaces was a subject of several papers by Yu. L. Daletskiĭ (see, e.g., [Dal60]) in connection with the path-integral representation (*Feynman-Kac* formula) for solutions of evolution equations. Although the first abstract result appeared in 1959 due to a celebrated paper by H. Trotter [Trot59], who extended the *Lie* product formula (1) to the case of strongly continuous contraction semigroups on a *Banach* space $\mathfrak{X}$. There the convergence of the *Lie-Trotter* product formula was proved in the *strong* operator topology on the Banach space of bounded operators $\mathcal{L}(\mathfrak{X})$.

Namely, let A and B be generators of strongly continuous contraction semigroups in a Banach space $\mathfrak{X}$. If the closure C of the operator sum $A + B$ on domain: dom $(A) \cap$ dom (B) is generator of a contraction semigroup, then limit

$$\text{s-}\lim_{n\to\infty} \left(e^{-tA/n} e^{-tB/n} \right)^n = e^{-tC} , \qquad C := \overline{A + B} , \tag{3}$$

holds for any $t \geq 0$ in the strong operator topology on $\mathcal{L}(\mathfrak{X})$.

In 1964, E. Nelson pointed out [Nel64] the importance of the *Trotter* product formula (3) for semigroups generated by *Schrödinger* operators, and for the proof of the *Feynman-Kac* formula. It is also to him belongs a simple proof of (3) in a separable Hilbert space $\mathfrak{H}$ where A and B are two non-negative self-adjoint operators such that the operator sum $A + B$ is also self-adjoint [Nel64]. It is striking that much later it was proven in [IT01] and [ITTZ01] that Nelson's conditions ensure convergence of the Trotter product formula (3) in the *operator-norm* topology ($\|\cdot\| - \lim_{n\to\infty}$) on $\mathcal{L}(\mathfrak{H})$, with error bound estimate for the rate of convergence which is (ultimately) *optimal*. This means that it coincides with the rate in (1) for matrices, and it cannot be improved without supplementary conditions, or by a simple symmetrisation of approximants as in (2).

A beautiful and elegant way to treat the Lie-Trotter product formulæ and to prove (3) in the framework of a general theory of strongly continuous contractions on Hilbert and Banach spaces is due to P. Chernoff [Cher68, Cher74]. There he generalised (3) by proposition to study on $\mathfrak{X}$ (or $\mathfrak{H}$) strongly continuous families of contractions $\{F(t)\}_{t\geq 0}$ and the product formula of the type

$$\text{s-}\lim_{n\to\infty} \left(F(t/n)\right)^n = G(t)\,, \quad t \geq 0\,, \tag{4}$$

which is now known as the *Chernoff* product formula. If the closure of the strong right derivative $F'(+0)$ is equal to (minus) generator: $(-C)$ of a contraction semigroup, the arguments in [Cher68] reveal that the strong limit (4) exists and $\{G(t) = e^{-tC}\}_{t\geq 0}$. The *Chernoff theory* and its generalisation for convergence in the operator-norm topology will be developed and systematically employed below in this book.

Further progress in the product formulæ approximations was achieved by T. Kato [Kato78]. In 1978, he obtained the following important results:

(a) Let $A \geq 0$ and $B \geq 0$ be non-negative self-adjoint operators in a separable Hilbert space $\mathfrak{H}$. Denote by $\mathfrak{H}_0$ the subspace

$$\mathfrak{H}_0 := \overline{\mathrm{dom}\,(A^{1/2}) \cap \mathrm{dom}\,(B^{1/2})}\,.$$

Note that it may happen that intersection of operator domains $\mathrm{dom}\,(A) \cap \mathrm{dom}\,(B) = \{0\}$, but the *form-sum* $C := A \dot{+} B$ is well-defined and is generator in the subspace $\mathfrak{H}_0$. Under these conditions, the Trotter product formula converges strongly to *degenerate* semigroup $\{e^{-tC}P_0\}_{t>0}$, *locally uniformly* away from zero. That is, one gets (*Kato* product formula):

$$\text{s-}\lim_{n\to\infty} \left(e^{-tA/n}e^{-tB/n}\right)^n = e^{-tC}P_0, \qquad t > 0, \tag{5}$$

uniformly in $t \in [a, b]$, for any $[a, b] \subset \mathbb{R}^+$. Here P_0 denotes the orthogonal projection from $\mathfrak{H}$ onto $\mathfrak{H}_0$ and $\mathbb{R}^+ := (0, +\infty)$.

(b) Further, T. Kato [Kato74, Kato78] discovered that the product formula (5) is valid not only for the exponential function e^{-x}, $x \geq 0$, but also for the whole class of Borel measurable functions f and g, which are defined on $\mathbb{R}_0^+ = [0, \infty)$

and satisfy the conditions

$$0 \leq f(x) \leq 1, \qquad f(0) = 1, \qquad f'(+0) = -1, \tag{6}$$

$$0 \leq g(x) \leq 1, \qquad g(0) = 1, \qquad g'(+0) = -1. \tag{7}$$

Kato proved that in this case one obtains instead (5) the product formula

$$\text{s-}\lim_{n\to\infty} (f(tA/n)g(tB/n))^n = \mathrm{e}^{-tC} P_0, \qquad t > 0, \tag{8}$$

converging locally uniformly in $\mathbb{R}_0^+$ away from zero. These product formulæ for pairs $f, g \in \mathcal{K}$ are known as the *Trotter-Kato product formulæ* for the set of *generic* Kato functions $\mathcal{K}$ satisfying (6), (7).

The next important step concerning the Trotter product formula is related to the following remarkable result by Dzh. L. Rogava in 1993 [Rog93] about convergence of (3) in the *operator-norm* topology on $\mathcal{L}(\mathfrak{H})$. More precisely:

Let A and B be two bounded from below self-adjoint operators in a separable Hilbert space $\mathfrak{H}$. If $\mathrm{dom}\,(A) \subset \mathrm{dom}\,(B)$ and operator $C = A + B$ is self-adjoint (cf. [Nel64]), then

$$\left\| \left(\mathrm{e}^{-tA/n}\mathrm{e}^{-tB/n}\right)^n - \mathrm{e}^{-tC} \right\| \leq \frac{c\,\ln(n)}{\sqrt{n}}, \qquad c > 0, \quad n > 1, \tag{9}$$

uniformly in $t \in [0, T]$, $0 < T < \infty$. This *operator-norm* estimate leads to convergence of the Trotter product formula (3) with the error bound estimate (9) for the rate. This convergence is uniform in $t \geq 0$ if the self-adjoint operators A and B are non-negative.

Ever since the discovery by Rogava [Rog93] that the Trotter product formula for strongly continuous semigroups may exhibit convergence in the *operator-norm* topology, a compelling question has emerged regarding the *optimal* operator-norm error bound estimate. This inquiry delves into the error bound estimate for the rate of convergence within the framework of Eqs. (9) and scrutinises its dependency on the pair of generators A and B. The question has become a focal point for both Trotter and Trotter-Kato product formulæ, captivating researches in Hilbert and Banach spaces.

The core focus of this book is to provide an exhaustive account of recent findings pertaining to operator-norm convergent Trotter and Trotter-Kato product formulæ, along with several of their generalisations. The dedication of the book to this exploration underscores its significance in advancing our understanding of these mathematical constructs.

It is noteworthy to mention that, in the realm of Hilbert spaces, Trotter-Kato product formulæ converging in the operator-norm topology have been acknowledged since 1988. However, during this period, there was an absence of estimates regarding the rate of convergence. This was particularly observed for a distinct class of *strongly* continuous at zero semigroups (C_0-semigroups), known as *Gibbs semigroups*, as documented in works such as [Zag88] and [NZ90a, NZ90b]. The established topology of *convergence* was stronger than the operator-norm topology, aligning with the *trace-norm* topology of continuity for Gibbs semigroups *away* from zero; for details, see [Zag19]. This observation extends also to the C_0-semigroups considered in this book, emphasising in particular their *operator-norm* continuity *away* from zero.

The principal thrust of this book is to communicate concrete results regarding the elevation (*uplifting*) of diverse *strongly* convergent Trotter-Kato product formulæ, encompassing those with time dependence, to a state of convergence in the *operator-norm* topology within both Banach and Hilbert spaces. Additionally, the book explores alternative options, including different spaces for semigroups and topologies of convergence of product formulæ, while outlining potential limits to this overarching programme. In essence, the book aims to serve as a comprehensive resource, shedding light on the intricacies of these mathematical concepts and their convergence behaviour in various settings.

2. Overview

Contents of the book is essentially based on original publications by the authors and their co-authors. According to particular subjects, we split the presentation into five parts: Parts I–V, which are accompanying by a few technical Appendices.

Part I contains some standard preliminaries about the C_0-semigroups and their generators. Besides we recall the abstract non-autonomous Cauchy problem (nACP) for linear operator-valued evolution equations. The corresponding product formulæ will be the main subject of Part IV. In the last section of the Part I, we introduce a class of *quasi-sectorial contractions* [CZ01b], [ArZag10], together with suitable (in the operator-norm continuity context) elements of the Chernoff theory including revisions for future purposes.

In Part II, we collect results on the Trotter-Kato product formulæ for *self-adjoint* semigroups. To this aim, the Chernoff theory of approximation is extended from the strong to the operator-norm topology. As a consequence, the Trotter-Kato product formulæ are established in the operator-norm topology first without and then with the error bound estimates. In the last section, the ITTZ Theorem [IT01], [ITTZ01] about the *optimal* operator-norm error bound estimate is proven. Another theorem about the optimal error bound estimate, which is due to the *fractional powers* conditions, [NZ99a], [Tam00], and [INZ04], concludes this part.

Part III is dedicated to extensions of results presented in the Part II to *non-self-adjoint* semigroups and also to certain results about the Trotter product formula in a Banach space. Although we developed non-self-adjoint improvement of the Chernoff theory [CZ01b], [Zag08], [Zag22c], there are no results about optimality of the operator-norm convergence rate for non-self-adjoint semigroups, cf. [CZ99], [CNZ01] and [CNZ02]. These results are partially summarised in [Ca10]. On the other hand, very little results are known about the operator-norm convergent Trotter product formula in Banach spaces [CZ01a], [Zag23]. They are more poor than the corresponding results in Hilbert spaces, and they are still essentially based on the useful properties of *holomorphic* semigroups.

A new result in a Banach space concerns the operator-norm convergence of the Trotter product formula for *contraction* semigroups, [NSZ18a], [Zag23]. It shows in particular that the rate of the operator-norm convergence of the Trotter product formula may be *arbitrary slow*, even if one of generators in formula is bounded; see Example in Sect. 10.3.

In Part IV, we present the results on product formulæ approximations (11.1.2), (11.1.3) of *solution operator* $\{U(t,s)\}_{0\leq s\leq t\leq T}$ (also known as *fundamental solution*, or *propagator*) corresponding to the abstract *non-autonomous* Cauchy problem (nACP) (11.1.1) in Banach, or Hilbert spaces. Note that in Banach space a direct study of product approximants (11.1.2) yields convergence (11.1.3) only in the *strong* operator topology (Sect. 11.1). So, we use a rather *round-about* way to estimate their convergence in the operator-norm topology. This is possible owing to the *Howland-Evans-Neidhardt* theory of nACP ([How74], [Ev76], [Nei79], [Nei81]) based on the *evolution semigroup* method in the Banach space L^p-setting; see (11.3.24). As a consequence, one gets estimates (11.5.68) and (11.5.79) for the the operator-norm rate of convergence.

The Hilbert space L^2-setting (Sect. 12.1) essentially follows the L^p-setting for Banach spaces, but with a better operator-norm estimate of the rate of convergence, Theorem 12.1.19. This improvement is due to the self-adjointness, which yields some number of straightforward estimates which are available in the Hilbert space L^2-setting. In Sect. 12.2, we present the *Ichinose-Tamura* approach to a direct analysis of product approximants for the nACP solution operator in a Hilbert space. It deserves to mention that for the *Lipschitz* condition (12.1.3), this approach yields for convergence rates in (12.1.73) and (12.1.74) the best of the actually known estimate: $O(\ln(n)/n)$, for the operator-norm error bound.

Part V contains two different subjects. The first one (Chap. 13) is a revision in Sect. 13.2 of the well-known strongly convergent Trotter product formula for unitary groups striving to uplift it to convergence to the operator-norm topology. In Sect. 13.2, it is shown that Trotters's conditions are neither sufficient, nor necessary for convergence of the unitary product formula in the operator-norm topology. Our results there on the limit of unitary Trotter product formula in the operator-norm topology are based on a certain *commutator* conditions. Searching for generalisations of Trotters's conditions, we present in Sect. 13.3 the unitary Trotter-Kato product formulæ, which converge only in the operator *locally convex* topology (L^2-topology) of the *Fréchet* space. The fact that we retreated into the corner of convergence in this topology (weaker than the strong operator topology) indicates that we are far from our ultimate aim of *uplifting* this result to the operator-norm topology.

The second subject concerns the *Zeno* product formula. Formally it corresponds to a unitary Trotter product formula when one of the factors is a *degenerate* (semi)*group* $\{P_t\}_{t\in\mathbb{R}}$ of projections: $P_t = P$, for $t \neq 0$, and $P_{t=0} = \mathbb{1}$. First we present the results on convergence of the Zeno product formula in the operator topology of the *Fréchet* space, Sect. 14.1. Then we consider extensions of this result to the *non-exponential* Zeno product formulæ, Sect. 14.2, for the *strong* as well as for the *operator-norm* convergences. In Sect. 14.3, we elucidate and improve the theory developed for the operator topology of the *Fréchet* space. Due to the *Exner-Ichinose* approach, one can show that the exponential Zeno product formula holds in the *strong* operator topology.

3. Guide to the Reader

Part I of this book is a consistent and quite detailed introduction into preliminaries for further reading. For non-specialists, it is indispensable for orientation before entering into the main subject of the book, which concerns a variety of the product formulæ approximations in different spaces and topologies.

The specialists may proceed quickly to Part II and to Parts III–V, which describe original concrete results on the Trotter-Kato product formulæ. The notes to each chapter allow to continue the reading of a more specialised literature. They also contain additional comments that escaped from the main text and that may be also useful for non-specialists.

Acknowledgements

The genesis of this book spans more than two decades, tracing its roots back to October 2002, when Hagen Neidhardt visited the Centre de Physique Théorique (CPT-Marseille). During that visit, the seed of an idea was planted, a concept that would mature over the years. We shared this idea with Takashi Ichinose (Kanazawa University), who joined us, providing invaluable support to breathe life into the project. As we embarked on this journey, numerous institutions have played a pivotal rôle in promoting our work.

- Gratitude is owed to the Weierstraß-Institut für Angewandte Analysis und Stochastik (WIAS-Berlin) for Professor Alexander Mielke's generous support during annual visits from 2012 to 2018.

The following institutions provided crucial support and hospitality accommodating our respective visits, for which we extend our heartfelt thanks:

- The Centre International de Rencontres Mathématiques (CIRM-Luminy) in 2013
- The Kanazawa University in 2010 and in 2014
- The Mittag-Leffler Mathematical Institute (Stockholm) in 2014 under Professor Ari Laptev's sponsorship
- The Research Institute for Mathematical Sciences (RIMS-Kyoto University) in 2015 with Professor Yoshitsugu Takei's assistance
- The Institut de Mathématiques de Marseille (UMR 7373) in 2015 and in 2017

We are also indebted to Mme Nathalie Granottier and Mme Fabienne Grosjean (Library of the Centre International de Rencontres Mathématiques-Luminy) for their kind assistance and help.

The welcoming environments of these institutions were not only conducive to our work but also facilitated encounters with colleagues who left an indelible mark on our project. Special acknowledgement goes to our co-authors, including Pavel Exner, Hideo Tamura, Hiroshi Tamura, Vincent Cachia, Arthur Stephan, Yury Arlinskiĭ, Satoshi Takanobu, Walter Wreszinski, Pierre-André Vuillermot and Bruno Iochum.

In addition to our co-authors, numerous colleagues influenced the content of this book. We express our gratitude to Robert A. Minlos, Tsuyoshi Ando, Paul R. Chernoff, E. Brain Davies, Barry Simon, Michail Sh. Birman, Bernard Helffer, Joachim Rehberg, Igor V. Volovich, Fëdor A.Sukochev, Bruno Nachtergaele, András Bátkai, Yury Tomilov, Alexander Gomilko, Vygantas Paulauskas, Jan Derezinski, Tom ter Elst, Valeri P. Frolov for their insightful discussions.

Hideo Tamura, Hiroshi Tamura and Pavel Exner deserve special mention for their review of different parts of the manuscript, providing invaluable feedback that significantly enhanced the text and reduced errors. Their contributions were instrumental in shaping and finalising the text.

A sincere thank you is extended to Birkhäuser Senior Editor Ms. Dorothy Mazlum, whose unwavering attention and support were pivotal throughout the publishing process.

The year 2019 brought the sombre news of the passing of our dear friend and co-author, Hagen Neidhardt. His demise on 23 March 2019, at the age of 68, was a profound loss. During his time in the hospital, Hagen expressed the hope that the book project would be completed.

We hope that the present work corresponds to endeavours in accomplishing the requests and ideas that together with Hagen we formulated many years ago. One such request was to dedicate this book to our wives.

Marseille, France
Kanazawa, Japan
December, 2023

Valentin A. Zagrebnov
Takashi Ichinose

Takashi Ichinose, Hagen Neidhardt, Valentin A.Zagrebnov at the Mittag-Leffler Mathematical Institute, July–August 2014

Contents

Part I
Preliminaries

Chapter 1
Semigroups and Their Generators

This chapter is a prelude. It contains a brief account of some basic notions and facts from the theory of strongly continuous semigroups of operators (C_0-semigroups) and their generators. We also introduce here notations and definitions indispensable in the following chapters. Further, we focus here on some special classes of semigroups important for entire of our further presentation. After definition of the strongly continuous exponential function, we consider contraction and quasi-bounded semigroups, emphasising the holomorphic semigroups.

The content of this chapter (except some details like perturbation theory based on the *product fromula*) is standard, and it can be found in many sources (see Notes to Chap. 1).

1.1 The Exponential Function

Let $\mathfrak{X}$ be a Banach space. For any bounded operator $A \in \mathcal{L}(\mathfrak{X})$, one can define for complex $t \in \mathbb{C}$ the exponential function $t \mapsto \mathrm{e}^{-tA} =: U_A(t)$ by the series

$$U_A(t) := \sum_{n=0}^{\infty} \frac{t^n}{n!}(-A)^n, \tag{1.1.1}$$

which is convergent in the operator-norm $\|\cdot\|$ on the Banach space $\mathcal{L}(\mathfrak{X})$ of linear-bounded operators. Therefore, for any $A \in \mathcal{L}(\mathfrak{X})$, the mapping $\mathbb{C} \ni t \mapsto U_A(t) \in$

V. A. Zagrebnov et al., *Trotter-Kato Product Formulæ*, Operator Theory: Advances and Applications 296, https://doi.org/10.1007/978-3-031-56720-9_1

$\mathcal{L}(\mathfrak{X})$ is entire $\|\cdot\|$-holomorphic operator-valued function on $\mathbb{C}$. The group property (*group law*)

$$U_A(t_1+t_2) = U_A(t_1)U_A(t_2), \quad t_1, t_2 \in \mathbb{C}, \tag{1.1.2}$$

is a direct consequence of formula (1.1.1) as well as equation

$$\|\cdot\|-\partial_t U_A(t) = (-A)U_A(t) = U_A(t)(-A), \tag{1.1.3}$$

where $\|\cdot\|-\partial_t$ means differentiation in the sense of the operator-norm on $\mathcal{L}(\mathfrak{X})$.

Now, let A be an unbounded operator with a dense in $\mathfrak{X}$ domain $\mathrm{dom}\,(A) \subset \mathfrak{X}$. Then, the existence of the exponential function $U_A(\cdot) : \mathcal{D} \to \mathcal{L}(\mathfrak{X})$ (1.1.3) for domain $\mathcal{D} \subset \mathbb{C}$ is much less obvious. One of the possibilities is to define $U_A(t)$ in a way alternative to (1.1.1), namely, by means of the classical *Euler formula* for exponential.

To this aim, suppose that $A \in \mathcal{C}(\mathfrak{X})$, where $\mathcal{C}(\mathfrak{X})$ denotes the set of *closed* linear operators in Banach space $\mathfrak{X}$, with resolvent $R_A(\zeta) := (A - \zeta\,\mathbb{1})^{-1}$ and *non-empty* resolvent set $\rho(A) := \{\zeta \in \mathbb{C} : R_A(\zeta) \in \mathcal{L}(\mathfrak{X})\}$. If the set $(-\infty, 0) \subset \rho(A)$, then one constructs a sequence of bounded *Euler approximants*:

$$\mathcal{E}_{A,n}(t) := \left(\mathbb{1} + \frac{t}{n}A\right)^{-n}, \quad n = 1, 2, \ldots, \qquad t \geq 0, \tag{1.1.4}$$

with intention to prove the limit (*Euler formula*)

$$U_A(t) = \lim_{n\to\infty}\left(\mathbb{1} + \frac{t}{n}A\right)^{-n}, \tag{1.1.5}$$

for $t \in \mathbb{R}_0^+ := \mathbb{R}^+ \cup \{0\}$, here $\mathbb{R}^+ := (0, +\infty)$, in one of the topologies on the set of bounded operators $\mathcal{L}(\mathfrak{X})$.

The following proposition provides a sufficient condition for that in the *strong operator* topology:

Proposition 1.1.1 *Let $A \in \mathcal{C}(\mathfrak{X})$ be a closed linear operator with dense domain* $\mathrm{dom}\,(A) \subset \mathfrak{X}$ *such that the following two conditions hold* :

(i) $\mathbb{R}^- := (-\infty, 0)$ *belongs to the resolvent set* $\rho(A)$.
(ii) $\|(A + \lambda\mathbb{1})^{-1}\| \leq \lambda^{-1}$ *for* $\lambda > 0$.

Then for $t \geq 0$,

(a) *The limit* (1.1.5) *exists in the strong operator topology to* (*exponential*) *function* $U_A(t)$, *such that* $\|U_A(t)\| \leq 1$, *locally uniformly in* $t \in \mathbb{R}_0^+$.
(b) *The mapping* $t \in \mathbb{R}^+ \mapsto U_A(t)x$ *is continuous for every* $x \in \mathfrak{X}$ (*i.e., strongly continuous*) *with the right limit* $\lim_{t\to+0}\|(U_A(t)\,x - x\| = 0$, *that is,* $U_A(0) = \mathbb{1}$.

(c) dom (A) *is invariant under* $U_A(t)$ *in the sense that* $U_A(t)(\mathrm{dom}\,(A)) \subseteq \mathrm{dom}\,(A)$ *for all* $t \geq 0$.
(d) *The semigroup law* $U_A(t+s) = U_A(t)U_A(s)$, *for all* $t, s \geq 0$.

Proof

(a) Owing to definition (1.1.4) and condition (ii), one infers that Euler approximants are bounded and $\|\mathcal{E}_{A,n}(t)\| \leq 1$ for $t \in \mathbb{R}_0^+$ and natural $n \in \mathbb{N}$. Then for $t > 0$, the operators

$$A\,\mathcal{E}_{A,n+1}(t) = \frac{n}{t}\,(\mathcal{E}_{A,n}(t) - \mathcal{E}_{A,n+1}(t))\,, \quad n \in \mathbb{N}\,, \tag{1.1.6}$$

are bounded. As a consequence, the functions $t \mapsto \mathcal{E}_{A,n}(t)$ are differentiable for $t > 0$ in the operator-norm topology on $\mathcal{L}(\mathfrak{X})$, and by (1.1.6) the operator-norm derivatives

$$\|\cdot\|-\partial_t\mathcal{E}_{A,n}(t) = (-A)\,\mathcal{E}_{A,n+1}(t) \in \mathcal{L}(\mathfrak{X})\,, \quad n \in \mathbb{N}\,, \tag{1.1.7}$$

are bounded for $t > 0$.

Now, we elucidate the continuity of functions $t \mapsto \mathcal{E}_{A,n}(t)$, at $t = 0$, where $\mathcal{E}_{A,n}(0) = \mathbb{1}$. To this end, we note that

$$\mathbb{1} - \mathcal{E}_{A,n}(t) = \frac{t}{n}\,A\sum_{k=0}^{n-1} C_{n-1}^k \left(\frac{t}{n}\,A\right)^k \mathcal{E}_{A,n}(t)\,, \quad n \in \mathbb{N}\,. \tag{1.1.8}$$

By iteration of (1.1.6) for $A^k\,\mathcal{E}_{A,n}(t)$ and $1 \leq k \leq n-1$, one concludes that operators

$$\left\| \frac{1}{n}\sum_{k=0}^{n-1} C_{n-1}^k \left(\frac{t}{n}\,A\right)^k \mathcal{E}_{A,n}(t) \right\| \leq M_n\,, \quad n \in \mathbb{N}\,, \tag{1.1.9}$$

are bounded for $t \in \mathbb{R}_0^+$. Given that operator A is unbounded, it is not necessarily that operator in the right-hand side of (1.1.8) tends to zero as $t \to +0$ in the operator-norm topology, although it is clear that in virtue of (1.1.9)

$$\|\,(\mathbb{1} - \mathcal{E}_{A,n}(t))\,x\,\| \leq t\,\|A\,x\|\,M_n\,, \quad x \in \mathrm{dom}\,(A)\,, \quad n \in \mathbb{N}\,. \tag{1.1.10}$$

Hence, $\mathbb{R}^+ \ni t \mapsto \mathcal{E}_{A,n}(t)$ is right-continuous at $t = 0$ on dom (A). Then because dom (A) is dense in $\mathfrak{X}$ and for any $n \in \mathbb{N}$ the family of operators $\{\mathbb{1} - \mathcal{E}_{A,n}(t)\}_{t\geq 0}$ is *uniformly* bounded, we can use Proposition A.1.6 (Appendix A.1) to extend the continuity in (1.1.10) from dom (A) to the whole space $\mathfrak{X}$: $\lim_{t\to+0}\|\,(\mathbb{1} - \mathcal{E}_{A,n}(t))\,x\,\| = 0$ for $x \in \mathfrak{X}$. Consequently, the function

$t \mapsto \mathcal{E}_{A,n}(t)$ is continuous at $t = 0$ in the *strong* operator sense

$$\text{s-}\lim_{t\to+0} \mathcal{E}_{A,n}(t) = \mathcal{E}_{A,n}(0) = \mathbb{1}. \qquad (1.1.11)$$

To proceed (with the proof of convergence in (1.1.5)), we note that (1.1.7) and (1.1.10) also yield

$$\mathcal{E}_{A,n}(t)\,x - \mathcal{E}_{A,k}(t)\,x = \lim_{\varepsilon\to+0} \int_{\varepsilon}^{t-\varepsilon} d\tau\, \partial_\tau \left[\mathcal{E}_{A,k}(t-\tau)\,\mathcal{E}_{A,n}(\tau)\right] x\,,$$
$$x \in \mathfrak{X}\,,\quad t > 0\,,$$

for $n, k \in \mathbb{N}$ and well-defined (*Bochner*) integral, Proposition 1.6.13. Then using (1.1.7), we obtain the following representation for any $x \in \mathfrak{X}$:

$$\mathcal{E}_{A,n}(t)\,x - \mathcal{E}_{A,k}(t)\,x = \qquad (1.1.12)$$
$$\lim_{\varepsilon\to+0} \int_{\varepsilon}^{t-\varepsilon} d\tau \left(\frac{\tau}{n} - \frac{t-\tau}{k}\right) A^2\,\mathcal{E}_{A,k+1}(t-\tau)\,\mathcal{E}_{A,n+1}(\tau)\,x\,.$$

If we let $x \in \text{dom}\,(A^2)$, then one may rearrange the integrand in (1.1.12) by commutations in this fashion:

$$\int_{\varepsilon}^{t-\varepsilon} d\tau \left(\frac{\tau}{n} - \frac{t-\tau}{k}\right) \mathcal{E}_{A,k+1}(t-\tau)\,\mathcal{E}_{A,n+1}(\tau)\;A^2 x\,. \qquad (1.1.13)$$

Owing to (1.1.7) and (1.1.11), the integrand in (1.1.13) is strongly continuous in $\tau \in [0,t]$. Since in addition $\|\mathcal{E}_{A,n}(t)\| \le 1$ for $t \in \mathbb{R}_0^+$, we obtain for any $n, k \in \mathbb{N}$ the estimate

$$\|\mathcal{E}_{A,n}(t)x - \mathcal{E}_{A,k}(t)x\| \le \frac{t^2}{2}\left(\frac{1}{n} + \frac{1}{k}\right)\|A^2 x\|\,,\quad x \in \text{dom}\,(A^2)\,. \qquad (1.1.14)$$

This means that $\{\mathcal{E}_{A,n}(t)x\}_{n\ge1}$ is a *Cauchy* sequence for $x \in \text{dom}\,(A^2)$. Moreover, condition (ii) also implies (see Propositions A.1.6, A.1.8 and Corollary A.1.9(a)) that $\text{dom}\,(A^2)$ is dense in $\mathfrak{X}$ and, in fact, is a *core* for A, Proposition 1.2.10. Then on account of the uniform operator-norm boundedness $\|\mathcal{E}_{A,n}(t) - \mathcal{E}_{A,k}(t)\| \le 2$, it follows by (1.1.14) and Corollary A.1.9(b) that $\{\mathcal{E}_{A,n}(t)x\}_{n\ge1}$ is a *Cauchy* sequence of contraction operators on the whole space $\mathfrak{X}$, which is locally uniformly convergent in $t \in \mathbb{R}_0^+$.

As a result, the limit (1.1.5) holds in the strong operator topology

$$U_A(t) := \text{s-}\lim_{n\to\infty} \mathcal{E}_{A,n}(t)\,,\quad t \ge 0\,, \qquad (1.1.15)$$

uniformly in $t \in [0,T]$ for any finite interval. This proves (a) and defines the family $\{U_A(t)\}_{t\ge0}$ of *contractions* $\{\|U_A(t)\| \le 1\}_{t\ge0}$, on the Banach space $\mathfrak{X}$.

(b) Since by (1.1.7) and (1.1.9) the mappings $t \mapsto \mathcal{E}_{A,n}(t)$, $n \in \mathbb{N}$, are strongly continuous for $t \in \mathbb{R}_0^+$ and seeing that convergence in (1.1.15) is uniform in t for any finite interval $0 \leq t \leq T$, the operator-valued function

$$t \mapsto U_A(t) : \mathbb{R}_0^+ \to \mathcal{L}(\mathfrak{X}) \tag{1.1.16}$$

is strongly continuous in $\mathbb{R}^+$ and also at $t = 0$ from the right: $U_A(+0) = \mathbb{1}$. This proves claim (b).

(c) On account of (1.1.8) and (1.1.12), it follows that

$$(\mathcal{E}_{A,n}(t) - \mathbb{1})x = -\int_0^t \mathrm{d}\tau \left(\mathbb{1} + \frac{\tau}{n}A\right)^{-1} \mathcal{E}_{A,n}(\tau)\, Ax, \quad x \in \operatorname{dom}(A). \tag{1.1.17}$$

Since

$$\operatorname{s-}\lim_{n\to\infty} \left(\mathbb{1} + \frac{\tau}{n}A\right)^{-1} \mathcal{E}_{A,n}(\tau) = U_A(\tau)$$

uniformly for any finite interval $0 \leq \tau \leq T$, (1.1.17) yields

$$(U_A(t) - \mathbb{1})x = -\int_0^t \mathrm{d}\tau\, U_A(\tau)\, Ax, \quad x \in \operatorname{dom}(A). \tag{1.1.18}$$

On the other hand, due to (1.1.16), the vector-valued family $\{U_A(\tau)(Ax)\}$ is continuous in τ for any $x \in \operatorname{dom}(A)$. Therefore, (1.1.18) shows that the vector-valued function $0 \leq t \mapsto U_A(t)x$ is differentiable in t for all $x \in \operatorname{dom}(A)$

$$\partial_t (U_A(t)x) = U_A(t)(-Ax), \quad t \geq 0,\ x \in \operatorname{dom}(A), \tag{1.1.19}$$

where by definition we consider in (1.1.19) at $t = 0$ the right derivative

$$\partial_t\, U_A(+0)\, x := \lim_{t\to+0} \frac{1}{t}(U_A(t) - \mathbb{1})x = -Ax\,, \quad x \in \operatorname{dom}(A)\,.$$

The last identity is due to the right continuity (1.1.16) and representation (1.1.18).

Note that in fact (1.1.8) gives more. Indeed, since

$$\left(\mathbb{1} + \frac{t}{n}A\right)^{-1} \mathfrak{X} \subseteq \operatorname{dom}(A)\,, \quad t > 0,$$

one has

$$\partial_t \mathcal{E}_{A,n}(t) = -\mathcal{E}_{A,n}(t)\, A \left(\mathbb{1} + \frac{t}{n}A\right)^{-1} = -A\, \mathcal{E}_{A,n}(t) \left(\mathbb{1} + \frac{t}{n}A\right)^{-1} \tag{1.1.20}$$

and

$$\lim_{n\to\infty} A\left(\mathbb{1} + \frac{t}{n}A\right)^{-1} x = \lim_{n\to\infty}\left(\mathbb{1} + \frac{t}{n}A\right)^{-1} Ax = Ax, \quad x \in \mathrm{dom}\,(A)$$

because $A \in \mathcal{C}(\mathfrak{X})$ is a closed operator. Hence, by virtue of (1.1.15) and by closeness of A, we get that for any $x \in \mathrm{dom}\,(A)$

$$\lim_{n\to\infty} \mathcal{E}_{A,n}(t)A\Big(\mathbb{1} + \frac{t}{n}A\Big)^{-1} x = U_A(t)\,A\,x =$$
$$= \lim_{n\to\infty} A\,\mathcal{E}_{A,n}(t)\Big(\mathbb{1} + \frac{t}{n}A\Big)^{-1} x = A\,U_A(t)\,x\,. \tag{1.1.21}$$

For that reason $U_A(t)x \in \mathrm{dom}\,(A)$, that proves (c).

(d) Note that by (1.1.21) the differential equation (1.1.19) takes the form

$$\partial_t(U_A(t)x) = -A\,(U_A(t)x), \quad t > 0, \;\; x \in \mathrm{dom}\,(A). \tag{1.1.22}$$

So, by virtue of (b), the *orbit* $\{U_A(t)\,x_0\}_{t\geq 0}$ of vector x_0 under U_A is a solution of the *autonomous* Cauchy problem (ACP)

$$\partial_t\, x(t) = -\,A\,x(t)\,, \qquad t \in \mathbb{R}_0^+, \tag{1.1.23}$$

provided the initial condition $x(t)|_{t=0} = x_0 \in \mathrm{dom}\,(A)$. Consequently, the orbit $\{x(t) := U_A(t)x_0\}_{t\geq 0}$ is differentiable including $t = 0$, where the right derivative is $(\partial_t x)(+0) = -\,A\,x(0)$.

Let $x(t)$ be a solution of (1.1.23) with initial condition $x_0 = x(0) \in \mathrm{dom}\,(A)$. This means that the vector-valued function $x(\cdot) : \mathbb{R}^+ \to \mathrm{dom}\,(A)$ is strongly differentiable, which means that

$$\lim_{\delta\to 0} \delta^{-1}(x(t+\delta) - x(t)) = \partial_t x(t)$$

exists, and is such that (1.1.23) holds for $t > 0$. By (1.1.19) and (1.1.23), we have

$$\partial_\tau(U_A(t-\tau)x(\tau)) = 0, \quad 0 < \tau < t,$$

with the one-side derivative at $\tau = +0$:

$$\lim_{\tau\to +0} \partial_\tau(U_A(t-\tau)x(\tau)) = 0.$$

Thus, for each $t > 0$ and arbitrary $0 \leq \tau \leq t$, one gets that

$$U_A(t-\tau)\,x(\tau) = U_A(t)\,x(0) = x(t) \in \mathrm{dom}\,(A). \tag{1.1.24}$$

This means that any solution of (1.1.23) has the unique form

$$x(t) = U_A(t)\, x_0, \qquad x_0 = x(0)\,, \tag{1.1.25}$$

that is, $U_A(t-\tau)U_A(\tau)x_0 = U_A(t)x_0$ for any $x_0 \in \mathrm{dom}\,(A)$. Since this holds on the dense set dom (A), we obtain the *functional equation* or *semigroup law*:

$$U_A(s)U_A(\tau) = U_A(s+\tau), \quad s, \tau \geq 0, \tag{1.1.26}$$

on $\mathfrak{X}$. This proves (d) and also motivates for $U_A(t)$, according to the classical Euler formula and the limit (1.1.5), the notation of the *exponential* function

$$U_A(t) := \mathrm{e}^{-tA} \in \mathcal{L}(\mathfrak{X}), \quad t \geq 0, \tag{1.1.27}$$

generated by the operator A. □

We showed above that conditions of Proposition 1.1.1 are *sufficient* for construction of a strongly continuous semigroup of contractions $\{U_A(t)\}_{t\geq 0}$. The converse assertion about *necessity* of these conditions and uniqueness of the construction we shall prove in Proposition 1.2.7 of Sect. 1.2.

1.2 Strongly Continuous and Contraction Semigroups

Proposition 1.1.1 motivates definition of the following fundamental notion for this book: the one-parameter *strongly continuous* semigroup or the C_0-semigroup on a Banach space $\mathfrak{X}$.

Definition 1.2.1 (C_0-semigroups) A family $\{U(t)\}_{t\geq 0}$ of bounded linear operators on $\mathfrak{X}$ is called a strongly continuous one-parameter C_0-semigroup if

(a) $U(0) = \mathbb{1}$
(b) $U(t)U(s) = U(t+s)$, for $t, s \in \mathbb{R}_0^+$ (*functional equation* or the *semigroup law*)
(c) The mapping $t \mapsto U(t)$ is strongly continuous from $\mathbb{R}^+$ into the Banach space $\mathcal{L}(\mathfrak{X})$ of bounded operators and strongly *right-continuous* at $t = 0$:

$$(s_0) \qquad \lim_{t\to+0} \|U(t)\,x - x\| = 0\,, \tag{1.2.1}$$

for every $x \in \mathfrak{X}$, that is, $\text{s-lim}_{t\to+0}\, U(t) = \mathbb{1}$ (*non-degenerate* semigroup) .

Remark 1.2.2 Here and in Sects. 1.4 and 1.5, we show that crucial for the properties of a semigroup is topology of the *right* continuity at $t = 0$. We denote it also as continuity at $t = 0$. On the other hand, it is known that (the *strongest*) topology of continuity of a semigroup away from $t = 0$ may be rather different from that at $t = 0$. Moreover, this topology may even *vary* with $t > 0$ from the *strong* topology to the *operator-norm* and then to the *trace-norm* topology (see examples and comments in Notes to Chap. 1 (Sect. 1.6)).

The strong continuity (1.2.1) makes the C_0-semigroups, in a certain sense, exceptional and the most important among the one-parameter operator semigroups. On account of this remark, we come back to Definition 1.2.1 for scrutinising the continuity condition (s_0) at $t = 0$. In fact, the Proposition 1.2.3, which follows below, says that conditions (a)–(c) are *equivalent* to (a) and (b) and condition (s_0) *only* at the origin $t = 0$. That is, conditions (a) and (b) and (s_0) are necessary and sufficient for the family $\{U(t)\}_{t\geq 0}$ be a C_0-semigroup.

Since (a)–(c) evidently imply (a), (b) and (s_0), we have to prove only the converse statement.

Proposition 1.2.3 *If a family of bounded operators* $\{U(t)\}_{t\geq 0}$ *satisfy* (a), (b) *and* (s_0) *in Definition* 1.2.1, *then*

(1) *There exist constants* $M \geq 1$ *and* $\beta \geq 0$ *such that*

$$\|U(t)\| \leq M\, e^{\beta t} \tag{1.2.2}$$

for all $t \geq 0$.

(2) *The orbit mappings* : $\{t \mapsto U(t)\, x\}_{x\in\mathfrak{X}}$ *are strongly continuous from* $\mathbb{R}_0^+$ *into* $\mathfrak{X}$ *for every* $x \in \mathfrak{X}$. *That is, the one-parameter semigroup* $\{U(t)\}_{t\geq 0}$ *satisfies condition* (c).

Proof

(1) On account of (s_0), one infers that for some $n^* \in \mathbb{N}$

$$M_{n^*} := \sup_{0\,\leq\, t\,\leq\, 1/n^*} \|U(t)\| < \infty\,.$$

Indeed, for otherwise, there would exist a sequence $\{t_n \in [0, 1/n]\}_{n\in\mathbb{N}}$ such that $\|U(t_n)\| > n$. Then it would follow that

$$\limsup_{n\to\infty} \|U(t_n)\, x\| = \infty\,,$$

for some $x \in \mathfrak{X}$, which contradicts to the limit (1.2.1). Note that owing to (a) one also gets that $M_{n^*} \geq 1$.

Now, we put $M := (M_{n^*})^{n^*} \geq 1$. Then by the functional equation (b) and definition of M_{n^*}, we obtain that $\|U(s)\| \leq \|U(s/n^*)\|^{n^*} \leq M$ for $s \in [0, 1]$. For arbitrary $t \in \mathbb{R}_0^+$, there exists $n \in \mathbb{N}$ such that $t = n + s$, where $s \in [0, 1]$. As a consequence of (b) and $M \geq 1$, we deduce (1.2.2) since

$$\|U(t)\| \leq \|U(1)\|^n\, \|U(s)\| \leq M^{n+1} \leq M\, \mathrm{e}^{\beta t}$$

holds for $\beta := \ln(M)$ and each $t \geq 0$.

(2) Let $t_0 > 0$. Then the right continuity at t_0 of the family of bounded operators $\{U(t)\}_{t\geq 0}$ follows from the right continuity (s_0) at zero by virtue of

$$\lim_{\delta\to+0} \|U(t_0+\delta)\,x - U(t_0)\,x\| \leq \|U(t_0)\| \lim_{\delta\to+0} \|U(\delta)\,x - x\| = 0\,, \quad x \in \mathfrak{X}\,.$$

For the proof of the left continuity at t_0, we note that by estimate (1.2.2) and again by condition (s_0)

$$\lim_{\delta\to+0} \|U(t_0-\delta)\,x - U(t_0)\,x\| \leq \lim_{\delta\to+0} \|U(t_0-\delta)\|\,\|x - U(\delta)\,x\| \leq$$
$$\lim_{\delta\to+0} M\,\mathrm{e}^{\beta\,(t_0-\delta)}\,\|x - U(\delta)\,x\| = 0\,, \quad x \in \mathfrak{X}\,.$$

Thus, the strong one-sided continuity (s_0): s-$\lim_{t\to+0} U(t) = U(0)$, together with non-degeneracy (a) : $U(0) = \mathbb{1}$, and the functional equation (b) imply that the orbit mappings: $\{t \mapsto U(t)\,x\}_{x\in\mathfrak{X}}$ are continuous from $\mathbb{R}_0^+$ into $\mathfrak{X}$ for every $x \in \mathfrak{X}$. That is, the one-parameter semigroup $\{U(t)\}_{t\geq 0}$ satisfies the condition (c). □

Remark 1.2.4

(a) A priori, a one-parameter C_0-semigroup $\{U(t)\}_{t\geq 0}$ is *not* continuous for $t > 0$ in the *operator-norm* topology, even though one has Example 1.6.5 of such behaviour in Sect. 1.6. On the other hand, a semigroup, which is operator-norm continuous at $t = +0$, is, in a certain sense, a *trivial* concept (see Sects. 1.4 and 1.10 (Notes to Sect. 1.2)).
(b) Instead, one gets that for a C_0-semigroup $\{U(t)\}_{t\geq 0}$ the family $\{\|U(t)\,x\|\}_{t\geq 0}$ is continuous for each $x \in \mathfrak{X}$. Then function $\mathbb{R}_0^+ \ni t \mapsto \|U(t)\|$ is *lower* semi-continuous and, as a consequence, is measurable. On that account, the family of operator-norms $\{\|U(t)\|\}_{t\geq 0}$ need *not* be continuous.
(c) Moreover, on account of the semigroup law, the measurable function $\mathbb{R}^+ \ni t \mapsto \ln(\|U(t)\|)$ is *subadditive*. As a consequence, we infer that there exists

$$\beta_0 := \inf_{t>0} \frac{1}{t} \ln(\|U(t)\|) = \lim_{t\to\infty} \frac{1}{t} \ln(\|U(t)\|)\,, \tag{1.2.3}$$

which is called the *type* of the C_0-semigroup $\{U(t)\}_{t\geq 0}$.

We note that Proposition 1.1.1 proves the existence of a special class of strongly continuous semigroups (1.1.15) for which $M = 1$ and $\beta = 0$, cf. (1.2.2) in Proposition 1.2.3. This class is generated by operators satisfying the *Hille-Yosida* conditions (i) and (ii) of Proposition 1.1.1.

Definition 1.2.5 An operator-valued function $t \mapsto U(t)$ is called a contraction C_0-semigroup if it is a strongly continuous semigroup and $\|U(t)\| \leq 1$ for $t \in \mathbb{R}_0^+$.

It is important that the *converse* statement of Proposition 1.1.1 is also true. To this aim, one needs to define another basic object.

Definition 1.2.6 The infinitesimal generator A of a strongly continuous semigroup $U(t)$ on a Banach space $\mathfrak{X}$ is the linear operator on domain dom (A) defined by

$$Ax = \lim_{t \to +0} \frac{1}{t} (\mathbb{1} - U(t))\, x\,, \quad x \in \mathrm{dom}\,(A), \tag{1.2.4}$$

where dom (A) is the set of $x \in \mathfrak{X}$ for which the limit in (1.2.4) exists.

Proposition 1.2.7 *Let* $\{U(t)\}_{t \geq 0}$ *be a contraction semigroup on a Banach space* $\mathfrak{X}$. *Then there exists a unique closed densely defined operator A (the semigroup generator) satisfying the conditions* (i) *and* (ii) *of Proposition* 1.1.1, *in other words, such that* (*cf.* (1.1.27))

$$U(t) = \mathrm{e}^{-tA}\,, \quad t \geq 0\,. \tag{1.2.5}$$

Proof Let $\{A_t := t^{-1}(\mathbb{1} - U(t))\}_{t>0} \subset \mathcal{L}(\mathfrak{X})$ be family of bounded operators, and let domain

$$D := \{x \in \mathfrak{X} : \lim_{t \to +0} A_t x \ \text{ exists}\}.$$

By these definitions and by (1.1.22) for $\{U(t)\}_{t \geq 0}$ in the proposition, we can define for all $x \in D$ the linear operator

$$Ax := \lim_{t \to +0} A_t x = -\partial_t\, (U(t)\, x)\, \big|_{t=+0}\,, \quad x \in \mathrm{dom}\,(A) := D\,. \tag{1.2.6}$$

To verify that the closure $\overline{\mathrm{dom}\,(A)} = \mathfrak{X}$, we define for any $x \in \mathfrak{X}$ the *strong* Bochner integral

$$x(s) := \int_0^s \mathrm{d}\tau\, U(\tau)\, x \ \in \mathfrak{X}\,. \tag{1.2.7}$$

Then owing to (1.2.7), the bounded operator $A_t := t^{-1}(\mathbb{1} - U(t))$ gets the form

$$A_t\, x(s) = \frac{1}{t} \int_0^t \mathrm{d}\tau\, U(\tau)\, x - \frac{1}{t} \int_s^{s+t} \mathrm{d}\tau\, U(\tau)\, x\,, \quad t > 0\,. \tag{1.2.8}$$

Since family $\{U(t)\}_{t>0}$ is strongly continuous, by virtue of (1.2.6), (1.2.7) and (1.2.8), one gets

$$Ax(s) = \text{s-}\lim_{t \to +0} A_t x(s) = x - U(s)\, x\,, \quad x \in \mathfrak{X}\,. \tag{1.2.9}$$

Consequently, $s^{-1}x(s) \in \mathrm{dom}\,(A)$ for each $x \in \mathfrak{X}$ and $s > 0$. Hence, seeing that by the strong continuity of semigroup in (1.2.7) we infer

$$\lim_{s\downarrow 0} s^{-1}x(s) = x\,, \quad x \in \mathfrak{X}\,, \tag{1.2.10}$$

the operator A is densely defined.

Note that on $\mathfrak{X}$ for any $\tau > 0$ and $t \geq 0$ we have $A_\tau\, U(t)\, x = U(t)\, A_\tau\, x$. If now $x \in \mathrm{dom}\,(A)$, then by (1.1.22) and limit (1.2.6), one infers that

$$\partial_t U(t)x = -\,AU(t)x = -\,U(t)\,Ax\,, \quad x \in \mathrm{dom}\,(A)\,. \tag{1.2.11}$$

On that account, $\mathrm{dom}\,(A)$ is invariant under the semigroup $\{U(t)\}_{t\geq 0}$:

$$t \mapsto U(t) : \mathrm{dom}\,(A) \to \mathrm{dom}\,(A)\,, \quad t \in \mathbb{R}_0^+\,. \tag{1.2.12}$$

Now, we can prove that operator A is closed. Let $x_n \in \mathrm{dom}\,(A)$ and let $x_n \to x$ together with $Ax_n \to y$ for $n \to \infty$. Then by (1.2.6) and (1.2.11), we obtain

$$\lim_{t\to +0} A_t x = \lim_{t\to +0}\lim_{n\to\infty} \frac{1}{t}(\mathbb{1} - U(t))x_n = \lim_{t\to +0}\lim_{n\to\infty} \frac{1}{t}\int_0^t d\tau\, U(\tau)\, Ax_n =$$
$$\lim_{t\to +0} \frac{1}{t}\int_0^t d\tau\, U(\tau)\, y = y\,.$$

As a result, we infer that $x \in D = \mathrm{dom}\,(A)$ and $Ax = y$, that is, operator A is closed.

To verify (i) and (ii) of Proposition 1.1.1, we calculate the *Laplace* transform of the semigroup $U(t)$

$$\hat{U}(z) := \lim_{T\to\infty}\int_0^T dt\, e^{-zt}U(t) = \int_0^\infty dt\, e^{-zt}U(t)\,. \tag{1.2.13}$$

The integral in the right-hand side of (1.2.13) is an improper strong *Riemann* integral, which is defined as the strongly convergent limit of bounded operators $\{\hat{U}_{z,T} = \int_0^T dt\, e^{-zt}U(t)\}_{T>0}$ for $\mathrm{Re}\, z > 0$. Here, the proper integrals are well-defined because the integrand is a strongly continuous function of t. Since $\|U(t)\| \leq 1$, (1.2.13) defines a bounded linear operator with $\|\hat{U}(z)\| \leq (\mathrm{Re}\, z)^{-1}$. Now, let $x \in \mathrm{dom}\,(A)$, then by virtue of (1.2.11)

$$\partial_t (e^{-zt}U(t)x) = -e^{-zt}U(t)\,(z\mathbb{1} + A)\,x\,, \quad \mathrm{Re}\, z > 0\,.$$

Integration of this equality gives

$$x = \int_0^\infty dt\, e^{-zt}\, U(t)\, (z\mathbb{1} + A)\, x\,,$$

or, with $w := (z\mathbb{1} + A)x$, this becomes

$$(z\mathbb{1} + A)^{-1} w = \int_0^\infty dt\, e^{-zt}\, U(t)\, w\,, \quad \operatorname{Re} z > 0\,. \tag{1.2.14}$$

As far as $\|\hat{U}(z)\| \le (\operatorname{Re} z)^{-1}$, the equality (1.2.14), which means that

$$\hat{U}(z) = (z\mathbb{1} + A)^{-1} = R_A(-z)\,, \tag{1.2.15}$$

coincides with the resolvent of the operator A at the point $\zeta = -z$. Hence, the set $\operatorname{Re}\zeta < 0$ belongs to $\rho(A)$, and we get the estimate (ii) of Proposition 1.1.1.

Suppose that $U(t)^{(1)}$ and $U(t)^{(2)}$ are two different contraction semigroups, i.e., $U_{t_0}^{(1)} \neq U_{t_0}^{(2)}$ for some $t_0 > 0$. Then there exist a linear functional $\phi \in \mathfrak{X}^*$ and a vector $w \in \mathfrak{X}$ such that $\phi(U_{t_0}^{(1)} w) \neq \phi(U_{t_0}^{(2)} w)$. Due to (1.2.14), this implies that $\phi((z\mathbb{1} + A_1)^{-1} w) \neq \phi((z\mathbb{1} + A_2)^{-1} w)$ for some z with $\operatorname{Re} z > 0$. Thus, $R_\zeta(A_1) \neq R_\zeta(A_2)$, and so $A_1 \neq A_2$. Conversely, two different generators correspond, by (1.2.14), to different semigroups.

It remains to prove (1.2.5). Since the operator A, constructed above, satisfies the conditions of Proposition 1.1.1, the contraction semigroup $U_A(t) = e^{-tA}$ for $t \ge 0$ exists. The Laplace transform of this semigroup, calculated as in (1.2.13) and (1.2.14), gives

$$\int_0^\infty dt\, e^{-zt} U_A(t) = (z\,\mathbb{1} + A)^{-1}\,. \tag{1.2.16}$$

Therefore, (1.2.5) follows from the uniqueness that we proved above. □

We recall that the concept of a *core* (core $(A) \subset$ dom (A)) for operator A is often useful if it is difficult to identify precisely the domain dom $(A) \subset \mathfrak{X}$. In the context of a strongly continuous semigroup $\{U_A(t)\}_{t\ge 0}$ this means that generator A is entirely defined if one knows its restriction to a core.

Definition 1.2.8 We say that $\mathcal{C} \subset$ dom (A) is a *core* for operator A (and we denote it as $\mathcal{C} =$ core (A)) if for all $f \in$ dom (A) there exists a sequence $\{f_n\}_{n\ge 1} \subset \mathcal{C}$ such that

$$\lim_{n\to\infty} f_n = f\,, \quad \lim_{n\to\infty} A\, f_n = A\, f\,, \tag{1.2.17}$$

or equivalently if subspace $\mathcal{C}$ is dense in dom (A) for the graph-norm

$$\|f\|_A := \|f\| + \|Af\| , \quad f \in \text{dom}\,(A) , \tag{1.2.18}$$

particularly, the closure in norm (1.2.18) of restriction $A \restriction \mathcal{C}$ restores the operator A, that is, $\overline{A \restriction \mathcal{C}} = A$.

It is clear that core (A) is not unique, as well as that domain dom (A) equipped with the graph-norm (1.2.18) is a *complete* space $(\text{dom}\,(A), \|\cdot\|_A)$.

Proposition 1.2.9 (Nelson Theorem) *Let A be generator of a strongly continuous semigroup $\{U_A(t)\}_{t\geq 0}$ on $\mathfrak{X}$. If subset $\mathcal{D} \subseteq$ dom (A) is dense in $\mathfrak{X}$ and invariant under the semigroup, that is, $U_A(t) : \mathcal{D} \to \mathcal{D}$ for any $t \geq 0$, then $\mathcal{D} =$ core (A).*

Proof If $f \in$ dom (A), then given that $\mathcal{D} \subseteq$ dom (A) is dense in $\mathfrak{X}$, there exists a sequence $\{f_n\}_{n\geq 1} \subset \mathcal{D}$ such that $\lim_{n\to\infty} \|f - f_n\| = 0$. Continuing, we have to use invariance of $\mathcal{D}$, and according to Definition 1.2.8, establish for $\{f_n\}_{n\geq 1}$ the second limit in (1.2.17).

To this aim, we introduce auxiliary sequence $\{U_A(\tau)\, f_n\}_{n\geq 1}$, which for any $\tau \geq 0$ belongs to $\mathcal{D}$ by invariance of the subspace $\mathcal{D}$. Note that because of (1.2.11) the mapping: $\tau \mapsto U_A(\tau)\, f_n \in \mathcal{D}$ is continuous in the graph-norm (1.2.18) for each $f_n \in \mathcal{D}$, $n \in \mathbb{N}$. As a consequence, for each $n \in \mathbb{N}$, the integral

$$f_n(t) := \int_0^t d\tau \, U_A(\tau)\, f_n \, \in \overline{\mathcal{D}}^A , \tag{1.2.19}$$

that is, it belongs to the $\|\cdot\|_A$-closure of $\mathcal{D}$. Then by (1.2.7), (1.2.9) and (1.2.19), we obtain for the sequence $\{f_n\}_{n\geq 1} \subset \mathcal{D}$ that

$$\lim_{n\to\infty} \Big\| \int_0^t d\tau \, U_A(\tau)\, f_n - \int_0^t d\tau \, U_A(\tau)\, f \Big\|_A = \lim_{n\to\infty} \Big\| \int_0^t d\tau \, U_A(\tau)\, (f_n - f) \Big\| +$$
$$\lim_{n\to\infty} \|f_n - U_A(t)\, f_n - f + U_A(t)\, f\| = 0$$

for every $t > 0$. So, by virtue of (1.2.18) and (1.2.19),

$$f(t) := \int_0^t d\tau \, U_A(\tau)\, f \, \in \overline{\mathcal{D}}^A . \tag{1.2.20}$$

Now, using (1.2.9) again, one gets for $f \in$ dom (A) and $\{t^{-1}\, f(t)\}_{t>0}$

$$\lim_{t\to+0} \Big\|t^{-1} \int_0^t d\tau \, U_A(\tau)\, f - f \Big\|_A = \lim_{t\to+0} \Big\|t^{-1} \int_0^t d\tau \, U_A(\tau)\, f - f \Big\| +$$
$$\lim_{t\to+0} \|t^{-1}\,(f - U_A(t)\, f) - A\, f\| = 0 .$$

Then on account of (1.2.18) and (1.2.20), this limit shows that $f \in \overline{\mathcal{D}}^A$, and therefore $\overline{\mathcal{D}}^A = \mathrm{dom}\,(A)$ as required. □

On account of Proposition 1.1.1(b), domain $\mathrm{dom}\,(A)$ is invariant under semigroup $\{U_A(t)\}_{t\geq 0}$ in the sense that $U_A(t)(\mathrm{dom}\,(A)) \subseteq \mathrm{dom}\,(A)$. Hence, by Proposition 1.2.9 $\mathrm{dom}\,(A) = \mathrm{core}\,(A)$ and $\mathrm{dom}\,(A)$ is a *trivial* example of a core for generator A of contraction semigroup.

The *triviality of domain* problem for the square of unbounded operators in Hilbert and Banach spaces warns that $\mathrm{dom}\,(A^2) = \{0\}$ may happen for a nice densely defined and closed operator A. However, it is not the case if A is *generator* of a C_0-semigroup (see Appendix A.1, Sect. A.1.2). In fact, this is a source of more examples of cores for A given by domains $\mathrm{dom}\,(A^n)$ of powers of the generator A.

Proposition 1.2.10 *Let A be generator of strongly continuous contraction semigroup $\{U_A(t)\}_{t\geq 0}$ on $\mathfrak{X}$. Then domains $\mathrm{dom}\,(A^n)$ of the powers: $A^n = A\,A^{n-1}$, $n \in \mathbb{N}$, of the generator A are cores of the operator A.*

Proof By virtue of Corollary A.1.9(a) domain

$$\mathrm{dom}\,(A^2) = \{x \in \mathrm{dom}\,(A) : A\,x \in \mathrm{dom}\,(A)\} \subseteq \mathrm{dom}\,(A)\,, \tag{1.2.21}$$

is dense in $\mathfrak{X}$. Then by invariance of $\mathrm{dom}\,(A)$ (Proposition 1.1.1(b)) and Corollary A.1.9(a), one infers the invariance of $\mathrm{dom}\,(A^2)$:

$$\begin{aligned} &U_A(t)\,(\mathrm{dom}\,(A^2)) = U_A(t)\,(A + \lambda\,\mathbb{1})^{-1}(\mathrm{dom}\,(A)) = \qquad (1.2.22)\\ &(A + \lambda\,\mathbb{1})^{-1}(U_A(t)\,(\mathrm{dom}\,(A))) \subseteq (A + \lambda\,\mathbb{1})^{-1}(\mathrm{dom}\,(A)) = \mathrm{dom}\,(A^2)\,, \quad \lambda > 0\,. \end{aligned}$$

By Proposition 1.2.9, this implies that subset $\mathrm{dom}\,(A^2) = \mathrm{core}\,(A)$.

Now, iterating (1.2.21) and (1.2.22), we obtain for

$$\mathrm{dom}\,(A^n) = \{x \in \mathrm{dom}\,(A) : A\,x \in \mathrm{dom}\,(A^{n-1})\} \subseteq \mathrm{dom}\,(A)\,, \quad n \in \mathbb{N}\,, \tag{1.2.23}$$

dense by Corollary A.1.9(a) in $\mathfrak{X}$ (if it holds for $\mathrm{dom}\,(A^{n-1})$), that

$$U_A(t)\,(\mathrm{dom}\,(A^n)) = U_A(t)\,(A + \lambda\,\mathbb{1})^{-1}(\mathrm{dom}\,(A^{n-1})) \subseteq \mathrm{dom}\,(A^n)\,, \quad \lambda > 0\,, \tag{1.2.24}$$

if it is true for $\mathrm{dom}\,(A^{n-1})$. As a result, by Proposition 1.2.9 and (1.2.23) (1.2.24), we conclude that subset $\mathrm{dom}\,(A^n) = \mathrm{core}\,(A)$ for $n \in \mathbb{N}$. □

1.3 Quasi-bounded Semigroups

Propositions 1.1.1 and 1.2.7 show that generator A plays a central rôle in characterisation of strongly continuous semigroups. In fact, the proof of the Proposition 1.2.7 shows that the existence of the closed densely defined generator for a semigroup

$\{U(t)\}_{t\geq 0}$ is ensured by its strong continuity alone, i.e., it does not require that $U(t)$ is a contraction. As a consequence, in order that an operator A generates a strongly continuous semigroup, it is not necessary that the generator satisfies condition (ii) of Proposition 1.1.1.

Let $\{U(t)\}_{t\geq 0}$ be a strongly continuous semigroup and $\tau > 0$. Since for any $f \in \mathfrak{X}$ the vector-valued function $t \mapsto U(t)f : [0, \tau] \to \mathfrak{X}$ is continuous and $[0, \tau]$ is compact, the set $\{\|U(t)f\| : f \in \mathfrak{X}\}_{0\leq t\leq \tau}$ is bounded. Then applying two times the *uniform boundedness principle* and taking into account $\|U(0)\| = 1$, we infer that there exists an $M \geq 1$ such that $\|U(t)\| \leq M$ for $t \in [0, \tau]$. As a consequence, by the semigroup properties, one gets the estimate

$$\|U(t)\| = \|(U(\tau))^n U(s)\| \leq M e^{\beta t} \tag{1.3.1}$$

for any $t = n\tau + s \in \mathbb{R}_0^+$, $n \in \mathbb{N}_0 = \{0, 1, 2, \ldots\}$ and $\beta = \tau^{-1}\ln(M) \geq 0$. Therefore, C_0-semigroups are exponentially bounded and (1.3.1), together with Proposition 1.2.3 (1), suggest that $\beta \geq 0$.

But taking into account Remark 1.2.4 (c), we see that it is not true since β in (1.2.2) and (1.3.1) is only an evident *upper bound*, which does not very much depend on the value of M. Instead of $\{U(t)\}_{t\geq 0}$, we consider a *rescaled* C_0-semigroup $\{U^{(\alpha)}(t) := e^{\alpha t}\, U(t)\}_{t\geq 0}$. Then for $\alpha > 0$, the estimate (1.3.1) must be modified to $\|U^{(\alpha)}(t)\| \leq M e^{\beta+\alpha}$, whereas (1.3.1) for the same β is still valid when $\alpha < 0$. In both cases, we have the same M. To scrutinise dependence of the estimate (1.3.1) on (M, β) we use observation in Remark 1.2.4 (c).

Definition 1.3.1 The infimum β_0 (1.2.2) of the set of numbers $\{\beta\} \subset \mathbb{R}$, such that there is an M satisfying the estimate (1.3.1), is called the *type* of the strongly continuous semigroup $\{U(t)\}_{t\geq 0}$. Slightly abusing notations we shall denote below the infimum in (1.2.2) by the same letter β .

If $\beta = 0$, we call $\{U(t)\}_{t\geq 0}$ a bounded semigroup. If $\beta > 0$, we call it quasi-bounded of the *type* $\mathcal{G}(M, \beta)$ (for short, of the *type* β), and we denote the corresponding *class* of generators by the same $\mathcal{G}(M, \beta)$ since it does not produce any confusion. Semigroups of the type $\mathcal{G}(1, 0)$ are called *contraction* semigroups.

Remark 1.3.2 (Rescaling and Bounded Semigroups) Let $\{U(t)\}_{t\geq 0}$ with generator $A \in \mathcal{G}(M, \beta)$. Then by virtue of Definition 1.3.1 the *rescaled* C_0-semigroup $\{U^{(\alpha)}(t)\}_{t\geq 0} \in U(M, \beta + \alpha)$ for any $\alpha \in \mathbb{R}$. In particular, this entails that by *rescaling* a quasi-bounded semigroup $\{U(t)\}_{t\geq 0}$ may be transformed into a *bounded* semigroup $\{U^{(\alpha)}(t)\}_{t\geq 0} \in U(M, \beta + \alpha = 0)$. Moreover, the *type* of semigroup $\{U^{(\alpha)}(t)\}_{t\geq 0}$ may be even *negative*: $\beta + \alpha < 0$. As a result, $\lim_{t\to\infty} \|U^{(\alpha)}(t)\| = 0$, that is, the semigroup is *quasi-nilpotent.*

By Definition 1.3.1, the class $\mathcal{G}(1, 0)$ corresponds to generators of contraction semigroups. Let generator $B \in \mathcal{G}(1, 0)$, then operator $(B - \alpha\mathbb{1}) \in \mathcal{G}(1, \alpha)$. As a consequence, $\|U_{B-\alpha\mathbb{1}}(t)\| = \|U_B(t)\|\, e^{\alpha t} \leq e^{\alpha t}$.

Remark 1.3.3 (Equivalent Norms and Contraction Semigroups)

(a) It is also useful to put $M = 1$ and to introduce the notion of *quasi-contraction* semigroups of the type $\mathcal{G}(1, \beta)$. If $\{U_A(t)\}_{t\geq 0}$ is of the type $\mathcal{G}(1, \beta)$, then similar to the case of quasi-bounded semigroups the *rescaling*: $\{e^{-\beta t} U_A(t)\}_{t\geq 0}$ gives a contraction semigroup of the type $\mathcal{G}(1, 0)$. Note that $A \in \mathcal{G}(1, \beta)$ implies that generator A is quasi-m-accretive (see Proposition 1.3.4 for $M = 1$ and Corollary 1.8.11).

(b) To reach $M = 1$, we may introduce on a Banach space $(\mathfrak{X}, \|\cdot\|)$ an equivalent norm $|||\cdot|||$, such that a bounded C_0-semigroup of the type $\mathcal{G}(M, 0)$ becomes a contraction semigroup on $\mathfrak{X}$ endowed with norm $|||\cdot|||$. To this aim, we recall that if $\{U(t)\}_{t\geq 0}$ of the type $\mathcal{G}(M, 0)$ is a bounded strongly continuous semigroup on $(\mathfrak{X}, \|\cdot\|)$, then the norm

$$|||x||| := \sup_{\tau\geq 0} \|U(\tau)x\|, \qquad x \in \mathfrak{X}, \tag{1.3.2}$$

is equivalent to the norm $\|\cdot\|$ on $\mathfrak{X}$ and consequently $\{U(t)\}_{t\geq 0}$ is a contraction semigroup of the type $\mathcal{G}(1, 0)$ on $(\mathfrak{X}, |||\cdot|||)$.

Proposition 1.3.4 *Let A be a closed linear operator in a Banach space $\mathfrak{X}$ with $\overline{\mathrm{dom}\,(A)} = \mathfrak{X}$. Then, A is the generator of a strongly continuous quasi-bounded semigroup $U_A(t)$ of the type β, that is, $A \in \mathcal{G}(M, \beta)$, if and only if the following conditions are satisfied*:

(i) *There is a real β so that $\{z \in \mathbb{C} : \mathrm{Re}\, z < (-\beta)\}$ belongs to the resolvent set $\rho(A)$.*
(ii) *There exists an M such that $\|(z\mathbb{1} + A)^{-1}\| \leq M(\mathrm{Re}\,(z - \beta))^{-1}$, for $\mathrm{Re}\, z > \beta$.*
(iii) *One has*

$$\left\|(\lambda\mathbb{1} + A)^{-m}\right\| \leq \frac{M}{(\lambda - \beta)^m} \tag{1.3.3}$$

for all $\lambda > \beta$ and $m \in \mathbb{N}$.

Proof We prove here only sufficiency. The necessity follows *verbatim* along the arguments in the proof of Proposition 1.2.7 for semigroups of the type $\mathcal{G}(1, 0)$.

Note that without loss of generality we may assume that $\beta = 0$. If $\beta \neq 0$, then we can replace operator A by $A' := A + \beta\mathbb{1}$.

Letting $\beta = 0$, then (1.3.3) takes the form

$$\left\|(\lambda\mathbb{1} + A)^{-m}\right\| \leq \frac{M}{\lambda^m}, \quad \lambda > 0, \quad m \in \mathbb{N}. \tag{1.3.4}$$

Our strategy is to construct on the space $\mathfrak{X}$ a new (but equivalent to $\|\cdot\|$) norm to reduce the conditions (1.3.4) and *a fortiori* the proof to the case of contraction semigroups.

To this aim, we introduce first the norm

$$\|x\|_\mu := \sup_{m\in\mathbb{N}_0} \mu^m \left\|(\mu\mathbb{1} + A)^{-m}x\right\|, \quad x \in \mathfrak{X}, \quad \mu > 0. \tag{1.3.5}$$

This norm has the following properties:

(a) $\|x\| \le \|x\|_\mu \le M\|x\|$, which means that $\|\cdot\|$ and $\|\cdot\|_\mu$ are equivalent.
(b) $\left\|\mu\,(\mu\mathbb{1} + A)^{-1}\right\|_\mu \le 1$.
(c) $\left\|\lambda^n(\lambda\mathbb{1} + A)^{-n}\right\|_\mu \le 1$ for all $0 < \lambda \le \mu$ and $n \in \mathbb{N}$.
(d) $\|x\|_\lambda \le \|x\|_\mu$, $x \in \mathfrak{X}$, for all $0 < \lambda \le \mu$.

The property (a) follows from definition (1.3.5) and estimate (1.3.4), whereas (b) follows from definition of the operator-norm with respect norm (1.3.5). To prove (c), we use the resolvent equation to show that for any $x \in \mathfrak{X}$ one has

$$\begin{aligned} v := (\lambda\mathbb{1} + A)^{-1}x &= (\mu\mathbb{1} + A)^{-1}x + (\mu - \lambda)\,(\mu\mathbb{1} + A)^{-1}\,(\lambda\mathbb{1} + A)^{-1}x \\ &= (\mu\mathbb{1} + A)^{-1}\,(x + (\lambda - \mu)\,v)\,, \quad \lambda, \mu \in \mathbb{R}^+\,, \end{aligned} \tag{1.3.6}$$

which by definition of norm (1.3.5) and by (b) yields

$$\|v\|_\mu \le \frac{1}{\mu}\|x\|_\mu + \frac{\mu - \lambda}{\mu}\|v\|_\mu\,, \quad 0 < \lambda \le \mu\,. \tag{1.3.7}$$

Hence, $\|\lambda(\lambda\mathbb{1} + A)^{-1}x\|_\mu \le \|x\|_\mu$, which implies (c).

Next, by (a) and (c), we obtain $\left\|\lambda^n(\lambda\mathbb{1} + A)^{-n}x\right\| \le \left\|\lambda^n(\lambda\mathbb{1} + A)^{-n}x\right\|_\mu \le \|x\|_\mu$. Then (d) follows if one takes $\sup_{n\in\mathbb{N}_0}$ in the left-hand side of these inequalities.

Now, we introduce another norm, which is equivalent to $\|\cdot\|$:

$$|||x||| := \sup_{\mu>0} \|x\|_\mu\,. \quad x \in \mathfrak{X}\,. \tag{1.3.8}$$

By definition (1.3.8) and by (a) and (c), one obtains the following properties:

(e) $\|x\| \le |||x||| \le M\|x\|$ (norm equivalence).
(f) $|||\lambda(\lambda\mathbb{1} + A)^{-1}||| \le 1$, $\lambda > 0$.

Note that by (f) the operator A satisfies (ii) of Proposition 1.1.1 for contraction semigroups in the equivalent norm (1.3.8). Therefore, by applying this proposition, one completes the proof for the quasi-bounded case conditioned by (1.3.4). □

1.4 Norm Continuity

Here, we consider another continuity condition for semigroups. Suppose that instead of the strong operator continuity of Definition 1.2.1 (c), one has a semigroup $U_A(t)$, which is *operator-norm* continuous for $t \to +0$ in the sense that

$$(n_0) \qquad \lim_{t\to+0} \|U_A(t) - \mathbb{1}\| = 0. \tag{1.4.1}$$

Since the operator-norm continuity (n_0) implies the strong (s_0)-continuity (1.2.1) and, by consequence, the quasi-boundedness (1.3.1) of semigroup $U_A(t)$, we also obtain

$$\lim_{\delta\to 0} \|U_A(t+\delta) - U_A(t)\| \le \lim_{\delta\to 0} M\, e^{\beta(t+|\delta|)} \|U_A(|\delta|) - \mathbb{1}\| = 0\,, \quad t > 0\,. \tag{1.4.2}$$

This yields the operator-norm continuity of the semigroup $U_A(t)$ for any $t > 0$.

If A is a bounded generator, then the semigroup $U_A(t)$ is well-defined by the operator-norm convergent series (1.1.1) for any $t \in \mathbb{C}$. It is obviously operator-norm continuous on $\mathbb{R}_0^+$ in the sense of (1.4.1) and (1.4.2). Note that the converse is also true.

Proposition 1.4.1 *A semigroup* $\{U_A(t)\}_{t\ge 0}$ *is norm continuous on* $\mathbb{R}_0^+$ *if and only if its generator* $A \in \mathcal{L}(\mathfrak{X})$.

Proof By virtue of the above remark and by Propositions 1.1.1 and 1.2.1, we only have to prove that (1.4.1) implies $A \in \mathcal{L}(\mathfrak{X})$.

On account of

$$\left\| \int_0^\tau ds\, (U_A(s) - \mathbb{1}) \right\| \le \tau \sup_{0\le s\le\tau} \|U_A(s) - \mathbb{1}\| \tag{1.4.3}$$

and taking $\tau > 0$ small enough, $\tau < \tau_0$, one gets

$$\left\| \tau^{-1} \int_0^\tau ds\, U_A(s) - \mathbb{1} \right\| < 1.$$

Therefore, the bounded operator $L_\tau := \tau^{-1} \int_0^\tau ds\, U_A(s)$, $0 < \tau < \tau_0$, is invertible. Note that for $0 < \tau < \tau_0$ and for any $x \in \mathrm{dom}\,(A)$ one gets

$$(U_A(\tau) - \mathbb{1})x = -\int_0^\tau ds\, U_A(s)\, A\, x = -\tau\, L_\tau\, A\, x\,. \tag{1.4.4}$$

Then by invertibility of L_τ we obtain that

$$A\, x = \frac{\mathbb{1} - U_A(\tau)}{\tau\, L_\tau} x\,, \quad x \in \mathrm{dom}\,(A)\,, \quad \tau \in (0, \tau_0). \tag{1.4.5}$$

Now, we remind that generator of the semigroups $\{U_A(t)\}_{t\geq 0}$ must be a closed operator. Therefore, since the closure of the right-hand side of (1.4.5) is a bounded operator, one gets that $\operatorname{dom}(A) = \mathfrak{X}$ and hence $A \in \mathcal{L}(\mathfrak{X})$. □

It is appropriate to mention here a special class of semigroups, which is of particular importance in the context of this section and below in the context of the holomorphic semigroups.

It is known (see Propositions 1.1.1(b) and 1.2.7) that any strongly continuous semigroup $U_A(t)$ on a Banach space $\mathfrak{X}$ maps $\operatorname{dom}(A)$ into itself and that any dense invariant subspace $\mathcal{C} \subseteq \operatorname{dom}(A)$ is a *core* for the generator A (see Propositions 1.2.9). Now, we assume a bit more, namely, that $U_A(t)$ maps $\mathfrak{X}$ to $\operatorname{dom}(A)$ for $t > 0$.

Proposition 1.4.2 *Let $\{U_A(t)\}_{t\geq 0}$ be a strongly continuous semigroup on $\mathfrak{X}$ such that $U_A(t > 0) : \mathfrak{X} \to \operatorname{dom}(A)$. Then $t \mapsto U_A(t)$ is an infinitely operator-norm differentiable function on $\mathbb{R}^+$.*

Proof Let $\delta > 0$. Owing to assumptions of the proposition, the operator $A\, U_A(s)$ is closed with $\operatorname{dom}(A\, U_A(\delta)) = \mathfrak{X}$, so it is bounded: $\|A\, U_A(\delta)\| \leq C_\delta$.

Now, by (1.1.18), one has

$$U_A(t)f - f = -\int_0^t d\tau\, U_A(\tau)\, Af$$

for all $f \in \operatorname{dom}(A)$ and $t > 0$. Therefore,

$$(U_A(t+\delta) - U_A(\delta))\, x = -\int_0^t d\tau\, U_A(\tau)\, A\, U_A(\delta)\, x\,, \quad \delta > 0\,, \quad x \in \mathfrak{X}\,. \tag{1.4.6}$$

Since any strongly continuous semigroup is exponentially bounded (1.3.1), we have $\|U_A(t)\| \leq M\, e^{\beta t}$ for some $M > 0$ and $\beta \geq 0$ for all $t \geq 0$. Then, (1.4.6) yields the estimate

$$\|U_A(t+\delta) - U_A(\delta)\| \leq C_\delta \int_0^t d\tau\, M\, e^{\beta\tau}. \tag{1.4.7}$$

Thus, the semigroup $\{U_A(t)\}_{t\geq 0}$ is operator-norm continuous at $t = \delta$ for all $\delta > 0$. Moreover, by (1.4.6), (1.4.7) and by the closeness of the generator A, we obtain

$$\begin{aligned}
&\left\|\frac{1}{t}\,(U_A(t+\delta) - U_A(\delta)) + A\, U_A(\delta)\right\| \\
&= \left\|\frac{1}{t}\int_0^t d\tau\ (U_A(\tau+\delta/2) - U_A(\delta/2))\, A\, U_A(\delta/2)\right\| \\
&\leq (C_{\delta/2})^2 \int_0^t ds\, M\, e^{\beta s},
\end{aligned}$$

which means that $U_A(t)$ is operator-norm differentiable for all $t > 0$ and $\| \cdot \|-\partial_t U_A(t) = -A\, U_A(t)$. For $0 < \delta < t$, we may rewrite this as

$$\| \cdot \|-\partial_t U_A(t) = U_A(t-\delta)\,(-AU_A(\delta)) \tag{1.4.8}$$

and hence differentiate (1.4.8) repeatedly in the operator-norm topology. □

A knowledge of semigroups generated by bounded operators motivates another construction of strongly continuous semigroups than that in Sects. 1.1 and 1.2. The idea is to approximate them by sequences of the norm-continuous semigroups. Since the resolvent of generator of a strongly continuous semigroup plays a key rôle (see Proposition 1.1.1 and (1.2.16)) in their construction and characterisation, we establish first the following general convergence property:

Proposition 1.4.3 *Let A be a closed linear operator with a dense domain* dom $A \subseteq \mathfrak{X}$. *If there exist a real β and an $M > 0$ such that $(-\infty, -\beta] \subset \rho(A)$ and $\|\lambda(A + \lambda\mathbb{1})^{-1}\| \leq M$ for all $\lambda \geq \beta$, then*

(i) $\text{s-lim}_{\lambda\to\infty} \lambda R_A(-\lambda) = \mathbb{1}$.
(ii) $\lim_{\lambda\to\infty} \lambda A\, R_A(-\lambda)x = \lim_{\lambda\to\infty} \lambda R_A(-\lambda)\, Ax = Ax\,, \quad x \in \text{dom}\,(A)$.

Proof

(i) If $x \in \text{dom}\,(A)$, then $\lambda R_A(-\lambda)x = -R_A(-\lambda)Ax + x$. By conditions of the proposition, we have estimate $\|R_A(-\lambda)Ax\| \leq \lambda^{-1}M\ \|Ax\|$. For that reason, (i) holds for $x \in \text{dom}\,(A)$. This result can be extended to any $x \in \mathfrak{X}$ owing to Proposition A.1.6, because by conditions of the proposition the operators $\lambda R_A(-\lambda)$ are uniformly bounded by M for all $\lambda \geq \beta$ and domain dom (A) is dense in $\mathfrak{X}$.
(ii) The limit (ii) is a straightforward consequence of (i).

□

The statement (ii) suggests which kind of bounded operators could be chosen to approximate the unbounded generator A. The next assertion claims that under conditions of Proposition 1.1.1 one can use for construction of a C_0-semigroup, besides the *Euler* approximants (1.1.4), also the *semigroup* approximants (1.1.1), for bounded generators $\{A_n \in \mathcal{L}(\mathfrak{X})\}_{n\geq 1}$. Since in Sect. 1.3 we showed that the results for the case of quasi-bounded semigroups can be deduced from those for contraction semigroups, the subsequent proof is given only for the latter.

Proposition 1.4.4 (Yosida Approximation) *Let $A \in \mathcal{C}(\mathfrak{X})$ be a closed linear operator with dense domain* dom $(A) \subset \mathfrak{X}$ *such that $(-\infty, 0) \subset \rho(A)$ and $\|\lambda R_A(-\lambda)\| \leq 1$ for $\lambda > 0$, then operator A generates a strongly continuous contraction semigroup $\{U_A(t)\}_{t\geq 0}$.*

Proof In view of Proposition 1.4.3, we define a sequence of bounded operators (*Yosida approximants* for generator A):

$$A_n := nAR_A(-n) = n\mathbb{1} - n^2 R_A(-n), \quad n \in \mathbb{N}\,, \tag{1.4.9}$$

which, by (ii) of this statement, converges to A on $\mathrm{dom}\,(A)$. Consider then the corresponding sequence of the operator-norm-continuous semigroups, cf. (1.1.1),

$$U_{A_n}(t) := \mathrm{e}^{-tA_n}, \quad t \geq 0,\ n \in \mathbb{N}. \tag{1.4.10}$$

By (1.4.9), each $\{U_{A_n}(t)\}_{t\geq 0}$ is a contraction semigroup,

$$\left\| U_{A_n}(t) \right\| \leq \mathrm{e}^{-t\,n} \mathrm{e}^{t\,\| n^2 R_A(-n)\|} \leq 1.$$

Therefore, by virtue of the *uniform boundedness principle*, it is sufficient to prove the convergence of this approximating sequence of semigroups on the dense set $\mathrm{dom}\,(A)$ (see Proposition A.1.6). To this aim, we define, for $0 \leq s \leq t$ and $m, n \in \mathbb{N}$, the operator-valued functions

$$s \mapsto U_{A_m}(t-s)\, U_{A_n}(s)\, x\,, \quad x \in \mathrm{dom}\,(A)\,. \tag{1.4.11}$$

Elements of the semigroups $\{U_{A_n}(t)\}_{t\geq 0}$ commute for all $n \in \mathbb{N}$, and they are norm continuous. Then, by virtue of definition (1.4.11), one gets

$$\begin{aligned} U_{A_n}(t)x - U_{A_m}(t)x &= \int_0^t \mathrm{d}s\, \partial_s \left(U_{A_m}(t-s) U_{A_n}(s) x \right) \\ &= \int_0^t \mathrm{d}s\, U_{A_m}(t-s) U_{A_n}(s)\, (A_m - A_n)\, x\,. \end{aligned}$$

This yields the estimate

$$\left\| U_{A_n}(t)x - U_{A_m}(t)x \right\| \leq t\, \|(A_n - A_m)\, x\| \tag{1.4.12}$$

for any $x \in \mathfrak{X}$. Since, by Proposition 1.4.3(ii), $\{A_n x\}_{n\geq 1}$ is a Cauchy sequence for each $x \in \mathrm{dom}\,(A)$, Eq. (1.4.12) implies that $\{U_{A_n}(t)u\}_{n\geq 1}$ converges uniformly on each interval $[0, T]$. Hence, by the uniform boundedness of the sequence $\{U_{A_n}(t)\}_{n\geq 1}$, the limit

$$U(t)\, x := \lim_{n\to\infty} U_{A_n}(t)\, x \tag{1.4.13}$$

exists for each $x \in \mathfrak{X}$ and for each $t \geq 0$ (see Proposition A.1.6).

On account of (1.4.13), the limit $\{U(t)\}_{t\geq 0}$ inherits the semigroup property of $\{U_{A_n}(t)\}_{t\geq 0}$. Hence, it is a contraction semigroup, i.e., $\|U(t)\| \leq 1$. By the estimate

$$\|(U(t) - \mathbb{1})\, x\| \leq \left\| \left(U(t) - U_{A_n}(t)\right) x \right\| + \left\| \left(U_{A_n}(t) - \mathbb{1}\right) x \right\|$$

and by the uniform convergence of $\{U_{A_n}(t)x\}_{n\geq 1}$ on each interval $[0, T]$ for any $x \in \mathrm{dom}\,(A)$, we obtain

$$\lim_{t\to+0} U(t)x = x\,, \quad x \in \mathrm{dom}\,(A). \tag{1.4.14}$$

Then again the uniform boundedness of $\|U(t)\| \leq 1$ allows us to extend (1.4.14) to $\mathfrak{X}$. Thus, the semigroup $\{U(t)\}_{t\geq 0}$ is strongly continuous.

It remains to prove that the operator A generates this semigroup. Suppose that B is the generator of $\{U(t)\}_{t\geq 0}$. Fix $u \in \mathrm{dom}\,(A)$. Since the sequence $\{U_{A_n}(t)x\}_{n\geq 1}$ converges uniformly on each interval $[0, T]$, we get

$$\begin{aligned}(\mathbb{1} - U(t))x &= x - \lim_{n\to\infty} U_{A_n}(t)x \\ &= \lim_{n\to\infty} \int_0^t \mathrm{d}s\, U_{A_n}(s)\, A_n x = \int_0^t \mathrm{d}s\, U(s)\, Ax\,,\end{aligned} \tag{1.4.15}$$

by virtue of (1.4.12) and (1.4.13) and Proposition 1.4.3(ii). Then, (1.4.15) implies that

$$Ax = Bx\,, \quad x \in \mathrm{dom}\,(A), \tag{1.4.16}$$

or $\mathrm{dom}\,(A) \subseteq \mathrm{dom}\,(B)$. Since $\{U(t)\}_{t\geq 0}$ is a strongly continuous contraction semigroup, its Laplace transform (1.2.16) exists for the set $\{z : \mathrm{Re}\, z > 0\}$. Hence, $(-\infty, 0) \subset \rho(B)$. Moreover, by (1.4.16), we get $R_A(\zeta) = R_B(\zeta)$ for any $\zeta \in (-\infty, 0)$, which implies equality $A = B$ with $\mathrm{dom}\,(A) = \mathrm{dom}\,(B)$. □

For comments on the Yosida construction of C_0-semigroups using *approximations by semigroups* (Proposition 1.4.4), see Notes to Chap. 1 (Notes to Sect. 1.4).

1.5 Weak Continuity and Adjoint Semigroups

We discover a surprising result if instead of the strong (or the norm) continuity of a semigroup for $t \to +0$ we assume in Definition 1.2.1 (c) a continuity in the *weak* operator topology.

Recall that the set of bounded linear functionals $\mathfrak{X}^* := \mathcal{L}(\mathfrak{X}, \mathbb{C})$ on $\mathfrak{X}$ is a dual of the Banach space $\mathfrak{X}$. Note that $\mathfrak{X}^*$ is in turn a Banach space. By $\sigma(\mathfrak{X}, \mathfrak{X}^*)$, we denote the weak topology on $\mathfrak{X}$ generated by functionals $\phi \in \mathfrak{X}^*$. Suppose that a non-degenerate semigroup $\{U(t)\}_{t\geq 0}$ is *weakly* continuous in the sense that the mappings

$$\mathbb{R}_0^+ \ni t \mapsto \langle U(t)x, \phi\rangle \in \mathbb{C} \tag{1.5.1}$$

are continuous for each $x \in \mathfrak{X}$, $\phi \in \mathfrak{X}^*$, including the right continuity for $t \to +0$, cf. Definition 1.2.1. As it is indicated in Remark 1.2.2, a decisive for the semigroup definition and properties is topology of the right continuity at $t = 0$.

Definition 1.5.1 A semigroup $\{U(t)\}_{t\geq 0}$ on Banach space $\mathfrak{X}$ is called *weakly continuous* (or continuous in the *weak operator topology)* if instead of assumption *(c)* in Definition 1.2.1 the mapping: $t \mapsto U(t)x \in \mathfrak{X}$ is a *weakly* continuous vector-valued function of the parameter $t \in \mathbb{R}^+$ for every $x \in \mathfrak{X}$ and also weakly right-continuous at $t = 0$. We denote the latter as

$$(w_0) \qquad \text{w-}\lim_{t\to+0} U(t) = \mathbb{1}\,, \tag{1.5.2}$$

which is equivalent to $\lim_{t\to+0} \langle U(t)x, \phi\rangle = \langle x, \phi\rangle$ for each $x \in \mathfrak{X}$ and $\phi \in \mathfrak{X}^*$.

It is clear that the strong continuity in Definition 1.2.1 (c) implies (1.5.1) and (1.5.2). In a sense unexpected is that the opposite is also true.

We consider first the case of the Hilbert space setting. Then we get the following assertion:

Proposition 1.5.2 *Every weakly continuous semigroup $\{U(t)\}_{t\geq 0}$ on a Hilbert space $\mathfrak{H}$ is also strongly continuous, that is, semigroup satisfies conditions* (a), (b) *and* (c) *of Definition* 1.2.1.

Proof Seeing that by Proposition 1.2.3 the condition (c) is equivalent to condition (s_0) (1.2.1), it is sufficient to establish equivalence of conditions (w_0) and (s_0). Since the statement $(s_0) \Rightarrow (w_0)$ follows directly from definitions, it rests only to prove the assertion $(s_0) \Leftarrow (w_0)$.

Assume that the mapping $t \mapsto U(t)$ is *contraction* for $t \geq 0$ and satisfies condition (w_0). Then for any $u \in \mathfrak{H}$

$$\lim_{t\to+0} \|U(t)\,u - u\|^2 = \lim_{t\to 0} \{\|U(t)\,u\|^2 + \|u\|^2 - 2\,\mathrm{Re}\,(U(t)\,u, u)\} \leq$$
$$\lim_{t\to+0} \{2\,\|u\|^2 - 2\,\mathrm{Re}\,(U(t)\,u, u)\} = 0\,. \tag{1.5.3}$$

Hence, the mapping $t \mapsto U(t)$ also satisfies condition s_0. Consequently, a weakly continuous (Definition 1.5.1) *contraction* semigroup $\{U(t)\}_{t\geq 0}$ is also a C_0-semigroup.

Now, let $\{U(t)\}_{t\geq 0}$ satisfy conditions of Definition 1.2.1. Then following the arguments similar to that of Proposition 1.2.3 (1), one gets after application twice the *uniform boundedness principle* that semigroup $\{U(t)\}_{t\geq 0}$ is quasi-bounded:

$$\|U(t)\| \leq M\,\mathrm{e}^{\beta t}\,, \quad t \geq 0\,, \tag{1.5.4}$$

cf. (1.2.2). After *rescaling* (Remark 1.3.2), the semigroup $\{U^{(-\beta)}(t) = \mathrm{e}^{-\beta t}\,U(t)\}_{t\geq 0}$ is bounded and satisfies (w_0). Then, owing to Remark 1.3.3 on the Hilbert space $(\mathfrak{H}, \|\cdot\|)$, there is a norm $|||\cdot|||$, which is *equivalent* to the original norm $\|\cdot\| = (\cdot\,,\,\cdot)_{\mathfrak{H}}^{1/2}$, such that semigroup $\{U^{(-\beta)}(t)\}_{t\geq 0}$ is *contraction* on $(\mathfrak{H}, |||\cdot|||)$ and still satisfies (w_0).

Applying (1.5.3) to semigroup $\{U^{(-\beta)}(t)\}_{t\geq 0}$ on the space $(\mathfrak{H}, |||\cdot|||)$, we obtain that $\{U^{(-\beta)}(t)\}_{t\geq 0}$ is *strongly continuous* on $(\mathfrak{H}, |||\cdot|||)$ and by equivalence of norms also on the space $(\mathfrak{H}, \|\cdot\|)$. Since the *rescaling* does not affect continuity, the weakly continuous semigroup $\{U(t)\}_{t\geq 0}$ with property (w_0) is *strongly* continuous on $\mathfrak{H}$.

□

Seeing that adjoint semigroup $\{U^*(t)\}_{t\geq 0}$ is defined by $(u, U^*(t)\, v) := (U(t)\, u, v)$, for any $u, v \in \mathfrak{H}$, we infer the next assertion.

Corollary 1.5.3 *If semigroup $\{U(t)\}_{t\geq 0}$ on a Hilbert space $\mathfrak{H}$ is strongly continuous, then so is for adjoint semigroup $\{U^*(t)\}_{t\geq 0}$.*

Indeed, since $\{U(t)\}_{t\geq 0}$ is strongly continuous, the semigroup $\{U^*(t)\}_{t\geq 0}$ is weakly continuous by definition of adjoint operators: $U^*(t),\ t \geq 0$ and consequently strongly continuous by Proposition 1.5.2.

A similar to Proposition 1.5.2 (but less elementary) assertion also holds in Banach space $\mathfrak{X}$.

Proposition 1.5.4 *A semigroup $\{U_A(t)\}_{t\geq 0}$ on a Banach space $\mathfrak{X}$ is strongly continuous if and only if it is weakly continuous in the sense of Definition* 1.5.1.

Proof We set for $x \in \mathfrak{X}$

$$p_g[x] := \sup_{s\in[0,1]} |(U(s)x, g)|\,, \qquad g \in \mathcal{B}_1(\mathfrak{X}^*) := \{g \in \mathfrak{X}^* : \|g\| \leq 1\}.$$

Notice that for $x', x'' \in \mathfrak{X}$ one gets

$$p_g[x' + x''] \leq p_g[x'] + p_g[x''] \quad \text{and} \quad p_g[-x] = p_g[x],$$

$x \in \mathfrak{X}$ and $g \in \mathcal{B}_1(\mathfrak{X}^*)$. Since $\{U(s)\}_{s\in\mathbb{R}_0^+}$ is weakly continuous, one gets $p_g[x] < \infty$ for each $x \in \mathfrak{X}$ and $g \in \mathfrak{X}^*$. By the *uniform boundedness principle* , we find that the set $\{p_g[x]\}_{g\in\mathcal{B}_1(\mathfrak{X}^*)}$ is uniformly bounded for $x \in \mathcal{B}_1(\mathfrak{X}) := \{x \in \mathfrak{X} : \|x\| \leq 1\}$, i.e., there is a constant $C > 0$ such that $p_g[x] \leq C$ for $x \in \mathcal{B}_1(\mathfrak{X})$ and $g \in \mathcal{B}_1(\mathfrak{X}^*)$. Hence, $\|U(s)\| \leq C$ for $s \in [0, 1]$.

Let $\phi(\cdot) : \mathbb{R} \to \mathfrak{X}$ be a continuous function such that $\text{supp}(\phi) \subset (0, 1)$. We consider the element

$$f = \int_0^1 ds\, \phi(s)U(s)x\,, \quad x \in \mathfrak{X},$$

$$\phi \in \widetilde{C}([0, 1]) := \{\phi \in C([0, 1]) : \text{supp}(\phi) \subset (0, 1)\}.$$

The integral exists since $\|U(s)\| \leq C$ for $s \in [0, 1]$. We have

$$U(t)f = U(t) \int_0^1 ds\, \phi(s)U(s)x = \int_t^{1+t} ds\, \phi(s-t)U(s)x\,, \quad t \in \mathbb{R}_0^+.$$

Obviously, we have $\lim_{t\to+0} U(t)f = f$. Let us show that the set

$$\mathcal{D} := \left\{ f = \int_0^1 ds\, \phi(s)U(s)x : \phi \in \widetilde{C}([0, 1]), \quad x \in \mathfrak{X} \right\}$$

is dense in $\mathfrak{X}$. If $\mathcal{D}$ is not dense, then there is an element $h \in \mathfrak{X}^*$ such that for any $f \in \mathcal{D}$ one has $(f, h) = 0$. Hence,

$$0 = (f, h) = \int_0^1 ds\, \phi(s)(U(s)x, h)\,, \quad \phi \in \widetilde{C}([0, 1]).$$

Since $\phi(\cdot)$ is arbitrary, we get $(U(s)x, h) = 0$ for $s \in [0, 1]$. By the weak continuity, we immediately get $0 = (x, h)$ for $x \in \mathfrak{X}$. Hence, $h = 0$ which shows that $\mathcal{D}$ is dense in $\mathfrak{X}$. Using estimate $\|U(s)\| \leq C$ for $s \in [0, 1]$, we find that semigroup $\{U(s)\}_{s\in[0,1]}$ is strongly continuous at zero. □

Remark 1.5.5

(1) One should not confuse the *weak operator* topology on $\mathcal{L}(\mathfrak{X})$, Definition 1.5.1, with the *weak Banach space* topology $\sigma(\mathcal{L}(\mathfrak{X}), \mathcal{L}(\mathfrak{X})^*)$, on bounded operators $\mathcal{L}(\mathfrak{X})$ generated by the dual of $\mathcal{L}(\mathfrak{X})$ space $\mathcal{L}(\mathfrak{X})^*$, which is stronger.
(2) The strong operator and the weak Banach space topologies are stronger than the weak operator topology on $\mathcal{L}(\mathfrak{X})$, but the uniform boundedness principle is still valid in this topology on $\mathcal{L}(\mathfrak{X})$.

Note that the adjoint semigroup $\{U_A^*(t)\}_{t\geq 0}$ on the dual space $\mathfrak{X}^*$ is in general neither strongly continuous nor continuous in the weak Banach space topology $\sigma(\mathfrak{X}^*, \mathfrak{X}^{**})$. It is however always *weak*-continuous* in the sense that the mappings

$$t \mapsto \Gamma_{x,\phi}(t) := (x, U_A^*(t)\phi) = (U_A(t)x, \phi) \tag{1.5.5}$$

are continuous for all $x \in \mathfrak{X}$, $\phi \in \mathfrak{X}^*$. which means in the weak*-topology $\sigma(\mathfrak{X}^*, \mathfrak{X})$.

If the Banach space $\mathfrak{X}^*$ is *reflexive*, then the weak and weak*-topologies on it coincide. Consequently, by (1.5.5), the adjoint semigroup $\{U_A^*(t)\}_{t\geq 0}$ is weakly and, hence by Proposition 1.5.4, strongly continuous on $\mathfrak{X}^*$.

Note that this remark bolsters the statement of Proposition 1.5.2 since the Hilbert space $\mathfrak{H}$ is reflexive, cf. Remark 13.1.2.

1.6 Special Classes of Semigroups

The analysis in Sects. 1.4 and 1.5 shows that to produce a variety of *special classes* of semigroups one either modifies (relaxes or reinforces) in Definition 1.2.1 conditions (a) and (c), that is, conditions *at* the origin $t = 0$, or (for producing *subclasses* of C_0-semigroups) changes conditions *away* from the origin $t = 0$. Finally, one can also combine the both of them.

First, we consider a C_0-continuity *breaking*, which is of a different nature than modifications of the strong continuity at the origin $t = 0$ presented in Sect. 1.4 and in Sect. 1.5, cf. (a), (b) and (c) in Definition 1.2.1.

Definition 1.6.1 (Degenerate Semigroups) A family $\{U(t)\}_{t\geq 0}$ of linear-bounded operators on a Banach space $\mathfrak{X}$ is called a *degenerate* one-parameter semigroup if
(b') $U(t)U(s) = U(t+s)$ for $t, s \in \mathbb{R}^+$,
(c') $t \mapsto U(t)f \in \mathfrak{X}$ is a continuous function of parameter $t \in \mathbb{R}^+$ for every $f \in \mathfrak{X}$.

Proposition 1.6.2 *Let $\{U(t)\}_{t>0}$ be a degenerate semigroup on $\mathfrak{X}$ such that*

(i) $\sup_{0<t\leq 1} \|U(t)\| < \infty$.
(ii) *The strong operator limit* s-$\lim_{t\to +0} U(t) =: U(+0) \in \mathcal{L}(\mathfrak{X})$, *exists.*

Then it is called the continuous degenerate semigroup, and the following assertions hold :

(a) *The semigroup $\{U(t)\}_{t\geq 0}$ is quasi-bounded, that is, there are constants $M \geq 1$ and $\beta \in \mathbb{R}$ such that $\|U(t)\| \leq M\, e^{\beta t}$ for $t \geq 0$.*
(b) *The limit $U(+0)$ is a projection operator mapping $\mathfrak{X}$ onto subspace of the strong continuity for degenerate semigroup $\{U(t)\}_{t\geq 0}$:*

$$\mathfrak{X}_0 := \{f \in \mathfrak{X} : \lim_{t\to +0} U(t)f = f\} \subseteq \mathfrak{X}\,. \tag{1.6.1}$$

(c) *The subspace* (1.6.1) *is $U(t)$-invariant, and restriction $\{U(t) \upharpoonright \mathfrak{X}_0\}_{t\geq 0}$ defines on $\mathfrak{X}_0 \subseteq \mathfrak{X}$ a C_0-semigroup $\{U_0(t)\}_{t\geq 0}$, that is, $U_0(+0)\,\mathfrak{X}_0 = \mathfrak{X}_0$.*
   If $\mathfrak{X}_0 = \mathfrak{X}$, then $\{U_0(t)\}_{t\geq 0}$ is a C_0-semigroup on $\mathfrak{X}$. In this case, $U_0(+0) = U(+0) = \mathbb{1}$.
(d) *The subspace $\mathfrak{X}_0^c := (\mathbb{1} - U(+0))\,\mathfrak{X}$ is also $U(t)$-invariant and restriction of the continuous degenerate semigroup $\{U(t)\}_{t>0}$ onto $\mathfrak{X}_0^c$ is trivial : $\{U(t) \upharpoonright \mathfrak{X}_0^c = 0 \cdot \mathbb{1}\}_{t\geq 0}$.*

Proof

(a) The proof follows *verbatim* the proof for C_0-semigroups in Sect. 1.3.
(b) Because of (b') and (ii), one gets that

$$\begin{aligned} U(+0) &= \text{s-}\lim_{t_1,\, t_2\to +0} U(t_1 + t_2) \\ &= \text{s-}\lim_{t_1\to +0} U(t_1)\ \text{s-}\lim_{t_2\to +0} U(t_2) = (U(+0))^2\,. \end{aligned}$$

As a consequence, $U(+0)$ is a projection operator and the subspace $U(+0)\,\mathfrak{X}$ coincides with $\mathfrak{X}_0$ defined in (1.6.1).

(c) If $f \in \mathfrak{X}_0$, then by (b') and (ii), we obtain for $s \geq 0$

$$U(+0)(U(s)\,f) = \text{s-}\lim_{t\to+0} U(t)\,(U(s)\,f) = \text{s-}\lim_{t\to+0} U(t+s)\,f =$$
$$U(s)\;\text{s-}\lim_{t\to+0} U(t)\,f = U(s)\,f\,.$$

So, by (ii) and definition (1.6.1), one infers that $U(s)\,f \in \mathfrak{X}_0$. The invariant subspace $\mathfrak{X}_0$ is called the *space of strong continuity* for degenerate semigroup since its restriction onto $\mathfrak{X}_0 : \{U_0(t) := U(t) \restriction \mathfrak{X}_0\}_{t\geq 0}$ is strongly continuous for $t \geq 0$. Then owing to (1.6.1) $\{U_0(t)\}_{t\geq 0}$ is a C_0-semigroup on $\mathfrak{X}_0$.
If $\mathfrak{X}_0 = \mathfrak{X}$, then on account of (1.6.1), $\{U(t)\}_{t\geq 0}$ is a C_0-semigroup on $\mathfrak{X}$.

(d) The proof follows from (c) and $\mathfrak{X}_0 \cap \mathfrak{X}_0^c = U(+0)\,(\mathbb{1} - U(+0))\,\mathfrak{X} = \{0\}$, which yields the $U(t)$-invariance and triviality: $\{U(t) \restriction \mathfrak{X}_0^c = 0 \cdot \mathbb{1}\}_{t\geq 0}$.

□

In Definition 1.6.1 of degenerate semigroups the condition, Definition 1.2.1 (a), for C_0-continuity at the origin $t = 0$, namely, $U(t = 0) = \mathbb{1}$, is completely missed. Note that in Proposition 1.6.2 we consider the case of the strongly *continuous* degenerate semigroup $\{U(t)\}_{t\geq 0}$, where breaking of the C_0-continuity at the origin $t = 0$ is expressed by (a *mild*) condition $\text{s-lim}_{t\to+0}\, U(t) = U(+0) \neq \mathbb{1}$.

Corollary 1.6.3 *If $P_0 \neq \mathbb{1}$ is a projection operator mapping $\mathfrak{X}$ onto the subspace $\mathfrak{X}_0$ and $\{S(t)\}_{t\geq 0}$ is a quasi-bounded C_0-semigroup on $\mathfrak{X}$, then the families*

$$U_+(t) := P_0\,S(t) \;\text{ and }\; U_-(t) := S(t)\,P_0\,, \quad t \geq 0\,, \tag{1.6.2}$$

with composition laws, respectively,: $U_+(t_1) \circ U_+(t_2) = U_+(t_1 + t_2) = P_0\,S(t_1 + t_2)$ and $U_-(t_1) \circ U_-(t_2) = U_-(t_1 + t_2) = S(t_1 + t_2)\,P_0$, are continuous degenerate (quasi-bounded) semigroups $\{U_\pm(t)\}_{t>0}$ on $\mathfrak{X}$ such that $U_\pm(+0) = P_0$.

On account of Corollary 1.6.3, the simplest non-trivial example of degenerate (contraction) semigroup on $\mathfrak{X}$ is the *constant* degenerate semigroup $\{U(t) := P_0\}_{t\geq 0}$, where P_0 is a projection operator mapping $\mathfrak{X}$ onto $\mathfrak{X}_0 = P_0\,\mathfrak{X} \subset \mathfrak{X}$.

Remark 1.6.4 We evoke that the limits of the Trotter-Kato product formulae (away from zero) are often *degenerate* semigroups. For example, in Chaps. 5 and 8, one discovers (continuous) degenerate semigroups on Hilbert space $\mathfrak{H}$ in the following contexts :

(1) Let $P_0 : \mathfrak{H} \to \mathfrak{H}_0$ be an orthogonal projection from $\mathfrak{H}$ onto $\mathfrak{H}_0$, and let $\{e^{-t\,H}\}_{t\geq 0}$ be a (quasi-bounded) C_0-semigroup on the subspace $\mathfrak{H}_0$ with domain of generator $\operatorname{dom}(H) \subset \mathfrak{H}_0$. Then by Definition 1.6.1, the bounded operator family $\{W(t) := \mathrm{e}^{-t\,\widetilde{H}}\,P_0\}_{t\geq 0}$ for $\widetilde{H} := P_0\,H\,P_0$ is a continuous *degenerate* semigroup on $\mathfrak{H}$ for $W(+0) = P_0$ and with the subspace $\mathfrak{H}_0$ as the *space of strong continuity* for $\{W(t)\}_{t\geq 0}$, cf. (1.6.1) and Corollary 1.6.3.

(2) In Chap. 14, we consider Trotter product formulæ that involve the *constant* degenerate *semigroup* $\{U(t) := P\}_{t\in\mathbb{R}}$, for mapping $\mathfrak{H}$ onto $\mathfrak{H}_0 = P\,\mathfrak{H} \subset \mathfrak{H}$ and a unitary *group* $\{\,e^{-itC}\,\}_{t\in\mathbb{R}}$. Namely, they are product formulæ of the type

$$\left(P\,e^{-itC/n}\,P\right)^n, \quad n \in \mathbb{N}, \tag{1.6.3}$$

where C is a self-adjoint operator with dom $(C) \subseteq \mathfrak{H}$.

For more comments and details, see Notes to Chap. 1 (Notes to Sect. 1.6).

After presented above (in Sects. 1.4 and 1.5) penultimate versions of *continuity* at $t = +0$ and after the last example of a kind of *discontinuity* $U(+0) \neq \mathbb{1}$, for continuous degenerate semigroups at the origin $t = 0$, yet another source of variety of semigroups is possible and related to their *continuity* for $t \in \mathbb{R}^+$.

For instance, one of the most important classes is due to a *holomorphic* extension of semigroups of *bounded* operators from $\mathbb{R}^+$ to subsets of the right half-plane $\mathbb{C}_+$ (see Sects. 1.8 and 1.9). Besides that, other examples arise from different *constraints* imposed on the values $\{U(t)\}_{t>0}$ of a semigroup. This can be a topology of the *image* of mappings, $t \mapsto U(t)$, *away* from zero. The extreme example is the value of operators $U(t)$ for $t > 0$ in the case of the *nilpotent* semigroups (see Chaps. 11 and 12).

Example 1.6.5 (Self-adjoint C_0-semigroups) Let $A \geq 0$ be a non-negative self-adjoint operator in a Hilbert space $\mathfrak{H}$. Then it satisfies conditions of Proposition 1.1.1 and consequently (cf. Proposition 1.2.7) is generator of a contraction C_0-semigroup $\{U_A(t) = e^{-tA}\}_{t\geq 0}$, which is in turn a family of bounded self-adjoint operators. That is, $t \mapsto U_A(t) \in \mathcal{L}^{sa}(\mathfrak{H}) \subset \mathcal{L}(\mathfrak{H})$.

Seeing that the *spectral theorem* for A provides a canonical *functional calculus* for (unbounded) self-adjoint operators, we obtain representation:

$$U_A(t) = \int_{\mathbb{R}_0^+} E_A(d\lambda)\, e^{-t\lambda}. \tag{1.6.4}$$

Here, $E_A(d\lambda)$ is a unique projection-valued *spectral measure*, associated with operator A, such that it has in turn the spectral representation: $A = \int_{\mathbb{R}_0^+} E_A(d\lambda)\,\lambda$.

Now, let $t > 0$ and $t + \sigma > 0$. Then on account of (1.6.4),

$$\begin{aligned} \|U_A(t+\sigma) - U_A(t)\| &= \left\| \int_{\mathbb{R}_0^+} E_A(d\lambda)\, e^{-t\lambda}(e^{-\sigma\lambda} - 1) \right\| \leq \\ \sup_{\lambda\in\mathbb{R}_0^+} |\, e^{-t\lambda}(e^{-\sigma\lambda} - 1)\,| &= \frac{\sigma}{t+\sigma}\left(1 + \frac{\sigma}{t}\right)^{-t/\sigma}, \end{aligned} \tag{1.6.5}$$

where for the first inequality in (1.6.5) we used that the spectral family $\{E_A(\lambda)\}_{\lambda\geq 0}$ is a *resolution of the identity*.

The estimate (1.6.5) yields that a *strongly* continuous semigroup $\{U_A(t)\}_{t\geq 0}$ with values in the subset $\mathcal{L}^{sa}(\mathfrak{H})$ of self-adjoint operators is (immediately) *operator-norm* continuous for $t > 0$.

The following examples show that a stronger regularity of the image of the mapping $t \mapsto U(t)$, *away* from zero, may also provide a stronger topology of *continuity* of the C_0-semigroup. Besides that, this continuity may start *eventually* with delay, $t \geq t_0$, for some $t_0 > 0$.

To enhance a topology of the image of mappings (to improve the regularity), one can suppose, for example, that a C_0-semigroup $\{U_A(t)\}_{t\geq 0}$, for generator A, is a subset of *compact* operators $\mathcal{C}_\infty(\mathfrak{H})$ on a Hilbert space $\mathfrak{H}$, as long as $t > 0$. So, $\{U_A(t)\}_{t\geq 0}$ is the *immediately* compact C_0-semigroup. Then similar to the self-adjoint case (Example 1.6.5), the strong continuity of $\{U_A(t)\}_{t\geq 0}$ is *uplifted* to the operator-norm continuity, although now it might happen with delay ($t \geq t_0$) for some $t_0 > 0$.

Proposition 1.6.6 *Let $\{U_A(t)\}_{t\geq 0}$ be a (quasi-bounded) C_0-semigroup such that operator $U_A(t_0)$ is compact for some $t_0 > 0$. Then $\{U_A(t)\}_{t\geq t_0} \subset \mathcal{C}_\infty(\mathfrak{H})$ and the mapping $t \mapsto U_A(t)$ is operator-norm continuous in (t_0, ∞), including the operator-norm right continuity at $t = t_0 + 0$.*

Correspondingly, any immediately compact C_0-semigroup $\{U_A(t)\}_{t\geq 0}$ is (immediately) operator-norm continuous for $t > 0$.

Proof Since $\mathcal{C}_\infty(\mathfrak{H})$ is a two-sided $*$-*ideal* in $\mathcal{L}(\mathfrak{H})$, by virtue of the semigroup law, one infers that $U_A(t_0)\,U_A(\tau) = U_A(t_0+\tau) \in \mathcal{C}_\infty(\mathfrak{H})$, for $\tau \geq 0$. This proves the first part of the claim: $\{U_A(t)\}_{t\geq t_0} \subset \mathcal{C}_\infty(\mathfrak{H})$.

Now, let $B_1 \subset \mathfrak{H}$ be the unit ball in the Hilbert space $\mathfrak{H}$, and let

$$\Delta(\tau) := \|U_A(t_0+\tau) - U_A(t_0)\| = \sup_{u\in B_1} \|\,(U_A(\tau) - \mathbb{1})\,U_A(t_0)\,u\,\|\,, \quad \tau \geq 0\,. \tag{1.6.6}$$

Suppose that $\lim_{\tau\to 0} \Delta(\tau) > \varepsilon > 0$. Then on account of (1.6.6), there exist $\tau_n \to 0$ $(n \to \infty)$ and a sequence $\{u_n\}_{n\geq 1} \subset B_1$ such that

$$\varepsilon < \|\,(U_A(\tau_n) - \mathbb{1})\,U_A(t_0)\,u_n\,\|\,, \quad n \geq 1\,. \tag{1.6.7}$$

Since operator $U_A(t_0)$ is compact, the sequence $\{U_A(t_0)\,u_n\}_{n\geq 1}$ contains a convergent subsequence: $\lim_{n_k\to\infty} U_A(t_0)\,u_{n_k} =: w$. As a consequence, the estimate (1.6.7) yields

$$\begin{aligned}\varepsilon < \|\,(U_A(\tau_n) - \mathbb{1})\,U_A(t_0)\,u_n\,\| \leq \|\,(U_A(\tau_{n_k}) - \mathbb{1})\,w\,\| + \\ +\|(U_A(\tau_{n_k}) - \mathbb{1})\,(U_A(t_0)\,u_{n_k} - w)\,\|\,, \quad n_k \geq 1\,.\end{aligned} \tag{1.6.8}$$

Owing to quasi-boundedness of the semigroup, one gets $\|(U_A(\tau_{n_k}) - \mathbb{1})\| \leq M\,\mathrm{e}^{\beta\,t_0} + 1$. Then (1.6.8) contradicts for $\tau_n \to 0$ to the strong continuity of

$\{U_A(t)\}_{t\geq 0}$ and to convergence to the limit w. Therefore, by virtue of (1.6.6), the semigroup $\{U_A(t)\}_{t\geq 0}$ is operator-norm continuous at $t = t_0 + 0$. Since by the first part of the claim this semigroup is compact for any $t > t_0$, we obtain the proof of the second part of the claim for all $t \geq t_0$ *verbatim* the arguments for $t = t_0 + 0$.

Seeing that immediately compact C_0-semigroup $\{U_A(t)\}_{t\geq 0}$ is such that $U_A(t) \in \mathcal{C}_\infty(\mathfrak{H})$ for any $t \in \mathbb{R}^+$, one gets the proof of assertion by the same line of reasoning as our arguments above for $t \geq t_0$. □

We note that although it seems plausible the relationship between *compactness* of a C_0-semigroup with generator A and *compactness* of resolvent $\{R_A(z)\}_{z\in\rho(A)}$ is not straightforward. For example, a resolvent corresponding to (eventually) compact C_0-semigroup is not necessarily compact, cf. Example 1.10.3 in Notes to Chap. 1 (Notes to Sect. 1.6).

The following theorem of *characterisation* the compact semigroups serves to elucidate this relationship:

Proposition 1.6.7 *A C_0-semigroup $\{U_A(t)\}_{t\geq 0}$ on $\mathfrak{H}$ with generator A is immediately compact if and only if*

(*i*) *The mapping $t \mapsto U_A(t) \in \mathcal{L}(\mathfrak{H})$ is operator-norm continuous in $\mathbb{R}^+$*
(*ii*) *The resolvent $R_A(z) = (A - z\mathbb{1})^{-1}$ is compact for some (and thus for any) $z \in \rho(A)$.*

Proof The proof follows *mutatis mutandis* the arguments in the proof of a similar assertion for the Gibbs semigroups (see Definition 1.6.8 and Theorem 1.6.14). Note that, in contrast to the Gibbs semigroups, for the case of compact semigroups the problem of invertibility of generator A can be easily handled by means of the shift, $A + \alpha\mathbb{1}$, for $\alpha \in \mathbb{R}$. □

The next example concerns the C_0-semigroups such that $\{U_A(t)\}_{t\geq t_0} \subset \mathcal{C}_1(\mathfrak{H})$, where $t_0 > 0$ and $\mathcal{C}_1(\mathfrak{H}) \subset \mathcal{C}_\infty(\mathfrak{H})$ is a two-sided $*$-*ideal* in the algebra of *compact* operators $\mathcal{C}_\infty(\mathfrak{H}) \subset \mathcal{L}(\mathfrak{H})$. This ideal is endowed with the norm topology:

$$\|K\|_p := \sum_{n=1}^{\infty} \lambda_n^p(\sqrt{K^*K}), \quad K \in \mathcal{C}_p(\mathfrak{H}), \ p \geq 1, \tag{1.6.9}$$

for $p = 1$. Here, $\{\lambda_n(C)\}_{n\geq 1}$ denotes eigenvalues of compact operator $C \in \mathcal{C}_\infty(\mathfrak{H})$. We recall that for any $p \geq 1$ the $*$-*ideal* $\mathcal{C}_p(\mathfrak{H})$ is a Banach space with the norm (1.6.9) and that $\mathcal{C}_1(\mathfrak{H})$ are the *trace-class* operators.

Definition 1.6.8 A strongly continuous semigroup $\{G(t)\}_{t\geq 0}$ on a Hilbert space $\mathfrak{H}$ is called *immediately* Gibbs semigroup (or, for simplicity, Gibbs semigroup) if for *positive* $t \mapsto G(t) \in \mathfrak{H}_1(\mathfrak{H})$. We call a strongly continuous semigroup $\{G(t)\}_{t\geq 0}$ the *eventually* Gibbs semigroup if there exists a threshold $t_0 > 0$ such that $t \mapsto G(t) \in \mathcal{C}_1(\mathfrak{H})$ for $t \geq t_0$.

A semigroup $\{G(t)\}_{t\geq 0}$ is called the self-adjoint Gibbs semigroup if $G^*(t) = G(t)$ for $t \in \mathbb{R}_0^+$.

Example 1.6.9 Let $\{e_n\}_{n\geq 1}$ be an orthonormal basis in the Hilbert space $\mathfrak{H}$. Then the one-parameter family of operators $\{G(t)\}_{t\geq 0}$ defined for $u \in \mathfrak{H}$ by mappings

$$u \mapsto G(t)u := \sum_{n=1}^{\infty} \mathrm{e}^{-t\,\ln(n)}\,(u,\, e_n)\, e_n\,, \quad t \geq 0\,, \tag{1.6.10}$$

is a self-adjoint strongly continuous contraction semigroup on $\mathfrak{H}$ with unbounded self-adjoint generator $Au := \sum_{n=1}^{\infty} \ln(n)\,(u,\, e_n)\, e_n$.

By definition (1.6.10), the eigenvalues $\{\lambda_n(G(t)) = 1/(n+1)^t\}_{n\geq 1}$ and, consequently, the C_0-semigroup (1.6.10) are immediately compact: $\{G(t)\}_{t>0} \subset \mathcal{C}_\infty(\mathfrak{H})$. In addition, by (1.6.9) and (1.6.10), we infer that

$$\|G(t)\|_1 = \sum_{n=1}^{\infty} \frac{1}{n^t} < \infty\,, \quad \text{for} \quad t > 1\,. \tag{1.6.11}$$

Then by Definition 1.6.8 and by (1.6.10), (1.6.11), the immediately compact C_0-semigroup (1.6.10) is eventually Gibbs semigroup because $\{G(t)\}_{t>0} \subset \mathcal{C}_1(\mathfrak{H})$ for $t > t_0$, where $t_0 = 1$.

For analysis of continuity of C_0-semigroups with values in $\mathcal{C}_1$ or $\mathcal{C}_\infty$, one needs the following statement about *continuity of multiplication* on these ideals:

Proposition 1.6.10 *Suppose that* s-$\lim_{k\to\infty} B_k = B \in \mathcal{L}(\mathfrak{H})$ *and that* $\|\cdot\|_p -$ $\lim_{k\to\infty} C_k = C \in \mathcal{C}_p(\mathfrak{H})$ *for* $1 \leq p \leq \infty$. *Then*

$$\|\cdot\|_p - \lim_{k\to\infty} B_k C_k = B\,C\,. \tag{1.6.12}$$

If, in addition, s-$\lim_{k\to\infty} B_k^* = B^*$, *then also*

$$\|\cdot\|_p - \lim_{k\to\infty} C_k B_k = C\,B\,. \tag{1.6.13}$$

Note that here $\mathcal{C}_{p=\infty}(\mathfrak{H})$ *is the set of compact operators and* $\|\cdot\|_{p=\infty}$ *coincides with the operator-norm* $\|\cdot\|$ *on* $\mathcal{L}(\mathfrak{H})$.

Proof Writing

$$\|B_k C_k - B\,C\|_p \leq \|B_k\|\,\|C_k - C\|_p + \|(B_k - B)\,C\|_p \tag{1.6.14}$$

and using the uniform boundedness principle

$$\sup_{k\geq 1} \|B_k\| \leq L < \infty,$$

for a strongly convergent sequence $\{B_k\}_{k\geq 1}$, we see that it suffices to show that in (1.6.14) $\lim_{k\to\infty} \|(B_k - B)\,C\|_p = 0$.

Since $C \in \mathcal{C}_p(\mathfrak{H})$, where $1 \le p \le \infty$, we can find for C and any $\varepsilon > 0$ a finite-rank orthogonal projection $P \in \mathcal{K}(\mathfrak{H})$ such that $\|(\mathbb{1} - P)\, C\|_p < \varepsilon$. Then the last term in (1.6.14) can be estimated as

$$\|(B_k - B)\, C\|_p \le \|(B_k - B)\, P\| \, \|C\|_p + (\|B_k\| + \|B\|)\, \|(\mathbb{1} - P)\, C\|_p \,. \tag{1.6.15}$$

Then, since the strong convergence of $\{B_k\}_{k\ge 1}$ implies $\lim_{k\to\infty} \|(B_k - B)\, P\| = 0$, (1.6.15) gives

$$\limsup_{k\to\infty} \|(B_k - B)\, C\|_p \le (L + \|B\|)\, \varepsilon \,.$$

Since ε is arbitrary, the result (1.6.12) is proven owing to (1.6.14).

The proof of (1.6.13) follows along the same line of reasoning, with the property $\|B_k^*\|_p = \|B_k\|_p$ of the norm $\|\cdot\|_p$ $(1 \le p \le \infty)$, which is taken into account. □

Remark 1.6.11 (Strong Continuity of Multiplication) Suppose that $\text{s-}\lim_{k\to\infty} B_k = B \in \mathcal{L}(\mathfrak{H})$ and $\text{s-}\lim_{k\to\infty} C_k = C \in \mathcal{L}(\mathfrak{H})$. Then

$$\text{s-}\lim_{k\to\infty} B_k\, C_k = B\, C \quad \text{and} \quad \text{s-}\lim_{k\to\infty} C_k\, B_k = C\, B \,. \tag{1.6.16}$$

The proof (similar to that in Proposition 1.6.10) is based on the *uniform boundedness principle* for strongly convergent sequences of bounded operators.

Corollary 1.6.12

(a) *By Definition* 1.6.8 *and Proposition* 1.6.10, *we conclude that an immediately Gibbs semigroup* $\{G(t)\}_{t\ge 0}$ *is immediately trace-norm continuous for* $t > 0$, *whereas an eventually Gibbs semigroup (with threshold* $t_0 > 0$*) is trace-norm continuous only for* $t > t_0$ *(or* $t \ge t_0$*).*

(b) *Similarly, an immediately compact* C_0*-semigroup* $\{U(t)\}_{t\ge 0}$ *is immediately operator-norm continuous for* $t > 0$, *whereas an eventually compact* C_0*-semigroup (with a threshold* $t_0 > 0$*) is* $\|\cdot\|$*-continuous only for* $t > t_0$ *(or* $t \ge t_0$*).*

Proof

(a) Let threshold $t_0 > 0$. Then for any $t \ge t_0$, there is $\delta \in \mathbb{R}$ such that $(t - t_0)/2 + \delta > 0$. By the semigroup law, it follows that $G(t+\delta) = G((t-t_0)/2+\delta)\, G((t+t_0)/2)$, where $G((t - t_0)/2 + \delta) \in \mathcal{L}(\mathfrak{H})$ and $G((t + t_0)/2) \in \mathcal{C}_1(\mathfrak{H})$. Since the semigroup $\{G(t)\}_{t\ge 0}$ is strongly continuous, $\text{s-}\lim_{\delta\to 0} G((t - t_0)/2 + \delta) = G((t - t_0)/2)$. Hence, for $t \ge t_0$, one infers the trace-norm limit,

$$\|\cdot\|_1 - \lim_{\delta\to 0} G(t+\delta) = \|\cdot\|_1 - \lim_{\delta\to 0} (G((t - t_0)/2 + \delta)\, G((t + t_0)/2) =$$
$$\left(\text{s-}\lim_{\delta\to 0} G((t - t_0)/2 + \delta)\right) G((t + t_0)/2) = G(t) \,, \tag{1.6.17}$$

by virtue of the $\|\cdot\|_1$-continuity of multiplication in $\mathcal{C}_1(\mathfrak{H})$, Proposition 1.6.10.
If $t_0 = 0$, then proof of the immediate trace-norm continuity of the Gibbs semigroup $\{G(t)\}_{t\geq 0}$ for $t > 0$ is *verbatim* the above arguments and (1.6.17).

(b) The proof follows *mutatis mutandis* the line of reasoning in (a) for C_0-semigroup $\{U(t)\}_{t\geq 0}$ and for the operator-norm $\|\cdot\|$ in (1.6.17).

□

Next, we prove the extension of Proposition 1.6.7 for compact semigroups to the case of the trace-class ideal $\mathcal{C}_1(\mathfrak{H})$, that is, a *characterisation* theorem for the Gibbs semigroups. To this aim, we first remind a fundamental statement, which characterises the Bochner integrable functions.

Proposition 1.6.13 (Bochner Theorem) *Let $I \subset \mathbb{R}$ be an interval (bounded or unbounded) in $\mathbb{R}$ and $\mathfrak{X}$ be a Banach space. A vector-valued function $f : I \to \mathfrak{X}$ is Bochner integrable if and only if f is $\|\cdot\|_{\mathfrak{X}}$-measurable and the function $t \mapsto \|f(t)\|_{\mathfrak{X}}$ is Lebesgue integrable. Moreover, when f is $\|\cdot\|_{\mathfrak{X}}$-Bochner integrable, one has the estimate*

$$\left\| \int_I \mathrm{d}t\, f(t) \right\|_{\mathfrak{X}} \leq \int_I \mathrm{d}t\, \|f(t)\|_{\mathfrak{X}} . \tag{1.6.18}$$

In particular, this function f is measurable in the topology of $\mathfrak{X}$.

Theorem 1.6.14 *Let $\{\mathrm{e}^{-tA}\}_{t\geq 0}$ be a C_0-semigroup on Hilbert space $\mathfrak{H}$ with invertible generator A, that is,* $\ker(A) = \{0\}$. *If $p \geq 1$, then the following assertions are equivalent:*

(a) The C_0-semigroup $\{\mathrm{e}^{-tA}\}_{t\geq 0}$ is immediately Gibbs semigroup $\{G_A(t)\}_{t\geq 0}$, with asymptotic $\|G_A(t)\|_1 = O(t^{-p})$ for $t \downarrow 0$.

(b) For any $q \in \mathbb{N}$, such that $q > p$,

 (i) The mapping $\mathbb{R}^+ \ni t \mapsto \mathrm{e}^{-tA}$ is $\|\cdot\|_q$-continuous.

 (ii) The resolvent family $\{R_A(z)\}_{z\in\rho(A)}$ is defined using the Laplace transform of $\{\mathrm{e}^{-tA}\}_{t\geq 0}$ (cf. (1.2.13), (1.2.15)) *by means of $\|\cdot\|_q$-Bochner integral and $(R_A(z))^q \in \mathcal{C}_1(\mathfrak{H})$.*

Proof *(a)* $\Longrightarrow$ *(b)*: Suppose that C_0-semigroup $\{e^{-tA}\}_{t\geq 0}$ is an immediately Gibbs semigroup. Then it is trace-norm continuous for $t \in \mathbb{R}^+$ and thus immediately continuous in topology of the ideal $\mathcal{C}_q(\mathfrak{H}) \supset \mathcal{C}_1(\mathfrak{H})$ for $q > p \geq 1$. This proves (i).

To prove (ii), we note that Gibbs semigroup $\{\mathrm{e}^{-tA}\}_{t\geq 0}$ is strongly continuous, quasi-bounded (Definition 1.3.1), and compact (for $t > 0$) since $\mathcal{C}_1(\mathfrak{H}) \subset \mathcal{C}_\infty(\mathfrak{H})$. Consequently, there exists β such that $(-\infty, -\beta) \subset \rho(A)$, and choosing $\lambda > \beta$ (i.e., $-\lambda \in \rho(A)$), we obtain $\|\mathrm{e}^{-\lambda t}\,\mathrm{e}^{-tA}\| \leq M\,\mathrm{e}^{(\beta-\lambda)t}$, cf. (1.3.1). Then by virtue of Proposition 1.6.13, the resolvent of generator A at $z = -\lambda$ has the operator-norm Bochner integral representation (*Laplace* transform),

$$R_A(-\lambda) = \int_0^\infty \mathrm{d}t\, \mathrm{e}^{-\lambda t}\,\mathrm{e}^{-tA}, \quad \beta < \lambda , \tag{1.6.19}$$

which by compactness of the family $\{e^{-tA}\}_{t>0}$ and by construction of the integral yields in the left-hand side of (1.6.19) a compact operator. On account of the resolvent identity, the fact $R_A(-\lambda) \in \mathcal{C}_\infty(\mathfrak{H})$ for some $-\lambda \in \rho(A)$ implies that $R_A(z)$ is compact for any $z \in \rho(A)$.

Due to representation (1.6.19) and the *Fubini* theorem, we infer that powers $q \in \mathbb{N}$ of the resolvent have representation,

$$(R_A(z))^q = \Gamma(q)^{-1} \int_0^\infty dt\, t^{q-1}\, e^{-t\,(A-z\mathbb{1})}\,, \quad \beta < \mathrm{Re}\,(-z)\,, \tag{1.6.20}$$

which similar to (1.6.19) is the operator-norm convergent Bochner integral. Note that the Gibbs semigroup is $\|\cdot\|_1$-continuous (so $\|\cdot\|_1$-measurable) on a closed $I \subset \mathbb{R}^+$, cf. Proposition 1.6.13. Next, by conditions $q > p$ and $\|e^{-tA}\|_1 = O(t^{-p})$ for $t \downarrow 0$, the function $\|t^{q-1}\, e^{-t\,(A-z\,\mathbb{1})}\|_1$ is integrable in the (positive) vicinity of zero. Moreover, by estimate,

$$\|e^{-t\,(A-z\,\mathbb{1})}\|_1 \le \|e^{-T\,(A-z\,\mathbb{1})}\|_1\, M\, e^{(t-T)\,(\beta+\mathrm{Re}\,z)} \le M_{z,T}\, e^{-t\,|\beta+\mathrm{Re}\,z\,|}\,,$$
$$\mathrm{Re}\,z < -\beta\,,$$

for $t > T$, where $T > 0$, this function is also integrable at infinity. Hence, the right-hand side of (1.6.20) is the $\|\cdot\|_1$-Bochner integral, and by Proposition 1.6.13 for $\mathfrak{X} = \mathcal{C}_1(\mathfrak{H})$, together with (1.6.18), one obtains estimate

$$\|(R_A(z))^q\|_1 \le \Gamma(q)^{-1} \int_0^\infty dt\, t^{q-1}\, \|e^{-tA}\|_1\, e^{t\,\mathrm{Re}\,(z)} < \infty\,, \quad \beta < \mathrm{Re}\,(-z)\,. \tag{1.6.21}$$

This proves that $(R_A(z))^q \in \mathcal{C}_1(\mathfrak{H})$, for $z \in \rho(A)$, and hence the assertion (ii).

$(b) \implies (a)$: The converse is much harder since we have to construct a Gibbs semigroup $\{G_A(t)\}_{t\ge 0}$ starting essentially from condition (ii) for resolvent, cf. (1.6.20).

By condition of theorem, the set $\{U_A(t) := e^{-tA}\}_{t\ge 0}$ is a (quasi-bounded) C_0-semigroup, which by virtue of (i) is also $\mathcal{C}_q(\mathfrak{H})$-continuous in $\mathbb{R}^+$. Hence, there exists β such that $(-\infty, -\beta) \subset \rho(A)$ and $\{e^{-tA}\}_{t\in\mathbb{R}^+}$ is by $\mathcal{C}_q(\mathfrak{H}) \subset \mathcal{L}(\mathfrak{H})$ the operator-norm continuous *family* of operators. To proceed, we fix $\lambda > \beta$ and define by the operator-norm Bochner integral (Proposition 1.6.13 for $\mathfrak{X} = \mathcal{L}(\mathfrak{H})$) a family $\{F(t)\}_{t\ge 0}$ of bounded operators:

$$F(t) := \int_0^t d\tau\, e^{-\lambda\tau}\, U_A(\tau)\,, \quad t \ge 0\,. \tag{1.6.22}$$

Now, integrating equation

$$\partial_t (e^{-\lambda t}\, U_A(t)\, u) = -\, e^{-\lambda t}\, U_A(t)\, (\lambda \mathbb{1} + A)\, u\,, \quad u \in \mathrm{dom}\,(A)\,,$$

we obtain

$$(\mathbb{1} - \mathrm{e}^{-\lambda t} U_A(t))\, u = \int_0^t \mathrm{d}\tau\, \mathrm{e}^{-\lambda \tau} U_A(\tau)\, (\lambda\, \mathbb{1} + A)\, u$$

or, for $w := (\lambda \mathbb{1} + A)\, u$, the equation

$$(\mathbb{1} - e^{-\lambda t} U_A(t))\, (A + \lambda \mathbb{1})^{-1}\, w = \int_0^t \mathrm{d}\tau\, e^{-\lambda \tau} U_A(\tau)\, w\,. \tag{1.6.23}$$

By virtue of $-\lambda \in \rho(A)$, the range ran $(\lambda\, \mathbb{1} + A) = \mathfrak{H}$. Therefore, Eq. (1.6.23) yields that the family of operators (1.6.22) belongs to the ideal $\mathcal{C}_q(\mathfrak{H})$,

$$F(t) = (\mathbb{1} - \mathrm{e}^{-\lambda t} U_A(t))\, (A + \lambda\, \mathbb{1})^{-1} \in \mathcal{C}_q(\mathfrak{H})\,, \quad t \geq 0\,, \tag{1.6.24}$$

because by condition (ii) the resolvent $R_{-\lambda}(A) \in \mathcal{C}_q(\mathfrak{H})$ and $F(0) = 0 \cdot \mathbb{1}$. As a consequence,

$$F(t + \delta) - F(t) = \int_t^{t+\delta} \mathrm{d}\tau\, e^{-\lambda \tau} U_A(\tau) \in \mathcal{C}_q(\mathfrak{H})\,, \quad t > 0\,,\ \delta > 0\,. \tag{1.6.25}$$

On account of condition (i), the $\|\cdot\|_q$-continuity of the family $\{U_A(\tau)\}_{\tau > 0}$ yields, for $t > 0$ and $\delta > 0$, the estimate

$$\left\| \frac{1}{\delta} \int_t^{t+\delta} \mathrm{d}\tau\, U_A(\tau) - U_A(t) \right\|_q \leq \frac{1}{\delta} \int_t^{t+\delta} \mathrm{d}\tau\, \|U_A(\tau) - U_A(t)\,\|_q \leq \varepsilon_t(\delta)\,, \tag{1.6.26}$$

where $\varepsilon_t(\delta) := \sup_{\tau \in [t, t+\delta]} \|U_A(\tau) - U_A(t)\|_q$ and $\lim_{\delta \downarrow 0} \varepsilon_t(\delta) = 0$. Therefore, (1.6.26) implies that

$$U_A(t) = \|\cdot\|_q - \lim_{\delta \downarrow 0} \frac{1}{\delta} \int_t^{t+\delta} \mathrm{d}\tau\, U_A(\tau)\,, \quad t > 0\,. \tag{1.6.27}$$

Note that by invertibility of generator A (ker $(A) = \{0\}$) and by virtue of (1.6.24) and (1.6.25) for $\lambda = 0$ the integral in (1.6.27) coincides with operator $F(t + \delta) - F(t)$, which belongs to the ideal $\mathcal{C}_q(\mathfrak{H})$. Because the limit (1.6.27) exists and also holds in the $\|\cdot\|_q$-norm for the $\|\cdot\|_q$-valued integrals, in the left-hand side of (1.6.27), the operator $U_A(t) \in \mathcal{C}_q(\mathfrak{H})$ for any $t > 0$. By the reason of semigroup law, $U_A(t) = (U_A(t/q))^q$, and the properties of ideals $\{\mathcal{C}_q(\mathfrak{H})\}_{1 \leq q < \infty}$, this implies that $U_A(t) \in \mathcal{C}_1(\mathfrak{H})$ for $t > 0$ and consequently the C_0-semigroup $\{U_A(t) =: G_A(t)\}_{t \geq 0}$ is a Gibbs semigroup.

Seeing that by (1.6.27) the family $\{U_A(t)\}_{t > 0}$ belongs to $\mathcal{C}_q(\mathfrak{H})$ and that it is $\mathcal{C}_q(\mathfrak{H})$-continuous function by condition (i), the integrand in (1.6.22) is $\mathcal{C}_q(\mathfrak{H})$-measurable. Then for $\beta < \lambda$, we can construct the representation of resolvent

$R_A(z = -\lambda)$ as the $\mathcal{C}_q(\mathfrak{H})$-Bochner integral (Laplace transform). To this end, it is sufficient to take in (1.6.22) and (1.6.24) the $\|\cdot\|_q$-$\lim_{t\to\infty} F(t)$, which then can be extended to $z \in \rho(A)$. As a result, we obtain that

$$\|R_A(z)\|_q = \left\| \int_0^\infty dt\, e^{t\,\mathrm{Re}\,z}\, e^{-tA} \right\|_q , \quad \mathrm{Re}\,z < -\beta , \tag{1.6.28}$$

where the left-hand side of (1.6.28) is finite by condition (ii). By the same condition, similarly to (1.6.20) and (1.6.21), we also obtain the $\mathcal{C}_1(\mathfrak{H})$-Bochner integral for the power q of Laplace transform, which by the *Fubini* theorem gives

$$\|(R_A(z))^q\|_1 = \left\| \Gamma(q)^{-1} \int_0^\infty dt\, t^{q-1}\, e^{-tA}\, e^{t\,\mathrm{Re}\,z} \right\|_1 < \infty , \quad \text{when } \mathrm{Re}\,z < -\beta . \tag{1.6.29}$$

Since by Proposition 1.6.13 the integrand in (1.6.29) is $\|\cdot\|_1$-integrable, one gets estimate

$$\Gamma(q)^{-1} \int_0^\infty dt\, t^{q-1}\, \|e^{-tA}\|_1\, e^{t\,\mathrm{Re}\,z} < \infty , \quad \text{when } \mathrm{Re}\,z < -\beta , \tag{1.6.30}$$

which implies the asymptotic $\|e^{-tA} = G_A(t)\|_1 = O(t^{-p})$, for $t \downarrow 0$, where $p < q$. □

Corollary 1.6.15

(a) *The statement of Theorem* 1.6.14 *also holds if* $\ker(A) \neq \{0\}$, *but* $\dim(\ker(A))$ *is finite. The problem of invertibility of generator A can be easily handled then by means of the shift* $A + \alpha\,\mathbb{1}$, *for* $\alpha > 0$.
(b) *The statement of Theorem* 1.6.14 *holds for any real* $q > p$, *where* $p \geq 1$. *For the proof, see Notes to Chap.* 1 (*Notes to Sect.* 1.6).

Notice that Example 1.6.9 does not satisfy conditions of Theorem 1.6.14 since $\ker(A) \neq \{0\}$. But on account of $\dim(\ker(A)) = 1$, this problem can be mended as it is indicated in Corollary 1.6.15 (a). For a generalisation of Theorem 1.6.14 which covers the case of eventually Gibbs semigroups for $t > t_0$, as in Example 1.6.9, one needs more arguments. They follow *verbatim* the line of reasoning in Theorem 1.6.14 with conditions, which are "shifted" to $t_0 > 0$.

Remark 1.6.16 A warning similar to that before Proposition 1.6.7 is also relevant to Theorem 1.6.14.

- Indeed, a C_0-semigroup with generator A, which has a trace-class resolvent $\{R_A(z)\}_{z\in\rho(A)}$ on the resolvent set $\rho(A)$, cf. Theorem 1.6.14 (b)-(ii), is not necessarily a Gibbs semigroup. For example, it is the case when $\dim(\ker(A)) = \infty$.
- On the other hand, a resolvent of an eventually Gibbs semigroup is not necessarily even compact. It is the case when semigroup is *nilpotent*.

- In fact, to avoid these discrepancies or triviality, one needs certain supplementary conditions. In our Theorem 1.6.14, they are expressed, for example, by conditions in (a) for Gibbs semigroups or in (b)-(ii) for C_0-semigroups.
- We also note that Theorem 1.6.14 provides a construction of *a priori* non-holomorphic Gibbs semigroups, which is the alternative to construction of Gibbs semigroups with aid of the *p-generators* (see Example 1.10.8(5)).

For illustrations of these remarks, we refer to Examples 1.10.3, 1.10.8 and 1.10.9 in Notes to Chap. 1 (Notes to Sects. 1.6 and 1.7).

On account of Examples 1.6.5 and 1.6.9 (or in general by Corollary 1.6.12), we infer that a C_0-semigroup $\{U(t)\}_{t\geq 0}$ may be *immediately* or *eventually* operator-norm, or trace-norm, continuous by virtue of distinct properties of regularity of the *image* of the mapping $t \mapsto U(t)$, away from zero. The causes may be rather different, a self-adjointness, a compactness, or a triviality of the image, as in the case of a nilpotent semigroup that we are going to consider below.

The next is a simple example of the semigroup with many *eventual* properties. We use this class of semigroups in a few places of this book below, for example, in Chap. 2 and in Chaps. 11 and 12.

Example 1.6.17 (Nilpotent C_0-semigroup) Let the Banach space $\mathfrak{X} = L^p([0, 1])$. Then for any $f \in \mathfrak{X}$ and $t \geq 0$, the equation

$$(U_0(t)f)(s) = \begin{cases} f(s-t) & \text{for } \{t < 1\} \text{ and } s \in (t, 1]\,, \\ 0 & \text{for } s \notin (t, 1] \text{ or } \{t \geq 1\}\,, \end{cases}$$

defines on $\mathfrak{X}$ a strongly continuous *nilpotent* semigroup $\{U_0(t)\}_{t\geq 0}$. Below (cf. (2.1.24)), we also use for $\{U_0(t)\}_{t\geq 0}$ the following representation:

$$(U_0(t)f)(s) = \chi_{[0,1]}(s-t)f(s-t), \quad s \in [0, 1]\,, \tag{1.6.31}$$

for $f \in \mathfrak{X}$ and *characteristic* function $\chi_{[0,1]}(\cdot)$ of the interval $[0, 1] \subset \mathbb{R}_0^+$.

Note that for $t \geq 1$ the *image* of the mappings $t \mapsto U_0(t) = 0 \cdot \mathbb{1}$, which is a *trivial* compact operator. Therefore, the nilpotent semigroup (1.6.31) is eventually compact, and as a consequence (see Corollary 1.6.12 (b)), it is also *eventually* operator-norm continuous for the threshold $t_0 = 1$. (However, it is valid without this reference.) For $t < 1$, the semigroup $\{U_0(t)\}_{t\geq 0}$ is only strongly continuous.

By virtue of (1.2.2) and Definition 1.3.1, one might to say that the *type* of the nilpotent C_0-semigroup $\{U_0(t)\}_{t\geq 0}$, such that $U_0(t \geq t_0) = 0$, is $\beta = -\infty$, whereas for *quasi-nilpotent* semigroup (see Remark 1.3.2), that is, for $t_0 = \infty$, the negative *type* may be bounded from below.

In the next section, we present an important special class of immediately operator-norm continuous C_0-semigroups, which are *holomorphic* semigroups.

1.7 Bounded Holomorphic Semigroups

Instead of the semigroup construction described in Sects. 1.1–1.4, we can use a suitable functional calculus to express the exponential function. One of the ways is to define semigroup $\{U_A(t) = \mathrm{e}^{-tA}\}_{t \geq 0}$ via the *Cauchy* integral as it is used in the *Riesz-Dunford* formula for operator-valued exponential function:

$$U_A(t) = \frac{1}{2\pi i} \int_\Gamma dz\, \mathrm{e}^{-tz} (z\mathbb{1} - A)^{-1}, \quad \Gamma \subset \rho(A)\,. \tag{1.7.1}$$

This formula is obviously valid if $A \in \mathcal{L}(\mathfrak{X})$ and if Γ is a positively oriented contour enclosing the spectrum $\sigma(A) := \mathbb{C} \setminus \rho(A)$ of the operator A in its interior.

In fact, the representation (1.7.1) is a way to construct semigroups if one assumes slightly more on the generator A than in Sect. 1.1 or in Sect. 1.3.

Definition 1.7.1 Let $\theta \in (0, \pi/2]$, and denote by S_θ the open sector $S_\theta := \{z \in \mathbb{C} : z \neq 0 \text{ and } |\arg z| < \theta\}$. A family of operators $\{U(z)\}_{z \in S_\theta}$ on a Banach space $\mathfrak{X}$ is called a holomorphic semigroup of semi-angle θ if

(i) $U(z + z') = U(z)\,U(z')$, for all $z,\ z' \in S_\theta$.
(ii) The mapping $z \to U(z)$ is analytic in S_θ.
(iii) If $x \in \mathfrak{X}$ and $\varepsilon \in (0, \theta)$, then $\lim_{z \to 0} U(z)\,x = x$, provided $z \in S_{\theta - \varepsilon}$.
If, in addition,
(iv) $\|U(z)\| \leq N_\varepsilon$ for some $N_\varepsilon < \infty$ in $\overline{S}_{\theta-\varepsilon}$ for every $\varepsilon \in (0, \theta)$ and $U(0) = \mathbb{1}$, we call $\{U(z)\}_{z \in S_\theta \cup \{0\}}$ a *bounded* holomorphic semigroup of semi-angle $\theta \in (0, \pi/2]$.

Remark 1.7.2 Notice that when we consider operator-valued holomorphic functions with maps in $\mathcal{L}(\mathfrak{X})$, there is no distinction between uniform (*operator-norm*), strong, and weak holomorphy (cf. Definition 1.7.1(ii)). See Proposition 1.10.5 and Corollary 1.10.6 in Notes to Chap. 1 (Notes to Sect. 1.7).

We also define generator A of the semigroup from Definition 1.7.1 by the use of the standard limit,

$$Ax = \lim_{t \to +0} \frac{1}{t} (\mathbb{1} - U(t))\,x\,, \quad x \in \mathrm{dom}\,(A)\,, \tag{1.7.2}$$

where $\mathrm{dom}\,(A)$ is the set of $x \in \mathfrak{X}$ for which the limit (1.7.2) *exists*.

Owing to these definitions, one can deduce the existence and properties of generators of holomorphic semigroups.

Remark 1.7.3 By (i) and (ii), for each $\varphi \in (-\theta, \theta)$, the family $\{U_A(z = r\,\mathrm{e}^{i\varphi})\}_{r \geq 0}$ is a bounded strongly continuous semigroup of a parameter $r \in \mathbb{R}_0^+$ with generator $\mathrm{e}^{i\varphi} A$. Then, by Proposition 1.3.4, the spectrum of this generator $\sigma(\mathrm{e}^{i\varphi}A) \subset \overline{S}_{\pi/2}$.

Since this is true for all $|\varphi| < \theta$, the spectrum of A belongs to the closed sector $\overline{S}_{\pi/2-\theta}$:

$$\sigma(A) \subset \overline{S}_{\pi/2-\theta} = \left\{ z \in \mathbb{C} : z = 0 \text{ and } |\arg z| \leq \frac{\pi}{2} - \theta \right\}. \tag{1.7.3}$$

By Definition 1.7.1(iv), for all $\varepsilon \in (0, \theta)$ and $|\varphi| \leq \theta' = \theta - \varepsilon/2$, the semigroup $\{U_{r\,e^{i\varphi}}(A)\}_{r \geq 0}$ is uniformly bounded by M'_ε. Therefore, by Proposition 1.3.4, one gets the estimate $\|(z\mathbb{1} + e^{i\varphi}A)^{-1}\| \leq M'_\varepsilon (\operatorname{Re} z)^{-1}$ for all z with $\operatorname{Re} z > 0$. For $|\arg z| \leq \pi/2 - \varepsilon/2$, this estimate implies

$$\left\| (z\,e^{-i\varphi}\mathbb{1} + A)^{-1} \right\| \leq \frac{M'_\varepsilon}{|z| \sin(\varepsilon/2)} .$$

As a consequence, generator A of the holomorphic semigroup in Definition 1.7.1 satisfies for some $\varepsilon \in (0, \theta)$ the condition

$$\left\| (A + \zeta\,\mathbb{1})^{-1} \right\| \leq \frac{M_\varepsilon}{|\zeta|}, \qquad \zeta \in \overline{S}_{\pi/2+\theta-\varepsilon} \setminus \{0\}, \tag{1.7.4}$$

for $M_\varepsilon > 0$, which is independent of ζ. By $\mathscr{H}(\theta, 0)$, we denote the *class* of generators with the property (1.7.4).

Moreover, as it will be shown in the next proposition, the conditions (1.7.3) and (1.7.4) are also sufficient.

Proposition 1.7.4 *A closed operator A on a Banach space $\mathfrak{X}$ is the generator of a bounded holomorphic semigroup $\{U(t)\}_{t \in S_\theta}$ of semi-angle, $0 < \theta < \pi/2$, if and only if $A \in \mathscr{H}(\theta, 0)$.*

Proof The necessity has been already proved, cf. Remark 1.7.3. To prove the sufficiency, we use the *Riesz-Dunford* integral formula (1.7.1). By virtue of (1.7.3) and (1.7.4), the integral (1.7.1) (and those below) converges absolutely and uniformly for $t > 0$ if Γ is chosen as a contour in $\rho(A)$ running from infinity with $\arg z = \pi/2 - \theta + \varepsilon$ and to infinity with $\arg z = -(\pi/2 - \theta + \varepsilon)$ within the domain $D_{\theta-\varepsilon} = \mathbb{C} \setminus \overline{S}_{\pi/2-\theta+\varepsilon}$.

To verify the semigroup property, we alternatively define $U_A(s)$ by (1.7.1) for the contour of integration $\Gamma' \subset D_{\theta-\varepsilon}$, which is similar to Γ but shifted by a small amount to the *left*. Then

$$\begin{aligned}
U_A(t)U_A(s) &= \frac{1}{(2\pi i)^2} \int_\Gamma dz \int_{\Gamma'} dz'\, e^{-tz-sz'} (z\mathbb{1} - A)^{-1} (z'\mathbb{1} - A)^{-1} \\
&= \frac{1}{(2\pi i)^2} \int_\Gamma dz \int_{\Gamma'} dz'\, e^{-tz-sz'} \left[(z\mathbb{1} - A)^{-1} - (z'\mathbb{1} - A)^{-1} \right] (z' - z)^{-1} \\
&= \frac{1}{(2\pi i)^2} \int_\Gamma dz\, e^{-tz} (z\mathbb{1} - A)^{-1} \int_{\Gamma'} dz'\, e^{-sz'} (z' - z)^{-1}
\end{aligned}$$

$$+ \frac{1}{(2\pi i)^2} \int_{\Gamma'} dz' \, e^{-sz'} (z'\mathbb{1} - A)^{-1} \int_{\Gamma} dz \, e^{-tz} (z - z')^{-1}$$
$$= \frac{1}{2\pi i} \int_{\Gamma} dz \, e^{-(t+s)z} (z\mathbb{1} - A)^{-1} = U_A(t+s),$$

where we have used the resolvent equation

$$R_A(z)\, R_A(z') = (R_A(z) - R_A(z'))\,(z - z')^{-1}$$

and properties of the Cauchy integral.

Note that if $|\arg t| < \theta$, the integral (1.7.1) can be made absolutely and uniformly convergent by deforming $\Gamma \subset D_{\theta-\varepsilon}$ to ensure that $|\arg(t\,z)| < \pi/2$ for $z \in \Gamma$ and $|z| \to \infty$. Hence, the semigroup $U_A(t)$ can be extended to complex t if $|\arg t| < \theta$. Since (1.7.1) is operator-norm differentiable under the integral sign, it follows that $U_A(t)$ is $\|\cdot\|$-holomorphic in an open sector $S_\theta = \{t \in \mathbb{C} : |\arg t| < \theta\}$. Within this sector, the Cauchy theorem yields

$$\|\cdot\|-\partial_t U_A(t) = \frac{1}{2\pi i} \int_{\Gamma} dz \, e^{-tz} \, (-z)\,(z\mathbb{1} - A)^{-1} \tag{1.7.5}$$
$$= \frac{(-1)}{2\pi i} \int_{\Gamma} dz \, e^{-tz} \left[\mathbb{1} + A\,(z\mathbb{1} - A)^{-1}\right]$$
$$= (-A)\,\frac{1}{2\pi i} \int_{\Gamma} dz \, e^{-tz} (z\mathbb{1} - A)^{-1} = (-A)\,U_A(t) \in \mathcal{L}(\mathfrak{X}),$$

where closeness of the operator A justifies that A can be taken out of the integral. As a consequence, this yields

$$U_A(t)A \subset AU_A(t) = -\partial_t U_A(t) \in \mathcal{L}(\mathfrak{X}), \quad t \in S_\theta. \tag{1.7.6}$$

Letting $z' = z\,t$, then Eq. (1.7.1) transforms into

$$U_A(t) = \frac{1}{2\pi i t} \int_{\Gamma'} dz' \, e^{-z'} \left(\frac{z'}{t}\mathbb{1} - A\right)^{-1}, \tag{1.7.7}$$

where the contour Γ' may be taken within the domain D_θ. For $z' \in \Gamma'$, $\varepsilon > 0$ and $t \in \overline{S}_{\theta-\varepsilon}$, we have $\|(z'/t - A)^{-1}\| \le M_\varepsilon\,|t/z'|$. Moreover, the condition $\Gamma' \subset D_\theta$ ensures the convergence of the integral (1.7.7) (see estimate (1.7.4)). Therefore,

$$\|U_A(t)\| \le M_\varepsilon \, \frac{1}{2\pi} \int_{\Gamma'} |dz'|\,|z'|^{-1}\,|e^{-z'}| = M'. \tag{1.7.8}$$

So, the semigroup $\{U_A(t)\}_{t\geq 0}$ is uniformly bounded in the closed sector $\overline{S}_{\theta'}$ for $\theta' < \theta$. Similarly, we obtain the estimate, cf. (1.7.5),

$$\|\partial_t U_A(t)\| = \left\| \frac{1}{2\pi i t^2} \int_{\Gamma'} \mathrm{d}z'\, z'\, \mathrm{e}^{-z'} \left(\frac{z'}{t}\mathbb{1} - A\right)^{-1} \right\|$$

$$\leq \frac{M_\varepsilon}{|t|} \int_{\Gamma'} |\mathrm{d}z'|\, |\mathrm{e}^{-z'}| := \frac{M_1'}{|t|}, \quad t \in \overline{S}_{\theta' < \theta} \setminus \{0\}. \tag{1.7.9}$$

To verify the strong continuity of $U_A(t)$, for $t \to +0$, in the sector $S_{\theta-\varepsilon}$, we use again the representation (1.7.7),

$$U_A(t) - \mathbb{1} = \frac{1}{2\pi i} \int_{\Gamma'} \mathrm{d}z' \, \frac{\mathrm{e}^{-z'}}{z'} \left(\frac{z'}{t}\mathbb{1} - A\right)^{-1} A.$$

Then, due to (1.7.4), for any $u \in \mathrm{dom}\,(A)$, we get

$$\|(U_A(t) - \mathbb{1})u\| \leq |t|\, M_\varepsilon\, \|Au\| \int_{\Gamma'} |\mathrm{d}z'|\, |z'|^{-2}\, |\mathrm{e}^{-z'}|, \quad t \in \overline{S}_{\theta-\varepsilon}. \tag{1.7.10}$$

Since $\overline{\mathrm{dom}\,(A)} = \mathfrak{X}$ and $\|U_A(t)\| \leq M'$ for $t \in \overline{S}_{\theta-\varepsilon}$ and $\varepsilon > 0$, it follows that

$$\text{s-}\lim_{t\to+0} U_A(t) = U_A(t)\big|_{t=0} = \mathbb{1}, \quad t \in \overline{S}_{\theta-\varepsilon},$$

which finishes the proof. □

Corollary 1.7.5 *If $A \in \mathscr{H}(\theta, 0)$, then a calculation similar to that in* (1.7.9) *yields for the family* $\{U_A(t)\}_{t>0}$:

$$\|\partial_t^n U_A(t)\| = \|A^n U_A(t)\|$$

$$= \left\| \frac{1}{2\pi i t^{n+1}} \int_{\Gamma'} \mathrm{d}z'\, (z')^n \mathrm{e}^{-z'} \left(\frac{z'}{t}\mathbb{1} - A\right)^{-1} \right\| \leq \frac{M_n'}{t^n}. \tag{1.7.11}$$

The next corollary gives an alternative characterisation of the bounded holomorphic semigroups and their operator-norm continuity and differentiability (see Proposition 1.4.2).

Corollary 1.7.6 *Let $\{U_A(z)\}_{z\in S_\theta}$ be a bounded strongly (thence equivalently in the operator-norm, see Remark* 1.7.2 *) holomorphic semigroup on $\mathfrak{X}$ in the sector S_θ with $\theta < \pi/2$, then*

$$U_A(z) \,:\, \mathfrak{X} \to dom\,(A), \quad z \in S_\theta. \tag{1.7.12}$$

Moreover, for any $\varepsilon \in (0, \theta)$, *there exists a constant* $C_\varepsilon > 0$ *such that*

$$\|A\, U_A(z)\| \leq \frac{C_\varepsilon}{|z|} \tag{1.7.13}$$

for all $z \in S_{\theta-\varepsilon}$. *The estimate* (1.7.13) *implies that* $U_A(z)$ *is* $\|\cdot\|$*-continuous in* S_θ.

Proof The mapping (1.7.12) follows from (1.7.5) to (1.7.6). They also imply that the semigroup $\{U_A(z)\}_{z\in S_\theta}$ is holomorphic in the operator-norm topology (the $\|\cdot\|$-holomorphic) in the open sector S_θ. □

Now, we present sufficient conditions for extension of a C_0-semigroup to holomorphic semigroup in some open sector.

Proposition 1.7.7 *Let* $\{U_A(t)\}_{t\geq 0}$ *be a strongly continuous semigroup on* $\mathfrak{X}$ *with generator* A *such that* $\mathfrak{X} \ni u \mapsto U_A(t)\, u \in \mathrm{dom}\,(A)$, $\|U_A(t)\| \leq M$ *and* $\|A\, U_A(t)\| \leq M_1/t$, *for all* $t > 0$. *Then there exists angle* $\theta > 0$, *which depends on* M_1, *such that* $\{U_A(t)\}_{t\geq 0}$ *may be extended to a* $\|\cdot\|$*-holomorphic in the open sector* S_θ*-bounded semigroup* $\{U_A(z)\}_{z\in S_\theta\cup\{0\}}$. *Moreover, the converse of this statement is also true.*

Proof By Proposition 1.4.2, we infer that semigroup $\{U_A(t)\}_{t\geq 0}$ is an operator-norm infinitely differentiable function on $\mathbb{R}^+$ with $\|\cdot\|$–$\partial_t^n U_A(t) = (-A\, U_A(t/n))^n$; see (1.4.8). As a result, by conditions of the proposition, one gets estimate

$$\|\partial_t^n U_A(t)\| \leq (n\, M_1/\, t)^n . \tag{1.7.14}$$

Seeing that, the C^∞-function $t \mapsto U_A(t)$ can be analytically continued for $eM_1 > 1$ and each $t > 0$ to disc $D_t := \{z \in \mathbb{C}_+ : |z - t| < (eM_1)^{-1}\, t\}$ by setting

$$U_A(z) = \sum_{n=0}^{\infty} \frac{(z-t)^n}{n!}\, \partial_t^n U_A(t)\,, \quad z \in D_t\,. \tag{1.7.15}$$

The union of these discs, $\bigcup_{t>0} D_t =: S_\theta$, is a sector in $\mathbb{C}_+$ with semi-angle

$$\theta := \arcsin\,(eM_1)^{-1} < \pi/2\,. \tag{1.7.16}$$

In order to verify that the operator-valued function (1.7.15) satisfies conditions (i) of Definition 1.7.1, we first remark that by construction of $\{U_A(z)\}_{z\in S_\theta}$, its restriction to $\mathbb{R}^+$ coincides with $U_A(t)$. Then semigroup relation $U_A(z_1)\, U_A(z_2) = U_A(z_1 + z_2)$ for $z_1,\ z_2,\ z_1 + z_2 \in S_\theta$ holds by analytic continuation of this relation from the case, where both z_1 and z_2 are real and positive. This conclusion can be extended to $\{U_A(z)\}_{z\in S_\theta\cup\{0\}}$ if construction (1.7.15) is complemented by the following condition: $U_A(z = 0) = \mathbb{1}$.

Now, we show that family $\{U_A(z)\}_{z \in S_\theta \cup \{0\}}$ satisfies conditions (ii) of Definition 1.7.1. To this aim, let $\theta' < \theta$ and $z \in S_{\theta'} \subset S_\theta$. Then there exists (a unique) $t > 0$ such that $z = |z|\, e^{i\phi}$ for $|\phi| < \theta$ belongs to the disc $D'_t \subset S_{\theta'}$ of radius

$$r'_t = t\, \sin(\theta') = |z - t| + |z|\, \sin(\theta' - \phi)\,, \tag{1.7.17}$$

centered at t. Since z also belongs to (the concentric to D'_t) disc D_t of radius $r_t = t\, \sin(\theta) > r'_t$, by (1.7.16) and (1.7.17), we obtain that

$$|z - t| \leq t\, \sin(\theta') < t\, \sin(\theta) = t\, (eM_1)^{-1}\,. \tag{1.7.18}$$

Hence, by conditions of proposition and on account of (1.7.14), (1.7.15) and (1.7.18), one gets using $n!e^n > n^n$ the estimates

$$\|U_A(z)\| \leq \|U_A(t)\| + \sum_{n=1}^{\infty} \frac{|z-t|^n}{n!}\, \|\partial_t^n U_A(t)\| \leq M + \sum_{n=1}^{\infty} (\sin(\theta'))^n \frac{1}{n!}\, (n\, M_1)^n$$

$$\leq M + \sum_{n=1}^{\infty} (\sin(\theta')/\sin(\theta))^n =: M_\theta(\theta')\,, \qquad z \in \overline{D'}_{t>0}\,. \tag{1.7.19}$$

So, (1.7.19) proves (uniform) boundedness of the family $\{U_A(z)\}_{z \in \overline{S}_{\theta'}}$ and it remains only to look at its strong continuity for $z \to 0$.

To this aim, let $u \in \mathrm{dom}\,(A)$ and $z \in \overline{S}_{\theta'}$. Then on account of (1.4.8) and (1.7.15),

$$(U_A(z) - U_A(t))\, u = \sum_{n=1}^{\infty} \frac{(z-t)^{n-1}}{(n-1)!}\, \partial_t^{n-1} U_A(t)\, (-A\, u)\, \frac{(z-t)}{n}\,,$$

that provides by boundedness (1.7.19) and $u \in \mathrm{dom}\,(A)$ the estimate

$$\|(U_A(z) - U_A(t))\, u\| \leq \|U_A(z)\|\, \|Au\|\, |z-t| \leq |z-t|\, M_\theta(\theta')\, \|Au\|\,. \tag{1.7.20}$$

As a consequence of (1.7.20),

$$\|(U_A(z) - \mathbb{1})\, u\| \leq \|(U_A(z) - U_A(t))\, u\| + \|(U_A(t) - \mathbb{1})\, u\| \leq$$

$$|z - t|\, M_\theta(\theta')\, \|Au\| + \|(U_A(t) - \mathbb{1})\, u\|\,, \quad u \in \mathrm{dom}\,(A)\,, \tag{1.7.21}$$

in particular, for $t \geq 0$ satisfying equation (1.7.18), which provides that $t = \alpha\, |z|$. For that reason, the estimate (1.7.21) and the strong continuity of semigroup $\{U_A(t)\}_{t \geq 0}$ yield

$$\lim_{z \to 0} \|(U_A(z) - \mathbb{1})\, u\| = 0\,, \quad z \in \overline{S}_{\theta'}\,, \quad u \in \mathrm{dom}\,(A)\,. \tag{1.7.22}$$

Since operator A is generator of the C_0-semigroup, $\mathrm{dom}\,(A)$ is dense in $\mathfrak{X}$; see Propositions 1.2.7. Then by the operator-norm boundedness, $\|(U_A(z) - \mathbb{1})\| \leq (M_\theta(\theta') + 1)$, (1.7.19), and by Proposition A.1.6, the convergence in (1.7.22) holds for all $u \in \mathfrak{X}$. Consequently, the semigroup $\{U_A(z)\}_{z \in S_\theta \cup \{0\}}$ (1.7.15) satisfies both conditions (i) and (ii) of Definition 1.7.1.

The converse of this statement stems from Proposition 1.7.4; see (1.7.5) and (1.7.9). □

Remark 1.7.8 For $t > 0$, the holomorphic semigroups have a stronger form of continuity than the operator-norm continuity. Since the family $\{U_A(t)\}_{t>0}$ can be differentiated any number of times, one gets by (1.7.11)

$$U_A(t)u \in \bigcap_{n=1}^{\infty} \mathrm{dom}\,(A^n) \qquad \text{and} \qquad \|A^n U_A(t)\| \leq M_n t^{-n}, \tag{1.7.23}$$

for all $t > 0$. Therefore, one gets the estimate

$$\begin{aligned} \|A^n(U_A(t) - U_A(s))\| &= \left\| \int_s^t \mathrm{d}\tau\, A^n \partial_\tau U_A(\tau) \right\| \\ &\leq \frac{M_{n+1}}{n} \left(\frac{1}{s^n} - \frac{1}{t^n} \right), \end{aligned} \tag{1.7.24}$$

for all $0 < s \leq t$ and $n > 0$. Corollary 1.7.6 treats the case $n = 0$.

For discussion of the semigroup analyticity *versus* differentiability, see Remark 1.10.7 and Example 1.10.8 in Sect. 1.10 (Notes to Sect. 1.7).

The notion of a *bounded* holomorphic semigroup can be generalised in the same way as in Sect. 1.3 to strongly continuous *quasi-bounded* semigroups. This means that we can relax condition (ii) of Definition 1.7.1.

Definition 1.7.9 A strongly continuous semigroup $\{U_A(z)\}_{z \in S_\theta}$ on a Banach space $\mathfrak{X}$ is called a holomorphic semigroup of semi-angle $\theta \in (0, \pi/2]$ if $U_A(z)$ satisfies all the properties of a *bounded* holomorphic semigroup of semi-angle θ except that it is not required to be uniformly bounded in the sector $\overline{S}_{\theta'}$ for $\theta' < \theta$.

Definition 1.7.10 It is appropriate to mention here that a certain *restriction* of holomorphic semigroups is also useful; see, for example, Chap. 10. The $\{U_A(z)\}_{z \in S_\theta}$ is a *holomorphic contraction* semigroup if its restriction $\{U_A(t)\}_{t \geq 0}$ on $\mathbb{R}_0^+$ is a contraction C_0-semigroup. We denote by $\mathscr{H}_c(\theta, 0) \subset \mathscr{H}(\theta, 0)$ the corresponding *subclass* of generators of holomorphic semigroups. Then

$$\mathscr{H}_c(\theta, 0) \subseteq \mathscr{H}(\theta, 0) \cap \mathscr{G}(1, 0)\,, \tag{1.7.25}$$

cf. definition (1.7.4) of $\mathscr{H}(\theta, 0)$ and characteristic of contraction semigroups $\mathscr{G}(1, 0)$ by Proposition 1.3.4 (1.3.3) for $M = 1$ and $\beta = 0$.

Remark 1.7.11 To show that intersection in (1.7.25) is non-empty, it is sufficient to consider in Proposition 1.7.7 the analytic extension of contraction semigroup for $M = 1$ and $e\,M_1 > 1$. Then result is a holomorphic contraction semigroup in the sense of Definition 1.7.10. In Sect. 1.8.2, we shall discuss this concept in the case of a Hilbert space. It turns out that the property to be contracted for holomorphic semigroup $\{U_A(z)\}_{z\in S_\theta}$ on a Hilbert space $\mathfrak{H}$ persists in a complex vicinity of $\mathbb{R}_0^+$.

Proceeding in the same way as in Sect. 1.3, we can obtain an exponential estimate on the growth of $U_A(z)$ in the sector $S_{\theta'}$. If $U_A(z)$ is a holomorphic semigroup of semi-angle θ, then $\|U_A(z)\,u\|$ is bounded in domain $D_{\theta',\tau} = \{z \in \overline{S}_{\theta'} : |z| \leq \tau,\ \tau > 0\}$ for any $u \in \mathfrak{X}$. Hence, by the *uniform boundedness principle*, $\|U_A(z)\| \leq M$ for $z \in D_{\theta',\tau}$. For any $z \in \overline{S}_{\theta'}$, by the semigroup property, Definition 1.7.1 (i), we have $U_A(z) = (U_A(z'))^n U_A(\zeta)$, where $z = n\,z' + \zeta$ and $z', \zeta \in D_{\theta',\tau}$. Therefore, one concludes that there are $M,\ \beta > 0$ such that

$$\|U_A(z)\| \leq M\,\mathrm{e}^{\beta|z|}, \quad z \in \overline{S}_{\theta'}\,. \tag{1.7.26}$$

These M and β are in fact θ'-dependent, i.e., the semigroup $\tilde{U}_A(z) := \mathrm{e}^{-\beta z} U_A(z)$ is a bounded holomorphic semigroup of semi-angle θ'.

The above line of reasoning motivates a generalisation of Proposition 1.7.4, which is in the spirit of Proposition 1.3.4.

Proposition 1.7.12 *A closed operator A on a Banach space $\mathfrak{X}$ is the generator of a quasi-bounded holomorphic semigroup of semi-angle $\theta \in (0, \pi/2]$ if and only if*

(i) $\rho(A) = \mathbb{C} \setminus \{z \in \mathbb{C} : z + \beta \in \overline{S}_{\pi/2-\theta}\}$ *for some* $\beta \in \mathbb{R}$.
(ii) *For any* $0 < \varepsilon \leq \pi/2 + \theta$, *there exists a constant* $M_\varepsilon > 0$ *such that*

$$\|(z\mathbb{1} + \beta\mathbb{1} + A)^{-1}\| \leq \frac{M_\varepsilon}{|z|}, \quad z \in \overline{S}_{\pi/2+\theta-\varepsilon} \setminus \{0\}. \tag{1.7.27}$$

We denote this class of generators by $\mathscr{H}(\theta, \beta)$. We comment that $A \in \mathscr{H}(\theta, \beta)$ implies $A \in \mathscr{G}(M, \beta)$ (1.3.3) for $M = M_\varepsilon$, but the converse is not true.

Proof The operator $A + \beta\mathbb{1}$ satisfies the conditions of Proposition 1.7.4. Therefore, $U_{A+\beta\mathbb{1}}(z) = \mathrm{e}^{-\beta z}\,U_A(z)$ is a bounded holomorphic semigroup of semi-angle $\theta' = \theta - \varepsilon$, which yields the assertion. □

Remark 1.7.13 To make evident that condition (1.7.27) is stronger than (1.3.3), consider, for example, the case $\beta = 0$. Note that since $t \mapsto U_A(t)$ is strongly continuous for $t \geq 0$ and differentiable (1.7.6) for $t > 0$, the $A \in \mathscr{H}(\theta, 0)$ implies the existence of the Laplace transform

$$\hat{U}_\lambda(A) = \int_0^\infty \mathrm{d}t\,\mathrm{e}^{-\lambda t}\,U_A(t) = (A + \lambda\mathbb{1})^{-1} \in \mathcal{L}(\mathfrak{X}),$$

for $\lambda > 0$; see (1.2.13)–(1.2.15). Then, using the *Fubini* theorem, we obtain

$$(\lambda \mathbb{1} + A)^{-n} = \int_0^\infty dt_1 \int_0^\infty dt_2 \ldots \int_0^\infty dt_n \, e^{-\lambda(t_1+\cdots+t_n)} U_A(t_1 + \cdots + t_n)$$
$$= \frac{1}{(n-1)!} \int_0^\infty d\tau \, \tau^{n-1} e^{-\lambda\tau} U_A(\tau) \, . \tag{1.7.28}$$

Since by (1.7.8) the holomorphic semigroup is uniformly bounded $\|U_A(\tau)\| \leq M'$ in a smaller sector with $\theta' < \theta$, the inversion formula (1.7.28) gives the estimate (1.3.4) for $M = M'$. Hence, we conclude that $A \in \mathscr{G}(M', 0)$.

Note that the arguments in Remark 1.7.3 show the converse. Namely, if $e^{i\varphi} A \in \mathscr{G}(M'_\varepsilon, 0)$ for any $\varphi \in (-\theta, \theta)$, then $A \in \mathscr{H}(\theta, 0)$; see (1.7.4).

We comment that condition $A \in \mathscr{H}(\theta, \beta)$ (see (1.7.3) or (1.7.4)), guarantees that for $z = t > 0$ the holomorphic semigroup $\{U_A(z)\}_{z \in S_\theta}$ is operator-norm continuous and that $\mathfrak{X} \ni u \mapsto U_A(t) u \in \mathrm{dom}\,(A)$ (see Sect. 1.4 and Corollary 1.7.6). This is in contrast to the strongly continuous semigroups, for which one has a restricted invariance: if $u \in \mathrm{dom}\,(A)$, then $u \mapsto U_A(t) u \in \mathrm{dom}\,(A)$, for $t \geq 0$; see Sect. 1.3.

Remark 1.7.14 The condition (1.7.4) also ensures that for $t > 0$ the mapping $t \mapsto U_A(t)$ is infinitely operator-norm differentiable, cf. Corollary 1.7.5. In Sect. 1.10 (Notes to Sect. 1.7), we give Example 1.10.8 of semigroup, which for $t > 0$ is *infinitely* differentiable (and consequently $\mathfrak{X} \ni u \mapsto U_A(t) u \in \mathrm{dom}\,(A)$), but it is *not* holomorphic since the sectorial condition (1.7.3) does not hold.

1.8 Holomorphic Semigroups on Hilbert Spaces

1.8.1 Operator Numerical Range

In this section, we look closer on holomorphic semigroups with generators of the type $\mathscr{H}(\theta, \beta)$ in a separable complex Hilbert space $\mathfrak{H}$.

To this aim, we introduce, first of all, the notion of the operator *numerical range*. Let A be a densely defined in $\mathfrak{H}$ linear operator with domain $\mathrm{dom}(A) \subset \mathfrak{H}$.

Definition 1.8.1 The set of complex numbers,

$$W(A) := \{(Au, u) \in \mathbb{C} : u \in \mathrm{dom}(A) \, , \ \|u\| = 1\}, \tag{1.8.1}$$

is called the *numerical range* of the operator A.

Remark 1.8.2

(a) It is known (*Töplitz-Hausdorff* theorem) that the set $W(A)$ is convex and in general it is neither open nor closed, even for a closed operator A.

(b) Let $\Delta(A) := \mathbb{C} \setminus \overline{W(A)}$ be complement of the numerical range closure in the complex plane. Then $\Delta(A)$ is either a connected open set, or it consists of two components $\Delta_1(A)$ and $\Delta_2(A)$. The latter possibility occurs whenever $\overline{W(A)}$ is a strip bounded by two parallel straight lines, including a degenerate situation, when those lines coincide.

We denote by $\operatorname{def}(A) := \dim((\operatorname{ran}(A))^{\perp})$ the *deficiency* (or *defect*) of a closed operator A in $\mathfrak{H}$.

Proposition 1.8.3 *Let A be a closed operator in $\mathfrak{H}$. Then for any $\zeta \in \Delta(A)$, the operator $A - \zeta \mathbb{1}$ is injective with a closed range $\operatorname{ran}(A - \zeta \mathbb{1})$ such that*

$$\|(A - \zeta \mathbb{1})\, w\| \geq \operatorname{dist}(\zeta, \overline{W(A)})\, \|w\|, \quad w \in \operatorname{dom}(A). \tag{1.8.2}$$

If, in addition, for each of those ζ, the deficiency $\operatorname{def}(A - \zeta \mathbb{1})$ *is zero, then $\Delta(A) \subset \rho(A)$ or, equivalently, $\sigma(A) \subset \overline{W(A)}$. Moreover,*

$$\|R_A(\zeta)\| \leq \frac{1}{\operatorname{dist}(\zeta, \overline{W(A)})}. \tag{1.8.3}$$

Proof For any $u \in \operatorname{dom}(A)$ with $\|u\| = 1$, one has

$$|(Au, u) - \zeta| = |((A - \zeta \mathbb{1})\, u, u)| \leq \|(A - \zeta \mathbb{1})\, u\|, \quad \zeta \in \mathbb{C}. \tag{1.8.4}$$

Then, if $\zeta \in \Delta(A) = \mathbb{C} \setminus \overline{W(A)}$ so that $\operatorname{dist}(\zeta, \overline{W(A)}) = \delta > 0$, the estimate (1.8.4) gives $\|(A - \zeta \mathbb{1})u\| \geq \delta$. This implies that the operator $A - \zeta \mathbb{1}$ is an injection, i.e., $\ker(A - \zeta \mathbb{1}) = \{0\}$. Moreover, since A is closed, the set $\operatorname{ran}(A - \zeta \mathbb{1})$ is also closed, i.e., it is a subspace. If the dimension of its orthogonal complement $(\operatorname{ran}(A - \zeta \mathbb{1}))^{\perp}$ is zero, then $\operatorname{ran}(A - \zeta \mathbb{1}) = \mathfrak{H}$. Thus, the inverse operator $(A - \zeta \mathbb{1})^{-1} = R_A(\zeta)$ has domain $\mathfrak{H}$ and $\|R_A(\zeta)\| \leq \delta^{-1}$ by virtue of (1.8.4). This proves that $\zeta \in \rho(A)$, or $\Delta(A) \subset \rho(A)$, and the estimate (1.8.4). □

In fact, in Proposition 1.8.3, it is sufficient that $\operatorname{def}(A - \zeta \mathbb{1}) = 0$ for only one $\zeta \in \Delta(A)$. To make it short, we put below $A_\zeta := A - \zeta \mathbb{1}$ for $\zeta \in \mathbb{C}$.

Lemma 1.8.4 *The deficiency* $\operatorname{def}(A_\zeta)$ *is constant for $\zeta \in \Delta(A)$, except for the abovementioned special case, in which the* $\operatorname{def}(A_\zeta)$ *is constant in both $\Delta_1(A)$ and $\Delta_2(A)$. These constants are called the defects of a A in $\Delta(A)$, $\Delta_1(A)$ and $\Delta_2(A)$, respectively.*

Proof Let $\zeta \in \Delta(A)$ so that $\operatorname{dist}(\zeta, \overline{W(A)}) = \delta > 0$. We put $d_\zeta := \operatorname{def}(A_\zeta)$. For $a < 1$ and $\zeta' \in \mathbb{C}$ such that $|\zeta' - \zeta| \leq a\,\delta$, we have by (1.8.4) that

$$a\, \|A_\zeta u\| \geq \|(\zeta - \zeta')\, u\|, \quad u \in \operatorname{dom}(A). \tag{1.8.5}$$

Notice that the operator $A_{\zeta'}$ is closed and $A_{\zeta'} = A_\zeta + (\zeta - \zeta')\, \mathbb{1}$.

Suppose that $d_{\zeta'} < d_\zeta$. Then there exists a $\varphi \in \mathfrak{H} \ominus \operatorname{ran} A_\zeta$, $\varphi \neq 0$, such that $\varphi \perp (\mathfrak{H} \ominus \operatorname{ran} A_{\zeta'})$. This means that $\varphi \in \operatorname{ran} A_{\zeta'}$ or $\varphi = A_{\zeta'} v$ for some $v \in \operatorname{dom}(A)$. Since $\varphi \perp \operatorname{ran} A_\zeta$, one has $(\varphi, A_\zeta v) = 0$ and, hence,

$$(A_\zeta v, A_\zeta v) = -(\zeta - \zeta')(v, A_\zeta v). \tag{1.8.6}$$

Suppose now that $d_{\zeta'} > d_\zeta$. Then, similarly, one can find $\psi \in \mathfrak{H} \ominus \operatorname{ran} A_{\zeta'}$, $\psi \neq 0$, such that $\psi \in \operatorname{ran} A_\zeta$, i.e., $\psi = A_\zeta w$ for some $w \in \operatorname{dom}(A)$. Since $\psi \perp \operatorname{ran} A_{\zeta'}$, we get

$$(A_\zeta w, A_\zeta w) = -(A_\zeta w, (\zeta - \zeta') w). \tag{1.8.7}$$

Both (1.8.6) and (1.8.7) are impossible for $a < 1$, because, by (1.8.5), one gets

$$\|A_\zeta v\|^2 \leq \|(\zeta - \zeta') v\| \, \|A_\zeta v\| \leq a \, \|A_\zeta v\|^2,$$

$$\|A_\zeta w\|^2 \leq \|(\zeta - \zeta') w\| \, \|A_\zeta w\| \leq a \, \|A_\zeta w\|^2.$$

Therefore, we must have $d_{\zeta'} = d_\zeta$ for all ζ' in the disc $D_{r=\delta}(\zeta) := \{z \in \mathbb{C} : |z - \zeta| < \delta\} \subset \Delta(A)$.

By the same reasoning, one can extend the equality $d_{\zeta''} = d_\zeta$ to all ζ'' in the disc $D_{r\leq\delta}(z) \subset \Delta(A)$ for any center $z \in D_{r=\delta}(\zeta)$. Covering $\Delta(A)$ by these intersecting discs $D_{r\leq\delta}(z)$, we get that the deficiency of A is constant on $\Delta(A)$. The proof for the case with two components $\Delta_1(A)$ and $\Delta_2(A)$ is similar. □

The next statement gives a relation between the spectrum $\sigma(A)$ and the numerical range $W(A)$ in case of a bounded operator A.

Proposition 1.8.5 *If $A \in \mathcal{L}(\mathfrak{H})$, then*

$$\sigma(A) \subset \overline{W(A)}. \tag{1.8.8}$$

Proof Since for $u \in \mathfrak{H}$, $\|u\| = 1$, one has $|(Au, u)| \leq \|A\|$, we get $\overline{W(A)} \subseteq \overline{D}_{r=\|A\|}(z = 0)$. Hence, $\Delta(A)$ is a connected open set containing the exterior of the disc $\overline{D}_{r=\|A\|}(z = 0)$. Since $A \in \mathcal{L}(\mathfrak{H})$, this exterior belongs to the resolvent set $\rho(A)$, that is, the resolvent $R_A(\zeta) \in \mathcal{L}(\mathfrak{H})$, or $\operatorname{def}(A_\zeta) = 0$ for $\zeta \in \mathbb{C} \setminus \overline{D}_{\|A\|}(z = 0) \subset \Delta(A)$. By Lemma 1.8.4, the same must be true for all $\zeta \in \Delta(A) = \mathbb{C} \setminus \overline{W(A)}$. Since the operator A is closed, the conditions of Proposition 1.8.3 are satisfied, and this implies that $\Delta(A) \subset \rho(A)$, which is equivalent to (1.8.8). □

For unbounded operator A, the relation between spectrum and numerical range is more complicated. For example, it may happen that $\sigma(A)$ is not contained in $\overline{W(A)}$, but for a closed operator A the *essential* spectrum $\sigma_{ess}(A)$ is always a subset of $\overline{W(A)}$.

Proposition 1.8.6 *If A is unbounded densely defined in Hilbert space $\mathfrak{H}$ closed operator, then $\sigma_{ess}(A) \subseteq \overline{W(A)}$.*

Proof Recall that the essential spectrum of A is the set

$$\sigma_{ess}(A) := \tag{1.8.9}$$
$$\{\lambda \in \mathbb{C} : \dim(\ker(A - \lambda\mathbb{1})) = \infty \quad \text{or} \quad \operatorname{ran}(A - \lambda\mathbb{1}) \neq \overline{\operatorname{ran}(A - \lambda\mathbb{1})}\} ;$$

see Appendix B.1 (Sect. B.1.2)) for definition of different components of the spectrum $\sigma(A)$. Suppose that $\mathbb{C} \setminus \overline{W(A)} = \Delta(A)$ (see Remark 1.8.2). Then on account of Proposition 1.8.3, one gets for $\lambda \in \Delta(A)$ that $\ker(A - \lambda\mathbb{1}) = \{0\}$ and $\operatorname{ran}(A - \lambda\mathbb{1}) = \overline{\operatorname{ran}(A - \lambda\mathbb{1})}$, which gives a contradiction if $\lambda \in \sigma_{ess}(A)$. □

1.8.2 Holomorphic Contraction Semigroups

The notion of numerical range is very useful for classification of generators of contraction and holomorphic semigroups on a Hilbert space.

Definition 1.8.7 Operator A on a Hilbert space $\mathfrak{H}$ is said to be *accretive* if $W(A) \subset \overline{\mathbb{C}}_+ = \{z \in \mathbb{C} : \operatorname{Re} z \geq 0\}$. If, in addition, A is closed, then by the stability of the deficiency, Lemma 1.8.4, $\operatorname{def}(A_z) = \dim((\operatorname{ran}(A_z))^\perp)$ is constant for $\operatorname{Re} z < 0$. If $\operatorname{def}(A_z) = 0$ for $\operatorname{Re} z < 0$, then by Proposition 1.8.3, one obtains that $\mathbb{C}_+ = \{z \in \mathbb{C} : \operatorname{Re} z > 0\} \subset \rho(-A)$ with

$$(A + z\mathbb{1})^{-1} \in \mathcal{L}(\mathfrak{H}) \quad \text{and} \quad \|(A + z\mathbb{1})^{-1}\| \leq \frac{1}{\operatorname{Re} z}, \qquad \operatorname{Re} z > 0. \tag{1.8.10}$$

A closed operator A satisfying (1.8.10) is called *m-accretive*.

Note that m-accretive operator A is *maximal* accretive in the sense that A has no *proper* accretive extensions. Indeed, let A' be an accretive extension of A. Then, since the corresponding extension for resolvent is bounded and such that $(A' + z\mathbb{1})^{-1} = (A + z\mathbb{1})^{-1} \in \mathcal{L}(\mathfrak{H})$ for $\operatorname{Re} z > 0$, one infers that $A = A'$.

Remark 1.8.8 In Definition 1.8.7, we did not require $\operatorname{dom}(A)$ to be dense in $\mathfrak{H}$. In fact, an m-accretive operator A is necessarily densely defined. Indeed, since $\operatorname{dom}(A) = \operatorname{ran}(A+z\mathbb{1})^{-1}$ for $\operatorname{Re} z > 0$, we have to show that $((A+z\mathbb{1})^{-1}u, v) = 0$ for all $u \in \mathfrak{H}$ implies $v = 0$. Let $u = v$ and $w = (A + z\mathbb{1})^{-1}v$. Then,

$$0 = |(w, (A + z\mathbb{1})w)| \geq \operatorname{Re}(Aw, w) + \operatorname{Re} z\, \|w\|^2 \geq \operatorname{Re} z\, \|w\|^2,$$

where $\operatorname{Re} z > 0$, and hence, $w = 0$ and $v = 0$.

Operator A is called *quasi-accretive* if there is a $\gamma \in \mathbb{R}$ such that $A + \gamma\mathbb{1}$ is accretive. In the same way, we say that A is *quasi-m-accretive* if $A + \gamma\,\mathbb{1}$ is m-accretive for some $\gamma \in \mathbb{R}$.

Definition 1.8.9 An accretive operator A is called *sectorial* with *semi-angle* $\alpha \in [0, \pi/2)$ if

$$W(A) \subseteq \overline{S}_\alpha := \{z \in \mathbb{C} : z = 0 \text{ and } |\arg z| \le \alpha\}\,. \tag{1.8.11}$$

If in addition A is m-accretive, then it is called *m-sectorial*. If A is quasi-m-accretive and

$$W(A) \subseteq \overline{S}_{\alpha,\gamma} := \{z \in \mathbb{C} : z = \gamma \text{ and } |\arg(z - \gamma)| \le \alpha\}\,, \tag{1.8.12}$$

for some $\gamma \in \mathbb{R}$, then it is called *quasi-m-sectorial* with *vertex* γ and *semi-angle* α.

Owing to notions of m-accretive and quasi-m-sectorial operators, we can identify generators of *contraction* or, more generally, *quasi-bounded* and *holomorphic* semigroups.

Proposition 1.8.10 *Operator $A \in \mathscr{G}(1, 0)$ in $\mathfrak{H}$ is generator of a contraction semigroup $\{U_A(t)\}_{t\ge 0}$ on $\mathfrak{H}$ if and only if A is m-accretive.*

Proof If the operator A is m-accretive, then, by Definition 1.8.7 and Remark 1.8.8, it is densely defined and closed, and its properties (1.8.10) imply the *Hille-Yosida* conditions of Proposition 1.1.1 for the Euler formula construction of a C_0-semigroup $\{U_A(t)\}_{t\ge 0}$. Hence, by Definition 1.2.5, this semigroup is a contraction semigroup. The converse follows on account of the Proposition 1.1.1. □

Corollary 1.8.11 *Operator $A \in \mathscr{G}(1, \beta)$ (see Definition* 1.3.1*) if and only if it is quasi-m-accretive for $\gamma = \beta$.*

Remark 1.8.12 Let A be the generator of a quasi-bounded holomorphic semigroup with semi-angle $\theta \in (0, \pi/2]$, that is, $A \in \mathscr{H}(\theta, \beta)$. In Sect. 1.7 (see (1.7.3), (1.7.4), and Proposition 1.7.12), we have shown (see Remark 1.7.3 and Propositions 1.7.4) that $A \in \mathscr{H}(\theta, 0)$ is equivalent to the existence, for each $\varepsilon > 0$, of a constant M'_ε such that $e^{i\varphi}A \in \mathscr{G}(M'_\varepsilon, 0)$ for any angle $|\varphi| \le \theta - \varepsilon/2$.

Remark 1.8.12 and Definition 1.8.9 of m-sectorial operators motivate a convenient criterion for operator A to be a generator of a *holomorphic contraction* semigroup on $\mathfrak{H}$ in a *sector* that includes $\mathbb{R}^+$. We remind here Remark 1.7.3, together with Definition 1.7.10 and inclusion (1.7.25) for the restricted case of contractions on $\mathbb{R}^+_0$.

Proposition 1.8.13 *Let A be an m-sectorial operator in Hilbert space $\mathfrak{H}$ with semi-angle $\alpha \in [0, \pi/2)$. Then $A \in \mathscr{H}_c(\theta, 0)$ is generator of a holomorphic contraction semigroup $\{U_A(z)\}_{z\in S_\theta}$ for $\theta = \pi/2 - \alpha$.*

Proof According to Remark 1.8.12, it suffices to prove that $e^{i\varphi}A \in \mathscr{G}(M = 1, 0)$ for any $|\varphi| \le \pi/2 - \alpha$. Since by (1.8.11) and (1.8.12) the numerical range $W(A) \subset \overline{S}_{\alpha,\gamma=0}$, one obtains $W(e^{i\varphi}A) = e^{i\varphi}W(A) \subset \overline{\mathbb{C}}_+$ for $|\varphi| \le \pi/2 - \alpha$. Thus, $\rho(e^{i\varphi}A) \supset \mathbb{C} \setminus \overline{W(e^{i\varphi}A)} \supset \mathbb{C}_- = \{z \in \mathbb{C} : \mathrm{Re}\, z < 0\}$ by Proposition 1.8.3. Therefore, if $\mathrm{Re}\,\zeta < 0$, then by (1.8.3),

$$\|(-\zeta \mathbb{1} + e^{i\varphi}A)^{-1}\| \le \frac{1}{\mathrm{dist}\,(\zeta, \overline{\mathbb{C}}_+)}. \tag{1.8.13}$$

Since $\mathrm{dist}\,(\zeta, \overline{\mathbb{C}}_+)$ is the distance of ζ from the imaginary axis, (1.8.13) together with (1.8.10) and Corollary 1.8.11 (for $\gamma = 0$) implies that $e^{i\varphi}A \in \mathscr{G}(M = 1, 0)$ for $|\varphi| \le \pi/2 - \alpha$. Therefore, $\|U_A(z)\| = \|U_{e^{i\varphi}A}(t)\| \le 1$, where $z = e^{i\varphi}t$, for $t \ge 0$. □

See Sect. 10.1.2 for some examples.

Corollary 1.8.14 *Let A be a non-negative self-adjoint operator in $\mathfrak{H}$, that is, the numerical range $W(A) \subset \mathbb{R}_0^+$. Then $A \in \mathscr{H}_c(\pi/2, 0)$ is generator of a holomorphic contraction self-adjoint semigroup $\{U_A(t)\}_{t\ge0}$ on $\mathfrak{H}$.*

Remark 1.8.15 The self-adjoint semigroup $\{U_A(t)\}_{t\ge0}$ is a particular case of the semigroup on a Hilbert space $\mathfrak{H}$ generated by a *normal* operator.

(a) Let generator A be a normal operator. This means that densely defined closed generator A satisfies $AA^* = A^*A$, where by the *von Neumann theorem* the left- and right-hand sides are self-adjoint operators with $\mathrm{dom}\,(AA^*) = \mathrm{dom}\,(A^*A) =: D$ such that D is a *core* for both operators A and A^*, especially $\mathrm{dom}\,(A) = \mathrm{dom}\,(A^*)$.
(b) Recall that similar to the self-adjoint semigroups, the *spectral theorem* for normal operators states that there exists a *standard* measure space (Ω, μ) and a measurable function $q : \Omega \to \mathbb{C}$ such that $\{U_A(t)\}_{t\ge0}$ is unitarity equivalent to the *multiplication operators* $\{U_q(t)\}_{t\ge0}$ on the space $L^2(\Omega, \mu)$. This means that there is a *unitary isomorphism* $\mathcal{U} : \mathfrak{H} \to L^2(\Omega, \mu)$, taking (bijectively) the graph of $U_A(t)$ onto the graph of $U_q(t)$ for each $t \ge 0$.
(c) This isomorphism to multiplication operators allows explicitly to compare the analytic properties of semigroups generated by normal operators with those for holomorphic self-adjoint semigroups. For more details, see Remark 1.10.7, Examples 1.10.8 and 1.10.9 in Notes to Sect. 1.8 (Sect. 1.8.2).

1.9 Semigroup Perturbations and Approximations

By semigroup perturbations, we mean *modifications* of the C_0-semigroup generator A in such a way that they still conserve the property to be generators of the corresponding perturbed C_0-semigroups. More precisely, let A be generator of a

C_0-semigroup. We study, in a Banach space $\mathfrak{X}$ or in a Hilbert space $\mathfrak{H}$, the problem of *stability*, namely, whether operator "$A+B$" (after closure, extensions, etc.) is still able to yield a generator of C_0-semigroup after perturbation of A by an operator B.

This problem is easy to solve for a *quasi-contraction* semigroup with generator $A \in \mathcal{G}(1, \beta)$ (Remark 1.3.3) if perturbation is a *bounded* linear operator $B \in \mathcal{L}(\mathfrak{X})$. We use this example to illustrate two methods of construction of the semigroup perturbations. The *first* one is based on perturbation of the resolvent of generator A and then checking semigroup criteria provided the resolvent properties (see (1.3.3) and (1.7.4)). The *second* method is ground on the construction of a perturbed semigroup by iterations that produce a convergent *Dyson-Phillips* series and then completed by the proof of the semigroup properties of the corresponding sum.

The *third* method is much closer to the main subject of this book (the *product formulæ*), and we present it in the Sect. 1.9.6.

1.9.1 Resolvent- and Semigroup-Based Methods

The first step in the resolvent-based method is the following assertion:

Proposition 1.9.1 *Let $A \in \mathcal{G}(1, \beta)$ be generator of the quasi-contraction semigroup and $B \in \mathcal{L}(\mathfrak{X})$. Then operator $H := A + B$ with* $\mathrm{dom}\,(H) = \mathrm{dom}\,(A)$ *is closed, and it is in turn a generator of quasi-contraction semigroup*: $H \in \mathcal{G}(1, \beta + \|B\|)$.

Proof Since the *closeness* of the operator A is stable with respect to bounded perturbations, by virtue of Proposition 1.3.4, one has only to estimate $\|R_H^m(-\lambda)\|$ for $m = 1$. From the second *Neumann* series for the resolvent $R_H(-\zeta)$, one gets

$$(H + \zeta\mathbb{1})^{-1} = \sum_{k=0}^{\infty}(A + \zeta\mathbb{1})^{-1}[-B\,(A + \zeta\mathbb{1})^{-1}]^k. \tag{1.9.1}$$

Seeing that $A \in \mathcal{G}(M = 1, \beta)$ implies that $\|\,(A + \zeta\mathbb{1})^{-1}\| \leq (\mathrm{Re}\,\zeta - \beta)^{-1}$ for $\mathrm{Re}\,\zeta > \beta$, we obtain from (1.9.1) for $\|B\|\,(\mathrm{Re}\,\zeta - \beta)^{-1} < 1$ that

$$\|\,(H + \zeta\mathbb{1})^{-1}\| \leq (\mathrm{Re}\,\zeta - \beta - \|B\|)^{-1}. \tag{1.9.2}$$

By Definition 1.3.1, the estimate (1.9.2) yields $H \in \mathcal{G}(1, \beta + \|B\|)$. □

Note that for extension of this statement to the case of a quasi-bounded semigroup $\{U_A(t)\}_{t\geq 0}$ for generator $A \in \mathcal{G}(M, \beta)$, where $M \geq 1$, one needs the estimates (1.3.3) for $m \geq 1$ (see Proposition 1.3.4). This line of reasoning is complicated because of certain combinatorial computations. An alternative is to introduce on the space $\mathfrak{X}$ the *equivalent norm* $(\mathfrak{X}, |||\cdot|||)$ (Remark 1.3.3), in which semigroup $\{U_A(t)\}_{t\geq 0}$ becomes quasi-contraction, $A \in \mathscr{G}(1,\beta)$, and then again apply Proposition 1.9.1.

Remark 1.9.2 The method that we used in Proposition 1.9.1 proves directly the existence of generator of the perturbed semigroup $\{U_{A+B}(t) = \mathrm{e}^{-t(A+B)}\}_{t\geq 0}$ via corresponding resolvent estimates (1.3.3). For that reason, it is called the *resolvent-based* method.

Now, we scrutinise the very same problem using the traditional perturbation theory approach based on iterations.

Proposition 1.9.3 *Let $A \in \mathscr{G}(M,\beta)$ be generator of a quasi-bounded semigroup $\{U_A(t)\}_{t\geq 0}$ on the Banach space $\mathfrak{X}$. If $B \in \mathcal{L}(\mathfrak{X})$, then operator $H := A + B$ with* $\mathrm{dom}\,(H) = \mathrm{dom}\,(A)$ *is closed, and $H \in \mathscr{G}(M, \beta + M\|B\|)$ is generator of the quasi-bounded semigroup $\{U_{A+B}(t)\}_{t\geq 0}$.*

Proof For any $t \geq 0$, we define the families of operators by

$$S_0(t) = U_A(t)\,, \tag{1.9.3}$$

$$S_n(t) = -\int_0^t ds\; S_{n-1}(t-s)\, B\, U_A(s)\,, \quad n \in \mathbb{N}\,. \tag{1.9.4}$$

Here, operators $S_{n\geq 1}(t)$ are the n-fold operator-norm convergent *Bochner* integrals

$$S_n(t) = \int_0^t ds_1 \int_0^{s_1} ds_2 \ldots \int_0^{s_{n-1}} ds_n\; \times \tag{1.9.5}$$
$$\times\, U_A(t-s_1)(-B)U_A(s_1-s_2)\ldots U_A(s_{n-1}-s_n)(-B)U_A(s_n)\,.$$

In virtue of $A \in \mathscr{G}(M,\beta)$ and definitions (1.9.3)–(1.9.5), we obtain for the operator-norm convergent *Dyson-Phillips* series $U(t) := \sum_{n\geq 0} S_n(t)$ the estimate

$$\|U(t)\| \leq M\mathrm{e}^{\beta t} \sum_{n=0}^{\infty} \frac{1}{n!}\,(t\, M\,\|B\|)^n\,. \tag{1.9.6}$$

The proof for any $t \geq 0$ and $s \geq 0$ of the semigroup composition law, $U(t)\,U(s) = U(t+s)$, can be checked by the term-by-term multiplication of the corresponding series and rearranging integrals (see Proposition 1.9.27 and (1.9.68)–(1.9.70)).

For any $x \in \mathfrak{X}$, we get from (1.9.3) to (1.9.5)

$$\lim_{t\to+0} \|U(t)\,x - x\| \leq \lim_{t\to+0} \left\{ \|U_A(t)\,x - x\| + M\mathrm{e}^{\beta t} \sum_{n=1}^{\infty} (t\, M\,\|B\|)^n/n! \right\} = 0\,.$$

Hence, $\{U(t)\}_{t\geq 0}$ is a C_0-semigroup. Now, we calculate generator H of this semigroup. To this aim, we consider the estimate

$$\left\| \frac{1}{t}(U(t) - \mathbb{1})x + (A+B)x \right\| = \tag{1.9.7}$$

$$\left\| \frac{1}{t}(U_A(t) - \mathbb{1})x + Ax + \frac{1}{t}S_1(t)x + Bx + \frac{1}{t}\sum_{n\geq 2} S_n(t)\,x \right\| \leq$$

$$\left\| \frac{1}{t}(U_A(t) - \mathbb{1})x + Ax \right\| + \left\| \frac{1}{t}S_1(t)x + Bx \right\| + \left\| \frac{1}{t}\sum_{n\geq 2} S_n(t)\,x \right\| .$$

By definition of the C_0-semigroup $\{U_A(t)\}_{t\geq 0}$, the limit

$$\lim_{t\to +0} \left\| \frac{1}{t}(U_A(t) - \mathbb{1})x + Ax \right\| = 0 \tag{1.9.8}$$

if and only if $x \in \mathrm{dom}\,(A)$. Since $B \in \mathcal{L}(\mathfrak{X})$, for $n = 1$ and any $x \in \mathfrak{X}$, the integrand in (1.9.4)

$$s \mapsto -U_A(t-s)\,B\,U_A(s)\,x$$

is bounded continuous function from $[0, t]$ to $\mathfrak{X}$. Therefore, owing to (1.9.4),

$$\lim_{t\to +0} \left\| \frac{1}{t}S_{n=1}(t)x + Bx \right\| = 0\,, \quad x \in \mathfrak{X}. \tag{1.9.9}$$

Finally, by the estimate similar to (1.9.6), we obtain

$$\lim_{t\to +0} \left\| \frac{1}{t}\sum_{n\geq 2} S_n(t)\,x \right\| = 0\,, \quad x \in \mathfrak{X}. \tag{1.9.10}$$

Therefore, on account of (1.9.7)–(1.9.10), one infers that generator $H = A + B$ on domain $\mathrm{dom}\,(H) = \mathrm{dom}\,(A)$, that is, $\{U(t) =: U_{A+B}(t)\}_{t\geq 0}$. □

Remark 1.9.4 The arguments that we used in Proposition 1.9.3 are less direct than in Proposition 1.9.1. The existence of the perturbed semigroup $\{U_{A+B}(t) = \mathrm{e}^{-t(A+B)}\}_{t\geq 0}$ with generator $H = A + B$ follows from the proof of the operator-norm convergent *Dyson-Phillips* series

$$U_H(t) = \sum_{n\geq 0} S_n(t). \tag{1.9.11}$$

This way of constructing a perturbed semigroup is called the *semigroup-based* method.

Corollary 1.9.5 *If* $A \in \mathscr{G}(M, \beta)$ *and* $\{z\, B\}_{z \in \mathbb{C}} \subset \mathcal{L}(\mathfrak{X})$ *is a complex family of bounded perturbations, then* $z \mapsto U_{A+z\,B}(t)$ *is an entire operator-valued function of* $z \in \mathbb{C}$ *for every* $t > 0$.

The proof of this claim is an immediate consequence the operator-norm convergent Dyson-Phillips series (1.9.11).

Note that in the framework of the semigroup-based method using the iterations (1.9.3)–(1.9.5) and series (1.9.11) we deduce identities

$$U_{A+B}(t) = U_A(t) - \int_0^t ds\, U_{A+B}(t-s)\, B\, U_A(s) \tag{1.9.12}$$

$$= U_A(t) - \int_0^t ds\, U_A(t-s)\, B\, U_{A+B}(s)\,. \tag{1.9.13}$$

In turn, they can be considered as integral equations for the perturbed semigroup $\{U_{A+B}(t)\}_{t\geq 0}$ and the Dyson-Phillips series (1.9.11) as their solution by iterations.

1.9.2 Perturbations of Holomorphic Semigroups

In contrast to bounded perturbations (Sect. 1.9.1), it is difficult in general to perturb generator A of a C_0-semigroup by an unbounded operator B without destroying the property of $A + B$ being a generator. For example, the operator $A + B$ need not to be a generator in $\mathscr{G}(M', \beta')$ even if B is a *relatively* A*-bounded* perturbation of $A \in \mathscr{G}(M, \beta)$,

Definition 1.9.6 Let A with domain dom $(A) \subset \mathfrak{X}$ be a linear operator in a Banach space $\mathfrak{X}$. A linear operator B with dom $(B) \subset \mathfrak{X}$ is called *(relatively)* A*-bounded* if dom $(A) \subseteq$ dom (B) and if there exist $a \geq 0$ and $b \geq 0$ such that

$$\|Bu\| \leq a\,\|Au\| + b\,\|u\|\,, \quad \text{for all} \quad u \in \mathrm{dom}\,(A)\,. \tag{1.9.14}$$

The *relative bound* of B with respect of A or simply the A*-bound* of B is

$$a_0 := \inf\{a \geq 0 : \exists\, b \geq 0 \ \text{ such that } (1.9.14) \text{ holds}\}\,. \tag{1.9.15}$$

We call B the *Kato-small* operator with respect to A if its A-bound $a_0 < 1$.

Note that the value of a may significantly vary in (1.9.14) with the value of b (see Example 1.9.8 of the *infinitesimally small* relative bound $a = 0^+$).

Definition 1.9.7

(1) Let A be a closed operator in $\mathfrak{X}$. We denote by $\mathcal{P}_a(A)$ the class of operators B satisfying (1.9.14) with relative bound $a > 0$ and by $\mathcal{P}_1(A) := \{\mathcal{P}_a(A) : 0 < a < 1\}$. Note that $\mathcal{P}_a(A) \subseteq \mathcal{P}_{a'}(A)$ for $a \leq a'$.
(2) The class $\mathcal{P}_1(A)$ is the set of the *Kato-small* perturbations of the operator A. If $a = 0$, then (1.9.14) defines for $u \in \mathfrak{X}$ the class $\mathcal{P}_0(A) = \mathcal{L}(\mathfrak{X})$, which coincides the set of *bounded* operators on $\mathfrak{X}$.
(3) We also define the class of (*unbounded*) operators:

$$\mathcal{P}_{0^+}(A) := \lim_{b \to +0} \bigcap_{0<\epsilon<b} \mathcal{P}_\epsilon(A) .$$

It is the class of *infinitesimally small* perturbations (1.9.14) of the operator A. As a consequence, we obtain $\mathcal{P}_0 \subset \mathcal{P}_{0^+}(A) \subset \mathcal{P}_a(A)$, for $a > 0$.

Example 1.9.8 Let $A \in \mathscr{G}(1, 0)$ be the generator of contraction semigroup $\{U_A(t)\}_{t \geq 0}$ that without loss of generality we suppose to be invertible. The fractional power A^α, $0 < \alpha < 1$, of generator can be expressed by the integral

$$A^\alpha u = \frac{1}{\Gamma(-\alpha)} \int_0^\infty dt\, t^{-\alpha-1}(U_A(t) - \mathbb{1})\, u \,, \quad u \in \operatorname{dom}(A), \tag{1.9.16}$$

where t^α is chosen to be positive for $t > 0$ and $\Gamma(-\alpha) < 0$ is the Gamma-function. Since for any $u \in \operatorname{dom}(A)$ and $0 < \alpha < 1$ the integral (1.9.16) is convergent, $\operatorname{dom}(A) \subseteq \operatorname{dom}(A^\alpha)$. Then by (1.9.16) and by the closeness of A, we obtain for each $\alpha \in (0, 1)$ and for all $\mu > 0$ the estimate

$$\left\| A^\alpha (A + \mu \mathbb{1})^{-1} \right\| \leq \frac{C_\alpha}{\mu^{1-\alpha}}, \quad C_\alpha := \frac{2}{\Gamma(-\alpha)\, \alpha(1-\alpha)} . \tag{1.9.17}$$

Let unbounded closed perturbation operator B be such that $\operatorname{dom}(A^\alpha) \subseteq \operatorname{dom}(B)$ for some $\alpha \in (0, 1)$. Since the operators A^α and B are closed, this inclusion is equivalent to A^α-boundedness of B, that is, to $\|BA^{-\alpha}\| \leq d$. Then by (1.9.17), one gets for any $\eta > 0$

$$\|B(A + \eta\mathbb{1})^{-1} x\| \leq \|BA^{-\alpha}\| \, \|A^\alpha (A + \eta\mathbb{1})^{-1} x\| \leq \frac{C_\alpha d}{\eta^{1-\alpha}} \|x\| , \quad x \in \mathfrak{X} .$$

Seeing that $u = (A + \eta\mathbb{1})^{-1} x \in \operatorname{dom}(A)$, this estimate yields (1.9.14), where

$$\|Bu\| \leq C_\alpha\, \eta^\alpha d \,\|u\| \, + \frac{C_\alpha\, d}{\eta^{1-\alpha}} \|Au\| , \quad u \in \mathfrak{X} , \tag{1.9.18}$$

for $a(\eta) := C_\alpha\, d/\eta^{1-\alpha}$, as the A-bound of B, and for $b(\eta) := C_\alpha\, d\, \eta^\alpha$. Seeing that for large η the value of the relative bound $a(\eta)$ can be turned to the *infinitesimally small* one but not to *zero*, the unbounded perturbation $B \in \mathcal{P}_{0^+}(A)\backslash\mathcal{P}_0(A)$, where $\mathcal{P}_0(A) = \mathcal{L}(\mathfrak{X})$ is the set of bounded operators.

We shall see more unbounded *infinitesimally* small operators from the class $\mathcal{P}_{0^+}$ throughout this book in what follows.

Proofs of Propositions 1.9.1 and 1.9.3 demonstrate that extension of the arguments even to *infinitesimally* small unbounded perturbations $\mathcal{P}_{0^+}(A)$ needs certain precautions. Another way to make the class of perturbations wider is to put more conditions on the generator A.

Description of allowed perturbations of generators $A \in \mathscr{G}(M, \beta)$ for quasi-bounded semigroups is postponed to the next section. If, however, A is assumed to be the generator of a *holomorphic* semigroup, a wider class of perturbations is permitted. Below, we use the *resolvent-based* method (Remark 1.9.2) to treat unbounded *Kato-small* perturbations of the class $\mathcal{P}_1(A)$ where unperturbed semigroup is holomorphic with generator $A \in \mathscr{H}(\theta, \beta)$.

Theorem 1.9.9 *Let $A \in \mathscr{H}(\theta, \beta)$ be generator of a quasi-bounded holomorphic semigroup of semi-angle $\theta \in (0, \pi/2]$. If perturbation $B \in \mathcal{P}_a(A)$, for $b \geq 0$ and $a < \delta \leq 1$, then operator $H = A + B \in \mathscr{H}(\theta - \varepsilon, \gamma)$ for some $\varepsilon \in (0, \theta)$ and $\gamma > 0$. If, in particular, $\beta = 0$ and $b = 0$, then $H \in \mathscr{H}(\theta - \varepsilon, 0)$ is generator of a bounded holomorphic semigroup.*

Proof Without loss of generality, we may assume $\beta = 0$ (see Proposition 1.7.12). Let $A \in \mathscr{H}(\theta, 0)$. Then $B \in \mathcal{P}_a(A)$ implies that

$$\|B(A + \zeta\mathbb{1})^{-1}\| \leq b\|(A + \zeta\mathbb{1})^{-1}\| + a\|A(A + \zeta\mathbb{1})^{-1}\|. \tag{1.9.19}$$

If $|\arg\zeta| \leq \pi/2 + \theta - \varepsilon$, for $\varepsilon \in (0, \theta)$, then by (1.7.4),

$$\|(A + \zeta\mathbb{1})^{-1}\| \leq \frac{M_\varepsilon}{|\zeta|},$$

and consequently,

$$\|A(A + \zeta\mathbb{1})^{-1}\| = \|\mathbb{1} - \zeta(A + \zeta\mathbb{1})^{-1}\| \leq 1 + M_\varepsilon.$$

Hence, (1.9.19) yields the estimate,

$$\|B(A + \zeta\mathbb{1})^{-1}\| \leq b\, M_\varepsilon|\zeta|^{-1} + a\,(1 + M_\varepsilon), \tag{1.9.20}$$

which implies convergence of the corresponding *Neumann* series for $(H + \zeta\mathbb{1})^{-1}$ as soon as the sum in the right-hand side of (1.9.20) is smaller than 1. The norm estimate of this series provides

$$\|(H+\zeta \mathbb{1})^{-1}\| \le \frac{M_\varepsilon |\zeta|^{-1}}{1 - b\, M_\varepsilon |\zeta|^{-1} - a\,(1+M_\varepsilon)}$$
$$= \frac{M_\varepsilon (1 - a\,(1+M_\varepsilon))^{-1}}{|\zeta| - b\, M_\varepsilon (1 - a\,(1+M_\varepsilon))^{-1}}. \tag{1.9.21}$$

As a consequence, if $a < (1+M_\varepsilon)^{-1}$, then (1.9.21) implies that

$$\|(H+\zeta \mathbb{1})^{-1}\| \le \frac{M'}{|\zeta - \gamma|}, \tag{1.9.22}$$

for $|\arg(\zeta-\gamma)| \le \pi/2+\theta-\varepsilon$, $M' := M'(a, b, M_\varepsilon) > 0$, where $\gamma := \gamma(a, b, M_\varepsilon)$. From estimate (1.9.21), we also deduce that if $b = 0$, then $\gamma = 0$, that is, generator $H \in \mathscr{H}(\theta - \varepsilon, 0)$. □

Theorem 1.9.10 *Let A be an m-sectorial operator in a Hilbert pace $\mathfrak{H}$ with semi-angle $\alpha \in [0, \pi/2)$. Define*

$$D_R := \{\kappa \in \mathbb{C} : |\kappa| < R\}, \qquad \text{and} \qquad R(a, \varepsilon) := \frac{\sin\varepsilon}{a\,(1+\sin\varepsilon)}. \tag{1.9.23}$$

If $B \in \mathcal{P}_a(A)$ for some $b < 1$, then for any $0 < \varepsilon < \theta := \pi/2 - \alpha$ we obtain that

(i) *For some $\gamma \ge 0$, the family of generators $\{H(\kappa) := A + \kappa\, B\}_{\kappa \in D_{R(a,\varepsilon)}}$ belongs to the class $\mathscr{H}(\theta - \varepsilon, \gamma)$.*
(ii) *For any $z \in S_{\theta-\varepsilon}$, the family $\{U_{H(\kappa)}(z)\}_{\kappa \in D_{R(a,\varepsilon)}}$ of holomorphic semigroups is also holomorphic in the operator-norm topology for $\kappa \in D_{R(a,\varepsilon)}$.*

Proof

(i) By virtue of Proposition 1.8.13, $A \in \mathscr{H}_c(\pi/2-\alpha, 0)$, i.e., it is the generator of a contraction holomorphic semigroup $\{U_A(z)\}_{z \in S_\theta}$ on $\mathfrak{H}$. Since A is m-sectorial, by Proposition 1.8.3, one gets the following estimates on its resolvent $R_A(-\zeta)$,

$$\|R_A(-\zeta)\| \le \frac{1}{\operatorname{dist}(\zeta, \overline{W(-A)})}$$

$$= \begin{cases} |\zeta|^{-1} & \text{if } |\arg\zeta| \le \theta, \\ \left[|\zeta| \sin(\pi/2+\theta \mp \arg\zeta)\right]^{-1} & \text{if } \theta < \pm\arg\zeta < \pi/2+\theta. \end{cases} \tag{1.9.24}$$

Now, we can use Theorem 1.9.9 with the value of M_ε refined by (1.9.24). Indeed, by (1.9.19) and (1.9.24), one gets

$$\|B(A+\zeta\mathbb{1})^{-1}\| \leq \begin{cases} b|\zeta|^{-1}+2a & \text{if } |\arg\zeta|\leq\theta, \\ b|\zeta|^{-1}M_{\mp}(\zeta)+a(1+M_{\mp}(\zeta)) & \text{if } \theta<\pm\arg\zeta<\pi/2+\theta, \end{cases} \tag{1.9.25}$$

where $M_{\mp}(\zeta) := \sin(\pi/2+\theta\mp\arg\zeta)$. Now, the condition of the $\|\cdot\|$-convergence of the *Neumann* series,

$$R_{H(\kappa)}(-\zeta) = R_A(-\zeta)\sum_{n=0}^{\infty}(\kappa\, B\, R_A(-\zeta))^n, \tag{1.9.26}$$

takes the form $\|\kappa\, B\, R_A(-\zeta)\| < 1$. This condition, together with the norm estimate of (1.9.26),

$$\|R_{H(\kappa)}(-\zeta)\| \leq \|R_A(-\zeta)\|\,[1-\|\kappa\, B\, R_A(-\zeta)\|]^{-1},$$

and the estimates (1.9.25), shows that the resolvent set $\rho(-H(\kappa))$ contains the domain

$$\begin{aligned}\mathcal{D}_{\kappa,a} :=& \left\{\zeta\in\mathbb{C} : |\arg\zeta|\leq\theta, |\zeta|>\frac{|\kappa|b}{1-2|\kappa|a}\right\}\\ &\bigcup\left\{\zeta\in\mathbb{C} : \theta<\arg\zeta<\frac{\pi}{2}+\theta,\right.\\ &\qquad\qquad \left. b|\zeta|^{-1}M_-(\zeta)+a(1+M_-(\zeta))<|\kappa|^{-1}\right\}\\ &\bigcup\left\{\zeta\in\mathbb{C} : \theta<-\arg\zeta<\frac{\pi}{2}+\theta,\right.\\ &\qquad\qquad \left. b|\zeta|^{-1}M_+(\zeta)+a(1+M_+(\zeta))<|\kappa|^{-1}\right\}.\end{aligned} \tag{1.9.27}$$

By virtue of the first line in (1.9.27), we must restrict κ to the disc

$$D_{(2a)^{-1}} := \{\kappa\in\mathbb{C} : |\kappa|<(2a)^{-1}\}. \tag{1.9.28}$$

By the second and the third part in (1.9.27), the asymptotic slopes of the border line of the domain $\mathcal{D}_{\kappa,a}$, when $|\zeta|\to\infty$, are determined by the inequalities

$$\begin{cases} 1>|\kappa|a\,(1+1/\cos(\arg\zeta-\theta)) & \text{if } \theta<\arg\zeta<\pi/2+\theta, \\ 1>|\kappa|a\,(1+1/\cos(\arg\zeta+\theta)) & \text{if } \theta<-\arg\zeta<\pi/2+\theta. \end{cases} \tag{1.9.29}$$

For any $0 < \varepsilon < \theta$, the slope asymptotics are

$$\arg \zeta \xrightarrow[|\zeta| \to \infty]{} \pm\left[\left(\frac{\pi}{2} + \theta\right) - \varepsilon\right], \tag{1.9.30}$$

in the sectors $\theta < \pm \arg \zeta < \pi/2 + \theta$, and the inequalities (1.9.29) give the allowed values of $|\kappa|$ in these sectors,

$$|\kappa| < \frac{\sin \varepsilon}{a(1 + \sin \varepsilon)} = R(a, \varepsilon),$$

see (1.9.23). Notice that the border line of the domain $\mathcal{D}_{\kappa,a}$ is continuous at $\arg \zeta = \pm\theta$, see (1.9.27). Since for any $0 < \varepsilon \leq \pi/2$ we have $R(a, \varepsilon) \leq (2b)^{-1}$, we put

$$\gamma(\theta, a, b, \varepsilon) := \frac{b R(a, \varepsilon)}{1 - 2a R(a, \varepsilon)} (\cos \theta)^{-1}. \tag{1.9.31}$$

Then by (1.9.27)–(1.9.31), there exists a $\gamma \geq \gamma(\theta, a, b, \varepsilon) \geq 0$ such that, for all $\kappa \in D_{R(a,\varepsilon)}$, the complement of $\mathcal{D}_{\kappa,a}$ belongs to the sector

$$- \; S_{\frac{\pi}{2}-\theta+\varepsilon, \, -\gamma} = \left\{z \in \mathbb{C} \, : \, |\arg(-z - \gamma)| < \frac{\pi}{2} - \theta + \varepsilon\right\},$$

for all $|\kappa| < R(a, \varepsilon)$. Since $\mathcal{D}_{\kappa,a} \subset \rho(-H(\kappa))$, we get

$$W(H(\kappa)) \subset \mathbb{C} \setminus \rho(H(\kappa)) \subset S_{\frac{\pi}{2}-\theta+\varepsilon,-\gamma},$$

that is, $\{H(\kappa)\}_{\kappa \in D_{R(a,\varepsilon)}}$ are quasi-m-sectorial operators with vertex $-\gamma$, see Definition 1.8.9. Similar to (1.9.22), this implies

$$\|R_{-\zeta}(H(\kappa))\| \leq \frac{M'_\varepsilon}{|\zeta - \gamma|}, \quad \kappa \in D_{R(a,\varepsilon)}.$$

Therefore, by Propositions 1.7.12 and 1.8.13, the operators $\{H(\kappa)\}_{\kappa \in D_{R(a,\varepsilon)}}$ are the generators of the family of quasi-bounded holomorphic semigroups

$$\{U_{H(\kappa)}(z)\}_{z \in S_{\theta-\varepsilon}}, \quad \kappa \in D_{R(a,\varepsilon)}. \tag{1.9.32}$$

(ii) To prove that the family (1.9.32) is $\|\cdot\|$-holomorphic in $\kappa \in D_{R(a,\varepsilon)}$, consider the *Riesz-Dunford* formula (1.7.1),

$$U_{H(\kappa)}(z) = \frac{1}{2\pi i} \int_\Gamma d\zeta \; e^{z\zeta} \; (\zeta \mathbb{1} + H(\kappa))^{-1}, \tag{1.9.33}$$

where the positively oriented contour Γ is chosen in such a way that (a) $\Gamma \subset \mathcal{D}_{\kappa,a}$ for all $\kappa \in D_{R(a,\varepsilon)}$, (b) the asymptotic slopes are

$$\arg \zeta \underset{\substack{|\zeta|\to\infty \\ \zeta\in\Gamma}}{\longrightarrow} \pm\left[\left(\frac{\pi}{2}+\theta\right)-\varepsilon\right],$$

as in (1.9.30). Since, for this choice of Γ, the integral (1.9.33) is absolutely $\|\cdot\|$-convergent for any z in the open sector $S_{\theta-\varepsilon}$, see Proposition 1.7.4, the resolvent identity gives

$$U_{H(\kappa+\delta)}(z) - U_{H(\kappa)}(z) = \\ = \frac{1}{2\pi i}\int_\Gamma d\zeta \; e^{z\zeta}\; R_{H(\kappa+\delta)}(-\zeta)\,(-\delta B)\, R_{H(\kappa)}(-\zeta)\,, \tag{1.9.34}$$

for $\kappa, \kappa+\delta \in D_{R(a,\varepsilon)}$. Since $B \in \mathcal{P}_a(A)$, the representation (1.9.34) and the estimates (1.9.24)–(1.9.25) imply that, for any $z \in S_{\theta-\varepsilon}$, the operator-valued function $\kappa \to U_{H(\kappa)}(z)$ is $\|\cdot\|$-differentiable in $\kappa \in D_{R(a,\varepsilon)}$. Hence, for any $z \in S_{\theta-\varepsilon}$, the family $\{U_{H(\kappa)}(z)\}_{\kappa\in D_{R(a,\varepsilon)}}$ is $\|\cdot\|$-holomorphic in the κ.

□

Of particular interest is the case of a *self-adjoint* generator $A \geq 0$, namely, the case of m-sectorial operator A with semi-angle $\alpha = 0$, see Proposition 1.8.13. Then, the estimates of Theorem 1.9.10 can be improved significantly.

Theorem 1.9.11 *Let $A \geq 0$ be a densely defined self-adjoint operator in $\mathfrak{H}$. If $B \in \mathcal{P}_a(A)$ for some $a < 1$, then it holds that:*

(i) *For any $|\kappa| < a^{-1}$ generators $H(\kappa) \in \mathscr{H}(\theta(\kappa, a), \gamma(\kappa, a, b))$ with*

$$\operatorname{ctg}\theta(\kappa, a) = \frac{|\kappa|\,a}{\sqrt{1-|\kappa|^2 a^2}}\,, \tag{1.9.35}$$

and

$$\gamma(\kappa, a, b) \leq \frac{b\,|\kappa|}{(1-a\,|\kappa|)}\,. \tag{1.9.36}$$

(ii) *For any $r < a^{-1}$, the family $\{U_{H(\kappa)}(z)\}_{z\in S_{\theta(r,a)}}$ of holomorphic semigroups is $\|\cdot\|$-holomorphic in variable $\kappa \in D_r = \{\kappa \in \mathbb{C} : |\kappa| < r\}$.*

Proof

(i) By Proposition 1.8.13, $A \in \mathscr{H}_c(\pi/2, 0)$. Hence, $\{U_A(z)\}_{z\in S_{\pi/2}}$ is a contraction holomorphic semigroup in the open sector $S_{\theta=\pi/2}$. Since A is self-adjoint, by

Proposition 1.8.3 we get

$$\|R_A(-\zeta)\| \leq \begin{cases} |\zeta|^{-1} & \text{for } |\arg\zeta| \leq \pi/2, \\ |\mathrm{Im}\ \zeta|^{-1} & \text{for } \pi/2 < \pm\arg\zeta < \pi. \end{cases} \tag{1.9.37}$$

Now, we refine the estimate (1.9.25). Indeed, since $A = A^*$, by the spectral representation

$$\begin{aligned} \|A(A+\zeta)^{-1}\| &= \left\| \int_0^\infty \mathrm{d}E_\lambda(A)\, \frac{\lambda}{\lambda+\zeta} \right\| \\ &\leq \sup_{\lambda\geq 0} \left\| \frac{\lambda}{\lambda+\zeta} \right\| \\ &= \begin{cases} 1 & \text{if } |\arg\zeta| \leq \pi/2, \\ |\zeta||\mathrm{Im}\ \zeta|^{-1} & \text{if } \pi/2 < \pm\arg\zeta < \pi. \end{cases} \end{aligned} \tag{1.9.38}$$

Then, by virtue of (1.9.19), (1.9.37) and (1.9.38), the estimate (1.9.25) takes the form

$$\begin{aligned} &\|B(A+\zeta\mathbb{1})^{-1}\| \\ &\quad \leq \begin{cases} b\,|\zeta|^{-1} + a & \text{if } |\arg\zeta| \leq \pi/2, \\ b\,|\mathrm{Im}\ \zeta|^{-1} + a\,|\zeta||\mathrm{Im}\ \zeta|^{-1} & \text{if } \pi/2 < \pm\arg\zeta < \pi. \end{cases} \end{aligned} \tag{1.9.39}$$

Therefore, the condition $\|\kappa\, B\, R_A(-\zeta)\| < 1$ for the $\|\cdot\|$-convergence of the *Neumann* series $R_{H(\kappa)}(-\zeta)$ (see (1.9.26)), and the estimates (1.9.37) and (1.9.39) show that the resolvent set $\rho(-H(\kappa))$ contains the domain

$$\begin{aligned} \mathcal{M}_{\kappa,a} := &\left\{\zeta \in \mathbb{C} : |\arg\zeta| \leq \frac{\pi}{2},\ |\zeta| > \frac{|\kappa|b}{1-|\kappa|a}\right\} \\ \bigcup &\left\{\zeta \in \mathbb{C} : \frac{\pi}{2} \leq |\arg\zeta| \leq \pi,\right. \\ &\left. |\mathrm{Im}\,\zeta| > \frac{|\kappa|b}{1-|\kappa|^2a^2} + \frac{|\kappa|a}{\sqrt{1-|\kappa|^2a^2}}\left[(\mathrm{Re}\,\zeta)^2 + \frac{|\kappa|^2b^2}{1-|\kappa|^2a^2}\right]^{1/2}\right\} \end{aligned} \tag{1.9.40}$$

By virtue of (1.9.40), for any $\kappa \in D_{a^{-1}}$, the asymptotic slopes of the border line of $\mathcal{M}_{\kappa,a}$, when $|\zeta| \to \infty$, are equal to

$$\lim_{|\zeta|\to\infty} \frac{\mathrm{Im}\,\zeta}{|\mathrm{Re}\,\zeta|} = \pm\mathrm{tg}\left(\frac{\pi}{2} + \theta(\kappa,a)\right) = \pm\frac{|\kappa|a}{\sqrt{1-|\kappa|^2a^2}}, \tag{1.9.41}$$

for, respectively, $\pi/2 < \pm \arg \zeta < \pi$, which proves (1.9.35).
Since $\mathcal{M}_{\kappa,b} \subset \rho(-H(\kappa))$, by (1.9.40) and (1.9.41), we get

$$W(H(\kappa)) \subset S_{\pi/2-\theta(\kappa,a),\ -\gamma(\kappa,a,b)}\ , \tag{1.9.42}$$

where $\gamma(\kappa, a, b) \leq |\kappa|\, b\,(1-|\kappa|\, a)^{-1}$ for $|\arg \zeta| \leq \pi/2$, which proves (1.9.36). Hence, $H(\kappa) \in \mathscr{H}(\theta(\kappa, a), \gamma(\kappa, a, b))$ by Propositions 1.7.12 and 1.8.13. As a consequence, $\{U_{H(\kappa)}(z)\}_{z \in S_{\theta(\kappa,b)}}$ is a quasi-bounded holomorphic semigroup for any $\kappa \in D_{a^{-1}}$.

(ii) For a fixed $r < a^{-1}$, there is a contour $\Gamma \subset \bigcap_{\kappa \in D_r} \mathcal{M}_{\kappa,a}$ such that, for any $\kappa \in D_r$, one gets the *Riesz-Dunford* integral formula (1.9.33) for the family of holomorphic semigroups $\{U_{H(\kappa)}(z)\}_{z \in S_{\theta(r)}}$ with $S_{\theta(r)} := \bigcap_{\kappa \in D_r} S_{\theta(\kappa,a)}$. Now, we can proceed as in the proof of Theorem 1.9.10 (ii). Taking into account (1.9.34), we obtain that for any $z \in S_{\theta(r)}$, the mapping $\kappa \mapsto U_{H(\kappa)}(z)$ is $\|\cdot\|$-holomorphic in $\kappa \in D_r$.

□

Remark 1.9.12 Formally, the self-adjoint case $A = A^* \geq 0$ in Theorem 1.9.10 corresponds to an m-sectorial operator with $\alpha = 0$. Hence, we may put $\theta = \pi/2$ in the proof. We then see that the estimate (1.9.25) is less refine than (1.9.39). This entails all other modifications, in particular $\pi/2 - \varepsilon$ instead of $\theta(r)$, where ε is defined by the condition $R(a, \varepsilon) < (2a)^{-1}$.

Remark 1.9.13 If $b = 0$, then because of (1.9.31) or (1.9.36), we get $\gamma(\theta, a, b = 0, \varepsilon) = 0$ or $\gamma(\kappa, a, b = 0,) = 0$. So, in both cases, the operator $H(\kappa)$ is the generator of a *bounded* holomorphic semigroup. The $\gamma(\theta, a, b = 0, \varepsilon) = 0$ does not guarantee that $H(\kappa)$ satisfies conditions of Proposition 1.8.13 for *contraction* holomorphic semigroups.

Remark 1.9.14 If perturbation is infinitesimally small, $B \in \mathcal{P}_{0^+}(A)$, then the proof of Theorem 1.9.10 implies that $R(0, \varepsilon) = \infty$ (1.9.23) for arbitrary small $\varepsilon > 0$. Hence, $H(\kappa) \in \mathscr{H}(\theta, \gamma)$ for any $\kappa \in \mathbb{C}$. For the self-adjoint case, we get a similar result, $H(\kappa) \in \mathscr{H}(\pi/2, 0)$ for $\kappa \in \mathbb{C}$ but only if $b = 0$.

The next assertion elucidates the *Kato-small* perturbation of a self-adjoint generator by *symmetric* operators, which is known as the *Kato-Rellich* theorem about *stability* of self-adjointness.

Corollary 1.9.15 *Let* $A \geq -\gamma_A \mathbb{1}$ *be a self-adjoint operator in* $\mathfrak{H}$, *and let* $B \in \mathcal{P}_a(A)$, *for some* $a < 1$, *be a symmetric operator. Then,*

(a) *Operator* $H = A + B \geq -(\gamma_A + b/(1-a))\mathbb{1}$ *on* dom $(H) =$ dom (A) *is self-adjoint and* $H \in \mathscr{G}(1, \beta)$ *for* $\beta = (\gamma_A + b/(1-a))$.
(b) *One has*

$$|(Bu, u)| \leq a'(Au, u) + b'\|u\|^2, \quad u \in \mathrm{dom}\,(A), \tag{1.9.43}$$

for any $a' > a$ *and some* $a' > 0$.

Proof

(a) By Theorem 1.9.11 (see (1.9.36) and (1.9.40)), we get that for any real κ, $|\kappa| < a^{-1}$, the operator $H(\kappa)$ is self-adjoint (the self-adjoint *stability*), that is, $\theta(\kappa, a) = \pi/2$, and semibounded from below,

$$H(\kappa) \geq (-\gamma_A - \gamma(\kappa, a, b))\,\mathbb{1} = -\Big(\gamma_A + \frac{b|\kappa|}{1 - a|\kappa|}\Big)\,\mathbb{1}. \tag{1.9.44}$$

Since $a < 1$, we may put $\kappa = 1$ in (1.9.44). The last part of this assertion follows from Proposition 1.9.17.

(b) The inequality (1.9.44) is equivalent to

$$-\,\kappa(Bu, u) \leq \Big(\gamma_A + \frac{b|\kappa|}{1 - a|\kappa|}\Big)\|u\|^2 + (Au, u), \quad u \in \operatorname{dom}(A)\,,$$

which implies (1.9.43) for $\kappa = -(a')^{-1}$.

□

Corollary 1.9.16 *Let $A \geq 0$ be a self-adjoint operator in $\mathfrak{H}$. Let B be symmetric with* $\operatorname{dom}(A) \subset \operatorname{dom} B$ *and* $\|Bu\| \leq \|Au\|$, $u \in \operatorname{dom}(A)$. *Then,*

$$|(Bu, u)| \leq (Au, u), \quad u \in \operatorname{dom}(A). \tag{1.9.45}$$

Proof For any $\kappa \in (-1, 1)$, $\kappa B \in \mathcal{P}_1(A)$. Hence, by (1.9.44), one gets $A + \kappa B \geq 0$, which implies (1.9.45) by letting $\kappa \to \pm 1$. □

1.9.3 Perturbations of Quasi-bounded Semigroups

Another class of unbounded admissible perturbations results from analysis of stability of quasi-bounded semigroup with generator $A \in \mathscr{G}(M, \beta)$. To treat the unbounded *Kato-small* perturbations, we use in this section the *resolvent-based* method (Remark 1.9.2). First, we consider supplementary conditions on A and perturbation $B \in \mathcal{P}_1(A)$ (see Definition 1.9.7) in a Hilbert space $\mathfrak{H}$.

Proposition 1.9.17 *Let $A \in \mathscr{G}(1, 0)$ be generator of a contraction C_0-semigroup on $\mathfrak{H}$. If $B \in \mathcal{P}_1(A)$ is an accretive operator, then $H = A + B$ with* $\operatorname{dom}(H) = \operatorname{dom}(A)$ *is generator of contraction semigroup $\{e^{-tH}\}_{t \geq 0}$ on $\mathfrak{H}$, that is, $H \in \mathscr{G}(1, 0)$.*

Proof By Proposition 1.8.10, it is necessary to prove that H is an m-accretive operator in $\mathfrak{H}$. Then according to Definition 1.8.7, we have to check first that the operator $H = A + B$ with $\mathrm{dom}\,(H) = \mathrm{dom}\,(A)$ is closed. Indeed, since $B \in \mathcal{P}_1(A)$, we obtain by (1.9.14)

$$-\,b\|u\| + (1-a)\|Au\| \leq \|Hu\| \leq b\|u\| + (1+a)\|Au\|, \quad u \in \mathrm{dom}\,(A), \;\; a < 1. \tag{1.9.46}$$

Because of the inequalities (1.9.46), the operator H is closed if A is closed.

Next, by Proposition 1.8.10, the operator A is m-accretive. Since B is accretive, we get that

$$\mathrm{Re}\,(u, Hu) = \mathrm{Re}\,(u, Au) + \mathrm{Re}\,(u, Bu) \geq 0, \quad u \in \mathrm{dom}\,(A).$$

Therefore, H is a closed accretive operator with $W(H) \subseteq \overline{\mathbb{C}}_+$. For $\zeta \in \mathbb{C}_+$,

$$\|(H + \zeta\mathbb{1})u\| \geq \mathrm{Re}\,(u, (H+\zeta)u) \geq \mathrm{Re}\,\zeta\,\|u\|^2, \quad u \in \mathrm{dom}\,(A)\,,$$

which implies that $\mathrm{def}\,(H + \zeta\mathbb{1}) = 0$ and $\|(H + \zeta\mathbb{1})^{-1}\| \leq (\mathrm{Re}\,\zeta)^{-1}$. Then by Definition 1.8.7, the operator H is m-accretive. Consequently, it is a generator of contraction semigroup $\{U_H(t)\}_{t\geq 0}$ on $\mathfrak{H}$. □

Remark 1.9.18 Consider operator $H(\kappa) := A + \kappa B$ with $\mathrm{dom}\,(H(\kappa)) = \mathrm{dom}\,(A) \subset \mathfrak{H}$, where A and B are the same as in Proposition 1.9.17. Then one cannot guarantee the result of this proposition for $\kappa \in \mathbb{C}$, even if $|\kappa|$ is small. However, if A is a generator of a *holomorphic* semigroup and $B \in \mathcal{P}_{a<1}(A)$, there is a complete theory of this semigroup perturbed by κB, for $\kappa \in \mathbb{C}$ (see Theorem 1.9.9).

We focus now on similar claims in a Banach space $\mathfrak{X}$ and recall that in general there is no guarantee that $(A + B) \in \mathscr{G}(M', \beta')$ if generator $A \in \mathscr{G}(M, \beta)$, even if B is Kato-small (Definition 1.9.6), that is, (1.9.14) for $a < 1$, with respect to A. A well-known positive result in this direction concerns the case when *both* A and B are generators of contraction semigroups.

Proposition 1.9.19 *Let $A \in \mathscr{G}(1, \beta)$ and $B \in \mathscr{G}(1, \beta')$ in a Banach space $\mathfrak{X}$. If B is small with respect to A with A-bound $a < 1/2$, then $(A + B) \in \mathscr{G}(1, \beta + \beta')$.*

We return to this case at the end of this section (Proposition 1.9.24) and also in discussion of the *product formula-based* method for construction the semigroup perturbations (see Sect. 1.9.6).

Note that in a Hilbert space $\mathfrak{H}$ the relative A-bound in Proposition 1.9.19 can be extended to $a < 1$. However, to relax the restriction $a < 1/2$ to $a < 1$ in a Banach space $\mathfrak{X}$, we consider in Proposition 1.9.19 the case of *accretive* perturbations B of generator A.

Definition 1.9.20 Operator B in a Banach space $\mathfrak{X}$ is called *accretive* (then operator $(-B)$ is called *dissipative*) if $\|(B + \lambda\mathbb{1})\,x\| \geq \lambda\,\|x\|$ for all $x \in \mathrm{dom}\,(B)$ and any $\lambda > 0$.

Now, to proceed, we have to recall that for any $x \in \mathfrak{X}$ one can define a *duality* set of element x by

$$\{x\}^* := \{\phi \in \mathfrak{X}^* : \langle x, \phi \rangle = \|x\|_{\mathfrak{X}}^2 = \|\phi\|_{\mathfrak{X}^*}^2 = 1\} , \tag{1.9.47}$$

where suitable ϕ always exists by the *Hahn-Banach* theorem. ϕ is unique if one considers a Hilbert space $\mathfrak{H}$. Let $\mathcal{B} := \{(x, \phi) \in \mathfrak{X} \times \mathfrak{X}^* : x \in \mathrm{dom}\,(B),\ \phi \in \{x\}^*\}$. Then by virtue of Definition 1.9.20, the operator B is accretive if and only if $\mathrm{Re}\,\langle Bx, \phi\rangle \geq 0$ for all pairs $(x, \phi) \in \mathcal{B}$.

For description of accretive operators in $\mathfrak{H}$, see Definition 1.8.7. Then one gets a criterion for generator $B \in \mathscr{G}(1, 0)$ of contraction C_0-semigroup in a Banach space $\mathfrak{X}$, which is similar to Proposition 1.8.10 in a Hilbert space.

Proposition 1.9.21 *A density defined in $\mathfrak{X}$ operator B belongs to the class $\mathscr{G}(1, 0)$ if and only if it is accretive and the range $\mathrm{ran}\,(B + \lambda\mathbb{1}) = \mathfrak{X}$ for $\lambda > 0$.*

Proof If B is accretive, then for $(x, \phi) \in \mathcal{B}$, one gets by Definition 1.9.20

$$\|(B + z\,\mathbb{1})\,x\| \geq |(Bx + z\,\mathbb{1}, \phi)| \geq \mathrm{Re}\,(Bx, \phi) + \mathrm{Re}\,z\,(x, \phi) \geq \mathrm{Re}\,z\ \|x\|.$$

Given that $\mathrm{ran}\,(B+\lambda\mathbb{1}) = \mathfrak{X}$ for $\lambda > 0$, the operator $B+z\,\mathbb{1}$ is a one-to-one mapping with range equal to $\mathfrak{X}$ for $\mathrm{Re}\,z > 0$. As a consequence,

$$\|(B + z\mathbb{1})^{-1}\| \leq \frac{1}{\mathrm{Re}\,z}, \quad \mathrm{Re}\,z > 0\,, \tag{1.9.48}$$

that (similar to (1.8.10)) yields the *Hille-Yosida* conditions for existence on $\mathfrak{X}$ of a contraction C_0-semigroup with generator $A \in \mathscr{G}(1, 0)$.

Conversely, let $\{U_B(t) = \mathrm{e}^{-tB}\}_{t\geq 0}$ be a contraction semigroup on a Banach space $\mathfrak{X}$. Then by Proposition 1.2.7, the closed densely defined operator B is accretive with $\mathrm{ran}\,(B + \lambda\mathbb{1}) = \mathfrak{X}$ for $\lambda > 0$ with estimate (1.9.48). □

Now, we are in position to improve Proposition 1.9.19 to the case of the Kato-small perturbations, that is, to the relative bound $a < 1$. Moreover, we do not assume that $B \in \mathscr{G}(1, 0)$, but only that it is *accretive*, cf. Proposition 1.9.21.

Proposition 1.9.22 *If in $\mathfrak{X}$ the operator $A \in \mathscr{G}(1, 0)$ and B is an accretive operator with A-bound $a < 1$, then $(A + B) \in \mathscr{G}(1, 0)$.*

Proof

(1) By Proposition 1.9.21, operator A is accretive. Then condition A-bound $a < 1$ implies that $\mathrm{Re}\,((A + B)x, \phi) > 0$ for all pairs $(x, \phi) \in \mathcal{A}$, i.e., densely defined closed operator $A + B$ on $\mathrm{dom}\,(A)$ is also accretive. To prove that $(A + B) \in \mathscr{G}(1, 0)$, it is sufficient by Proposition 1.9.21 to show that $\mathrm{ran}\,(A+B+\lambda\mathbb{1}) = \mathfrak{X}$ for $\lambda > 0$.

For $\varepsilon \in (0, 1)$, we consider an auxiliary accretive operator $\varepsilon\,B$. Note that $A \in \mathscr{G}(1, 0)$ yields the representation $\mathrm{ran}\,(A+\varepsilon\,B+\lambda\mathbb{1}) = (\mathbb{1}+\varepsilon\,B(A+\lambda\mathbb{1})^{-1})\mathfrak{X}$

for $\lambda > 0$. Then for $\mathrm{ran}\,(A + \varepsilon B + \lambda \mathbb{1}) = \mathfrak{X}$, it is sufficient to show that $\|\varepsilon B(A + \lambda\mathbb{1})^{-1}\| < 1$. By (1.9.14), we obtain for $x \in \mathfrak{X}$ the estimate

$$\|\varepsilon B(A + \lambda\mathbb{1})^{-1} x\| \leq \varepsilon b\|(A + \lambda\mathbb{1})^{-1}x\| + \varepsilon a\|A(A + \lambda\mathbb{1})^{-1}x\| \leq$$

$$\varepsilon b\|(A + \lambda\mathbb{1})^{-1}x\| + \varepsilon a \lambda\|(A + \lambda\mathbb{1})^{-1}x\| + \varepsilon a \|x\| \leq (\varepsilon b\lambda^{-1} + 2\varepsilon a) \|x\| .$$

Therefore, there exist $\lambda_0 > 0$ and $\varepsilon < 1/2$ such that $\|\varepsilon B(A + \lambda\mathbb{1})^{-1}\| < 1$ if $\lambda > \lambda_0$. This proves that $\mathrm{ran}\,(A + \varepsilon B + \lambda\mathbb{1}) = \mathfrak{X}$ for $\varepsilon < 1/2$ and $\lambda > \lambda_0$, cf. Proposition 1.9.19. As a consequence, $C_\varepsilon \in \mathscr{G}(1, 0)$, where $C_\varepsilon := (A + \varepsilon B)$.

(2) For extension of the above result to $\varepsilon = 1$, we consider a perturbation of generator $C_\varepsilon \in \mathscr{G}(1, 0)$. By (1.9.14), we obtain

$$\|Bu\| \leq b\|u\| + a\|C_\varepsilon - \varepsilon Bu\| \leq b\|u\| + a\|C_\varepsilon\| + a \|Bu\|, \quad u \in \mathrm{dom}\,(A),$$

which yields for $m \in \mathbb{N}$ the estimate

$$\left\|\frac{1}{m}Bu\right\| \leq \frac{b}{m(1-a)}\|u\| + \frac{a}{m(1-a)}\|C_\varepsilon u\| . \tag{1.9.49}$$

If $a/m(1 - a) < 1/2$, then the statement in (1) ensures that perturbed generator $(C_\varepsilon + B/m) \in \mathscr{G}(1, 0)$. Let $\varepsilon = 1/m$ be such that $1/m < \{(1-a)/2a \wedge 1/2\}$. Then iterating (1.9.49) m-times, we obtain that $(A + B) \in \mathscr{G}(1, 0)$.

□

The next assertion extends the relative bound for A-bounded operator B to $a = 1$.

Proposition 1.9.23 *Let in $\mathfrak{X}$ the operator $A \in \mathscr{G}(1, 0)$ and B be an accretive operator, which satisfies (1.9.14) for $b < \infty$ and $a = 1$. Suppose also that B^* has a dense domain in $\mathfrak{X}$. Then $\overline{(A + B)} \in \mathscr{G}(1, 0)$, where $\overline{(A + B)}$ is the closure of operator sum on dom (A).*

Note that hypothesis on the adjoint operator B^* is not needed if $\mathfrak{X}$ is *reflexive*.

Proof Since the operator $A + B$ is accretive and hence closable with the accretive closure, by Proposition 1.9.21, we have to check only that the range $\mathrm{ran}\,(A+B+\lambda\mathbb{1})$ is dense in $\mathfrak{X}$.

Suppose that $\phi \in \mathfrak{X}^*$ is such that $\mathrm{ran}\,(A + B + \lambda\mathbb{1}) = \ker\,(\phi)$. Then one can find element $y \in \mathfrak{X}$ with $\|y\| = \|\phi\|$ and $\langle y, \phi\rangle \geq \|\phi\|^2/2$.

Note that by the perturbation theory for A-bond < 1 (Proposition 1.9.22) we have $(A + t B) \in \mathscr{G}(1, 0)$ if $0 \leq t < 1$. Hence, for each t, there is an element $x(t) \in \mathrm{dom}\,(A + tB) = \mathrm{dom}\,(A)$, such that $y = (A + t B + \lambda\mathbb{1})\, x(t)$.

Since $(A + t B)$ is accretive, one gets $\|x(t)\| \leq \|y\|$. Then by (1.9.14), we obtain

$$\|Bx(t)\| \leq b\|x(t)\| + \|A x(t)\| \leq b\|x(t)\| + \|(A + tB)\, x(t)\| + \|t B x(t)\|$$

$$= b\|x(t)\| + \|x(t) - y\| + \|t B x(t)\|$$

and finally

$$\|(1-t)Bx(t)\| \leq (2+b)\|y\| . \tag{1.9.50}$$

Note that $\{x(t)\}_{0\leq t<1}$ is bounded by $\|y\|$. Therefore, it has a subnet $\{x(t')\}_{t'}$ that $w^* - \lim_{t'\to 1} x(t') = x^{**}$, where $x^{**} \in \mathfrak{X}^{**}$.

To show that the weak limit $\text{w}-\lim_{t'\to 1}(1-t')Bx(t') = 0$, let $\varphi \in \text{dom}\,(B^*)$. Then

$$\begin{aligned}\lim_{t'\to 1}((1-t')Bx(t'), \varphi) &= \lim_{t'\to 1}(1-t')(x(t'), B^*\varphi) \\ &= \lim_{t'\to 1}(1-t')(x^{**}, B^*\varphi) = 0 .\end{aligned}$$

Since dom (B^*) is dense in $\mathfrak{X}^*$, by the uniform boundedness (1.9.50), one gets that the last limit holds for all elements of $\mathfrak{X}^*$ and in particular for ϕ.

On the hand, since we supposed that ker $(\phi) =$ ran $(A + B + \lambda\mathbb{1})$, one gets

$$\begin{aligned}&\langle y, \phi\rangle = \langle (A + t\,B + \lambda\mathbb{1})\,x(t), \phi\rangle = \\ &\langle (A + B + \lambda\mathbb{1})\,x(t) + (1-t)Bx(t), \phi\rangle = \langle (1-t)Bx(t), \phi\rangle .\end{aligned}$$

This implies $\|\phi\|^2/2 \leq \langle y, \phi\rangle = \lim_{t'\to 1}\langle (1-t')Bx(t'), \phi\rangle$ and proves that $\phi = 0$. Hence, the range ran $(A + B + \lambda\mathbb{1})$ is dense in $\mathfrak{X}$. □

Now, we return to Proposition 1.9.19 and show that the *resolvent-based* method may give results for perturbations of contraction semigroups *without* hypothesis about relative boundedness.

Proposition 1.9.24 *Let A and B be generators of contraction semigroups on a Banach space $\mathfrak{X}$, that is, they belong to $\mathscr{G}(1,0)$. Let $D =$* dom $(A) \cap$ dom (B) *be dense in $\mathfrak{X}$, and let there be $\lambda_0 > 0$ such that* ran $(A + B + \lambda\mathbb{1})$ *is also dense for $\lambda > \lambda_0$. If operator $A + B$ on domain D is closable, then the closure $H := \overline{A + B} \in \mathscr{G}(1,0)$.*

Proof Given that A and B belong to $\mathscr{G}(1,0)$, we have $\|(A + B + \lambda\mathbb{1})u\| \geq \lambda\|u\|$ for $u \in D$. Then by assumption that ran $(A+B+\lambda\mathbb{1})$ is dense, for $\lambda > \lambda_0$, and that operator $A + B$ on the dense domain D is closable, one gets $\|(H + \lambda\mathbb{1})^{-1}\| \leq \lambda^{-1}$. By the first Neumann series, we extend the last estimate to $\lambda > 0$, which yields $H \in \mathscr{G}(1,0)$. □

Since A and B are both generators of C_0-semigroups, we revise Proposition 1.9.24 in discussion of the *product formula-based* method for construction of semigroup perturbations in Sect. 1.9.6.

1.9.4 The Miyadera-Voigt Perturbation Theorem

In the following result, the construction of perturbed quasi-bounded semigroup is owing to the *semigroup-based* method (see Remark 1.9.4). Recall that within this method a perturbed semigroup is presented by the *Dyson-Phillips* series (1.9.11), which is an iterative (*mild*) solution of the one of integral equations (1.9.12) or (1.9.13).

The next preparatory lemma introduces the integral equation for perturbed semigroup in the framework of the *Miyadera-Voigt* theory.

Lemma 1.9.25 *Let* $\{U_A(t)\}_{t\geq 0}$ *be a* C_0*-semigroup on* $\mathfrak{X}$ *with generator* $A \in \mathcal{G}(M_A, \beta_A)$*, and let* B *with* dom (B) *be a linear operator in* $\mathfrak{X}$*:*

(1) *Let there exists a dense subset* $\mathcal{D} \subset \mathfrak{X}$ *such that* $\mathcal{D} \subset \text{dom}\,(A) \cap \text{dom}\,(B)$ *and that* $U_A(t)\mathcal{D} \subset \mathcal{D}$ *for* $t \in [0, \infty)$.
(2) *Let for each* $u \in \mathcal{D}$ *and* $t \in [0, \infty)$ *the function* $t \mapsto B\,U_A(t)u$ *be continuous.*
(3) *If* $\{U_H(t)\}_{t\geq 0}$ *is a* C_0*-semigroup with generator* H*, then the following conditions are equivalent* :

(A) *Generator* H *is a closed extension of operator* $(A + B) \restriction \mathcal{D}$*.*
(B) *For all* $u \in \mathcal{D}$*, the semigroup* $\{U_H(t)\}_{t\geq 0}$ *satisfies integral equation*:

$$U_H(t)u = U_A(t)u - \int_0^t ds\, U_H(t-s)\, B\, U_A(s)\, u\,. \tag{1.9.51}$$

(C) *There exists* $\alpha > 0$ *such that* (1.9.51) *is valid for all* $u \in \mathcal{D}$ *and* $t \in [0, \alpha)$*.*

Proof (A)⇒(B) By virtue of conditions (2) and (3) of lemma for each $u \in \mathcal{D}$ and for $s \in [0, t]$, the function $s \mapsto U_H(t-s)\, U_A(s)\, u$ is continuously differentiable since by (1) and (A) one gets $\partial_s U_H(t-s)\, U_A(s)u = U_H(t-s)\, B\, U_A(s)$. Then by integration, we obtain

$$U_A(t)\, u - U_H(t)\, u = \int_0^t ds\, U_H(t-s)\, B\, U_A(s)\, u\,,$$

that is, Eq. (1.9.51).
(B)⇒(C) By (1.9.51), this assertion is trivial.
(C)⇒(A) For $u \in \mathcal{D}$ and $0 \leq s \leq t < \alpha$, we have

$$\begin{aligned}\|U_H(t-s)\, B\, U_A(s)\, u - B\, u\| \leq & \sup_{0\leq s\leq t<\alpha} \|U_H(t-s)\|\, \|B\, U_A(s)u - Bu\| \\ & + \|U_H(t-s)\, B\, u - Bu\|.\end{aligned}$$

Therefore, by conditions (2) and (3) of lemma, $\lim_{t\to+0} \|U_H(t-s)\, B\, U_A(s)\, u - B\, u\| = 0$ uniformly for $0 \le s \le t < \alpha$. This yields for any $u \in \mathcal{D}$

$$\lim_{t\to+0} \left\| \frac{1}{t} \int_0^t ds\, U_H(t-s)\, B\, U_A(s)\, u - B\, u \right\| = 0. \tag{1.9.52}$$

As a consequence, by (1.9.51), together with definition of C_0-semigroups and the limit (1.9.52), we obtain

$$Hu = \lim_{t\to+0} \frac{1}{t}(\mathbb{1} - U_H(t))\, u = \lim_{t\to+0} \frac{1}{t}(\mathbb{1} - U_A(t))\, u + \frac{1}{t} \int_0^t ds\, U_H(t-s)\, B\, U_A(s)\, u = Au + Bu\,, \quad u \in \mathcal{D}. \tag{1.9.53}$$

For that reason, $u \in \mathrm{dom}\,(H)$, and restriction $H \restriction \mathcal{D} = A \restriction \mathcal{D} + B \restriction \mathcal{D}$. Since H is generator of a C_0-semigroup and subset $\mathcal{D} \subset \mathfrak{X}$ is dense, the operator $H \restriction \mathcal{D} = (A+B) \restriction \mathcal{D}$ in (1.9.53) is closable. For that reason, the generator $H = ((A+B) \restriction \mathcal{D})_{\mathrm{ext}}$ is a closed extension of the right-hand side of (1.9.53). □

Remark 1.9.26 Note that conditions (1) and (2) of Lemma 1.9.25 are weaker than condition of A-boundedness of operator B, Definition 1.9.6. Indeed, if A and B satisfy (1.9.14), that is, $\mathcal{D} = \mathrm{dom}\,(A)$, then we get estimate

$$\begin{aligned} \|B\, U_A(s)\, u - B\, U_A(s')\, u\| \le\ & a\, \|U_A(s)\, u - U_A(s')\, u\| + \\ & + b\, \|A\, U_A(s)\, u - A\, U_A(s')\, u\|, \quad u \in \mathrm{dom}\,(A)\,. \end{aligned} \tag{1.9.54}$$

For any $u \in \mathrm{dom}\,(A)$, there is $x \in \mathfrak{X}$ such that $u = (A + \lambda\mathbb{1})^{-1}x$ and $y = A(A + \lambda\mathbb{1})^{-1}x \in \mathfrak{X}$, for $\lambda > \beta_A$. Therefore, by $A\, U_A(s)\, u = U_A(s)\, y$ and by the strong continuity of semigroup $\{U_A(t)\}_{t\ge0}$, we obtain from estimate (1.9.54)

$$\lim_{s'\to s} \|B\, U_A(s)\, u - B\, U_A(s')\, u\| = 0\,, \quad u \in \mathcal{D}. \tag{1.9.55}$$

The next assertion is the *Miyadera-Voigt* theorem that gives the construction of perturbed quasi-bounded semigroups within the semigroup-based method.

Proposition 1.9.27 (Miyadera-Voigt Theorem) *Let $\{U_A(t)\}_{t\ge0}$ be a semigroup acting on a Banach space $\mathfrak{X}$ with generator $A \in \mathcal{G}(M_A, \beta_A)$. Let B be densely defined in $\mathfrak{X}$ linear operator, and assume there is a dense subset $\mathcal{D} \subset \mathfrak{X}$ such that*

(i) *For $\mathcal{D} \subset \mathrm{dom}\,(A) \cap \mathrm{dom}\,(B)$, one has $U_A(t)\mathcal{D} \subset \mathcal{D}$, for $t \in [0, \infty)$, and for all $u \in \mathcal{D}$, the function $t \mapsto B\, U_A(t)\, u$ is continuous.*
(ii) *There are constants $\alpha \in (0, \infty]$ and $\beta \in [0, 1)$ such that*

$$\int_0^\alpha dt\, \mathrm{e}^{-\beta_A t}\, \|B\, U_A(t)\, u\,\| \le \beta\, \|u\| \tag{1.9.56}$$

holds for all $u \in \mathcal{D}$. Then

(a) *There exists a (unique) semigroup $\{U_H(t)\}_{t\geq 0}$ with generator $H = ((A + B) \upharpoonright \mathcal{D})_{\mathrm{ext}}$, which is a closed extension of operator $(A + B) \upharpoonright \mathcal{D}$.*
(b) *The perturbed semigroup $\{U_H(t)\}_{t\geq 0}$ is quasi-bounded, and the following estimate holds:*

$$\|U_H(t)\| \leq \frac{M_A}{1-\beta}\left(\frac{M_A}{1-\beta}\right)^{t/\alpha} \mathrm{e}^{\beta_A t}\,. \tag{1.9.57}$$

Thus, H is generator of quasi-bounded semigroup, that is, $H \in \mathcal{G}(M_H, \beta_H)$, for

$$M_H := M_A/(1-\beta) \quad \text{and} \quad \beta_H := \beta_A + (1/\alpha)\ln(M_A/(1-\beta))\,. \tag{1.9.58}$$

(c) *The semigroup $\{U_H(t)\}_{t\geq 0}$ is unique. In fact, dom(H) = dom(A), and generator H is the closure of restriction $(A + B) \upharpoonright \mathcal{D}$. Moreover, the restriction $B \upharpoonright \mathcal{D}$ of operator B is $A \upharpoonright \mathcal{D}$-bounded, and it can be extended uniquely to an A-bounded operator $\widehat{B}$ with domain dom$(\widehat{B})$ = dom(A). By this extension, it holds that $H = A + \widehat{B}$, cf.* (a).

Proof
(a) Let $\alpha > 0$. We define by recursion a sequence of the families of (bounded) operator-valued functions $\{[0,\alpha) \ni t \mapsto U_n(t)\}_{n\in\mathbb{N}_0}$, where $U_0(t) := U_A(t)$ and

$$\begin{aligned} U_{n+1}(t)\,u :&= \int_0^t ds\, U_n(t-s)\,(-B)\,U_A(s)\,u = \\ &= \int_0^t ds\, U_n(s)\,(-B)\,U_A(t-s)\,u\,, \quad u \in \mathcal{D}\,, \quad t \in [0,\alpha)\,. \end{aligned} \tag{1.9.59}$$

We show first that for each $n \in \mathbb{N}_0$ the family $\{U_n(t)\}_{t\in[0,\alpha)} \subset \mathcal{L}(\mathfrak{X})$ and that it is strongly continuous. Note that it true for $\{U_0(t)\}_{t\in[0,\alpha)}$ since it is quasi-bounded semigroup. Consider $n = 1$

$$U_1(t)\,u = \int_0^t ds\, U_0(t-s)\,(-B)\,U_A(s)\,u\,, \quad u \in \mathcal{D}\,, \quad t \in [0,\alpha)\,. \tag{1.9.60}$$

By condition (i), the integrand is continuous function, then Bochner integral (1.9.60) is well-defined, and by $A \in \mathcal{G}(M_A, \beta_A)$ together with condition (ii), one gets estimate

$$\|U_1(t)\,u\| \leq M_A \mathrm{e}^{\beta_A t} \int_0^t ds\, \mathrm{e}^{-\beta_A s}\,\|B\,U_A(s)\,u\| \leq M_A \mathrm{e}^{\beta_A t}\,\beta\|u\|\,, \tag{1.9.61}$$

for $u \in \mathcal{D}$ and $t \in [0, \alpha)$. Since $\overline{\mathcal{D}} = \mathfrak{X}$, the estimate (1.9.61) yields

$$\|U_1(t)\| \leq M_A e^{\beta_A t} \beta, \quad t \in [0, \alpha). \tag{1.9.62}$$

To prove the strong continuity of $\{U_1(t)\}_{t\in[0,\alpha)}$, we use (1.9.60) and estimate

$$\|U_1(t+\delta)\, u - U_1(t)\, u\| \leq \int_0^t ds\, \|(U_0(t+\delta-s) - U_0(t-s))\, B\, U_A(s)\, u\|$$
$$+ \int_t^{t+\delta} ds\, \|U_0(t+\delta-s)\, B\, U_A(s)\, u\|, \quad u \in \mathcal{D}. \tag{1.9.63}$$

for $t \in [0, \alpha)$ and $t + \delta \in [0, \alpha)$. By condition (i), the function $[0, t] \ni s \mapsto B\, U_A(s)u \in \mathfrak{X}$ is continuous. Then function $s \mapsto (U_0(t+\delta-s) - U_0(t-s))\, B\, U_A(s)\, u$ for $u \in \mathcal{D}$ is also continuous and

$$R_\alpha(u) := \sup_{s\in[0,\alpha]} \|B\, U_A(s)\, u\| \leq \infty. \tag{1.9.64}$$

The strong continuity of quasi-bounded semigroup $\{U_0(s)\}_{s\in[0,\alpha)}$ then implies

$$\lim_{\delta\to 0} \int_0^t ds\, \|(U_0(t+\delta-s) - U_0(t-s))\, B\, U_A(s)\, u\| = 0. \tag{1.9.65}$$

By similar arguments, cf. (1.9.64), we obtain

$$\lim_{\delta\to 0} \int_t^{t+\delta} ds\, \|U_0(t+\delta-s)\, B\, U_A(s)\, u\| \leq \tag{1.9.66}$$
$$\leq \lim_{\delta\to 0} M_A e^{\beta_A(t+\delta)} \int_t^{t+\delta} ds\, e^{-\beta_A s} R_\alpha(u) = 0.$$

Thus, (1.9.65) and (1.9.66) prove continuity in (1.9.63) for any $u \in \mathcal{D}$. Since $\mathcal{D}$ is dense, for any $x \in \mathfrak{X}$ and $\varepsilon(x) > 0$, there exists $u(x) \in \mathcal{D}$ such that $\|x - u(x)\| < \varepsilon(x)$, and we have estimate:

$$\|U_1(t+\delta)\, x - U_1(t)\, x\| \leq \|U_1(t+\delta)\, u(x) - U_1(t)\, u(x)\| +$$
$$+ \|(U_1(t+\delta) - U_1(t))(x - u(x))\|.$$

By virtue of (1.9.63), (1.9.65), (1.9.66), and (1.9.62) for arbitrary small $\varepsilon(x)$, this estimate yields the *strong* continuity of *quasi-bounded* family $\{U_1(t)\}_{t\in[0,\alpha)}$.

Suppose now that family $\{U_n(t)\}_{t\in[0,\alpha)}$ is strongly continuous and quasi-bounded: $\|U_n(t)\| \leq M_A e^{\beta_A t}\, \beta^n$, cf. (1.9.62). By (1.9.59) and (1.9.61), we obtain

$$\|U_{n+1}(t)\, u\| \leq \int_0^t ds\, \|U_n(t-s)\|\, \|(-B)\, U_A(s)\, u\| \leq \tag{1.9.67}$$
$$\leq M_A e^{\beta_A t}\, \beta^n \int_0^t ds\, e^{-\beta_A s}\, \|B\, U_A(s)\, u\| \leq M_A e^{\beta_A t}\, \beta^{n+1}\, \|u\|\,,$$
$$u \in \mathcal{D}\,, t \in [0,\alpha)\,.$$

Then by the same arguments as for $n = 1$, we deduce that the family $\{U_{n+1}(t)\}_{t\in[0,\alpha)}$ is also strongly continuous and quasi-bounded: $\|U_{n+1}(t)\| \leq M_A e^{\beta_A t}\, \beta^{n+1}$. So, this assertion is valid by induction for any family $\{U_n(t)\}_{t\in[0,\alpha)}$, $n \in \mathbb{N}_0$.

Following the semigroup-based method, we introduce the *Dyson-Phillips* series

$$U(t) := \sum_{n=0}^{\infty} U_n(t)\,, \quad t \in [0,\alpha)\,. \tag{1.9.68}$$

Then estimate $\|U_n(t)\| \leq M_A e^{\beta_A t}\, \beta^n$ implies that series (1.9.68) converges in the operator-norm topology and the family $\{U(t)\}_{t\in[0,\alpha)}$ is quasi-bounded:

$$\|U(t)\| \leq \sum_{n=0}^{\infty} \|U_n(t)\| \leq \frac{M_A}{1-\beta}\, e^{\beta_A t}\,, \quad t \in [0,\alpha)\,. \tag{1.9.69}$$

Note that (1.9.69) implies the uniform in $t \in [a,b] \subset [0,\alpha)$ convergence of the series (1.9.68) in the *strong* operator topology. Therefore, the established above strong continuity of the terms $\{U_n(t)\}_{t\in[0,\alpha)}$, $n \in \mathbb{N}_0$, provides the *strong* continuity of the quasi-bounded family $\{U(t)\}_{t\in[0,\alpha)}$.

Next, we prove that this family satisfies the semigroup *composition* law, $U(t)U(s) = U(t+s)$, when all arguments belong to $[0,\alpha)$. To this aim, we show by induction that

$$\sum_{k=0}^{n} U_k(t)\, U_{n-k}(s) = U_n(t+s)\,. \tag{1.9.70}$$

For $n = 0$, Eq. (1.9.70) is obvious since $U_0(t) = U_A(t)$. Let (1.9.70) be true for $n \in \mathbb{N}$. Then by (1.9.59), we obtain for $u \in \mathcal{D}$

$$\sum_{k=0}^{n+1} U_k(t)\, U_{n+1-k}(s)u =$$
$$\sum_{k=0}^{n} U_k(t) \int_0^s d\tau\, U_{n-k}(s-\tau)\, (-B)\, U_A(\tau)\, u$$

$$+\int_0^t d\tau\, U_n(t-\tau)\,(-B)\,U_A(\tau)\,U_A(s)\,u =$$

$$\int_0^s d\tau\, U_n(t+s-\tau)\,(-B)\,U_A(\tau)\,u + \int_0^t d\tau\, U_n(t-\tau)\,(-B)\,U_A(s+\tau)\,u =$$

$$\int_0^{t+s} d\tau\, U_n(t+s-\tau)\,(-B)\,U_A(\tau)\,u = U_{n+1}(t+s)\,u\,.$$

Since $\overline{\mathcal{D}} = \mathfrak{X}$, this relation proves (1.9.70) for $n+1$, and as a result,

$$\begin{aligned} U(t)U(s) &= \sum_{m=0}^{\infty} U_m(t) \sum_{n=0}^{\infty} U_n(s) = \sum_{m=0}^{\infty}\sum_{k=0}^{m} U_k(t)U_{m-k}(s) \\ &= \sum_{m=0}^{\infty} U_m(t+s) = U(t+s)\,. \end{aligned}$$

For extension of the quasi-bounded family $\{U(t)\}_{t\in[0,\alpha)}$ (1.9.69) to any $t \in \mathbb{R}_0^+$, let $t \geq 0$. Then there exists $n \in \mathbb{N}$ such that $t/n < \alpha$, and we define $\{\widehat{U}(t)\}_{t\in\mathbb{R}_0^+}$ by $\widehat{U}(t) := U(t/n)^n$. The family $\{\widehat{U}(t)\}_{t\geq 0}$ inherits the semigroup properties and strong continuity of the family $\{U(t)\}_{t\in[0,\alpha)}$, and, consequently, it is a C_0-semigroup.

Let H be generator of this semigroup. Then we denote $\{\widehat{U}(t)\}_{t\in\mathbb{R}_0^+}$ by $\{U_H(t)\}_{t\geq 0}$. By virtue of construction (1.9.59) and (1.9.68), this semigroup satisfies equation (1.9.51). Then by Lemma 1.9.25, (B) $\Rightarrow$ (A), generator $H = ((A+B) \restriction \mathcal{D})_{\mathrm{ext}}$ is a closed extension of operator $(A+B) \restriction \mathcal{D} = H \restriction \mathcal{D}$ (see (1.9.53)). Since generator H is *an* extension, at this stage, we have no proof of *uniqueness* of the semigroup $\{U_H(t)\}_{t\geq 0}$.

(b) By construction of semigroup $\{U_H(t)\}_{t\geq 0}$, one has for any $t \geq 0$ the representation $U_H(t) = U(t/n)^n$, where $n \in \mathbb{N}$ is such that $t/n < \alpha$. Then we can take $n^* \in \mathbb{N}$, which satisfies conditions $t/n^* < \alpha \leq t/(n^*-1)$, that is, $n^* \leq 1 + t/\alpha$. As a consequence, by (1.9.69), we obtain the estimate

$$\|U_H(t)\| = \|U(t/n^*)^{n^*}\| \leq \left(\frac{M_A}{1-\beta}\, \mathrm{e}^{\beta_A t/n^*}\right)^{n^*} \leq \left(\frac{M_A}{1-\beta}\right) \mathrm{e}^{\widehat{\beta}_A t}\,, \qquad (1.9.71)$$

which coincides with (1.9.57) since $\widehat{\beta}_A := \beta_A + (1/\alpha)\ln(M_A/(1-\beta))$.

(c) For the proof of uniqueness of semigroup $\{U_H(t)\}_{t\geq 0}$, we scrutinise the structure of generator H. To this purpose, we compare it to the generator A to establish uniqueness of the operator H constructed by means of $\{\widehat{U}(t)\}_{t\in\mathbb{R}_0^+}$.

To this aim, we use for $\eta > \widehat{\beta}_A$ and $u \in \mathcal{D}$ the representation

$$(A + \eta \mathbb{1})^{-1} u - (H + \eta \mathbb{1})^{-1} u = \tag{1.9.72}$$
$$= \int_0^\infty ds \, \mathrm{e}^{-\eta s} \, U_H(s) \int_0^\infty dt \, \mathrm{e}^{-\eta t} \, B \, U_A(t) \, u \, .$$

To justify the right-hand side of (1.9.72), we estimate the last integral:

$$\int_0^\infty dt \, \mathrm{e}^{-\eta t} \, \|B \, U_A(t) \, u\| =$$
$$= \int_0^\alpha dt \, \mathrm{e}^{-\eta t} \, \|B \, U_A(t) \, u\| + \sum_{k=1}^{\infty} \int_{k\alpha}^{(k+1)\alpha} dt \, \mathrm{e}^{-\eta t} \, \|B \, U_A(t) \, u\| \leq$$
$$\leq \int_0^\alpha dt \, \mathrm{e}^{-\beta_A t} \, \|B \, U_A(t) \, u\| + \sum_{k=1}^{\infty} \mathrm{e}^{-\eta \alpha k} \int_0^\alpha dt \, \mathrm{e}^{-\beta_A t} \, \|B \, U_A(t) \, U_A(\alpha k) \, u\| \leq$$
$$\leq \beta \, \|u\| + \sum_{k=1}^{\infty} \mathrm{e}^{-\eta \alpha k} \beta \, M_A \, \mathrm{e}^{\beta_A \alpha k} \, \|u\| = \beta(\eta) \, \|u\| \, ,$$

where $\beta(\eta) := \beta(1 + M_A/(\mathrm{e}^{(\eta - \beta_A)\alpha} - 1))$. Since $\mathcal{D}$ is dense, this estimate implies that operator

$$B(\eta) := \int_0^\infty dt \, \mathrm{e}^{-\eta t} \, B \, U_A(t) \in \mathcal{L}(\mathfrak{X}) \tag{1.9.73}$$

is bounded and its norm $\|B(\eta)\| \leq \beta(\eta)$.

Definition (1.9.73) and representation (1.9.72) yield

$$(A + \eta \mathbb{1})^{-1} = (H + \eta \mathbb{1})^{-1} (\mathbb{1} + B(\eta)) \, . \tag{1.9.74}$$

Since for large $\eta > \widehat{\beta}_A$ (as well as for large α or for small β) the value of parameter $\beta(\eta) < 1$, one can make $\|B(\eta)\| \leq 1$ and operator $\mathbb{1} + B(\eta)$ injective. Then by virtue of (1.9.74), we deduce that domain

$$\mathrm{dom}\,(A) = \mathrm{ran}\,((A + \eta \mathbb{1})^{-1}) = \mathrm{ran}\,((H + \eta \mathbb{1})^{-1}) = \mathrm{dom}\,(H) \tag{1.9.75}$$

but also that the operator $(H + \eta \mathbb{1})(A + \eta \mathbb{1})^{-1} \in \mathcal{L}(\mathfrak{X})$, as well as the operator $(A + \eta \mathbb{1})(H + \eta \mathbb{1})^{-1} \in \mathcal{L}(\mathfrak{X})$, is bounded. Thus, in consistency with (1.9.75), operator H is A-bounded and A is H-bounded,

$$b_- \|u\| + a_- \|A \, u\| \leq b \|u\| + a \|H \, u\| \leq b_+ \|u\| + a_+ \|A \, u\| \, , \quad u \in \mathrm{dom}\,(A) \, , \tag{1.9.76}$$

for some $a_{\mp} > 0$, $b_{\mp} > 0$ and $a \geq 0$, $b > 0$. Then, (1.9.75) and (1.9.76) yield that the graph-norm $\|\cdot\|_A$ of A and the graph-norm $\|\cdot\|_H$ of H are equivalent.

Recall that if a dense $\mathcal{D} \subset \operatorname{dom}(A)$ is invariant under the semigroup $\{U_A(t)\}_{t\geq 0}$: $U_A(t)\mathcal{D} \subset \mathcal{D}$, for $t \in [0, \infty)$ (see condition (i)), then $\mathcal{D} = \operatorname{core}(A)$ by the *Nelson* theorem (see Proposition 1.2.9).

This means that the closure $\overline{A \restriction \mathcal{D}} = A$ or, equivalently, that subset $\mathcal{D}$ is dense in $\operatorname{dom}(A)$ for the graph-norm $\|\cdot\|_A$. Because the graph-norms $\|\cdot\|_A$ and $\|\cdot\|_H$ on $\mathfrak{X}_T^p$ are equivalent and $\mathcal{D} \subset \operatorname{dom}(H)$ by (1.9.75), we infer that $\mathcal{D}$ is also a core for generator H, that is, $H = \overline{H \restriction \mathcal{D}}$. Consequently, the generator H is uniquely defined as the minimal extension by the closure. So, $\overline{H \restriction \mathcal{D}} = \overline{(A+B) \restriction \mathcal{D}}$ coincides with the closed extension $H = ((A + B) \restriction \mathcal{D})_{\text{ext}}$ that we have already found in the part (a) of the proof. Then, a one-to-one correspondence between C_0-semigroups and their generators proves uniqueness of semigroup $\{U_H(t)\}_{t\geq 0}$.

Let $\widehat{B} := H - A$ be defined on $\operatorname{dom}(\widehat{B}) = \operatorname{dom}(A)$. Then $\widehat{B}$ is closed and A-bounded, cf. (1.9.76). Similarly, we conclude that operator $\widehat{B} \restriction \mathcal{D} = (H - A) \restriction \mathcal{D} = B \restriction \mathcal{D}$ is $A \restriction \mathcal{D}$-bounded, that is, for some $a_0 > 0$, $b_0 > 0$:

$$\|B\,u\| \leq a_0\|A\,u\| + b_0\|u\|\,, \quad u \in \mathcal{D}\,. \tag{1.9.77}$$

Since A is closed and $\mathcal{D} = \operatorname{core}(A)$, the inequality (1.9.77) implies that operator $B \restriction \mathcal{D}$ is closable and that it has a unique closed A-bounded extension $(B \restriction \mathcal{D})_{\text{ext}}$ for $\operatorname{dom}(A)$, which coincides with $\widehat{B}$. □

Corollary 1.9.28 *Let in addition to conditions of Proposition* 1.9.27 *the operator B be closed. Then B is A-bounded, and operator $H := A + B$, on* $\operatorname{dom}(H) = \operatorname{dom}(A)$, *is generator of the C_0-semigroup* $\{U_H(t)\}_{t\geq 0}$.

Proof Let $u \in \operatorname{dom}(A)$. Since $\mathcal{D} = \operatorname{core}(A)$, there exists a sequence $\{u_n\}_{n\geq 1} \subset \mathcal{D}$ such that $\lim_{n\to\infty} u_n = u$ and $\lim_{n\to\infty} Au_n = Au$. By definition of $\widehat{B} := H - A$ on $\operatorname{dom}(\widehat{B}) = \operatorname{dom}(A)$, the operator $\widehat{B}$ is A-bounded, cf. (1.9.76). This implies for $\lim_{n\to\infty} u_n = u$:

$$\lim_{n\to\infty} Bu_n = \lim_{n\to\infty} \widehat{B}u_n =: \widehat{B}u = w \in \mathfrak{X}\,, \quad \{u_n\}_{n\geq 1} \subset \mathcal{D}\,. \tag{1.9.78}$$

Given that operator B is closed, the limit (1.9.78) yields that $u \in \operatorname{dom}(B)$ and $Bu = w$, i.e., $\operatorname{dom}(B) \supset \operatorname{dom}(A)$. Since operators A and B are closed, $\operatorname{dom}(B) \supset \operatorname{dom}(A)$ implies that B is A-bounded. Moreover, by (1.9.78) and $Bu = w$, we also obtain $B \restriction \operatorname{dom}(A) = \widehat{B} \restriction \operatorname{dom}(A)$. Then by definition of $\widehat{B}$, the operator $H = A + B$ is closed on domain $\operatorname{dom}(H) = \operatorname{dom}(A)$, and, consequently, it is a generator of the C_0-semigroup $\{U_H(t)\}_{t\geq 0}$. □

1.9.5 Approximation of Semigroups and Generators

In the framework of general approximation problem, we recall here relations between convergence of semigroups and generators.

Lemma 1.9.29 *Let $K, L \in \mathscr{G}(M, \beta)$. Then one gets for* $\operatorname{Re}\zeta < (-\beta)$ *the estimate*

$$\|R_K(\zeta)\,(U_K(t) - U_L(t))R_L(\zeta)\| \leq t\,M^2\,\mathrm{e}^{\beta\,t}\,\|R_K(\zeta) - R_L(\zeta)\|\,. \tag{1.9.79}$$

Proof Since the functions $U_K(t)\,R_K(\zeta)$ and $U_L(t)R_L(\zeta)$ are strongly differentiable in $t > 0$, we obtain that for $0 < s < t$

$$\partial_s(U_K(t-s)\,R_K(\zeta)\,U_L(s)R_L(\zeta)) = U_K(t-s)(R_\zeta(L) - R_\zeta(K))U_L(s)\,. \tag{1.9.80}$$

Then integration of (1.9.80) over s yields

$$R_K(\zeta)\,(U_L(t) - U_K(t))\,R_L(\zeta) = \int_0^t ds\;U_K(t-s)\,(R_L(\zeta) - R_K(\zeta))\,U_L(s)\,. \tag{1.9.81}$$

Hence, (1.9.79) follows from (1.9.81) by virtue of estimates, $\|U_K(t-s)\| \leq M\,\mathrm{e}^{\beta\,(t-s)}$, for $t \geq s$, and $\|U_L(s)\| \leq M\,\mathrm{e}^{\beta\,s}$. □

Proposition 1.9.30 (Trotter-Kato-Neveu Theorem) *Let $\{U_A(t)\}_{t\geq 0}$ and $\{U_{A_n}(t)\}_{t\geq 0}$, for $n \in \mathbb{N}$, be quasi-bounded semigroups of the type $\mathscr{G}(M, \beta)$ on $\mathfrak{X}$ such that*

$$\|U_{A_n}(t)\| \leq M\mathrm{e}^{\beta\,t}\,, \quad \text{for all } t \geq 0 \text{ and } n \in \mathbb{N}\,. \tag{1.9.82}$$

Then one has

$$\lim_{n\to\infty}\sup_{t\in[0,\tau]}\|U_{A_n}(t)x - U_A(t)x\| = 0\,, \tag{1.9.83}$$

for all $x \in \mathfrak{X}$ and for any finite $\tau > 0$, if and only if

$$\lim_{n\to\infty} R_{A_n}(\zeta)\,x = R_A(\zeta)\,x\,, \tag{1.9.84}$$

for all $x \in \mathfrak{X}$ and for some $\operatorname{Re}\zeta < (-\beta)$.

Proof Without loss of generality, we can reduce the proof to bounded semigroups, $M \geq 1$, of the type $\beta = 0$, that is, for generators from $\mathscr{G}(M, 0)$. Then $\|U_{A_n}(t-s)\| \leq M$, for $t \geq s$, and (1.9.81) yields

$$\|R_{A_n}(\zeta)(U_A(t) - U_{A_n}(t))\,R_A(\zeta)\,x\| \leq M\int_0^t ds\;\|(R_A(\zeta) - R_{A_n}(\zeta))\,U_A(s)\,x\|, \tag{1.9.85}$$

for any $x \in \mathfrak{X}$.

Suppose that one has (1.9.84); then for each s and $n \to \infty$, the integrand in (1.9.85) converges to zero. Since all generators $\{A_n\}_{n\geq 1}$, $A \in \mathscr{G}(M, 0)$, the estimate $\|(R_A(\zeta)) - R_{A_n}(\zeta))\, U_A(s)\, x\| \leq 2\, M^2\, (-\mathrm{Re}\,\zeta)^{-1}\, \|x\|$ holds for $\mathrm{Re}\,\zeta < 0$. As a consequence, by (1.9.84), the pointwise convergence of the integrand implies the convergence to zero of the integral in the right-hand side of (1.9.85) which is *uniform* in $t \in [0, \tau]$ for any finite $\tau > 0$. Then on account of the mapping $R_A(\zeta) : \mathfrak{X} \to \mathrm{dom}\,(A)$, this convergence yields for the left-hand side of (1.9.85):

$$\lim_{n\to\infty} \|\, R_{A_n}(\zeta)\, (U_A(t) - U_{A_n}(t))\, x\| = 0\,, \quad x \in \mathrm{dom}\,(A)\,, \tag{1.9.86}$$

uniformly in $t \in [0, \tau]$. Since domain $\mathrm{dom}\,(A)$ is dense in $\mathfrak{X}$ and $\|R_{A_n}(\zeta)(U_A(t) - U_{A_n}(t))\| \leq 2\, M^2\, (-\mathrm{Re}\,\zeta)^{-1}$, the limit (1.9.86) can be extended to all $x \in \mathfrak{X}$ (see Proposition A.1.6). Now, we remark that by (1.9.84) and by condition of quasi-boundedness of semigroup one also infers that

$$\lim_{n\to\infty} \|U_{A_n}(t)\, (R_A(\zeta) - R_{A_n}(\zeta))\, x\| = 0\,, \tag{1.9.87}$$

$$\lim_{n\to\infty} \|(R_A(\zeta) - R_{A_n}(\zeta))\, U_A(t)\, x\| = 0\,. \tag{1.9.88}$$

for all $x \in \mathfrak{X}$. From (1.9.86) to (1.9.88), we obtain

$$\lim_{n\to\infty} \|\, (U_A(t) - U_{A_n}(t))\, R_A(\zeta)\, x\| = 0\,, \quad x \in \mathfrak{X}\,. \tag{1.9.89}$$

Seeing that $R_A(\zeta) : \mathfrak{X} \to \mathrm{dom}\,(A)$, by the same arguments (Proposition A.1.6) as for extension of (1.9.86) to $\mathfrak{X}$, we deduce from (1.9.89) that

$$\lim_{n\to\infty} \|\, (U_A(t) - U_{A_n}(t))\, x\| = 0\,, \quad x \in \mathfrak{X}\,. \tag{1.9.90}$$

Note that by virtue of $\|U_{A_n}(t)\| \leq M$ (1.9.82) the convergence in (1.9.87) is uniform in $t \in [0, \tau]$ for any finite $\tau > 0$. To prove the same for convergence in (1.9.88), we comment that for any $x \in \mathfrak{X}$ the set $\{U_A(t)\, x\}_{t\in[0,\tau]} \subset \mathfrak{X}$ is a compact subset of $\mathfrak{X}$. Then the strong limit (1.9.84) implies on the compact set $\{U_A(t)\, x\}_{t\in[0,\tau]}$ a uniform pointwise convergence, that is, convergence which is uniform in $t \in [0, \tau]$ (see Proposition A.1.7 and Notes to Chap. 1 (Notes to Sects. 1.9 and 1.9.5)). As a result, the limit (1.9.90) inherits the uniformity in $t \in [0, \tau]$ from (1.9.86) to (1.9.88). This proves (1.9.83).

Conversely, let (1.9.83) be true for $t \in [0, \tau]$. Then by the semigroup property, it is also true for any $t \geq 0$. Consequently, by the *dominated convergence* theorem (for the Bochner integral), we obtain the strong limit:

$$\lim_{n\to\infty} \left\| \int_0^{\infty} ds\, e^{\zeta t}\, (U_{A_n}(s) - U_A(s))\, x \right\| = 0\,, \quad \mathrm{Re}\,\zeta < 0\,,\ x \in \mathfrak{X}\,.$$

Then by the Laplace transform (1.2.16), this yields (1.9.84) for $\mathrm{Re}\,\zeta < 0$. □

Remark 1.9.31 It is often useful to have another way to control the strong resolvent convergence (1.9.84). Let $D \subset \mathrm{dom}\,(A)$ be a *core* for $A \in \mathcal{G}(M,0)$, denoted by $D := \mathrm{core}(A)$. Suppose that generators $\{A_n\}_{n\geq 1} \subset \mathcal{G}(M,0)$ convergence on a core(A) in the sense that for each $u \in D$ there exists $\{u_n \in \mathrm{dom}\,(A_n)\}_{n\geq 1}$ such that

$$\lim_{n\to\infty} u_n = u \text{ and } \lim_{n\to\infty} A_n u_n = Au \,. \tag{1.9.91}$$

If $\mathrm{Re}\,\zeta < 0$, we have $\|R_A(\zeta)\| \leq M/(-\mathrm{Re}\,\zeta)$ and $\|R_{A_n}(\zeta)\| \leq M/(-\mathrm{Re}\,\zeta)$ for all $n \geq 1$. For $u \in D$, we put $y := (A - \zeta\mathbb{1})u$. Then by assumption, there exists $\{u_n \in \mathrm{dom}\,(A_n)\}_{n\geq 1}$ such that $\lim_{n\to\infty} u_n = u$ and hence $\lim_{n\to\infty}(A_n - \zeta\mathbb{1})u_n = y$.

Let $y_n := (A - \zeta\mathbb{1})u_n$. Then one obtains the estimate

$$\begin{aligned}\|R_{A_n}(\zeta)y - R_A(\zeta)y\| &\leq \|R_{A_n}(\zeta)y - R_{A_n}(\zeta)y_n\| + \|R_{A_n}(\zeta)y_n - R_A(\zeta)y\| \\ &\leq \|R_{A_n}(\zeta)\|\,\|y - y_n\| + \|x_n - x\|\,.\end{aligned}$$

Since $\overline{\mathrm{ran}\,(A - \zeta\mathbb{1})} = \mathfrak{X}$ for $\mathrm{Re}\,\zeta < 0$, this estimate proves that the convergence on the core(A) (1.9.91) implies the convergence of generators $\{A_n\}_{n\geq 1}$ in the *strong resolvent* sense (1.9.84).

We remind here that the sequence $\{A_n\}_{n\geq 1}$ converges to A in the *uniform resolvent* sense if

$$\lim_{n\to\infty} \|R_{A_n}(\zeta) - R_A(\zeta)\| = 0\,,$$

for any $\zeta \in \rho(A)$, where $\rho(A)$ is the resolvent set of the operator A.

In Proposition 1.9.30, the limit operator A was *assumed* to be a generator of semigroup $U_A(t)$. The next statement (which is also often called the *Trotter-Kato-Neveu* theorem) aims to prove that it indeed does but under certain additional conditions.

Proposition 1.9.32 *Let $\{U_{A_n}(t)\}_{t\geq 0}$, for $n \in \mathbb{N}$, be quasi-bounded semigroups of the type $\mathcal{G}(M,\beta)$ (1.9.82). If for some* $\mathrm{Re}\,\zeta_0 < (-\beta)$ *the resolvents $\{R_{A_n}(\zeta_0)\}_{n\geq 1}$ converge strongly to a bounded operator $R(\zeta_0) \in \mathcal{L}(\mathfrak{X})$ with the range* $\mathrm{ran}(R(\zeta_0))$ *dense in the space $\mathfrak{X}$, then the sequence of semigroups $\{U_{A_n}(t)\}_{t\geq 0}$, $n \in \mathbb{N}$, converges for all $x \in \mathfrak{X}$ uniformly on* $[0,\tau]$,

$$\lim_{n\to\infty} \sup_{t\in[0,\tau]} \|U_{A_n}(t)x - U_A(t)x\| = 0\,, \tag{1.9.92}$$

to a strongly continuous semigroup quasi-bounded semigroup $\{U_A(t)\}_{t\geq 0}$ with generator $A \in \mathcal{G}(M,\beta)$ such that $R_A(\zeta) = R(\zeta)$ for all $\mathrm{Re}\,\zeta < (-\beta)$.

Proof For simplicity, we suppose again that $\beta = 0$. By virtue of the strong convergence of $\{R_{A_n}(\zeta_0)\}_{n\geq 1}$, the limit

$$R(\zeta)x := \lim_{n\to\infty} R_{A_n}(\zeta)x \ , \ x \in \mathfrak{X} \ , \tag{1.9.93}$$

exists for all $\zeta \in \mathbb{C}_-$ with negative real part, and it is a *pseudo-resolvent* (see Appendix B.1 (Sect. B.1.3)) such that $\|\zeta R(\zeta)\| \leq M$ when $\mathrm{Re}\,\zeta < 0$. The limit (1.9.91) implies $R(\zeta)^k x = \lim_{n\to\infty} R_{A_n}(\zeta)^k x$. Then we also obtain $\|\zeta^k R(\zeta)^k\| \leq M$, for all $k \in \mathbb{N}$ and $\zeta \in \mathbb{C}_-$. Since by conditions the range $\mathrm{ran}(R(\zeta))$ is dense in $\mathfrak{X}$ for $\zeta = \zeta_0$ (and hence for all $\zeta \in \mathbb{C}_-$), the $\ker(R(\zeta)) = \{0\}$ for all of these ζ, that is, the limit (1.9.93), is, in fact, a *resolvent*. Then there exists a densely defined closed operator A with $\mathrm{dom}\,(A)$ such that for $\zeta \in \mathbb{C}_-$ one has $R(\zeta) = R_A(\zeta)$ (Appendix B.1, Sect. B.1.3) and the resolvent $R_A(\zeta)$ satisfies the so-called *Hille-Yosida* estimate (1.3.4)

$$\|\zeta^k R_A(\zeta)^k\| \leq M \ , \ k \in \mathbb{N} \, . \tag{1.9.94}$$

Hence, by Proposition 1.3.4, the operator $A \in \mathscr{G}(M, \beta = 0)$ generates a bounded strongly continuous semigroup $\{U_A(t)\}_{t\geq 0}$.

Finally, we use the convergence of resolvents (1.9.93) and $R(\zeta) = R_A(\zeta)$ to deduce (1.9.83) from Proposition 1.9.30 . □

1.9.6 Method Based on the Product Formulæ

This approach, which is close to the main topic of our book, presents an alternative to methods developed in Sect. 1.9.1 and then in Sects. 1.9.2–1.9.4. The most important difference is that *a priori* it does not presume any *hierarchy* between generators A and B involved into construction of resulting perturbed (or approximated) semigroup with generator H.

Therefore, this method allows to treat perturbations of *quasi-bounded* semigroups *without* hypothesis about relative boundedness of generators A and B, cf. Propositions 1.9.23 and 1.9.24. We shall also see (Sect. 5.2, Remark 5.2.6) that the method based on the *product formulae* is able to treat a *singular* case when

$$\mathrm{dom}\,(A) \cap \mathrm{dom}\,(B) = \{0\}\, .$$

In the simplest circumstances, the construction of semigroup with generator $H = A + B$ is due to the *product formula* approximation, which for finite *square matrices* A and B dates back to *Sophus Lie* (1875) (see Notes to Chap. 1 (Notes to Sects. 1.9, 1.9.6)):

$$\mathrm{e}^{-t(A+B)} = \tau - \lim_{n\to\infty} (\mathrm{e}^{-tA/n}\mathrm{e}^{-tB/n})^n \ , \quad t \geq 0\, . \tag{1.9.95}$$

Here, τ denotes the topology of convergence. For example, it may be the matrix-norm topology induced by the vector-norm topology on the (corresponding) finite-dimensional Euclidian space.

We note that in contrast to the *resolvent*- and *semigroup*-based methods (Sect. 1.9.1) the full power of the *product formulæ*-based method is owing to condition that *both* operators A and B *must* be generators of C_0-semigroups.

In fact, a recent progress in the product formula-based method is motivated by the discovery of the *operator-norm* convergence of the product formula (1.9.95) and the *Chernoff* approximation (or product) formula. Recall that the *operator-norm* convergence of the product formulæ is the central topic of the present book.

For the first acquaintance and for warming up, we investigate in this section the extension of (1.9.95) to the case of the strong operator topology, $\tau = s$, on a Banach space $\mathfrak{X}$, when A and B are generators of the corresponding C_0-semigroups. In contrast to the perturbation theory (Sect. 1.9.1), there is no subordination between A and B, that is, neither of these operators should be *relatively* bounded to the other.

To this aim, we begin with *Chernoff estimate*, which is coming back to the historically first *Chernoff* $\sqrt{n}$-Lemma. The scheme of the proof will be useful later when we use it for proving the convergence in (1.9.95) in the operator-norm topology $\tau = \|\cdot\|$.

Lemma 1.9.33 **($\sqrt{n}$ -Lemma)** *Let linear operator C be a contraction on Banach space $\mathfrak{X}$. Then $\{e^{t\,(C-\mathbb{1})}\}_{t\geq 0}$ is a norm-continuous contraction semigroup on $\mathfrak{X}$, and one has the estimate*

$$\|(C^n - \mathrm{e}^{n(C-\mathbb{1})})x\| \leq \sqrt{n}\, \|(C - \mathbb{1})x\| \,, \tag{1.9.96}$$

for all $x \in \mathfrak{X}$ and natural $n \in \mathbb{N}$.

Proof To prove the estimate (1.9.96), we apply representation

$$C^n - e^{n(C-\mathbb{1})} = \mathrm{e}^{-n} \sum_{m=0}^{\infty} \frac{n^m}{m!} (C^n - C^m) \,. \tag{1.9.97}$$

Then to proceed, we implement a *telescopic* inequality for $\|C\| \leq 1$:

$$\|(C^n - C^m)\, x\| \leq \left\|(C^{|n-m|} - \mathbb{1})\, x\right\| \leq |m-n|\; \|(C-\mathbb{1})\, x\| \,, \quad x \in \mathfrak{X} \,. \tag{1.9.98}$$

On account of (1.9.97), (1.9.98), and the *Cauchy-Bouniakovski-Schwarz* inequality, one deduces estimate

$$\begin{aligned} \|(C^n - \mathrm{e}^{n(C-\mathbb{1})})x\| &\leq \|(C - \mathbb{1})x\|\, e^{-n} \sum_{m=0}^{\infty} \frac{n^m}{m!}\, |m-n| \leq \\ &\left\{\sum_{m=0}^{\infty} \mathrm{e}^{-n} \frac{n^m}{m!}\, (m-n)^2\right\}^{1/2} \|(C - \mathbb{1})x\| \,, \quad x \in \mathfrak{X} \,. \end{aligned} \tag{1.9.99}$$

Seeing that the sum in the right-hand side of inequality (1.9.99) is equal to n, we obtain (1.9.96). □

Now, on account of estimate (1.9.96), we obtain the *Chernoff* product formula (theorem), thanks to the $\sqrt{n}$ -Lemma.

Theorem 1.9.34 (Chernoff Product Theorem) *Let $\Phi : t \mapsto \Phi(t)$ be a mapping from $\mathbb{R}_0^+$ to contractions on $\mathfrak{X}$ such that $\Phi(0) = \mathbb{1}$. Let $\{U_A(t)\}_{t\geq 0}$ be a contraction C_0-semigroup, and let $D \subset$ dom (A) be a core for the generator A. If $\Phi(t)$ has on $D =$ core (A) a strong right derivative $\Phi'(+0)$ at $t = 0$, such that*

$$\Phi'(+0)\, x := \lim_{t\to+0} \frac{1}{t}(\Phi(t) - \mathbb{1})\, x = -\, Ax\,, \tag{1.9.100}$$

for all $x \in D$ and the closure : $\Phi'(+0) = -\, A$, then

$$\lim_{n\to\infty} (\Phi(t/n))^n\, x = U_A(t)\, x\,, \tag{1.9.101}$$

for all $t \in \mathbb{R}_0^+$ and $x \in \mathfrak{X}$.

Proof Consider bounded approximations $\{A_n(s)\}_{n\geq 1}$ of generator A:

$$A_n(s) := \frac{\mathbb{1} - \Phi(s/n)}{s/n}\,, \qquad s \in \mathbb{R}^+\,, \qquad n \in \mathbb{N}\,. \tag{1.9.102}$$

These operators are m-accretive (Definition 1.8.7): $\|(A_n(s)+\zeta\mathbb{1})^{-1}\| \leq (\mathrm{Re}\,(\zeta))^{-1}$, for $\mathrm{Re}\,(\zeta) > 0$ and for any $n \in \mathbb{N}$. By $\|\Phi(t)\| \leq 1$ together with (1.9.102), we obtain $\|\, e^{-t\, A_n(s)}\| \leq 1$ but also

$$\lim_{n\to\infty} A_n(s)\, x = A\, x\,, \tag{1.9.103}$$

for all $x \in D$ and any $s \in \mathbb{R}^+$, cf. (1.9.100). Then, given that $D =$ core (A), by virtue of the *Trotter-Kato-Neveu* theorem (Proposition 1.9.30 and Remark 1.9.31), one gets

$$\lim_{n\to\infty} e^{-t\, A_n(s)}\, x = U_A(t)\, x\,, \quad x \in \mathfrak{X}\,, \quad t \in \mathbb{R}_0^+\,, \tag{1.9.104}$$

which is strong and locally uniform in t and in s convergence (1.9.83) of contraction approximants $\{e^{-t\, A_n(s)}\}_{n\geq 1}$ for $t \in [0, t_0]$ and $s \in (0, s_0]$.

Now, by Lemma 1.9.33 (1.9.96) for contractions $C := \Phi(t/n)$, we obtain

$$\begin{aligned} &\|(\Phi(t/n))^n\, x - e^{-t\, A_n(t)}\, x\| = \|((\Phi(t/n))^n - e^{\,n(\Phi(t/n)-\mathbb{1})})\, x\| \\ &\leq \sqrt{n}\, \|(\mathbb{1} - \Phi(t/n))\, x\,\|\,, \quad x \in \mathfrak{X}\,. \end{aligned} \tag{1.9.105}$$

Since by (1.9.103) one gets for any $x \in D$ and uniformly in $t \in [0, t_0]$

$$\lim_{n\to\infty} \sqrt{n} \, \| (\mathbb{1} - \Phi(t/n)) \, x \, \| = \lim_{n\to\infty} \frac{t}{\sqrt{n}} \, \|A_n(t) \, x\| = 0 \,, \tag{1.9.106}$$

the Eqs. (1.9.105) and (1.9.106) provide

$$\lim_{n\to\infty} \| \, (\Phi(t/n))^n \, x - \mathrm{e}^{-t\, A_n(t)} \, x\| = 0, \qquad x \in D \,. \tag{1.9.107}$$

Then (1.9.104) and (1.9.107) yield uniformly in $t \in [0, t_0]$ the limit:

$$\lim_{n\to\infty} (\Phi(t/n))^n \, x = U_A(t) \, x \,, \quad x \in D \,. \tag{1.9.108}$$

Note that by density of $D = \mathrm{core}\,(A)$ and uniform in $x \in \mathfrak{X}$ bound $\| \, (\Phi(t/n))^n \, x - \mathrm{e}^{-t\, A_n(t)} x\| \le 2\, \|x\|$ the convergence in (1.9.107) can be extended to all $x \in \mathfrak{X}$ (see Proposition A.1.6 in Appendix A (Sect. A.1.2)). As a consequence, (1.9.108) holds for all $x \in \mathfrak{X}$, which yield the proof of (1.9.101). □

We remark that for $t \in \mathbb{R}_0^+$ the sequence $\{(\Phi(t/n))^n\}_{n\in\mathbb{N}}$, in the *Chernoff* product formula (1.9.101), is a family of the generic *Chernoff product approximants* for contraction C_0-semigroup $\{U_A(t)\}_{t\ge 0}$ in the *strong* operator topology.

Proposition 1.9.35 (Lie-Trotter Product Formula) *Let $A \in \mathscr{G}(1, 0)$ and $B \in \mathscr{G}(1, 0)$ be generators of contraction C_0-semigroups on $\mathfrak{X}$. Consider the operator*

$$\widehat{C} x := Ax + Bx \,, \quad x \in D = \mathrm{dom}\,(A) \cap \mathrm{dom}\,(B) \,, \tag{1.9.109}$$

and assume that D and the map $(\lambda\mathbb{1} + A + B)\, D$ are dense in $\mathfrak{X}$ for $\lambda > 0$. Then the closure of the sum (1.9.109): $C := \overline{(A + B) \restriction D}$ generates a contraction C_0-semigroup $\{U_C(t) = \mathrm{e}^{-tC}\}_{t\ge 0}$, which can be approximated on $\mathfrak{X}$ in the strong operator topology $\tau = s$ (see (1.9.95)) by the Lie-Trotter product formula,

$$\mathrm{e}^{-tC} \, x = \lim_{n\to\infty} (\mathrm{e}^{-tA/n} \mathrm{e}^{-tB/n})^n \, x \,, \quad x \in \mathfrak{X} \,, \tag{1.9.110}$$

locally uniformly for $t \in \mathbb{R}_0^+$.

Proof We follow the *Chernoff product* Theorem 1.9.34 and define contraction $t \mapsto V_{A,B}(t)$, for $t \in \mathbb{R}_0^+$ with $V_{A,B}(0) = \mathbb{1}$, by

$$V_{A,B}(t) := \mathrm{e}^{-tA} \mathrm{e}^{-tB} \,. \tag{1.9.111}$$

Note that if $x \in D$, then the right derivative at $t = 0$

$$\partial_t V_{A,B}(+0) \, x = \lim_{t\to +0} \frac{1}{t} \, (V_{A,B}(t) - \mathbb{1}) \, x = -\, (A + B) \, x \tag{1.9.112}$$

and the closure $\partial_t V_{A,B}(+0) = -\, \overline{(A + B) \restriction D} = C$.

Then we are in position to apply Theorem 1.9.34 for the *Lie-Trotter product approximants* given by $\{(\Phi(t/n))^n = (V_{A,B}(t/n))^n\}_{n\in\mathbb{N}}$. In virtue of (1.9.111) and (1.9.112), this yields (1.9.110) for the closed operator $C := \overline{(A+B) \upharpoonright D}$, which is such that $C \in \mathscr{G}(1,0)$. □

Corollary 1.9.36 *Extension of the strongly convergent Lie-Trotter product formula* (1.9.110) *to quasi-contraction semigroups when $A \in \mathscr{G}(1,\beta_A)$ and $B \in \mathscr{G}(1,\beta_B)$ follows verbatim. To this end, one notes that $(A+\beta_A \mathbb{1}) \in \mathscr{G}(1,0)$ and $(B+\beta_B \mathbb{1}) \in \mathscr{G}(1,0)$ and then applies Proposition* 1.9.35.

Extension of Proposition 1.9.35 to the general case of quasi-bounded semigroups needs a supplementary the *Trotter stability* condition.

Definition 1.9.37 The pair $\{A, B\}$ of generators of quasi-bounded C_0-semigroups is called *Trotter-stable* if

$$\|(\mathrm{e}^{-tA/n}\mathrm{e}^{-tB/n})^n\|_{\mathcal{L}(\mathfrak{X})} \leq M\,\mathrm{e}^{\beta t}\,, \qquad n \in \mathbb{N}\,, \quad t \in \mathbb{R}_0^+\,, \tag{1.9.113}$$

uniformly for $M \geq 1$ and some $\beta \in \mathbb{R}$.

Note that condition (1.9.113) is trivially satisfied for any pair of the *quasi-contraction* C_0-semigroups from $\mathscr{G}(1,\beta_A)$, $\mathscr{G}(1,\beta_B)$ (see Remark 1.3.3). Subsequently, the following assertion will be useful for *quasi-bounded* semigroups with generators from $\mathscr{G}(M_A,\beta_A)$, $\mathscr{G}(M_B,\beta_B)$.

Lemma 1.9.38 *Let A and B be generators of quasi-bounded C_0-semigroups on the Banach space $\mathfrak{X}$. Let $T(t) := \mathrm{e}^{-t\,A}\mathrm{e}^{-t\,B}$, for $t \in \mathbb{R}_0^+$:*

(i) *If the pair $\{A, B\}$ of generators is Trotter-stable for $\beta = 0$, then*

$$\|T(t/m)^n\|_{\mathcal{L}(\mathfrak{X})} \leq M\,, \tag{1.9.114}$$

for any $m \in \mathbb{N}$, $n \in \mathbb{N}$, and $t \geq 0$. In particular, $\|T(t)^n\|_{\mathcal{L}(\mathfrak{X})} \leq M$ for any $n \in \mathbb{N}$.

(ii) *If the pair $\{A, B\}$ is Trotter-stable, then the pair of generators $\{B, A\}$ is also Trotter-stable.*

Proof

(i) Replacing in (1.9.113) the argument t by $t\,n/m$, we obtain the proof of (1.9.114). For $m = 1$, one gets the additional statement.
(ii) Replacing in (1.9.113) the argument t by $(t\,n)/(n+1)$, we obtain

$$\|T(t/(n+1))^n\| \leq M\mathrm{e}^{\beta t\,n/(n+1)}.$$

Since $A \in \mathcal{G}(M_A, \beta_A)$ and $B \in \mathcal{G}(M_B, \beta_B)$, this estimate yields

$$\begin{aligned}&\|\mathrm{e}^{-tB/(n+1)}T(t/(n+1))^n \mathrm{e}^{-tA/(n+1)}\| \\ &\quad \leq M_B\, \mathrm{e}^{\beta_B t/(n+1)} M\, \mathrm{e}^{\beta tn/(n+1)} M_A\, \mathrm{e}^{\beta_A t/(n+1)}.\end{aligned}$$

Setting $M' := M_A M_B M$ and $\beta' := 2\max\{\beta_A, \beta_B, \beta\}$, we get

$$\|\,\mathrm{e}^{-tB/(n+1)}\mathrm{e}^{-tA/(n+1)} \cdots \mathrm{e}^{-tB/(n+1)}\mathrm{e}^{-tA/(n+1)}\| \leq M \mathrm{e}^{\beta' t}, \quad n \in \mathbb{N}, \tag{1.9.115}$$

for $t \geq 0$. Consequently, the pair of generators $\{B, A\}$ is Trotter-stable.

□

To proceed further with the *Trotter stability* condition and then *Trotter product* formula, we need certain preliminaries. Recall that in general it is possible to endow the Banach space $\mathfrak{X}$ with a *new norm* such that one of the operators, A or B, becomes a generator of a quasi-contraction semigroups on $\mathfrak{X}$, cf. Remark 1.3.3.

Indeed, if $A \in \mathscr{G}(M_A, \beta_A)$ is generator of quasi-bounded C_0-semigroup, one has

$$\|\mathrm{e}^{-\sigma A}x\| \leq M_A\, \mathrm{e}^{\beta_A \sigma}\|x\|, \quad \sigma \geq 0, \quad x \in \mathfrak{X}, \tag{1.9.116}$$

for $M_A \geq 1$ and $\beta_A \in \mathbb{R}$. Now, let the new norm of any $x \in \mathfrak{X}$ be defined by

$$|||x||| := \sup_{\sigma > 0}\, \mathrm{e}^{-\beta_A \sigma}\, \|\mathrm{e}^{-\sigma A}x\|. \tag{1.9.117}$$

Then we obtain

$$\left|\left|\left|\mathrm{e}^{-\tau A}x\right|\right|\right| = \mathrm{e}^{\beta_A \tau} \sup_{\sigma > 0}\, \mathrm{e}^{-\beta_A\{\sigma+\tau\}}\, \|\mathrm{e}^{-(\sigma+\tau)A}x\| \leq \mathrm{e}^{\beta_A \tau}\, |||x|||. \tag{1.9.118}$$

This observation show that in the Banach space $(\mathfrak{X}, |||\cdot|||)$ endowed with the norm $|||\cdot|||$ the semigroup $\{\mathrm{e}^{-\sigma A}\}_{\sigma \geq 0}$ belongs to the class $\mathscr{G}(1, \beta_A)$ of quasi-contraction semigroups. Moreover, since by (1.9.116) and by definition (1.9.117) one gets

$$\|x\| \leq |||x||| \leq M_A\, \|x\|,$$

the norm $|||\cdot|||$ is *equivalent* to $\|\cdot\|$.

The same reasoning can be applied to the semigroup $\{\mathrm{e}^{-\sigma B}\}_{\sigma \geq 0}$, but in general, it is *impossible* to find an equivalent norm such that *both* semigroups become quasi-contraction.

Definition 1.9.39 Let $A \in \mathscr{G}(M_A, \beta_A)$ and $B \in \mathscr{G}(M_B, \beta_B)$ be generators of quasi-bounded C_0-semigroups on $\mathfrak{X}$. The pair $\{A, B\}$ is called *renormalisable* with constants β_A and β_B if for any finite sequences $\{\tau_k\}_{k=1}^N$, $\tau_k \geq 0$, and $\{\sigma_k\}_{k=1}^N$, $\sigma_k \geq 0$, $n \in \mathbb{N}$ one has

$$\sup_{\substack{\tau_1 \geq 0, \ldots, \tau_n \geq 0 \\ \sigma_1 \geq 0, \ldots, \sigma_n \geq 0 \\ n \in \mathbb{N}}} \mathrm{e}^{-\beta_A \sum \tau_k} \mathrm{e}^{-\beta_B \sum \sigma_k} \|\mathrm{e}^{-\tau_1 A} \mathrm{e}^{-\sigma_1 B} \cdots \mathrm{e}^{-\tau_n A} \mathrm{e}^{-\sigma_n B} x\| < \infty \tag{1.9.119}$$

for each $x \in \mathfrak{X}$.

Lemma 1.9.40 *Let $A \in \mathscr{G}(M_A, \beta_A)$ and $B \in \mathscr{G}(M_B, \beta_B)$ be generators of quasi-bounded C_0-semigroups in Banach space $\{\mathfrak{X}, \|\cdot\|\}$. Then on $\mathfrak{X}$, there exists an equivalent norm $|||\cdot|||$ such that on the Banach space $\{\mathfrak{X}, |||\cdot|||\}$ both semigroups are quasi-contraction : $A \in \mathscr{G}(1, \beta_A)$ and $B \in \mathscr{G}(1, \beta_B)$, if and only if the pair $\{A, B\}$ is renormalisable with constants β_A and β_B.*

Proof Let the pair $\{A, B\}$ be renormalisable with constants β_A and β_B. On the Banach space $\mathfrak{X}$, we define for $x \in \mathfrak{X}$, cf. (1.9.119), a new norm by

$$|||x||| := \sup_{\substack{\tau_1 \geq 0, \ldots, \tau_n \geq 0 \\ \sigma_1 \geq 0, \ldots, \sigma_n \geq 0 \\ n \in \mathbb{N}}} \mathrm{e}^{-\beta_A \sum \tau_k} \mathrm{e}^{-\beta_B \sum \sigma_k} \|\mathrm{e}^{-\tau_1 A} \mathrm{e}^{-\sigma_1 B} \cdots \mathrm{e}^{-\tau_n A} \mathrm{e}^{-\sigma_n B} x\| . \tag{1.9.120}$$

Then by virtue of (1.9.116), one gets $\|x\| \leq |||x|||$. On the other hand, by (1.9.119) and by the *uniform boundedness principle* , we find that the value of

$$M := \sup_{\substack{\tau_1 \geq 0, \ldots, \tau_n \geq 0 \\ \sigma_1 \geq 0, \ldots, \sigma_n \geq 0 \\ n \in \mathbb{N},\ \|x\| \leq 1}} \mathrm{e}^{-\beta_A \sum \tau_k} \mathrm{e}^{-\beta_B \sum \sigma_k} \|\mathrm{e}^{-\tau_1 A} \mathrm{e}^{-\sigma_1 B} \cdots \mathrm{e}^{-\tau_n A} \mathrm{e}^{-\sigma_n B} x\| , \tag{1.9.121}$$

is finite. By (1.9.120) and (1.9.121), this yields $|||x||| \leq M \|x\|$, $x \in \mathfrak{X}$. Hence, the norms $\|\cdot\|$ and $|||\cdot|||$ are equivalent.

Consider semigroup $\{e^{-\sigma B}\}_{\sigma \geq 0}$ on the Banach space $(\mathfrak{X}, |||\cdot|||)$. By definition (1.9.120) and following arguments in (1.9.118), we obtain

$$
\begin{aligned}
\left|\left|\left| e^{-\sigma B} x \right|\right|\right| &:= \\
&\sup_{\substack{\{\tau_k \geq 0, \sigma_k \geq 0\}_{1 \leq k \leq n} \\ n \in \mathbb{N}}} e^{-\beta_A \sum \tau_k} e^{-\beta_B \sum \sigma_k} \| e^{-\tau_1 A} e^{-\sigma_1 B} \cdots e^{-\tau_n A} \\
&\quad \times e^{-\sigma_n B} e^{-\sigma B} x \| \\
&\leq e^{\beta_B \sigma} |||x||| , \quad x \in \mathfrak{X} .
\end{aligned} \tag{1.9.122}
$$

Similarly (cf. Lemma 1.9.38), one also gets the estimate

$$
\left|\left|\left| e^{-\tau A} x \right|\right|\right| \leq e^{\beta_A \tau} \, |||x||| , \quad x \in \mathfrak{X} . \tag{1.9.123}
$$

Therefore, in the Banach space $(\mathfrak{X}, |||\cdot|||)$, the generators A and B belong, respectively, to $\mathscr{G}(1, \beta_A)$ and $\mathscr{G}(1, \beta_B)$.

Conversely, suppose that on a Banach space $\{\mathfrak{X}, \|\cdot\|\}$ there is an equivalent norm $|||\cdot|||$

$$
c_1 \|x\| \leq |||x||| \leq c_2 \|x\| , \quad c_2 \geq c_1 > 0 , \quad x \in \mathfrak{X} , \tag{1.9.124}
$$

such that $A \in \mathscr{G}(1, \beta_A)$ and $B \in \mathscr{G}(1, \beta_B)$ in $\{\mathfrak{X}, |||\cdot|||\}$. Then

$$
\begin{aligned}
& e^{-\beta_A \sum \tau_k} e^{-\beta_B \sum \sigma_k} \| e^{-\tau_1 A} e^{-\sigma_1 B} \cdots e^{-\tau_n A} e^{-\sigma_n B} x \| \\
& \leq (c_1)^{-1} e^{-\beta_A \sum \tau_k} e^{-\beta_B \sum \sigma_k} \left|\left|\left| e^{-\tau_1 A} e^{-\sigma_1 B} \cdots e^{-\tau_n A} e^{-\sigma_n B} x \right|\right|\right| \\
& \leq (c_1)^{-1} \, |||x||| \leq (c_1)^{-1} c_2 \, \|x\| , \quad x \in \mathfrak{X} .
\end{aligned} \tag{1.9.125}
$$

For that reason, (1.9.125) proves (1.9.119) (or (1.9.121)), i.e., that the pair $\{A, B\}$ is renormalisable with constants β_A and β_B. □

Proposition 1.9.41 (Trotter Product Formula) *Let $A \in \mathscr{G}(M_A, \beta_A)$ and $B \in \mathscr{G}(M_B, \beta_B)$ be generators of quasi-bounded semigroups on $\mathfrak{X}$. Suppose that these semigroups satisfy condition of the Trotter stability* (1.9.113) *and that operator $A + B$ on a dense domain $D :=$* dom $(A) \cap$ dom (B) *is closable with a dense set $(\lambda \mathbb{1} + A + B)\, D$, for $\lambda > \beta$, and the closure*

$$
C := \overline{A + B} \, . \tag{1.9.126}
$$

Then operator (1.9.126) *generates a quasi-bounded C_0-semigroup $\{U_C(t) = e^{-tC}\}_{t \geq 0}$ for $C \in \mathscr{G}(M, \beta)$, which can be approximated on $\mathfrak{X}$ in the strong operator* *topology* $\tau = s$, (1.9.95), *by the Trotter product formula*

$$e^{-tC}x = \lim_{n\to\infty}(e^{-tA/n}e^{-tB/n})^n x\,, \quad x \in \mathfrak{X}\,, \tag{1.9.127}$$

locally uniformly for $t \in \mathbb{R}_0^+$.

Proof By virtue of the *Trotter stability* condition (1.9.113) and Definition 1.9.39, the pair of generators $\{A, B\}$ is renormalisable in the Banach space $\{\mathfrak{X}, \|\cdot\|\}$ with constants β_A and β_B. Then by Lemma 1.9.40, there exists on $\mathfrak{X}$ an equivalent norm $|||\cdot|||$ such that on the Banach space $(\mathfrak{X}, |||\cdot|||)$ both semigroups $\{U_A(t)\}_{t\geq 0}$ and $\{U_B(t)\}_{t\geq 0}$ are quasi-contraction: $A \in \mathscr{G}(1, \beta_A)$ and $B \in \mathscr{G}(1, \beta_B)$. So, by conditions of the proposition and by results of Corollary 1.9.36 together with Proposition 1.9.35, we obtain for $C = \overline{A + B}$, (1.9.126), the strongly convergent *Lie-Trotter* product formula

$$\lim_{n\to\infty} \left|\left|\left|(U_A(t/n)\,U_B(t/n))^n x - U_C(t)\,x\right|\right|\right| = 0\,, \quad x \in (\mathfrak{X}, |||\cdot|||)\,, \tag{1.9.128}$$

for $t \in \mathbb{R}_0^+$ on the Banach space $(\mathfrak{X}, |||\cdot|||)$.

As a consequence, we deduce that by equivalence of norms (1.9.124) formula (1.9.128) implies the Trotter product formula (1.9.127) for quasi-bounded semigroups on the Banach space $(\mathfrak{X}, \|\cdot\|)$. □

We note that the supplementary *Trotter stability* condition (1.9.113) is indispensable for product formulæ of quasi-bounded semigroups in Sect. 3.1.2, as well as in Sect. 11.4 (see comments in Notes to Chap. 1 (Notes to Sects. 2.1 and 1.9.6). A similar stability condition is also indispensable for product approximants of the *solution operator* in extension of the Trotter product formula to *non-autonomous* linear equations, see Definitions 2.1.36 and 2.2.27).

On the other hand, extensions of the strongly convergent product formula (1.9.110) to the operator-norm topology, $\tau = \|\cdot\|$, or to the trace-norm topology $\tau = \|\cdot\|_1$ for the Gibbs semigroups, are non-trivial problem. These are in fact a relatively recent results, see Notes to Chap. 1, that is a *principal* subject of this book.

1.10 Notes to Chap. 1

The material of this chapter is standard. Some comments are collected in the notes to corresponding sections and/or subsections. For the rest we provide some popular references for further reading and few historical remarks.

Notes to Sect. 1.1

One of the first definitions of the *exponential* function in the *infinite dimensional* spaces appeared probably in the paper [Gram10]. Here we follow the Kato book

[Kato80, Chapter IX]. A brief history of the exponential function (by T. Hahn and C. Parazzoli) can be found in [EN00, Chapter VII].

In construction of the operator-valued *exponential* function (Proposition 1.1.1) we followed the method based on the well-known approximations by the *Euler formula* (1.1.5). This approach was initiated by E. Hille [Hille48] (see [HiPh57, Theorem 12.3.1, Ch.XII, Sec.12.3]) with a suggestion by T. Kato which yields an elegant proof of convergence of the *Euler approximants* (1.1.4) due to Proposition A.1.8 and Corollary A.1.9.

We note that in the very same year K. Yosida [Yos48], [Yos80, Chapter IX], proposed a clever alternative to the Euler formula construction of C_0-semigroups. It was approximation involved the operator-valued *exponential* functions generated by *bounded* operators called the *Yosida approximants* (1.4.9). We present this construction of strongly continuous contraction semigroups in Proposition 1.4.4.

For more comments on the Yosida construction and Proposition 1.4.4, see Notes to Sect. 1.4.

Notes to Sect. 1.2

The right continuity assumption at $t = 0$ is crucial for a *non-trivial* semigroup theory. That is, the *strong* continuity for $t \to +0$ seems to be, in this respect, the most appropriate restriction to make this theory so significant, see discussions in [HiPh57, Chapter X], [Dav80, Chapter 1] and [EN00, Chapter II].

Several other types of continuity at the origin $t = 0$ have been studied in the book of E. Hille and R. S. Phillips [HiPh57], the great classic on the one-parameter semigroup theory. Their conclusion, see [HiPh57, Chapter X (Sections 10.5 and 10.6)], was that to obtain interesting structure theorems the semigroups should be (right)-continuous at the origin in a some particular sense. A variety of these conditions were produced in [HiPh57, Chapter X (Sections 10.6)]: *The basic classes of semigroups*. Then in virtue of Theorem 10.6.4 there, any of the basic classes reduces to the "class C_0" under natural condition of the semigroup operator-norm boundedness for t in some neighborhood of zero. This property makes the C_0-semigroups, in a certain sense, *exceptional* and the most important among the one-parameter operator semigroups.

By Proposition 1.2.3 the *strong continuity* at the origin for $t \to +0$ and conditions (a), (b) extend the strong continuity *away* from $t = 0$ and, as a consequence, to the whole $\mathbb{R}_0^+$. It is worth mentioning here that the *strong measurability* of the family $\{U(t)\}_{t\geq 0}$, and conditions (a), (b) imply the strong continuity of the semigroup $\{U(t)\}_{t>0}$ away from $t = 0$, see [HiPh57, Theorem 10.2.3], or [Dav80, Proposition 1.25]. Incidentally, there exists a number of examples showing that in this case one *cannot* prove the strong continuity for $t \to +0$. The behaviour of semigroup $\{U(t)\}_{t\geq 0}$ as $t \to +0$ may be *highly singular* although $U(0) = \mathbb{1}$, if it is not a C_0-semigroup. For instance, in this limit one may even get that $\lim_{t\to+0} \|U(t)\| = \infty$, see [Dav80, Example 1.26]. This bolsters the *exceptional* status of the C_0-semigroups.

On the other hand, the condition of the strong measurability is convenient and useful for certain applications. For the *Bochner integral* see Sect. E.1.

In this connection we also note (see [HiPh57, Theorem 3.5.3]) that a vector-valued *weakly* measurable function with *separable* image is *strongly* measurable. Then a one-sided weak continuity implies strong measurability. As a consequence, on account of [HiPh57, Theorem 10.2.3], one concludes that the *weak* one-sided continuity of semigroup $\{U(t)\}_{t>0}$ yields the *strong* continuity of this semigroup *away* from $t = 0$. We scrutinise consequences of the *weak* continuity at the origin: $t = 0$, in Sect. 1.5, cf. also Notes to Sect. 1.5.

The next *example*, which illustrates Remark 1.2.4, is due to [EN00, Chapter I, 5.7 Examples (iii)]. It shows that although $\|U(0)\| = 1$, it is sometimes *not* possible to put in (1.2.2) $M = 1$. Moreover, since the operator-norm

$$A \mapsto \|A\| , \quad A \in \mathcal{L}(\mathfrak{X}) ,$$

is *not* a continuous function if the space $\mathcal{L}(\mathfrak{X})$ is endowed with the *strong* operator topology, the family of norms $\{\|U(t)\|\}_{t\geq 0}$ corresponding to a C_0-semigroup may be *discontinuous*. For more examples of strongly continuous (semi)groups that manifest discontinuity of function: $t \mapsto \|U(t)\|$, such that $\|U(t \neq 0)\| > \|U(0)\| = 1$, we refer to [Dav07, Example 6.2.3 and Theorem 6.3.8].

For more comments on nuances conserving C_0-semigroups with other types of continuity, when $t > 0$, see Notes to Sects. 1.4–1.7. Therefore, a variety of continuity for $t > 0$ produces *special classes* of semigroups (see [Dav07, Chapter 7], [EN00, Chapter II, Section 4], and [Zag19a]), which are complementary to *basic classes* of semigroups (cf. [HiPh57, Sections 10.6]) defined by conditions for $t \geq 0$ in a neighborhood of zero. We discuss the continuity *away* from $t = 0$ in Sect. 1.6. For references, see Notes to Sect. 1.6.

The Nelson *invariance* theorem (Proposition 1.2.9) is due to [Nel59]. It is important for characterisation of generator. Our proof follows essentially the line of reasoning from [Dav07, Theorem 6.1.18].

The *triviality of domain* problem for the square of unbounded operators in Hilbert and Banach spaces $\mathrm{dom}\,(A^2) = \{0\}$ appears in the context of construction of the C_0-semigroups using the *Euler formula* (see Proposition 1.1.1, (1.1.14) and Proposition A.1.8, Corollary A.1.9). Although it is not difficult to produce simple examples of unbounded operators with the properly $\mathrm{dom}\,(A^2) = \{0\}$ (see, for example, [Mor22, Question 18.2.5]), in virtue of Proposition A.1.8 and Corollary A.1.9, this problem has no solution if A is generator of a C_0-semigroup.

Note that the triviality of domain problem has an independent interest. It becomes substantial under certain restrictions on the operator A (as above for $A \in \mathscr{G}(M, \beta)$) or on the involved spaces. In the case of a separable Hilbert space, we refer to [ArZag15] and [ArTr20] for a quite exhaustive list of results concerning this and related problems of operator domains.

We use in Proposition 1.2.10 the Nelson theorem (Proposition 1.2.9) and Corollary A.1.9(a) to demonstrate that under condition that operator A is generator of a C_0-semigroup the *triviality of domain* problem becomes non-trivial. The square of this generator rests *densely* defined and $\mathrm{dom}\,(A^n) = \mathrm{core}\,(A)$ for any $n \in \mathbb{N}$. For density of intersection $\bigcap_{n\in\mathbb{N}} \mathrm{dom}\,(A^n)$, see [EN00, Chapter II, 1.8 Proposition].

Notes to Sect. 1.3

The material of this section is a part of any standard course on the strongly continuous one-parameter semigroups like [Dav80], or the exhaustive treatise [EN00]. To complete the references that have been already quoted above, we only add here the books: [Yos80], [SKre67], as well as [Paz83], [Gol85] and [EN06].

Proposition 1.1.1 for $\beta = 0$ and $M = 1$ is known as the *Hille* theorem for contraction semigroups, while the general case for semigroups of the type $\mathscr{G}(M, \beta)$ is often called the *Hille-Yosida-Phillips* theorem. Our proof of these theorems is based on the *Euler approximants* (1.1.4). Another popular way to construct the C_0-semigroups is to uilise the so-called *Yosida approximants* (1.4.9) and Notes to Sect. 1.4.

Although for a quasi-bounded C_0-semigroup $\{U(t)\}_{t\geq 0}$ one has s-$\lim_{t\to+0} U(t) = U(0) = \mathbb{1}$, it is not necessarily true that $\lim_{t\to+0} \|U(t)\| = 1$, cf. (1.3.1). In Notes to Sect. 1.2, we mentioned few examples even with *discontinuity*: $\lim_{t\to+0} \|U(t)\| > 1$ and $\|U(0)\| = 1$. On that account, to control in (1.3.1) the parameter $M \geq 1$: $\|U(t)\| \leq M\,\mathrm{e}^{\beta t}$, we have to use a *rescaling* as in Remark 1.3.2.

For rescaling and for equivalent norms (Remarks 1.3.2 and 1.3.3), see [Nei82, Lemma 5.1] and [EN00, 3.10 Lemma].

Notes to Sect. 1.4

Proposition 1.4.1 states that semigroups which are norm continuous at $t = 0$, are in a certain sense trivial. Notice that it is not the case of the so-called *immediately* norm continuous, or *eventually* norm continuous semigroups (see discussion in [EN00, Chapter II. Section 4.c]). In the latter case, the semigroup is norm continuous for all $t > t_0 > 0$, whereas for the former case $t_0 = 0$. Although such semigroups are known since [HiPh57, Theorem 16.4.2], the complete characterisation of generators of these semigroups is still an open question.

Proposition 1.4.4 is the *Yosida theorem* [Yos48], which is an alternative to the *Euler formula* approach to construction of C_0-semigroups by the *Euler approximants* (1.1.4), as it was proposed by E.Hille, cf. Proposition 1.1.1 and Notes to Sect. 1.1. For construction in Proposition 1.4.4 a C_0-semigroup with generator A , we used the *Yosida approximants* (1.4.10) that produce semigroups with bounded generators (1.4.9), the so-called Yosida approximants of generator A.

We remark that conclusion of the proof of Proposition 1.4.4 stems from some accurate analysis of the *second* resolvent identity:

$$R_B(\zeta) - R_A(\zeta) = R_B(\zeta)\,(A - B)\,R_A(\zeta)\,, \;\; \mathrm{dom}\,(A) \subseteq \mathrm{dom}\,(B)\,, \;\; \zeta \in (-\infty, 0)\,, \tag{1.10.1}$$

see, for example, [EN00, Chapter II, 3.5 Generation Theorem].

Notes to Sect. 1.5

The material of this section is standard, and the surprising observations concerning the equivalence between *weakly* and *strongly* continuous semigroups are well-known (see [HiPh57, Chapter X], Section 10.6. *The basic classes of semigroups*). To accent the impact of this observation on the semigroup theory, we quote [Dav07]:

> *As far as we know, the main importance of our next theorem (Theorem 6.2.6 - our ref.) is to rule out a possible generalization of the notion of one-parameter semigroup.*

In our presentation and the proof of Proposition 1.5.4, we follow [Dav07, Theorem 6.2.6] and [EN00, 5.8 Theorem]. For completeness, we also recall some general assertions related to this observation (see [HiPh57, Theorem 10.2.3]).

Theorem 1.10.1 *If semigroup* $\{U(t)\}_{t\geq 0}$ *is (locally) strongly measurable, then it is strongly continuous on* $\mathbb{R}^+$.

Seeing that for any vector-valued function the weak continuity implies the strong measurability, one gets the following:

Corollary 1.10.2 *A weak one-sided continuity of semigroup* $\{U(t)\}_{t\geq 0}$ *at* $t = +0$ *implies the strong continuity on* $\mathbb{R}^+$.

Notes to Sect. 1.6

The notion of *degenerate* semigroups was, in fact, introduced in [HiPh57, Section 10.5] in the framework of discussion the semigroup discontinuity (*non-continuity*) at the origin $t = 0$.

Then this notion was explicitly reformulated for the case of *continuous* degenerate semigroups by W. Arendt in [Are01, Section 2]:

> *As the strongly continuous mappings having the semigroup properties but with the strong limit at the origin, which may be different from the identity.*

Our Definition 1.6.1 and Proposition 1.6.2 are motivated by Hille and Phillips [HiPh57, Theorem 10.5.5], Arendt [Are01, Section 2] and Haase [Haa06, Appendix A.8].

The main result of the paper [Are01, Section 4] is Trotter-Kato-Neveu-type approximation Theorem 4.2 for continuous degenerate semigroups. It elucidates how the strong convergence of these semigroups can be deduced from strong convergence of the associated *pseudoresolvents*.

We note that degenerate semigroups in Trotter-Kato product formulæ (Remark 1.6.4) are strong, or operator-norm, limits of *non-degenerate* product approximants (see, for example, Remark 5.2.6 or Theorem 8.3.4. For that reason, one needs another (than in [Are01, Section 4]) version of the Trotter-Kato-Neveu-type approximation theorem (see Theorem 8.3.1 and Remark 8.3.2). This problem was scrutinised by T. Kato in [Kato78, Sections 2 and 3] and in [Zag80a, Theorem 3.1 and Corollary 3.1].

Some special classes of semigroups have their origin from C_0-semigroups specified by their properties *away from zero*: $t = 0$. For more details about compact, differentiable, holomorphic and other types of semigroups, we refer to [EN00, Chapter II, Section 4], [Zag19a] and also Examples 1.10.3, 1.10.8 and 1.10.9.

The Gibbs semigroups were considered in details in [Zag19a]. A characterisation of the Gibbs semigroups by Theorem 1.6.14 is due to [ZI24] (see also a recent paper [IZ23, Section 5] for results on the trace asymptotic). For the Bochner integral and proof of Theorem 1.6.13, one can consult, for example, [ABHN01, Theorem 1.1.4].

A general case of semigroups in *symmetrically normed* ideals and in the *Dixmier* ideal was studied in [Zag19a, Chapters 6 and 7].

For completeness, we prove here Corollary 1.6.15 about extension of $q \in \mathbb{N}$ to $q \in \mathbb{R}^+$. By conditions of Theorem 1.6.14, we have that $\{U_A(t)\}_{t\geq 0}$ with $A \in \mathscr{G}(M, \beta)$ is a quasi-bounded C_0-semigroup and hence $\|U_A(t)\| \leq M\, e^{\beta t}$ (1.3.1).

Proof The key point is to show that fractional power of a resolvent can be defined by the right-hand side of representation (1.6.20) for any real $q > p$. First, we introduce the function

$$F_z : (p, \infty) \ni q \mapsto \Gamma(q)^{-1} \int_0^{\infty} dt\, t^{q-1}\, e^{-t\,(A - z\mathbb{1})}\,, \quad \operatorname{Re} z < -\beta\,, \tag{1.10.2}$$

where $t \mapsto t^{q-1} := e^{(q-1)\,\ln(t)}$ is the first branch of the fractional power function corresponding to the branch $t \mapsto \ln(t)$ of logarithm on $\mathbb{R}^+$. On account of (1.6.29) and (1.6.30), the function (1.10.2) is well-defined and $\|\cdot\|_1$-continuous on $(p, +\infty)$.

For integer $q > p$, definitions (1.6.20) and (1.10.2) yield $R_A(z)^q = F_z(q)$ with composition law for integers q_1, q_2 : $R_A(z)^{q_1} R_A(z)^{q_2} = R_A(z)^{q_1+q_2}$. To ensure that representation (1.10.2) defines the *fractional powers* of the resolvent, one has to show that F_z satisfies the *composition law*:

$$F_z(q_1)\, F_z(q_2) = F_z(q_1 + q_2)\,, \quad \operatorname{Re} z < -\beta\,, \text{ for any real } q_i > p,\ i = 1, 2\,. \tag{1.10.3}$$

After change of variables, $t_1 = x^2$, $t_2 = y^2$ and then $x = r\cos\theta$, $y = r\sin\theta$, we obtain

$$F_z(q_1)\, F_z(q_2) = \Gamma(q_1)^{-1}\Gamma(q_2)^{-1} \int_0^{\infty} dt_1 \int_0^{\infty} dt_2\, t_1^{q_1-1}\, t_2^{q_2-1}\, e^{-(t_1+t_2)\,(A-z\mathbb{1})} \tag{1.10.4}$$

$$= \frac{2}{\Gamma(q_1)\Gamma(q_2)} \int_0^{\infty} dr^2\, r^{2\,(q_1+q_2)-2}\, e^{-r^2\,(A-z\mathbb{1})} \int_0^{\pi/2} d\theta\, (\cos\theta)^{2q_1-1}\, (\sin\theta)^{2q_2-1}.$$

The last θ-integral is equal to $B(q_1, q_2)/2$, where $B(q_1, q_2) = \Gamma(q_1)\,\Gamma(q_2)/\Gamma(q_1+q_2)$ is the *Euler* Beta-function. Consequently, (1.10.4) and (1.10.2) prove (1.10.3).

□

Example 1.10.3 We consider here examples/counterexamples for illustration assertions in Proposition 1.6.7 and in Theorem 1.6.14. Note that examples (a)–(d) for a Gibbs semigroup (Theorem 1.6.14) can be translated *mutatis mutandis* for compact semigroups (Proposition 1.6.7).

(a) Consider the Hilbert space $\mathfrak{H} = l^2(\mathbb{C})$. We define in $l^2(\mathbb{C})$ a linear densely defined multiplication operator

$$M_q : u \mapsto q \cdot u = \{q_k\, u_k\}_{k\geq 1}\,, \quad q_k \in \mathbb{C}\,, \tag{1.10.5}$$

with domain $\mathrm{dom}\,(M_q) = \{u \in l^2(\mathbb{C}) : M_q u \in l^2(\mathbb{C})\}$ for a given vector $q = \{q_k\}_{k\geq 1}$.

(b) By definition (1.10.5), the spectrum $\sigma(M_q)$ of M_q coincides with the set of eigenvalues $\{\lambda_k(M_q) = q_k\}_{k\geq 1}$. If $\sigma(M_q) \subset \mathbb{C}_+$, then the closed operator M_q generates the C_0-semigroup $\{U_q(t) = \mathrm{e}^{-t\, M_q}\}_{t\geq 0}$ on the Hilbert space $l^2(\mathbb{C})$:

$$\mathrm{e}^{-t\, M_q}\, u := \mathrm{e}^{-t\, q} \cdot u = \{\mathrm{e}^{-t\, q_k}\, u_k\}_{k\geq 1}\,, \quad t \geq 0\,, \quad u \in l^2(\mathbb{C})\,. \tag{1.10.6}$$

(c) For the next example in $\mathfrak{H} = l^2(\mathbb{C})$, we suppose that the closed densely defined multiplication operator M_{q_2} corresponds to

$$q_2 = \{q_k = i\, k^2\}_{k\geq 1}\,, \quad k \in \mathbb{N}\,, \tag{1.10.7}$$

with spectrum $\sigma(M_{q_2}) = \{i\, k^2\}_{k\geq 1}$ (see (b)).

(i) Since $\|(\mu\, \mathbb{1} + M_{q_2})^{-1}\| \leq |\mathrm{Re}\,\mu|^{-1}$ for $\mu \in \mathbb{C}$, the operator M_{q_2} satisfies the hypotheses of the Hille-Yosida theorem. Therefore, $\{U_{q_2}(t) := \mathrm{e}^{-t\, M_{q_\beta}}\}_{t\geq 0}$ is a *strongly* continuous contraction semigroup. Note that for $t \gtrless 0$ it gives two branches of a unitary group, which is *nowhere* continuous in *operator-norm* topology for $t \in \mathbb{R}$ since by the spectral representation for the branch $\{U_{q_2}(t)\}_{t\geq 0}$ generated by self-adjoint operator (M_{q_2}/i) one gets

$$\|U_{q_2}(t)\, u - u\|^2 = 2 \int_{\sigma(M_{q_2}/i)} (dE(\lambda)\, u,\, u)\, (1 - \cos(t\, \lambda))\,. \tag{1.10.8}$$

Owing to representation (1.10.8), we obtain for $u \in l^2(\mathbb{C})$ and any $t > 0$ the estimate from below: $\sup_{\|u\|=1} \|U_{q_2}(t)\, u - u\| \geq \sqrt{2}\,$.

(ii) The resolvent $R_{M_{q_2}}(-\mu)$, for $-\mu \in \rho(M_{q_2})$, is a bounded operator with discrete spectrum $\sigma(R_{M_{q_2}}(-\mu)) = \{(\mu + i\, k^2)^{-1}\}_{k\geq 1}$. For $\mu_0 \in \mathbb{C}$ with $\mathrm{Im}\,\mu_0 = 0$, the *singular values* $\{s_k(R_{M_{q_2}}(\mu_0))\}_{k\geq 1}$ of operator $R_{M_{q_2}}(\mu_0)$, that is, the eigenvalues $\{\lambda_k(|R_{M_{q2}}(\mu_0)|)\}_{k\geq 1}$ of operator $|R_{M_{q2}}(\mu_0)| := (R^*_{M_{q2}}(\mu_0)R_{M_{q2}}(\mu_0))^{1/2}$, are

$$s_k(R_{M_{q2}}(\mu_0)) = \lambda_k(|R_{M_{q2}}(\mu_0)|) = (\mu_0^2 + k^4)^{-1/2}\,, \quad k \in \mathbb{N}\,. \tag{1.10.9}$$

Consequently, $R_{M_{q2}}(\mu_0) \in \mathcal{C}_1(\mathfrak{H})$, which by the first resolvent equation yields $R_{M_{q2}}(\mu) \in \mathcal{C}_1(\mathfrak{H})$ for any $\mu \in \rho(M_{q2})$.

(iii) As a result, although the resolvent of generator M_{q2} consists of trace-norm operators, the semigroup $\{U_{q2}(t)\}_{t\geq 0}$ is not a Gibbs semigroup since by (i) it is only strongly continuous. Notice that for this semigroup the Bochner integral representation (Laplace transform, cf. (1.6.19))

$$R_A(-\lambda) = \int_0^\infty dt\, e^{-\lambda t}\, e^{-tA}\,, \quad \operatorname{Re}\lambda > 0\,, \tag{1.10.10}$$

is defined only in the *strong operator topology*, which implies that the left-hand side of (1.10.10) is simply a bounded operator.

On that account, the condition (b)(i) in Theorem 1.6.14 serves for improving topology in definition (1.10.10) of the Bochner integral with the aim to exclude this kind of discrepancy.

(d) Now, let $\mathfrak{H} = L^2(0, 1)$. For $t \geq 0$ and $f \in \mathfrak{H}$, define $U(t) \in \mathcal{L}(\mathfrak{H})$ by

$$(U(t)f)(s) = \begin{cases} f(s+t) & \text{if } 0 \leq s+t \leq 1\,, \\ 0 & \text{if } s+t > 1\,. \end{cases}$$

Then $\{U(t)\}_{t\geq 0}$ is a C_0-semigroup, which is nilpotent and hence eventually (compact) Gibbs semigroup with $\operatorname{Tr} U(t > 1) = 0$. Since the Laplace transform (1.6.19) defines the resolvent of generator of this semigroup by a strongly convergent Bochner integral, we obtain

$$R(-z) = \int_0^1 dt\, e^{-zt}\, e^{-tA}\,, \quad z \in \mathbb{C}\,. \tag{1.10.11}$$

Therefore, the resolvent (1.10.11) is an entire function of z and consequently not in the von Neumann-Schatten ideals since the spectrum of generator is *empty*.

Remark 1.10.4 We note that theory of continuous semigroups on the *locally convex* vector spaces is somewhat *murky*. For references, we suggest [Yos80, Capter IX, Section 13] and some papers ([Koma64], [Komu68], [Ouc73], [Dem74], [Cho85], [LeWu11]) for understanding the origin, evolution and certain progress in this direction. We do not consider this class of semigroups in the present book.

On the other hand, in Chap. 13 and in Chap. 14, we study *convergence* of the product Trotter and Trotter-Kato formulæ in topology of the locally convex vector space, which is a complete, metrisable *Fréchet* space $L^2_{\mathrm{loc}}(\mathbb{R}, \mathfrak{H})$. Note that approximants in these formulæ are families which are continuous in the *strong* operator topology on $\mathcal{L}(\mathfrak{H})$ for Hilbert space $\mathfrak{H}$.

Notes to Sect. 1.7

Holomorphic semigroups arise from parabolic partial differential equations or abstract linear evolution equations in normed spaces (see Chap. 2). They represent an important subclass of the strongly continuous semigroups, which are also immediately norm continuous. In this section, we followed essentially [Dav80], [Kato80], [RS75] and [EN00, Chapter II, Section 4].

Notice that when we consider holomorphic operator-valued function with a map in $\mathcal{L}(\mathfrak{X})$, there is no distinction among uniform (*operator-norm*), strong and weak holomorphy (cf. Definition 1.7.1(ii)), which is similar to the case of vector-valued functions on Banach space $\mathfrak{X}$. See Proposition 1.10.5 and Corollary 1.10.6 or, for example, [Kato80, Chapter III, Theorem 3.12] and [RS80, Theorem VI.4]. We (often) tacitly refer to this observation in few places of the book. So, we present the proof of this useful assertion below.

Proposition 1.10.5 *Let $\Delta \ni z \mapsto V(z) \in \mathcal{L}(\mathfrak{X})$ be bounded operator-valued function in domain $\Delta \subset \mathbb{C}$ of the complex plane. Let for any $x \in \mathfrak{X}$ the function $\Delta \ni z \mapsto \langle V(z)\, x\,,\, \phi\rangle$ be holomorphic for each $\phi \in \mathfrak{X}^*$ (weak holomorphy). Then $z \mapsto V(z)$ is strongly holomorphic, that is, strongly differentiable in $z \in \Delta$,*

$$\partial_z V(z)\, x = \lim_{z' \to z} \frac{1}{z' - z}\, (V(z') - V(z))\, x\,, \quad z,\, z' \in \Delta\,, \tag{1.10.12}$$

for any $x \in \mathfrak{X}$.

Proof First, we note that for a Banach space $\mathfrak{X}$ and its adjoint $\mathfrak{X}^*$ one has for the vector norm of $x \in \mathfrak{X}$:

$$\|x\| = \sup_{\phi \in \mathfrak{X}^* \,:\, \|\phi\| \leq 1} |\langle x, \phi\rangle|\,. \tag{1.10.13}$$

Then on account of (1.10.13), the weak convergence $\langle x_n, \phi\rangle \to \langle x, \phi\rangle$ for $n \to \infty$ uniformly for $\|\phi\| \leq 1$ yields

$$\lim_{n\to\infty} \|x_n - x\| = \lim_{n\to\infty} \sup_{\phi \in \mathfrak{X}^* \,:\, \|\phi\| \leq 1} |\langle x_n - x\,,\, \phi\rangle| = 0\,. \tag{1.10.14}$$

Now, we use the *Cauchy* integral formula for holomorphic function $z \mapsto \langle V(z)\, x\,,\, \phi\rangle$ and positively oriented contour $\Gamma \subset \Delta$ enclosing $z \in \Delta$ for equation:

$$\langle V(z)\, x\,,\, \phi\rangle = \frac{1}{2\pi i} \int_\Gamma d\zeta\, \frac{\langle V(\zeta)\, x\,,\, \phi\rangle}{\zeta - z}\,. \tag{1.10.15}$$

Owing to (1.10.15), one infers for $z', z \in \Delta$ enclosed by Γ the representation:

$$\frac{1}{z'-z}\,\langle V(z')\,x\,-\,V(z)\,x\,,\,\phi\rangle - \partial_z\langle V(z)\,x\,,\,\phi\rangle = \tag{1.10.16}$$
$$= \frac{z'-z}{2\pi i}\int_\Gamma d\zeta\,\frac{\langle V(\zeta)\,x\,,\,\phi\rangle}{(\zeta - z')\,(\zeta - z)^2}\,.$$

Given that $\langle V(\zeta)\,x\,,\,\phi\rangle$ is weakly continuous in ζ, it is bounded on the contour Γ, that is, $|\langle V(\zeta)\,x\,,\,\phi\rangle| \le M\,\|x\|\,\|\phi\|$ for $\zeta \in \Gamma$. Then since integral in (1.10.16) converges, this representation provides estimate

$$\left|\frac{1}{z'-z}\,\langle V(z')\,x - V(z)\,x\,,\,\phi\rangle - \partial_z\langle V(z)\,x\,,\,\phi\rangle\right| \le |z'-z|\,\widetilde{M}\,\|x\|\,\|\phi\|\,, \tag{1.10.17}$$

for any $x \in \mathfrak{X}$ and uniformly for $\|\phi\| \le 1$. Because of (1.10.14), the estimate (1.10.17) proves strong differentiability (1.10.14) and hence the strong holomorphy of function $z \mapsto V(z)$ on Δ. □

Corollary 1.10.6 *If operator-valued function $z \mapsto V(z) \in \mathcal{L}(\mathfrak{X})$ is strongly holomorphic in $\Delta \subset \mathbb{C}$, then $z \mapsto V(z)$ is holomorphic operator-valued function in the operator-norm topology on $\mathcal{L}(\mathfrak{X})$. Following* (1.10.16), *we obtain for this function with values in Banach space $\mathcal{L}(\mathfrak{X})$ representation*

$$\left\{\frac{1}{z'-z}(V(z') - V(z)) - \partial_z V(z)\right\}\,x = \tag{1.10.18}$$
$$= \frac{z'-z}{2\pi i}\int_\Gamma d\zeta\,\frac{V(\zeta)\,x}{(\zeta - z')\,(\zeta - z)^2}\,,\qquad x \in \mathfrak{X}.$$

Then verbatim the proof of (1.10.17) *representation* (1.10.18) *yields estimate in the norm of $\mathfrak{X}$,*

$$\left\|\left\{\frac{1}{z'-z}\,(V(z') - V(z)) - \partial_z\langle V(z)\right\}\,x\right\| \le |z'-z|\,\widetilde{M}\,\|x\|\,, \tag{1.10.19}$$

for any $x \in \mathfrak{X}$. Now, taking $\sup_{\|x\|=1}$ in (1.10.19), *we obtain the operator-norm estimate saying that operator-valued function $z \mapsto V(z)$ is differentiable in operator-norm. Hence, it is holomorphic in the operator-norm topology.*

Remark 1.10.7 For illustration of different types of the C_0-semigroups *continuity* (cf. also Sect. 1.6), we consider below some examples of semigroups generated on a Banach space $\mathfrak{X}$ by *multiplication operators*, whereas for a general setting that involves the multiplication operators (see Chap. 2).

(a) To this aim, let $\mathfrak{X} = c_0$ be the Banach space of *null sequences* of complex numbers, this means such that $c_0 = \{f = \{f_k\}_{k\geq 1} \in c_\infty : \lim_{k\to\infty} f_k = 0\}$, endowed with the norm $\|f\|_\infty := \sup_{k\geq 1} |f_k|$. We define in c_0 a linear densely defined multiplication operator

$$M_q : f \mapsto q \cdot f = \{q_k \, f_k\}_{k\geq 1}, \quad q_k \in \mathbb{C}, \tag{1.10.20}$$

with domain $\operatorname{dom}(M_q) = \{f \in c_0 : M_q f \in c_0\}$ for a given vector $q = \{q_k\}_{k\geq 1}$.

(b) By definition (1.10.20), the spectrum $\sigma(M_q)$ of operator M_q coincides with the set $\{q_k\}_{k\geq 1}$. If $\sigma(M_q) \subset \mathbb{C}_+$, then by Proposition 1.1.1, the closed operator M_q is generator of the C_0-semigroup $\{U_q(t)\}_{t\geq 0}$:

$$U_q(t) f := \mathrm{e}^{-t q} \cdot f, \quad f \in c_0. \tag{1.10.21}$$

(c) As the first illustration, we consider the case of *holomorphic* semigroups generated on Banach space c_0 by multiplication operators (1.10.20). To this end, let

$$q_\alpha := \{q_k(\alpha) = \mathrm{e}^{i\,\alpha}\, k\}_{k\geq 1}, \quad |\alpha| < \pi/2. \tag{1.10.22}$$

Then unbounded operator M_{q_α} is generator of C_0-semigroup $\{U_{q_\alpha}(t) = \mathrm{e}^{-t\, M_{q_\alpha}}\}_{t\geq 0}$ such that the operator-norm

$$\|M_{q_\alpha} U_{q_\alpha}(t)\| = \sup_{k\geq 1} |\, k\, \mathrm{e}^{-t\, k\, \cos\alpha} \,| \leq \frac{M(\alpha)}{t}, \quad M(\alpha) := (e\, \cos\alpha)^{-1}. \tag{1.10.23}$$

As a consequence, by Proposition 1.7.7 and (1.7.16), the semigroup $\{U_{q_\alpha}(t)\}_{t\geq 0}$ may be analytically continued to a bounded $\|\cdot\|$-holomorphic semigroup $\{U_{q_\alpha}(z)\}_{z\in S_\theta}$ in the open sector S_θ, where $\theta = \arcsin(e\, M(\alpha))^{-1}$. That is, $\theta = \pi/2 - \alpha$.

(d) Next, we consider the case when semigroup $\{U_q(t)\}_{t\geq 0}$ generated on Banach space c_0 by multiplication operator (1.10.20) is the operator-norm *differentiable*. This condition is weaker than (1.10.23) and has the form $U_q(t > 0) : c_0 \to \operatorname{dom}(M_q)$. That is, for given q and any $t > 0$, it yields

$$\|M_q U_q(t)\| = \sup_{f\in c_0 : \|f\|_\infty = 1} \|M_q U_q(t)\, f\|_\infty = \sup_{k\geq 1} |\, q_k\, \mathrm{e}^{-t\, q_k} \,| \tag{1.10.24}$$

$$\leq \sup_{k\geq 1} |\, \mathrm{Re}(q_k)\, \mathrm{e}^{-t\, \mathrm{Re}(q_k)} \,| + \sup_{k\geq 1} |\, \mathrm{Im}(q_k)\, \mathrm{e}^{-t\, \mathrm{Re}(q_k)} \,|.$$

Because $\sigma(M_q) \subset \mathbb{C}_+$, the estimate (1.10.24) implies the condition of the semigroup differentiability for $t > 0$: there exists $C > 0$ such that the spectrum

$$\sigma(M_q) \subset \{q \in \mathbb{C}_+ : |\operatorname{Im}(q_k)| \leq C\, \mathrm{e}^{t \operatorname{Re}(q_k)}\,, \text{ for any } t > 0 \text{ and all } k \geq 1\}. \tag{1.10.25}$$

Note that if semigroup $\{U_q(t)\}_{t\geq 0}$ is the operator-norm *eventually differentiable* for $t > t_0 > 0$ then condition (1.10.25) gets the form

$$\sigma(M_q) \subset \{q \in \mathbb{C}_+ : |\operatorname{Im}(q_k)| \leq C\, \mathrm{e}^{t_0 \operatorname{Re}(q_k)}\,, \text{ for all } k \geq 1\}. \tag{1.10.26}$$

Example 1.10.8

(1) Let $q_{t_0} = \{q_k = k + i\, \mathrm{e}^{t_0 k}\}_{k\geq 1}$. Then by virtue of (1.10.25) and (1.10.26), the C_0-semigroup $\{U_{q_{t_0}}(t)\}_{t\geq 0}$ generated on Banach space c_0 by multiplication operator $M_{q_{t_0}}$ is the operator-norm differentiable for $t > t_0$, but not for $0 \leq t < t_0$. Therefore, the semigroup $\{U_{q_{t_0}}(t)\}_{t\geq 0}$ is *eventually* differentiable.

(2) On the other hand, for $t_0 = 0$, we infer that by (1.10.25) the semigroup $\{U_{q_{t_0=0}}(t)\}_{t\geq 0}$ is *immediately* differentiable. Moreover, seeing that for $t_0 = 0$ the spectrum (cf. Remark 1.10.7(c)),

$$\sigma(M_{q_{t_0=0}}) = \{k + i\}_{k\geq 1} \subset \overline{S}_\alpha\,, \tag{1.10.27}$$

for a closed sector with semi-angle $\pi/4 \leq \alpha < \pi/2$ (see condition (1.7.3)), we conclude by Definition 1.7.1 that semigroup $\{U_{q_{t_0=0}}(t)\}_{t\geq 0}$ is *holomorphic* in open sector S_θ for $\theta = \pi/2 - \alpha \leq \pi/4$.

(3) Let $q = \{q_k = k + i\, k^2\}_{k\geq 1}$. Note that for any $t > 0$ there exists $C > 0$ such that $k^2 < C\, \mathrm{e}^{t k}$ for $k \geq 1$. Then on account of (1.10.25) the C_0-semigroup $\{U_q(t)\}_{t\geq 0}$ generated on the Banach space c_0 by multiplication operator M_q is *immediately* and *infinitely* differentiable.

(4) On the contrary, since $\sigma(M_q) = \{q_k = k + i\, k^2\}_{k\geq 1}$ does *not* belong to any sector S_θ with semi-angle $\theta < \pi/2$ (cf. condition (1.7.3)), the semigroup $\{U_q(t)\}_{t\geq 0}$ is *not* holomorphic. Or the other way around, by virtue of (1.10.23), we obtain that

$$\|M_q U_q(t)\| = \sup_{k\geq 1} |\, k^2\, \mathrm{e}^{-t k}\,| \leq \frac{M}{t^2}\,, \quad M > 0\,, \tag{1.10.28}$$

which bolsters by Proposition 1.7.7 the same conclusion about the (infinitely differentiable) semigroup $\{U_q(t)\}_{t\geq 0}$, which has no analytic extension from the real $\mathbb{R}^+$.

(5) On the Hilbert space $\mathfrak{H} = l^2(\mathbb{C})$, the semigroup $\{U_q(t) = \mathrm{e}^{-t\,M_q}\}_{t\geq 0}$ is the immediately Gibbs semigroup since it is strongly continuous and the trace-norm $\|U_q(t)\|_1 < \infty$ (cf. Sect. 1.6). Moreover, owing to the trace-norm estimates,

$$\|(M_q)^n U_q(t)\|_1 = \sum_{k\geq 1}\{(k^2 + k^4)^{n/2}\mathrm{e}^{-t\,k}\} \leq \frac{M_n}{t^{2n}}\,, \quad n \in \mathbb{N}\,, \tag{1.10.29}$$

this semigroup is infinitely trace-norm differentiable.

We note that generator M_q and family $\{\mathrm{e}^{-t\,M_q}\}_{t>0}$ satisfy conditions of Theorem 1.6.14. As a consequence, the Gibbs semigroup $\{U_q(t)\}_{t\geq 0}$ could be reconstructed due to this theorem, but not via p-generator ([Zag19, Section 4.3]), which provides $\|\cdot\|_1$-holomorphic semigroups.

Again, this is because of estimate (1.10.29)) for $n = 1$ and Proposition 1.7.7 that infinitely differentiable Gibbs semigroup $\{U_q(t)\}_{t\geq 0}$ has no analytic extension in the trace-norm topology from the real $\mathbb{R}^+$.

We remark that examples above are also valid for illustrations of the similar properties in the Hilbert space $\mathfrak{H} = l^2(\mathbb{C})$ in the framework of a general setting for the C_0-semigroups with *normal* generators. See Notes to Sect. 1.8.

Notes to Sect. 1.8

Section 1.8.1

The *numerical range* of a generator seems a very natural means for characterisation the holomorphic semigroups on a Hilbert space. Our presentation here was motivated by Kato [Kato80, Chapter V, §3.2, §3.10] and Schmüdgen [Schm12, Chapter 3, Section 3.3].

Section 1.8.2

A convenient *criterion* that operator $A \in \mathscr{H}_c(\theta, 0)$, in other words, a generator of the *holomorphic contraction* semigroup $\{U_A(z)\}_{z\in S_\theta}$ on $\mathfrak{H}$ (Proposition 1.8.13), is due to T. Kato [Kato80, Ch.IX, Theorem 1.24].

Note that in review [ArtE12] one can find more about holomorphic contraction semigroups as well as about a new *form method* for construction of their generations.

According to Remark 1.8.15 and on account of similarly to Example 1.10.8, we infer the properties of semigroups generated by normal operators along the following example.

Example 1.10.9

(a) We consider the Hilbert space $L^2(\Omega, \mu) = l^2(\mathbb{C})$ and define in $l^2(\mathbb{C})$ a linear densely defined multiplication operator

$$M_q : u \mapsto q \cdot u = \{q_k\, u_k\}_{k\geq 1}\,, \quad q_k \in \mathbb{C}\,, \tag{1.10.30}$$

with domain $\text{dom}(M_q) = \{u \in l^2(\mathbb{C}) : M_q u \in l^2(\mathbb{C})\}$ for a given vector $q = \{q_k\}_{k\geq 1}$.

(b) By definition (1.10.30), the spectrum $\sigma(M_q)$ of operator M_q coincides with the set $\{q_k\}_{k\geq 1}$. If $\sigma(M_q) \subset \mathbb{C}_+$, then by Proposition 1.1.1, the closed operator M_q is generator of the C_0-semigroup $\{U_q(t)\}_{t\geq 0}$ on the Hilbert space $l^2(\mathbb{C})$:

$$U_q(t)\,u := \mathrm{e}^{-t\,q} \cdot u\,, \quad t \geq 0\,, \quad u \in l^2(\mathbb{C})\,. \tag{1.10.31}$$

(c) Let C_0-semigroup is generated on $l^2(\mathbb{C})$ by multiplication operator (1.10.30) defined by

$$q_\alpha := \{q_k(\alpha) = \mathrm{e}^{i\,\alpha}\,k\,\}_{k\geq 1}\,, \quad |\alpha| < \pi/2\,. \tag{1.10.32}$$

Then unbounded operator M_{q_α} is generator of C_0-semigroup $\{U_{q_\alpha}(t) = \mathrm{e}^{-t\,M_{q_\alpha}}\}_{t\geq 0}$ such that for $\|u\| = 1$ and $t > 0$ one gets

$$\|M_{q_\alpha}U_{q_\alpha}(t)\,u\|^2 = \sum_{k\geq 1} \mathrm{e}^{-2\,k\,t\,\cos\alpha}\,k^2\,|u_k|^2 \leq \frac{M(\alpha)^2}{t^2}\,,$$

$$M(\alpha) := (e\,\cos\alpha)^{-1}\,. \tag{1.10.33}$$

As a consequence of (1.10.33), the operator-norm $\|M_{q_\alpha}U_{q_\alpha}(t)\| \leq M(\alpha)/t$. Then by Proposition 1.7.7 and (1.7.16), the semigroup $\{U_{q_\alpha}(t)\}_{t\geq 0}$ may be analytically continued to a bounded $\|\cdot\|$-holomorphic semigroup $\{U_{q_\alpha}(z)\}_{z\in S_\theta}$ in the open sector S_θ, where $\theta = \arcsin{(e\,M(\alpha))^{-1}}$. That is, $\theta = \pi/2 - \alpha$.

Note that by (1.10.30) and (1.10.32) the numerical range $W(M_{q_\alpha}) = \{\mathrm{e}^{i\,\alpha}\,k\,\}_{k\geq 1}$, where $|\alpha| < \pi/2$, cf. (1.8.1), belongs to the sector $\overline{S}_\alpha \subset \mathbb{C}_+$.

(d) Now, let

$$q_\beta := \{q_k(\beta) = k + i\,k^\beta\,\}_{k\geq 1}\,, \quad \beta \geq 0\,, \tag{1.10.34}$$

that is, the numerical range $W(M_{q_\beta}) = \{k+i\,k^\beta\,\}_{k\geq 1}$. Then unbounded operator M_{q_β} is generator of C_0-semigroup $\{U_{q_\beta}(t) = \mathrm{e}^{-t\,M_{q_\beta}}\}_{t\geq 0}$ such that for $u \in l^2(\mathbb{C})$ we infer

$$\|M_{q_\beta}U_{q_\beta}(t)\,u\|^2 = \sum_{k\geq 1} \mathrm{e}^{-2\,k\,t}\,(k^2 + k^{2\beta})\,|u_k|^2\,. \tag{1.10.35}$$

(i) If $\beta \leq 1$, then numerical range $W(M_{q_\beta}) \subset \overline{S}_{\alpha=\pi/4}$ and (1.10.35) yields for some $M > 0$ that operator-norm

$$\|M_{q_\beta}U_{q_\beta}(t)\| \leq \frac{M}{t}\,, \quad t > 0\,.$$

Consequently, by Proposition 1.7.7 and (1.7.16), the semigroup $\{U_{q_\beta}(t)\}_{t\geq 0}$ allows the analytic continuation to a bounded $\|\cdot\|$-holomorphic semigroup $\{U_{q_\alpha}(z)\}_{z\in S_\theta}$ in the open sector $S_{\theta=\pi/2-\alpha}$, where $\alpha = \pi/4$.

(ii) If $\beta > 1$, then numerical range $W(M_{q_\beta})$ does not belong to any sector $S_{\alpha<\pi/2}$ and (1.10.35) yields that operator-norm

$$\|M_{q_\beta} U_{q_\beta}(t)\| \leq \frac{M(\beta)}{t^\beta}, \quad t > 0.$$

for $M(\beta) > 0$. Hence, although for any $n \in \mathbb{N}$ and $t > 0$ owing to (1.10.35) one has $\|M^n_{q_\beta} U_{q_\beta}(t)\| < \infty$ (i.e., semigroup $\{U_{q_\beta}(t)\}_{t\geq 0}$ is infinitely differentiable for $t > 0$), the Proposition 1.7.7 does not allow any analytic continuation of this semigroup for the case $\beta > 1$.

Notes to Sect. 1.9

The results of this section can be found in monographs [HiPh57], [Kato80] and [Dav80], as well as in some relatively recent books [EN00], [Dav07] and [Bob16]. We note that the latter book is particularly dedicated to the subject of the present section and, in contrast to other treatises, contains a number of instructive and useful examples for the reader from the pure and applied mathematics.

Section 1.9.1

Our classification of different methods for perturbation of C_0-semigroups is motivated by the books [Dav07, Sections 11.4 and 11.5] and [Zag19a, Sections 1.7].

For a direct extension of Proposition 1.9.1 to quasi-bounded C_0-semigroups (without application of the *equivalent norms* method, Remark 1.3.3), see [HiPh57, Theorem 13.2.1].

Section 1.9.2

Besides the abstract but important for other chapters of our book, Example 1.9.8 (cf. [Yos65]), one can find instructive examples of perturbations with the relative bound $a = 0^+$ in [Kato80, Chapter V, §5.2, §5.3] and in [Amr09, Proposition 2.49].

In the proofs of Theorem 1.9.9, Theorem 1.9.10 and Theorem 1.9.11, we essentially follow the line of reasoning developed in [Mais71] and [Zag89].

Section 1.9.3

For accretive generators in Banach space, the Proposition 1.9.21 is known as the *Lumer-Phillips* theorem [LuPh61]. The Proposition 1.9.23 is due to Chernoff [Cher72]. For more details (including Proposition 1.9.24), see [Kato80, Section IX].

Section 1.9.4

We present the *Miyadera-Voigt* perturbation theory following basically the paper by J. Voigt [Voi77] and [EN00, Section III.3c].

Section 1.9.5

There are several versions of Propositions 1.9.30 and 1.9.32, known collectively as the *Trotter-Kato-Neveu* convergence theorem [Cher72, §1] or the *Trotter-Kato* approximation theorem [EN00, Chapter III, Section 4].

H. Trotter [Trot58] proved Propositions 1.9.30 within a general semigroup set-up. For *contractions*, this statement has been proved independently by J. Neveu [Nev58]. In paper [Kato59], T. Kato pointed out (see Propositions 1.9.32) the necessity of a more careful treatment of the *pseudo-resolvents* appeared in [Trot58] as a technical tool. For details, one can consult, for example, [Kato80, Chapter IX, §5.2], or [EN00, Chapter III, Section 4].

We warn the reader against a possible confusion between the *Trotter-Kato* approximations (convergence) and the *Trotter-Kato product* approximations (convergence) (see Sect. 1.9.6). However, for the first of these notions, we shall systematically use in this book the name *Trotter-Kato-Neveu* approximation (convergence) theorem, whereas for the second the name *Trotter-Kato* product formulæ (approximations). Note that the *Trotter* product formula involves only *exponential* Kato functions.

Section 1.9.6

The proof of the Chernoff $\sqrt{n}$ -Lemma (Lemma 1.9.33) is due to [Cher68, Lemma 2]. For more Chernoff's estimates and applications, see Sect. 3.1.2. Theorem 1.9.34, which is also known as the *Chernoff product theorem* (about Chernoff product formula) appeared for the first time in [Cher68, Theorem and Corollary]. Together with the Chernoff estimates, it is one of the most important tools in the Trotter-Kato theory [Trot59], [Kato74], [Kato78] of the C_0-semigroup product approximations that we shall use systematically throughout this book (see Parts II–V).

Two fundamental statements, Propositions 1.9.35 and 1.9.41, are formulated as a general setting in the framework of a Banach space. We shall show in this book that in a Hilbert space the control of the product formulæ is more accurate and the corresponding results may be extended to include the *Trotter-Kato* product formulæ (see Parts II and III).

Here, two key points, which are both due to T. Kato (1978), are important to mention as a *part* of the programme for this book:

(1) A relaxation of condition on *intersection* up to $\mathrm{dom}\,(A) \cap \mathrm{dom}\,(B) = \{0\}$
(2) A generalisation of product formulæ for *non-exponential* (Kato) functions

The third point is in many aspects due to results by the authors. It consists of *uplifting* (improvement) of convergence the product formulæ from the *strong* to the *operator-norm* topology.

For completeness and for the reader convenience, we recall (see Proposition 1.10.10) a *direct* proof of the product formula, which does *not use* the Chernoff estimate and which is due to *Sophus Lie*. In fact, it is instructive since the *telescopic representation* used there was applied in [NZ90a] and [NZ90b], for the first proof of extension the Trotter-Kato product formulæ beyond the well-known *strong* operator convergence. We use this representation in some of our arguments below.

In fact, it is worthy to note that in fact there are *essentially two* methods to prove the product formulæ:

- The historically first, which dates back to S. Lie (1875), is due to the *telescopic representation*.
- The second one is due to P. Chernoff (1968), which is based on the *Chernoff* $\sqrt{n}$ *-Lemma* and the *Trotter-Kato-Neveu* approximation (convergence) theorem.

Since the observation by S. Lie for finite matrices, it is known that the semigroup $\{U_{A+B}(t)\}_{t\geq 0}$ generated by the sum $A+B$ of two *bounded* operators A and B on a Banach space $\mathfrak{X}$ can be approximated in the *operator-norm* topology in terms of the products of semigroups $\{U_A(t)\}_{t\geq 0}$ and $\{U_B(t)\}_{t\geq 0}$.

Proposition 1.10.10 (S. Lie Product Formulæ) *Let* $A, B \in \mathcal{L}(\mathfrak{X})$. *Then for any* $t \geq 0$ *and* $n \in \mathbb{N}$, *we obtain*

(a) *The product (operator-norm) approximations of semigroup* $\{U_{A+B}(t)\}_{t\geq 0}$

$$\left\|(U_A(t/n)\,U_B(t/n))^n - U_{A+B}(t)\right\| \leq \frac{c}{n}\,, \tag{1.10.36}$$

where $c = c(t, \|A\|, \|B\|) > 0$

(b) *The symmetric product (operator-norm) approximations of semigroup* $\{U_{A+B}(t)\}_{t\geq 0}$

$$\left\|(U_A(t/2n)\,U_B(t/n)\,U_A(t/2n))^n - U_{A+B}(t)\right\| \leq \frac{c_{\text{sym}}}{n^2}\,, \tag{1.10.37}$$

where $c_{\text{sym}} = c_{\text{sym}}(t, \|A\|, \|B\|) > 0$.

Proof

(a) Since A, B and $A+B$ are bounded, we can use for exponents the operator-norm convergent series. Let $V_{A,B}(t) := U_A(t)\,U_B(t)$, cf. (1.9.111). Then for *Lie product approximants* $\{V_{A,B}(t/n)^n\}_{n\geq 1}$, one uses for the difference between approximants and semigroup the *telescopic representation*,

$$\begin{aligned} &V_{A,B}(t/n)^n - U^n_{A+B}(t/n) \\ &\qquad = \sum_{m=0}^{n-1} V_{A,B}(t/n)^{n-m-1}\left(V_{A,B}(t/n) - U_{A+B}(t/n)\right) U^m_{A+B}(t/n)\,, \end{aligned} \tag{1.10.38}$$

which can be estimated as follows:

$$\left\| V_{A,B}(t/n)^n - U_{A+B}^n(t/n) \right\| \tag{1.10.39}$$

$$\leq n \left(\max_{n\geq 1} \left\{ \|V_{A,B}(t/n)\|,\ \|U_{A+B}(t/n)\| \right\} \right)^n \|V_{A,B}(t/n) - U_{A+B}(t/n)\|.$$

Using exponential series, one gets the estimates

$$\|V_{A,B}(t/n)\| \leq \mathrm{e}^{t(\|A\|+\|B\|)/n} \qquad \text{and} \qquad \|U_{A+B}(t/n)\| \leq \mathrm{e}^{t(\|A\|+\|B\|)/n}. \tag{1.10.40}$$

Then (1.10.39) yields

$$\left\| V_{A,B}(t/n)^n - U_{A+B}^n(t/n) \right\| \leq n\, \mathrm{e}^{t(\|A\|+\|B\|)} \|V_{A,B}(t/n) - U_{A+B}(t/n)\| . \tag{1.10.41}$$

Given that

$$\begin{aligned}
&\left\| V_{A,B}(t/n) - U_{A+B}(t/n) \right\| = \Bigg\| \sum_{k=0}^{\infty} \frac{1}{k!}\Big(-\frac{t}{n}A\Big)^k \sum_{\ell=0}^{\infty} \frac{1}{\ell!}\Big(-\frac{t}{n}B\Big)^\ell \\
&- \sum_{s=0}^{\infty} \frac{1}{s!}\Big[-\frac{t}{n}(A+B)\Big]^s \Bigg\| \\
&\leq \sum_{k=2}^{\infty} \frac{1}{k!}\Big(\frac{t}{n}\|A\|\Big)^k \sum_{\ell=2}^{\infty} \frac{1}{\ell!}\Big(\frac{t}{n}\|B\|\Big)^\ell + \sum_{k=3}^{\infty} \frac{1}{k!}\Big(\frac{t}{n}\|A\|\Big)^k + \sum_{\ell=3}^{\infty} \frac{1}{\ell!}\Big(\frac{t}{n}\|B\|\Big)^\ell \\
&+ \frac{t}{n}\|A\| \sum_{\ell=2}^{\infty} \frac{1}{\ell!}\Big(\frac{t}{n}\|B\|\Big)^\ell + \frac{t}{n}\|B\| \sum_{k=2}^{\infty} \frac{1}{k!}\Big(\frac{t}{n}\|A\|\Big)^k \\
&+ \sum_{s=3}^{\infty} \frac{1}{s!}\Big[\frac{t}{n}(\|A\|+\|B\|)\Big]^s \\
&+ \frac{t^2}{2n^2}\|AB - BA\| =: \frac{1}{n^2} M_n(t, \|A\|, \|B\|),
\end{aligned} \tag{1.10.42}$$

the inequality (1.10.41) yields the estimate (1.10.36), where c is defined as the *maximum* over $n \geq 1$ of the right-hand side of (1.10.42) multiplied by $n^2 \mathrm{e}^{t(\|A\|+\|B\|)}$. Then $c = \mathrm{e}^{t(\|A\|+\|B\|)} M(t, \|A\|, \|B\|)$, where $M(t, \|A\|, \|B\|) := \max_{n\geq 1} M_n(t, \|A\|, \|B\|)$.

(b) Let $W_{A,B}(t) := U_B(t/2)U_A(t)U_B(t/2)$. Then $\{W_{A,B}(t/n)^n\}_{n\geq 1}$ is the family of *symmetric* Lie product *approximants*. Now arguing as in (a), we obtain

$$\left\| W_{A,B}(t/n)^n - U_{A+B}^n(t/n) \right\| \leq n\, \mathrm{e}^{\,t(\|A\|+\|B\|)} \left\| W_{A,B}(t/n) - U_{A+B}(t/n) \right\| . \tag{1.10.43}$$

Since the symmetrisation eliminates the last term in (1.10.42), we get

$$\begin{aligned} &\left\| W_{A,B}(t/n) - U_{A+B}(t/n) \right\| = \\ &\Big\| \sum_{k_1=0}^{\infty} \frac{1}{k_1!}\Big(-\frac{t}{2n}B\Big)^{k_1} \sum_{\ell=0}^{\infty} \frac{1}{\ell!}\Big(-\frac{t}{n}A\Big)^{\ell} \sum_{k_2=0}^{\infty} \frac{1}{k_2!}\Big(-\frac{t}{2n}B\Big)^{k_2} \\ &- \sum_{s=0}^{\infty} \frac{1}{s!}\Big[-\frac{t}{n}(A+B)\Big]^{s} \Big\| \leq \frac{1}{n^3} L(t, \|A\|, \|B\|) , \end{aligned} \tag{1.10.44}$$

where L is determined in a similar way as the right-hand side of (1.10.42). The estimate (1.10.44) provides for $c_{\mathrm{sym}} = \mathrm{e}^{\,t(\|A\|+\|B\|)} L(t, \|A\|, \|B\|)$ the *rate* of convergence (1.10.37), which is better than (1.10.36).

□

We note that *symmetrisation* is in fact a simplest modification of the original Lie-Trotter product formula (1.10.36), which elucidates in this book different aspects of the related convergence problems. As we see in Proposition 1.10.10, these modifications might be instructive even for matrices. There is a long series of papers by M. Suzuki collected in [Suz96], where the reader can find more sophisticated, useful (e.g., *higher-order*) decompositions of exponential product formulæ for matrices or bounded operators. They are out of the scope of the present book.

The rest of this subsection concerning the strongly convergent Lie-Trotter product formula is standard ()see, for example, [Dav80, Chapter 3.4] or a detailed treatise [EN00, Part III, Chapter 5]).

We remark that the *renormalisation* Lemma 1.9.40, which is addressed to check the *Trotter stability* condition (Definition 1.9.37), is due to Hagen Neidhardt [Nei82, Lemma 5.1]. We also comment that this supplementary condition (Definition 1.9.37) is not at all innocent or harmless to be neglected (see discussion in [Trot59, 4.The norm condition] and Corollary 1.10.12). To this aim, we reformulate the result of Proposition 1.9.41 as follows:

Proposition 1.10.11 (Trotter Product Formula) *Let* $\{U_A(t) = \mathrm{e}^{-tA}\}_{t\geq 0}$ *and* $\{U_B(t) = \mathrm{e}^{-tB}\}_{t\geq 0}$ *be* C_0*-semigroups on a Banach space* $\mathfrak{X}$ *such that*

(a) *The Trotter stability condition holds .*
(b) *The closure* $C := \overline{A+B}$ *is generator of a* C_0*-semigroup* $\{U_C(t) = \mathrm{e}^{-tC}\}_{t\geq 0}$.

Then $U_C(t)\,x = \lim_{n\to\infty}(U_A(t/n)U_B(t/n))^n x$ *locally uniformly in* $t \in \mathbb{R}_0^+$ *for each fixed* $x \in \mathfrak{X}$.

The question arises whether only (b) is enough for convergence of the Trotter product formula to $\{U_C(t)\}_{t\geq 0}$. In [KW00], [CG12] the authors provided examples of A and B that satisfy (b), but not (a); and as a result, the convergence of the Trotter formula was violated. For that reason, to obtain the strong convergence of the Trotter product formula to $\{U_C(t)\}_{t\geq 0}$, the *Trotter stability* condition is *necessary*.

Corollary 1.10.12 *If the sum (or its closure) of two generators of* C_0*-semigroups is again a generator of a* C_0*-semigroup, the corresponding semigroup, in general, can not be given by the Trotter product formula.*

On account of elementary example of construction the semigroup approximants (Proposition 1.10.10) and discussion of the Trotter stability (Proposition 1.10.11), it is appropriate to mention yet another approach approximating semigroup. It is widely used in the numerical solution of evolution equations the *operator splitting* method motivated by the difference schemes (see [Lax02, Section 34.3] for formulation of the main idea (the *Lax equivalence theorem*) and review in [Str69]).

For example, the simplest form of the operator splitting for solution operator (semigroup) $\{U(t)\}_{t\geq 0}$ is called the *sequential splitting*. It defines the approximants of solution operator U_{n+1} on the $(n+1)$-step of iteration as $U_0(\tau) = \mathbb{1}$ and

$$U_{n+1}(\tau) = S(\tau)\, U_n(\tau)\,, \quad \tau = t/(n+1)\,, \quad t \in \mathbb{R}_0^+\,, \quad n \in \mathbb{N}_0\,, \tag{1.10.45}$$

where $S(\tau)$ is the splitting, which selects variety of *difference schemes*. It is clear that convergence for $n \to \infty$ of the difference scheme corresponding to sequential splitting (1.10.45) is determined by the smoothness of the splitting, which is called *consistency*, [Lax02, Section 34.3, (30)], and by *stability*, which controls the growing of $U_n(\tau)$, [Lax02, Section 34.3, (32)].

Applying the splitting scheme (1.10.45) to the problem of approximation of semigroup $\{U_{A+B}(t)\}_{t\geq 0}$ in Proposition 1.10.10, we see that there are (at least) two splitting available: $S(\tau) = V_{A,B}(\tau)$ (non-symmetric) and $S(\tau) = W_{A,B}(\tau)$ (symmetric). Then by (1.10.40) and (1.10.45), both splitting schemes are *exponentially stable*, cf. Definition 1.9.37.

On the other hand, the *consistency* is "measured" by Faragó and Havasi [FH07, Definition 2.6]

$$\Delta_S(x,\tau) = \|U_{A+B}(\tau)x - S(\tau)x\|\,, \quad x \in \mathcal{D} \subseteq \mathfrak{X}\,, \tag{1.10.46}$$

where $\mathcal{D}$ is a *dense* subset in $\mathfrak{X}$. Then for $\tau \to 0$, one obtains for non-symmetric splitting $\Delta_V(x,\tau) = O(\tau^2)$ (consistency of order $p=1$), whereas for symmetric splitting $\Delta_W(x,\tau) = O(\tau^3)$ (consistency of order $p=2$), here $\mathcal{D} = \mathfrak{X}$.

If $A+B$ is unbounded operator, then a *dense* $\mathcal{D} \subset \mathfrak{X}$, and it is defined via domains of powers of this operator (see [FH07, Proposition 4.1]).

Chapter 2
Linear Evolution Equations

Since in this book we also study the product formulæ approximations of the *solution operator* for the abstract *non-autonomous Cauchy problem* (nACP), this chapter collects preliminaries which are indispensable for this purpose.

For construction of the corresponding approximants and for the proof of their convergence, we broadly follow in this book the *Howland-Evans-Neidhardt* (HEN) method, which relates (within the L^p-setting) the *solution operator* to a suitable *evolution semigroup* in a certain extended Banach space of *evolution orbits*. Although for solution of the nACP via product formulæ it sounds as a rather *roundabout way*, this method allows to apply a full power of the Trotter product approximation theory to construct the *evolution semigroup* and then to translate the result to approximation of solution operator.

The main notations, definitions and results concerning the linear evolution equations in a Banach space and about the corresponding solution operator are recollected first in the framework of a relatively restricted (smooth) C-setting (Sect. 2.1). Here, we introduce the key notion of the *evolution semigroups* in the space of *orbits* and the corresponding HEN *formula* (representation) for solution operator of the original nACP.

To advance (in Sect. 2.2) with a less restrictive and more appropriate for the product formulæ approximations L^p-setting, we use a similarity and results, which are due to the C-setting. The idea of construction first the evolution semigroups for proving the HEN representation within the C-setting and then to generalise these assertions to the L^p-setting is due to [Nei79, Nei81]. Note that the proof of uniqueness in identification of evolution generator and evolution operator in the HEN representation owing to the *dense cross-section* property of orbits is also due to [Nei79, Nei81].

V. A. Zagrebnov et al., *Trotter-Kato Product Formulæ*, Operator Theory: Advances and Applications 296, https://doi.org/10.1007/978-3-031-56720-9_2

2.1 Linear Evolution Equations in C-Setting

2.1.1 Abstract Non-autonomous Cauchy Problem

Note that for a *closed* operator $A \in \mathcal{C}(\mathfrak{X})$ with dom (A) in a Banach space $\mathfrak{X}$ one can consider *abstract*, that is, vector-valued linear *autonomous* Cauchy problem (ACP) defined by initial condition $x \in \mathfrak{X}$ and equation

$$\begin{aligned} \partial_t u(t) + Au(t) = 0\,, \quad &\text{for} \quad t \in [s, T] \subset \mathbb{R}_0^+\,, \\ u(s) = x\,, \quad &\text{for} \quad 0 \le s < T \;\; \text{and}\; x \in \mathfrak{X}. \end{aligned} \tag{2.1.1}$$

Remark 2.1.1 The notion of *solution* of the abstract non-autonomous Cauchy problem needs elucidations that we start by comments concerning solutions of the ACP:

(a) Recall that continuous, for $t \in [s, T]$, function $t \mapsto u(t) \in \mathrm{dom}\,(A)$ is a unique *classical* solution of ACP (2.1.1) if it is strongly continuously differentiable in t, that is, $u \in C([s, T], \mathrm{dom}\,(A)) \cap C^1([s, T], \mathfrak{X})$, and satisfies (2.1.1) for initial condition $u(s) = x \in \mathrm{dom}\,(A)$ with *one-side* derivatives in equation (2.1.1) at $t = s$ and at $t = T$.
Note that for the *classical* solution of ACP (2.1.1) it is crustal that *initial* condition $u(s) \in \mathrm{dom}\,(A)$. As a result, the right derivative of the solution u at $t = s$ exists and satisfies equation (2.1.1). For example, this is the case when a unique solution $\{u(t) = U_A(t - s)\, u(s) \in \mathrm{dom}\,(A)\}_{t \in [s,T]}$ of ACP (2.1.1) is generated by a C_0-semigroup $\{U_A(t)\}_{t \ge 0}$ (see (1.1.22) and (1.1.23) in Proposition 1.1.1).

(b) For general case of ACP (2.1.1), when $x \in \mathfrak{X}$, one has to modify the concept of *solution* by requiring differentiability *only* for $t > s$. The continuous, for $t \in [s, T]$, function $t \mapsto u(t) \in \mathfrak{X}$ is called a *strict* solution of ACP (2.1.1) if $u \in C([s, T], \mathfrak{X}) \cap C^1((s, T], \mathfrak{X})$ and satisfies (2.1.1) in $(s, T]$ for initial condition $\lim_{t \to s} u(t) = x \in \mathfrak{X}$.
Note that for the *strict* solution of ACP (2.1.1) the initial condition $u(s) = x \in \mathfrak{X}$ is satisfied because of continuity $u \in C([s, T], \mathfrak{X})$ and the limit $x = \lim_{t \to s} u(t)$. For example, this is the case when for any $x \in \mathfrak{X}$ a unique solution $\{u(t) = U_A(t - s)\, x \in \mathrm{dom}\,(A)\}_{t \in (s,T]}$ of ACP (2.1.1) in $(s, T]$ is generated by a holomorphic semigroup $\{U_A(t)\}_{t \ge 0}$ (see (1.7.6)). By the strong continuity, this semigroup ensures also the initial condition $\lim_{t \to s} U_A(t - s)\, x = x$.

(c) Recall that there exists another concept of the ACP for the case of initial conditions $x \in \mathfrak{X}$. It completely renounces differentiability and substitutes the differential equation in (2.1.1) by an integral equation. Then continuous, for

$t \in [s, T]$, function $t \mapsto u(t) \in \mathfrak{X}$, i.e. $u \in C([s, T], \mathfrak{X})$, is called a *mild* solution of ACP (2.1.1) if

$$\int_s^t d\tau\, u(\tau) \in \operatorname{dom}(A)\,, \quad \text{for} \quad t \in [s, T]\,; \tag{2.1.2}$$

$$\text{it satisfies equation } u(t) = x - A \int_s^t d\tau\, u(\tau)\,, \text{ for } t \in [s, T]\,, \tag{2.1.3}$$

and, given that $A \in \mathcal{C}(\mathfrak{X})$, also the initial condition $x = \lim_{t \to s} u(t)$.

A mild solution of ACP (2.1.1) is *weaker* than the classical or the strict solution. Let A be generator of a C_0-semigroup $\{U_A(t)\}_{t \ge 0}$. Note that the *orbit* of any $x \in \mathfrak{X}$ under this semigroup, $\{u(t) = U_A(t-s)\,x\}_{t \in [s,T]}$, is continuous. Then this continuity and the C_0-semigroup property of $\{U_A(t)\}_{t \ge 0}$ ensure the existence of limits:

$$\lim_{\varepsilon \to +0} \frac{1}{\varepsilon}(U_A(\varepsilon) - \mathbb{1}) \int_s^t d\tau\, u(\tau) = \lim_{\varepsilon \to +0} \left(\frac{1}{\varepsilon} \int_t^{t+\varepsilon} d\tau\, u(\tau) - \frac{1}{\varepsilon} \int_0^{\varepsilon} d\tau\, u(\tau) \right),$$

which yield (2.1.2) and (2.1.3). To check the uniqueness, we insert in (2.1.3) solution, $w(\tau) = u_1(\tau) - u_2(\tau)$, for initial condition $u_1(s) - u_2(s) = 0$, and use identity

$$\partial_\eta \left(U_A(t - \eta) \int_s^\eta d\tau\, w(\tau) \right) = 0\,, \quad s < \eta < t\,,$$

which gives after integration over interval $[s, t]$ that $\int_s^t d\tau\; w(\tau) = 0$ for any $t \in (s, T]$. Since $w(s) = 0$, this yields $w(t) = 0$ for any $t \in (s, T]$, that is, the uniqueness.

We note that converse to the above assertion does not hold. On that account, the *existence* and *uniqueness* of the mild solution of the ACP (2.1.1) should be supplemented with an additional ingredient: the *continuous dependence* on the initial data. Then this *triad* ensures a *well-posedness* of the ACP, which yields a link of equation (2.1.1) with the C_0-semigroup setting for the ACP. See Notes to Chap. 2 (Sect. 2.1). On account of Remark 2.1.1 (c), it is clear that a mild solution of ACP coincides with a classical solution of this problem if and only if the solution $u \in C^1([s, T], \mathfrak{X})$.

To study the non-autonomous case, we evoke that if there is a mapping that generates orbits $\{u(t) = U(t, s)\; x\}_{t \in [s,T]}$ with initial conditions $u(s) = x \in \mathfrak{X}$ in (2.1.1), then the family $\{U(t, s)\}_{t \in [s,T]}$ is called a *solution operator* of ACP. If operator A in ACP (2.1.1) generates a C_0-semigroup, then by Proposition 1.1.1 the solution operator coincides with this semigroup $\{U(t, s) = U_A(t - s)\}_{t \in [s,T]}$ and readily yields (see (1.1.22), (1.1.23)), a *classical* solution of the ACP.

Now, we embark on elucidation of the abstract *non-autonomous* Cauchy problem (nACP). It corresponds to a linear *evolution equation* defined by the family of operators $\{A(t)\}_{t\in[0,T]}$ in a Banach space $\mathfrak{X}$:

$$\begin{aligned} \partial_t u(t) + A(t)u(t) &= 0\,, \quad \text{for} \quad t \in [s, T] \subset \mathbb{R}^+\,, \\ u(s) &= x\,, \quad \text{for} \quad 0 \le s < T \text{ and } x \in \mathfrak{X}. \end{aligned} \tag{2.1.4}$$

Definition 2.1.2 Let $\{A(t) \in \mathcal{C}(\mathfrak{X})\}_{t\in[0,T]}$ be a family of closed linear operators with dense domains $\{\mathrm{dom}\,(A(t))\}_{t\in[0,T]} \subset \mathfrak{X}$. Then for $s \in [0, T)$ and $x \in \mathfrak{X}$, a continuous function $u[s, T] \longrightarrow \mathfrak{X}$, is called the *classical* solution of nACP (2.1.4) if

(a) $u(t) \in \mathrm{dom}\,(A(t))$ for $s \le t \le T$ and $u(s) = x$,
(b) $u \in C^1([s, T], \mathfrak{X})$,
(c) $\partial_t u(t) + A(t)u(t) = 0$ for $t \in [s, T]$,
with convention that $\partial_t u(t)$ is the *one-side* derivative at the end-points of the interval.

For initial condition $u(s) = x$ of a classical solution of the nACP, we denote by $\{u(t|s, x)\}_{t\in[s,T]}$ the *orbit* of x under evolution (2.1.4) on the interval $[s, T]$. See Remark 2.1.1 (a) and (b) to distinguish the classical and the strict solution cases.

Definition 2.1.3 Let $\{A(t)\}_{t\in[0,T]}$ be a family of linear, closed, and densely defined operators in a Banach space $\mathfrak{X}$. The nACP (2.1.4) is called (classically) *well-posed* if for any $s \in [0, T)$ there is a *regularity* set $\mathfrak{Y}(s) \subset \mathfrak{X}$ satisfying the following conditions:

(i) (*Existence*) For each $s \in [0, T)$, the set

$$\mathfrak{Y}(s) := \{x \in \mathfrak{X} : \text{there exists a classical solution} \{u(t|s, x)\}_{t\in[s,T]} \text{ of nACP}\}$$

is dense in $\mathfrak{X}$ and $\mathfrak{Y}(s) \subseteq \mathrm{dom}\,(A(s))$.
(ii) (*Uniqueness*) For every $x \in \mathfrak{Y}(s)$, the solution $\{u(t|s, x)\}_{t\in[s,T]}$ is unique.
(iii) (*Continuous dependence on data*) The solution depends continuously on the initial data s and x, i.e. if $s_n \longrightarrow s$, for $s_n, s \in [0, T)$, and $x_n \longrightarrow x$, for $x_n \in \mathfrak{Y}(s_n)$, $x \in \mathfrak{Y}(s)$, when $n \to \infty$, then

$$\lim_{n\to\infty} \sup_{t\in(0,T]} \|\widehat{u}\,(t|s_n, x_n) - \widehat{u}\,(t|s, x)\| = 0,$$

where for $r \in (0, T)$ we define

$$\widehat{u}\,(t|r, x) := \begin{cases} u(t|r, x) & t \in [r, T] \\ x & t \in (0, r) \end{cases} .$$

(iv) If in addition there exist $M \geq 1$ and $\beta \in \mathbb{R}$ such that

$$\|u(t|s,x)\| \leq M e^{\beta(t-s)} \|x\| \,, \tag{2.1.5}$$

for $x \in \mathfrak{Y}(s)$ and $t \geq s$, then the nACP (2.1.4) is called (classically) *well-posed* with *exponentially* bounded solution.

Similarly to ACP (2.1.1) the important notion related to nACP (2.1.4) is the *solution operator*, which is also known as *fundamental solution*, *evolution family*, or *propagator*.

Definition 2.1.4 We call the family of bounded operators $\{U(t,s)\}_{(t,s)\in\Delta}$ defined in a Banach space $\mathfrak{X}$ on the triangle domain

$$\Delta := \{(t,s) \in [0,T] \times [0,T] : 0 \leq s \leq t \leq T\},$$

by (strongly continuous) *evolution family* if the following conditions are satisfied:

(a) The function $\Delta \ni (t,s) \longrightarrow U(t,s)$ is strongly continuous.
(b) For every $t \in [0,T]$, one has $U(t,t) = \mathbb{1}$ and the Chapman-Kolmogorov *composition* law

$$U(t,r)\,U(r,s) = U(t,s)\,, \tag{2.1.6}$$

for $(t,r,s) \in [0,T] \times [0,T] \times [0,T]\,,\ 0 \leq s \leq r \leq t \leq T$.
(c) $\sup_{(t,s)\in\Delta} \|U(t,s)\|_{\mathcal{L}(\mathfrak{X})} < \infty$.

Definition 2.1.5 Let $\{A(t)\}_{t\in[0,T]}$ be a family of linear, closed, and densely defined operators in a Banach space $\mathfrak{X}$. We say that a strongly continuous evolution family $\{U(t,s)\}_{(t,s)\in\Delta}$ solves the nACP (2.1.4) (by this reason, it is called the *solution operator*) if for every $s \in [0,T)$ the set

$$\mathfrak{Y}_U(s) := \left\{ x \in \mathrm{dom}\,(A(s)) : \begin{array}{l} u(t|s,x) := U(t,s)x, \quad t \in [s,T], \\ \text{a classical solution of nACP} \end{array} \right\}$$

is a *regularity* set $\mathfrak{Y}(s)$, cf. Definition 2.1.3.

Few remarks are now in order to illuminate the Definitions 2.1.3–2.1.5.

Remark 2.1.6 Let $s < r < T$. Then by Definitions 2.1.3–2.1.5, the solution of a well-posed nACP (2.1.4) with regularity set $\mathfrak{Y}(s)$ can be rewritten for $x \in \mathfrak{Y}(s)$ and $s < r \leq t < T$ as

$$u(t) = U(t,r)\,U(r,s)\,x = U(t,r)\,u(r|s,x)\,. \tag{2.1.7}$$

Here, $u(r|s,x) := U(r,s)\,x$ is taken in (2.1.7) as initial condition for $t = r$. By the reason of uniqueness of the classical solution (i.e. of the orbit $\{U(t,s)\,x\}_{t\in[s,T]}$ of the initial point x), the representation (2.1.7) implies that $U(t,s)\,x\big|_{t=r} = u(r|s,x)$ must belong to the set of initial conditions for (classically) *well-posed* nACP (2.1.4) on the interval $[r,T]$. According to Definition 2.1.3, this set is in turn a regularity set

$$\mathfrak{Y}(r) := \{U(r,s)\,x : \text{there exists a classical solution} \\ \times \{u(t|r, u(r|s,x))\}_{t\in[r,T]} \text{ of nACP}\}.$$

Consequently, by definition of $\mathfrak{Y}(r)$, together with Definition 2.1.3 (i), we obtain

$$U(r,s)\,\mathfrak{Y}(s) \subseteq \mathfrak{Y}(r) \subseteq \mathrm{dom}\,(A(r))\,, \quad s < r \le t < T. \tag{2.1.8}$$

Remark 2.1.7 Note that in Definitions 2.1.3 and 2.1.5 it is important to distinguish for every $s \in (0,T)$ between the operator *domain* $\mathrm{dom}\,(A(s))$ and the *regularity set* $\mathfrak{Y}_U(s)$. One can construct example (see Notes to Chap. 2, Sect. 2.1.1) of a well-posed nACP such that $U(t,s)\,\mathfrak{Y}(s) \subseteq \mathfrak{Y}(t)$, whereas $U(t,s)\,\mathrm{dom}\,(A(s)) \nsubseteq \mathrm{dom}\,(A(t))$ for $t > s$. Consequently, if nACP is well-posed, it may happen that it *cannot* be solved on the whole $\mathrm{dom}\,(A(t))$, for a $t \in (s,T]$. This is contrary to the ACP, where the C_0-semigroup properties (Proposition 1.1.1 (b)) yield $U_A(t-s)\,\mathrm{dom}\,(A) \subseteq \mathrm{dom}\,(A)$ for $\mathfrak{Y}_{U_A}(s) = \mathrm{dom}\,(A)$ and any $t \ge s \ge 0$.

Another remark is that in contrast to the ACP the regularity set $\mathfrak{Y}(\cdot)$ and hence the well-posedness are a subtle matter for nACP. It is known that if nACP is well-posed for a family of unbounded operators $\{A(t)\}_{t\in[0,T]}$ on $\mathfrak{X}$ and perturbation $\{B(t)\}_{t\in[0,T]}$ is a uniformly *bounded* family of operators on $\mathfrak{X}$, then the nACP for the family $\{A(t)+B(t)\}_{t\in[0,T]}$ is, in general, *not* well-defined.

For illustration of this phenomenon, see example below and also Example 2.2.1 that manifests the corresponding impact on restrictions of the C-setting.

Example 2.1.8 Let non-negative continuous function $a : [0,T] \to \mathbb{R}_0^+$ and $t_0 \in (0,T)$ be such that $a(t > t_0) > 0$ and $a(t \le t_0) = 0$. Let A be unbounded generator of a C_0-semigroup and B be a bounded operator on a Banach space $\mathfrak{X}$. Then for the nACP corresponding to $A(t) := a(t)\,A + B$ the regularity set $\mathfrak{Y}(s > t_0) = \mathrm{dom}\,(A)$, whereas to ensure that the nACP for $\{A(t)\}_{t\in[0,T]}$ is well-posed the regularity set for $s < t_0$ is defined by condition

$$\mathfrak{Y}(s) = \{x \in \mathfrak{X} : \mathrm{e}^{(s-t_0)\,B}\,x \in \mathrm{dom}\,(A)\}\,.$$

In fact, choosing appropriately the operator B, one can make this subspace quite arbitral. See Notes to Chap. 2 (Sects. 2.1, 2.1.1).

According to Remark 2.1.7, *a priori*, the nACP (2.1.4) does not possess a universal (time-independent) regularity set, which is similar to $\mathrm{dom}\,(A)$ in the case of ACP. The next assertion relates the well-posedness of the nACP with the existence of solution operator for this problem.

Proposition 2.1.9 *Let $\{A(t)\}_{t\in[0,T]}$ be a family of closed linear operators densely defined in a Banach space $\mathfrak{X}$. The following assertions are equivalent:*

(a) *The nACP* (2.1.4) *is (classically) well-posed.*
(b) *There exists a unique solution operator $\{U(t,s)\}_{(t,s)\in\Delta}$ for the nACP* (2.1.4).

Proof (a) $\Rightarrow$ (b): By Definition 2.1.3 (i), the existence of classical solution implies the existence of satisfying the co-cycle property mappings, $U(t,s) : \mathfrak{Y}(s) \to \mathfrak{Y}(t)$, $(t,s) \in \Delta$, on dense regularity sets. Then by Definition 2.1.3 (ii) the family $\{U(t,s)\}_{(t,s)\in\Delta}$ is unique and by continuity (Definition 2.1.3 (iii)) can be extended from regularity sets to the strongly continuous solution operator $\{U(t,s)\}_{(t,s)\in\Delta}$ for the nACP (2.1.4) defined by Definition 2.1.5.
(b) $\Rightarrow$ (a): By Definition 2.1.5, the existence of unique solution operator implies the existence of uniquely defined classical solutions with initial conditions on dense regularity sets, which proves the properties in Definition 2.1.3 (i) and (ii). Next, on account of the strong continuity of evolution family $\{U(t,s)\}_{(t,s)\in\Delta}$, we infer the properties (iii) and (iv) indicated in Definition 2.1.3. □

Example 2.1.10 Examples below give sufficient conditions of well-posedness for two principal cases: *hyperbolic* nACP and *parabolic* nACP:

(1) It was shown that if $\{A(t)\}_{t\in[0,T]}$ is a family of generators of contraction semigroups (hyperbolic case), i.e. $A(t) \in \mathcal{G}(1,0)$, $t \in [0,T]$, such that regularity set $\mathfrak{Y} := \mathrm{dom}\,(A(t))$ is independent of $t \in [0,T]$ and the mapping $A(\cdot)x : [0,T] \longrightarrow \mathfrak{X}$ is continuous for $x \in \mathfrak{Y}$, then the nACP (2.1.4) is well-posed. See Notes to Chap. 2 (Sects. 2.1, 2.1.1).
(2) Let $\{A(t)\}_{t\in[0,T]}$ be a family of generators of holomorphic contraction semigroups (parabolic case), that is, $A(t) \in \mathscr{H}(\theta,0)$, $\theta \in (0,\pi/2]$, such that for $\varepsilon \in (0,\theta)$ there exists a constant $M_\varepsilon > 0$ such that the estimate

$$\|(z\,\mathbb{1} + A(t))^{-1}\| \le \frac{M_\varepsilon}{|z|}, \quad z \in \overline{S}_{\pi-\theta-\varepsilon}\,, \quad t \in [0,T], \tag{2.1.9}$$

holds. If $0 \in \rho(A(t))$, for $t \in [0,T]$, and there are constants $L > 0$ and $\mu, \nu \in (0,1]$, with $\mu + \nu > 1$, such that the estimate

$$\left\| A(t)(A(t) + z\mathbb{1})^{-1}\left(A(t)^{-1} - A(s)^{-1}\right)\right\| \le L|t-s|^{\mu}|z|^{-\nu}\,, \tag{2.1.10}$$

for $(t,s) \in [0,T] \times [0,T]$, then the nACP (2.1.4) is well-posed for regularity set $\mathfrak{Y}(s) = \mathrm{dom}\,(A(s))$, $s \in [0,T)$, cf. Remark 2.1.7.
For further comments and details, see Notes to Chap. 2 (Sects. 2.1, 2.1.1).

Definition 2.1.11 The nACP in a normed space $E = \mathfrak{X}$ (or $\mathfrak{H}$)

$$\begin{aligned} \partial_t u(t) + A(t)u(t) &= 0\,, \quad \text{for} \quad t \in [s,T] \subset \mathbb{R}^+\,, \\ u(s) &= x\,, \quad \text{for} \quad s \in [0,T) \text{ and } x \in E. \end{aligned} \tag{2.1.11}$$

is said to be of *parabolic-type* (or simply *parabolic*) if operator $A(t)$ for each $t \in [0, T]$ is generator of a holomorphic semigroup on E, that is, $A(t) \in \mathscr{H}(\theta, \beta)$ for $\theta \in (0, \pi/2]$ and $\beta \geq 0$.

For the parabolic case, we formulate two assertions that will be useful in our analysis of the product formula approximation for solution operator in Sects. 11.1 and 12.1. To this end, we note that differential nACP (2.1.4) can be reformulated on the orbits $\{u(t) = U(t, s)\, x\}_{t\in[s,T]}$ of $x \in \mathfrak{Y}(s)$ as relation for solution operator:

$$\begin{aligned} \partial_t U(t, s)\, x + A(t)U(t, s)\, x = 0\,, &\quad \text{for} \quad t \in [s, T] \subset \mathbb{R}^+\,, \\ U(s, s)\, x = x\,, &\quad \text{for} \quad s \in [0, T) \text{ and } x \in \mathfrak{Y}(s). \end{aligned} \tag{2.1.12}$$

Holomorphic semigroups generated by $\{A(t)\}_{t\in[0,T]}$ make analytic properties of solution operator $U(t, s)_{(t,s)\in\Delta}$ for parabolic case similar to analytic properties of solution operator for ACP, which is the holomorphic semigroup with generator $A(t) = A$, cf. (2.1.1) and Remark 2.1.1 (b).

Proposition 2.1.12 *Under assumptions* (2.1.9) and (2.1.10) *and invariance of domain* dom$(A(t))$ $=$ D, *a unique solution operator* $\{U(t, s)\}_{(t,s)\in\Delta}$ *for the parabolic-type nACP* (2.1.12) *exits and has the following properties*:

(a) *For* $(t, s) \in \Delta_>$, *where* $\Delta_> := \{(t, s) \in \Delta : t > s\}$, *the range* ran $U(t, s) \subset D$.
(b) *It is a strongly continuously differentiable with respect to* $t(> s)$ (*strict, see Remark* 2.1.1(b)) *solution of the problem*:

$$\begin{aligned} \partial_t U(t, s) + A(t)U(t, s) = 0\,, &\quad (t, s) \in \Delta_>\,, \\ \lim_{t\to s} U(t, s) = \mathbb{1}\,. & \end{aligned} \tag{2.1.13}$$

(c) *The following estimates hold*:

$$\|\partial_t U(t, s)\| = \|A(t)U(t, s)\| \leq C(t - s)^{-1} \text{ and } \|A(t)U(t, s)A(s)^{-1}\| \leq C. \tag{2.1.14}$$

(d) *For each* $t \in (0, T]$ *and each* $u \in D$, *the function* $s \mapsto U(t, s)u$ *is differentiable for* $s \in [0, t]$ *and* (*seeing that* $\partial_s(U(t, s)U(s, r)) = 0$) *satisfies equation*

$$\begin{aligned} \partial_s U(t, s)u - U(t, s)\, A(s)u = 0\,, &\quad (t, s) \in \Delta_>\,, \\ \lim_{s\to t} U(t, s) = \mathbb{1}\,. & \end{aligned} \tag{2.1.15}$$

(e) *The following estimates hold*:

$$\|\partial_s U(t, s)\| = \|\overline{U(t, s)A(s)}\| \leq C(t - s)^{-1} \text{ and } \|\overline{A(t)^{-1}U(t, s)A(s)}\| \leq C. \tag{2.1.16}$$

Here, $\overline{U(t,s)A(s)}$ and $\overline{A(t)^{-1}U(t,s)A(s)}$ are bounded extensions (the closure) of densely defined for $0 \le s < t \le T$ operators $U(t,s)A(s)$ and $A(t)^{-1}U(t,s)A(s)$ with domain D. We also make use of one-side derivatives at the end-points of intervals.

For the case, when domain dom $(A(t))$ *varies* with time dom $(A(t)) = D(t)$, one needs more assumptions, cf. Sect. 11.1:

(H1) The function $t \mapsto A^{-1}(t)$ is continuously differentiable in the *operator-norm* topology on $\mathcal{L}(\mathfrak{X})$, and there exist constants $a_1 \in \mathbb{R}^+$, $\widehat{a_1} \in (0, 1]$ such that the Hölder continuity estimate

$$\left\| \partial_t A^{-1}(t) - \partial_s A^{-1}(s) \right\| \le a_1 \, |t - s|^{\widehat{a_1}} \tag{2.1.17}$$

is valid for every $s, t \in [0, T]$.

(H2) There exist constants $a_2 \in \mathbb{R}^+, \widehat{a_2} \in (0, 1]$ such that the inequality

$$\| \partial_t (\lambda \mathbb{1} + A(t))^{-1} \| \le a_2 \, |\lambda|^{-\widehat{a_2}} \tag{2.1.18}$$

holds for every $t \in [0, T]$ and every $\lambda \in \overline{S}_{\pi-\theta-\varepsilon} \setminus \{0\}$.

Proposition 2.1.13 *Under assumptions* (2.1.9) *and* (H1) and (H2), *a unique solution operator $\{U(t,s)\}_{(t,s)\in\Delta}$ for the parabolic-type nACP* (2.1.12) *exists and has the following properties*:

(a) *For $0 \le s < t \le T$, the range* ran $U(t,s) \subset$ dom $(A(t))$.
(b) *For $s < t$, the function $(t,s) \mapsto U(t,s)$ is differentiable with respect to t and s in the operator-norm topology and satisfies equations*

$$\partial_t U(t,s) + A(t)U(t,s) = 0 \text{ and } \partial_s U(t,s) - \overline{U(t,s)\,A(s)} = 0. \tag{2.1.19}$$

(c) *The following estimates hold for $s < t$*:

$$\begin{aligned} \|\partial_t U(t,s)\| &= \|A(t)U(t,s)\| \le C(t-s)^{-1}, \\ \|\partial_s U(t,s)\| &= \|\overline{U(t,s)\,A(s)}\| \le C(t-s)^{-1}. \end{aligned} \tag{2.1.20}$$

Here, by $\overline{U(t,s)\,A(s)}$, we denote the closure, which gives a bounded extension of the densely defined product, cf. the notes at the end of Proposition 2.1.12.

About details and references regarding Propositions 2.1.12 and 2.1.13, see Notes to Chap. 2 (Sects. 2.1, 2.1.1). For further discussion of supplementary assumptions and results about nACP (2.1.12) for the case of *time-dependent* domain, we refer to Sect. 11.1.

We conclude this subsection by more comments about peculiarities of the nACP solution operator $\{U(t,s)\}_{t \geq s > 0}$ versus the ACP solution operator $\{U_A(t) = e^{-tA}\}_{t \geq 0}$. They are a continuation and extension of Remark 2.1.1 and Remark 2.1.7.

Remark 2.1.14 We recall that the functional equation, or semigroup law (Definition 1.2.1 (b)), implies exponential boundedness of $\{U_A(t-s)\}_{t \geq s \geq 0}$ and makes a weakly continuous semigroup strongly continuous (see Proposition 1.5.4). The Chapman-Kolmogorov composition law, Definition 2.1.4 (b), which is weaker than semigroup law, has no such consequences:

(a) If $\Delta^+ := \mathbb{R}^+ \times \mathbb{R}^+$, then constraint in Definition 2.1.4 (c) is not superfluous and must be modified to (2.1.5)

$$(c') \quad \|U(t,s)\| \leq M\, e^{\beta(t-s)}, \text{ for some } M \geq 1, \beta \geq 0, \text{ and } t \geq s > 0. \tag{2.1.21}$$

To bolster a necessity of (2.1.21), we note that the family $\{U(t,s) := e^{(t^2-s^2)}\}_{t \geq s \geq 0}$ on the Banach space $X = \mathbb{C}$ satisfies the *composition law*, but it is *not* exponentially bounded.

It is known that solution operator $\{U(t,s)\}_{t \geq s > 0}$ is exponentially bounded if and only if $\sup_{\{s \leq t \leq s+\delta\} \wedge \{s \geq 0\}} \|U(t,s)\| \leq M_\delta$ for some $\delta > 0$ and $M_\delta > 0$ (see Notes to Chap. 2 (Sect. 2.1)).

(b) Next, let us consider a family $\{U(t,s) := Q(t)Q(s)^{-1}\}_{t \geq s > 0}$, where $\{Q(t)\}_{t \in \mathbb{R}_0^+}$ are bounded invertible operators which are *weakly* continuous in $\mathbb{R}^+$ and strongly continuous at $t = s = +0$. Then family $\{U(t,s)\}_{t \geq s > 0}$ satisfies the Chapman-Kolmogorov composition law, but it is *not* strongly continuous away from zero. For that reason, Definition 2.1.4 (a) is *not* superfluous.

(c) Following the same line of reasoning as above, let us consider a continuous positive function $f : \mathbb{R}_0^+ \to [a,b] \subset \mathbb{R}^+$, and let $\{U(t,s) := f(t)/f(s)\}_{t \geq s \geq 0}$. Then it is a bounded continuous family on the Banach space $X = \mathbb{C}$ satisfying the Chapman-Kolmogorov composition law which may be *nowhere differentiable* if the function f possesses this property.

(d) We continue with the case of *similarity* between ACP and nACP solution operators. It is known that an nACP solution operator $\{U(t,s)\}_{t \geq s \geq 0}$ is operator-norm continuous if and only if there exists a uniformly operator-norm continuous invertible family $\{Q(t)\}_{t \geq 0}$, continuous at $t = +0$, such that it defines $U(t,s) := Q(t)Q(s)^{-1}$ for $t \geq s \geq 0$. Because generator A of operator-norm continuous semigroup is a bounded operator (Proposition 1.4.1), on account of the above arguments, we infer the same formula for the ACP solution operator: $\{U_A(t-s) = U_A(t)U_A(s)^{-1}\}_{t \geq s \geq 0}$.

Since *a priori* there is no guarantee of differentiability of strongly continuous solution operator for (parabolic) nACP, one can relax the framework of this problem to look at *mild* solutions of nACP (2.1.12). To this aim, we introduce and survey in the next sections the *Howland-Evans-Neidhardt* approach to analysis of the nACP based on the evolution semigroups method. The advantage is that evolution semigroups are always differentiable on domains of their generators. This allows to establish the well-posedness of the corresponding nACP in terms of evolution semigroups.

Another, and historically the first case, when the mild solutions of nACP (2.1.12) was helpful, dates back to the *Sobolebskiĭ-Tanabe-Kato* approach to the abstract nACP for parabolic type evolution equations in the *normed spaces*. We shall use this classical method in Sect. 12.2 to prove, following the *Ichinose-Tamura* approach, the operator-norm convergence of product approximants to solution operator of the nACP in a Hilbert space.

2.1.2 Cauchy Problem and Evolution Semigroups in C-Setting

Consider the Banach space $\mathfrak{X}_T^C := C_0([0,T],\mathfrak{X})$ consisting of all continuous functions $f : [0,T] \longrightarrow \mathfrak{X}$ such that $f(0) = 0$ and endowed with the norm topology:

$$\|f\|_{\mathfrak{X}_T^C} = \sup_{t\in[0,T]} \|f(t)\|_{\mathfrak{X}} . \tag{2.1.22}$$

Note that the set-up (2.1.1) in the Banach space $\mathfrak{X}_T^C := C_0([0,T],\mathfrak{X})$ is called the *C-setting*.

We remind that any solution operator $\{U(t,s)\}_{(t,s)\in\Delta}$ on $\mathfrak{X}$ (Definition 2.1.4 and Remark 2.1.14 (2.1.21)) induces on the Banach space $\mathfrak{X}_T^C$ a strongly continuous (nilpotent) semigroup $\{\mathcal{U}(\tau)\}_{\tau\in\mathbb{R}_0^+}$, which is defined by formula:

$$(\mathcal{U}(\tau)f)(t) = \begin{cases} U(t,t-\tau)f(t-\tau), & \text{if } \{\tau < T\} \text{ and } t \in (\tau,T], \\ 0, & \text{if } t \notin (\tau,T] \text{ or } \{\tau \geq T\}. \end{cases} \tag{2.1.23}$$

We call (2.1.23) the *Howland-Evans-Neidhardt* (HEN) formula, or representation (see Appendix F.1), and we use for (2.1.23) the following expression:

$$(\mathcal{U}(\tau)f)(t) := U(t,t-\tau)\chi_{[0,T]}(t-\tau)f(t-\tau), \quad t \in [0,T], \tag{2.1.24}$$

for any $f \in \mathfrak{X}_T^C$ and characteristic function $\chi_{[0,T]}(\cdot)$ of the interval $[0,T] \subset \mathbb{R}_0^+$.

Definition 2.1.15 A strongly continuous semigroup $\{\mathcal{U}(\tau)\}_{\tau\in\mathbb{R}_0^+}$ defined on Banach space $\mathfrak{X}_T^C = C_0([0,T],\mathfrak{X})$ is called the *evolution semigroup* if there is a solution operator $\{U(t,s)\}_{(t,s)\in\Delta}$ (Definition 2.1.5) such that the representation (2.1.24) is valid.

Let $\phi \in C([0, T])$. We consider a (*scalar*) multiplication operator $\mathcal{M}(\phi)$ (cf. Notes to Chap. 1 (Sect. 1.7)):

$$(\mathcal{M}(\phi) f)(t) := \phi(t) f(t), \quad t \in [0, T], \quad f \in \mathfrak{X}_T^C = C_0([0, T], \mathfrak{X}). \tag{2.1.25}$$

Then for each $\phi \in C([0, T])$ the operator $\mathcal{M}(\phi) \in \mathcal{L}(\mathfrak{X}_T^C)$ is bounded, and the norm $\|\mathcal{M}(\phi)\|_{\mathcal{L}(C_0([0,T],\mathfrak{X}))} = \|\phi\|_{C([0,T])}$.

Definition 2.1.16 A densely defined in $\mathfrak{X}_T^C$ operator $\mathcal{K}$ is called the *evolution pre-generator* if for $\phi \in C^1([0, T])$ and $f \in \mathrm{dom}\,(\mathcal{K})$, $\phi'(t) := \partial_t \phi(t)$ one has

$$\begin{aligned} &\mathcal{M}(\phi)(\mathrm{dom}\,(\mathcal{K})) \subseteq \mathrm{dom}\,(\mathcal{K}), \\ &\mathcal{K}\mathcal{M}(\phi) f - \mathcal{M}(\phi)\mathcal{K} f = \mathcal{M}(\phi') f . \end{aligned} \tag{2.1.26}$$

On account of Definition 2.1.16, one infers the following criterion identifying generator of evolution semigroup (2.1.24), $\{\mathcal{U}(\tau) = \mathrm{e}^{-\tau\mathcal{K}}\}_{\tau\in\mathbb{R}_0^+}$, which is called the *evolution generator* $\mathcal{K}$ in $\mathfrak{X}_T^C$.

Theorem 2.1.17 *Let operator $\mathcal{K}$ be evolution pre-generator in Banach space $\mathfrak{X}_T^C$. Then $\mathcal{K}$ is evolution generator of semigroup* (2.1.24) *on $\mathfrak{X}_T^C$ if and only if operator $\mathcal{K}$ is generator of a strongly continuous on $\mathfrak{X}_T^C$ semigroup $\{\mathrm{e}^{-\tau\mathcal{K}}\}_{\tau\geq 0}$.*

For the proof and remarks, see Appendix F.1 (Sect. F.1.4., Theorem F.1.7).

Theorem 2.1.18 *Between the set of all evolution semigroups $\{\mathcal{U}(\tau) = \mathrm{e}^{-\tau\mathcal{K}}\}_{\tau\in\mathbb{R}_0^+}$ on $\mathfrak{X}_T^C$ (equivalently, all evolution generators $\mathcal{K}$ defined by Theorem* 2.1.17*) and the set of all strongly continuous quasi-bounded solution operators $\{U(t, s)\}_{(t,s)\in\Delta}$ on $\mathfrak{X}$, there exists a one-to-one correspondence such that the Howland-Evans-Neidhardt representation*

$$(\mathrm{e}^{-\tau\mathcal{K}} f)(t) = U(t, t-\tau)\chi_{[0,T]}(t-\tau) f(t-\tau), \quad \tau \geq 0, \tag{2.1.27}$$

holds for all $f \in \mathfrak{X}_T^C$ and $t \in [0, T]$, cf. (2.1.24).

Owing to the bound (2.1.21) *and the nilpotence* (2.1.27), *one gets estimate*

$$\|\mathrm{e}^{-\tau\mathcal{K}} f\|_{\mathfrak{X}_T^C} \leq M\,\mathrm{e}^{\beta T}\, \|f\|_{\mathfrak{X}_T^C}, \quad f \in \mathfrak{X}_T^C, \quad \tau \geq 0. \tag{2.1.28}$$

Therefore, generator $\mathcal{K}$ of the evolution semigroup belongs to the class $\mathscr{G}(M\,\mathrm{e}^{\beta T}, 0)$.

For the proof and comments, we suggest to consult Appendix F.1 (Theorems F.1.1, F.1.5 and F.1.6).

Corollary 2.1.19 *Let $\mathcal{K}$ be the generator of evolution semigroup $\{\mathcal{U}(\tau)\}_{\tau\in\mathbb{R}_0^+}$ on $\mathfrak{X}_T^C$, and let $\{U(t, s)\}_{(t,s)\in\Delta}$ be the solution operator such that the representation* (2.1.24) *holds. Then given that this semigroup is nilpotent, equation* (2.1.27) *yields for any $f \in \mathfrak{X}_T^C$ and the unit operator $\mathcal{I} f \mapsto f$, the following representation:*

$$((\mathcal{K} + \xi\, \mathcal{I})^{-1} f)(t) = \int_0^t ds\, e^{-\xi(t-s)} U(t,s) f(s), \quad t \in [0,T], \quad \xi \in \mathbb{C}. \tag{2.1.29}$$

Consequently, owing to (2.1.21), (2.1.22) *and* (2.1.29), *we obtain inequality*

$$\|(\mathcal{K} + \xi\, \mathcal{I})^{-1} f\|_{\mathfrak{X}_T^C} \leq T M\, e^{(\beta+|\xi|)T}\, \|f\|_{\mathfrak{X}_T^C}, \quad \xi \in \mathbb{C}. \quad f \in \mathfrak{X}_T^C. \tag{2.1.30}$$

Therefore, the resolvent set $\rho(\mathcal{K}) = \mathbb{C}$ and operator $\mathcal{K}$ is invertible.

Finally, we are going to characterise the domain dom $(\mathcal{K})$ of generator of an evolution semigroup.

Lemma 2.1.20 *Let $\mathcal{K}$ be the generator of an evolution semigroup $\{\mathcal{U}(\tau)\}_{\tau \in \mathbb{R}_0^+}$ on $C_0([0,T], \mathfrak{X})$, and let $\{U(t,s)\}_{(t,s)\in\Delta}$ be a solution operator such that* (2.1.24) *holds. Then $f \in$ dom $(\mathcal{K})$ if and only if for each $t \in (0,T]$ one has $U(t,\cdot)f(\cdot) \in C_0^1([0,t], \mathfrak{X})$ together with*

$$\partial_s \left(U(t,s) f(s)\right) = U(t,s) g(s), \quad f \in \text{dom}\,(\mathcal{K}), \quad s \in [0,t], \tag{2.1.31}$$

for some $g \in C_0([0,T], \mathfrak{X})$ and right derivative $\partial_s U(t,+0)$ at $s = 0$.

Proof Let $f \in$ dom $(\mathcal{K})$. Then

$$e^{-\tau\mathcal{K}} f = f - \int_0^\tau dr\, e^{-r\mathcal{K}} g, \tag{2.1.32}$$

where $g := \mathcal{K} f$. Using (2.1.24), we obtain

$$U(t, t-\tau)\chi_{[0,T]}(t-\tau) f(t-\tau) =$$
$$\varphi(t) - \int_0^\tau dr\, U(t, t-r)\chi_{[0,T]}(t-r) g(t-r),$$

$t \in [0,T]$. Change of variables to $\alpha := t - r$, $\alpha \in [0,t]$ and $s := t - \tau$, $s \in [0,t]$ yields

$$U(t,s) f(s) = f(t) - \int_s^t d\alpha\, U(t,\alpha) g(\alpha), \quad (t,s) \in \Delta. \tag{2.1.33}$$

Hence, $U(t,\cdot) f(\cdot) \in C_0^1([0,t], \mathfrak{X})$ for each $t \in (0,T]$ and (2.1.31) is valid.

Conversely, if (2.1.31) is satisfied, then we obtain (2.1.33) which yields (2.1.32)

$$\mathrm{e}^{-\tau\mathcal{K}} f = f - \int_0^\tau dr\, \mathrm{e}^{-r\mathcal{K}} \mathcal{K} f, \quad \tau \geq 0.$$

Hence, $f \in \mathrm{dom}\,(\mathcal{K})$. This completes the proof. □

We introduce in the space $\mathfrak{X}_T^C := C_0([0, T], \mathfrak{X})$ endowed with topology (2.1.22) the closed operator D_0 defined by

$$(D_0 f)(t) := \partial_t f(t), \quad f \in \mathrm{dom}\,(D_0) := C_0^1([0, T], \mathfrak{X}). \tag{2.1.34}$$

The operator (2.1.34) generates on $\mathfrak{X}_T^C$ a strongly continuous (*nilpotent*) right-shift *contraction* semigroup $\{\mathcal{S}_R(\tau) =: \mathrm{e}^{-\tau D_0}\}_{\tau\in\mathbb{R}_0^+}$ of automorphisms $\mathcal{S}_R(\tau)$: $L^p([0, T], \mathfrak{X}) \to L^p([0, T], \mathfrak{X})$, $p \geq 1$, such that

$$(\mathcal{S}_R(\tau) f)(t) = \chi_{[0,T]}(t-\tau) f(t-\tau), \quad f \in \mathfrak{X}_T^C, \quad \tau \geq 0\,,\ t \in [0, T]. \tag{2.1.35}$$

That is, for $\tau \in \mathbb{R}_0^+$, one obtains automorphisms $\mathcal{S}_R(\tau) : C_0([0, T], \mathfrak{X}) \to C_0([0, T], \mathfrak{X})$ on the subspace $C_0([0, T], \mathfrak{X}) \subset L^p([0, T], \mathfrak{X})$.

We remark that (2.1.35) is an example of nilpotent evolution semigroup (2.1.27), when the solution operator is trivial: $\{U(t, s) = \mathbb{1}\}_{(t,s)\in\Delta}$. Then similar to (2.1.28)–(2.1.30), we establish that *evolution generator* $D_0 \in \mathscr{G}(1, 0)$, with empty spectrum, $\sigma(D_0) = \emptyset$, that is, the operator D_0 is invertible.

Definition 2.1.21 With family $\{A(t)\}_{t\in[0,T]}$ of (in general) unbounded linear operators in $\mathfrak{X}$, one can associate in the Banach space $\mathfrak{X}_T^C = C_0([0, T], \mathfrak{X})$ a *multiplication* operator $\mathcal{A}$ (cf. Notes to Sect. 1.7), which is defined by

$$\begin{aligned} &(\mathcal{A} f)(t) := A(t) f(t), \quad f \in \mathrm{dom}\,(\mathcal{A}), \quad t \in (0, T]\,, \\ &\mathrm{dom}\,(\mathcal{A}) := \{f \in \mathfrak{X}_T^C : f(t) \in \mathrm{dom}\,(A(t)) \text{ and } \mathcal{A} f \in \mathfrak{X}_T^C\}\,. \end{aligned} \tag{2.1.36}$$

Note that $\mathrm{dom}\,(\mathcal{A})$ is not *a priory* dense in $\mathfrak{X}_T^C$ since it is restricted by condition that functions $t \mapsto A(t) f(t) \in C_0([0, T], \mathfrak{X})$ are continuous for any $f \in \mathrm{dom}\,(\mathcal{A})$. This may reduce its domain to $\mathrm{dom}\,(\mathcal{A}) = \{0\}$ even for family $\{A(t)\}_{t\in[0,T]}$ of bounded operators (see Example 2.2.1).

Lemma 2.1.22 *If $\{A(t)\}_{t\in[0,T]}$ is a family of closed operators, then the induced multiplication operator $\mathcal{A}$ is also closed.*

Proof Let $f_n \to f$ be a sequence in $\mathfrak{X}_T^C$ with $f_n \in \mathrm{dom}\,(\mathcal{A})$ and $\mathcal{A} f_n \to g \in C_0([0, T], \mathfrak{X})$. Then by definition (2.1.36), it holds pointwise for any $t \in [0, T]$ that

$$f_n(t) \in \mathrm{dom}\,(A(t)), \quad f_n(t) \to f(t), \quad A(t) f_n(t) \to g(t) \in \mathfrak{X}.$$

Since the operators $\{A(t)\}_{t\in[0,T]}$ are closed, we conclude that $f(t) \in \operatorname{dom}(A(t))$ and $A(t)f(t) = g(t)$ for any $t \in [0, T]$. This yields that $f \in \operatorname{dom}(\mathcal{A})$ and $(\mathcal{A}f)(t) = A(t)f(t) = g(t)$. □

Using the scalar multiplication operators (2.1.25) on the space $\mathfrak{X}_T^C$, we supplement Definition 2.1.21 with criterium to discern a general *bounded* multiplication operator on $\mathfrak{X}_T^C$.

Definition 2.1.23 We say that operator $\mathcal{C} \in \mathcal{L}(\mathfrak{X}_T^C)$ is bounded multiplication operator on $\mathfrak{X}_T^C$ if it commutes

$$\mathcal{C}\,\mathcal{M}(\phi) = \mathcal{M}(\phi)\,\mathcal{C}\,, \tag{2.1.37}$$

with all scalar multiplication operators $\{\mathcal{M}(\phi)\}_{\phi\in C([0,T]}$.

Proposition 2.1.24 *For every bounded multiplication operator $\mathcal{C}$ on $\mathfrak{X}_T^C$, there exists a unique family $\{C(t)\}_{t\in[0,T]}$ of bounded linear operators $t \mapsto C(t) \in \mathcal{L}(\mathfrak{X})$, such that*

(i) $(\mathcal{C}f)(t) = C(t)f(t)$, *for any* $f \in \mathfrak{X}_T^C$ *and* $t \in [0, T]$.
(ii) *Uniformly bounded:* $\sup_{t\in[0,T]} \|C(t)\|_{\mathcal{L}(\mathfrak{X})} < \infty$.
(iii) *The operator-norm of bounded multiplication operator $\mathcal{C}$ is given*

$$\|\mathcal{C}\|_{\mathcal{L}(\mathfrak{X}_T^C)} = \sup_{t\in[0,T]} \|C(t)\|_{\mathcal{L}(\mathfrak{X})}\,. \tag{2.1.38}$$

(iv) *Function $t \mapsto C(t)$ is strongly continuous on $\mathfrak{X}$ for $t \in [0, T]$.*

See Notes to Chap. 2 (Sect. 2.1.2) and Appendix E.3.1.

We draw attention that in contrast to Definition 2.1.23, for a family of *unbounded* operators $\{A(t)\}_{t\in[0,T]}$ in $\mathfrak{X}$, we define multiplication operator $\mathcal{A}$ following Definition 2.1.21, that is, taking into account continuity of the function $t \mapsto A(t)f(t)$, $f \in \operatorname{dom}(\mathcal{A})$ (see (2.1.36)). Assumption of continuity is too restrictive in the L^p-setting, and it may be relaxed to the strong measurability. For more comments and proofs related to multiplication operators and Definition 2.1.21 and Definition 2.1.23, we refer to Appendix E.3.

Further, we proceed with nACP (2.1.4) for a family of unbounded in $\mathfrak{X}$ operators $\{A(t)\}_{t\in[0,T]}$. To this aim, we consider in $C_0([0, T], \mathfrak{X})$ the operator

$$\widetilde{\mathcal{K}}f := D_0 f + \mathcal{A}f, \quad f \in \operatorname{dom}(\widetilde{\mathcal{K}}) \subseteq \operatorname{dom}(D_0) \cap \operatorname{dom}(\mathcal{A}). \tag{2.1.39}$$

A priori, operator $\widetilde{\mathcal{K}}$ is not an *evolution generator* of perturbed by multiplication operator $\mathcal{A}$ evolution semigroup (2.1.35). However, one has the following assertion:

Lemma 2.1.25 *Let $\mathfrak{D} \subseteq \operatorname{dom}(\widetilde{\mathcal{K}})$ be a dense subset in $C_0([0, T], \mathfrak{X})$ such that $\mathcal{M}(\phi)\mathfrak{D} \subseteq \mathfrak{D}$ for $\phi \in C^1([0, T])$. Then $\widetilde{\mathcal{K}} \restriction \mathfrak{D}$ is an evolution pre-generator in $C_0([0, T], \mathfrak{X})$. In particular, operator $\widetilde{\mathcal{K}}$ (2.1.39) itself is the evolution pre-generator in $C_0([0, T], \mathfrak{X})$.*

Proof By conditions, the operator $\widetilde{\mathcal{K}} \restriction \mathfrak{D}$ is densely defined in $C_0([0,T],\mathfrak{X})$ and $\mathcal{M}(\phi)\mathfrak{D} \subseteq \mathfrak{D}$ for $\phi \in C^1([0,T])$. Since $\mathcal{A}$ is induced multiplication operator, by virtue of definition (2.1.36), it commutes with the operator $\mathcal{M}(\phi)$ for $\phi \in C^1([0,T])$. So, the relation $\widetilde{\mathcal{K}}\mathcal{M}(\phi)f - \mathcal{M}(\phi)\widetilde{\mathcal{K}}f = \mathcal{M}(\phi')f$ for $f \in \mathfrak{D}$ follows from (2.1.39) and the *Leibnitz rule*. Then by Definition 2.1.16, the operator $\widetilde{\mathcal{K}} \restriction \mathfrak{D}$ is an evolution pre-generator in $C_0([0,T],\mathfrak{X})$ as well as the $\widetilde{\mathcal{K}}$. □

Definition 2.1.26 Let $\{A(t)\}_{t\in[0,T]}$ be a family of linear operators in $\mathfrak{X}$, and let $\widetilde{\mathcal{K}}$ be operator defined by (2.1.39). Further, let $\mathfrak{D} \subseteq \mathrm{dom}\,(\widetilde{\mathcal{K}})$ be a dense linear subset:

(i) The nACP (2.1.4) (or (2.1.12)) is called *C-solvable* with respect to $\mathfrak{D}$ if the operator $\widetilde{\mathcal{K}} \restriction \mathfrak{D}$ admits a closed extension $\mathcal{K}$ in $C_0([0,T],\mathfrak{X})$, which generates evolution semigroup.
(ii) The nACP (2.1.4) (or (2.1.12)) is called *uniquely C-solvable* with respect to $\mathfrak{D}$ if the operator $\widetilde{\mathcal{K}} \restriction \mathfrak{D}$ admits only *one* closed extension in $C_0([0,T],\mathfrak{X})$, which generates an evolution semigroup.
(iii) The nACP (2.1.4) (or (2.1.12)) is called *strictly C-solvable* with respect to $\mathfrak{D}$ if the operator $\widetilde{\mathcal{K}} \restriction \mathfrak{D}$ admits a closed extension $\mathcal{K}$ in $C_0([0,T],\mathfrak{X})$, which generates the evolution semigroup such that

$$e^{-\tau\mathcal{K}}\mathfrak{D} \subseteq \mathfrak{D}, \quad \tau \in \mathbb{R}_0^+, \tag{2.1.40}$$

and $\mathcal{A}\,e^{-\tau\mathcal{K}}f$ is continuous in $\tau \in \mathbb{R}_0^+$ for $f \in \mathfrak{D}$.

Here, it is worthy of note that if the nACP (2.1.4) is strictly C-solvable, then it is also uniquely C-solvable since (2.1.40) implies that $\mathfrak{D}$ is a *core* for generator $\mathcal{K}$. We shall return to this remark in the proof of Theorem 2.1.34 (ii).

Furthermore, we note that for conditions above the existence of a dense linear subset $\mathfrak{D} \subseteq \mathrm{dom}\,(\widetilde{\mathcal{K}})$ is already sufficient. Indeed, then there is in $C_0([0,T],\mathfrak{X})$ another dense linear subset

$$\mathfrak{D}' := \bigcup_{\phi\in C^1([0,T])} \mathcal{M}(\phi)\mathfrak{D},$$

which has the properties $\mathcal{M}(\phi)\mathfrak{D}' \subseteq \mathfrak{D}'$, for $\phi \in C^1([0,T])$, and $\mathfrak{D}' \subseteq \mathrm{dom}\,(\widetilde{\mathcal{K}})$.

Theorem 2.1.27 *Let $\{A(t)\}_{t\in[0,T]}$ be a family of linear operators in $\mathfrak{X}$. The nACP (2.1.4) (or (2.1.12)) is C-solvable with respect to $\mathfrak{D}$ if and only if there is a solution operator $\{U(t,s)\}_{(t,s)\in\Delta}$ such that for each $f \in \mathfrak{D}$ the derivative $[\partial_s U(t,s)]f(s)$, $(t,s) \in \Delta$ exists and the non-autonomous backward evolution equation*

$$[\partial_s U(t,s)]f(s) = U(t,s)A(s)f(s), \quad (t,s) \in \Delta, \tag{2.1.41}$$

is satisfied.

Proof Assume that $\widetilde{\mathcal{K}} \restriction \mathfrak{D}$ admits an extension $\mathcal{K}$ which generates an evolution semigroup. Then there is a solution operator $\{U(t,s)\}_{(t,s)\in\Delta}$ such that the semigroup $\{\mathcal{U}(\tau)\}_{\tau\in\mathbb{R}_0^+}$ generated by $\mathcal{K}$ admits the representation (2.1.24). If $f \in \mathfrak{D}$, then $f \in \mathrm{dom}\,(\mathcal{K})$. Let $f \in \mathfrak{D}$. Then from (2.1.33), (2.1.34) and (2.1.39), we find

$$\begin{aligned}[\partial_s U(t,s)]f(s) &= -U(t,s)f'(s) + U(t,s)(\mathcal{K}f)(s)\\ &= -U(t,s)f'(s) + U(t,s)(\widetilde{K}f)(s) = U(t,s)A(s)f(s), \quad (t,s)\in\Delta.\end{aligned} \tag{2.1.42}$$

Conversely, from (2.1.42) and $\widetilde{\mathcal{K}} = D_0 + \mathcal{A}$, (2.1.39), we obtain

$$\partial_s[U(t,s)f(s)] = U(t,s)f'(s) + U(t,s)A(s)f(s) = U(t,s)(\widetilde{\mathcal{K}}f)(s),$$

$(t,s)\in\Delta$. Therefore,

$$U(t,s)f(s) = f(t) - \int_s^t dr\; U(t,r)g(r)\,, \tag{2.1.43}$$

where $g := \widetilde{\mathcal{K}}f$. Let $\mathcal{K}$ be the generator of the evolution semigroup $\{\mathcal{U}(\tau)\}_{\tau\in\mathbb{R}_0^+}$ induced by (2.1.24). From Lemma 2.1.20 and (2.1.43), we get that $f \in \mathrm{dom}\,(\mathcal{K})$. This shows that $\mathcal{K}$ is an extension of $\widetilde{\mathcal{K}} \restriction \mathfrak{D}$. □

Remark 2.1.28 Note that a starting point for solution of nACP (2.1.4) is strongly continuous evolution family and solution operator (Definitions 2.1.4 and 2.1.5), associated with a strongly continuous evolution semigroup (2.1.24). In general, no differentiability holds for the evolution family and solution operator, that we do need for well-posedness (Definition 2.1.3) of the nACP. On the other hand, the evolution semigroup $\{\mathcal{U}(\tau)\}_{\tau\in\mathbb{R}_0^+}$ is always differentiable on domain of its generator $\mathcal{K}$. By Definition 2.1.3, the well-posedness means that the regularity set is dense or, in virtue of (2.1.24), that there are *sufficiently many* differentiable orbits generated by the evolution semigroup.

For characterisation of the well-posedness of nACP by the properties of the evolution semigroup, we introduce the notion of a dense *cross-section* $[\mathfrak{D}]_{\mathfrak{X}_T^C}$ of the set $\mathfrak{D} \subset \mathfrak{X}_T^C = C_0([0,T],\mathfrak{X})$ in $\mathfrak{X}$. Similar to density in $\mathfrak{X}$ of the *regularity set* (see Definition 2.1.3(i)), dense cross-section in $\mathfrak{X}$ indicates that in $C_0([0,T],\mathfrak{X})$ there are sufficiently many trajectories $\{f(t)\}_{t\in[0,T]} \in \mathfrak{D}$.

Definition 2.1.29 We say that cross-section $[\mathfrak{D}]_{\mathfrak{X}_T^C}$ of the set $\mathfrak{D} \subset \mathfrak{X}_T^C = C_0([0,T],\mathfrak{X})$ is dense in $\mathfrak{X}$ if for every $t \in (0,T]$ the *cross-section* of $\mathfrak{D}$ at t

$$[\mathfrak{D}]_t := \{x \in \mathfrak{X} : \exists f \in \mathfrak{D} \text{ such that } f(t) = x\} \tag{2.1.44}$$

is *dense* in $\mathfrak{X}$ or, equivalently, if one has for the closure in $\mathfrak{X}$ $\overline{[\mathfrak{D}]_{\mathfrak{X}_T^C}} = \mathfrak{X}$, where we denote $[\mathfrak{D}]_{\mathfrak{X}_T^C} := \bigcap_{t\in(0,T]} [\mathfrak{D}]_t$.

Remark 2.1.30 If set $\mathfrak{D}$ is dense in $\mathfrak{X}_T^C = C_0([0,T],\mathfrak{X})$, then it has a dense cross-section $[\mathfrak{D}]_{\mathfrak{X}_T^C}$ in $\mathfrak{X}$. Indeed, let $x_0 \in \mathfrak{X}$ and $t \in (0,T]$. Then there exists $f_0 \in \mathfrak{X}_T^C$ such that $f_0(t) = x_0$. Given that $\mathfrak{D}$ is dense in $\mathfrak{X}_T^C$, there is a sequence $\{f_n\}_{n\geq 1} \subset \mathfrak{D}$, which converges in the $\|\cdot\|_{\mathfrak{X}_T^C}$-norm (2.1.22) to f_0. Consequently,

$$\lim_{n\to\infty} \|f_n - f_0\|_{\mathfrak{X}_T^C} \geq \lim_{n\to\infty} \|f_n(t) - f_0(t)\|_{\mathfrak{X}} = 0\,.$$

These limits prove the assertion, seeing that $\{f_n(t)\}_{n\geq 1} \subset [\mathfrak{D}]_t$ and $f_0(t) = x_0$.

On the other hand, by (2.1.44), we obtain for unbounded multiplication operator $\mathcal{A}$ (2.1.36) that

$$[\mathrm{dom}\,(\mathcal{A})]_t \subset \mathrm{dom}\,(A(t)) \quad \text{for} \quad t \in (0,T]. \tag{2.1.45}$$

On account of (2.1.45), the following lemma provides a criterion ensuring that induced multiplication operator $\mathcal{A}$ is densely defined in $\mathfrak{X}_T^C = C_0([0,T],\mathfrak{X})$.

Lemma 2.1.31 *Let $\mathfrak{D}$ be a linear set in $\mathfrak{X}_T^C$. Then $\mathfrak{D}$ is dense in the Banach space $\mathfrak{X}_T^C$ if and only if it has the following properties:*

(i) *$\mathfrak{D}$ has dense cross-section in $\mathfrak{X}$ in the sense of Definition* 2.1.29.
(ii) *For every $\varphi \in C^1([0,T])$, one has $\mathcal{M}(\varphi)\mathfrak{D} \subseteq \mathfrak{D}$.*

Proof Let us assume that $\mathfrak{D}$ is not dense in $\mathfrak{X}_T^C$. Then there is a functional $\mu \in C_0([0,T],\mathfrak{X})^*$ such that $\mu(f) = 0$ for $f \in \mathfrak{D}$. By the *Hahn-Banach* theorem , there is an extension $\tilde{\mu} \in C([0,T],\mathfrak{X})^*$ of μ such that $\tilde{\mu} \restriction C_0([0,T],\mathfrak{X}) = \mu$ and $\|\tilde{\mu}\| = \|\mu\|$ such that $\tilde{\mu}(f) = 0$ for $f \in \mathfrak{D}$. Owing to Proposition E.2.4, there is a measure $m \in \mathsf{rcabv}([0,T],\mathfrak{X}^*)$ such that

$$\tau(m)(f) = \int_{[0,T]} (f, dm), \quad f \in \mathfrak{X}_T^C.$$

By Lemma E.2.6, there is a w^*-measurable function $\phi : [0,T] \to \mathfrak{X}^*$ such that

$$0 = \mu(f) = \int_{[0,T]} (f, dm) = \int_{[0,T]} (f(t), \phi(t)) d\|m\|(t), \quad f \in \mathfrak{D}.$$

Since $\mathcal{M}(\varphi)\mathfrak{D} \subset \mathfrak{D}$ for $\varphi \in C^1([0,T])$, it yields

$$0 = \int_{[0,T]} (\varphi(t)f(t), \phi(t)) d\|m\|(t) = \int_{[0,T]} \varphi(t)(f(t), \phi(t)) d\|m\|(t), \quad f \in \mathfrak{D}.$$

The function $t \mapsto (f(t), \phi(t))$ belongs to $L^1(\|m\|)$. On account of $\varphi \in C^1([0, T])$ is arbitrary, we conclude that

$$0 = \langle f(t), \phi(t) \rangle$$

for a.a. $t \in [0, T]$ with respect to $\|m\|$. Since $\mathfrak{D}$ has dense cross-section, we have $\phi(t) = 0$ for a.a. $t \in [0, T]$ with respect to $\|m\|$. Hence, $\mu = 0$ yields that the linear set $\mathfrak{D}$ is dense in $\mathfrak{X}_T^C$.

For proof of the converse statement, see Remark 2.1.30. □

For applying Lemma 2.1.31 as a criterion for multiplication operator $\mathcal{A}$, one takes as a subset $\mathfrak{D} \subset \mathrm{dom}\,(\mathcal{A})$.

For a quasi-bounded C_0-semigroup with generator, which corresponds to induced multiplication operator, there is a following assertion:

Theorem 2.1.32 *Let $\{A(t)\}_{t \in [0,T]}$ be a family of generators in $\mathfrak{X}$ such that for all $t \in [0, T]$ it holds $A(t) \in \mathcal{G}(M, \beta)$, for some $M \geq 1$ and $\beta \in \mathbb{R}$. If the function $[0, T] \ni t \mapsto (A(t) + \xi \mathbb{1})^{-1} x$ is continuous for each $x \in \mathfrak{X}$ and $\xi > \beta$, then the induced multiplication operator $\mathcal{A}$ in $\mathfrak{X}_T^C$ is generator of a C_0-semigroup on $\mathfrak{X}_T^C$ and $\mathcal{A} \in \mathcal{G}(M, \beta)$. The corresponding semigroup is given by*

$$(\mathrm{e}^{-\tau \mathcal{A}} f)(t) = \mathrm{e}^{-\tau A(t)} f(t), \quad f \in \mathfrak{X}_T^C, \quad t \in [0, T].$$

In particular, we obtain that $\| \mathrm{e}^{-\tau \mathcal{A}} \|_{\mathcal{L}(\mathfrak{X}_T^C)} = \sup_{t \in [0,T]} \|\mathrm{e}^{-\tau A(t)}\|_{\mathcal{L}(\mathfrak{X})}$.

Proof Since by Lemma 2.1.22 the induced multiplication operator $\mathcal{A}$ is closed, for the first claim, it rests to prove that $\mathcal{A}$ is densely defined.

Note that for any $\varphi \in C^1([0, T])$ one gets $\mathcal{M}(\varphi)\mathrm{dom}\,(\mathcal{A}) \subseteq \mathrm{dom}\,(\mathcal{A})$. So, by Lemma 2.1.31, it remains only to show that $\mathrm{dom}\,(\mathcal{A})$ has dense cross-section in $\mathfrak{X}$.

For every $t_0 \in (0, T]$, there is a function $\varphi \in C^1([0, T])$ such that $\varphi(0) = 0$ and $\varphi(t_0) = 1$. We set

$$f_\xi(t) := \xi\, \varphi(t)\, (A(t) + \xi \mathbb{1})^{-1} x, \quad \xi > \beta, \quad x \in \mathfrak{X}. \tag{2.1.46}$$

Then, definition (2.1.46) and assumptions of the theorem yield $f_\xi(t) \in \mathrm{dom}\,(A(t))$, $f_\xi \in C_0([0, T], \mathfrak{X})$ and

$$(\mathcal{A} f_\xi)(t) = \xi\, \varphi(t)\, x - \xi^2\, \varphi(t)\, (A(t) + \xi \mathbb{1})^{-1} x \in \mathfrak{X}.$$

Hence, $f_\xi \in \mathrm{dom}\,(\mathcal{A})$. Since $A(t_0)$ is generator of C_0-semigroup, by (2.1.46), we get for any $x \in \mathfrak{X}$:

$$f_\xi(t_0) = \xi\, (A(t_0) + \xi \mathbb{1})^{-1} x \to x, \quad \text{for } \xi \to \infty.$$

Or putting it differently, for every $t_0 \in (0, T]$, the cross-section $[\mathrm{dom}\,(\mathcal{A})](t_0)$ is dense in $\mathfrak{X}$. Then by Lemma 2.1.31 $\mathrm{dom}\,(\mathcal{A})$ is dense in $\mathfrak{X}$.

For any $t \in [0, T]$, the operators $\{A(t)\}_{t\in[0,T]}$ belong to the same class $\mathcal{G}(M, \beta)$. Thus, uniformly in $t \in [0, T]$, we get for any $k \in \mathbb{N}$

$$\| (A(t) + \lambda \mathbb{1})^{-k} \|_{\mathcal{L}(\mathfrak{X})} \leq \frac{M}{|\lambda - \beta|^k}, \quad \lambda > \beta, \quad k \in \mathbb{N}.$$

Seeing that $\mathcal{A}$ is a multiplication operator, we obtain for any $t \in [0, T]$ and $f \in \mathfrak{X}_T^C$

$$\|((\mathcal{A}+\lambda\mathcal{I})^{-k} f)(t)\|_{\mathfrak{X}} = \|(A(t)+\lambda\mathbb{1})^{-k}(f(t))\|_{\mathfrak{X}} \leq \frac{M}{|\lambda - \beta|^k} \|f(t)\|_{\mathfrak{X}}, \quad \lambda > \beta .$$

Taking the supremum on $[0, T]$, one gets

$$\|(\mathcal{A} + \lambda\mathcal{I})^{-k} f\|_{\mathfrak{X}_T^C} \leq \frac{M}{|\lambda - \beta|^k} \|f\|_{\mathfrak{X}_T^C}, \quad \lambda > \beta, \quad k \in \mathbb{N}.$$

Then by the *Hille-Yosida* theorem (Proposition 1.1.1), the closed densely defined operator $\mathcal{A}$ is generator $\mathcal{A} \in \mathcal{G}(M, \beta)$ in $\mathfrak{X}_T^C$. The corresponding quasi-bounded C_0-semigroup (Proposition 1.3.4) is given by the *Euler formula*:

$$\mathrm{e}^{-\tau\mathcal{A}} f = \lim_{n\to\infty} \left(\mathcal{I} + \frac{\tau}{n}\mathcal{A}\right)^{-n} f, \quad f \in \mathfrak{X}_T^C .$$

Given that for any $n \in \mathbb{N}$ one has for multiplication operator $\mathcal{A}$

$$\left(\left(\mathcal{I} + \frac{\tau}{n}\mathcal{A}\right)^{-n} f\right)(t) = \left(\mathbb{1} + \frac{\tau}{n}A(t)\right)^{-n} f(t),$$

this formula yields for any $f \in \mathfrak{X}_T^C$

$$\begin{aligned}(\mathrm{e}^{-\tau\mathcal{A}} f)(t) &= \lim_{n\to\infty} \left(\left(\mathcal{I} + \frac{\tau}{n}\mathcal{A}\right)^{-n} f\right)(t) = \lim_{n\to\infty} \left(\mathbb{1} + \frac{\tau}{n}A(t)\right)^{-n} f(t) \\ &= \mathrm{e}^{-\tau A(t)} f(t) .\end{aligned}$$

Then the last claim of the theorem follows from (2.1.38) of Proposition 2.1.24. □

Similar to quasi-boundedness, we can establish in the space $\mathfrak{X}_T^C$ the inheritance of the semigroup analyticity.

Theorem 2.1.33 *Let $\{A(t)\}_{t\in[0,T]}$ be a family of generators in $\mathfrak{X}$ such that $A(t) \in \mathscr{H}(\theta, 0)$ for some $\theta \in (0, \pi/2]$ and for all $t \in [0, T]$. If the function $[0, T] \ni t \mapsto (A(t) + \xi\mathbb{1})^{-1} x$ is continuous for each $x \in \mathfrak{X}$ and $\xi > 0$, then the induced in $\mathfrak{X}_T^C$ multiplication operator $\mathcal{A}$ is generator of a holomorphic in the open sector $S_\theta \subset \mathbb{C}_+$ semigroup $\{\mathcal{U}_{\mathcal{A}}(\tau) := \mathrm{e}^{-\tau\mathcal{A}}\}_{\tau\in S_\theta}$ on $\mathfrak{X}_T^C$, that is, $\mathcal{A} \in \mathscr{H}(\theta, 0)$.*

Proof Seeing that by Lemma 2.1.22 and Theorem 2.1.32 the induced multiplication operator $\mathcal{A}$ is closed and densely defined, it rests only to check that $\mathcal{A}$ satisfies for some $\varepsilon \in (0, \theta)$ the condition

$$\left\|(\zeta\,\mathcal{I} + \mathcal{A})^{-1}\right\|_{\mathcal{L}(\mathfrak{X}_T^C)} \le \frac{M_\varepsilon}{|\zeta|}, \quad \zeta \in \overline{S}_{\pi/2+\theta-\varepsilon} \setminus \{0\}. \tag{2.1.47}$$

Here, $M_\varepsilon > 0$, which is independent of ζ, and the closed sector $\overline{S}_{\pi/2+\theta-\varepsilon}$ (see (1.7.4).

To this aim, we recall that generator $A(t) \in \mathscr{H}(\theta, 0)$ satisfies condition (1.7.4) for any $t \in [0, T]$. Consequently, one gets estimate

$$\left\|(\zeta\,\mathcal{I} + \mathcal{A})^{-1} f\right\|_{\mathfrak{X}_T^C} = \sup_{t\in[0,T]} \left\|(\zeta\,\mathbb{1} + A(t))^{-1} f(t)\right\|_{\mathfrak{X}} \le$$

$$\le \sup_{t\in[0,T]} \frac{M_\varepsilon}{|\zeta|} \|f(t)\|_{\mathfrak{X}} = \frac{M_\varepsilon}{|\zeta|} \|f\|_{\mathfrak{X}_T^C}, \quad \zeta \in \overline{S}_{\pi/2+\theta-\varepsilon} \setminus \{0\}, \tag{2.1.48}$$

for any $f \in \mathfrak{X}_T^C$. Then (2.1.47) follows from (2.1.48) after taking the operation,

$$\sup_{\{f\in\mathfrak{X}_T^C : \|f\|_{\mathfrak{X}_T^C}=1\}} \|\cdot\|_{\mathfrak{X}_T^C},$$

on the left- and right-hand sides of (2.1.48). □

After these preliminaries, we return to the problem of the well-posedness of nACP versus its C-solubility, cf. Definition 2.1.26.

Theorem 2.1.34 *Let* $\{A(t)\}_{t\in[0,T]}$ *be a family of linear operators in* $\mathfrak{X}$. *The following assertions are equivalent:*

(i) *The nACP* (2.1.4) (*or* (2.1.12)) *is well-posed.*
(ii) *The nACP* (2.1.4) (*or* (2.1.12)) *is strictly C-solvable.*

Proof (i) $\Rightarrow$ (ii): If the nACP is well-posed, then by Proposition 2.1.9, there is a unique solution operator $\{U(t,s)\}_{(t,s)\in\Delta}$ solving the nACP (2.1.4) (or (2.1.12)).

By equation (2.1.27), one associates with $\{U(t,s)\}_{(t,s)\in\Delta}$ an evolution semigroup $\{\mathcal{U}(\tau)\}_{\tau\in\mathbb{R}_0^+}$ generated by $\mathcal{K}$. Let the linear hull

$$\mathfrak{L} := \operatorname{span}\left\{\varphi(t)U(t,s)y : \begin{array}{ll} \varphi(\cdot) \in C_0^1([0,T]), & \operatorname{supp}(\varphi) \subseteq [s,T] \\ y \in \mathfrak{Y}(s), & (t,s) \in \Delta \end{array}\right\},$$

where $\mathfrak{Y}(s)$ is a regularity set (Definition 2.1.5) for solution operator $\{U(t,s)\}_{(t,s)\in\Delta}$.

We check that the set $\mathfrak{L}$ is dense in the space $\mathfrak{X}_T^C = C_0([0,T],\mathfrak{X})$. To this end, we show that $\mathfrak{L}$ has dense cross-section in $\mathfrak{X}$ (see Definition 2.1.29 and Lemma 2.1.31). Let $t_0 \in (0, T]$. If the cross-section $[\mathfrak{L}]_{t_0}$ (2.1.44) is not dense in $\mathfrak{X}$, then there is a functional $\mu \in \mathfrak{X}^*$ such that $\mu(f(t_0)) = 0$ for any $f \in \mathfrak{L}$. In particular, one has $\mu(U(t_0, s)y_s) = 0$ for any $s \in (0, t_0)$ and $y_s \in \mathfrak{Y}(s)$. For $y \in \mathfrak{Y}(t_0)$, there are sequences $\{s_n\}_{n\in\mathbb{N}}$, $s_n \in (0, t_0)$ and $\{y_{s_n}\}_{n\in\mathbb{N}}$ such that

$$\lim_{n\to\infty} s_n = t_0 \quad \text{and} \quad \lim_{n\to\infty} y_{s_n} = y.$$

Consequently,

$$0 = \lim_{n\to\infty} \mu(U(t_0, s_n)y_{s_n}) = \mu(y).$$

Because $\mathfrak{Y}(t_0)$ is dense in $\mathfrak{X}$, we obtain $\mu = 0$. Then on account of $\mathcal{M}(\varphi)\mathfrak{L} \subseteq \mathfrak{L}$ for any $\varphi \in C^1([0, T])$, we conclude that $\mathfrak{L}$ is dense in $\mathfrak{X}^C_T$ applying Lemma 2.1.31.

Furthermore, $\mathcal{U}(\tau)\mathfrak{L} \subseteq \mathfrak{L}$ holds for any $\tau \in \mathbb{R}^+_0$. Indeed, by HEN formula (2.1.24), one has

$$\begin{aligned}(\mathcal{U}(\tau)\varphi(\cdot)U(\cdot, s)y)(t) &= \chi_{[0,T]}(t-\tau)U(t, t-\tau)\varphi(t-\tau)U(t-\tau, s)y \\ &= \varphi(t-\tau)U(t, s)y.\end{aligned}$$

Seeing that $\mathrm{supp}(\varphi(\cdot - \tau)) \subseteq [s+\tau, T] \subseteq [s, T]$, we get $\varphi(\cdot - \tau)U(\cdot, s)y \in \mathfrak{L}$, $\tau \in \mathbb{R}^+_0$. Let $\mathfrak{D} := \mathcal{K}^{-1}\mathfrak{L}$, where $\mathcal{K}^{-1}$ exists since $\rho(\mathcal{K}) = \mathbb{C}$, Corollary 2.1.19. Then $\mathfrak{D} \subseteq \mathrm{dom}\,(\mathcal{K})$, which yields $\mathcal{U}(\tau)\mathfrak{D} \subseteq \mathfrak{D}$, $\tau \in \mathbb{R}^+_0$. Since $\mathfrak{L}$ is dense, the set $\mathfrak{D}$ is also dense in $\mathfrak{X}^C_T$. Then the last inclusion implies that $\mathfrak{D}$ is a core for generator $\mathcal{K}$. That is, the closure $\overline{\mathcal{K} \restriction \mathfrak{D}} = \mathcal{K}$.

It remains to show that $\mathfrak{D} \subseteq \mathrm{dom}\,(\widetilde{\mathcal{K}}) \subseteq \mathrm{dom}\,(D_0) \cap \mathrm{dom}\,(\mathcal{A})$ (see (2.1.39)). To this aim, we define $f(t) = \varphi(t)\,U(t, s)\,y$ for $y \in \mathfrak{Y}(s)$ and $g = \mathcal{K}^{-1}f \in \mathfrak{D}$. Then by (2.1.24), one gets the representation

$$g(t) = \int_0^t dr\, U(t, r)\,\varphi(r)\,U(r, s)\,y = \widehat{\varphi}(t)\,U(t, s)\,y \in \mathfrak{L}, \quad t \in [0, T],$$

where

$$\widehat{\varphi}(t) := \int_0^t dr\, \varphi(r), \quad t \in [0, T].$$

Note that $y \in \mathfrak{Y}(s)$ yields $U(\cdot, s)y \in C^1([s, T], \mathfrak{X})$ and then $g(\cdot) = \widehat{\varphi}(\cdot)\,U(\cdot, s)\,y \in C^1_0([0, T], \mathfrak{X}) = \mathrm{dom}\,(D_0)$. Since by condition the evolution equation (2.1.4) (or (2.1.12)) is well-posed, we also have $U(t, s)y \in \mathrm{dom}\,(A(t))$, $t \in [s, T]$ and $A(\cdot)U(\cdot, s)y \in C([s, T], \mathfrak{X})$. Hence, $\widehat{\varphi}(\cdot)\,A(\cdot)\,U(\cdot, s)\,y \in \mathfrak{X}^C_T$, that is, $g \in \mathrm{dom}\,(\mathcal{A})$ and consequently $g \in \mathrm{dom}\,(D_0) \cap \mathrm{dom}\,(\mathcal{A}) = \mathrm{dom}\,(\widetilde{\mathcal{K}})$.

Finally, we show that mapping $\mathcal{A}\,\mathcal{U}(\cdot)f : \mathbb{R}_0^+ \longrightarrow \mathfrak{X}_T^C$, for $f \in \mathfrak{D}$, is continuous. Let $f(t) = \varphi(t)\, U(t,s)\, y$, for $(t,s) \in \Delta$ and $y \in \mathfrak{Y}(s)$, where $\varphi \in C^1([0,T])$ with $\operatorname{supp}(\varphi) \in [s,T]$. Then

$$
\begin{aligned}
(\mathcal{U}(\tau)f)(t) &= U(t, t-\tau)\, \chi_{[0,T]}(t-\tau)\, \varphi(t-\tau)\, U(t-\tau, s)\, y \\
&= \varphi(t-\tau)\, U(t,s)\, y, \quad (t,s) \in \Delta.
\end{aligned}
$$

As a result,

$$
(\mathcal{A}\, e^{-\tau\mathcal{K}} f)(t) = \varphi(t-\tau)\, A(t)\, U(t,s)\, y\,, \quad t \in [0,T], \quad \tau \in \mathbb{R}_0^+. \tag{2.1.49}
$$

Since the nACP (2.1.4) is well-posed and $y \in \mathfrak{Y}(s)$, the right-hand side of (2.1.49) yields $A(\cdot)\, U(\cdot, s)\, y \in C([s,T], \mathfrak{X})$. So, function $\mathcal{A}\; \mathcal{U}(\cdot)f : \mathbb{R}_0^+ \longrightarrow \mathfrak{X}_T^C$ is continuous.

(ii) $\Rightarrow$ (i): To prove this claim, we show first that condition (ii) implies the existence and uniqueness of solution operator $\{U(t,s)\}_{(t,s)\in\Delta}$ for the nACP (2.1.4) (or (2.1.12)). Then, completing the proof, we make use of the assertion (ii) $\Rightarrow$ (i), of Proposition 2.1.9.

To this end, we introduce on the space $L^p([0,T], \mathfrak{X})$ a family of the *left-shift* automorphisms $\{\mathcal{S}_L(\tau)\}_{\tau \geq 0}$ such that

$$
(\mathcal{S}_L(\tau)\, f)(t) := \chi_{[0,T]}(t+\tau)\; f(t+\tau),\;\; f \in C_T([0,T], \mathfrak{X})\,,\;\; \tau \geq 0\,,\;\; t \in [0,T]\,, \tag{2.1.50}
$$

where $C_T([0,T], \mathfrak{X}) := \{f \in C([0,T], \mathfrak{X}) : f(T) = 0\}$ is the Banach space endowed with the standard sup-norm. The family $\{\mathcal{S}_L(\tau)\}_{\tau\geq 0}$ forms on $C_T([0,T], \mathfrak{X})$ a (*nilpotent*) contraction C_0-semigroup, cf. Sect. F.1, 2. Generator D_T of this semigroup is defined as follows:

$$
\begin{aligned}
f \in \operatorname{dom}(D_T) &:= \{f \in C^1([0,T], \mathfrak{X}) : f(T) = \partial_t f(T-0) = 0\}, \\
(D_T f)(t) &:= -\partial_t f(t), \quad t \in [0,T].
\end{aligned}
$$

On account of assumption (ii) (see Definition 2.1.26 (iii)), there exists a dense set $\mathfrak{D} \subset \mathfrak{X}_T^C$ and operator $\widetilde{\mathcal{K}} = D_0 + \mathcal{A}$, (2.1.39), such that restriction $\widetilde{\mathcal{K}} \restriction \mathfrak{D}$ is closable. Its closure $\mathcal{K}$ generates the evolution semigroup $\{e^{-\tau\mathcal{K}}\}_{\tau\in\mathbb{R}_0^+}$, with property $e^{-\tau\mathcal{K}}\mathfrak{D} \subseteq \mathfrak{D}$, and $\mathcal{A}\, e^{-\tau\mathcal{K}} f$ is continuous in $\tau \in \mathbb{R}_0^+$ for $f \in \mathfrak{D}$.

For construction of associated evolution family $\{U(t,s)\}_{(t,s)\in\Delta}$, let $f \in \mathfrak{D}$. Then derivative $\partial_\tau(\mathcal{S}_L(\tau)\mathcal{M}(\phi)\, e^{-\tau\mathcal{K}})f$, $\tau \in [0,T]$, exists for any $\phi \in C^1([0,T])$, such that $\phi'(T) = \phi(T) = 0$, and we obtain for $\tau \in [0,T]$

$$
\partial_\tau(\mathcal{S}_L(\tau)\mathcal{M}(\phi) e^{-\tau\mathcal{K}})f = -\,\mathcal{S}_L(\tau)(D_T\mathcal{M}(\phi) + \mathcal{M}(\phi)\mathcal{K})\, e^{-\tau\mathcal{K}} f. \tag{2.1.51}
$$

Note that as a consequence of (2.1.34), (2.1.39) and Definition 2.1.16 one gets

$$(D_T\mathcal{M}(\phi) + \mathcal{M}(\phi)\mathcal{K})\, g = \mathcal{M}(\phi)\mathcal{A}\, g - \mathcal{M}(\phi')\, g, \quad \text{for} \quad g \in \mathfrak{D},$$

which yields for $g = \mathcal{U}(\tau)f$ and $\phi'(\tau) := \partial_\tau\phi(\tau)$:

$$\begin{aligned}&(\mathcal{K}_T\mathcal{M}(\phi) + \mathcal{M}(\phi)\mathcal{K})\, \mathcal{U}(\tau)f \\ &\quad = \mathcal{M}(\phi)\mathcal{A}\,\mathcal{U}(\tau)f - \mathcal{M}(\phi')\mathcal{U}(\tau)f, \quad f \in \mathfrak{D}, \quad \tau \in \mathbb{R}_0^+.\end{aligned}$$

Since by assumptions (ii) the function $\mathcal{A}\,\mathcal{U}(\cdot)f : \mathbb{R}_0^+ \longrightarrow \mathfrak{X}_T^C$ is continuous, the last formula and (2.1.51) imply

$$\begin{aligned}&\mathcal{S}_L(\tau)\mathcal{M}(\phi)\,\mathrm{e}^{-\tau\mathcal{K}} f = \\ &= \mathcal{M}(\phi)f - \int_0^\tau d\tau'\, \mathcal{S}_L(\tau')\mathcal{M}(\phi)\mathcal{A}\,\mathcal{U}(\tau')f + \int_0^\tau d\tau'\, \mathcal{S}_L(\tau')\mathcal{M}(\phi')\,\mathrm{e}^{-\tau'\mathcal{K}} f.\end{aligned}$$

Because of the HEN formula (2.1.27), one gets for the first integrand,

$$\begin{aligned}&(\mathcal{S}_L(\tau)\mathcal{M}(\phi)\mathcal{A}\,\mathcal{U}(\tau)f)(t) \\ &\quad = \phi(t+\tau)A(t+\tau)U(t+\tau,t)\chi_{[0,T]}(t+\tau)f(t), \quad t \in [0,T],\end{aligned}$$

and similarly for the second,

$$(\mathcal{S}_L(\tau)\mathcal{M}(\phi')\,\mathrm{e}^{-\tau\mathcal{K}} f)(t) = \phi'(t+\tau)U(t+\tau,t)\chi_{[0,T]}(t+\tau)f(t), \quad t \in [0,T],$$

as well as

$$(\mathcal{S}_L(\tau)\mathcal{M}(\phi)\,\mathrm{e}^{-\tau\mathcal{K}} f)(t) = \phi(t+\tau)U(t+\tau,t)\chi_{[0,T]}(t+\tau)f(t), \quad t \in [0,T],$$

given that

$$(\mathcal{M}(\phi)f)(t) = \phi(t)f(t), \quad t \in [0,T].$$

Summarising, we obtain for $t \in [0,T]$ and $f \in \mathfrak{D}$

$$\begin{aligned}&\phi(t+\tau)U(t+\tau,t)\chi_{[0,T]}(t+\tau)f(t) - \phi(t)f(t) = \\ &\int_0^\tau d\tau'\, \phi'(t+\tau')U(t+\tau',t)\,\chi_{[0,T]}(t+\tau')f(t) - \\ &\int_0^\tau d\tau'\, \phi(t+\tau')A(t+\tau')U(t+\tau',t)\,\chi_{[0,T]}(t+\tau')f(t),\end{aligned}$$

or for arguments $(t, s) \in \Delta$

$$\phi(t)U(t,s)f(s) - \phi(s)f(s) =$$
$$\int_s^t dr\ \phi'(r)U(r,s)f(s) - \int_s^t dr\ \phi(r)A(r)U(r,s)f(s).$$

Choosing function $\phi \in C^1([0,T])$ with properties $\mathrm{supp}(\phi) \supset [t,s]$ and $\phi(r) = 1$ for $r \in [t,s]$, we obtain that evolution family $\{U(t,s)\}_{(t,s)\in\Delta}$ satisfies the (*mild*) evolution equation:

$$U(t,s)f(s) - f(s) = -\int_s^t dr\ A(r)U(r,s)f(s). \tag{2.1.52}$$

Since function $\mathcal{A}\ \mathcal{U}(\cdot)f : \mathbb{R}_0^+ \longrightarrow \mathfrak{X}_T^C$ is continuous, the function $A(\cdot)U(\cdot,s)f(s) : [s,T] \longrightarrow \mathfrak{X}$, is also continuous. Then (2.1.52) entails that for $f \in \mathfrak{D}$ the continuous orbit $\{t \mapsto u(t|s, f(s)) = U(t,s)f(s)\}_{t\in[s,T)} \subset C^1([s,T],\mathfrak{X})$ of f under solution operator $\{U(t,s)\}_{(t,s)\in\Delta}$ provides a *classical solution* of the nACP (2.1.4) (or (2.1.12)) (see Definition 2.1.2).

Note that definition of $\mathfrak{D}$ (cf. Lemma 2.1.25) and Definition 2.1.3 of regularity set $\mathfrak{Y}_U(s)$ yield that cross-section $[\mathfrak{D}]_s \subseteq \mathfrak{Y}_U(s)$ for any $s \in (0,T)$. Given that by (2.1.52) the cross-section $[\mathfrak{D}]_s$ (see (2.1.44)) is dense for each $s \in (0,T)$, we obtain that regularity set $\mathfrak{Y}_U(s)$ is also dense in $\mathfrak{X}$. Then by Definition 2.1.5, the family $\{U(t,s)\}_{(t,s)\in\Delta}$ is a solution operator for nACP (2.1.4) (or for (2.1.12)).

Now, by virtue of Proposition 2.1.9, it will follow that the nACP (2.1.4) (or (2.1.12)) is (classically) well-posed if $\{U(t,s)\}_{(t,s)\in\Delta}$ is unique. Indeed, the latter comes from the fact that $e^{-\tau\mathcal{K}}\mathfrak{D} \subseteq \mathfrak{D}$, for $\tau \geq 0$, which implies that $\mathfrak{D}$ is a *core* for the semigroup generator $\mathcal{K}$. □

2.1.3 Cauchy Problem and Product Formulæ in C-Setting

Let $\{A(t)\}_{t\in[0,T]}$ be a family of linear operators such that $A(t) \in \mathcal{G}(M,\beta)$ for $t \in [0,T]$. Assume that for each $x \in \mathfrak{X}$ and $\xi > \beta$ the function

$$[0,T] \ni t \mapsto (A(t) + \xi\mathbb{1})^{-1}x \in C([0,T],\mathfrak{X})\,, \tag{2.1.53}$$

is strongly continuous. Then by Theorem 2.1.32, the multiplication operator $\mathcal{A}$ is densely defined and generates a strongly continuous quasi-bounded evolution semigroup $\{\mathcal{U}_{\mathcal{A}}(\tau)\}_{\tau\in\mathbb{R}_0^+}$ given by

$$(\mathcal{U}_{\mathcal{A}}(\tau)f)(t) = (e^{-\tau\mathcal{A}}f)(t) = e^{-\tau A(t)}f(t), \quad f \in \mathfrak{X}_T^C\,, \quad t \in [0,T]\,, \tag{2.1.54}$$

that is, $\mathcal{A} \in \mathcal{G}(M,\beta)$.

Theorem 2.1.35 *Let $\{A(t)\}_{t\in[0,T]}$ be a family of linear operators such that $A(t) \in \mathcal{G}(1,0)$ for each $t \in [0,T]$ and condition (2.1.53) be satisfied for $\beta = 0$. If the nACP (2.1.4) is well-posed, then for each $x \in \mathfrak{X}$, the strictly* C*-solution operator $\{U(t,s)\}_{(t,s)\in\Delta}$ admits the following product approximation* à la Lie-Trotter,

$$\lim_{n\to\infty} \sup_{(t,s)\in\Delta_\varepsilon} \|(U(t,s) - U_n(t,s))x\| = 0, \tag{2.1.55}$$

for any $x \in \mathfrak{X}$ and $\varepsilon \in (0,T]$, where the product

$$U_n(t,s) := \prod_{k=n-1}^{0} \mathrm{e}^{-\frac{(t-s)}{n} A\left(s+k\frac{t-s}{n}\right)}, \quad (t,s) \in \Delta\,, \tag{2.1.56}$$

and $\Delta_\varepsilon := \{(t,s) \in \Delta : 0 < \varepsilon \le s \le t \le T\}$.

Proof Condition (2.1.54) for $M = 1$ and $\beta = 0$ implies that multiplication operator $\mathcal{A} \in \mathcal{G}(1,0)$. By Theorem 2.1.34, (i) $\Rightarrow$ (ii), there is a dense subset $\mathfrak{D} \subseteq \mathrm{dom}\,(\widetilde{\mathcal{K}})$ such that $\mathcal{K} = \overline{\widetilde{\mathcal{K}} \restriction \mathfrak{D}}$ is the generator of an evolution semigroup. To show that the closed operator $\mathcal{K} \in \mathcal{G}(1,0)$, we use estimates

$$\begin{aligned}&\frac{1}{\tau}\left\| f - \mathrm{e}^{-\xi\tau}\mathrm{e}^{-\tau D_0}\mathrm{e}^{-\tau\mathcal{A}} f\right\| \ge \\ &\frac{1}{\tau}\left|\, \|f\| - \mathrm{e}^{-\xi\tau}\left\|\mathrm{e}^{-\tau D_0}\mathrm{e}^{-\tau\mathcal{A}} f\right\|\,\right| \ge \frac{1}{\tau}\left(1 - \mathrm{e}^{-\xi\tau}\right)\|f\|, \quad \tau > 0\,,\end{aligned} \tag{2.1.57}$$

for $f \in \mathfrak{D}$ and $\xi > 0$. Then for the limit $\tau \to +0$, we obtain by (2.1.39) and (2.1.57)

$$\|D_0 f + \mathcal{A} f + \xi f\| = \|\widetilde{\mathcal{K}} f + \xi f\| \ge \xi \|f\|, \quad f \in \mathfrak{D}.$$

Since operator $\mathcal{K} = \overline{\widetilde{\mathcal{K}} \restriction \mathfrak{D}}$, we also obtain that $\|(\mathcal{K}+\xi)f\| \ge \xi\|f\|$, $f \in \mathrm{dom}\,(\mathcal{K})$. Hence, densely defined closed operator $\mathcal{K}$ is accretive:

$$\left\|(\mathcal{K} + \xi\,\mathcal{I})^{-1}\right\| \le \frac{1}{\xi}\,, \quad \xi > 0 \tag{2.1.58}$$

(see Definition 1.9.20), with the resolvent set $\rho(\mathcal{K})$ that contains $(-\infty, 0)$. Then by Proposition 1.1.1, one gets $\mathcal{K} \in \mathcal{G}(1,0)$.

Since $\mathfrak{D} = \mathrm{core}\,(\mathcal{K})$, we get by the Lie-Trotter product formula for C_0-semigroups, cf. Proposition 1.9.35 and (1.9.110), the limit

$$\lim_{n\to\infty} \sup_{\tau\in[0,T]} \left\|\mathrm{e}^{-\tau\mathcal{K}} f - \left(\mathrm{e}^{-\tau D_0/n}\mathrm{e}^{-\tau\mathcal{A}/n}\right)^n f\right\| = 0, \quad f \in C_0([0,T],\mathfrak{X})\,. \tag{2.1.59}$$

Note that by definitions of semigroup $\{e^{-\tau D_0}\}_{\tau\geq 0}$, (2.1.35) and semigroup $\{e^{-\tau\mathcal{A}}\}_{\tau\geq 0}$, (2.1.54), it follows that

$$(e^{-\tau D_0/n}\, e^{-\tau\mathcal{A}/n}\, f)(t) = \chi_{[0,T]}(t-\tau/n)\, e^{-\tau A(t-\tau/n)}\, f(t-\tau/n)\,. \tag{2.1.60}$$

Then n-times iteration of (2.1.60) yields

$$((e^{-\tau D_0/n}\, e^{-\tau\mathcal{A}/n})^n\, f)(t) = \prod_{j=1}^{n} e^{-\tau\, A(t-j\,\tau/n)}\, \chi_{[0,T]}(t-\tau)\, f(t-\tau)\,, \tag{2.1.61}$$

where the product in the right-hand side of (2.1.61) coincides with (2.1.56) after substitutions: $s = t-\tau$ and $k = n-j$.

Let $\{U(t,s)\}_{(t,s)\in\Delta}$ be solution operator which by formula (2.1.27) corresponds to generator $\mathcal{K}$. By virtue of (2.1.27) and (2.1.56), we obtain

$$\lim_{n\to\infty}\ \sup_{\tau\in[0,T]}\ \sup_{t\in[0,T]} \left\| (U(t,t-\tau) - U_n(t,t-\tau))\, \chi_{[0,T]}(t-\tau)\, f(t-\tau)\right\| = 0.$$

As a consequence, if $f(t) := (\phi(t)\otimes x)$, where $\phi\in C_0([0,T])$ and $x\in\mathfrak{X}$ (see Remark 2.2.3), then

$$\lim_{n\to\infty}\ \sup_{(t,s)\in\Delta} \phi(s)\, \|(U(t,s) - U_n(t,s))\,x\| = 0\,. \tag{2.1.62}$$

Let ϕ be such that $\phi(s) = 1$ for $s\in[\varepsilon,T]$. Then (2.1.62) yields (2.1.55). □

Note that assumption $A(t)\in\mathcal{G}(1,0)$, $t\in[0,T]$, is not always satisfied. To extend the result to $A(t)\in\mathcal{G}(M,\beta)$, $t\in[0,T]$, one has to introduce a *stability condition*. (Compare with the Trotter stability condition (1.9.113) and Definition 1.9.39 for renormalisable pairs $\{A,B\}$ of operators.)

Definition 2.1.36 The family $\{A(t)\}_{t\in[0,T]}$ of generators $A(t)\in\mathcal{G}(M,\beta)$ is called *stable* if there are constants $M\geq 1$ and $\beta\in\mathbb{R}$ such that for any sequences $\{t_k\}_{k=1}^n$, $0 < t_1\leq t_2\leq\ldots\leq t_n\leq T$ and $\{\tau_k\}_{k=1}^n$, $0\leq\tau_k$, the operator-norm estimate

$$\left\| e^{-\tau_n A(t_n)} e^{-\tau_{n-1} A(t_{n-1})}\cdots e^{-\tau_1 A(t_1)}\right\|_{\mathcal{L}(\mathfrak{X})} \leq M e^{\beta\,(\tau_n+\tau_{n-1}+\cdots+\tau_1)} \tag{2.1.63}$$

holds for any $n\in\mathbb{N}$.

Lemma 2.1.37 *Let $\{A(t)\}_{t\in[0,T]}$ a family of generators such that $A(t)\in\mathcal{G}(M,\beta)$ and the condition (2.1.53) is satisfied. The family $\{A(t)\}_{t\in[0,T]}$ is stable if and only if there are constants $M\geq 1$ and $\beta\in\mathbb{R}$ such that for any $n\in\mathbb{N}$ and any sequences $\{\sigma_k\}_{k=1}^n$, $0\leq\sigma_k$ and $\{\tau_k\}_{k=1}^n$, $0\leq\tau_k$, the estimate*

$$\left\| e^{-\sigma_n D_0} e^{-\tau_n\mathcal{A}} e^{-\sigma_{n-1} D_0} e^{-\tau_{n-1}\mathcal{A}}\cdots e^{-\sigma_1 D_0} e^{-\tau_1\mathcal{A}}\right\|_{\mathcal{L}(\mathfrak{X}_T^C)} \leq M e^{\beta\,(\tau_n+\tau_{n-1}+\cdots+\tau_1)}\,, \tag{2.1.64}$$

is valid in the operator-norm topology of bounded operators $\mathcal{L}(\mathfrak{X}_T^C)$ on the Banach space $\mathfrak{X}_T^C$ (2.1.22).

Proof By definition of D_0, we get

$$
\begin{aligned}
&\mathrm{e}^{-\sigma_n D_0}\mathrm{e}^{-\tau_n \mathcal{A}}\mathrm{e}^{-\sigma_{n-1} D_0}\mathrm{e}^{-\tau_{n-1}\mathcal{A}}\cdots\mathrm{e}^{-\sigma_1 D_0}\mathrm{e}^{-\tau_1\mathcal{A}} f)(t) = \\
&\qquad \chi_{[0,T]}(t-\sigma_n-\sigma_{n-1}-\cdots-\sigma_1)\mathrm{e}^{-\tau_n A(t-\sigma_n)}\mathrm{e}^{-\tau_{n-1}A(t-\sigma_n-\sigma_{n-1})}\times \\
&\qquad \times\cdots\mathrm{e}^{-\tau_1 A(t-\sigma_n-\sigma_{n-1}-\cdots-\sigma_1)} f(t-\sigma_n-\sigma_{n-1}-\cdots-\sigma_1)\,, \quad t\in[0,T],
\end{aligned}
$$

$f \in C_0([0,T], \mathfrak{X})$. Then by (2.1.64), we obtain

$$
\begin{aligned}
\sup_{t\in[\sigma_1+\cdots+\sigma_n,T]} &\Big\| \mathrm{e}^{-\tau_n A(t-\sigma_n)}\mathrm{e}^{-\tau_{n-1}A(t-\sigma_n-\sigma_{n-1})}\times \\
&\cdots\mathrm{e}^{-\tau_1 A(t-\sigma_n-\sigma_{n-1}-\cdots-\sigma_1)} f(t-\sigma_n-\sigma_{n-1}-\cdots-\sigma_1)\Big\| \\
&\le M\,\mathrm{e}^{\beta(\tau_1+\cdots\tau_n)}\sup_{t\in[0,T]}\|f(t)\|\,,
\end{aligned}
$$

which yields

$$
\left\|\mathrm{e}^{-\tau_n A(t-\sigma_n)}\mathrm{e}^{-\tau_{n-1}A(t-\sigma_n-\sigma_{n-1})}\cdots\mathrm{e}^{-\tau_1 A(t-\sigma_n-\sigma_{n-1}-\cdots-\sigma_1)}\right\|_{\mathcal{L}(\mathfrak{X})} \le M\mathrm{e}^{\beta(\tau_1+\cdots\tau_n)},
$$

for $t \in (\sigma_1 + \cdots \sigma_{n-1} + \sigma_n, T]$. Setting

$$
\begin{aligned}
t_n &:= t-\sigma_n, \\
t_{n-1} &:= t-\sigma_n-\sigma_{n-1} = t_n - \sigma_{n-1}, \\
&\;\vdots \\
t_1 &= t-\sigma_n-\sigma_{n-1}-\ldots-\sigma_1 = t_2-\sigma_1\,,
\end{aligned}
$$

we get $0 < t_1 \le \ldots \le t_{n-1} \ldots \le t_n$ and (2.1.63).

Conversely, assume that (2.1.63) is satisfied. Then setting

$$
\begin{aligned}
\sigma_n &:= t-t_n, \\
\sigma_{n-1} &:= t_n - t_{n-1}, \\
&\;\vdots \\
\sigma_1 &= t_2 - t_1\,,
\end{aligned}
$$

we get from (2.1.63) that

$$\begin{aligned}\sup_{t\in[\sigma_1+\cdots+\sigma_n,T]} & \Big\| \mathrm{e}^{-\tau_n A(t-\sigma_n)} \mathrm{e}^{-\tau_{n-1} A(t-\sigma_n-\sigma_{n-1})} \times \\ & \cdots \mathrm{e}^{-\tau_1 A(t-\sigma_n-\sigma_{n-1}-\cdots-\sigma_1)} f(t-\sigma_n-\sigma_{n-1}-\cdots-\sigma_1)\Big\| \\ & \le M\,\mathrm{e}^{\beta(\tau_1+\cdots\tau_n)} \sup_{t\in(\sigma_1+\cdots+\sigma_n,T]} \|f(t-\sigma_n-\sigma_{n-1}-\cdots-\sigma_1)\| \\ & \le M\,\mathrm{e}^{\beta(\tau_1+\cdots\tau_n)} \sup_{t\in[0,T]} \|f(t)\|. \end{aligned} \tag{2.1.65}$$

Therefore, (2.1.65) yields (2.1.64). □

Theorem 2.1.38 *Let* $\{A(t)\}_{t\in[0,T]}$ *be a family of generators* $A(t) \in \mathcal{G}(M, \beta)$, *and let condition* (2.1.53) *be satisfied. If the* $nACP$ (2.1.4) *is well-posed and the family* $\{A(t)\}_{t\in[0,T]}$ *is stable, then corresponding solution operator* $\{U(t,s)\}_{(t,s)\in\Delta}$, *which ensures a strict solution, admits the product approximation* (2.1.55).

Proof By Lemma 2.1.37, the condition (2.1.63) is equivalent to (2.1.64). Let $|||\cdot|||$ denote a (*renormalised*) norm on the Banach space $\mathfrak{X}_T^C = C_0([0,T],\mathfrak{X})$. For any $f \in \mathfrak{X}_T^C$, we set

$$\begin{aligned} |||f||| := & \sup_{\substack{\sigma_k\ge 0,\ \tau_k\ge 0\\ k=1,2,\ldots,n}} \mathrm{e}^{-\beta\sum_{k=1}^n \tau_k} \\ & \times \left\| \mathrm{e}^{-\sigma_n D_0} \mathrm{e}^{-\tau_n \mathcal{A}} \mathrm{e}^{-\sigma_{n-1} D_0} \mathrm{e}^{-\tau_{n-1}\mathcal{A}} \cdots \mathrm{e}^{-\sigma_1 D_0} \mathrm{e}^{-\tau_1 \mathcal{A}} f \right\|_{\mathcal{L}(\mathfrak{X}_T^C)} . \end{aligned}$$

Then inequality (2.1.64) yields the equivalence of the norms $\|\cdot\|$ and $|||\cdot|||$,

$$\|f\| \le |||f||| \le M\|f\|, \quad f \in C_0([0,T],\mathfrak{X}),$$

as well as the estimates (see Lemma 1.9.40)

$$\left|\left|\left| \mathrm{e}^{-\sigma D_0} f \right|\right|\right| \le |||f||| \quad \text{and} \quad \left|\left|\left| \mathrm{e}^{-\tau \mathcal{A}} f \right|\right|\right| \le |||f|||, \quad \sigma \ge 0,\ \ \tau \ge 0 .$$

Consequently, $D_0 \in \mathcal{G}(1,0)$ and $\mathcal{A} \in \mathcal{G}(1,0)$. Since the norms $\|\cdot\|$ and $|||\cdot|||$ are equivalent, we obtain that $\mathcal{K} \in \mathcal{G}(1,0)$. As a result,

$$\text{s-}\lim_{n\to\infty} \sup_{\tau\in[0,T]} \left|\left|\left| \mathrm{e}^{-\tau\mathcal{K}} f - \left(\mathrm{e}^{-\tau D_0/n} \mathrm{e}^{-\tau\mathcal{A}/n}\right)^n f \right|\right|\right| = 0, \quad f \in \mathfrak{X}_T^C,$$

which yields the limit (2.1.55). □

2.2 Linear Evolution Equations in L^p-Setting

2.2.1 *Evolution Semigroups and Their Generators in L^p-Setting*

It turns out that to handle nACP in the C-setting (Definition 2.1.16) is a too subtle and restraining problem. Note, for example, that evolution generator in Banach space $\mathfrak{X}_T^C = C_0([0,T],\mathfrak{X})$ is *unstable* even under bounded perturbations.

Example 2.2.1 Let

$$A(t) := \begin{cases} A & t \in [0, T/2)\,, \\ A + B & t \in [T/2, T]\,, \end{cases}$$

where A is the generator of a strongly continuous semigroup on $\mathfrak{X}$ and B is bounded on $\mathfrak{X}$ operator such that $\ker(B) = \{0\}$. Then multiplication operator $\mathcal{A}$ associated with the family $\{A(t)\}_{t\in[0,T]}$ in Banach space $\mathfrak{X}_T^C$ with norm (2.1.22) is not densely defined. Indeed, let $f \in \mathrm{dom}(\mathcal{A})$. Then since $\ker(B) = \{0\}$, the vector $\mathcal{A}f \in \mathfrak{X}_T^C$ if and only if $f(T/2) = 0$. But the set $\{f \in \mathfrak{X}_T^C : f(T/2) = 0\}$ is not dense in the Banach space $\mathfrak{X}_T^C$.

So, it makes sense to look for a larger setting. A possible candidate is the so-called L^p-*setting* that replaces Banach space $C_0([0,T],\mathfrak{X})$ by the Banach space $\mathfrak{X}_T^p := L^p([0,T],\mathfrak{X})$ of p-summable functions on the interval $[0,T]$ with values in $\mathfrak{X}$, endowed with the norm

$$\|f\|_{\mathfrak{X}_T^p} := \left(\int_0^T dt\, \|f(t)\|_{\mathfrak{X}}^p\right)^{1/p}, \quad \text{for } f \in \mathfrak{X}\,. \tag{2.2.1}$$

Here, $\|\cdot\|_{\mathfrak{X}}$ denotes the norm on the Banach space $\mathfrak{X}$, for which (as above) we systematically use (if there is no confusion) the notation $\|\cdot\|$. We denote by $\|\cdot\|_{\mathcal{L}(\mathfrak{X}_T^p)}$ the operator norm

$$\|C\|_{\mathcal{L}(\mathfrak{X}_T^p)} := \sup_{\{f\in\mathfrak{X}_T^p:\, \|f\|_{\mathfrak{X}_T^p}=1\}} \|Cf\|_{\mathfrak{X}_T^p}\,, \quad \text{for } C \in \mathcal{L}(\mathfrak{X}_T^p)\,, \tag{2.2.2}$$

of bounded on $\mathfrak{X}_T^p$ set of operators $\mathcal{L}(\mathfrak{X}_T^p)$.

Similar to (2.1.24), any strongly continuous on $\mathfrak{X}$ solution operator $\{U(t,s)\}_{(t,s)\in\Delta}$ induces on the Banach space $\mathfrak{X}_T^p$ by the *Howland-Evans-Neidhardt* (HEN) formula a strongly continuous nilpotent semigroup $\{\mathcal{U}_p(\tau)\}_{\tau\in\mathbb{R}_0^+}$

$$(\mathcal{U}_p(\tau)f)(t) := U(t, t-\tau)\, \chi_{[0,T]}(t-\tau) f(t-\tau), \quad t \in [0,T], \quad p \in [1,\infty), \tag{2.2.3}$$

for any $f \in \mathfrak{X}_T^p$ (see Corollary F.1.2 in Appendices F.1 and F.2).

Definition 2.2.2 A strongly continuous semigroup $\{\mathcal{U}_p(\tau)\}_{\tau\in\mathbb{R}_0^+}$ defined on Banach space $\mathfrak{X}_T^p = L^p([0,T],\mathfrak{X})$, $p \in [1,\infty)$, is called *evolution semigroup* if there is solution operator $\{U(t,s)\}_{(t,s)\in\Delta}$ (Definition 2.1.5) such that the HEN representation (2.2.3) is valid *almost everywhere* (a.e.) on $[0,T]$.

Remark 2.2.3 We note that solution operator $\{U(t,s)\}_{(t,s)\in\Delta}$ can be obtained from the action of the evolution semigroup $\{\mathcal{U}_p(\tau)\}_{\tau\in\mathbb{R}_0^+}$ on the linear space $\mathfrak{X}_{T\otimes}$ of *constant* functions. Indeed, by formula (2.2.3), one gets

$$U(t,s)\,x = (\mathcal{U}_p(t-s)\,(\mathbb{1}_T\otimes x))(t)\,, \quad 0\le s\le t\le T, \quad x\in\mathfrak{X}\,, \tag{2.2.4}$$

where $t\mapsto \mathbb{1}_T(t) := \chi_{[0,T]}(t)$ and vector $(\mathbb{1}_T\otimes x)(t) := \chi_{[0,T]}(t)\,x$ belongs to the space

$$\mathfrak{X}_{T\otimes} = \{f\in\mathfrak{X}_T^p : (z\,\mathbb{1}_T\otimes x),\ \ z\in\mathbb{C},\ \ x\in\mathfrak{X}\}\,. \tag{2.2.5}$$

We also note that (2.2.5) is a subspace of

$$\mathfrak{X}_{T\otimes}^p = \{f\in\mathfrak{X}_T^p : f = (\varphi\otimes x),\ \ \varphi\in L^p([0,T]),\ \ x\in\mathfrak{X}\}\,. \tag{2.2.6}$$

which is, in turn, a subspace of $\mathfrak{X}_T^p = L^p([0,T],\mathfrak{X})$. Here, $(\varphi\otimes x)(t) := \varphi(t)\,x$.

Let $\phi(\cdot)\in L^\infty([0,T])$. Then similar to C-setting, we introduce first a *scalar* multiplication operator $\mathcal{M}_p(\phi)$:

$$(\mathcal{M}_p(\phi)f)(t) := \phi(t)f(t), \quad t\in[0,T], \quad f\in\mathfrak{X}_T^p, \quad p\in[1,\infty). \tag{2.2.7}$$

Obviously, $\mathcal{M}_p(\phi)$ is a bounded operator on $\mathfrak{X}_T^p$ for each $\phi\in L^\infty([0,T])$. Similar to C-setting (cf. Definition 2.1.23, Proposition 2.1.24), the scalar multiplication operators (2.2.7) allow, *mutatis mutandis*, the extension to Banach space $L^p([0,T],\mathfrak{X})$ of the notion of *general* bounded multiplication operator.

We note that Banach space $C_0([0,T],\mathfrak{X})$ can be regarded as *continuously* and *densely* embedded in Banach space $L^p([0,T],\mathfrak{X})$:

$$\begin{aligned}&\iota_p : C_0([0,T],\mathfrak{X}) \longrightarrow L^p([0,T],\mathfrak{X})\,, \quad p\in[1,\infty)\,,\\ &\|f\|_{\mathfrak{X}_T^p} \le T^{1/p}\,\|f\|_{\mathfrak{X}_T^C}\,, \quad f\in\mathfrak{X}_T^C\,,\end{aligned} \tag{2.2.8}$$

Here, ι_p is the *embedding* operator for $C_0([0,T],\mathfrak{X}) \overset{d}{\hookrightarrow} L^p([0,T],\mathfrak{X})$. Then $\ker(\iota_p) = \{0\}$ and also

$$\mathcal{M}_p(\phi)\,\iota_p = \iota_p\,\mathcal{M}(\phi), \quad \text{for } \phi\in C([0,T])\,, \ \ p\in[1,\infty)\,. \tag{2.2.9}$$

Compare Definition (2.1.25) of $\mathcal{M}(\phi)$ in the C-setting.

The embedding (2.2.8) yields a plain indication concerning the contact of the L^p-setting with the C-setting developed in preceding Sect. 2.1. The following assertion justifies this approach to the L^p-setting.

Theorem 2.2.4 *For every strongly continuous evolution semigroup* $\{\mathcal{U}(\tau)\}_{\tau\in\mathbb{R}_0^+}$ *(see* (2.1.24)*) on the Banach space* $C_0([0,T],\mathfrak{X})$ *there exists a unique strongly continuous evolution semigroup* $\{\mathcal{U}_p(\tau)\}_{\tau\in\mathbb{R}_0^+}$ *on the Banach space* $L^p([0,T],\mathfrak{X})$ *with the property*

$$(\mathcal{U}_p(\tau)\,f)(t) = (\mathcal{U}(\tau)\,f)(t)\,, \quad f \in C_0([0,T],\mathfrak{X})\,, \quad \text{for a.a. } t \in [0,T]\,. \tag{2.2.10}$$

For more details and for the proof of Theorem 2.2.4, we refer to Appendix F.2, Theorem F.2.1.

Note that the L^p-setting admits a natural extension from $\mathfrak{X}_T^C = C_0([0,T],\mathfrak{X})$ to the Banach space $\mathfrak{X}_T^p = L^p([0,T],\mathfrak{X})$ of the concept of a dense *cross-section* $[\mathfrak{D}]_{\mathfrak{X}_T^p}$, cf. Definition 2.1.29. We also denote the cross-section of $\mathfrak{D}$ in $\mathfrak{X}$ at $t \in [0,T]$ by $[\mathfrak{D}]_t$.

Definition 2.2.5 We say that *cross-section* $[\mathfrak{D}]_{\mathfrak{X}_T^p}$ of the set $\mathfrak{D} \subset \mathfrak{X}_T^p$ is dense in the space $\mathfrak{X}$ for some $p \in [1,\infty]$ if

(a) $\mathfrak{D} \subseteq \mathfrak{X}_T^p \cap \mathfrak{X}_T^C$.
(b) $\mathfrak{D}$, *viewed* as a subset of $\mathfrak{X}_T^C$, that is, the set $\mathfrak{D} \cap \mathfrak{X}_T^C$, has a *dense* cross-section $[\mathfrak{D} \cap \mathfrak{X}_T^C]_{\mathfrak{X}_T^C}$ in the sense of Definition 2.1.29.

Similar to the C-setting (cf. Lemma 2.1.31), the following assertion holds also in the L^p-setting but only in one direction.

Lemma 2.2.6 *Let* $\mathfrak{D} \subseteq \mathfrak{X}_T^p \cap \mathfrak{X}_T^C$, $p \in [1,\infty)$ *be a linear set such that* $\mathcal{M}_p(\phi)\mathfrak{D} \subseteq \mathfrak{D}$ *holds for every* ϕ *from the Sobolev space* $W^{1,\infty}([0,T])$*. If domain* $\mathfrak{D}$ *has a dense cross-section* $[\mathfrak{D}]_{\mathfrak{X}_T^p}$ *in* $\mathfrak{X}$*, then* $\mathfrak{D}$ *is a dense set in Banach space* $\mathfrak{X}_T^p$*.*

Proof Let $\mu \in \mathfrak{X}_T^{p*}$ be any bounded linear functional on $\mathfrak{X}_T^p$ such that

$$\langle f, \mu\rangle = 0, \quad f \in \mathfrak{D}.$$

We use now characterisation of the dual space $\mathfrak{X}_T^{p*}$ from Proposition E.2.7. Hence, there is a w^*-measurable function $\Psi : [0,T] \to \mathfrak{X}^*$ such that

$$0 = \langle f, \mu\rangle = \int_{[0,T]} \langle f(t), \Psi(t)\rangle\, dt, \text{ for } f \in \mathfrak{D}. \tag{2.2.11}$$

Since $\mathcal{M}_p(\phi)\mathfrak{D} \subset \mathfrak{D}$, for $\phi \in W^{1,\infty}([0,T])$, it follows that

$$0 = \int_{[0,T]} dt\, \langle \phi(t) f(t), \Psi(t)\rangle = \int_{[0,T]} dt\, \phi(t)\, \langle f(t), \Psi(t)\rangle\,.$$

The function $t \mapsto \langle f(t), \phi(t)\rangle$ is in $L^1([0,T])$. Since $\phi \in W^{1,\infty}([0,T])$ is arbitrary, we conclude that $0 = \langle f(t), \Psi(t)\rangle$, for a.a. $t \in [0,T]$. By condition, $\mathfrak{D}$ has a dense cross-section in $\mathfrak{X}$. Then on account of Definition 2.2.5, this implies $\Psi(t) = 0$ for a.a. $t \in [0,T]$, and by (2.2.11), it follows that $\mu = 0$. □

Remark 2.2.7

(a) In contrast to the C-setting (see Remark 2.1.30 and Lemma 2.1.31), the converse of Lemma 2.2.6 is not true. That is, let $\mathfrak{D} \subseteq \mathfrak{X}_T^p \cap \mathfrak{X}_T^C$, $p \in [1,\infty)$ be a linear set such that $\mathcal{M}_p(\phi)\mathfrak{D} \subseteq \mathfrak{D}$ for each $\phi \in W^{1,\infty}([0,T])$. If $\mathfrak{D}$ is dense in $\mathfrak{X}_T^p$, then in general, it does not have dense cross-section $[\mathfrak{D}]_{\mathfrak{X}_T^p}$.

(b) If by condition of Lemma 2.2.6 the cross-section $[\mathfrak{D}]_{\mathfrak{X}_T^p}$ is dense, then by Definition 2.2.5 (b), the set $\mathfrak{D}$ (considered as a subset of $\mathfrak{X}_T^C$, that is, as the set $\mathfrak{D} \cap \mathfrak{X}_T^C$) has dense cross-section $[\mathfrak{D} \cap \mathfrak{X}_T^C]_{\mathfrak{X}_T^C}$ and, for that reason, satisfies conditions of Lemma 2.1.31. As a consequence, the set $\mathfrak{D} \cap \mathfrak{X}_T^C$ is dense in the Banach space $\mathfrak{X}_T^C$.

Similar to the C-setting, it is useful to introduce within the L^p-setting a notion of the evolution pre-generator, which is slightly modified for taking into account Remark 2.2.7 (a).

Definition 2.2.8 Operator $\mathcal{K}_p$ in $\mathfrak{X}_T^p = L^p([0,T],\mathfrak{X})$ is called the *evolution pre-generator* for $p \in [1,\infty)$ if

(i) $\mathrm{dom}\,(\mathcal{K}_p) \subset \mathfrak{X}_T^p \cap \mathfrak{X}_T^C$.

(ii) $\mathcal{M}_p(\phi)(\mathrm{dom}\,(\mathcal{K}_p)) \subset \mathrm{dom}\,(\mathcal{K}_p)$ for $\phi \in W^{1,\infty}([0,T])$ and

$$\mathcal{K}_p\mathcal{M}_p(\phi)f - \mathcal{M}_p(\phi)\mathcal{K}_p f = \mathcal{M}_p(\phi')f, \quad f \in \mathrm{dom}\,(\mathcal{K}_p),$$

where $\phi' = \partial_t\phi$.

(iii) $\mathrm{dom}\,(\mathcal{K}_p)$ has a dense cross-section $[\mathrm{dom}\,(\mathcal{K}_p)]_{\mathfrak{X}_T^p}$ in $\mathfrak{X}$ (see Definition 2.2.5).

Remark 2.2.9

(a) Note that $\mathrm{dom}\,(\mathcal{K}_p)$ of evolution pre-generator is always dense in $\mathfrak{X}_T^p$. Indeed, this claim follows from Definition 2.2.8 and Lemma 2.2.6. So, conditions (i)–(iii) are stronger than those in Definition 2.1.16 of evolution pre-generator in the C-setting.

(b) However, *a priori*, it is unclear whether the closure of evolution pre-generator $\mathcal{K}_p$ exists, as well as whether it gives an evolution pre-generator or evolution generator.

(c) In the C-setting, this question is solved by Theorems 2.1.17 and 2.1.18, and by arguments in Appendix F, Sect. F.1, which prove *uniqueness* of the evolution semigroup in this setting. In the L^p-setting, this result is valid only with supplementary condition (iii) in Definition 2.2.8. Note that *a priori* the evolution semigroups $\{\mathcal{U}_p(\tau)\}_{\tau\in\mathbb{R}_0^+}$, defined by extension (2.2.10) and that defined below by evolution generator from Theorem 2.2.10 together with the

HEN representation (2.2.12), are *different* (see comments in Sects. F.1 and F.2 of Appendix F).

We continue denoting the evolution semigroup on $\mathfrak{X}_T^p$ by the unique $\{\mathcal{U}_p(\tau)\}_{\tau\in\mathbb{R}_0^+}$ as far as it does not produce any confusion, cf. Corollaries F.2.7 and F.2.8.

(d) Besides a general criterion for *extension* of evolution pre-generator to evolution generator and for its *uniqueness* (see Theorems 2.2.10 and 2.2.11 and 2.2.22), we shall study in Sects. 11.2 and 11.3 some special case of evolution equations.

(e) We also note that by the same token as in C-setting (see Definition 2.1.16 and Theorem 2.1.17), we shall prove in Theorem 2.2.10 that evolution pre-generator $\mathcal{K}_p$ is, in fact, an evolution generator if $\mathcal{K}_p$ generates on $\mathfrak{X}_T^p$ a C_0-semigroup. On the other hand, an important difference is that in the L^p-setting, we add in Definition 2.2.8 of evolution pre-generator $\mathcal{K}_p$ the condition (iii), which ensures the *one-to-one* correspondence of evolution semigroups and solution operators in the *Howland-Evans-Neidhardt* representation (formula) (see Theorem 2.2.11) and finally a desirable uniqueness.

Owing to Definition 2.2.8, we obtain the following criterion for generators of evolution semigroups (i.e. for evolution generators) in $\mathfrak{X}_T^p$, cf. Theorem 2.1.17 for the C-setting.

Theorem 2.2.10 *Let operator $\mathcal{K}_p$ be evolution pre-generator in Banach space $\mathfrak{X}_T^p$. Then $\mathcal{K}_p$ is evolution generator of semigroup $\{\mathcal{U}_p(\tau)\}_{\tau\in\mathbb{R}_0^+}$ on $\mathfrak{X}_T^p$ if and only if operator $\mathcal{K}_p$ is generator of a strongly continuous on $\mathfrak{X}_T^p$ semigroup, that is, $\{\mathcal{U}_p(\tau) = \mathrm{e}^{-\tau\,\mathcal{K}_p}\}_{\tau\in\mathbb{R}_0^+}$.*

For the proof, see Appendix F.2 (Sect. F.2.3), as well as Notes to Chap. 2 (Notes to Sects. 2.2, 2.2.1) for more comments and remarks.

To formulate the next assertion, we recall that evolution semigroup $\{\mathcal{U}_p(\tau)\}_{\tau\in\mathbb{R}_0^+}$ has the HEN representation (2.2.3) for strongly continuous on $\mathfrak{X}$ solution operator $\{U(t,s)\}_{(t,s)\in\Delta}$ of the nACP (2.1.12).

Theorem 2.2.11 *We claim that between the set of all quasi-bounded evolution semigroups $\{\mathcal{U}_p(\tau) = \mathrm{e}^{-\tau\mathcal{K}_p}\}_{\tau\in\mathbb{R}_0^+}$ on $\mathfrak{X}_T^p$ (equivalently, all evolution generators $\mathcal{K}_p \in \mathcal{G}(\mathfrak{X}_T^p, M, \beta)$ defined by Theorem* 2.2.10) *and the set of all strongly continuous quasi-bounded solution operators $\{U(t,s)\}_{(t,s)\in\Delta}$ on $\mathfrak{X} : \|U(t,s)\| \leq M\,\mathrm{e}^{\beta(t-s)}$, $M \geq 1$, $\beta \geq 0$ (Definition* 2.1.4 *and Remark* 2.1.14), *there exists a one-to-one correspondence such that the Howland-Evans-Neidhardt representation*

$$(\mathrm{e}^{-\tau\mathcal{K}_p}\, f)(t) = U(t,t-\tau)\;\chi_{[0,T]}(t-\tau) f(t-\tau)\,, \quad \tau \geq 0\,, \tag{2.2.12}$$

holds for all $f \in \mathfrak{X}_T^p$ and for almost all (a.a.) $t \in [0,T]$. *Besides, the semigroup $\{\mathcal{U}_p(\tau) = \mathrm{e}^{-\tau\mathcal{K}_p}\}_{\tau\in\mathbb{R}_0^+}$ is nilpotent on the space $\mathfrak{X}_T^p$, that is, $\mathcal{U}_p(\tau)|_{\tau>T} =$*

0. *Therefore, by definition* (2.2.1) *and representation* (2.2.12) *for quasi-bounded solution operator, one gets estimate*

$$\| e^{-\tau \mathcal{K}_p} f \|_{\mathfrak{X}_T^p} \le M\, e^{\beta T} \, \| f \|_{\mathfrak{X}_T^p} , \quad \tau \ge 0 .$$

As a consequence of the nilpotence on $\mathfrak{X}_T^p$, the evolution semigroup is bounded,

$$\| e^{-\tau \mathcal{K}_p} \|_{\mathcal{L}(\mathfrak{X}_T^p)} \le M_T , \quad \tau \ge 0 , \tag{2.2.13}$$

where $M_T := M\, e^{\beta T}$.

For the proof, see Appendix F.2 (Sect. F.2.3) and Notes to Chap. 2 (Notes to Sects. 2.2, 2.2.1) for comments.

Corollary 2.2.12 *Let $\mathcal{K}_p$ be generator of evolution semigroup $\{\mathcal{U}_p(\tau) = e^{-\tau \mathcal{K}_p}\}_{\tau \in \mathbb{R}_0^+}$ on $\mathfrak{X}_T^p$, and let $\{U(t,s)\}_{(t,s)\in \Delta}$ be the corresponding solution operator. Then by formula* (2.2.12) *and quasi-boundedness $\|U(t,s)\| \le M\, e^{\beta(t-s)}$, one gets for $f \in \mathfrak{X}_T^p$* a.a. $t \in [0,T]$ *and for any $\xi > \beta$ the following representation:*

$$\begin{aligned} ((\mathcal{K}_p + \xi\, \mathcal{I}_p)^{-1} f)(t) &= \int_0^\infty d\tau\, e^{-\xi \tau}\, U(t, t-\tau)\, \chi_{[0,T]}(t-\tau) f(t-\tau) \\ &= \int_0^t ds\, e^{-\xi (t-s)}\, U(t,s)\, f(s) , \end{aligned} \tag{2.2.14}$$

cf. (F.2.23) *in Appendix* F.2 (*Sect.* F.2.3). *As a consequence, the representation* (2.2.14) *together with definition* (2.2.1) *and with bound* : $\|U(t,s)\| \le M\, e^{\beta(t-s)}$ *by the Hölder inequality provide estimates*

$$\begin{aligned} &\big\| (\mathcal{K}_p + \xi\, \mathcal{I}_p)^{-1} f(t) \big\|_{\mathfrak{X}} \le \int_0^\infty d\tau\, e^{-(\xi-\beta)\tau}\, M\, \chi_{[0,T]}(t-\tau)\, \| f(t-\tau) \|_{\mathfrak{X}} \le \\ &\le \int_0^\infty d\tau\, \left(e^{-q(\xi-\beta)\tau}\, M^q \right)^{1/q} \chi_{[0,T]}(t-\tau) \left(\big\| f(t-\tau) \big\|_{\mathfrak{X}}^p \right)^{1/p} \le \\ &\le \left(\int_0^\infty d\tau\, e^{-q(\xi-\beta)\tau}\, M^q \right)^{1/q} \left(\int_0^\infty d\tau\, \chi_{[0,T]}(t-\tau)\, \big\| f(t-\tau) \big\|_{\mathfrak{X}}^p \right)^{1/p} \le \\ &\le \frac{M}{(q(\xi-\beta))^{1/q}} \left(\int_0^t ds\, \big\| f(s) \big\|_{\mathfrak{X}}^p \right)^{1/p} \le \frac{M}{((\xi-\beta)\, q)^{1/q}} \, \| f \|_{\mathfrak{X}_T^p} , \end{aligned} \tag{2.2.15}$$

for $\{q : 1/p + 1/q = 1,\ p \ge 1\}$ and for any $f \in \mathfrak{X}_T^p$.

Note that on account of representation (2.2.14) *one has*

$$(\mathcal{K}_p + \xi\, \mathcal{I}_p)^{-1} : \mathfrak{X}_T^C \to \mathfrak{X}_T^C , \quad \xi > \beta .$$

Then inequality (2.2.15) *yields estimate*

$$\begin{aligned} \left\|(\mathcal{K}_p+\xi\,\mathcal{I}_p)^{-1}\,f\right\|_{\mathfrak{X}_T^C} &= \sup_{t\in[0,T]}\left\|(\mathcal{K}_p+\xi\,\mathcal{I}_p)^{-1}\,f(t)\right\|_{\mathfrak{X}} \le \\ &\le \frac{M}{(q\,(\xi-\beta))^{1/q}}\,\left\|f\right\|_{\mathfrak{X}_T^p}, \quad f\in\mathfrak{X}_T^C\,. \end{aligned} \tag{2.2.16}$$

We also mention another argument that uses representation (2.2.14). The Hölder inequality and the fact that evolution semigroup $\{\mathcal{U}_p(\tau)=\mathrm{e}^{-\tau\mathcal{K}_p}\}_{\tau\in\mathbb{R}_0^+}$ on the space $\mathfrak{X}_T^p$ is *nilpotent* provide estimates

$$\begin{aligned} \left\|(\mathcal{K}_p+\xi\,\mathcal{I}_p)^{-1}\,f(t)\right\|_{\mathfrak{X}} &\le \int_0^T d\tau\ \mathrm{e}^{(|\xi|+\beta)\,\tau}\,M\,\left\|f(\tau)\right\|_{\mathfrak{X}} \le \\ &\le \left(\int_0^T d\tau\,(\mathrm{e}^{(|\xi|+\beta)\,\tau}\,M)^q\right)^{1/q}\left(\int_0^T d\tau\,\left\|f(\tau)\right\|_{\mathfrak{X}}^p\right)^{1/p} \le \\ &\le T^{1/q}\,M\,\mathrm{e}^{(|\xi|+\beta)\,T}\,\left\|f\right\|_{\mathfrak{X}_T^p}, \quad 1/q+1/p=1\,, \quad \xi\in\mathbb{C}\,. \end{aligned} \tag{2.2.17}$$

Then (2.2.1) and (2.2.17) yield

$$\begin{aligned} \sup_{\{f\in\mathfrak{X}_T^p:\ \|f\|_{\mathfrak{X}_T^p}=1\}} &\left\|(\mathcal{K}_p+\xi\,\mathcal{I}_p)^{-1}\,f\right\|_{\mathfrak{X}_T^p} = \\ &= \left\|(\mathcal{K}_p+\xi\,\mathcal{I}_p)^{-1}\right\|_{\mathcal{L}(\mathfrak{X}_T^p)} \le T\,M\,\mathrm{e}^{(|\xi|+\beta)\,T}\,. \end{aligned} \tag{2.2.18}$$

This means that resolvent of operator $\mathcal{K}_p$ is bounded on the space $\mathfrak{X}_T^p$ for all $\xi\in\mathbb{C}$. That is, generator of (nilpotent) semigroup $\{\mathcal{U}_p(\tau)\}_{\tau\in\mathbb{R}_0^+}$ is invertible and the resolvent set $\rho(\mathcal{K}_p)=\mathbb{C}$, cf. the case of C-setting (2.1.28)–(2.1.30).

2.2.2 Cauchy Problem and Evolution Semigroups in L^p-Setting

Similar to the C-setting, an important example of the evolution generator in the Banach space $\mathfrak{X}_T^p:=L^p([0,T],\mathfrak{X})$ for $p\in[1,\infty)$ is the *differentiation* operator

$$(D_{0p}f)(t):=\partial_t f(t), \quad \mathrm{dom}\,(D_{0p}):=\{f\in W^{1,p}([0,T],\mathfrak{X}) : f(0)=0\}. \tag{2.2.19}$$

in the *Sobolev* space $W^{1,p}([0,T],\mathfrak{X})$ of the vector-valued *Bochner* p-integrable mappings from $[0,T]$ to $\mathfrak{X}$.

Remark 2.2.13 Note that on the space $\mathfrak{X}_T^p$ the operator (2.2.19) is plainly a generator of strongly continuous *nilpotent* right-shift *contraction* semigroup $\{\mathcal{S}_R(\tau) := e^{-\tau D_{0p}}\}_{\tau \geq 0}$ of automorphisms $\mathcal{S}_R(\tau) : \mathfrak{X}_T^p \to \mathfrak{X}_T^p$, such that

$$(\mathcal{S}_R(\tau) f)(t) = \chi_{[0,T]}(t - \tau) f(t - \tau), \quad t \in [0, T], \quad f \in \mathfrak{X}_T^p. \tag{2.2.20}$$

Indeed, according to (2.2.12) and (2.2.20), the solution operator corresponding to this shift-semigroup is *identity* $U(t, s) = \mathbb{1}$, for $0 < s \leq t \leq T$, and $e^{-\tau D_{0p}} = 0$, for $\tau > T$. As a consequence, $D_{0p} \in \mathcal{G}(1, 0)$. Similar to the result of Corollary 2.2.12, the operator D_{0p} is *invertible* since the resolvent set $\rho(D_{0p}) = \mathbb{C}$.

Definition 2.2.14 With a given operator family $\{A(t)\}_{t\in[0,T]}$ in $\mathfrak{X}$, we may associate induced multiplication operator $\mathcal{A}_p$, $p \in [1, \infty]$, in $\mathfrak{X}_T^p$ (cf. Sect. 2.1.2 (Definition 2.1.21)), which is defined by

$$\begin{aligned} &(\mathcal{A}_p f)(t) := A(t) f(t), \quad f \in \operatorname{dom}(\mathcal{A}_p), \quad \text{for a.a. } t \in (0, T], \\ &\operatorname{dom}(\mathcal{A}_p) := \{f \in \mathfrak{X}_T^p : f(t) \in \operatorname{dom}(A(t)),\ t \mapsto A(t) f(t) \in \mathfrak{X}_T^p\}. \end{aligned} \tag{2.2.21}$$

Then similar to Lemma 2.1.22 for C-setting, the following assertion holds:

Lemma 2.2.15 *If $\{A(t)\}_{t\in[0,T]}$ is a family of closed linear operators in $\mathfrak{X}$, then the induced multiplication operator $\mathcal{A}_p$ in $\mathfrak{X}_T^p$, $p \in [1, \infty]$, is also closed.*

Proof Let the sequence $\{f_n\}_{n\geq 1} \subset \operatorname{dom}(\mathcal{A}_p)$ be such that limits $\lim_{n\to\infty} f_n = f$ and $\lim_{n\to\infty} \mathcal{A}_p f_n = g$ exist in the $\mathfrak{X}_T^p$-topology (2.2.1). This implies that by a diagonal procedure one can find a subsequence $\{f_{n_k}\}_{k\geq 1}$ such that $\lim_{n_k\to\infty} f_{n_k}(t) = f(t)$ and $\lim_{n_k\to\infty} (\mathcal{A}_p f_{n_k})(t) = g(t) \in \mathfrak{X}$ for a.a. $t \in [0, T]$. Since for any $t \in [0, T]$ the operator $A(t)$ is closed in $\mathfrak{X}$, we conclude that $f(t) \in \operatorname{dom}(A(t))$ and $A(t) f(t) = g(t)$ for almost all $t \in [0, T]$. On the other hand, since $g \in \mathfrak{X}_T^p$, it follows that $f \in \operatorname{dom}(\mathcal{A}_p)$ and that $(\mathcal{A}_p f)(t) = g(t)$ almost everywhere in $[0, T]$. The latter proves that operator $\mathcal{A}_p$ is closed. □

Now, it is worth noting that if $\{A(t)\}_{t\in[0,T]}$ is a family of closed densely defined linear operators in $\mathfrak{X}$, then the induced closed multiplication operator $\mathcal{A}_p$ in $\mathfrak{X}_T^p$, $p \in [1, \infty]$, is *not* necessarily densely defined in the space $\mathfrak{X}_T^p$. For that reason, $\operatorname{dom}(\mathcal{A}_p)$ must meet some additional conditions (cf. Lemma 2.2.6). To this aim, one can use, for example, the C-setting in $\mathfrak{X}_T^C$, together with embedding (2.2.8), that is, $\iota_p : \operatorname{dom}(\mathcal{A}) \to \operatorname{dom}(\mathcal{A}_p)$ and

$$\iota_p \mathcal{A} f = \mathcal{A}_p \iota_p f, \quad f \in \operatorname{dom}(\mathcal{A}) \subset \mathfrak{X}_T^C, \quad p \in [1, \infty),$$

for the cross-section analysis of $\operatorname{dom}(\mathcal{A}_p)$ in $\mathfrak{X}$ (see Definition 2.2.5).

For bounded operators $\mathcal{L}(\mathfrak{X}_T^p)$ in the L^p-setting, we have a statement, which is similar to Proposition 2.1.24 in the C-setting.

Proposition 2.2.16 *Let $\{C(t)\}_{t\in[0,T]} \subset \mathcal{L}(\mathfrak{X})$ be a strongly measurable family of bounded linear operators on $\mathfrak{X}$, i.e. such that function $t \mapsto C(t)x$ is measurable for each $x \in \mathfrak{X}$. If $\operatorname{ess\,sup}_{t\in[0,T]} \|C(t)\|_{\mathcal{L}(\mathfrak{X})} < \infty$, then the induced multiplication operator $\mathcal{C}_p$ on the Banach space $\mathfrak{X}_T^p$ is also bounded and one has the norm equality*

$$\left\|\mathcal{C}_p\right\|_{\mathcal{L}(\mathfrak{X}_T^p)} = \operatorname*{ess\,sup}_{t\in[0,T]} \left\|C(t)\right\|_{\mathcal{L}(\mathfrak{X})}, \quad p \in [1,\infty), \tag{2.2.22}$$

where $\|\cdot\|_{\mathcal{L}(\mathfrak{X}_T^p)}$ denotes the operator norm on the Banach space $\mathcal{L}(\mathfrak{X}_T^p)$.

For comments and proof, we refer to Theorem E.3.12 in Appendix E.3.

Similar to the C-setting (Theorem 2.1.32) for the C_0-semigroup generators, which correspond to induced multiplication operators in $\mathfrak{X}_T^p = L^p([0,T],\mathfrak{X})$, we have the following statement:

Theorem 2.2.17 *Let closed operators $\{A(t)\}_{t\in[0,T]} \subset \mathcal{C}(\mathfrak{X})$ be a family of generators in $\mathfrak{X}$ such that $A(t) \in \mathcal{G}(M,\beta)$, $M \geq 1$, $\beta \in \mathbb{R}$ for a.a. $t \in [0,T]$. If the function $[0,T] \ni t \mapsto (A(t) + \xi\,\mathbb{1})^{-1} x \in \mathfrak{X}$ is strongly measurable for $\xi > \beta$, $x \in \mathfrak{X}$, then for each $p \in [1,\infty)$ the induced multiplication operator $\mathcal{A}_p$ in $\mathfrak{X}_T^p$ is generator of a quasi-bounded C_0-semigroup such that $\mathcal{A}_p \in \mathcal{G}(M,\beta)$ and*

$$(\mathrm{e}^{-\tau\mathcal{A}_p} f)(t) = \mathrm{e}^{-\tau A(t)} f(t) \quad \text{for } f \in \mathfrak{X}_T^p \text{ and } \text{ a.a. } t \in [0,T].$$

Moreover, $\|\,\mathrm{e}^{-\tau\mathcal{A}_p}\|_{\mathcal{L}(\mathfrak{X}_T^p)} = \operatorname{ess\,sup}_{t\in[0,T]} \|\mathrm{e}^{-\tau A(t)}\|_{\mathcal{L}(\mathfrak{X})}$ *for $\tau \in \mathbb{R}_0^+$.*

Proof Since Lemma 2.2.15 ensures that operator $\mathcal{A}_p$ is closed, we continue by proving that this operator is densely defined in $\mathfrak{X}_T^p$.

Let $\mathfrak{I} \subset [0,T]$ be a Borel set, $\xi > \beta$ and $x \in \mathfrak{X}$. We set

$$f_{\mathfrak{I},x,\xi}(t) := \xi\,(A(t) + \xi\,\mathbb{1})^{-1}\,\chi_{\mathfrak{I}}(t)\,x, \quad t \in [0,T]. \tag{2.2.23}$$

Then (2.2.23) yields that $f_{\mathfrak{I},x,\xi}(t) \in \operatorname{dom}(A(t))$, for a.a. $t \in [0,T]$, and that also $f_{\mathfrak{I},x,\xi} \in \mathfrak{X}_T^p$ by condition of the strong measurability of function $t \mapsto (A(t) + \xi\,\mathbb{1})^{-1}x$. Owing to $(\chi_{\mathfrak{I}}\,x)(t) = \chi_{\mathfrak{I}}(t)\,x \in \mathfrak{X}$, that is, $\chi_{\mathfrak{I}}\,x \in \mathfrak{X}_T^p$, by the same argument as for (2.2.23), the function

$$t \mapsto A(t) f_{\mathfrak{I},x,\xi}(t) = \xi\,\chi_{\mathfrak{I}}(t)\,x - \xi^2\,(A(t) + \xi\,\mathbb{1})^{-1}\,\chi_{\mathfrak{I}}(t)\,x, \quad t \in [0,T],$$

has values in $\mathfrak{X}_T^p = L^p([0,T],\mathfrak{X})$, $p \in [1,\infty)$. This implies that $f_{\mathfrak{I},x,\xi} \in \operatorname{dom}(\mathcal{A}_p)$.

Given that for a.a. $t \in [0,T]$ the operators $A(t)$ are generators of quasi-bounded C_0-semigroups on $\mathfrak{X}$, by the *Yosida approximation* argument (Proposition 1.4.3 (i)), we deduce from (2.2.23) that for any $\mathfrak{I} \subset [0,T]$ the limit

$$\lim_{\xi\to\infty} f_{\mathfrak{I},x,\xi}(t) = \chi_{\mathfrak{I}}(t)\,x,$$

for $x \in \mathfrak{X}$ and a.a. $t \in \mathfrak{I}$. Thereby, for any $\chi_{\mathfrak{I}}\, x \in \mathfrak{X}^p_T$, there exists a sequence of elements $\{f_{\mathfrak{I},\,x,\,\xi} \in \mathrm{dom}\,(\mathcal{A}_p)\}_{\xi>\beta}$ such that

$$\lim_{\xi\to\infty} \big\| f_{\mathfrak{I},\,x,\,\xi} - \chi_{\mathfrak{I}}\, x \,\big\|_{\mathfrak{X}^p_T} = 0\,. \tag{2.2.24}$$

Due to the fact that the subspace of *simple* (i.e. *finitely valued*) functions

$$s_n := \sum_{j=1}^{n} \chi_{\mathfrak{I}_{j,n}}\, x_{j,n}\,, \quad n \in \mathbb{N}, \quad \mathfrak{I}_{j,n} \subset [0,T] \text{ and } x_{j,n} \in \mathfrak{X}\,,$$

is dense in the Banach space $\mathfrak{X}^p_T = L^p([0,T],\mathfrak{X})$, $p \in [1,\infty)$ (and defines the *Bochner integral*, cf. Appendix E.1), we get for any $\varepsilon > 0$ approximation

$$\big\| f - s_n(f) \big\|_{\mathfrak{X}^p_T} < \varepsilon\,, \quad \text{for } n > N_\varepsilon\,, \tag{2.2.25}$$

of any $f \in \mathfrak{X}^p_T$ by appropriate sequence of simple functions

$$s_n(f) := \sum_{j=1}^{n} \chi_{\mathfrak{I}_{j,n}}\, f_{j,n}\,, \quad n \in \mathbb{N}. \tag{2.2.26}$$

On account of (2.2.24) and (2.2.26), we also obtain for any $n > N_\varepsilon$ the limit

$$\lim_{\xi\to\infty} \Big\| \sum_{j=1}^{n} f_{\mathfrak{I}_{j,n},\, f_{j,n},\, \xi} - s_n(f) \Big\|_{\mathfrak{X}^p_T} = 0\,. \tag{2.2.27}$$

Therefore, (2.2.23), (2.2.25) and (2.2.27) imply that $\mathrm{dom}\,(\mathcal{A}_p)$ is dense in $\mathfrak{X}^p_T$.

The rest of the proof can be completed *verbatim* the arguments in the proof of Theorem 2.1.32, including Proposition 2.2.16 for the last claim. □

Similar to the quasi-boundedness, the next theorem establishes in the space $\mathfrak{X}^p_T$ the inheritance of the semigroup analyticity, cf. Theorem 2.1.33.

Theorem 2.2.18 *Let $\{A(t)\}_{t\in[0,T]}$ be a family of generators in $\mathfrak{X}$ such that $A(t) \in \mathscr{H}(\theta,0)$ for some $\theta \in (0,\pi/2]$ and for all $t \in [0,T]$. If the function $[0,T] \ni t \mapsto (A(t) + \xi\,\mathbb{1})^{-1} x$ is measurable for each $x \in \mathfrak{X}$ and $\xi > 0$, then the induced in $\mathfrak{X}^p_T$ multiplication operator $\mathcal{A}_p$ is generator of a holomorphic bounded in the closed sector $\overline{S}_{\theta-\varepsilon} \subset \mathbb{C}_+$ ($\varepsilon < \theta$) semigroup $\{\mathcal{U}_{\mathcal{A}_p}(\tau) := \mathrm{e}^{-\tau\mathcal{A}_p}\}_{\tau\in S_\theta}$ on $\mathfrak{X}^p_T$, that is, in turn $\mathcal{A}_p \in \mathscr{H}(\theta,0)$.*

Proof Owing to the fact that by Lemma 2.2.15 and Theorem 2.2.17 the induced multiplication operator $\mathcal{A}_p$ is closed and densely defined, one has only to check that resolvent of $\mathcal{A}_p$ satisfies for some $\varepsilon \in (0, \theta)$ the condition

$$\left\| (\mathcal{A}_p + \zeta\, \mathcal{I})^{-1} \right\|_{\mathcal{L}(\mathfrak{X}_T^p)} \leq \frac{M_\varepsilon}{|\zeta|}, \quad \zeta \in \overline{S}_{\pi/2+\theta-\varepsilon} \setminus \{0\}\,, \tag{2.2.28}$$

which defines a holomorphic semigroup, cf. (1.7.4). Here, $M_\varepsilon > 0$, which is independent of ζ, and the closed sector $\overline{S}_{\pi/2+\theta-\varepsilon}$ (see (1.7.4)).

To this end, we recall that for each $t \in [0, T]$ generator $A(t) \in \mathscr{H}(\theta, 0)$ satisfies condition (1.7.4). Consequently, one gets estimate

$$\left\| (A(t) + \zeta\, \mathbb{1})^{-1} \right\|_{\mathcal{L}(\mathfrak{X})} \leq \frac{M_\varepsilon}{|\zeta|}, \quad \zeta \in \overline{S}_{\pi/2+\theta-\varepsilon} \setminus \{0\}\,, \quad t \in [0, T]\,. \tag{2.2.29}$$

Then condition (2.2.28) follows from (2.2.29) by virtue of equality (2.2.22) in Proposition 2.2.16. □

Now, we consider a candidate for evolution pre-generator (Definition 2.2.8):

$$\widetilde{\mathcal{K}}_p := D_{0p} + \mathcal{A}_p, \quad \mathrm{dom}\,(\widetilde{\mathcal{K}}_p) = \mathrm{dom}\,(D_{0p}) \cap \mathrm{dom}\,(\mathcal{A}_p)\,. \tag{2.2.30}$$

Here, $\mathcal{A}_p$ is multiplication operator induced by the family $\{A(t)\}_{t \in [0,T]}$ (see Definition 2.2.14) and $\mathrm{dom}\,(\widetilde{\mathcal{K}}_p) \subset \mathfrak{X}_T^p \cap \mathfrak{X}_T^C$ by definition of operator D_{0p} (2.2.19).

Lemma 2.2.19 *Let $\mathfrak{D} \subseteq \mathrm{dom}\,(\widetilde{\mathcal{K}}_p)$ be a linear subset in $\mathfrak{X}_T^p$ for some $p \in [1, \infty)$ with dense cross-section in $\mathfrak{X}$ and such that $\mathcal{M}_p(\phi)\mathfrak{D} \subseteq \mathfrak{D}$ for $\phi \in W^{1,\infty}([0, T])$. Then restriction $\widetilde{\mathcal{K}}_p \restriction \mathfrak{D}$ of operator (2.2.30) is an evolution pre-generator in $\mathfrak{X}_T^p$. In particular, operator $\widetilde{\mathcal{K}}_p$ itself is an evolution pre-generator.*

Proof In fact, one has only to check conditions (i), (ii) and (iii) of Definition 2.2.8.

(i) Note that $\mathfrak{D} \subseteq \mathrm{dom}\,(\widetilde{\mathcal{K}}_p)$ and definition (2.2.30) yield $\mathfrak{D} \subset \mathfrak{X}_T^p \cap \mathfrak{X}_T^C$.

(ii) The property $\mathcal{M}_p(\phi)\mathfrak{D} \subseteq \mathfrak{D}$ for $\phi \in W^{1,\infty}([0, T])$ is due to condition of lemma. Seeing that $\mathcal{A}_p$ in (2.2.30) is induced multiplication operator, it commutes with operator $\mathcal{M}_p(\phi)$. Then by virtue of (2.2.30), the relation

$$\widetilde{\mathcal{K}}_p \mathcal{M}_p(\phi) f - \mathcal{M}_p(\phi) \widetilde{\mathcal{K}}_p f = \mathcal{M}_p(\phi') f\,, \quad f \in \mathfrak{D}\,,$$

follows from the *Leibnitz rule*.

(iii) Given that domain $\mathfrak{D}$ has a dense cross-section in the Banach space $\mathfrak{X}$, the conditions of Lemma 2.2.6 are satisfied, from where we infer that $\mathfrak{D}$ is dense in the space $\mathfrak{X}_T^p$. Hence, the proof holds for densely defined operator $\widetilde{\mathcal{K}}_p \restriction \mathfrak{D}$.

Finally, since the line of reasoning of this proof is evidently valid if domain $\mathfrak{D} = \mathrm{dom}\,(\widetilde{\mathcal{K}}_p)$, it yields the same conclusion for operator $\widetilde{\mathcal{K}}_p$. □

Definition 2.2.20 Let $\{A(t)\}_{t\in[0,T]}$ be a family of linear operators in $\mathfrak{X}$, and let $\widetilde{\mathcal{K}}_p$ be operator defined by (2.2.30) in $\mathfrak{X}_T^p$ for $p \in [1,\infty)$. Let $\mathfrak{D} \subseteq \mathrm{dom}\,(\widetilde{\mathcal{K}}_p)$ be a linear subset which dense cross-section in $\mathfrak{X}$ is such that $\mathcal{M}_p(\phi)\mathfrak{D} \subseteq \mathfrak{D}$ for $\phi \in W^{1,\infty}([0,T])$:

(i) The nACP (2.1.4) (or (2.1.12)) is called *L^p-solvable* with respect to $\mathfrak{D}$ if the evolution pre-generator $\widetilde{\mathcal{K}}_p \restriction \mathfrak{D}$ admits an extension $\mathcal{K}_p$ in $\mathfrak{X}_T^p$, which is an evolution generator.
(ii) The nACP (2.1.4) (or (2.1.12)) is called *uniquely* L^p-solvable with respect to $\mathfrak{D}$ if the pre-generator $\widetilde{\mathcal{K}}_p \restriction \mathfrak{D}$ admits only *one* extension in $\mathfrak{X}_T^p$, which is an evolution generator.
(iii) The nACP (2.1.4) (or (2.1.12)) is called *strictly* L^p-solvable with respect to $\mathfrak{D}$ if $\widetilde{\mathcal{K}}_p \restriction \mathfrak{D}$ can be extended to an evolution generator $\mathcal{K}_p$ in $\mathfrak{X}_T^p$ such that

$$e^{-\tau\mathcal{K}_p}\mathfrak{D} \subseteq \mathfrak{D}, \quad \tau \in \mathbb{R}_0^+, \tag{2.2.31}$$

and $\tau \mapsto \mathcal{A}_p e^{-\tau\mathcal{K}_p} f$ is Bochner integrable on $\mathbb{R}_0^+$ for $f \in \mathfrak{D}$.

Remark 2.2.21 Similar to the C-setting (cf. Definition 2.1.26), it is worth mentioning that if the nACP (2.1.4) is strictly L^p-solvable, then it is also uniquely L^p-solvable since (2.2.31) implies that $\mathfrak{D}$ is a core for generator $\mathcal{K}_p$. Then nACP (2.1.4) is well-posed in the L^p-setting, cf. Theorem 2.1.34. We shall return to remark about Definition 2.2.20 (ii) and (iii), e.g. Theorems 2.2.22, 2.2.26 and Sect. 11.2. In general, it is possible that the evolution pre-generator $\widetilde{\mathcal{K}}_p \restriction \mathfrak{D}$ has *several* extensions, which yield different evolution generators, that is, the nACP (2.1.4) (or (2.1.12)) has more than one solution (see Definition 2.2.20 (i)).

Theorem 2.2.22 *Let $\{A(t)\}_{t\in[0,T]}$ be a family of linear operators in $\mathfrak{X}$ and operator $\widetilde{\mathcal{K}}_p$ be defined by* (2.2.30)*. Let $\mathfrak{D} \subseteq \mathrm{dom}\,(\widetilde{\mathcal{K}}_p)$ be a linear subset such that restriction $\widetilde{\mathcal{K}}_p \restriction \mathfrak{D}$ is an evolution pre-generator in the Banach space $\mathfrak{X}_T^p$.*

If operator $\widetilde{\mathcal{K}}_p \restriction \mathfrak{D}$ is closable and its closure $\mathcal{K}_p := \overline{\widetilde{\mathcal{K}}_p \restriction \mathfrak{D}}$ is generator of a C_0-semigroup on $\mathfrak{X}_T^p$, then nACP (2.1.4) *(or* (2.1.12)*) is uniquely L^p-solvable with respect to $\mathfrak{D} := \mathrm{dom}\,(\widetilde{\mathcal{K}}_p)$ (see Definition* 2.2.20 (ii)*).*

Proof First, we prove that operator $\mathcal{K}_p$ is a *pre-generator* (see Definition 2.2.8). To this aim, we note that by definition (2.2.30) domain $\mathrm{dom}\,(\widetilde{\mathcal{K}}_p) \subset \mathfrak{X}_T^C \cap \mathfrak{X}_T^p$. Then definition of operator $\mathcal{K}_p := \overline{\widetilde{\mathcal{K}}_p \restriction \mathfrak{D}}$ implies that $\mathrm{dom}\,(\mathcal{K}_p) \subset \mathfrak{X}_T^C \cap \mathfrak{X}_T^p$ and, therefore, proves the condition Definition 2.2.8 (i).

Since $\widetilde{\mathcal{K}}_p \restriction \mathfrak{D}$ is evolution pre-generator, we infer by Remark 2.2.9 that $\mathfrak{D}$ is dense in $\mathfrak{X}_T^p$. Let $f \in \mathrm{dom}\,(\mathcal{K}_p)$. Then, by definition of $\mathfrak{D}$ and operators $\widetilde{\mathcal{K}}_p$ and $\mathcal{K}_p$, there is a sequence $f_n \in \mathfrak{D}$ such that $f_n \to f$ and $\widetilde{\mathcal{K}}_p f_n \to \mathcal{K}_p f$. Let $\phi \in W^{1,\infty}([0,T])$. Given that $\widetilde{\mathcal{K}}_p \restriction \mathfrak{D}$ is evolution pre-generator, by Definition 2.2.8 (ii), we obtain

$$\widetilde{\mathcal{K}}_p\mathcal{M}_p(\phi)f_n = \mathcal{M}_p(\phi)\widetilde{\mathcal{K}}_p f_n + \mathcal{M}_p(\phi')f_n\,.$$

Here, the right-hand side converges to $\mathcal{M}_p(\phi)\mathcal{K}_p f + \mathcal{M}_p(\phi')f$, and therefore, we conclude that $\mathcal{M}_p(\phi)f \in \operatorname{dom}(\mathcal{K}_p)$, i.e. $\mathcal{M}_p(\phi)(\operatorname{dom}(\mathcal{K}_p)) \subset \operatorname{dom}(\mathcal{K}_p)$, as well as that $\mathcal{K}_p\mathcal{M}_p(\phi)f = \mathcal{M}_p(\phi)\mathcal{K}_p f + \mathcal{M}_p(\phi')f$. Hence, the operator $\mathcal{K}_p$ satisfies condition (ii) in Definition 2.2.8.

Next, because $\widetilde{\mathcal{K}}_p \upharpoonright \mathfrak{D}$ is evolution pre-generator and since by definition of $\mathcal{K}_p$ the subset $\mathfrak{D}$ is a core for $\mathcal{K}_p$, domain $\operatorname{dom}(\mathcal{K}_p)$ has a dense cross-section in $\mathfrak{X}$, cf. Definition 2.2.8 (iii). As a consequence, operator $\mathcal{K}_p$ is the evolution pre-generator.

Second, we use that $\mathcal{K}_p$ is generator of a C_0-semigroup on $\mathfrak{X}_T^p$ and Theorem 2.2.10 to deduce that operator $\mathcal{K}_p$ is the evolution generator.

Third, it remains to show that evolution generator $\mathcal{K}_p$ constructed as closure $\mathcal{K}_p := \overline{\widetilde{\mathcal{K}}_p \upharpoonright \mathfrak{D}}$ is unique. Let $\mathcal{K}_p$ and $\mathcal{K}'_p$ be two different extensions of $\widetilde{\mathcal{K}}_p$, which are evolution generators. Since $\mathcal{K}_p$ is the closure of $\widetilde{\mathcal{K}}_p$ and extension $\mathcal{K}'_p$ is closed, we obtain that $\operatorname{dom}(\mathcal{K}_p) \subseteq \operatorname{dom}(\mathcal{K}'_p)$ and $\mathcal{K}'_p \upharpoonright \operatorname{dom}(\mathcal{K}_p) = \mathcal{K}_p$. For any $f \in \operatorname{dom}(\mathcal{K}_p)$, we have equation

$$\partial_s \{\mathrm{e}^{-(\tau-s)\mathcal{K}'_p}\,\mathrm{e}^{-s\mathcal{K}_p} f\} = \mathrm{e}^{-(\tau-s)\mathcal{K}'_p}(\mathcal{K}'_p - \mathcal{K}_p)\,\mathrm{e}^{-s\mathcal{K}_p} f = 0.$$

Hence, the function $s \mapsto \mathrm{e}^{-(\tau-s)\mathcal{K}'_p}\,\mathrm{e}^{-sK_p}$ is a constant operator. So, the semigroup generated by $\mathcal{K}'_p$ must be the same as the one generated by $\mathcal{K}_p$. Thus, $\mathcal{K}_p = \mathcal{K}'_p$ and the claim follows: □ □

As a consequence, Theorem 2.2.22 suggests for examination a unique solution of nACP the following *strategy*:

- First, one has to associate with nACP (2.1.4) (or (2.1.12)) a pre-generator $\widetilde{\mathcal{K}}_p$ for some $p \in [1, \infty)$ (see Definition 2.2.8).
- If it can be shown that $\widetilde{\mathcal{K}}_p$ is closable and its closure $\mathcal{K}_p$ generates a C_0-semigroup, then $\mathcal{K}_p$ is evolution generator, and the nACP (2.1.4) (cf. (2.1.12)) is uniquely, or strictly, L^p-solvable.

On the other hand, it is worth noting that the problem of L^p-solvability of nACP (2.1.4) is also related to the L^p-maximal (parabolic) regularity ($\mathcal{MR}_p$).

Definition 2.2.23 Let $\mathfrak{X}$ and $\mathfrak{Y}$ be Banach spaces such that $\mathfrak{Y}$ is densely and continuous embedded in $\mathfrak{X}$, that is, $\mathfrak{Y} \overset{d}{\hookrightarrow} \mathfrak{X}$. Further, let $\{A(t)\}_{t\in[0,T]}$ be a family of linear operators in $\mathfrak{X}$ such that $\operatorname{dom}(A(t)) = \mathfrak{Y}$ for a.a. $t \in [0, T]$ and restriction

$$A(\cdot) \upharpoonright \mathfrak{Y} : [0, T] \longrightarrow \mathcal{L}(\mathfrak{Y}, \mathfrak{X}),$$

is *strongly measurable* and *essentially* bounded, that is, $A(\cdot) \in L^\infty([0, T], \mathcal{L}(\mathfrak{Y}, \mathfrak{X}))$. The family $\{A(t)\}_{t\in[0,T]}$ is called *L^p-maximal regular* for some $p \in (1, \infty)$, if for every $g \in \mathfrak{X}_T^p$ there exists a unique $f \in W^{1,p}([0, T], \mathfrak{X}) \cap L^p([0, T], \mathfrak{Y})$ satisfying $f(0) = 0$, such that equation $D_{0p} f + \mathcal{A}_p f = g$ holds.

We denote this by $\{A(t)\}_{t\in[0,T]} \in \mathcal{MR}_p([0,T])$. Let $A \in \mathcal{L}(\mathfrak{Y}, \mathfrak{X})$. Then A is called L^p-maximal regular if the *constant* family $\{A(t) = A\}_{t\in[0,T]}$ is L^p-maximal regular.

Remark 2.2.24

(a) If A is L^p-maximal regular, then A regarded as operator in $\mathfrak{X}$ is a generator of holomorphic semigroup. The converse is, in general, not true, However, if $\mathfrak{H}$ is a Hilbert space, then A is L^p-maximal regular if and only if A is a generator of holomorphic semigroup.
(b) If $A(\cdot) \in C_0([0,T], \mathcal{L}(\mathfrak{Y}, \mathfrak{X}))$ and $\{A(t)\}_{t\in[0,T]}$ is L^p-maximal regular, then for each fixed $t \in [0,T]$, the operator $A(t)$ is L^p-maximal regular. In particular, this yields that $A(t)$ is a generator of a holomorphic semigroup for each $t \in [0,T]$.
(c) For linear operator $\widetilde{\mathcal{K}}_p$ defined by (2.2.30), the family $\{A(t)\}_{t\in[0,T]}$ is L^p-maximal regular if ran $(\widetilde{\mathcal{K}}_p) = \mathfrak{X}_T^p$ and $\widetilde{\mathcal{K}}_p$ is an evolution pre-generator. But it remains an open question whether the operator $\widetilde{\mathcal{K}}_p$ is generator of a semigroup.

Hypotheses 2.2.25 Let $\mathfrak{Y}$ be Banach space which is densely and continuously embedded into $\mathfrak{X}$, that is, $\mathfrak{Y} \overset{d}{\hookrightarrow} \mathfrak{X}$:

(H1) The family $\{A(t)\}_{t\in[0,T]}$ consists of holomorphic semigroup generators such that dom $(A(t)) = \mathfrak{Y}$ (in the graph norm $\|\cdot\|_{\mathfrak{Y}}$ of $A(t)$) and $A(\cdot) \in C([0,T], \mathcal{L}(\mathfrak{Y}, \mathfrak{X}))$.
(H2) For each $t \in [0,T]$, the operator $A(t)$ is L^p-maximal regular for some $p \in (1,\infty)$.

It is known (see Notes to Chap. 2 (Notes to Sects. 2.2, 2.2.2)) that under the hypotheses (H1) and (H2) the nACP (2.1.4) is *strictly* L^p-solvable (Definition 2.2.20 (iii)) for $\widetilde{\mathcal{K}}_p$ with respect to

$$\mathfrak{D} := \text{dom}\,(\widetilde{\mathcal{K}}_p) = W^{1,p}([0,T], \mathfrak{X}) \cap L^p([0,T], \mathfrak{Y})\,.$$

Moreover, its extension $\mathcal{K}_p$, which generates the evolution semigroup, coincides with $\widetilde{\mathcal{K}}_p$, that is, evolution pre-generator $\widetilde{\mathcal{K}}_p$ itself is an evolution generator.

Furthermore, if $\{A(t)\}_{t\in[0,T]}$ is L^p-maximal regular for some $p \in (1,\infty)$ and, in addition, $\{A(t)\}_{t\in[0,T]}$ is a family of generators of contraction semigroups on $\mathfrak{X}$, then nACP (2.1.4) is strictly L^p-solvable with respect to $\mathfrak{D}$, cf. Definition 2.2.20 (iii). Indeed, this follows from the estimate (cf. (2.1.57) and (2.1.58))

$$\|(\widetilde{\mathcal{K}}_p + \xi\,\mathcal{I})f\| = \|(D_{0p} + \mathcal{A}_p + \xi\,\mathcal{I})f\| \geq \xi\,\|f\|, \quad f \in \text{dom}\,(\widetilde{\mathcal{K}}_p), \quad \xi > 0.$$

Consequently, by virtue of L^p-maximal regularity, ran $(\widetilde{\mathcal{K}}_p) = \mathfrak{X}_T^p$, or ran $(\widetilde{\mathcal{K}}_p + \xi\,\mathcal{I}) = \mathfrak{X}_T^p$ for $\xi > 0$, and hence we deduce that $\widetilde{\mathcal{K}}_p$ is generator of a contraction semigroup, that is, $\widetilde{\mathcal{K}}_p \in \mathscr{G}(1,0)$.

2.2.3 Cauchy Problem and Product Formulæ in L^p-Setting

Recall that if $\{A(t)\}_{t\in[0,T]}$ is a family of linear operators such that $A(t) \in \mathcal{G}(M, \beta)$ for $t \in [0, T]$ and if we assume that for each $x \in \mathfrak{X}$ the vector-valued function

$$[0, T] \ni t \mapsto (A(t) + \xi\, \mathbb{1})^{-1} x \in \mathfrak{X}, \quad \xi > \beta, \tag{2.2.32}$$

is *strongly* measurable, then for $p \in [1, \infty)$, corresponding to $\{A(t)\}_{t\in[0,T]}$ multiplication operator $\mathcal{A}_p$ in $\mathfrak{X}^p_T$ (2.2.21) is generator of C_0-semigroup quasi-bounded semigroup $\{\mathcal{U}_{\mathcal{A}_p}(\tau)\}_{\tau\geq 0}$, that is, $\mathcal{A}_p \in \mathcal{G}(M, \beta)$. On that account

$$(\mathcal{U}_{\mathcal{A}_p}(\tau) f)(t) := (\mathrm{e}^{-\tau \mathcal{A}_p} f)(t) = \mathrm{e}^{-\tau A(t)} f(t), \quad t \in [0, T], \quad \tau \in \mathbb{R}^+_0, \quad f \in \mathfrak{X}^p_T,$$

see Theorem 2.2.17.

In this subsection, we collect some preliminaries about the product formulæ approximations for nACP in the L^p-setting, which will be considered in a full generality in Part IV: time-dependent product formulæ.

Theorem 2.2.26 *Let $\{A(t)\}_{t\in[0,T]}$ be a family of linear operators such that $A(t) \in \mathcal{G}(1, 0)$ for each $t \in [0, T]$ and satisfies condition* (2.2.32) *for $\beta = 0$. Further, let $\mathfrak{D} \subseteq \mathrm{dom}\,(\widetilde{\mathcal{K}}_p)$ be a linear subset such that $\widetilde{\mathcal{K}}_p \restriction \mathfrak{D}$ is an evolution pre-generator in $\mathfrak{X}^p_T$ for some $p \in [1, \infty)$. If restriction $\widetilde{\mathcal{K}}_p \restriction \mathfrak{D}$ is closable and the closure $\mathcal{K}_p$ is a generator, then the nACP* (2.1.4) (*or* (2.1.12)) *is uniquely L^p-solvable with respect to $\mathfrak{D}$ and for each $x \in \mathfrak{X}$ the corresponding solution operator $\{U(t, s)\}_{(t,s)\in\Delta}$ admits the following product approximation* :

$$\lim_{n\to\infty} \sup_{\tau\in[0,T]} \int_0^{T-\tau} ds\, \|(U_n(s + \tau, s) - U(s + \tau, s))\, x\|^p_{\mathfrak{X}} = 0, \tag{2.2.33}$$

where approximants $\{U_n(t, s)\}_{(t,s)\in\Delta}$, $n \in \mathbb{N}$, are given by the product formula (2.1.56).

Proof Condition (2.2.32) for $\beta = 0$ implies that multiplication operator $\mathcal{A}_p \in \mathcal{G}(1, 0)$. By Theorem 2.2.22, the closure $\mathcal{K}_p$ of restriction $\widetilde{\mathcal{K}}_p \restriction \mathfrak{D}$ is evolution generator, which generates a contraction semigroup in $\mathfrak{X}^p_T$. Then by the Lie-Trotter product formula for C_0-semigroups on $\mathfrak{X}^p_T$ (Proposition 1.9.35, (1.9.110)), we obtain

$$\lim_{n\to\infty} \sup_{\tau\in[0,T]} \left\| \left(\mathrm{e}^{-\tau D_{0p}/n} \mathrm{e}^{-\tau \mathcal{A}_p/n}\right)^n f - \mathrm{e}^{-\tau \mathcal{K}_p} f \right\|_{\mathfrak{X}^p_T} = 0 \tag{2.2.34}$$

for $f \in \mathfrak{X}^p_T$. Note that by (2.1.56) together with definitions of multiplication and shift operators in $\mathfrak{X}^p_T$ one gets representation

$$\left(\left(\mathrm{e}^{-\tau D_{0p}/n} \mathrm{e}^{-\tau \mathcal{A}_p/n}\right)^n f\right)(t) = U_n(t, t - \tau)\chi_{[0,T]}(t - \tau) f(t - \tau), \tag{2.2.35}$$

for a.a. $t \in [0, T]$, cf. (2.1.60), (2.1.61). As a consequence of (2.2.35) and the HEN representation (2.2.12), one infers that

$$\left\| \left(e^{-\tau D_{0p}/n} e^{-\tau \mathcal{A}_p/n}\right)^n f - e^{-\tau \mathcal{K}_p} f \right\|^p_{\mathfrak{X}^p_T} = \\ = \int_0^T dt \left\| (U_n(t, t-\tau) - U(t, t-\tau)) \chi_{[0,T]}(t-\tau) f(t-\tau) \right\|^p_{\mathfrak{X}} .$$

Setting $t = s + \tau$, we obtain

$$\left\| \left(e^{-\tau D_{0p}/n} e^{-\tau \mathcal{A}_p/n}\right)^n f - e^{-\tau \mathcal{K}_p} f \right\|^p_{\mathfrak{X}^p_T} = \\ = \int_0^{T-\tau} ds \left\| (U_n(s+\tau, s) - U(s+\tau, s)) f(s) \right\|^p_{\mathfrak{X}} .$$

Then for $f(t) = (\mathbb{1}_T \otimes x)(t)$, where $x \in \mathfrak{X}$ (see Remark 2.2.3), one gets (2.2.33).

□

The assumption $A(t) \in \mathcal{G}(1, 0)$, $t \in [0, T]$, is not always satisfied. So, to extend the result (2.2.33) to generators of quasi-bounded semigroups, $A(t) \in \mathcal{G}(M, \beta)$, $t \in [0, T]$, one has to introduce a certain stability condition, cf. Definition 2.1.36, or Definition 1.9.37 of the Trotter stability condition (1.9.113), and also Definition 1.9.39 of renormalisable pair of generators.

Definition 2.2.27 The family $\{A(t)\}_{t\in[0,T]}$ of generators $A(t) \in \mathcal{G}(M, \beta)$ is called almost every (a.e.) *stable* if there are constants $M \geq 1$ and $\beta \in \mathbb{R}$ such that for any sequence $\{\tau_k \geq 0\}_{k=1}^n$, $n \in \mathbb{N}$ the estimate

$$\left\| e^{-\tau_n A(t_n)} e^{-\tau_{n-1} A(t_{n-1})} \cdots e^{-\tau_1 A(t_1)} \right\|_{\mathcal{L}(\mathfrak{X})} \leq M e^{\beta(\tau_n + \tau_{n-1} + \ldots + \tau_1)} \tag{2.2.36}$$

holds for a.a. $(t_1, t_2, \ldots, t_n) \in \Delta_n$, where

$$\Delta_n := \{(t_1, t_2, \ldots, t_n) \in [0, T]^n : 0 < t_1 \leq t_2 \leq \ldots \leq t_n \leq T\}, \quad n \in \mathbb{N}.$$

Lemma 2.2.28 *Let $\{A(t)\}_{t\in[0,T]}$ be a family of generators such that $A(t) \in \mathcal{G}(M, \beta)$ and the condition* (2.2.32) *be satisfied. The family $\{A(t)\}_{t\in[0,T]}$ is a.e. stable* (2.2.36) *if and only if there are constants $M \geq 1$ and $\beta \in \mathbb{R}$ such that for any sequence $\{\tau_k \geq 0\}_{k=1}^n$ and any sequences $\{\sigma_k \geq 0\}_{k=1}^n$, the estimate*

$$\left\| e^{-\sigma_n D_{0p}} e^{-\tau_n \mathcal{A}_p} e^{-\sigma_{n-1} D_{0p}} e^{-\tau_{n-1} \mathcal{A}_p} \cdots e^{-\sigma_1 D_{0p}} e^{-\tau_1 \mathcal{A}_p} \right\|_{\mathcal{L}(\mathfrak{X}^p_T)} \leq \\ \leq M e^{\beta(\tau_n + \tau_{n-1} + \ldots + \tau_1)} \tag{2.2.37}$$

is valid.

Proof Recall that condition (2.2.32) implies that multiplication operator $\mathcal{A}_p \in \mathcal{G}(M,\beta)$. For any $f \in \mathfrak{X}_T^p$, we have identity

$$
\begin{aligned}
&(\mathrm{e}^{-\sigma_n D_{0p}}\mathrm{e}^{-\tau_n\mathcal{A}}\mathrm{e}^{-\sigma_{n-1}D_{0p}}\mathrm{e}^{-\tau_{n-1}\mathcal{A}}\cdots\mathrm{e}^{-\sigma_1 D_{0p}}\mathrm{e}^{-\tau_1\mathcal{A}}f)(t)=\\
&\qquad \chi_{[0,T]}(t-\sigma_n-\sigma_{n-1}-\ldots-\sigma_1)\mathrm{e}^{-\tau_n A(t-\sigma_n)}\mathrm{e}^{-\tau_{n-1}A(t-\sigma_n-\sigma_{n-1})}\times\\
&\qquad \times\cdots\mathrm{e}^{-\tau_1 A(t-\sigma_n-\sigma_{n-1}-\ldots-\sigma_1)}f(t-\sigma_n-\sigma_{n-1}-\ldots-\sigma_1),\quad t\in[0,T].
\end{aligned}
$$

If $\{\tau_k \geq 0\}_{k=1}^n$ are fixed, (2.2.36) yields the estimate

$$
\begin{aligned}
&\Big\|\mathrm{e}^{-\tau_n A(t-\sigma_n)}\mathrm{e}^{-\tau_{n-1}A(t-\sigma_n-\sigma_{n-1})}\times\\
&\qquad\cdots\mathrm{e}^{-\tau_1 A(t-\sigma_n-\sigma_{n-1}-\ldots-\sigma_1)}f(t-\sigma_n-\sigma_{n-1}-\ldots-\sigma_1)\Big\|_{\mathfrak{X}}\\
&\qquad\leq M\mathrm{e}^{\beta(\tau_1+\ldots+\tau_n)}\|f(t-\sigma_n-\sigma_{n-1}-\ldots-\sigma_1)\|_{\mathfrak{X}}
\end{aligned}
$$

for a.a. $(\sigma_1,\sigma_2,\ldots,\sigma_n)\in[0,T]^n$, such that $0\leq\sigma_1+\sigma_2+\ldots+\sigma_n<T$, and for a.a. $t\in[\sigma_1+\ldots+\sigma_{n-1}+\sigma_n,T]$. On that account, we obtain

$$
\begin{aligned}
&\Big\|\mathrm{e}^{-\sigma_n D_{0p}}\mathrm{e}^{-\tau_n\mathcal{A}}\mathrm{e}^{-\sigma_{n-1}D_{0p}}\mathrm{e}^{-\tau_{n-1}\mathcal{A}}\cdots\mathrm{e}^{-\sigma_1 D_{0p}}\mathrm{e}^{-\tau_1\mathcal{A}}f\Big\|_{\mathfrak{X}_T^p}\\
&\qquad\leq M\mathrm{e}^{\beta(\tau_1+\ldots+\tau_n)}\|f\|_{\mathfrak{X}_T^p},\quad f\in\mathfrak{X}_T^p,
\end{aligned}
\tag{2.2.38}
$$

for $\{\tau_k \geq 0\}_{k=1}^n$ and for a.a. $(\sigma_1,\sigma_2,\ldots,\sigma_n)\in[0,T]^n$. Since the left-hand side of this inequality is continuous in σ_k, we get the estimate for any sequences $\{\tau_k \geq 0\}_{k=1}^n$ and $\{\sigma_k \geq 0\}_{k=1}^n$.

Conversely, assume that (2.2.37) is satisfied. We remind that similar to definition (2.1.50) the automorphisms $\mathcal{S}_L(\tau):\mathfrak{X}_T^p\to\mathfrak{X}_T^p$, $\tau\geq 0$, form on the space the $\mathfrak{X}_T^p$ a *left-shift* contraction C_0-semigroup $\{\mathcal{S}_L(\tau)\}_{\tau\geq 0}$ such that

$$
(\mathcal{S}_L(\tau)f)(t):=\chi_{[0,T]}(t+\tau)\,f(t+\tau),\quad f\in\mathfrak{X}_T^p,\quad \tau\geq 0,\ t\in[0,T].
\tag{2.2.39}
$$

Note that operator family

$$
\begin{aligned}
\mathcal{T}_p(\sigma_1+\sigma_2+\ldots\sigma_n) &:= \mathcal{S}_L(\sigma_1+\sigma_2+\ldots+\sigma_n)\mathrm{e}^{-\sigma_n D_{0p}}\mathrm{e}^{-\tau_n\mathcal{A}_p}\\
&\quad\times\mathrm{e}^{-\sigma_{n-1}D_{0p}}\mathrm{e}^{-\tau_{n-1}\mathcal{A}_p}\cdots\mathrm{e}^{-\sigma_1 D_{0p}}\mathrm{e}^{-\tau_1\mathcal{A}_p}
\end{aligned}
$$

is a multiplication operator on $\mathfrak{X}_T^p$ for $\{\tau_k \geq 0\}_{k=1}^n$ and $\{\sigma_k \geq 0\}_{k=1}^n$. Indeed, we have

$$(\mathcal{T}_p(\sigma_1+\sigma_2+\ldots\sigma_n)f)(t) = \mathrm{e}^{-\tau_n A(t+\sigma_1+\sigma_2+\ldots+\sigma_{n-1})}\times$$
$$\mathrm{e}^{-\tau_{n-1} A(t+\sigma_1+\sigma_2+\cdots\sigma_{n-2})}\cdots\mathrm{e}^{-\tau_1 A(t)}\,\chi_{[0,T]}(t+\sigma_1+\sigma_2+\ldots+\sigma_n)f(t),$$

for a.a. $t \in [0, T]$ and for fixed $\{\tau_k \geq 0\}_{k=1}^n$ and $\{\sigma_k \geq 0\}_{k=1}^n$. Since (2.2.38) yields

$$\begin{aligned}
&\| \mathcal{T}_p(\sigma_1+\sigma_2\ldots+\sigma_n)\| \\
&\leq \| \mathrm{e}^{-\sigma_n D_0 p}\mathrm{e}^{-\tau_n \mathcal{A}_p}\mathrm{e}^{-\sigma_{n-1} D_0 p}\mathrm{e}^{-\tau_{n-1}\mathcal{A}_p}\cdots\mathrm{e}^{-\sigma_1 D_0 p}\mathrm{e}^{-\tau_1\mathcal{A}_p}\| \\
&\leq M\,\mathrm{e}^{\tau_1+\tau_2+\ldots+\tau_n},
\end{aligned}$$

we find

$$\begin{aligned}
\operatorname*{ess\,sup}_{t\in[0,T-\sigma_1-\sigma_2-\ldots-\sigma_n]} &\|\mathrm{e}^{-\tau_n A(t+\sigma_1+\sigma_2+\ldots+\sigma_{n-1})}\mathrm{e}^{-\tau_{n-1} A(t+\sigma_1+\sigma_2+\ldots+\sigma_{n-2})}\cdots\mathrm{e}^{-\tau_1 A(t)}\| \\
&\leq M\,\mathrm{e}^{\tau_1+\tau_2+\ldots+\tau_n},
\end{aligned}$$

for fixed $\{\tau_k \geq 0\}_{k=1}^n$ and $\{\sigma_k \geq 0\}_{k=1}^n$ such that $\sigma_1+\sigma_2+\ldots+\sigma_n < T$. Then setting

$$\begin{aligned}
t_n &:= t+\sigma_1+\sigma_2+\ldots+\sigma_{n-1},\\
t_{n-1} &:= t+\sigma_1+\sigma_2+\ldots+\sigma_{n-2},\\
&\ \ \vdots\\
t_1 &:= t\,,
\end{aligned}$$

we obtain (2.2.36) for a.a. $(t_1, t_2, \ldots, t_n) \in \Delta_n$. □

Corollary 2.2.29 *Stability in the sense* (2.2.36) *(see Definition* 2.2.27*) is equivalent to* (2.2.37)*. This means that the pair* $\{D_{0p}, \mathcal{A}_p\}$ *of generators in* $\mathfrak{X}_T^p$ *is, as a consequence, Trotter-stable, cf. Definition* 1.9.37 *and* (1.9.113).

Theorem 2.2.30 *Let* $\{A(t)\}_{t\in[0,T]}$ *be a family of generators of quasi-bounded semigroups on* $\mathfrak{X}$ *such that it is stable in the sense of Definition* 2.2.27 *and satisfies condition* (2.2.32) *for* $\beta \geq 0$*. Further, let* $\mathfrak{D} \subseteq \mathrm{dom}\,(\widetilde{\mathcal{K}}_p)$ *be a linear subset such that* $\widetilde{\mathcal{K}}_p \restriction \mathfrak{D}$ *is evolution pre-generator in* $\mathfrak{X}_T^p$ *for some* $p \in [1, \infty)$*. If restriction* $\widetilde{\mathcal{K}}_p \restriction \mathfrak{D}$ *is closable and the closure* $\mathcal{K}_p$ *is a generator, then the nACP* (2.1.4) *(or* (2.1.12)*) is uniquely* L^p*-solvable with respect to* $\mathfrak{D}$*, and the corresponding solution operator* $\{U(t, s)\}_{(t,s)\in\Delta}$ *admits the product approximation* (2.2.33) *for each* $x \in \mathfrak{X}$ *in the case when* $A(t) \in \mathcal{G}(M, \beta)$*,* $t \in [0, T]$.

Proof Condition (2.2.32) for $\beta \geq 0$ implies that $\mathcal{A}_p \in \mathcal{G}(M, \beta)$. By Lemma 2.2.28, the condition (2.2.36) is equivalent to (2.2.37). Let $|||\cdot|||$ denote a *norm* on the Banach space $\mathfrak{X}_T^p = L^p([0, T], \mathfrak{X})$, which is defined as follows: For any $f \in \mathfrak{X}_T^p$, we set

$$|||f||| := \sup_{\substack{\sigma_k \geq 0,\ \tau_k \geq 0 \\ k=1,2,\ldots,n}} e^{-\beta \sum_{k=1}^{n} \tau_k} \times \left\| e^{-\sigma_n D_{0p}} e^{-\tau_n \mathcal{A}_p} e^{-\sigma_{n-1} D_{0p}} e^{-\tau_{n-1} \mathcal{A}_p} \cdots e^{-\sigma_1 D_{0p}} e^{-\tau_1 \mathcal{A}_p} f \right\| .$$

Then (2.2.38) yields the *equivalence* of the norms $|||\cdot|||$ and $\|\cdot\|$

$$\|f\| \leq |||f||| \leq M\|f\|, \quad f \in \mathfrak{X}_T^p , \tag{2.2.40}$$

as well as the estimates (see Lemma 1.9.40)

$$\left|\left|\left| e^{-\sigma D_{0p}} f \right|\right|\right| \leq |||f||| \quad \text{and} \quad \left|\left|\left| e^{-\tau \mathcal{A}_p} f \right|\right|\right| \leq |||f||| , \quad \sigma \geq 0, \ \tau \geq 0 .$$

As a consequence, $D_{0p} \in \mathcal{G}(1, 0)$ and $\mathcal{A}_p \in \mathcal{G}(1, 0)$ in this norm. Hence, by the Lie-Trotter product formula (Proposition 1.9.35) for C_0-semigroups (2.2.34), we obtain

$$\lim_{n \to \infty} \sup_{\tau \in [0,T]} \left|\left|\left| \left(e^{-\tau D_{0p}/n} e^{-\tau \mathcal{A}_p/n} \right)^n f - e^{-\tau \mathcal{K}_p} f \right|\right|\right| = 0, \quad f \in \mathfrak{X}_T^p ,$$

which on account of (2.2.40) yields the limit (2.2.34), which by Theorem 2.2.26 gives the product approximation (2.2.33) for $\mathcal{A}_p \in \mathcal{G}(M, \beta)$. □

2.3 Notes to Chap. 2

Notes to Sect. 2.1

Non-autonomous Cauchy problem (nACP) and construction of the *solution operator* (or *fundamental solution* of nACP) for *parabolic* and *hyperbolic* types of time-dependent generators have a long history and large literature, essentially staring in papers by T. Kato [Kato53, Kato56, Kato61, KaTa62] and by P. E. Sobolevskiĭ [KKS56, Sob57, Sob58, Sob58, Sob61, Sob61a, Sob62]. The approach of these papers was developed and then summarised in [Tan60, Tan67, Sob61a]. Now, it is known as the *Tanabe-Sobolevskiĭ* method (for solution of the parabolic-type nACP in Banach spaces, see [Yos80, Chapter XIV, Section 5]). Ideas of this method were surveyed in two detailed and well-written books [SKre67] and [Tan79].

Note that simultaneously there was an important progress in studying of the *hyperbolic*-type nACP. The first advance was due to T. Kato [Kato53] and then it was improved in [Kato70, Kato73]. This initiated researches of the Schrödiger-type equations with *time-dependent* domain of the nACP generator, which before was considered by T. Kato [Kato61] for parabolic case. (For the proof of product formula approximation of the nACP solution operator in this case, see Sect. 11.1 and [Far67b, VWZ09].) We also mention here the paper [Kis64], as well as a certain elucidation and generalisation of this paper in [NZ09, Theorem 5.5], where we used the *evolution semigroups* method for the Schrödiger-type nACP. Applying this method we also improved for parabolic case [NZ09, Theorem 4.6] the result of [Kato61]: the existence of *solution operator* was proven in [NZ09] under *weaker* assumptions.

We introduced the *evolution semigroups* method in C-setting (Sect. 2.1) and in L^p-setting (Sect. 2.2) within the framework of the *strong* operator topology. Then in Sect. 11.5, we shall uplift the L^p-setting for the proof of the product formulæ approximations of the *parabolic* nACP solution operator in the *operator-norm* topology.

Since in our book there are no results about product formulæ approximations of solution operator for the *hyperbolic*-type nACP, we do not discuss this case in Chap. 2 and in the rest of the book. We shall show in Parts II–IV that removing the *holomorphic*, or a tightly related to it, the *parabolic* conditions, produces a serious obstacle for proving the Trotter-Kato product formulae, as well as the product formula approximations of the nACP solution operator, in the *operator-norm* topology. This is what happens and produces the problems in the *hyperbolic* case.

Section 2.1.1

This introduction is standard (see, for example, [SKre67, Tan75, Fat83], or a short review [Schn00]). In Remarks 2.1.7 and 2.1.14 and in Example 2.1.8, we used [Nick96] as well as Examples 3.5 and 3.6 from [NaNi02].

Example 2.1.10 (1) is due to [Kato70], cf. also [Kato73]. For Example 2.1.10 (2), we refer to [EN00, Chapter VI, Sec.9.6] or to original papers by [Sob58, Yagi90](1.2). For more details, we suggest [SKre67, Chapter II, §3–§5].

For the proof of Proposition 2.1.12 (with comments on the properties (a)–(e)) as well as of Proposition 2.1.13, see [Tan79, Theorem 5.3.3, (5.90), (5.91)] and also remarks in [Kato61, Tan61, KaTa62, Tan79, Yos80, VWZ09].

Section 2.1.2

Here, we essentially follow ideas formulated in papers by [How74, Ev76, Nei79, Nei81, Nei81a, NZ09] and [NSZ20].

For the first time, the *Howland-Evans-Neidhardt* (HEN) formula (or representation), which relates unitary *solution operator* (propagator) for nACP and corresponding unitary *evolution group* in the L^2-setting over a separable Hilbert space appeared in [How74]. Then this observation was extended in [Ev76] to bounded measurable propagators and corresponding strongly continuous bounded evolution groups in C and L^p settings over separable Banach spaces.

In paper [Nei81] (based on his thesis [Nei79]), these results were generalised in both settings over arbitrary (nonseparable) Banach space $\mathfrak{X}$ and strongly continuous *evolution semigroup*. In contrast to predecessors, the paper [Nei81] began with an abstract definition of the *evolution semigroups* and then proved a *reconstruction* theorem saying that between them and the strongly continuous propagators (*solution operator*) on $\mathfrak{X}$ there exists a one-to-one correspondence such that the HEN formula (2.1.24) holds. For exact formulations and proofs, we refer to Appendix F.

The notion of the evolution semigroup *pre-generator* was introduced in [NSZ20, Definition 3.1], but a version of Theorem 2.1.17 was proven much earlier in [Nei81, Theorem 2.8].

Proposition 2.1.24 is due to [Ev76, Proposition 1.3] and [Nei81, Proposition 3.2] (see also Appendix E.3.1).

Section 2.1.3

It is a preliminary introduction into the problem of the *product formulæ* approximations of the nACP solution operator within the C-setting, Theorem 2.1.35.

We discuss the *stability* condition and the *renormalised* norm (see Definition 2.1.36 and Theorem 2.1.38) following [Nei82] and [NZ09].

Notes to Sect. 2.2

Section 2.2.1

To advance with the L^p-setting, we use here certain similarity and results due to the C-setting. The idea to construct first the evolution semigroups, next to prove the HEN formula within the C-setting and then to generalise these assertions to the L^p-setting (see Theorem 2.2.4) is due to [Nei81, Theorem 3.4 and Corollary 3.9].

Unique identification of evolution generator and evolution operator in the HEN representation (Theorems 2.2.10 and 2.2.11) because of the *dense cross-section* property is also due to [Nei81]. To this aim, he introduced in [Nei79] and [Nei81, Definition 4.8] a certain supplementary class of C_{00} *evolution semigroups*. Here, we call this class: the semigroups with *evolution generators* (see Definition 2.2.8 and Theorem 2.2.10). They are an L^p-dense cross-section extension of the pre-generator concept introduced in Sect. 2.1.2 within the C-setting, cf. Definition 2.1.16 and Theorem 2.1.17.

Section 2.2.2

Here, we translate to L^p-setting essential statements from the C-setting. Note that assertion in Proposition 2.2.16 is much less trivial than Proposition 2.1.24 in C-setting (see Appendix E.3).

The estimates in the Proof of Theorem 2.2.22 are due to [Nei81, Corollary 4.7]. Definition 2.2.23 of the L^p-*maximal* (*parabolic*) *regularity* ($\mathcal{MR}_p$) is identical to that in [ACFP07, Definition 2.1].

The assertion formulated just after Hypothesis 2.2.25 is, in fact, a theorem from [PrSchn01, Theorem 2.5] about solvability and maximal regularity ($\mathcal{MR}_p$) of parabolic evolution equations.

Section 2.2.3

These results are translation to L^p-setting of assertions from Sect. 2.1.3 concerning the C-setting.

We collect here preliminaries about the *product formulæ approximations* for nACP in the L^p-setting, which we shall make use of in Part IV: time-dependent product formulæ.
For completeness and to make a contact with discussion of the *well-posedness*, see, as an illustration, Example 2.1.10. We quote below a result from [NSZ20, §3.1], where the notion of the *correctly posed* nACP was introduced.

- To this aim, we precise first the notion of the *solution operator* of the problem (2.1.4) versus the *evolution family* $\{U(t,s)\}_{(t,s)\in\Delta}$ (Definition 2.1.4) on the Banach space $\mathfrak{X}$.

Definition 2.3.1 ([NSZ20, Definition 3.5])

(i) The evolution nACP (2.1.4) is called *correctly posed* on $[0, T]$ if operator $\widetilde{\mathcal{K}}$ is an evolution *pre-generator* (Definition 2.1.16).
(ii) An evolution family $\{U(t,s)\}_{(t,s)\in\Delta}$ is called a *solution operator* of the correctly posed evolution problem (2.1.4) if the corresponding *evolution generator* $\mathcal{K}$ is an operator extension of pre-generator $\widetilde{\mathcal{K}}$, that is, $\widetilde{\mathcal{K}} \subseteq \mathcal{K}$.
(iii) The evolution problem (2.1.4) has a unique solution operator if pre-generator $\widetilde{\mathcal{K}}$ admits *only one* extension which is the evolution generator.

Remark 2.3.2

(a) In general, it is a nontrivial question: whether an evolution pre-generator admits several extensions which are evolution generators.
However, if this is the case, then nACP (2.1.4) has more than *one* solution operators.
(b) Our Definition 2.3.1 (i) of the *correctly posed* nACP is a *weak* property. For example, the property of well-posedness developed in Sects. 2.1.1 and 2.1.2 (see also [Nick96, Nick97]) implies the correct posedness, which was introduced in [NSZ20, Definition 3.5].

- It is, overall, a nontrivial problem to describe extensions of evolution pre-generator $\widetilde{\mathcal{K}}$ that give evolution generators. However, there are some special cases, when it is easy to guarantee the existence of such extension and, moreover, to show that it is unique. We quote here a result from [Nick97, Theorem 2.9], about well-posedness by means of evolution semigroup.

The main result relates the properties of domain dom $(\mathcal{K})$ to the *well-posedness* of the non-autonomous Cauchy problem (2.1.4). Roughly speaking, nACP is well-posed if there are *sufficiently many* orbits $\{e^{-\tau\mathcal{K}} f\}_{\tau\geq 0}$ for evolution semigroup staying in the space of differentiable functions $C_0^1([0,t], \mathfrak{X})$ (see representation (2.1.27) and Theorem 2.1.34 about *dense* cross-sections).

Proposition 2.3.3 (**[Nick97]**) *Let* $\mathfrak{X}$ *be a Banach space and* $\{A(t), \operatorname{dom}(A(t))\}_{t\in[0,T]}$ *be a family of closed linear operators in* $\mathfrak{X}$. *The following assertions are equivalent.*

(i) *The nACP* (2.1.4) *for the family* $\{A(t)\}_{t\in[0,T]}$ *is well-posed (with exponentially bounded solutions).*
(ii) *There exists a unique evolution semigroup* $\{\mathcal{U}(\tau)\}_{\tau\geq 0}$ *with generator* $\{\mathcal{K}, \operatorname{dom}(\mathcal{K})\}$ *which has invariant core* $\mathcal{D} \subseteq C_0^1([0,t], \mathfrak{X}) \cap \operatorname{dom}(\mathcal{K})$ *such that*

$$(\mathcal{K}f - \partial_t f)(t) = (\mathcal{A}f)(t) = A(t)f(t),$$

for $f \in \mathcal{D}$, *cf. Definition* (2.1.39).

Chapter 3
Quasi-Sectorial Contractions and Operator-Norm Convergence

Quasi-sectorial contractions play important rôle throughout this book. We need this type of mappings in different parts of the text, but mainly in the Part III: Trotter-Kato product formula for non-self-adjoint semigroups. For quasi-sectorial contractions we can improve the original *Chernoff* $\sqrt{n}$ - Lemma up to the $(\sqrt[3]{n})^{-1}$-Lemma estimate in the operator-norm topology, as well as to uplift to this topology the convergence of the corresponding product formulæ.

The next Sect. 3.1 is an introduction and a collection of basic results including the main theorem about relation between *sectorial* generators and *quasi-sectorial* contractions on a Hilbert space. In Sects. 3.2 we present some accurate results about relations between the *numerical range* of generators and approximations in the *operator-norm* topology of contraction holomorphic semigroups by the *Euler* approximants.

For references and comments on quasi-sectorial contractions and the operator-norm Chernoff-type estimates see Notes to Chap. 3, Sect. 3.3.

3.1 Sectorial Generators

3.1.1 Quasi-Sectorial Contractions

Let C be contraction operator on a Hilbert space $\mathfrak{H}$ with $\|C\| < 1$. Then its numerical range $W(C) = \{(Cu, u) \in \mathbb{C} : u \in \mathfrak{H}, \|u\| = 1\}$ (Definition 1.8.1) is a (proper) subset of the open unit disc $\mathbb{D} \subset \mathbb{C}$. A *quasi-sectorial* contraction is a deformation of of C in such a way that $W(C)$ may have the *"angle"* (in contrast to the *"tangent"*) contact with boundary $\partial\mathbb{D}$ at only one point $z = 1$.

The notion of *quasi-sectorial* contractions was introduced to scrutinise the non-self-adjoint semigroups with *sectorial* generators, which is the subject of this section. Therefore, we pass to formal definitions and results.

V. A. Zagrebnov et al., *Trotter-Kato Product Formulæ*, Operator Theory: Advances and Applications 296, https://doi.org/10.1007/978-3-031-56720-9_3

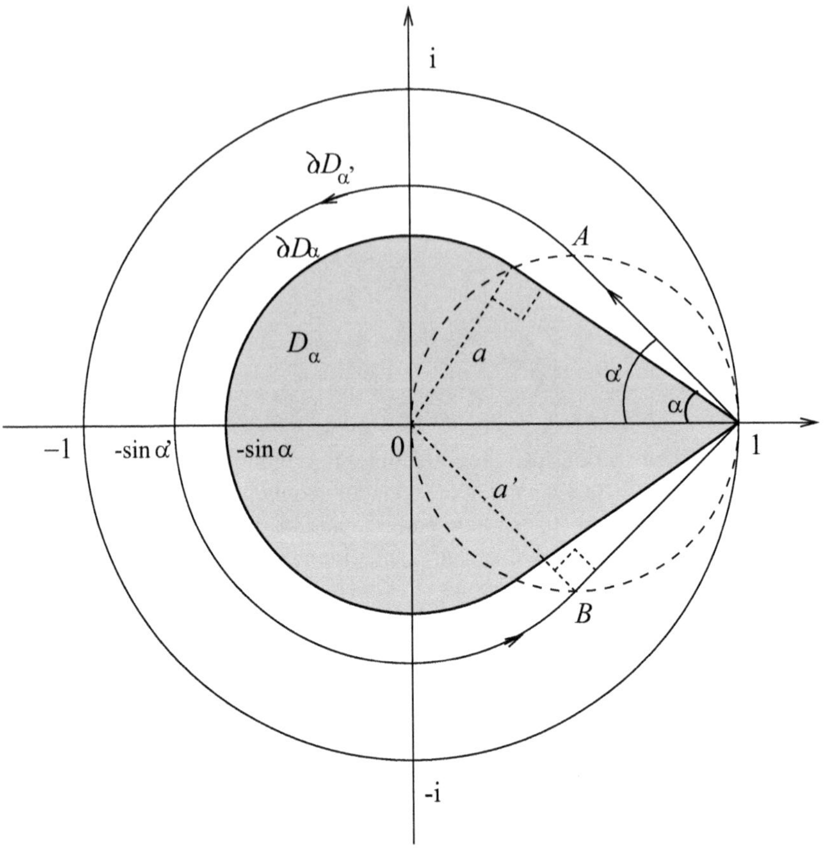

Fig. 3.1 Illustration of the set $D_\alpha \subset \mathbb{C}$ (shaded domain) with boundary ∂D_α for $a = \sin\alpha$, as well as of our choice of the contour $\partial D_{\alpha'}$ in the resolvent set $\rho(C)$, where $a' = \sin\alpha' > a$. The contour consists of two segments of tangent straight lines $(1, A)$ and $(1, B)$ and the arc (A, B) of radius a'. The dotted circle corresponds to the set of tangency points for the different values of $\alpha \in (0, \pi/2)$

Definition 3.1.1 For $\alpha \in [0, \pi/2)$ we define in the complex plane $\mathbb{C}$ *closed* domain:

$$D_\alpha := \{z \in \mathbb{C} : |z| \leq \sin\alpha\} \cup \{z \in \mathbb{C} : |\arg(1 - z)| \leq \alpha \text{ and } |z - 1| \leq \cos\alpha\}. \tag{3.1.1}$$

As a consequence, domain D_α is a closed convex subset of the (open) unit disc $\mathbb{D} \subset \mathbb{C}$, with the "angle" one-point contact of boundary $\partial\mathbb{D}$ at $z = 1$, see Fig. 3.1.

On account of definition (3.1.1) we obtain that $D_\alpha \subset D_{\alpha'>\alpha}$ for $\alpha' \in (0, \pi/2)$.

Definition 3.1.2 (Quasi-Sectorial Contractions) Contraction operator $C \in \mathcal{L}(\mathfrak{H})$ on a Hilbert space $\mathfrak{H}$ is called *quasi-sectorial* contraction with semi-angle $\alpha \in (0, \pi/2)$ with respect to the vertex at $z = 1$, if its numerical range $W(C) \subseteq D_\alpha$.

We comment that $D_\alpha \subset D_{\pi/2} := \overline{\mathbb{D}}$ and recall that a generic *contraction* operator $\|C\| \leq 1$ satisfies a weaker condition: $W(C) \subseteq \overline{\mathbb{D}}$, see Sect. 1.8.1, and therefore, it may have a *"tangent"* contact with boundary $\partial\mathbb{D}$ at $z = 1$.

Note that if operator C is a quasi-sectorial contraction, then $\mathbb{1} - C$ is, by Definition 1.8.9, an m-sectorial operator with semi-angle α. The limit cases: $\alpha = 0$ and $\alpha = \pi/2$ for $\mathbb{1} - C$ correspond, respectively, to non-negative self-adjoint and to m-accretive (Definition 1.8.7) bounded operators.

We claim that resolvent of an m-sectorial operator A, with semi-angle $\alpha \in (0, \alpha_0]$, for some $\alpha_0 < \pi/2$ yields the simplest nontrivial example of a quasi-sectorial contraction if $\alpha_0 = \pi/4$.

Lemma 3.1.3 *Let A be m-sectorial operator with semi-angle $\alpha \in (0, \pi/4]$. Then $\{F(t) := (\mathbb{1} + tA)^{-1}\}_{t>0}$ is a family of quasi-sectorial contractions which numerical ranges $W(F(t)) \subseteq D_\alpha$ for any $t \geq 0$.*

Proof By condition on A and Definition 1.8.9 the numerical range

$$W(A) = \{(Au, u) \in \mathbb{C} : u \in \mathrm{dom}(A)\,, \;\; \|u\| = 1\} \subseteq \overline{S}_\alpha\,. \tag{3.1.2}$$

Here S_α is open sector $S_\alpha = \{z \in \mathbb{C} : z \neq 0\,,\, |\arg(z)| < \alpha < \pi/2\}$. Then in virtue of Proposition 1.8.3 we obtain from (1.8.3) the estimate:

$$\|F(t)\| = \frac{1}{t}\left\|\frac{\mathbb{1}}{1/t\,\mathbb{1} + A}\right\| \leq \frac{1}{t\,\mathrm{dist}(1/t\,,\, -\overline{S}_\alpha)} = 1\,, \quad t > 0\,. \tag{3.1.3}$$

As a consequence, the family of operators $\{F(t)\}_{t>0}$ consists of contractions with numerical range $W(F(t)) \subseteq \overline{\mathbb{D}}$.

Next, on account of (3.1.2) one gets $(F(t)u, u) = (v_t, v_t) + t(v_t, A\,v_t) \in \overline{S}_\alpha$ for $v_t := F(t)\,u$ and $u \in \mathfrak{H}$, $\|u\| = 1$. This means that for any $t > 0$ the numerical range $\{W(F(t)) \subseteq \overline{S}_\alpha$. Similarly, one finds that $((\mathbb{1} - F(t))u, u) = t(A\,v_t, v_t) + t^2(A\,v_t, A\,v_t) \in \overline{S}_\alpha$, that is, $W(\mathbb{1} - F(t)) \subseteq \overline{S}_\alpha$, or $W(F(t)) \subseteq (1 - \overline{S}_\alpha)$. Finally, since by (3.1.2) one has $W(\mathbb{1}) = 1$ and $W(0\,\mathbb{1}) = 0$, for any $t \geq 0$ and $\alpha \leq \pi/3$ we obtain:

$$W(F(t)) \subseteq (\overline{S}_\alpha \cap (1 - \overline{S}_\alpha)) \subseteq \overline{\mathbb{D}}\,. \tag{3.1.4}$$

On the other hand, by Definition 3.1.1 one infers that inclusion:

$$(\overline{S}_\alpha \cap (1 - \overline{S}_\alpha)) \subseteq D_\alpha\,, t \geq 0_{,},$$

is possible only if $\alpha \leq \pi/4$, see see Fig. 3.1, $\alpha' = \pi/4$. Then in virtue of (3.1.4) for these values of α the operators $\{F(t)\}_{t\geq 0}$ are quasi-sectorial contractions with numerical ranges in D_α , with the common point $z = 1$. □

The following statement establishes a relation between *sectorial* generators and *quasi-sectorial* contraction semigroups.

Theorem 3.1.4 *Let A be an m-sectorial operator with semi-angle $\alpha \in (0, \pi/2)$. Then $\{e^{-tA}\}_{t\geq 0}$ is a holomorphic in sector $S_{\pi/2-\alpha}$ and quasi-sectorial contraction semigroup with numerical range $W(e^{-tA}) \subseteq D_\alpha$ for $t \geq 0$.*

For the proof of this theorem we need some preliminary statements.

Note that holomorphic extension of the semigroup, which is claimed in Theorem 3.1.4, results from Proposition 1.8.13, where it is also proved that this semigroup is a contraction. The rest of the proof is based on a series of lemmata and on the following *global Kato mapping* theorem that control localisation of the operator numerical range.

Proposition 3.1.5 *Let $f(z)$ be a rational function on the complex plane $\mathbb{C}$, with $f(\infty) = \infty$. Let for some compact and convex set $E' \subset \mathbb{C}$ the inverse function $f^{-1} : E' \mapsto E \supseteq K$, where K is a convex kernel of E, that is, a subset of E such that E is star-shaped relative to any $z \in K$. If C is operator with the numerical range $W(C) \subseteq K$, then $W(f(C)) \subseteq E'$.*

See Notes to Chap. 3 (Notes to Sects. 3.1, 3.1.1) and Appendix C (Sect. C.2).

We remark that for a convex set E the *maximal* kernel K coincides with E, although for nonconvex E the kernel K may be *empty*, see (C.2.1). In the latter case the Proposition 3.1.5 is of no use. The assertion is called *global* because the behaviour of function f in the whole plane $\mathbb{C}$ is essential. One can find comments on some *local Kato mapping* theorems for the operator numerical range in Appendix C (Sect. C.2).

Lemma 3.1.6 *Let $f_n(z) = z^n$ be complex functions, for $z \in \mathbb{C}$ and $n \in \mathbb{N}$. Then the sets $f_n(D_\alpha)$ are convex and domains $f_n(D_\alpha) \subseteq D_\alpha$ for any $n \in \mathbb{N}$, if $\alpha < \pi/2$.*

Lemma 3.1.7 (Euler Formula) *Let A be an m-sectorial operator, that is, with semi-angle less than $\pi/2$ and vertex at zero. Then for $t \geq 0$ one gets the strong limit*

$$\text{s-}\lim_{n\to\infty} (\mathbb{1} + tA/n)^{-n} = e^{-tA} , \tag{3.1.5}$$

for $\alpha \in (0, \pi/4]$. Here $\{e^{-tA}\}_{t\geq 0}$ is holomorphic contraction semigroup in sector $S_{\pi/4}$.

The last part of this statement can be proven independently of the Euler formula (3.1.5), see Proposition 1.8.13. We need the explicit construction (3.1.5) of semigroup for description of numerical range $\{W(e^{-tA})\}_{t>0}$.

Remark 3.1.8 The rest of the section is reserved for the proofs. These proofs are instructive and straightforward, but the upper bound: $\alpha = \pi/4$, in Lemma 3.1.7, is far from the optimal $\alpha < \pi/2$ as in Lemma 3.1.6. Note that restriction $\alpha \leq \pi/4$ is because of Lemma 3.1.3 and remark that $(\overline{S}_\alpha \cap (1 - \overline{S}_\alpha)) \not\subseteq D_\alpha$ for $\alpha > \pi/4$, see Definition 3.1.1 and Fig. 3.1. We remove this bound and restore the optimal $\pi/2$ in the next Sect. 3.2.

Proof of Lemma 3.1.6 First we check that the set D_α is closed with respect to *multiplication*: that is $z_1, z_2 \in D_\alpha$ implies $z_1 \cdot z_2 \in D_\alpha$.

Since D_α is convex, for any $z \in D_\alpha$ one has representation $z = r\,\zeta$, where $0 \leq r < 1$ and $\zeta \in \partial D_\alpha$. Hence, it is sufficient to check this assertion only for the corresponding boundary points: $\zeta_1, \zeta_2 \in \partial D_\alpha$.

By Definition 3.1.1 the boundary $\partial D_\alpha = \Gamma(d_1, d_2) \cup [1, d_1] \cup [1, d_2]$, where the points: $d_1 := a \sin\alpha + i\, a\cos\alpha$ and $d_2 := a\sin\alpha - i\, a\cos\alpha$, see Fig. 3.1. Since the circle arc $\Gamma(d_1, d_2)$ has radius $a = \sin\alpha < 1$, then $|\zeta_1\zeta_2| < a$ if at least one the factors ζ_1, or ζ_2, belongs to this arc, that is, the product $\zeta_1 \cdot \zeta_2$ and consequently $z_1 \cdot z_2$ belong to the set D_α.

Now let at least one of factors: ζ_1, or ζ_2, belongs to the union of two intervals $[1, d_1] \cup [1, d_2]$. For the sake of definiteness suppose that $\zeta_1 \in [1, d_1]$. Then we fix arbitrary $\zeta_2 \in \partial D_\alpha$ and note that the function

$$g : \zeta \mapsto \zeta \cdot \zeta_2\,, \; \zeta \in [1, d_1] \tag{3.1.6}$$

maps the interval $[1, d_1]$ onto the interval $[\zeta_2, d_1\,\zeta_2]$. Since $|d_1| = a < 1$, the both end-points of this interval are in the closure $\overline{D}_\alpha$. Then convexity of D_α implies that $[\zeta_2, d_1\,\zeta_2] \subset \overline{D}_\alpha$. Hence, $\zeta_1 \cdot \zeta_2 \in \overline{D}_\alpha$. Therefore, by virtue of $z = r\,\zeta$ and $0 \leq r < 1$ one gets that the corresponding to $\zeta_1 \cdot \zeta_2$ product $z_1 \cdot z_2 \in D_\alpha$.

If $z \in D_\alpha$, then the above arguments yield that $f_2(z) = z^2 \in D_\alpha$ and by induction $f_n(z) \in D_\alpha$, which implies $f_n(D_\alpha) \subseteq D_\alpha$. □

Proof of Lemma 3.1.7 Since operator A is m-sectorial, the holomorphic and contraction properties of semigroup $\{e^{-zA}\}_{z \in S_{\pi/2-\alpha}}$ follow from Proposition 1.8.13.

On account of Lemma 3.1.3 we also obtain that

$$\|(\mathbb{1} + t\,A)^{-1}\| \leq 1\,, \quad t \geq 0\,, \tag{3.1.7}$$

as well as that $\{(\mathbb{1} + t\,A)^{-1}\}_{t\geq 0}$ is for $\alpha \in (0, \pi/4]$ a family of quasi-sectorial contractions. Because of (3.1.7) the m-sectorial operator A is also *accretive* (1.8.10). Then standard arguments used for construction of C_0-semigroups (Proposition 1.1.1 and (1.1.4)) yield due to (3.1.7) for approximants $\{(\mathbb{1} + t\,A/n)^{-n}\}_{t\geq 0, n\in\mathbb{N}}$ the Euler formula (3.1.5) that converges in the strong operator topology to contraction semigroup $\{e^{-tA}\}_{t\geq 0}$, which was established in (1.1.15). □

Now we show that limit (3.1.5) yields in fact a quasi-sectorial contraction. Note that by a simple argument in Lemma 3.1.3 $\{(\mathbb{1} + t\,A)^{-1}\}_{t\geq 0}$ is a family of quasi-sectorial contractions with numerical range in D_α, for $\alpha \in (0, \pi/4]$. A heritage of this property by approximants $\{(\mathbb{1} + t\,A/n)^{-n}\}_{t\geq 0, n\in\mathbb{N}}$ and also by the limit $\{e^{-t\,A}\}_{t\geq 0}$ is due to the reasoning below. The proof will be based on the *numerical range mapping* arguments due to Proposition 3.1.5 and on Lemma 3.1.6.

Proof of Theorem 3.1.4 According to Lemma 3.1.6 and following Proposition 3.1.5, for $z \in \mathbb{C}$ and the rational function $f_2 : z \mapsto z^2$, we take $E' := f_2(D_\alpha) \subseteq D_\alpha$ as a compact convex set in $\mathbb{C}$. Then the inverse function $f_2^{-1} : z \mapsto \pm\sqrt{z}$ yields $E := f_2^{-1}(E') = D_\alpha \cup (-D_\alpha)$, which is, in turn, *convex* set. As a consequence, the convex *kernels* K of E exist and we take $K = D_\alpha$. Note that the *maximal* kernel $K = E$. Hence, by the Kato mapping theorem (Proposition 3.1.5) we conclude that if for operator C the numerical range $W(C) \subseteq K = D_\alpha$, then one gets

$$W(f_2(C)) \subseteq E' = f_2(D_\alpha) \subseteq D_\alpha = K\,, \tag{3.1.8}$$

where the inclusion is due to Lemma 3.1.6.

Now, let contraction $C_1 := (\mathbb{1} + t\,A/2)^{-1} = F(t/2)$. Then by Lemma 3.1.3 we have $W(C_1) \subseteq D_\alpha$ for any $t \geq 0$. Since the convex $D_\alpha \subset E = D_\alpha \cup (-D_\alpha)$, as above we can choose $K = D_\alpha$ as as a kernel of E and infer that $W(F(t/2)) \subseteq K$. Then by the *Kato mapping theorem* (Proposition 3.1.5) and Lemma 3.1.6 applied as in (3.1.8) to operator C_1 we obtain:

$$W(f_2(C_1)) = W(F(t/2)^2) \subseteq E' = f_2(D_\alpha) \subseteq D_\alpha = K\,, \quad t \geq 0\,. \tag{3.1.9}$$

That is, $W(F(t/2)) \subseteq K$ implies $W(F(t/2)^2) \subseteq K$ for $K = D_\alpha$.

Similarly, this *numerical range mapping* arguments can be applied to contraction $C_2 := F(t/4)^2$. Indeed, since (3.1.9) is valid for any $t \geq 0$, it is also true after substitution: $t \mapsto t/2$, in the argument of $F(\cdot)$. On account of (3.1.9) one obtains that $W(C_2) \subseteq K$. Then again as in (3.1.8) the Proposition 3.1.5 together with Lemma 3.1.6 provide:

$$W(f_2(C_2) = W(F(t/4)^4)) \subseteq E' = f_2(D_\alpha) \subseteq D_\alpha = K\,, \quad t \geq 0\,. \tag{3.1.10}$$

Owing to (3.1.9) and (3.1.10), we obtain by induction: $W(F(t/2^n)^{2^n}) \subseteq D_\alpha$, for any $n \in \mathbb{N}$. Then on account of Lemma 3.1.7 the strong limit (3.1.5) yields

$$\lim_{n\to\infty} ((\mathbb{1} + t\,A/2^n)^{-2^n} u, u) = (e^{-t\,A} u, u) \in D_\alpha\,, \quad t \geq 0\,, \tag{3.1.11}$$

for unit vectors $u \in \mathfrak{H}$. As a consequence, in virtue of (3.1.2) and (3.1.11), the numerical range of semigroup $\{e^{-t\,A}\}_{t\geq 0}$ for each $t \geq 0$ belongs to the set D_α: $W(e^{-t\,A}) \subseteq D_\alpha$, (3.1.1). This proves that semigroup generated by m-sectorial operator A with semi-angle $\alpha \in (0, \pi/2)$ is a quasi-sectorial contraction semigroup, which is holomorphic thanks to Proposition 1.8.13. □

As we indicated in Remark 3.1.8 certain results of the present Sect. 3.1 are undeservedly limited by $\alpha \leq \pi/4$. In Sect. 3.2 (see Remark 3.2.15) we shall extend Lemma 3.1.3 to semi-angle $\alpha \in (0, \pi/2)$ merely by some more refined arguments, which will be also useful for other applications.

3.1.2 *Chernoff Estimate: Revisited*

As we see, the Definition 3.1.2 of *quasi-sectorial* contractions C is quite restrictive comparing to the notion of *general* contractions, which demands *only* $W(C) \subseteq \overline{\mathbb{D}}$ for the unit disc $\mathbb{D} \subset \mathbb{C}$. Recall that in the latter case the assertion of Lemma 1.9.33, known as the *Chernoff* $\sqrt{n}$-Lemma, and the corresponding estimate (1.9.96), known as the *Chernoff estimate*, are valid.

Seeing that the Chernoff estimate will be useful further as one of the instrument for *uplifting* (improving) the Chernoff (1.9.101) and Trotter-Kato (1.9.110), (1.9.127) product formulæ from the *strong* to the *operator-norm* convergence, we scrutinise this estimate below.

The next n-Lemma yields a straightforward revision of the Chernoff estimate (1.9.96). In fact, this revision is still restricted to convergence in the *strong* operator topology on a Banach space $\mathfrak{X}$.

Lemma 3.1.9 **(n-Lemma)** *Let $C \in \mathcal{L}(\mathfrak{X})$ be a contraction on Banach space $\mathfrak{X}$. Then $\{e^{t(C-\mathbb{1})}\}_{t\geq 0}$ is a norm-continuous contraction semigroup on $\mathfrak{X}$ and the following estimate*

$$\|(C^n - \mathrm{e}^{n(C-\mathbb{1})})\,x\| \leq \frac{n}{2}\left(\|(C-\mathbb{1})^2\,x\| + \frac{e^2}{3}\,\|(C-\mathbb{1})^3\,x\|\right) \tag{3.1.12}$$

holds for all $n \in \mathbb{N}$ and $x \in \mathfrak{X}$.

Proof Since operator C is bounded and $\|C\| \leq 1$, the operator $(\mathbb{1} - C)$ is generator of a norm-continuous contraction semigroup since:

$$\|\,\mathrm{e}^{-t\,(\mathbb{1}-C)}\,\| \leq \mathrm{e}^{-t}\left\|\sum_{m=0}^{\infty}\frac{t^m}{m!}C^m\right\| \leq 1\,, \quad t \geq 0\,. \tag{3.1.13}$$

To prove inequality (3.1.12) we use the *telescopic* representation:

$$C^n - e^{n(C-\mathbb{1})} = \sum_{k=0}^{n-1} C^{n-k-1}\,(C - \mathrm{e}^{(C-\mathbb{1})})\,\mathrm{e}^{\,k(C-\mathbb{1})}\,. \tag{3.1.14}$$

To proceed we exploit that operator $C \in \mathcal{L}(\mathfrak{X})$ is bounded and therefore

$$C - e^{(C-\mathbb{1})} = -\frac{1}{2}(\mathbb{1} - C)^2 - (\mathbb{1} - C)^3 \sum_{m=3}^{\infty} \frac{(-1)^m}{m!} (\mathbb{1} - C)^{m-3}, \tag{3.1.15}$$

for the operator-norm convergent series. Hence, owing to $\|C\| \leq 1$ one gets estimate

$$\Big\| \sum_{m=3}^{\infty} \frac{1}{m!} (\mathbb{1} - C)^{m-3} \Big\| \leq \frac{1}{6}\, e^{\|\mathbb{1}-C\|} \leq \frac{e^2}{6}. \tag{3.1.16}$$

Then on account of (3.1.14)–(3.1.16) and contraction semigroup in (3.1.13) we obtain inequality (3.1.12). □

Now similarly to Theorem 1.9.34 we obtain for estimate (3.1.12) the Chernoff product theorem, which is owing to the n-Lemma.

Theorem 3.1.10 *Let $\Phi : t \mapsto \Phi(t)$ be a function from $\mathbb{R}_0^+$ to contractions on $\mathfrak{X}$ such that $\Phi(0) = \mathbb{1}$, which satisfies conditions of Theorem* 1.9.34, *that is, closed right derivative $\Phi'(+0)\,x := -A\,x$ for all $x \in D = \mathrm{core}\,(A)$, where A is generator of contraction C_0-semigroup. Then*

$$\lim_{n\to\infty} \|([\Phi(t/n)]^n - e^{n(\Phi(t/n)-\mathbb{1})})\, x\| = 0, \quad x \in \mathfrak{X}, \tag{3.1.17}$$

and as a result one gets the Chernoff product formula (cf. (1.9.101))

$$\lim_{n\to\infty} [\Phi(t/n)]^n\, x = e^{-t\,A}\, x, \tag{3.1.18}$$

for all $t \in \mathbb{R}_0^+$ and $x \in \mathfrak{X}$.

Proof If $C := \Phi(t/n)$, then on account of Lemma 3.1.9 we obtain by (3.1.12) for $t > 0$ the estimate

$$\|([\Phi(t/n)]^n - e^{n(\Phi(t/n)-\mathbb{1})})\, x\| \leq \tag{3.1.19}$$
$$\frac{t^2}{2n} \left\{ \Big\| \frac{n^2}{t^2} (\mathbb{1} - \Phi(t/n))^2\, x \Big\| + \frac{2\, e^2}{3} \Big\| \frac{n^2}{t^2} (\mathbb{1} - \Phi(t/n))^2\, x \Big\| \right\}, \quad x \in \mathfrak{X}.$$

Note that by condition of theorem (cf. (1.9.103)) for any $t \in \mathbb{R}^+$ we infer on the dense set $D = \mathrm{core}(A)$:

$$\lim_{n\to\infty} \frac{n}{t}\, (\mathbb{1} - \Phi(t/n))\, x = A\, x, \quad x \in D. \tag{3.1.20}$$

Seeing that generator A of contraction C_0-semigroup is *accretive*, the range of resolvent: $\mathrm{ran}((A + \zeta\mathbb{1})^{-1}) = \mathfrak{X}$, for $\mathrm{Re}(\zeta) > 0$. As a consequence (see

Corollary A.1.9(a)) domain dom $(A^2) \subset$ dom (A) is *dense* in $\mathfrak{X}$. Moreover, by Proposition 1.2.10 one infers that dom (A^2) is a *core* for A. On that account, the limit (3.1.20) provides for any $t > 0$

$$\lim_{n\to\infty} (A_n(t))^2\, x = A^2\, x\,, \quad x \in D\,, \tag{3.1.21}$$

where $A_n(t) := (t/n)^{-1}\,(\mathbb{1} - \Phi(t/n))$. Then in virtue of estimate (3.1.19) and limit (3.1.21) we obtain

$$\lim_{n\to\infty} \|([\Phi(t/n)]^n - \mathrm{e}^{-t\,A_n(t)})\, x\,\| = 0\,, \quad x \in D\,. \tag{3.1.22}$$

Due to estimate: $\|[\Phi(t/n)]^n - \mathrm{e}^{-t\,A_n(t)}\,\| \leq 2$, the topology of the *pointwise* convergence (3.1.22) on a *dense* subset $D \subset \mathfrak{X}$ coincides with the *strong* operator topology, see Proposition A.1.6. So, the limit (3.1.22) can be extended to $x \in \mathfrak{X}$ (cf. Corollary A.1.9(b)), that is, to (3.1.17).

Finally, given that D = core(A), because of (3.1.20) and Remark 1.9.31 the *Trotter-Kato-Neveu* theorem (Proposition 1.9.30) yields the limit

$$\lim_{n\to\infty} \mathrm{e}^{-t\,A_n(t)}\, x = \mathrm{e}^{-t\,A}\, x\,, \quad x \in \mathfrak{X}, \quad t \in \mathbb{R}^+\,, \tag{3.1.23}$$

cf. (1.9.104). Then owing to (3.1.22) and (3.1.23) for $x \in \mathfrak{X}$, we deduce the *Chernoff product formula* (3.1.18). □

We note that $\sqrt{n}$-Lemma 1.9.33 and n-Lemma 3.1.9 give the Chernoff estimates (1.9.96) and (3.1.12) with growing (for $n \to \infty$) upper bounds proportional to $\sqrt{n}$ and correspondingly to n. The next revision of the Chernoff estimate ($\sqrt[3]{n}$-Lemma) has the same property, but with the *smaller* rate of growing, which is proportional to $\sqrt[3]{n}$. Moreover, a revised scheme for the proof of this rate will be used then for *uplifting* the convergence of the Chernoff and the Lie-Trotter product formulæ to the *operator-norm* topology in the case of *quasi-sectorial* contractions.

Lemma 3.1.11 **($\sqrt[3]{n}$ -Lemma)** *Let C be a contraction on a Banach space $\mathfrak{X}$. Then $\{\mathrm{e}^{t(C-\mathbb{1}}\}_{t\geq 0}$ is a norm-continuous contraction semigroup on $\mathfrak{X}$ and one infers the estimate*

$$\|(C^n - \mathrm{e}^{n\,(C-\mathbb{1})})\, x\| \leq \frac{n}{\epsilon_n^2}\, 2\,\|x\| + \epsilon_n\,\|(\mathbb{1} - C)\, x\,\|\,, \quad n \in \mathbb{N}\,, \tag{3.1.24}$$

for all $x \in \mathfrak{X}$ and $\epsilon_n > 0$. For the optimal (minimiser) value of the parameter ϵ_n :

$$\epsilon_n^* := \left(\frac{4\,n\,\|x\|}{\|(\mathbb{1} - C)\, x\,\|}\right)^{1/3}, \tag{3.1.25}$$

in the right-hand side of (3.1.24) *one gets the estimate*

$$\|(C^n - \mathrm{e}^{n\,(C-\mathbb{1})})\| \leq \frac{3}{2}\,\sqrt[3]{n}\,\|2\,(\mathbb{1}-C)\,\|^{2/3}\,, \tag{3.1.26}$$

which is the claim of the $\sqrt[3]{n}$ *-Lemma.*

Proof Since operator C is bounded and $\|C\| \leq 1$, the operator $(\mathbb{1} - C)$ is generator of a norm-continuous and contraction semigroup:

$$\|\,\mathrm{e}^{\,t(C-\mathbb{1})}\,\| \leq \mathrm{e}^{-t}\left\|\sum_{m=0}^{\infty}\frac{t^m}{m!}C^m\right\| \leq 1\,.$$

For proving the estimate (3.1.24) we use representation

$$C^n - \mathrm{e}^{n(C-\mathbb{1})} = \mathrm{e}^{-n}\sum_{m=0}^{\infty}\frac{n^m}{m!}(C^n - C^m)\,. \tag{3.1.27}$$

Then we split the sum (3.1.27) into two parts: the *central* part for $|m-n| \leq \epsilon_n$ and the *tails* for $|m-n| > \epsilon_n$. Optimisation of the *splitting* parameter ϵ_n in (3.1.24) yields the best estimate and thus the optimal value of $\epsilon_n \in \mathbb{R}$.

For evaluation the tails we use the *Tchebychëv inequality*. Let $X_n \in \mathbb{N}_0$ be the *Poisson random variable* with the *rate* parameter n, that is, with the probability distribution $\mathbb{P}\{X_n = m\} = n^m \mathrm{e}^{-n}/m!$. Then one gets for the expectation: $\mathbb{E}(X_n) = n$, and for the variance: $\mathrm{Var}(X_n) := \mathbb{E}((X_n - \mathbb{E}(X_n))^2) = n$. That being so, the Tchebychëv inequality yields

$$\mathbb{P}\{|X_n - \mathbb{E}(X_n)| > \epsilon\} \leq \frac{\mathrm{Var}(X_n)}{\epsilon_n^2}\,, \quad \text{for any } \epsilon_n > 0. \tag{3.1.28}$$

Note that although for any $x \in \mathfrak{X}$ there is an evident bound: $\|(C^n - C^m)\,x\| \leq 2\,\|x\|$, for estimating (3.1.27) we shall also use below the *telescopic* inequality:

$$\begin{aligned}\|(C^n - C^m)\,x\| &= \|C^{n-k}(C^k - C^{m-n+k})\,x\| \\ &\leq |m-n|\;\|C^{n-k}(\mathbb{1}-C)\,x\|, \quad k = 0, 1, \ldots, n\,.\end{aligned} \tag{3.1.29}$$

Then by $\|C\| \leq 1$ and by the *Tchebychëv inequality* (3.1.28) we can estimate in (3.1.27) the operator-norm of tails:

$$\begin{aligned}\mathrm{e}^{-n}\sum_{|m-n|>\epsilon_n}\frac{n^m}{m!}\|(C^n - C^m)\,x\| &\leq \mathrm{e}^{-n}\sum_{|m-n|>\epsilon_n}\frac{n^m}{m!}\cdot 2\,\|x\| \\ = \mathbb{P}\{|X_n - \mathbb{E}(X_n)| > \epsilon_n\}\cdot 2\,\|x\| &\leq \frac{n}{\epsilon_n^2}\,2\,\|x\|\,, \quad x \in \mathfrak{X}\,.\end{aligned} \tag{3.1.30}$$

To evaluate the central part of the sum (3.1.27), when $|m-n| \leq \epsilon_n$, note that in virtue of (3.1.29) one gets the estimate

$$\begin{aligned} \|(C^n - C^m)x\| &\leq |m-n| \; \|C^{n-[\epsilon_n]} (\mathbb{1} - C)x\| \\ &\leq \epsilon_n \; \|(\mathbb{1} - C)\, x\| \, , \end{aligned} \tag{3.1.31}$$

which contains the difference $(\mathbb{1} - C)\, x$. Then we obtain for the central part of the sum (3.1.27) estimate

$$\mathrm{e}^{-n} \sum_{|m-n| \leq \epsilon_n} \frac{n^m}{m!} \; \|(C^n - C^m)\, x\| \leq \epsilon_n \; \|(\mathbb{1} - C)\, x\| \, , \quad x \in \mathfrak{X} \, , \tag{3.1.32}$$

for $n \in \mathbb{N}$. As a consequence, estimate (3.1.32), together with (3.1.30), yields (3.1.24) for all $x \in \mathfrak{X}$ and $\epsilon_n > 0$.

Minimising the right-hand side of (3.1.24) with respect to parameter $\epsilon_n > 0$ we workout out its optimal value (3.1.25) and the upper bound:

$$\frac{n}{\epsilon_n^{*2}} \, 2\, \|x\| + \epsilon_n^* \; \|(\mathbb{1} - C)\, x\,\| = \frac{3}{2} \, \sqrt[3]{n} \, (4\, \|x\|)^{1/3} \; \|(\mathbb{1} - C)\, x\,\|^{2/3} \, , \tag{3.1.33}$$

for all $x \in \mathfrak{X}$ and $n \in \mathbb{N}$. Therefore, (3.1.24) and (3.1.33) yield estimate (3.1.26), which is the $\sqrt[3]{n}$ -Lemma. □

Now similarly to Theorem 3.1.10 we obtain for estimate (3.1.26) the Chernoff product theorem, which is owing to the $\sqrt[3]{n}$ -Lemma.

Theorem 3.1.12 *Let $\Phi : t \mapsto \Phi(t)$ be a function from $\mathbb{R}_0^+$ to contractions on $\mathfrak{X}$ such that $\Phi(0) = \mathbb{1}$. Let $\{U_A(t) = \mathrm{e}^{-tA}\}_{t \geq 0}$ be a contraction C_0-semigroup, and let domain $D \subset \mathrm{dom}(A)$ be a core for the related generator A.*

If the function $\Phi(t)$ has a strong right derivative $\Phi'(+0)$ at $t = 0$ (that is, $\Phi'(+0)x$ exists for any $x \in \mathrm{dom}(\Phi'(+0))$) and if

$$\Phi'(+0)\, x := \lim_{t \to +0} \frac{1}{t} (\Phi(t) - \mathbb{1})\, x = -\, A\, x \, , \tag{3.1.34}$$

for all $x \in D$, then

$$\lim_{n \to \infty} [\Phi(t/n)]^n \, x = U_A(t)\, x \, , \tag{3.1.35}$$

for all $t \in \mathbb{R}_0^+$ and $x \in \mathfrak{X}$.

Proof Consider the bounded approximations $\{A_n(s)\}_{n \geq 1}$ of generator A:

$$A_n(s) := \frac{\mathbb{1} - \Phi(s/n)}{s/n} \, , \quad s \in \mathbb{R}^+ \, , \quad n \in \mathbb{N} \, . \tag{3.1.36}$$

Note that these operators are m-accretive: $\|(A_n(s) + \zeta\, \mathbb{1})^{-1}\| \leq (\mathrm{Re}(\zeta))^{-1}$ for $\mathrm{Re}(\zeta) > 0$ and for any $n \in \mathbb{N}$. By $\|\Phi(t)\| \leq 1$, together with (3.1.36) and condition (3.1.34), we obtain $\|e^{-t\,A_n(s)}\| \leq 1$, and also

$$\lim_{n\to\infty} A_n(s)\, x = A\, x\,, \tag{3.1.37}$$

for all $x \in D$ and any $s \in \mathbb{R}^+$. Then, given that $D = \mathrm{core}(A)$, by virtue of the *Trotter-Kato-Neveu* theorem (Proposition 1.9.30 and Remark 1.9.31) one gets

$$\lim_{n\to\infty} \mathrm{e}^{-t\,A_n(s)}\, x = U_A(t)\, x\,, \quad x \in \mathfrak{X}, \quad t \in \mathbb{R}_0^+\,, \quad s \in \mathbb{R}^+\,. \tag{3.1.38}$$

So, the limit (3.1.38) yields a strong and uniform, in $t \in [0, \tau]$ and in $s \in (0, s_0]$, convergence of contraction approximants $\{e^{-t\,A_n(s)}\}_{n\geq 1}$ to semigroup $\{U_A(t)\}_{t\geq 0}$.

Now, by Lemma 3.1.11 (3.1.26) we obtain for contraction $C := \Phi(t/n)$:

$$\begin{aligned} &\|[\Phi(t/n)]^n\, x - \mathrm{e}^{-t\,A_n(t)}\, x\| = \|([\Phi(t/n)]^n - \mathrm{e}^{n(\Phi(t/n)-\mathbb{1})})\, x\| \\ &\leq \frac{3}{2}\sqrt[3]{n}\,(4\,\|x\|)^{1/3}\, \|(\mathbb{1} - \Phi(t/n))\, x\,\|^{2/3}\,, \quad x \in \mathfrak{X}\,. \end{aligned} \tag{3.1.39}$$

Since by (3.1.37) one gets for any $x \in D$ and $t \in \mathbb{R}_0^+$

$$\lim_{n\to\infty} \sqrt[3]{n}\, \|\,(\mathbb{1} - \Phi(t/n))\, x\,\|^{2/3} = \lim_{n\to\infty} t^{2/3}\, n^{-1/3}\, \|A_n(t)\, x\|^{2/3} = 0\,, \tag{3.1.40}$$

estimate (3.1.39) and limit (3.1.40) provide uniformly on $[0, t_0]$

$$\lim_{n\to\infty} \|\,[\Phi(t/n)]^n\, x - \mathrm{e}^{-t\,A_n(t)}\, x\| = 0, \quad x \in D\,. \tag{3.1.41}$$

Then (3.1.38) and (3.1.41) yield uniformly in $t \in [0, t_0]$ the limit:

$$\lim_{n\to\infty} [\Phi(t/n)]^n\, x = U_A(t)\, x\,, \quad x \in D\,. \tag{3.1.42}$$

On account of Proposition A.1.6 , by density of D and by estimate $\|\,[\Phi(t/n)]^n - \mathrm{e}^{-t\,A_n(t)}\| \leq 2$ the convergence in (3.1.41) can be extended to all $x \in \mathfrak{X}$. As a consequence, the limit (3.1.41), when being extended to $x \in \mathfrak{X}$, and limit (3.1.38) yield the extension of (3.1.42) to (3.1.35) for $t \in \mathbb{R}_0^+$. □

3.1.3 Operator-Norm Estimates and Convergence

The Chernoff $\sqrt{n}$ -estimate (1.9.96) for contractions due to the $\sqrt{n}$ -Lemma 1.9.33 can be drastically improved if the contractions are *quasi-sectorial.*

In fact, the idea of the proof of the $\sqrt[3]{n}$ -Lemma 3.1.11 allows to go beyond the $\sqrt[3]{n}$-estimate if the contraction in this lemma is quasi-sectorial. This improvement is based on the following observation:

Theorem 3.1.13 *If C is a quasi-sectorial contraction on a Hilbert space $\mathfrak{H}$ with semi-angle $0 \leq \alpha < \pi/2$, that is, the numerical range $W(C)$ is a subset of the domain D_α (Definition 3.1.1), then*

$$\|C^n(\mathbb{1} - C)\| \leq \frac{K}{n+1}, \quad n \in \mathbb{N}, \tag{3.1.43}$$

for some $K > 0$.

The property (3.1.43) yields that the quasi-sectorial contractions belong to the class of, so-called, *Ritt operators*. We postpone the proof of the estimate (3.1.43) to Sect. 8.2 (Lemma 8.2.1). Here we mention only (Lemma 3.1.14) how this estimate allows us to go beyond the $\sqrt[3]{n}$ -Lemma 3.1.11 to the operator-norm estimate that we call the *operator-norm* $(\sqrt[3]{n})^{-1}$-Lemma.

Lemma 3.1.14 (Operator-Norm $(\sqrt[3]{n})^{-1}$-Lemma) *Let C be a quasi-sectorial contraction on a Hilbert space $\mathfrak{H}$ with semi-angle $0 < \alpha < \pi/2$, that is, $W(C) \subset D_\alpha$. Then $\{e^{t(C-\mathbb{1})}\}_{t\geq 0}$ is a norm-continuous quasi-sectorial contraction semigroup on $\mathfrak{H}$ and there exist $n_0 \in \mathbb{N}$ and $M > 0$ such that the operator-norm estimate*

$$\| C^n - e^{n(C-\mathbb{1})} \| \leq \frac{M}{\sqrt[3]{n}} \tag{3.1.44}$$

holds for $n > n_0$, where $n \in \mathbb{N}$.

Proof Note that with help of the inequality (3.1.43) one can improve the estimate (3.1.24) in $\sqrt[3]{n}$ -Lemma 3.1.11. Let the splitting parameter in the sum (3.1.24) be $\epsilon_n := n^{\delta+1/2}$ ($\delta < 1/2$). Then on account of (3.1.43) one gets, instead of (3.1.31) inequalities:

$$\|C^n - C^m\| \leq |m-n| \, \| C^{n-[\epsilon_n]} (\mathbb{1} - C)\| \leq \epsilon_n \frac{K}{n - [\epsilon_n] + 1}, \quad n, m \in \mathbb{N}.$$

As a consequence, for $\delta < 1/2$ there exists $n_0 \in \mathbb{N}$ such that above inequalities, together with (3.1.30), give instead of (3.1.24) (or (1.9.96)) the operator-norm estimate

$$\left\| C^n - e^{n(C-\mathbb{1})} \right\| \leq \frac{2}{n^{2\delta}} + \frac{2K}{n^{1/2-\delta}}, \quad n > n_0. \tag{3.1.45}$$

Then operator-norm estimate $M/\sqrt[3]{n}$ in (3.1.44) stems owing to the optimal choice: $\delta = 1/6$, which minimises the right-hand side of (3.1.45) and yields (3.1.44). □

Remark 3.1.15

(a) As a consequence, in the case of quasi-sectorial contractions we find, instead of divergent, for $n \to \infty$, estimates: (1.9.96), (3.1.12) and (3.1.26), the estimate (3.1.44), which converges for $n \to \infty$ to zero in the operator-norm topology. Note that the rate $O(1/\sqrt[3]{n})$ of this convergence is obtained with help of the Poisson representation and the Tchebychëv inequality in the spirit of the proof of the Lemma 3.1.11 due to the splitting and improvement of (3.1.31) by (3.1.43). We comment that this rate is *not optimal*, see proof of the $(\sqrt[3]{n})^{-1}$-Theorem 8.2.2 in Sect. 8.2.

(b) As a matter of fact, the estimate $M/\sqrt[3]{n}$ can be improved by a more refined arguments and/or more conditions, see Notes to Chap. 3 (Sect. 3.1.3). For example, one can easily obtain the *ultimate* optimal estimate $O(n^{-1})$ in (3.1.44) for a particular case of self-adjoint (that is quasi-sectorial for $\alpha = 0$) contractions with help of the spectral calculus for self-adjoint operators.

(c) Now we consider application of $(\sqrt[3]{n})^{-1}$-Lemma 3.1.14 for proving the *Dunford-Segal* approximation of holomorphic quasi-sectorial contraction semigroups in the operator-norm topology.
Let A be an m-sectorial operator in Hilbert space $\mathfrak{H}$, with semi-angle $\alpha \in (0, \pi/2)$. Then arguments of Theorem 3.1.4 and Lemma 3.1.14 can be applied, *mutatis mutandis*, to the *Dunford-Segal* approximants $\{e^{-n(\mathbb{1}-\mathrm{e}^{-tA/n})}\}_{n\geq 1}$ for $t \geq 0$. Indeed, on account of Theorem 3.1.4 one gets that $\{e^{-tA/n}\}_{n\geq 1}$, for $t \geq 0$ and for each $n \in \mathbb{N}$, is a holomorphic quasi-sectorial contraction. If we put $C := \mathrm{e}^{-tA/n}$, then by $(\sqrt[3]{n})^{-1}$-Lemma 3.1.14 the operator-norm estimate (3.1.44) implies that

$$\left\| \mathrm{e}^{-tA} - \mathrm{e}^{-n(\mathbb{1}-\mathrm{e}^{-tA/n})} \right\| \leq \frac{M_{DS}}{n^{1/3}}, \quad t \geq 0, \quad n \in \mathbb{N}, \tag{3.1.46}$$

for some $M_{DS} > 0$. This yields on a Hilbert space $\mathfrak{H}$ the operator-norm approximation (3.1.46) of the quasi-sectorial contraction semigroup $\{e^{-tA}\}_{t\geq 0}$ by the Dunford-Segal approximants $\{e^{-n(\mathbb{1}-\mathrm{e}^{-tA/n})}\}_{n\geq 1}$ with a (*non-optimal*) estimate of the rate of convergence $O(1/n^{1/3})$.

(d) For other applications of quasi-sectorial contractions that generalise the Chernoff approximation theory for semigroups with m-sectorial generators see Chaps. 4 and 8.

A direct application in Remark 3.1.15 (c) of the operator-norm $(\sqrt[3]{n})^{-1}$-Lemma 3.1.14 is a particular case of the following necessary and sufficient conditions for the operator-norm approximations of quasi-sectorial contractions, *without* estimates.

Theorem 3.1.16 (Product Formula for Quasi-Sectorial Contractions) *Let $\{\Phi(s)\}_{s\geq 0}$ be a family of uniformly quasi-sectorial contractions on a Hilbert space $\mathfrak{H}$ such that there exists $0 < \alpha < \pi/2$ and $W(\Phi(s)) \subseteq D_\alpha$, for $s > 0$. Let*

$$X(s) := (\mathbb{1} - \Phi(s))/s\,, \quad s > 0\,, \tag{3.1.47}$$

and let X_0 be a closed operator with non-empty resolvent set, defined in $\mathfrak{H}$. Then the family $\{X(s)\}_{s>0}$ converges, when $s \to +0$, in the uniform resolvent sense (i.e., in the operator-norm topology) to the operator X_0 if and only if

$$\lim_{n\to\infty} \left\| \Phi(t/n)^n - \mathrm{e}^{-tX_0} \right\| = 0\,, \quad t \geq 0\,. \tag{3.1.48}$$

Proof In virtue of Theorem 3.1.13 and Lemma 3.1.14 the proof of this assertion follows *verbatim* the proof of Theorem 8.3.4. □

Corollary 3.1.17 (Operator-Norm Euler Formula) *If A is an m-sectorial operator in Hilbert space $\mathfrak{H}$, with semi-angle $\alpha \in (0, \pi/4]$, then*

$$\lim_{n\to\infty} \left\| (\mathbb{1} + tA/n)^{-n} - \mathrm{e}^{-tA} \right\| = 0\,, \quad t \geq 0\,. \tag{3.1.49}$$

Proof Since by condition the numerical range $W(A) \subset S_\alpha$, on account of Lemma 3.1.3 and Lemma 3.1.7, $\{\Phi(t) := (\mathbb{1} + tA)^{-1}\}_{t\geq 0}$ is a family of quasi-sectorial contractions with $W(\Phi(t)) \subset D_\alpha$ for $\alpha \in (0, \pi/4]$ and $t \geq 0$.

Let $X(s) := (\mathbb{1} - \Phi(s))/s,\ s > 0$, and $X_0 := A$. Then on account of estimate:

$$\left\| \frac{A}{\zeta\mathbb{1} + A + s\zeta A} \cdot \frac{A}{\zeta\mathbb{1} + A} \right\| \leq$$

$$\frac{1}{|1+s\zeta|}\left(1 + \frac{|\zeta|\,|1+s\zeta|^{-1}}{\mathrm{dist}\,(\zeta\,(1+s\zeta)^{-1}, -S_\alpha)}\right)\left(1 + \frac{|\zeta|}{\mathrm{dist}(\zeta, -S_\alpha)}\right), \quad \zeta \in S_{\pi-\alpha}\,,$$

the family $\{X(s)\}_{s>0}$ converges, when $s \to +0$, to operator X_0 in the uniform resolvent sense:

$$\|(\zeta\mathbb{1} + X(s))^{-1} - (\zeta\mathbb{1} + X_0)^{-1}\| = \frac{s\,|\zeta|^2}{|1+s\zeta|}\left\| \frac{A}{\zeta\mathbb{1} + A + s\zeta A} \cdot \frac{A}{\zeta\mathbb{1} + A} \right\| = O(s)\,. \tag{3.1.50}$$

Then owing to Theorem 3.1.16, (3.1.48), the limit (3.1.50) implies

$$\lim_{n\to\infty} \left\| \Phi(t/n)^n - e^{-tX_0} \right\| = 0\,, \quad t \geq 0\,, \tag{3.1.51}$$

which coincides with the Euler formula (3.1.49). □

Corollary 3.1.18 (Operator-Norm (Modified) Yosida Approximation) *For the same conditions as in Corollary* 3.1.17 *we introduce contraction* $C := (\mathbb{1} + tA/n)^{-1}$. *Then operator-norm estimate* (3.1.44) *yields*

$$\left\|(\mathbb{1} + tA/n)^{-n} - \mathrm{e}^{-tA/(\mathbb{1}+tA/n)}\right\| \leq \frac{M_Y}{n^{1/3}}, \quad t \geq 0, \quad n \in \mathbb{N}, \tag{3.1.52}$$

uniformly in $t \in [0, t_0]$, *for* $t_0 > 0$. *Bounded generators* $\{A_n(t) := A(\mathbb{1} + tA/n)^{-1}\}_{n\geq 1}$ *represent the (modified) Yosida approximates* $\{\mathrm{e}^{-tA_n(t)}\}_{n\geq 1}$. *Then because of* (3.1.49) *and* (3.1.52) *the estimate*

$$\left\|\mathrm{e}^{-tA} - \mathrm{e}^{-tA_n(t)}\right\| \leq \left\|\mathrm{e}^{-tA} - (\mathbb{1} + tA/n)^{-n}\right\| + \left\|(\mathbb{1} + tA/n)^{-n} - \mathrm{e}^{-tA_n(t)}\right\| \tag{3.1.53}$$

yields a (modified) Yosida approximation of holomorphic quasi-sectorial contraction semigroup $\{\mathrm{e}^{-tA}\}_{t\geq 0}$ *in the operator-norm topology.*

We remind that in contrast to the *modified* a well-known Yosida approximants of generator A are defined by $\{A_n := A_n(t = 1)\}_{n\geq 1}$.

Remark 3.1.19

(a) For assessment of the operator-norm rate of convergence in (3.1.49) one can use the inequality

$$\left\|\Phi(t/n)^n - e^{-tX_0}\right\| \leq \tag{3.1.54}$$
$$\|\Phi(t/n)^n - e^{-tX(t/n)}\| + \|e^{-tX(t/n)} - \mathrm{e}^{-tX_0}\|.$$

Note that on account of (3.1.54) the estimate of the rate of convergence in (3.1.51) is determined by a competition of *two* terms in the right-hand side of this inequality. The first term corresponds to a Chernoff-type estimate for quasi-sectorial contractions, which is controlled by the operator-norm $(\sqrt[3]{n})^{-1}$-Lemma 3.1.14 with $C = \Phi(t/n)$ and $X(s) = (\mathbb{1} - \Phi(s))/s$, $s > 0$. The second one is the *Trotter-Kato-Neveu* term $\|e^{-tX(t/n)} - \mathrm{e}^{-tX_0}\|$. So, to advance with operator-norm estimate in (3.1.54) one needs *uplifting* (improving) of the Trotter-Kato-Neveu theorem (Proposition 1.9.30) to operator-norm topology with an explicit estimate.

We comment here that the same observation holds for strongly convergent Chernoff product formula in Theorem 3.1.10 (see (3.1.22) and (3.1.23)), as well as in Theorem 3.1.12 (see (3.1.38) and (3.1.41)). But (in contrast to operator-norm estimates) one benefits there of existence of the Trotter-Kato-Neveu theorem in the strong operator topology, see Proposition 1.9.30.

Although in full generality we shall study the operator-norm uplifting of the Trotter-Kato-Neveu theorem in Chap. 8, here we take an advantage of (3.1.50) to deduce operator-norm estimate of the *Trotter-Kato-Neveu* term in (3.1.54).

Note that by conditions of Corollary 3.1.17 operators $\{X(s)\}_{s>0}$ and X_0 are uniformly m-sectorial with numerical ranges in sector S_α. Then contractions generated by $\{X(s)\}_{s>0}$, and X_0 can be expressed via the *Riesz-Dunford* integral formula (1.7.1), which yields for $t > 0$ the operator-norm estimate:

$$\left\| e^{-tX(s)} - e^{-tX_0} \right\| \leq \frac{1}{2\pi} \int_\Gamma |d\zeta| \, |e^{t\zeta}| \left\| (\zeta \mathbb{1} + X(s))^{-1} - (\zeta \mathbb{1} + X_0)^{-1} \right\| . \tag{3.1.55}$$

Here Γ is a positively oriented closed (at infinity) contour around the sector: $-S_\alpha$.

As a result, in virtue of (3.1.50) and (3.1.55) we infer for $s = t/n$ the estimate

$$\left\| e^{-tX(t/n)} - e^{-tX_0} \right\| \leq \frac{M_{TKN}}{n}, \quad t > 0 . \tag{3.1.56}$$

Thus, the rate of convergence in the Euler formula (3.1.51) is determined by the operator-norm $(\sqrt[3]{n})^{-1}$-Lemma 3.1.14, with asymptotic $O(1/\sqrt[3]{n})$ for $n \to \infty$.

(b) In contrast to (3.1.54), the *Dunford-Segal* formula (3.1.46) is an example of the *one-term* approximation. One can use this advantage to check by the spectral calculus that for self-adjoint generator A ($\alpha = 0$) the *optimal* (best possible) estimate of the rate of the operator-norm convergence in (3.1.46) is of the order $O(1/n)$.

Note that the same conclusion about the rate of convergence is valid also for the self-adjoint *Euler formula* (3.1.49) and the self-adjoint (modified) *Yosida approximation* (3.1.53).

(c) In Sect. 8.5 (Sect. 8.5.1) we revise Corollary 3.1.17 and show that following a less *round-about* way than two-terms estimate (3.1.54), one can prove for m-sectorial A the operator-norm Euler formula with estimate of the rate of convergence:

$$\left\| e^{-tA} - (\mathbb{1} + tA/n)^{-n} \right\| \leq O(\ln(n)/n), \quad t \geq 0, \quad n \in \mathbb{N} .$$

For the proof see Sect. 8.5, Theorem 8.5.1.

The optimal rate $O(1/n)$ of the operator-norm estimate for convergence rate in (3.1.46) and in (3.1.49), (3.1.53) for quasi-sectorial contractions, will be established for $\alpha \in (0, \pi/2)$ in Sect. 3.2, cf. Remark 3.1.19(b) for self-adjoint case.

The Chernoff product formula, studied in Sect. 3.1.2 (cf. Theorem 3.1.10 and Theorem 3.1.12 on a Banach space) and the operator-norm product formula for quasi-sectorial contractions on a Hilbert space in Theorem 3.1.16, will be also a subject of Chap. 4 (self-adjoint case, $\alpha = 0$) with some optimal estimates and of Chap. 8 (for non-self-adjoint case, $\alpha \neq 0$) without optimal estimates.

As we mentioned in Remark 3.1.8 the extension to $\alpha \in (0, \pi/2)$ needs a revision of elementary Lemma 3.1.3 and also some new estimates that we collect in the next section. To this aim we shall refine first of all our arguments concerning localisation of the *numerical range* for quasi-sectorial contractions.

3.2 Quasi-Sectorial Contractions: Revisited

In this section we present a method that allows to give a more precise description of localisation the numerical range for quasi-sectorial contractions. This method allows to extend results of Sect. 3.1 to semigroups for m-sectorial generators with semi-angle $\alpha \in (0, \pi/2)$ along with improvement of estimates, as those for convergence rate in the operator-norm topology for *Euler*, or *Dunford-Segal* and *Yosida* approximants of holomorphic C_0-semigroups.

3.2.1 *Operators of the Class* $C_{\mathfrak{H}}(\alpha)$

Definition 3.2.1 Let $\alpha \in (0, \pi/2)$. The subset of bounded operators $\mathcal{L}(\mathfrak{H})$, which is defined by

$$C_{\mathfrak{H}}(\alpha) := \{T \in \mathcal{L}(\mathfrak{H}) : \|T \ \sin\alpha + i\,\mathbb{1}\cos\alpha\| \leq 1 \text{ and } \|T \ \sin\alpha - i\,\mathbb{1}\cos\alpha\| \leq 1\}, \tag{3.2.1}$$

is called the class $C_{\mathfrak{H}}(\alpha)$.

It is clear that as intersection of two closed operator balls: $C^{+}_{\mathfrak{H}}(\alpha) \cap C^{-}_{\mathfrak{H}}(\alpha)$ corresponding to $\pm$ in (3.2.1), the class $C_{\mathfrak{H}}(\alpha)$ is a convex and closed set with respect to the strong operator topology. Moreover, owing to (3.2.1) and $\|T\| = \|T^*\|$, one immediately infers that there are equivalences:

$$T \in C_{\mathfrak{H}}(\alpha) \iff -T \in C_{\mathfrak{H}}(\alpha) \iff T^* \in C_{\mathfrak{H}}(\alpha)\,, \tag{3.2.2}$$

and that condition (3.2.1) is equivalent to the following criterion:

$$(\|f\|^2 - \|Tf\|^2)\,\mathrm{tg}\alpha \geq 2\,|\mathrm{Im}\,(Tf, f)|\,, \tag{3.2.3}$$

for all $f \in \mathfrak{H}$.

The inequality (3.2.3) implies that operator T is a *contraction*. Note that together with Definition 1.8.1 of the *numerical range* inequality (3.2.3) also proves that $T \in C_{\mathfrak{H}}(\alpha)$ is equivalent to the statement that $(\mathbb{1} - T^*)(\mathbb{1} + T)$ is a bounded m-sectorial operator with semi-angle $\alpha \in (0, \pi/2)$.

According to (3.2.3) it is natural to identify $C_{\mathfrak{H}}(0)$ with the set of all *self-adjoint* contractions, whereas the *limit case*: $C_{\mathfrak{H}}(\pi/2)$, with the set of general contractions

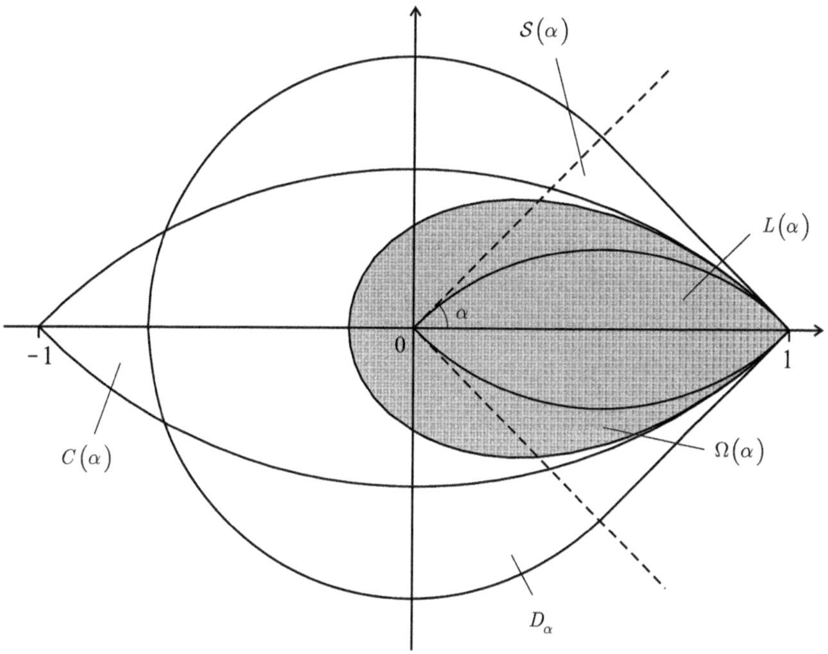

Fig. 3.2 Illustration of subsets of the complex plane involved in Sects. 3.1 and 3.2. The subset D_α is defined in (3.1.1), see also Fig. 3.1. Domain $C(\alpha)$ is described by equation (3.2.4). The shaded domains $L(\alpha)$ and $\Omega(\alpha)$ are defined, respectively, in (3.2.13) and (3.2.17), see also relations between them in (3.2.38). Open sector $S(\alpha) \equiv S_\alpha$ is defined by (1.8.11)

on $\mathfrak{H}$, see Sect. 1.8.1. Relation between class $C_{\mathfrak{H}}(\alpha)$ and quasi-sectorial contractions with numerical range D_α (3.1.1) is not straightforward, see Remark 3.2.3 and Fig. 3.2.

Definition 3.2.2 Now for $\alpha \in (0, \pi/2)$ we define by

$$C(\alpha) := \{z \in \mathbb{C} : |z \sin\alpha \pm i \cos\alpha| \leq 1\} \tag{3.2.4}$$
$$= \left\{z \in \mathbb{C} : (1 - |z|^2)\,\mathrm{tg}\alpha \geq 2\,|\mathrm{Im}\, z|\right\},$$

the family of subsets of the complex plane $\mathbb{C}$, cf. definitions (3.2.1), (3.2.3).

Then, similarly to (3.2.1), each set $C(\alpha)$ is, in turn, an intersection of two closed discs of the complex plane, and $C(\alpha)$ is contained in the closed unit disc $\overline{\mathbb{D}} = \{z \in \mathbb{C} : |z| \leq 1\}$. Definition (3.2.4) implies that $C(0) = [-1, 1]$.

Remark 3.2.3 On account of definitions (3.2.1)–(3.2.4) we obtain that $T \in C_{\mathfrak{H}}(\alpha)$ implies for numerical range: $W(T) \subseteq C(\alpha)$. Owing to (3.2.4) we obtain, after parametrisation: $\tau \mapsto z(\tau) = r_\alpha(\tau)\, e^{i\tau}$ for $\tau \in [-\pi, \pi]$, equation for the boundary $\partial C(\alpha)$:

$$1 - r_\alpha^2(\tau) = 2\,\mathrm{ctg}(\alpha)\, r_\alpha(\tau)\, |\sin\tau| \,. \tag{3.2.5}$$

By virtue of (3.2.5) one gets $\sup_{\tau\in[-\pi,\pi]} |\mathrm{Im}\, z(\tau)| = r_\alpha(\pi/2) = \mathrm{tg}(\alpha/2)$. Then we obtain

$$\|\mathrm{Im}\, T\| \leq \mathrm{tg}(\alpha/2)\,, \quad \mathrm{Im}\, T := (T - T^*)/2i\,. \tag{3.2.6}$$

We also find that $z(\tau = \pm\pi) = -1$, that is, $C(\alpha) \cap D_\alpha \neq C(\alpha)$ (see Fig. 3.2). Hence, operator T is (in general) *not* quasi-sectorial.

Note that the operators from the class $C_{\mathfrak{H}}(\alpha)$ have the following useful properties:

Proposition 3.2.4 *Let T_1 and T_2 belong to the class $C_{\mathfrak{H}}(\alpha)$. Then the operator*

$$(T_1T_2 + T_2T_1)/2 \in C_{\mathfrak{H}}(\alpha)\,. \tag{3.2.7}$$

If T_1 and T_2 are two commuting operators from the class $C_{\mathfrak{H}}(\alpha)$, then the product $T = T_1\, T_2$ also belongs to $C_{\mathfrak{H}}(\alpha)$.

Remark 3.2.5 The set of operators $C(\alpha)$ inherits a similar property: for each $\alpha \in (0, \pi/2)$ it forms a multiplicative semigroup of complex numbers in the plane $\mathbb{C}$.

The next statement establishes a connection between the set of contractions $C_{\mathfrak{H}}(\alpha)$ and the m-sectorial operators.

Proposition 3.2.6 *If A is m-sectorial operator with semi-angle α, then*

$$T = (\mathbb{1} - A)\,(\mathbb{1} + A)^{-1} \tag{3.2.8}$$

belongs to the class $C_{\mathfrak{H}}(\alpha)$. Conversely, if $T \in C_{\mathfrak{H}}(\alpha)$, and $\ker(\mathbb{1} + T) = \{0\}$, *then $A = (\mathbb{1} - T)(\mathbb{1} + T)^{-1}$ is m-sectorial operator.*

This statement follows directly from Definition 3.2.1 and the Kato numerical range mapping theorem (Proposition 3.1.5), see Notes to Chap. 3 (Notes to Sects. 3.2, 3.2.1).

Now let A be m-sectorial operator with semi-angle α and let $\lambda > 0$. Then λA is also m-sectorial. By Proposition 3.2.6 the operator $U(\lambda) = (\mathbb{1} - \lambda A)(\mathbb{1} + \lambda A)^{-1}$ belongs to the class $C_{\mathfrak{H}}(\alpha)$. Let

$$F(\lambda) := (\mathbb{1} + \lambda A)^{-1} = \frac{1}{2}(U(\lambda) + \mathbb{1}). \tag{3.2.9}$$

Since operators $U(\lambda)$ and $\mathbb{1}$ belong to the class $C_{\mathfrak{H}}(\alpha)$ and since this class is a convex set, the operator $F(\lambda) \in C_{\mathfrak{H}}(\alpha)$. Therefore, by Remark 3.2.3 one obtains that $W(F(\lambda)) \subseteq C(\alpha)$.

Remark 3.2.7 Following the arguments of Sect. 3.1.1 we can now to localise the numerical range $W(F(\lambda))$ even more precisely. Since for any $u \in \mathfrak{H}$ we have:

$$(F(\lambda)u, u) = (w_\lambda, w_\lambda) + \lambda(v_\lambda, A\, v_\lambda) \in \overline{S}_\alpha\,, \tag{3.2.10}$$

where $w_\lambda := F(\lambda)u$, it follows that

$$W(F(\lambda)) \subseteq \overline{S}_\alpha \cap C(\alpha)\,, \tag{3.2.11}$$

for $\lambda > 0$ and $\alpha \in (0, \pi/2)$. Moreover, since $U(\lambda) = 2F(\lambda) - \mathbb{1} \in C_{\mathfrak{H}}(\alpha)$, see (3.2.9), we find that

$$\|(F(\lambda) - \mathbb{1}/2)\sin\alpha \pm i\mathbb{1}(\cos\alpha)/2\| \leq 1/2\,. \tag{3.2.12}$$

For the numerical range this implies:

$$W(F(\lambda)) \subseteq L(\alpha) := \{\zeta \in \mathbb{C} : |(\zeta - 1/2)\sin\alpha \pm (i\cos\alpha)/2| \leq 1/2\}\,, \tag{3.2.13}$$

for $\lambda > 0$. Remark that $L(\alpha) \subset \overline{S}_\alpha \cap C(\alpha)$, see Fig. 3.2.

Theorem 3.2.8 *If A is m-sectorial operator in a Hilbert space $\mathfrak{H}$, then the corresponding semigroup $T(t) = \exp(-tA) \in C_{\mathfrak{H}}(\alpha)$, for all $t \geq 0$. Conversely, let $T(t) = \exp(-tA)$ for $t \geq 0$ be a C_0-semigroup on a Hilbert space $\mathfrak{H}$. If $T(t) \in C_{\mathfrak{H}}(\alpha)$ for positive t in a neighbourhood of $t = 0$, then the generator A is an m-sectorial operator with semi-angle α.*

Remark 3.2.9 The proof of Theorem 3.2.8 is based on Propositions 3.2.4, 3.2.6, and on the Euler approximation (Corollary 3.1.17). The functional calculus for contractions on Hilbert spaces allows to establish the following :

If T is a contraction in a Hilbert space $\mathfrak{H}$, if $\ker(\mathbb{1} + T) = \{0\}$ and operator $S := (\mathbb{1} - T)(\mathbb{1} + T)^{-1}$, then for C_0-semigroup of contractions $T(t) = \exp(-tS)$ the equality $T(t) = f_t(T)$ holds for $t \geq 0$, where $f_t(z) := \exp(-t\,(1-z)\,(1+z)^{-1})$.

Moreover, the next statement is true : if ϕ is analytic in the unit disc $\mathbb{D} = \{z \in \mathbb{C} : |z| < 1\}$, $\phi(z) = \overline{\phi(\bar{z})}$ and $|\phi(z)| < 1$ for $z \in \mathbb{D}$, then one has implication: $T \in C_{\mathfrak{H}}(\alpha) \Rightarrow \phi(T) \in C_{\mathfrak{H}}(\alpha)$.

We also remark that if $\dim \mathfrak{H} \geq 2$ and $T \in \mathcal{L}(\mathfrak{H})$, then inclusion $W(T) \subseteq C(\alpha)$ does not imply $T \in C_{\mathfrak{H}}(\alpha)$. On the other hand, there exists a bounded operator $T \in \mathcal{L}(\mathfrak{H})$ such that $W(T) \subseteq \overline{\mathbb{D}}$ and $1 < \|T\| \leq 2$. Now, let $\|T\| = a > 1$ and let $b \in (1, a)$. Then one gets

$$W(b^{-1}T) \subseteq \overline{\mathbb{D}}_{b^{-1}} = \{z \in \mathbb{C} : |z| \leq b^{-1}\} \subset \mathbb{D}.$$

Clearly, $\mathbb{D}_{b^{-1}} \subset C(\alpha_b)$, where $\mathrm{tg}(\alpha_b/2) = b^{-1}$ (3.2.6), and $\|b^{-1}T\| > \|a^{-1}T\| = 1$.

Nevertheless, the following statement takes place.

Theorem 3.2.10 *Let $\alpha \in (0, \pi/2)$ and let $T(t) = \exp(-tA)$, $t \geq 0$ be a C_0-semigroup. Suppose that*

$$W(T(t)) \subset 1 - S_\alpha = \{z \in \mathbb{C} : (1 - z) \in S_\alpha\}, \tag{3.2.14}$$

for $\{t > 0\}$ in a neighbourhood of $t = 0$. Then the generator A is an m-sectorial operator with semi-angle α and hence $T(t) \in C_{\mathfrak{H}}(\alpha)$ for all $t \geq 0$.

Proof For a small $t > 0$ and for any $u \in \mathrm{dom}\,(A)$, $\|u\| = 1$, one gets

$$\frac{1}{t}\,(u - T(t)\,u,\, u) = \frac{1 - (T(t)\,u,\, u)}{t} \in t^{-1} S_\alpha \subset S_\alpha\,. \tag{3.2.15}$$

Letting $t \to +0$ yields $(Au, u) \in S_\alpha$. Hence, the operator A is m-sectorial and by Theorem 3.2.8 we obtain: $T(t) \in C_{\mathfrak{H}}(\alpha)$ for all $t \geq 0$. □

Therefore, since $C(\alpha) \subset 1 - S_\alpha$, by Theorems 3.2.8 and 3.2.10 we find that the *three* equivalent conditions: $\exp(-tA) \in C_{\mathfrak{H}}(\alpha)$, $W(\exp(-tA)) \subseteq C(\alpha)$, and also $W(\exp(-tA)) \subset 1 - S_\alpha$, *completely* characterise m-sectorial generators A.

3.2.2 Numerical Range of Quasi-Sectorial Contraction Semigroups

From Theorem 3.2.8 it follows, in particular, that for m-sectorial generator A, with semi-angle $\alpha \in (0, \pi/2)$, the numerical range of the corresponding contraction semigroup

$$W(\mathrm{e}^{-tA}) \subseteq C(\alpha) \quad \text{for all} \quad t \geq 0\,. \tag{3.2.16}$$

But as we warranted in Remark 3.2.3 this does not imply that the semigroup is a quasi-sectorial contraction.

Recall that the conformal mapping : $z \mapsto z^2$ (Theorem 3.1.4) and the Kato *numerical range mapping* theorem (Proposition 3.1.5) play a special rôle in analysis of quasi-sectorial contractions and the problem of localisation of numerical ranges.

Definition 3.2.11 Let $\alpha \in (0, \pi/2)$. We define in the complex plane $\mathbb{C}$ domain

$$\Omega(\alpha) := \left\{z^2 \in \mathbb{C} : z \in C(\alpha)\right\}. \tag{3.2.17}$$

So, if $f(z) = z^2$, then $\Omega(\alpha) = f(C(\alpha))$. Since (see Remark 3.2.5) $C(\alpha)$ is a *multiplicative semigroup*, we obtain that $\Omega(\alpha) \subseteq C(\alpha)$, that is, the subset $\Omega(\alpha)$

is in turn a multiplicative semigroup. From (3.2.4) and Proposition 3.2.4 it follows then that for any $\alpha \in (0, \pi/2)$ the set (3.2.11) has the representation:

$$\begin{aligned} \Omega(\alpha) &= \left\{z \in \mathbb{C} : |\sqrt{z}\sin\alpha \pm i\cos\alpha| \leq 1\right\} \\ &= \left\{z \in \mathbb{C} : 2\,|\mathrm{Im}\sqrt{z}| \leq (1 - |z|)\,\mathrm{tg}\alpha\right\}, \end{aligned} \tag{3.2.18}$$

with the limiting cases: $\Omega(\alpha = 0) = [0, 1]$ and $\Omega(\alpha = \pi/2) = \overline{\mathbb{D}}$.

Lemma 3.2.12 *The set $\Omega(\alpha)$ (3.2.17) is convex and $\Omega(\alpha) \subseteq D_\alpha$.*

Proof Let $C_+(\alpha) := \{z \in C(\alpha) : \mathrm{Im}\, z \geq 0\}$. Then clearly: $\Omega(\alpha) = f(C_+(\alpha))$, where $f(z) := z^2$. Denote

$$\Gamma(\alpha) = \partial C_+(\alpha) \setminus (-1, 1). \tag{3.2.19}$$

Then $\partial\Omega(\alpha) = f(\Gamma(\alpha))$. Since

$$\Gamma(\alpha) = \left\{z : z = \frac{\mathrm{e}^{it} - i\cos\alpha}{\sin\alpha},\ t \in \left[\frac{\pi}{2} - \alpha, \frac{\pi}{2} + \alpha\right]\right\}, \tag{3.2.20}$$

the boundary $\partial\Omega(\alpha)$ can be parametrised as follows:

$$\partial\Omega(\alpha) = \left\{z = \zeta(t) := \frac{(\mathrm{e}^{it} - i\cos\alpha)^2}{\sin^2\alpha},\ t \in \left[\frac{\pi}{2} - \alpha, \frac{\pi}{2} + \alpha\right]\right\}. \tag{3.2.21}$$

Put $x = x(t) := \mathrm{Re}\,\zeta(t)$, $y = y(t) := \mathrm{Im}\,\zeta(t)$. Then, since one has

$$\frac{d^2y}{dx^2} = \frac{y''(t)x'(t) - y'(t)x''(t)}{(x'(t))^3}, \tag{3.2.22}$$

we obtain

$$\frac{d^2y}{dx^2} = -\frac{\mathrm{Im}\,(\zeta'(t)\overline{\zeta''(t)})}{(\mathrm{Re}\,\zeta'(t))^3}. \tag{3.2.23}$$

Further, by explicit calculations we get:

$$\zeta'(t) = \frac{2i\mathrm{e}^{it}(\mathrm{e}^{it} - i\cos\alpha)}{\sin^2\alpha},\quad \zeta''(t) = \frac{2i\mathrm{e}^{it}(2i\mathrm{e}^{it} + \cos\alpha)}{\sin^2\alpha}, \tag{3.2.24}$$

$$\begin{aligned} -\mathrm{Im}\,(\zeta'(t)\overline{\zeta''(t)}) &= \frac{4(2 + \cos^2\alpha - 3\cos\alpha\sin t)}{\sin^4\alpha}, \\ \mathrm{Re}\,\zeta'(t) &= \frac{2\cos t(\cos\alpha - 2\sin t)}{\sin^2\alpha}, \ \ldots . \end{aligned} \tag{3.2.25}$$

Let $t \in [\pi/2 - \alpha, \pi/2 + \alpha]$. Then by the estimates

$$\begin{aligned}&\cos^2\alpha - 3\cos\alpha \sin t + 2 \\ &\qquad \geq \cos^2\alpha - 3\cos\alpha + 2 = (2-\cos\alpha)(1-\cos\alpha) > 0,\end{aligned} \tag{3.2.26}$$

we obtain $-\mathrm{Im}\,(\zeta'(t)\overline{\zeta''(t)}) > 0$.

Since $\cos\alpha - 2\sin t \leq -\cos\alpha < 0$, then for $t \in [\pi/2 - \alpha, \pi/2 + \alpha]$, we find that

$$\begin{aligned}&\frac{d^2y}{dx^2} < 0 \quad \text{for} \quad t \in \left[\frac{\pi}{2} - \alpha, \frac{\pi}{2}\right), \\ &\frac{d^2y}{dx^2} > 0 \quad \text{for} \quad t \in \left(\frac{\pi}{2}, \frac{\pi}{2} + \alpha\right],\end{aligned} \tag{3.2.27}$$

which implies that the set $\Omega(\alpha)$ is convex, see Fig. 3.2.

Seeing that the map: $z \mapsto z^2$ is conformal and $\Omega(\alpha)$ is convex, the proof of the second part of the lemma follows from the estimate

$$-\,[\mathrm{tg}(\alpha/2)]^2 \leq \mathrm{Re}\,\zeta(t)\,, \tag{3.2.28}$$

where $[\mathrm{tg}(\alpha/2)]^2 \leq \sin\alpha$ for $\alpha \in (0, \pi/2)$, see Definition 3.1.1 and (3.2.21). □

Now we remark that by virtue of relation

$$y(t) = \frac{\sin 2t - 2\cos\alpha\cos t}{\sin^2\alpha}\,, \tag{3.2.29}$$

one obtains

$$\max_{z \in \Omega(\alpha)} |\mathrm{Im}\, z| = \frac{\sin 2\gamma - 2\cos\alpha\sin\gamma}{\sin^2\alpha}\,, \tag{3.2.30}$$

where $\gamma := \gamma(\alpha) \in (\pi/4, \pi/2)$ satisfies the equation:

$$\sin\gamma = \frac{\cos\alpha + \sqrt{\cos^2\alpha + 8}}{4}\,, \quad \alpha \in (0, \pi/2)\,. \tag{3.2.31}$$

Since $\Omega(\alpha) \subset C(\alpha)$, by virtue of (3.2.6) the number in the right-hand side of (3.2.30) is less than $\mathrm{tg}(\alpha/2)$.

Lemma 3.2.13 *Let $\alpha \in (0, \pi/2)$ and let $T \in C_{\mathfrak{H}}(\alpha)$. Then the numerical range $W(T^{2n}) \subseteq \Omega(\alpha)$ for all natural numbers $n \in \mathbb{N}$.*

Proof We apply the Kato *numerical range mapping* theorem (Proposition 3.1.5) for the rational function $f : z \mapsto z^2$, with $E' = \Omega(\alpha)$. Then by Definition 3.2.11 we have $E := f^{-1}(\Omega(\alpha)) = C(\alpha)$, see (3.2.17). Since $C(\alpha)$ is a convex set, its maximal convex kernel K *coincides* with $C(\alpha)$. By virtue of Proposition 3.2.4 from $T \in C_{\mathfrak{H}}(\alpha)$ it follows that for all $n \in \mathbb{N}$ one gets $T^n \in C_{\mathfrak{H}}(\alpha)$. Then by Remark 3.2.3 we obtain $W(T^n) \subseteq C(\alpha) = K$. Applying now the Kato numerical range mapping theorem again for function $f(z) = z^2$ we obtain $W(T^{2n}) \subseteq \Omega(\alpha)$. □

Now we are in position to prove the main statement of this subsection about *localisation* of numerical range of quasi-sectorial contraction semigroups.

Theorem 3.2.14 *Let A be an m-sectorial operator with semi-angle α. Then for any* $\alpha \in (0, \pi/2)$

$$W(\mathrm{e}^{-tA}) \subseteq \Omega(\alpha)\,, \quad t > 0\,. \tag{3.2.32}$$

In particular, the family $\{\mathrm{e}^{-tA}\}_{t \geq 0}$ *is a holomorphic quasi-sectorial contraction semigroup.*

Proof The holomorphic property comes from Proposition 1.8.13. Now, by Theorem 3.2.8 the operators $T(t) = \mathrm{e}^{-tA}$ belong to the class $C_{\mathfrak{H}}(\alpha)$ for all $t > 0$ and $T(t) = T^2(t/2)$. Then by Lemma 3.2.13 we obtain $W(T(t)) \subseteq \Omega(\alpha)$. Since Lemma 3.2.12 implies $\Omega(\alpha) \subset D_\alpha$, the operators $T(t)$ are quasi-sectorial contractions. □

Remark 3.2.15 Since (3.2.32) yields $W(\exp(-tA)) \subset D_\alpha$, the first statement of Theorem 3.2.14 is an extension of Theorem 3.1.4 (proven only for $\alpha \in (0, \pi/4)$) to the *whole* class of the m-sectorial generators with semi-angle $\alpha \in (0, \pi/2)$.

Taking into account the *multiplicative semigroup* property of the set $\Omega(\alpha)$ and Theorem 3.2.14 we obtain as a corollary the following inequalities:

$$\left| \sin\alpha \left(\prod_{k=1}^{n} (\exp(-t_k A_k) u_k, u_k) \right)^{1/2} \pm i \cos\alpha \right| \leq 1\,, \tag{3.2.33}$$

for arbitrary m-sectorial operators $A_1, \ldots A_n$, any non-negative numbers $t_1, \ldots, t_n$, and arbitrary normalised vectors $u_1, \ldots, u_n$ from $\mathfrak{H}$. By the semigroup property one has : $\exp(-t_1 A)\exp(-t_2 A) = \exp(-(t_1 + t_2)A)$. Hence, we also get

$$|\sin\alpha \sqrt{(\exp(-t_1 A)u, \exp(-t_2 A^*)u)} \pm i \cos\alpha| \leq 1\,, \tag{3.2.34}$$

for arbitrary $t_1, t_2 \geq 0$ and for any $u \in \mathfrak{H}$, $\|u\| = 1$.

Notice that since $\mathrm{Re}\, z \geq -\mathrm{tg}^2(\alpha/2)$ for all $z \in \Omega(\alpha)$ and $W(\exp(-tA)) \subseteq \Omega(\alpha)$, we obtain the estimate from below:

$$\mathrm{Re}\, \exp(-tA) \geq -\mathrm{tg}^2(\alpha/2)\, \mathbb{1}\,,\ t \geq 0\,. \tag{3.2.35}$$

On the other hand the inclusion $W(\exp(-tA)) \subseteq \Omega(\alpha)$ also implies (see (3.2.30)) that

$$\|\mathrm{Im}\, \exp(-tA)\| \leq \frac{\sin 2\gamma - 2\cos\alpha \sin\gamma}{\sin^2\alpha} < \mathrm{tg}(\alpha/2)\,,\ t \geq 0\,, \tag{3.2.36}$$

where $\gamma = \gamma(\alpha)$ satisfies (3.2.31). By virtue of relation $\mathrm{Re}\, T^2 = (\mathrm{Re}\, T)^2 - (\mathrm{Im}\, T)^2$ (valid for any bounded operator T) one has $\mathrm{Re}\, \exp(-tA) = (\mathrm{Re}\, \exp(-tA/2))^2 - (\mathrm{Im}\, \exp(-tA/2))^2$. Therefore, by (3.2.36) we obtain another estimate from below:

$$\mathrm{Re}\, \exp(-tA) \geq -\left(\frac{\sin 2\gamma - 2\cos\alpha \sin\gamma}{\sin^2\alpha}\right)^2 \mathbb{1} > -\mathrm{tg}^2(\alpha/2)\, \mathbb{1}\,,\ t \geq 0\,. \tag{3.2.37}$$

Remark 3.2.16 Let D_α be numerical set of a quasi-sectorial contraction, see (3.1.1). Then one finds (Remark 3.2.3) that D_α is not a subset of $C(\alpha)$. In addition, because of (3.2.13) and (3.2.18) one obtains: $L(\alpha) \subset \Omega(\alpha)$, where $L(\alpha)$ is defined by (3.2.13). So, for all $\alpha \in (0, \pi/2)$ and $F(\lambda) = (\mathbb{1} + \lambda A)^{-1}$, $\lambda > 0$, we have the following inclusions, see Fig. 3.2:

$$\begin{aligned} &L(\alpha) \subset \Omega(\alpha) \subset D_\alpha\,, \\ &W(F(\lambda)) \subseteq L(\alpha)\,, \\ &W(F^n(\lambda)) \subseteq C(\alpha) \quad \text{and} \quad W(F^{2n}(\lambda)) \subseteq \Omega(\alpha)\,, \end{aligned} \tag{3.2.38}$$

for all $n \in \mathbb{N}$. The inclusion: $W(\mathrm{e}^{-tA}) \subseteq \Omega(\alpha)$, $t > 0$ (3.2.32), is the key observation for the proof of the optimal rate $O(1/n)$ of the operator-norm convergence for $\alpha \in (0, \pi/2)$.

3.2.3 *Operator-Norm Convergence of the Euler Formula*

According to Remark 3.1.19 (c) the Euler approximation of a quasi-sectorial contraction semigroup on $\mathfrak{H}$ the rate of convergence:

$$\left\| \mathrm{e}^{-tA} - (\mathbb{1} + tA/n)^{-n} \right\| \leq O(\ln(n)/n)\,, \quad t \geq 0\,, \quad n \in \mathbb{N}\,, \tag{3.2.39}$$

for m-sectorial generator A, cf. Corollary 3.1.17 for $\alpha \in (0, \pi/4]$.

The results of Sects. 3.2.1, 3.2.2 and in particular Theorem 3.2.14 about localisation of numerical range allow us:

(i) To extend (3.2.39) to a general case of m-sectorial generators with $\alpha \in (0, \pi/2)$.

(ii) To take into account the *α-dependence* of the optimal operator-norm estimate $O(1/n)$ for the convergence rate instead of (3.2.39).

Theorem 3.2.17 *Let A be an m-sectorial operator with semi-angle $\alpha \in (0, \pi/2)$ defined in a Hilbert space $\mathfrak{H}$. Then*

$$\left\| \mathrm{e}^{-tA} - (\mathbb{1} + tA/n)^{-n} \right\| \leq \frac{K_\alpha}{n\,(\cos\alpha)^2}\,, \quad t \geq 0\,,\ n \in \mathbb{N}\,, \tag{3.2.40}$$

where

$$\frac{\pi \sin\alpha}{2\alpha} \leq K_\alpha \leq \min\left(\frac{\pi - \alpha}{\alpha}, K_0\right) \quad \text{and} \quad \frac{\pi}{2} \leq K_0 \leq 2 + \frac{2}{\sqrt{3}}\,. \tag{3.2.41}$$

For the proof we shall use a result, related to *extension* of the *von Neumann* theory of the *spectral sets* (for a particular generalisation of *von Neumann's inequality*, see Notes to Chap. 3 (Notes to Sects. 3.2, 3.2.3)). For our purposes we reformulate this extension of the von Neumann inequality in the following form:

Proposition 3.2.18 (Crouzeix-Delyon Theorem) *Let S_α be a (convex) open sector in the complex plane with vertex at $z = 0$ and with semi-angle $\alpha \in (0, \pi/2)$. Then for any bounded operator $B \in \mathcal{L}(\mathfrak{H})$ with numerical range such that $W(B) \subseteq \overline{S}_\alpha$ there is the best (i.e., optimal) α-dependent constant $K_\alpha > 0$ such that the inequality*

$$\|r(B)\| \leq K_\alpha \sup_{z \in S_\alpha} |r(z)|\,, \tag{3.2.42}$$

holds for all rational functions $r : \mathbb{C} \to \mathbb{C}$, bounded in S_α. Furthermore, K_α is a continuous and decreasing function of $\alpha \in (0, \pi/2)$ satisfying the estimates (3.2.41).

The inequality (3.2.42) *is still valid if r is a holomorphic function in S_α, which is bounded on the closure $\overline{S}_\alpha$.*

Proof of Theorem 3.2.17 for a (linear) Sector $S_\alpha = \{z \in \mathbb{C} : |\arg(z)| < \alpha < \pi/2\}$ Let

$$g_n(w) := \exp(-w) - (1 + w/n)^{-n}\,, \quad w \in S_\alpha\,. \tag{3.2.43}$$

Since $\{g_n(w)\}_{n \geq 1}$ are holomorphic in S_α and bounded on $\overline{S}_\alpha$, then

$$\sup_{w \in S_\alpha} |g_n(w)| = \sup_{w \in \partial S_\alpha} |g_n(w)|\,. \tag{3.2.44}$$

We remark that

$$\partial S_\alpha = \{x\ \exp(-i\alpha),\ x \in \mathbb{R}^+\} \cup \{x\ \exp(i\alpha),\ x \in \mathbb{R}^+\}\ . \tag{3.2.45}$$

Therefore, we have to estimate the value of

$$\sup_{x\in\mathbb{R}^+} |g_n(x \exp(\pm i\alpha))|\ . \tag{3.2.46}$$

To this end we use the representation

$$g_n(x \exp(i\alpha)) = -\int_0^x ds\ \partial_s \left((1 + s\,\mathrm{e}^{\,i\alpha}/n)^{-n} \exp(-(x-s)\,\mathrm{e}^{\,i\alpha})\right)\ , \tag{3.2.47}$$

where

$$\begin{aligned}&\partial_s \left((1 + s\,\mathrm{e}^{\,i\alpha}/n)^{-n} \exp(-(x-s)\,\mathrm{e}^{\,i\alpha})\right) = \\ &(s\,\mathrm{e}^{\,2i\alpha}/n)\ (1 + s\,\mathrm{e}^{\,i\alpha}/n)^{-(n+1)} \exp(-(x-s)\,\mathrm{e}^{\,i\alpha})\ .\end{aligned} \tag{3.2.48}$$

By straightforward estimates:

$$\begin{aligned}&\left|1 + s\,\mathrm{e}^{\,i\alpha}/n\right|^{n+1} = \left(1 + s^2/n^2 + 2s\ \cos\alpha/n\right)^{(n+1)/2} \geq (1 + (s\ \cos\alpha)/n)^{n+1} \\ &\quad \geq 1 + ((n+1)\,s\ \cos\alpha)/n \geq 1 + s\ \cos\alpha\ ,\end{aligned} \tag{3.2.49}$$

we infer that

$$s\ |1 + (s\cos\alpha)/n|^{\ -(n+1)} \leq s/(1 + s\cos\alpha) \leq 1/\cos\alpha\ . \tag{3.2.50}$$

Hence, we obtain as an upper bound:

$$\begin{aligned}|g_n(x \exp(i\alpha))| &\leq \int_0^x ds\ \frac{\exp(-(x-s)\cos\alpha)}{n\cos\alpha} \\ &\leq \frac{1}{n\ (\cos\alpha)^2}\ (1 - \exp(-x\cos\alpha)) \leq \frac{1}{n\ (\cos\alpha)^2}\ .\end{aligned} \tag{3.2.51}$$

Similarly one gets the estimate:

$$|g_n(x \exp(-i\alpha))| \leq \frac{1}{n\ (\cos\alpha)^2}\ . \tag{3.2.52}$$

Therefore, by (3.2.44) we obtain that

$$\sup_{w \in S_\alpha} |g_n(w)| \leq \frac{1}{n\,(\cos\alpha)^2} \,. \tag{3.2.53}$$

Now applying Proposition 3.2.18 to $R(z) = g_n(z)$, by estimates (3.2.41)–(3.2.53) we find evaluation

$$\left\| \mathrm{e}^{-tA} - (\mathbb{1} + tA/n)^{-n} \right\| = \|g_n(tA)\| \leq \frac{K_\alpha}{n\,(\cos\alpha)^2} \,, \quad t \geq 0\,, \ \ n \in \mathbb{N}\,, \tag{3.2.54}$$

which completes the proof. □

Similarly, using the *Crouzeix-Delyon* theorem (Proposition 3.2.18), one can improve to the optimal value the rate of convergence of the *Dunford-Segal* approximants (3.1.46) and also establish the optimal rate for convergence of the *Yosida* approximants for $\{A_n = A_n(t = 1)\}_{n \geq 1}$, cf. Corollary 3.1.18.

Proposition 3.2.19 *Let A be an m-sectorial operator with semi-angle $\alpha \in (0, \pi/2)$, which is generator of C_0-semigroup $\{\mathrm{e}^{-tA}\}_{t \geq 0}$ defined in a Hilbert space $\mathfrak{H}$.*

(a) Then for K_α given by (3.2.41) *one gets for the Dunford-Segal approximants $\{U_n^{DS}(t) := \mathrm{e}^{-n(\mathbb{1} - \mathrm{e}^{-tA/n})}\}_{n \geq 1}$ estimates*

$$\left\| \mathrm{e}^{-tA} - U_n^{DS}(t) \right\| \leq \frac{e^{-1} K_\alpha}{t\,n\,(\cos\alpha)^2} \,, \quad t > 0\,, \quad n \in \mathbb{N}\,. \tag{3.2.55}$$

(b) Correspondingly, for bounded generators $\{A_n := A\,(\mathbb{1} + A/n)^{-1}\}_{n \geq 1}$ (Yosida approximation of generator), cf. (3.1.52), (3.1.53), *and for Yosida approximants $\{\mathrm{e}^{-tA_n}\}_{n \geq 1}$ one infers estimates*

$$\left\| \mathrm{e}^{-tA} - \mathrm{e}^{-tA_n} \right\| \leq \frac{2\,e^{-1} K_\alpha}{t\,n\,(\cos\alpha)^2} \,, \quad t > 0\,, \quad n \in \mathbb{N}\,. \tag{3.2.56}$$

Here again K_α is given by (3.2.41).

In spite of a certain similarity, the proof of Proposition 3.2.19 is quite different from that of Theorem 3.2.17, see Notes to Chap. 3 (Notes to Sects. 3.2, 3.2.3).

3.3 Notes to Chap. 3

Notes to Sect. 3.1

We note that relation between m-sectorial generators and holomorphic contraction semigroups is studied in great generality in a Banach space in the fundamental treatise about sectorial operators [Haa06, Capter 3.4].

In this chapter we consider a similar relation in a Hilbert space with a focus on the *numerical range* description of the quasi-sectorial contractions. Essentially we follow here the ideas from the papers [CZ01b, Zag08, ArZag10, Arl12, Zag17] and [Zag20, Zag22a, Zag22c, MoRo23].

Section 3.1.1

The *quasi-sectorial* contractions have been introduced in [CZ01b, Section 2] and [Zag08, Section 2] with the aim to improve the $\sqrt{n}$-Chernoff estimate [Cher68, Lemma 2] known for generic contractions, see Lemma 1.9.33.

Example in Lemma 3.1.3 is due to [CZ01b]. Our proof of Theorem 3.1.4 (see [CZ01b, Theorem 2.1] and [Zag08, Theorem 2.1]) is based on Lemma 3.1.3 and the *global* Kato *numerical range mapping theorem* [Kato65, Theorem 1]. For convenience it is reformulated in Proposition 3.1.5 and presented in more details in Appendix C (Sect. C.2).

We point out that $E(\subset \mathbb{C})$ is *star-shaped* relative to the *point* $z \in E$ if for any $\zeta \in E$ the interval $[z, \zeta] \subset E$. Collection $K \subset E$ of *all* of such points is called a *kernel* of E, see (C.2.1). It is known (see references in [Kato65]) that *kernel* K is a *convex* subset of E with some evident (and less evident) properties that we discuss in Appendix C (Sect. C.2).

Section 3.1.2

The revisited Chernoff estimates in the strong operator topology expressed by n-Lemma 3.1.9, and by $\sqrt[3]{n}$-Lemma 3.1.11 as well as the Chernoff product formulæ in Theorem 3.1.10 and Theorem 3.1.12 in Hilbert and Banach spaces, are due to [Zag17, Lemma 2.1], [Zag22a, Lemma 3.1, Lemma 2.1], [Zag22c, Lemma 2.1, Theorem 3.1] and [MoRo23, Lemma 4.2].

The standard proof of the Chernoff $\sqrt{n}$-Lemma 1.9.33 is well-known, see [Cher74, Theorem 1.1] and, e.g., [Dav80, Chapter 3.4]. The proof, which is based on the *Poisson representation* and the *Tchebychëv inequality* (see, e.g., [Sin92, Lecture 1.2]), was proposed in [CZ01b, Section 3]. We use them also in the proof of $\sqrt[3]{n}$-Lemma 3.1.11. This approach allows to consider extension of the Chernoff estimates to the operator-norm topology in Sect. 3.1.3.

Section 3.1.3

The *Ritt operators* appeared for the first time in paper [Ri53]. We use that *quasi-sectorial* contractions are Ritt's operators, Theorem 3.1.13, see [CZ01b, Lemma 3.1]. This allows to improve for *quasi-sectorial* contractions the $\sqrt[3]{n}$-Lemma 3.1.11 ([Zag22c, Lemma 2.1]) and to deduce in a Hilbert space $\mathfrak{H}$ the $(\sqrt[3]{n})^{-1}$-Lemma 3.1.14, which, instead the *bound* $\sqrt{n}$ (1.9.96)

(Chernoff $\sqrt{n}$-Lemma 1.9.33), yields a new operator-norm *estimate* of the order $O(\sqrt[3]{n})^{-1}$ (3.1.44), see [Zag17, Theorem 3.1] and [Zag20, Theorem 4.3], [Zag22c, Theorem 4.1].

The *Dunford-Segal* approximation in the strong operator topology appeared in [DS46]. For approximations in the strong operator topology and extensions to Banach spaces see [GoTo14, Theorem 1.1(b)].

We mentioned in Remark 3.1.19(b) the *one-term* advantage of the *Dunford-Segal* approximations and our estimate (3.1.44) to produce the operator-norm convergent estimate (3.1.46). Note that the rate of convergence due to (3.1.46) is not optimal.

The proofs in Corollary 3.1.17 of the *Euler formula* and in Corollary 3.1.18 of the (modified) *Yosida approximation* involve the two-terms estimate (3.1.54). Then the proof of the operator-norm convergence in (3.1.49) and as a consequence in (3.1.53) is heavily limited by the *Trotter-Kato-Neveu* term $\| e^{-t\,X(t/n)} - \mathrm{e}^{-t\,X_0} \|$ in (3.1.54), even in the case without estimate of the rate of convergence.

In Remark 3.1.17(a) we show that the limit in (3.1.50) and the m-sectorial properties of generator A, which yields (3.1.55), allow to obtain operator-norm estimate (3.1.56) proving convergence of the *Trotter-Kato-Neveu* term in the two-terms estimate (3.1.54).

In our presentation of this subsection we partially follow the paper [CZ01b, Section 4 and Section 5].

The optimal rate of convergence and extension to m-sectorial generators with semi-angle $\alpha \in (0, \pi/2)$ for the *Dunford-Segal*, *Euler* (Corollary 3.1.17) and *Yosida* (Corollary 3.1.18) approximations are presented in [Arl12, Theorem 3.1, Theorem 3.4] and [ArZag10, Theorem 4.1]. We consider these results in the next Sect. 3.2.

Notes to Sect. 3.2

This section is based on the *Arlinskiĭ–Tsekanovsiĭ* theory [ArTs84, ArTs88], which allows to study the contraction C_0-semigroups on a Hilbert space through an elegant link with certain multiplication semigroups on the complex plane. For further development see [Arl87, Arl91, Arl05].

Applications to quasi-sectorial contractions and proof of convergence the product formulæ in the operator-norm topology are motivated by results from [CZ01b, Zag08, ArZag10], and [Arl12].

Section 3.2.1

Operators of the class $C_{\mathfrak{H}(\alpha)}$ were introduced in [Arl87, Arl91]. In this subsection we follow [Arl87] and [ArZag10]. For the proofs of Proposition 3.2.4 and Proposition 3.2.6 see [Arl87]. Details concerning Theorem 3.2.8 and Remark 3.2.9 one can find in [ArZag10, Section 2].

Section 3.2.2

The approach developed in [Arl87, Arl91] allows to extend the results in [CZ01b, Zag08]. The first step (Lemma 3.2.12), which yields an accurate localisation of the image of a key conformal mapping, was motivated by [Arl05]. It is due to [Zag08] and [ArZag10, Lemma 3.2]. Then from Theorem 3.2.14 it follows, in particular, that for m-sectorial generator A the localisation of numerical range of the corresponding contraction holomorphic semigroup is defined by (3.2.32).

Theorem 3.2.14 is an extension of Theorem 3.1.4 (proven in [Zag08] only for $\alpha \in [0, \pi/4)$) to the *whole* class of the m-sectorial generators with semi-angle $\alpha \in [0, \pi/2)$. In this subsection we essentially follow [ArZag10].

Section 3.2.3

A (*non-optimal*) rate of convergence $O(\ln(n)/n)$ for the Euler approximants (3.2.39) was established in [CZ01b, Theorem 5.1]. A new rate of convergence is due to an accurate localisation of the numerical range for quasi-sectorial contractions (Sect. 3.2.2) and to the proof of Theorem 3.2.17.

In Theorem 3.2.17, [ArZag10, Theorem 4.1], we first discovered for m-sectorial generators the *α-dependence* of the *optimal* error bound estimates: $O(1/(n(\cos\alpha)^2))$, for the operator-norm convergence rate of the Euler approximants, see (3.2.40), (3.2.41).

Second, we use the results of [CrDe03, BeCr06] that yield in Proposition 3.2.18 the best α-dependent constant $K_\alpha > 0$ for the *Crouzeix-Delyon* inequality (3.2.42) in sector S_α. For more details see, for example, [Haa06, Sections 7.1.3–7.1.5].

Notice that for $\alpha = \pi/2$ one gets $K_\alpha = 1$ and half-plane $S_{\alpha=\pi/2} = \mathbb{C}_+$. As a consequence the Proposition 3.2.18 yields the *von Neumann theorem* [vN51], which in particular asserts that the half-planes $\Omega = \mathbb{C}_\pm$ are the *spectral sets* for operator B.

Recall (cf. [BCD06]) that Ω is said to be a *K-spectral set* (respectively a *spectral set*) for operators A if the inequality $\|r(A)\| \leq K \sup_\Omega |r(z)|$ (respectively $\|r(A)\| \leq \sup_\Omega |r(z)|$) holds for all bounded *rational functions*: $z \mapsto r(z)$, on the spectrum $\sigma(A) \subseteq \overline{W(A)} \subseteq \Omega$. Then, by virtue of (3.2.42), sector $\Omega = S_{\alpha<\pi/2}$ is a *K_α-spectral set* for any bounded operator B, whereas $\Omega = S_{\alpha=\pi/2}$ is a *spectral set* for all B.

A classical result in this direction is a part of the *von Neumann theorem* [vN51] on the closed unit disc: $\Omega = \overline{\mathbb{D}}$.

Proposition 3.3.1 (von Neumann Theorem) *Let A be a contraction acting on a Hilbert space $\mathfrak{H}$. Then*

$$\|p(A)\| \leq \sup_{z\in\overline{\mathbb{D}}} |p(z)| \tag{3.3.1}$$

holds for all polynômials $p : \mathbb{C} \to \mathbb{C}$, where $\overline{\mathbb{D}} \subset \mathbb{C}$ is the closed unit disc.

The *von Neumann inequality* (3.3.1) reads also as: the closed unit disc $\overline{\mathbb{D}}$ is a *spectral set* for all operators A such that $W(A) \subseteq \overline{\mathbb{D}}$.

We remark that deformations of disc $\mathbb{D}$ to, for example, *parabola–*, or *lenticular–shaped*, domains Ω imply modifications of inequality (3.3.1) and determine the corresponding K-spectral sets, see [Cr04, BeCr06]. A substitution in (3.3.1) the $\Omega = \overline{\mathbb{D}}$ by $\Omega = W(A)$ also modifies the *spectral set* $\overline{\mathbb{D}}$ into a *K-spectral set*, with a value of coefficient K, which is a source of the long standing *Crouzeix conjecture* [PPW20].

Proposition 3.2.19 [Arl12, Theorem 3.1, Theorem 3.4] is a nontrivial extension of Theorem 3.2.17 for the operator-norm Dunford-Segal and Yosida approximations with a similar to Euler formula α*-dependence* of error bound estimates for optimal rates of convergence in the operator-norm, see (3.2.55) and (3.2.56). For the proof of Proposition 3.2.19 we refer to [Arl12, Section 3].

In this subsection we basically follow [ArZag10] and [Arl12].

Part II
Trotter-Kato Product Formulæ for Self-Adjoint Semigroups

Chapter 4
Product Approximations of Self-Adjoint Semigroups

In this chapter, we proceed with a systematic analysis of the main for this book subject: the *product formulæ approximations* in the *operator-norm* topology for strongly continuous semigroups. This concept has been already introduced in Preliminaries, Sect. 1.9.6. The current chapter concerns the actually most developed case of the Trotter-Kato product formulæ: the formulæ for *self-adjoint* C_0-semigroups.

Although the self-adjointness of the involved semigroups allows to apply certain direct methods (for instance, the *spectral calculus*) for proving important estimates and convergence of the product approximants even with operator-norm error bounds, we use this chapter for developing a more universal tool, which is known as the *Chernoff product formula*, cf. Sect. 3.1.2.

4.1 Strong Operator Topology: Chernoff Product Formula

Let $F(\cdot) : \mathbb{R}_0^+ \longrightarrow \mathcal{L}(\mathfrak{H})$ be a (strongly) measurable family of non-negative self-adjoint contractions $F(t) \leq \mathbb{1}$ on a Hilbert space $\mathfrak{H}$ such that $F(0) = \mathbb{1}$. We set

$$S(\tau) := \frac{\mathbb{1} - F(\tau)}{\tau}, \quad \tau > 0. \tag{4.1.1}$$

Then for each $\tau > 0$, bounded operator $S(\tau)$ is self-adjoint and positive.

Let H with a dense domain dom (H) be self-adjoint operator in $\mathfrak{H}$. Then by the *Trotter-Kato-Neveu* convergence theorem (see Proposition 1.9.30) we obtain that

$$\text{s-}\lim_{\tau \to +0} (\mathbb{1} + S(\tau))^{-1} = (\mathbb{1} + H)^{-1}, \tag{4.1.2}$$

V. A. Zagrebnov et al., *Trotter-Kato Product Formulæ*, Operator Theory: Advances and Applications 296, https://doi.org/10.1007/978-3-031-56720-9_4

holds in the strong operator topology if and only if

$$\text{s-}\lim_{\tau\to+0} \mathrm{e}^{-tS(\tau)} = \mathrm{e}^{-tH}, \tag{4.1.3}$$

uniformly in $t \in \mathcal{I}$ for any bounded closed interval $\mathcal{I}$ in $\mathbb{R}^+ = \mathbb{R}_0^+ \setminus \{0\}$. Here, operator $(-H)$ is the closure of the strong (right) derivative $F'(+0)$.

Remark that we shall say in this case that the convergence in (4.1.2) and (4.1.3) holds *locally uniformly* in $t \in \mathbb{R}_0^+$ *away from zero*, locally uniformly $t \in \mathbb{R}^+$.

Setting $\tau = t/\eta$, for $\eta \geq 1$, we get, in particular, that

$$\text{s-}\lim_{\eta\to+\infty} \mathrm{e}^{-tS(t/\eta)} = \mathrm{e}^{-tH}, \tag{4.1.4}$$

locally uniformly in $t \in \mathbb{R}_0^+$ away from zero.

To proceed, we need two elementary technical lemmata, which will be regularly applied below owing to self-adjointness.

Lemma 4.1.1 *For* $\lambda \in [0, 1]$ *and* $r \geq 1$, *one has*

$$0 \leq \mathrm{e}^{-r(1-\lambda)} - \lambda^r \leq \frac{1}{r}. \tag{4.1.5}$$

Proof We show first that $h(\lambda) := \mathrm{e}^{(1-\lambda)}\lambda \leq 1$, $\lambda \in [0, 1]$. Note that

$$h'(\lambda) = -\mathrm{e}^{(1-\lambda)}\lambda + \mathrm{e}^{(1-\lambda)} = (1-\lambda)\,\mathrm{e}^{(1-\lambda)} \geq 0, \quad \lambda \in [0, 1].$$

Hence $0 = h(0) \leq h(\lambda) \leq h(1) = 1$, $\lambda \in [0, 1]$. In particular, $0 \leq h(\lambda)^r = \mathrm{e}^{r(1-\lambda)}\lambda^r \leq 1$, which yields $0 \leq \mathrm{e}^{-r(1-\lambda)} - \lambda^r$ for $\lambda \in [0, 1]$, $r \geq 1$.

To prove the right-hand side of (4.1.5), we verify the inequality

$$1 \leq \frac{\mathrm{e}^{rx}}{r} + \left(\mathrm{e}^{x}(1-x)\right)^r, \quad x \in [0, 1], \quad r \geq 1. \tag{4.1.6}$$

To this aim let $g(x) := \mathrm{e}^{rx}/r + (\mathrm{e}^{x}(1-x))^r$, $x \in [0, 1]$, $r \geq 1$. Then

$$g'(x) = \mathrm{e}^{rx} - r\mathrm{e}^{rx}(1-x)^{r-1}x = \mathrm{e}^{rx}k(x), \quad x \in [0, 1], \quad r \geq 1,$$

where $k(x) := 1 - r(1-x)^{r-1}x$, $x \in [0, 1]$, $r \geq 1$. Obviously, one has $k(x) \leq 1$ and $k(0) = k(1) = 1$. Now, since

$$k'(x) = r(1-x)^{r-2}(rx-1), \quad x \in [0, 1), \quad r \geq 1,$$

we find that $k(x)$ has a minimum at $x = 1/r$. Then

$$k(1/r) = 1 - (1 - 1/r)^{r-1} \geq 0, \quad r \geq 1,$$

or $0 \le k(x) \le 1$, $x \in [0, 1]$, $r \ge 1$. Hence $g'(x) \ge 0$, $x \in [0, 1]$, $r \ge 1$. Therefore, the function $g(x)$ is nondecreasing that yields $1 \le g(0) = 1/r + 1 \le g(x)$, $x \in [0, 1]$, $r \ge 1$. Hence (4.1.6) is proven. Setting $x := 1 - \lambda$ we get from (4.1.6) that

$$1 \le \frac{\mathrm{e}^{\,r(1-\lambda)}}{r} + \mathrm{e}^{\,r(1-\lambda)}\lambda^r, \quad \lambda \in [0, 1], \quad r \ge 1.$$

Multiplying this inequality by $\mathrm{e}^{\,-r(1-\lambda)}$, we find

$$\mathrm{e}^{\,-r(1-\lambda)} \le \frac{1}{r} + \lambda^r, \quad \lambda \in [0, 1], \quad r \ge 1,$$

which yields the right-side inequality of (4.1.5). □

Lemma 4.1.2 *Let* $\{u_n\}_{n\ge1}$ *be a weakly convergent sequence of vectors* $\text{w-}\lim_{n\to\infty} u_n = u(\neq 0)$ *in a Hilbert space* $\mathfrak{H}$. *If, in addition, one has* $\lim_{n\to\infty} \|u_n\| = \|u\|$, *then* $\text{s-}\lim_{n\to\infty} u_n = u$.

Proof Note that

$$\|u_n - u\|^2 = \|u_n\|^2 + \|u\|^2 - 2\,\mathrm{Re}\,(u_n, u).$$

Then by conditions of the lemma, this yields $\text{s-}\lim_{n\to\infty} u_n = u$. □

Now we are in position to prove the *Chernoff product theorem* for self-adjoint semigroups in the *strong operator* topology in a Hilbert space $\mathfrak{H}$. For the case of a Banach space, see Theorem 1.9.34 (Sect. 1.9.6).

Theorem 4.1.3 *Let* $F(\cdot) : \mathbb{R}_0^+ \longrightarrow \mathcal{L}(\mathfrak{H})$ *be a strongly (Borel) measurable family of non-negative self-adjoint contractions such that* $F(0) = \mathbb{1}$ *and let* H *be a self-adjoint operator in* $\mathfrak{H}$. *The convergence*

$$\text{s-}\lim_{\eta\to+\infty} F(t/\eta)^{\eta} = \mathrm{e}^{\,-tH}, \tag{4.1.7}$$

holds locally uniformly in $t \in \mathbb{R}_0^+$ *away from zero if and only if the condition* (4.1.2) *is satisfied.*

Proof Let us assume that condition (4.1.2) is satisfied. We are going to show that

$$\text{s-}\lim_{\eta\to+\infty} \left(F(t/\eta)^{\eta} - \mathrm{e}^{\,-tS(t/\eta)}\right) = 0, \tag{4.1.8}$$

uniformly in $t \ge 0$. Using the functional calculus for self-adjoint operators, we get

$$F(\tau)^r - \mathrm{e}^{\,-r(\mathbb{1}-F(\tau))} = \int_{[0,1]} E_{F(\tau)}(d\lambda)\left(\lambda^r - \mathrm{e}^{\,-r(1-\lambda)}\right), \quad \tau > 0, \quad r \ge 1.$$

Then inequality (4.1.5) yields the estimate

$$\left\|\left(F(\tau)^r - \mathrm{e}^{-r(\mathbb{1}-F(\tau))}\right) u\right\| \leq r^{-1}\,\|u\|, \quad r \geq 1, \quad u \in \mathfrak{H}. \tag{4.1.9}$$

Hence, setting $\tau = t/\eta$, $\eta \geq 1$ and $r = t/\tau = \eta \geq 1$, we obtain that

$$\left\|\left(F(t/\eta)^\eta - \mathrm{e}^{-tS(t/\eta)}\right) u\right\| \leq \eta^{-1}\|u\|, \quad \eta \geq 1, \quad u \in \mathfrak{H}.$$

This proves the limit

$$\text{s-}\lim_{\eta\to+\infty}\left(F(t/\eta)^\eta - \mathrm{e}^{-tS(t/\eta)}\right) = 0, \tag{4.1.10}$$

locally uniformly in t away from zero. Since

$$F(t/\eta)^\eta - \mathrm{e}^{-tH} = F(t/\eta)^\eta - \mathrm{e}^{-tS(t/\eta)} + \mathrm{e}^{-tS(t/\eta)} - \mathrm{e}^{-tH}, \quad \eta \geq 1, \quad t \geq 0,$$

then using (4.1.4) and (4.1.10), we deduce (4.1.7).

Conversely, assume that (4.1.7) is satisfied. Using representation

$$\mathrm{e}^{-tS(t/\eta)} - \mathrm{e}^{-tH} = F(t/\eta)^\eta - \mathrm{e}^{-tH} + \mathrm{e}^{-tS(t/\eta)} - F(t/\eta)^\eta,$$

we get from (4.1.7) and (4.1.10) that (4.1.4) holds locally uniformly in t away from zero. Setting $\tau := t/\eta$, we verify that convergence in (4.1.3) holds locally uniformly in t away from zero. Then by Trotter-Kato-Neveu convergence theorem (Proposition 1.9.30), the limit (4.1.3) implies (4.1.2). □

We note that in particular the relation (4.1.7) yields that the semigroup approximation formula

$$\text{s-}\lim_{n\to\infty} F(t/n)^n = \mathrm{e}^{-tH}, \tag{4.1.11}$$

holds locally uniformly in t away from zero. We also remark that formula (4.1.11) follows for a particular case of *self-adjoint* semigroups from *Chernoff product formula* (1.9.101) established in Theorem 1.9.34.

Lemma 4.1.4 *Let $K(\cdot) : \mathbb{R}^+ \to \mathcal{L}(\mathfrak{H})$ be a strongly measurable family of non-negative self-adjoint operators and let H be non-negative self-adjoint operator. If*

$$\text{w-}\lim_{\tau\to+0}(\lambda\mathbb{1} + K(\tau))^{-1} = (\lambda\mathbb{1} + H)^{-1}, \tag{4.1.12}$$

for each $\lambda \geq 1$, then

$$\text{s-}\lim_{\tau\to+0}(\lambda\mathbb{1} + K(\tau))^{-1} = (\lambda\mathbb{1} + H)^{-1}. \tag{4.1.13}$$

Proof From (4.1.12) we get

$$\lim_{\tau\to+0} \|(\lambda\mathbb{1} + K(\tau))^{-1/2}u\|^2 = \|(\lambda\mathbb{1} + H)^{-1/2}u\|^2, \quad u \in \mathfrak{H}, \tag{4.1.14}$$

for $\lambda \geq 1$. Since

$$(\mathbb{1} + K(\tau))^{-1/2} = \frac{1}{\pi}\int_0^\infty dt\, \frac{1}{\sqrt{t}}\,(t\,\mathbb{1} + \mathbb{1} + K(\tau))^{-1}\,, \tag{4.1.15}$$

Equation (4.1.12) also yields w$-\lim_{\tau\to+0}(\mathbb{1} + K(\tau))^{-1/2} = (\mathbb{1} + H)^{-1/2}$. This, together with (4.1.14) and the uplifting Lemma 4.1.2, imply s-$\lim_{\tau\to+0}(\mathbb{1} + K(\tau))^{-1/2} = (\mathbb{1} + H)^{-1/2}$. Then seeing that product of bounded operators is strongly continuous, we obtain the strong limit (4.1.13). □

Taking into account Lemma 4.1.4, the conditions of Theorem 4.1.3 can be now relaxed as follows.

Corollary 4.1.5 *Let $F(\cdot) : \mathbb{R}_0^+ \longrightarrow \mathcal{L}(\mathfrak{H})$ be a strongly measurable family of non-negative contractions such that $F(0) = \mathbb{1}$ and let H be a self-adjoint operator in $\mathfrak{H}$. The statement* (4.1.7) *is valid if and only if the condition*

$$\text{w}-\lim_{\tau\to+0} (\lambda\mathbb{1} + S(\tau))^{-1} = (\lambda\mathbb{1} + H)^{-1}\,, \tag{4.1.16}$$

is satisfied for each $\lambda \geq 1$.

Proof If (4.1.16) is valid, then applying Lemma 4.1.4 we verify (4.1.2). Using Theorem 4.1.3, we prove (4.1.7).

To prove the converse assertion, we note that by Theorem 4.1.3 the statement (4.1.7) yields (4.1.2). However, from (4.1.2) one gets

$$\text{s-}\lim_{\tau\to+0} (\lambda\mathbb{1} + S(\tau))^{-1} = (\lambda\mathbb{1} + H)^{-1}\,, \tag{4.1.17}$$

for $\lambda \geq 1$, which implies (4.1.16). □

4.2 Extension of the Chernoff Theory to Operator-Norm Topology

A natural question arises: whether Theorem 4.1.3 can be *uplifted* (improved) to convergence in a stronger topology, which is the operator-norm topology.

First, we note that in contrast to quasi-sectorial contractions, for a self-adjoint contraction C, the estimate (3.1.44) in the $(\sqrt[3]{n})^{-1}$-Theorem (Remark 3.1.15), as well as the estimate (1.9.96) in the Chernoff $\sqrt{n}$-Lemma 1.9.33, can be drastically improved. Namely, by functional calculus for self-adjoint contraction C and by

Lemma 4.1.1, we obtain the estimate

$$\|C^n - e^{n(C-\mathbb{1})}\| = \left\| \int_0^1 dE_C(\lambda)\,(\lambda^n - e^{n(\lambda-1)}) \right\| \leq \frac{1}{n}, \qquad n \in \mathbb{N}, \tag{4.2.1}$$

Then similarly to the case of the strong operator topology (Sect. 1.9.5), the next step in the program for uplifting it to the operator-norm topology must involve a generalisation of the *Trotter-Kato-Neveu* approximation theorem, cf. Proposition 1.9.30 for the strong operator topology.

Given that, we proceed with the following lemma.

Lemma 4.2.1 *Let K and L be non-negative self-adjoint operators in a Hilbert space $\mathfrak{H}$. Then*

$$\|e^{-K} - e^{-L}\| \leq C\|(\mathbb{1} + K)^{-1} - (\mathbb{1} + L)^{-1}\| \tag{4.2.2}$$

with a constant $C > 0$ independent of operators K and L.

Proof By the *Riesz-Dunford* integral formula, one obtains for the difference of exponentials the representation

$$e^{-K} - e^{-L} = \frac{1}{2\pi i}\int_\Gamma dz\, e^{-z}\left((z-K)^{-1} - (z-L)^{-1}\right), \tag{4.2.3}$$

where the contour Γ is a union of two branches: $\Gamma = \Gamma_0 \cup \Gamma_\infty$, with

$$\begin{aligned} \Gamma_0 &= \{z \in \mathbb{C} : z = e^{i\varphi}, \quad \pi/4 \leq \varphi \leq 2\pi - \pi/4\}, \\ \Gamma_\infty &= \{z \in \mathbb{C} : z = re^{\pm i\pi/4}, \quad r \geq 1\}. \end{aligned} \tag{4.2.4}$$

From (4.2.3) we find the representation

$$\begin{aligned} e^{-K} - e^{-L} = \frac{1}{2\pi i}\int_\Gamma dz\, e^{-z}(\mathbb{1} + K)(z-K)^{-1} \cdot \\ \cdot \left((\mathbb{1}+L)^{-1} - (\mathbb{1}+K)^{-1}\right)(\mathbb{1}+L)(z-L)^{-1}. \end{aligned} \tag{4.2.5}$$

Since

$$(\mathbb{1} + K)(z-K)^{-1} = -\mathbb{1} + (1+z)(z-K)^{-1},$$

one gets the estimate

$$\left\|(\mathbb{1} + K)(z-K)^{-1}\right\| \leq 1 + \frac{1+|z|}{\operatorname{dist}(z, \mathbb{R}^+)}.$$

Setting

$$C_\Gamma := \sup_{z\in\Gamma} \frac{1+|z|}{\mathrm{dist}\,(z,\mathbb{R}^+)} < \infty\,,$$

we find

$$\sup_{z\in\Gamma} \left\| (\mathbb{1}+K)(z-K)^{-1} \right\| \leq (1+C_\Gamma)\,, \tag{4.2.6}$$

where the constant C_Γ depends only on contour Γ but not on the operator K. Similarly, from (4.2.6) one also gets

$$\sup_{z\in\Gamma} \left\| (\mathbb{1}+L)(z-L)^{-1} \right\| \leq (1+C_\Gamma)\,.$$

Using these estimates, we find from (4.2.5) that

$$\left\| \mathrm{e}^{-K} - \mathrm{e}^{-L} \right\| \leq C \left\| (\mathbb{1}+K)^{-1} - (\mathbb{1}+L)^{-1} \right\|$$

with coefficient

$$C := \frac{1}{2\pi}\,(1+C_\Gamma)^2 \int_\Gamma |dz|\, |\mathrm{e}^{-z}|\,,$$

which depends only on the contour Γ. □

The first step to the proof of the operator-norm convergence of Chernoff approximation formula (4.1.7) would be *uplifting* of the strong convergence in (4.1.2) to the operator-norm convergence. To study consequences of this uplifting, we consider the following assertion.

Lemma 4.2.2 *Let $F(\cdot) : \mathbb{R}_0^+ \longrightarrow \mathcal{L}(\mathfrak{H})$ be a measurable family of non-negative self-adjoint contractions such that $F(0) = \mathbb{1}$. Let self-adjoint family $\{S(\tau)\}_{\tau>0}$ be defined by* (4.1.1)–(4.1.2)*, where H is non-negative self-adjoint operator in $\mathfrak{H}$. Then condition*

$$\lim_{\tau\to+0} \|(\mathbb{1}+S(\tau))^{-1} - (\mathbb{1}+H)^{-1}\| = 0. \tag{4.2.7}$$

is satisfied if and only if

$$\lim_{\eta\to\infty} \sup_{t\in\mathcal{I}} \|(\mathbb{1}+tS(t/\eta))^{-1} - (\mathbb{1}+tH)^{-1}\| = 0\,, \tag{4.2.8}$$

for any closed away from zero interval $\mathcal{I} \subset \mathbb{R}_0^+$.

Proof A direct computation shows that

$$
\begin{aligned}
&(\mathbb{1} + tS(\tau))^{-1} - (\mathbb{1} + tH)^{-1}\\
&= t(\mathbb{1} + S(\tau))(\mathbb{1} + tS(\tau))^{-1}((\mathbb{1} + S(\tau))^{-1} - (\mathbb{1} + H)^{-1})(\mathbb{1} + H)(\mathbb{1} + tH)^{-1} .
\end{aligned}
$$

Here we used that if $t > 0$ and $\tau > 0$, then for self-adjoint $S(\tau)$ the closure

$$
\overline{(\mathbb{1} + t\,S(\tau))^{-1}(\mathbb{1} + S(\tau))} = (\mathbb{1} + S(\tau))(\mathbb{1} + t\,S(\tau))^{-1} .
$$

For these values of arguments, we obviously have

$$
\begin{aligned}
&\|(\mathbb{1} + S(\tau))(\mathbb{1} + t\,S(\tau))^{-1}\| \leq (1 + 2/t)\,,\\
&\|(\mathbb{1} + H)(\mathbb{1} + t\,H)^{-1}\| \leq (1 + 2/t)\,.
\end{aligned}
$$

If $\mathcal{I} \subset \mathbb{R}_0^+$ is a closed interval away from zero, that is, $\mathcal{I} \subset [a, b]$, $0 < a < b < \infty$, then one gets

$$
\|(\mathbb{1} + tS(\tau))^{-1} - (\mathbb{1} + tH)^{-1}\| \leq b(1 + 2/a)^2\|(\mathbb{1} + S(\tau))^{-1} - (\mathbb{1} + H)^{-1}\|,
$$

for $t \in \mathcal{I}$ and $\tau > 0$. Setting $\tau = t/\eta$, we find

$$
\|(\mathbb{1} + tS(t/\eta))^{-1} - (\mathbb{1} + tH)^{-1}\| \leq b(1 + 2/a)^2\|(\mathbb{1} + S(t/\eta))^{-1} - (\mathbb{1} + H)^{-1}\|. \tag{4.2.9}
$$

Since by (4.2.7) we obtain for the right-hand side of (4.2.9)

$$
\lim_{\eta\to\infty} \sup_{t\in\mathcal{I}} \|(\mathbb{1} + S(t/\eta))^{-1} - (\mathbb{1} + H)^{-1}\| = 0,
$$

the estimate (4.2.9) yields (4.2.8). The converse is straightforward. □

Lemma 4.2.2 allows to prove the first result about operator-norm convergence of the Chernoff product formula (4.1.7).

Theorem 4.2.3 *Let $F(\cdot) : \mathbb{R}_0^+ \longrightarrow \mathcal{L}(\mathfrak{H})$ be a strongly measurable family of non-negative self-adjoint contractions such that $F(0) = \mathbb{1}$. Let self-adjoint family $\{S(\tau)\}_{\tau>0}$ be defined by* (4.1.1)–(4.1.2), *where H is a positive self-adjoint operator in $\mathfrak{H}$. Then we have*

$$
\lim_{\eta\to\infty} \sup_{t\in\mathcal{I}} \|F(t/\eta)^{\eta} - \mathrm{e}^{-tH}\| = 0, \tag{4.2.10}
$$

for any closed away from zero interval $\mathcal{I} \subset \mathbb{R}_0^+$ if and only if the family $\{S(\tau)\}_{\tau>0}$ satisfies condition (4.2.7).

Proof Obviously, for $t > 0$ and $\eta > 0$, we have the estimate

$$\| F(t/\eta)^{\eta} - e^{-tH} \| \le \| F(t/\eta)^{\eta} - e^{-t\,S(t/\eta)} \| + \| e^{-t\,S(t/\eta)} - e^{-tH} \| \,. \tag{4.2.11}$$

By the functional calculus for self-adjoint contraction $C := F(t/\eta)$ and for $S(t/\eta) = \eta\,(\mathbb{1} - C)/t$, Lemma 4.1.1 yields estimate (4.2.1) which improves (1.9.96) in the Chernoff $\sqrt{n}$-Lemma for general contractions. Then (4.2.1) and (4.2.11) imply

$$\| F(t/\eta)^{\eta} - e^{-tH} \| \le \frac{1}{\eta} + \| e^{-t\,S(t/\eta)} - e^{-tH} \|, \quad t > 0, \quad \eta > 0\,. \tag{4.2.12}$$

Note that by Lemma 4.2.1, there is a constant $C > 0$ such that

$$\| e^{-t\,S(t/\eta)} - e^{-tH} \| \le C\, \| (\mathbb{1} + t\,S(t/\eta))^{-1} - (\mathbb{1} + tH)^{-1} \|, \tag{4.2.13}$$

for $t > 0$ and $\eta > 0$. Inserting the estimate (4.2.13) into (4.2.12), we obtain

$$\| F(t/\eta)^{\eta} - e^{-tH} \| \le \frac{1}{\eta} + C\, \| (\mathbb{1} + t\,S(t/\eta))^{-1} - (\mathbb{1} + t\,H)^{-1} \|, \tag{4.2.14}$$

Then using Lemma 4.2.2, we get (4.2.10).

Conversely, let us assume (4.2.10). Then obviously, we have the estimate

$$\| e^{-t\,S(t/\eta)} - e^{-tH} \| \le \| F(t/\eta)^{\eta} - e^{-tH} \| + \| F(t/\eta)^{\eta} - e^{-t\,S(t/\eta)} \| \,,$$

for $t > 0$ and $\eta > 0$. Applying to the last term the spectral representation and Lemma 4.1.1, we find

$$\| e^{-t\,S(t/\eta)} - e^{-tH} \| \le \| F(t/\eta)^{\eta} - e^{-tH} \| + \frac{1}{\eta}, \quad t > 0, \quad \eta > 0. \tag{4.2.15}$$

Then by assumption (4.2.10), the estimate (4.2.15) yields

$$\lim_{\eta\to\infty} \sup_{t\in\mathcal{I}} \| e^{-tS(t/\eta)} - e^{-tH} \| = 0\,,$$

for any closed away from zero interval $\mathcal{I} \subset \mathbb{R}_0^+$. As a consequence, the limit $\lim_{\eta\to\infty} \| e^{\eta\,(F(t/\eta) - \mathbb{1})} - e^{-tH} \| = 0$, or equivalently $\lim_{\tau\to+0} \| e^{-t\,S(\tau)} - e^{-tH} \| = 0$, holds for any $t > 0$. Then by virtue of representation,

$$(\mathbb{1} + S(\tau))^{-1} - (\mathbb{1} + H)^{-1} = \int_{\mathbb{R}^+} ds\, e^{-s} \left(e^{-s\,S(\tau)} - e^{-sH} \right), \tag{4.2.16}$$

we infer estimate

$$\|(\mathbb{1} + S(\tau))^{-1} - (\mathbb{1} + H)^{-1}\| \leq \int_0^\infty ds\, \mathrm{e}^{-s} \left\| \mathrm{e}^{-s\,S(\tau)} - \mathrm{e}^{-s\,H} \right\| . \tag{4.2.17}$$

Note that by conditions of the theorem

$$e^{-s} \left\| e^{-sS(\tau)} - \mathrm{e}^{-sH} \right\| \leq 2\,\mathrm{e}^{-s} \in L^1(\mathbb{R}_0^+) .$$

Then $\lim_{\tau\to+0}$ in the right-hand side of (4.2.17) is zero by the *Lebesgue* dominated convergence theorem that yields (4.2.7). □

Extension of Chernoff product Formula (4.2.10) to *any* bounded interval $\mathcal{I} \subset \mathbb{R}_0^+$ needs a *uniform* operator-norm generalisation of the *Trotter-Kato-Neveu* theorem that we present below; see Theorem 4.2.6.

Lemma 4.2.4 *Let $X \geq 0$ be a bounded self-adjoint operator on $\mathfrak{H}$. If $\{X_\nu\}_{\nu>0}$ is a sequence of non-negative self-adjoint operators such that* $\lim_{\nu\to\infty} \|X_\nu - X\| = 0$, *then for any continuous function $g(\cdot) : [0, \|X\|] \longrightarrow \mathbb{R}$, one gets limit*

$$\lim_{\nu\to\infty} \|g(X_\nu) - g(X)\| = 0,$$

in the operator-norm topology.

Proof Note that $\lim_{\nu\to\infty} \|X_\nu - X\| = 0$ implies $\lim_{\nu\to\infty} \|X_\nu^n - X^n\| = 0$ for $n \in \mathbb{N}$, and estimate $\|X_\nu\| \leq \|X\| + \delta$ for any $\delta > 0$, where $\nu > \nu(\delta)$. Then

$$\lim_{\nu\to\infty} \|p(X_\nu) - p(X)\| = 0,$$

for any polynomials $p(\cdot) : [0, \|X\|] \to \mathbb{R}$.

By the *Weierstrass* theorem, polynomials are dense in the set of continuous functions $C_\infty([0, \|X\|])$ on the interval $[0, \|X\|]$. Thus, given $\varepsilon > 0$ we can find polynomial $p(\cdot)$ such that $\|g - p\|_\infty = \sup_{x\in[0,\|X\|]} |g(x) - p(x)| < \varepsilon/3$. Then by the spectral representation for operators X and X_ν, we obtain

$$\|g(X) - p(X)\| < \varepsilon/3 ,$$
$$\|g(X_\nu) - p(X_\nu)\| < \varepsilon/3 .$$

Now taking $\nu > \nu(\delta)$ large enough, one gets $\|p(X_\nu) - p(X)\| < \varepsilon/3$, which consequently gives for any $\varepsilon > 0$ the estimate $\|g(X_\nu) - g(X)\| < \varepsilon$, and, therefore, proves the lemma. □

Corollary 4.2.5 *Let $F(\cdot) : \mathbb{R}_0^+ \longrightarrow \mathcal{L}(\mathfrak{H})$ be a measurable family of non-negative self-adjoint contractions such that $F(0) = \mathbb{1}$. Let self-adjoint family $\{S(\tau)\}_{\tau>0}$ be defined by* (4.1.1)–(4.1.2), *where H is a non-negative self-adjoint operator in $\mathfrak{H}$. If*

$$\lim_{\eta\to\infty} \|F(t_0/\eta)^\eta - \mathrm{e}^{-t_0 H}\| = 0 , \tag{4.2.18}$$

holds for some $t_0 > 0$, *then the operator-norm Chernoff approximation formula* (4.2.10) *holds for any closed interval* $\mathcal{I} \subset \mathbb{R}_0^+$ *away from zero.*

Proof We use the representation

$$F(t/\eta)^{\eta} = \left(F(t_0/\nu)^{\nu}\right)^{t/t_0}, \quad \nu := \frac{t_0}{t}\,\eta\,, \quad t > 0\,. \tag{4.2.19}$$

Setting $X := \mathrm{e}^{-t_0 H}$ and $X_\nu := F(t_0/\nu)^\nu$, we have by assumption (4.2.18) that $\lim_{\nu\to\infty} \|X_\nu - X\| = 0$. Setting $g(x) := x^{t/t_0}$, $x \geq 0$, and $g(x) := 0$, $x < 0$, by Lemma 4.2.4, we obtain

$$\lim_{\nu\to\infty} \left\|X_\nu^{t/t_0} - X^{t/t_0}\right\| = 0.$$

Then by virtue of representation (4.2.19), it follows that $\lim_{\eta\to\infty} \|F(t/\eta)^\eta - \mathrm{e}^{-tH}\| = 0$ holds for any $t \in \mathbb{R}_0^+$. Proceeding now similar to above arguments for (4.2.17), we find (4.2.7). Finally, applying Theorem 4.2.3, we obtain (4.2.10) for any *closed* interval $\mathcal{I} \subset \mathbb{R}_0^+$ away from zero. □

Seeing that by definition of $\tau \mapsto t\,S(\tau)$ the limit (4.2.8) is also valid for $t = 0$, a question arises: what happens if the condition (4.2.8) is satisfied for *any* bounded interval $\mathcal{I} \subset \mathbb{R}_0^+$?

Theorem 4.2.6 *Let* $F(\cdot) : \mathbb{R}_0^+ \longrightarrow \mathcal{L}(\mathfrak{H})$ *be a measurable family of non-negative self-adjoint contractions such that* $F(0) = \mathbb{1}$. *Let self-adjoint family* $\{S(\tau)\}_{\tau>0}$ *be defined by* (4.1.1)–(4.1.2), *where* H *is a positive self-adjoint operator in* $\mathfrak{H}$. *Then the convergence*

$$\lim_{\tau\to+0} \sup_{t\in\mathcal{I}} \|\,\mathrm{e}^{-t\,S(\tau)} - \mathrm{e}^{-t\,H}\,\| = 0\,, \tag{4.2.20}$$

holds for any bounded interval $\mathcal{I} \subset \mathbb{R}_0^+$ *if and only if the condition*

$$\lim_{\tau\to+0} \sup_{t\in\mathcal{I}} \|(\mathbb{1} + t\,S(\tau))^{-1} - (\mathbb{1} + t\,H)^{-1}\| = 0\,, \tag{4.2.21}$$

is valid for any bounded interval $\mathcal{I} \subset \mathbb{R}_0^+$.

Proof By conditions of the theorem, we obtain from (4.2.13) the estimate

$$\sup_{t\in\mathcal{I}} \|\mathrm{e}^{-tS(\tau)} - \mathrm{e}^{-tH}\| \leq C \sup_{t\in\mathcal{I}} \|(\mathbb{1} + t\,S(\tau))^{-1} - (\mathbb{1} + t\,H)^{-1}\|\,,$$

for $\tau > 0$ and for any bounded interval $\mathcal{I} \subset \mathbb{R}_0^+$. This estimate and condition (4.2.21) imply for $\tau := t/\eta$ the convergence in (4.2.20).

Conversely, assume (4.2.20). Note that by representation (4.2.16) one gets for $\tau > 0$ and $t \geq 0$:

$$(\mathbb{1} + tS(\tau))^{-1} - (\mathbb{1} + tH)^{-1} = \int_0^\infty ds\, e^{-s} \left(e^{-st\,S(\tau)} - e^{-st\,H}\right).$$

This yields estimate

$$\left\| (\mathbb{1} + tS(\tau))^{-1} - (\mathbb{1} + tH)^{-1} \right\| \leq \int_0^\infty ds\, e^{-s} \left\| e^{-st\,S(\tau)} - e^{-st\,H} \right\|,$$

for $\tau > 0$ and $t \geq 0$.

Now let $0 < \varepsilon \leq 1$ and let $N_\varepsilon \geq -\ln(\varepsilon/2)$. Then one infers that

$$\int_{N_\varepsilon}^\infty ds\, e^{-s} \left\| e^{-st\,S(t/\eta)} - e^{-st\,H} \right\| \leq \varepsilon,$$

for $\tau > 0$ and $t \geq 0$. As a consequence,

$$\left\| (\mathbb{1} + tS(\tau))^{-1} - (\mathbb{1} + tH)^{-1} \right\| \leq \int_0^{N_\varepsilon} ds\, e^{-s} \left\| e^{-stS(\tau)} - e^{-stH} \right\| + \varepsilon,$$

for $\tau > 0$ and $t \geq 0$. On that account,

$$\begin{aligned} &\sup_{t\in\mathcal{I}} \left\| (\mathbb{1} + t\,S(\tau))^{-1} - (\mathbb{1} + t\,H)^{-1} \right\| \\ &\leq \sup_{t\in\mathcal{I}\,\wedge\, s\in[0,N_\varepsilon]} \left\| e^{-st\,S(\tau)} - e^{-st\,H} \right\| + \varepsilon, \end{aligned}$$

for $\tau > 0$ and for any bounded interval $\mathcal{I} \subset \mathbb{R}_0^+$. Applying now (4.2.20), we get

$$\lim_{\tau\to+0} \sup_{t\in\mathcal{I}} \left\| (\mathbb{1} + tS(\tau))^{-1} - (\mathbb{1} + tH)^{-1} \right\| \leq \varepsilon,$$

for any bounded interval $\mathcal{I} \subset \mathbb{R}^+$ and for any $\varepsilon > 0$. This completes the proof of condition (4.2.21). □

Now we are in position to improve Theorem 4.2.3 for the *operator-norm* Chernoff product formula. We relax the condition that intervals $\mathcal{I}$ are *away* from zero.

Theorem 4.2.7 *Let $F(\cdot) : \mathbb{R}_0^+ \longrightarrow \mathcal{L}(\mathfrak{H})$ be a strongly measurable family of non-negative self-adjoint contractions such that $F(0) = \mathbb{1}$. Let self-adjoint family $\{S(\tau)\}_{\tau>0}$ be defined by* (4.1.1)–(4.1.2)*, where H is positive self-adjoint operator in $\mathfrak{H}$. Then*

$$\lim_{\eta\to\infty} \sup_{t\in\mathcal{I}} \|F(t/\eta)^\eta - \mathrm{e}^{-tH}\| = 0, \tag{4.2.22}$$

for any bounded interval $\mathcal{I} \subset \mathbb{R}_0^+$ if and only if

$$\lim_{\eta\to\infty} \sup_{t\in\mathcal{I}} \|(\mathbb{1} + tS(t/\eta))^{-1} - (\mathbb{1} + tH)^{-1}\| = 0, \tag{4.2.23}$$

is satisfied for any bounded interval $\mathcal{I} \subset \mathbb{R}_0^+$, that is, locally uniformly on $\mathbb{R}_0^+$.

Proof By (4.2.14) and by assumption (4.2.23), we obtain the limit (4.2.22). Conversely, by estimate (4.2.15) and assumption (4.2.22), we obtain (4.2.20) for any bounded interval $\mathcal{I} \subset \mathbb{R}_0^+$. Then applying Theorem 4.2.6, we deduce (4.2.23). □

4.3 Operator-Norm Approximations and Rate of Convergence

Theorem 4.2.6 permits a further improvement. In particular, a rate of the operator-norm convergence can be established.

Theorem 4.3.1 *Let $F(\cdot) : \mathbb{R}_0^+ \longrightarrow \mathcal{L}(\mathfrak{H})$ be a measurable family of non-negative self-adjoint contractions such that $F(0) = \mathbb{1}$. Let self-adjoint family $\{S(\tau)\}_{\tau>0}$ be defined by* (4.1.1)–(4.1.2)*, where H is non-negative self-adjoint operator in $\mathfrak{H}$.*

(i) *If $\rho \in (0, 1]$ and there is a constant $M_\rho > 0$ such that the estimate*

$$\|(\mathbb{1} + tS(\tau))^{-1} - (\mathbb{1} + tH)^{-1}\| \leq M_\rho \left(\frac{\tau}{t}\right)^\rho, \tag{4.3.1}$$

holds for $\tau, t \in (0, 1]$ and $0 < \tau \leq t$, then there is a constant $C_\rho > 0$ such that the estimate

$$\|F(\tau)^{t/\tau} - \mathrm{e}^{-tH}\| \leq C_\rho \left(\frac{\tau}{t}\right)^\rho, \tag{4.3.2}$$

is valid for $\tau, t \in (0, 1]$ with $0 < \tau \leq t$.

(ii) *If $\rho \in (0, 1)$ and there is a constant C_ρ such that* (4.3.2) *holds, then there is a constant $M_\rho > 0$ such that the estimate* (4.3.1) *is valid.*

Proof

(i) By Lemma 4.2.1, there is a constant $C > 0$ such that

$$\| e^{-tS(\tau)} - e^{-tH} \| \leq C \, \|(\mathbb{1} + tS(\tau))^{-1} - (\mathbb{1} + tH)^{-1}\| , \qquad (4.3.3)$$

for $\tau, t > 0$. Owing to (4.3.1), we obtain

$$\left\| e^{-tS(\tau)} - e^{-tH} \right\| \leq C \, M_\rho \left(\frac{\tau}{t}\right)^\rho ,$$

for $\tau, t \in (0, 1]$ with $0 < \tau \leq 1$. Then by definition (4.1.1), estimate (4.1.5) and spectral representations for functions of $F(\tau)$, we infer that

$$\left\| F(\tau)^{t/\tau} - e^{-tS(\tau)} \right\| \leq \frac{\tau}{t}, \quad 0 < \tau \leq t \leq 1 . \qquad (4.3.4)$$

As a consequence, the estimate

$$\left\| F(\tau)^{t/\tau} - e^{-tH} \right\| \leq \left\| F(\tau)^{t/\tau} - e^{-tS(\tau)} \right\| + \left\| e^{-tS(\tau)} - e^{-tH} \right\| , \qquad (4.3.5)$$

yields

$$\left\| F(\tau)^{t/\tau} - e^{-tH} \right\| \leq \frac{\tau}{t} + C \, M_\rho \left(\frac{\tau}{t}\right)^\rho , \quad 0 < \tau \leq t \leq 1 . \qquad (4.3.6)$$

Since for $0 < \tau \leq t$ such that $\tau, t \in (0, 1]$, and for $\rho \in [0, 1]$ one has $\tau/t \leq (\tau/t)^\rho$, from (4.3.6) we obtain

$$\left\| F(\tau)^{t/\tau} - e^{-tH} \right\| \leq (1 + C \, M_\rho) \left(\frac{\tau}{t}\right)^\rho . \qquad (4.3.7)$$

Thus, setting $C_\rho := 1 + C \, M_\rho$, we prove (4.3.2) for $\rho \in (0, 1]$.

(ii) To prove (4.3.1), we note that

$$(\mathbb{1}+tS(\tau))^{-1} - (\mathbb{1}+tH)^{-1} = \int_0^\infty dx \, e^{-x} \left(e^{-x\,t\,S(\tau)} - e^{-x\,t\,H}\right), \quad \tau, t > 0 ,$$

and use the representation

$$(\mathbb{1} + tS(\tau))^{-1} - (\mathbb{1} + tH)^{-1} = \sum_{n=0}^{\infty} \int_n^{n+1} dx \, e^{-x} \left(e^{-x\,t\,S(\tau)} - e^{-x\,t\,H}\right).$$

By the substitution $x = y + n$, it yields

$$(\mathbb{1} + tS(\tau))^{-1} - (\mathbb{1} + tH)^{-1} =$$

$$\sum_{n=0}^{\infty} e^{-n} \int_0^1 dy \, e^{-y} \left(e^{-(y+n)\,t\,S(\tau)} - e^{-(y+n)\,t\,H}\right),$$

for $\tau, t > 0$. Then seeing that

$$e^{-(y+n)t\,S(\tau)} - e^{-(y+n)t\,H} = \left(e^{-nt\,S(\tau)} - e^{-nt\,H}\right)e^{-yt\,S(\tau)} + e^{-nt\,H}\left(e^{-yt\,S(\tau)} - e^{-yt\,H}\right),$$

and

$$e^{-ntS(\tau)} - e^{-ntH} = \sum_{k=0}^{n-1} e^{-kt\,S(\tau)}\left(e^{-t\,S(\tau)} - e^{-t\,H}\right)e^{-(n-k-1)t\,H},$$

we deduce equation

$$(\mathbb{1} + tS(\tau))^{-1} - (\mathbb{1} + tH)^{-1} = \sum_{n=0}^{\infty} e^{-n}\left\{\left(\sum_{k=0}^{n-1} e^{-ktS(\tau)}\left(e^{-tS(\tau)} - e^{-tH}\right)e^{-(n-k-1)t\,H}\right) \times \int_0^1 dy\, e^{-y}\, e^{-ytS(\tau)} + e^{-ntH}\int_0^1 dy\, e^{-y}\left(e^{-yt\,S(\tau)} - e^{-yt\,H}\right)\right\}.$$

As a result, we obtain the estimate

$$\|(\mathbb{1} + tS(\tau))^{-1} - (\mathbb{1} + tH)^{-1}\| \leq \sum_{n=0}^{\infty} e^{-n}\left\{n\,\|e^{-tS(\tau)} - e^{-tH}\| + \int_0^1 dy\, e^{-y}\,\|e^{-yt\,S(\tau)} - e^{-yt\,H}\|\right\}, \tag{4.3.8}$$

where $\tau, t > 0$.

Given that $\rho \in (0, 1)$, the assumption (4.3.2) and estimate (4.3.4) yield

$$\|e^{-tS(\tau)} - e^{-tH}\| \leq (1 + C_\rho)\left(\frac{\tau}{t}\right)^\rho, \tag{4.3.9}$$

for $0 < \tau \leq t \leq 1$. Then to treat the last term in (4.3.8), we use decomposition

$$\int_0^1 dy\, e^{-y}\,\|e^{-yt\,S(\tau)} - e^{-yt\,H}\| = \int_{\tau/t}^1 dy\, e^{-y}\,\|e^{-yt\,S(\tau)} - e^{-yt\,H}\| + \int_0^{\tau/t} dy\, e^{-y}\,\|e^{-yt\,S(\tau)} - e^{-yt\,H}\|. \tag{4.3.10}$$

Since on account of (4.3.9) and $\tau \leq t\, y$, one has

$$\|\mathrm{e}^{-y\,t\,S(\tau)} - \mathrm{e}^{-y\,t\,H}\| \leq (1 + C_\rho) \left(\frac{\tau}{y\,t}\right)^\rho ,$$

for the first term in the right-hand side of (4.3.10), we obtain estimate

$$\int_{\tau/t}^{1} dy\, \mathrm{e}^{-y}\, \|\mathrm{e}^{-y\,t\,S(\tau)} - \mathrm{e}^{-y\,t\,H}\| \leq (1 + C_\rho) \int_0^1 dy\, \mathrm{e}^{-y}\, y^{-\rho} \left(\frac{\tau}{t}\right)^\rho , \tag{4.3.11}$$

where $0 < \tau \leq t \leq 1$.

Next, because of $\rho < 1$ and $\tau/t \leq 1$, we deduce for the second term in the right-hand side of (4.3.10) the estimate

$$\int_0^{\tau/t} dy\, \mathrm{e}^{-y}\, \|\mathrm{e}^{-y\,t\,S(\tau)} - \mathrm{e}^{-y\,t\,H}\| \leq 2 \left(\frac{\tau}{t}\right)^\rho , \quad 0 < \tau \leq t\,. \tag{4.3.12}$$

Taking into account (4.3.11) and (4.3.12), we infer from (4.3.10) the estimate

$$\int_0^1 dy\, \mathrm{e}^{-y}\, \|\mathrm{e}^{-y\,t\,S(\tau)} - \mathrm{e}^{-y\,t\,H}\| \leq \left((1 + C_\rho) \int_0^1 dy\, \mathrm{e}^{-y}\, y^{-\rho} + 2\right) \left(\frac{\tau}{t}\right)^\rho , \tag{4.3.13}$$

for $\tau, t \in (0, 1]$ with $0 < \tau \leq t$.

Finally, using (4.3.9) and (4.3.13), we get for (4.3.8) the estimate

$$\|(\mathbb{1} + tS(\tau))^{-1} - (\mathbb{1} + tH)^{-1}\| \leq \sum_{n=0}^{\infty} \mathrm{e}^{-n} \left\{n\, (1 + C_\rho) + (1 + C_\rho) \int_0^1 dy\, \mathrm{e}^{-y}\, y^{-\rho} + 2\right\} \left(\frac{\tau}{t}\right)^\rho ,$$

where $0 < \tau \leq t \leq 1$. Then setting

$$M_\rho := \sum_{n=0}^{\infty} \mathrm{e}^{-n} \left\{n\, (1 + C_\rho) + (1 + C_\rho) \int_0^1 dy\, \mathrm{e}^{-y} y^{-\rho} + 2\right\},$$

we obtain estimate (4.3.1). □

Remark 4.3.2

(a) In Theorem 4.3.1, (i) it is shown that for $\rho = 1$ the condition (4.3.1) implies (4.3.2). But it is *unclear* whether the assertion is also true for converse, given that Theorem 4.3.1(ii) does not cover this case.
(b) If we put $\tau = t/n$, the estimate (4.3.2) transforms into

$$\sup_{t \in [0,1]} \|F(t/n)^n - \mathrm{e}^{-tH}\| \leq C_\rho \frac{1}{n^\rho} , \quad n \geq 1\,. \tag{4.3.14}$$

This inequality yields the convergence rate estimate for $\rho \in (0, 1]$ on the unit interval $[0, 1]$ under *local* conditions: $0 < \tau \le t \le 1$ in (4.3.1).

Theorem 4.3.1(i) and (4.3.14) motivate sufficient conditions for the *Chernoff product formula* with estimate of the rate of operator-norm convergence on *any* bounded interval $\mathcal{I} \subset \mathbb{R}_0^+$.

Corollary 4.3.3 *Let $F(\cdot) : \mathbb{R}_0^+ \longrightarrow \mathcal{L}(\mathfrak{H})$ be a measurable family of non-negative self-adjoint contractions such that $F(0) = \mathbb{1}$. Let self-adjoint family $\{S(\tau)\}_{\tau>0}$ be defined by* (4.1.1)–(4.1.2)*, where H is a non-negative self-adjoint operator in $\mathfrak{H}$.*

If for some $\rho \in (0, 1]$, there is a constant $M_\rho > 0$ such that the estimate (4.3.1) *holds for $\tau, t \in (0, 1]$ and $0 < \tau \le t$, then for any bounded interval $\mathcal{I} \subset \mathbb{R}_0^+$, there exists a constant $C_\rho^{\mathcal{I}} > 0$ such that the estimate*

$$\sup_{t\in\mathcal{I}} \|F(t/\eta)^\eta - \mathrm{e}^{-tH}\| \le C_\rho^{\mathcal{I}} \frac{1}{\eta^\rho}, \tag{4.3.15}$$

holds for $\eta \ge 1$.

Proof Let $N \in \mathbb{N}$ such that $\mathcal{I} \subseteq [0, N]$. Then representation

$$F(t/N\eta)^{N\eta} - \mathrm{e}^{-tH} \\ = \sum_{k=0}^{N-1} \mathrm{e}^{-k\,t\,H/N}(F(t/N\eta)^\eta - \mathrm{e}^{-t\,H/N})\, F(t/N\eta)^{(N-1-k)\eta},$$

yields the estimate

$$\|F(t/N\eta)^{N\eta} - \mathrm{e}^{-t\,H}\| \le N\, \|F(t/N\eta)^\eta - \mathrm{e}^{-t\,H/N}\|.$$

Let $t \in \mathcal{I}$, $t' := t/N$ and $\tau' := t'/\eta$, $\eta \ge 1$. Then $0 < \tau' \le t' \le 1$. Applying Theorem 4.3.1, we find that

$$\|F(t/N\eta)^\eta - \mathrm{e}^{-tH/N}\| = \|F(\tau')^{t'/\tau'} - \mathrm{e}^{-t'H}\| \le C_\rho \left(\frac{\tau'}{t'}\right)^\rho.$$

This implies for $0 < \tau' \le t' \le 1$, that is, for $t \le N$ that

$$\|F(t/N\eta)^{N\eta} - \mathrm{e}^{-tH}\| \le C_\rho\, N \left(\frac{\tau'}{t'}\right)^\rho.$$

If $\eta' := N\eta \ge 1$, then we obtain the estimate

$$\|F(t/\eta')^{\eta'} - \mathrm{e}^{-tH}\| \le C_\rho\, N^{1+\rho} \left(\frac{1}{\eta'}\right)^\rho, \quad t \in [0, N]. \tag{4.3.16}$$

Setting $C_\rho^{[0,N]} := C_\rho N^{1+\rho}$, we prove corollary for $\mathcal{I} = [0, N]$. Because for any bounded interval $\mathcal{I}$, one can always find a $N \in \mathbb{N}$ such that $\mathcal{I} \subseteq [0, N]$; this completes the proof. □

In Corollary 4.3.3 the condition on t is restricted to $t \in (0, 1]$, and estimate (4.3.14) does not allow infinite intervals. To extend (4.3.15) to $\mathcal{I} \subseteq \mathbb{R}_0^+$, one needs a *global* condition for domains $0 < \tau \leq t < \infty$.

Theorem 4.3.4 *Let $F(\cdot) : \mathbb{R}_0^+ \longrightarrow \mathcal{L}(\mathfrak{H})$ be a measurable family of non-negative self-adjoint contractions such that $F(0) = \mathbb{1}$. Let self-adjoint family $\{S(\tau)\}_{\tau>0}$ be defined by* (4.1.1)–(4.1.2), *where H is a non-negative self-adjoint operator in $\mathfrak{H}$.*

If for some $\rho \in (0, 1]$ there is a constant $M_\rho > 0$ such that the estimate (4.3.1) *holds for $0 < \tau \leq t < \infty$, then there is a constant $C_\rho^{\mathbb{R}^+} > 0$ such that the estimate*

$$\sup_{t\in\mathbb{R}^+} \|F(t/\eta)^\eta - \mathrm{e}^{-tH}\| \leq C_\rho^{\mathbb{R}^+} \frac{1}{\eta^\rho}, \tag{4.3.17}$$

holds for $\eta \geq 1$.

Proof It is clear that arguments leading from (4.3.3) to estimate (4.3.7) are still valid if we assume $0 < \tau \leq t < \infty$. Then setting $\tau := t/\eta$, $\eta \geq 1$, we deduce from (4.3.7)

$$\|F(t/\eta)^\eta - \mathrm{e}^{-tH}\| \leq C_\rho^{\mathbb{R}^+} \frac{1}{\eta^\rho}, \quad \eta \geq 1,$$

where $C_\rho^{\mathbb{R}^+} := 1 + C\, M_\rho$ and consequently the proof of (4.3.17). □

Remark 4.3.5 Note that in Theorem 4.2.7, we established the self-adjoint operator-norm convergent extension of the Chernoff product formula for any bounded interval $\mathcal{I} \subset \mathbb{R}_0^+$ under *t-dependent* condition (4.2.23), which is necessary and sufficient. We also note that Corollary 4.3.3 and Theorem 4.3.4 prove self-adjoint operator-norm Chernoff product formula in $\mathbb{R}_0^+$ with estimate of the rate of convergence. They are also based on *t-dependent* fractional power condition (4.3.1), which is necessary and sufficient for $\rho \in (0, 1)$.

Our next remark is that if $\rho = 1$, then t-dependence in assumption (4.3.1) can be eliminated. The assertion below extends to $\mathbb{R}_0^+$ the established in Theorem 4.2.3 operator-norm convergence of the self-adjoint Chernoff product formula for any bounded interval $\mathcal{I} \subset \mathbb{R}_0^+$. This yields operator-norm version of the Chernoff product formula in the *original* formulation (cf. Theorem 4.2.3), but now with the operator-norm estimate of the convergence rate as in the ITTZ-Theorem 7.1.9 (Chap. 7).

To this aim, the *local* condition $0 < \tau \leq 1$ is yet sufficient.

Theorem 4.3.6 *Let* $F(\cdot) : \mathbb{R}_0^+ \longrightarrow \mathcal{L}(\mathfrak{H})$ *be a measurable family of non-negative self-adjoint contractions such that* $F(0) = \mathbb{1}$. *Let self-adjoint family* $\{S(\tau)\}_{\tau>0}$ *be defined by* (4.1.1)–(4.1.2), *where* H *is a non-negative self-adjoint operator in* $\mathfrak{H}$.

If there is a constant $M_1 > 0$ *such that the estimate*

$$\|(\mathbb{1} + S(\tau))^{-1} - (\mathbb{1} + H)^{-1}\| \leq M_1 \tau \,, \tag{4.3.18}$$

holds for $\tau \in (0, 1]$, *then for any bounded interval* $\mathcal{I} \subset \mathbb{R}_0^+$, *there is a constant* $C_1^{\mathcal{I}} > 0$ *such that the estimate*

$$\sup_{t \in \mathcal{I}} \|F(t/\eta)^\eta - \mathrm{e}^{-tH}\| \leq C_1^{\mathcal{I}} \frac{1}{\eta} \,, \tag{4.3.19}$$

holds for $\eta \geq 1$.

Proof The obvious identity

$$\begin{aligned}(\mathbb{1} + tS(\tau))^{-1} - (\mathbb{1} + tH)^{-1} = \\ t\,(\mathbb{1} + S(\tau))(\mathbb{1} + tS(\tau))^{-1}[(\mathbb{1} + S(\tau))^{-1} - (\mathbb{1} + H)^{-1}]\,(\mathbb{1} + H)\,(\mathbb{1} + tH)^{-1} \,,\end{aligned}$$

yields for $t > 0$ the estimate

$$\begin{aligned}&\|(\mathbb{1} + tS(\tau))^{-1} - (\mathbb{1} + tH)^{-1}\| \\ &\leq M_1 \tau\, t\, \|\,(\mathbb{1} + S(\tau))\,(\mathbb{1} + tS(\tau))^{-1}\|\, \|(\mathbb{1} + H)(\mathbb{1} + tH)^{-1}\| \,,\end{aligned} \tag{4.3.20}$$

where we used (4.3.18).

Therefore (4.3.20) together with estimates $\|(\mathbb{1} + S(\tau))\,(\mathbb{1} + tS(\tau))^{-1}\| \leq 1/t$ and $\|(\mathbb{1} + H)\,(\mathbb{1} + tH)^{-1}\| \leq 1/t$, for $t \leq 1$, imply the estimate (4.3.1) for $0 < \tau \leq t \leq 1$ and $\rho = 1$. Finally, applying Corollary 4.3.3, we extend the proof to any bounded interval $\mathcal{I} \subset \mathbb{R}_0^+$. □

Now we extend Theorem 4.3.6, for the case $\rho = 1$, to any infinite interval $\mathcal{I} = \mathbb{R}_0^+$. To this end one needs additional assumptions (cf. Appendix D for a similar assumption for the Kato function), which also include a *global* conditions similar to $0 < \tau \leq t < \infty$. This would allow us to use the result of Theorem 4.3.4.

Theorem 4.3.7 *Let in addition to conditions of Theorem* 4.3.6 *operator* $H \geq \mu \mathbb{1}$, $\mu > 0$. *Moreover, we assume that there exist* $\varepsilon \in (0, 1)$ *and* $\delta_\varepsilon \in (0, 1)$ *such that*

$$0 \leq F(\tau) \leq (1 - \delta_\varepsilon)\mathbb{1}, \quad \tau \geq \varepsilon, \tag{4.3.21}$$

is valid. If there is a constant $M_1 > 0$ *such that the estimate* (4.3.18) *holds for any* $\tau \in (0, \varepsilon)$, *then there exists constant* $C_1^{\mathbb{R}^+} > 0$ *such that estimate* (4.3.19) *is valid for infinite interval* $\mathcal{I} = \mathbb{R}_0^+$.

Proof Since (4.3.18) implies the resolvent-norm convergence of $\{S(\tau)\}_{\tau>0}$, when $\tau \downarrow 0$, and since $H \geq \mu\mathbb{1}$, there exists $0 < \mu' \leq \mu$ such that $S(\tau) \geq \mu'\mathbb{1}$ for $\tau \in (0, \varepsilon)$ where ε is *sufficiently* small.

On the other hand, (4.3.20) yields that for $t > 0$

$$\begin{aligned} &\|(\mathbb{1} + t\,S(\tau))^{-1} - (\mathbb{1} + t\,H)^{-1}\| = \\ &\leq M_1\,\frac{\tau}{t}\,\|(\mathbb{1} + S(\tau))\,(\mathbb{1}/t + S(\tau))^{-1}\|\,\|(\mathbb{1} + H)(\mathbb{1}/t + H)^{-1}\|\,. \end{aligned} \tag{4.3.22}$$

Since $S(\tau) \geq \mu'\mathbb{1}$, $\tau \in (0, \varepsilon)$, and $H \geq \mu I$, for $t > 0$, we obtain estimates

$$\begin{aligned} \|(\mathbb{1} + S(\tau))\,(\mathbb{1}/t + S(\tau))^{-1}\| &\leq \frac{1+\mu'}{\mu'}\,, \\ \|(\mathbb{1} + H)\,(\mathbb{1}/t + H)^{-1}\| &\leq \frac{1+\mu}{\mu}\,. \end{aligned} \tag{4.3.23}$$

By (4.3.22) these estimates give

$$\|(\mathbb{1} + t\,S(\tau))^{-1} - (\mathbb{1} + t\,H)^{-1}\| \leq M_1^{\mathbb{R}^+}\,\frac{\tau}{t}\,, \tag{4.3.24}$$

for $\tau \in (0, \varepsilon)$ and $t > 0$. Here $M_1^{\mathbb{R}^+} := M_1\,(1+\mu')(1+\mu)/\mu'\mu$.

Note that if $\tau/t \leq 1$, then we can use inequalities (4.1.5) and spectral representation for $F(\tau) \geq 0$ to obtain estimate

$$\left\|F(\tau)^{t/\tau} - \mathrm{e}^{-tS(\tau)}\right\| \leq \frac{\tau}{t}\,, \quad 0 < \tau \leq t < \infty,$$

cf. (4.3.4). Therefore, (4.3.3), (4.3.5) and (4.3.24), which is valid for $\tau \in (0, \varepsilon)$, allow to extend the result (4.3.19) of Theorem 4.3.6 to the case $0 < \tau \leq t < \infty$, for $\tau = t/\eta$, $\eta \geq 1$. This yields

$$\|F(t/\eta)^{\eta} - \mathrm{e}^{-tH}\| \leq \widehat{C}_1^{\mathbb{R}^+}\,\frac{1}{\eta}\,, \tag{4.3.25}$$

where $\widehat{C}_1^{\mathbb{R}^+} := 1 + C\,M_1^{\mathbb{R}^+}$ and $t \in [0, \varepsilon\eta)$.

Now let $t \geq \varepsilon\eta$. Then by assumption (4.3.21) we have

$$\|F(t/\eta)^{\eta}\| \leq (1-\delta_\varepsilon)^{\eta} = \mathrm{e}^{\eta\,\ln(1-\delta_\varepsilon)}, \quad t \geq \eta\varepsilon\,. \tag{4.3.26}$$

Note that $H \geq \mu I$ implies $\|\mathrm{e}^{-tH}\| \leq \mathrm{e}^{-\eta\,\varepsilon\,\mu}$ for $t \geq \eta\,\varepsilon$. This together with (4.3.25) and (4.3.26) yield the estimate

$$\|F(t/\eta)^{\eta} - \mathrm{e}^{-tH}\| \leq \widehat{C}_1^{\mathbb{R}^+}\,\frac{1}{\eta} + \mathrm{e}^{\eta\,\ln(1-\delta_\varepsilon)} + \mathrm{e}^{-\eta\,\varepsilon\,\mu}\,,$$

for *appropriately* small $\varepsilon > 0$, cf. (4.3.24) and for *any* $t \geq 0$.

Since $\widetilde{C}_1 := \sup_{\eta \geq 1} \eta(\mathrm{e}^{\eta \ln(1-\delta_\varepsilon)} + \mathrm{e}^{-\eta\,\varepsilon\,\mu}) < \infty$, there exists constant $C_1^{\mathbb{R}^+} := \widehat{C}_1^{\mathbb{R}^+} + \widetilde{C}_1$ such that (4.3.19) is valid for $\eta \geq 1$ and infinite interval $\mathcal{I} = \mathbb{R}_0^+$ □

We return to Theorems 4.3.6 and 4.3.7 in Chap. 7 for discussion *optimality* of the operator-norm estimate (4.3.19) established due to the *uplifted* (improved) Chernoff product formula.

4.4 Notes to Chap. 4

Notes to Sect. 4.1

Equivalence of condition (4.1.2)–(4.1.3) and to existence of the closure H of the strong derivative $F'(+0)$ is part of the *Trotter-Kato-Neveu* theorem (Proposition 1.9.30) and the Chernoff product Theorem 1.9.34. Note that in (4.1.2)–(4.1.3), the limit generator H is assumed to be known. The theorem, [Kato80, Theorem 2.17], gives sufficient conditions which ensure existence of the generator H, cf. also [EN00, 4.9 Second Trotter-Kato Approximation Theorem].

We note that owing to *self-adjointness* of the family $\{F(t)\}_{t\geq 0}$ the estimate (4.1.9) for $\eta = n$ ($n \in \mathbb{N}$) is stronger than the $\sqrt{n}$-estimate for general contractions in Sect. 1.9.6; see Lemma 1.9.33. Since estimate (4.1.9) holds in the operator-norm, the strong convergence in (4.1.7) is determined by the strong convergence in the *Trotter-Kato-Neveu* theorem (Proposition 1.9.30).

For further comments, we refer to [Zag20, Section 1] and [Zag22c].

Notes to Sect. 4.2

The first result about *uplifting* (improvement) the Chernoff product formula to the *operator-norm* topology without estimate of the operator-norm error-bound (see Theorem 4.2.3, (4.2.10)) is due to [NZ99b, Theorem 2.2 and Theorem 2.3]. This uplifting was owing to the proof of the operator-norm version of the *Trotter-Kato-Neveu* theorem [NZ99b, Theorem 2.3] for the uniform operator-norm convergence on the closed intervals in $\mathbb{R}_0^+$ away from zero.

Relaxation of this condition to any bounded intervals in $\mathbb{R}_0^+$ follows from [Zag20, Theorem 2.4 and Theorem 2.5]; see Theorem 4.2.6 for extension of the *Trotter-Kato-Neveu* theorem and Theorem 4.2.7 for extension of the *Chernoff* product formula. These assertions permit to work out the operator-norm estimates for the rate of convergence.

Notes to Sect. 4.3

Seeing that operator-norm estimate (4.1.9) has optimal rate $O(1/n)$, the operator-norm rate of convergence for Chernoff product formula (4.2.22) is (again) limited by the rate of convergence in the *Trotter-Kato-Neveu* theorem. The optimal operator-norm rate in the *Trotter-Kato-Neveu* theorem (see (4.3.1) and (4.3.2)) was obtained in [INZ04, Theorem 2.1]; see furthermore [IT01, Lemma 2.1] as well as [Zag20, Theorem 3.6, Theorem 3.7] for more comments on the self-adjoint case and $F \in$

$\widehat{\mathcal{K}}_\beta$ (4.3.21), where $\widehat{\mathcal{K}}_\beta$ is a class of the Kato functions; see [Zag19a, Appendix C.1].

Here we would like to note that in Theorem 4.3.6 the rate of convergence (4.3.19), for $\eta = n$ ($n \in \mathbb{N}$), is a part of the ITTZ-Theorem 7.1.9 (Chap. 7) about optimal error bound $O(1/n)$. It was proved in [ITTZ01, Theorem 1, Corollary 1] along the idea presented in [IT01] by T.Ichinose and Hideo Tamura, cf. Appendix D.

For extension of the operator-norm Chernoff product formula (4.3.19) with optimal rate $O(1/n)$ to convergence on unbounded intervals, one has to add in Theorem 4.3.7 [Zag20, Theorem 3.7] the *Ichinose-Tamura* condition (4.3.21). It concerns a definition of the class of admissible for Theorem 4.3.7 *Kato functions* $\widehat{\mathcal{K}}_\beta$; see [Zag19a, Definition C.3]. These condition and class $\widehat{\mathcal{K}}_\beta$ were introduced and studied by T. Ichinose and Hideo Tamura; see [IT01, Section 4].

Chapter 5
Trotter-Kato Product Formulæ: Strong Operator Topology

After general product approximations *à la Chernoff* developed in Chap. 4, we start here with the first concrete applications for a couple of semigroups generated on a Hilbert space $\mathfrak{H}$ by self-adjoint operators A, B.

To begin we consider a well-known case of the *Trotter product formula*, where A, B are generators of self-adjoint contraction C_0-semigroups involved into the Trotter approximants or in their symmetrised (i.e. self-adjoint) versions. A sufficient condition for convergence of these approximants in the *strong* operator topology is that operator $A + B$ should be *essentially* self-adjoint. Next, we extend this assertion to the case where the *form-sum* $A \dotplus B$ of (non-negative) generators A and B is *self-adjoint*. For that reason, we prove the *Kato product theorem*, that is, the Kato version of the strongly convergent self-adjoint Trotter product theorem.

To this aim, we introduce a fundamental for this book notion of the (generic) *Kato functions* and the corresponding *Trotter-Kato product formulæ*. We prove in this chapter the Kato product theorem about strong operator convergence of the Trotter-Kato product formulæ for the whole class of generic Kato functions and the self-adjoint form-sums: $A \dotplus B$.

These statements cover the both cases of the *self-adjoint* as well as *non-self-adjoint* Trotter-Kato *approximants* for self-adjoint semigroups.

5.1 Trotter and Kato Product Formulæ

In the case of self-adjoint semigroups, the *Trotter product theorem* about strong operator convergence of *Trotter* product formula reads as follows.

V. A. Zagrebnov et al., *Trotter-Kato Product Formulæ*, Operator Theory: Advances and Applications 296, https://doi.org/10.1007/978-3-031-56720-9_5

Theorem 5.1.1 (Trotter Product Theorem) *Let A and B be two non-negative self-adjoint operators in a Hilbert space $\mathfrak{H}$. Let operator $A+B$ be densely defined and essentially self-adjoint on domain* $\mathrm{dom}\,(A) \cap \mathrm{dom}\,(B)$ *. Then the Trotter product formula*

$$\mathrm{s-}\lim_{n\to\infty} \left(\mathrm{e}^{-tA/n}\mathrm{e}^{-tB/n}\right)^n = \mathrm{e}^{-tH}\,, \quad t \in \mathbb{R}_0^+\,, \tag{5.1.1}$$

converges strongly to C_0-semigroup with non-negative self-adjoint generator $H = \overline{A+B}$ for contraction Trotter product approximants (cf. (1.9.111))

$$V_{A,B}(t/n) := \left(\mathrm{e}^{-tA/n}\mathrm{e}^{-tB/n}\right)^n\,, \quad n \in \mathbb{N}\,. \tag{5.1.2}$$

The convergence holds uniformly in $t \in \mathcal{I}$ for any bounded interval of $\mathcal{I} \subset \mathbb{R}_0^+$, that is, locally uniformly on $\mathbb{R}_0^+$.

Proof Let $W_{B,A}(\tau) := \mathrm{e}^{-\tau A/2}\mathrm{e}^{-\tau B}\mathrm{e}^{-\tau A/2}$ for $\tau \geq 0$, cf. Definition 1.10.43. Note that symmetric approximants $\{W_{B,A}(t/n)\}_{n\geq 1}$ are self-adjoint contractions for $t \geq 0$. Then for $S_W(\tau) := (\mathbb{1} - W_{B,A}(\tau))/\tau$ and $\tau > 0$, one infers on $\mathfrak{H}$ the representation

$$S_W(\tau)\,u = \mathrm{e}^{-\tau A/2}\mathrm{e}^{-\tau B}\,\frac{\mathbb{1}-\mathrm{e}^{-\tau A/2}}{\tau}\,u + \mathrm{e}^{-\tau A/2}\,\frac{\mathbb{1}-\mathrm{e}^{-\tau B}}{\tau}\,u + \frac{\mathbb{1}-\mathrm{e}^{-\tau A/2}}{\tau}\,u\,. \tag{5.1.3}$$

If $u \in \mathrm{dom}\,(A) \cap \mathrm{dom}\,(B)$, then we obtain

$$\lim_{\tau\to+0} \frac{\mathbb{1}-\mathrm{e}^{-\tau A/2}}{\tau}\,u = \frac{1}{2}\,A\,u \quad \text{and} \quad \lim_{\tau\to+0} \frac{\mathbb{1}-\mathrm{e}^{-\tau B}}{\tau}\,u = B\,u\,. \tag{5.1.4}$$

Owing to (5.1.3) and (5.1.4), this yields $\lim_{\tau\to+0} S_W(\tau)\,u = H'\,u$ for operator $H' := A+B$ on domain $\mathrm{dom}\,(H') := \mathrm{dom}\,(A) \cap \mathrm{dom}\,(B)$.

Given that H' on $\mathrm{dom}\,(H')$ is essentially self-adjoint, the set $\mathcal{D} := \mathrm{ran}\,(\mathbb{1}+H') = (\mathbb{1}+H')\,\mathrm{dom}\,(H')$ is dense in $\mathfrak{H}$. If $v \in \mathcal{D}$, then using

$$(\mathbb{1}+S_W(\tau))^{-1}v - (\mathbb{1}+H')^{-1}v = -(\mathbb{1}+S_W(\tau))^{-1}(S_W(\tau)-H')(\mathbb{1}+H')^{-1}v\,, \tag{5.1.5}$$

we obtain $\lim_{\tau\to+0}(\mathbb{1}+S_W(\tau))^{-1}v = (\mathbb{1}+H)^{-1}v$. Since $\mathcal{D}$ is dense in $\mathfrak{H}$ and the operator norm $\|(\mathbb{1}+S_W(\tau))^{-1} - (\mathbb{1}+H)^{-1}\| \leq 2$, for $\tau > 0$, the pointwise limit on $\mathcal{D}$ implies $\mathrm{s-}\lim_{\tau\to+0}(\mathbb{1}+S_W(\tau))^{-1} = (\mathbb{1}+H)^{-1}$, for the self-adjoint closure $H := \overline{H'}$; see Proposition A.1.6 in Appendix A.1.

As a consequence, seeing that by condition of the theorem $W_{B,A}(\cdot) : \mathbb{R}_0^+ \to \mathcal{L}(\mathfrak{H})$ is a strongly measurable mapping to positive self-adjoint contractions such that $W_{B,A}(0) = \mathbb{1}$, we can apply Theorem 4.1.3 (see also Corollary 4.1.5) about the Chernoff product formula for the case of self-adjoint approximants. Hence, we obtain for symmetric (self-adjoint) approximants the Trotter product formula

$$\text{s-}\lim_{n\to+\infty}(\mathrm{e}^{-\tau A/2}\mathrm{e}^{-\tau B}\mathrm{e}^{-\tau A/2})^n = \mathrm{e}^{-tH}\,, \quad t \in \mathbb{R}_0^+\,. \tag{5.1.6}$$

convergent in the strong operator topology locally uniformly in $t \in \mathbb{R}_0^+$.

For proving the Trotter product formula for contraction non-self-adjoint approximants (5.1.2), we use identity

$$\mathrm{e}^{-tA/2n}\left(W_{B,A}(t/n)\right)^n = \left(V_{A,B}(t/n)\right)^n \mathrm{e}^{-tA/2n}\,, \quad n \in \mathbb{N}\,. \tag{5.1.7}$$

Then $\text{s}-\lim_{n\to+\infty} \mathrm{e}^{-tA/2n} = \mathbb{1}$ and (5.1.6) together with the *strong continuity* of the product in the left-hand side of (5.1.7) yield (5.1.1). □

Corollary 5.1.2 *Seeing that* $\left(V_{A,B}(t/n)\right)^n \mathrm{e}^{-tA/n} = \mathrm{e}^{-tA/n}\left(V_{B,A}(t/n)\right)^n$, *the limit* (5.1.1) *implies*

$$\mathrm{e}^{-tH} = \text{s}-\lim_{n\to\infty}\left(\mathrm{e}^{-tB/n}\mathrm{e}^{-tA/n}\right)^n. \tag{5.1.8}$$

Similarly to (5.1.7), *one infers*

$$\mathrm{e}^{-tB/2n}\left(W_{A,B}(t/n)\right)^n = \left(V_{B,A}(t/n)\right)^n \mathrm{e}^{-tB/2n}\,, \quad n \in \mathbb{N}\,. \tag{5.1.9}$$

Then the strong continuity of the product (5.1.8) *and* (5.1.9) *yields the limit*

$$\mathrm{e}^{-tH} = \text{s}-\lim_{n\to+\infty}\left(\mathrm{e}^{-tB/2n}\mathrm{e}^{-tA/n}\mathrm{e}^{-tB/2n}\right)^n. \tag{5.1.10}$$

The limits (5.1.6) *and* (5.1.10) *are called the symmetric (or symmetrised) Trotter product formulæ.*

It is appropriate to remind here that Theorem 5.1.1 can be substantially strengthened in such a way the it covers the case when the sum $A + B$ of two non-negative self-adjoint operators is "ill-defined". For example, $\mathrm{dom}\,(A) \cap \mathrm{dom}\,(B) = \{0\}$.

Let A and B be non-negative self-adjoint operators in $\mathfrak{H}$. Recall that in general, the operator sum of operators $A + B$ may *not* be densely defined on $\mathrm{dom}\,(A) \cap \mathrm{dom}\,(B)$. On the other hand, the *form-sum* $A \dotplus B$ of these operators may be still well-defined. To this aim with help of the *second representation theorem* for sesquilinear forms (see Sect. 9.1), we introduce densely defined, closed, symmetric, non-negative sesquiliniear forms $\mathfrak{t}_A$ and $\mathfrak{t}_B$, associated, respectively, with operators A and B:

$$\begin{aligned} \mathfrak{t}_A[u,v] &:= (A^{1/2}u, A^{1/2}v), \quad u, v \in \mathrm{dom}\,(\mathfrak{t}_A) := \mathrm{dom}\,(A^{1/2})\,, \\ \mathfrak{t}_B[u,v] &:= (B^{1/2}u, B^{1/2}v), \quad u, v \in \mathrm{dom}\,(\mathfrak{t}_B) := \mathrm{dom}\,(B^{1/2})\,. \end{aligned} \tag{5.1.11}$$

In the following, we assume (the opposite case and *degenerate* semigroups we shall consider separately) that

$$\overline{\mathrm{dom}\,(\mathfrak{t}_A) \cap \mathrm{dom}\,(\mathfrak{t}_B)} = \mathfrak{H} \tag{5.1.12}$$

and we define in $\mathfrak{H}$ the sesquilinear form

$$\mathfrak{t}_H[u,v] := \mathfrak{t}_A[u,v] + \mathfrak{t}_B[u,v], \quad u, v \in \mathrm{dom}\,(\mathfrak{t}_H) := \mathrm{dom}\,(\mathfrak{t}_A) \cap \mathrm{dom}\,(\mathfrak{t}_B). \tag{5.1.13}$$

Since the set of the closed forms is a *linear space*, the form $\mathfrak{t}_H$ is closed, symmetric non-negative and densely defined by virtue of (5.1.12). Then by the *first representation theorem*, one can associate with form (5.1.13) a non-negative self-adjoint operator H such that $\mathfrak{t}_H[u,v] = (Hu, v)$ for $u \in \mathrm{dom}\,(H)$ and $v \in \mathrm{dom}\,(\mathfrak{t}_H)$.

Definition 5.1.3 The self-adjoint operator H constructed above is called the *form-sum* of self-adjoint operators A, B. We denote it by $H := A \dotplus B$.

For comments on representation theorems, see Notes to Chap. 5 (Notes to Sects. 5.1) and 9.1.

To elucidate this definition, we prove a lemma about strong operator convergence, which will be also necessary in the following.

Lemma 5.1.4 *Let A and B be two non-negative self-adjoint operators and let operator*

$$H_\tau := A + B\,\mathrm{e}^{-\tau B}, \quad \mathrm{dom}\,(H_\tau) = \mathrm{dom}\,(A), \quad \tau > 0.$$

If the set dom$(A^{1/2}) \cap \mathrm{dom}\,(B^{1/2})$ is dense in $\mathfrak{H}$ and operator H is the form-sum $A \dotplus B$ of operators A, B, then

$$\text{s-}\lim_{\tau \to +0} (\lambda \mathbb{1} + H_\tau)^{-1} = (\lambda \mathbb{1} + H)^{-1}, \quad \lambda > 0. \tag{5.1.14}$$

Proof By $\mathfrak{t}_{H_\tau}$ we denote the sesquilinear form associated with operator H_τ, i.e.

$$\mathfrak{t}_{H_\tau}[u,v] = \mathfrak{t}_A[u,v] + \mathfrak{t}_B[\mathrm{e}^{-\tau B/2}u, \mathrm{e}^{-\tau B/2}v], \quad u, v \in \mathrm{dom}\,(\mathfrak{t}_{H_\tau}) := \mathrm{dom}\,(\mathfrak{t}_A).$$

Then by the spectral calculus for operator B, one checks that the sequence $\{\mathfrak{t}_{H_\tau}\}_{\tau>0}$ is monotonously nondecreasing for $\tau \to +0$. Moreover, one has

$$0 \le \mathfrak{t}_{H_\tau}[u,u] \le \mathfrak{t}_H[u,u], \quad u \in \mathrm{dom}\,(\mathfrak{t}_H), \quad \tau > 0.$$

Now, to proceed, we use the following line of reasoning. Let

$$\mathrm{dom}\,(\mathrm{t}_\infty) := \left\{ u \in \bigcap_{\tau>0} \mathrm{dom}\,(\mathrm{t}_{H_\tau}) : \sup_{\tau>0} \mathrm{t}_{H_\tau}[u,u] < \infty \right\},$$

where $\mathrm{t}_\infty[u,v] := \lim_{\tau\to+0} \mathrm{t}_{H_\tau}[u,v]$, $u, v \in \mathrm{dom}\,(\mathrm{t}_\infty)$. The sesquilinear form t_∞ is densely defined, symmetric and closed. Notice that $\bigcap_{\tau>0} \mathrm{dom}\,(\mathrm{t}_{H_\tau}) = \mathrm{dom}\,(\mathrm{t}_A)$. Hence, we infer

$$\mathrm{dom}\,(\mathrm{t}_\infty) = \{u \in \mathrm{dom}\,(\mathrm{t}_A) : \sup_{\tau>0} \mathrm{t}_B[u,u] = \sup_{\tau>0} \|\sqrt{B}\mathrm{e}^{-\tau B/2}u\|^2 < \infty\}.$$

As a consequence, $u \in \mathrm{dom}\,(\mathrm{t}_\infty)$ if and only if $u \in \mathrm{dom}\,(\mathrm{t}_A) \cap \mathrm{dom}\,(\mathrm{t}_B)$, which yields $\mathrm{dom}\,(\mathrm{t}_\infty) = \mathrm{dom}\,(\mathrm{t}_H)$. Since $\lim_{\tau\to+0} \mathrm{t}_{H_\tau}[u,v] = \mathrm{t}_H[u,v]$, $u, v \in \mathrm{dom}\,(\mathrm{t}_H)$, one obtains equality: $\mathrm{t}_\infty = \mathrm{t}_H \geq 0$.

Seeing that monotone convergence of closed, symmetric, non-negative sesquilinear forms yields convergence of the corresponding to these forms non-negative self-adjoint operators in the strong resolvent sense, we obtain the proof of (5.1.14) for $\lambda > 0$; see Notes to Chap. 5 (Notes to Sect. 5.1). □

The next assertion extends the strong convergence of Trotter product approximants $\{V_{A,B}(t/n) = \left(\mathrm{e}^{-tA/n}\mathrm{e}^{-tB/n}\right)^n\}_{n\geq 1}$ (5.1.2) to the case when the sum of generators A and B is well-defined as the form-sum $A \dotplus B$. Note that for the proof, we again implement Theorem 4.1.3 about the Chernoff product formula for the case of self-adjoint approximants and $\lambda \geq 1$ (4.1.16), cf. Corollary 4.1.5.

Theorem 5.1.5 (Kato Product Formula) *Let A and B be two non-negative self-adjoint operators in the Hilbert space $\mathfrak{H}$. If domain $dom(A^{1/2}) \cap \mathrm{dom}\,(B^{1/2})$ is dense in $\mathfrak{H}$, then the Trotter product formula*

$$\text{s-}\lim_{n\to\infty} \left(\mathrm{e}^{-tA/n}\mathrm{e}^{-tB/n}\right)^n = \mathrm{e}^{-tH}, \tag{5.1.15}$$

converges in the strong operator topology uniformly for $t \in \mathcal{I}$ for any bounded interval $\mathcal{I} \subset \mathbb{R}_0^+$. Here self-adjoint operator $H := A \dotplus B$.

Proof We start by symmetric (and consequently self-adjoint), strongly continuous product approximants

$$W_{B,A,}(t/n) = \left(\mathrm{e}^{-tA/2n}\mathrm{e}^{-tB/n}\mathrm{e}^{-tA/2n}\right)^n, \quad n \geq 1, \quad t \in \mathbb{R}_0^+, \tag{5.1.16}$$

and $S_W(\tau) := (\mathbb{1} - W_{B,A,}(\tau))/\tau$ for $\tau > 0$. Our aim is to verify conditions of the Chernoff product formula Theorem 4.1.3 (cf. Corollary 4.1.5) for self-adjoint, positive (strongly measurable) family of contractions $\{W_{B,A}(\tau)\}_{\tau\geq 0}$. To this avail, we shall check (4.1.16).

To this aim, we first note that

$$\lambda\, e^{\tau A} + \frac{e^{\tau A} - I}{\tau} + \frac{\mathbb{1} - e^{-\tau B}}{\tau} \geq \lambda\, \mathbb{1} + A + \frac{\mathbb{1} - e^{-\tau B}}{\tau} \geq \lambda\, \mathbb{1} + A + e^{-\tau B} B \geq \lambda\, \mathbb{1}\,,$$

for $\lambda \geq 1$ and $\tau > 0$. Hence

$$0 \leq \left(\lambda e^{\tau A} + \frac{e^{\tau A} - I}{\tau} + \frac{\mathbb{1} - e^{-\tau B}}{\tau}\right)^{-1} \leq \left(\lambda \mathbb{1} + A + e^{-\tau B} B\right)^{-1},$$

which yields

$$e^{-\tau A/2}(\lambda \mathbb{1} + S_W(\tau))^{-1} e^{-\tau A/2} \leq \left(\lambda \mathbb{1} + A + e^{-\tau B} B\right)^{-1}, \quad \lambda \geq 1, \quad \tau > 0. \tag{5.1.17}$$

Since by Lemma 5.1.4, one has for $H = A \dot{+} B$

$$\text{s-}\lim_{\tau \to +0} \left(\lambda \mathbb{1} + A + e^{-\tau B} B\right)^{-1} = (\lambda \mathbb{1} + H)^{-1}, \quad \lambda \geq 1\,, \tag{5.1.18}$$

we use this limit to identify generator H in (5.1.15). On account of (5.1.17) and (5.1.18), we obtain

$$\begin{aligned} &\limsup_{\tau \to +0}\, ((\lambda \mathbb{1} + S_W(\tau))^{-1} e^{-\tau A/2} u, e^{-\tau A/2} u) \\ &\leq ((\lambda \mathbb{1} + H)^{-1} u, u), \quad \lambda \geq 1, \quad u \in \mathfrak{H}\,. \end{aligned}$$

Now, using $\text{s}-\lim_{\tau \to +0} e^{-\tau A/2} = \mathbb{1}$, this provides estimate (*upper* bound)

$$\limsup_{\tau \to +0} ((\lambda \mathbb{1} + S_W(\tau))^{-1} u, u) \leq ((\lambda \mathbb{1} + H)^{-1} u, u), \quad \lambda \geq 1, \quad u \in \mathfrak{H}. \tag{5.1.19}$$

To check (4.1.16) we have to establish in (5.1.19) also the estimate from *below*, cf. (5.1.23). Note that (5.1.16) yields representation

$$\mathbb{1} - W_{B,A}(\tau) = \mathbb{1} - e^{-\tau A} + \left(Q(\tau)^* - \sqrt{\mathbb{1} - e^{-\tau B}}\right)\left(Q(\tau) - \sqrt{\mathbb{1} - e^{-\tau B}}\right), \quad \tau \geq 0\,,$$

where $Q(\tau)$ is given by

$$Q(\tau) := \sqrt{\mathbb{1} - e^{-\tau B}}\,\left(\mathbb{1} - e^{-\tau A/2}\right), \quad \tau \geq 0\,. \tag{5.1.20}$$

Therefore, we find

$$((\mathbb{1}-W_{B,A}(\tau))u,u) \le ((\mathbb{1}-\mathrm{e}^{-\tau A})u,u) + \|Q(\tau)u\|^2 + 2\,\|Q(\tau)u\|\;\|\sqrt{\mathbb{1}-\mathrm{e}^{-\tau B}}\,u\| + \|\sqrt{\mathbb{1}-\mathrm{e}^{-\tau B}}\,u\|^2,$$

for $u \in \mathfrak{H}$. Seeing that for each $\varepsilon > 0$

$$2\,\|Q(\tau)u\|\;\|\sqrt{\mathbb{1}-\mathrm{e}^{-\tau B}}u\|^2 \le \frac{1}{\varepsilon^2}\|Q(\tau)u\|^2 + \varepsilon^2\|\sqrt{\mathbb{1}-\mathrm{e}^{-\tau B}}u\|, \quad u \in \mathfrak{H},$$

one gets

$$((\mathbb{1}-W_{B,A}(\tau))u,u) \le ((\mathbb{1}-\mathrm{e}^{-\tau A})u,u) + (1+\varepsilon^2)((\mathbb{1}-\mathrm{e}^{-\tau B})u,u) + \left(1+\frac{1}{\varepsilon^2}\right)\|Q(\tau)u\|^2,$$

or equivalently

$$((\mathbb{1}-W_{B,A}(\tau))u,u) \le (1+\varepsilon^2)\Big[((\mathbb{1}-\mathrm{e}^{-\tau A})u,u) + ((\mathbb{1}-\mathrm{e}^{-\tau B})u,u)\Big] + \left(1+\frac{1}{\varepsilon^2}\right)\|Q(\tau)u\|^2.$$

As a consequence, for $S_W(\tau) = (\mathbb{1} - W_{B,A}(\tau))/\tau$, we obtain for $u \in \mathfrak{H}$ estimate

$$(S_W(\tau)u,u) \le (1+\varepsilon^2)\left[\left(\frac{\mathbb{1}-\mathrm{e}^{-\tau A}}{\tau}u,u\right) + \left(\frac{\mathbb{1}-\mathrm{e}^{-\tau B}}{\tau}u,u\right)\right] + \left(1+\frac{1}{\varepsilon^2}\right)\frac{1}{\tau}\|Q(\tau)u\|^2.$$

Therefore, given that $(\tau^{-1}(\mathbb{1}-\mathrm{e}^{-\tau A})u,\,u) \le (Au,u)$ for $u \in \operatorname{dom}(A^{1/2})$, as well as $(\tau^{-1}(\mathbb{1}-\mathrm{e}^{-\tau B})u,\,u) \le (Bu,u)$ for $u \in \operatorname{dom}(B^{1/2})$, we infer for $H = A \dot{+} B$:

$$\lambda\,\mathbb{1} + S_W(\tau) \le \lambda\,\mathbb{1} + (1+\varepsilon^2)\,H + \left(1+\frac{1}{\varepsilon^2}\right)\frac{1}{\tau}Q(\tau)^*Q(\tau), \quad \tau > 0, \quad \varepsilon > 0.$$

As a result, we obtain for $u \in \mathfrak{H}$ the estimate

$$\left([\lambda\,\mathbb{1} + (1+\varepsilon^2)\,H + \left(1+\frac{1}{\varepsilon^2}\right)\frac{1}{\tau}Q(\tau)^*Q(\tau)]^{-1}u,\,u\right) \le ((\lambda\,\mathbb{1} + S_W(\tau))^{-1}u,\,u), \tag{5.1.21}$$

where $\tau > 0$, $\varepsilon > 0$ and $\lambda \geq 1$. Now setting

$$Z(\tau) := \frac{1}{\sqrt{\tau}} Q(\tau) (\lambda \mathbb{1} + (1+\varepsilon^2) H)^{-1/2}, \quad \tau > 0, \quad \varepsilon > 0, \quad \lambda \geq 1, \tag{5.1.22}$$

we infer from (5.1.21) and (5.1.22) estimate

$$(\lambda \mathbb{1} + (1+\varepsilon^2) H)^{-1/2} \left(\mathbb{1} + \left(1 + \frac{1}{\varepsilon^2}\right) Z(\tau)^* Z(\tau)\right)^{-1} (\lambda \mathbb{1} + (1+\varepsilon^2) H)^{-1/2}$$
$$\leq (\lambda \mathbb{1} + S_W(\tau))^{-1}. \tag{5.1.23}$$

Next, we study (for small $\tau > 0$) the product $Z(\tau)^* Z(\tau)$ in (5.1.23). On account of definitions (5.1.20) and (5.1.22) and positivity of operators A, B, one concludes that

$$Z(\tau)^* Z(\tau) \leq$$
$$(\lambda \mathbb{1} + (1+\varepsilon^2) H)^{-1/2} \left(\frac{\mathbb{1} - \mathrm{e}^{-\tau A/2}}{\tau}\right) (\mathbb{1} - \mathrm{e}^{-\tau A/2}) (\lambda \mathbb{1} + (1+\varepsilon^2) H)^{-1/2} \leq$$
$$\frac{1}{2} (\lambda \mathbb{1} + (1+\varepsilon^2) H)^{-1/2} \sqrt{A}\, (\mathbb{1} - \mathrm{e}^{-\tau A/2}) \sqrt{A}\, (\lambda \mathbb{1} + (1+\varepsilon^2) H)^{-1/2}, \tag{5.1.24}$$

for $\tau > 0$, $\varepsilon > 0$ and $\lambda \geq 1$. Now let

$$Z_0 := \frac{1}{\sqrt{2}} \sqrt{A}\, (\lambda \mathbb{1} + (1+\varepsilon^2) H)^{-1/2}.$$

Then owing to $H := A \dot{+} B$ it follows that $\|Z_0\| \leq 1/\sqrt{2}$. So, by virtue of (5.1.24) we obtain

$$Z(\tau)^* Z(\tau) \leq Z_0^* (\mathbb{1} - \mathrm{e}^{-\tau A/2}) Z_0,$$

which implies $\text{s-}\lim_{\tau \to +0} Z(\tau)^* Z(\tau) = 0$ for $\varepsilon > 0$ and $\lambda \geq 1$. Therefore, by definition (5.1.22) and estimate (5.1.23), we obtain

$$\text{s-}\lim_{\tau \to +0} \left(\lambda \mathbb{1} + (1+\varepsilon^2) H + \left(1 + \frac{1}{\varepsilon^2}\right) \frac{1}{\tau} Q(\tau)^* Q(\tau)\right)^{-1}$$
$$= \left(\lambda \mathbb{1} + (1+\varepsilon^2) H\right)^{-1}.$$

Then application of this limit in (5.1.21) yields for $u \in \mathfrak{H}$

$$\left(\left(\lambda \mathbb{1} + (1+\varepsilon^2)\, H\right)^{-1} u, u\right) \leq \liminf_{\tau \to +0}((\lambda \mathbb{1} + S_W(\tau))^{-1} u, u),$$

where $\varepsilon > 0$, $\lambda \geq 1$. Since $\varepsilon > 0$ is arbitrary, then we obtain for the limit in the right-hand side the promised (cf. (5.1.19)) *lower* bound:

$$((\lambda \mathbb{1} + H)^{-1} u, u) \leq \liminf_{\tau \to +0}((\lambda \mathbb{1} + S_W(\tau))^{-1} u, u).$$

Together with the *upper* bound (5.1.19), this proves the limit

$$\lim_{\tau \to +0}((\lambda \mathbb{1} + S_W(\tau))^{-1} u, u) = ((\lambda \mathbb{1} + H)^{-1} u, u), \quad u \in \mathfrak{H},$$

which in turn yields that $\text{w}-\lim_{\tau \to +0}(\lambda \mathbb{1} + S_W(\tau))^{-1} = (\lambda \mathbb{1} + H)^{-1}$ for $\lambda \geq 1$, cf. (4.1.16). Applying now the Chernoff product formula (Corollary 4.1.5), we obtain the proof of the Kato product formula

$$\mathrm{e}^{-tH} = \text{s}-\lim_{n \to \infty} \left(\mathrm{e}^{-tA/2n} \mathrm{e}^{-tB/n} \mathrm{e}^{-tA/2n}\right)^n, \tag{5.1.25}$$

locally uniformly in $t \in \mathbb{R}_0^+$ for symmetric self-adjoint (measurable) approximants (5.1.16) corresponding to the family of self-adjoint strongly continuous contractions $\{W_{B,A}(t)\}_{t \geq 0}$.

To complete the proof by product formula for non-symmetric approximants $\{V_{A,B}(t/n)\}_{n \geq 1}$, (5.1.15), one has to use (5.1.25) and to mimic the arguments in Corollary 5.1.2. □

5.2 Kato Functions and Trotter-Kato Product Formulæ

In this section, we introduce a fundamental for this book notion of (generic) *Kato functions* of class $\mathcal{K}$. It is the first step to generalisation of the Trotter product formula to *non-exponential* approximants for corresponding strongly convergent *Trotter-Kato product formulæ*.

We prove in this section the *Kato product* theorem about strong convergence of Trotter-Kato product formulæ for the whole class of generic Kato functions and for *self-adjoint* form-sums $A \dot{+} B$ as generators of the limiting C_0-semigroups.

Definition 5.2.1 (Kato Functions of Class $\mathcal{K}$) Let $f : \mathbb{R}_0^+ \to \mathbb{R}_0^+$ and $g : \mathbb{R}_0^+ \to \mathbb{R}_0^+$ be Borel measurable functions such that the conditions

$$0 \leq f(x) \leq 1 \quad \text{and} \quad 0 \leq g(x) \leq 1, \quad x \in \mathbb{R}_0^+, \tag{5.2.1}$$

and $f(0) = g(0) = 1$ together with

$$\lim_{x\to+0} \frac{1-f(x)}{x} = 1 \quad \text{and} \quad \lim_{x\to+0} \frac{1-g(x)}{x} = 1\,, \tag{5.2.2}$$

are satisfied. Borel measurable functions satisfying the assumptions (5.2.1) and (5.2.2) are called in the following the Kato functions of class $\mathcal{K}$ or *generic* Kato functions.

Note that for a non-negative self-adjoint operator C and $f \in \mathcal{K}$, the operator-valued function $C \mapsto f(C)$ is defined by the spectral representation:

$$f(C) = \int_{\mathbb{R}_0^+} dE_C(\lambda)\, f(\lambda)\,.$$

Here $dE_C(\lambda)$ denotes the projection-valued spectral measure attributed to self-adjoint operator C.

Below we define for $\tau \in \mathbb{R}_0^+$ and for non-negative self-adjoint operators A and B functions:

$$\begin{aligned} \tau &\mapsto F(\tau) := g(\tau B/2) f(\tau A) g(\tau B/2)\,, \\ \tau &\mapsto G(\tau) := f(\tau A/2) g(\tau B) f(\tau A/2)\,, \end{aligned} \tag{5.2.3}$$

as well as

$$\begin{aligned} A(\tau) &:= \tau^{-1}[\mathbb{1} - f(\tau A)]\,, \quad B(\tau) := \tau^{-1}[\mathbb{1} - g(\tau B)]\,, \\ S_F(\tau) &:= \tau^{-1}[\mathbb{1} - F(\tau)]\,, \quad S_G(\tau) := \tau^{-1}[\mathbb{1} - G(\tau)]\,. \end{aligned} \tag{5.2.4}$$

Then by Definition 5.2.1 and spectral representations for (5.2.4), one gets

$$0 \le \tau A(\tau) = \mathbb{1} - f(\tau A) \le \mathbb{1}\,, \quad 0 \le \tau B(\tau) = \mathbb{1} - g(\tau B) \le \mathbb{1}\,, \quad \tau \ge 0. \tag{5.2.5}$$

We also set

$$K(\tau) := A(\tau) + B(\tau/2) - \frac{\tau}{4} B(\tau/2)^2, \quad \tau > 0. \tag{5.2.6}$$

Then owing to representation

$$K(\tau) = A(\tau) + B(\tau/2)^{1/2} \Big(\frac{\mathbb{1} + g(\tau B/2)}{2}\Big) B(\tau/2)^{1/2}, \quad \tau > 0, \tag{5.2.7}$$

and (5.2.5) we find that $K(\tau) \geq A(\tau)$ for $\tau > 0$. Furthermore, a straightforward computation shows that by (5.2.3) and (5.2.4), one gets equation

$$S_F(\tau) = K(\tau) + W(\tau), \quad \tau > 0, \tag{5.2.8}$$

where we define for $\tau > 0$

$$W(\tau) := \frac{\tau^2}{4} B(\tau/2)A(\tau)B(\tau/2) - \frac{\tau}{2}\left(A(\tau)B(\tau/2) + B(\tau/2)A(\tau)\right). \tag{5.2.9}$$

Lemma 5.2.2 *Let A and B be non-negative self-adjoint operators in $\mathfrak{H}$ such that domain* $\operatorname{dom}(A^{1/2}) \cap \operatorname{dom}(B^{1/2})$ *is dense in $\mathfrak{H}$ and let $H = A \dot{+} B$. If f and g are Kato-functions of class $\mathcal{K}$, then*

$$\text{s-}\lim_{\tau\to+0} (\mathbb{1} + K(\tau))^{-1} = (\mathbb{1} + H)^{-1}. \tag{5.2.10}$$

Proof The proof is divided into three steps.

(1) First of all we are going to show (cf. (5.1.19)) that

$$\limsup_{\tau\to+0} \left((\mu\mathbb{1} + K(\tau))^{-1}u, u\right) \leq \left((\mu\mathbb{1} + H)^{-1}u, u\right), \quad u \in \mathfrak{H}, \quad \mu \geq 1. \tag{5.2.11}$$

Let $\Delta_r = [0, r)$, $r > 0$. We set

$$K_{r,s}(\tau) := A(\tau)E_A(\Delta_r) + B(\tau/2)^{1/2}\Big(\frac{\mathbb{1} + g(\tau B/2)}{2}\Big)B(\tau/2)^{1/2}E_B(\Delta_s) \tag{5.2.12}$$

$r, s > 0$ and $E_A(\cdot)$, $E_B(\cdot)$ are corresponding to A, B spectral measures. Then we obtain

$$K_{r,s}(\tau) \leq K(\tau), \quad \tau > 0, \quad r, s > 0, \tag{5.2.13}$$

which yields

$$(\mu\mathbb{1} + K(\tau))^{-1} \leq (\mu\mathbb{1} + K_{r,s}(\tau))^{-1}, \quad \tau > 0, \quad r, s > 0, \quad \mu \geq 1. \tag{5.2.14}$$

Setting

$$H_{r,s} := AE_A(\Delta_r) + BE_B(\Delta_s), \quad r, s > 0, \tag{5.2.15}$$

one verifies that

$$\text{s-}\lim_{\tau\to+\infty} (\mu\mathbb{1}+K_{r,s}(\tau))^{-1} = (\mu\mathbb{1}+H_{r,s})^{-1}, \quad r,s>0, \quad \mu\geq 1. \tag{5.2.16}$$

Let $H_s := A + BE_B(\Delta_s)$, $s \in \mathbb{R}^+$. Then on account of definition (5.2.15)

$$\text{s-}\lim_{r\to\infty} (\mu\mathbb{1}+H_{r,s})^{-1} = (\mu\mathbb{1}+H_s)^{-1}, \quad s>0, \quad \mu\geq 1, \tag{5.2.17}$$

as well as by (5.2.14) together with (5.2.16), (5.2.17), one infers that

$$\limsup_{\tau\to+0} \left((\mu\mathbb{1}+K(\tau))^{-1}u, u\right) \leq \left((\mu\mathbb{1}+H_s)^{-1}u, u\right), \quad \mu\geq 1, \quad s>0, \tag{5.2.18}$$

$u \in \mathfrak{H}$. The sesquilinear form $\mathfrak{t}_{H_s}$ associated with positive operator H_s is given by

$$\begin{aligned} \mathfrak{t}_{H_s}[u,v] =& \mathfrak{t}_A[u,v] + \mathfrak{t}_B[E_B(\Delta_s)u, E_B(\Delta_s)u], \\ & u,v \in \text{dom}\,(\mathfrak{t}_{H_s}) = \text{dom}\,(\mathfrak{t}_A) = \text{dom}\,(A^{1/2}), \quad s>0. \end{aligned} \tag{5.2.19}$$

By definition of H_s, the sequence $\{\mathfrak{t}_{H_s}\}_{s\in\mathbb{R}^+}$ is *monotone nondecreasing* and such that

$$\text{s}-text-\lim_{s\to\infty} \mathfrak{t}_{H_s}[u,u] = \mathfrak{t}_H[u,u], \quad u \in \text{dom}\,(\mathfrak{t}_H). \tag{5.2.20}$$

Then we get $\text{s}-\lim_{s\to\infty}(\mu\mathbb{1}+H_s)^{-1} = (\mu\mathbb{1}+H)^{-1}$, $\mu \geq 1$, which because of (5.2.18) yields (5.2.11), cf. the proof in Lemma 5.1.4 and Notes to Chap. 5 (Notes to Sect. 5.1).

(2) Next we show that

$$\left((\mu\mathbb{1}+H)^{-1}u, u\right) \leq \liminf_{\tau\to+0} \left((\mu\mathbb{1}+K(\tau))^{-1}u, u\right), \quad u\in\mathfrak{H}, \quad \mu\geq 1. \tag{5.2.21}$$

Given that f and g are Kato functions, by reason of (5.2.1) and (5.2.2), there is $\delta > 0$ such that

$$\frac{1-f(x)}{x} \leq 1 \quad \text{and} \quad \frac{1-g(x)}{x} \leq 1, \quad x\in[0,\delta). \tag{5.2.22}$$

Let $\delta_\tau := [0, \delta/\tau)$. In this case, we find

$$A(\tau)E_A(\delta_\tau) \leq AE_A(\delta_\tau) \quad \text{and} \quad B(\tau)E_B(\delta_\tau) \leq BE_B(\delta_\tau), \quad \tau>0. \tag{5.2.23}$$

Hence,

$$A(\tau)E_A(\delta_\tau) + B(\tau/2)E_B(\delta_{\tau/2}) \le AE_A(\delta_\tau) + BE_B(\delta_{\tau/2}) \le H, \quad \tau > 0. \tag{5.2.24}$$

Using (5.2.24) we obtain

$$\begin{aligned} K(\tau) &\le A(\tau) + B(\tau/2) \le \\ &A(\tau)E_A(\delta_\tau) + B(\tau/2)E_B(\delta_{\tau/2}) + A(\tau)E_A(\mathbb{R}_0^+ \setminus \delta_\tau) + B(\tau/2)E_B(\mathbb{R}_0^+ \setminus \delta_{\tau/2}) \\ &\le H + A(\tau)E_A(\mathbb{R}_0^+ \setminus \delta_\tau) + B(\tau/2)E_B(\mathbb{R}_0^+ \setminus \delta_{\tau/2}), \quad \tau > 0. \end{aligned} \tag{5.2.25}$$

From (5.2.25) it follows that

$$(\mu\mathbb{1} + H + A(\tau)E_A(\mathbb{R}_0^+ \setminus \delta_\tau) + B(\tau/2)E_B(\mathbb{R}_0^+ \setminus \delta_{\tau/2}))^{-1} \le (\mu\mathbb{1} + K(\tau))^{-1}, \tag{5.2.26}$$

$\tau > 0$, $\mu \ge 1$. Setting

$$\begin{aligned} &\Xi_H(\tau) := \\ &(\mu\mathbb{1} + H)^{-1/2} \left(A(\tau)E_A(\mathbb{R}_0^+ \setminus \delta_\tau) + B(\tau/2)E_B(\mathbb{R}_0^+ \setminus \delta_{\tau/2})\right)(\mu\mathbb{1} + H)^{-1/2}, \end{aligned} \tag{5.2.27}$$

$\tau > 0$, we find

$$(\mu\mathbb{1} + H)^{-1/2}(\mathbb{1} + \Xi_H(\tau))^{-1}(\mu\mathbb{1} + H)^{-1/2} \le (\mu\mathbb{1} + K(\tau))^{-1}, \quad \tau > 0. \tag{5.2.28}$$

Notice that

$$\begin{aligned} &(\mu\mathbb{1}+H)^{-1/2}A(\tau)E_A(\mathbb{R}^+ \setminus \delta_\tau)(\mu\mathbb{1} + H)^{-1/2} = \\ &\qquad Z_0^* \left(\frac{\mathbb{1} - f(\tau A)}{\tau A}\right)^{1/2} E_A(\mathbb{R}^+ \setminus \delta_\tau) \left(\frac{\mathbb{1} - f(\tau A)}{\tau A}\right)^{1/2} Z_0\,, \end{aligned} \tag{5.2.29}$$

for $\tau > 0$ and $\mu \ge 1$, where $Z_0 := A^{1/2}(\mu\mathbb{1} + H)^{-1/2}$. Moreover, for $\mu \ge 1$ one has $\|Z_0\| = \|A^{1/2}(\mu\mathbb{1} + H)^{-1/2}\| \le 1$ and we obtain

$$\text{s-}\lim_{\tau\to+0} (\mu\mathbb{1} + H)^{-1/2}A(\tau)\,E_A(\mathbb{R}_0^+ \setminus \delta_\tau)\,(\mu\mathbb{1} + H)^{-1/2} = 0, \quad \mu \ge 1. \tag{5.2.30}$$

Similarly, for $\mu \geq 1$, we prove

$$\text{s-}\lim_{\tau \to +0} (\mu \mathbb{1} + H)^{-1/2} B(\tau/2)\, E_B(\mathbb{R}_0^+ \setminus \delta_{\tau/2})\, (\mu \mathbb{1} + H)^{-1/2} = 0. \tag{5.2.31}$$

Summing up (5.2.27)–(5.2.31), we find $\text{s}-\lim_{\tau \to +0}(\mathbb{1} + \Xi_H(\tau))^{-1} = \mathbb{1}$. Hence

$$\text{s-}\lim_{\tau \to +0} (\mu \mathbb{1} + H)^{-1/2} (\mathbb{1} + \Xi_H(\tau))^{-1} (\mu \mathbb{1} + H)^{-1/2} = (\mu \mathbb{1} + H)^{-1},$$

for any $\mu \geq 1$, which due to (5.2.28) yields (5.2.21).

(3) From (5.2.11) and (5.2.21) we derive

$$\text{w-}\lim_{\tau \to +0} (\mu \mathbb{1} + K(\tau))^{-1} = (\mu \mathbb{1} + H)^{-1}, \quad \mu \geq 1\,. \tag{5.2.32}$$

Then applying Lemma 4.1.4, we obtain (5.2.10). □

For $\tau > 0$ we introduce the operator-valued function:

$$\begin{aligned} Q(\tau) :=\; & \frac{\tau^2}{4} (\mathbb{1} + K(\tau))^{-1/2} B(\tau/2) A(\tau) B(\tau/2) (\mathbb{1} + K(\tau))^{-1/2} \\ & - \frac{\tau}{2} (\mathbb{1} + K(\tau))^{-1/2} (A(\tau) B(\tau/2) + B(\tau/2) A(\tau)) (\mathbb{1} + K(\tau))^{-1/2}\,. \end{aligned} \tag{5.2.33}$$

A straightforward calculation shows that because of definitions (5.2.3) and (5.2.4),

$$\mathbb{1} + S_F(\tau) = (\mathbb{1} + K(\tau))^{1/2} (\mathbb{1} + Q(\tau)) (\mathbb{1} + K(\tau))^{1/2}, \quad \tau > 0, \tag{5.2.34}$$

Lemma 5.2.3 *Let A and B be non-negative self-adjoint operators in $\mathfrak{H}$. If f, g are Kato functions of class $\mathcal{K}$, then $\mathbb{1} + Q(\tau)$ is boundedly invertible for each $\tau > 0$ and*

$$\|(\mathbb{1} + Q(\tau))^{-1}\| \leq q_0 := \left(\frac{\sqrt{17} + 1}{\sqrt{17} - 1} \right)^2, \quad \tau > 0. \tag{5.2.35}$$

Moreover, one has $\text{s}-\lim_{\tau \to +0} Q(\tau) = 0$.

Proof We set

$$\begin{aligned} A_K(\tau) &:= A(\tau)^{1/2} (\mathbb{1} + K(\tau))^{-1/2}, \\ B_K(\tau) &:= ((\mathbb{1} + g(\tau B/2))/2)^{1/2} B(\tau/2)^{1/2} (\mathbb{1} + K(\tau))^{-1/2}\,. \end{aligned} \tag{5.2.36}$$

Note that $\|A_K(\tau)\| \leq 1$ and $\|B_K(\tau)\| \leq 1$ for $\tau > 0$.

Thanks to definition (5.2.33), one gets representation

$$
\begin{aligned}
Q(\tau) =&(\mathbb{1} + K(\tau))^{-1/2}(\tau B(\tau/2)/2 - r)A(\tau)(\tau B(\tau/2)2 - r)(\mathbb{1} + K(\tau))^{-1/2} \\
&- ((1-r)\tau/2)(\mathbb{1} + K(\tau))^{-1/2}[A(\tau)B(\tau/2) + B(\tau/2)A(\tau)] \\
&\quad \times (\mathbb{1} + K(\tau))^{-1/2} \\
&- r^2(\mathbb{1} + K(\tau))^{-1/2}A(\tau)(\mathbb{1} + K(\tau))^{-1/2},
\end{aligned}
$$

which yields the estimate

$$
\begin{aligned}
&Q(\tau) \geq \\
&\quad - ((1-r)\tau/2)(\mathbb{1} + K(\tau))^{-1/2}\,[A(\tau)B(\tau/2) + B(\tau/2)A(\tau)]\,(\mathbb{1} + K(\tau))^{-1/2} \\
&\quad - r^2(\mathbb{1} + K(\tau))^{-1/2}A(\tau)(\mathbb{1} + K(\tau))^{-1/2},
\end{aligned} \tag{5.2.37}
$$

where $r \in (0, 1)$ is a constant, which will be determined later. Notice that by (5.2.36)

$$
A(\tau)(\mathbb{1} + K(\tau))^{-1/2} = \frac{1}{\tau^{1/2}}(\mathbb{1} - f(\tau A))^{1/2}A_K(\tau) \quad \tau > 0\,, \tag{5.2.38}
$$

and that it can be estimated as

$$
\|A(\tau)(\mathbb{1} + K(\tau))^{-1/2}u\| \leq \frac{1}{\tau^{1/2}}\|A_K(\tau)u\|, \quad u \in \mathfrak{H}, \quad \tau > 0\,. \tag{5.2.39}
$$

Further, by virtue of (5.2.36), one infers

$$
B(\tau/2)(\mathbb{1} + K(\tau))^{-1/2} = \frac{1}{\tau^{1/2}}\left(\frac{\mathbb{1} - g(\tau B/2)}{\mathbb{1} + g(\tau B/2)}\right)^{1/2} B_K(\tau/2), \quad \tau > 0\,. \tag{5.2.40}
$$

Then we get inequality

$$
\|B(\tau/2)(\mathbb{1} + K(\tau))^{-1/2}u\| \leq \frac{1}{\tau^{1/2}}\|B_K(\tau)u\|, \quad u \in \mathfrak{H}, \quad \tau > 0. \tag{5.2.41}
$$

As a result by (5.2.37), (5.2.39) and (5.2.41), we obtain for $u \in \mathfrak{H}$,

$$
\begin{aligned}
(Q(\tau)u, u) &\geq -(1-r)\|A_K(\tau)u\|\;\|B_K(\tau)u\| - r^2\|A_K(\tau)u\|^2 \\
&\geq -((1-r)p/2 + r^2)\|A_K(\tau)u\|^2 - ((1-r)/2p)\|(B_K(\tau)u\|^2,
\end{aligned} \tag{5.2.42}
$$

$\tau > 0$ and $p > 0$.

Recall that $\{A_K(\tau)\}_{\tau>0}$ and $\{B_K(\tau)\}_{\tau>0}$ are contractions (5.2.36). Then choosing p such that the equation is satisfied $(1-r)p/2 + r^2 = (1-r)/2p$, that is, $p := \big(-r^2 + \sqrt{r^4 + (1-r)^2}\big)/(1-r)$, we get

$$(Q(\tau)u, u) \geq -\beta(r)\big(\|A_K(\tau)u\|^2 + \|B_K(\tau)u\|^2\big) = -2\,\beta(r)\,\|u\|^2\,, \qquad (5.2.43)$$

here $\beta(r) = (r^2 + \sqrt{r^4 + (1-r)^2})/2$. The function $\beta(r)$ attains its minimum at $r_0 := (\sqrt{17}-1)/8 = 2/(\sqrt{17}+1)$, where $\beta(r_0) = (2\sqrt{17})/(\sqrt{17}+1)^2$ and consequently $(Q(\tau)u, u) \geq -4\sqrt{17}\|u\|^2/(\sqrt{17}+1)^2$. Therefore,

$$((1 + Q(\tau))u, u) \geq \left(\frac{\sqrt{17}-1}{\sqrt{17}+1}\right)^2 \|u\|^2, \quad u \in \mathfrak{H}\,,$$

which yields (5.2.35).

Finally we prove that $\text{s}-\lim_{\tau\to+0} Q(\tau) = 0$. To this aim, we introduce a family of mappings

$$X(\tau) : \mathfrak{H} \longrightarrow \widehat{\mathfrak{H}} := \mathfrak{H} \oplus \mathfrak{H} \oplus \mathfrak{H}\,, \quad \tau > 0\,, \qquad (5.2.44)$$

where $\widehat{\mathfrak{H}}$ is direct sum of Hilbert spaces $\{\mathfrak{H}\}$, which is equipped with a standard scalar product. Here mapping $X(\tau)$ is such that for any $u \in \mathfrak{H}$

$$X(\tau)\,u := \big((\mathbb{1} + K(\tau))^{-1/2}\,u,\ A_K(\tau)\,u,\ B_K(\tau)\,u\big) \in \widehat{\mathfrak{H}}.\quad \tau > 0\,, \qquad (5.2.45)$$

Because of definitions (5.2.6) and (5.2.36) and definitions (5.2.44) and (5.2.45), the operator-valued vector function

$$\tau \mapsto X(\tau) \in \mathcal{L}(\widehat{\mathfrak{H}})\,, \quad \tau > 0\,, \qquad (5.2.46)$$

is bounded. Moreover, representation (5.2.7) and straightforward calculation yield

$$\|X(\tau)u\|^2_{\widehat{\mathfrak{H}}} = \|u\|^2, \quad u \in \mathfrak{H},\ \tau > 0\,. \qquad (5.2.47)$$

On account of Lemma 5.2.2, we infer that similarly to $\{K(\tau)\}_{\tau>0}$ the limit (5.2.10) also holds for families $\{A(\tau)\}_{\tau>0}$ and $\{B(\tau)\}_{\tau>0}$; see (5.2.4), if f and g are Kato-functions. Then by definitions (5.2.36) and $\|A_K(\tau)\| \leq 1$, $\|B_K(\tau)\| \leq 1$, we conclude that

$$\text{w-}\lim_{\tau\to+0} A_K(\tau) = A^{1/2}(\mathbb{1} + H)^{-1/2} \ \text{and}\ \text{w}-\lim_{\tau\to+0} B_K(\tau) = B^{1/2}(\mathbb{1} + H)^{-1/2}\,. \qquad (5.2.48)$$

So, by virtue of (5.2.45) and by limits (5.2.10) and (5.2.46), one infers the *weak* operator convergence: $\text{w}-\lim_{\tau\to+0} X(\tau) = X$ on the Hilbert space $\widehat{\mathfrak{H}}$ to bounded operator

$$X := \left((\mathbb{1}+H)^{-1/2},\ A^{1/2}(\mathbb{1}+H)^{-1/2},\ B^{1/2}(\mathbb{1}+H)^{-1/2}\right) \in \mathcal{L}(\widehat{\mathfrak{H}}) .$$

Since $\|Xu\|^2_{\widehat{\mathfrak{H}}} = \|u\|^2$, $u \in \mathfrak{H}$, by (5.2.47) and Lemma 4.1.2, the weak convergence indicated above uplifts to the *strong* operator convergence: $\text{s}-\lim_{\tau\to+0} X(\tau) = X$, on $\widehat{\mathfrak{H}}$. As a consequence, (5.2.48) is also improved to

$$\text{s-}\lim_{\tau\to+0} A_K(\tau) = A^{1/2}(\mathbb{1}+H)^{-1/2} \text{ and } \text{s}-\lim_{\tau\to+0} B_K(\tau) = B^{1/2}(\mathbb{1}+H)^{-1/2} . \tag{5.2.49}$$

As a result, on account of (5.2.38) and (5.2.40), it follows from (5.2.49) for Kato functions $f, g \in \mathcal{K}$ that

$$\begin{aligned} &\text{s-}\lim_{\tau\to+0} \sqrt{\tau}\, A(\tau)\,(\mathbb{1}+K(\tau))^{-1/2} = 0 , \\ &\text{s-}\lim_{\tau\to+0} \sqrt{\tau}\, B(\tau/2)\,(\mathbb{1}+K(\tau))^{-1/2} = 0 . \end{aligned} \tag{5.2.50}$$

As a consequence, by definitions (5.2.33) and (5.2.5) and (5.2.50), we obtain the limit $\text{s}-\lim_{\tau\to+0} Q(\tau) = 0$. □

It is worthwhile mentioning here that from Lemma 5.2.3 and Eq. (5.2.34), we deduce the representation

$$(\mathbb{1}+S_F(\tau))^{-1} = (\mathbb{1}+K(\tau))^{-1/2}(\mathbb{1}+Q(\tau))^{-1}(\mathbb{1}+K(\tau))^{-1/2} , \quad \tau > 0 . \tag{5.2.51}$$

Similarly to Trotter product formulæ (5.1.1) and (5.1.8) and to their symmetrised versions (5.1.6) and (5.1.10), the (*non-exponential*) Trotter-Kato product formulæ correspond to approximants generated by monomials and fractional powers of Kato functions. Then counterparts of the Trotter approximants (5.1.1) and (5.1.8), as well as of (5.1.6) and(5.1.10), are the *Trotter-Kato approximants*

$$\{F_{fg}(t/n)^n := (f(tA/n)g(tB/n))^n\}_{n\geq 1} , \quad t \in \mathbb{R}_0^+ , \ n \in \mathbb{N} , \tag{5.2.52}$$

for the non-self-adjoint case, and

$$\{F(t/\eta)^\eta := (g(tB/2\eta) f(tA/\eta) g(tB/2\eta))^\eta\}_{\eta>0} , \tag{5.2.53}$$

$$\{G(t/\eta)^\eta := (f(tA/2\eta) g(tB/\eta) f(tA/2\eta))^\eta\}_{\eta>0} , \tag{5.2.54}$$

for the self-adjoint case, where $\eta \in \mathbb{R}^+$. Hence, the *exponential* Trotter-Kato approximants coincide with the Trotter approximants. To this end, compare, for example, (5.2.52) with $V_{A,B}$, or (5.2.53), (5.2.54) with $W_{A,B}$, $W_{B,A}$, in Theorem 5.1.1.

Definition 5.2.4 The product formulæ generated by approximants (5.2.52) and (5.2.53), (5.2.54), or other monomials and fractional powers of Kato functions we call the *Trotter-Kato product formulæ*. Their variety is evidently larger than the exponential product formulæ.

Now we are in position to prove the Trotter-Kato product formulæ in the strong operator topology for approximants (5.2.53) and (5.2.54).

Theorem 5.2.5 (Kato Product Theorem) *Let A and B be non-negative self-adjoint operators in $\mathfrak{H}$ such that* $\mathrm{dom}\,(A^{1/2}) \cap \mathrm{dom}\,(B^{1/2})$ *be dense in $\mathfrak{H}$ and self-adjoint operator $H = A \dot{+} B$ by defined by the form-sum. Further, let f and g be Kato functions from the class $\mathcal{K}$. Then the Trotter-Kato product formula for approximants $\{(F(t/\eta))^\eta\}_{\eta>0}$ and $\{(G(t/\eta))^\eta\}_{\eta>0}$*

$$e^{-tH} = \mathrm{s-}\lim_{\eta\to\infty} F(t/\eta)^\eta = \mathrm{s-}\lim_{\eta\to\infty} G(t/\eta)^\eta\,, \tag{5.2.55}$$

and the Trotter-Kato product formula for approximants $\{f(tA/n)g(tB/n)\}_{n\geq 1}$ (and evidently $\{g(tB/n)f(tA/n)\}_{n\geq 1}$)

$$e^{-tH} = \mathrm{s-}\lim_{n\to\infty} F_{fg}(t/n)^n = \mathrm{s-}\lim_{n\to\infty} F_{gf}(t/n)^n\,, \tag{5.2.56}$$

hold in the strong operator topology locally uniformly away from zero, that is, uniformly in $t \in \mathcal{I}$ for any bounded interval $\mathcal{I} \subset \mathbb{R}^+$.

Proof Since by (5.2.53), $F(\tau) = g(\tau B/2)f(\tau A)g(\tau B/2)$, $\tau \geq 0$, we have to show, following Theorem 4.1.3, that

$$\mathrm{s-}\lim_{\tau\to+0} (\mathbb{1} + S_F(\tau))^{-1} = (\mathbb{1} + H)^{-1}\,, \tag{5.2.57}$$

where $S_F(\tau)$ is given by (5.2.3) and (5.2.4).

Indeed, from Lemma 5.2.3, we find $\mathrm{s-}\lim_{\tau\to+0}(\mathbb{1} + Q(\tau))^{-1} = \mathbb{1}$, and from Lemma 5.2.2, we get $\mathrm{s-}\lim_{\tau\to+0}(\mathbb{1} + K(\tau))^{-1/2} = (\mathbb{1} + H)^{-1/2}$. Then thanks to representation (5.2.51) we obtain (5.2.57), which in turn yields (5.2.55). Note that *mutatis mutandis* we obtain by (5.2.54) the proof of (5.2.55) also for $\{G(t/\eta)^\eta\}_{\eta>0}$.

To prove (5.2.56) we set $\widehat{f}(\lambda) := f(2\lambda)^{1/2}$, $\lambda \in \mathbb{R}^+$. By Definition 5.2.1, it is clear that $\widehat{f} \in \mathcal{K}$. Let

$$\widehat{G}(\tau) := \widehat{f}(\tau A/2)g(\tau B)\widehat{f}(\tau A/2) = f(\tau A)^{1/2}g(\tau B)f(\tau A)^{1/2}, \quad \tau \geq 0\,. \tag{5.2.58}$$

For $\{\widehat{G}(t/\eta)^{\eta}\}_{\eta>0}$ the first part of Theorem 5.2.5 (where $\eta = n \in \mathbb{N}$) and the identity (5.2.58) yield that the strong limit

$$e^{-tH} u = \lim_{n\to\infty} \left(f(tA/n)^{1/2} g(tB/n) f(tA/n)^{1/2}\right)^n u\,, \tag{5.2.59}$$

exists uniformly in $t \in \mathcal{I}$ on any fixed vector $u \in \mathfrak{H}$. Then the representation

$$(f(tA/n)g(tB/n))^n f(tA/n)^{1/2} = f(tA/n)^{1/2}\, \widehat{G}(t/n)^n, \quad t \geq 0, \quad n \in \mathbb{R}\,, \tag{5.2.60}$$

together with (5.2.59) and $\text{s}-\lim_{n\to\infty} f(tA/n)^{1/2} = \mathbb{1}$ give for the right-hand side of (5.2.60)

$$\lim_{n\to\infty} F_{fg}(t/n)^n f(tA/n)^{1/2}\, u = e^{-tH} u\,, \tag{5.2.61}$$

uniformly in $t \in \mathcal{I}$ on any fixed vector $u \in \mathfrak{H}$. Since $\lim_{n\to\infty} f(tA/n)^{1/2} u = u$ uniformly in $t \in \mathcal{I}$ for any fixed $u \in \mathfrak{H}$, the limit (5.2.61) implies (5.2.56). □

Remark 5.2.6

(a) Seeing that exponential function $f(x) = e^{-x}$, $x \in \mathbb{R}_0^+$, is obviously an example of (generic) Kato function. Theorem 5.1.5 follows from Theorem 5.2.5.
(b) Besides the exponential function, there are many other generic Kato functions $\mathcal{K}$; see Definition 5.2.1. Typical examples are as follows: $f(x) = (1+x)^{-1}$, $f(x) = (1 + x/a)^{-a}$, $a > 0$, $f(x) = 1 - |\sin(x)|$, etc. See Notes to Chap. 5 (Notes to Sect. 5.2).
(c) We note that if $f \in \mathcal{K}$ is Kato function, then $f_\delta(x) := f(x)\, \chi_{[0,\delta]}(x)$ for $x \in \mathbb{R}_0^+$ and $\delta > 0$ is also a Kato function. It is also useful to remark that if function $f \in \mathcal{K}$, then $\widehat{f}(x) := f(x/m)^m$, $x \in \mathbb{R}_0^+$, is in turn a generic Kato function for any $m \in \mathbb{N}$.
(d) In contrast to the Trotter product formula (see (5.1.1), Theorem 5.1.1), the proof of the Trotter-Kato product formulæ (5.2.55) and (5.2.56) covers not only the *singular* case

$$\text{dom}\,(A) \cap \text{dom}\,(B) = \{0\}\,,$$

but also a case of nontrivial intersections $(\text{dom}\,(\sqrt{A}) \cap \text{dom}\,(\sqrt{B})) \subset \mathfrak{H}$. Suppose that $\text{dom}\,(\sqrt{A}) \cap \text{dom}\,(\sqrt{B})$ is *not* dense in Hilbert space $\mathfrak{H}$. Then the closure $\mathfrak{H}_0 := \overline{\text{dom}\,(\sqrt{A}) \cap \text{dom}\,(\sqrt{B})}$ is a subspace in $\mathfrak{H}$. By Definition 5.1.3 the form-sum $H = A \dotplus B$ exists, but now only on the subspace $\mathfrak{H}_0$.
(e) In this case the Kato Theorem 5.2.5 still remains true, but now in the following sense: Let P_0 be the orthogonal projection onto $\mathfrak{H}_0$. Then the product formulæ (5.2.55) and (5.2.56) take the form

$$\text{s-}\lim_{\eta\to\infty} (g(tB/2\eta) f(tA/\eta) g(tB/2\eta))^{\eta} = e^{-tH} P_0\,, \tag{5.2.62}$$

and

$$\text{s-}\lim_{n\to\infty} (f(tA/n)g(tB/n))^n = \mathrm{e}^{-tH} P_0 , \tag{5.2.63}$$

locally uniformly in $t \in \mathbb{R}_0^+$ away from zero for a *degenerate* semigroup $\{\mathrm{e}^{-tH} P_0\}_{t\geq 0}$ (Definition 1.6.1 and Remark 1.6.4) on $\mathfrak{H}$ in the right-hand side of (5.2.62) and (5.2.63).

For more comments, see Notes to Chap. 5 (Notes to Sect. 5.2) and Proposition 8.1.2, Sect. 8.3 together with Sect. 8.5.2.

5.3 Notes to Chap. 5

Notes to Sect. 5.1

Historically, the first result about approximation of semigroups by the *exponential* product formula in the *strong* operator topology dates back to H.Trotter (1959) [Trot59] and to Yu.L.Daletskiĭ (1960) [Dal60].

Theorem 5.1.1 is the famous Trotter product theorem about strongly convergent product formula on a Hilbert space. In [Trot59, Theorem 2] H.Trotter proved it in a general setting of a Banach space; see Sect. 1.9.6 (Proposition 1.9.35) and Chap. 10.

For details on *representation* theorems and construction of the *form-sum* in Definition 5.1.3, we refer to Sect. 9.1.2 and to [Kato80, Chapter VI, §2].

The proof of Lemma 5.1.4 is motivated by [Si78, Theorem 3.1] (see [RS80, Theorem S.14]) about *monotone convergence* of forms and of the corresponding operators in the strong resolvent sense.

Generalisation of the strongly convergent Trotter product formula for the case where generator of the semigroup in the limit is defined as a form-sums is due to T.Kato [Kato74, Theorem 1, Theorem 2] and [Kato78, Theorem, p.186]. These two papers plaid important rôle in motivations of the whole project related to the present book since (together with the form-sum for generators) they also introduced a *non-exponential* versions of the product formulæ and a link with degenerate semigroups.

On that account, Kato's version of the Lie-Trotter product formula in Theorem 5.1.5 and non-exponential product formulæ in the Kato product Theorem 5.2.5 we call the *Trotter-Kato product formulæ*.

In our arguments for the proof of Theorem 5.1.5, we used the *Chernoff product formula* (4.1.7), Corollary 4.1.5, and results due to B.Simon [Si78, Theorem 3.1] on the monotone convergence of sesquilinear forms associated with operators.

Notes to Sect. 5.2

Definition 5.2.1 of *generic* Kato function appeared in [Kato74]. For other types of the Kato functions, see [Zag19, Appendix C].

We recall that Theorem 5.2.5 for self-adjoint semigroups and strong operator convergence dates back to T.Kato [Kato74, Theorem 1, Theorem 2] and [Kato78,

Theorem, p.186]. It is valid even when the strong limit yields a *degenerate* semigroup defined by projection onto *subspace* $\mathfrak{H}_0 \subset \mathfrak{H}$, with generator given by a form-sum of operators; see Definition 1.6.1 and Remark 1.6.4. In Theorem 5.2.5 we study a nondegenerate case when the limit semigroup $\{e^{-tH}\}_{t\geq 0}$ is defined on $\mathfrak{H}$. The limits (5.2.55) and (5.2.56) are known now as the *Trotter-Kato product formulæ*. The formulæ of type (5.2.55) (and (5.2.59)) we call the *symmetric* (or *symmetrised*) Trotter-Kato product formulae.

A difference of parametrisation in Theorem 5.2.5 between the limits (5.2.55) and (5.2.56) is owing to *self-adjointness* and *non-self-adjointness* of the corresponding approximants. For the self-adjoint approximants (5.2.55), it is possible (and sometimes convenient) to use the *continuous* parametrisation, $\eta > 0$, seeing that these approximants can be defined by means of the spectral calculus. This is not the case for non-self-adjoint approximants (5.2.56), where only *product* of factors is well-defined. In proof of Theorem 5.2.5, we used the line of reasoning developed for nondegenerate semigroups in [INZ04, Sections 2 and 3].

We note that assertion about the case $\mathfrak{H}_0 \subset \mathfrak{H}$

$$\mathfrak{H}_0 := \overline{\mathrm{dom}\,(\sqrt{A}) \cap \mathrm{dom}\,(\sqrt{B})} \subset \mathfrak{H}\,,$$

mentioned in Remark 5.2.6 (d), is again due to [Kato74, Theorem 1, Theorem 2] and [Kato78, Theorem, p.186].

What concerning the proof of Theorem 5.2.5 for a *degenerate* limit semigroup $\{e^{-tH} P_0\}_{t\geq 0}$ defined by (5.2.62), (5.2.63), we follow here [NZ99b, Theorem 2.2, Theorem 2.3], cf. Proposition 8.1.2 in Sect. 8.1.

Chapter 6
Trotter-Kato Product Formulæ: Operator-Norm Topology

After strong convergence (Chap. 5), we elucidate in this and the next chapter one of the main topics of the present book. It concerns different results on convergence of the Trotter and the Trotter-Kato product formulæ in the *operator-norm* topology with, and without, operator-norm error-bound estimates of the *rate* of convergence. Similarly to Chap. 5, we focus on the case where generators A and B for a couple of C_0-semigroups are non-negative self-adjoint operators in a separable Hilbert space $\mathfrak{H}$. The principal results are (relatively) recent, and they essentially are in many aspects attributable to the authors of this book; see Notes to Chap. 6.

In Chap. 6 we focus on the operator-norm convergence of Trotter-Kato product formulæ for *self-adjoint* C_0-semigroups *without* operator-norm estimates of the rate of convergence. We shall study these estimates in Chap. 7.

6.1 Operator-Norm Convergence

Let A and B be two densely defined non-negative self-adjoint operators in a Hilbert space $\mathfrak{H}$. Further, let $f(\cdot)$ and $g(\cdot)$ be the Kato functions of class $\mathcal{K}$ (Definition 5.2.1). We consider self-adjoint contraction functions $\widehat{F}(\cdot) : \mathbb{R}_0^+ \longrightarrow \mathcal{L}(\mathfrak{H})$ and $\widehat{G}(\cdot) : \mathbb{R}_0^+ \longrightarrow \mathcal{L}(\mathfrak{H})$ defined by

$$\widehat{F}(\tau) := g(\tau B)^{1/2} f(\tau A) g(\tau B)^{1/2}\,, \quad \text{and} \quad \widehat{G}(\tau) := f(\tau A)^{1/2} g(\tau B) f(\tau A)^{1/2}\,, \tag{6.1.1}$$

$\tau \geq 0$, cf. definitions in (5.2.3) and (5.2.58). We also set on $\mathbb{R}^+$:

$$S_{\widehat{F}}(\tau) := \frac{1}{\tau}\,(\mathbb{1} - \widehat{F}(\tau)) \quad \text{and} \quad S_{\widehat{G}}(\tau) := \frac{1}{\tau}\,(\mathbb{1} - \widehat{G}(\tau)), \quad \tau > 0\,, \tag{6.1.2}$$

compare with definitions in (5.2.4).

V. A. Zagrebnov et al., *Trotter-Kato Product Formulæ*, Operator Theory: Advances and Applications 296, https://doi.org/10.1007/978-3-031-56720-9_6

Remark 6.1.1 Note that the Trotter-Kato approximants defined in (5.2.53) and (5.2.54) and $\{\widehat{F}(\tau/\eta)^{\eta}\}_{\eta>0}$, $\{\widehat{G}(\tau/\eta)^{\eta}\}_{\eta>0}$ defined by (6.1.1) are *equivalent* in the sense that they produce the same strong limit $\{e^{-tH}\}_{t\geq 0}$ as in (5.2.55). To this aim let $\widehat{g}(\lambda) := g(2\lambda)^{1/2}$ and $\widehat{f}(\lambda) := f(2\lambda)^{1/2}$, for $\lambda \in \mathbb{R}^+$. It is clear that by Definition 5.2.1 the functions $\widehat{g}, \widehat{f} \in \mathcal{K}$. Then contraction functions (6.1.1)

$$\widehat{F}(\tau) = \widehat{g}(\tau B/2) f(\tau A) \widehat{g}(\tau B/2)\,, \quad \tau \geq 0\,, \tag{6.1.3}$$

$$\widehat{G}(\tau) = \widehat{f}(\tau A/2) g(\tau B) \widehat{f}(\tau A/2)\,, \quad \tau \geq 0\,, \tag{6.1.4}$$

are equivalent to contractions $\{F(t)\}_{t\geq 0}$, $\{G(t)\}_{t\geq 0}$ in (5.2.53), (5.2.54).

Lemma 6.1.2 *Let A and B non-negative self-adjoint operators in $\mathfrak{H}$ such that* $\operatorname{dom}(A^{1/2}) \cap \operatorname{dom}(B^{1/2})$ *is dense in $\mathfrak{H}$ and let $H := A \dotplus B$ be the form-sum of two operators. Further, let $f(\cdot)$ and $g(\cdot)$ be Kato functions of the class $\mathcal{K}$ and let $\mathcal{I} \subset \mathbb{R}^+$ be a closed interval. Then there is an equivalence* :

$$\lim_{\eta\to\infty} \sup_{t\in\mathcal{I}} \|\widehat{F}(t/\eta)^{\eta} - e^{-tH}\| = 0 \Longleftrightarrow \lim_{\eta\to\infty} \sup_{t\in\mathcal{I}} \|\widehat{G}(t/\eta)^{\eta} - e^{-tH}\| = 0\,. \tag{6.1.5}$$

Proof We start by proving the implication:

$$\lim_{\eta\to\infty} \sup_{t\in\mathcal{I}} \|\widehat{F}(t/\eta)^{\eta} - e^{-tH}\| = 0 \Longrightarrow \lim_{\eta\to\infty} \sup_{t\in\mathcal{I}} \|\widehat{G}(t/\eta)^{\eta} - e^{-tH}\| = 0\,.$$

To this aim, use the identity

$$\widehat{F}(\tau) g(\tau B)^{1/2} f(\tau A)^{1/2} = g(\tau B)^{1/2} f(\tau A)^{1/2} \widehat{G}(\tau)\,, \quad \tau \geq 0\,.$$

Hence

$$\widehat{F}(\tau)^{\eta} g(\tau B)^{1/2} f(\tau A)^{1/2} = g(\tau B)^{1/2} f(\tau A)^{1/2} \widehat{G}(\tau)^{\eta}\,, \quad \tau \geq 0\,, \quad \eta \geq 1.$$

or equivalently

$$f(tA/\eta)^{1/2} g(tB/\eta)^{1/2} \widehat{F}(t/\eta)^{\eta} g(tB/\eta)^{1/2} f(tA/\eta)^{1/2} = \widehat{G}(t/\eta)^{\eta+1}$$

for $t \geq 0$ and $\eta \geq 1$. Then one gets

$$\begin{aligned}\widehat{G}(t/\eta)^{\eta} &= \widehat{G}(t/\eta)^{\eta}(\mathbb{1} - \widehat{G}(t/\eta)) + \\ &\quad f(tA/\eta)^{1/2} g(tB/\eta)^{1/2} \widehat{F}(t/\eta)^{\eta} g(tB/\eta)^{1/2} f(tA/\eta)^{1/2}\,,\end{aligned}$$

which leads to representation

$$
\begin{aligned}
&\widehat{G}(t/\eta)^{\eta} - \mathrm{e}^{-tH} = \widehat{G}(t/\eta)^{\eta}(\mathbb{1} - \widehat{G}(t/\eta)) + \\
&\quad f(tA/\eta)^{1/2} g(tB/\eta)^{1/2}[\widehat{F}(t/\eta)^{\eta} - \mathrm{e}^{-tH}] g(tB/\eta)^{1/2} f(tA/\eta)^{1/2} + \\
&\quad f(tA/\eta)^{1/2}[g(tB/\eta)^{1/2} - \mathbb{1}]\mathrm{e}^{-tH} g(tB/\eta)^{1/2} f(tA/\eta)^{1/2} + \\
&\quad [f(tA/\eta)^{1/2} - \mathbb{1}]\mathrm{e}^{-tH} g(tB/\eta)^{1/2} f(tA/\eta)^{1/2} + \\
&\quad \mathrm{e}^{-tH}[g(tB/\eta)^{1/2} - \mathbb{1}] f(tA/\eta)^{1/2} + \mathrm{e}^{-tH}[f(tA/\eta)^{1/2} - \mathbb{1}]\,.
\end{aligned}
$$

This representation readily implies the estimate

$$
\begin{aligned}
\|\widehat{G}(t/\eta)^{\eta} - \mathrm{e}^{-tH}\| &\leq \|\widehat{G}(t/\eta)^{\eta}(\mathbb{1} - \widehat{G}(t/\eta))\| + \|\widehat{F}(t/\eta)^{\eta} - \mathrm{e}^{-tH}\| \\
&\quad + 2\,\|\,[\mathbb{1} - g(tB/\eta)^{1/2}]\,\mathrm{e}^{-tH}\| + 2\,\|\,[\mathbb{1} - f(tA/\eta)^{1/2}]\,\mathrm{e}^{-tH}\|\,.
\end{aligned}
\tag{6.1.6}
$$

Now by the *spectral* theorem for self-adjoint operators, we obtain estimate

$$
\|\widehat{G}(t/\eta)^{\eta}(\mathbb{1} - \widehat{G}(t/\eta))\| \leq \frac{1}{1+\eta}, \qquad \eta \geq 1, \quad t \geq 0\,. \tag{6.1.7}
$$

Seeing that $(1 - f(x)^{1/2})/x^{1/2} \leq \sqrt{(1 - f(x))/x}$, for $x > 0$, one infers the inequality

$$
\sup_{x>1} \frac{1 - f(x)^{1/2}}{x^{1/2}} \leq \sqrt{\sup_{x>1} \frac{1 - f(x)}{x}} \leq \sqrt{\gamma[f]}\,,
$$

where we denoted $\gamma[f] := \sup_{x>0}(1 - f(x))/x < \infty$.

Next we use for $u \in \mathfrak{H}$ the equation

$$
\begin{aligned}
&[\mathbb{1} - f(tA/\eta)^{1/2}]\,\mathrm{e}^{-tH} u = \\
&\eta^{-1/2}[\mathbb{1} - f(tA/\eta)^{1/2}]\,(tA/\eta)^{-1/2}(tA)^{1/2}(\mathbb{1} + tH)^{-1/2}(\mathbb{1} + tH)^{1/2}\,\mathrm{e}^{-tH} u\,,
\end{aligned}
$$

Then owing to the *spectral* theorem for operator H and to $0 \leq (1+x)^{1/2}\mathrm{e}^{-x} \leq 1$, for $x \geq 0$, one gets inequalities

$$
\begin{aligned}
&\|\,[\mathbb{1} - f(tA/\eta)^{1/2}]\,\mathrm{e}^{-tH} u\,\| \\
&\quad \leq \eta^{-1/2}\sqrt{\gamma[f]}\,\|\,(\mathbb{1} + tH)^{1/2}\mathrm{e}^{-tH}\|\,\|u\| \leq \eta^{-1/2}\sqrt{\gamma[f]}\,\|u\|\,,
\end{aligned}
$$

for $h \in \mathfrak{H}$ and $t \geq 0$. As a result we obtain the operator-norm estimate

$$
\|\,(\mathbb{1} - f(tA/\eta)^{1/2})\,\mathrm{e}^{-tH}\| \leq \eta^{-1/2}\sqrt{\gamma[f]}\,, \tag{6.1.8}
$$

for any $t \geq 0$ and $\eta \geq 1$. Similarly, one also proves that

$$\| [\mathbb{1} - g(tB/\eta)^{1/2}] \mathrm{e}^{-tH} \| \leq \eta^{-1/2} \sqrt{\gamma[g]} , \tag{6.1.9}$$

for $t \geq 0$ and $\eta \geq 1$. Therefore, by virtue of (6.1.6), (6.1.8), and (6.1.9) one gets

$$\begin{aligned} &\sup_{t \in \mathcal{I}} \| \widehat{G}(t/\eta)^{\eta} - \mathrm{e}^{-tH} \| \\ &\quad \leq \sup_{t \in \mathcal{I}} \| \widehat{F}(t/\eta)^{\eta} - \mathrm{e}^{-tH} \| + \frac{1}{1+\eta} + \frac{2}{\eta^{1/2}} (\sqrt{\gamma[f]} + \sqrt{\gamma[g]}) . \end{aligned} \tag{6.1.10}$$

Now if $\lim_{\eta \to \infty} \sup_{t \in \mathcal{I}} \| \widehat{F}(t/\eta)^{\eta} - \mathrm{e}^{-tH} \| = 0$, then the estimate (6.1.10) implies that also $\lim_{\eta \to \infty} \sup_{t \in \mathcal{I}} \| \widehat{G}(t/\eta)^{\eta} - \mathrm{e}^{-tH} \| = 0$.

The converse statement one proves *verbatim* the line of reasoning we developed above. □

Remark 6.1.3 We note that in general, the self-adjoint operator $H := A \dot{+} B$ (form-sum (9.1.8)) differs from the self-adjoint *Friedrichs* extension $H_F := (A + B)_F$ defined by a closure of the sesquilinear form constructed from non-negative operator $A + B$; see Sect. 6.5, Examples, 1.

Corollary 6.1.4 *Let the assumptions of Lemma* 6.1.2 *be satisfied. If* $\mathcal{I} \subset \mathbb{R}^+$ *is a closed interval, then we obtain equivalences*:

$$\begin{aligned} &\lim_{\eta \to \infty} \sup_{t \in \mathcal{I}} \| (f(tA/\eta)^{1/2} g(tB/\eta)\, f(tA/\eta)^{1/2})^{\eta} - \mathrm{e}^{-tH} \| = 0 \\ &\Longleftrightarrow \lim_{n \to \infty} \sup_{t \in \mathcal{I}} \| (f(tA/n) g(tB/n))^{n} - \mathrm{e}^{-tH} \| = 0 , \\ &\lim_{\eta \to \infty} \sup_{t \in \mathcal{I}} \| (g(tB/\eta)^{1/2} f(tA/\eta)\, g(tB/\eta)^{1/2})^{\eta} - \mathrm{e}^{-tH} \| = 0 \\ &\Longleftrightarrow \lim_{n \to \infty} \sup_{t \in \mathcal{I}} \| (g(tB/n) f(tA/n))^{n} - \mathrm{e}^{-tH} \| = 0 , \\ &\lim_{n \to \infty} \sup_{t \in \mathcal{I}} \| (f(tA/n) g(tB/n))^{n} - \mathrm{e}^{-tH} \| = 0 \\ &\Longleftrightarrow \lim_{n \to \infty} \sup_{t \in \mathcal{I}} \| (g(tB/n) f(tA/n))^{n} - \mathrm{e}^{-tH} \| = 0 . \end{aligned} \tag{6.1.11}$$

Proof We start by the proof that

$$\begin{aligned} &\lim_{\eta \to \infty} \sup_{t \in \mathcal{I}} \| (f(tA/\eta)^{1/2} g(tB/\eta)\, f(tA/\eta)^{1/2})^{\eta} - \mathrm{e}^{-tH} \| = 0 \\ &\Longrightarrow \lim_{n \to \infty} \sup_{t \in \mathcal{I}} \| (f(tA/n) g(tB/n))^{n} - \mathrm{e}^{-tH} \| = 0 . \end{aligned} \tag{6.1.12}$$

Since by definition (6.1.1) for $\eta = n$ one has

$$
\begin{aligned}
&(f(tA/n)g(tB/n))^n \\
&\quad = f(tA/n)^{1/2}\widehat{G}(t/n)^{n-1} f(tA/n)^{1/2} g(tB/n), \quad n = 1, 2\dots,
\end{aligned}
$$

we obtain the representation

$$
\begin{aligned}
&(f(tA/n)g(tB/n))^n - \mathrm{e}^{-tH} = \\
&\quad f(tA/n)^{1/2}\widehat{G}(t/n)^{n-1}(\mathbb{1} - \widehat{G}(t/n)) f(tA/n)^{1/2} g(tB/n) + \\
&\quad f(tA/n)^{1/2}(\widehat{G}(t/n)^n - \mathrm{e}^{-tH}) f(tA/n)^{1/2} g(tB/n) + \\
&\quad [f(tA/n)^{1/2} - \mathbb{1}]\mathrm{e}^{-tH} f(tA/n)^{1/2} g(tB/n) + \\
&\quad \mathrm{e}^{-tH}[f(tA/n)^{1/2} - \mathbb{1}]g(tB/n) + \mathrm{e}^{-tH}[g(tB/n)^{1/2} - \mathbb{1}][g(tB/n)^{1/2} + \mathbb{1}],
\end{aligned}
$$

which, in turn, gives the estimate

$$
\begin{aligned}
&\|(f(tA/n)g(tB/n))^n - \mathrm{e}^{-tH}\| \le \\
&\qquad \|\widehat{G}(t/n)^{n-1}(\mathbb{1} - \widehat{G}(t/n))\| + \|\widehat{G}(t/n)^n - \mathrm{e}^{-tH}\| + \\
&\qquad 2\,\|(\mathbb{1} - f(tA/n)^{1/2})\,\mathrm{e}^{-tH}\| + 2\,\|(\mathbb{1} - g(tB/n)^{1/2})\,\mathrm{e}^{-tH}\|\,.
\end{aligned}
$$

As a result, we can use (6.1.7) and (6.1.8) and (6.1.9) to deduce for $\eta = n$ the estimate

$$
\begin{aligned}
&\sup_{t\in\mathcal{I}} \|(f(tA/n)g(tB/n))^n - \mathrm{e}^{-tH}\| \le \\
&\qquad \sup_{t\in\mathcal{I}} \|\widehat{G}(t/n)^n - \mathrm{e}^{-tH}\| + \frac{1}{n} + \frac{2}{n^{1/2}}\left(\sqrt{\gamma[f]} + \sqrt{\gamma[g]}\right),
\end{aligned}
$$

which proves (6.1.12).

The converse to (6.1.12) follows for $\eta = n$ from representation

$$
\begin{aligned}
&\widehat{G}(t/n)^{n+1} \\
&\quad = f(tA/n)^{1/2} g(tB/n)(f(tA/n)g(tB/n))^n f(tA/n)^{1/2}, \quad n = 1, 2, \dots,
\end{aligned}
$$

along the similar line of reasoning as the proof of (6.1.12).

The proof of the second equivalence in (6.1.11) follows by the same arguments as for equivalence corresponding to (6.1.12).

The last equivalence of the operator-norm convergent Trotter-Kato product formulæ is due to (6.1.5) and the two first equivalences in (6.1.11). □

Remark 6.1.5 In the following, we say that the Trotter-Kato product formula converges *locally uniformly* in the *operator-norm* topology if the limits in (6.1.11) exists uniformly on the closed intervals $\mathcal{I} \subset \mathbb{R}_0^+$ or uniformly on the closed intervals $\mathcal{I} \subset \mathbb{R}^+$. In the latter case, we also say that there is a locally uniform convergence *away* from zero, that is, in $\mathbb{R}^+$.

Let us assume that $f(x) > 0$ for $x \in \mathbb{R}_0^+$. In this case we have $\ker(f(\tau A)) = \{0\}$ for any $\tau \geq 0$. Hence the closure $\overline{\operatorname{ran}(f(\tau A))} = \mathfrak{H}$ for any $\tau \geq 0$. Under this assumption, the operator $f(\tau A)^{-1}$ is a self-adjoint operator on $\mathfrak{H}$ for any $\tau \geq 0$. Note that $f(\tau A)^{-1} \geq \mathbb{1}$ for $\tau \geq 0$. To proceed, we set the operator family

$$\tau \mapsto M(\tau) := \frac{1}{\tau}(f(\tau A)^{-1} - \mathbb{1}) + \frac{1}{\tau}(\mathbb{1} - g(\tau B)), \quad \tau \in \mathbb{R}^+, \tag{6.1.13}$$

with domain $\operatorname{dom}(M(\tau)) := \operatorname{ran}(f(\tau A))$.

Lemma 6.1.6 *Let A and B be non-negative self-adjoint operators in the separable Hilbert space $\mathfrak{H}$ such that* $\operatorname{dom}(A^{1/2}) \cap \operatorname{dom}(B^{1/2})$ *is dense in $\mathfrak{H}$ and let $H := A \dot{+} B$. Further, let $f(\cdot)$ and $g(\cdot)$ be Kato functions of the class $\mathcal{K}$. If $f(x) > 0$ for $x \in \mathbb{R}_0^+$, then the following equivalences hold :*

$$\text{s-}\lim_{\tau \to +0} (f(\tau A)^{-1} + M(\tau))^{-1} = (\mathbb{1} + H)^{-1} \tag{6.1.14}$$

$$\iff \text{s-}\lim_{\tau \to +0} (\mathbb{1} + M(\tau))^{-1} = (\mathbb{1} + H)^{-1}, \tag{6.1.15}$$

$$\lim_{\tau \to +0} \|(f(\tau A)^{-1} + M(\tau))^{-1} - (\mathbb{1} + H)^{-1}\| = 0 \tag{6.1.16}$$

$$\iff \lim_{\tau \to +0} \|(\mathbb{1} + M(\tau))^{-1} - (\mathbb{1} + H)^{-1}\| = 0, \tag{6.1.17}$$

$$\lim_{\eta \to \infty} \sup_{t \in \mathcal{I}} \|(f(tA/\eta)^{-1} + tM(t/\eta))^{-1} - (\mathbb{1} + tH)^{-1}\| = 0 \tag{6.1.18}$$

$$\iff \lim_{\eta \to \infty} \sup_{t \in \mathcal{I}} \|(\mathbb{1} + tM(t/\eta))^{-1} - (\mathbb{1} + tH)^{-1}\| = 0, \tag{6.1.19}$$

where $\mathcal{I}$ is a closed interval of $\mathbb{R}^+$.

Proof (6.1.18) $\Longrightarrow$ (6.1.19): To prove (6.1.19) we note that

$$f(\tau A)^{-1} + tM(\tau) = (t + \tau)M(\tau) + g(\tau B), \quad \tau > 0, \quad t \in \mathbb{R}_0^+,$$

which gives

$$(f(\tau A)^{-1} + tM(\tau))^{-1} = ((t + \tau)M(\tau) + g(\tau B))^{-1}, \ \tau > 0, \ t \in \mathbb{R}_0^+. \tag{6.1.20}$$

Thus, one also obtains

$$
\begin{aligned}
&(\mathbb{1}+tM(\tau))^{-1} - ((t+\tau)M(\tau)+g(\tau B))^{-1} = \qquad (6.1.21)\\
&\quad (\mathbb{1}+tM(\tau))^{-1}(g(\tau B)-\mathbb{1})\Big(((t+\tau)M(\tau)+g(\tau B))^{-1} - (\mathbb{1}+tH)^{-1}\Big)\\
&+(\mathbb{1}+tM(\tau))^{-1}(g(\tau B)-\mathbb{1})(\mathbb{1}+tH)^{-1}\\
&+\frac{\tau M(\tau)}{\mathbb{1}+tM(\tau)}((t+\tau)M(\tau)+g(\tau B))^{-1}, \quad \tau>0, \quad t\in\mathbb{R}_0^+ .
\end{aligned}
$$

Now, using (6.1.20), we get

$$
\begin{aligned}
&(\mathbb{1}+tM(\tau))^{-1} - \left(f(\tau A)^{-1}+tM(\tau)\right)^{-1} = \qquad (6.1.22)\\
&\quad (\mathbb{1}+tM(\tau))^{-1}(g(\tau B)-\mathbb{1})\left(\left(f(\tau A)^{-1}+tM(\tau)\right)^{-1} - (\mathbb{1}+tH)^{-1}\right)\\
&+(\mathbb{1}+tM(\tau))^{-1}(g(\tau B)-\mathbb{1})(\mathbb{1}+tH)^{-1}\\
&+\frac{\tau M(\tau)}{\mathbb{1}+tM(\tau)}\left(f(\tau A)^{-1}+tM(\tau)\right)^{-1}, \quad \tau>0, \quad t\in\mathbb{R}_0^+ .
\end{aligned}
$$

Thereupon we find the estimate

$$
\begin{aligned}
&\left\|(\mathbb{1}+tM(\tau))^{-1} - \left(f(\tau A)^{-1}+tM(\tau)\right)^{-1}\right\| \qquad (6.1.23)\\
&\le \left\|\left(f(\tau A)^{-1}+tM(\tau)\right)^{-1} - (\mathbb{1}+tH)^{-1}\right\| + \left\|(g(\tau B)-\mathbb{1})(\mathbb{1}+tH)^{-1}\right\| +\\
&\left\|\frac{\tau M(\tau)}{\mathbb{1}+tM(\tau)}\left(f(\tau A)^{-1}+tM(\tau)\right)^{-1}\right\|, \quad \tau>0, \quad t\in\mathbb{R}_0^+ .
\end{aligned}
$$

Now taking into account identity

$$
(\mathbb{1}-g(\tau B))(\mathbb{1}+tH)^{-1} = \sqrt{\frac{\tau}{t}}\,\frac{\mathbb{1}-g(\tau B)}{\sqrt{\tau B}}\left(\sqrt{tB}\,(\mathbb{1}+tH)^{-1/2}\right)(\mathbb{1}+tH)^{-1/2},
$$

for $\tau > 0$ and $t \in \mathbb{R}^+$, we obtain the estimate

$$
\left\|(\mathbb{1}-g(\tau B))(\mathbb{1}+tH)^{-1}\right\| \le \sqrt{\frac{\tau}{t}}\left\|\frac{\mathbb{1}-g(\tau B)}{\sqrt{\tau B}}\right\|\left\|\sqrt{tB}\,(\mathbb{1}+tH)^{-1/2}\right\| .
$$

Given that $\sup_{x>0} 1 - g(x)/\sqrt{x} \le \sup_{x>0} \sqrt{1 - g(x)/x} \le \sqrt{\gamma[g]}$, where $\gamma[g] := \sup_{x>0}(1 - g(x))/x < \infty$, and on account of $\left\| \sqrt{tB}\, (\mathbb{1} + tH)^{-1/2} \right\| \le 1$, for $t \in \mathbb{R}^+$, we infer

$$\left\| (\mathbb{1} - g(\tau B))\, (\mathbb{1} + H)^{-1} \right\| \le \sqrt{\gamma[g]}\, \sqrt{\frac{\tau}{t}}\,, \quad \tau > 0, \quad t > 0\,. \tag{6.1.24}$$

Further, from $\mathbb{1} \le f(\tau A)^{-1} + tM(\tau)$, $\tau > 0$, $t \in \mathbb{R}^+$, one gets

$$0 \le \left(f(\tau A)^{-1} + tM(\tau) \right)^{-1} \le \mathbb{1}, \quad \tau > 0, \quad t > 0,$$

which yields

$$\left\| \left(f(\tau A)^{-1} + tM(\tau) \right)^{-1} \right\| \le 1, \quad \tau > 0, \quad t > 0. \tag{6.1.25}$$

Next, by virtue of $0 \le \tau M(\tau)/(\mathbb{1} + tM(\tau)) \le (\tau/t)\, \mathbb{1}$, for $\tau > 0$ and $t > 0$, we find

$$\left\| \frac{\tau M(\tau)}{\mathbb{1} + tM(\tau)} \right\| \le \frac{\tau}{t}\,, \quad \tau > 0, \quad t > 0. \tag{6.1.26}$$

Taking into account the estimates (6.1.25) and (6.1.26), we obtain the estimate

$$\left\| \frac{\tau M(\tau)}{\mathbb{1} + tM(\tau)} \left(f(\tau A)^{-1} + tM(\tau) \right)^{-1} \right\| \le \frac{\tau}{t}\,, \quad \tau > 0, \quad t > 0. \tag{6.1.27}$$

Now inserting the estimates (6.1.24) and (6.1.27) into (6.1.23), we infer that

$$\begin{aligned} &\left\| (\mathbb{1} + tM(\tau))^{-1} - \left(f(\tau A)^{-1} + tM(\tau) \right)^{-1} \right\| \le \\ &\left\| \left(f(\tau A)^{-1} + tM(\tau) \right)^{-1} - (\mathbb{1} + tH)^{-1} \right\| + \sqrt{\gamma[g]}\, \sqrt{\frac{\tau}{t}} + \frac{\tau}{t}\,, \end{aligned}$$

for $\tau > 0$, $t > 0$. Therefore, by

$$\begin{aligned} &\|(\mathbb{1} + tM(\tau))^{-1} - (\mathbb{1} + tH)^{-1}\| \le \\ &\|(\mathbb{1} + tM(\tau))^{-1} - \left(f(\tau A)^{-1} + tM(\tau) \right)^{-1}\| + \\ &\| \left(f(\tau A)^{-1} + tM(\tau) \right)^{-1} - (\mathbb{1} + tH)^{-1}\|, \quad \tau > 0, \quad t > 0\,, \end{aligned}$$

we derive the estimate

$$\|(\mathbb{1}+tM(\tau))^{-1}-(\mathbb{1}+tH)^{-1}\| \leq$$
$$2\,\|\left(f(\tau A)^{-1}+tM(\tau)\right)^{-1}-(\mathbb{1}+tH)^{-1}\|+\sqrt{\gamma[g]}\sqrt{\frac{\tau}{t}}+\frac{\tau}{t}, \tag{6.1.28}$$

where $\tau > 0, t > 0$. Then for $\tau := t/\eta$ and $\eta > 0, t \in \mathbb{R}^+$, one gets

$$\begin{aligned}&\|(\mathbb{1}+tM(t/\eta))^{-1}-(\mathbb{1}+tH)^{-1}\|\\ &\leq 2\left\|\left(f(tA/\eta)^{-1}+tM(t/\eta)\right)^{-1}-(\mathbb{1}+tH)^{-1}\right\|+\sqrt{\gamma[g]}\sqrt{\frac{1}{\eta}}+\frac{1}{\eta}.\end{aligned}$$

Therefore, if $\lim_{\eta\to\infty}\sup_{t\in\mathcal{I}}\left\|\left(f(tA/\eta)^{-1}+tM(t/\eta)\right)^{-1}-(\mathbb{1}+tH)^{-1}\right\|=0$ for interval $\mathcal{I}\subset\mathbb{R}^+$ (cf. (6.1.18)), then (6.1.19) follows.

(6.1.18) $\Longleftarrow$ (6.1.19): Similarly to (6.1.22) we obtain the representation

$$\begin{aligned}&(\mathbb{1}+tM(\tau))^{-1}-\left(f(\tau A)^{-1}+tM(\tau)\right)^{-1} \\ &=\left(f(\tau A)^{-1}+tM(\tau)\right)^{-1}(g(\tau B)-\mathbb{1})\left((\mathbb{1}+tM(\tau))^{-1}-(\mathbb{1}+tH)^{-1}\right)\\ &+\left(f(\tau A)^{-1}+tM(\tau)\right)^{-1}(g(\tau B)-\mathbb{1})(\mathbb{1}+tH)^{-1}\\ &+\left(f(\tau A)^{-1}+tM(\tau)\right)^{-1}\frac{\tau M(\tau)}{\mathbb{1}+tM(\tau)},\quad \tau>0,\quad t\in\mathbb{R}^+.\end{aligned} \tag{6.1.29}$$

As a consequence, by (6.1.24)–(6.1.26), we infer the estimate

$$\begin{aligned}&\left\|\left(f(\tau A)^{-1}+tM(\tau)\right)^{-1}-(\mathbb{1}+tM(\tau))^{-1}\right\|\\ &\leq\left\|(\mathbb{1}+tM(\tau))^{-1}-(\mathbb{1}+tH)^{-1}\right\|+\sqrt{\gamma[g]}\sqrt{\frac{\tau}{t}}+\frac{\tau}{t},\quad \tau>0,\quad t>0.\end{aligned}$$

Now taking into account the estimate

$$\begin{aligned}&\|(f(\tau A)^{-1}+tM(\tau))^{-1}-(\mathbb{1}+tH)^{-1}\|\leq\\ &\|\left(f(\tau A)^{-1}+tM(\tau)\right)^{-1}-(\mathbb{1}+tM(\tau))^{-1}\|+\\ &\|(\mathbb{1}+tM(\tau))^{-1}-(\mathbb{1}+tH)^{-1}\|,\quad \eta>0,\quad t>0,\end{aligned}$$

we deduce that

$$\left\|(f(\tau A)^{-1}+tM(\tau))^{-1}-(\mathbb{1}+tH)^{-1}\right\| \leq \tag{6.1.30}$$
$$2\left\|(\mathbb{1}+tM(\tau))^{-1}-(\mathbb{1}+tH)^{-1}\right\|+\sqrt{\gamma[g]}\sqrt{\frac{\tau}{t}}+\frac{\tau}{t}, \quad \eta>0, \quad t>0,$$

As a result, for $\tau := t/\eta$, $t > 0$, $\eta > 0$ we obtain

$$\left\|(f(tA/\eta)^{-1}+tM(t/\eta))^{-1}-(\mathbb{1}+tH)^{-1}\right\| \leq$$
$$2\left\|(\mathbb{1}+tM(t/\eta))^{-1}-(\mathbb{1}+tH)^{-1}\right\|+\sqrt{\gamma[g]}\sqrt{\frac{1}{\eta}}+\frac{1}{\eta}, \quad \eta>0, \quad t>0.$$

That is, if $\lim_{\eta\to\infty}\sup_{t\in\mathcal{I}}\left\|(\mathbb{1}+tM(t/\eta))^{-1}-(\mathbb{1}+tH)^{-1}\right\|=0$ for some interval $\mathcal{I}\subset\mathbb{R}^+$ (cf. (6.1.19)), then one gets (6.1.18).

(6.1.16) $\Longrightarrow$ (6.1.17): To prove (6.1.17) we use (6.1.28) for $t = 1$. Then we get

$$\left\|(\mathbb{1}+M(\tau))^{-1}-(\mathbb{1}+H)^{-1}\right\| \leq$$
$$2\left\|(f(\tau A)^{-1}+M(\tau))^{-1}-(\mathbb{1}+H)^{-1}\right\|+\sqrt{\gamma[g]}\,\sqrt{\tau}+\tau, \quad \tau>0.$$

Therefore, if $\lim_{\tau\to+0}\left\|\left(f(\tau A)^{-1}+M(\tau)\right)^{-1}-(\mathbb{1}+H)^{-1}\right\|=0$, then (6.1.17) follows.

(6.1.16) $\Longleftarrow$ (6.1.17): Conversely, using (6.1.30) and setting $t = 1$, we obtain

$$\left\|(f(\tau A)^{-1}+M(\tau))^{-1}-(\mathbb{1}+H)^{-1}\right\| \leq$$
$$2\left\|(\mathbb{1}+M(\tau))^{-1}-(\mathbb{1}+H)^{-1}\right\|+\sqrt{\gamma[g]}\sqrt{\tau}+\tau, \quad \eta>0,$$

which yields (6.1.16).

(6.1.14) $\Longrightarrow$ (6.1.15): To show (6.1.15), we use (6.1.22) for $t = 1$. As a consequence, we get

$$(\mathbb{1}+M(\tau))^{-1}-\left(f(\tau A)^{-1}+M(\tau)\right)^{-1}=$$
$$=(\mathbb{1}+M(\tau))^{-1}(g(\tau B)-\mathbb{1})\left(\left(f(\tau A)^{-1}+M(\tau)\right)^{-1}-(\mathbb{1}+H)^{-1}\right)$$
$$+(\mathbb{1}+M(\tau))^{-1}(g(\tau B)-\mathbb{1})(\mathbb{1}+H)^{-1}$$
$$+\frac{\tau M(\tau)}{\mathbb{1}+M(\tau)}\left(f(\tau A)^{-1}+M(\tau)\right)^{-1}, \quad \tau>0.$$

Then one checks that $\text{s-lim}_{\tau\to+0}\left(f(\tau A)^{-1}+M(\tau)\right)^{-1}=(\mathbb{1}+H)^{-1}$ yields

$$\text{s-}\lim_{\tau\to+0}\left((\mathbb{1}+M(\tau))^{-1}-\left(f(\tau A)^{-1}+M(\tau)\right)^{-1}\right)=0\,,$$

which proves (6.1.15).

(6.1.14) $\Longleftarrow$ (6.1.15): To prove the converse, we use (6.1.29) for $t=1$. Hence

$$\begin{aligned}
&(\mathbb{1}+M(\tau))^{-1}-\left(f(\tau A)^{-1}+M(\tau)\right)^{-1}=\\
&=\left(f(\tau A)^{-1}+M(\tau)\right)^{-1}(g(\tau B)-\mathbb{1})\left((\mathbb{1}+M(\tau))^{-1}-(\mathbb{1}+H)^{-1}\right)\\
&+\left(f(\tau A)^{-1}+M(\tau)\right)^{-1}(g(\tau B)-\mathbb{1})\,(\mathbb{1}+H)^{-1}\\
&+\left(f(\tau A)^{-1}+M(\tau)\right)^{-1}\frac{\tau M(\tau)}{\mathbb{1}+M(\tau)}\,,\qquad \tau>0\,.
\end{aligned}$$

On that account from $\text{s-lim}_{\tau\to+0}(\mathbb{1}+M(\tau))^{-1}=(\mathbb{1}+H)^{-1}$ (6.1.15), we obtain

$$\text{s-}\lim_{\tau\to+0}\left((\mathbb{1}+M(\tau))^{-1}-\left(f(\tau A)^{-1}+M(\tau)\right)^{-1}\right)=0\,,$$

which readily provides (6.1.14). □

Theorem 6.1.7 *Let A and B be non-negative self-adjoint operators in the separable Hilbert space $\mathfrak{H}$ and let $f(\cdot)$ and $g(\cdot)$ be the Kato functions of the class $\mathcal{K}$ such that $f(x)>0$ for $x\in\mathbb{R}_0^+$. Further, let $H:=A\dotplus B$. Then the following equivalences are valid :*

$$\text{s-}\lim_{\tau\to+0}(\mathbb{1}+S_{\widehat{G}}(\tau))^{-1}=(\mathbb{1}+H)^{-1} \tag{6.1.31}$$

$$\Longleftrightarrow\ \text{s-}\lim_{\tau\to+0}(\mathbb{1}+M(\tau))^{-1}=(\mathbb{1}+H)^{-1}, \tag{6.1.32}$$

$$\lim_{\tau\to+0}\|(\mathbb{1}+S_{\widehat{G}}(\tau))^{-1}-(\mathbb{1}+H)^{-1}\|=0 \tag{6.1.33}$$

$$\Longleftrightarrow\ \lim_{\tau\to+0}\|(\mathbb{1}+M(\tau))^{-1}-(\mathbb{1}+H)^{-1}\|=0, \tag{6.1.34}$$

$$\lim_{\eta\to\infty}\sup_{t\in\mathcal{I}}\|(\mathbb{1}+tS_{\widehat{G}}(t/\eta))^{-1}-(\mathbb{1}+tH)^{-1}\|=0 \tag{6.1.35}$$

$$\Longleftrightarrow\ \lim_{\eta\to\infty}\sup_{t\in\mathcal{I}}\|(\mathbb{1}+tM(t/\eta))^{-1}-(\mathbb{1}+tH)^{-1}\|=0\,, \tag{6.1.36}$$

where $\mathcal{I}$ is a closed interval of $\mathbb{R}^+$.

Proof (6.1.35) $\Longrightarrow$ (6.1.36): First we consider identity

$$
\begin{aligned}
&f(\tau A)^{1/2}(\mathbb{1}+tS_{\widehat{G}}(\tau))^{-1}f(\tau A)^{1/2}-(\mathbb{1}+tH)= \\
&f(\tau A)^{1/2}\left((\mathbb{1}+tS_{\widehat{G}}(\tau))^{-1}-(\mathbb{1}+tH)^{-1}\right)f(\tau A)^{1/2}+ \\
&(f(\tau A)^{1/2}-\mathbb{1})(\mathbb{1}+tH)^{-1}f(\tau A)^{1/2}+(\mathbb{1}+tH)^{-1}(f(\tau A)^{1/2}-\mathbb{1})
\end{aligned}
\tag{6.1.37}
$$

which, for $\tau > 0$, $t \in \mathbb{R}^+$, gives estimate

$$
\begin{aligned}
&\|f(\tau A)^{1/2}(\mathbb{1}+tS_{\widehat{G}}(\tau))^{-1}f(\tau A)^{1/2}-(\mathbb{1}+tH)^{-1}\| \leq \\
&\left\|(\mathbb{1}+tS_{\widehat{G}}(\tau))^{-1}-(\mathbb{1}+tH)^{-1}\right\|+2\left\|(\mathbb{1}-f(\tau A)^{1/2})(\mathbb{1}+tH)^{-1}\right\|.
\end{aligned}
\tag{6.1.38}
$$

Then using equation

$$
\begin{aligned}
&(\mathbb{1}-f(\tau A)^{1/2})\,(\mathbb{1}+tH)^{-1}= \\
&\frac{\mathbb{1}-f(\tau A)^{1/2}}{\sqrt{\tau A}}\sqrt{\tau A}(\mathbb{1}+tH)^{-1/2}(\mathbb{1}+tH)^{-1/2},
\end{aligned}
$$

together with inequality

$$
\left\|\frac{\mathbb{1}-f(\tau A)^{1/2}}{\sqrt{\tau A}}\right\| \leq \sqrt{\gamma[f]}, \quad \tau > 0,
$$

and

$$
\left\|\sqrt{\tau A}\,(\mathbb{1}+tH)^{-1/2}\right\| \leq \sqrt{\frac{\tau}{t}}, \quad \tau > 0, \quad t > 0.
$$

we obtain for $\tau > 0$, $t \in \mathbb{R}^+$ the estimate

$$
\begin{aligned}
&\|(\mathbb{1}-f(\tau A)^{1/2})\,(\mathbb{1}+tH)^{-1}\| \leq \\
&\left\|\frac{\mathbb{1}-f(\tau A)^{1/2}}{\sqrt{\tau A}}\right\|\,\left\|\sqrt{\tau A}(\mathbb{1}+tH)^{-1/2}\right\|\,\left\|(\mathbb{1}+tH)^{-1/2}\right\| \leq \sqrt{\gamma[f]}\sqrt{\frac{\tau}{t}}.
\end{aligned}
\tag{6.1.39}
$$

Inserting (6.1.39) into (6.1.38), one gets

$$
\begin{aligned}
&\|f(\tau A)^{1/2}(\mathbb{1}+tS_{\widehat{G}}(\tau))^{-1}f(\tau A)^{1/2}-(\mathbb{1}+tH)^{-1}\| \leq \\
&\left\|(\mathbb{1}+tS_{\widehat{G}}(\tau))^{-1}-(\mathbb{1}+tH)^{-1}\right\|+2\sqrt{\gamma[f]}\sqrt{\frac{\tau}{t}}, \quad \tau > 0,\ t > 0.
\end{aligned}
$$

As a consequence, seeing that by virtue of (6.1.13)

$$f(\tau A)^{1/2}(\mathbb{1}+tS_{\widehat{G}}(\tau))^{-1}f(\tau A)^{1/2}=(f(\tau A)^{-1}+t\,M(\tau))^{-1}\,,\quad \tau>0\,,\,t>0\,, \tag{6.1.40}$$

we get the estimate

$$\begin{gathered}\|(f(\tau A)^{-1}+t\,M(\tau))^{-1}-(\mathbb{1}+tH)^{-1}\|\le\\ \left\|(\mathbb{1}+tS_{\widehat{G}}(\tau))^{-1}-(\mathbb{1}+tH)^{-1}\right\|+2\sqrt{\gamma[f]}\sqrt{\frac{\tau}{t}}\,,\quad \tau>0,\ t>0\,.\end{gathered} \tag{6.1.41}$$

Then setting in (6.1.41) $\tau:=t/\eta$, for $\eta>0$, we obtain

$$\begin{gathered}\|(f(tA/\eta)^{-1}+tM(t/\eta))^{-1}-(\mathbb{1}+tH)^{-1}\|\le\\ \left\|(\mathbb{1}+tS_{\widehat{G}}(t/\eta))^{-1}-(\mathbb{1}+tH)^{-1}\right\|+2\sqrt{\gamma[f]}\sqrt{\frac{1}{\eta}}\,.\end{gathered} \tag{6.1.42}$$

Now, if $\lim_{\eta\to\infty}\sup_{t\in\mathcal{I}}\left\|(\mathbb{1}+tS_{\widehat{G}}(t/\eta))^{-1}-(\mathbb{1}+tH)^{-1}\right\|=0$ holds for some interval $\mathcal{I}\subset\mathbb{R}^+$ (cf. (6.1.35)), then by (6.1.42) we obtain the limit (6.1.18), which by Lemma 6.1.6 yields (6.1.19), and consequently (6.1.36).

(6.1.35) $\Longleftarrow$ (6.1.36): To prove the converse, we consider

$$\begin{aligned}&\widehat{G}(\tau)\,(\mathbb{1}+t\,S_{\widehat{G}}(\tau))^{-1}\widehat{G}(\tau)-(\mathbb{1}+tH)^{-1}=\\ &f(\tau A)^{1/2}g(\tau B)\left(f(\tau A)^{1/2}(\mathbb{1}+tS_{\widehat{G}}(\tau))^{-1}f(\tau A)^{1/2}-(\mathbb{1}+tH)^{-1}\right)\\ &\times g(\tau B)f(\tau A)^{1/2}+f(\tau A)^{1/2}(g(\tau B)-\mathbb{1})(\mathbb{1}+tH)^{-1}g(\tau B)f(\tau A)^{1/2}+\\ &(f(\tau A)^{1/2}-\mathbb{1})(\mathbb{1}+tH)^{-1}g(\tau B)f(\tau A)^{1/2}+\\ &(\mathbb{1}+tH)^{-1}(g(\tau B)-\mathbb{1})f(\tau A)^{1/2}+(\mathbb{1}+tH)^{-1}(f(\tau A)^{1/2}-\mathbb{1})\,,\end{aligned} \tag{6.1.43}$$

for $\tau>0$ and $t\in\mathbb{R}^+$. As a result, we infer estimate

$$\begin{aligned}&\left\|\widehat{G}(\tau)\,(\mathbb{1}+t\,S_{\widehat{G}}(\tau))^{-1}\widehat{G}(\tau)-(\mathbb{1}+tH)^{-1}\right\|\le\\ &\left\|f(\tau A)^{1/2}\,(\mathbb{1}+tS_{\widehat{G}}(\tau))^{-1}f(\tau A)^{1/2}-(\mathbb{1}+tH)^{-1}\right\|+\\ &2\left\|(g(\tau B)-\mathbb{1})(\mathbb{1}+t\,H)^{-1}\right\|+2\left\|(f(\tau A)^{1/2}-\mathbb{1})\,(\mathbb{1}+t\,H)^{-1}\right\|\,.\end{aligned}$$

Then inequalities (6.1.24) and (6.1.39) and Eq. (6.1.40) yield for $\tau > 0$ and $t \in \mathbb{R}^+$ the estimate

$$\begin{aligned}
&\left\|\widehat{G}(\tau)(\mathbb{1} + tS_{\widehat{G}}(\tau))^{-1}\widehat{G}(\tau) - (\mathbb{1} + tH)^{-1}\right\| \leq \\
&\left\|(f(\tau A)^{-1} + t\,M(\tau))^{-1} - (\mathbb{1} + t\,H)^{-1}\right\| + 2\,(\sqrt{\gamma[f]} + \sqrt{\gamma[g]})\sqrt{\frac{\tau}{t}}\,.
\end{aligned} \tag{6.1.44}$$

Next, we have identify

$$\begin{aligned}
&(\mathbb{1} + tS_{\widehat{G}}(\tau))^{-1} - \widehat{G}(\tau)(\mathbb{1} + t\,S_{\widehat{G}}(\tau))^{-1}\widehat{G}(\tau) = \\
&\tau S_{\widehat{G}}(\tau)(\mathbb{1} + tS_{\widehat{G}}(\tau))^{-1}\widehat{G}(\tau) + \tau S_{\widehat{G}}(\tau)(\mathbb{1} + tS_{\widehat{G}}(\tau))^{-1}, \quad \tau > 0, \quad t > 0,
\end{aligned}$$

which leads to estimate

$$\left\|(\mathbb{1}+tS_{\widehat{G}}(\tau))^{-1} - \widehat{G}(\tau)\,(\mathbb{1}+t\,S_{\widehat{G}}(\tau))^{-1}\widehat{G}(\tau)\right\| \leq 2\,\frac{\tau}{t}\,\left\|t\,S_{\widehat{G}}(\tau)\,(\mathbb{1} + t\,S_{\widehat{G}}(\tau))^{-1}\right\|\,.$$

Since $\left\|t\,S_{\widehat{G}}(\tau)\,(\mathbb{1} + tS_{\widehat{G}}(\tau))^{-1}\right\| \leq 1$, for $\tau > 0$ and $t \in \mathbb{R}_0^+$, one gets

$$\left\|(\mathbb{1} + tS_{\widehat{G}}(\tau))^{-1} - \widehat{G}(\tau)(\mathbb{1} + tS_{\widehat{G}}(\tau))^{-1}\widehat{G}(\tau)\right\| \leq 2\,\frac{\tau}{t}\,, \quad \tau > 0, \quad t > 0. \tag{6.1.45}$$

Now we take the estimate

$$\begin{aligned}
&\left\|(\mathbb{1} + tS_{\widehat{G}}(\tau))^{-1} - (\mathbb{1} + tH)^{-1}\right\| \leq \\
&\left\|(\mathbb{1} + tS_{\widehat{G}}(\tau))^{-1} - \widehat{G}(\tau)(\mathbb{1} + tS_{\widehat{G}}(\tau))^{-1}\widehat{G}(\tau)\right\| + \\
&\left\|\widehat{G}(\tau)(\mathbb{1} + tS_{\widehat{G}}(\tau))^{-1}\widehat{G}(\tau) - (\mathbb{1} + tH)^{-1}\right\|, \quad \tau > 0, \quad t > 0\,.
\end{aligned}$$

Inserting into this estimate bounds (6.1.44) and (6.1.45), we obtain

$$\begin{aligned}
&\left\|(\mathbb{1} + tS_{\widehat{G}}(\tau))^{-1} - (\mathbb{1} + tH)^{-1}\right\| \leq \\
&\left\|(f(\tau A)^{-1} + tM(\tau))^{-1} - (\mathbb{1} + tH)^{-1}\right\| + 2\,(\sqrt{\gamma[f]} + \sqrt{\gamma[g]})\sqrt{\frac{\tau}{t}} + \frac{2\tau}{t}\,,
\end{aligned} \tag{6.1.46}$$

for $\tau > 0$ and $t > 0$. Now setting $\tau := t/\eta$ for $\eta > 0$ and $t \in \mathbb{R}^+$, we deduce that

$$\begin{aligned}
&\left\|(\mathbb{1} + tS_{\widehat{G}}(t/\eta))^{-1} - (\mathbb{1} + tH)^{-1}\right\| \leq \\
&\left\|(f(tA/\eta)^{-1} + tM(t/\eta))^{-1} - (\mathbb{1} + tH)^{-1}\right\| + 2\,(\sqrt{\gamma[f]} + \sqrt{\gamma[g]})\sqrt{\frac{1}{\eta}} + \frac{2}{\eta}\,,
\end{aligned} \tag{6.1.47}$$

Consequently, let $\lim_{\eta\to\infty}\sup_{t\in\mathcal{I}}\|(\mathbb{1}+t\,M(t/\eta))^{-1}-(\mathbb{1}+tH)^{-1}\|=0$ in (6.1.36) for some interval $\mathcal{I}\subseteq\mathbb{R}^+$. Seeing that (6.1.36) coincides with (6.1.19), by Lemma 6.1.6 one gets the limit (6.1.18), which on account of (6.1.47) yields in turn the limit (6.1.35). That is, (6.1.35) follows from (6.1.36).

(6.1.33) $\Longrightarrow$ (6.1.34): Setting $t=1$ we get from (6.1.41) that

$$\begin{aligned}&\|(f(\tau A)^{-1}+M(\tau))^{-1}-(\mathbb{1}+H)^{-1}\|\le\\&\qquad\left\|(\mathbb{1}+S_{\widehat{G}}(\tau))^{-1}-(\mathbb{1}+H)^{-1}\right\|+2\sqrt{\gamma[f]}\sqrt{\tau}\,,\quad \tau>0.\end{aligned}\tag{6.1.48}$$

If $\lim_{\tau\to+0}\|(\mathbb{1}+S_{\widehat{G}}(\tau))^{-1}-(\mathbb{1}+H)^{-1}\|=0$ (6.1.33), then by virtue of (6.1.48) the limit (6.1.16) in Lemma 6.1.6 is valid. Hence, due to equivalence of (6.1.16) and (6.1.17), we obtain the proof of (6.1.34) since it is identical to expression (6.1.17).

(6.1.33) $\Longleftarrow$ (6.1.34): To prove the converse, we note that from (6.1.46) for $t=1$ one gets for $\tau>0$:

$$\begin{aligned}&\left\|(\mathbb{1}+S_{\widehat{G}}(\tau))^{-1}-(\mathbb{1}+H)^{-1}\right\|\le\\&\left\|(f(\tau A)^{-1}+M(\tau))^{-1}-(\mathbb{1}+H)^{-1}\right\|+2(\sqrt{\gamma[f]}+\sqrt{\gamma[g]})\sqrt{\tau}+2\tau\,.\end{aligned}\tag{6.1.49}$$

If $\lim_{\tau\to+0}\|(\mathbb{1}+M(\tau))^{-1}-(\mathbb{1}+H)^{-1}\|=0$ (cf. (6.1.34)), then by virtue of Lemma 6.1.6 the limit (6.1.16) is valid. So, estimate (6.1.49) yields the limit (6.1.33) and thus proves the assertion.

(6.1.31) $\Longrightarrow$ (6.1.32): We consider (6.1.37) for $t=1$ that gives

$$\begin{aligned}&f(\tau A)^{1/2}(\mathbb{1}+S_{\widehat{G}}(\tau))^{-1}f(\tau A)^{1/2}-(\mathbb{1}+H)^{-1}=\\&f(\tau A)^{1/2}\left((\mathbb{1}+S_{\widehat{G}}(\tau))^{-1}-(\mathbb{1}+H)^{-1}\right)f(\tau A)^{1/2}+\\&(f(\tau A)^{1/2}-\mathbb{1})(\mathbb{1}+H)^{-1}f(\tau A)^{1/2}+(\mathbb{1}+H)^{-1}(f(\tau A)^{1/2}-\mathbb{1})\,.\end{aligned}$$

Then using (6.1.40) we obtain equation

$$\begin{aligned}&(f(\tau A)^{-1}+M(\tau))^{-1}-(\mathbb{1}+H)^{-1}=\\&f(\tau A)^{1/2}\left((\mathbb{1}+S_{\widehat{G}}(\tau))^{-1}-(\mathbb{1}+H)^{-1}\right)f(\tau A)^{1/2}+\\&(f(\tau A)^{1/2}-\mathbb{1})(\mathbb{1}+H)^{-1}f(\tau A)^{1/2}+(\mathbb{1}+H)^{-1}(f(\tau A)^{1/2}-\mathbb{1})\,.\end{aligned}\tag{6.1.50}$$

Now, if $\text{s}-\lim_{\tau\to+0}(\mathbb{1}+S_{\widehat{G}}(\tau))^{-1}=(\mathbb{1}+H)^{-1}=0$ (cf. (6.1.31)), then by (6.1.39) (for $t=1$) and (6.1.50), the strong limit (6.1.14) holds. By Lemma 6.1.6 this implies the limit (6.1.15) and proves the assertion since (6.1.15) coincides with the strong limit in (6.1.32).

(6.1.31) $\Longleftarrow$ (6.1.32): To prove the converse we set $t = 1$ to obtain from (6.1.43)

$$\widehat{G}(\tau)(\mathbb{1} + S_{\widehat{G}}(\tau))^{-1}\widehat{G}(\tau) - (\mathbb{1} + H)^{-1} = \tag{6.1.51}$$

$$f(\tau A)^{1/2}g(\tau B)\left(f(\tau A)^{1/2}(\mathbb{1} + S_{\widehat{G}}(\tau))^{-1}f(\tau A)^{1/2} - (\mathbb{1} + H)^{-1}\right)$$

$$\times\, g(\tau B)f(\tau A)^{1/2} + f(\tau A)^{1/2}(g(\tau B) - \mathbb{1})(\mathbb{1} + H)^{-1}g(\tau B)f(\tau A)^{1/2} +$$

$$(f(\tau A)^{1/2} - \mathbb{1})(\mathbb{1} + H)^{-1}g(\tau B)f(\tau A)^{1/2} +$$

$$(\mathbb{1} + H)^{-1}(g(\tau B) - \mathbb{1})f(\tau A)^{1/2} + (\mathbb{1} + H)^{-1}(f(\tau A)^{1/2} - \mathbb{1}), \quad \tau > 0.$$

Then on account of (6.1.40), we deduce formula

$$\widehat{G}(\tau)(\mathbb{1} + S_{\widehat{G}}(\tau))^{-1}\widehat{G}(\tau) - (\mathbb{1} + H)^{-1} = \tag{6.1.52}$$

$$f(\tau A)^{1/2}g(\tau B)\left((f(\tau A)^{-1} + M(\tau))^{-1} - (\mathbb{1} + H)^{-1}\right)g(\tau B)f(\tau A)^{1/2} +$$

$$f(\tau A)^{1/2}(g(\tau B) - \mathbb{1})(\mathbb{1} + H)^{-1}g(\tau B)f(\tau A)^{1/2} +$$

$$(f(\tau A)^{1/2} - \mathbb{1})(\mathbb{1} + H)^{-1}g(\tau B)f(\tau A)^{1/2} +$$

$$(\mathbb{1} + H)^{-1}(g(\tau B) - \mathbb{1})f(\tau A)^{1/2} + (\mathbb{1} + H)^{-1}(f(\tau A)^{1/2} - \mathbb{1}), \quad \tau > 0.$$

Therefore, if $\text{s}-\lim_{\tau\to+0}(\mathbb{1} + M(\tau))^{-1} = (\mathbb{1} + H)^{-1}$ (cf. (6.1.32)), then by Lemma 6.1.6 the strong limit (6.1.14) holds. Using formula (6.1.52) for the limit $\tau \to +0$ in (6.1.14), as well as for estimates (6.1.24) and (6.1.39), we obtain

$$\text{s}-\lim_{\tau\to+0}\widehat{G}(\tau)(\mathbb{1} + S_{\widehat{G}}(\tau))^{-1}\widehat{G}(\tau) = (\mathbb{1} + H)^{-1}. \tag{6.1.53}$$

Owing to (6.1.53), the strong limit of the right-hand side of (6.1.51) for $\tau \to +0$ is zero. Then because of (6.1.24) and (6.1.39) and by the strong continuity of $f(\tau A)$, $g(\tau B)$ at $\tau = +0$, we obtain from the limit of the right-hand side of (6.1.51) that $\text{s}-\lim_{\tau\to+0}(\mathbb{1} + S_{\widehat{G}}(\tau))^{-1} = (\mathbb{1} + H)^{-1}$. That is, (6.1.32) implies (6.1.31). □

Corollary 6.1.8 *Let A and B be non-negative self-adjoint operators in the separable Hilbert space $\mathfrak{H}$ and let $f(\cdot)$ and $g(\cdot)$ be Kato functions of the class $\mathcal{K}$ such that $f(x) > 0$ for $x \in \mathbb{R}_0^+$. Further, let $H := A \dotplus B$. Then one has*

$$\text{s-}\lim_{\eta\to\infty}\left(f(tA/\eta)^{1/2}g(tB/\eta)f(tA/\eta)^{1/2}\right)^{\eta} = \mathrm{e}^{-tH} \tag{6.1.54}$$

uniformly in $t \in \mathcal{I}$ for any closed interval $\mathcal{I} \subset \mathbb{R}^+$

$$\Longleftrightarrow \text{s-}\lim_{\tau\to+0}(\mathbb{1} + M(\tau))^{-1} = (\mathbb{1} + H)^{-1}, \tag{6.1.55}$$

$$\lim_{\eta\to\infty}\sup_{t\in\mathcal{I}}\left\|\left(f(tA/\eta)^{1/2}g(tB/\eta)f(tA/\eta)^{1/2}\right)^{\eta} - \mathrm{e}^{-tH}\right\| = 0 \tag{6.1.56}$$

for any closed interval $\mathcal{I} \subset \mathbb{R}^+$

$$\Longleftrightarrow \lim_{\tau \to +0} \|(\mathbb{1} + M(\tau))^{-1} - (\mathbb{1} + H)^{-1}\| = 0\,, \tag{6.1.57}$$

$$\lim_{\eta \to \infty} \sup_{t \in \mathcal{I}} \left\| \left(f(tA/\eta)^{1/2} g(tB/\eta) f(tA/\eta)^{1/2} \right)^{\eta} - \mathrm{e}^{-tH} \right\| = 0 \tag{6.1.58}$$

for any closed interval $\mathcal{I} \subset \mathbb{R}_0^+$

$$\Longleftrightarrow \lim_{\eta \to \infty} \sup_{t \in \mathcal{I}} \|(\mathbb{1} + tM(t/\eta))^{-1} - (\mathbb{1} + tH)^{-1}\| = 0 \tag{6.1.59}$$

for any closed interval $\mathcal{I} \subset \mathbb{R}_0^+$.

Proof Recall that $\widehat{G}(\tau) = f(\tau A)^{1/2} g(\tau B) f(\tau A)^{1/2}$, $\tau \in \mathbb{R}_0^+$; see (6.1.1).

(6.1.54) $\Longleftrightarrow$ (6.1.55): Theorem 4.1.3 yields that (6.1.54) is equivalent to (6.1.31). Using Theorem 6.1.7, we obtain the equivalence of (6.1.31) to (6.1.55).

(6.1.56) $\Longleftrightarrow$ (6.1.57): By Theorem 4.2.3, we get that (6.1.56) is equivalent to (4.2.7). Then by Theorem 6.1.7 (6.1.34)-(6.1.34) we infer the equivalence of (4.2.7) to (6.1.57).

(6.1.58) $\Longleftrightarrow$ (6.1.59): Theorem 4.2.7 establishes the equivalence of the limit (6.1.58) and (4.2.23). Applying Theorem 6.1.7 (6.1.35)-(6.1.36), we infer the equivalence of (4.2.23) and (6.1.59). □

6.2 Resolvent Compactness

The aim of this section is to show that under a supplementary condition of the *resolvent compactness* on at least one of generators of C_0-semigroups involved into the Trotter-Kato approximants, we can prove their operator-norm convergence. This improvement of the strong operator convergence holds uniformly on closed intervals in $\mathbb{R}_0^+$ *away* from zero.

For this purpose we first need the following technical lemma.

Lemma 6.2.1 *Let* $\{Y(\tau)\}_{\tau>0}$ *be a family of bounded non-negative self-adjoint operators on* $\mathfrak{H}$ *and let* $X \in \mathcal{C}_\infty(\mathfrak{H})$ *be a non-negative compact operator on* $\mathfrak{H}$ *such that*

$$0 \le Y(\tau) \le X, \qquad \tau > 0. \tag{6.2.1}$$

If we suppose that

$$\text{w-}\lim_{\tau \to +0} Y(\tau) = Y\,, \tag{6.2.2}$$

then $Y \in \mathcal{C}_\infty(\mathfrak{H})$ *and moreover one gets*

$$\lim_{\tau \to +0} \|Y(\tau) - Y\| = 0\,, \tag{6.2.3}$$

that is, the weak operator convergence (6.2.2) *can be uplifted to the operator-norm convergence* (6.2.3).

Proof Note that the square-roots of self-adjoint operators $Y(\tau)$, X are well-defined. Then in virtue of the *Heinz inequality* (9.4.65), we obtain

$$0 \le \sqrt{Y(\tau)} \le \sqrt{X}\,, \qquad \tau > 0\,, \tag{6.2.4}$$

see Notes to Chap. 9 (Notes to Sects. 9.4 and 9.4.4). On account of the *canonical representation* of compact operators (Proposition 6.6.1), one gets that $\sqrt{X} \in \mathcal{C}_\infty(\mathfrak{H})$, with singular values $\{s_n(\sqrt{X}) = \sqrt{s_n(X)}\}_{n\ge1}$. Let vectors $\{u_n\}_{n\ge1} \subset \mathfrak{H}$ be such that $\text{w}-\lim_{n\to\infty} u_n = 0$. Then by compactness of $\sqrt{X}$ and due to (6.2.1), one gets

$$0 \le \lim_{n\to\infty} \|\sqrt{Y(\tau)}\, u_n\|^2 \le \lim_{n\to\infty} \|\sqrt{X} u_n\|^2 = 0\,, \qquad \tau > 0\,. \tag{6.2.5}$$

That is, as a consequence of (6.2.5), the operator $\sqrt{Y(\tau)} \in \mathcal{C}_\infty(\mathfrak{H})$, for $\tau > 0$, with singular values $\{s_n(\sqrt{Y(\tau)}) = \sqrt{s_n(Y(\tau))}\}_{n\ge1}$. Next, by the *Weyl-Courant minimax principle* determining the singular values of compact operators and by (6.2.4), we infer that

$$s_n(\sqrt{Y(\tau)}) \le s_n(\sqrt{X})\,, \qquad n \ge 1\,. \tag{6.2.6}$$

Then owing to (6.2.6) and to the *canonical* representation (6.6.1) of compact operators $\{\sqrt{Y(\tau)}\,\}_{\tau>0}$ and $\sqrt{X}$, there exists on $\mathfrak{H}$ a family of *contractions* $\{\Gamma(\tau)\}_{t>0}$, such that $\Gamma(\tau) : \overline{\text{ran}\,(\sqrt{X})} \longrightarrow \mathfrak{H}$, and

$$\sqrt{Y(\tau)} = \Gamma(\tau)\sqrt{X}, \qquad \tau > 0\,. \tag{6.2.7}$$

Furthermore, by (6.2.1) and (6.2.2), one gets $0 \le Y \le X$ which by the *Heinz inequality* (9.4.65) yields

$$0 \le \sqrt{Y} \le \sqrt{X}\,. \tag{6.2.8}$$

Then similarly to (6.2.5), we infer that operator $\sqrt{Y}$ and consequently Y are compact. Hence, the weak limit (6.2.2) of $\{Y(\tau) \in \mathcal{C}_\infty(\mathfrak{H})\,\}_{\tau>0}$ and (6.2.1) prove that $Y \in \mathcal{C}_\infty(\mathfrak{H})$.

Moreover, on account of (6.2.8) and *verbatim* the arguments, which have provided representation (6.2.7), we deduce that there exists a *contraction* Γ : $\overline{\mathrm{ran}\,(\sqrt{X})} \longrightarrow \mathfrak{H}$, such that

$$\sqrt{Y} = \Gamma \sqrt{X}\,. \tag{6.2.9}$$

To proceed with the proof of (6.2.3), we use (6.2.7) and (6.2.9) to introduce operators

$$Z(\tau) = \Gamma^*(\tau)\Gamma(\tau) : \overline{\mathrm{ran}\,(\sqrt{X})} \to \overline{\mathrm{ran}\,(\sqrt{X})}\,,$$
$$Z = \Gamma^*\Gamma : \overline{\mathrm{ran}\,(\sqrt{X})} \to \overline{\mathrm{ran}\,(\sqrt{X})}\,.$$

Since by (6.2.7) and (6.2.9) we obtain $Y(\tau) = \sqrt{X}\Gamma^*(\tau)\,\Gamma(\tau)\sqrt{X}$ and $Y = \sqrt{X}\Gamma^*\,\Gamma\sqrt{X}$, the weak operator limit (6.2.2) yields for any $u, v \in \mathfrak{H}$:

$$\lim_{\tau\to+0}((Y(\tau) - Y)u, v) = \lim_{\tau\to+0}((Z(\tau) - Z)\sqrt{X}u, \sqrt{X}v) = 0\,. \tag{6.2.10}$$

Seeing that by canonical form of $\sqrt{X} \in \mathcal{C}_\infty(\mathfrak{H})$ one gets $\overline{\mathrm{ran}\,(\sqrt{X})} = \mathfrak{H}$, the limit (6.2.10) bears

$$\text{w-}\lim_{\tau\to+0}(Z(\tau) - Z) = 0. \tag{6.2.11}$$

On account of representation

$$Y(\tau) - Y = \sqrt{X}\,(Z(\tau) - Z)\sqrt{X}\,, \qquad \tau > 0\,,$$

the proof of (6.2.3) reduces to demonstration of the operator-norm limit

$$\lim_{\tau\to+0}\|\sqrt{X}\,(Z(\tau) - Z)\sqrt{X}\,\| = 0\,, \qquad \tau > 0\,.$$

The latter follows from (6.2.11) thanks to suitable finite-rank approximants for the compact operator $\sqrt{X}$, since finite-rank operators are dense in the set of compact operators $\mathcal{C}_\infty(\mathfrak{H})$ in the operator-norm topology. □

Now we introduce more details related to description of generic Kato function.

Definition 6.2.2 Let $f \in \mathcal{K}$ be a (generic) Kato function of class $\mathcal{K}$ (Definition 5.2.1). We define

$$f_0(x) := \begin{cases} (1 + x\,\varphi_0(x))^{-1}\,, & x > 0, \\ 1\,, & x = 0, \end{cases} \tag{6.2.12}$$

where we set

$$\varphi_0(x) := \inf_{0<s\leq x} \frac{1-f(s)}{s\, f(s)}, \quad x > 0. \tag{6.2.13}$$

Here we have agreed to set $1/f(x) := \infty$, if $f(x) = 0$.

Taking as example the exponential Kato function: $f(x) = \mathrm{e}^{-x}$, $x \geq 0$, one obtains that $\varphi_0(x) = 1$ and $f_0(x) = (1+x)^{-1}$.

Lemma 6.2.3 *Let $f \in \mathcal{K}$ be a Kato function. Then f_0 is also a Kato function of class $\mathcal{K}$ such that $f(x) \leq f_0(x)$ and*

$$\frac{1}{1+x} \leq f_0(x), \quad x \in \mathbb{R}_0^+. \tag{6.2.14}$$

Proof Since $\varphi_0(x) \geq 0$, $x > 0$, one gets $0 \leq f_0(x) \leq 1$, $x \in \mathbb{R}_0^+$. Notice that

$$\frac{1-f_0(x)}{x} = \frac{\varphi_0(x)}{1+x\,\varphi_0(x)}, \quad x > 0.$$

Owing to $\lim_{x\to+0} \varphi_0(x) = 1$, we get $\lim_{x\to+0}(1-f_0(x))/x = 1$, which shows that $f_0 \in \mathcal{K}$; see Definition 5.2.1.

On account of $\varphi_0(x) \leq (1-f(x))/(x\,f(x))$, $x \in \mathbb{R}^+$, we find that $f(x) \leq f_0(x)$, $x \in \mathbb{R}_0^+$. By virtue of $\varphi_0(x) \leq 1$, $x \geq 0$, we deduce that

$$\frac{1}{1+x\,\varphi_0(x)} \geq \frac{1}{1+x}, \quad x \in \mathbb{R}_0^+, \tag{6.2.15}$$

which proves (6.2.14). □

Theorem 6.2.4 *Let A and B be non-negative self-adjoint operators in $\mathfrak{H}$ such that $\mathrm{dom}\,(A^{1/2}) \cap \mathrm{dom}\,(B^{1/2})$ is dense in $\mathfrak{H}$ and let $H := A \dotplus B$. Further, let $f(\cdot)$ and $g(\cdot)$ be Kato functions of class $\mathcal{K}$ such that $f(x) > 0$ for each $x \in \mathbb{R}_0^+$. If the condition $f_0(\tau_0 A) \in \mathcal{C}_\infty(\mathfrak{H})$ holds for some $\tau_0 > 0$, then the Trotter-Kato formula converges (cf. Remark 6.1.5) in the operator-norm topology uniformly on any closed interval $\mathcal{I} \subset \mathbb{R}_0^+$ away from zero and, moreover, $(\mathbb{1}+H)^{-1} \in \mathcal{C}_\infty(\mathfrak{H})$.*

Proof By Theorems 4.1.3 and 5.2.5, we obtain $\mathrm{s}-\lim_{\tau\to+0}(\mathbb{1}+S_{\widehat{G}}(\tau))^{-1} = (\mathbb{1}+H)^{-1}$. Applying Theorem 6.1.7 we get $\mathrm{s}-\lim_{\tau\to+0} Y(\tau) = Y$, where

$$Y(\tau) := (\mathbb{1}+M(\tau))^{-1}, \quad \tau > 0, \quad \text{and} \quad Y := (\mathbb{1}+H)^{-1}.$$

By definitions (6.2.12), one infers

$$\varphi_0(x) \geq \varphi_0(y), \qquad 0 < x \leq y, \tag{6.2.16}$$

and also

$$0 \le f(x) \le f_0(x) \le 1, \qquad x > 0. \tag{6.2.17}$$

Therefore, we get

$$f(\tau A) \le f_0(\tau A), \qquad \tau \ge 0.$$

Now, since $f_0(x) > 0$, for $x \ge 0$, we find

$$f_0(\tau A)^{-1} \le f(\tau A)^{-1}, \qquad \tau > 0,$$

which yields

$$\frac{1}{\tau}\left[f_0(\tau A)^{-1} - \mathbb{1}\right] \le \frac{1}{\tau}\left[f(\tau A)^{-1} - \mathbb{1}\right], \qquad \tau > 0.$$

By (6.1.13) we finally infer

$$\frac{1}{\tau}\left[f_0(\tau A)^{-1} - \mathbb{1}\right] \le M(\tau), \qquad \tau > 0,$$

or equivalently, cf. (6.2.12),

$$A\varphi_0(\tau A) \le M(\tau), \qquad \tau > 0.$$

As a result one gets inequality

$$(\mathbb{1} + M(\tau))^{-1} \le (\mathbb{1} + A\varphi_0(\tau A))^{-1}, \quad \tau > 0.$$

Since $\varphi_0(x)$ is a non-increasing function (6.2.16), we have

$$\varphi_0(\tau A) \ge \varphi_0(\tau_0 A), \qquad 0 < \tau \le \tau_0,$$

which yields

$$(\mathbb{1} + A\varphi_0(\tau A))^{-1} \le (\mathbb{1} + A\varphi_0(\tau_0 A))^{-1}, \qquad 0 < \tau \le \tau_0.$$

As a consequence

$$Y(\tau) = (\mathbb{1} + M(\tau))^{-1} \le (\mathbb{1} + A\varphi_0(\tau_0 A))^{-1}, \qquad 0 < \tau \le \tau_0\,.$$

On the other hand, by definition (6.2.12), we get from $f_0(\tau_0 A) \in \mathcal{C}_\infty(\mathfrak{H})$ that

$$(\mathbb{1} + \tau_0 A\varphi(\tau_0 A))^{-1} \in \mathcal{C}_\infty(\mathfrak{H}).$$

Then by virtue of identity

$$(\mathbb{1} + A\varphi_0(\tau_0 A))^{-1} = (\mathbb{1} + \tau_0 A\varphi_0(\tau_0 A))^{-1} +$$
$$(\tau_0 - 1)A\varphi_0(\tau_0 A)(\mathbb{1} + A\varphi_0(\tau_0 A))^{-1}(\mathbb{1} + \tau_0 A\varphi_0(\tau_0 A))^{-1}$$

we obtain $(\mathbb{1} + A\varphi_0(\tau_0 A))^{-1} \in \mathcal{C}_\infty(\mathfrak{H})$ which allows to identify $X := (\mathbb{1} + A\varphi_0(\tau_0 A))^{-1} \in \mathcal{C}_\infty(\mathfrak{H})$ in Lemma 6.2.1.

Now, applying Lemma 6.2.1, we obtain $\lim_{\tau\to+0} \|Y(\tau) - Y\| = 0$, which shows that

$$\lim_{\tau\to+0} \|(\mathbb{1} + M(\tau))^{-1} - (\mathbb{1} + H)^{-1}\| = 0\,, \tag{6.2.18}$$

and, moreover, that $Y = (\mathbb{1} + H)^{-1} \in \mathcal{C}_\infty(\mathfrak{H})$. Finally, applying to (6.2.18) Corollary 6.1.8, (6.1.56)–(6.1.57), we conclude that the Trotter-Kato product formula converges in the operator-norm topology uniformly on any closed interval $\mathcal{I} \subset \mathbb{R}_0^+$ away from zero. □

Remark 6.2.5 We note that assumption $f_0(\tau_0 A) \in \mathcal{C}_\infty(\mathfrak{H})$ yields $(\mathbb{1} + A)^{-1} \in \mathcal{C}_\infty(\mathfrak{H})$. Indeed, from (6.2.14) we get $(\mathbb{1} + \tau_0 A)^{-1} \in \mathcal{C}_\infty(\mathfrak{H})$, $\tau_0 > 0$, which yields $(\mathbb{1} + A)^{-1} \in \mathcal{C}_\infty(\mathfrak{H})$. The converse is in general *not* true. To elucidate this subtle point, we suggest Example 6.6.3 in Notes to Chap. 6 (Notes to Sect. 6.2(3)).

Definition 6.2.6 Let $f \in \mathcal{K}$ be a (generic) Kato function of class $\mathcal{K}$. The Kato function $f \in \mathcal{K}_r$ is called *regular* ($\mathcal{K}_r \subset \mathcal{K}$) if its transformation f_0 defined by (6.2.12) has the property

$$(\mathbb{1} + A)^{-1} \in \mathcal{C}_\infty(\mathfrak{H}) \Longleftrightarrow f_0(A) \in \mathcal{C}_\infty(\mathfrak{H})\,, \tag{6.2.19}$$

for any non-negative self-adjoint operator A with compact resolvent.

Remark that $f(x) = e^{-x}$ and $f(x) = 1/(1 + x)$, $x \in \mathbb{R}_0^+$, are *regular* Kato functions, whereas Kato function $f(x) = 1 - |\sin(x)|$, $x \in \mathbb{R}_0^+$, is *not* regular.

Then on account of Theorem 6.2.4, we surmise that the Trotter-Kato product formula converges in the operator norm uniformly on any bounded interval away from zero for *any regular* Kato function $f(\cdot)$, *any generic* Kato function $g(\cdot)$ and *any* non-negative self-adjoint operator B if the resolvent of operator A is *compact*.

It turns out that these assumptions are not only *sufficient* but also *necessary*. To prove this assertion, we need the following lemma.

Lemma 6.2.7 *Let $f(x) = (1 + x)^{-1}$, $x \in \mathbb{R}_0^+$, and let $g(\cdot)$ be a generic Kato function. Further, let D be a bounded positive self-adjoint operator such that $\|D\| < 1$. If the Trotter-Kato product formula converges in the operator norm topology uniformly for any bounded interval away from zero for a couple $\{A, B\}$ of non-negative self-adjoint operators A and B, then the same holds for the couple $\{A + D, B\}$.*

Proof We are going to verify the condition (6.1.57) of Corollary 6.1.8. Since the Trotter-Kato product formula converges in the operator-norm topology for self-adjoint operators A and B, by (6.1.57), one has

$$\lim_{\tau \to +0} \|(\mathbb{1} + M(\tau))^{-1} - (\mathbb{1} + H)^{-1}\| = 0, \tag{6.2.20}$$

where, instead of (6.1.13), we introduce corresponding to $f(x) = (1+x)^{-1}$ formula

$$M(\tau) := A + \frac{1}{\tau}[\mathbb{1} - g(\tau B)], \qquad \tau > 0. \tag{6.2.21}$$

The lemma will be proven if we show that (6.2.20) implies

$$\lim_{\tau \to +0} \|(\mathbb{1} + M(\tau) + D)^{-1} - (\mathbb{1} + H + D)^{-1}\| = 0. \tag{6.2.22}$$

Since $\|D\| < 1$, one has the representations

$$(\mathbb{1} + M(\tau) + D)^{-1} = (\mathbb{1} + M(\tau))^{-1} \sum_{k=0}^{\infty} (-1)^k (D\,(\mathbb{1} + M(\tau))^{-1})^k, \tag{6.2.23}$$

and

$$(\mathbb{1} + H + D)^{-1} = (\mathbb{1} + H)^{-1} \sum_{k=0}^{\infty} (-1)^k (D\,(\mathbb{1} + H)^{-1})^k.$$

Since $\|D\| < 1$, for any $\epsilon > 0$, there is a k_0 such that for $k \geq k_0$, one has

$$\|(\mathbb{1} + M(\tau))^{-1} \sum_{k=k_0}^{\infty} (-1)^k (D\,(\mathbb{1} + M(\tau))^{-1})^k\| \leq \epsilon,$$

and

$$\|(\mathbb{1} + H)^{-1} \sum_{k=k_0}^{\infty} (-1)^k (D_0\,(\mathbb{1} + H)^{-1})^k\| \leq \epsilon,$$

uniformly in $\tau > 0$. Moreover, by (6.2.20) there is $\delta > 0$ such that one has

$$\|(\mathbb{1}+M(t))^{-1}(D(\mathbb{1}+M(\tau))^{-1})^k - (\mathbb{1}+H)^{-1}(D(\mathbb{1}+H)^{-1})^k\| \leq \frac{\epsilon}{k_0}, \tag{6.2.24}$$

for $k = 0, 1, \ldots, k_0 - 1$ and $\tau \in (0, \delta]$. Summing up (6.2.23) - (6.2.24), we get

$$\|(\mathbb{1} + M(\tau) + D)^{-1} - (\mathbb{1} + H + D)^{-1}\| \leq 3\epsilon,$$

for $\tau \in (0, \delta]$. Therefore, the limit (6.2.22) is proven. □

Theorem 6.2.8 *Let A be a non-negative self-adjoint operator. The Trotter-Kato product formula converges in the operator norm topology uniformly on any bounded interval of $\mathbb{R}_0^+$ away from zero for any regular Kato function $f(\cdot) \in \mathcal{K}_r$, any Kato function $g(\cdot) \in \mathcal{K}$, as well as any non-negative self-adjoint operator B, if and only if one has $(\mathbb{1} + A)^{-1} \in \mathcal{C}_\infty(\mathfrak{H})$.*

Proof If $(\mathbb{1} + A)^{-1} \in \mathcal{C}_\infty(\mathfrak{H})$, then by Definition 6.2.6 one has $f_0(A) \in \mathcal{C}_\infty(\mathfrak{H})$. Applying Theorem 6.2.4, we verify the sufficient part of the theorem.

Conversely, assume that the Trotter-Kato product formula converges in the operator-norm for *any* regular Kato function $f(\cdot)$, for *any* (generic) Kato function $g(\cdot)$ and for *any* $B \geq 0$. In particular, this holds for $f(x) = (1 + x)^{-1}$. Therefore, one gets that (6.2.20) is valid for *each* Kato function $g(\cdot)$ and *each* $B \geq 0$, where $M(t)$ is given by (6.2.21).

Here we remind that by the *Weyl-von Neumann* theorem, for a self-adjoint operator A and for any $\epsilon > 0$, there exists a self-adjoint operator V with the *Hilbert-Schmidt* norm $\|V\|_2 := (\mathrm{Tr}\,(V^* V))^{1/2} \leq \epsilon$ such that operator $A + V$ has a *pure point* spectrum $\sigma_{pp}(A + V)$.

Now let $\epsilon < 1/2$. Setting $D = \frac{1}{2}\mathbb{1} + V$, we construct a non-negative self-adjoint operator such that $\|D\| < 1$ and the spectrum of $A + D$ is purely point. Note that $\sigma_{ess}(A + D) = \sigma_{ess}(A + \frac{1}{2}\mathbb{1}) = \frac{1}{2} + \sigma_{ess}(A)$.

If $(\mathbb{1} + A)^{-1} \in \mathcal{C}_\infty(\mathfrak{H})$ is not satisfied, then $\sigma_{ess}(A) \neq \emptyset$. As a consequence, $\sigma_{ess}(A + D) \neq \emptyset$. Let $\mu \in \sigma_{ess}(A + D)$. Since A is non-negative, one gets $\mu \geq \frac{1}{2}$. Then there is an orthonormal sequence of eigenfunctions $\{\varphi_n\}_{n=1}^\infty$ and a sequence of eigenvalues $\{\mu_n\}_{n=1}^\infty$ such that

$$(A + D)\,\varphi_n = \mu_n \varphi_n, \qquad n = 1, 2, \ldots, \tag{6.2.25}$$

and

$$\lim_{n\to\infty} \mu_n = \mu\,.$$

By $\mathfrak{K}$ we denote the closed linear span of $\{\varphi_n\}_{n=1}^\infty$. Setting

$$B\varphi_n = 2^{n+1}\pi\,\varphi_n\,, \quad \pi \in \mathbb{R}_0^+\,, \quad n \in \mathbb{N}\,, \tag{6.2.26}$$

and

$$Bf = 0, \qquad f \in \mathfrak{H} \ominus \mathfrak{K},$$

we define a non-negative self-adjoint operator B. Let $g(x) = 1 - |\sin(x)|$, $x \in \mathbb{R}_0^+$, which is (*nonregular*) Kato function. Obviously, we have

$$(\mathbb{1} - g(\tau_n B))\,\varphi_n = |\sin(\tau_n B)|\,\varphi_n = |\sin(\tau_n 2^{n+1}\pi)|\,\varphi_n = 0\,, \tag{6.2.27}$$

for $\tau_n := 2^{-n}$, $n \in \mathbb{N}$. Therefore, by (6.2.21), (6.2.25) and (6.2.27) one gets

$$(\mathbb{1} + M(t_n) + D)\,\varphi_n = (1 + \mu_n)\,\varphi_n\,, \qquad n \in \mathbb{N}.$$

Further, we have

$$(\mathbb{1} + H + D)\,\varphi_n = (1 + \mu_n + 2^{n+1}\pi)\,\varphi_n\,, \quad \pi \in \mathbb{R}_0^+\,, \quad n \in \mathbb{N}.$$

Therefore,

$$\|(\mathbb{1}+M(\tau_n)+D)^{-1}\varphi_n - (\mathbb{1}+H+D)^{-1}\varphi_n\| \le |(1+\mu_n)^{-1} - (1+\mu_n+2^{n+1}\pi)^{-1}|\,,$$

$n \in \mathbb{N}$. By (6.2.26) we find

$$\lim_{n\to\infty} \|(\mathbb{1} + M(\tau_n) + D)^{-1}\varphi_n - (\mathbb{1} + H + D)^{-1}\varphi_n\| = (1 + \mu)^{-1}\,. \qquad (6.2.28)$$

Now taking into account (6.2.22), we obtain

$$\lim_{n\to\infty} \|(\mathbb{1} + M(t_n) + D)^{-1}\varphi_n - (\mathbb{1} + H + D)^{-1}\varphi_n\| = 0\,,$$

which contradicts (6.2.28). As a consequence, the assumption $\sigma_{ess}(A) \neq \emptyset$ is false, and on that account, we infer $(\mathbb{1} + A)^{-1} \in \mathcal{C}_\infty(\mathfrak{H})$. □

6.3 Compactness of the Resolvent Product

In this section we consider a condition of compactness, which slightly generalises the condition $(\mathbb{1} + A)^{-1} \in \mathcal{C}_\infty(\mathfrak{H})$ of Theorem 6.2.8. On the other hand, we also modify a class of acceptable for the main assertion Kato functions from $\mathcal{K}_r$ to the class $\mathcal{K}_\sigma$; see definition in (6.3.1).

Since the product of resolvents is involved in the assumption, we call these new conditions the *resolvents product compactness*. To proceed we start with a technical lemma.

Lemma 6.3.1 *Let A and B be two self-adjoint operator. If f is a Kato function from the class $\mathcal{K}_\sigma$, which satisfies the conditions $f(x) > 0$, $x \in \mathbb{R}_0^+$, and*

$$\sigma[f] := \sup_{x>0} \frac{x f(x)}{1 - f(x)} < \infty\,, \qquad (6.3.1)$$

then one has

$$\lim_{\tau\to+0}\left\|\left(\mathbb{1}+\frac{1}{\tau}(f(\tau A)^{-1}-\mathbb{1})\right)^{-1}-(\mathbb{1}+A)^{-1}\right\|=0\,, \tag{6.3.2}$$

$$\lim_{\tau\to+0}\left\|\left(\mathbb{1}+\frac{1}{\tau}(f(\tau A)^{-1}-\mathbb{1})\right)^{-1/2}-(\mathbb{1}+A)^{-1/2}\right\|=0\,. \tag{6.3.3}$$

If $g(\cdot)$ is also a Kato function from the class $\mathcal{K}_\sigma$, then

$$\lim_{\tau\to+0}\left\|\left(\mathbb{1}+\frac{1}{\tau}(\mathbb{1}-g(\tau B))\right)^{-1}-(\mathbb{1}+B)^{-1}\right\|=0\,, \tag{6.3.4}$$

$$\lim_{\tau\to+0}\left\|\left(\mathbb{1}+\frac{1}{\tau}(\mathbb{1}-g(\tau B))\right)^{-1/2}-(\mathbb{1}+B)^{-1/2}\right\|=0\,. \tag{6.3.5}$$

Proof We start with the proof of (6.3.2). Note that

$$\left(1+\frac{1}{-}\tau(f(\tau x)^{-1}-1)\right)^{-1}-(1+x)^{-1}$$
$$=\left(1+x\,\frac{f(s)^{-1}-1)}{s}\right)^{-1}\left(1-\frac{f(s)^{-1}-1}{s}\right)\frac{x}{1+x}\,,$$

where $s:=\tau x$, $\tau>0$, $x\in\mathbb{R}_0^+$. Consequently

$$\left(1+\frac{1}{\tau}\,(f(\tau x)^{-1}-1)\right)^{-1}-(1+x)^{-1}$$
$$=\left(x+\frac{sf(s)}{1-f(s)}\right)^{-1}\left(\frac{sf(s)}{1-f(s)}-1\right)\frac{x}{1+x}\,.$$

This, together with (6.3.1), yields the estimate

$$\left|\left(1+\frac{1}{\tau}\,(f(\tau x)^{-1}-1)\right)^{-1}-(1+x)^{-1}\right|\le(1+\sigma[f])\,\frac{1}{1+x}\,, \tag{6.3.6}$$

for $x\in\mathbb{R}_0^+$ and $\tau>0$. Let $N>0$, then we define

$$L_N^f(\tau):=\sup_{x\ge N}\left|\left(1+\frac{1}{\tau}\,(f(\tau x)^{-1}-1)\right)^{-1}-(1+x)^{-1}\right|\,,$$

and

$$M_N^f(\tau) := \sup_{0<x<N} \left| \left(1 + \frac{1}{\tau}(f(\tau x)^{-1} - 1)\right)^{-1} - (1+x)^{-1} \right| .$$

As a result, one gets

$$\begin{aligned} M_\infty^f(\tau) &:= \sup_{x>0} \left| \left(\mathbb{1} + \frac{1}{\tau}\,(f(\tau x)^{-1} - 1)\right)^{-1} - (\mathbb{1}+x)^{-1} \right| \\ &= \max_{N>0} \{L_N^f(\tau), M_N^f(\tau)\} , \end{aligned}$$

for $\tau > 0$. From (6.3.6) we find that $L_N^f(\tau) \leq (1 + \sigma[f])(1+N)$ for $\tau > 0$. Seeing that $\lim_{s\to+0} sf(s)/(1-f(s)) = 1$, for any fixed $N > 0$, we get $\lim_{\tau\to+0} M_N^f(\tau) = 0$. Both observations immediately yield that $\lim_{\tau\to+0} M_\infty^f(\tau) = 0$. Since

$$M_\infty^f(\tau) := \left\| \left(\mathbb{1} + \frac{1}{\tau}(f(\tau A)^{-1} - 1)\right)^{-1} - (\mathbb{1}+A)^{-1} \right\| ,$$

we obtain the proof of (6.3.2).

The convergence (6.3.3) follows from (6.3.2) using the estimate $|\sqrt{a} - \sqrt{b}\,| \leq \sqrt{|a-b|}$, $a, b \in \mathbb{R}_0^+$.

Next, we consider

$$\begin{aligned} &\left(1 + \frac{1-g(\tau x)}{\tau}\right)^{-1} - (1+x)^{-1} \\ &\qquad = \left(1 + x\,\frac{1-g(s)}{s}\right)^{-1} \left(1 - \frac{1-g(s)}{s}\right) \frac{x}{1+x} , \end{aligned}$$

where $s := \tau x$, $\tau > 0$, $x \in \mathbb{R}_0^+$. This gives

$$\begin{aligned} &\left(1 + \frac{1-g(\tau x)}{\tau}\right)^{-1} - (1+x)^{-1} \\ &\qquad = \left(\frac{s}{1-g(s)} + x\right)^{-1} \left(\frac{s}{1-g(s)} - 1\right) \frac{x}{1+x} , \end{aligned}$$

and as a consequence

$$\left(1+\frac{1-g(\tau x)}{\tau}\right)^{-1}-(1+x)^{-1}$$
$$=\left(\frac{1}{x}\frac{s}{1-g(s)}+1\right)^{-1}\left(\frac{sg(s)}{1-g(s)}-1\right)\frac{1}{1+x}+\left(\frac{1}{x}\frac{s}{1-g(s)}+1\right)^{-1}\frac{s}{1+x},$$

which leads to estimate

$$\left|\left(1+\frac{1-g(\tau x)}{\tau}\right)^{-1}-(1+x)^{-1}\right|\leq\left|\frac{sg(s)}{1-g(s)}-1\right|\frac{1}{1+x}+\frac{s}{1+x}.$$

As above we set for $N>0$

$$L_N^g(\tau):=\sup_{x\geq N}\left|\left(1+\frac{1-g(\tau x)}{\tau}\right)^{-1}-(1+x)^{-1}\right|$$

and

$$M_N^g(\tau):=\sup_{0<x<N}\left|\left(1+\frac{1-g(\tau x)}{\tau}\right)^{-1}-(1+x)^{-1}\right|.$$

Notice that $M_\infty^g:=\max_{N>0}\{L_N^g(\tau),M_N^g(\tau)\}$. Then we get the estimates

$$L_N^g(\tau)\leq\frac{1+\sigma[g]}{1+N}+\tau,\quad \tau>0,$$

and

$$M_N^g(\tau)\leq\sup_{0<x<N}\left|\frac{\tau\, x\, g(\tau x)}{1-g(\tau x)}-1\right|+\tau,\quad \tau>0.$$

Since $\limsup_{\tau\to+0}L_N^g(\tau)\leq(1+\sigma[g])/(1+N)$ and $\limsup_{\tau\to+0}M_N^g(\tau)=0$, one gets

$$\limsup_{\tau\to+0}M_\infty^g(\tau)\leq\frac{1+\sigma[g]}{1+N},$$

where $N>0$ is *arbitrary*. Therefore, $\limsup_{\tau\to+0}M_\infty^g(\tau)=0$ and consequently by virtue of

$$M_\infty^g(\tau)=\left\|\left(\mathbb{1}+\frac{1-g(\tau B)}{\tau}\right)^{-1}-(\mathbb{1}+B)^{-1}\right\|,$$

we finally obtain (6.3.4). Then convergence (6.3.5) follows from (6.3.4) by the same argument as the proof of (6.3.3). □

Remark 6.3.2 Note that generic Kato functions $f(x) = e^{-x}$ and $f(x) = 1/(1+x)$, $x \in \mathbb{R}_0^+$, belong to the class $\mathcal{K}_\sigma$. Seeing that condition (6.3.1) implies $\sup_{x>0} x\, f(x) < \infty$, which is *stronger* than condition (5.2.1), we infer that $\mathcal{K}_\sigma \subset \mathcal{K}$. In fact, there are inclusions: $\mathcal{K}_\sigma \subset \mathcal{K}_r \subset \mathcal{K}$; see Definition 6.2.6 and also Lemma 6.6.5 in Notes to Chap. 6 (Notes to Sect. 6.3).

Lemma 6.3.3 *Let $A \geq 0$ and $B \geq 0$ be non-negative self-adjoint operators. Then one has the equivalence:*

$$(\mathbb{1}+A)^{-1}(\mathbb{1}+B)^{-1} \in \mathcal{C}_\infty(\mathfrak{H}) \Longleftrightarrow (\mathbb{1}+A)^{-1/2}(\mathbb{1}+B)^{-1/2} \in \mathcal{C}_\infty(\mathfrak{H}).$$

Proof Let $(\mathbb{1}+A)^{-1}(\mathbb{1}+B)^{-1} \in \mathcal{C}_\infty(\mathfrak{H})$, then

$$(\mathbb{1}+B)^{-1}(\mathbb{1}+A)^{-1/2}(\mathbb{1}+A)^{-1/2}(\mathbb{1}+B)^{-1} \in \mathcal{C}_\infty(\mathfrak{H}).$$

If $X := (\mathbb{1}+A)^{-1/2}(\mathbb{1}+B)^{-1}$, then $X^*X \in \mathcal{C}_\infty(\mathfrak{H})$ that yields $X \in \mathcal{C}_\infty(\mathfrak{H})$. Hence, $X^* = (\mathbb{1}+B)^{-1}(\mathbb{1}+A)^{-1/2} \in \mathcal{C}_\infty(\mathfrak{H})$. Therefore,

$$(\mathbb{1}+A)^{-1/2}(\mathbb{1}+B)^{-1}(\mathbb{1}+A)^{-1/2} \in \mathcal{C}_\infty(\mathfrak{H}).$$

Setting $Y := (\mathbb{1}+B)^{-1/2}(\mathbb{1}+A)^{-1/2}$ we get $Y^*Y \in \mathcal{C}_\infty(\mathfrak{H})$, which yields $Y \in \mathcal{C}_\infty(\mathfrak{H})$.

The converse assertion is immediate since operators $(\mathbb{1}+A)^{-1/2}$, $(\mathbb{1}+B)^{-1/2}$ are bounded and $(\mathbb{1}+A)^{-1}(\mathbb{1}+B)^{-1} = (\mathbb{1}+A)^{-1/2}[(\mathbb{1}+A)^{-1/2}(\mathbb{1}+B)^{-1/2}](\mathbb{1}+B)^{-1/2}$. □

We note that although the set of compact operators $\mathcal{C}_\infty(\mathfrak{H})$ is a two-sided *ideal* in C^*-algebra $\mathcal{L}(\mathfrak{H})$ of bounded operators, compactness of the product $(K\,L) \in \mathcal{C}_\infty(\mathfrak{H})$ does not imply compactness of any of factors.

Lemma 6.3.4 *Let $A \geq 0$ and $B \geq 0$ be non-negative self-adjoint operators such that* dom $(A^{1/2}) \cap$ dom $(B^{1/2})$ *is dense in $\mathfrak{H}$ and let $H := A \dot{+} B$. If $(\mathbb{1}+A)^{-1}(\mathbb{1}+B)^{-1} \in \mathcal{C}_\infty(\mathfrak{H})$, then $(\mathbb{1}+H)^{-1} \in \mathcal{C}_\infty(\mathfrak{H})$.*

Proof First of all the conditions of lemma imply that

$$(\mathbb{1}+H)^{-1} \leq (\mathbb{1}+A)^{-1} \quad \text{and} \quad (\mathbb{1}+H)^{-1} \leq (\mathbb{1}+B)^{-1}.$$

Then there are two contractions $\Gamma \in \mathcal{L}(\mathfrak{H})$ and $\Pi \in \mathcal{L}(\mathfrak{H})$ such that

$$(\mathbb{1}+H)^{-1/2} = \Gamma\,(\mathbb{1}+A)^{-1/2} \quad \text{and} \quad (\mathbb{1}+H)^{-1/2} = \Pi\,(\mathbb{1}+B)^{-1/2}. \tag{6.3.7}$$

As a result,

$$(\mathbb{1} + H)^{-1} = \Gamma\,(\mathbb{1} + A)^{-1/2}(\mathbb{1} + B)^{-1/2}\Pi^* .$$

Then by Lemma 6.3.3, we get $(\mathbb{1} + H)^{-1} \in \mathcal{C}_\infty(\mathfrak{H})$. □

Theorem 6.3.5 *Let* A *and* B *be non-negative self-adjoint operators such that the set* $\mathrm{dom}\,(A^{1/2}) \cap \mathrm{dom}(B^{1/2})$ *is dense in* $\mathfrak{H}$. *Let* f *and* g *be Kato functions such that* $f, g \in \mathcal{K}_\sigma$. *If*

$$(\mathbb{1} + A)^{-1}(\mathbb{1} + B)^{-1} \in \mathcal{C}_\infty(\mathfrak{H}), \tag{6.3.8}$$

then the Trotter-Kato product formula converges in the operator-norm topology uniformly on any closed interval away from zero (i.e. locally uniformly in $\mathbb{R}^+$*) and* $(\mathbb{1} + H)^{-1} \in \mathcal{C}_\infty(\mathfrak{H})$.

Proof By Lemma 6.3.4 we immediately get that $(\mathbb{1} + H)^{-1} \in \mathcal{C}_\infty(\mathfrak{H})$. Note that (6.1.13) implies

$$(\mathbb{1} + M(\tau))^{-1} \leq \left(\mathbb{1} + \frac{f(\tau A)^{-1} - \mathbb{1}}{\tau}\right)^{-1}, \qquad \tau > 0\,.$$

Hence, there is a family of contractions $\{\Gamma(\tau)\}_{\tau>0}$ such that

$$(\mathbb{1} + M(\tau))^{-1/2} = \Gamma(\tau)\left(\mathbb{1} + \frac{f(\tau A)^{-1} - \mathbb{1}}{\tau}\right)^{-1/2}, \qquad \tau > 0\,. \tag{6.3.9}$$

Again, by (6.1.13) we also obtain

$$(\mathbb{1} + M(\tau))^{-1} \leq \left(\mathbb{1} + \frac{\mathbb{1} - g(\tau B)}{\tau}\right)^{-1}, \qquad \tau > 0.$$

Therefore, there exists a family of contractions $\{\Pi(\tau)\}_{\tau>0}$ such that

$$(\mathbb{1} + M(\tau))^{-1/2} = \Pi(\tau)\left(\mathbb{1} + \frac{\mathbb{1} - g(\tau B)}{\tau}\right)^{-1/2}, \qquad \tau > 0. \tag{6.3.10}$$

Now, by Theorem 6.1.7, we obtain

$$\operatorname*{s-lim}_{\tau\to+0} (\mathbb{1} + M(\tau))^{-1/2} = (\mathbb{1} + H)^{-1/2}.$$

Furthermore, one has

$$\operatorname*{s-lim}_{\tau\to+0} \left(\mathbb{1} + \frac{f(\tau A)^{-1} - \mathbb{1}}{\tau}\right)^{-1/2} = (\mathbb{1} + A)^{-1/2}\,, \tag{6.3.11}$$

and

$$\text{s-}\lim_{\tau\to+0}\left(\mathbb{1}+\frac{\mathbb{1}-g(\tau B)}{\tau}\right)^{-1/2}=(\mathbb{1}+B)^{-1/2}. \tag{6.3.12}$$

Using the representation

$$(\mathbb{1}+M(\tau))^{-1/2}=\Gamma(\tau)\left(\left(\mathbb{1}+\frac{f(\tau A)^{-1}-\mathbb{1}}{\tau}\right)^{-1/2}-(\mathbb{1}+A)^{-1/2}\right)$$
$$+\Gamma(\tau)(\mathbb{1}+A)^{-1/2},\quad \tau>0,$$

as well as (6.3.11), we obtain $\text{s-lim}_{\tau\to+0}\,\Gamma(\tau)=\Gamma$ where Γ is defined by (6.3.7). Similarly, we prove $\text{s-lim}_{\tau\to+0}\,\Pi(\tau)=\Pi$. Further, from (6.3.9) and (6.3.10) we find the representation

$$(\mathbb{1}+M(\tau))^{-1}=$$
$$\Gamma(\tau)\left(\mathbb{1}+\frac{f(\tau A)^{-1}-I}{\tau}\right)^{-1/2}\left(\mathbb{1}+\frac{\mathbb{1}-g(\tau B)}{\tau}\right)^{-1/2}\Pi(\tau)^*,\quad \tau>0,$$

which leads to

$$(\mathbb{1}+M(\tau))^{-1}-\Gamma(\tau)(\mathbb{1}+A)^{-1/2}(\mathbb{1}+B)^{-1/2}\Pi(\tau)^*$$
$$=\Gamma(\tau)\left(\left(\mathbb{1}+\frac{f(\tau A)^{-1}-I}{\tau}\right)^{-1/2}-(\mathbb{1}+A)^{-1/2}\right)$$
$$\times\left(\mathbb{1}+\frac{\mathbb{1}-g(\tau B)}{\tau}\right)^{-1/2}\Pi(\tau)^*$$
$$+\Gamma(\tau)(\mathbb{1}+A)^{-1/2}\left(\left(\mathbb{1}+\frac{\mathbb{1}-g(\tau B)}{\tau}\right)^{-1/2}-(\mathbb{1}+B)^{-1/2}\right)\Pi(\tau)^*,$$

for $\tau>0$. Hence, one gets the estimate

$$\left\|(\mathbb{1}+M(\tau))^{-1}-\Gamma(\tau)(\mathbb{1}+A)^{-1/2}(\mathbb{1}+B)^{-1/2}\Pi(\tau)^*\right\|$$
$$\le\left\|\left(\mathbb{1}+\frac{f(\tau A)^{-1}-I}{\tau}\right)^{-1/2}-(\mathbb{1}+A)^{-1/2}\right\|$$
$$+\left\|\left(\mathbb{1}+\frac{\mathbb{1}-g(\tau B)}{\tau}\right)^{-1/2}-(\mathbb{1}+B)^{-1/2}\right\|.$$

Therefore, by Lemma 6.3.4, we infer

$$\lim_{\tau\to+0}\left\|(\mathbb{1}+M(\tau))^{-1}-\Gamma(\tau)(\mathbb{1}+A)^{-1/2}(\mathbb{1}+B)^{-1/2}\Pi(\tau)^*\right\|=0\,. \qquad (6.3.13)$$

Next we use the representation

$$\begin{aligned}\Gamma(\tau)(\mathbb{1}+A)^{-1/2}&(\mathbb{1}+B)^{-1/2}\Pi(\tau)^*-(\mathbb{1}+H)^{-1}\\ &=\Gamma(\tau)(\mathbb{1}+A)^{-1/2}(\mathbb{1}+B)^{-1/2}\Pi(\tau)^*\\ &\quad-\Gamma(\mathbb{1}+A)^{-1/2}(\mathbb{1}+B)^{-1/2}\Pi^*\\ &=(\Gamma(\tau)-\Gamma)(\mathbb{1}+A)^{-1/2}(\mathbb{1}+B)^{-1/2}\Pi(\tau)^*\\ &\quad+\Gamma(\mathbb{1}+A)^{-1/2}(\mathbb{1}+B)^{-1/2}(\Pi(\tau)^*-\Pi^*)\,,\end{aligned}$$

which yields the estimate

$$\begin{aligned}&\left\|\Gamma(\tau)(\mathbb{1}+A)^{-1/2}(\mathbb{1}+B)^{-1/2}\Pi(\tau)^*-(\mathbb{1}+H)^{-1}\right\|\\ &\le\left\|(\Gamma(\tau)-\Gamma)(\mathbb{1}+A)^{-1/2}(\mathbb{1}+B)^{-1/2}\right\|\\ &+\left\|(\mathbb{1}+A)^{-1/2}(\mathbb{1}+B)^{-1/2}(\Pi(\tau)^*-\Pi^*)\right\|\,.\end{aligned}$$

Since by Lemma 6.3.4 the operator $(\mathbb{1}+A)^{-1/2}(\mathbb{1}+B)^{-1/2}$ is compact, and owing to the limits $\text{s-lim}_{\tau\to+0}(\Gamma(\tau)-\Gamma))=0$, $\text{s-lim}_{\tau\to+0}(\Pi(\tau)-\Pi))=0$, we get from Lemma 6.4.3 that

$$\lim_{\tau\to+0}\left\|(\Gamma(\tau)-\Gamma))(\mathbb{1}+A)^{-1/2}(\mathbb{1}+B)^{-1/2}\right\|=0\,,$$

and

$$\lim_{\tau\to+0}\left\|(\mathbb{1}+A)^{-1/2}(\mathbb{1}+B)^{-1/2}(\Pi(\tau)^*-\Pi^*)\right\|=0.$$

Summarising, we obtain that

$$\lim_{\tau\to+0}\left\|\Gamma(\tau)(\mathbb{1}+A)^{-1/2}(\mathbb{1}+B)^{-1/2}\Pi(\tau)^*-(\mathbb{1}+H)^{-1}\right\|=0\,. \qquad (6.3.14)$$

Ultimately, by (6.3.13) and (6.3.14), we infer the locally uniform in $\mathbb{R}^+$ limit (6.1.57). Then applying Corollary 6.1.8 we conclude that the Trotter-Kato product formula (6.1.56) converges in the operator-norm topology uniformly *away* from zero, that is, locally uniformly in $\mathbb{R}^+$. □

6.4 Relative Compactness

In this section we consider the case when $\mathrm{dom}\,(A^{1/2}) \subseteq \mathrm{dom}\,(B^{1/2})$. This assumption and definition of H yield $\mathrm{dom}\,(H^{1/2}) = \mathrm{dom}\,(A^{1/2})$. For discussion of values other than $\alpha = 1/2$, see Sect. 7.4.

Below we shall use the notation $A(\tau) := \tau^{-1}[\mathbb{1} - f(\tau A)]$, cf. (5.2.4), and we make precise the class of the Kato functions $\widehat{\mathcal{K}} \subset \mathcal{K}$, which are acceptable for the main assertion of this section.

Definition 6.4.1 A generic Kato function f (Definition 5.2.1) belongs to the class $\widehat{\mathcal{K}} \subset \mathcal{K}$ if the condition

$$\sup_{x>0} \frac{1}{x} \left| \frac{x}{1 - f(x)} - 1 \right| = 1\,, \tag{6.4.1}$$

is satisfied for $f \in \mathcal{K}$.

We note that (6.4.1) implies $f(x) \leq 1/(1+x)$ for $x \in \mathbb{R}_0^+$. As a consequence, typical examples of $f \in \widehat{\mathcal{K}}$ are $f(x) = \mathrm{e}^{-x}$, $f(x) = (1 + x/a)^{-a}$, $a > 0$, but *not* the function $f(x) = 1 - |\sin(x)|$.

Lemma 6.4.2 *Let A be a non-negative self-adjoint operator and let Kato function $f \in \widehat{\mathcal{K}}$. Then*

$$\left\|(\mathbb{1} + tA(\tau))^{-1} - (\mathbb{1} + tA)^{-1}\right\| \leq \frac{\tau}{t}, \quad t > 0, \quad \tau > 0. \tag{6.4.2}$$

and

$$\left\|(\mathbb{1} + tA(\tau))^{-1/2} - (\mathbb{1} + tA)^{-1/2}\right\| \leq \left(\frac{\tau}{t}\right)^{1/2}, \quad t > 0, \quad \tau > 0. \tag{6.4.3}$$

Here $A(\tau) = (\mathbb{1} - f(\tau A))/\tau$ for $\tau > 0$.

Proof To proceed we use identity

$$\left(1 + t\frac{1 - f(\tau x)}{\tau}\right)^{-1} - (1 + tx)^{-1}$$
$$= \left(1 + t\frac{1 - f(\tau x)}{\tau}\right)^{-1} \left(1 - \frac{1 - f(\tau x)}{\tau x}\right) \frac{tx}{1 + tx}\,.$$

Let $s := \tau x$ and $y := tx$ for $x > 0$, $t > 0$ and $\tau > 0$. With these notations, we obtain

$$\left(1 + y\frac{1 - f(s)}{s}\right)^{-1} - (1 + y)^{-1} =$$
$$\left(\frac{s}{1 - f(s)} + y\right)^{-1} \left(\frac{s}{1 - f(s)} - 1\right) \frac{y}{1 + y}, \quad s > 0, \quad y > 0.$$

Then using condition (6.4.1), one gets the estimates

$$\left|\left(1+y\frac{1-f(s)}{s}\right)^{-1}-(1+y)^{-1}\right|\leq\frac{s}{y},\quad s>0,\quad y>0,$$

or

$$\left|\left(1+t\frac{1-f(\tau x)}{\tau}\right)^{-1}-(1+tx)^{-1}\right|\leq\frac{\tau}{t},\quad x>0,\quad t>0,\quad \tau>0. \tag{6.4.4}$$

Seeing that for non-negative self-adjoint A, one infers

$$\begin{aligned}&\sup_{x>0}\left|\left(1+t\frac{1-f(\tau x)}{\tau}\right)^{-1}-(1+tx)^{-1}\right|\\ &=\left\|\left(1+t\frac{1-f(\tau A)}{\tau}\right)^{-1}-(1+tA)^{-1}\right\|,\end{aligned}$$

$t>0$, $\tau>0$, we immediately obtain (6.4.2).

Then using inequality $|\sqrt{a}-\sqrt{b}\,|\leq\sqrt{|a-b|}$, $a,b>0$, we also deduce from (6.4.4)

$$\left|\left(1+t\frac{1-f(\tau x)}{\tau}\right)^{-1/2}-(1+tx)^{-1/2}\right|\leq\sqrt{\frac{\tau}{t}}\,,$$

for $x>0, t>0, \tau>0$, which proves (6.4.3). □

Lemma 6.4.3 *Let $\{\mathfrak{B}(t,\tau)\}_{t>0,\tau>0}$ and $\{\mathfrak{B}_0(t)\}_{t>0}$ be families of bounded operators such that the conditions*

$$C_{\mathfrak{B}}:=\sup_{t,\tau>0}\|\mathfrak{B}(t,\tau)\|\quad\text{and}\quad C_{\mathfrak{B}_0}:=\sup_{t>0}\|\mathfrak{B}_0(t)\|\,, \tag{6.4.5}$$

be satisfied. Further, let $C\in\mathcal{C}_\infty(\mathfrak{H})$ be a compact operator. If

$$\lim_{\tau\to+0}\sup_{t\in\mathcal{I}}\|\mathfrak{B}(t,\tau)\psi-\mathfrak{B}_0(t)\psi\|=0 \tag{6.4.6}$$

for any $\psi\in\mathfrak{H}$ and any bounded interval $\mathcal{I}\subset\mathbb{R}^+$, then

$$\lim_{\tau\to+0}\sup_{t\in\mathcal{I}}\|(\mathfrak{B}(t,\tau)-\mathfrak{B}_0(t))\,C\|=0 \tag{6.4.7}$$

$$\lim_{\tau\to+0}\sup_{t\in\mathcal{I}}\|C\,(\mathfrak{B}^*(t,\tau)-\mathfrak{B}_0^*(t))\|=0. \tag{6.4.8}$$

Proof Let C be one-rank operator: $C = (\cdot\,,\,\phi)\,\psi$, for $\phi, \psi \in \mathfrak{H}$. Then we obtain

$$(\mathfrak{B}(t,\tau) - \mathfrak{B}_0(t))\, C = (\cdot\,,\,\phi)\, (\mathfrak{B}(t,\tau) - \mathfrak{B}_0(t))\, \psi\,,$$

which in turn gives estimate

$$\|(\mathfrak{B}(t,\tau) - \mathfrak{B}_0(t))\, C\| \leq \|\phi\|\; \|(\mathfrak{B}(t,\tau) - \mathfrak{B}_0(t))\, \psi\|.$$

Using (6.4.6), we obtain for this operator (6.4.7). Now, if C is a finite sum of one-rank operators, then the result (6.4.7) remains true. Since finite-rank operators are dense in the set of compact operators $\mathcal{C}_\infty(\mathfrak{H})$ in the operator-norm topology, the result (6.4.7) extends to any compact operator C.

Because adjoint operator C^* is also compact, using (6.4.7), we also infer that $\lim_{\tau\to+0} \sup_{t\in\mathcal{I}} \|(\mathfrak{B}(t,\tau) - \mathfrak{B}_0(t))\, C^*\| = 0$. Then seeing that $\|(\mathfrak{B}(t,\tau) - \mathfrak{B}_0(t))\, C^*\| = \|C\, (\mathfrak{B}^*(t,\tau) - \mathfrak{B}_0^*(t))\|$, we obtain (6.4.8). □

Theorem 6.4.4 *Let A and B be non-negative self-adjoint operators such that*

$$\mathrm{dom}\,(A^{1/2}) \subseteq \mathrm{dom}\,(B^{1/2})\,. \tag{6.4.9}$$

Further, let f and g be Kato functions. If the condition (6.4.1) *is satisfied, that is, $f, g \in \widehat{\mathcal{K}}$, and also*

$$B^{1/2}(\mathbb{1} + A)^{-1/2} \in \mathcal{C}_\infty(\mathfrak{H})\,,$$

then the Trotter-Kato product formula (6.1.58) *converges in the operator-norm topology uniformly on closed intervals $\mathcal{I} \subset \mathbb{R}_0^+$ (Remark* 6.1.5*).*

Proof For family $\{f(\tau A)^{1/2}\, g(\tau B)\, f(\tau A)^{1/2}\}_{\tau\geq 0}$, cf. (6.1.58), we use representation

$$\mathbb{1} + t\, S(\tau) = \mathbb{1} + t\,\frac{\mathbb{1} - f(\tau A)}{\tau} + t f(\tau A)^{1/2} \frac{\mathbb{1} - g(\tau B)}{\tau} f(\tau A)^{1/2}\,, \tag{6.4.10}$$

for (4.2.23) in Theorem 4.2.7 (cf.(6.1.59)). Here $\tau > 0$ and $t \geq 0$.

Implying $A(\tau) := (\mathbb{1} - f(\tau A))/\tau$ and $B(\tau) := (\mathbb{1} - g(\tau B))/\tau$, $\tau > 0$, we find

$$\mathbb{1}{+}tS(\tau) = \tag{6.4.11}$$

$$(\mathbb{1} + tA(\tau))^{1/2} \left(\mathbb{1} + t\sqrt{\frac{f(\tau A)}{\mathbb{1} + tA(\tau)}} B(\tau) \sqrt{\frac{f(\tau A)}{\mathbb{1} + tA(\tau)}}\right) (\mathbb{1} + tA(\tau))^{1/2}\,,$$

$\tau > 0$, $t \geq 0$. We set

$$X(t,\tau) := \sqrt{t}\,\sqrt{B(\tau)}\sqrt{\frac{f(\tau A)}{\mathbb{1} + tA(\tau)}}, \quad \tau > 0 \quad t \geq 0\,.$$

Taking into account that

$$X(t,\tau) =$$
$$\sqrt{t}\sqrt{\frac{\mathbb{1}-g(\tau B)}{\tau B}}\sqrt{B}(\mathbb{1}+A)^{-1/2}\sqrt{\frac{\mathbb{1}+A}{\mathbb{1}+\tau A}}\sqrt{\frac{1}{\mathbb{1}+tA(\tau)}}\sqrt{(\mathbb{1}+\tau A)f(\tau A)},$$

and also

$$\mathbb{1}+tS(\tau) = (\mathbb{1}+tA(\tau))^{1/2}\left(\mathbb{1}+X(t,\tau)^*X(t,\tau)\right)(\mathbb{1}+tA(\tau))^{1/2},$$

we deduce

$$(\mathbb{1}+tS(\tau))^{-1} = (\mathbb{1}+tA(\tau))^{-1/2}\left(\mathbb{1}+X(t,\tau)^*X(t,\tau)\right)^{-1}(\mathbb{1}+tA(\tau))^{-1/2}, \tag{6.4.12}$$

$\tau > 0$ and $t \geq 0$. Condition (6.4.1) implies that $f \in \widehat{\mathcal{K}}$, i.e. $f(x) \leq 1/(1+x)$, $x \geq 0$, which yields $\left\|\sqrt{(\mathbb{1}+\tau A)f(\tau A)}\right\| \leq 1$ for any $\tau > 0$.

Now we set

$$Z(\tau) := \sqrt{\frac{\mathbb{1}-g(\tau B)}{\tau B}},$$
$$C := \sqrt{B}(\mathbb{1}+A)^{-1/2},$$
$$Y(t,\tau) := \sqrt{t}\sqrt{\frac{\mathbb{1}+A}{\mathbb{1}+\tau A}}\sqrt{\frac{1}{\mathbb{1}+tA(\tau)}}\sqrt{(\mathbb{1}+\tau A)f(\tau A)},$$

for $\tau > 0$, $t \geq 0$. Note that $X(t,\tau) := Z(\tau)\,C\,Y(t,\tau)$, $\tau > 0$ and $t \geq 0$. Further, we set

$$Y_0(t) := \sqrt{t}\sqrt{\frac{\mathbb{1}+A}{\mathbb{1}+tA}},$$
$$Y_1(t,\tau) := \sqrt{t}\sqrt{\frac{\mathbb{1}+A}{\mathbb{1}+\tau A}}\left(\sqrt{\frac{1}{\mathbb{1}+tA(\tau)}} - \sqrt{\frac{1}{\mathbb{1}+tA}}\right)\sqrt{(\mathbb{1}+\tau A)f(\tau A)},$$
$$Y_2(t,\tau) := \sqrt{t}\sqrt{\frac{\mathbb{1}+A}{\mathbb{1}+tA}}\left(\sqrt{f(\tau A)}-\mathbb{1}\right),$$

$\tau > 0$, $t \geq 0$. Note that $Y(t,\tau) = Y_0(t) + Y_1(t,\tau) + Y_2(t,\tau)$. Using the estimate (6.4.3), we infer that

$$\|Y_1(t,\tau)\| \leq 1, \quad 0 < \tau \leq 1, \quad t \geq 0. \tag{6.4.13}$$

Moreover, we also get

$$\|Y_2(t,\tau)\| \le \sup\{1, \sqrt{t}\}, \quad \tau > 0, \quad t \ge 0. \tag{6.4.14}$$

On account of estimates (6.4.13) and (6.4.14), one obtains

$$\lim_{\tau\to+0} \sup_{t\in\mathcal{I}} \|Y_1(t,\tau)\psi\| = 0 \quad \text{and} \quad \lim_{\tau\to+0} \sup_{t\in\mathcal{I}} \|Y_2(t,\tau)\psi\| = 0, \tag{6.4.15}$$

for each $\psi \in \mathfrak{H}$ and any closed interval $\mathcal{I} \subset \mathbb{R}_0^+$. Let

$$\begin{aligned} X_0(t,\tau) &:= (Z(\tau) - \mathbb{1})\, C\, Y_0(t), \\ X_1(t,\tau) &:= Z(\tau)\, C\, Y_1(t,\tau), \\ X_2(t,\tau) &:= Z(\tau)\, C\, Y_2(t,\tau), \end{aligned}$$

for $\tau > 0$, $t \ge 0$. Notice that

$$X(t,\tau) - C\, Y_0(t) = X_0(t,\tau) + X_1(t,\tau) + X_2(t,\tau), \quad \tau > 0, \quad t \ge 0. \tag{6.4.16}$$

Since $g \in \mathcal{K}$, we have

$$\gamma[g] := \sup_{x>0} \frac{1 - g(x)}{x} < \infty.$$

Then $\|Z(\tau)\| \le \sqrt{\gamma[g]}$ for $\tau > 0$. By virtue of (6.4.15) and Lemma 6.4.3, we get that

$$\lim_{\tau\to+0} \sup_{t\in\mathcal{I}} \|X_1(t,\tau)\| = 0, \quad \text{and} \quad \lim_{\tau\to+0} \sup_{t\in\mathcal{I}} \|X_2(t,\tau)\| = 0, \tag{6.4.17}$$

for any closed interval $\mathcal{I} \subset \mathbb{R}^+$. Since $\text{s-}\lim_{\tau\to+0}(Z(\tau) - \mathbb{1}) = 0$ and $\|Y_0(t)\| \le \sup\{1, \sqrt{t}\}$ for $t \ge 0$, we find by Lemma 6.4.3 that

$$\lim_{\tau\to+0} \sup_{t\in\mathcal{I}} \|X_0(t,\tau)\| = 0, \tag{6.4.18}$$

for any closed interval $\mathcal{I} \subset \mathbb{R}_0^+$. Finally, on account of (6.4.16), (6.4.17) and (6.4.18) we obtain

$$\lim_{\tau\to+0} \sup_{t\in\mathcal{I}} \|X(t,\tau) - C\, Y_0(t)\| = 0, \quad \mathcal{I} \subset \mathbb{R}_0^+.$$

This yields

$$\lim_{\tau\to+0}\sup_{t\in\mathcal{I}}\|X(t,\tau)^*X(t,\tau)-Y_0(t)\,C^*C\,Y_0(t)\|=0\,,\quad \mathcal{I}\subset\mathbb{R}_0^+\,. \tag{6.4.19}$$

From (6.4.11), we get the representation

$$\begin{aligned}
&(\mathbb{1}+tS(\tau))^{-1}-(\mathbb{1}+tA)^{-1/2}\left(\mathbb{1}+Y_0(t)\,C^*C\,Y_0(t)\right)^{-1}(\mathbb{1}+tA)^{-1/2}\\
&=\left((\mathbb{1}+tA(\tau))^{-1/2}-(\mathbb{1}+tA)^{-1/2}\right)\left(\mathbb{1}+X(t,\tau)^*X(t,\tau)\right)^{-1}(\mathbb{1}+tA(\tau))^{-1/2}\\
&+(\mathbb{1}+tA)^{-1/2}\left(\mathbb{1}+X(t,\tau)^*X(t,\tau)\right)^{-1}(\left((\mathbb{1}+tA(\tau))^{-1/2}-(\mathbb{1}+tA)^{-1/2}\right)\\
&+(\mathbb{1}+tA)^{-1/2}\left(\left(\mathbb{1}+X(t,\tau)^*X(t,\tau)\right)^{-1}-\left(\mathbb{1}+Y_0(t)\,C^*C\,Y_0(t)\right)^{-1}\right)\\
&\quad\times(\mathbb{1}+tA)^{-1/2}\,,
\end{aligned}$$

for $\tau>0$, $t\geq 0$, which as a consequence gives the estimate

$$\begin{aligned}
&\left\|(\mathbb{1}+tS(\tau))^{-1}-(\mathbb{1}+tA)^{-1/2}\left(\mathbb{1}+Y_0(t)\,C^*C\,Y_0(t)\right)^{-1}(\mathbb{1}+tA)^{-1/2}\right\|\\
&\leq 2\left\|(\mathbb{1}+tA(\tau))^{-1/2}-(\mathbb{1}+tA)^{-1/2}\right\|+\|X(t,\tau)^*X(t,\tau)-Y_0(t)\,C^*C\,Y_0(t)\|\,.
\end{aligned} \tag{6.4.20}$$

To proceed we set $\tau=t/\eta$, $t\geq 0$ and $\eta\geq 1$. Then (6.4.3) yields the estimate

$$\left\|(\mathbb{1}+tA(t/\eta))^{-1/2}-(\mathbb{1}+tA)^{-1/2}\right\|\leq\frac{1}{\sqrt{\eta}}\,,\quad t\geq 0,\quad \eta\geq 1\,.$$

From (6.4.19), we get

$$\lim_{\eta\to\infty}\sup_{t\in\mathcal{I}}\|X(t,t/\eta)^*X(t,t/\eta)-Y_0(t)\,C^*C\,Y_0(t)\|=0\,,$$

for any bounded interval $\mathcal{I}\subset\mathbb{R}_0^+$. Hence from (6.4.20) we find

$$\begin{aligned}
&\lim_{\eta\to\infty}\sup_{t\in\mathcal{I}}\Big\|(\mathbb{1}+tS(t/\eta))^{-1}-(\mathbb{1}+tA)^{-1/2}\left(\mathbb{1}+Y_0(t)\,C^*C\,Y_0(t)\right)^{-1}\\
&\qquad\qquad\times(\mathbb{1}+tA)^{-1/2}\Big\|\\
&=0\,,
\end{aligned}$$

for any bounded interval $\mathcal{I}$ of $\mathbb{R}_0^+$. Let

$$(\mathbb{1} + tH)^{-1} := (\mathbb{1} + tA)^{-1/2} \left(\mathbb{1} + Y_0(t)\, C^*C\, Y_0(t)\right)^{-1} (\mathbb{1} + tA)^{-1/2}, \quad t \geq 0,$$

then we obtain

$$\lim_{\eta\to\infty} \sup_{t\in\mathcal{I}} \left\| (\mathbb{1} + tS(t/\eta))^{-1} - (\mathbb{1} + tH)^{-1} \right\| = 0, \tag{6.4.21}$$

where $H = A \dot{+} B$ is the form-sum of operators; see (6.4.9), for bounded intervals $\mathcal{I}$ of $\mathbb{R}_0^+$

Now applying to (6.4.21) Theorem 4.2.7, we infer that for Kato functions of the class $\widehat{\mathcal{K}}$, the operator-norm product formula

$$\lim_{\eta\to\infty} \sup_{t\in\mathcal{I}} \|F(t/\eta)^\eta - \mathrm{e}^{-tH}\| = 0, \tag{6.4.22}$$

holds uniformly for on bounded interval $\mathcal{I} \subset \mathbb{R}_0^+$ for $F(t/\eta) = \mathbb{1} - t\, S(t/\eta)$. Since by representation (6.4.10) one has

$$F(t) = \mathbb{1} - t\, S(t) = f(tA)^{1/2}\, g(tB)\, f(tA)^{1/2}, \tag{6.4.23}$$

on account of equivalence of (6.4.21) and (6.1.59) established in Corollary 6.1.8, the limits (6.4.22) and (6.4.23) yield the Trotter-Kato product formula (6.1.58), which converges in the operator-norm topology locally uniformly on $\mathbb{R}_0^+$. □

Given that $f, g \in \widehat{\mathcal{K}}$, the proof for other symmetrised and non-symmetrised Trotter-Kato product formulæ (6.1.11) follows owing to equivalence established in Corollary 6.1.4.

6.5 Examples

1. Let Ω be an open bounded set of $\mathbb{R}^d$, $d \in \mathbb{N}$. Let H_D be Laplacian operator in the Hilbert space $\mathfrak{H} = L^2(\Omega, dx)$ with *Dirichlet* boundary conditions, which is defined as follows. We denote by $C_0^\infty(\Omega)$ the C^∞-functions of compact support.

   One considers the *minimal* operator $H_{\min}$, which is the closure of operator

$$(\widetilde{H}u)(x) := (-\Delta u)(x), \quad u \in \mathrm{dom}\,(\widetilde{H}) = C_0^\infty(\Omega).$$

   The operator $H_{\min}$ is symmetric and non-negative, and for the adjoint operator, one gets $\mathrm{dom}\,(H_{\min}^*) = \{u \in \mathfrak{H} : -\Delta u \in \mathfrak{H}\}$.

The *Friedrichs* extension H_F of operator $H_{\min}$ is obtained by associating with $H_{\min}$ the sesquilinear form

$$\mathfrak{t}[u,v] := (H_{\min}u, v) = (\nabla u, \nabla v), \quad u, v \in \mathrm{dom}\,(\mathfrak{t}) := C_0^\infty(\Omega).$$

Closing this form, we obtain a non-negative closed symmetric sesquilinear form $\mathfrak{t}_F$ with domain $\mathrm{dom}\,(\mathfrak{t}_F)$ which is $W_0^{1,2}(\Omega)$. By the sesquilinear form *representation theorem* (Sect. 9.1.2) for non-negative closed symmetric forms, there is a unique non-negative self-adjoint operator H_F such that

$$\mathfrak{t}_F[u,v] = (H_F u, v), \quad u \in \mathrm{dom}\,(H_F), \quad v \in \mathrm{dom}\,(\mathfrak{t}_F) = W_0^{1,2}(\Omega), \tag{6.5.1}$$

where $\mathrm{dom}\,(H_F) = \mathrm{dom}\,(H_{\min}^*) \cap W_0^{1,2}(\Omega)$. Then operator H_F is the Laplacian operator on Ω with *Dirichlet* boundary conditions. So, we denote it by $H_D := H_F$ and its associated closed quadratic form by $\mathfrak{t}_{H_D} := \mathfrak{t}_F$. Since Ω is a bounded open set of $\mathbb{R}^d$, the operator $(\mathbb{1} + H_D)^{-1} \in \mathcal{C}_\infty(\mathfrak{H})$.

Further, let $V(\cdot) : \Omega \longrightarrow \mathbb{R}$ be any non-negative measurable function such that $V(\cdot) \in L^1_{\mathrm{loc}}(\Omega, dx)$ (absolutely integrable on any compact subset of Ω). We introduce the multiplication operator V,

$$(Vu)(x) := V(x)u(x),$$
$$u \in \mathrm{dom}\,(V) := \{u \in L^2(\Omega, dx) : V(x)u(x) \in L^2(\Omega, dx)\}.$$

The operator is self-adjoint and non-negative. Notice that $V \geq 0$. The corresponding sesquilinear form $\mathfrak{t}_V$ is given by

$$\begin{aligned} \mathfrak{t}_V[u,v] &= \int_\Omega dx\, V(x)\, u(x)\, \overline{v(x)}, \quad u, v \in \mathrm{dom}\,(\mathfrak{t}_V)\,, \\ \mathrm{dom}\,(\mathfrak{t}_V) &= \{u \in L^2(\Omega, dx) : \int_\Omega dx\, |\sqrt{V(x)}\, u(x)|^2 < \infty\}. \end{aligned} \tag{6.5.2}$$

On account of (6.5.1) and (6.5.2), one infers that intersection $\mathrm{dom}\,(\mathfrak{t}_D) \cap \mathrm{dom}\,(\mathfrak{t}_V)$ is dense in $L^2(\Omega, dx)$.

Notice that it may happen that $\mathrm{dom}\,(H_D) \cap \mathrm{dom}\,(V) = \{0\}$, which makes the operator sum $H_{DV} := H_D + V$ trivially zero. This is what we observe for H_D and $V(\cdot) \in L^1_{\mathrm{loc}}(\Omega)$; see Example 7.4.12(d). However, the intersection $\mathrm{dom}\,(\mathfrak{t}_{H_D}) \cap \mathrm{dom}\,(\mathfrak{t}_V) \supseteq C_0^\infty(\Omega)$ is always dense in $L^2(\Omega, dx)$. Then, let us consider the form-sum $H := H_D \dotplus V$.

Applying Theorem 6.2.4, we find that for any *regular* Kato function $f \in \mathcal{K}_r$ (Definition 6.2.6) and any Kato function $g \in \mathcal{K}$, one infers

$$\lim_{n\to\infty} \sup_{t\in\mathcal{I}} \left\| (f(tH_D/n)g(tV/n))^n - \mathrm{e}^{-tH} \right\| = 0\,, \tag{6.5.3}$$

for any bounded interval $\mathcal{I} \subseteq \mathbb{R}_0^+$ *away* from zero. Since $f(x) = \mathrm{e}^{-x}$, $x \in \mathbb{R}^+$, is a *regular* Kato function, we get

$$\lim_{n\to\infty} \sup_{t\in\mathcal{I}} \left\| \left(\mathrm{e}^{-tH_D/n}\mathrm{e}^{-tV/n}\right)^n - \mathrm{e}^{-tH} \right\| = 0\,, \tag{6.5.4}$$

uniformly on any closed interval $\mathcal{I} \subset \mathbb{R}_0^+$ *away* from zero.

Now, if we define the non-negative multiplication operator V as

$$V(\cdot) \in \begin{cases} L^1(\Omega, dx), & d = 1, \\ L^r(\Omega, dx), \quad r > d/2, & d = 2, 3, \ldots, \end{cases} \tag{6.5.5}$$

then $\sqrt{V}\,(\mathbb{1} + H_D)^{-1/2} \in \mathcal{C}_\infty(\mathfrak{H})$. Indeed, we have the *compact embeddings*

$$W_0^{1,2}(\Omega) \hookrightarrow \begin{cases} C_b^0(\Omega), & & d = 1, \\ L^q(\Omega, dx), & 1 \le q < \infty, & d = 2, \\ L^q(\Omega, dx), & 1 \le q < 2d/d-2, & d \ge 3, \end{cases}$$

where $C_b^0(\Omega)$ is the Banach space of continuous functions $u(\cdot)$ on Ω such that $\sup_{x\in\Omega} |u(x)| < \infty$.

If $V(\cdot) \in L^1(\Omega, dx)$ and $u(\cdot) \in C_b^0(\Omega)$, then

$$\|\sqrt{V}u\|^2 = \int_\Omega dx\, V(x)\, |u(x)|^2 \le (\sup_{x\in\Omega} |u(x)|)^2 \int_\Omega dx\, V(x) \, < \infty. \tag{6.5.6}$$

Hence, $\mathrm{dom}\,(\sqrt{V}) \supseteq C_b^0(\Omega)$ and $\sqrt{V}$ can be considered as a bounded operator acting from the Banach space $C_b^0(\Omega)$ into $L^2(\Omega, dx)$. Let us denote this operator by $T_0 := \sqrt{V} : C_b^0(\Omega) \longrightarrow L^2(\Omega, dx)$. Since $\mathrm{ran}\,((\mathbb{1} + H_D)^{-1/2}) = W_0^{1,2}(\Omega)$, the operator $(\mathbb{1} + H_D)^{-1/2}$ can be considered as an invertible operator acting from $L^2(\Omega, dx)$ into $W_0^{1,2}(\Omega)$. Doing so, we denote it by $R = (\mathbb{1} + H_D)^{-1/2} : L^2(\Omega, dx) \longrightarrow W_0^{1,2}(\Omega)$.

If $d = 1$, then $W^{1,2}(\Omega) \hookrightarrow C_b^0(\Omega)$ and the corresponding *embedding operator* $J_{C_b^0} : W_0^{1,2}(\Omega) \longrightarrow C_b^0(\Omega)$ is compact. Hence the representation

$$\sqrt{V}\,(\mathbb{1} + H_D)^{-1/2} = T_0 J_{C_b^0} R\,,$$

holds and $\sqrt{V}\,(\mathbb{1} + H_D)^{-1/2} \in \mathcal{C}_\infty(\mathfrak{H})$.

If $V(\cdot) \in L^r(\Omega, dx)$ for some $r > 1$, then by the *Hölder* inequality, we get

$$\|\sqrt{V}u\|^2 = \int_\Omega dx\, V(x)|u(x)|^2 \leq \left(\int_\Omega dx\, V(x)^r\right)^{1/r} \left(\int_\Omega dx\, |u(x)|^s\right)^{2/s} < \infty$$

for $u(\cdot) \in L^s(\Omega, dx)$, where $s := 2r/(r-1) \in (2, \infty)$. Hence, $\mathrm{dom}\,(\sqrt{V}) \supseteq L^s(\Omega, dx)$ and $T_s := \sqrt{V} : L^s(\Omega, dx) \longrightarrow L^2(\Omega, dx)$ are bounded operators.

If $d = 2$ and $r > 1$, then $W^{1,2}(\Omega) \hookrightarrow L^s(\Omega, dx)$ and the embedding operator $J_{L^2(\Omega,dx)} : W_0^{1,2}(\Omega) \longrightarrow L^s(\Omega, dx)$ are compact. Therefore, using the representation $\sqrt{V}\,(\mathbb{1} + H_D)^{-1/2} = T_s J_{L^2(gO,dx)} R$, we find $\sqrt{V}\,(\mathbb{1} + H_D)^{-1/2} \in \mathcal{C}_\infty(\mathfrak{H})$.

If $d \geq 3$ and $r > d/2$, then $W_0^{1,2}(\Omega) \hookrightarrow L^s(\Omega, dx)$, $s = 2r/(r-1)$, provided $s \in [1, 2d/(d-2))$. However, this is satisfied if $r > d/2$. Since the embedding operator $J_{L^2(\Omega,dx)} : W_0^{1,2}(\Omega) \longrightarrow L^s(\Omega, dx)$ is compact, we obtain that $\sqrt{V}\,(\mathbb{1} + H_D)^{-1/2} \in \mathcal{C}_\infty(\mathfrak{H})$.

Therefore, we have shown that under assumption (6.5.5) for any dimension $d \in \mathbb{N}$ the condition of *relative compactness* $\sqrt{V}\,(\mathbb{1} + H_D)^{-1/2} \in \mathcal{C}_\infty(\mathfrak{H})$ is satisfied. Then using Theorem 6.4.4 we deduce for Kato functions f and g (such that $f \in \widehat{\mathcal{K}}$ satisfies in addition the condition (6.4.1)) that for our example of generators H_D and V the Trotter-Kato product formulæ (6.1.11) converge locally uniformly on $\mathbb{R}_0^+$, that is, the *proviso*"away from zero" can be dropped. Since the Kato function $f(x) = \mathrm{e}^{-x}$, $x \in \mathbb{R}_0^+$, satisfies condition (6.4.1), we find that the convergence (6.5.4) holds uniformly for any bounded interval of $\mathbb{R}_0^+$.

2. Let us consider the Hilbert space $\mathfrak{H} = L^2(\mathbb{R}^d, dx)$, $d \in \mathbb{N}$ and Laplacian operator $H_0 := -\Delta$ defined on $\mathrm{dom}\,(H_0) = W^{2,2}(\mathbb{R}^d)$, $d \in \mathbb{N}$. Further, let $V(\cdot) \in L^1_{\mathrm{loc}}(\mathbb{R}^d, dx)$ be a non-negative function such that

$$\lim_{|x|\to\infty} V(x) = \infty.$$

From this condition, one infers that family of multiplication operators $\{Q_\sigma\}_{\sigma>0}$, where

$$Q_\sigma := (\mathbb{1} + V)^{-1}\, \chi_{\{x:\, |x| \geq \sigma\}}\,,$$

satisfies $\lim_{\sigma\to\infty} \|Q_\sigma\| = 0$. Because for any $\sigma > 0$ the operator $\chi_{\{x:|x|<\sigma\}}(\mathbb{1} + H_0)^{-1} \in \mathcal{C}_\infty(\mathfrak{H})$ is compact, one gets that the *product* $(\mathbb{1} + V)^{-1}(\mathbb{1} + H_0)^{-1} \in \mathcal{C}_\infty(\mathfrak{H})$ is also a compact operator. If f and g are Kato functions from the class $\mathcal{K}_\sigma$, then by Theorem 6.3.5, we obtain that

$$\lim_{n\to\infty} \sup_{t\in\mathcal{I}} \left\| (f(tH_0/n)g(tV/n))^n - \mathrm{e}^{-tH} \right\| = 0\,, \tag{6.5.7}$$

locally uniformly in $t \in \mathbb{R}^+$ (i.e. away from zero), for the form-sum $H := H_0 \dot{+} V$. Seeing that exponential functions $f(x) = g(x) = e^{-x}$, $x \in \mathbb{R}_0^+$ satisfy $\sigma[f] < \infty$ and $\sigma[g] < \infty$ (see (6.3.1)), that is, $f, g \in \mathcal{K}_\sigma$, we get

$$\lim_{n\to\infty} \sup_{t\in\mathcal{I}} \left\| \left(e^{-tH_0/n} e^{-tV/n}\right)^n - e^{-tH} \right\| = 0, \tag{6.5.8}$$

for any bounded interval $\mathcal{I} \subset \mathbb{R}_0^+$, away from zero.

3. Here we consider again the Hilbert space $\mathfrak{H} = L^2(\mathbb{R}^d, dx)$, $d \in \mathbb{N}$ and Laplacian operator $H_0 = -\Delta$ defined on dom $(H_0) = W^{2,2}(\mathbb{R}^d)$, $d \in \mathbb{N}$. Let $V(\cdot) : \mathbb{R}^d \longrightarrow \mathbb{R}$ be a non-negative measurable function such that

$$V(\cdot) \in \begin{cases} L^1_{\text{loc}}(\mathbb{R}^d), & d = 1, \\ L^p_{\text{loc}}(\mathbb{R}^d), \quad p > d/2, & d = 2, 3, \ldots, \end{cases} \tag{6.5.9}$$

and suppose that there exists $a > 0$ such that

$$V(x)\big|_{|x|>a} = 0, \quad x \in \mathbb{R}^d, \quad d \in \mathbb{N}. \tag{6.5.10}$$

Then by considerations from above (cf. part 2.), one infers that (6.5.9) implies

$$\sqrt{V}\, \chi_{\{x:\, |x| \leq a\}}\, (\mathbb{1} + H_0)^{-1/2} \in \mathcal{C}_\infty(\mathfrak{H}), \quad a > 0,$$

and that on account of (6.5.10)

$$\sqrt{V} \chi_{\{x:\, |x| > a\}}\, (\mathbb{1} + H_0)^{-1/2} = 0. \tag{6.5.11}$$

As a consequence, operator $\sqrt{V}(\mathbb{1} + H_0)^{-1/2} \in \mathcal{C}_\infty(\mathfrak{H})$, for $d \in \mathbb{N}$. Then using Theorem 6.4.4, we find that for Kato functions f and g such that in addition f satisfies the condition (6.4.1) (i.e. $f \in \widehat{\mathcal{K}}$), the convergence (6.5.7) takes place for any bounded interval $\mathcal{I} \subset \mathbb{R}_0^+$, for $H := H_0 \dot{+} V$. In particular, one has (6.5.8) for any bounded interval $\mathcal{I} \subset \mathbb{R}_0^+$.

6.6 Notes to Chap. 6

The principal results of this chapter are (relatively) recent, and they are essentially due to the authors of this book. Our interest to the *operator-norm* convergent product formulæ for self-adjoint semigroups was triggered by a pioneering paper on this subject due to Dzh.L.Rogava [Rog93], bolstered then by T.Ichinose, H.Tamura [IT97] and [Zag95], as well as by some preliminary results by V.A.Zagrebnov [Zag88] and by H.Neidhardt, V.A.Zagrebnov [NZ90a], [NZ90b] on the Trotter-

Kato product formulæ for Gibbs semigroups converging in (stronger than *operator-norm*) the *trace-norm* topology.

In Chap. 6 we present a list of *sufficient* conditions that ensure operator-norm convergence of the Trotter-Kato approximants for *self-adjoint* semigroups. Besides a general setting, they also concern the resolvent *compactness* properties of the involved self-adjoint semigroup generators.

Notes to Sect. 6.1

In this section we present *uplifting* of the approximation theory for the Trotter-Kato approximants to the *operator-norm* convergence, which was developed in Chap. 5 for the strong operator convergence. Here we study the operator-norm convergence *without* estimates of the *rate* of convergence.

The principle result presented in this section is formulated in four statements about *equivalences*: Corollary 6.1.4, Lemma 6.1.6, Theorem 6.1.7 and Corollary 6.1.8. They involve the Trotter-Kato approximants, resolvents and important family of *auxiliary* operators (6.1.13), which serve as principle instruments for analysis in the following sections. Essentially we were motivated here by results in [NZ99b, Section 2], [INZ04, Sections 2 and 3] and [Zag20, Section 2].

We delay examination of the *rate* of the operator-norm error-bound estimate of convergence in the Trotter-Kato product formulæ to Chap. 7.

Notes to Sect. 6.2

(1) We note that Lemma 6.2.1 has a number of variations [Si79, Theorem 2.16], [Zag80, Proposition 1], [Zag03a], [Zag03b, Proposition 2.56], [Si05, Theorem 2.16]. Here we were inspired by [NZ99b, Lemma 3.1] and [Zag19a, Proposition 2.65].

For the *Heinz inequality*, which concerns the fractional powers $\alpha \in (0, 1)$ of positive bounded self-adjoint operators, we refer to Notes to Chap. 9 (Notes to Section 9.5, Subsection 9.4.4) and [BS87, Theorem 3, Chapter 10, §3] or [RS80, Theorem VI.17].

We recall that for a compact operator $A \in \mathcal{C}_\infty(\mathfrak{H})$, the *canonical representation* has a similar importance as the spectral representation for *self-adjoint* operators; see, e.g. [BS87, Theorem 1, Chapter 11, §1], and in particular for our construction of contractions (6.2.7) and (6.3.9) and (6.3.10).

Proposition 6.6.1 (Canonical Form for Compact Operators) *If* $A \in \mathcal{C}_\infty(\mathfrak{H})$, *then* A^*A *is compact and self-adjoint, and there exists a (not necessarily complete) orthonormal set* $\{\varphi_n\}_{n\geq 1}$ *such that* $A^*A\varphi_n = \lambda_n(A^*A)\varphi_n$ *with* $\lambda_n(A^*A) \neq 0$. *For any* $u \in \mathcal{H}$, *one has* $A^*Au = \sum_{n\geq 1} \lambda_n(A^*A)u_n\varphi_n$ *with* $u_n = (u, \varphi_n)$. *Since* $\ker(A^*A) = [\varphi_1, \varphi_2, \ldots]^\perp$ *and seeing that* $A^*A \geq 0$, *each* $\lambda_n(A^*A) > 0$ *with* $\lim_{n\to\infty} \lambda_n(A^*A) = 0$. *Let* $s_n(A) := \sqrt{\lambda_n(A^*A)}$ *and set* $\psi_n = s_n(A)^{-1}A\varphi_n$, *then the* $\{\psi_n\}_{n\geq 1}$ *are orthonormal, and we obtain the canonical representation of the operator* A:

$$Au = \sum_{n\geq 1} s_n(A)(u, \varphi_n)\psi_n, \qquad A^*u = \sum_{n\geq 1} s_n(A)(u, \psi_n)\varphi_n, \quad u \in \mathfrak{H}. \qquad (6.6.1)$$

Remark 6.6.2 The canonical representation (6.6.1) entails:

(a) That the orthonormal sets $\{\varphi_n\}_{n\geq 1}$ and $\{\psi_n\}_{n\geq 1}$ are related by $\psi_n = s_n(A)^{-1}A\varphi_n$ and that they are complete in $\overline{\text{ran}\, A}$ and in $\overline{\text{ran}\, A^*}$, respectively.
(b) That the numbers $\{s_n(A)\}_{n\geq 1}$, called the singular values of A, are the eigenvalues $\{\lambda_n(|A|)\}_{n\geq 1}$ of the self-adjoint operator $|A| = \sqrt{A^*A}$, with corresponding eigenvectors $\{\varphi_n\}_{n\geq 1}$. Therefore, if $A = U|A|$ is the *polar decomposition* of A, then $\lambda_n(|A|) = s_n(A)$ and $\psi_n = U\varphi_n$.
(c) That if $A = A^* \in \mathcal{C}_\infty(\mathfrak{H})$, $B = B^* \in \mathcal{C}_\infty(\mathfrak{H})$, then by the *Weyl-Courant minimax principle* (see, e.g. [RS78, Theorem XIII.1]) $s_n(A) \leq s_n(B)$. So, on account of Proposition 6.6.1 for operators A and B, we obtain due to canonical representation (6.6.1)

$$Au = \sum_{n\geq 1} s_n(A)(u, \varphi_n)\, \varphi_n\,, \qquad Bu = \sum_{n\geq 1} s_n(B)(u, \phi_n)\, \phi_n\,, \quad u \in \mathfrak{H}, \qquad (6.6.2)$$

that there exists a *contraction* $\varGamma_{AB} : \overline{\text{ran}\, B} \to \mathfrak{H}$, such that $A = \varGamma_{AB}B$.

(2) The condition $f(x) > 0$, $x \in \mathbb{R}^+$ in Theorem 6.2.4 can be removed. This makes the arguments more complicated.

(3) The condition $f_0(\tau A) \in \mathcal{C}_\infty(\mathfrak{H})$ for some $\tau > 0$ is stronger than $(\mathbb{1} + A)^{-1} \in \mathcal{C}_\infty(\mathfrak{H})$. Notice that $0 \leq \varphi_0(x) \leq 1$, $x \in \mathbb{R}^+$, implies $(\mathbb{1} + A)^{-1} \in \mathcal{C}_\infty(\mathfrak{H})$.

Indeed, for $x \in \mathbb{R}^+$, we have $1+x\varphi_0(x) \leq 1+x$, which yields $(1+x)^{-1} \leq f_0(x)$. As a consequence, $f_0(\tau A) \in \mathcal{C}_\infty(\mathfrak{H})$ implies

$$(\mathbb{1} + \tau A)^{-1} \in \mathcal{C}_\infty(\mathfrak{H})\,, \quad \tau > 0\,.$$

The converse is certainly *not* true; see Remark 6.2.5. This results from the following example.

Example 6.6.3 Let $\{\varDelta_n\}_{n=0}^\infty$ be a partition of $\mathbb{R}_+$ given by

$$\varDelta_0 = [0, 1), \qquad \varDelta_n := [n^n, (n+1)^{n+1}), \qquad n = 1, 2, \ldots.$$

We define a Borel function $f(\cdot)$ setting

$$\varphi(x) := \begin{cases} 1 & : \quad x \in \varDelta_0 \\ n^{-n} & : \quad x \in \varDelta_n, \qquad n = 1, 2, \ldots \end{cases}$$

and

$$f(x) := (1 + x\varphi(x))^{-1}, \qquad x \in \mathbb{R}_+.$$

A straightforward computation shows that by definition (6.2.12) that $f_0(x) = f(x)$, $x \in \mathbb{R}^+$. Let A be a non-negative self-adjoint operator with simple discrete spectrum given by $\{\lambda_n\}_{n=0}^{\infty}$, $\lambda_n = n^n$. Since $\lim_{n\to\infty} \lambda_n = +\infty$, the resolvent of A is compact. Further, we have

$$f_0(\lambda_n) = f(\lambda_n) = \left(1 + n^n n^{-n}\right)^{-1} = \frac{1}{2}, \qquad n = 1, 2, \dots ,$$

which yields $\lim_{n\to\infty} f_0(\lambda_n) = 1/2$. Therefore, the essential spectrum of $f_0(A)$ contains the point $\{1/2\}$, which is impossible if $f_0(A)$ is compact. Note that if the spectrum of A is given by $\lambda_n := (n+1)^{n+1} - 1$, then $\lim_{n\to\infty} f_0(\lambda_n) = 0$. On that account $f_0(A) \in \mathcal{C}_\infty(\mathfrak{H})$.

(4) Seeing that Definition 6.2.6 of the *regular* Kato functions $\mathcal{K}_r$ is *roundabout*, a natural question arises about another characterisation of the class $\mathcal{K}_r$; for more details, see [Zag19, Appendix C.3].

Let $f(\cdot)$ be a Kato function and let f_0 be defined by (6.2.12). We set

$$b(x) := \sup_{0\le s\le x} sf(s) \quad \text{and} \quad r(x) := \sup_{s\in[x,\infty)} f(s)\,, \qquad x \in \mathbb{R}_0^+ .$$

Then we note that $b(\cdot)$ is non-decreasing and $0 \le b(x) \le x$, for $x \in \mathbb{R}_0^+$, while $r(x)$ is non-increasing such that $0 \le r(x) \le 1$, for $x \in \mathbb{R}_0^+$.

Theorem 6.6.4 *The Kato function $f \in \mathcal{K}$ is regular (i.e. $f \in \mathcal{K}_r$) if and only if*

$$\lim_{x\to+\infty} \frac{b(x)}{x} = 0\,, \tag{6.6.3}$$

and

$$0 \le r(x) < 1\,, \tag{6.6.4}$$

for $x \in \mathbb{R}^+$.

We skip the proof and refer to [NZ99b, Lemma 3.4].

5. About the *Weyl-von Neumann* theorem and the corresponding proof, see, for example, [Kato80, Chapter X, Theorem 2.1].

For the *von Neumann-Schatten* ideals (including the *Hilbert-Schmidt* operators), see, e.g. [BS87, Chapter 11], or [Zag19, Chapter 2].

Notes to Sect. 6.3

We note that on account of Remark 6.2.5, the condition $f_0(\tau_0 A) \in \mathcal{C}_\infty(\mathfrak{H})$ for some $\tau_0 > 0$ yields $(\mathbb{1} + A)^{-1} \in \mathcal{C}_\infty(\mathfrak{H})$. As a consequence, if $f_0(\tau_0 A) \in \mathcal{C}_\infty(\mathfrak{H})$ for some $\tau > 0$ is satisfied , then the product $(\mathbb{1} + A)^{-1}(\mathbb{1} + B)^{-1}$ is compact, and Theorem 6.3.5 yields Theorem 6.2.4. However, there is a slight difference since Theorem 6.3.5 holds for Kato functions f and g satisfying $\sigma[f] < \infty$ and $\sigma[g] <$

∞, whereas Theorem 6.2.4 holds for arbitrary Kato functions such that $f_0(\tau_0 A)$ is compact for some $\tau_0 > 0$.

Let us show that the class of functions for which Theorem 6.2.4 holds is, in fact, larger. Indeed, it turns out that Kato functions $\mathcal{K}$ satisfying $\sigma[f] < \infty$ are regular.

Lemma 6.6.5 *If $f \in \mathcal{K}$ is a Kato function such that $\sigma[f] < \infty$, then it is a regular Kato function, that is, $f \in \mathcal{K}_\sigma \subset \mathcal{K}_r$.*

Proof If $\sigma := \sigma[f] < \infty$, then

$$f(x) \leq \frac{\sigma}{\sigma + x}, \qquad x \in \mathbb{R}_0^+.$$

Consequently $r(x) := \sup_{s\in[x,\infty)} f(s) < 1$, for $x > 0$. Moreover, $b(x) = \sup_{0\leq s\leq x} sf(s) < \infty$. Hence $\lim_{x\to\infty} b(x)/x = 0$. Then applying Theorem 6.6.4, we obtain $f \in \mathcal{K}_r$. □

We remind here that if $f \in \mathcal{K}_r$, then owing to Definition 6.2.6, one obtains the equivalence: $f_0(A) \in \mathcal{C}_\infty(\mathfrak{H}) \Longleftrightarrow (\mathbb{1} + A)^{-1} \in \mathcal{C}_\infty(\mathfrak{H})$.

In this section we partially follow [NZ99b, Section 5.]. More about *hierarchy* of the Kato functions one can find in [Zag19, Appendix C].

Notes to Sect. 6.4

The results on relative compactness presented here are stronger than those in [NZ99a, Section 7] and [NZ99b, Theorem 4.1]. There it was proven that under assumption of the fractional relative compactness for $\alpha = 1/2$, the Trotter-Kato product formula converges in the operator-norm topology uniformly on any (bounded) interval in $\mathbb{R}_0^+$, *away* from zero (cf. Remark 6.1.5).

In this section we refined our arguments developed in [NZ99b]. As a consequence, we prove in Theorem 6.4.4 that requirement "away from zero" may be omitted.

The conjecture in [NZ99b, Section 6 (5.)] that Theorem 6.4.4 remains true under assumption that operator $B^{1/2}$ is relatively $A^{1/2}$-bounded with the *relative bound* equals to *zero* is (up to our knowledge) an open question. In this connection, we mention paper [BiHr00] for arguments on this question.

Notes to Sect. 6.5

For quadratic forms and construction of the Friedrichs extension, see, for example, [Kato80, Chapter VI, §2.3], [Schm12, Section 10.6].

To bolster our abstract results from Sects. 6.2–6.4, we present here a variety of examples (with different compactness properties), which are sufficient to ensure the operator-norm convergence of the Trotter-Kato approximants for self-adjoint semigroups.

Chapter 7
Trotter-Kato Product Formulae: Operator-Norm Topology and Error Bounds

In this chapter, we present two fundamental results about the operator-norm *optimal* error bound estimates for convergence rate of the Trotter-Kato product formulæ for C_0-semigroups with noncommuting self-adjoint generators A and B.

First, the ITTZ-Theorem (Sect. 7.1) shows that *self-adjointness* of the *sum* of operators $A + B$, is a crucial condition for the operator-norm convergence of the Trotter-Kato product formulæ with the *optimal* error bound estimate for the rate of convergence.

The second statement (INZ-Theorem, Sect. 7.4) concerns the case when it is the *form-sum* of operators: $A \dot{+} B$, which is self-adjoint. In Sect. 7.4 we also establish a *dependence* of the *optimal* operator-norm error bound estimate for convergence rate of the Trotter-Kato product formulæ on relations between *fractional* powers of the self-adjoint generators A and B.

7.1 Trotter-Kato Product Formulæ and the Rate of Convergence

In this section, we improve results of the previous chapter about the operator-norm convergence of the Trotter-Kato product formulæ by indicating the error bound estimates for the rate of convergence. We shall call them *optimal* if the corresponding rate is the best possible for a given conditions.

V. A. Zagrebnov et al., *Trotter-Kato Product Formulæ*, Operator Theory: Advances and Applications 296, https://doi.org/10.1007/978-3-031-56720-9_7

Definition 7.1.1 For (generic) Kato function $f \in \mathcal{K}$, we define three functionals:

$$\gamma[f] := \sup_{x>0} \frac{1 - f(x)}{x}, \tag{7.1.1}$$

$$\varkappa[f] := \sup_{x>0} \frac{1}{x} \left| \frac{1 - f(x)}{x} - 1 \right|, \tag{7.1.2}$$

and

$$\omega[f] := \sup_{x>0} \frac{1}{\sqrt{1 - f(x)}} \left| \frac{1 - f(x)}{x} - 1 \right|. \tag{7.1.3}$$

A priori, some of them may be *unbounded.*

Note that for *generic* Kato functions (class $\mathcal{K}$, Definition 5.2.1), one always has $\gamma[f] < \infty$. However, the functional (7.1.2) for $f \in \mathcal{K}$ may be infinite. That being so, the condition $\varkappa[f] < \infty$ controls the limit $x \to 0$ and defines a proper *subclass* of the Kato functions $\mathcal{K}$. Similarly, the condition $\omega[f] < \infty$ is nontrivial too; see Remark D.0.2. It controls behaviour of the "tail" of function $f \in \mathcal{K}$ "away from zero", namely, for $x \geq \delta$ and arbitrary $\delta > 0$, Lemma D.0.1. This condition also selects a proper *subclass* of the Kato functions $\mathcal{K}$; see Appendix D.

Let A and B be non-negative self-adjoint operators in a Hilbert space $\mathfrak{H}$. We denote by $\{E_A(\lambda)\}_{\lambda \geq 0}$ and $\{E_B((\lambda)\}_{\lambda \geq 0}\}$ the projection-valued (*spectral*) families associated with these operators and by $E_A(d\lambda)$ and $E_B((d\lambda)$ the corresponding *spectral measures.*

Then $P_A := E_A((0, \infty))$ and $P_B = E_B((0, \infty))$ are spectral projections corresponding to the interval $(0, \infty)$. We note that these projections are non-trivial ($\neq \mathbb{1}$) if ker (A) and ker (B) are non-empty. We also comment that operators $A^{-1} P_A$ and $B^{-1} P_B$ are well-defined on $\mathfrak{H}$.

Lemma 7.1.2 *Let A and B be non-negative self-adjoint operators in a Hilbert space $\mathfrak{H}$. If f and g are Kato functions from $\mathcal{K}$ satisfying $\varkappa[f] < \infty$ and $\varkappa[g] < \infty$ (see* (7.1.2)*), then*

$$\|A(\tau)\, A^{-1} P_A\| \leq \gamma[f], \quad \tau > 0, \tag{7.1.4}$$

$$\|(A(\tau) - A)\, A^{-2} P_A\| \leq \varkappa[f]\, \tau, \quad \tau > 0, \tag{7.1.5}$$

$$\left\| \left(B(\tau/2) - B - \frac{\tau}{4} B(\tau/2)^2 \right) B^{-2} P_B \right\| \leq \frac{1}{2} (\varkappa[g] + \frac{1}{2} \gamma[g]^2)\, \tau, \quad \tau > 0\,, \tag{7.1.6}$$

here $A(\tau) = (\mathbb{1} - f(\tau A))/\tau$ *and* $B(\tau) = (\mathbb{1} - g(\tau B))/\tau$, *cf.* (5.2.4).

Proof By (5.2.4) and definition of projector P_A, one readily gets

$$\left\| A(\tau)\, A^{-1} P_A \right\| = \sup_{x>0} \left| \frac{1 - f(\tau x)}{\tau x} \right| = \sup_{x>0} \left| \frac{1 - f(x)}{x} \right| = \gamma[f]\,,$$

as well as

$$\begin{aligned} \left\| (A(\tau) - A)\, A^{-2} P_A \right\| &= \sup_{x>0} \frac{1}{x^2} \left| \frac{1 - f(\tau x)}{\tau} - x \right| \\ &= \sup_{x>0} \frac{1}{x} \left| \frac{1 - f(x)}{x} - 1 \right| = \varkappa[f]\,. \end{aligned}$$

Finally, we note the

$$\begin{aligned} &\left\| \left(B(\tau/2) - B - \frac{\tau}{4} B(\tau/2)^2 \right) B^{-2} P_B \right\| \\ &\leq \left\| (B(\tau/2) - B)\, B^{-2} P_B \right\| + \frac{\tau}{4} \left\| B(\tau/2)^2\, B^{-2} P_B \right\| \,, \end{aligned}$$

which yields

$$\left\| \left(B(\tau/2) - B - \frac{\tau}{4}\, B(\tau/2)^2 \right) B^{-2} P_B \right\| \leq \frac{1}{2} \left(\varkappa[g] + \frac{1}{2} \gamma[g]^2 \right) \tau\,,$$

that finishes the proof of (7.1.4)–(7.1.6). □

Remark 7.1.3 (The key assumption) Note that in the following we *assume* that

- operators A and B are self-adjoint and *non-negative*,
- operator $H := A + B$ on $\mathrm{dom}\,(H) := \mathrm{dom}\,(A) \cap \mathrm{dom}\,(B)$ is *self-adjoint.*

These relatively strong assumptions have several implications and finally yields the *main* result of Sect. 7.2 (ITTZ-Theorem 7.1.9), namely, the *optimal* error bound for convergence of the Trotter-Kato product formula. For discussion and comments, see Remark 7.1.14.

As the first consequence of this key assumption, one can readily deduce that operators $A(\mathbb{1} + H)^{-1}$ and $B(\mathbb{1} + H)^{-1}$ are bounded. Then we set

$$C_H^A := \|A(\mathbb{1} + H)^{-1}\| \quad \text{and} \quad C_H^B := \|B(\mathbb{1} + H)^{-1}\|. \tag{7.1.7}$$

Lemma 7.1.4 *Let A and B be non-negative self-adjoint operators in a Hilbert space such that operator $H := A+B$ is self-adjoint. If $f, g \in \mathcal{K}$ are Kato functions satisfying $\varkappa[f] < \infty$ and $\varkappa[g] < \infty$ (see* (7.1.2)*), then*

$$\|(\mathbb{1} + H)^{-1}(S(\tau) - H)(\mathbb{1} + H)^{-1}\| \leq C_{H,H}^{S-H}\, \tau\,, \quad \tau > 0\,. \tag{7.1.8}$$

Here $F(\tau) = g(\tau B/2) f(\tau A) g(\tau B/2)$ *and* $S(\tau) = (\mathbb{1} - F(\tau))/\tau$, *cf.* (5.2.3) *and* (5.2.4).

Proof Notice that $S(\tau) = K(\tau) + W(\tau)$, where $K(\tau)$ and $W(\tau)$ are given by (5.2.6) and (5.2.9), respectively. Hence

$$
\begin{aligned}
\|(\mathbb{1}+H)^{-1}(S(\tau) - H)(\mathbb{1} + H)^{-1}\| \leq \\
\|(\mathbb{1} + H)^{-1}(K(\tau) - H)(\mathbb{1} + H)^{-1}\| + \qquad (7.1.9)\\
\|(\mathbb{1} + H)^{-1} W(\tau)(\mathbb{1} + H)^{-1}\|, \qquad (7.1.10)
\end{aligned}
$$

$\tau > 0$. We use the representation

$$
\begin{aligned}
(\mathbb{1}+H)^{-1}(K(\tau) - H)(\mathbb{1} + H)^{-1} &= (\mathbb{1} + H)^{-1}(A(\tau) - A)(\mathbb{1} + H)^{-1} \\
&+ (\mathbb{1} + H)^{-1}(B(\tau/2) - B - \frac{\tau}{4} B(\tau/2)^2)(\mathbb{1} + H)^{-1} ,
\end{aligned}
$$

which yields the estimate

$$
\begin{aligned}
\|(\mathbb{1} + H)^{-1}(K(\tau) - H)(\mathbb{1} + H)^{-1}\| \leq \\
\|(\mathbb{1} + H)^{-1}(A(\tau) - A)(\mathbb{1} + H)^{-1}\| + \qquad (7.1.11)\\
\|(\mathbb{1} + H)^{-1}(B(\tau/2) - B - \frac{\tau}{4} B(\tau/2)^2)(\mathbb{1} + H)^{-1}\| ,
\end{aligned}
$$

$\tau > 0$. Using the relation

$$
\begin{aligned}
&(\mathbb{1} + H)^{-1}(A(\tau) - A)(\mathbb{1} + H)^{-1} \\
&\qquad = (\mathbb{1} + H)^{-1} A \, (A(\tau) - A) A^{-2} P_A \, A(\mathbb{1} + H)^{-1} ,
\end{aligned}
$$

we find

$$
\|(\mathbb{1} + H)^{-1}(A(\tau) - A)(\mathbb{1} + H)^{-1}\| \leq (C_H^A)^2 \left\|(A(\tau) - A)A^{-2} P_A\right\| , \quad \tau > 0.
$$

Applying (7.1.5), we obtain the estimate

$$
\|(\mathbb{1} + H)^{-1}(A(\tau) - A)(\mathbb{1} + H)^{-1}\| \leq (C_H^A)^2 \, \varkappa[f] \, \tau, \quad \tau > 0. \qquad (7.1.12)
$$

Similarly we prove that

$$
\begin{aligned}
&\|(\mathbb{1} + H)^{-1}(B(\tau/2) - B - \frac{\tau}{4} B(\tau/2)^2)(\mathbb{1} + H)^{-1}\| \\
&\qquad \leq \frac{(C_H^B)^2}{2} \left(\varkappa[g] + \frac{1}{2}\gamma[g]^2\right) \tau, \quad \tau > 0.
\end{aligned} \qquad (7.1.13)
$$

Hence, inserting (7.1.12) and (7.1.13) into (7.1.11), we obtain

$$\|(\mathbb{1}+H)^{-1}(K(\tau)-H)(\mathbb{1}+H)^{-1}\| \leq \left((C_H^A)^2\varkappa[f] + \frac{(C_H^B)^2}{2}\left(\varkappa[g]+\frac{1}{2}\gamma[g]^2\right)\right)\tau, \quad \tau>0.$$

Setting $C_{H,H}^{K-H} := (C_H^A)^2\,\varkappa[f] + (C_H^B)^2\,(\varkappa[g]+\frac{1}{2}\gamma[g]^2)/2$, we prove

$$\|(\mathbb{1}+H)^{-1}(K(\tau)-H)(\mathbb{1}+H)^{-1}\| \leq C_{H,H}^{K-H}\,\tau, \quad \tau>0. \tag{7.1.14}$$

Note that from (5.2.9), we infer the estimate

$$\|(1+H)^{-1}W(\tau)(\mathbb{1}+H)^{-1}\|$$
$$\leq \frac{\tau^2}{4}\,\|(\mathbb{1}+H)^{-1}B(\tau/2)A(\tau)B(\tau/2)(\mathbb{1}+H)^{-1}\| + \tag{7.1.15}$$
$$\tau\,\|(\mathbb{1}+H)^{-1}A(\tau)\|\,\|(B(\tau/2)(\mathbb{1}+H)^{-1})\| \tag{7.1.16}$$

To estimate (7.1.15), we note that

$$\frac{\tau^2}{4}\,\|(\mathbb{1}+H)^{-1}B(\tau/2)A(\tau)B(\tau/2)(\mathbb{1}+H)^{-1}\| \leq \frac{\tau}{4}\|(\mathbb{1}+H)^{-1}B(\tau/2)\|\,\|B(\tau/2)(\mathbb{1}+H)^{-1}\|, \quad \tau>0,$$

which leads to

$$\frac{\tau^2}{4}\,\|(\mathbb{1}+H)^{-1}B(\tau/2)A(\tau)B(\tau/2)(\mathbb{1}+H)^{-1}\| \leq \frac{\tau}{4}(C_H^B)^2\|B(\tau/2)B^{-1}P_B\|\,\|B(\tau/2)B^{-1}P_B\|, \quad \tau>0.$$

Note that P_B is the orthogonal projection onto $\ker(B)^{\perp}$. Using definition (7.1.1), we immediately obtain

$$\frac{\tau^2}{4}\,\|(\mathbb{1}+H)^{-1}B(\tau/2)A(\tau)B(\tau/2)(\mathbb{1}+H)^{-1}\| \leq \frac{\tau}{4}\,(C_H^B)^2\gamma[g]^2, \quad \tau>0. \tag{7.1.17}$$

Further we have

$$\tau\,\|(\mathbb{1}+H)^{-1}A(\tau)\|\,\|(B(\tau/2)(\mathbb{1}+H)^{-1})\|$$
$$\leq \tau\,C_H^A C_H^B\|A(\tau)A^{-1}P_A\|\,\|(B(\tau/2)B^{-1}P_B\| \leq \tau)\,C_H^A C_H^B\gamma[f]\gamma[g], \quad \tau>0. \tag{7.1.18}$$

Inserting (7.1.17) and (7.1.18) into (7.1.15) and (7.1.16), we obtain

$$\|(1+H)^{-1}W(\tau)(\mathbb{1}+H)^{-1}\| \leq \left(\frac{1}{4}(C_H^B)^2\gamma[g]^2 + C_H^A C_H^B \gamma[f]\gamma[g]\right)\tau, \quad \tau > 0.$$

Then setting $C_{H,H}^W := (C_H^B)^2\gamma[g]^2/4 + C_H^A C_H^B \gamma[f]\gamma[g]$ we prove the estimate

$$\|(1+H)^{-1}W(\tau)(\mathbb{1}+H)^{-1}\| \leq C_{H,H}^W\,\tau\,, \quad \tau > 0\,. \tag{7.1.19}$$

Inserting (7.1.14) and (7.1.19) into (7.1.9) and (7.1.10), we obtain

$$\|(\mathbb{1}+H)^{-1}(S(\tau)-H)(\mathbb{1}+H)^{-1}\| \leq (C_{H,H}^{K-H} + C_{H,H}^W)\,\tau\,, \quad \tau > 0. \tag{7.1.20}$$

As a consequence, definition $C_{H,H}^{S-H} := C_{H,H}^{K-H} + C_{H,H}^W$ yields the estimate (7.1.8). □

Lemma 7.1.5 *Let A and B be non-negative self-adjoint operators in a Hilbert space. If $f, g \in \mathcal{K}$ are Kato functions satisfying $\omega[f] < \infty$ and $\omega[g] < \infty$ (see (7.1.3)), then*

$$\|(\mathbb{1}+A(\tau))^{-1/2}(A(\tau)-A)\,A^{-1}P_A\| \leq \omega[f]\sqrt{\tau}, \quad \tau > 0, \tag{7.1.21}$$

and

$$\|(\mathbb{1}+B(\tau/2))^{-1/2}\left(B(\tau/2) - B - \frac{\tau}{4}\,B(\tau/2)^2\right)B^{-1}P_B\| \leq \frac{1}{\sqrt{2}}(\omega[g] + \frac{1}{2}\gamma[g])\,\sqrt{\tau}, \quad \tau > 0. \tag{7.1.22}$$

Proof Notice that by (7.1.3), we get

$$\left\|(\mathbb{1}+A(\tau))^{-1/2}(A(\tau)-A)\,A^{-1}P_A\right\| = \sup_{x>0}\left|\frac{1}{\sqrt{1+(1-f(\tau x))/\tau}}\left(\frac{1-f(\tau x)}{\tau} - x\right)\frac{1}{x}\right|, \tag{7.1.23}$$

which for $\tau > 0$ yields

$$\left\|(\mathbb{1}+A(\tau))^{-1/2}(A(\tau)-A)\,A^{-1}P_A\right\| \leq \sqrt{\tau}\,\sup_{x>0}\left|\frac{1}{\sqrt{1-f(x)}}\left(\frac{1-f(x)}{\tau} - 1\right)\right| = \omega[f]\,\sqrt{\tau}\,, \tag{7.1.24}$$

and consequently the proof of (7.1.21).

On the other hand, we have inequality

$$\begin{aligned} &\left\|(\mathbb{1} + B(\tau/2))^{-1/2}\left(B(\tau/2) - B - \frac{\tau}{4}B(\tau/2)^2\right)B^{-1}P_B\right\| \\ &\leq \left\|(\mathbb{1} + B(\tau/2))^{-1/2}\,(B(\tau/2) - B)\,B^{-1}P_B\right\| \\ &+ \frac{\tau}{4}\left\|(\mathbb{1} + B(\tau/2))^{-1/2}B(\tau/2)^2B^{-1}P_B\right\| . \end{aligned} \tag{7.1.25}$$

Now, similarly to (7.1.23) and (7.1.24), we obtain estimate

$$\left\|(\mathbb{1} + B(\tau/2))^{-1/2}\,(B(\tau/2) - B)\,B^{-1}P_B\right\| \leq \frac{1}{\sqrt{2}}\,\omega[g]\,\sqrt{\tau} \tag{7.1.26}$$

and by (7.1.3), together with definition $B(\tau) = (\mathbb{1} - g(\tau B))/\tau$, the representation

$$\begin{aligned} &\frac{\tau}{4}\left\|(\mathbb{1} + B(\tau/2))^{-1/2}B(\tau/2)^2B^{-1}P_B\right\| \\ &= \frac{\tau}{4}\sup_{x>0}\left|\frac{1}{\sqrt{1 + (1 - g(\tau x/2))/\tau/2}}\,\frac{1 - g(\tau x/2)}{\tau/2}\,\frac{1 - g(\tau x/2)}{\tau/2}\,\frac{1}{x}\right| , \end{aligned}$$

which yields

$$\begin{aligned} &\frac{\tau}{4}\left\|(\mathbb{1} + B(\tau/2))^{-1/2}B(\tau/2)^2B^{-1}P_B\right\| \\ &\leq \frac{1}{2\sqrt{2}}\sqrt{\tau}\,\sup_{x>0}\left|\sqrt{1 - g(\tau x/2)}\,\frac{1 - g(\tau x/2)}{\tau x/2}\right| \leq \frac{1}{2\sqrt{2}}\,\gamma[g]\,\sqrt{\tau} . \end{aligned} \tag{7.1.27}$$

From (7.1.25), (7.1.26) and (7.1.27), we obtain (7.1.22). □

Next we prove the following important technical lemma.

Lemma 7.1.6 *Let A and B be non-negative self-adjoint operators in a Hilbert space such that $H := A + B$ is self-adjoint. If f and g are Kato functions from $\mathcal{K}$, then there are constants $C_H^S > 0$ and $C_K^S > 0$ such that the estimates*

$$\|S(\tau)\,(\mathbb{1} + H)^{-1}\| \leq C_H^S, \quad \tau > 0, \tag{7.1.28}$$

and

$$\|(\mathbb{1} + K(\tau))^{-1/2}\sqrt{S(\tau)}\| \leq 2, \quad \tau > 0. \tag{7.1.29}$$

are valid for $S(\tau) = \tau^{-1}[\mathbb{1} - g(\tau B/2)f(\tau A)g(\tau B/2)]$.

Proof By $S(\tau) = K(\tau) + W(\tau)$, where $K(\tau)$ and $W(\tau)$ are given by (5.2.6) and (5.2.9), respectively (cf. (5.2.8)), we get

$$\|S(\tau)(\mathbb{1} + H)^{-1}\| \le \|K(\tau)(\mathbb{1} + H)^{-1}\| + \|W(\tau)(\mathbb{1} + H)^{-1}\|, \tau > 0.$$

By (5.2.7) we find the estimate

$$\|K(\tau)(\mathbb{1} + H)^{-1}\| \le \|A(\tau)(\mathbb{1} + H)^{-1}\| + \|B(\tau/2)(\mathbb{1} + H)^{-1}\|, \quad \tau > 0. \tag{7.1.30}$$

Since $H = A + B$ we have $\mathrm{dom}\,(H) \subseteq \mathrm{dom}\,(A)$. Hence $C_H^A := \|A(\mathbb{1} + H)^{-1}\| < \infty$, (7.1.7). Therefore

$$\|A(\tau)(\mathbb{1} + H)^{-1}\| \le \|A(\tau)A^{-1}P_A A(\mathbb{1} + H)^{-1}\| \le C_H^A \|A(\tau)A^{-1}P_A\|,$$

$\tau > 0$, where P_A is the orthogonal projection onto the subspace $\ker(A)^{\perp}$. Using definition of operator $A(\tau)$ and functional (7.1.1), we find

$$\|A(\tau)(\mathbb{1} + H)^{-1}\| \le \|A(\tau)A^{-1}P_A A(\mathbb{1} + H)^{-1}\| \le C_H^A \gamma[f], \quad \tau > 0. \tag{7.1.31}$$

Similarly we have $\mathrm{dom}\,(H) \subseteq \mathrm{dom}\,(B)$, which yields

$$\|B(\tau/2)(\mathbb{1} + H)^{-1}\| \le \|B(\tau/2)B^{-1}P_B B(\mathbb{1} + H)^{-1}\| \le C_H^B \gamma[g], \quad \tau > 0, \tag{7.1.32}$$

where $C_H^B := \|B(\mathbb{1} + H)^{-1}\|$, (7.1.7). Inserting (7.1.31) and (7.1.32) into (7.1.30), we find

$$\|K(\tau)(\mathbb{1} + H)^{-1}\| \le C_H^A \gamma[f] + C_H^B \gamma[g], \quad \tau > 0.$$

Setting $C_H^S := C_H^A \gamma[f] + C_H^B \gamma[g]$ we prove (7.1.28).

To prove (7.1.29) we note that

$$(\mathbb{1}+K(\tau))^{-1/2} S(\tau)(\mathbb{1} + K(\tau))^{-1/2} = (\mathbb{1} + K(\tau))^{-1/2} K(\tau)(\mathbb{1} + K(\tau))^{-1/2}$$
$$+ (\mathbb{1} + K(\tau))^{-1/2} W(\tau)(\mathbb{1} + K(\tau))^{-1/2}, \quad \tau > 0,$$

which yields the estimate

$$\begin{aligned} &\|(\mathbb{1}+K(\tau))^{-1/2} S(\tau)(\mathbb{1} + K(\tau))^{-1/2}\| \\ &\quad \le 1 + \|(\mathbb{1} + K(\tau))^{-1/2} W(\tau)(\mathbb{1} + K(\tau))^{-1/2}\|, \quad \tau > 0. \end{aligned} \tag{7.1.33}$$

By definition of $W(\tau)$ (5.2.9), we get the estimate

$$\|(\mathbb{1}+K(\tau))^{-1/2}W(\tau)(\mathbb{1}+K(\tau))^{-1/2}\| \leq$$
$$\frac{\tau^2}{4}\,\|(\mathbb{1}+K(\tau))^{-1/2}B(\tau/2)A(\tau)B(\tau/2)(\mathbb{1}+K(\tau))^{-1/2}\|+$$
$$\tau\,\|(\mathbb{1}+K(\tau))^{-1/2}A(\tau)\|\,\|B(\tau/2)(\mathbb{1}+K(\tau))^{-1/2}\|, \quad \tau>0,$$

which on account of $f, g \in \mathcal{K}$ (thus, $\|\sqrt{A(\tau)}\| = \sup_{x>0}\sqrt{(1-f(\tau x))/\tau} \leq \tau^{-1/2}$, and similarly $\|\sqrt{B(\tau/2)}\| \leq \sqrt{2/\tau}$) leads to inequality

$$\|(\mathbb{1}+K(\tau))^{-1/2}W(\tau)(\mathbb{1}+K(\tau))^{-1/2}\| \leq$$
$$\frac{\tau^2}{4}\|(\mathbb{1}+K(\tau))^{-1/2}B(\tau/2)A(\tau)B(\tau/2)(\mathbb{1}+K(\tau))^{-1/2}\|+$$
$$\sqrt{2}\,\|(\mathbb{1}+K(\tau))^{-1/2}\sqrt{A(\tau)}\|\,\|\sqrt{B(\tau/2)}(\mathbb{1}+K(\tau))^{-1/2}\|, \quad \tau>0. \tag{7.1.34}$$

Now we note that by definition of operators $A(\tau)$ and $B(\tau)$, one gets

$$\frac{\tau^2}{4}\|(\mathbb{1}+K(\tau))^{-1/2}B(\tau/2)A(\tau)B(\tau/2)(\mathbb{1}+K(\tau))^{-1/2}\| \leq \tag{7.1.35}$$
$$\frac{1}{2}\|(\mathbb{1}+K(\tau))^{-1/2}\sqrt{B(\tau/2)}\|^2 = \frac{1}{2}\|\sqrt{B(\tau/2)}(\mathbb{1}+K(\tau))^{-1}\sqrt{B(\tau/2)}\|\,,$$

for $\tau > 0$. On account of (5.2.7), we have

$$\mathbb{1}+\frac{1}{2}B(\tau/2) \leq \mathbb{1}+K(\tau), \quad \tau>0,$$

which yields

$$(\mathbb{1}+K(\tau))^{-1} \leq (\mathbb{1}+\frac{1}{2}B(\tau/2))^{-1} \leq 2(\mathbb{1}+B(\tau/2))^{-1}, \quad \tau>0.$$

Therefore,

$$(\mathbb{1}+K(\tau))^{-1} \leq (\mathbb{1}+\frac{1}{2}B(\tau/2))^{-1} \leq 2(\mathbb{1}+B(\tau/2))^{-1}, \quad \tau>0.$$

This leads for $\tau > 0$ to inequality

$$\frac{1}{2}\sqrt{B(\tau/2)}(\mathbb{1}+K(\tau))^{-1}\sqrt{B(\tau/2)} \leq \sqrt{B(\tau/2)}(\mathbb{1}+B(\tau/2))^{-1}\sqrt{B(\tau/2)}\,,$$

which shows that

$$\frac{1}{2}\sqrt{B(\tau/2)}(\mathbb{1} + K(\tau))^{-1}\sqrt{B(\tau/2)} \leq \mathbb{1}, \quad \tau > 0. \tag{7.1.36}$$

As a consequence, by virtue of (7.1.35), we obtain

$$\frac{\tau^2}{4}\|(\mathbb{1} + K(\tau))^{-1/2}B(\tau/2)A(\tau)B(\tau/2)(\mathbb{1} + K(\tau))^{-1/2}\| \leq 1, \quad \tau > 0. \tag{7.1.37}$$

Next, from (7.1.36) we also infer that

$$\|\sqrt{B(\tau/2)}(\mathbb{1} + K(\tau))^{-1}\sqrt{B(\tau/2)}\| = \|(\mathbb{1} + K(\tau))^{-1/2}\sqrt{B(\tau/2)}\|^2 \leq 2,$$

for $\tau > 0$. As a result

$$\|(\mathbb{1} + K(\tau))^{-1/2}\sqrt{B(\tau/2)}\| \leq \sqrt{2}, \quad \tau > 0. \tag{7.1.38}$$

and similarly we prove

$$\|(\mathbb{1} + K(\tau))^{-1/2}\sqrt{A(\tau)}\| \leq 1, \quad \tau > 0. \tag{7.1.39}$$

Inserting (7.1.37), (7.1.38) and (7.1.39) into (7.1.34), we obtain

$$\|(\mathbb{1} + K(\tau))^{-1/2}W(\tau)(\mathbb{1} + K(\tau))^{-1/2}\| \leq 3, \quad \tau > 0. \tag{7.1.40}$$

Finally, by (7.1.33) and (7.1.40), we deduce estimate

$$\|(\mathbb{1} + K(\tau))^{-1/2}S(\tau)(\mathbb{1} + K(\tau))^{-1/2}\| \leq 4, \quad \tau > 0,$$

which proves (7.1.29). □

Corollary 7.1.7 *Let A and B be non-negative self-adjoint operators in a Hilbert space such that $H := A + B$ is self-adjoint. If $f, g \in \mathcal{K}$ are Kato functions, then there is a constant $C^W_{K,H} > 0$ such that the estimate*

$$\|(\mathbb{1} + K(\tau))^{-1/2}W(\tau)(\mathbb{1} + H)^{-1}\| \leq C^W_{K,H}\sqrt{\tau}, \quad \tau > 0, \tag{7.1.41}$$

holds for $C^W_{K,H} := C^A_H\,\gamma[f] + C^B_H\,\gamma[g]$, where $C^A_H = \|A\,(\mathbb{1} + H)^{-1}\|$ and $C^B_H = \|B\,(\mathbb{1} + H)^{-1}\|$.

Proof Using definition of $W(\tau)$ (5.2.9), we infer the inequality

$$\|(\mathbb{1}+K(\tau))^{-1/2}W(\tau)(\mathbb{1}+H)^{-1}\| \leq$$
$$\frac{\tau^2}{4}\|(\mathbb{1}+K(\tau))^{-1/2}B(\tau/2)A(\tau)B(\tau/2)(\mathbb{1}+H)^{-1}\| + \tag{7.1.42}$$
$$\frac{\tau}{2}\|(\mathbb{1}+K(\tau))^{-1/2}A(\tau)\| \, \|B(\tau/2)(\mathbb{1}+H)^{-1}\| + \tag{7.1.43}$$
$$\frac{\tau}{2}\|(\mathbb{1}+K(\tau))^{-1/2}B(\tau/2)\| \, \|A(\tau)(\mathbb{1}+H)^{-1}\|, \quad \tau > 0, \tag{7.1.44}$$

To estimate (7.1.42), we note that by $\|\sqrt{B(\tau/2)}\| \leq, \sqrt{2/\tau}$ one gets

$$\frac{\tau^2}{4}\|(\mathbb{1}+K(\tau))^{-1/2}B(\tau/2)A(\tau)B(\tau/2)(\mathbb{1}+H)^{-1}\|$$
$$\leq \frac{\sqrt{\tau}}{2\sqrt{2}}\|(\mathbb{1}+K(\tau))^{-1/2}\sqrt{B(\tau/2)}\| \, \|B(\tau/2)(\mathbb{1}+H)^{-1}\|, \quad \tau > 0.$$

Therefore, due to (7.1.32) and (7.1.38), we find for $\tau > 0$

$$\frac{\tau^2}{4}\|(\mathbb{1}+K(\tau))^{-1/2}B(\tau/2)A(\tau)B(\tau/2)(\mathbb{1}+H)^{-1}\| \leq \frac{\sqrt{\tau}}{2}C_H^B\gamma[g], \tag{7.1.45}$$

which gives estimate for (7.1.42).

Now, on account of inequality (5.2.39), one infers

$$\begin{aligned}&\frac{\tau}{2}\|(\mathbb{1}+K(\tau))^{-1/2}A(\tau)\| \, \|B(\tau/2)(\mathbb{1}+H)^{-1}\| \\ &\leq \frac{\sqrt{\tau}}{2}\|(\mathbb{1}+K(\tau))^{-1/2}\sqrt{A(\tau)}\| \, \|B(\tau/2)(\mathbb{1}+H)^{-1}\|, \quad \tau > 0.\end{aligned} \tag{7.1.46}$$

Then using (7.1.32) and (7.1.39), we deduce that

$$\frac{\tau}{2}\|(\mathbb{1}+K(\tau))^{-1/2}A(\tau)\| \, \|B(\tau/2)(\mathbb{1}+H)^{-1}\| \leq \frac{\sqrt{\tau}}{2}C_H^B\gamma[g]. \quad \tau > 0. \tag{7.1.47}$$

Hence, (7.1.46) and (7.1.47) yield estimate of (7.1.43).

Similarly to (7.1.46) we deduce the inequality

$$\frac{\tau}{2}\|(\mathbb{1}+K(\tau))^{-1/2}B(\tau/2)\| \, \|A(\tau)(\mathbb{1}+H)^{-1}\|$$
$$\leq \frac{\sqrt{\tau}}{\sqrt{2}}\|(\mathbb{1}+K(\tau))^{-1/2}\sqrt{B(\tau/2)}\| \, \|A(\tau)(\mathbb{1}+H)^{-1}\|, \quad \tau > 0,$$

which by (7.1.31) and (7.1.38) implies estimate

$$\frac{\tau}{2}\|(\mathbb{1}+K(\tau))^{-1/2}B(\tau/2)\|\,\|A(\tau)(\mathbb{1}+H)^{-1}\| \leq \sqrt{\tau}C_H^A\gamma[f], \quad \tau > 0, \tag{7.1.48}$$

and, as a consequence, the estimate of (7.1.44).

Now, collecting (7.1.45), (7.1.47) and (7.1.48), one gets

$$\|(\mathbb{1}+K(\tau))^{-1/2}W(\tau)(\mathbb{1}+H)^{-1}\| \leq (C_H^B\gamma[g]+C_H^A\gamma[f])\sqrt{\tau}, \quad \tau > 0. \tag{7.1.49}$$

Then setting $C_{K,H}^W := C_H^A\gamma[f]+C_H^B\gamma[g]$, we prove (7.1.41). □

Lemma 7.1.8 *Let A and B be non-negative self-adjoint operators in a Hilbert space such that $H := A+B$ is self-adjoint. If f and g are Kato functions $\mathcal{K}$ satisfying $\omega[f] < \infty$, $\omega[g] < \infty$, (7.1.3), then there is a constant $C_{K,H}^{S-H} > 0$ such that*

$$\|(\mathbb{1}+K(\tau))^{-1/2}(S(\tau)-H)(\mathbb{1}+H)^{-1}\| \leq C_{K,H}^{S-H}\sqrt{\tau}, \tag{7.1.50}$$

holds for $\tau > 0$.

Proof From $S(\tau) = K(\tau)+W(\tau)$, we infer inequality

$$\begin{aligned}
&\|(\mathbb{1}+K(\tau))^{-1/2}(S(\tau)-H)(\mathbb{1}+H)^{-1}\| \leq \\
&\quad\|(\mathbb{1}+K(\tau))^{-1/2}(K(\tau)-H)(\mathbb{1}+H)^{-1}\| + &(7.1.51)\\
&\quad\|(\mathbb{1}+K(\tau))^{-1/2}W(\tau)(\mathbb{1}+H)^{-1}\|, \quad \tau > 0. &(7.1.52)
\end{aligned}$$

By definition (5.2.6) we obtain for (7.1.51) and $\tau > 0$

$$\begin{aligned}
&\|(\mathbb{1}+K(\tau))^{-1/2}(K(\tau)-H)(\mathbb{1}+H)^{-1}\| \\
\leq\ &\|(\mathbb{1}+K(\tau))^{-1/2}(A(\tau)-A)(\mathbb{1}+H)^{-1}\| + &(7.1.53)\\
&\left\|(\mathbb{1}+K(\tau))^{-1/2}\left(B(\tau/2)-B-\frac{\tau}{2}B(\tau/2)^2\right)(\mathbb{1}+H)^{-1}\right\|. &(7.1.54)
\end{aligned}$$

Notice that

$$\begin{aligned}
&\|(\mathbb{1}+K(\tau))^{-1/2}(A(\tau)-A)(\mathbb{1}+H)^{-1}\| \\
&\qquad\leq C_H^A\|(\mathbb{1}+K(\tau))^{-1/2}(A(\tau)-A)A^{-1}P_A\|, \quad \tau > 0,
\end{aligned}$$

where $C_H^A := A(\mathbb{1} + H)^{-1}$. Then using $\mathbb{1} + A(\tau) \leq \mathbb{1} + K(\tau)$, which yields $(\mathbb{1} + K(\tau))^{-1} \leq (\mathbb{1} + A(\tau))^{-1}$, we get

$$\|(\mathbb{1} + K(\tau))^{-1/2}(A(\tau) - A)\,(\mathbb{1} + H)^{-1}\| \\ \leq C_H^A \|(\mathbb{1} + A(\tau))^{-1/2}(A(\tau) - A)A^{-1}P_A\|, \quad \tau > 0,$$

where P_A is the orthogonal projection on $\ker(A)^{\perp}$. Owing to (7.1.21) we find

$$\|(\mathbb{1} + K(\tau))^{-1/2}(A(\tau) - A)\,(\mathbb{1} + H)^{-1}\| \leq C_H^A\, \omega[f]\, \sqrt{\tau}, \quad \tau > 0\,, \tag{7.1.55}$$

which provides estimate of (7.1.53).

To estimate (7.1.54) we apply (7.1.38). Then

$$\left\|(\mathbb{1} + K(\tau))^{-1/2}\left(B(\tau/2) - B - \frac{\tau}{4}B(\tau/2)^2\right)(\mathbb{1} + H)^{-1}\right\| \\ \leq \sqrt{2}\,C_H^B \left\|(\mathbb{1} + B(\tau/2))^{-1/2}\left(B(\tau/2) - B - \frac{\tau}{4}B(\tau/2)^2\right)B^{-1}P_B\right\|,$$

where $C_H^B := \|B(\mathbb{1} + H)^{-1}\|$. Together with (7.1.22), this gives

$$\left\|(\mathbb{1} + K(\tau))^{-1/2}\left(B(\tau/2) - B - \frac{\tau}{4}B(\tau/2)\right)(\mathbb{1} + H)^{-1}\right\| \\ \leq C_H^B\left(\omega[g] + \frac{1}{2}\gamma[g]\right)\sqrt{\tau}, \quad \tau > 0. \tag{7.1.56}$$

Inserting (7.1.55) and (7.1.56) into (7.1.53) and (7.1.54), one gets

$$\|(\mathbb{1} + K(\tau))^{-1/2}(K(\tau) - H)(\mathbb{1} + H)^{-1}\| \\ \leq (C_H^A\omega[f] + C_H^B(\omega[g] + \tfrac{1}{2}\gamma[g]))\,\sqrt{\tau}, \quad \tau > 0.$$

Then setting $C_{K,H}^{K-H} := C_H^A\omega[f] + C_H^B(\omega[g] + \frac{1}{2}\gamma[g])$, we obtain

$$\|(\mathbb{1} + K(\tau))^{-1/2}(K(\tau) - H)(\mathbb{1} + H)^{-1}\| \leq C_{K,H}^{K-H}\,\sqrt{\tau}\,, \quad \tau > 0\,, \tag{7.1.57}$$

that is, the estimate of the term (7.1.51).

To estimate the next term (7.1.52), we apply Corollary 7.1.7 (7.1.41).

Finally, inserting (7.1.57) into (7.1.51) and (7.1.41) into (7.1.52), we infer that

$$\|(\mathbb{1} + K(\tau))^{-1/2}(S(\tau) - H)(\mathbb{1} + H)^{-1}\| \leq (C_{K,H}^{K-H} + C_{K,H}^{W})\,\sqrt{\tau}\,, \quad \tau > 0,$$

which proves (7.1.50) for $C_{K,H}^{S-H} := C_{K,H}^{K-H} + C_{K,H}^{W}$. □

Now we are in position to prove the *operator-norm* convergence of the Trotter-Kato product formulæ for *symmetrised* (thence, self-adjoint) approximants:

$$\{(g(tB/2\eta)f(tA/\eta)g(tB/2\eta))^{\eta}\}_{\eta>0}\,,\ \{\left(g(tB/\eta)^{1/2}f(tA/\eta)g(tB/\eta)^{1/2}\right)^{\eta}\}_{\eta>0}\,,$$
$$\{(f(tA/2\eta)g(tB/\eta)f(tA/2\eta))^{\eta}\}_{\eta>0}\,,\ \{\left(f(tA/\eta)^{1/2}g(tB/\eta)f(tA/\eta)^{1/2}\right)^{\eta}\}_{\eta>0}\,, \tag{7.1.58}$$

together with estimate of error bounds for the *rate* of operator-norm convergence.

Theorem 7.1.9 (ITTZ-Theorem) *Let A and B be non-negative self-adjoint operators in $\mathfrak{H}$ and let $f, g \in \mathcal{K}$ be Kato functions satisfying conditions $\varkappa[f]+\varkappa[g] < \infty$ and $\omega[f] + \omega[g] < \infty$. If the sum $H := A + B$ is self-adjoint operator on* $\operatorname{dom}(H) = \operatorname{dom}(A) \cap \operatorname{dom}(B)$, *then there exists a constant $C_1^{\mathcal{I}} > 0$ such that on each compact interval $\mathcal{I} \subset \mathbb{R}_0^+$, one obtains estimates* :

$$\sup_{t\in\mathcal{I}}\left\|(g(tB/2\eta)f(tA/\eta)g(tB/2\eta))^{\eta} - \mathrm{e}^{-tH}\right\| \le C_1^{\mathcal{I}}\,\frac{1}{\eta}\,, \quad \eta > 0\,, \tag{7.1.59}$$

$$\sup_{t\in\mathcal{I}}\left\|\left(g(tB/\eta)^{1/2}f(tA/\eta)g(tB/\eta)^{1/2}\right)^{\eta} - \mathrm{e}^{-tH}\right\| \le C_1^{\mathcal{I}}\,\frac{1}{\eta}\,, \quad \eta > 0\,, \tag{7.1.60}$$

$$\sup_{t\in\mathcal{I}}\left\|(f(tA/2\eta)g(tB/\eta)f(tA/2\eta))^{\eta} - \mathrm{e}^{-tH}\right\| \le C_1^{\mathcal{I}}\,\frac{1}{\eta}\,, \quad \eta > 0\,, \tag{7.1.61}$$

$$\sup_{t\in\mathcal{I}}\left\|\left(f(tA/\eta)^{1/2}g(tB/\eta)f(tA/\eta)^{1/2}\right)^{\eta} - \mathrm{e}^{-tH}\right\| \le C_1^{\mathcal{I}}\,\frac{1}{\eta}\,, \quad \eta > 0\,. \tag{7.1.62}$$

On that account, for $\eta \to \infty$, the operator-norm convergence of symmetrised Trotter-Kato product approximants (7.1.58) *to semigroup $\{\mathrm{e}^{-tH}\}_{t\ge0}$ is uniform on each compact interval $\mathcal{I} \subset \mathbb{R}_0^+$, that is (cf. Remark* 6.1.5*), locally uniformly on $\mathbb{R}_0^+$.*

Proof To this aim we have to verify condition (4.3.18) of Theorem 4.3.6, namely, that there is a constant $M_1 > 0$ such that

$$\|(\mathbb{1} + S(\tau))^{-1} - (\mathbb{1} + H)^{-1}\| \le M_1\tau, \quad \tau > 0\,, \tag{7.1.63}$$

for $S(\tau) = \tau^{-1}[\mathbb{1} - g(\tau B/2)f(\tau A)g(\tau B/2)]$.

Note that by the second resolvent formula, we obtain for $\tau > 0$ identities

$$(\mathbb{1} + S(\tau))^{-1} = (\mathbb{1} + H)^{-1} + (\mathbb{1} + S(\tau))^{-1}(H - S(\tau))(\mathbb{1} + H)^{-1}\,, \tag{7.1.64}$$

and

$$(\mathbb{1} + S(\tau))^{-1} = (\mathbb{1} + H)^{-1} + (\mathbb{1} + H)^{-1}(H - S(\tau))(\mathbb{1} + S(\tau))^{-1} . \qquad (7.1.65)$$

We call (7.1.64) the *first* iteration of expression for operator $(\mathbb{1} + S(\tau))^{-1}$. Then inserting (7.1.65) into (7.1.64) (or (7.1.64) into (7.1.65)), one infers the *second* iteration of expression (7.1.64) for operator $(\mathbb{1} + S(\tau))^{-1}$:

$$(\mathbb{1} + S(\tau))^{-1} = (\mathbb{1} + H)^{-1} + (\mathbb{1} + H)^{-1}(H - S(\tau))(\mathbb{1} + H)^{-1} + \\ (\mathbb{1} + H)^{-1}(H - S(\tau))(\mathbb{1} + S(\tau))^{-1}(H - S(\tau))(\mathbb{1} + H)^{-1} . \qquad (7.1.66)$$

It yields the estimate

$$\|(\mathbb{1}+S(\tau))^{-1} - (1 + H)^{-1}\| \leq \|(\mathbb{1} + H)^{-1}(H - S(\tau))(\mathbb{1} + H)^{-1}\| \\ + \|(\mathbb{1} + H)^{-1}(H - S(\tau))(\mathbb{1} + S(\tau))^{-1}(H - S(\tau))(\mathbb{1} + H)^{-1}\| . \qquad (7.1.67)$$

By Lemma 7.1.4 there exists a constant $C_{H,H}^{S-H} > 0$ such that for the right-hand side of (7.1.67), the estimate

$$\|(\mathbb{1} + H)^{-1}(H - S(\tau))(\mathbb{1} + H)^{-1}\| \leq C_{H,H}^{S-H}\, \tau , \quad \tau > 0 , \qquad (7.1.68)$$

is valid for the first term. To estimate the second term, we use the representation (5.2.34), which provides

$$(\mathbb{1} + S(\tau))^{-1} = (\mathbb{1} + K(\tau))^{-1/2}(\mathbb{1} + Q(\tau))^{-1}(\mathbb{1} + K(\tau))^{-1/2}, \quad \tau > 0.$$

As a consequence,

$$\begin{aligned}(\mathbb{1} + H)^{-1}&(H - S(\tau))(1 + S(\tau))^{-1}(H - S(\tau))(\mathbb{1} + H)^{-1} \\ &=(\mathbb{1} + H)^{-1}(H - S(\tau))(\mathbb{1} + K(\tau))^{-1/2}(\mathbb{1} + Q(\tau))^{-1} \\ &\quad \times (\mathbb{1} + K(\tau))^{-1/2}(H - S(\tau))(1 + H)^{-1} .\end{aligned}$$

Next, by (5.2.35), we obtain

$$\begin{aligned}\|(\mathbb{1} + H)^{-1}&(H - S(\tau))(\mathbb{1} + S(\tau))^{-1}(H - S(\tau))(\mathbb{1} + H)^{-1}\| \\ &\leq q_0\, \|(\mathbb{1} + K(\tau))^{-1/2}(H - S(\tau))(1 + H)^{-1}\|^2, \quad \tau > 0.\end{aligned}$$

Then using Lemma 7.1.8, we get

$$\|(\mathbb{1}+H)^{-1}(H-S(\tau))(\mathbb{1}+S(\tau))^{-1}(H-S(\tau))(\mathbb{1}+H)^{-1}\| \leq q_0\,(C_{K,H}^{S-H})^2\,\tau, \tag{7.1.69}$$

$\tau > 0$. Inserting (7.1.68) and (7.1.69) into (7.1.67), we finally infer the proof of (7.1.63) for $M_1 := C_{H,H}^{S-H} + q_0\,(C_{K,H}^{S-H})^2$. Now applying Theorem 4.3.6 for $F(t) = g(\tau B/2)f(\tau A)g(\tau B/2)$, we obtain the proof of the estimate (7.1.59).

For demonstration of (7.1.60), we introduce function (cf. Remark 6.1.3)

$$\widehat{g}(x) := \sqrt{g(2x)}\,, \quad x \in \mathbb{R}_0^+\,. \tag{7.1.70}$$

Notice that $\widehat{g}(\cdot)$ is again a generic Kato function: $\widehat{g} \in \mathcal{K}$, Definition 5.2.1. Furthermore, by (7.1.2) we have

$$\varkappa[\widehat{g}] = \sup_{x>0} \frac{1}{x}\left|\frac{1-\widehat{g}(x)}{x} - 1\right| = \sup_{x>0} \frac{1}{x}\left|\frac{1-\sqrt{g(2x)}}{x} - 1\right|. \tag{7.1.71}$$

Seeing that

$$\frac{1-\sqrt{g(2x)}}{x} - 1 = \left(\frac{1-g(2x)}{2x} - 1\right)\frac{2}{1+\sqrt{g(2x)}} + \frac{1-g(2x)}{(1+\sqrt{g(2x)})^2}\,,$$

by virtue of (7.1.1), (7.1.2) and (7.1.71), we obtain the estimate

$$\varkappa[\widehat{g}] = \sup_{x>0} \frac{1}{x}\left|\frac{1-\widehat{g}(x)}{x} - 1\right| \leq 4\,\varkappa[g] + 2\,\gamma[g]\,.$$

Further, by definition (7.1.3), we have

$$\omega[\widehat{g}] = \sup_{x>0} \frac{1}{\sqrt{1-\widehat{g}(x)}}\left|\frac{1-\widehat{g}(x)}{x} - 1\right| = \sup_{x>0} \frac{1}{\sqrt{1-\sqrt{g(2x)}}}\left|\frac{1-\sqrt{g(2x)}}{x} - 1\right|, \tag{7.1.72}$$

where

$$\begin{aligned}
&\frac{1}{\sqrt{1-\sqrt{g(2x)}}}\left(\frac{1-\sqrt{g(2x)}}{x} - 1\right)\\
&= \frac{\sqrt{1+\sqrt{g(2x)}}}{\sqrt{1-g(2x)}}\left(\left(\frac{1-g(2x)}{2x} - 1\right)\frac{2}{1+\sqrt{g(2x)}} + \frac{2}{1+\sqrt{g(2x)}} - 1\right).
\end{aligned}$$

On account of

$$\frac{\sqrt{1+\sqrt{g(2x)}}}{\sqrt{1-g(2x)}}\left(\frac{2}{1+\sqrt{g(2x)}} - 1\right) = \frac{\sqrt{1-\sqrt{g(2x)}}}{1+\sqrt{g(2x)}} \leq 1\,,$$

and due to definition (7.1.72), which yields

$$\sup_{x>0} \frac{\sqrt{1+\sqrt{g(2x)}}}{\sqrt{1-g(2x)}} \left(\frac{1-g(2x)}{2x} - 1\right) \frac{2}{1+\sqrt{g(2x)}} \leq 4\,\omega[g]\,,$$

we deduce that $\omega[\widehat{g}] \leq 4\omega[g]+1$. As a consequence, if $\varkappa[g] < \infty$ and $\omega[g] < \infty$, then $\varkappa[\widehat{g}] < \infty$ and $\omega[\widehat{g}] < \infty$.

Therefore, setting in (4.3.19): $\widehat{F}(\tau) := \widehat{g}(\tau B/2) f(\tau A)\widehat{g}(\tau B/2)$, $\tau \geq 0$, we find for each compact interval $\mathcal{I}$ of $\mathbb{R}_0^+$ a constant $\widetilde{C}_1^{\mathcal{I}}$ such that

$$\sup_{\eta\in\mathcal{I}} \left\| (\widehat{g}(tB/2\eta) f(tA/\eta)\widehat{g}(tB/2\eta))^{\eta} - \mathrm{e}^{-tH} \right\| \leq \widetilde{C}_1^{\mathcal{I}} \frac{1}{\eta}, \quad \eta > 0\,.$$

As a result, given that $\widehat{g}(\tau B/2) = g(\tau B)^{1/2}$, $\tau \geq 0$ (see (7.1.70)), this inequality yields the estimate (7.1.60).

In conclusion, two other estimates, (7.1.61) and (7.1.62), follow from (7.1.59) and (7.1.60) due to the fact that the pairs $\{A, f\}$ and $\{B, g\}$ enter into the theorem in a totally symmetric way. □

Remark 7.1.10 It is worth elucidating now the *iterations* in the ITTZ-Theorem 7.1.9 (see expressions (7.1.64), (7.1.66)) for operator $(\mathbb{1}+S(\tau))^{-1}$, which serve for evaluation of the operator difference in the left-hand side of (7.1.67). To this aim we introduce operators

$$\begin{aligned}
\Delta(\tau) &:= (\mathbb{1} + S(\tau))^{-1} - (\mathbb{1} + H)^{-1}\,,\\
\Gamma(\tau) &:= (\mathbb{1} + H)^{-1}(H - S(\tau))\,(\mathbb{1} + H)^{-1}\,,\\
\Phi(\tau) &:= (H - S(\tau))\,(\mathbb{1} + H)^{-1}\,.
\end{aligned}$$

Then expressions (7.1.64) and (7.1.66) yield equation for the operator difference $\Delta(\tau)$ in question:

$$\Delta(\tau) = \Gamma(\tau) + \Delta(\tau)\,\Phi(\tau)\,, \quad \tau > 0\,. \tag{7.1.73}$$

Given that by conditions of ITTZ-Theorem 7.1.9 one has $S(\tau) = \tau^{-1}(\mathbb{1} - F(\tau))$ and $F(\tau) = g(\tau B/2) f(\tau A) g(\tau B/2)$, for $f, g \in \mathcal{K}$ (7.1.1), we obtain for operator $H = A + B$ that

$$\lim_{\tau\to+0} S(\tau)\,u = (A + B)\,u\,, \quad u \in \mathrm{dom}\,(H)\,. \tag{7.1.74}$$

As a consequence, (7.1.74) implies that $\{\Phi(\tau)\}_{0<\tau<T}$, for $T > 0$, is a family of bounded operators such that the strong limit

$$\text{s-}\lim_{\tau\to+0} \Phi(\tau) = 0\,. \tag{7.1.75}$$

Then by the *uniform boundedness principle* and Corollary A.1.5 (Appendix A.1.2)

$$\sup_{\tau \in (0,T)} \|\Phi(\tau)\| =: m(T) . \tag{7.1.76}$$

On account of Proposition A.1.4 and (7.1.75) and (7.1.76), there exists $T > 0$ such that $m(T) < 1$. So, in virtue of (7.1.73) and (7.1.76), we obtain

$$\Delta(\tau) = \Gamma(\tau)\,(\mathbb{1} - \Phi(\tau))^{-1}, \quad \tau \in (0, T) .$$

Consequently, one gets for difference $\Delta(\tau)$ representation

$$\begin{aligned}&(\mathbb{1} + S(\tau))^{-1} - (\mathbb{1} + H)^{-1} = \\ &(\mathbb{1} + H)^{-1}(H - S(\tau))\,(\mathbb{1} + H)^{-1}[\mathbb{1} - (H - S(\tau))\,(\mathbb{1} + H)^{-1}]^{-1} .\end{aligned} \tag{7.1.77}$$

Owing to (7.1.77), the proof of estimate (7.1.63) reduces to the operator-norm upper bound for small $\tau > 0$ of the operator $\Gamma(\tau)$; see (7.1.68).

Now we prove the *operator-norm* convergence of the Trotter-Kato product formulæ for *non-symmetric* (thence, *non-self-adjoint*) approximants

$$(g(tB/n)f(tA/n))^n = ((g(tB/n)f(tA/n))^n)^* , \quad n \in \mathbb{N}, \quad t \in \mathbb{R}_0^+ , \tag{7.1.78}$$

with estimates of the *rate* of convergence.

Lemma 7.1.11 *Let A and B be non-negative self-adjoint operators with domains* dom (A) *and* dom (B) *in a Hilbert space $\mathfrak{H}$. Let the operator sum $H := A + B$ be a self-adjoint operator. If f and g are Kato functions from $\mathcal{K}$, then the estimates*

$$\|(\mathbb{1} - f(\tau A))\,\mathrm{e}^{-tH}\| \le C_H^A\,\gamma[f]\,\frac{\tau}{t} \quad \text{and} \quad \|(\mathbb{1} - g(\tau B))\,\mathrm{e}^{-tH}\| \le C_H^B\,\gamma[g]\,\frac{\tau}{t} , \tag{7.1.79}$$

hold for $\tau > 0$ and $t > 0$.

Proof Since dom $(H) \subseteq$ dom (A), we use definitions $A(\tau) = (\mathbb{1} - f(\tau A))/\tau$ and $A^{-1}P_A$, to infer representation

$$(\mathbb{1} - f(\tau A))\,\mathrm{e}^{-tH} = \tau\, A(\tau)\, A^{-1} P_A\, A\,(\mathbb{1}+H)^{-1}(\mathbb{1}+H)\,\mathrm{e}^{-tH}, \quad \tau > 0, \quad t > 0 ,$$

which by (7.1.4) yields the estimate

$$\left\|(\mathbb{1} - f(\tau A))\,\mathrm{e}^{-tH}\right\| \le C_H^A\,\gamma[f]\left\|(\mathbb{1} + H)\,\mathrm{e}^{-tH}\right\|\tau, \quad \tau > 0, \quad t > 0 .$$

Since for a self-adjoint operator $H \geq 0$ one has $\|(\mathbb{1} + H)\, \mathrm{e}^{-tH}\| \leq t^{-1}$, $t > 0$, this estimate yields in (7.1.79) the first inequality. The proof of the second inequality in (7.1.79) is similar. □

Corollary 7.1.12 *Let A and B be non-negative self-adjoint operators in $\mathfrak{H}$ and let f and g be Kato functions satisfying $\varkappa[f] + \varkappa[g] < \infty$ and $\omega[f] + \omega[g] < \infty$. If the sum $H := A + B$ is a self-adjoint operator, then for each compact interval $\mathcal{I} \subset \mathbb{R}_0^+$ there exists a constant $\widetilde{C}_1^{\mathcal{I}} > 0$ such that*

$$\sup_{t\in\mathcal{I}} \left\| (f(tA/n)g(tB/n))^n - \mathrm{e}^{-tH} \right\| \leq \widetilde{C}_1^{\mathcal{I}} \frac{1}{n}, \quad n \in \mathbb{N}, \tag{7.1.80}$$

$$\sup_{t\in\mathcal{I}} \left\| (g(tB/n)f(tA/n))^n - \mathrm{e}^{-tH} \right\| \leq \widetilde{C}_1^{\mathcal{I}} \frac{1}{n}. \quad n \in \mathbb{N}, \tag{7.1.81}$$

For $n \to \infty$ these approximants converge in the operator norm topology to semigroup $\{\mathrm{e}^{-tH}\}_{t\geq 0}$ with the rate $O(1/n)$ for each compact t-interval $\mathcal{I} \subset \mathbb{R}_0^+$, that is, (cf. Remark 6.1.5*), locally uniformly on $\mathbb{R}_0^+$.*

Proof To this aim we consider the family $\widehat{G}(\tau) = f(\tau A)^{1/2} g(\tau B) f(\tau A)^{1/2}$, $\tau \geq 0$; see (7.1.82). Since $\widehat{G}(\tau)$ are self-adjoint contractions, one gets

$$\|\widehat{G}(\tau)^{n-1}(\mathbb{1} - \widehat{G}(\tau))\| \leq \frac{1}{n}, \quad n \in \mathbb{N}, \quad \tau \geq 0\,. \tag{7.1.82}$$

Using

$$\widehat{G}(t/n)^{n-1} - \mathrm{e}^{-tH} = \widehat{G}(t/n)^{n-1}(\mathbb{1} - \widehat{G}(t/n)) + \widehat{G}(t/n)^n - \mathrm{e}^{-tH}\,, \quad n \in \mathbb{N}\,,$$

we obtain the estimate

$$\|\widehat{G}(t/n)^{n-1} - \mathrm{e}^{-tH}\| \leq \|\widehat{G}(t/n)^{n-1}(\mathbb{1} - \widehat{G}(t/n))\| + \|\widehat{G}(t/n)^n - \mathrm{e}^{-tH}\|\,, \quad t \in \mathbb{R}_0^+\,.$$

Then from Theorem 7.1.9, (7.1.61) and (7.1.82) it follows that

$$\sup_{t\in\mathcal{I}} \|\widehat{G}(t/n)^{n-1} - \mathrm{e}^{-tH}\| \leq (1 + C_1^{\mathcal{I}})\, \frac{1}{n}, \quad n \in \mathbb{N}\,, \quad t \in \mathbb{R}_0^+\,. \tag{7.1.83}$$

Note that for $n \in \mathbb{N}$ and $t \geq 0$

$$g(tB/n) f(tA/n)^{1/2} \widehat{G}(t/n)^{n-1} f(tA/n)^{1/2} = (g(tB/n) f(tA/n))^n\,.$$

Then using representation

$$
\begin{aligned}
&(g(tB/n)f(tA/n))^n - \mathrm{e}^{-tH} = \\
&g(tB/n)f(tA/n)^{1/2}(\widehat{G}(t/n)^{n-1} - \mathrm{e}^{-tH})f(tA/n)^{1/2} + \\
&g(tB/n)f(tA/n)^{1/2}\mathrm{e}^{-tH}(f(tA/n)^{1/2} - \mathbb{1}) + \\
&g(tB/n)\,(f(tA/n)^{1/2} - \mathbb{1})\,\mathrm{e}^{-tH} + (g(tB/n) - \mathbb{1})\,\mathrm{e}^{-tH}\,,
\end{aligned}
$$

we get for $n \in \mathbb{N}$ and $t \geq 0$ inequality

$$
\begin{aligned}
\|(g(tB/n)f(tA/n))^n - \mathrm{e}^{-tH}\| &\leq \|\,\widehat{G}(t/n)^{n-1} - \mathrm{e}^{-tH}\| \\
&+ 2\,\|(\mathbb{1} - f(tA/n)^{1/2})\,\mathrm{e}^{-tH}\| + \|(\mathbb{1} - g(tB/n))\,\mathrm{e}^{-tH}\|\,.
\end{aligned}
$$

Therefore $\|(\mathbb{1} - f(tA/n)^{1/2})\,\mathrm{e}^{-tH}\| \leq \|(\mathbb{1} - f(tA/n))\,\mathrm{e}^{-tH}\|$ implies

$$
\begin{aligned}
\|(g(tB/n)f(tA/n))^n - \mathrm{e}^{-tH}\| &\leq \|\widehat{G}(t/n)^{n-1} - \mathrm{e}^{-tH}\| \\
&+ 2\,\|(\mathbb{1} - f(tA/n))\,\mathrm{e}^{-tH}\| + \|(\mathbb{1} - g(tB/n))\,\mathrm{e}^{-tH}\|\,.
\end{aligned}
$$

Implementing in this inequality (7.1.83) and (7.1.79), for $\tau = t/n$, we deduce the locally uniform in $\mathbb{R}_0^+$ estimate

$$
\|(g(tB/n)f(tA/n))^n - \mathrm{e}^{-tH}\| \leq (1 + C_1^{\mathcal{I}} + 2C_H^A\gamma[f] + C_H^B\gamma[g])\,\frac{1}{n}\,,
$$

for $n \in \mathbb{N}$. Setting $\widetilde{C}_1^{\mathcal{I}} := 1 + C_1^{\mathcal{I}} + 2C_H^A\gamma[f] + C_H^B\gamma[g]$, we prove (7.1.81). The proof of (7.1.80) is similar. □

Corollary 7.1.13 *Let A and B be non-negative self-adjoint operators. If the operator sum $H := A + B$ is self-adjoint, then for each compact interval $\mathcal{I} \subset \mathbb{R}_0^+$, there is a constant $C^{\mathcal{I}} > 0$ such that*

$$
\begin{aligned}
\sup_{t\in\mathcal{I}} \|(\mathrm{e}^{-tB/2n}\mathrm{e}^{-tA/n}\mathrm{e}^{-tB/2n})^n - \mathrm{e}^{-tH}\| &\leq C^{\mathcal{I}}\,\frac{1}{n}\,, \quad n \in \mathbb{N}, \\
\sup_{t\in\mathcal{I}} \|(\mathrm{e}^{-tA/n}\mathrm{e}^{-tB/n})^n - \mathrm{e}^{-tH}\| &\leq C^{\mathcal{I}}\,\frac{1}{n}\,, \quad n \in \mathbb{N}\,.
\end{aligned}
\tag{7.1.84}
$$

If self-adjoint generator H is strictly positive, $H \geq \eta\,\mathbb{1}$, $\eta > 0$, then in Theorem 7.1.9, *Corollary* 7.1.12 *and* (7.1.84), *the interval $\mathcal{I} = \mathbb{R}_0^+$.*

Proof Note that the strict positivity of H implies that at least one of operators A and B is (strictly) positive. Then the proof follows verbatim the arguments in Theorem 7.1.9 and Corollary 7.1.12. To this aim we note that since generator H is (strictly) positive, one can apply Theorem 4.3.7. □

Remark 7.1.14 (a) We note here that unless the sum of *non-negative* self-adjoint operators A and B, i.e. $A + B$, is *self-adjoint* on $\mathrm{dom}\,(A) \cap \mathrm{dom}\,(B)$, the *operator-norm* convergence of the Trotter product formula does *not* always hold, even though the sum is *essentially* self-adjoint there and B is A-form-bounded with relative bound less than 1. However, it *may* be true under certain complementary conditions; see, for example, the fractional power conditions in Sect. 7.4.

This fact has been pointed out by *Hiroshi Tamura* for exponential Kato functions by means of a *counterexample*; see Notes to Chap. 7 (Sect. 7.5) and Example H.0.5 in Appendix H. Seeing that, the condition of the Trotter product theorem (Theorem 5.1.1) is *not* sufficient for the *operator-norm* convergence of the Trotter product formula. While on the contrary, the result of Rogava's theorem (Proposition 7.5.1) with the rate of convergence, $O\,(\ln(n)\,/\sqrt{n})$, or improvements, $O\,(\ln(n)^2/n)$ and $O\,(\ln(n)/n)$ (see Notes to Chap. 7 (Notes to Sect. 7.1)), is a corollary of the ITTZ-Theorem 7.1.9 (Corollary 7.1.13), which provides the *optimal* rate of convergence, $O\,(1/n)$, instead of all these preliminary estimates.

(b) We also note that condition of non-negativity of operators A and B is not superfluous either. In fact it yields the *Trotter stability* condition (Definition 1.9.37 for $M = 1$ and $\beta = 0$). Remember that without stability, even the strong operator convergence of the Trotter product formula cannot be guaranteed; see Proposition 1.10.11 and Corollary 1.10.12.

7.2 Optimality of Error Bound Estimate and ITTZ-Theorem

In this section, we prove that the operator-norm error bound of order $O(1/n)$ established in the ITTZ-Theorem 7.1.9 is *optimal*, that is, a posteriori the best possible under conditions of this theorem. To this end it is sufficient to provide an example satisfying conditions of the ITTZ-Theorem 7.1.9 and such that the operator-norm estimate of the error bound for the Trotter-Kato product formula from *below* has the same order $O(1/n)$ as that from *above*.

Seeing that optimal operator-norm estimate may depend on symmetrisation of approximants (cf. *S.Lie* product formulæ, Proposition 1.10.10), we consider also the operator-norm estimate of the error bound from *below* for the Trotter-Kato product formulæ in Corollary 7.1.12.

To this aim we start by examination of the exponential Kato functions, $f(s) = g(s) = \mathrm{e}^{-s}$, and we study first *non-symmetric* and next the *symmetric* product formulæ.

The Non-symmetric Product Formula

To begin, let A and B be bounded operators, cf. the *Lie* product formula in Notes to Chap. 1 (Notes to Sects. 1.9, 1.9.6), Proposition 1.10.10. Then by the *Baker-Campbell-Hausdorff* formula, we obtain for small $|t|$ representation

$$\mathrm{e}^{-tA}\mathrm{e}^{-tB} = \exp\big[-t(A+B)+t^2\,[A,B]/2-t^3\,[A-B,[A,B]/2]/6+O_p(|t|^4)\big], \tag{7.2.1}$$

with $[A, B] = AB - BA$. In this representation and below the $O_p(|t|^k)$, for $k > 0$, denotes a *bounded operator* with norm of order $O(|t|^k)$. Let

$$N(t) := (\mathrm{e}^{-tA}\mathrm{e}^{-tB})^{1/t} = \exp\{-(A+B) + t[A, B]/2 + O_p(|t|^2)\}, \tag{7.2.2}$$

where we understand that $N(0) = \mathrm{e}^{-(A+B)}$. Then by (7.2.1) and (7.2.2)

$$\begin{aligned} E_1(A; B) &:= \frac{d}{dt}N(t)|_{t=0} \\ &= \frac{1}{2}\sum_{k=1}^{\infty}\frac{(-1)^{k-1}}{k!}\sum_{j=1}^{k}(A+B)^{j-1}[A, B](A+B)^{k-j}, \end{aligned} \tag{7.2.3}$$

where the right-hand side is norm-convergent. It can be a nonzero operator with bound $2^{-1}\|[A, B]\|\,\mathrm{e}^{\|A+B\|}$, if A and B do not commute.

It follows that

$$N(t) = \mathrm{e}^{-(A+B)} + tE_1(A; B) + O_p(t^2), \tag{7.2.4}$$

so that with $t = 1/n$

$$(\mathrm{e}^{-A/n}\mathrm{e}^{-B/n})^n = N(1/n)^n = \mathrm{e}^{-(A+B)} + n^{-1}E_1(A; B) + O_p(n^{-2}). \tag{7.2.5}$$

Thus we have seen that, in the non-symmetric case, the error bound $O(n^{-1})$ is *optimal* because it is a direct consequence of *non-commutativity* of (bounded) operators A and B. See also (1.10.42) in Proposition 1.10.10 for the Lie product formula.

The Symmetric Product Formula

Owing to (7.2.1), we infer that

$$\mathrm{e}^{-tB/2}\mathrm{e}^{-tA}\mathrm{e}^{-tB/2} = \exp\big[-t(A+B) - \tfrac{t^3}{24}[2A+B, [A, B]] + O_p(|t|^4)\big]. \tag{7.2.6}$$

Similarly it follows that

$$\begin{aligned} L(t) &:= (\mathrm{e}^{-tB/2}\mathrm{e}^{-tA}\mathrm{e}^{-tB/2})^{1/t} \\ &= \exp\big[-(A+B) - t^2[2A+B, [A, B]]/24 + O_p(|t|^3)\big]. \end{aligned} \tag{7.2.7}$$

Notice that $L(0) = \mathrm{e}^{-(A+B)}$ and have $dL(t)/dt|_{t=0} = 0$, so that with $t = 1/n$

$$(\mathrm{e}^{-B/2n}\mathrm{e}^{-A/n}\mathrm{e}^{-B/2n})^n = L(1/n)^n = \mathrm{e}^{-(A+B)} + O_p(n^{-2}), \quad n \in \mathbb{N}. \tag{7.2.8}$$

Therefore, in the symmetric case, still due to *non-commutativity* of (bounded) operators A and B, the *optimal* error bound is of the order $O(n^{-2})$, cf. estimate (1.10.44) for the symmetrised Lie product formula in Proposition 1.10.10.

We note that in this case the manipulation with non-commutativity in (1.10.44), or in (7.2.8), is more subtle and that it does not survive for *unbounded* operators A and B. In fact, it turns out that the optimal error bound is just of the order $O(n^{-1})$.

In the following, we shall show that in a Hilbert space $\mathfrak{H}$, there exist unbounded non-negative self-adjoint operators A and B, satisfying the ITTZ-Theorem 7.1.9, such that the operator sum $A + B$ is self-adjoint on $\mathrm{dom}\,(A) \cap \mathrm{dom}\,(B)$ and the estimate from below

$$\| \mathrm{e}^{-t(A+B)} - (\mathrm{e}^{-tB/2n}\mathrm{e}^{-tA/n}\mathrm{e}^{-tB/2n})^n \| \geq L(t)\, n^{-1}\,, \quad n \in \mathbb{N}\,, \quad t > 0\,,$$

holds for a *positive* continuous function $L(t)$ defined on $\mathbb{R}^+$, which is independent of $n \in \mathbb{N}$.

Example 7.2.1 (cf. Appendix H) Let us consider the Hilbert space $\mathfrak{H}$ defined by

$$\mathfrak{H} := \bigoplus_{k\in\mathbb{N}} \mathfrak{H}_k, \quad \mathfrak{H}_k := \mathbb{C}^2, \tag{7.2.9}$$

where $(\cdot,\cdot)_k$ is the scalar product in $\mathfrak{H}_k$, $k \in \mathbb{N}$, and the scalar product in $\mathfrak{H}$ is given by $(\xi,\eta) = \sum_{k=1}^{\infty}(\xi_k,\eta_k)_k$, for vectors $\xi = \oplus_{k\in\mathbb{N}}\xi_k$ and $\eta = \oplus_{k\in\mathbb{N}}\eta_k$ from $\mathfrak{H}$. Let S, T and E be matrices

$$S = \begin{pmatrix} 1 & 0 \\ 0 & -1 \end{pmatrix}, \quad T = \begin{pmatrix} 0 & 1 \\ 1 & 0 \end{pmatrix} \quad E := \mathbb{1}_{\mathbb{C}^2} = \begin{pmatrix} 1 & 0 \\ 0 & 1 \end{pmatrix}.$$

Notice that

$$ST + TS = 0, \qquad S^2 = T^2 = E\,. \tag{7.2.10}$$

For each $k \in \mathbb{N}$, one defines on $\mathfrak{H}_k$ two bounded non-negative self-adjoint operators

$$A_k = k\,(S + E), \qquad B_k = k\,(S\cos(\theta_k) + T\sin(\theta_k) + E)\,, \tag{7.2.11}$$

where the parameters $\theta_k \in (0, \pi/2]$ are chosen in such a way that

$$\cos\theta_k = 1 - \varepsilon_k, \quad \varepsilon_k = 1/(2k^2)\,. \tag{7.2.12}$$

We set $H_k := A_k + B_k$ for $k \in \mathbb{N}$.

Let us mention some useful relations. One can check that

$$A_k^2 = 2\,k\,A_k \quad \text{and} \quad B_k^2 = 2\,k\,B_k\,, \quad k \in \mathbb{N}\,.$$

Moreover, we find

$$A_k B_k + B_k A_k = E + A_k^2 + B_k^2 , \quad k \in \mathbb{N}. \tag{7.2.13}$$

In particular, one has $2\,k\, H_k = A_k^2 + B_k^2$, $k \in \mathbb{N}$, which yields $H_k \geq 0$, $k \in \mathbb{N}$.
Let us consider in $\mathfrak{H}$ operators

$$A := \oplus_{k\in\mathbb{N}} A_k \quad \text{and} \quad B := \oplus_{k\in\mathbb{N}} B_k , \tag{7.2.14}$$

with domains

$$\begin{aligned} \mathrm{dom}\,(A) &:= \{\xi = \oplus_{k\in\mathbb{N}} \xi_k : \sum_k \|A_k \xi_k\|^2 < \infty\}, \\ \mathrm{dom}\,(B) &:= \{\xi = \oplus_{k\in\mathbb{N}} \xi_k : \sum_k \|B_k \xi_k\|^2 < \infty\}. \end{aligned} \tag{7.2.15}$$

Then on account of (7.2.11) and (7.2.15), the operators A and B are *unbounded*, non-negative and self-adjoint.

Theorem 7.2.2 *Let operators A and B be given by* (7.2.11)–(7.2.15). *Then the operator sum*

$$H := A + B = \oplus_{k\in\mathbb{N}} (A_k + B_k) , \quad \mathrm{dom}\,(H) := \mathrm{dom}\,(A) \cap \mathrm{dom}\,(B) , \tag{7.2.16}$$

is non-negative and self-adjoint.

Proof Owing to the facts that the sum of symmetric and non-negative operators is symmetric and non-negative the operator H (7.2.16) is symmetric and non-negative.
Now let $\xi = \{\xi_k\}_{k\in\mathbb{N}} \in \mathrm{dom}\,(H) = \mathrm{dom}\,(A) \cap \mathrm{dom}\,(B)$. Using (7.2.13) we infer that

$$\begin{aligned} \|H\xi\|^2 &= \sum_{k\in\mathbb{N}} \|H_k \xi_k\|^2 \\ &= \sum_{k\in\mathbb{N}} \left\{ \|A_k \xi_k\|^2 + ((A_k B_k + B_k A_k)\xi_k, \xi_k) + \|B_k \xi_k\|^2 \right\} \\ &= \sum_{k\in\mathbb{N}} \left\{ \|\xi_k\|^2 + 2\|A_k \xi_k\|^2 + 2\|B_k \xi_k\|^2 \right\} . \end{aligned}$$

As a result $\|H\xi\|^2 = \|\xi\|^2 + 2(\|A\xi\|^2 + \|B\xi\|^2)$, for $\xi \in \mathrm{dom}\,(H)$. Hence the operator H is closed.

Since H is non-negative, for its self-adjointness it is sufficient to check that the operator range, $\mathrm{ran}\,(\mathbb{1} + H) = \mathfrak{H}$. To this end assume that there is an element $\eta \in \mathfrak{H}$ such that $((\mathbb{1} + H)\,\xi, \eta) = 0$, for every $\xi \in \mathrm{dom}\,(H)$. Then by definition (7.2.16), we get $((\mathbb{1} + H_k)\,\xi_k, \eta_k) = 0$ for any $k \in \mathbb{N}$ and $\xi_k \in \mathfrak{H}_k$. Consequently, by self-adjointness of operators (7.2.11), one obtains $\eta_k = 0$ for any $k \in \mathbb{N}_k$, which yields $\eta = 0$. Hence, the operator H is self-adjoint. □

Theorem 7.2.3 *Let A and B be given by* (7.2.11)-(7.2.15). *Then there exists in* $\mathbb{R}^+$ *a positive bounded continuous function* $t \mapsto L(t)$, *such that the estimate from below*

$$\| \mathrm{e}^{-t(A+B)} - (\mathrm{e}^{-tB/2n}\mathrm{e}^{-tA/n}\mathrm{e}^{-tB/2n})^n \| \geq L(t)\, n^{-1}\,, \tag{7.2.17}$$

holds for every $t > 0$ *and* $n \in \mathbb{N}$.

Proof Note that by definition of Hilbert space (7.2.9), we infer inequalities:

$$\begin{aligned} &\| \mathrm{e}^{-tH} - (\mathrm{e}^{-tB/2n}\mathrm{e}^{-tA/n}\mathrm{e}^{-tB/2n})^n \|_{\mathfrak{H}} \\ &\geq \| \mathrm{e}^{-tH_n} - (\mathrm{e}^{-tB_n/2n}\mathrm{e}^{-tA_n/n}\mathrm{e}^{-tB_n/2n})^n \|_{\mathfrak{H}_n} \\ &\geq \frac{1}{2}\left|\mathrm{Tr}_{\,\mathfrak{H}_n}\left(\mathrm{e}^{-tH_n} - (\mathrm{e}^{-tB_n/2n}\mathrm{e}^{-tA_n/n}\mathrm{e}^{-tB_n/2n})^n\right)\right|, \quad n \in \mathbb{N}\,. \end{aligned} \tag{7.2.18}$$

Here the norm in the *first term* means the operator norm of bounded operators on $\mathfrak{H}$ and in the *second term* the operator norm on $\mathfrak{H}_n = \mathbb{C}^2$, and the trace in the *third term* is that one over the 2×2 matrices. Here $H_n = A_n + B_n$; see definitions (7.2.11) and (7.2.16) for $n \in \mathbb{N}$.

For a later use, let us note that for $\Lambda(\theta) := S\cos(\theta) + T\sin(\theta)$, the following formulæ

$$\mathrm{e}^{-s\Lambda(\theta)} = E\,\mathrm{ch}\,(s) - \Lambda(\theta)\,\mathrm{sh}\,(s), \tag{7.2.19}$$

$$\mathrm{Tr}_{\,\mathbb{C}^2}\,(\mathrm{e}^{-s\Lambda(\theta)}) = 2\,\mathrm{ch}\,(s), \tag{7.2.20}$$

hold for $s \in \mathbb{R}^+$. To get the first formula in (7.2.19), we expand $\mathrm{e}^{-s\Lambda}$ into a series and use the property $\Lambda^2 = E$, which is a consequence of (7.2.10). The second formula follows from the first one and $\mathrm{Tr}_{\,\mathbb{C}^2}(\Lambda) = 0$.

On account of (7.2.11), we obtain that $H_n = 2n\,E + 2n\,\cos(\theta_n/2)\,\Lambda(\theta_n/2)$. Then in virtue of (7.2.13), (7.2.19) and (7.2.20)

$$\mathrm{Tr}_{\,\mathfrak{H}_n}\,(\mathrm{e}^{-tH_n}) = 2\,\mathrm{e}^{-2nt}\mathrm{ch}\,(2nt\,\cos(\theta_n/2)), \quad n \in \mathbb{N}\,. \tag{7.2.21}$$

On the other hand, from (7.2.19) and (7.2.20), we also obtain that

$$\begin{aligned}\operatorname{Tr}_{\mathfrak{H}_n}\left(\left(\mathrm{e}^{-tA_n/2n}\mathrm{e}^{-tB_n/n}\mathrm{e}^{-tA_n/2n}\right)^n\right) &= \mathrm{e}^{-2nt}\operatorname{Tr}_{\mathfrak{H}_n}\left(\left(\mathrm{e}^{-tS/2}\mathrm{e}^{-t(S\cos\theta_n+T\sin\theta_n)}\mathrm{e}^{-tS/2}\right)^n\right)\\ &= \mathrm{e}^{-2nt}\operatorname{Tr}_{\mathfrak{H}_n}\left(\left(a_n(t)\,E - b_n(t)\,S - c_n(t)\,T\right)^n\right),\end{aligned} \tag{7.2.22}$$

where a_n, b_n and c_n are positive numbers defined by

$$\begin{aligned} a_n(t) &:= \operatorname{ch}^2(t) + \operatorname{sh}^2(t)\cos\theta_n = \operatorname{ch}(2t) - \varepsilon_n \operatorname{sh}^2(t),\\ b_n(t) &:= \operatorname{sh}(t)\operatorname{ch}(t)(1+\cos(\theta_n)),\\ c_n(t) &:= \operatorname{sh}(t)\sin(\theta_n).\end{aligned} \tag{7.2.23}$$

Since the identity $a_n^2(t) - b_n^2(t) - c_n^2(t) = 1$ holds, there exist positive numbers $K_n(t)$ and $\Theta_n(t)$ such that

$$\begin{aligned} a_n(t) &= \operatorname{ch}(K_n(t)),\\ b_n(t) &= \operatorname{sh}(K_n(t))\cos(\Theta_n(t)),\\ c_n(t) &= \operatorname{sh}(K_n(t))\sin(\Theta_n(t)).\end{aligned} \tag{7.2.24}$$

Setting $s := n\,K_n(t)$ and $\theta := \Theta_n(t)$ in (7.2.19), (7.2.20), one gets

$$\begin{aligned}\operatorname{Tr}_{\mathfrak{H}_n}\left(\left(\mathrm{e}^{-tB_n/2n}\mathrm{e}^{-tA_n/n}\mathrm{e}^{-tB_n/2n}\right)^n\right) &= \mathrm{e}^{-2nt}\operatorname{Tr}_{\mathfrak{H}_n}\left(\left(\mathrm{e}^{-K_n(t)(S\cos(\Theta_n(t))+T\sin(\Theta_n(t)))}\right)^n\right) = 2\mathrm{e}^{-2nt}\operatorname{ch}(n\,K_n(t)).\end{aligned} \tag{7.2.25}$$

Now, on account of (7.2.12) for θ_n, ε_n, we can introduce positive numbers $\{\delta_n\}_{n\geq 1}$ such that

$$\delta_n = 2 - 2\cos(\theta_n/2) \quad \text{and} \quad \varepsilon_n = 2\delta_n - \delta_n^2/2\,, \quad n\in\mathbb{N}\,. \tag{7.2.26}$$

Note that $\varepsilon_n/2 \le \delta_n \le \varepsilon_n$. By (7.2.23) and (7.2.24), we deduce estimate

$$\begin{aligned}
&\operatorname{ch}(K_n(t)) - \operatorname{ch}((2-\delta_n)\,t) \\
&\quad = (1-\varepsilon_n/2)\operatorname{ch}(2t) + \varepsilon_n/2 - \operatorname{ch}((2-\delta_n)\,t) \\
&\quad = \int_0^t ds\,(t-s)\,\partial_s^2\,[(1-\varepsilon_n/2)\operatorname{ch}(2s) + \varepsilon_n/2 - \operatorname{ch}((2-\delta_n)s)] \\
&\quad = \int_0^t ds\,(t-s)\,[4(1-\varepsilon_n/2)\operatorname{ch}(2s) - (2-\delta_n)^2\operatorname{ch}((2-\delta_n)s)] \\
&\quad \ge (2\delta_n - 2\delta_n^2 + \delta_n^3/2)\int_0^t ds\,(t-s)\,(\operatorname{ch}(2s)-1)\,.
\end{aligned} \tag{7.2.27}$$

Here the last step is due to the convexity of the function $s \mapsto \operatorname{ch} s$, which yields

$$\operatorname{ch}\big((1-\delta_n/2)\,2s + (\delta_n/2)\,0\big) \le (1-\delta_n/2)\operatorname{ch} 2s + \delta_n/2\,.$$

Then from (7.2.12) and (7.2.26), we obtain for the coefficient in the lower bound (7.2.27) the estimate

$$2\delta_n - 2\delta_n^2 + \delta_n^3/2 = \delta_n(2-\varepsilon_n) \ge \varepsilon_n(1-\varepsilon_n/2) \ge 3/(8n^2)\,, \quad n \in \mathbb{N}\,. \tag{7.2.28}$$

and by integration

$$\int_0^t ds\,(t-s)\,(\operatorname{ch}(2s)-1) = (\operatorname{ch}(2t) - 1 - 2t^2)/4\,. \tag{7.2.29}$$

Next we are going to use the *Cauchy mean value* theorem saying that for two real-valued functions f and g, which are continuous on the closed interval $[a, b]$ and differentiable in (a, b), there exists c such that

$$(f(b) - f(a))\,g'(c) = (g(b) - g(a))\,f'(c)\,, \quad a < c < b\,. \tag{7.2.30}$$

To this end, we note that (7.2.27) yields $b := K_n(t) > a := (2-\delta_n)\,t$ for $t > 0$ and apply (7.2.30) to the functions $f(x) := \operatorname{ch}(n\,x)$ and $g(x) := \operatorname{ch}(x)$ for $x \in [a, b]$ and $n \in \mathbb{N}$. Then by (7.2.30) there exists M_n with $(2-\delta_n)\,t < M_n(t) < K_n(t)$ for which

$$\operatorname{ch}(nK_n(t)) - \operatorname{ch}(n(2-\delta_n)\,t) = \frac{n\operatorname{sh}(nM_n(t))}{\operatorname{sh}(M_n(t))}\big(\operatorname{ch}(K_n(t)) - \operatorname{ch}((2-\delta_n)\,t)\,. \tag{7.2.31}$$

To proceed with lower bound in the estimate (7.2.18), we use (7.2.21), (7.2.25) and identity (7.2.31) that yields

$$\begin{aligned}
&\tfrac{1}{2}\mathrm{Tr}\left(\left(\mathrm{e}^{-tB_n/2n}\mathrm{e}^{-tA_n/n}\mathrm{e}^{-tB_n/2n}\right)^n - \mathrm{e}^{-t(A_n+B_n)}\right)\\
&\quad = \mathrm{e}^{-2nt}\big(\mathrm{ch}\,(nK_n(t)) - \mathrm{ch}\,(n(2-\delta_n)t)\big)\\
&\quad = \mathrm{e}^{-2nt}\, n\frac{\mathrm{sh}\ (nM_n(t))}{\mathrm{sh}\ (M_n(t))}\big(\mathrm{ch}\,(K_n(t)) - \mathrm{ch}\,((2-\delta_n)t)\big)\\
&\quad \geq \mathrm{e}^{-2nt}\mathrm{e}^{(n-1)M_n(t)}\,(\mathrm{ch}\,(2t) - 1 - 2t^2)/32n\,, \quad n\in\mathbb{N}\,.
\end{aligned} \tag{7.2.32}$$

Here the last inequality is due to (7.2.27)–(7.2.29) and to estimate $\mathrm{sh}\ (ns)/\mathrm{sh}\ (s) \geq \mathrm{e}^{(n-1)s}$. Seeing that

$$(n-1)\,M_n(t) \geq (n-1)\,(2-\delta_n)t \geq (n-1)\,(2-1/2n^2)\,t \geq 2\,(n-1)\,t - t/8\,, \tag{7.2.33}$$

by virtue of (7.2.32) and (7.2.33), we infer the proof of the estimate from below (7.2.17) for $L(t) := \mathrm{e}^{-17t/8}\,(\mathrm{ch}\,(2t) - 1 - 2t^2)/32$. □

Remark 7.2.4 By Theorem 7.2.2 the operators A and B involved in Theorem 7.2.3 satisfy the conditions of the ITTZ-Theorem 7.1.9, that is, the operator-norm bound for the rate of convergence in (7.2.17) must be of the order $O(1/n)$. The operator-norm estimate of the *same* order $O(1/n)$ for the *lower* bound for the rate of convergence in (7.2.17) says that under conditions of the ITTZ-Theorem 7.1.9, this rate of convergence is *optimal* for *unbounded* generators A and/or B.

Indeed, we remind that for *bounded* generators, the operator-norm bound of difference in (7.2.17) has the same order $O(1/n^2)$ as in the *S.Lie* product formula for matrices; see (1.10.37). Therefore, in the case of *unbounded* generators, the symmetrisation of approximants in the ITTZ-setting (7.2.17) does not improve the operator-norm rate of operator-norm convergence of the corresponding product formula; see Corollary 7.1.13, (7.1.84).

For that reason, similar to the *S.Lie* product formula for matrices, we prove optimality of the operator-norm estimate in Corollary 7.1.12 for the case of non-symmetric approximants.

Theorem 7.2.5 *Let A and B be given by* (7.2.11)–(7.2.15). *Then there exists in $\mathbb{R}^+$ a positive bounded continuous function $t \mapsto M(t)$, on $\mathbb{R}^+$, such that the estimate from below* :

$$\|\,\mathrm{e}^{-t(A+B)} - (\mathrm{e}^{-tA/n}\mathrm{e}^{-tB/n})^n\| \geq M(t)\,n^{-1}\,, \tag{7.2.34}$$

holds for every $t > 0$ and $n \in \mathbb{N}$.

Proof The proof mimics the line of reasoning of the ITTZ-Theorem 7.1.9 and follows mutatis mutandis the arguments in the proof of Lemma H.0.7 for $\alpha = 1$. □

For improving the rate of operator-norm convergence of product formulæ in the case of *unbounded* generators, one has to modify conditions on operators A and B; see Notes to Chap. 7 (Sects. 7.2 and 7.4). In Notes to Chap. 7 (Remarks 7.5.2 and 7.5.3), we scrutinise subtlety (*nuances*) of conditions ensuring the operator-norm convergence of the product formulæ with *optimal* rates.

7.3 Product Formulæ for More Than Two Generators

In this section, we present a generalisation of certain results from Sect. 7.2 for the case of the sum of $m > 2$ non-negative self-adjoint generators $A_1, A_2, \ldots, A_m$ of C_0-semigroups on $\mathfrak{H}$.

Note that by virtue of Theorem 5.1.1 (*Trotter* theorem) for *essentially* self-adjoint on $\cap_{l=1}^{m} \mathrm{dom}\,(A_l)$ operator $T_m := A_1 + A_2 + \ldots + A_m$, or on account of Theorem 5.1.5 (*Kato* theorem) for dense intersection $\cap_{l=1}^{m} \mathrm{dom}\,(A_l^{1/2})$ and the self-adjoint operator $H_m = A_1 \dotplus A_2 \dotplus \cdots \dotplus A_m$, defined as a *form-sum*, the product formula

$$\mathrm{s-}\lim_{n\to\infty} (\mathrm{e}^{-tA_1/n} \mathrm{e}^{-tA_2/n} \cdots \mathrm{e}^{-tA_m/n})^n = \mathrm{e}^{-tH_m} , \tag{7.3.1}$$

holds in the strong operator topology locally uniformly in $t \in \mathcal{I}$ for any bounded interval $\mathcal{I} \subset \mathbb{R}_0^+$. Here generator H_m either is the closure, $H_m = \overline{T}_m$, or it coincides with H_m defined as the form-sum.

To make a contact with the Trotter-Kato product formulæ, let $f_l \in \mathcal{K}$, $l \in \mathbb{N}$, and

$$\begin{aligned} F_1(\tau) &:= f_1(\tau A_1), \\ F_l(\tau) &:= f_l(\tau A_l/2) \cdots f_2(\tau A_2/2) f_1(\tau A_1) f_2(\tau A_2/2) \cdots f_l(\tau A_l), \end{aligned}$$

for $l \geq 2$ and $\tau \geq 0$. Then we also define

$$S_l(\tau) := \frac{\mathbb{1} - F_l(\tau)}{\tau}, \quad \tau > 0, \quad l = 1, 2, \ldots, m,$$

and we set

$$A_l(\tau) := \frac{\mathbb{1} - f_l(\tau A_l)}{\tau}, \quad \tau > 0, \quad l = 1, 2, \ldots, m.$$

Note that $S_1(\tau) = A_1(\tau)$, for $\tau > 0$ and $l = 1$. If $A_1 = A$, $A_2 = B$, $f_1 = f$, $f_2 = g$, then $F_2(\tau) = F(\tau)$, $\tau \geq 0$, cf. (5.2.3), as well as $S_2(\tau) = S(\tau)$ and $A_1(\tau) = A(\tau)$, $\tau > 0$, cf. (4.1.1) and (5.2.4).

Furthermore, we have equation:

$$F_l(\tau) = f_l(\tau A_l/2) F_{l-1}(\tau) f_l(\tau A_l/2)\,, \quad \tau \geq 0, \quad l = 2, 3, \ldots, m.$$

Then one can check that

$$S_l(\tau) = K_l(\tau) + W_l(\tau), \quad \tau > 0, \tag{7.3.2}$$

where

$$K_l(\tau) = S_{l-1}(\tau) + A_l(\tau/2) - \frac{\tau}{4} A_l(\tau/2)^2, \quad \tau > 0, \tag{7.3.3}$$

and

$$W_l(\tau) = \frac{\tau^2}{4} A_l(\tau/2) S_{l-1}(\tau) A_l(\tau/2) - \frac{\tau}{2}\left(S_{l-1}(\tau) A_l(\tau/2) + A_l(\tau/2) S_{l-1}(\tau)\right), \tag{7.3.4}$$

$\tau > 0, l = 2, 3, \ldots, m$. Notice that

$$K_l(\tau) = S_{l-1}(\tau) + A_l(\tau/2)^{1/2}\, \frac{\mathbb{1} + f_l(\tau A_l/2)}{2}\, A_l(\tau/2)^{1/2}, \quad \tau > 0, \tag{7.3.5}$$

$l = 2, 3, \ldots, m$. Moreover, we define

$$S_{K_l}(\tau) := S_{l-1}(\tau)^{1/2} (\mathbb{1} + K_l(\tau))^{-1/2}, \quad \tau > 0,$$

$$A_{K_l}(\tau) := \left(\frac{\mathbb{1} + f_l(\tau A_l/2)}{2}\right)^{1/2} A_l(\tau/2)^{1/2} (\mathbb{1} + K_l(\tau))^{-1/2}, \quad \tau > 0,$$

$l = 2, 3, \ldots, m$. As a consequence, one gets $\|S_{K_l}(\tau)\| \leq 1$ and $\|A_{K_l}(\tau)\| \leq 1$ for $\tau > 0, l = 2, 3, \ldots, m$. Finally, we set

$$Q_l(\tau) := (\mathbb{1} + K_l(\tau))^{-1/2}\, W_l(\tau)\, (\mathbb{1} + K_l(\tau))^{-1/2}\,, \tag{7.3.6}$$

for $\tau > 0$ and $l = 2, 3, \ldots, m$.

The main result of this section is Theorem 7.3.9. It elucidates and generalises the ITTZ-Theorem 7.1.9, for example, of more than two generators. This is the case when $A_1, \ldots, A_m$, for $m \in \mathbb{N}$, are non-negative self-adjoint operators and operator sum

$$H_m = A_1 + A_2 + \ldots + A_m\,, \quad \mathrm{dom}\,(H_m) = \bigcap_{l=1}^{m} \mathrm{dom}\,(A_l)\,, \tag{7.3.7}$$

is a self-adjoint operator on domain $\mathrm{dom}\,(H_m)$ in a Hilbert space $\mathfrak{H}$, for any $m \in \mathbb{N}$.

Lemma 7.3.1 *Let $A_1, \ldots, A_m$ be non-negative self-adjoint operators in $\mathfrak{H}$. If $f_1, \ldots, f_m$ are Kato functions, then $\mathbb{1} + Q_l(\tau)$ is boundedly invertible for each $\tau > 0$ and*

$$\|(\mathbb{1} + Q_l(\tau))^{-1}\| \leq q_0 := \left(\frac{\sqrt{17}+1}{\sqrt{17}-1}\right)^2, \quad \tau > 0, \tag{7.3.8}$$

for $l = 2, 3, \ldots, m$.

Proof On account of (7.3.6), we obtain that

$$\begin{aligned} Q_l(\tau) =& (\mathbb{1} + K_l(\tau))^{-1/2}(\tfrac{\tau}{2}A_l(\tau/2) - r)S_{l-1}(\tau)(\tfrac{\tau}{2}A_l(\tau/2) - r)(\mathbb{1} + K_l(\tau))^{-1/2} \\ & - \tfrac{(1-r)\tau}{2}(\mathbb{1} + K_l(\tau))^{-1/2}(S_{l-1}(\tau)A_l(\tau/2) + A_l(\tau/2)S_{l-1}(\tau))(\mathbb{1} + K_l(\tau))^{-1/2} \\ & - r^2(\mathbb{1} + K_l(\tau))^{-1/2}S_{l-1}(\tau)(\mathbb{1} + K_l(\tau))^{-1/2}, \end{aligned}$$

for $l = 2, 3, \ldots, m$, which yields inequality

$$\begin{aligned} & Q_l(\tau) \geq \\ & - \tfrac{(1-r)\tau}{2}(\mathbb{1} + K_l(\tau))^{-1/2}(S_{l-1}(\tau)A_l(\tau/2) + A_l(\tau/2)S_{l-1}(\tau))(\mathbb{1} + K_l(\tau))^{-1/2} \\ & - r^2\,(\mathbb{1} + K_l(\tau))^{-1/2}\,S_{l-1}(\tau)\,(\mathbb{1} + K_l(\tau))^{-1/2}, \end{aligned}$$

where $r \in (0, 1)$ is a *constant* that will be determined later. Notice that

$$S_{l-1}(\tau)\,(\mathbb{1} + K_l(\tau))^{-1/2} = \frac{1}{\tau^{1/2}}\,(\mathbb{1} - F_{l-1}(\tau))^{1/2}\,S_{K_l}(\tau), \quad \tau > 0, \tag{7.3.9}$$

$l = 2, 3, \ldots, m$, which can be estimated by

$$\|S_{l-1}(\tau)\,(\mathbb{1} + K_l(\tau))^{-1/2}u\| \leq \frac{1}{\tau^{1/2}}\,\|S_{K_l}(\tau)\,u\|, \quad u \in \mathfrak{H}, \quad \tau > 0,$$

$l = 2, 3, \ldots, m$. Further, using

$$A_l(\tau/2)\,(\mathbb{1} + K_l(\tau))^{-1/2} = \frac{1}{\tau^{1/2}}\left(\frac{\mathbb{1} - f_l(\tau A_l/2)}{\mathbb{1} + f_1(\tau A_l/2)}\right)^{1/2} A_{K_l}(\tau/2), \quad \tau > 0,$$

$l = 2, 3, \ldots, m$, we get the estimate

$$\|A_l(\tau/2)\,(\mathbb{1} + K_l(\tau))^{-1/2}u\| \leq \frac{1}{\tau^{1/2}}\,\|A_{K_l}(\tau)u\|, \quad u \in \mathfrak{H}, \quad \tau > 0,$$

$l = 2, 3, \ldots, m$. Hence, we have for $u \in \mathfrak{H}$,

$$\begin{aligned}(Q_l(\tau)u, u) &\geq -(1-r)\,\|S_{K_l}(\tau)u\|\,\|A_{K_l}(\tau)u\| - r^2\|A_{K_l}(\tau)u\|^2\\ &\geq -((1-r)p/2 + r^2)\,\|S_{K_l}(\tau)u\|^2 - (1-r)\,\|(A_{K_l}(\tau)\,u\|^2/2p)\,,\end{aligned}$$

$\tau > 0$, $p > 0$ and $l = 2, 3, \ldots, m$. Starting from this point we follow the proof of Lemma 5.2.3 (below (5.2.42)) using that S_{K_l} and A_{K_l} are contractions for $l = 2, 3, \ldots, m$. Following this line of reasoning we obtain the proof of estimate (7.3.8). □

If $A_1, \ldots, A_m$ are non-negative self-adjoint operators, we set

$$H_l = A_1 + A_2 + \cdots + A_l\,, \quad l = 2, 3, \ldots, m\,. \tag{7.3.10}$$

In the following, we also assume that for any $l = 2, 3, \ldots, m$, the sum H_l is a densely defined *self-adjoint* operator. The sums $\{H_l\}_{l>1}$ are called the *partial operator sums*.

Lemma 7.3.2 *Let $A_1, \ldots, A_m$ be non-negative self-adjoint operators in a Hilbert space $\mathfrak{H}$ such that the partial operator sums H_l are densely defined for any $l = 2, 3, \ldots, m$. Further, let $f_1, \cdots, f_m$ are Kato functions. If for some $l = 3, \ldots, m$ one has that $H_l = H_{l-1} + A_l$ as well as there are constants $C^{S_{l-1}}_{H_{l-1}} > 0$ and $C^{S_{l-1}}_{K_{l-1}} > 0$ such that the estimates*

$$\|S_{l-1}(\tau)\,(\mathbb{1} + H_{l-1})^{-1}\| \leq C^{S_{l-1}}_{H_{l-1}}, \quad \tau > 0, \tag{7.3.11}$$

and

$$\left\|(\mathbb{1} + K_{l-1}(\tau))^{-1/2}\sqrt{S_{l-1}(\tau)}\right\| \leq C^{S_{l-1}}_{K_{l-1}}, \quad \tau > 0, \tag{7.3.12}$$

are valid, then there are constants $C^{S_l}_{H_l} > 0$ and $C^{S_l}_{K_l} > 0$ such that the estimates

$$\|S_l(\tau)\,(\mathbb{1} + H_l)^{-1}\| \leq C^{S_l}_{H_l}, \quad \tau > 0, \tag{7.3.13}$$

and

$$\left\|(\mathbb{1} + K_l(\tau))^{-1/2}\sqrt{S_l(\tau)}\right\| \leq C^{S_l}_{K_l}, \quad \tau > 0. \tag{7.3.14}$$

Proof From (7.3.2), (7.3.3) and (7.3.5), we get

$$S_l(\tau) = S_{l-1}(\tau) + \frac{\mathbb{1} + f_l(\tau A_l/2)}{2} A_l(\tau/2) + W_l(\tau), \quad \tau > 0.$$

Hence, we find the estimate

$$\begin{aligned}\|S_l(\tau)(\mathbb{1}+H_l)^{-1}\| \le \|S_{l-1}(\tau)(\mathbb{1}+H_l)^{-1}\|+ \\ \|A_l(\tau/2)(\mathbb{1}+H_l)^{-1}\|+\|W_l(\tau)(\mathbb{1}+H_l)^{-1}\|, \quad \tau>0.\end{aligned} \tag{7.3.15}$$

Since $H_l = H_{l-1} + A_l$, one has $\mathrm{dom}\,(H_l) \subseteq \mathrm{dom}\,(H_{l-1})$ which yields that the constant $C_{H_l}^{H_{l-1}} := \|(\mathbb{1}+H_{l-1})(\mathbb{1}+H_l)^{-1}\|$ is well defined and finite. Hence, using the assumption (7.3.11), we get the estimate

$$\|S_{l-1}(\tau)\,(\mathbb{1}+H_l)^{-1}\| \le C_{H_l}^{H_{l-1}}\|S_{l-1}(\tau)(\mathbb{1}+H_{l-1})^{-1}\| \le C_{H_l}^{H_{l-1}}C_{H_{l-1}}^{S_{l-1}}, \tag{7.3.16}$$

for $\tau > 0$. Using again $H_l = H_{l-1} + A_l$, we get $\mathrm{dom}\,(H_l) \subseteq \mathrm{dom}\,(A_l)$ which implies $C_{H_l}^{A_l} := \|A_l(\mathbb{1}+H_l)^{-1}\| < \infty$. Hence, we have

$$\|A_l(\tau/2)(\mathbb{1}+H_l)^{-1}\| \le C_{H_l}^{A_l}\|A_l(\tau/2)A_l^{-1}P_{A_l}\| \le C_{H_l}^{A_l}\gamma[f_l], \quad \tau>0, \tag{7.3.17}$$

where P_{A_l} is the orthogonal projection onto $\ker(A_l)^{\perp}$.

To estimate $\|W_l(\tau)(\mathbb{1}+H_l)^{-1}\|$, we use formula (7.3.4). Hence, we get

$$\begin{aligned}W_l(\mathbb{1}+H_l)^{-1} &= \frac{\tau^2}{4}A_l(\tau/2)S_{l-1}(\tau)A_l(\tau/2)(\mathbb{1}+H_l)^{-1} \\ &- \frac{\tau}{2}S_{l-1}(\tau)A_l(\tau/2)(\mathbb{1}+H_l)^{-1} - \frac{\tau}{2}A_l(\tau/2)S_{l-1}(\tau)(\mathbb{1}+H_l)^{-1}\end{aligned}$$

which gives the estimate

$$\|W_l(\mathbb{1}+H_l)^{-1}\| \le \|A_l(\tau/2)(\mathbb{1}+H_l)^{-1}\|+\|S_{l-1}(\tau)(\mathbb{1}+H_l)^{-1}\|, \quad \tau>0.$$

Using the estimates (7.3.17) and (7.3.16), we find

$$\|W_l(\mathbb{1}+H_l)^{-1}\| \le C_{H_l}^{A_l}\gamma[f_l]+C_{H_l}^{H_{l-1}}C_{H_{l-1}}^{S_{l-1}}, \quad \tau>0. \tag{7.3.18}$$

Inserting the estimates (7.3.16), (7.3.17) and (7.3.18) into (7.3.15) we obtain

$$\|S_l(\tau)(\mathbb{1}+H_l)^{-1}\| \le C_{H_l}^{S_l}, \quad \tau>0,$$

where

$$C_{H_l}^{S_l} := C_{H_l}^{H_{l-1}}C_{H_{l-1}}^{S_{l-1}} + 2C_{H_l}^{A_l}\gamma[f_l] + C_{H_l}^{H_{l-1}}C_{H_{l-1}}^{S_{l-1}}$$

which proves (7.3.13).

To prove (7.3.14) we note that

$$(\mathbb{1} + K_l(\tau))^{-1/2} S_l(\tau)(\mathbb{1} + K_l(\tau))^{-1/2} =$$
$$(\mathbb{1} + K_l(\tau))^{-1/2} K_l(\tau)(\mathbb{1} + K_l(\tau))^{-1/2} +$$
$$(\mathbb{1} + K_l(\tau))^{-1/2} W_l(\tau)(\mathbb{1} + K_l(\tau))^{-1/2},$$

$\tau > 0$, where we have used (7.3.2). Hence

$$\|(\mathbb{1} + K_l(\tau))^{-1/2} S_l(\tau)(\mathbb{1} + K_l(\tau))^{-1/2}\| \leq$$
$$1 + \|(\mathbb{1} + K_l(\tau))^{-1/2} W_l(\tau)(\mathbb{1} + K_l(\tau))^{-1/2}\|, \quad \tau > 0. \tag{7.3.19}$$

Using (7.3.4) we get

$$\|(\mathbb{1}+K_l(\tau))^{-1/2} W_l(\tau)(\mathbb{1} + K_l(\tau))^{-1/2}\| \leq$$
$$\frac{1}{2}\|\sqrt{A_l(\tau/2)}(\mathbb{1} + K_l(\tau))^{-1/2}\|^2 + \tag{7.3.20}$$
$$\sqrt{2}\|\sqrt{A_l(\tau/2)}(\mathbb{1} + K_l(\tau))^{-1/2}\| \, \|\sqrt{S_{l-1}(\tau)}(\mathbb{1} + K_l(\tau))^{-1/2}\|,$$

$\tau > 0$. We note

$$\mathbb{1} + \frac{1}{2} A_l(\tau/2) \leq \mathbb{1} + K_l(\tau), \quad \tau > 0,$$

which yields

$$(\mathbb{1} + K_l(\tau))^{-1} \leq (\mathbb{1} + \frac{1}{2} A_l(\tau/2))^{-1} \leq 2(\mathbb{1} + A_l(\tau/2))^{-1}, \quad \tau > 0.$$

Hence, we get

$$\sqrt{A_l(\tau/2)}(\mathbb{1} + K_l(\tau))^{-1}\sqrt{A_l(\tau/2)} \leq$$
$$2\sqrt{A_l(\tau/2)}(\mathbb{1} + A_l(\tau/2))^{-1}\sqrt{A_l(\tau/2)} \leq 2\mathbb{1}, \quad \tau > 0.$$

This yields the estimate

$$\|(\mathbb{1} + K_l(\tau))^{-1/2}\sqrt{A_l(\tau/2)}\| \leq \|\sqrt{A_l(\tau/2)}(\mathbb{1} + K_l(\tau))^{-1/2}\| \leq \sqrt{2}, \tag{7.3.21}$$

$\tau > 0$. Next we notice that

$$\begin{aligned}\mathbb{1} + K_l(\tau) &\geq \mathbb{1} + K_{l-1}(\tau) + W_{l-1}(\tau)\\ &= \sqrt{\mathbb{1} + K_{l-1}(\tau)}(\mathbb{1} + Q_{l-1}(\tau))\sqrt{\mathbb{1} + K_{l-1}(\tau)},\end{aligned}$$

$\tau > 0$. As a result

$$(\mathbb{1} + K_l(\tau))^{-1} \leq (\mathbb{1} + K_{l-1}(\tau))^{-1/2}(\mathbb{1} + Q_{l-1}(\tau))^{-1}(\mathbb{1} + K_{l-1}(\tau))^{-1/2}$$

for $\tau > 0$. Using Lemma 7.3.1 we get

$$0 \leq (\mathbb{1} + Q_{l-1}(\tau))^{-1} \leq q_0\mathbb{1}, \quad \tau > 0,$$

which leads to

$$(\mathbb{1} + K_l(\tau))^{-1} \leq q_0(\mathbb{1} + K_{l-1}(\tau))^{-1}, \quad \tau > 0.$$

Consequently,

$$\begin{aligned}&\sqrt{S_{l-1}(\tau)}(\mathbb{1} + K_l(\tau))^{-1}\sqrt{S_{l-1}(\tau)} \leq\\ &\qquad q_0\sqrt{S_{l-1}(\tau)}(\mathbb{1} + K_{l-1}(\tau))^{-1}\sqrt{S_{l-1}(\tau)}, \quad \tau > 0,\end{aligned}$$

which yields

$$\|(\mathbb{1} + K_l(\tau))^{-1/2}\sqrt{S_{l-1}(\tau)}\| \leq \sqrt{q_0}\|(\mathbb{1} + K_{l-1}(\tau))^{-1/2}\sqrt{S_{l-1}(\tau)}\|,$$

$\tau > 0$, or

$$\|\sqrt{S_{l-1}(\tau)}(\mathbb{1} + K_l(\tau))^{-1/2}\| \leq \sqrt{q_0}\|\sqrt{S_{l-1}(\tau)}(\mathbb{1} + K_{l-1}(\tau))^{-1/2}\|, \tag{7.3.22}$$

Inserting (7.3.21) and (7.3.22) into (7.3.20), we obtain

$$\begin{aligned}&\|(\mathbb{1}+K_l(\tau))^{-1/2}W_l(\tau)(\mathbb{1} + K_l(\tau))^{-1/2}\| \leq\\ &\qquad 1 + 2\sqrt{q_0}\ \|\sqrt{S_{l-1}(\tau)}(\mathbb{1} + K_{l-1}(\tau))^{-1/2}\|, \quad \tau > 0.\end{aligned} \tag{7.3.23}$$

Using now the assumption (7.3.12), we find from (7.3.23) that

$$\|(\mathbb{1}+K_l(\tau))^{-1/2}W_l(\tau)(\mathbb{1}+K_l(\tau))^{-1/2}\| \leq 1+2\sqrt{q_0}\ C^{S_{l-1}}_{K_{l-1}}, \quad \tau > 0. \tag{7.3.24}$$

If we put (7.3.24) in (7.3.19), we get

$$\|(\mathbb{1} + K_l(\tau))^{-1/2}S_l(\tau)(\mathbb{1} + K_l(\tau))^{-1/2}\| \leq 2(1 + \sqrt{q_0}\ C^{S_{l-1}}_{K_{l-1}}), \quad \tau > 0,$$

which yields

$$\|\sqrt{S_l(\tau)}(\mathbb{1} + K_l(\tau))^{-1/2}\| \leq \sqrt{2}\sqrt{1 + \sqrt{q_0}\, C^{S_{l-1}}_{K_{l-1}}}, \quad \tau > 0.$$

Finally, setting

$$C^{S_l}_{K_l} := \sqrt{2}\sqrt{1 + \sqrt{q_0}\, C^{S_{l-1}}_{K_{l-1}}} \tag{7.3.25}$$

we prove (7.3.14). □

Corollary 7.3.3 *Let $A_1, \ldots, A_m$ be non-negative self-adjoint operators in a Hilbert space such that the partial operator sums H_l are densely defined for any $l = 2, 3, \ldots, m$. Further, let $f_1, \cdots, f_m$ are Kato functions from $\mathcal{K}$. If for some $l = 3, \ldots, m$ one has $H_l = H_{l-1} + A_l$ as well as the estimate* (7.3.11) *and* (7.3.12) *are satisfied, then there is a constant $C^{W_l}_{K_l,H_l} > 0$ such that*

$$\left\|(\mathbb{1} + K_l(\tau))^{-1/2} W_l(\tau)(\mathbb{1} + H_l)^{-1}\right\| \leq C^{W_l}_{K_l,H_l}\sqrt{\tau}, \tag{7.3.26}$$

is valid for $\tau > 0$.

Proof Using formula (7.3.4) we obtain

$$\begin{aligned} W_l(\tau) := &\frac{\tau^2}{4}(\mathbb{1} + K_l(\tau))^{-1/2} A_l(\tau/2) S_{l-1}(\tau) A_l(\tau/2)(\mathbb{1} + H_l)^{-1} \\ &- \frac{\tau}{2}(\mathbb{1} + K_l(\tau))^{-1/2} S_{l-1}(\tau) A_l(\tau/2)(\mathbb{1} + H_l)^{-1} \\ &- \frac{\tau}{2}(\mathbb{1} + K_l(\tau))^{-1/2} A_l(\tau/2) S_{l-1}(\tau)(\mathbb{1} + H_l)^{-1}, \end{aligned}$$

$\tau > 0$, which yields the estimate

$$\|W_l(\tau)\| \leq \frac{\tau^2}{4}\|(\mathbb{1} + K_l(\tau))^{-1/2} A_l(\tau/2) S_{l-1}(\tau) A_l(\tau/2)(\mathbb{1} + H_l)^{-1}\| \tag{7.3.27}$$

$$+ \|\frac{\tau}{2}(\mathbb{1} + K_l(\tau))^{-1/2} S_{l-1}(\tau) A_l(\tau/2)(\mathbb{1} + H_l)^{-1}\| \tag{7.3.28}$$

$$+ \|\frac{\tau}{2}(\mathbb{1} + K_l(\tau))^{-1/2} A_l(\tau/2) S_{l-1}(\tau)(\mathbb{1} + H_l)^{-1}\|, \tag{7.3.29}$$

To estimate the first term (7.3.27), we use inequality

$$\begin{aligned} &\frac{\tau^2}{4}\|(\mathbb{1} + K_l(\tau))^{-1/2} A_l(\tau/2) S_{l-1}(\tau) A_l(\tau/2)(\mathbb{1} + H_l)^{-1}\| \\ &\quad \leq \frac{\sqrt{\tau}}{2\sqrt{2}}\|(\mathbb{1} + K_l(\tau))^{-1/2}\sqrt{A_l(\tau/2)}\|\, \|A_l(\tau/2)(\mathbb{1} + H_l)^{-1}\|, \quad \tau > 0. \end{aligned}$$

By (7.3.21) and (7.3.17), we obtain

$$\frac{\tau^2}{4}\|(\mathbb{1} + K_l(\tau))^{-1/2} A_l(\tau/2) S_{l-1}(\tau) A_l(\tau/2)(\mathbb{1} + H_l)^{-1}\| \le \frac{\sqrt{\tau}}{2} C^{A_l}_{H_l} \gamma[f_l], \tag{7.3.30}$$

$\tau > 0$. To estimate the second term (7.3.28), we note that

$$\frac{\tau}{2}\|(\mathbb{1} + K_l(\tau))^{-1/2} S_{l-1}(\tau) A_l(\tau/2)(\mathbb{1} + H_l)^{-1}\|$$
$$\le \frac{\sqrt{\tau}}{2}\|(\mathbb{1} + K_l(\tau))^{-1/2}\sqrt{S_{l-1}(\tau)}\|\ \|A_l(\tau/2)(\mathbb{1} + H_l)^{-1}\|, \quad \tau > 0.$$

Using (7.3.22) we get

$$\frac{\tau}{2}\|(\mathbb{1} + K_l(\tau))^{-1/2} S_{l-1}(\tau) A_l(\tau/2)(\mathbb{1} + H_l)^{-1}\|$$
$$\le \sqrt{q_0}\frac{\sqrt{\tau}}{2}\|(\mathbb{1} + K_{l-1}(\tau))^{-1/2}\sqrt{S_{l-1}(\tau)}\|\ \|A_l(\tau/2)(\mathbb{1} + H_l)^{-1}\|, \tag{7.3.31}$$

$\tau > 0$. So, by assumption (7.3.12) and estimate (7.3.17), we obtain

$$\frac{\tau}{2}\|(\mathbb{1} + K_l(\tau))^{-1/2} S_{l-1}(\tau) A_l(\tau/2)(\mathbb{1} + H_l)^{-1}\| \le \sqrt{q_0}\frac{\sqrt{\tau}}{2} C^{S_{l-1}}_{K_{l-1}} C^{A_l}_{H_l} \gamma[f_l], \tag{7.3.32}$$

$\tau > 0$. To estimate the third term (7.3.29), we use

$$\frac{\tau}{2}\|(\mathbb{1} + K_l(\tau))^{-1/2} A_l(\tau/2) S_{l-1}(\tau)(\mathbb{1} + H_l)^{-1}\|$$
$$\le \frac{\sqrt{\tau}}{\sqrt{2}}\|(\mathbb{1} + K_l(\tau))^{-1/2}\sqrt{A_l(\tau/2)}\|\ \|S_{l-1}(\tau)(\mathbb{1} + H_l)^{-1}\|, \quad \tau > 0.$$

Taking into account (7.3.21) and (7.3.16), we get

$$\frac{\tau}{2}\|(\mathbb{1} + K_l(\tau))^{-1/2} A_l(\tau/2) S_{l-1}(\tau)(\mathbb{1} + H_l)^{-1}\| \le \sqrt{\tau}\, C^{H_{l-1}}_{H_l} C^{S_{l-1}}_{H_{l-1}}, \tag{7.3.33}$$

$\tau > 0$. Inserting the estimates (7.3.30), (7.3.32) and (7.3.33) into (7.3.27)–(7.3.29), we find

$$\|W_l(\tau)\| \le \left(\frac{1}{2} + \sqrt{q_0}\, C^{S_{l-1}}_{K_{l-1}}\right) C^{A_l}_{H_l} \gamma[f_l]\, \sqrt{\tau}, \quad \tau > 0.$$

Definition $C^{W_l}_{K_l,H_l} := (1/2 + \sqrt{q_0}\, C^{S_{l-1}}_{K_{l-1}})\, C^{A_l}_{H_l}\, \gamma[f_l]$ finalises the prove of (7.3.26).

□

Corollary 7.3.4 *Let A_1 and A_2 be self-adjoint operators in a Hilbert space $\mathfrak{H}$ such that $H_2 := A_1 + A_2$ is self-adjoint. If f_1 and f_2 are Kato functions from $\mathcal{K}$, then there are constants $C^{S_2}_{H_2} > 0$ and $C^{S_2}_{K_2} > 0$ such that the estimates (7.3.11) and (7.3.4) are valid for $l = 3$ and $\tau > 0$.*

Proof Setting $A := A_1$ and $B := A_2$ as well as $f := f_1$ and $g := f_2$, we see that $F(\tau) = F_2(\tau)$, cf. (5.2.3), $S_2(\tau) = S(\tau)$, cf. (4.1.1), $A(\tau) = A_1(\tau)$, $A_2(\tau) = B(\tau)$, cf. (5.2.4), $K(\tau) := K_2(\tau)$, cf. (5.2.6) and $W(\tau) = W_2(\tau)$, cf. (5.2.9); the proof follows from Lemma 7.1.6. □

Lemma 7.3.5 *Let $A_1, \dots, A_m$ be non-negative self-adjoint operators in a Hilbert space $\mathfrak{H}$ such that the partial operator sums H_l are densely defined for any $l = 2, 3, \dots, m$. Let f_l, $l = 1, 2, \dots, m$, be Kato functions. If for certain $l = 3, \dots, m$ one has $\varkappa[f_l] < \infty$, (7.1.2), and $H_l = H_{l-1} + A_l$, as well as there are constants $C^{S_{l-1}}_{H_{l-1}} > 0$ and $C^{S_{l-1}-H_{l-1}}_{H_{l-1},H_{l-1}} > 0$ such that the conditions (7.3.11) and*

$$\|(1+H_{l-1})^{-1}(S_{l-1} - H_{l-1})(\mathbb{1} + H_{l-1})^{-1}\| \le C^{S_{l-1}-H_{l-1}}_{H_{l-1},H_{l-1}}\,\tau, \quad \tau > 0, \tag{7.3.34}$$

are satisfied, then there is a constant $C^{S_l-H_l}_{H_{l,l}} > 0$ such that

$$\|(1 + H_l)^{-1}(S_l - H_l)(\mathbb{1} + H_l)^{-1}\| \le C^{S_l-H_l}_{H_l,H_l}\,\tau, \quad \tau > 0, \tag{7.3.35}$$

holds.

Proof Using (7.3.2) and (7.3.5) as well as $H_l = H_{l-1}$, we get the estimate

$$\begin{aligned}
&\|(\mathbb{1} + H_l)^{-1}(S_l(\tau) - H_l)(\mathbb{1} + H_l)^{-1}\| \le \\
&\qquad \|(\mathbb{1} + H_l)^{-1}(S_{l-1}(\tau) - H_{l-1})(\mathbb{1} + H_l)^{-1}\| + &(7.3.36)\\
&\qquad \|(\mathbb{1} + H_l)^{-1}(A_l(\tau/2) - A_l)(\mathbb{1} + H_l)^{-1}\| + &(7.3.37)\\
&\qquad \|(\mathbb{1} + H_l)^{-1}W_l(\tau)(\mathbb{1} + H_l)^{-1}\| &(7.3.38)
\end{aligned}$$

$\tau > 0$. Since $H_l = H_{l-1} + A_l$, we get $\mathrm{dom}\,(H_l) \subseteq \mathrm{dom}\,(H_{l-1})$, and the constant $C^{H_{l-1}}_{H_l} := \|(\mathbb{1} + H_{l-1})(\mathbb{1} + H_l)\|$ is well defined and finite. Hence,

$$\begin{aligned}
&\|(\mathbb{1} + H_l)^{-1}(S_{l-1}(\tau) - H_{l-1})(\mathbb{1} + H_l)^{-1}\| \le \\
&\qquad (C^{H_{l-1}}_{H_l})^2\|(\mathbb{1} + H_{l-1})^{-1}(S_{l-1}(\tau) - H_{l-1})(\mathbb{1} + H_{l-1})^{-1}\|, \quad \tau > 0.
\end{aligned}$$

Using the assumption (7.3.34), we estimate (7.3.36) by

$$\|(\mathbb{1} + H_l)^{-1}(S_{l-1}(\tau) - H_{l-1})(\mathbb{1} + H_l)^{-1}\| \le (C^{H_{l-1}}_{H_l})^2 C^{S_{l-1}-H_{l-1}}_{H_{l-1},H_{l-1}}\,\tau, \tag{7.3.39}$$

$\tau > 0$. Since $H_l = H_{l-1} + A_l$, we get $\operatorname{dom}(H_l) \subseteq \operatorname{dom}(A_{l-1})$, and the constant $C_{H_l}^{A_{l-1}} := \|A_{l-1}(\mathbb{1} + H_l)\|$ is well defined and finite. Hence

$$\|(\mathbb{1} + H_l)^{-1}(A_l(\tau/2) - A_l)(\mathbb{1} + H_l)^{-1}\| \leq \\ (C_{H_l}^{A_l})^2 \|(A_l(\tau/2) - A_l)A_l^{-2} P_{A_l}\|, \quad \tau > 0.$$

Using Lemma 7.1.2, in particular (7.1.5), we obtain

$$\|(\mathbb{1} + H_l)^{-1}(A_l(\tau/2) - A_l)(\mathbb{1} + H_l)^{-1}\| \leq (C_{H_l}^{A_l})^2 \varkappa[f_l]\tau, \quad \tau > 0. \tag{7.3.40}$$

To estimate (7.3.38) we use (7.3.4). First we observe that

$$\|(\mathbb{1} + H_l)^{-1} W_l(\tau)(\mathbb{1} + H_l)^{-1}\| \leq$$

$$\frac{\tau^2}{4}\|(\mathbb{1} + H_l)^{-1} A_l(\tau/2) S_{l-1}(\tau) A_l(\tau/2)(\mathbb{1} + H_l)^{-1}\| + \tag{7.3.41}$$

$$\frac{\tau}{2}\|(\mathbb{1} + H_l)^{-1} S_{l-1}(\tau) A_l(\tau/2)(\mathbb{1} + H_l)^{-1}\| + \tag{7.3.42}$$

$$\frac{\tau}{2}\|(\mathbb{1} + H_l)^{-1} A_l(\tau/2) S_{l-1}(\tau)(\mathbb{1} + H_l)^{-1}\|, \tag{7.3.43}$$

$\tau > 0$. It is clear that

$$\frac{\tau^2}{4}\|(\mathbb{1} + H_l)^{-1} A_l(\tau/2) S_{l-1}(\tau) A_l(\tau/2)(\mathbb{1} + H_l)^{-1}\| \leq \\ \frac{\tau}{4}(C_{H_l}^{A_l})^2 \|A_l(\tau/2) A_l^{-1} P_{A_l}\|^2, \quad \tau > 0.$$

By Lemma 7.1.2, we infer

$$\frac{\tau^2}{4}\|(\mathbb{1} + H_l)^{-1} A_l(\tau/2) S_{l-1}(\tau) A_l(\tau/2)(\mathbb{1} + H_l)^{-1}\| \leq \frac{\tau}{4}(C_{H_l}^{A_l})^2 (\gamma[f_l])^2, \tag{7.3.44}$$

$\tau > 0$. Moreover, there is inequality

$$\frac{\tau}{2}\|(\mathbb{1} + H_l)^{-1} S_{l-1}(\tau) A_l(\tau/2)(\mathbb{1} + H_l)^{-1}\| \leq \\ \frac{\tau}{2}\|S_{l-1}(\tau)(\mathbb{1} + H_l)^{-1}\| \, \|A_l(\tau/2)(\mathbb{1} + H_l)^{-1}\|, \quad \tau > 0,$$

which yields

$$\frac{\tau}{2}\|(\mathbb{1} + H_l)^{-1} S_{l-1}(\tau) A_l(\tau/2)(\mathbb{1} + H_l)^{-1}\| \leq$$
$$\frac{\tau}{2} C_{H_l}^{H_{l-1}} C_{H_l}^{A_l} \|S_{l-1}(\tau)(\mathbb{1} + H_{l-1})^{-1}\| \, \|A_l(\tau/2) A_l^{-1}\|, \quad \tau > 0\,.$$

By assumption (7.3.11) and on account of Lemma 7.1.2, we obtain

$$\frac{\tau}{2}\|(\mathbb{1} + H_l)^{-1} S_{l-1}(\tau) A_l(\tau/2)(\mathbb{1} + H_l)^{-1}\| \leq \frac{\tau}{2} C_{H_l}^{H_{l-1}} C_{H_l}^{A_l} C_{H_{l-1}}^{S_{l-1}} \gamma[f_l], \qquad (7.3.45)$$

$\tau > 0$. Similarly one infers

$$\frac{\tau}{2}\|(\mathbb{1} + H_l)^{-1} A_l(\tau/2) S_{l-1}(\tau)(\mathbb{1} + H_l)^{-1}\| \leq \frac{\tau}{2} C_{H_l}^{H_{l-1}} C_{H_l}^{A_l} C_{H_{l-1}}^{S_{l-1}} \gamma[f_l], \qquad (7.3.46)$$

$\tau > 0$. Inserting (7.3.44), (7.3.45) and (7.3.46) into (7.3.41), (7.3.42) and (7.3.43), we deduce

$$\|(\mathbb{1} + H_l)^{-1} W_l(\tau)(\mathbb{1} + H_l)^{-1}\| \leq$$
$$\left(\frac{1}{4}(C_{H_l}^{A_l})^2(\gamma[f_l])^2 + C_{H_l}^{H_{l-1}} C_{H_l}^{A_l} C_{H_{l-1}}^{S_{l-1}} \gamma[f_l]\right)\tau, \quad \tau > 0.$$

Then setting $C_{H_l,H_l}^{W_l} := \gamma[f_l](\frac{1}{4} C_{H_l}^{A_l} \gamma[f_l] + C_{H_l}^{H_{l-1}} C_{H_{l-1}}^{S_{l-1}}) C_{H_l}^{A_l}$,we establish

$$\|(\mathbb{1} + H_l)^{-1} W_l(\tau)(\mathbb{1} + H_l)^{-1}\| \leq C_{H_l,H_l}^{W_l}\tau, \quad \tau > 0. \qquad (7.3.47)$$

Inserting (7.3.39), (7.3.40) and (7.3.41) into (7.3.36), (7.3.37) and (7.3.38), we get

$$\|(\mathbb{1} + H_l)^{-1}(S_l(\tau) - H_l)(\mathbb{1} + H_l)^{-1}\| \leq$$
$$\left((C_{H_l}^{H_{l-1}})^2 C_{H_{l-1},H_{l-1}}^{S_{l-1}-H_{l-1}} + (C_{H_l}^{A_l})^2 \varkappa[f_l] + C_{H_l,H_l}^{W_l}\right)\tau, \quad \tau > 0.$$

Now setting

$$C_{H_l,H_l}^{S_l - H_l} := (C_{H_l}^{H_{l-1}})^2 C_{H_{l-1},H_{l-1}}^{S_{l-1}-H_{l-1}} + (C_{H_l}^{A_l})^2 \varkappa[f_l] + C_{H_l,H_l}^{W_l}$$

we obtain (7.3.35). □

Corollary 7.3.6 *Let A_1 and A_2 be self-adjoint operators in a Hilbert space $\mathfrak{H}$ such that $H_2 := A_1 + A_2$ is self-adjoint. Further, let f_1 and f_2 be Kato functions from $\mathcal{K}$. If $\varkappa[f_1] < \infty$ and $\varkappa[f_2] < \infty$, (7.1.2), then there exists a constants $C^{S_2}_{H_2,H_2} > 0$ such that the estimate*

$$\|(\mathbb{1} + H_2)^{-1}(S_2(\tau) - H_2)(\mathbb{1} + H_2)^{-1}\| \leq C^{S_2}_{H_2,H_2}\,\tau\,,$$

is valid for $\tau > 0$.

Proof Following the proof of Corollary 7.3.4, we see that the result is a consequence of Lemma 7.1.4. □

Lemma 7.3.7 *Let $A_1, \ldots, A_m$ be non-negative self-adjoint operators in a Hilbert space such that the partial operator sums H_l are densely defined for any $l = 2, 3, \ldots, m$. Further, let $f_1, \cdots, f_m$ are Kato functions from $\mathcal{K}$. Further let the estimates (7.3.11) and (7.3.12) be satisfied. If $\omega[f_l] < \infty$ and $H_l = H_{l-1} + A_l$ as well as there is a constant $C^{K_{l-1}}_{K_{l-1},H_{l-1}} > 0$ such that the estimate*

$$\|(\mathbb{1}+K_{l-1}(\tau))^{-1/2}(S_{l-1}(\tau)-H_{l-1})(\mathbb{1}+H_{l-1})^{-1}\| \leq C^{S_{l-1}-H_{l-1}}_{K_{l-1},H_{l-1}}\sqrt{\tau}, \tag{7.3.48}$$

is valid for $\tau > 0$, then there is a constant $C^{S_l-H_l}_{K_l,H_l} > 0$ such that the estimate

$$\|(\mathbb{1} + K_l(\tau))^{-1/2}(S_l(\tau) - H_l)(\mathbb{1} + H_l)^{-1}\| \leq C^{S_l-H_l}_{K_l,H_l}\sqrt{\tau}, \tag{7.3.49}$$

holds for $\tau > 0$.

Proof Using that $S_l(\tau) = K_l(\tau) + W_l(\tau)$ and formula (7.3.3), we find for the left-hand side of (7.3.49) the estimate

$$\begin{aligned}
&\|(\mathbb{1} + K_l(\tau))^{-1/2}(S_l(\tau) - H_l)(\mathbb{1} + H_l)^{-1}\| \leq \\
&\quad \|(\mathbb{1} + K_l(\tau))^{-1/2}(S_{l-1}(\tau) - H_{l-1})(\mathbb{1} + H_l)^{-1}\| + && (7.3.50)\\
&\quad \|(\mathbb{1} + K_l(\tau))^{-1/2}(A_l(\tau/2) - \tfrac{\tau}{4}A_l(\tau/2)^2 - A_l)(\mathbb{1} + H_l)^{-1}\| + && (7.3.51)\\
&\quad \|(\mathbb{1} + K_l(\tau))^{-1/2}W_l(\mathbb{1} + H_l)^{-1}\|, \quad \tau > 0. && (7.3.52)
\end{aligned}$$

We start by upper bound for (7.3.50). Since $H_l = H_{l-1}+A_l$, we have dom $(H_l) \subseteq$ dom (H_{l-1}). In particular,

$$C^{H_{l-1}}_{H_l} := \|(\mathbb{1} + H_{l-1})(\mathbb{1} + H_l)^{-1}\| < \infty. \tag{7.3.53}$$

Then we note that by (7.3.2), (7.3.3) and (7.3.6)

$$\begin{aligned}\mathbb{1} + K_l(\tau) &\geq \mathbb{1} + K_{l-1}(\tau) + W_{l-1}(\tau) \\ &= \sqrt{\mathbb{1} + K_{l-1}(\tau)}(\mathbb{1} + Q_{l-1}(\tau))\sqrt{\mathbb{1} + K_{l-1}(\tau)},\end{aligned}$$

for $\tau > 0$, and as a consequence

$$(\mathbb{1} + K_l(\tau))^{-1} \leq (\mathbb{1} + K_{l-1}(\tau))^{-1/2}(\mathbb{1} + Q_{l-1}(\tau))^{-1}(\mathbb{1} + K_{l-1}(\tau))^{-1/2}, \tag{7.3.54}$$

for $\tau > 0$. Since using Lemma 7.3.1, we get

$$0 \leq (\mathbb{1} + Q_{l-1}(\tau))^{-1} \leq q_0 \mathbb{1}, \quad \tau > 0, \tag{7.3.55}$$

by virtue of (7.3.54) and (7.3.55), it follows that

$$(\mathbb{1} + K_l(\tau))^{-1} \leq q_0(\mathbb{1} + K_{l-1}(\tau))^{-1}, \quad \tau > 0. \tag{7.3.56}$$

Thanks to (7.3.53) and (7.3.56), we find the upper bound for (7.3.50) in the form

$$\begin{aligned}&\|(\mathbb{1} + K_l(\tau))^{-1/2}(S_{l-1}(\tau) - H_{l-1})(\mathbb{1} + H_l)^{-1}\| \leq \\ &\sqrt{q_0}\, C_{H_l}^{H_{l-1}} \|(\mathbb{1} + K_{l-1}(\tau))^{-1/2}(S_{l-1}(\tau) - H_{l-1})(\mathbb{1} + H_{l-1})^{-1}\|, \quad \tau > 0.\end{aligned}$$

As a result, by assumption (7.3.48) we obtain for (7.3.50) the estimate

$$\begin{aligned}&\|(\mathbb{1} + K_l(\tau))^{-1/2}(S_{l-1}(\tau) - H_{l-1})(\mathbb{1} + H_l)^{-1}\| \leq \\ &\qquad \sqrt{q_0}\, C_{H_l}^{H_{l-1}}\, C_{K_{l-1},H_{l-1}}^{S_{l-1}-H_{l-1}} \sqrt{\tau}, \quad \tau > 0.\end{aligned} \tag{7.3.57}$$

To estimate (7.3.51) we note that by (7.3.5), one has

$$\mathbb{1} + \frac{1}{2}A_l(\tau/2) \leq \mathbb{1} + K_l(\tau), \quad \tau > 0,$$

which yields $(\mathbb{1} + K_l(\tau))^{-1} \leq 2\,(\mathbb{1} + A_l(\tau/2))^{-1}$, for $\tau > 0$. Then one gets for (7.3.51) an upper bound

$$\begin{aligned}&\left\|(\mathbb{1} + K_l(\tau))^{-1/2}\left(A_l(\tau/2) - A_l - \frac{\tau}{2}A_l(\tau/2)^2\right)(\mathbb{1} + H_l)^{-1}\right\| \\ &\leq \sqrt{2}\left\|(\mathbb{1} + A_l(\tau/2))^{-1/2}\left(A_l(\tau/2) - A_l - \frac{\tau}{2}A_l(\tau/2)^2\right)(\mathbb{1} + H_l)^{-1}\right\|,\end{aligned}$$

$\tau > 0$. Since $\operatorname{dom}(A_l) \subseteq \operatorname{dom}(H_l)$, we find that

$$\left\|(\mathbb{1} + K_l(\tau))^{-1/2}\left(A_l(\tau/2) - A_l - \frac{\tau}{2}A_l(\tau/2)^2\right)(\mathbb{1} + H_l)^{-1}\right\|$$
$$\leq \sqrt{2}\, a_l \left\|(\mathbb{1} + A_l(\tau/2))^{-1/2}\left(A_l(\tau/2) - A_l - \frac{\tau}{2}A_l(\tau/2)^2\right)A_l^{-1}P_{A_l}\right\|,$$

$\tau > 0$, where $a_l := \|A_l(\mathbb{1} + H_l)\|$ and P_{A_l} denotes the orthogonal projection onto $\ker(A_l)^\perp$. Finally, using Lemma 7.1.5, in particular estimate (7.1.22), we get

$$\left\|(\mathbb{1} + K_l(\tau))^{-1/2}\left(A_l(\tau/2) - A_l - \frac{\tau}{2}A_l(\tau/2)^2\right)(\mathbb{1} + H_l)^{-1}\right\|$$
$$\leq (\omega[f_l] + \frac{1}{2}\gamma[f_l])\sqrt{\tau}, \quad \tau > 0. \tag{7.3.58}$$

By Corollary 7.3.3, there is a constant $C^{W_l}_{K_l,H_l} > 0$ such that the (7.3.26) is valid and yields an upper bound for (7.3.52).

Inserting now the estimates (7.3.57), (7.3.58) together with (7.3.26) into (7.3.50)–(7.3.52), we obtain the upper bound

$$\|(\mathbb{1}+K_l(\tau))^{-1/2}(S_l(\tau) - H_l)(\mathbb{1} + H_l)^{-1}\| \leq$$
$$\left(\sqrt{q_0}\, C^{H_{l-1}}_{H_l} C^{S_{l-1}-H_{l-1}}_{K_{l-1},H_{l-1}} + (\omega[f_l] + \frac{1}{2}\gamma[f_l]) + C^{W_l}_{K_l,H_l}\right)\sqrt{\tau}, \quad \tau > 0.$$

Then by setting

$$C^{S_l-H_l}_{K_l,H_l} := \sqrt{q_0}\, C^{H_{l-1}}_{H_l}\, C^{S_{l-1}-H_{l-1}}_{K_{l-1},H_{l-1}} + (\omega[f_l] + \frac{1}{2}\gamma[f_l]) + C^{W_l}_{K_l,H_l},$$

we finish the proof of (7.3.49). □

Corollary 7.3.8 *Let A_1 and A_2 be self-adjoint operators in a Hilbert space $\mathfrak{H}$ such that $H_2 := A_1 + A_2$ is self-adjoint. Further, let f_1 and f_2 be Kato functions from $\mathcal{K}$. If $\omega[f_1] < \infty$ and $\omega[f_2] < \infty$, (7.1.3), then there is a constant $C^{S_2-H_2}_{K_2,H_2} > 0$ such that the estimate*

$$\|(\mathbb{1} + K_2(\tau))^{-1/2}(S_2(\tau) - H_2)(\mathbb{1} + H_2)^{-1}\| \leq C^{S_2-H_2}_{K_2,H_2}\,\tau, \quad \tau > 0,$$

is valid.

Proof Following verbatim the proof of Corollary 7.3.4, we infer that the result is a consequence of Lemma 7.1.8. □

Theorem 7.3.9 *Let $A_1, \cdots, A_m$ are m non-negative self-adjoint operators in a Hilbert space $\mathfrak{H}$. Further, let $f_1, f_2, \ldots, f_m$ be Kato functions from $\mathcal{K}$ satisfying* (7.1.2), (7.1.3):

$$\sum_{j=1}^{m} \varkappa[f_j] < \infty \quad \textit{and} \quad \sum_{j=1}^{m} \omega[f_j] < \infty .$$

If all partial sums $H_l := A_1 + A_2 + \cdots + A_l$, for $l = 2, 3, \ldots, m$, are self-adjoint operators, then for each $l = 2, 3, \ldots, m$ and each bounded interval $\mathcal{I} \subset \mathbb{R}^+$, there is a constant $C_{H_l}^{\mathcal{I}} > 0$ such that the estimate

$$\sup_{t\in\mathcal{I}} \left\| \left(f_l(tA_l/2\eta)\cdots f_2(tA_2/2\eta) f_1(tA_1/\eta) f_2(tA_2/2\eta)\cdots f_l(tA_l/2\eta)\right)^\eta - \mathrm{e}^{-tH_l} \right\|$$
$$\leq \frac{C_{H_l}^{\mathcal{I}}}{\eta}, \quad \eta > 0, \quad l = 2, 3, \ldots, m, \tag{7.3.59}$$

holds locally uniformly in $t \in \mathbb{R}^+$.

Proof Let $l = 2$. In this case we set $A := A_1$, $B := A_2$, $f := f_1$ and $g = f_2$. The proof immediately follows from Theorem 7.1.9. By Corollary 7.3.4 we know that there are constants $C_{H_2}^{S_2} > 0$ and $C_{K_2}^{S_2} > 0$ such that

$$\|S_2(\tau)(\mathbb{1} + H_2)^{-1}\| \leq C_{H_2}^{S_2} \quad \text{and} \quad \|(\mathbb{1} + K_2)^{-1/2}\sqrt{S_2(\tau)}\| \leq C_{K_2}^{S_2} ,$$

are satisfied for $\tau > 0$. Applying Lemma 7.3.2, we get that for each $l = 2, 3, \ldots, m$ there are constants $C_{H_l}^{S_l} > 0$ and $C_{K_l}^{S_l} > 0$ such that

$$\|S_l(\tau)(\mathbb{1} + H_l)^{-1}\| \leq C_{H_l}^{S_l} \quad \text{and} \quad \|(\mathbb{1} + K_l)^{-1/2}\sqrt{S_l(\tau)}\| \leq C_{K_l}^{S_l} ,$$

holds for $\tau > 0$. Further, from Corollary 7.3.6 and Lemma 7.3.5, we get that for each $l = 2, 3, \ldots, m$, there is a constant $C^{S_l - H_l} > 0$ such that the estimate (7.3.35) is valid for $\tau > 0$. Moreover, from Corollary 7.3.8 and Lemma 7.3.7 for each $l = 2, 3, \ldots, m$, there is a constant $C_{K_l,H_l}^{S_l - H_l} > 0$ such that the estimate (7.3.49) holds for $\tau > 0$. From the resolvent formulæ

$$(\mathbb{1} + S_l(\tau))^{-1} - (\mathbb{1} + H_l)^{-1} = (\mathbb{1} + S_l(gt))^{-1}(H_l - S_l(\tau))(\mathbb{1} + H_l)^{-1} ,$$

and

$$(\mathbb{1} + S_l(\tau))^{-1} - (\mathbb{1} + H_l)^{-1} = (\mathbb{1} + H_l)^{-1}(H_l - S_l(\tau))(\mathbb{1} + S_l(\tau))^{-1} ,$$

we deduce a representation

$$(\mathbb{1} + S_l(\tau))^{-1} - (\mathbb{1} + H_l)^{-1} = (\mathbb{1} + H_l)^{-1}(H_l - S_l(\tau))(\mathbb{1} + H_l)^{-1} + \\ (\mathbb{1} + H_l)^{-1}(H_l - S_l(\tau))(\mathbb{1} + K_l(\tau))^{-1}(H_l - S_l(\tau))(\mathbb{1} + H_l)^{-1},$$

which yields the estimate

$$\|(\mathbb{1} + S_l(\tau))^{-1} - (\mathbb{1} + H_l)^{-1}\| \leq \|(\mathbb{1} + H_l)^{-1}(H_l - S_l(\tau))(\mathbb{1} + H_l)^{-1}\| + \\ \|(\mathbb{1} + H_l)^{-1}(H_l - S_l(\tau))(\mathbb{1} + S_l(\tau))^{-1}(H_l - S_l(\tau))(\mathbb{1} + H_l)^{-1}\| \tag{7.3.60}$$

On the other hand, thanks to $S_l(\tau) = K_l(\tau) + W_l(\tau)$ (7.3.2) and notation $Q_l(\tau) = (\mathbb{1} + K_l(\tau))^{-1/2} W_l(\tau)(\mathbb{1} + K_l(\tau))^{-1/2}$, $\tau > 0$, we infer the identity

$$(\mathbb{1}+H_l)^{-1}(H_l - S_l(\tau))(\mathbb{1} + S_l(\tau))^{-1}(H_l - S_l(\tau))(\mathbb{1} + H_l)^{-1} = \\ (\mathbb{1} + H_l)^{-1}(H_l - S_l(\tau))(\mathbb{1} + K_l(\tau))^{-1/2}(\mathbb{1} + Q_l(\tau))^{-1} \times \\ (\mathbb{1} + K_l(\tau))^{-1/2}(H_l - S_l(\tau))(\mathbb{1} + H_l)^{-1},$$

for $\tau > 0$. Then taking into account Lemma 7.3.1 and (7.3.60), we deduce inequality

$$\|(\mathbb{1}+H_l)^{-1}(H_l - S_l(\tau))(\mathbb{1} + S_l(\tau))^{-1}(H_l - S_l(\tau))(\mathbb{1} + H_l)^{-1}\| \leq \\ q_0 \, \|(\mathbb{1} + K_l(\tau))^{-1/2}(H_l - S_l(\tau))(\mathbb{1} + H_l)^{-1}\|^2, \quad \tau > 0.$$

Note that estimates (7.3.35) and (7.3.49) yield for $\tau > 0$

$$\|(\mathbb{1} + S_l(\tau))^{-1} - (\mathbb{1} + H_l)^{-1}\| \leq \left(C^{S_l - H_l}_{H_l, H_l} + q_0 \, (C^{S_l - H_l}_{K_l, H_l})^2\right) \tau .$$

Consequently, setting $M_{H_l} := C^{S_l - H_l}_{H_l, H_l} + q_0 (C^{S_l - H_l}_{K_l, H_l})^2$, we prove that

$$\|(\mathbb{1} + S_l(\tau))^{-1} - (\mathbb{1} + H_l)^{-1}\| \leq M_{H_l} \, \tau, \quad \tau > 0.$$

Finally, applying Theorem 4.3.6, we get that for each $l = 2, 3, \ldots, m$ and each bounded interval $\mathcal{I} \subseteq \mathbb{R}^+$, there exists a constant $C^{\mathcal{I}}_{H_l} > 0$ such that the estimate (7.3.59) holds for $\tau > 0$. □

Remark 7.3.10

(i) We note that from (7.3.59) for each *permutation* of the self-adjoint operators $A_1, A_2, \ldots, A_l$, $l = 2, 3, \ldots, m$, there exists a particular constant $C^{\mathcal{I}}_{H_l} > 0$ such that the estimate of type (7.3.59) holds, cf. Theorem 7.1.9. Remark that dependence of this constant on permutation persists because of the fact that the sum $H_l = A_1 + \cdots + A_l$ is not *associative*.

(ii) Note that in contrast to Corollary 7.1.12, the estimate (7.3.59) does *not* imply estimates

$$\sup_{t\in\mathcal{I}} \left\| (f_k(tA_k/n)\cdots f(tA_2/n)f(tA_1/n))^n - \mathrm{e}^{-itH} \right\| \leq \widetilde{C}^{\mathcal{I}}_{H_l}\,\frac{1}{n}\,,$$

$$k = 2, 3, \ldots, m\,,$$

for *non-symmetric* product formula and some constants $\widetilde{C}^{\mathcal{I}}_{H_l}$.

7.4 Fractional Power Conditions on Generators

In contrast to the setup of Sect. 7.1, when the sum of generators $A + B$ is self-adjoint, we study here certain (fractional power) conditions for self-adjointness of generators given by the *form-sum*: $A \dot{+} B$. We also remind here a setup formulated in the beginning of Sect. 7.3.

The aim of this section is to describe how the *optimal* operator-norm error bound estimate for convergence rate of the Trotter-Kato product formulæ depends on relation between fractional powers of self-adjoint generators A and B of C_0-semigroups involved into product. The main result is the INZ-Theorem 7.4.7 for *symmetric* (i.e. self-adjoint) approximants with the *optimal* rate of convergence; see Remark 7.4.8. These results are extended to *non-symmetric* (i.e. non-self-adjoint) approximants in Theorem 7.4.10.

7.4.1 Auxiliary Estimates

The list of lemmata collected below clarifies the strategy of proving the operator-norm error estimates.

Lemma 7.4.1 *Let A and B be non-negative self-adjoint operators. If f and g are Kato functions of the class $\mathcal{K}$, then there are upper bounds*

$$\|\sqrt{A(\tau)}\,(\mathbb{1} + K(\tau))^{-1/2}\| \leq 1, \quad \tau > 0\,, \tag{7.4.1}$$

and

$$\|\sqrt{B(\tau/2)}\,(\mathbb{1} + K(\tau))^{-1/2}\| \leq \sqrt{2}, \quad \tau > 0\,, \tag{7.4.2}$$

where operators $\{\}K(\tau)\}_{\tau>0}$ *are defined by* (5.2.4) *and* (5.2.7).

Proof We have

$$\|\sqrt{A(\tau)}(\mathbb{1}+K(\tau))^{-1/2}\|^2 = \|\sqrt{A(\tau)}(\mathbb{1}+K(\tau))^{-1}\sqrt{A(\tau)}\|, \quad \tau > 0.$$

Since (5.2.7) yields $K(\tau) \geq A(\tau)$, we find that

$$\|\sqrt{A(\tau)}\,(\mathbb{1}+K(\tau))^{-1}\sqrt{A(\tau)}\| \leq 1\,, \quad \tau > 0\,,$$

which proves (7.4.1). Seeing that (5.2.7) also implies $K(\tau) \geq B(\tau)/2$, we similarly obtain (7.4.2). □

In the following for any (generic) Kato function $h \in \mathcal{K}$, we define

$$m_h(x) := \sup_{y\in[x,\infty)} h(y), \quad x > 0\,. \tag{7.4.3}$$

Lemma 7.4.2 *Let A and B be non-negative self-adjoint operators. If f and g are Kato functions of the class $\mathcal{K}$ and $m_f(x) < 1$ for $x > 0$, then there are constants $C > 0$ and $C_0 > 0$ such that for $u \in \mathfrak{H}$ and $\tau > 0$*

$$\|(\mathbb{1}+S(\tau))^{-1/2}u\| \leq C\,\|(\mathbb{1}+A)^{-1/2}u\| + C_0\,\|u\|\,\tau^{1/2}\,, \tag{7.4.4}$$

where for $S(\tau) := (\mathbb{1} - F(\tau))/\tau$ we use the representation (5.2.8).

Proof Using Lemma 5.2.3 (see (5.2.34) and (5.2.35)), we obtain for $u \in \mathfrak{H}$ and $\tau > 0$

$$\|(\mathbb{1}+S(\tau))^{-1/2}u\| \leq q_0\,\|(\mathbb{1}+K(\tau))^{-1/2}\|\,,$$

where q_0 is defined in (5.2.35) or (7.3.8). Seeing that by (5.2.7): $K(\tau) \geq A(\tau/2)/2$, we find

$$\|(\mathbb{1}+S(\tau))^{-1/2}u\| \leq \sqrt{2q_0}\,\|(\mathbb{1}+B(\tau/2))^{-1/2}u\|\,,$$

for $u \in \mathfrak{H}$ and $\tau > 0$.

Note that for the Kato function $f \in \mathcal{K}$, there is a $\delta > 0$ such that $1 - f(x) \geq x/2$ for $x \in (0, \delta)$, which yields estimate

$$1 - f(x) \geq \frac{x}{2}\,\chi_{(0,\delta)}(x) + (1 - m_f(\delta))\,\chi_{[\delta,\infty)}(x)\,, \quad x > 0\,,$$

where $\chi_{(0,\delta)}(\cdot)$ and $\chi_{[\delta,\infty)}(\cdot)$ are the characteristic functions of the intervals $(0, \delta)$ and $[\delta, \infty)$, respectively. Hence, for $m_f(\delta))$ (7.4.3) we find

$$A(\tau/2) \geq \frac{1}{2}\,A\,E_A([0, 2\delta/\tau)) + 2\,\frac{1 - m_f(\delta)}{\tau}\,E_A([2\delta/\tau, \infty))\,, \quad \tau > 0\,,$$

where $E_A(\cdot)$ is the spectral measure of A. As a result, we obtain

$$\|(\mathbb{1} + S(\tau))^{-1/2}u\| \leq 2\sqrt{q_0}\, \|(\mathbb{1} + A)^{-1/2} E_A([0, 2\delta/\tau))u\| + \\ + \frac{2\sqrt{q_0}}{\sqrt{1 - m_f(\delta)}}\, \|E_A([2\delta/\tau, \infty))u\|\, \tau^{1/2}, \tag{7.4.5}$$

for $u \in \mathfrak{H}$ and $\tau > 0$. Setting $C := 2\sqrt{q_0}$ and $C_0 := 2\sqrt{q_0}\,/\sqrt{1 - m_f(\delta)}$, one gets the proof of the estimate (7.4.4). □

Definition 7.4.3 In the following, we set (cf. Definition 7.1.1)

$$\gamma_r[h] := \sup_{x>0} \frac{1 - h(x)}{x^r}, \quad r \in (0, 1], \tag{7.4.6}$$

and similarly to (7.1.2) (cf. (D.0.1)), we introduce

$$\varkappa_s[h] := \sup_{x>0} \frac{|1 - h(x) - x|}{x^s}, \quad s \in [1, 2], \tag{7.4.7}$$

where $h \in \mathcal{K}$ is a generic Kato function, Definition 5.2.1.

Lemma 7.4.4 *Let A and B be non-negative self-adjoint operators such that their self-adjoint form-sum $H := A \dot{+} B$ satisfies*

$$\operatorname{dom}(H^\alpha) \subseteq \operatorname{dom}(A^\alpha) \cap \operatorname{dom}(B^\alpha), \tag{7.4.8}$$

for some $\alpha \in (1/2, 1)$. Further, let f and g be Kato functions such that $\gamma_r[f] < \infty$ and $\gamma_r[g] < \infty$ for $r \in (0, \alpha]$.

(i) *If $p \in [1/2, \alpha]$, $q \in [1/2, \alpha]$ and $\varkappa_s[f] < \infty$, $\varkappa_s[g] < \infty$, for $s \in [1, 2]$, then there is a constant $D_{p,q} > 0$ such that*

$$\|(\mathbb{1} + H)^{-p}(H - S(\tau))(\mathbb{1} + H)^{-q}\| \leq D_{p,q}\, \tau^{p+q-1}, \quad \tau > 0. \tag{7.4.9}$$

(ii) *There exists also a constant $D_\alpha > 0$ such that for $\tau > 0$*

$$\|S(\tau)\,(\mathbb{1} + H)^{-\alpha}\| \leq D_\alpha\, \tau^{\alpha-1}, \quad \alpha \in (1/2, 1). \tag{7.4.10}$$

Here $S(\tau) := (\mathbb{1} - F(\tau))/\tau$, where $F(\tau) = g(\tau B/2) f(\tau A) g(\tau B/2)$ (5.2.3), for $\tau > 0$.

Proof

(i) First we note that

$$\begin{aligned}\mathbb{1} - F(\tau) = \mathbb{1} - g(\tau B/2)^2 + (\mathbb{1} - f(\tau A)) + \\ (\mathbb{1} - g(\tau B/2))\,(\mathbb{1} - f(\tau A))\,(\mathbb{1} - g(\tau B/2)) - \\ (\mathbb{1} - g(\tau B/2))\,(\mathbb{1} - f(\tau A)) - (\mathbb{1} - f(\tau A))\,(\mathbb{1} - g(\tau B/2))\,,\end{aligned} \tag{7.4.11}$$

Then by definitions of H and $S(\tau)$, we find

$$\begin{aligned}&(\mathbb{1} + H)^{-p}(H - S(\tau))(\mathbb{1} + H)^{-q} = \\ &\quad (\mathbb{1} + H)^{-p}\left(B - \frac{\mathbb{1} - g(\tau B/2)^2}{\tau}\right)(\mathbb{1} + H)^{-q} + \\ &\quad (\mathbb{1} + H)^{-p}\left(A - \frac{\mathbb{1} - f(\tau A)}{\tau}\right)(\mathbb{1} + H)^{-q} - \\ &\quad (\mathbb{1} + H)^{-p}(\mathbb{1} - g(\tau B/2))(\mathbb{1} - f(\tau A))(\mathbb{1} - g(\tau B/2))(\mathbb{1} + H)^{-q}\,\tau^{-1} + \\ &\quad (\mathbb{1} + H)^{-p}(\mathbb{1} - g(\tau B/2))(\mathbb{1} - f(\tau A))(\mathbb{1} + H)^{-q}\,\tau^{-1} + \\ &\quad (\mathbb{1} + H)^{-p}(\mathbb{1} - f(\tau A))(\mathbb{1} - g(\tau B/2))(\mathbb{1} + H)^{-q}\,\tau^{-1}\,,\end{aligned}$$

which gives for $\tau > 0$ the inequality

$$\|(\mathbb{1} + H)^{-p}(H - S(\tau))(\mathbb{1} + H)^{-q}\| \leq$$

$$\left\|(\mathbb{1} + H)^{-p}\left(B - \frac{\mathbb{1} - g(\tau B/2)^2}{\tau}\right)(\mathbb{1} + H)^{-q}\right\| + \tag{7.4.12}$$

$$\left\|(\mathbb{1} + H)^{-p}\left(A - \frac{\mathbb{1} - f(\tau A)}{\tau}\right)(\mathbb{1} + H)^{-q}\right\| + \tag{7.4.13}$$

$$\|(\mathbb{1} - g(\tau B/2))(\mathbb{1} + H)^{-p}\|\;\|(\mathbb{1} - g(\tau B/2))(\mathbb{1} + H)^{-q}\|\,\tau^{-1} + \tag{7.4.14}$$

$$\|(\mathbb{1} - g(\tau B/2))(\mathbb{1} + H)^{-p}\|\;\|(\mathbb{1} - f(\tau A))(\mathbb{1} + H)^{-q}\|\,\tau^{-1} + \tag{7.4.15}$$

$$\|(\mathbb{1} - f(\tau A))(\mathbb{1} + H)^{-p}\|\;\|(\mathbb{1} - g(\tau B/2))(\mathbb{1} + H)^{-q}\|\,\tau^{-1}\,. \tag{7.4.16}$$

To estimate (7.4.12) we note that because of conditions (7.4.8) and $p, q \in [1/2, \alpha]$, the operators $(\mathbb{1}+A)^p(\mathbb{1}+H)^{-p}$ and $(\mathbb{1}+B)^p(\mathbb{1}+H)^{-p}$ are bounded. For that reason

$$\left\|(\mathbb{1}+H)^{-p}\left(B-\frac{\mathbb{1}-g(\tau B/2)^2}{\tau}\right)(\mathbb{1}+H)^{-q}\right\| \leq$$
$$\|(\mathbb{1}+B)^p(\mathbb{1}+H)^{-p}\|\;\|(\mathbb{1}+B)^q(\mathbb{1}+H)^{-q}\| \times$$
$$\times\left\|(\mathbb{1}+B)^{-p}\left(B-\frac{\mathbb{1}-g(\tau B/2)^2}{\tau}\right)(\mathbb{1}+B)^{-q}\right\|,$$

which by the *functional calculus* for self-adjoint operator B and by definition (7.4.6) yields

$$\left\|(\mathbb{1}+H)^{-p}\left(B-\frac{\mathbb{1}-g(\tau B/2)^2}{\tau}\right)(\mathbb{1}+H)^{-q}\right\| \leq$$
$$\|(\mathbb{1}+B)^p\,(\mathbb{1}+H)^{-p}\|\;\|(\mathbb{1}+B)^q\,(\mathbb{1}+H)^{-q}\| \times$$
$$\times\left\{\varkappa_{p+q}[g]+\frac{1}{2}\,\gamma_{(p+q)/2}[g]^2\right\}\,\tau^{p+q-1}, \quad \tau>0\,.$$

Similarly one estimates (7.4.13). This gives for $\tau > 0$ the inequality

$$\left\|(\mathbb{1}+H)^{-p}\left(A-\frac{\mathbb{1}-f(\tau A)}{\tau}\right)(\mathbb{1}+H)^{-q}\right\| \leq$$
$$\|(\mathbb{1}+A)^p(\mathbb{1}+H)^{-p}\|\;\|(\mathbb{1}+A)^q(\mathbb{1}+H)^{-q}\|\;\varkappa_{p+q}[f]\,\tau^{p+q-1}\,.$$

Seeing that by the *functional calculus* for self-adjoint operators A, B and by definition (7.4.6), one infers for $\tau > 0$ and $r \in (0, \alpha]$

$$\|(1-g(\tau B/2))(\mathbb{1}+H)^{-r}\| \leq 2^{-r}\,\|(\mathbb{1}+B)^r(\mathbb{1}+H)^{-r}\|\,\gamma_r[g]\,\tau^r\,, \tag{7.4.17}$$

and

$$\|(\mathbb{1}-f(\tau A))(\mathbb{1}+H)^{-r}\| \leq \|(\mathbb{1}+A)^r(\mathbb{1}+H)^{-r}\|\,\gamma_r[f]\,\tau^r\,, \tag{7.4.18}$$

(7.4.17) and (7.4.18) yield estimates for the terms (7.4.14)–(7.4.16).

Collecting now all of them for (7.4.12)–(7.4.16), we finally obtain the estimate

$$\|(\mathbb{1}+H)^{-p}(H-S(\tau))(\mathbb{1}+H)^{-q}\| \;\leq\; \|(\mathbb{1}+B)^p(\mathbb{1}+H)^{-p}\|\,\cdot$$
$$\|(\mathbb{1}+B)^q(\mathbb{1}+H)^{-q}\|\left\{\varkappa_{p+q}[g]+\frac{1}{2}\,\gamma_{(p+q)/2}[g]^2\right\}\tau^{p+q-1}+$$

$$\|(\mathbb{1}+A)^p(\mathbb{1}+H)^{-p}\|\,\|(\mathbb{1}+A)^q(\mathbb{1}+H)^{-q}\|\,\varkappa_{p+q}[f]\,\tau^{p+q-1}+$$
$$2^{-(p+q)}\,\gamma_p[g]\,\gamma_q[g]\,\|(\mathbb{1}+B)^p(\mathbb{1}+H)^{-p}\|$$
$$\|(\mathbb{1}+B)^q(\mathbb{1}+H)^{-q}\|\,\tau^{p+q-1}+$$
$$2^{-p}\,\gamma_p[g]\,\gamma_q[f]\,\|(\mathbb{1}+B)^p(\mathbb{1}+H)^{-p}\|\,\|(\mathbb{1}+A)^q(\mathbb{1}+H)^{-q}\|\,\tau^{p+q-1}+$$
$$2^{-q}\,\gamma_q[g]\,\gamma_p[f]\,\|(\mathbb{1}+B)^q(\mathbb{1}+H)^{-q}\|\,\|(\mathbb{1}+A)^p(\mathbb{1}+H)^{-p}\|\,\tau^{p+q-1}.$$

Now setting for $\tau > 0$

$$D_{p,q} := \|(\mathbb{1}+B)^p(\mathbb{1}+H)^{-p}\|\cdot$$
$$\|(\mathbb{1}+B)^q(\mathbb{1}+H)^{-q}\|\left\{\varkappa_{p+q}[g]+\frac{1}{2}\,\gamma_{(p+q)/2}[g]^2\right\}+$$
$$\|(\mathbb{1}+A)^p(\mathbb{1}+H)^{-p}\|\,\|(\mathbb{1}+A)^q(\mathbb{1}+H)^{-q}\|\,\varkappa_{p+q}[f]+$$
$$2^{-(p+q)}\,\gamma_p[g]\,\gamma_q[g]\,\|(\mathbb{1}+B)^p(\mathbb{1}+H)^{-p}\|\,\|(\mathbb{1}+B)^q(\mathbb{1}+H)^{-q}\|+$$
$$2^{-p}\,\gamma_p[g]\,\gamma_q[f]\,\|(\mathbb{1}+B)^p(\mathbb{1}+H)^{-p}\|\,\|(\mathbb{1}+A)^q(\mathbb{1}+H)^{-q}\|+$$
$$2^{-q}\,\gamma_q[g]\,\gamma_p[f]\,\|(\mathbb{1}+B)^q(\mathbb{1}+H)^{-q}\|\,\|(\mathbb{1}+A)^p(\mathbb{1}+H)^{-p}\|,$$

we obtain the proof of the estimate (7.4.9).

(ii) Using decomposition (7.4.11), we infer inequality

$$\left\|S(\tau)(\mathbb{1}+H)^{-\alpha}\right\| \le$$
$$4\left\|(\mathbb{1}-g(\tau B/2))(\mathbb{1}+H)^{-\alpha}\right\|\,\tau^{-1}+\|(\mathbb{1}-f(\tau A))(\mathbb{1}+H)^{-\alpha}\|\,\tau^{-1},$$

for $\tau > 0$. Then taking into account (7.4.17) and (7.4.18), we obtain

$$\left\|S(\tau)(\mathbb{1}+H)^{-\alpha}\right\| \le$$
$$2^{2-\alpha}\|(\mathbb{1}+B)^{\alpha}(\mathbb{1}+H)^{-\alpha}\|\,\gamma_{\alpha}[g]\,\tau^{\alpha-1}+$$
$$\|(\mathbb{1}+A)^{\alpha}(\mathbb{1}+H)^{-\alpha}\|\,\gamma_{\alpha}[f]\,\tau^{\alpha-1},$$

for $\tau > 0$. As a consequence, on account of definition

$$D_{\alpha} := 2^{2-\alpha}\|(\mathbb{1}+B)^{\alpha}(\mathbb{1}+H)^{-\alpha}\|\,\gamma_{\alpha}[g]$$
$$+\|(\mathbb{1}+A)^{\alpha}(\mathbb{1}+H)^{-\alpha}\|\,\gamma_{\alpha}[f],$$

one gets the proof of (7.4.10) for $\tau > 0$.

□

7.4.2 Error Bound Estimate and INZ-Theorem

In order to prove the main assertion about optimal operator-norm error bound estimate of the rate of convergence (INZ-Theorem 7.4.7), we need the following two lemmata.

Lemma 7.4.5 *Let A and B be non-negative self-adjoint operators such that for some $\alpha \in (1/2, 1)$ the condition* (7.4.8) *is satisfied. Further, let f and g be Kato functions from $\mathcal{K}$, which obey conditions of Lemma* 7.4.4. *If in addition one has that* $\mathrm{dom}\,(A^{1/2}) \subseteq \mathrm{dom}\,(B^{1/2})$ *and* $m_f(x) < 1$ (7.4.3), *for $x > 0$, then there is a constant $\mathcal{N}_{\alpha-1/2} > 0$ such that*

$$t\,\|(\mathbb{1} + t\,S(\tau))^{-1/2}(H - S(\tau))(\mathbb{1} + t\,H)^{-1}\| \leq \mathcal{N}_{\alpha-1/2}\left(\frac{\tau}{t}\right)^{\alpha-1/2}, \tag{7.4.19}$$

for $t, \tau \in (0, 1]$ with $0 < \tau \leq t$.

Proof By the functional calculus for self-adjoint operators $\{S(\tau) = \tau^{-1}(\mathbb{1} - F(\tau))\}_{\tau>0}$ and $t \in (0, 1]$ one obtains estimate $\|((\mathbb{1} + S(\tau))^{1/2})/(\mathbb{1} + t\,S(\tau))^{1/2}\| \leq t^{-1/2}$ and consequently

$$\begin{aligned} t\,\|(\mathbb{1} + tS(\tau))^{-1/2}(H - S(\tau))(\mathbb{1} + tH)^{-1}\| &\leq \\ t^{1/2}\,\|(\mathbb{1} + S(\tau))^{-1/2}&(H - S(\tau))(\mathbb{1} + tH)^{-1}\|\,, \end{aligned}$$

for $t \in (0, 1]$ and $\tau > 0$. Then by Lemma 7.4.2, we infer

$$\begin{aligned} t\,\|(\mathbb{1} + tS(\tau))^{-1/2}(H - S(\tau))(\mathbb{1} + tH)^{-1}\| &\leq \\ C\,\|(\mathbb{1} + B)^{-1/2}(H - S(\tau))(\mathbb{1} + tH)^{-1}\|\,t^{1/2} &+ \\ C_0\,\|(H - S(\tau))(\mathbb{1} + tH)^{-1}\|\,\tau^{1/2}\,t^{1/2}&, \end{aligned} \tag{7.4.20}$$

and, since $\mathrm{dom}\,(H^{1/2}) = \mathrm{dom}\,(B^{1/2})$, also

$$\begin{aligned} &\|(\mathbb{1} + B)^{-1/2}(H - S(\tau))(\mathbb{1} + tH)^{-1}\| \leq \\ &\quad \|(\mathbb{1} + H)^{1/2}(\mathbb{1} + B)^{-1/2}\|\,\|(\mathbb{1} + H)^{-1/2}(H - S(\tau))(\mathbb{1} + tH)^{-1}\|\,, \end{aligned}$$

for the same range: $t \in (0, 1]$ and $\tau > 0$.

Using now the functional calculus for the self-adjoint operator H, we obtain $\|((\mathbb{1} + H)^{\alpha})/(\mathbb{1} + t\,H)\| \leq t^{-\alpha}$, for $\alpha \in (1/2, 1)$, and $t \in (0, 1]$. Therefore,

$$\begin{aligned} &\|(\mathbb{1} + B)^{-1/2}(H - S(\tau)(\mathbb{1} + tH)^{-1}\| \leq \\ &\quad \|(\mathbb{1} + H)^{1/2}(\mathbb{1} + B)^{-1/2}\|\,\|(\mathbb{1} + H)^{-1/2}(H - S(\tau))(\mathbb{1} + H)^{-\alpha}\|\,t^{-\alpha}\,, \end{aligned}$$

for $t \in (0, 1]$ and $\tau > 0$. Then by Lemma 7.4.4, (i) there is a constant $D_{1/2,\alpha} > 0$ such that

$$\|(\mathbb{1} + B)^{-1/2}(H - S(\tau)(\mathbb{1} + tH)^{-1})\| \leq \\ D_{1/2,\alpha}\ \|(\mathbb{1} + H)^{1/2}(\mathbb{1} + B)^{-1/2}\|\ \tau^{\alpha-1/2}\ t^{-\alpha}\,,$$

which yields

$$\|(\mathbb{1} + B)^{-1/2}(H - S(\tau))(\mathbb{1} + tH)^{-1}\|\ t^{1/2} \leq \\ D_{1/2,\alpha}\ \|(\mathbb{1} + H)^{1/2}(\mathbb{1} + B)^{-1/2}\|\ \left(\frac{\tau}{t}\right)^{\alpha-1/2}, \tag{7.4.21}$$

for $t \in (0, 1]$ and $\tau > 0$.

To estimate the second term of the right-hand side of (7.4.20), we use again the *functional calculus* for H to infer that

$$\|(H - S(\tau))(\mathbb{1} + tH)^{-1}\| \leq \frac{1}{t} + \|S(\tau)(\mathbb{1} + H)^{-\alpha}\|\ t^{-\alpha}\,, \quad \alpha \in (1/2, 1)\,,$$

for $t \in (0, 1]$ and $\tau > 0$. By virtue of Lemma 7.4.4, (ii) there is a constant $D_\alpha > 0$ such that

$$\|S(\tau)(\mathbb{1} + tH)^{-1}\| \leq D_\alpha\ \tau^{\alpha-1}\ t^{-\alpha}\,, \quad \alpha \in (1/2, 1)\,,$$

for $t \in (0, 1]$ and $\tau > 0$, which yields

$$\|S(\tau)(\mathbb{1} + tH)^{-1}\|\ \tau^{1/2}\ t^{1/2} \leq D_\alpha\ \left(\frac{\tau}{t}\right)^{\alpha-1/2}. \tag{7.4.22}$$

As a result, we obtain

$$\|(H - S(\tau))(\mathbb{1} + tH)^{-1}\|\ \tau^{1/2}\ t^{1/2} \leq (1 + D_\alpha)\ \left(\frac{\tau}{t}\right)^{\alpha-1/2},$$

for $t, \tau \in (0, 1]$ and $0 < \tau \leq t$. Applying the estimates (7.4.21) and (7.4.22), we find from (7.4.20) that

$$t\|(\mathbb{1} + tS(\tau))^{-1/2}(H - S(\tau))(\mathbb{1} + tH)^{-1}\| \leq \\ \left\{C\ D_{1/2,\alpha}\ \|(\mathbb{1} + H)^{1/2}(\mathbb{1} + B)^{-1/2}\| + C_0(1 + D_\alpha)\right\} \left(\frac{\tau}{t}\right)^{\alpha-1/2},$$

for $t, \tau \in (0, 1]$ with $0 < \tau \leq t$. Owing to identification

$$\mathcal{N}_{\alpha-1/2} := C\ D_{1/2,\alpha}\ \|(\mathbb{1} + H)^{1/2}(\mathbb{1} + B)^{-1/2}\| + C_0(1 + D_\alpha)\,,$$

we obtain the estimate (7.4.19). □

Lemma 7.4.6 *Let A and B be non-negative self-adjoint operators such that for some $\alpha \in (1/2, 1)$ the condition (7.4.8) is satisfied. Further, let f and g be Kato functions of the class $\mathcal{K}$, which obey conditions of Lemma 7.4.4. Then there is a constant $L_{2\alpha-1} > 0$ such that*

$$t\,\|(\mathbb{1} + tH)^{-1}(H - S(\tau))(\mathbb{1} + tH)^{-1}\| \le L_{2\alpha-1}\left(\frac{\tau}{t}\right)^{2\alpha-1}, \tag{7.4.23}$$

for $t \in (0, 1]$ and $\tau > 0$.

Proof By the functional calculus for self-adjoint operator H, we get

$$\begin{aligned} t\,\|(\mathbb{1} + tH)^{-\alpha}(H - S(\tau))(\mathbb{1} + tH)^{-\alpha}\| \le \\ \|(\mathbb{1} + H)^{-\alpha}(H - S(\tau))(\mathbb{1} + H)^{-\alpha}\|\, t^{-(2\alpha-1)}, \end{aligned}$$

for $t \in (0, 1]$ and $\tau > 0$. Applying Lemma 7.4.4, we find a constant $L_{2\alpha-1} > 0$ such that

$$t\,\|(\mathbb{1} + tH)^{-\alpha}(H - S(\tau))(\mathbb{1} + tH)^{-\alpha}\| \le L_{2\alpha-1}\left(\frac{\tau}{t}\right)^{2\alpha-1},$$

for $t \in (0, 1]$ and $\tau > 0$, proving (7.4.23). □

Now we are in position for proving the main theorem about convergence of the Trotter-Kato product formula for approximants

$$\{F(t/n)^n = (g(tB/2n)f(tA/n)g(tB/2n))^n\}_{n\in\mathbb{N}}, \quad t \ge 0, \tag{7.4.24}$$

with (optimal) operator-norm error bound estimate corresponding to a fractional powers conditions.

Theorem 7.4.7 (INZ-Theorem) *Let A and B be non-negative self-adjoint operators and let self-adjoint operator $H := A \dot{+} B$ be their form-sum. Assume that for some $\alpha \in (1/2, 1)$ the condition (7.4.8) is satisfied. Further, let f and g be Kato functions of the class $\mathcal{K}$ obeying conditions of Lemma 7.4.4. If in addition one has that $\mathrm{dom}\,(A^{1/2}) \subseteq \mathrm{dom}\,(B^{1/2})$ and $m_f(x) < 1$ for $x > 0$, then for any finite interval $[0, T]$, there is a constant $C_{T,2\alpha-1} > 0$ such that*

$$\left\|(g(tB/2n)f(tA/n)g(tB/2n))^n - \mathrm{e}^{-tH}\right\| \le C_{T,2\alpha-1}\frac{1}{n^{2\alpha-1}}, \tag{7.4.25}$$

for $t \in [0, T]$ and $n \in \mathbb{N}$.

Proof On account of resolvent identities

$$\begin{aligned} (\mathbb{1} + tS(\tau))^{-1} - (\mathbb{1} + tH)^{-1} &= t\,(\mathbb{1} + tS(\tau))^{-1}(H - S(\tau))(\mathbb{1} + tH)^{-1} \\ &= t\,(\mathbb{1} + tH)^{-1}(H - S(\tau))(\mathbb{1} + tS(\tau))^{-1}, \end{aligned}$$

one infers that for $\tau, t > 0$

$$(\mathbb{1} + tS(\tau))^{-1} - (\mathbb{1} + tH)^{-1} = t(\mathbb{1} + tH)^{-1}(H - S(\tau))(\mathbb{1} + tH)^{-1} + \\ t^2(\mathbb{1} + tH)^{-1}(H - S(\tau))(\mathbb{1} + tS(\tau))^{-1}(H - S(\tau))(\mathbb{1} + tH)^{-1}.$$

As a consequence, we deduce estimate

$$\|(\mathbb{1} + tS(\tau))^{-1} - (\mathbb{1} + tH)^{-1}\| \leq \\ t\|(\mathbb{1} + tH)^{-1}(H - S(\tau))(\mathbb{1} + tH)^{-1}\| + \\ t^2\|(\mathbb{1} + tH)^{-1}(H - S(\tau))(\mathbb{1} + tS(\tau))^{-1}(H - S(\tau))(\mathbb{1} + tH)^{-1}\|,$$

for $\tau, t > 0$, which can be re-written as

$$\|(\mathbb{1} + tS(\tau))^{-1} - (\mathbb{1} + tH)^{-1}\| \leq \\ t\ \|(\mathbb{1} + tH)^{-1}(H - S(\tau))(\mathbb{1} + tH)^{-1}\| + \\ t^2\ \|(\mathbb{1} + tS(\tau))^{-1/2}(H - S(\tau))(\mathbb{1} + tH)^{-1}\|^2.$$

Taking into account Lemma 7.4.5 and Lemma 7.4.6, we obtain

$$\|(\mathbb{1} + tS(\tau))^{-1} - (\mathbb{1} + tH)^{-1}\| \leq L_{2\alpha-1}\left(\frac{\tau}{t}\right)^{2\alpha-1} + \mathcal{N}^2_{\alpha-1/2}\left(\frac{\tau}{t}\right)^{2\alpha-1},$$

for $t, \tau \in (0, 1]$ with $0 < \tau \leq t$. Setting $M_{2\alpha-1} := L_{2\alpha-1} + \mathcal{N}^2_{\alpha-1/2}$ one gets a constant such that estimate (4.3.1) (cf. Theorem 4.3.1) holds for $\alpha \in (1/2, 1)$ and $\tau, t \in (0, 1]$ with $0 < \tau \leq t$. Then by Theorem 4.3.1 there exists a constant $C_{2\alpha-1} > 0$ such (4.3.2) is valid for $\rho = 2\alpha - 1$ and $\tau, t \in (0, 1]$ and $0 < \tau \leq t$. Setting $C_{T=1,2\alpha-1} := C_{2\alpha-1}$ and $\tau = t/n$, $n \in \mathbb{N}$, we deduce from (4.3.2) the Trotter-Kato product formula (7.4.25) for approximants (7.4.24), where $t \in [0, 1]$.

To extend the result to $t \in [0, T]$, $0 < T < \infty$, we set $A_\theta := \theta\ A$ and $B_\theta := \theta\ B$, where $\theta > 0$. Then one has $H_\theta := A_\theta \dotplus B_\theta = \theta\ H$. If A and A are non-negative self-adjoint operators such that for some $\alpha \in (1/2, 1)$ condition (7.4.8) is satisfied, then we also have $\mathrm{dom}\,(H_\theta^\alpha) \subseteq \mathrm{dom}\,(A_\theta^\alpha) \cap \mathrm{dom}\,(B_\theta^\alpha)$ for $\vartheta > 0$. Similarly, the condition $\mathrm{dom}\,(A^{1/2}) \subseteq \mathrm{dom}\,(B^{1/2})$ implies $\mathrm{dom}\,(A_\theta^{1/2}) \subseteq \mathrm{dom}\,(B_\theta^{1/2})$ for $\theta > 0$. Hence, there is a constant $C_{2\alpha-1}(\theta) > 0$ such that

$$\left\|(g(tB_\theta/2n)f(tA_\theta/n)g(tB_\theta/2n))^n - \mathrm{e}^{-tH_\theta}\right\| \leq C_{2\alpha-1}(\theta)\,\frac{1}{n^{2\alpha-1}},$$

for $t \in [0, 1]$ and $n \in \mathbb{N}$, which yields

$$\left\|(g(tB/2n)f(tA/n)g(tB/2n))^n - \mathrm{e}^{-tH}\right\| \leq C_{2\alpha-1}(\theta)\,\frac{1}{n^{2\alpha-1}},$$

for $t \in [0, \theta]$ and $n \in \mathbb{N}$. Choosing $\theta = T$ and setting $C_{T,2\alpha-1} := C_{2\alpha-1}(T)$, we get that for any finite interval $[0, T]$, there is a constant $C_{T,2\alpha-1} > 0$ such that the Trotter-Kato product formula (7.4.25) holds for any $T > 0$. □

Remark 7.4.8 We note that for $\alpha \in (1/2, 1)$, the operator-norm error bound estimate $O(n^{-(2\alpha-1)})$ in the INZ-Theorem 7.4.7 cannot be improved; in other words, it is the *ultimate optimal* one, cf. Theorem 7.2.3 and Theorem 7.2.5 in Sect. 7.2. Whereas, for $\alpha \in (0, 1/2]$ the product formula (7.4.25) converges only in the strong operator topology. We return to comments on this point in Notes to Chap. 7 (Notes to Sects. 7.4, 7.4.3 Remarks: (i),(ii)). For the proof of *optimality* of the operator-norm convergence rate (7.4.25), we refer to Theorem H.0.2 and Remark H.0.9 in Appendix H, which cover also the non-symmetric version of (7.4.25).

It is worth mentioning here that the proofs of *optimality* of the operator-norm convergence rate:

- in Sect. 7.2 (Theorem 7.2.3 and Theorem 7.2.5) for the ITTZ-Theorem 7.1.9 and Corollary 7.1.12,
- then in INZ-Theorem 7.4.7, as well as in *predecessor* of the INZ-Theorem (see Theorem H.0.1 and Notes to Sects. 7.4, 7.4.3, Remarks: (i) and (ii))

are *all* based on the Hiroshi Tamura Example H.0.5.

7.4.3 *Related Product Approximants and Equivalence*

In this subsection, we show that INZ-Theorem 7.4.7 with estimate (7.4.25) holds not only for approximants (7.4.24) corresponding to the self-adjoint families, $F(t) = g(tB/2)f(tA)g(tB/2)$ and $G(t) = f(tA/2)g(tB)f(tA/2)$ (see (5.2.3)), but also to a certain similar (*self-adjoint* and *non-self-adjoint*) families

$$\begin{aligned}\widehat{F}(t) &:= g(tB)^{1/2}f(tA)g(tB)^{1/2},\ \widehat{G}(t) := f(tA)^{1/2}g(tB)f(tA)^{1/2},\\ F_{fg}(t) &:= f(tA)g(tB),\ F^*_{fg}(t) := g(tB)f(tA),\end{aligned} \tag{7.4.26}$$

where $t \geq 0$, cf. formulæ (5.2.52) and (5.2.58).

To this aim we first prove the following lemma about *self-adjoint* approximants $\{F(t/n)^n\}_{n\geq 1}$ (7.4.24).

Lemma 7.4.9 *Let A and B be non-negative self-adjoint operators such that for some $\alpha \in (1/2, 1)$, the condition* (7.4.8) *is satisfied. Further, let $f, g \in \mathcal{K}$ be Kato* *functions, which obey conditions of Lemma 7.4.4. Then there is a constant $C_F > 0$ such that*

$$\left\|(\mathbb{1} - F(t/n))\, e^{-tH}\right\| \leq C_F \, \frac{e^{t}}{n^{\alpha}}, \quad t \geq 0, \quad n \in \mathbb{N}, \tag{7.4.27}$$

where $F(t) = g(tB/2)\, f(tA)\, g(tB/2)$.

Proof We use the representation

$$\begin{aligned}(\mathbb{1} - F(t/n))\, e^{-tH} = \\ & g(tB/2n) f(tA/n)(\mathbb{1} - g(tB/2n))\, e^{-tH} + g(tB/2n) \\ & (\mathbb{1} - f(tA/n))\, e^{-tH} + \\ & (\mathbb{1} - g(tB/2n))\, e^{-tH},\end{aligned}$$

which yields the estimate

$$\begin{aligned}\|(\mathbb{1} - F(t/n))\, e^{-tH}\| \leq \\ & 2\, \|(\mathbb{1} - g(tB/2n))\, e^{-tH}\| + \|(\mathbb{1} - f(tA/n))\, e^{-tH}\|.\end{aligned}$$

Then owing to

$$\begin{aligned}(\mathbb{1} - g(tB/2n))\, e^{-tH} = \\ & (\mathbb{1} - g(tB/2n))\, (\mathbb{1} + B)^{-\alpha} (\mathbb{1} + B)^{\alpha} (\mathbb{1} + H)^{-\alpha} (\mathbb{1} + H)^{\alpha}\, e^{-t\,(\mathbb{1}+H)}\, e^{t},\end{aligned}$$

we find the estimate

$$\begin{aligned}\|(\mathbb{1} - g(tB/2n)) e^{-tH}\| \leq \\ & \|(\mathbb{1} - g(tB/2n))(\mathbb{1} + B)^{-\alpha}\|\, \|(\mathbb{1} + B)^{\alpha} (\mathbb{1} + H)^{-\alpha}\|\, \|(\mathbb{1} + H)^{\alpha}\, e^{-t\,(\mathbb{1}+H)}\|\, e^{t}.\end{aligned}$$

Now, seeing that by $\gamma_g(\alpha) < \infty$, (7.4.6) for $\alpha \in (1/2, 1)$

$$\begin{aligned}\|(\mathbb{1} - g(tB/2n))(\mathbb{1} + B)^{-\alpha}\| \leq \\ & \sup_{\lambda \geq 0} \frac{1 - g(t\lambda/2n)}{(1 + \lambda)^{\alpha}} \leq \sup_{\lambda > 0} \frac{1 - g(\lambda)}{\lambda^{\alpha}} \left(\frac{t}{2n}\right)^{\alpha} \leq \gamma_g(\alpha) \left(\frac{t}{2n}\right)^{\alpha},\end{aligned}$$

we obtain for $n \in \mathbb{N}$ and $t \geq 0$ inequality

$$\|(\mathbb{1} - g(tB/2n))\, e^{-tH}\| \leq 2^{-\alpha} \gamma_g(\alpha)\, b(\alpha)\, \|(\mathbb{1} + B)^{\alpha} (\mathbb{1} + H)^{-\alpha}\| \, \frac{e^{t}}{n^{\alpha}},$$

where $b(\alpha) := \sup_{\lambda > 0} \lambda^{\alpha}\, e^{-\lambda}$.

Similarly, one proves that

$$\|(\mathbb{1} - f(tA/n))\, \mathrm{e}^{-tH}\| \le \gamma_f(\alpha)\, b(\alpha)\, \|(\mathbb{1} + A)^{\alpha}(\mathbb{1} + H)^{-\alpha}\| \,\frac{\mathrm{e}^{t}}{n^{\alpha}},$$

for $n \in \mathbb{N}$ and $t \ge 0$. Then setting

$$C_F := \\ 2^{1-\alpha}\gamma_g(\alpha)\, b(\alpha)\, \|(\mathbb{1} + B)^{\alpha}(\mathbb{1} + H)^{-\alpha}\| + \gamma_f(\alpha)\, b(\alpha)\, \|(\mathbb{1} + A)^{\alpha}(\mathbb{1} + H)^{-\alpha}\|,$$

we obtain the estimate (7.4.27). □

Theorem 7.4.10 *Let A and B be non-negative self-adjoint operators such that for some $\alpha \in (1/2, 1)$ the condition* (7.4.8) *is satisfied and Kato functions f and g satisfy conditions of Lemma* 7.4.4. *If in addition one has* $\mathrm{dom}\,(A^{1/2}) \subseteq \mathrm{dom}\,(B^{1/2})$ *and $m_f(x) < 1$* (7.4.3), *for $x > 0$, then for any finite interval* $[0, T]$, *there are (Φ-dependent) constants $C^{\Phi}_{T,2\alpha-1} > 0$, such that besides $\{F(t)\}_{t\ge0}$ for any family $\{\Phi(t)\}_{t\ge0}$ from the list in* (7.4.26), *including $\{G(t) = f(tA/2)g(tB)f(tA/2)\}_{t\ge0}$, we obtain estimates*

$$\left\|\Phi(t/n)^{n} - \mathrm{e}^{-tH}\right\| \le C^{\Phi}_{T,2\alpha-1}\,\frac{1}{n^{2\alpha-1}}, \tag{7.4.28}$$

where $t \in [0, T]$ and $n \in \mathbb{N}$. The rate of convergence in (7.4.28), *for $n \to \infty$, is optimal in the sense of Remark* 7.4.8.

Proof To this aim, we set

$$\widehat{g}(x) := g(2x)^{1/2}, \quad x \ge 0.$$

Note that if $f \in \mathcal{K}$, then function $\widehat{f}$ is also a Kato function (Remark 5.2.6(c)), which satisfies $\varkappa_{2\alpha}[\widehat{g}] < \infty$ (7.4.7) and $m_{\widehat{g}}(x) < 1$ for $x > 0$. As a consequence, one gets

$$\widehat{g}(\tau B/2) = g(\tau B)^{1/2}, \quad \tau \ge 0. \tag{7.4.29}$$

If now we define

$$\widehat{F}(\tau) := \widehat{g}(\tau B/2) f(\tau A)\widehat{g}(\tau B/2), \quad \tau \ge 0,$$

then by INZ-Theorem 7.4.7, there is a constant $C^{\widehat{F}}_{T,2\alpha-1} > 0$ such that

$$\|\widehat{F}(t/n)^{n} - \mathrm{e}^{-tH}\| \le C^{\widehat{F}}_{T,2\alpha-1}\,\frac{1}{n^{2\alpha-1}}, \quad t \in [0, T] \quad n \in \mathbb{N}. \tag{7.4.30}$$

Since by (7.4.29) and (7.4.26) one has $\widehat{F}(\tau) = g(tB)^{1/2} f(tA) g(tB)^{1/2}$ for $\tau \geq 0$, the estimate (7.4.30) proves the assertion (7.4.28) for self-adjoint family: $\Phi = \widehat{F}$ and constant $C^{\widehat{F}}_{T,2\alpha-1}$.

Next we consider $\Phi = F^*_{fg}$. On account of (7.4.26), one has representation

$$F^*_{fg}(\tau)^n = g(\tau B)^{1/2} \widehat{F}(\tau)^{n-1} g(\tau B)^{1/2} f(\tau A)\,, \quad \tau \geq 0, \quad n \in \mathbb{N}\,,$$

which provides formula

$$\begin{aligned} F^*_{fg}(\tau)^n - \mathrm{e}^{-tH} = \\ & g(\tau B)^{1/2}\left(\widehat{F}(\tau)^{n-1}(\mathbb{1} - \widehat{F}(\tau))\right) g(\tau B)^{1/2} f(\tau A) + \\ & g(\tau B)^{1/2}\left(\widehat{F}(\tau)^n - \mathrm{e}^{-tH}\right) g(\tau B)^{1/2} f(\tau A) + \\ & \left(g(\tau B)^{1/2} - \mathbb{1}\right) \mathrm{e}^{-tH} g(\tau B)^{1/2} f(\tau A) + \\ & \mathrm{e}^{-tH}\left(g(\tau B)^{1/2} - \mathbb{1}\right) f(\tau A) + \mathrm{e}^{-tH}\left(f(\tau A) - \mathbb{1}\right)\,, \end{aligned}$$

where $\tau = t/n$, $t \geq 0$ and $n \in \mathbb{N}$. On that account, we obtain estimate

$$\begin{aligned} \|F^*_{fg}(\tau)^n - \mathrm{e}^{-tH}\| \leq \left\|\widehat{F}(\tau)^{n-1}(\mathbb{1} - \widehat{F}(\tau))\right\| + \\ \left\|\widehat{F}(\tau)^n - \mathrm{e}^{-tH}\right\| + 2\left\|(\mathbb{1} - g(\tau B)^{1/2})\,\mathrm{e}^{-tH}\right\| + \left\|(\mathbb{1} - f(\tau A))\,\mathrm{e}^{-tH}\right\|\,. \end{aligned} \tag{7.4.31}$$

On account of Lemma 7.4.9, we infer that

$$\left\|(\mathbb{1} - g(tB/n)^{1/2})\,\mathrm{e}^{-tH}\right\| \leq \gamma_g(\alpha)\, b(\alpha)\, \|(\mathbb{1} + B)^{\alpha}\, (\mathbb{1} + H)^{-\alpha}\|\, \frac{\mathrm{e}^t}{n^{\alpha}}\,, \tag{7.4.32}$$

for $t \geq 0$ and $n \in \mathbb{N}$, as well as

$$\left\|(\mathbb{1} - f(tA/n))\,\mathrm{e}^{-tH}\right\| \leq 2\,\gamma_f(\alpha)\, b(\alpha)\, \|(\mathbb{1} + A)^{\alpha} (\mathbb{1} + H)^{-\alpha}\|\, \frac{\mathrm{e}^t}{n^{\alpha}}\,. \tag{7.4.33}$$

Note that self-adjointness: $\widehat{F}(\tau) = (\widehat{F})^*(\tau)$, and $\|\widehat{F}(\tau)\| \leq 1$, for $\tau \geq 0$, yield estimate

$$\left\|\widehat{F}(\tau)^{n-1}(\mathbb{1} - \widehat{F}(\tau))\right\| \leq \frac{1}{n}\,, \quad n \in \mathbb{N}\,,\ \tau \geq 0\,. \tag{7.4.34}$$

Therefore, by virtue of (7.4.31) the estimate (7.4.30) and (7.4.32)–(7.4.34) prove the assertion (7.4.28) for $\Phi = F^*_{fg}$ and $C^{F^*_{fg}}_{T,2\alpha-1} > 0$, where $\alpha \in (1/2, 1)$.

Because the mapping $F_{fg} \to F^*_{fg}$ is continuous in the operator-norm topology on $\mathcal{L}(\mathfrak{H})$ and on account of $H = H^*$, we infer from

$$\|F^*_{fg}(\tau)^n - \mathrm{e}^{-tH}\| = \|F_{fg}(\tau)^n - \mathrm{e}^{-tH}\|\,, \quad \tau = t/n\,, \quad n \in \mathbb{N}\,,\ t \geq 0\,,$$

the proof of assertion (7.4.28) for non-self-adjoint family, $\Phi = F_{fg}$, and coefficient $C^{F_{fg}}_{T,2\alpha-1} = C^{F^*_{fg}}_{T,2\alpha-1}$.

Now we consider the self-adjoint case $\Phi = \widehat{G}$. To this aim we use representation, which involves the self-adjoint family $\widehat{F}$ (7.4.26):

$$\begin{aligned}
&\widehat{G}(\tau)^n - \mathrm{e}^{-tH} = \\
&f(\tau A)^{1/2} g(\tau B)^{1/2} \widehat{F}(\tau)^{n-1} g(\tau B)^{1/2} f(\tau A)^{1/2} - \mathrm{e}^{-tH} = \\
&f(\tau A)^{1/2} g(\tau B)^{1/2} \,(\widehat{F}(\tau)^{n-1} - \mathrm{e}^{-tH})\, g(\tau B)^{1/2} f(\tau A)^{1/2} + \\
&f(\tau A)^{1/2} g(\tau B)^{1/2} \mathrm{e}^{-tH} g(\tau B)^{1/2} f(\tau A)^{1/2} - \mathrm{e}^{-tH}\,, \quad \tau = t/n\,.
\end{aligned} \tag{7.4.35}$$

Then by (7.4.30) and (7.4.34), one gets for $\alpha \in (1/2, 1)$ the estimate

$$\begin{aligned}
&\|f(\tau A)^{1/2} g(\tau B)^{1/2} (\widehat{F}(\tau)^{n-1} - \mathrm{e}^{-tH}) g(\tau B)^{1/2} f(\tau A)^{1/2}\| \leq \\
&\|\widehat{F}(\tau)^n - \mathrm{e}^{-tH}\| + \|\widehat{F}(\tau)^{n-1}(\mathbb{1} - \widehat{F}(\tau))\| \leq \\
&C^{\widehat{F}}_{T,2\alpha-1} \frac{1}{n^{2\alpha-1}} + \frac{1}{n}\,, \qquad \tau = t/n\,, \quad n \in \mathbb{N}\,, \quad t \in [0, T]\,.
\end{aligned} \tag{7.4.36}$$

To estimate the operator-norm of the last line in (7.4.35), one has to mimic the proof of (7.4.32) and (7.4.33). Then there exists $K(\alpha) > 0$ such that for $\tau = t/n$

$$\|f(\tau A)^{1/2} g(\tau B)^{1/2} \mathrm{e}^{-tH} g(\tau B)^{1/2} f(\tau A)^{1/2} - \mathrm{e}^{-tH}\| \leq K(\alpha)\, \frac{\mathrm{e}^{t}}{n^{\alpha}}\,, \tag{7.4.37}$$

where $n \in \mathbb{N}$, $t \in [0, T]$ and $\alpha \in (1/2, 1)$. On account of (7.4.36) and (7.4.37), we deduce from (7.4.35) the estimate in the form (7.4.28). This proves the assertion for the family $\Phi = \widehat{G}$.

Finally, we consider the self-adjoint family $\Phi = G$. Recall that by Remark 5.2.6(c) if $f \in \mathcal{K}$, then $\widehat{f}(x) := f(x/2)^2$ is also a Kato function of the class $\mathcal{K}$. As a consequence, if in $G(t) = f(tA/2) g(tB) f(tA/2)$ we set $f(tA/2) = \widehat{f}(tA)^{1/2}$, then one can apply to the family

$$\widehat{G}(t) := \widehat{f}(tA)^{1/2} g(tB) \widehat{f}(tA)^{1/2}\,, \quad t \geq 0\,, \tag{7.4.38}$$

the arguments developed above for the case $\Phi = \widehat{G}$. As a result, they prove the assertion for $\Phi = \widehat{G}$, and on account of definition $f(tA/2) = \widehat{f}(tA)^{1/2}$, these arguments also prove the assertion (7.4.28) for the family $\Phi = G$.

For the proof of *optimality* of the operator-norm estimate (7.4.28) (and consequently of the convergence rate for $n \to \infty$), we refer to Theorem H.0.2 and Remark H.0.9 in Appendix H. □

7.4.4 Examples

Below we address to the reader examples illustrating some applications of the main statements of the present chapter. Here and in the Notes to Chap. 7 (Notes to Sect. 7.5), we focused on the question about the Trotter-Kato product formulæ operator-norm *rate* of convergence. Since the fractional power conditions are widely known and instructive for differential operators, we consider, instead of an abstract operator setting, two explicit examples of the Schrödinger C_0-semigroups in L^2-Hilbert spaces.

Example 7.4.11 Let $\mathfrak{H} := L^2(\Omega)$, where Ω is a bounded domain in $\mathbb{R}^l$, $l = 1, 2, \ldots$, with boundary $\partial\Omega$ of C^∞-class. By self-adjoint operator A, we denote the negative half-*Laplacian* with *Dirichlet* boundary conditions in $L^2(\Omega)$, that is, $A := -\Delta_D/2$ on domain given by means of the *Sobolev* spaces $W^{p,q}$ as

$$\mathrm{dom}\,(A) := W^{2,2}(\Omega) \cap \mathring{W}^{1,2}(\Omega)\,, \tag{7.4.39}$$

where $\mathring{W}^{1,2}(\Omega)$ is the closure of $C_0^\infty(\Omega)$ in the (Lebesgue) space $W^{1,2}(\Omega)$. Now, using the space

$$W^{2,2}_{B_D}(\Omega) := \{u \in W^{2,2}(\Omega) : B_D u|_{\partial\Omega} = 0\}\,,$$

where the *boundary* operator B_D is defined as operator of *restriction*

$$B_D u := u|_{\partial\Omega}\,,$$

for $u \in W^{2,2}(\Omega)$, we obtain for domain (7.4.39) of operator A the representation

$$\mathrm{dom}\,(A) = W^{2,2}_{B_D}(\Omega)\,.$$

Further, let self-adjoint operator B be the negative half-*Laplacian* with *Neumann* boundary conditions, that is, $B := -\Delta_N/2$ on domain

$$\mathrm{dom}\,(B) := W^{2,2}_{B_N}(\Omega)\,,$$

where

$$W^{2,2}_{B_N}(\Omega) := \{u \in W^{2,2}(\Omega) : B_N u|_{\partial\Omega} = 0\}\,.$$

Here the *boundary* operator B_N is given for $u \in W^{2,2}(\Omega)$ by equation

$$(B_N u)(x) = \frac{\partial}{\partial \nu(x)} u(x) , \quad x \in \partial\Omega ,$$

where $\nu(x)$ is the outer unit normal to the boundary $\partial\Omega$ at the point $x \in \partial\Omega$.

Note that on account of $\operatorname{dom}(A^{1/2}) = \mathring{W}^{1,2}(\Omega)$ and $\operatorname{dom}(B^{1/2}) = W^{1,2}(\Omega)$, we infer that

$$\operatorname{dom}(A^{1/2}) \subseteq \operatorname{dom}(B^{1/2}) . \tag{7.4.40}$$

As a consequence

$$\operatorname{dom}(H^{1/2}) = \operatorname{dom}(A^{1/2}) \cap \operatorname{dom}(B^{1/2}) = \mathring{W}^{1,2}(\Omega) ,$$

and the *form-sum* of operators A and B yields $H := A \dotplus B = 2\,A = -\Delta_D$. Hence, $\operatorname{dom}(H^{\alpha}) = \operatorname{dom}(A^{\alpha})$ for arbitrary $\alpha \in [0, 1]$.

Next we elucidate a structure of domains $\operatorname{dom}(A^{\alpha})$ and $\operatorname{dom}(B^{\alpha})$ for fractional powers $\alpha \in (1/2, 1)$:

$$\operatorname{dom}(A^{\alpha}) = W^{2\alpha,2}_{B_D}(\Omega), \quad \alpha \in (1/2, 1) , \tag{7.4.41}$$

and

$$\operatorname{dom}(B^{\alpha}) = \begin{cases} W^{2\alpha,2}(\Omega) , \ \alpha \in (1/2, 3/4), \\ W^{2\alpha,2}_{B_N}(\Omega), \ \alpha \in (3/4, 1). \end{cases} \tag{7.4.42}$$

Given that $W^{2\alpha,2}_{B_D}(\Omega) \subseteq W^{2\alpha,2}(\Omega)$, we obtain

$$\operatorname{dom}(H^{\alpha}) = \operatorname{dom}(A^{\alpha}) \subseteq \operatorname{dom}(B^{\alpha}) ,$$

for $\alpha \in (1/2, 3/4)$.

Now, applying Theorem 7.4.10 to operators A and B defined above, we conclude that

$$\left\| \left(\mathrm{e}^{-t\,(-\Delta_D)/2n}\, \mathrm{e}^{-t\,(-\Delta_N)/2n} \right)^n - \mathrm{e}^{-t\,(-\Delta_D)} \right\| = O(n^{-\kappa}) , \quad n \to \infty ,$$

uniformly in $t \in [0, T]$, for any $\kappa := 2\alpha - 1 < \kappa_0 := 1/2$, where $\alpha \in (1/2, 3/4)$.

(a) Note that if $\alpha = 3/4$, then it does *not* hold that $\operatorname{dom}(A^{3/4}) \subseteq \operatorname{dom}(B^{3/4})$. Hence $\operatorname{dom}(A^{3/4}) \cap \operatorname{dom}(B^{3/4})$ is a proper subset of $\operatorname{dom}(H^{3/4})$. This does not allow to apply Theorem 7.4.10 (or equivalently INZ-Theorem 7.4.7) since condition (7.4.8) $\operatorname{dom}(H^{\alpha}) \subseteq \operatorname{dom}(A^{\alpha}) \cap \operatorname{dom}(B^{\alpha})$, for $\alpha = 3/4$, is not satisfied.

(b) Now we consider the case $\alpha \in (3/4, 1)$. Then by (7.4.41) and (7.4.42)

$$\operatorname{dom}(A^{\alpha}) \cap \operatorname{dom}(B^{\alpha}) = W^{2\alpha,2}_{\{B_D,B_N\}}(\Omega) :=$$
$$\{u \in W^{2\alpha,2}(\Omega) : B_D u|_{\partial\Omega} = 0,\ B_N u|_{\partial\Omega} = 0\} \subseteq W^{2\alpha,2}_{B_D}(\Omega).$$

As a result, instead of (7.4.8) the intersection $\operatorname{dom}(A^{\alpha}) \cap \operatorname{dom}(B^{\alpha})$ is again a proper *subset* of $\operatorname{dom}(H^{\alpha})$, which does not allow to apply Theorem 7.4.10 either.

(c) If $\alpha = 1$, then $\operatorname{dom}(A) \cap \operatorname{dom}(B)$ is again a proper subset of $\operatorname{dom}(H)$, which yields $H \neq A + B$. As a consequence, the ITTZ-Theorem 7.1.9 is also not applicable in this case.

Resuming Example 7.4.11, (a)–(c), we conclude that in contrast to a predecessor of the INZ-Theorem 7.4.7, which is Theorem H.0.1, the fractional power condition (7.4.40) yields for $\alpha \in (1/2, 3/4)$ the optimal rate of the operator-norm convergence as in (7.4.28). Here, INZ-Theorem 7.4.7 and Theorem 7.4.10 are illustrated by Example 7.4.11 which covers the case when operator B^{α} is *not* relatively small with respect to A^{α} for $\alpha \in (1/2, 3/4)$, cf. Theorem H.0.1 (ii), but they are related according to a *mild* subordination $\operatorname{dom}(A^{1/2}) \subseteq \operatorname{dom}(B^{1/2})$, (7.4.40).

The next Example 7.4.12 serves for analysis and illustration of these two conditions on generators A and B (cf. (7.4.50) and (7.4.51)), which ensure the same optimal rate of the operator-norm convergence due to (7.4.28). See Notes to Chap. 7 (Notes to Sects. 7.4, 7.4.3, and 7.4.4) for comments and references.

Example 7.4.12 Here we continue to consider fractional conditions on generators for example of the Schrödinger semigroups. To this end, we set $\mathfrak{H} = L^2(\mathbb{R})$ and define self-adjoint operator

$$(Af)(x) := (-\,\partial_x^2 + \mathbb{1})\, f(x)\,, \qquad f \in \operatorname{dom}(A) = W^{2,2}(\mathbb{R})\,. \tag{7.4.43}$$

We have added the identity to the one-dimensional Laplace operator in order to guarantee the semi-boundedness: $A \geq \mathbb{1}$. By B_γ we denote (unbounded) in $L^2(\mathbb{R})$ non-negative self-adjoint multiplication operator

$$\operatorname{dom}(B_\gamma) = \left\{ f \in L^2(\mathbb{R}) : \frac{C}{|x-y|^{\gamma}}\, f(x) \in L^2(\mathbb{R}) \right\}, \tag{7.4.44}$$

$$(B_\gamma) f(x) = \frac{C}{|x-y|^{\gamma}}\, f(x), \qquad f \in \operatorname{dom}(B)\,, \tag{7.4.45}$$

where $0 < \gamma < 1$, $C > 0$ and $y \in \mathbb{R}$.

By means of the *Fourier* transform $\widehat{f}$ of the function $f \in \mathrm{dom}(A)$, we obtain a (pseudo-differential) representation

$$(A^{\beta} f)(x) = \frac{1}{\sqrt{2\pi}} \int_{\mathbb{R}} dk \, \mathrm{e}^{\,i\,kx} \, (k^2+1)^{\beta} \, \widehat{f}(k) \,. \tag{7.4.46}$$

Then owing to (7.4.46) and on account of the *Cauchy-Bunyakovsky-Schwarz* inequality, one gets estimate

$$\begin{aligned}\left(\int_{\mathbb{R}} dk \, |\widehat{f}(k)|\right)^2 &\le \int_{\mathbb{R}} dk \, (k^2+1)^{-2\beta} \int_{\mathbb{R}} dk \, (k^2+1)^{2\beta} \, |\widehat{f}(k)|^2 \\ &= \int_{\mathbb{R}} dk \, (k^2+1)^{-2\beta} \, \|A^{\beta} f\|^2 \,,\end{aligned}$$

which yields

$$|f(x)|^2 \le \frac{1}{2} \, \rho(\beta) \, \|A^{\beta} f\|^2, \qquad f \in \mathrm{dom}(A^{\beta}), \quad \beta > \frac{1}{4}, \quad x \in \mathbb{R}\,, \tag{7.4.47}$$

where

$$\rho(\beta) := \frac{1}{\pi} \int_{-\infty}^{+\infty} dt \, \frac{1}{(1+t^2)^{2\beta}} \,.$$

Then (7.4.47) implies

$$\int_{|x-y|\le 1} dx \, \frac{1}{|x-y|^{2\kappa}} \, |f(x)|^2 \le \frac{1}{1-2\kappa} \, \rho(\beta) \, \|A^{\beta} f\|^2 \,, \qquad f \in \mathrm{dom}(A^{\beta}),$$

for $\beta > 1/4$, $0 < \kappa < 1/2$ and $y \in \mathbb{R}$. Since $A \ge \mathbb{1}$ we get for $f \in \mathrm{dom}(A^{\beta})$:

$$\int_{|x-y|\ge 1} dx \, \frac{1}{|x-y|^{2\kappa}} \, |f(x)|^2 \le \int_{|x-y|\ge 1} dx \, |f(x)|^2 \le \|f\|^2 \le \|A^{\beta} f\|^2 \,. \tag{7.4.48}$$

(a) Let $0 < \gamma < 1/2$. Setting $\kappa = \gamma$ and $\beta = 1$, we obtain from (7.4.47)–(7.4.48)

$$\mathrm{dom}(A) \subseteq \mathrm{dom}(B_{\gamma}) \,, \tag{7.4.49}$$

and

$$\|B_{\gamma} f\| \le a \, \|Af\|, \qquad f \in \mathrm{dom}(A),$$

where $a := C\sqrt{1+\rho(1)/(1-2\gamma)}$.

As a consequence, if C is sufficiently small such that $a < 1$, the operator B is Kato-small relative to A that ensures the self-adjointness of operator $H := A + B_\gamma$ on $\mathrm{dom}\,(H) = \mathrm{dom}\,(A)$; see Corollary 1.9.15. Then self-adjoint generators A and B of C_0-semigroups satisfy conditions of the ITTZ-Theorem 7.1.9. As a consequence, the Trotter-Kato product formulæ (7.1.59)–(7.1.62) and (7.1.80) and (7.1.81) converge, for $n \to \infty$, in the operator norm to the self-adjoint C_0-semigroup $\{\mathrm{e}^{-tH}\}_{t\geq 0}$ with the (*optimal*) rate $O(1/n)$. See comments in Notes to Chap. 7 (Notes to Sect. 7.1 and to Sects. 7.4, 7.4.3, and 7.4.4).

(b) However, if $1/2 \leq \gamma < 1$ the condition (7.4.49) is *not valid*. To see this let $f \in \mathrm{dom}\,(A) = W^{2,2}(\mathbb{R})$ be such that $\inf_{|x-y|\leq 1} |f(x)| \geq \epsilon > 0$. Note that such elements always exist. Then

$$\int_{-\infty}^{+\infty} dx \, \frac{1}{|x-y|^{2\gamma}} \, |f(x)|^2 = +\infty .$$

Therefore, $\mathrm{dom}\,(A) \not\subseteq \mathrm{dom}\,(B_\gamma)$. Moreover, if $\beta > 1/4$ and $\alpha\,\gamma \geq 1/2$, then one always has that $\mathrm{dom}\,(A^\beta) \not\subseteq \mathrm{dom}\,(B_\gamma^\alpha)$. On the other hand, if $\beta > 1/4$ and $\alpha\,\gamma < 1/2$, then in virtue of (7.4.47) and (7.4.48), we obtain $\mathrm{dom}\,(A^\beta) \subset \mathrm{dom}\,(B_\gamma^\alpha)$ and $\|B_\gamma^\alpha f\| \leq a\,\|A^\beta f\|$, for $f \in \mathrm{dom}\,(A^\beta)$, where $a := C^\alpha \sqrt{1+\rho(\beta)/(1-2\alpha\gamma)}$

So, if $1/2 \leq \gamma < 1$, then we can find $\alpha \in (1/2, 1)$, such that $0 < \alpha\,\gamma < 1/2$. For $\beta > 1/4$ this allows to make $a < 1$ by adjustment of $C > 0$ to be sufficiently small. Now we take $\alpha = \beta \in (1/2, 1)$ and obtain

$$\mathrm{dom}\,(A^\alpha) \subset \mathrm{dom}\,(B_\gamma^\alpha)\,, \quad \|B_\gamma^\alpha f\| \leq a\,\|A^\alpha f\|\,, \quad a < 1\,,\; f \in \mathrm{dom}\,(A^\alpha)\,. \tag{7.4.50}$$

Applying to (7.4.50) the *Heinz* inequality for power $\theta < 1$ (Proposition 9.2.15), one obtains $\|(B_\gamma^\alpha)^\theta f\| \leq \|(a\,A^\alpha)^\theta f\|$, for $f \in \mathrm{dom}\,(A^{\alpha\theta}) \subset \mathrm{dom}\,((B_\gamma^\alpha)^\theta)$. Let $\alpha\,\theta = 1/2$. Then $a^{1/2\alpha} < 1$. Therefore, we established that the relative boundedness (or *smallness*) (7.4.50) yield the subordination (7.4.40) and

$$\mathrm{dom}\,(H^{1/2}) = \mathrm{dom}\,(A^{1/2}) \cap \mathrm{dom}\,(B_\gamma^{1/2}) = \mathrm{dom}\,(A^{1/2}) \subset \mathrm{dom}\,(B_\gamma^{1/2})\,. \tag{7.4.51}$$

As a consequence, (7.4.51) guarantee that owing to (7.4.43) and (7.4.45) the self-adjoint *form-sum* of operators, $H = A \dotplus B_\gamma$, is well-defined in $\mathfrak{H} = L^2(\mathbb{R})$, and the conditions of the Theorem 7.4.10 are satisfied.

Hence, the Trotter-Kato product formulæ (5.2.3) and (7.4.28) converge, for $n \to \infty$, in the operator norm topology with (*optimal*) rate $O(1/n^{2\alpha-1})$ to the self-adjoint C_0-semigroup $\{\mathrm{e}^{-tH}\}_{t\geq 0}$, where generator H is given by the *form-sum*: $H = A \dotplus B_\gamma$. We note that in the INZ-Theorem 7.4.7, the *smallness* condition $a^{1/2\alpha} < 1$ (Theorem H.0.1) is *relaxed* to subordination of operators $A^{1/2}$ and $B^{1/2}$, cf. (7.4.51). See comments in Notes to Chap. 7 (Notes to Sect. 7.1 and to Sects. 7.4, 7.4.3, Remarks: (i), (ii)).

(c) In fact, the example (b) can be generalised. Let $\{x_n\}_{n=1}^{\infty}$ be set of all *rational* numbers of $\mathbb{R}$. We set

$$q_\gamma(x) := \sum_{n=1}^{\infty} \frac{C_n}{|x - x_n|^\gamma}, \qquad 0 < \gamma < 1, \quad x \in \mathbb{R},$$

where $\{C_n\}_{n=1}^{\infty}$ is a sequence of positive numbers such that

$$S := \sum_{n=1}^{\infty} C_n < +\infty .$$

Since

$$\int_{x_0-1}^{x_0+1} dx \sum_{n=1}^{N} \frac{C_n}{|x - x_n|^\gamma} \leq \frac{2}{1-\gamma} \sum_{n=1}^{N} C_n, \qquad N = 1, 2, \ldots, \quad x_0 \in \mathbb{R},$$

one gets that $q_\gamma \in L^1_{loc}(\mathbb{R})$. Furthermore, we note that by the *Jensen* inequality, one deduces

$$\left(\sum_{n=1}^{N} \frac{C_n}{|x - x_n|^\gamma} \right)^\nu \leq S^{\nu-1} \sum_{n=1}^{N} \frac{C_n}{|x - x_n|^{\gamma\nu}}, \qquad N = 1, 2, \ldots, \quad x \in \mathbb{R}, \tag{7.4.52}$$

for $\nu \geq 1$. We set

$$(\widehat{B}_\gamma f)(x) := q_\gamma(x) f(x), \qquad f \in \mathrm{dom}\,(\widehat{B}_\gamma), \tag{7.4.53}$$

$$\mathrm{dom}\,(\widehat{B}_\gamma) := \{f \in L^2(\mathbb{R}) : q_\gamma(x) f(x) \in L^2(\mathbb{R})\}. \tag{7.4.54}$$

Let $0 < \gamma < 1/2$. Then by (7.4.47), (7.4.48) and (7.4.52) for $\nu = 2$, we infer that $\mathrm{dom}\,(A) \subseteq \mathrm{dom}\,(\widehat{B}_\gamma)$ and $\|\hat{B}_\gamma f\| \leq a\, \|Af\|$ for $f \in \mathrm{dom}\,(A)$, where

$$a := S \sqrt{1 + \frac{\rho(1)}{1 - 2\gamma}} .$$

(d) If $0 < \gamma < 1/2$ and S is small enough, then operator $H = A + \widehat{B}_\gamma$ is self-adjoint. Hence, self-adjoint generators A and $\widehat{B}_\gamma$ of C_0-semigroups satisfy conditions of the ITTZ-Theorem 7.1.9. As a consequence, the Trotter-Kato product formulæ (7.1.59)–(7.1.62) and (7.1.80), (7.1.81) converge, for $n \to \infty$, in the operator norm to the self-adjoint C_0-semigroup $\{e^{-tH}\}_{t\geq 0}$ with the (*optimal*) rate $O(1/n)$.

Whereas, if $1/2 \leq \gamma < 1$, then condition

$$\frac{C_n}{|x - x_n|^{\gamma}} f(x) \in L^2(\mathbb{R}), \qquad f \in \operatorname{dom}(A) = W^{2,2}(\mathbb{R}),$$

entails $f(x_n) = 0$. As a consequence, $f \in \operatorname{dom}(\widehat{B}_{\gamma}) \cap \operatorname{dom}(A)$ requires that $f(x) \equiv 0$. Therefore,

$$\operatorname{dom}(\widehat{B}_{\gamma}) \cap \operatorname{dom}(A) = \{0\}, \qquad \frac{1}{2} \leq \gamma < 1.$$

Furthermore, since

$$\frac{C_n^{\alpha}}{|x - x_n|^{\alpha\gamma}} \leq q_{\gamma}(x)^{\alpha}, \qquad n = 1, 2, \ldots, \qquad x \in \mathbb{R},$$

one gets that $\operatorname{dom}(\widehat{B}_{\gamma}^{\alpha}) \cap \operatorname{dom}(A^{\beta}) = \{0\}$ for $1/2 \leq \alpha\gamma$ and $\beta > 1/4$.

(e) However, if $0 < \alpha\,\gamma < 1/2$ and $1/2 < \alpha < 1$, $\beta > 1/4$, then by (7.4.47) and (7.4.48) and (7.4.52) for $\nu = 2\alpha$, we obtain that

$$\operatorname{dom}(A^{\beta}) \subseteq \operatorname{dom}(\widehat{B}_{\gamma}^{\alpha}), \tag{7.4.55}$$

and $\|\widehat{B}_{\gamma}^{\alpha} f\| \leq a\, \|A^{\beta} f\|$ for $f \in \operatorname{dom}(A^{\beta})$, where

$$a := S^{\alpha} \sqrt{1 + \frac{\rho(\beta)}{1 - 2\alpha\gamma}}.$$

As above in (b), if $1/2 \leq \gamma < 1$, we can find $\alpha \in (1/2, 1)$ such that $0 < \alpha\,\gamma < 1/2$. Again similarly to (b), if S is sufficiently *small*, then by Theorem 7.4.10 the Trotter-Kato product formulæ converge in operator norm with the (*optimal*) rate $O(1/n^{2\alpha-1})$ for $n \to \infty$, where $H = A \dotplus \hat{B}_{\gamma}$.

We recall that in the INZ-Theorem 7.4.7 and in Theorem 7.4.10, the *smallness* condition was relaxed. See comments in Notes to Chap. 7 (Notes to Sect. 7.1 and to Sects. 7.4, 7.4.3, and 7.4.4).

(f) Now summing up the considerations above, let us introduce the *critical* value

$$\alpha_0 := \frac{1}{2\gamma}, \qquad 0 < \gamma < 1.$$

On account of this definition, we find that $\alpha_0 > 1/2$ and moreover

- If $\alpha_0 > 1$, then $\mathrm{dom}\,(A) \subseteq \mathrm{dom}\,(B_\gamma)$ and $\mathrm{dom}\,(A) \subseteq \mathrm{dom}\,(\widehat{B}_\gamma)$. Therefore, see (a) and (c), generators A and B_γ as well as A and $\widehat{B}_\gamma$ satisfy conditions of the ITTZ-Theorem 7.1.9. In this case the Trotter-Kato product formulæ converge with the optimal operator-norm rate $O(1/n)$.
- If $1/2 < \alpha_0 \leq 1$, then for each $\alpha \in [\alpha_0, 1]$, we have $\mathrm{dom}\,(A^\beta) \not\subseteq \mathrm{dom}\,(B_\gamma^\alpha)$ in the example (b), whereas $\mathrm{dom}\,(A^\beta) \cap \mathrm{dom}\,(\widehat{B}_\gamma^\alpha) = \{0\}$ in the example (d) for $\beta > 1/4$.
- Finally, if $1/2 < \alpha < \alpha_0$, then Trotter-Kato product formulæ converge for the *both* examples, B_γ^α (7.4.45) and $\widehat{B}_\gamma^\alpha$, with the *optimal* convergence rate : $O(1/n^{2\alpha-1})$.

7.5 Notes to Chap. 7

Notes to Sect. 7.1

Historically the first results on the *operator-norm* convergence of the Trotter-Kato product formulæ trace back to publications by Zagrebnov [Zag88] and then by Neidhardt-Zagrebnov [NZ90a], [NZ90b]. These results were established for convergence in the *trace-norm* and, consequently, in the *operator-norm* topology on a separable Hilbert space $\mathfrak{H}$ for a special class of C_0-semigroups (*Gibbs semigroups* [Zag19]), for a certain subset of the Kato functions $f, g \in \mathcal{K}$ and a couple of self-adjoint positive generators A, B.

The next fundamental observation in a Hilbert space was due to Dzh.L. Rogava [Rog93]. He proved that standard (exponential) Trotter product formula for (not necessarily Gibbsian) C_0-semigroups converges on the space $\mathfrak{H}$ in the *operator-norm* topology of $\mathcal{L}(\mathfrak{H})$ with the rate of convergence of order $O\,(\ln(n)\,/\sqrt{n})$.

Proposition 7.5.1 (Rogava's theorem [Rog93]) *Let A and B be positive self-adjoint operators in a Hilbert space $\mathfrak{H}$. Suppose that operator $H := A + B$ is self-adjoint on domain* $\mathrm{dom}\,(A) \cap \mathrm{dom}\,(B)$. *If in addition* $\mathrm{dom}\,(A) \subset \mathrm{dom}\,(B)$, *then there is a constant $C > 0$ such that the operator-norm estimates*

$$\begin{aligned} \|(\mathrm{e}^{-tA/2n}\mathrm{e}^{-tB/n}\mathrm{e}^{-tA/2n})^n - \mathrm{e}^{-tH}\| &\leq C\ \ln(n)\,/\sqrt{n}\,, \quad n \in \mathbb{N}, \\ \|(\mathrm{e}^{-tA/n}\mathrm{e}^{-tB/n})^n - \mathrm{e}^{-tH}\| &\leq C\ \ln(n)\,/\sqrt{n}\,, \quad n \in \mathbb{N}\,, \end{aligned} \tag{7.5.1}$$

for symmetric and non-symmetric product formulæ hold for $t \in \mathbb{R}_0^+$.

Note that, apart the additional $\mathrm{dom}\,(A) \subset \mathrm{dom}\,(B)$, conditions of Rogava's theorem coincide with conditions of the ITTZ-Theorem.

Trying to improve the operator-norm *rate* of convergence in (7.5.1), Zagrebnov [Zag95] (cf. [Zag23]) proved for a couple A, B of self-adjoint positive generators of C_0-semigroups, where operator $B \in \mathcal{L}(\mathfrak{H})$ is *bounded*, that the rate of convergence of the Trotter product formulæ (7.5.1) has the upper operator-norm bound of the order $O\,(\ln(n)^2/n)$. Note that for these A and B, the conditions of the Rogava theorem are certainly satisfied. The method proposed in [Zag95] was particularly useful later for the proof of the operator-norm Trotter product formula in a *Banach* space; see [CZ01a, Theorem 3.6] and [Zag23, Section 3.2]. We also refer to Theorem 10.2.6 (10.2.13) and Notes to Chap. 10 (Notes to Sect. 10.2).

The case of *unbounded* self-adjoint positive operator B was a subject of the paper [IT97] by Takashi Ichinose and Hideo Tamura. To this aim they supposed that $\mathrm{dom}\,(A^\alpha) \subset \mathrm{dom}\,(B)$ for some $\alpha \in [0, 1)$ (the IT-*condition*) that yields for the estimate of the upper operator-norm bound of the rate of convergence the order $O(1/n^{2/(3+\alpha)})$. Note that the IT condition imposes the *infinitesimal* relative A-smallness of unbounded operator B (see Definition 1.9.7 and (1.9.18)), that is, $B \in \mathcal{P}_{0^+}(A)\backslash\mathcal{L}(\mathfrak{H})$, for $\alpha \in (0, 1)$; see [Yos65]. If $\alpha = 0$, the operator $B \in \mathcal{L}(\mathfrak{H})$.

After that Neidhardt and Zagrebnov [NZ98] amended this statement. They proved that if *unbounded* self-adjoint positive generator B is A-bounded *relative bound* less than *one* (Definition 1.9.6 and Definition 1.9.7), that is, $B \in \mathcal{P}_1(A)$, then the upper operator-norm bound of the rate of convergence has instead of (7.5.1) the order $O\,(\ln(n)/n)$ for the Trotter product formulæ (7.1.84) as well as for the Trotter-Kato product formulæ (7.1.62) and (7.1.80). Note that $\mathcal{P}_{0^+}(A) \subset \mathcal{P}_1(A)$ and that by *stability* of the self-adjointness with respect to a (*symmetric*) perturbation $B \in \mathcal{P}_1(A)$, [Kato80, Theorem 4.3, Chapter V, §4], the conditions of the Rogava theorem are clearly satisfied for the both cases of perturbations studied in [IT97] and in [NZ98].

An instructive for activity around Rogava's Theorem 7.5.1 was at that time an example (due to Helffer [Hel95]) of the C_0-semigroup generated by the Schrödinger operator $H = -\Delta/2 + V(x)$, $x \in \mathbb{R}^d$. Here $V(x)$ is a C^∞-*non-negative* potential growing, when $\|x\| \to \infty$, as $O(\|x\|^\rho)$ for $\rho = 2$. Then taking a self-adjoint $A := (-\Delta/2)_{s-a}$ and $B := V(\cdot)$, he proved that the *symmetrised* Trotter product formula in (7.5.1) converges with the upper operator-norm bound for the rate of convergence, which has the order $O(1/n)$. Notice that as soon as $\rho > 0$ the spectrum $\sigma(H)$ of the positive self-adjoint operator $H = A+B$ is purely discrete (e.g. [Si09]) and $\{\mathrm{e}^{-tH}\}_{t\geq 0}$ is a Gibbs semigroup. Moreover, if potential $V(x)$ satisfies certain conditions of smoothness, then for (Schrödinger) Gibbs semigroups, the bounds for the rate of convergence $O(1/n^a)$, $a = \min\{1, 2/\rho\}$ are *identical* in the *operator-norm* and in the *trace-norm* topologies, [DIT98], [IT98b]. Some abstract results establishing this identity (or *uplifting* of the operator-norm convergence to the trace norm convergence) for self-adjoint as well as for non-self-adjoint Gibbs semigroups and the corresponding Trotter-Kato product formulæ are owing to [NZ99d] and [CZ01c]. See also [Zag19, Section 5.4] and Notes to Chap. 10 (Notes to Sect. 10.2).

In Sect. 7.1 we focused on the operator-norm estimate of the rate of convergence of the Trotter-Kato product formulæ.

To this aim, we present the proofs of ITTZ-Theorem 7.1.9 and Corollary 7.1.12, Corollary 7.1.13. Essentially we follow results appeared for the first time in papers [IT01] and [ITTZ01]. The (ultimate) optimal operator-norm error bound estimate

$O(1/n)$ for the Trotter-Kato product formulæ was proved in [ITTZ01, Theorem 1, Corollary 1] by T. Ichinose, Hideo Tamura, Hiroshi Tamura and V. A. Zagrebnov along the ideas presented in [IT01] by T.Ichinose and Hideo Tamura, which initially provided for the Trotter-Kato product formulæ the operator-norm error bound estimate of order $O(n^{-1/2})$.

Remark 7.1.10 elucidates the key idea of *iterations* that we use for improving the operator-norm rate of the error bound from $O(n^{-1/2})$ [IT01, Section 3, (3.6)], which is based on the *first iteration* (7.1.64), to the rate $O(n^{-1})$, which is due to the *second iteration* (7.1.66) established in [ITTZ01, Lemma 1, (2.6)]. For more comments and remarks, we refer to [IT02, Section 3]. Note that this idea allows also to refine for our purposes the *operator-norm* version of the Chernoff theorem [Cher74, Theorem 1.1]. This refinement was proposed in [NZ99b, Theorem 2.2] and then developed in [IT01, Lemma 2.1].

In Appendix D, we clarify the importance of condition $\omega[f] < \infty$ for generalisation of the Trotter product formula to the Trotter-Kato product formulæ, see [IT01, Section 4] and [IT05]. We also refer to [Zag19, Appendix D] for review on a variety of the Kato functions.

Therefore, the ITTZ-Theorem 7.1.9 and Corollary 7.1.12 and Corollary 7.1.13 generalise and improve results owing to [Rog93, Theorem (I)], [NZ98, Theorem 2.7] or [Hel95]. Moreover, they also cover a number of other known abstracts, as well as related to semigroups with Schrödinger generators assertions about the convergent in the operator-norm topology product formulæ for couples A and B, when the sum of operators, $A + B$, is *self-adjoint*.

Remark 7.5.2 Note that the main condition in the ITTZ-Theorem 7.1.9, i.e. that the operator $C := A + B$ on $\mathrm{dom}\,(A) \cap \mathrm{dom}\,(B)$ rests *self-adjoint* if A and B are self-adjoint, is quite subtle, and it *cannot* be relaxed without supplementary settings. To this end we refer to the *Hiroshi Tamura* Example H.0.5 ([Tam00, Propositions 1–3]), in which the operator-norm convergence of the Trotter product formula does not hold, cf. (7.1.84)

$$\|(\mathrm{e}^{-tA/n}\mathrm{e}^{-tB/n})^n - \mathrm{e}^{-t\overline{C}}\| \geq D_t\,, \quad D_{t>0} > 0\,, \quad n \in \mathbb{N}\,, \tag{7.5.2}$$

even though the operator sum $C = A + B$ is *essentially* self-adjoint on $\mathrm{dom}\,(A) \cap \mathrm{dom}\,(B)$. Here the self-adjoint operator B may be even A-form bounded with a relative bound less than one, and $\overline{C}$ is the closure of operator $A+B$. See Appendix H, Lemma H.0.7, (H.0.17) and comments in Remark H.0.9, 2.

Of course, in the latter case, the product formula (7.5.2) still converges *strongly* according to the Trotter product theorem [Trot59, Lemma, (5)]; see also Theorem 5.1.1 and Corollary 5.1.2. Therefore, (7.5.2) proves *necessity* of self-adjointness of operator $A+B$ for the operator-norm convergence of the Trotter product formula.

We conclude by emphasising that unless the sum $A + B$ is self-adjoint on domain $\mathrm{dom}\,(A) \cap \mathrm{dom}\,(B)$, the operator-norm convergence of the Trotter product formula does *not* always hold, even though the sum is *essentially* self-adjoint there and B is A-form-bounded with relative bound less than one. However, it *does* under complementary, for example, *fractional* conditions, which relate generators, cf. Sect. 7.4. This fact has been pointed out by Hiroshi Tamura in [Tam00]; see Appendix H, Example H.0.5.

The question of optimality of the rate $O(1/n)$ should be studied separately; see Sect. 7.2, as well as Sect. 7.4 and Appendix H. They include the *fractional power* condition for $\alpha \in (1/2, 1)$ that modifies the optimal rate of the operator-norm convergence to $O(1/n^{2\alpha-1})$.

For classification of the Kato functions, see [Zag19, Appendix C]. The condition $\varkappa[f] < \infty$ (7.1.2) selects a proper subclass, $\widehat{\mathcal{K}}_{\beta=2} \subset \mathcal{K}$, of the set of generic Kato functions $\mathcal{K}$ [Zag19, Definition C.3].

Note that condition $\omega[f] < \infty$ is nontrivial. The functional (7.1.3) for $f \in \mathcal{K}$ may be infinite. In contrast to $\varkappa[f] < \infty$ (7.1.2), which is monitoring the limit $x \to 0$, condition $\omega[f] < \infty$ controls a behaviour of the "tail" of function $f \in \mathcal{K}$ "away from zero", namely, for $x \geq \delta$ and arbitrary $\delta > 0$; see Lemma D.0.1 and Remark D.0.2 in Appendix D. This condition also selects a proper *subclass* of the Kato functions $\mathcal{K}$.

In reference [Zag19] the corresponding condition that controls the "tail" of function $f \in \mathcal{K}$ yields the proper subclass $\mathcal{K}_\alpha \subset \mathcal{K}$, [Zag19, Definition C.1].

Notes to Sect. 7.2

The question about *optimality* is quite subtle (cf. Remark 7.2.4) and concerns the estimate of a *lower* bound for operator-norm convergent product fromulæ. So, in Sect. 7.2 we study it only for symmetric approximants and exponential Kato functions with generators A and B satisfying conditions of the ITTZ-Theorem 7.1.9, to compare the results with the Lie-Trotter product formula in Proposition 1.10.10.

In fact, for the proof of optimality of the rate of the operator-norm convergence for a given conditions on generators $\{A, B\}$ and on the symmetry of product approximants, it is sufficient to provide an example of the couple $\{A, B\}$ and generic Kato functions $\{f, g\}$ such that the order of the operator-norm *lower* bound for convergence of the Trotter-Kato product formula *coincides* with the order of the operator-norm estimate of the *upper* bound for the rate of convergence.

Since for *symmetric* Lie-Trotter product formula and for *bounded* operators $\{A, B\}$ the rate of convergence is $O(1/n^2)$ (1.10.37), to check optimality of the *slower* rate $O(1/n)$ (Corollary 7.1.13), one has to consider *unbounded* operators $\{A, B\}$. To this end we present in Example 7.2.1 (motivated by [Tam00, Section 2, Example], cf. Appendix H, Example H.0.5), a couple of self-adjoint A and B such that operator $A + B$ is also self-adjoint (i.e. satisfying the ITTZ-Theorem), and we prove in Theorem 7.2.3 for the bound of the rate of operator-norm convergence for *symmetric* Trotter formula the estimate of order $O(1/n)$ from *below* (7.2.17).

As a consequence the conclusion is twofold:

- In contrast to the case of bounded generators (Proposition 1.10.10), the symmetrisation of the product approximants does not improve the rate of convergence.

- Theorem 7.2.3 together with the ITTZ-Theorem 7.1.9 yield *optimality* of the rate of convergence of order $O(1/n)$ in the framework of conditions of the ITTZ-Theorem for the both cases of symmetric and non-symmetric approximants.

A subtlety of these conditions we scrutinise in Remark 7.5.2 and Remark 7.5.3.

In presentation of Sect. 7.2 we essentially follow [Tam00, Section 2, Example], and [IT01, Sections 1 and 2], [ITTZ01, Section 4].

Remark 7.5.3 The problem of *optimality* was thoroughly treated by Takashi Ichinose and Hideo Tamura in [IT04] for Schrödinger C_0-semigroups on $\mathcal{H} = L^2(\mathbb{R}^d)$. In [IT04] it is proved that for generators $A = -\Delta$ and $B = V$, where V is a non-negative function $\mathbb{R}^d \ni x \mapsto V(x)$ growing polynomially (sufficiently fast) at infinity and hence verifying the conditions of Corollary 7.1.13, the *symmetric* Trotter product formula converges in the operator norm with the rate $O(1/n^2)$, cf. with $O(1/n)$ in [Hel95].

This result advises against a *prejudice* that for *any* pair of noncommuting unbounded operators A and B satisfying the conditions of ITTZ-Theorem 7.1.9, the rate of convergence $O(1/n)$ is necessarily the *best* possible for symmetric Trotter product formula. In fact, if one imposes on A and B in ITTZ-Theorem some *additional* conditions, for example, *compactness* of the operator $(\mathbb{1} + A)^{-1}(\mathbb{1} + B)^{-1}$, cf. Sect. 6.3, then symmetrisation could improve the operator-norm bound of the rate of convergence up to the asymptotic $O(1/n^2)$ as in the *S. Lie* product formula !

Moreover, in [AzIch08], taking a simple example of the one-dimensional harmonic oscillator, $H = A+B$, with $A = -\partial_x^2$, $B = V(x) = x^2$ in $\mathcal{H} = L^2(\mathbb{R})$, it was shown by explicit calculations that the operator-norm estimates for the symmetric Trotter product formula from *above* and from *below* are of the order $O(1/n^2)$

$$\frac{c}{n^2} \leq \left\| \left(\mathrm{e}^{-tB/2n}\mathrm{e}^{-tA/n}\mathrm{e}^{-tB/2n}\right)^n - \mathrm{e}^{-tH} \right\| \leq \frac{C}{n^2}, \quad n \in \mathbb{N}, \tag{7.5.3}$$

for $0 < c < C$ and locally uniformly in $t \in \mathbb{R}$, i.e. *away* from zero. Hence, the rate of convergence $O(n^{-2})$ is (ultimate) optimal in this case. For more details, see review [IT09].

Notes to Sect. 7.3

In this section we develop in detail the proof of generalisation of the ITTZ-Theorem 7.1.9 to the case of more than *two* generators. At first this generalisation appeared in [ITTZ01, Theorem 2], and here we essentially followed ideas and the line of reasoning presented there.

We note that the proof of the ITTZ-Theorem 7.3.9 for more than two generators is not an immediate consequence of the ITTZ-Theorem 7.1.9 for two generators even in the case of the *symmetric* (i.e. self-adjoint) approximants (7.3.59). In contrast to that, a specificity of the case of more than two generators indicated in Remark 7.3.10 does not allow a direct extension to non-symmetric (i.e. non-self-adjoint) approximants.

Notes to Sect. 7.4

Sections 7.4.1 and 7.4.2

The idea to study the operator-norm convergence of the Trotter-Kato product formulæ for self-adjoint generators A and B, such that generator $A \dotplus B$ of the limiting C_0-semigroup is a *form-sum*, is dated back to [NZ99a]. This paper describes a dependence of the operator-norm rate of convergence through relations between fractional powers of generators A and B.

Then *Hiroshi Tamura* proved in [Tam00, Section 2, Example] that for $\alpha \in (1/2, 1)$ the error bound estimate in [NZ99a, Theorem 5.5], and thus in our INZ-Theorem 7.4.7, is *optimal*, whereas for $\alpha \in (0, 1/2]$ convergence holds only in the strong operator topology, cf. Remark 7.4.8 and Example H.0.5 in Appendix H.

In contrast to direct estimates used in the predecessor [NZ99a] of our paper [INZ04], we followed in the latter the operator-norm Chernoff theory for product approximations developed in [NZ99b, Theorem 2.3]. Therefore, we used an improved version of this theory ([INZ04, Theorem 2.1]) with operator-norm error bound estimates, cf. Chap. 4 (Sect. 4.2). The paper [INZ04, Theorem 1.1] improves [NZ99a, Theorem 5.5]

- by relaxing the *smallness* condition on B^α with respect to A^α, for $\alpha \in (1/2, 1)$, to a *milder* assumption $\mathrm{dom}\,(A^{1/2}) \subseteq \mathrm{dom}\,(B^{1/2})$ of subordination,
- by extending essentially the admissible class of Kato functions.

In our presentation, see INZ-Theorem 7.4.7, we mainly follow [INZ04].

Section 7.4.3

For establishing in Theorem 7.4.10 an equivalence of the statement in INZ-Theorem 7.4.7 to other families of approximants, we follow [INZ04] (Sections 5 and 6).

Section 7.4.4

Example 7.4.11 in Sect. 7.4.4 is taken from [INZ04, Section 6]. It serves for illustration of relaxation of a subordination corresponding to the *relative boundedness* condition for generators A and B :

$$\|B^\alpha u\| \leq a\|A^\alpha u\| + b\|u\|, \quad u \in \mathrm{dom}\,(A^\alpha), \tag{7.5.4}$$

for some $\alpha \in (1/2, 1)$ and $a \in (0, 1)$, $b > 0$. It is imposed in [NZ99a, Section 1, (1.18), (1.20)] to infer the optimal rate of the operator-norm convergence of the Trotter-Kato product formulæ [NZ99a, Theorem 5.5].

Notes about Example 7.4.12, [NZ99a, Section 6].

1. We show that the operator-norm error bound estimate depends on $1/2 < \alpha < 1$, which orders $\mathrm{dom}\,(A^\alpha)$ and $\mathrm{dom}\,(B^\alpha)$, or $\mathrm{dom}\,(H^\alpha)$ and $\mathrm{dom}\,(A^\alpha)$; see [NZ99a, Section 1] and comments in Appendix H.
2. Example 7.4.12 illustrates this dependence and, in particular, the limit case $\alpha = 1$ considered in [NZ98].

3. We would like to note that for $\alpha = 1/2$ the Trotter-Kato formula is not considered here. It needs a special analysis (cf. [NZ99b] and [Tam00]), which shows that the operator-norm convergence can be ensured if one imposes some supplementary conditions on A and B.
4. Examples coming from the Trotter product formula for Schrödinger semigroups show [DIT98] that one can get error bounds of order $O(n^{-1})$ even in the case when $\mathrm{dom}\,(A^{\alpha}) \not\subseteq \mathrm{dom}\,(B^{\alpha})$ for $1/2 < \alpha \leq 1$, which means that the present abstract setting is not exhaustive.

Part III
Trotter-Kato Product Formulæ for Non-Self-Adjoint Semigroups

Chapter 8
Operator-Norm Approximation Theory à la Chernoff

This chapter is devoted to development of operator-norm approximations of the product formulæ for *non-self-adjoint* semigroups. To this aim we use here an advance in the *Chernoff theory* for the operator-norm approximations launched in Chap. 3 for general along with *quasi-sectorial* contractions and expounded in Chap. 4 for the case of self-adjoint semigroups.

The subject of this chapter is a demonstration that in a Hilbert space it is possible to generalise the self-adjoint operator-norm Chernoff product formula (Chap. 4) to the case of non-self-adjoint, but *quasi-sectorial* contractions introduced in Chap. 3. The improved operator-norm approximation theory *à la Chernoff* for the non-self-adjoint semigroups consists of two items: an *operator-norm* extension of the Chernoff $\sqrt{n}$-Lemma to non-self-adjoint contractions and the *operator-norm* generalisation of the non-self-adjoint *Trotter-Kato-Neveu* approximation theorem.

8.1 Advance in the Chernoff Theory for Quasi-sectorial Contractions

We note that Chap. 4 was devoted for extension of the Chernoff approximation theory to the operator-norm topology for self-adjoint semigroups. The aim of this chapter is to show that in a Hilbert space one can generalise this theory, first of, all to non-self-adjoint *quasi-sectorial* contractions introduced in Chap. 3.

Recall that Chernoff has proved (see Theorem 1.9.34 in Sect. 1.9.6) the following product formula for C_0-semigroups, which converges on a Banach space $\mathfrak{X}$ in the strong operator topology.

Proposition 8.1.1 *Let $t \mapsto \Phi(t)$ be a strongly continuous function from $\mathbb{R}_0^+ = [0, \infty)$ to linear contractions on $\mathfrak{X}$ such that $\Phi(0) = \mathbb{1}$. Suppose that the right strong derivative $\Phi'(+0)$ exists on a dense domain in $\mathfrak{X}$ and its closure is (minus) generator: $(-C)$, of a strongly continuous contraction semigroup. Then product:*

V. A. Zagrebnov et al., *Trotter-Kato Product Formulæ*, Operator Theory: Advances and Applications 296, https://doi.org/10.1007/978-3-031-56720-9_8

$\Phi(t/n)^n$, converges for $n \to +\infty$ to this C_0-semigroup $\{e^{-tC}\}_{t\geq 0}$ in the strong operator topology on $\mathcal{L}(\mathfrak{X})$.

This result describes a method for approximations of general contraction semigroups. It leads in particular to the proof of the *strong* convergence for the Trotter product formula, Sect. 5.1. For *self-adjoint* families of contractions on a Hilbert space, there exists a generalisation of the above result to convergence in the *operator-norm* topology that we presented in Sect. 6.1. It can be resumed as the following statement for convergence of the *Chernoff* product approximants $\{\Phi(t/n)^n\}_{n\geq 1}$ to C_0-semigroups, including, for example, *degenerate* semigroups (Definition 1.6.1 and Proposition 1.6.2).

Proposition 8.1.2 *Let $\{\Phi(s)\}_{s\geq 0}$ be a family of self-adjoint contractions on a Hilbert space $\mathfrak{H}$, and let X_0 be a self-adjoint operator in the closed subspace $\mathfrak{H}_0 \subseteq \mathfrak{H}$. Let $X(s) := s^{-1}(\mathbb{1} - \Phi(s))$, $s > 0$. Then the sequence $\{\Phi(t/n)^n\}_{n\geq 1}$, $t \geq 0$, converges in the operator norm to $e^{-tX_0} P_0$ as $n \to +\infty$, locally-uniformly in $t \in \mathbb{R}$, if and only if the family $\{X(s)\}_{s>0}$ converges in the uniform resolvent (i.e. in the operator-norm) sense to X_0 as $s \to +0$. Here P_0 denotes the orthogonal projection from $\mathfrak{H}$ onto subspace $\mathfrak{H}_0$ for degenerate semigroup $\{e^{-tX_0} P_0\}_{t\geq 0}$.*

The aim of the present chapter is to show that condition of *self-adjointness* in Proposition 8.1.2 can be replaced by a much *weaker* condition on the corresponding numerical range of the non-self-adjoint family $\{\Phi(s)\}_{s\geq 0}$ (Theorem 8.3.4) and that this leads to the notion of *quasi-sectorial* contractions introduced in Chap. 3.

Consequently, the non-self-adjoint operator-norm results of Chap. 8 are a further development of Chap. 3. Hence, it will be our starting point, which also includes the proofs of assertions (Theorems 3.1.13 and 3.1.16) promised in Sect. 3.1.3 of Chap. 3 and indispensable for application below.

Let $\{\Phi(s)\}_{s\geq 0}$ be a family of quasi-sectorial contractions on a Hilbert space $\mathfrak{H}$ such that numerical range $W(\Phi(s)) \subseteq D_\alpha$ (see Definition 1.8.1 and (3.1.1)) for some $0 \leq \alpha < \pi/2$ and for all $s > 0$. Let $X(s) := (\mathbb{1} - \Phi(s))/s$, and let X_0 be a closed operator with a non-empty resolvent set in a subspace $\mathfrak{H}_0 \subseteq \mathfrak{H}$. Then the main result of this chapter may be formulated as the equivalence between two operator-norm limits:

$$\|(\zeta\, \mathbb{1} + X(s))^{-1} - (\zeta\, \mathbb{1} + X_0)^{-1} P_0\| \longrightarrow 0\,, \quad \text{as } s \to +0\,, \tag{8.1.1}$$

$$\|\Phi(t/n)^n - e^{-tX_0} P_0\| \longrightarrow 0\,, \quad \text{as } n \to \infty,\ t > 0\,, \tag{8.1.2}$$

for some (and hence for any) $\zeta \in S_{\pi-\alpha}$; see Theorem 8.3.4.

Our method is an operator-norm advance in the semigroup approximation theory *à la* Chernoff. It basically consists of *two* key steps. To this end we recall that if operators $\{\Phi(s)\}_{s\geq 0}$ are not only contraction, but in addition a quasi-sectorial contractions, then estimate in the Chernoff $\sqrt{n}$ -Lemma may be drastically improved. As a consequence we obtain a convergent operator-norm estimate; see Remark 3.1.15. We remind that Chernoff's original estimate (Lemma 1.9.33): $\|(C^n - e^{n(C-\mathbb{1})})\| \leq \sqrt{n}\, \|(C - \mathbb{1})\|$, does converge for large n.

Our *first* step is an improvement of the Chernoff $\sqrt{n}$ -Lemma to the operator-norm $1/n^{1/3}$-estimate for the non-self-adjoint, but quasi-sectorial contraction case. Motivated by the $(\sqrt[3]{n})^{-1}$-Lemma 3.1.14 and Remark 3.1.15 (a)-(b), we prove this statement in Theorem 8.2.2.

As the *second* step (see Sect. 8.3) we consider the operator-norm extension of the *Trotter-Kato-Neveu* convergence theorem (Proposition 1.9.30) for quasi-sectorial contractions. To this aim we consider uniformly m-sectorial families of generators for holomorphic semigroups. The Corollary 8.3.3 shows that convergence in the *uniform resolvent sense* (operator-norm topology) (8.1.1) (see Remark 1.9.31) is well-adapted to approximate semigroups in the operator-norm topology (8.1.2). Note that this topology is natural for holomorphic contraction semigroups since they are holomorphic in the operator-norm topology in some open sector S_ω, for $\omega < \pi/2$, and generators of these semigroups are exactly the m-sectorial operators with semi-angle $\alpha = \pi/2 - \omega$; see Proposition 1.8.13.

These two steps allow to prove in Theorem 8.3.4 the equivalence between two limits (8.1.1) and (8.1.2). The second limit there is an *operator-norm* generalisation of the Chernoff product formula for approximation of semigroups. Note that this formula was established before in the strong operator topology (Theorem 4.1.3) as well as in the operator-norm topology (Theorem 4.2.7), but only for the *self-adjoint* semigroups. Note that in the latter case, we also establish the rate of convergence (Theorem 4.3.6) for the Chernoff product formula (8.1.2).

8.2 Non-Self-Adjoint Refinement of the Chernoff $\sqrt{n}$ -Lemma

As we demonstrated in Sect. 4.2, the estimate in the Chernoff $\sqrt{n}$ -Lemma 1.9.33 (for general contractions) can be drastically improved up to the *optimal* estimate with bound $O(1/n)$:

$$\|C^n - e^{n(C-\mathbb{1})}\| \leq \frac{1}{n}, \quad n \in \mathbb{N}, \tag{8.2.1}$$

if one utilises in (8.2.1) the *functional calculus* for self-adjoint contraction C.

In Sects. 1.9.6 and 3.1.2, it is shown that the Chernoff $\sqrt{n}$-Lemma 1.9.33 and its revised version, the n-Lemma 3.1.9, are sufficient to prove convergence of the Chernoff product formula in the strong operator topology (Theorems 1.9.34, 3.1.10). As a consequence, using a *diverging* $\sqrt{n}$-estimate, or n-estimate (see (1.9.96), or (3.1.12)) owing to Proposition 1.9.35, one obtains the strongly convergent *Lie-Trotter* product formula for general contraction semigroups.

In Sect. 3.1.2 we improved the non-self-adjoint Chernoff $\sqrt{n}$ -Lemma 1.9.33 for contraction C to the $\sqrt[3]{n}$ -Lemma 3.1.11 (still for diverging estimate (3.1.26)) and established the strongly convergent Chernoff product formula (3.1.35) in Theorem 3.1.12.

On the other hand, we announced in Sect. 3.1.2 that for non-self-adjoint, and quasi-sectorial contraction C (Definition 3.1.2), thanks to Theorem 3.1.13, one can obtain the operator-norm *converging* estimate (8.2.1) with asymptotic $O(1/n^{1/3})$, for $n \to \infty$; see Lemma 3.1.11.

Note that the self-adjoint contractions are a particular case of the quasi-sectorial contractions corresponding to $\alpha = 0$ in Definition 3.1.2, and that in contrast to $O(1/n)$, the estimate $O(1/n^{1/3})$ is not optimal; see Notes to Chap. 3 (Notes to Sect. 3.3) and to Chap. 8 (Notes to Sect. 8.6).

We present the proof of this uplifted operator-norm converging estimate for quasi-sectorial contractions in the $(\sqrt[3]{n})^{-1}$-Theorem 8.2.2.

Lemma 8.2.1 *Let C be a quasi-sectorial contraction on a Hilbert space $\mathfrak{H}$, that is, there exists $0 \leq \alpha < \pi/2$ such that the numerical range $W(C)$ is a subset of the domain D_α* (3.1.1)*, see* Fig. 8.1. *Then for $\alpha < \alpha' < \pi/2$*

$$\|C^n(\mathbb{1} - C)\| \leq \frac{K_{\alpha,\alpha'}}{n+1}, \quad n \in \mathbb{N}, \tag{8.2.2}$$

where $K_{\alpha,\alpha'} > 0$ is given by (8.2.7).

Proof Since operator C is a quasi-sectorial contraction, the spectrum $\sigma(C)$ is a subset of the closure $\overline{W(C)}$ (Proposition 1.8.5) of the numerical range $W(C) \subseteq D_\alpha$ (3.1.1), where

$$D_\alpha = \{z \in \mathbb{C} : |z| \leq \sin\alpha\} \cup \{z \in \mathbb{C} : |\arg(1-z)| \leq \alpha \text{ and } |z-1| \leq \cos\alpha\}. \tag{8.2.3}$$

So, taking $\alpha < \alpha' < \pi/2$, one gets (see Fig. 8.1) that $D_{\alpha'} \supset D_\alpha$. Hence, contour $\partial D_{\alpha'}$ is outside of D_α, but inside of the unit disc $\mathbb{D}$, and all of them have only one common point at $z = 1$. Then the Riesz-Dunford functional calculus provides the following representation of the operator in (8.2.2):

$$C^n(\mathbb{1} - C) = \frac{1}{2\pi i}\int_{\partial D_{\alpha'}} dz\, \frac{z^n(1-z)}{z\mathbb{1} - C}. \tag{8.2.4}$$

If z belongs to resolvent set $\rho(C)$ of C, then evaluation of the norm of resolvent: $\|(C - z\mathbb{1})^{-1}\| \leq \text{dist}(z, \overline{W(C)})^{-1}$, yields $\|(z\mathbb{1} - C)^{-1}\| \leq \text{dist}(z, D_\alpha)^{-1}$ for $z \in \partial D_{\alpha'} \subset \rho(C)$ in (8.2.4). We consider the following parametrisation of the positively oriented contour $\partial D_{\alpha'}$:

- for the arc(A, B) with end-points at $A = e^{i(\pi/2-\alpha')}\sin\alpha'$ and at $B = e^{i(3\pi/2+\alpha')}\sin\alpha'$, we take $z(t) = e^{it}\sin\alpha'$ with $\pi/2 - \alpha' \leq t \leq 3\pi/2 + \alpha'$;
- for the straight lines $(1, A)$ and $(B, 1)$, we take correspondingly $z_-(s) = 1 - s\,e^{-i\alpha'}$ with $s \in [0, \cos\alpha']$ and $z_+(s) = 1 - s\,e^{+i\alpha'}$ with $s \in [\cos\alpha', 0]$.

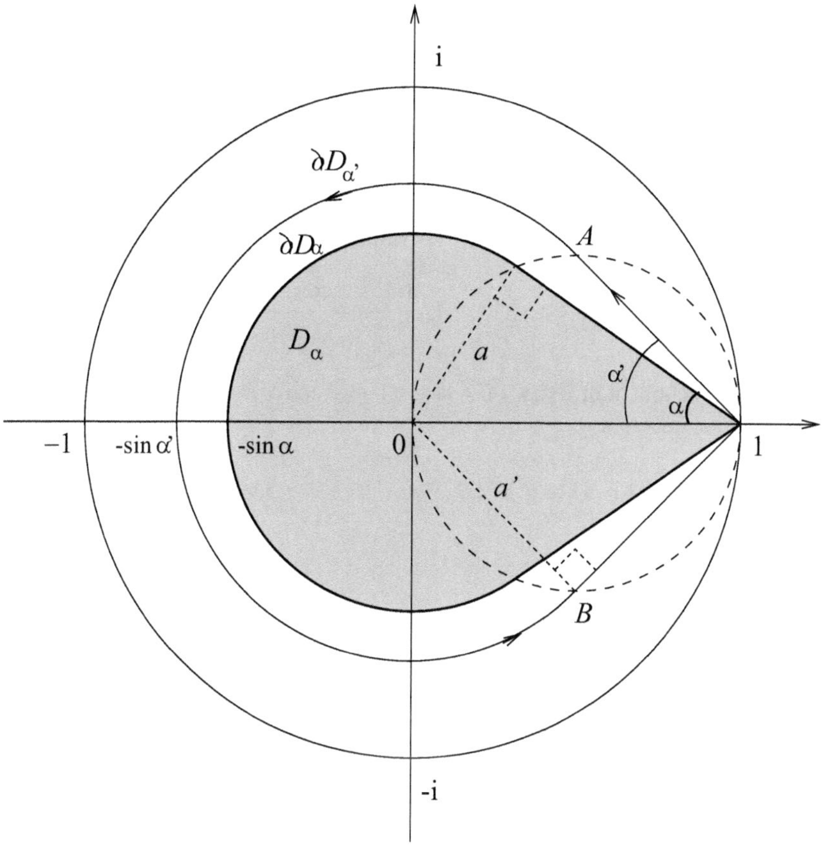

Fig. 8.1 Illustration of the set D_α (shaded domain inside of the unit disc $\mathbb{D}$) with boundary ∂D_α, where $a = \sin\alpha$, as well as of our choice of the contour $\partial D_{\alpha'}$ in the resolvent set $\rho(C)$, where $a' = \sin\alpha' > a$. The contour $\partial D_{\alpha'}$ consists of two segments of tangent straight lines $(1, A)$ and $(1, B)$ and the arc(A, B) of radius a'. The dotted circle corresponds to the set of tangency points for the different values of $\alpha \in [0, \pi/2)$

As a consequence, by definition of the (*shortest*) distance *from* $z \in \partial D_{\alpha'}$ *to* D_α, denoted as dist(z, D_α), we obtain:

- $\|(z\mathbb{1} - C)^{-1}\| \leq (\cos\alpha' \sin(\alpha' - \alpha))^{-1}$ for $|\arg z| \geq \pi/2 - \alpha'$, where we used that dist$(z, D_\alpha) \in [\cos\alpha' \sin(\alpha' - \alpha), \ (\sin\alpha' - \sin\alpha)]$ for $z \in$ arc(A, B), that is, $z \in \{e^{it}\sin\alpha'\}_{t\in[\pi/2-\alpha',\ 3\pi/2+\alpha']}$;
- $\|(z\mathbb{1} - C)^{-1}\| \leq (|1 - z|\sin(\alpha' - \alpha))^{-1}$ for $|\arg z| \leq \pi/2 - \alpha'$, that is, for $z \in \{1 - s\, e^{\mp i\alpha'}\}_{s\in[0,\ \cos\alpha']}$.

Then operator-norm estimate of the left-hand side in representation (8.2.4) takes the form

$$\begin{aligned}\|C^n(\mathbb{1}-C)\| &\le \frac{1}{2\pi}\int_{\frac{\pi}{2}-\alpha'}^{\frac{3\pi}{2}+\alpha'} dt\, \frac{|\sin\alpha'|^{n+1}|1-e^{it}\sin\alpha'|}{\cos\alpha' \sin(\alpha'-\alpha)} + \\ &+ \frac{1}{\pi}\int_0^{\cos\alpha'} ds\, \frac{|(1-e^{i\alpha'}s)^n e^{i\alpha'}s|}{s(\sin(\alpha'-\alpha))} \le \qquad (8.2.5)\\ &\le \frac{2(\sin\alpha')^{n+1}}{\cos\alpha' \sin(\alpha'-\alpha)} + \int_0^{\cos\alpha'} ds\, \frac{\left((1-s\cos\alpha')^2 + s^2(\sin\alpha')^2\right)^{n/2}}{\pi\sin(\alpha'-\alpha)} .\end{aligned}$$

Taking into account convexity of $s \mapsto (1 - s\cos\alpha')^2 + s^2(\sin\alpha')^2$, for $s \in [0, \cos\alpha']$, one gets that

$$(1 - s\cos\alpha')^2 + s^2(\sin\alpha')^2 \le 1 - s\cos\alpha' ,$$

which leads to inequality:

$$\begin{aligned}\int_0^{\cos\alpha'} ds\, \left((1-s\cos\alpha')^2 + s^2(\sin\alpha')^2\right)^{n/2} &\le \int_0^{\cos\alpha'} ds\, (1-s\cos\alpha')^{n/2} \\ &= \int_{(\sin\alpha')^2}^{1} du\, \frac{u^{n/2}}{\cos\alpha'} \le \frac{1-(\sin\alpha')^{n+2}}{(n/2+1)\cos\alpha'} .\end{aligned}$$

Therefore, by (8.2.5) we obtain the estimate

$$\begin{aligned}\|C^n(\mathbb{1}-C)\| &\le \frac{2\,(\sin\alpha')^{n+1}}{\cos\alpha'\,\sin(\alpha'-\alpha)} + 2\,\frac{1-(\sin\alpha')^{n+2}}{\pi\,(n+2)\,\cos\alpha'\,\sin(\alpha'-\alpha)} \le \\ &\frac{2}{(n+1)\,\cos\alpha'\,\sin(\alpha'-\alpha)}\left(\frac{1}{\pi} + (n+1)\,(\sin\alpha')^{n+1}\right) . \qquad (8.2.6)\end{aligned}$$

Now, after *optimisation* of the last factor in the right-hand side of (8.2.6) with respect to $n \in \mathbb{N}$, we infer (8.2.2) for

$$K_{\alpha,\alpha'} := \frac{2}{\cos\alpha'\sin(\alpha'-\alpha)}\left(\frac{1}{\pi} - \frac{1}{e\,\ln(\sin\alpha')}\right) , \qquad (8.2.7)$$

where $\alpha < \alpha' < \pi/2$. □

The next assertion is the *first* step of the programme outlined in Sect. 8.1. It is a rectification of the Chernoff $\sqrt{n}$ -Lemma 1.9.33.

Theorem 8.2.2 (**($(\sqrt[3]{n})^{-1}$-Theorem**) *Let C be a quasi-sectorial contraction on $\mathfrak{H}$ with numerical range $W(C) \subseteq D_\alpha$, $0 \le \alpha < \pi/2$. Then*

$$\left\| C^n - e^{n(C-\mathbb{1})} \right\| \le \frac{M_\alpha}{n^{1/3}}, \quad n = 1, 2, 3, \dots, \tag{8.2.8}$$

where $M_\alpha = 2\,K_\alpha + 2$ and $K_\alpha := \min_{\alpha' \in (\alpha, \pi/2)} K_{\alpha,\alpha'}$, is defined by (8.2.7).

Proof Since the operator C is a contraction, we have the representation:

$$C^n - e^{n(C-\mathbb{1})} = e^{-n} \sum_{m=0}^{\infty} \frac{n^m}{m!} (C^n - C^m), \quad n \in \mathbb{N}. \tag{8.2.9}$$

Let $\epsilon_n := n^{2/3}$, which corresponds to the optimal $\delta = 1/6$ in Remark 3.1.15 (a). Then similar to arguments in Lemma 3.1.11, we split the sum (8.2.9) into two parts: the *central* part for $|m - n| \le \epsilon_n$ and the *tails* corresponding to $|m - n| > \epsilon_n$.

To estimate the input of *tails* into sum (8.2.9), we use the *Tchebychëv* inequality, cf. Sect. 3.1.2. Recall that if we consider $\{X_n\}$ as the *Poisson random variable* of parameter n, i.e. that probability: $\mathbb{P}(X_n = m) = n^m e^{-n}/m!$, then expectation $\mathbb{E}(X_n) = n$ and the variance $\mathrm{Var}(X_n) = n$. Therefore, by the *Tchebychëv* inequality, one can estimate the probability of the event $\{|X_n - \mathbb{E}(X_n)| > \epsilon\}$ as

$$\mathbb{P}(|X_n - \mathbb{E}(X_n)| > \epsilon) \le \frac{\mathrm{Var}(X_n)}{\epsilon^2}, \quad \text{for any } \epsilon > 0.$$

Seeing that $\|C^n - C^m\| \le 2$, we can use the *Tchebychëv* inequality for the event $\{|X_n - n| > \epsilon_n\}$, that is, for evaluation of the *tail* parts of the operator norm of the sum (8.2.9). This yields the bound

$$e^{-n} \sum_{|m-n|>\epsilon_n} \frac{n^m}{m!} \|C^n - C^m\| \le \frac{2n}{\epsilon_n^2} = \frac{2}{n^{1/3}}, \quad n \in \mathbb{N}. \tag{8.2.10}$$

To estimate an input of the *central* part of the sum (8.2.9) for $|m - n| \le \epsilon_n$, we use the result of Lemma 8.2.1:

$$\|C^n - C^m\| = \left\| C^{n-[\epsilon_n]} \left(C^{[\epsilon_n]} - C^{m-n+[\epsilon_n]} \right) \right\|$$

$$\le |m - n| \left\| C^{n-[\epsilon_n]} (\mathbb{1} - C) \right\| \le \epsilon_n \frac{K\alpha}{n - [\epsilon_n] + 1} \le \frac{2K\alpha}{n^{1/3}},$$

for a corresponding $n_0 > 0$. Here $[x]$ denotes the integer part of $x \ge 0$. Then we obtain:

$$e^{-n} \sum_{|m-n|\le\epsilon_n} \frac{n^m}{m!} \|C^n - C^m\| \le \frac{2\,K\alpha}{n^{1/3}}, \quad n > n_0.$$

This estimate together with (8.2.10) yield (8.2.8). □

Remark 8.2.3 We note that if *no* estimate of the convergence rate is demanded, then the operator-norm convergence to zero of the difference (8.2.8) follows directly from the *Riesz-Dunford* representation of $C^n - \mathrm{e}^{n(C-\mathbb{1})}$ as integral along the contour $\partial D_{\alpha'}$:

$$C^n - \mathrm{e}^{n(C-\mathbb{1})} = \frac{1}{2\pi i}\int_{\partial D_{\alpha'}} dz\, \frac{z^n - \mathrm{e}^{n(z-1)}}{z - C}\,,$$

and the *Lebesgue* dominated convergence theorem applied to the right-hand side (in the third line) of inequalities:

$$\begin{aligned}\|C^n - \mathrm{e}^{n(C-\mathbb{1})}\| &\leq \frac{1}{2\pi}\int_{\frac{\pi}{2}-\alpha'}^{\frac{3\pi}{2}+\alpha'} dt\, \sin\alpha' \frac{|(\mathrm{e}^{it}\sin\alpha')^n - \exp\{n\,(\mathrm{e}^{it}\sin\alpha' - 1)\}|}{\cos\alpha'\ \sin(\alpha'-\alpha)} + \\ &+ \frac{1}{\pi}\int_0^{\cos\alpha'} ds\, \frac{|(1-\mathrm{e}^{-i\alpha'}s)^n - \exp(-n\,s\,\mathrm{e}^{-i\alpha'})|}{s\ \sin(\alpha'-\alpha)} \\ &\leq \sin\alpha'\, \frac{(\sin\alpha')^n + \mathrm{e}^{n(\sin\alpha'-1)}}{\cos\alpha'\ \sin(\alpha'-\alpha)} \\ &+ \int_0^{n\cos\alpha'} dr\, \frac{|(1-\mathrm{e}^{-i\alpha'}r/n)^n - \exp(-r\,\mathrm{e}^{-i\alpha'})|}{r\ \sin(\alpha'-\alpha)}\,.\end{aligned} \tag{8.2.11}$$

For parametrisation of integrands in (8.2.11), see notes in the proof of Lemma 8.2.1.

Remark 8.2.4 Recall that if the quasi-sectorial contraction C is self-adjoint ($\alpha = 0$), then one obtains instead of (8.2.2) and (8.2.8) the optimal estimates:

$$\|C^n(\mathbb{1} - C)\| \leq \frac{1}{n+1} \quad \text{and} \quad \|C^n - \mathrm{e}^{n(C-\mathbb{1})}\| \leq \frac{1}{n}\,, \tag{8.2.12}$$

directly from the spectral representation of C; see Sect. 4.2.

8.3 Approximation Theorem for the Operator-Norm Topology

In this section we present the *second* step of the programme outlined in Sect. 8.1. It is indispensable for the *operator-norm* generalisation of the Chernoff product formula in the case of non-self-adjoint semigroups. To this end we prove the operator-norm extension of the *Trotter-Kato-Neveu* theorem (Proposition 1.9.30) for the case of quasi-sectorial contractions.

For the self-adjoint case, the corresponding *Trotter-Kato-Neveu* theorem was obtained in Theorem 4.2.6 (Sect. 4.2). Below we prove it for non-self-adjoint, but

m-sectorial generators, that is, for quasi-sectorial contractions, see Theorems 3.1.4 and 3.2.8.

Theorem 8.3.1 *Let $\{X(s)\}_{s>0}$ be a family of m-sectorial operators in a Hilbert space $\mathfrak{H}$ with $W(X(s)) \subseteq S_\alpha$ for some $0 < \alpha < \pi/2$ and for all $s > 0$. Let X_0 be an m-sectorial operator defined in a closed subspace $\mathfrak{H}_0 \subseteq \mathfrak{H}$, with $W(X_0) \subseteq S_\alpha$. Then the two following assertions about operator-norm convergence are equivalent:*

$$(a)\ \lim_{s\to+0} \left\| (\zeta\mathbb{1} + X(s))^{-1} - (\zeta\mathbb{1} + X_0)^{-1}P_0 \right\| = 0\,,\ \ for\ \zeta \in S_{\pi-\alpha}\,,$$

$$(b)\ \lim_{s\to+0} \left\| \mathrm{e}^{-tX(s)} - \mathrm{e}^{-tX_0}P_0 \right\| = 0\,,\ \ for\ t > 0\,.$$

Here P_0 denotes the orthogonal projection from $\mathfrak{H}$ onto $\mathfrak{H}_0$.

Proof *(a) ⇒ (b)*
Since $\{X(s)\}_{s>0}$ are m-sectorial operators with numerical range $W(X(s)) \subseteq S_\alpha$, the *Riesz-Dunford* representation

$$\mathrm{e}^{-tX(s)} = \frac{1}{2\pi i}\int_\Gamma d\zeta\, \frac{\mathrm{e}^{t\zeta}}{\zeta\mathbb{1} + X(s)}\,,$$

defines a family of holomorphic semigroups $\{\mathrm{e}^{-tX(s)}\}_{s>0}$ with $t \in S_{\pi/2-\alpha}$. Here $\Gamma \subset S_{\pi-\alpha}$ is a positively oriented closed (at infinity) path around sector: $-S_\alpha$, that is, with the spectrum $\sigma(-X(s)) \subseteq \overline{W(-X(s))}$ in its interior (see Fig. 8.2). The same is true for the operator X_0:

$$\mathrm{e}^{-tX_0}P_0 = \frac{1}{2\pi i}\int_\Gamma d\zeta\, \frac{\mathrm{e}^{t\zeta}}{\zeta\mathbb{1} + X_0}P_0\,.$$

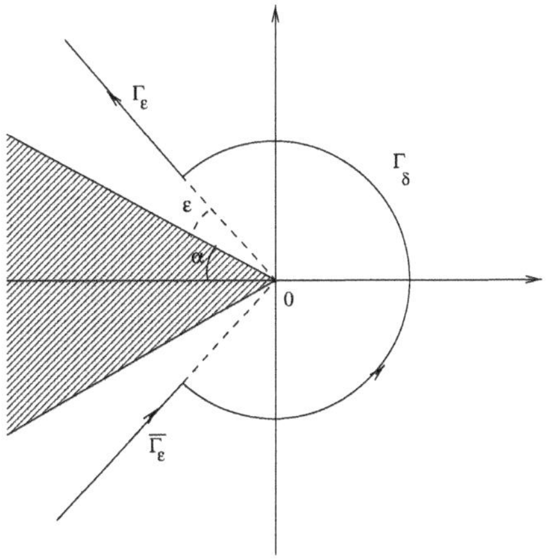

Fig. 8.2 Illustration of the path Γ in integrals

Here we set $\Gamma := \overline{\Gamma_\epsilon} \cup \Gamma_\delta \cup \Gamma_\epsilon$, where the arc $\Gamma_\delta := \{z \in \mathbb{C} : |z| = \delta > 0, \ |\arg z| \leq \pi - \alpha - \epsilon\}$ (for $0 < \epsilon < \pi/2 - \alpha$) and $\Gamma_\epsilon, \overline{\Gamma_\epsilon}$ are two conjugated straight lines, such that $\Gamma_\epsilon := \{z \in \mathbb{C} : \arg z = \pi - \alpha - \epsilon, \ |z| \geq \delta\}$. Then for $t > 0$ one gets

$$\left\| \mathrm{e}^{-tX(s)} - \mathrm{e}^{-tX_0} P_0 \right\| \leq \frac{1}{2\pi} \int_\Gamma |d\zeta| \, |\mathrm{e}^{t\zeta}| \left\| (\zeta \mathbb{1} + X(s))^{-1} - (\zeta \mathbb{1} + X_0)^{-1} P_0 \right\| . \tag{8.3.1}$$

Since operators $X(s)$ and X_0 are m-sectorial, by (1.8.3) one gets the estimate

$$\left\| (\zeta \mathbb{1} + X(s))^{-1} - (\zeta \mathbb{1} + X_0)^{-1} P_0 \right\| \leq \frac{2}{\mathrm{dist}(-S_\alpha, \zeta)} ,$$

which implies

$$\sup_{\zeta \in \Gamma} \left\| (\zeta \mathbb{1} + X(s))^{-1} - (\zeta \mathbb{1} + X_0)^{-1} P_0 \right\| \leq \frac{2}{\delta \sin \epsilon} . \tag{8.3.2}$$

Seeing that for the path $\Gamma = \overline{\Gamma_\epsilon} \cup \Gamma_\delta \cup \Gamma_\epsilon$, the integral $I_t = \int_\Gamma |d\zeta| \, |\mathrm{e}^{t\zeta}|$ converges for any $t > 0$, the condition (a) together with estimates (8.3.1), (8.3.2) and the *Lebesgue* dominated convergence theorem prove (b).

$(b) \Rightarrow (a)$

By the *Laplace* transform of semigroups: $\{\mathrm{e}^{-tX(s)}\}_{t \geq 0}$ and $\{\mathrm{e}^{-tX_0}\}_{t \geq 0}$, we obtain

$$(\zeta \mathbb{1} + X(s))^{-1} - (\zeta \mathbb{1} + X_0)^{-1} P_0 = \int_0^\infty dt \, \mathrm{e}^{-t\zeta} \left(\mathrm{e}^{-tX(s)} - \mathrm{e}^{-tX_0} P_0 \right) ,$$

for $\mathrm{Re}\, \zeta > 0$. Because generators $\{X(s)\}_{s>0}$ and X_0 are m-sectorial (with vertex at zero), the corresponding semigroups are contractions: $\| \mathrm{e}^{-tX(s)} \| \leq 1$, $\| \mathrm{e}^{-tX_0} P_0 \| \leq 1$, for $t \geq 0$ and $s > 0$. This entails for $\mathrm{Re}\, \zeta > 0$ the estimate:

$$\left\| (\zeta \mathbb{1} + X(s))^{-1} - (\zeta \mathbb{1} + X_0)^{-1} P_0 \right\| \leq \int_0^\infty dt \, \mathrm{e}^{-t \, \mathrm{Re}\, \zeta} \left\| \mathrm{e}^{-t X(s)} - \mathrm{e}^{-t X_0} P_0 \right\| ,$$

and on account of (b) the pointwise $\| \cdot \|$-convergence of integrand. Then the *Lebesgue* dominated convergence theorem ensures (a) for $\mathrm{Re}\, \zeta > 0$. Extension to the sector $S_{\pi - \alpha}$ follows in virtue of analyticity of resolvents in ζ. □

Remark 8.3.2 Since the functions in (a) and (b) are holomorphic and uniformly bounded, the *Vitali* theorem for operator-valued holomorphic functions allows to extend the pointwise convergence in (a) and in (b) to uniform convergence in arguments ζ and t. In particular, we can equivalently replace the conditions (a) and (b), respectively, by

$$(\tilde{a})\ \lim_{s\to+0}\left\|(\zeta\mathbb{1}+X(s))^{-1}-(\zeta\mathbb{1}+X_0)^{-1}P_0\right\|=0\,,$$

uniformly in ζ on compacts in $S_{\pi-\alpha}$;

$$(\tilde{b})\ \lim_{s\to+0}\left\|e^{-tX(s)}-e^{-tX_0}P_0\right\|=0\,,$$

uniformly in t on compacts in $S_{\pi/2-\alpha}$.

In fact, we can prove the equivalence in the *Trotter-Kato-Neveu* Theorem 8.3.1 under weaker conditions:

Corollary 8.3.3 *Let $\{X(s)\}_{s>0}$ be a family of m-sectorial operators with $W(X(s))\subseteq S_\alpha$ for some $0<\alpha<\pi/2$ and for all $s>0$. Let X_0 be a closed operator in $\mathfrak{H}_0\subseteq\mathfrak{H}$ with non-empty resolvent set. Then the following conditions are equivalent*:

$$(a')\ \lim_{s\to+0}\left\|(\zeta\mathbb{1}+X(s))^{-1}-(\zeta\mathbb{1}+X_0)^{-1}P_0\right\|=0\,,$$

for some $\zeta\in S_{\pi-\alpha}$,

$$(b')\ \lim_{s\to+0}\left\|e^{-tX(s)}-e^{-tX_0}P_0\right\|=0\,,$$

for t *in a subset* $\mathcal{T}$ of $(0,+\infty)$ *having a limit point,*

if X_0 is generator of a semigroup, especially, e^{-tX_0} *is well-defined at least for* $t\geq 0$.

Proof It is sufficient to check that the condition (a') implies (a) and $W(X_0)\subseteq S_\alpha$ for some $\alpha\in(0,\pi/2)$, as well as that the condition (b') implies (b) for the same operator X_0.

$(a')\Rightarrow(a)$ It is known, see Notes to Chap. 8 (Notes to Sect. 8.3), that condition (a') is sufficient to have the uniform (i.e. the operator-norm) resolvent convergence of operators $\{X(s)\}_{s>0}$. On the other hand, this convergence implies that (a') holds for any ζ in the resolvent set $\rho(X_0)$ of X_0. Since $\rho(X_0)$ is non-empty and open, it contains at least one limit point. Because the norms $\{\|(\zeta\mathbb{1}+X(s))^{-1}\|\}_{s>0}$ are uniformly bounded in s when (for any fixed $\epsilon>0$) $\zeta-\epsilon\in S_{\pi-\alpha}$, we can apply the *Vitali* theorem to the limit in (a'). This yields the pointwise convergence for any $\zeta\in S_{\pi-\alpha}$, and in particular (a) holds for operator X_0.

In order to show that X_0 is necessarily m-sectorial with $W(X_0)\subseteq S_\alpha$, we use that the operator-norm convergence in (a) implies:

$$\|(\zeta\mathbb{1}+X_0)^{-1}P_0\|=\lim_{s\to 0}\|(\zeta\mathbb{1}+X(s))^{-1}\|\leq\frac{1}{\operatorname{dist}(\zeta,-S_\alpha)}\,.$$

In the case $\mathrm{Re}\,\zeta > 0$, we obtain $\|(\zeta + X_0)^{-1}\|_0 \le 1/\mathrm{Re}\,\zeta$, where $\|\cdot\|_0$ is the operator-norm in the Hilbert space $\mathfrak{H}_0$. This estimate implies that X_0 generates a contraction semigroup and that X_0 is m-accretive. Similarly, one checks that $\mathrm{e}^{i\varphi} X_0$ is m-accretive for any $\alpha - \pi/2 \le \varphi \le \pi/2 - \alpha$, which proves that X_0 is m-sectorial with semi-angle α.

The implication $(b') \Rightarrow (b)$ is a direct consequence of the *Vitali* theorem. Indeed, by $\|\mathrm{e}^{-tX(s)}\| \le 1$, for any $t \in S_{\pi/2-\alpha}$ and $s > 0$, owing to (b') and the *Vitali* theorem, one infers that functions $\{\mathrm{e}^{-tX(s)}\}_{s>0}$ converge in the operator-norm topology, as $s \to +0$, for any $t \in S_{\pi/2-\alpha}$. This implies that the limit e^{-tX_0} is a holomorphic semigroup on $\mathfrak{H}_0$ with $\|\mathrm{e}^{-tX_0}\|_0 = \|\mathrm{e}^{-tX_0}P_0\| \le 1$ for $t \in S_{\pi/2-\alpha}$. Therefore, the operator X_0 in the subspace $\mathfrak{H}_0 = P_0\mathfrak{H}$ is m-sectorial with semi-angle α. □

It is worth recalling here that convergence in the *strong resolvent sense* (strong operator topology) is well-adapted to the problem of semigroup approximation in the strong topology. Similarly, the above results demonstrate that the convergence in the *uniform resolvent sense* is well-adapted for the semigroup approximation in the operator-norm topology. This topology is natural for holomorphic contraction semigroups, which are continuous (and analytic) in the operator-norm topology away from *zero*. We showed that due to analyticity the corresponding class of generators (the m-sectorial operators) is, in a certain sense, *stable* with respect to operator-norm convergence that we scrutinise also in Remark 8.3.5.

Now we are in position to prove as promised in Sect. 8.1 generalisation of Proposition 8.1.2 about the self-adjoint *Chernoff product formula* to the non-self-adjoint case of *quasi-sectorial* contractions, which converge to a degenerate semigroup, cf. Theorem 3.1.16.

Theorem 8.3.4 *Let $\{\Phi(s)\}_{s\ge 0}$ be a family of quasi-sectorial contractions on a Hilbert space $\mathfrak{H}$. Let there exists $\alpha \in (0, \pi/2)$ such that $W(\Phi(s)) \subseteq D_\alpha$ (8.2.3), for all $s > 0$. Let $X(s) := (\mathbb{1} - \Phi(s))/s$, and let X_0 be a closed operator with non-empty resolvent set, defined in a closed subspace $\mathfrak{H}_0 \subseteq \mathfrak{H}$. Then*

$$\lim_{n\to\infty} \left\| \Phi(t/n)^n - \mathrm{e}^{-tX_0} P_0 \right\| = 0\ ,\ t > 0\ , \tag{8.3.3}$$

if and only if the family $\{X(s)\}_{s>0}$ converges in the uniform resolvent sense to X_0, as $s \to +0$. Here P_0 denotes the orthogonal projection onto $\mathfrak{H}_0$.

Proof Sufficiency: Note that $W(\Phi(s)) \subseteq D_\alpha$, see (8.2.3), means that $W(X(s)) \subseteq S_\alpha$ for any $s > 0$. Thus, the convergence for $s \to +0$ of $\{X(s)\}_{s>0}$ to X_0 in the uniform resolvent sense coincides with condition (a) of Theorem 8.3.1. Then owing to

$$\begin{aligned} &\|\Phi(t/n)^n - \mathrm{e}^{-tX_0} P_0\| \le \\ &\qquad \|\Phi(t/n)^n - \mathrm{e}^{-n(\mathbb{1}-\Phi(t/n))}\| + \|\mathrm{e}^{-n(\mathbb{1}-\Phi(t/n))} - \mathrm{e}^{-tX_0} P_0\|\ , \end{aligned} \tag{8.3.4}$$

the assertion follows from the $(\sqrt[3]{n})^{-1}$-Theorem 8.2.2 for $C = \Phi(t/n)$ and the part $(a) \Rightarrow (b)$ of the *Trotter-Kato-Neveu* Theorem 8.3.1 for quasi-sectorial contractions. Necessity: On account of the operator-norm *Trotter-Kato-Neveu* Theorem 8.3.1, we have to estimate a difference

$$\|\mathrm{e}^{-tX(t/n)} - \mathrm{e}^{-tX_0} P_0\| \leq \\ \|\mathrm{e}^{-tX(t/n)} - \Phi(t/n)^n\| + \|\Phi(t/n)^n - \mathrm{e}^{-tX_0} P_0\| \,, \; t > 0 \,. \quad (8.3.5)$$

The first term in the right-hand side of (8.3.5) can be estimated by the $(\sqrt[3]{n})^{-1}$-Theorem 8.2.2 for $C = \Phi(t/n)$. The second term is supposed to tend to zero by (8.3.3). As a consequence, the assertion follows from the part $(b) \Rightarrow (a)$ of the *Trotter-Kato-Neveu* Theorem 8.3.1. □

Remark 8.3.5 As a matter of fact, in order to obtain (8.3.3), it is sufficient to have (a'). On the other hand, it is sufficient to have (8.3.3) for t in any subset of $(0, +\infty)$ having a limit point, in order to obtain the convergence (a). This, as well as observation that convergence (8.3.3) implies that $\{\mathrm{e}^{-tX_0}\}_{t\geq 0}$ is a holomorphic contraction semigroup for $t \in S_{\pi/2-\alpha}$, follows from Corollary 8.3.3.

8.4 Approximation Theorem for the Strong Operator Topology

In this section we evoke an alternative to the Chernoff approximation theory approach to study a *strongly* convergent product formulæ for quasi-sectorial contractions. In Chap. 9, we shall scrutinise this approach, called the *analytic extension* method for generators, in details; see Sect. 9.2.3. There we also uplift it from the *strong* to the *operator-norm* convergence in the framework of application to the Trotter product formula.

Here we demonstrate a power of the analytic extension method for sectorial generators (cf. Sect. 9.2.2) for strongly convergent Trotter product formula. This allows to generalise the Kato product Theorem 5.2.5 to non-self-adjoint (m-sectorial) generators of quasi-sectorial contraction semigroups.

Theorem 8.4.1 (Kato-Simon Product Theorem) *Let A and B be m-sectorial operators in $\mathfrak{H}$ such that $\mathfrak{H}_0 = \overline{\mathrm{dom}\,(A^{1/2}) \cap \mathrm{dom}\,(B^{1/2})}$ be a subspace in $\mathfrak{H}$ and let P_0 be the orthogonal projection onto the $\mathfrak{H}_0$. Let operator $H = A \dot{+} B$ defined in $\mathfrak{H}_0$ as a form-sum be self-adjoint. Then the Trotter product formula*

$$\mathrm{e}^{-tH} P_0 = \text{s-}\lim_{n\to\infty} (\mathrm{e}^{-tA/n}\mathrm{e}^{-tB/n})^n \,, \quad t > 0 \,, \qquad (8.4.1)$$

converges in the strong operator topology locally-uniformly in $t > 0$.

Proof The main idea of the *analytic extension* method for generators is to define for a pair of m-sectorial generators A and B in a Hilbert space $\mathfrak{H}$ two families of closed forms for $z \in \mathbb{C}$:

$$a(z) := \mathrm{Re}\, a + z\, \mathrm{Im}\, a, \quad \mathrm{dom}(a(z)) = \mathrm{dom}(a)\,, \tag{8.4.2}$$

$$b(z) := \mathrm{Re}\, b + z\, \mathrm{Im}\, b, \quad \mathrm{dom}(b(z)) = \mathrm{dom}(b)\,, \tag{8.4.3}$$

where a and b are *sectorial* sesquilinear forms associated with the m-sectorial operators A and B with vertices at zero.

Then by inspection of the linear complexifications: (8.4.2) and (8.4.3), one infers that there exists $m_{A,B} > 0$ such that the forms $a(z)$, $b(z)$ are sectorial with vertices at zero and holomorphic of the *type* (a) (Definition 9.1.10) for any z in the strip $\mathfrak{S}(m_{A,B}) := \{z \in \mathbb{C} \,:\, |\mathrm{Re}\, z| < m_{A,B}\}$.

As a consequence by the first representation theorem (Proposition 9.1.3), there exist m-sectorial operators $A(z)$ and $B(z)$ associated in the strip $\mathfrak{S}(m_{A,B})$ with the closed sectorial (a)-holomorphic forms $a(z)$ and $b(z)$, such that they produce a locally-uniformly sectorial holomorphic families: $\{A(z)\}_{\{z\in\mathfrak{S}(\epsilon_{A,B})\}}$ and $\{B(z)\}_{\{z\in\mathfrak{S}(\epsilon_{A,B})\}}$, of type (B) (Definition 9.1.12). Given that, $\{\mathrm{e}^{-tA(z)}\}_{\{z\in\mathfrak{S}(\epsilon_{A,B})\}}$ and $\{\mathrm{e}^{-tB(z)}\}_{\{z\in\mathfrak{S}(\epsilon_{A,B})\}}$ are holomorphic families of contraction semigroups for $t > 0$, which satisfy conditions of the Kato product Theorem 5.2.5 for self-adjoint generators, that is:

$$\mathrm{e}^{-tH(z)}\, P_0 = \text{s-}\lim_{n\to\infty} (\mathrm{e}^{-tA(z)/n}\mathrm{e}^{-tB(z)/n})^n\,,\ \mathrm{Im}\, z = 0\,,$$
$$\mathrm{Re}\, z = x \in (-m_{A,B}, m_{A,B})\,, \tag{8.4.4}$$

where $H(z) = A(z) \dot{+} B(z)$ is the form-sum operator.

For that reason, by the strong analyticity and the *Vitaly* theorem, it follows that (8.4.4) is true in the strip $\mathfrak{S}(m_{A,B})$, including $z = i$. Since by definitions (8.4.2) and (8.4.3) one has $A(z = i) = A$ and $B(z = i) = B$, this proves (8.4.1). □

Corollary 8.4.2 *Seeing that the Kato product Theorem* 5.2.5 *is valid, e.g. for symmetric approximants* (5.2.55), *we also obtain that*

$$\mathrm{e}^{-tH} = \text{s-}\lim_{n\to\infty} (\mathrm{e}^{-tB/2n}\mathrm{e}^{-tA/n}\mathrm{e}^{-tB/2n})^n = \text{s-}\lim_{n\to\infty} (\mathrm{e}^{-tA/2n}\mathrm{e}^{-tB/n}\mathrm{e}^{-tA/2n})^n, \tag{8.4.5}$$

for $t > 0$ *and the form-sum* $H = A \dot{+} B$.

In Chap. 9 (Sect. 9.2.3) we shall *uplift* the strong operator topology in the Kato-Simon product Theorem 8.4.1. We shall show that the *analytic extension* method allows also to improve the Trotter product formula up to convergence in the operator-norm topology.

In Sect. 8.5.2 we use Theorem 8.4.1 to demonstrate an application of the *Chernoff* product formula for proving convergence of a Trotter-Kato product formula in the operator-norm topology.

8.5 Applications

8.5.1 *Quasi-sectorial Euler Approximations*

Let A be an m-sectorial operator with semi-angle $0 < \alpha < \pi/2$. Then the operators $F(t) := (\mathbb{1} + tA)^{-1}$, $t \geq 0$ are quasi-sectorial contractions (Lemma 3.1.3), that is, $W(F(t)) \subseteq D_\alpha$ (8.2.3). Let $X(s) := (\mathbb{1} - F(s))/s$, $s > 0$, and $X_0 := A$. Then $X(s)$ converges, when $s \to +0$, to X_0 in the uniform resolvent sense with the asymptotic

$$\|(\zeta\mathbb{1} + X(s))^{-1} - (\zeta\mathbb{1} + X_0)^{-1}\| = \frac{s\,|\zeta|^2}{|1 + s\zeta|}\left\|\frac{A}{\zeta\mathbb{1} + A + s\zeta A}\cdot\frac{A}{\zeta\mathbb{1} + A}\right\| = O(s)\,, \tag{8.5.1}$$

for any $\zeta \in S_{\pi-\alpha}$, since we have the estimate:

$$\left\|\frac{A}{\zeta\mathbb{1} + A + s\zeta A}\cdot\frac{A}{\zeta\mathbb{1} + A}\right\| \leq$$

$$\frac{1}{|1+s\zeta|}\left(1 + \frac{|\zeta|\,|1+s\zeta|^{-1}}{\mathrm{dist}\left(\zeta\,(1+s\zeta)^{-1}, -S_\alpha\right)}\right)\left(1 + \frac{|\zeta|}{\mathrm{dist}(\zeta, -S_\alpha)}\right), \quad \zeta \in S_{\pi-\alpha}\,,$$

Therefore, the family $\{F(t)\}_{t\geq 0}$ satisfies conditions of Theorem 8.3.4. This gives the operator-norm approximation of the exponential function, i.e. the semigroup for m-sectorial generator, by the powers of resolvent (*Euler approximations*). The next statement yields the operator-norm estimate for the error bound in (8.3.3).

Theorem 8.5.1 *If A is an m-sectorial operator in a Hilbert space $\mathfrak{H}$, with semi-angle $\alpha \in (0, \pi/2)$, then*

$$\lim_{n\to\infty}\left\|(\mathbb{1} + tA/n)^{-n} - \mathrm{e}^{-tA}\right\| = 0\,, \quad t \in S_{\pi/2-\alpha}\,. \tag{8.5.2}$$

Moreover, uniformly in $t \geq 0$, one has the error-bound estimate:

$$\left\|(\mathbb{1} + tA/n)^{-n} - \mathrm{e}^{-tA}\right\| \leq O\left(\frac{\ln(n)}{n}\right)\,. \tag{8.5.3}$$

Proof The convergence (8.5.2) follows directly from Theorem 8.3.4. To obtain (8.5.3), we use the telescopic representation:

$$(\mathbb{1} + tA/n)^{-n} - \mathrm{e}^{-tA} = \sum_{m=0}^{n-1} (\mathbb{1} + tA/n)^{-(n-m-1)} \left((\mathbb{1} + tA/n)^{-1} - \mathrm{e}^{-tA/n}\right) \mathrm{e}^{-mt\,A/n}\,, \tag{8.5.4}$$

and the estimates from Lemmata 10.1.2–10.1.4 :

$$\|A^{-1}((\mathbb{1} + tA/n)^{-1} - \mathrm{e}^{-tA/n})\| \leq 2t/n\,,$$
$$\|A^{-1}((\mathbb{1} + tA/n)^{-1} - \mathrm{e}^{-tA/n})A^{-1}\| \leq 3/2(t/n)^2\,.$$

Since by Proposition 1.8.13 the operator A generates a holomorphic semigroup, owing to Corollary 1.7.5, one gets $\|A\,\mathrm{e}^{-\tau A}\| \leq C_A/\tau$ for some constant $C_A > 0$ and any $\tau > 0$. Finally, for any integer $k \geq 1$, we have the identity:

$$A\,(\mathbb{1} + tA/n)^{-k} = \frac{n}{t}\left(\mathbb{1} - (\mathbb{1} + tA/n)^{-1}\right)(\mathbb{1} + tA/n)^{-k+1}\,,$$

which leads to the estimate $\|(\mathbb{1}+tA/n)^{-k}\, A\| \leq n\, K/kt$ by Lemma 8.2.1, (8.2.12), for the quasi-sectorial contraction $C = (\mathbb{1} + tA/n)^{-1}$.

These estimates and (8.5.4) yield for $n \geq 3$:

$$\|(\mathbb{1} + tA/n)^{-n} - \mathrm{e}^{-tA}\| \leq \frac{nK}{(n-1)t}(2t/n) + (2t/n)\frac{C_A n}{(n-1)t} + \sum_{m=1}^{n-2} \frac{nK}{(n-m-1)\,t}\,\frac{3t^2}{2n^2}\,\frac{C_A n}{m\,t}\,.$$

Therefore, uniformly in $t \geq 0$, we obtain:

$$\|(\mathbb{1} + tA/n)^{-n} - \mathrm{e}^{-tA}\| \leq \frac{4(K + C_A)}{n} + 6\,KC_A\frac{\ln(n)}{n}\,, \tag{8.5.5}$$

which proves (8.5.3). □

Notice that in virtue of Theorem 3.2.17, the error-bound estimate (8.5.3) is not optimal.

8.5.2 *Operator-Norm Trotter-Kato Product Formulæ*

Let A be a non-negative self-adjoint operator, and let operator B be m-sectorial with $W(B) \subseteq S_\alpha$ for some angle $0 \leq \alpha < \pi/2$, in a Hilbert space $\mathfrak{H}$. Then it turns out that our Theorem 8.3.4 can be applied for the proof of operator-norm convergence for a particular class of the Trotter-Kato product formulæ.

Consider a family of operators

$$\Phi(s) := f(sA)^{1/2} g(sB) f(sA)^{1/2}, \quad s \geq 0, \tag{8.5.6}$$

where $f(t)$ and $g(t)$ are the *Kato functions*: either $(1+t)^{-1}$ or e^{-t}. Then in (8.5.6) the operators $f(sA)^{1/2}$ are self-adjoint and contraction: $\|f(sA)^{1/2}\| \leq 1$. In turn, by results of Sect. 3.1.1 (Lemma 3.1.3 and Theorem 3.1.4) for this choice of Kato functions and for $s > 0$, the operators $g(sB)$ are quasi-sectorial contractions with $W(g(sB)) \subset D_\alpha$, where D_α is defined by (3.1.1) (or (8.2.3)).

As a consequence one obtains for $s \geq 0$ that

$$(\Phi(s)\, u, u) = (f(sA)^{1/2} u, g(sB) f(sA)^{1/2} u) \in D_\alpha ,$$

as well as that

$$\begin{aligned}((\mathbb{1} - \Phi(s))\, u, u) = (u, (\mathbb{1} - f(sA))u) + \\ +(f(sA)^{1/2} u, (\mathbb{1} - g(sB)) f(sA)^{1/2} u) \in S_\alpha ,\end{aligned}$$

for any $u \in \mathfrak{H}$, $\|u\| = 1$. This proves that numerical range $W(\Phi(s)) \subseteq D_\alpha$, that is, $\{\Phi(s)\}_{s \geq 0}$ (8.5.6) are quasi-sectorial contractions on $\mathfrak{H}$, whereas $\{(\mathbb{1} - \Phi(s))\}_{s \geq 0}$ are (bounded) m-sectorial operators.

Since (8.5.6) is a family of quasi-sectorial contractions, Theorem 8.3.4 yields necessary and sufficient conditions for the operator-norm convergence of the approximants $\{\Phi(t/n)^n\}_{n \geq 1}$ as $n \to \infty$. However, we have to notify that we still have *no identification* (through operators A and B) of operator X_0, which is declared in Theorem 8.3.4. For elucidation of this point and for possible applications of Theorem 8.3.4, we consider two cases.

First, let non-negative self-adjoint generators A and B satisfy conditions of the Kato product Theorem 5.2.5. If $\overline{\mathrm{dom}\,(A^{1/2}) \cap \mathrm{dom}\,(B^{1/2})} \subseteq \mathfrak{H}$, then approximants $\{\Phi(t/n)^n\}_{n \geq 1}$ for Kato functions f and g (either $(1+t)^{-1}$ or e^{-t}) converge *strongly* to $\mathrm{e}^{-tH} P_0$, when $n \to \infty$. Here self-adjoint generator $H := A \dotplus B$ is the form-sum defined in the closure $\mathfrak{H}_0 := \overline{\mathrm{dom}\,(A^{1/2}) \cap \mathrm{dom}\,(B^{1/2})}$, and P_0 is the orthogonal projection of $\mathfrak{H}$ onto subspace $\mathfrak{H}_0$. Therefore, $X_0 = H$.

For non-self-adjoint case, when the both of operators A and B are m-sectorial (we consider self-adjoint operator A as a particular case $\alpha = 0$ of m-sectorial operator), we may still define $H := A \dotplus B$ as the form-sum of operators A and B in the closure $\mathfrak{H}_0$ of the common domain $\mathrm{dom}\,(a) \cap \mathrm{dom}\,(b)$ of the corresponding to these operators' sesquilinear forms a and b. If the Kato functions $f(t)$, $g(t)$ are e^{-t}, then in virtue of the Kato-Simon product Theorem 8.4.1 and Corollary 8.4.2 approximants $\{\Phi(t/n)^n\}_{n\geq 1}$ converge *strongly* to $\mathrm{e}^{-tH}P_0$ as $n \to \infty$, for $X_0 = H$, see (8.4.5).

Taking into account these observations and Theorem 8.3.4, we obtain the statement, where non-self-adjoint operator B is generator of a *quasi-sectorial* contraction semigroup.

Theorem 8.5.2 *Let A be a non-negative self-adjoint operator, and let B be an m-sectorial operator in $\mathfrak{H}$. Let the Kato functions f and g be exponetial. For $s > 0$ we put $X(s) = s^{-1}(\mathbb{1} - f(sA)^{1/2}g(sB)f(sA)^{1/2})$. Let $H = A \dot{+} B$ be the form-sum of A and B defined in the closure $\mathfrak{H}_0 = \overline{\mathrm{dom}\,(a) \cap \mathrm{dom}\,(b)}$, and let P_0 be the orthogonal projection onto $\mathfrak{H}_0$. Then the following assertions are equivalent:*

(i) $X(s)$ converges in the uniform resolvent sense to H as $s \to +0$,

$$(ii)\ \lim_{n\to\infty} \left\| \left(f(tA/n)^{1/2} g(tB/n) f(tA/n)^{1/2} \right)^n - \mathrm{e}^{-tH} P_0 \right\| = 0\,, \qquad (8.5.7)$$

locally-uniformly in $t > 0$ away from zero.

We comment that in contrast to Theorem 8.5.1, see (8.5.3), although similar to Theorem 8.3.4, Theorem 8.5.2 gives no operator-norm error-bound estimate for the rate of convergence in (8.5.7).

We shall return to this question in Sects. 9.3 and 9.4.

8.6 Notes to Chap. 8

Notes to Sect. 8.1

Operator-norm convergent product approximations of C_0-semigroups is one of the main topic of this book. After self-adjoint semigroups, in this section we present an approach suitable in a Hilbert space for the case of *not* necessarily self-adjoint semigroups. In particular, it covers the case of the *quasi-sectorial* contractions (Chap. 3). This approach in a Hilbert space was developed in paper [CZ01b], with further comments and remarks in [Zag08].

Notes to Sect. 8.2

The fact (Lemma 8.2.1) that *quasi-sectorial* contractions on a Hilbert space are the *Ritt* operators [Ri53] is due to [CZ01b, Lemma 3.1].

The non-self-adjoint improvement (Theorem 8.2.2) of the Chernoff $\sqrt{n}$ -Lemma for sectorial contractions, also known as the $(\sqrt[3]{n})^{-1}$-Theorem, see [Zag17, Section 3], [Zag20, Proposition 4.3], appeared for the first time in [CZ01b, Theorem 3.1]. Although this improvement is produced by means of a certain optimisation, cf. [Zag22a, Theorem 4.3], [Zag22c, Theorem 4.1], the probabilistic estimate (8.2.8), which involves the *Tchebychëv inequality*, is obviously not optimal.

As a consequence, for $n \to \infty$ the rate of the operator-norm convergence in Theorem 8.2.2 is also non-optimal. To find the optimal rate, one can try to improve our probabilistic arguments. This direction inspired the paper [Pau04], where with a more accurate bound for a tail of probability distribution, the estimate (8.2.8) has been improved to $O(\sqrt{\ln(n)/n})$; see [Pau04, Theorem 1] and comments in Notes to Sect. 8.3 below.

Recall that the $(\sqrt[3]{n})^{-1}$-Theorem 8.2.2 is the *first step* in validation of the operator-norm generalisation of the key for *à la Chernoff theory* the product formula (8.3.3) for non-self-adjoint, but *quasi-sectorial* contractions. In fact, it controls only one (the first) term in estimate (8.3.4).

Notes to Sect. 8.3

The second step in validation of the *operator-norm* generalisation of the product formula (8.3.3) for *quasi-sectorial* contractions is the proof operator-norm extension for quasi-sectorial contractions of the *Trotter-Kato-Neveu* theorem (Theorem 8.3.1); see [CZ01b, Lemma 4.1]. This extension controls the second term in the upper bound (8.3.4).

Altogether these terms ensure (see Theorem 8.3.4) the operator-norm generalisation of the Chernoff product formula (8.3.3) for *non-self-adjoint*, but quasi-sectorial contractions proved in [CZ01b, Theorem 4.1, Corollary 4.1]. We have to recognise that at this stage Theorem 8.3.4 does not give for the Chernoff approximation by formula (8.3.4) any information about the rate of operator-norm convergence since it is limited by result of the *Trotter-Kato-Neveu* theorem, cf. Applications in Sect. 8.5.

Note that by [Kato80, Ch.IV, §2.6, Theorem 2.25] the condition (a') (for the proof of Corollary 8.3.3) is sufficient to have the uniform resolvent convergence of operators $\{X(s)\}_{s>0}$ in the generalised sense. This convergence yields that (a') holds for any ζ in the resolvent set $\rho(X_0)$ of X_0, which allows to advance with the proof using analyticity of resolvents and the *Vitali* theorem. According to Remark 8.3.5, it is the analyticity of semigroups with m-sectorial generators that allows to make a "*bootstrap*" and to finish the proof of Theorem 8.3.4 *without* estimate of the error bound for the rate of convergence for $n \to \infty$.

Notes to Sect. 8.4

The *Kato-Simon* product Theorem 8.4.1 for the *strong* operator topology was presented as a private communication by B.Simon to the author in Addendum of the Kato paper [Kato78]. In [CZ99] it was shown that the *analytic extension* method announced in [Kato78] allows to improve the Trotter product formula for semigroups with m-sectorial generators up to convergence in the *operator-norm* topology.

In this subsection we proposed a sketch of the proof of the *Kato-Simon* product Theorem 8.4.1 for the strong operator topology. In Chap. 9 (Sect. 9.2.3) we present the results of [CZ99] with the corresponding detailed proof of the *analytic extension* method for m-sectorial generators and the Trotter product formula, which converges in the operator-norm topology.

Notes to Sect. 8.5

From the proof of Theorem 8.3.4, it follows that the operator-norm error-bound estimate in the Chernoff product formula (8.3.3) is defined by two terms; see (8.3.4). The $(\sqrt[3]{n})^{-1}$-Theorem 8.2.2 gives the estimate $O(1/n^{1/3})$ for the first of them. The estimate for the second term (the *Trotter-Kato-Neveu* Theorem 8.3.1) is more delicate. This term has a different nature, and it is related to the operator-norm error-bound for the convergence of resolvents and the estimate of the corresponding contour integral (8.3.1).

This poses questions about relation between operator-norm error-bounds for the Chernoff-type estimates (Sect. 8.2) and the corresponding rate of the operator-norm convergence for resolvents of quasi-sectorial contractions (Sect. 8.3) as well as about *optimality* of the splitting (8.3.4) of the estimate for the Chernoff product approximation into two terms.

Section 8.5.1

The application for the *Euler formula* approximants $\{F(t/n) := \Phi(t/n)\}_{n\geq 1}, t \geq 0$, is due to [CZ01b, Section 5.1] together with estimates proved in [CZ01a, Lemmata 2.6–2.8]. Here, to estimate the rate of the operator-norm convergence, we use instead of the Chernoff-type splitting (8.3.4) a *telescopic* representation (8.5.4).

Note that the estimate of the rate (8.5.5) is not optimal. In [BenPau04, Theorem 1.1], [Pau12, Theorem B] it was proven that for the Ritt-type contractions the optimal rate of the operator-norm convergence in Euler formula is $O(1/n)$. This statement was refined by a complement control of *sectoriality* for semigroups with m-sectorial generators; see [ArZag10, Theorem 4.1] and Theorem 3.2.17 in Sect. 3.2.3.

Section 8.5.2

In this book one of the important applications of the approximation theory *à la Chernoff* (Theorem 8.3.4) is demonstration of convergence of the Trotter-Kato product formulæ in the operator-norm topology.

For example, the paper [Cher68, Corollary] was directly addressed to this aim, but in the strong operator topology. In this subsection we demonstrated an application of Theorem 8.3.4 for uplifting (improving) the strong operator topology in the Kato product Theorem 5.2.5 and the Kato-Simon product Theorem 8.4.1 up to convergence in the operator-norm topology.

In this subsection we partially followed results from [CZ99], [CZ01b, Subsection 5.3], and thesis [Ca10, Sections 7.3 - 7.5].

Chapter 9
Product Formulæ for Non-Self-Adjoint Semigroups

We start this chapter by preliminaries (Sect. 9.1), which contain a short outline of the theory of *sesquilinear* forms and the *representation* theorems about corresponding operators. They are indispensable for our analysis and construction of the product formulæ for C_0-semigroups, which are based on idea of the *analytic extension* method that involves the *analytic families* (A) and (B) of m-sectorial generators. method

The results of this method in the proof of the operator-norm convergence of the Trotter product formula for non-self-adjoint semigroups with sectorial generators (*without* estimates of the bound for the operator-norm convergence rate) are presented in Sect. 9.2.

In Sects. 9.3 and 9.4, we utilise some results of Chap. 3 to estimate the *rate* of the operator-norm convergence for the corresponding Trotter product formula.

9.1 Preliminaries

9.1.1 Sectorial Forms and Operators

Let $\mathfrak{H}$ be a separable Hilbert space. We consider complex-valued *sesquilinear* forms $a[\cdot,\cdot]$ on $\mathfrak{H} \times \mathfrak{H}$ defined by the mapping $a : u, v \mapsto a[u,v] \in \mathbb{C}$ on domain $\mathrm{dom}(a) \subseteq \mathfrak{H}$. Here $u, v \in \mathrm{dom}(a)$ and the mapping is linear in u and semi-linear in v. The *adjoint* form is defined as $a^*[u,v] := \overline{a[v,u]}$. If $a^*[u,v] = a[u,v]$, the form is called *symmetric*.

With any form $a[\cdot,\cdot]$, one can associate its *real* and *imaginary* parts, which are two *complex*-valued symmetric forms

$$\mathrm{Re}\, a := \frac{1}{2}(a + a^*) \quad \text{and} \quad \mathrm{Im}\, a := \frac{1}{2i}(a - a^*) \tag{9.1.1}$$

V. A. Zagrebnov et al., *Trotter-Kato Product Formulæ*, Operator Theory: Advances and Applications 296, https://doi.org/10.1007/978-3-031-56720-9_9

Note that $a[u, v] = \text{Re}a[u, v] + i\ \text{Im}a[u, v]$, although in general $(\text{Re}a)[u, v] \neq \text{Re}(a[u, v])$ and $(\text{Im}a)[u, v] \neq \text{Im}(a[u, v])$.

One can also associate with $a[\cdot, \cdot]$ the *quadratic* form $a[u] := a[u, u]$, $u \in \text{dom}(a)$, which justifies the above notations (9.1.1) since we get $\text{Re}a[u] = \text{Re}(a[u])$ and $\text{Im}a[u] = \text{Im}(a[u])$.

The *numerical range* $W(a)$ of the form $a[\cdot, \cdot]$ is defined by

$$W(a) := \{a[u] \in \mathbb{C} : u \in \text{dom}(a)\,,\ \|u\| = 1\}\,. \tag{9.1.2}$$

The form is called *sectorial* with a *vertex* γ and a *semi-angle* θ if

$$W(a) \subseteq \overline{S}_{\theta,\gamma} := \{z \in \mathbb{C} : z = \gamma\,,\ |\arg(z - \gamma)| \leq \theta\,,\ 0 \leq \theta < \pi/2\,,\ \gamma \in \mathbb{R}\}\,, \tag{9.1.3}$$

for the closure of sector $S_{\theta,\gamma} := \{z \in \mathbb{C} : z \neq \gamma\,,\ |\arg(z - \gamma)| < \theta\,,\ 0 < \theta < \pi/2\,,\ \gamma \in \mathbb{R}\}$. The numbers γ and θ are not uniquely determined by the form $a[\cdot, \cdot]$ since relation (9.1.3) yields only that

$$\text{Re}a[u] \geq \gamma \quad \text{and} \quad |\,\text{Im}a[u]\,| \leq \text{tg}\theta\ (\text{Re}a[u] - \gamma)\,,\ u \in \text{dom}(a)\,. \tag{9.1.4}$$

Finally, if $\gamma = 0$, we identify $\overline{S}_\theta$ and $\overline{S}_{\theta,0}$.

The *sectorial* form $a[\cdot, \cdot]$ is said to be *closed* if the limits

$$\lim_{n\to\infty} u_n = u \ \text{ and } \lim_{n,m\to\infty} a[u_n - u_m] = 0\,,\ \text{ for } u_n, u_m \in \text{dom}(a)\,,$$

(denoted also as $a\text{-}\lim_{n\to\infty} u_n = u$) imply that

$$u \in \text{dom}(a) \ \text{ and } \lim_{n\to\infty} a[u_n - u] = 0\,.$$

A sectorial form is *closable* if and only if $a\text{-}\lim_{n\to\infty} u_n = 0$ implies $a[u_n] \to 0$.

Proposition 9.1.1 *Let a_1 and a_2 be sectorial form in $\mathfrak{H}$. Let $a := a_1 + a_2$ be form with domain* $\text{dom}(a) = \text{dom}(a_1) \cap \text{dom}(a_2)$. *Then, a is sectorial, and it is closed if the forms a_1, a_2 are closed. If a_1, a_2 are closable, then a is also closable and the closure $\widetilde{a} := (a_1 + a_2)^\sim \subset \widetilde{a}_1 + \widetilde{a}_2$.*

Let a be a *sectorial* form in $\mathfrak{H}$. Then a form b is said to be *relatively bounded* with respect to a, if $\text{dom}(a) \subset \text{dom}(b)$ and for non-negative α, β one has

$$|b[u]| \leq \alpha\ \|u\|^2 + \beta\ |a[u]|\,,\ \text{ for } u \in \text{dom}(a)\,. \tag{9.1.5}$$

Note that by virtue of (9.1.4) the *a-boundedness* of b (9.1.5) for the case $\text{Re}a > 0$ is equivalent to

$$|b[u]| \leq \alpha'\ \|u\|^2 + \beta'\ \text{Re}a[u]\,,\ \text{ for } u \in \text{dom}(a) = \text{dom}(\text{Re}a)\,.$$

If now $\beta' < 1$, then the form $a + b$ is sectorial, the form $a + b$ is closed if and only if a is closed, and finally $a+b$ is closable if and only if a is closable. In the last case domain of the closure of sum : $\mathrm{dom}(a + b)^{\sim} = \mathrm{dom}(\widetilde{a})$.

We remind (see Sect. 3.1) that similarly to (9.1.2), (9.1.3) the operator A in $\mathfrak{H}$ with *numerical range*

$$W(A) := \{(Au, u) : u \in \mathrm{dom}(A), \quad \|u\| = 1\} ,$$

is called *sectorial* with a vertex $\gamma \in \mathbb{R}$ and with a semi-angle θ, if

$$W(A) \subseteq \overline{S}_{\theta,\gamma} := \overline{\{z \in \mathbb{C} : z \neq \gamma\, ,\ \ |\arg(z - \gamma)| < \theta\, ,\ 0 \leq \theta < \pi/2\, ,\ \gamma \in \mathbb{R}\}}\, . \tag{9.1.6}$$

The operator A in (9.1.6) is called (quasi)-*m-sectorial,* if it is (quasi)-*m-accretive* (Definition 1.8.7) for $\gamma \in \mathbb{R}$. Then, A is closed, and the resolvent set $\rho(A)$ covers the exterior of $W(A)$.

We remind that by Definition 1.8.9 and (1.8.11), a sectorial m-accretive operator (i.e., vertex $\gamma = 0$) is called m-sectorial.

Remark 9.1.2 Let A be sectorial operator with $W(A) \subseteq \overline{S}_{\theta,\gamma}$. Then, the sesquilinear form generated by this operator:

$$a[u, v] := (Au, v) \quad \text{for} \quad \mathrm{dom}(a) = \mathrm{dom}(A)\, , \tag{9.1.7}$$

is a closable sectorial form with the vertex γ and semi-angle θ. The operator A is then called *form-closable.* If A is quasi-m-sectorial, that is, sectorial and quasi-m-accretive (Sect. 3.1.1), then the form (9.1.7) is closed. On account of (9.1.7) operator A and form a have the same numerical range. As a consequence, $W(a) = W(A) \subseteq \overline{S}_{\theta,\gamma}$. For m-sectorial (or sectorial m-accretive) operator A the vertex $\gamma = 0$ and we write: $W(A) \subseteq \overline{S}_{\theta}$.

Note that if operator A in (9.1.7) is self-adjoint, then form $a[u, v]$ is symmetric.

9.1.2 Representation Theorems

Relation between closed, sectorial sesquilinear forms and the corresponding operators is a subject of the *representation* theorem.

Proposition 9.1.3 (The *first* Representation Theorem) *Let $a[u, v]$ be a densely defined, closed, sectorial sesquilinear form in $\mathfrak{H}$. Then there exists an m-sectorial operator A_a associated with the form a such that:*

(i) $\mathrm{dom}(A_a) \subset \mathrm{dom}(a)$ *and* $a[u, v] = (A_a u, v)$ *for* $u \in \mathrm{dom}(A_a)$, $v \in \mathrm{dom}(a)$ *;*
(ii) $\mathrm{dom}(A_a)$ *is a core for the form* $a[\cdot, \cdot]$ *;*
(iii) The m-sectorial operator A is uniquely defined by the condition (i).

As a corollary of (i) one gets that if $A := A_a$, then the adjoint operator A^* is associated with the adjoint form a^*, that is, $A^* = A_{a^*}$. Together with Proposition 9.1.1, this representation theorem allows to define the *form-sum* of operators A_1 and A_2 in the form-sense:

$$\widetilde{A} = A_1 \dot{+} A_2 := A_{a_1+a_2} , \tag{9.1.8}$$

if the form $\widetilde{a} = a_1 + a_2$ satisfies the conditions of Proposition 9.1.3.

If operators $A_1 := A_{a_1}$ and $A_2 := A_{a_2}$ are m-sectorial, then corresponding sesquilinear forms a_1 and a_2 defined by (9.1.7) are closed. If domain $\mathcal{D} := \mathrm{dom}(a_1) \cap \mathrm{dom}(a_2)$ is dense in $\mathfrak{H}$, then the form-sum of these operators is well-defined and gives a densely defined operator $A_{a_1+a_2}$. Note that if the closure $\widetilde{\mathcal{D}} = \mathfrak{H}_0 \subset \mathfrak{H}$ is a proper subspace of $\mathfrak{H}$, then $A_1 \dot{+} A_2$ is operator with $\mathrm{dom}(A_1 \dot{+} A_2)$ dense in $\mathfrak{H}_0$.

This indicates that the closed sectorial forms yield a convenient tool for construction of the m-sectorial operators. First, because of the closed forms (in contrast to the closed operators) is a *vector space*. Second, due to the fact that (except a bounded case) there is the ease of *extending* such a form. Indeed, any closed symmetric form bounded from below admits a proper extension with the same properties. Consequently, there is *no* notion of the *maximal* symmetric, or sectorial, forms, whereas maximal symmetric and m-sectorial operators do exist, see Sect. 1.8.1.

Proposition 9.1.4 (The *second* Representation Theorem) *Let $a \geq 0$ be a densely defined, closed, symmetric form in $\mathfrak{H}$ and $A := A_a$ be associated with a self-adjoint operator. Then* $\mathrm{dom}(A^{1/2}) = \mathrm{dom}(a)$ *and*

$$a[u, v] = (A^{1/2}u, A^{1/2}v) , \quad u, v \in \mathrm{dom}(a) . \tag{9.1.9}$$

Note that the operator $G := A^{1/2}$, defined, e.g., by the *Riesz-Dunford* integral, is self-adjoint, non-negative and such that $G^2 = A$, with $\mathrm{dom}(A) = \mathrm{core}(G)$.

A remark related to this observation concerns a general form (*standard representation*) of an m-sectorial operator in $\mathfrak{H}$. Let a non-negative self-adjoint operator $A := A_a$ be defined by Proposition 9.1.4 and let $b[\cdot, \cdot]$ be a form relatively bounded with respect to a:

$$|b[u]| \leq \beta \, a[u] , \quad u \in \mathrm{dom}(a) .$$

Then there exists a bounded operator $B \in \mathcal{L}(\mathfrak{H})$ with $\|B\| \leq \varepsilon \beta$ ($\varepsilon = 1$, if c is a symmetric form, otherwise: $\varepsilon = 2$) such that

$$b[u, v] = (B\, Gu, Gv) , \quad G := A^{1/2} , \quad u, v \in \mathrm{dom}(a) = \mathrm{dom}(G) . \tag{9.1.10}$$

Let m-sectorial operator $A := A_a$ be defined by Proposition 9.1.3. Note that for the closed a the symmetric form $\phi := \mathrm{Re}\, a$ defined by (9.1.1) is also closed. Consequently, by Proposition 9.1.4, there exists self-adjoint operator

$$\mathrm{Re} A := A_{\mathrm{Re}\, a} , \tag{9.1.11}$$

which is called the *real part* of the operator A. Indeed, by definition (9.1.1) and by Proposition 9.1.3(i) one gets $\mathrm{Re} A^* = A_{\mathrm{Re}\, a^*}$, but also $\mathrm{Re}\, a^* = \mathrm{Re}\, a = (\mathrm{Re}\, a)^*$. This implies that $\mathrm{Re} A^* = \mathrm{Re} A = (\mathrm{Re} A)^*$.

We remark that relation $\mathrm{Re} A = (A + A^*)/2$ is correct for a bounded operator A, but it is not true in a general case of unbounded operators. On the other hand, if the forms a and $\mathrm{Re} a$ satisfy conditions of Proposition 9.1.3, then one gets $\mathrm{Re} A = (A \dotplus A^*)/2$ in the form-sense (9.1.8), that is $(A \dotplus A^*)/2 = A_{(a+a^*)/2}$, cf. (9.1.11).

Proposition 9.1.5 *Let A be an m-sectorial operator and with semi-angle θ* (9.1.6). *Then the real part of this operator is non-negative*: $\mathrm{Re} A \geq 0$, *and there exists a bounded symmetric operator $L \in \mathcal{L}(\mathfrak{H})$ such that $\|L\| \leq \mathrm{tg}\theta$ and the operator A has the following standard representation*

$$A = (\mathrm{Re}\, A)^{1/2}\, (\mathbb{1} + i\, L)\, (\mathrm{Re}\, A)^{1/2} , \tag{9.1.12}$$

where operator $\mathrm{Re}\, A$ *is defined by* (9.1.11).

Corollary 9.1.6 *Let A be m-sectorial operator. Then real part* $\mathrm{Re}\, A$ *has compact resolvent if and only if the resolvent of the operator A is compact.*

9.1.3 Holomorphic Families of Unbounded Operators

We recall in this subsection some indispensable information about analytic families of *unbounded closed* operators. We start with two Banach spaces $\mathfrak{X}$, $\mathfrak{Y}$ and denote by $\mathcal{L}(\mathfrak{X}, \mathfrak{Y})$ a set of bounded operators from $\mathfrak{X}$ to $\mathfrak{Y}$ and by $\mathcal{C}(\mathfrak{X}, \mathfrak{Y})$ a corresponding set of unbounded *closed* operators. Since in contrast to $\mathcal{L}(\mathfrak{X}, \mathfrak{Y})$ the set $\mathcal{C}(\mathfrak{X}, \mathfrak{Y})$ is *not* a vector space, one needs certain precaution (*generalisation*) to define analytic families of them.

Definition 9.1.7 A family $\{T(z)\}_{z\in D}$ of unbounded closed operators $\mathcal{C}(\mathfrak{X}, \mathfrak{Y})$, where D is an open subset of the complex plane: $D \subseteq \mathbb{C}$, defined in a neighbourhood of $z_0 \in D$ is *holomorphic* (in a *generalised* sense) at z_0 if

(a) there is a third Banach space $\mathfrak{Z}$;
(b) two families of bounded operators $U(z) \in \mathcal{L}(\mathfrak{Z}, \mathfrak{X})$ and $V(z) \in \mathcal{L}(\mathfrak{Z}, \mathfrak{Y})$;
(c) $U(z) : \mathfrak{Z} \to \mathrm{dom}(T(z))$ is surjective (i.e., *onto* and *one to one*);

(d) one has

$$T(z)\,U(z) = V(z)\ .$$

The family $\{T(z)\}_{z\in D}$ is holomorphic in D, if it is holomorphic at every $z \in D \subset \mathbb{C}$.

This definition generalises the case when $T(z) \in \mathcal{L}(\mathfrak{X}, \mathfrak{Y})$. There is a following resolvent characterisation of analyticity for family $\{T(z)\}_{z\in D}$ of unbounded operators.

Proposition 9.1.8 *Let for simplicity $T(z) \in \mathcal{C}(\mathfrak{X})$ be defined in a neighbourhood of $z_0 = 0$ and let $\zeta \in \rho(T(0))$ be in the resolvent set of $T(0)$. Then family $\{T(z)\}_{z\in D}$ is holomorphic at $z = 0$ if and only if $\zeta \in \rho(T(z))$ and the resolvent $R_{T(z)}(\zeta) = (T(z) - \zeta\mathbb{1})^{-1}$ is bounded-holomorphic for small $|z| < \delta$. In fact the mapping*

$$R_{(\cdot)}(T(\cdot)) : \{|z| < \delta\} \times \{\zeta \in \rho(T(0))\} \longrightarrow \mathcal{L}(\mathfrak{X})\ ,$$

is bounded-holomorphic in two variables $(z, \zeta,) \in \{|z| < \delta\} \times \{\zeta \in \rho(T(0))\}$.

There are several important special cases of holomorphic operator families, which can be selected on account of Definition 9.1.7.

Definition 9.1.9 (Holomorphic families of type (A)) A family of operators $T(z) \in \mathcal{C}(\mathfrak{X}, \mathfrak{Y})$ is said to be holomorphic family of *type* (A) in domain D if:

(i) $\mathrm{dom}(T(z)) = \mathcal{D}$ is independent of z ;
(ii) the vector-valued function $z \mapsto T(z)\,u$ is z-holomorphic in D for every $u \in \mathcal{D}$.

Another holomorphic families, which are useful concern sesquilinear forms and operators in a Hilbert space $\mathfrak{H}$.

Definition 9.1.10 (Holomorphic families of type (a)) A family $\{t(z)[\cdot,\cdot]\}_{z\in D}$ of sesquilinear forms is called a holomorphic family of *type* (a) in domain $D \subseteq \mathbb{C}$ if:

(i) each form $t(z)$ is sectorial and closed with domain $\mathrm{dom}(t(z)) = \mathcal{D}$ independent of z and dense in $\mathfrak{H}$;
(ii) $\{t(z)[u]\}_{z\in D}$ is holomorphic for each $u \in \mathcal{D}$.

Note that by polarisation identity, the (ii) implies that the family $\{t(z)[u, v]\}_{z\in D}$ is holomorphic for each fixed pair $u, v \in \mathcal{D}$.

Proposition 9.1.11 *Let $\{t(z)\}_{z\in D}$ be a holomorphic family of sesquilinear forms of type (a). Let m-sectorial operator $T(z) := T_{t(z)}$ be defined by Proposition* 9.1.3. *Then, operators $\{T(z)\}_{z\in D}$ are locally uniformly sectorial, and they form a holomorphic family in the sense of Definition* 9.1.7.

Definition 9.1.12 (Holomorphic families of type (B)) A holomorphic family of the m-sectorial operators $\{T(z) := T_{t(z)}\}_{z\in D}$ associated by Proposition 9.1.11 with the type (a) holomorphic family of sesquilinear forms is called a holomorphic family of *type* (B).

Remark 9.1.13 Similar to a holomorphic family of type (A), the m-sectorial operators $T(z)$ of holomorphic family of type (B) have compact resolvent either for all $z \in D$, or for no z.

If one restricts to *self-adjoint* families: $\{(T(z))^* = T(\overline{z})\}_{z \in D}$, of sectorial operators, then the families of type (B) form a wider class than those of type (A).

9.2 Operator-Norm Convergence Without Error Bound Estimate

As we discussed in Part II, there are two strategies to establish the operator-norm convergence of the Trotter–Kato product formulæ for *self-adjoint* semigroups. The historically first one was based on direct estimates of the error bounds, see Chap. 7. At the same time and then later in a more abstract form, another strategy has been found. It was based on some sufficient conditions for the operator-norm convergence of the Trotter–Kato product formulæ *without* estimate of the rate of convergence, see Chap. 6.

In this section, we develop a similar strategy for the case of *non-self-adjoint* C_0-semigroups on Hilbert spaces, although it is based on a very different idea of the *analytic extension* method, which involves *holomorphic families* of unbounded m-sectorial generators, cf. Sect. 8.4.

9.2.1 Sufficient Conditions for Convergence

Recall that in Sects. 6.2 and 6.3 or Sect. 7.4 we established two kinds of *sufficient* conditions that ensure the operator-norm convergence of the Trotter product formula for the self-adjoint semigroups without (or with) error bound estimates. They are based either on conditions of *compactness* (Theorems 6.3.5, 6.2.8):

Proposition 9.2.1 *Let A and B be non-negative self-adjoint operators in a Hilbert space $\mathfrak{H}$. If $(\mathbb{1} + A)^{-1}$, or $(\mathbb{1} + A)^{-1}(\mathbb{1} + B)^{-1}$ is compact, then the Trotter product approximants : $\{(e^{-tA/n}e^{-tB/n})^n\}_{n\geq 1}$, converge in the operator-norm to $e^{-tH}P$ locally uniformly for $t > 0$. Here $H = A \dotplus B$ is the form-sum of A and B, and P is the orthogonal projector on the closed subspace* $\overline{\mathrm{dom}(A^{1/2}) \cap \mathrm{dom}(B^{1/2})}$.

Either, they are based on condition of *subordination*: $\mathrm{dom}(A^{1/2}) \subset \mathrm{dom}(B^{1/2})$, and (7.4.8), see INZ-Theorem 7.4.7, or on the condition of *smallness* (7.5.4), see Theorem H.0.1:

Proposition 9.2.2 *Let A and B be non-negative densely defined self-adjoint operators in a Hilbert space $\mathfrak{H}$, with $A \geq \mathbb{1}$, $B \geq 0$, and* $\mathrm{dom}(A^\alpha) \subset \mathrm{dom}(B^\alpha)$ *for some* $1/2 < \alpha \leq 1$, $\|B^\alpha u\| \leq a\,\|A^\alpha u\|$, $u \in \mathrm{dom}(A^\alpha)$ *for* $a < 1$. *Then the Trotter product approximants* : $\{(e^{-tA/n}e^{-tB/n})^n\}_{n\geq 1}$, *converge in the operator-norm to* e^{-tH} *uniformly in* $t \geq 0$. *Here self-adjoint operator* $H = A \dot{+} B$ *is the form-sum of A and B.*

In the Part II, our main tool to prove the operator-norm convergence for the both cases was the standard *functional calculus* for self-adjoint operators. Below we shall exercise another strategy based on analyticity of semigroups and families of the unbounded m-sectorial generators.

Now let A and B be m-sectorial operators in a Hilbert space $\mathfrak{H}$ with domains $\mathrm{dom}(A)$ and $\mathrm{dom}(B)$. Let $S_\theta := \{z \in \mathbb{C} : z \neq 0, \ |\arg z| < \theta\}$ denote the open sector with semi-angle $0 < \theta < \pi/2$ in the complex plane $\mathbb{C}$, and let a and b be associated to operators A and B closed sectorial forms with non-negative vertices and semi-angles θ_A and θ_B. We recall that $\mathrm{Re}A$, $\mathrm{Re}B$ and $\mathrm{Im}A$, $\mathrm{Im}B$ denote respectively the real and imaginary parts of these operators, see Sect. 9.1.

First, we use the *analytic extension* method (Sect. 9.2.2) to prove the following non-self-adjoint generalisation of Proposition 9.2.1 under conditions reflecting the *compactness*.

Theorem 9.2.3 *Let A and B be m-sectorial operators with semi-angles $\theta_A < \pi/2$ and $\theta_B < \pi/2$ and with domains in a Hilbert space $\mathfrak{H}$. If operator* $(\mathbb{1} + A)^{-1}$, *or* $(\mathbb{1} + \mathrm{Re}A)^{-1/2}(\mathbb{1} + \mathrm{Re}B)^{-1/2}$ *are compact. Then the Trotter product approximants* $\{(e^{-tA/n}e^{-tB/n})^n\}_{n\geq 1}$ *converge in the operator-norm topology to* $e^{-tH}P$ *for any* $t \in S_{\theta^*}$. *Here* $H = A \dot{+} B$ *is the form-sum of A and B, P is the orthogonal projector on the closure:* $\overline{\mathrm{dom}(a) \cap \mathrm{dom}(b)}$, *and* $\theta^* < \theta = \pi/2 - \max\{\theta_A, \theta_B\}$. *The convergence is uniform in t on the compact subsets of the sector S_{θ^*}.*

Second, we use the *analytic extension* method (Sect. 9.2.2) to generalise to the non-self-adjoint case the Proposition 9.2.2 with assumptions reflecting the *smallness* of m-sectorial B with respect to m-sectorial A:

(i) $\mathrm{Re}A, \mathrm{Re}B \geq \gamma > 0$;

(ii) $\mathrm{dom}(\mathrm{Re}A) \subset \mathrm{dom}(\mathrm{Im}A)$;

(iii) $\mathrm{dom}(\mathrm{Re}B) \subset \mathrm{dom}(\mathrm{Im}B)$;

(iv) $\mathrm{dom}((\mathrm{Re}A)^\alpha) \subset \mathrm{dom}((\mathrm{Re}B)^\alpha)$ for some $\alpha \in (1/2, 1]$;

(v) $\|(\mathrm{Re}B)^\alpha u\| \leq c\,\|(\mathrm{Re}A)^\alpha u\|$, $u \in \mathrm{dom}((\mathrm{Re}A)^\alpha)$, $0 < c < 1$,

and for the same $\alpha \in (1/2, 1]$.

Theorem 9.2.4 *Let A and B be m-sectorial operators with semi-angles $\theta_A < \pi/2$ and $\theta_B < \pi/2$ and with domains in a Hilbert space $\mathfrak{H}$. If they satisfy assumptions (i)–(v), then for complex t in the open sector S_{θ^*}, the Trotter product approximants*

$\{(e^{-tA/n}e^{-tB/n})^n\}_{n\geq 1}$ converge in the operator-norm topology to e^{-tH}, uniformly on compact subsets of this sector. Here operator $H = A \dot{+} B$ is the form-sum of A and B, and $\theta^* < \theta = \pi/2 - \max\{\theta_A, \theta_B\}$.

Remark 9.2.5 Some comments concerning the assumptions (i)–(v) are in order.

(a) From (i) it follows that self-adjoint operators ReA and ReB are boundedly invertible, and thus, (ii), (iii) imply that there are positive constants λ, μ such that:

$$\|\mathrm{Im}Au\| \leq \lambda \, \|\mathrm{Re}Au\|, \quad u \in \mathrm{dom}(\mathrm{Re}A)\,, \tag{9.2.1}$$

$$\|\mathrm{Im}Bu\| \leq \mu \, \|\mathrm{Re}Bu\|, \quad u \in \mathrm{dom}(\mathrm{Re}B)\,. \tag{9.2.2}$$

(b) From (iv) it follows that $\mathrm{dom}(a) \subset \mathrm{dom}(b)$, thus $\overline{\mathrm{dom}(a) \cap \mathrm{dom}(b)} = \mathfrak{H}$ and the form-sum $A \dot{+} B$ is densely defined.
(c) To clarify assumption (ii) we recall a sufficient condition, which implies (ii). Let us consider the *standard* representation $A = (\mathrm{Re}\, A)^{1/2}\,(\mathbb{1} + i\, L)\,(\mathrm{Re}\, A)^{1/2}$, of the m-sectorial operator A. Here L is self-adjoint operator bounded by $\mathrm{tg}\theta_A$, (9.1.12). If

$$L\,(\,\mathrm{dom}((\mathrm{Re}\, A)^{1/2}) \cap \mathrm{ran}((\mathrm{Re}\, A)^{1/2})\,) \subset \mathrm{dom}((\mathrm{Re}\, A)^{1/2})\,,$$

where $\mathrm{ran}((\mathrm{Re}\, A)^{1/2})$ is the range of operator $(\mathrm{Re}\, A)^{1/2}$, then (ii) holds. A similar condition provides (iii) for operator B.

9.2.2 *Analytic Extension Method for Sectorial Generators*

We start with a pair of m-sectorial operators A and B in a Hilbert space $\mathfrak{H}$, which have the vertex $\gamma = 0$, and define for $z \in \mathbb{C}$ two families of closed forms:

$$a(z) := \mathrm{Re}\, a + z\, \mathrm{Im}\, a, \quad \mathrm{dom}(a(z)) = \mathrm{dom}(a)\,, \tag{9.2.3}$$

$$b(z) := \mathrm{Re}\, b + z\, \mathrm{Im}\, b, \quad \mathrm{dom}(b(z)) = \mathrm{dom}(b)\,. \tag{9.2.4}$$

Here a and b are *sectorial* sesquilinear forms associated with the m-sectorial operators A and B as in Remark 9.1.2 for vertex $\gamma = 0$.

Lemma 9.2.6 *There exists $m_{A,B} > 0$ such that the forms $a(z)$, $b(z)$ are sectorial with vertex $\gamma = 0$ and holomorphic of the type* (a) *for any z in the strip $\mathfrak{S}(m_{A,B}) := \{z \in \mathbb{C} : |\mathrm{Re}\, z| < m_{A,B}\}$, where $m_{A,B} := \min\{\mathrm{ctg}\theta_A, \mathrm{ctg}\theta_B\} > 0$.*

Proof Let $z = x + iy$ with $x, y \in \mathbb{R}$, and $u \in \mathrm{dom}(a)$. Since the form a is sectorial with vertex $\gamma = 0$, we have:

$$|x \,\mathrm{Im}\, a[u]| \leq |x| \,\mathrm{tg}\theta_A \,\mathrm{Re}\, a[u]\,, \quad u \in \mathrm{dom}(a)\,. \tag{9.2.5}$$

Hence $\mathrm{Re}\, a(z) = \mathrm{Re}\, a + x \,\mathrm{Im}\, a$ is positive for $|x| \,\mathrm{tg}\, \theta_A < 1$. Moreover, for such x:

$$\mathrm{Re}\, a(z)[u] \geq \mathrm{Re}\, a[u] - |\, x \,\mathrm{Im}\, a[u]\,| \geq (1 - |x| \,\mathrm{tg}\theta_A) \,\mathrm{Re}\, a[u]. \tag{9.2.6}$$

By (9.2.5) and (9.2.6) we obtain:

$$\begin{aligned} |\,\mathrm{Im}\, a(z)[u]\,| &= |\, y \,\mathrm{Im}\, a[u]\,| \leq |\, y\,| \,\mathrm{tg}\theta_A \,\mathrm{Re}\, a[u] \\ &\leq \frac{|\, y\,| \,\mathrm{tg}\theta_A}{1 - |\, x\,| \,\mathrm{tg}\theta_A} \mathrm{Re}\, a(z)[u]\,, \quad u \in \mathrm{dom}(a)\,, \end{aligned} \tag{9.2.7}$$

which proves that $a(z)$ is sectorial for any z in the strip $|\mathrm{Re}\, z| < \mathrm{ctg}\theta_A$. Similarly, $b(z)$ is sectorial for $|\mathrm{Re}\, z| < \mathrm{ctg}\theta_B$, and we can set $m_{A,B} := \min\{\mathrm{ctg}\theta_A, \mathrm{ctg}\theta_B\}$.

On the other hand, by (9.2.3) and (9.2.4), we also obtain that

$$\begin{aligned} \frac{d}{dz} a(z)[u] &= \mathrm{Im}\, a[u]\,, \quad u \in \mathrm{dom}(a)\,, \\ \frac{d}{dz} b(z)[u] &= \mathrm{Im}\, b[u]\,, \quad u \in \mathrm{dom}(b)\,. \end{aligned}$$

Hence, the forms $a(z)$ and $b(z)$ are holomorphic families of type (a) (see Sect. 9.1.3) for $z \in \mathbb{C}$ in the *strip* $\mathfrak{S}(m_{A,B}) := \{z \in \mathbb{C} : |\mathrm{Re}\, z| < m_{A,B}\}$. □

Corollary 9.2.7 *By the first representation theorem* (*Proposition* 9.1.3), *there exist* m*-sectorial operators* $A(z)$ *and* $B(z)$ *associated in the strip* $\mathfrak{S}(m_{A,B})$ *with the closed sectorial* (a)*-holomorphic forms* $a(z)$ (9.2.3) *and* $b(z)$ (9.2.4), *such that they produce a locally uniformly sectorial holomorphic families*: $\{A(z)\}_{\{z\in\mathfrak{S}(\epsilon_{A,B})\}}$ *and* $\{B(z)\}_{\{z\in\mathfrak{S}(\epsilon_{A,B})\}}$, *of type* (B). *Moreover, on account of definitions* (9.2.3) *and* (9.2.4), *one also gets* $A(z = i) = A$ *and* $B(z = i) = B$, *as well as,* $A(z = 0) = \mathrm{Re}\, A$ *and* $B(z = 0) = \mathrm{Re}\, B$, *cf. definition* (9.1.11)

To advance with presentation of the *analytic extension* method , we have to *select* in the strip $\mathfrak{S}(m_{A,B})$ a domain $(\mathfrak{D})$, which must contain a real interval $\mathfrak{I}$ with $z = 0$, the point $z = i$, as well as to be a finite subset of the strip. All together, this serves to ensure that holomorphic families $\{A(z)\}_{\{z\in\mathfrak{D}\}}$ and $\{B(z)\}_{\{z\in\mathfrak{D}\}}$ will be *uniformly* m-sectorial for some $\theta^*_{AB} < \pi/2$. This means that for all $z \in \mathfrak{D}$ the numerical ranges: $W(A(z))$ and $W(B(z))$, will belong to the sector $S_{\theta^*_{AB}}$.

To this aim, we note that owing to Remark 9.1.1, the numerical range $W(A(z)) = W(a(z))$ (as well as $W(B(z)) = W(b(z))$) and thus we may to use for analysis the inequality (9.2.7). This yields that numerical range $W(A(z))$ is sectorial with $\gamma = 0$ and that it is a subset of the sector $S_{\theta_A(z)}$ with semi-angle

$$\theta_A(z) = \text{arctg}\left\{\frac{|\,y\,|\ \text{tg}\theta_A}{1 - |\,x\,|\ \text{tg}\theta_A}\right\}, \qquad z = x + i\,y\,. \tag{9.2.8}$$

With help of (9.2.8), we select for z a suitable domain $\mathfrak{D}$.

Remark 9.2.8 Since we supposed that $m_{A,B} > 0$, there exists a positive $\delta_1 < m_{A,B}$, which defines a real interval $\mathfrak{I}_{A,B}(\delta_1) := [-(m_{A,B} - \delta_1), (m_{A,B} - \delta_1)]$. Next we define for some finite $\delta_2 > 0$ the interval $\mathfrak{I}_{Im}(\delta_2) := [-i\,(1+\delta_2), i\,(1+\delta_2)]$. As a consequence, domain $\mathfrak{D}_{A,B}(\delta_1, \delta_2) = \mathfrak{I}_{A,B}(\delta_1) \times \mathfrak{I}_{Im}(\delta_2) \subset \mathbb{C}$ satisfies conditions formulated above for a *suitable* finite domain $\mathfrak{D}$ because the real interval $\mathfrak{I}_{A,B}(\delta_1) \subset \mathfrak{D}_{A,B}(\delta_1, \delta_2)$ and $i \in \mathfrak{D}_{A,B}(\delta_1, \delta_2)$.

To check that on domain $\mathfrak{D}_{A,B}(\delta_1, \delta_2)$ the holomorphic families $\{A(z)\}_{\{z \in \mathfrak{D}_{A,B}(\delta_1,\delta_2)\}}$ and $\{B(z)\}_{\{z \in \mathfrak{D}_{A,B}(\delta_1,\delta_2)\}}$ are *uniformly* m-sectorial for some $\theta^*_{AB} < \pi/2$ we use (9.2.8).

The value of semi-angle (9.2.8) of the sector $S_{\theta_A(z)}$ on domain $\mathfrak{D}_{A,B}(\delta)$ is estimated from above by

$$\theta_A(z) \leq \text{arctg}\left\{\frac{(1 + \delta_2)\ \text{tg}\theta_A}{1 - (\,m_{A,B} - \delta_1\,)\ \text{tg}\theta_A}\right\} =: \theta^*_A < \pi/2\,. \tag{9.2.9}$$

Hence, for $z \in \mathfrak{D}_{A,B}(\delta_1, \delta_2)$ the numerical range $W(A(z)) \subset S_{\theta_A(z)} \subset S_{\theta^*_A}$ is uniformly sectorial for $\theta^*_A < \pi/2$. Similarly one infers that on domain $\mathfrak{D}_{A,B}(\delta)$ the estimate from above of semi-angle for the sector $S_{\theta_B(z)}$ is

$$\theta_B(z) \leq \text{arctg}\left\{\frac{(1 + \delta_2)\ \text{tg}\theta_B}{1 - (\,m_{A,B} - \delta_1\,)\ \text{tg}\theta_B}\right\} =: \theta^*_B < \pi/2\,. \tag{9.2.10}$$

Consequently, on domain $\mathfrak{D}_{A,B}(\delta)$ the numerical range $W(B(z)) \subset S_{\theta_B(z)} \subset S_{\theta^*_B}$ is uniformly sectorial for $\theta^*_B < \pi/2$.

Remark 9.2.9 Summarising (9.2.9) and (9.2.10), we obtain that holomorphic families $\{A(z)\}_{\{z \in \mathfrak{D}_{A,B}(\delta_1,\delta_2)\}}$ and $\{B(z)\}_{\{z \in \mathfrak{D}_{A,B}(\delta_1,\delta_2)\}}$ are uniformly m-sectorial for $\theta^*_{AB} := \max\{\theta^*_A, \theta^*_B\} < \pi/2$. This means: $W(A(z)) \subset S_{\theta^*_{AB}}$ and $W(B(z)) \subset S_{\theta^*_{AB}}$ for $z \in \mathfrak{D}_{A,B}(\delta_1, \delta_2)$ and $\theta^*_{AB} < \pi/2$. Seeing that by (9.2.9) and (9.2.10) one gets $\theta_A < \theta^*_A$ and $\theta_B < \theta^*_B$, we infer that $\theta_{AB} < \theta^*_{AB}$, where $\theta_{AB} := \max\{\theta_A, \theta_B\}$.

For continuation, we recall that if A is an m-sectorial operator with semi-angle $\theta_A < \pi/2$, then it generates on Hilbert space $\mathfrak{H}$ a holomorphic contraction semigroup $\{e^{-tA}\}_{t \in S_{\pi/2-\theta_A}}$, see Sect. 1.8.2. Then to proceed, we need a following general preparatory result.

Lemma 9.2.10 *Let $\{T(z)\}_{z \in \mathfrak{D}}$ be a holomorphic family of uniformly m-sectorial operators in Hilbert space $\mathfrak{H}$, such that for any $z \in \mathfrak{D} \subset \mathbb{C}$ the numerical range*

$W(T(z)) \subset S_{\theta_T^*}$, for open sector with semi-angle $\theta_T^* < \pi/2$. Then, the family of holomorphic contraction semigroups $\{e^{-tT(z)}\}_{t \in S_{\pi/2-\theta_T^*}}$, $z \in \mathfrak{D}$, is such that for each $t \in S_{\pi/2-\theta_T^*}$ the mapping : $\mathfrak{D} \ni z \mapsto e^{-tT(z)}$, is holomorphic in $\mathfrak{D}$ in the operator-norm topology.

Proof Since $\{T(z)\}_{z\in\mathfrak{D}}$ is a family of m-sectorial operators, by Proposition 1.8.13 for each $z \in \mathfrak{D}$ the operator $T(z)$ is a generator of the holomorphic contraction semigroup $\{\mathrm{e}^{-tT(z)}\}_{t \in S_{\pi/2-\theta_{T(z)}}}$. This proves the first part of assertion.

Given that family $\{T(z)\}_{z\in\mathfrak{D}}$ is uniformly sectorial, i.e., $W(T(z)) \subset S_{\theta_{T(z)}} \subset S_{\theta_T^*}$ for $z \in \mathfrak{D}$, we infer that family of holomorphic contraction semigroups $\{\mathrm{e}^{-tT(z)}\}_{t \in S_{\pi/2-\theta_T^*}}$ has the *Riesz-Dunford* integral representation (Proposition 1.7.4):

$$\mathrm{e}^{-tT(z)} = \frac{1}{2\pi i}\int_\Gamma d\zeta\, \frac{\mathrm{e}^{-t\zeta}}{\zeta\,\mathbb{1} - T(z)}\,, \quad z \in \mathfrak{D}\,, \tag{9.2.11}$$

where $\Gamma \subset \mathbb{C} \setminus S_{\theta_T^*}$ is an independent of $z \in \mathfrak{D}$ positively-oriented contour running from $\infty \cdot \mathrm{e}^{i\,(\theta_T^*+\varepsilon)}$ to $\infty \cdot \mathrm{e}^{-i\,(\theta_T^*+\varepsilon)}$ and enclosing the sector $S_{\theta_T^*}$ for $\theta_T^* + \varepsilon < \pi/2$.

Since $\{T(z)\}_{z\in\mathfrak{D}}$ is a holomorphic family, in virtue of Proposition 9.1.8 we obtain that resolvent:

$$z \mapsto (T(z) - \zeta\,\mathbb{1})^{-1}\,, \quad z \in \mathfrak{D}\,, \quad \zeta \in \mathbb{C} \setminus S_{\theta_T^*} \subset \rho(T(z))\,, \tag{9.2.12}$$

is bounded-holomorphic in $z \in \mathfrak{D}$ because ζ is in the resolvent set of $T(z)$. Thanks to the choice of Γ the convergence in (9.2.11) for integrand and for its derivatives is absolute and uniform in parameters t and z. So, (9.2.11) can be differentiated under the integral sign, which yields

$$\partial_z^n\, \mathrm{e}^{-tT(z)} = \frac{1}{2\pi i}\int_\Gamma d\zeta\, \mathrm{e}^{-t\zeta}\, \partial_z^n\, (\zeta\,\mathbb{1} - T(z))^{-1}\,, \quad z \in \mathfrak{D}\,. \tag{9.2.13}$$

On account of bounded analyticity (9.2.12) we obtain for $z \in \mathfrak{D}$ and $n \in \mathbb{N}$ that

$$\partial_z^n\, (\zeta\,\mathbb{1} - T(z))^{-1} = \frac{n!}{2\pi i}\int_{C_r} dz'\, (z'-z)^{-n-1}\, (\zeta\,\mathbb{1} - T(z'))^{-1}\,, \quad \zeta \in \Gamma\,,$$

where a circle $C_r := \{z' \in \mathfrak{D} : |z'-z| = r\}$, as well as the operator-norm estimates

$$\|\partial_z^n\, (\zeta\,\mathbb{1} - T(z))^{-1}\| \leq \frac{M_{\Gamma,r}\, n!}{|\zeta|\, r^n} \quad n \in \mathbb{N}\,, \quad \zeta \in \Gamma\,,\ z' \in C_r\,, \tag{9.2.14}$$

with constant $M_{\Gamma,r} = \sup_{\zeta\in\Gamma,\ z'\in C_r} \|\zeta\,(\zeta\,\mathbb{1} - T(z'))^{-1}\|$.

As a result, owing to representation (9.2.11) and (9.2.13), (9.2.14)

$$z \mapsto e^{-t\,T(z)}\,, \quad t \in S_{\pi/2-\theta_T^*}\,, \tag{9.2.15}$$

is a bounded-holomorphic (in the operator norm) function in a disc $\mathcal{D}_{r(t)} := \{z' \in \mathfrak{D} : |z' - z| < r(t) < r\}$ as the Taylor expansion in neighbourhood $\mathcal{D}_{r(t)}$ of $z \in \mathfrak{D}$ with coefficients holomorphic in $t \neq 0$, cf. (1.7.8). By analytic continuation this function can be extended to $\mathfrak{D}$, where it is contraction by the first part of assertion.

□

Remark 9.2.11

(a) We remark that the conclusions of Lemma 9.2.10 are true for any *type* of holomorphic family $\{T(z)\}_{z\in\mathfrak{D}}$ provided that it is uniformly m-sectorial: $W(T(z)) \subset S_{\theta_T^*}$, for some open sector with semi-angle $\theta_T^* < \pi/2$. Then on account of the resolvent analyticity (9.2.12) the holomorphic family of type (A) is equivalent to holomorphic family of type (B).
(b) Note that for bounded holomorphic operator-values functions on Hilbert (or on Banach) space there is no distinction between *uniform* (operator-norm), *strong* and *weak* analyticities. See Proposition 1.10.5 and Corollary 1.10.6 in Notes to Chap. 1 (Notes to Sect. 1.7). It is for this reason that we declare results in Lemma 9.2.10 in only the operator-norm topology.
(c) The simplest holomorphic family of type (A) is a *linear* perturbation: $T(z) = A + z\,B$, $z \in \mathfrak{D}$, of generator A of *contraction holomorphic* semigroup on a Hilbert space. For variety of perturbations B this problem is scrutinised in Sect. 1.9.2.

Corollary 9.2.12 *Now we apply the result of Lemma* 9.2.10 *to the case of uniformly m-sectorial for $z \in \mathfrak{D}_{A,B}(\delta_1,\delta_2)$ holomorphic families $\{A(z)\}_{\{z\in\mathfrak{D}_{A,B}(\delta_1,\delta_2)\}}$ and $\{B(z)\}_{\{z\in\mathfrak{D}_{A,B}(\delta_1,\delta_2)\}}$, such that (as in Remark* 9.2.9) *$W(A(z)) \subset S_{\theta^*}$ and $W(B(z)) \subset S_{\theta^*}$ for $z \in \mathfrak{D}_{A,B}(\delta_1,\delta_2)$ and $\theta^* < \pi/2$.*

Then two families of holomorphic contraction semigroups: $\{e^{-tA(z)}\}_{t\in S_{\pi/2-\theta^*}}$ and $\{e^{-tB(z)}\}_{t\in S_{\pi/2-\theta^*}}$, are also holomorphic in variable $z \in \mathfrak{D}_{A,B}(\delta_1,\delta_2)$ in the operator-norm topology.

The following proposition (*Vitali* theorem) is a key of the *analytic extension* method that ensures the proof of convergence the Trotter product formula for sectorial generators. We formulate this proposition here in a suitable for us form, cf. Notes to Chap. 9 (Notes to Sect. 9.2).

Proposition 9.2.13 (Vitali Theorem) *Let $\{f_n\}_{n\geq 1}$ be a family of holomorphic vector-valued functions $f_n : \mathbb{C} \to \mathfrak{X}$, in Banach space $\mathfrak{X}$ with a common domain $D \subset \mathbb{C}$. Suppose that this family is bounded: $\|f_n(z)\| \leq M$ uniformly in $n \in \mathbb{N}$ and in $z \in K \subset D$ for any compact $K \subset D$. If the sequence $\{f_n(z)\}_{n\geq 1}$ converges to $f(z)$ for all z from a subset $D_0 \subset D$ having at least one accumulation point in D, then $f_n(z)$ converges to $f(z)$ for any $z \in D$ uniformly on compact subsets of domain D, and the limit function $D \ni z \mapsto f(z)$ is holomorphic in D.*

Remark 9.2.14 Let $t > 0$ be fixed. Then, the *analytic extension* method for proving the operator-norm convergence of the Trotter product formula consists of application of the *Vitali* theorem to the following family of functions:

$$z \mapsto f_n(z) := \left(e^{-tA(z)/n}e^{-tB(z)/n}\right)^n, \quad z \in D \subset \mathbb{C}, \ n = 1, 2, \ldots, \tag{9.2.16}$$

with values in the Banach space of bounded operators $\mathcal{L}(\mathfrak{H})$ on $\mathfrak{H}$. This method consists of the following steps:

(1) By Lemma 9.2.6, Corollary 9.2.7, Lemma 9.2.10, and Corollary 9.2.12 functions (9.2.16) are holomorphic in $\mathfrak{D}_{A,B}(\delta_1, \delta_2)$, and $\|f_n(z)\| \leq 1$ for any $z \in \mathfrak{D}_{A,B}(\delta_1, \delta_2)$.
(2) Then we show that under conditions of Theorems 9.2.3 (or Theorem 9.2.4), the results of Proposition 9.2.1 (or Proposition 9.2.2) about the operator-norm convergence of the Trotter product formula can be applied to self-adjoint operators $A(x)$ and $B(x)$, where arguments $x \in \mathfrak{I}_{A,B}(\delta_1)$ are real and small to ensure the positivity of $A(x)$ and $B(x)$.
(3) By virtue of analytic continuation and the *Vitali* theorem this operator-norm convergence can be extended from $\mathfrak{I}_{A,B}(\delta_1)$ to uniform operator-norm convergence in domain $\mathfrak{D}_{A,B}(\delta_1, \delta_2) \subset \mathbb{C}$.
(4) Since the uniform limit is in turn a holomorphic function and $z = i \in \mathfrak{D}_{A,B}(\delta_1, \delta_2)$, we obtain the operator-norm convergent Trotter product formula (9.2.16) for initial m-sectorial generators $A \equiv A(z = i)$ and $B \equiv B(z = i)$ for all $t > 0$, as well as for $t \in S_{\pi/2-\theta^*_{AB}}$ (Remark 9.2.9).

Corollary 9.2.12 summarises the step (1) in Remark 9.2.14. The rest of these steps we shall scrutinise in the next Sect. 9.2.3.

9.2.3 Operator-Norm Convergence of the Trotter Product Formula

In this subsection we develop in details the proofs of the steps (2)–(4) outlined in Remark 9.2.14.

Proof of Theorem 9.2.3 Let $(\mathbb{1} + A)^{-1}$ be compact. Then, seeing that operator A is m-sectorial, $(\mathbb{1} + \mathrm{Re}A)^{-1}$ is also compact by Corollary 9.1.6. Here $\mathrm{Re}A = A(z = 0)$, see Corollary 9.2.7. Recall that if a type (B) holomorphic in $D \subset \mathbb{C}$ family of m-sectorial operators $\{A(z)\}_{z \in D}$ has compact resolvent for one $z_0 \in D$, then it is true for all $z \in D$, see Remark 9.1.13.

Since by Lemma 9.2.6 and by Corollary 9.2.7 the $\{\mathbb{1} + A(z)\}_{z \in \mathfrak{D}_{A,B}(\delta_1, \delta_2)}$ is the type (B) holomorphic family of invertible m-sectorial operators and $(\mathbb{1} + A(0))^{-1}$ is compact, on account of Remark 9.1.13 the operator $(\mathbb{1} + A(z))^{-1}$ is compact for any $z \in \mathfrak{D}_{A,B}(\delta_1, \delta_2)$. In particular, it is true for any $z = x$ from the real interval

$\mathfrak{I}_{A,B}(\delta_1)$, defined in Remark 9.2.8. Then for each $x \in \mathfrak{I}_{A,B}(\delta_1)$ the self-adjoint generators $A(x)$ and $B(x)$ (9.1.11) satisfy conditions of Proposition 9.2.1. This means that we have the operator-norm convergence of the Trotter approximants (9.2.16):

$$f_n(x) = \left(e^{-tA(x)/n} e^{-tB(x)/n}\right)^n , \quad x \in \mathfrak{I}_{A,B}(\delta_1) \subset \mathbb{C} , \quad n = 1, 2, \ldots , \tag{9.2.17}$$

where $t > 0$ and $\|f_n(x)\| \leq 1$, uniformly in $x \in \mathfrak{I}_{A,B}(\delta_1)$ to $f(x) = e^{-t\,(A(x)\dot{+}B(x))}\, P$.

Now following the steps (2) and (3) in Remark 9.2.14 we note that owing to Corollary 9.2.12, the approximants (9.2.17) have analytic extensions $\{f_n(z)\}_{n\geq 1}$ to domain $\mathfrak{D}_{A,B}(\delta_1, \delta_2)$, which preserves the property of contraction in this domain. Then applying to this sequence of approximants the *Vitali* theorem (Proposition 9.2.13) one obtains a uniform on compact $\mathfrak{D}_{A,B}(\delta_1, \delta_2)$ operator-norm convergence

$$\|\cdot\| - \lim_{n\to\infty} f_n(z) = f(z) , \quad z \in \mathfrak{D}_{A,B}(\delta_1, \delta_2) \subset \mathbb{C} , \quad t > 0 , \tag{9.2.18}$$

to holomorphic (by the *Weierstrass* theorem about uniform convergence of holomorphic functions) function $f(z) = e^{-t\,(A(z)\dot{+}B(z))}\, P$, with $\|f(z)\| \leq 1$.

Finally, step (4) in Remark 9.2.14, on account of $i \in \mathfrak{D}_{A,B}(\delta_1, \delta_2)$ and $A(i) = A$, $B(i) = B$, one obtains the proof of the first part of Theorem 9.2.3 for $t > 0$.

Proof of the second part of Theorem 9.2.3, when $(\mathbb{1} + \mathrm{Re}A)^{-1/2}(\mathbb{1} + \mathrm{Re}B)^{-1/2}$ is compact, follows *mutatis mutandis* the arguments of the first part. More comments one can find in Notes to Chap. 9 (Notes to Sect. 9.2). □

To continue with the proof of Theorem 9.2.4, we note that according to Lemma 9.2.6 and Corollary 9.2.7 to exploit the *analytic extension* method for sectorial generators $A(z)$ and $B(z)$ the parameter $x = \mathrm{Re}z$ must be restricted: $|x| < m_{A,B}$. This allows the analytic extension of sectorial families $\{A(x)\}_{\{x\,:\,|x|<m_{A,B}\}}$, $\{B(x)\}_{\{x\,:\,|x|<m_{A,B}\}}$ to the strip $\mathfrak{S}(m_{A,B}) := \{z \in \mathbb{C} \;:\; |\mathrm{Re}\, z| < m_{A,B}\}$. On the other hand, in contrast to Theorem 9.2.3, we have to choose in addition the value of $|x|$ in such a way that conditions of *smallness* in Proposition 9.2.2 and Theorem 9.2.4 must be satisfied.

To this aim, we start by a technical Lemma 9.2.16, which in turn is based on the following result known as the *Löwner-Heinz-Kato* theorem:

Proposition 9.2.15 *Let A and B be positive self-adjoint operators in a Hilbert space $\mathfrak{H}$, and suppose that* $\mathrm{dom}(A) \subset \mathrm{dom}(B)$. *If, in addition, there is a number $M > 0$ such that $\|Bu\| \leq M\|Au\|$ for all $u \in$* $\mathrm{dom}(A)$, *then for all α satisfying $0 < \alpha < 1$, we have* $\mathrm{dom}(A^\alpha) \subset \mathrm{dom}(B^\alpha)$ *and $\|B^\alpha u\| \leq M^\alpha \|A^\alpha u\|$ for all $u \in$* $\mathrm{dom}(A^\alpha)$.

Lemma 9.2.16 *If A and B satisfy the assumptions (i)–(v), then for any real x with $|x| < \min\{m_{A,B}, 1/\lambda\}$ one has:*

$$\|B(x)^{\alpha}u\| \leq \left(\frac{1+\mu\,|x|}{1-\lambda\,|x|}\right)^{\alpha} c\,\|A(x)^{\alpha}u\|\,, \quad u \in \mathrm{dom}(A(x)^{\alpha}). \tag{9.2.19}$$

Here λ and μ are defined by (9.2.1) *and* (9.2.2)*, whereas $c < 1$ and $1/2 < \alpha \leq 1$ are the same as in condition (v).*

Proof By (9.2.2), which follows from (iii), we have inequality:

$$\|B(x)\,u\| \leq \|\mathrm{Re}B\,u\| + |x|\,\|\mathrm{Im}B\,u\| \leq (1+\mu\,|x|)\,\|\mathrm{Re}B\,u\|\,,$$

for any $u \in \mathrm{dom}(\mathrm{Re}B) \subset \mathrm{dom}(B(x))$. Since positive $B(x)$ is self-adjoint, this inequality implies by the *Löwner-Heinz-Kato* theorem that

$$\|B(x)^{\alpha}u\| \leq (1+\mu|x|)^{\alpha}\,\|(\mathrm{Re}B)^{\alpha}u\|\,, \quad u \in \mathrm{dom}((\mathrm{Re}B)^{\alpha}). \tag{9.2.20}$$

In virtue of (9.2.1) and the *Kato* perturbation theory for self-adjoint operators (Theorem 1.9.11 and Corollary 1.9.15) the sum of two terms: $\mathrm{Re}(A) + x\,\mathrm{Im}(A)$, is self-adjoint on domain $\mathrm{dom}(\mathrm{Re}A)$ due to the *relative smallness*, for $|x| < 1/\lambda$, of perturbation: $\|x\,\mathrm{Im}(A)\,u\| \leq \|x\lambda\,\mathrm{Re}(A)\,u\|$, for $u \in \mathrm{dom}(\mathrm{Re}A)$. Seeing that by construction (cf. (9.1.8) and Corollary 9.2.7) the operator $A(x)$ is self-adjoint form-sum of these two terms, where $x\,\mathrm{Im}(A)$ is a small perturbation, we obtain $\mathrm{dom}(A(x)) = \mathrm{dom}(\mathrm{Re}A)$. Then by (9.2.1) one infers for any u in this domain estimate

$$\|\mathrm{Re}A\,u\| \leq \|A(x)u\| + |x|\,\|\mathrm{Im}A\,u\| \leq \|A(x)u\| + \lambda\,|x|\,\|\mathrm{Re}A\,u\|\,, \quad u \in \mathrm{dom}\,(\mathrm{Re}A).$$

Hence, due to $|x| < 1/\lambda$, we get

$$\|\mathrm{Re}A\,u\| \leq \frac{1}{1-\lambda\,|x|}\,\|A(x)u\|\,, \quad u \in \mathrm{dom}\,(\mathrm{Re}A)\,,$$

and consequently, by the *Löwner-Heinz-Kato* theorem (Proposition 9.2.15), the inequality:

$$\|(\mathrm{Re}A)^{\alpha}u\| \leq \frac{1}{(1-\lambda\,|x|)^{\alpha}}\,\|A(x)^{\alpha}u\|\,, \quad u \in \mathrm{dom}(A(x)^{\alpha})\,. \tag{9.2.21}$$

Finally the inequalities (9.2.20) and (9.2.21), together with assumption (v) yield the announced estimate (9.2.19). □

Proof of Theorem 9.2.4 Since by condition (v) on the fractional powers of operators A and B one has $c < 1$, there exists $\eta_{AB}(\lambda) < \min\{(m_{A,B}, 1/\lambda\}$ such that for $\alpha \in (1/2, 1]$ one has

$$c\left(\frac{1+\mu|x|}{1-\lambda|x|}\right)^{\alpha} < 1\,, \quad x \in \mathcal{I}_{AB}(\lambda) := [-\eta_{AB}(\lambda), \eta_{AB}(\lambda)]\,. \tag{9.2.22}$$

Therefore, if $x \in \mathcal{I}_{AB}(\lambda)$, then estimates (9.2.19) and (9.2.22) ensure that conditions of Proposition 9.2.2 are satisfied for the families of self-adjoint operators $A(x)$ and $B(x)$. This means that for $n \to \infty$ we have the operator-norm convergence of the Trotter approximants

$$f_n(x) = \left(\mathrm{e}^{-tA(x)/n}\mathrm{e}^{-tB(x)/n}\right)^n\,, \quad x \in \mathcal{I}_{AB}(\lambda)\,, \quad n = 1, 2, \ldots\,, \tag{9.2.23}$$

uniformly in $x \in \mathcal{I}_{AB}(\lambda)$ to $f(x) = \mathrm{e}^{-t\,(A(x)\dot{+}B(x))}$, where $t > 0$ and $\|f_n(x)\| \leq 1$.

Let $\mathcal{I}(\delta) := [-i\,\delta,\ i\,\delta]$, for $\delta > 0$. We select in the strip $\mathfrak{S}(m_{A,B})$ the closed domain $\mathfrak{D}_{A,B}(\lambda,\delta) := \mathcal{I}_{A,B}(\lambda) \times \mathcal{I}(\delta)$. Then by estimates (9.2.9) and (9.2.10) the holomorphic families $\{A(z)\}_{\{z\in\mathfrak{D}_{A,B}(\lambda,\delta)}$ and $\{B(z)\}_{\{z\in\mathfrak{D}_{A,B}(\lambda,\delta)\}}$ are uniformly m-sectorial on any compact of $\mathfrak{D}_{A,B}(\lambda,\delta)$ for some semi-angle $\theta^* < \pi/2$. As a consequence, in virtue of Corollary 9.2.12, they generate two families of holomorphic contraction semigroups: $\{\mathrm{e}^{-tA(z)}\}_{t>0}$ and $\{\mathrm{e}^{-tB(z)}\}_{t>0}$, which are also operator-norm holomorphic in variable $z \in \mathfrak{D}_{A,B}(\lambda,\delta)$.

Thereby the Trotter approximants (9.2.23) have analytic extensions $\{f_n(z)\}_{n\geq 1}$ to domain $\mathfrak{D}_{A,B}(\lambda,\delta)$, which preserve the property of contraction in this domain. Then, applying to this sequence of approximants the *Vitali* theorem (Proposition 9.2.13), we infer a uniform on compact $\mathfrak{D}_{A,B}(\lambda,\delta)$ operator-norm convergence

$$\|\cdot\| - \lim_{n\to\infty} f_n(z) = f(z)\,, \quad z \in \mathfrak{D}_{A,B}(\lambda,\delta) \subset \mathbb{C}\,, \quad t > 0\,, \tag{9.2.24}$$

to holomorphic function $f(z) = \mathrm{e}^{-t\,(A(z)\dot{+}B(z))}$, with $\|f(z)\| \leq 1$. Finally, seeing that $i \in \mathfrak{D}_{A,B}(\lambda,\delta)$ and also $A(i) = A$, $B(i) = B$, we conclude the proof of Theorem 9.2.4 for $t > 0$. □

Corollary 9.2.17 *Let families $\{A(z)\}_{z\in\mathfrak{D}}$ and $\{B(z)\}_{z\in\mathfrak{D}}$ be uniformly m-sectorial for some $\theta^*_{AB} < \pi/2$, cf. Remark* 9.2.9. *This means that for all $z \in \mathfrak{D}$ the numerical ranges: $W(A(z))$ and $W(B(z))$, belong to the sector $S_{\theta^*_{AB}}$. Then they generate two families of holomorphic contraction semigroups: $\{e^{-tA(z)}\}_{t\in S_{\theta^*}}$ and $\{e^{-tB(z)}\}_{t\in S_{\theta^*}}$, enumerated by $z \in \mathfrak{D}$ and also a holomorphic in $t \in S_{\theta^*}$ contraction family of the Trotter approximants*:

$$t \mapsto \left\{\left(e^{-tA(z)/n}e^{-tB(z)/n}\right)^n\right\}_{n\geq 1}\,, \quad z \in \mathfrak{D}\,, \tag{9.2.25}$$

where we denote $\theta^* = \pi/2 - \theta^*_{AB}$. *Now, if for* $z \in \mathfrak{D}$ *the Trotter product formula converges in the operator-norm topology for* $t > 0$:

$$\|\cdot\| - \lim_{n\to\infty} \left(e^{-tA(z)/n} e^{-tB(z)/n}\right)^n = e^{-tH(z)}, \quad z \in \mathfrak{D}, \tag{9.2.26}$$

then by the Vitali theorem (*Proposition* 9.2.13), *the convergence in* (9.2.26) *holds in the operator-norm topology to holomorphic function* : $S_{\theta^*} \ni t \mapsto e^{-tH(z)}$, *locally uniformly in* $t \in S_{\theta^*}$. *Owing to Remark* 9.2.9 *and Corollary* 9.2.12, *this reasoning proves extension of Theorems* 9.2.3 *and* 9.2.4 *from* $t > 0$ *to the sector* S_{θ^*}.

9.3 Error Bound Estimate for Operator-Norm Convergence

9.3.1 Accretive Perturbations of Self-Adjoint Semigroups

In Theorem 9.2.4 (Sect. 9.2), we established the operator-norm convergence of the Trotter product formula for non-self-adjoint semigroups with m-sectorial generators A and B with a certain *smallness* subordination between them. Note that the both of corresponding semigroups were holomorphic contractions.

To control the *rate* of the operator-norm convergence, we impose in this section a more *restrictive* condition on generator A but we *relax* conditions of Theorem 9.2.4 on generator B, cf. Sect. 8.5.2.

Hypothesis (H1) We suppose that $A = A^*$ is a semibounded from below densely defined self-adjoint operator in a Hilbert space $\mathfrak{H}$, which generates a C_0-semigroup $\{U_A(t) = \mathrm{e}^{-tA}\}_{t\geq 0}$. Without loss of generality, we suppose that $A \geq \mathbb{1}$.

Then $\{U_A(t)\}_{t\geq 0}$ can be extended to contraction holomorphic semigroup of semi-angle $\theta = \pi/2$ such that $\|U_A(t \geq 0)\| \leq 1$. Indeed, by Definition 1.7.1 and by estimate (1.7.4) one gets that generator $A \in \mathscr{H}_c(\theta = \pi/2, 0)$. Consequently, the semigroup has a holomorphic extension $\{U_A(z)\}_{z\in S_\theta}$ from $t \geq 0$ into the open complex half-plane: $S_{\theta=\pi/2} = \{z \in \mathbb{C} : z \neq 0 \text{ and } |\arg(z)| < \pi/2\}$.

Hypothesis (H2) About the generator B we suppose that:

(i) domain dom $(B) \supseteq$ dom (A) and there exists $a > 0$ such that

$$\|Bu\| \leq a\,\|Au\|, \quad u \in \text{dom}\,(A),$$

(ii) domain dom $(B^*) \supseteq$ dom (A) and there exists $a_* > 0$ such that

$$\|B^*u\| \leq a_*\,\|Au\|, \quad u \in \text{dom}\,(A),$$

(iii) B is a m-accretive (Definition 1.8.7) operator in $\mathfrak{H}$, that is one has

$$\|(B + z\mathbb{1})^{-1}\| \leq \frac{1}{\mathrm{Re}\, z}, \qquad \mathrm{Re}\, z > 0,$$

(iv) the parameters $a, a_* < 1$.

By H2 *(iii)* we obtain that $\{U_B(t) = \mathrm{e}^{-tB}\}_{t\geq 0}$ is a (non-self-adjoint) contraction semigroup, with $W(B) \subset \overline{S}_{\theta_B}$, but B is *not* necessarily *sectorial* because the angle θ_B may be equal to $\pi/2$, that is, $\overline{S}_{\pi/2} = \overline{\mathbb{C}}_+ := \{z \in \mathbb{C} : \mathrm{Re}\, z \geq 0\}$.

Theorem 9.3.1 *Let self-adjoint operator $A \geq \mathbb{1}$ and let B be m-accretive operator satisfying condition* (H2) *(i) for $a < 1$. Then operator $H := A + B$ is boundedly invertible generator of a contraction holomorphic semigroup of semi-angle $\theta =$* arccos(a) *(see (iv)), that is, $H \in \mathscr{H}_c(\theta, 0)$.*

Proof Note that by *(i)* and $a < 1$ (cf. (H2) *(iv)*) the operator B is *Kato-small* (Definition 1.9.7) with respect to A. Then the operator sum $H := A + B$ is densely defined and closed operator on dom (A).

Owing to *(iii)* the operator B is accretive. Then, given that $A \geq \mathbb{1}$ is self-adjoint and positive, the numerical range $W(A + B) \subset \overline{\mathbb{C}}_+$. This implies that the operator H is also *accretive*, Definition 1.8.7.

By *(i)* and *(iv)*, one gets the representation $H = (\mathbb{1} + BA^{-1})\, A$ with $\|BA^{-1}\| \leq a < 1$. Hence, the operator H is boundedly invertible. This ensures that operator $\lambda\mathbb{1} + H$ is also boundedly invertible for $\lambda > 0$. Then by virtue of (1.8.10) operator H is m-accretive, i.e., (see Proposition 1.8.10) it generates a *contraction* semigroup.

To verify that $H \in \mathscr{H}_c(\theta, 0)$, that is, a generator of holomorphic contraction semigroup (Definition 1.7.1) we show that for $\omega := \theta - \varepsilon$ and *any* positive ε $(< \theta)$

$$\left\|(\zeta\mathbb{1} + H)^{-1}\right\| \leq \frac{M_\varepsilon}{|\zeta|}, \qquad \zeta \in \overline{S}_{\pi/2+\theta-\varepsilon} \setminus \{0\}, \tag{9.3.1}$$

where $M_\varepsilon > 0$ is independent of ζ and $\theta = \arccos(a) < \pi/2$.

To this aim, we use the representation

$$(H + z\mathbb{1})f = \{\mathbb{1} + BA^{-1}A(A + z\mathbb{1})^{-1}\}(A + z\mathbb{1})f\ , \ \ f \in \mathrm{dom}\,(H) = \mathrm{dom}\,(A)\,. \tag{9.3.2}$$

Since

$$\|A(A + z\mathbb{1})^{-1}\| \leq \sup_{\lambda\geq 1}\left|\frac{\lambda}{\lambda + z}\right| \leq \sup_{\lambda\geq 0}\left|\frac{\lambda}{\lambda + z}\right|,$$

a straightforward computation shows that

$$\|A(A + z\mathbb{1})^{-1}\| \leq \begin{cases} 1 & \text{for } \mathrm{Re} z \geq 0 \\ |z|/|\mathrm{Im}\, z| & \text{for } \mathrm{Re} z < 0\,,\ \mathrm{Im}\, z \neq 0 \end{cases} \tag{9.3.3}$$

Let $z \in S_{\pi/2+\omega}$ and Re$z < 0$, then $|z|/|\text{Im}z| \leq 1/\cos\omega$. Hence (9.3.3) leads to

$$\|A\,(A + z\mathbb{1})^{-1}\| \leq \frac{1}{\cos\omega}\,, \tag{9.3.4}$$

for $z \in S_{\pi/2+\omega}$ and Re$z < 0$. If Re$z \geq 0$, then (9.3.3) yields the inequalities

$$\|A\,(A + z\mathbb{1})^{-1}\| \leq 1 \leq \frac{1}{\cos\omega}\,.$$

As a result, the estimate (9.3.4) holds for any $z \in S_{\pi/2+\omega}$.

Next, using the condition H2*(i)* we find that

$$\|B\,A^{-1}A(A + z\mathbb{1})^{-1}\| \leq \|B\,A^{-1}\|\,\|A(A + z\mathbb{1})^{-1}\| \leq \frac{a}{\cos\omega} < 1\,,$$

for $z \in S_{\pi/2+\omega}$. Hence, the operator $\mathbb{1} + B\,(A + z\mathbb{1})^{-1}$ is invertible for any $z \in S_{\pi/2+\omega}$ and we have the estimate

$$\|(\mathbb{1} + B\,(A + z\mathbb{1})^{-1})^{-1}\| \leq \frac{\cos\omega}{\cos\omega - a}\,. \tag{9.3.5}$$

Then, from (9.3.2) and (9.3.5) we obtain for $z \in S_{\pi/2+\omega} \subseteq \rho(-H)$ the representation $(H + z\mathbb{1})^{-1} = (A + z\mathbb{1})^{-1}(\mathbb{1} + B\,(A + z\mathbb{1})^{-1})^{-1}$, and the estimate

$$\|(H + z\mathbb{1})^{-1}\| \leq \frac{\cos\omega}{\cos\omega - a}\,\|(A + z\mathbb{1})^{-1}\|\,. \tag{9.3.6}$$

Now, since generator A is self-adjoint, we get

$$\|(z\mathbb{1} + A)^{-1}\| \leq \frac{1}{\cos\omega}\,\frac{1}{|z|}\,, \tag{9.3.7}$$

for any $z \in S_{\pi/2+\omega}$. Thus (9.3.6) and (9.3.7) yield

$$\|(z\mathbb{1} + H)^{-1}\| \leq \frac{M_\varepsilon}{|z|}\,,$$

for all $z \in S_{\pi/2+\omega}$. Recall that $\omega = \theta - \varepsilon$, for any $0 < \varepsilon < \theta$, and thus

$$M_\varepsilon := \frac{1}{\cos(\theta - \varepsilon) - a}\,, \tag{9.3.8}$$

cf. (9.3.1). This proves that operator $H \in \mathscr{H}_c(\theta, 0)$ is generator of holomorphic contraction semigroup with the semi-angle $\theta = \arccos(a)$. □

Note that since $\sigma(H) \subseteq \overline{S}_{\pi/2-\theta}$ implies $\sigma(H^*) \subseteq \overline{S}_{\pi/2-\theta}$ for spectra of opearators H and H^*, the characterisation (9.3.1) entails that operator H^* is also generator of a holomorphic contraction semigroup.

9.3.2 Preliminary Estimates

To advance with estimate of the operator-norm error bound for the rate of convergence of the Trotter product approximants, we need some preliminary estimates. Below we denote by $\mathbb{D} := \{z \in \mathbb{C} : |z| \leq 1\}$ and $\mathbb{D}_{1/2} := \{z \in \mathbb{C} : |z - 1/2| \leq 1/2\}$, two discs in the complex plane $\mathbb{C}$.

Lemma 9.3.2 *Let self-adjoint $A \geq \mathbb{1}$. Then, for any $\tau > 0$, one gets the estimate*

$$\|(\mathbb{1} - e^{-\tau A})(z - e^{-\tau A})^{-1}\| \leq \begin{cases} |z|^{-1}\,, & z \in \mathbb{D} \setminus \mathrm{D}_{1/2}\,, \\ |1 - z|/|\mathrm{Im}\, z|\,, & z \in \mathrm{D}_{1/2} \setminus [0, 1]\,. \end{cases} \tag{9.3.9}$$

Proof By the *functional calculus* for self-adjoint operator A one gets inequality

$$\|(\mathbb{1} - \mathrm{e}^{-\tau A})\,(z - \mathrm{e}^{-\tau A})^{-1}\| \leq \sup_{\lambda \geq 1} \left| \frac{1 - \mathrm{e}^{-\tau\lambda}}{z - \mathrm{e}^{-\tau\lambda}} \right| \leq \sup_{\lambda \geq 0} \left| \frac{1 - \mathrm{e}^{-\lambda}}{z - \mathrm{e}^{-\lambda}} \right| .$$

Let $z = x + iy$ and

$$m_z(\lambda) := \frac{1 - \mathrm{e}^{-\lambda}}{|z - \mathrm{e}^{-\lambda}|} = \frac{1 - \mathrm{e}^{-\lambda}}{\sqrt{(x - \mathrm{e}^{-\lambda})^2 + y^2}}\,, \quad \lambda \geq 0\,. \tag{9.3.10}$$

If $z \in \mathbb{D} \setminus \mathbb{D}_{1/2}$, then for $\lambda \geq 0$ derivative $m_z'(\lambda) \geq 0$. Therefore $m_z(\lambda)$ is a non-decreasing function such that

$$\sup_{\lambda \geq 0} m_z(\lambda) = \lim_{\lambda \to +\infty} m_z(\lambda) = \frac{1}{|z|}\,,$$

which proves the first estimate in (9.3.9).

If $z \in \mathbb{D}_{1/2} \setminus [0, 1]$, then $\lambda \mapsto m_z(\lambda)$ attains its maximum at λ_0 such that

$$\mathrm{e}^{-\lambda_0} = \left[\frac{1}{4} - (x - 1/2)^2 - y^2\right](1 - x)^{-1}\,. \tag{9.3.11}$$

Inserting (9.3.11) into (9.3.10) we obtain $m_z(\lambda_0) = |1 - z|/|\mathrm{Im} z|$, which yields

$$0 \leq m_z(\lambda) \leq \frac{|1 - z|}{|\mathrm{Im} z|}\,. \tag{9.3.12}$$

The inequality (9.3.12) proves the second estimate in (9.3.9). □

Note that since for $A \geq \mathbb{1}$ and for m-accretive B the operator

$$T(\tau) := \mathrm{e}^{-\tau A}\,\mathrm{e}^{-\tau B}, \qquad \tau > 0, \tag{9.3.13}$$

is a contraction, its spectrum $\sigma(T(\tau)) \subseteq \mathbb{D}$, $\tau \geq 0$. However, for operator B, which is *Kato-small* relative to A, we can show more.

To this aim, we introduce a family of closed convex sets in $\mathbb{C}$, which are defined for $r \in [0, 1]$ by $\Sigma_r := \Sigma_r^0 \cup \Sigma_r^1$. Here

$$\Sigma_r^0 := \{z \in \mathbb{D}_{1/2} : |\mathrm{Im} z| \leq r|1 - z|\},$$

and

$$\Sigma_r^1 := \{z \in \mathbb{D} \setminus \mathbb{D}_{1/2} : |z| \leq r\},$$

see Fig. 9.1. We remark that the family Σ_r increases with r in the sense that for $r < r'$ one gets $\Sigma_r \subset \Sigma_{r'}$. On the other hand, we denote by $R_r := \mathbb{D} \setminus \Sigma_r$ a *decreasing* family of domains $\mathbb{D} \setminus \Sigma_r$.

Theorem 9.3.3 *Let self-adjoint $A \geq \mathbb{1}$ and let B be m-accretive operator satisfying the condition* H2(*ii*) *with $a_* < 1$. Then, $\sigma(T(\tau)) \subseteq \Sigma_{a_*}$ for each $\tau > 0$. Moreover, for any $r \in (a_*, 1)$, one has the estimate*

$$\|(z\mathbb{1} - T(\tau))^{-1}\| \leq \frac{r}{r - a_*}\,\|(z\mathbb{1} - e^{-\tau A})^{-1}\|, \quad z \in R_r := \mathbb{D} \setminus \Sigma_r. \tag{9.3.14}$$

Proof One has to prove that each point $z \in \mathbb{D} \setminus \Sigma_{a_*} = R_{a_*}$ belongs to the resolvent set $\rho(T(\tau))$ of $T(\tau)$ for any $\tau > 0$. To this end we use the representation

$$z\mathbb{1} - T(\tau) = z\mathbb{1} - \mathrm{e}^{-\tau A} + \mathrm{e}^{-\tau A}(\mathbb{1} - \mathrm{e}^{-\tau B}). \tag{9.3.15}$$

If $\tau > 0$, then operator $(\mathbb{1} - \mathrm{e}^{-\tau A})$ is invertible. Let

$$X(\tau, z) := (\mathbb{1} - \mathrm{e}^{-\tau A})\,(z - \mathrm{e}^{-\tau A})^{-1}, \tag{9.3.16}$$

and

$$Y(\tau) := (\mathbb{1} - \mathrm{e}^{-\tau A})^{-1}\mathrm{e}^{-\tau A}(\mathbb{1} - \mathrm{e}^{-\tau B}). \tag{9.3.17}$$

Then, by (9.3.15)–(9.3.17), one gets the representation

$$z\mathbb{1} - T(\tau) = (z\mathbb{1} - \mathrm{e}^{-\tau A})\,\{\mathbb{1} + X(\tau, z)\,Y(\tau)\}. \tag{9.3.18}$$

For each $f \in \mathrm{dom}\,(B)$ we have

$$(\mathbb{1} - \mathrm{e}^{-\tau B})\, f = \int_0^\tau ds\, B\, \mathrm{e}^{-sB} f\,.$$

Hence, we infer

$$A^{-1}(\mathbb{1} - \mathrm{e}^{-\tau B})\, f = \int_0^\tau ds\, A^{-1}\, B\, \mathrm{e}^{-sB}\, f\,. \tag{9.3.19}$$

By condition H2*(ii)* the representation (9.3.19) leads to the estimate $\|A^{-1}(\mathbb{1} - \mathrm{e}^{-\tau B})\| \leq \tau\, a_*$. Then using for (9.3.17) the representation

$$Y(\tau) = (\mathbb{1} - \mathrm{e}^{-\tau A})^{-1}\, A\, \mathrm{e}^{-\tau A}\, A^{-1}(\mathbb{1} - \mathrm{e}^{-\tau B})\,,$$

we obtain

$$\|Y(\tau)\| \leq \tau\, a_*\, \|(\mathbb{1} - \mathrm{e}^{-\tau A})^{-1} A\, \mathrm{e}^{-\tau A}\|\,. \tag{9.3.20}$$

By the *functional calculus* for self-adjoint operators one verifies that

$$\tau\, \|(\mathbb{1} - \mathrm{e}^{-\tau A})^{-1} A\, \mathrm{e}^{-\tau A}\| \leq 1\,, \tag{9.3.21}$$

for each $\tau > 0$, and by consequence the estimates (9.3.20), (9.3.21) give

$$\|Y(\tau)\| \leq a_*\,. \tag{9.3.22}$$

Recall that $r > a_*$. To estimate $X(\tau, z)$ (9.3.16), we consider two cases:

(I) Let $z \in \mathbb{D}_{1/2} \cap R_r$. Then, $|\mathrm{Im}\, z| > r|1 - z|$. Since $z \in \mathbb{D}_{1/2}$, by Lemma 9.3.2 we find that $\|X(\tau, z)\| \leq |1 - z|/|\mathrm{Im} z|$. Hence one gets

$$\|X(\tau, z)\| < \frac{1}{r}\,, \tag{9.3.23}$$

for $z \in R_r \cap \mathbb{D}_{1/2}$ and $\tau > 0$.

(II) Let $z \in R_r \setminus \mathbb{D}_{1/2}$. Since $|z| > r$, by Lemma 9.3.2 we find

$$\|X(\tau, z)\| \leq \frac{1}{|z|} < \frac{1}{r}\,. \tag{9.3.24}$$

Therefore, we see that the estimate

$$\|X(\tau, z)\| < \frac{1}{r}\,, \tag{9.3.25}$$

cf. (9.3.23), (9.3.24), is valid for any $z \in R_r$ and $\tau > 0$. Moreover, by (9.3.22) and (9.3.25) we obtain $\|X(\tau, z)Y(\tau)\| \leq \|X(\tau, z)\| \, \|Y(\tau)\| < (a_*/r) \leq 1$, for $z \in R_r$, $\tau > 0$. This yields that the operator $\{\mathbb{1} + Y(\tau)X(\tau, z)\}$ is invertible for $z \in R_{a_*} = \bigcup_{r>a_*} R_r$ and $\tau > 0$, with the following estimate:

$$\|(\mathbb{1} + X(\tau, z)\, Y(\tau, z))^{-1}\| \leq \frac{1}{1 - a_*/r}, \qquad \tau > 0,$$

for $z \in R_r$ and $r > a_*$.

Since the operator $(z\mathbb{1} - \mathrm{e}^{-\tau A})$ is invertible for $z \in R_{a_*}$ and $\tau > 0$, we obtain from the representation (9.3.18) that the operator $z\mathbb{1} - T(\tau)$ is invertible for $z \in R_{a_*}$, $\tau > 0$, and thus announced estimate (9.3.14) is valid for $z \in R_r$ and $r \in (a_*, 1)$.

□

Let us define the closed contour $\Gamma_r = \partial\Sigma_r$, which is the boundary of the set Σ_r. Outside the disc $\mathbb{D}_{1/2}$, it coincides with the arc of radius r with the centre at 0, whereas inside $\mathbb{D}_{1/2}$, this curve consists of two segments of tangents to the preceding arc passing the point $(1, 0)$ (see Fig. 9.1).

Theorem 9.3.4 *Let self-adjoint $A \geq \mathbb{1}$, and let B be an m-accretive operator obeying condition* H2*(ii) with $a_* < 1$. Then, there is a constant $K_{AB} > 0$ such*

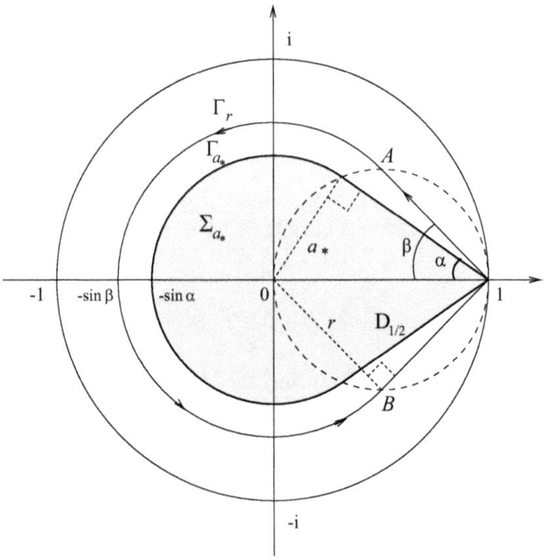

Fig. 9.1 Illustration of the set Σ_{a_*} (shaded domain) with boundary $\Gamma_{a_*} = \partial\Sigma_{a_*}$, where $a_* = \sin\alpha$, as well as of our choice of the contour Γ_r in the resolvent set $\rho(T(\tau))$, where $r = \sin\beta > a_*$. The contour Γ_r consists of the two segments of tangent straight lines $(1, A)$ and $(1, B)$ and the arc (A, B) of radius r. The dotted circle $\partial\mathbb{D}_{1/2}$ corresponds to the set of tangent points for the different values of $r \in [0, 1]$

that

$$\|T(\tau)^k(\mathbb{1} - T(\tau))\| \le \frac{K_{AB}}{k+1}, \quad k = 0, 1, 2, \ldots, \tag{9.3.26}$$

holds for operator $T(\tau)$ (9.3.13) *and any* $\tau \ge 0$.

Proof By Theorem 9.3.3, one gets that for any $r > a_*$ and $\tau > 0$, the contour Γ_r lies in $\rho(T(\tau))$ (see Fig. 9.1), the resolvent set of $T(\tau)$. Using the *Riesz–Dunford calculus*, we get the representation

$$T(\tau)^k(\mathbb{1} - T(\tau)) = \frac{1}{2\pi i}\int_{\Gamma_r} dz\, z^k\,(1 - z)\,(z\mathbb{1} - T(\tau))^{-1}\,. \tag{9.3.27}$$

We estimate the norm of the integral (9.3.27) with help of inequality (9.3.14):

$$\left\| \frac{1}{2\pi i}\int_{\Gamma_r} dz\, z^k(1 - z)\,(z\mathbb{1} - T(\tau))^{-1} \right\| \le \tag{9.3.28}$$

$$\frac{1}{2\pi}\int_{\Gamma_r} |dz|\, |z^k\,(1 - z)|\, \|\,(z\mathbb{1} - \mathrm{e}^{-\tau A})^{-1}\|\, \left(\frac{r}{r - a_*}\right).$$

Since the integral on the right-hand side of (9.3.28) is invariant with respect to conjugation $z \to \bar{z}$, it is sufficient to estimate the integral only on the path Γ_r^+, which is the branch of the contour Γ_r with $\mathrm{Im} z > 0$. This branch consists of the segment $(1, A)$ and the arc $(\pi/2 - \beta, \pi)$ of radius r, see Fig. 9.1.

We parameterise the segment $(1, A)$ by

$$z = 1 - s\mathrm{e}^{-i\beta}\,,\ 0 \le s \le \cos\beta\,,\ \text{ with } r = \sin\beta\,. \tag{9.3.29}$$

Since for (9.3.29) and self-adjoint operator $\mathrm{e}^{-\tau A}$ this gives $\|(z\mathbb{1} - \mathrm{e}^{-\tau A})^{-1}\| \le (\mathrm{Im} z)^{-1} \le (s\,\sin\beta)^{-1}$, we obtain

$$\int_{(1,A)} |dz|\, |z^k\,(1 - z)|\, \|(z\mathbb{1} - \mathrm{e}^{-\tau A})^{-1}\| \le \int_0^{\cos\beta} ds\, |1 - s\,\mathrm{e}^{-i\beta}\,|^k\, \frac{1}{\sin\beta}\,. \tag{9.3.30}$$

Notice that for $0 \le s \le \cos\beta$, one gets by convexity that

$$|1 - s\,\mathrm{e}^{-i\beta}| \le 1 - s\,\frac{1 - \sin\beta}{\cos\beta}\,.$$

Therefore, we obtain for (9.3.30) the estimate

$$\int_{(1,A)} |dz|\, |z^k(1 - z)|\, \|(z\mathbb{1} - \mathrm{e}^{-\tau A})^{-1}\| \le \frac{\cos\beta}{\sin\beta\,(1 - \sin\beta)}\,\frac{1 - (\sin\beta)^k}{k+1}\,. \tag{9.3.31}$$

Now for z that belongs to the arc $(\pi/2 - \beta, \pi)$ of radius r, and the self-adjoint operator $e^{-\tau A}$ we have: $\|(z\mathbb{1} - e^{-\tau A})^{-1}\| \leq (r\cos\beta)^{-1}$. As a consequence, one finds for (9.3.31):

$$\int_{\Gamma_r^+\setminus(1,A)} |dz|\, |z^k(1-z)|\, \|(z\mathbb{1} - e^{-\tau A})^{-1}\| \leq \tag{9.3.32}$$

$$(\pi/2+\beta)\, r\, \frac{2r^k}{r\cos\beta} \leq \frac{2\pi}{\cos\beta}\, \frac{1/r}{e\, \ln(1/r)}\, \frac{1}{k+1}\, .$$

In the last inequality, we used that $\sup_{k\geq 0}(k+1)\, e^{-k\ln x} = x/(e\ln x)$ for $x > 1$.

Finally, combining (9.3.28), (9.3.31) and (9.3.32) we obtain

$$\left\| \frac{1}{2\pi i}\int_{\Gamma_r} dz\, z^k(1-z)\,(z\mathbb{1} - T(\tau))^{-1} \right\| \leq$$

$$2\left(\frac{\cos\beta}{2\pi\sin\beta\,(1-\sin\beta)} + \frac{1/r}{e\cos\beta\,\ln(1/r)}\right)\frac{r}{r-a_*}\,\frac{1}{k+1}\, ,$$

which gives the announced estimate (9.3.26) for

$$K_{AB} := \left(\frac{r\,\cos\beta}{2\pi\sin\beta\,(1-\sin\beta)} + \frac{1}{e\cos\beta\,\ln(1/r)}\right)\frac{2}{r-a_*}\, , \tag{9.3.33}$$

and thus finishes the proof. □

Let $T(\tau) = e^{-\tau A}e^{-\tau B}$ be replaced by $F(\tau) = e^{-\tau B}e^{-\tau A}$. Then, Theorems 9.3.3 and 9.3.4 remain valid if B satisfies, instead of H2(ii), the condition H2(i) with $a < 1$. For that one has to consider the adjoint family $F(\tau)^* = e^{-\tau A}e^{-\tau B^*}$. Then the proofs of the estimates corresponding to (9.3.14), (9.3.26) carry *verbatim* by a simple replacement of parameter a_* by a.

Note that Theorem 9.3.4 is an adaptation to quasi-sectorial contraction $C = T(\tau)$ of a general Lemma 8.2.1. We comment that expressions (8.2.7) and (9.3.33) have different forms and that K_{AB} is a function of condition H2(*ii*).

9.3.3 Operator-Norm Error Bound Estimate

First we recall the following property of holomorphic semigroups (Sect. 1.8):

Proposition 9.3.5 *Let H be a generator of a bounded holomorphic semigroup. Then ran $(e^{-tH}) \subseteq$ dom (H), and we have the estimate:*

$$\left\| He^{-tH} \right\| \leq \frac{C_H}{t}\, , \quad \text{for} \quad C_H > 0\, ,\; t > 0\, . \tag{9.3.34}$$

Lemma 9.3.6 *Let A be a self-adjoint operator, $A \geq \mathbb{1}$. Let operator B be m-accretive and satisfying (ii) with $a_* < 1$. Then, operator $\mathbb{1} - T(\tau)$, cf. (9.3.13), is boundedly invertible for each $\tau > 0$ and*

$$\left\|(\mathbb{1} - T(\tau))^{-1}u\right\| \leq \frac{1}{1-a_*}\left(\|u\| + \frac{1}{\tau}\,\|A^{-1}u\|\right), \qquad u \in \mathfrak{H}, \tag{9.3.35}$$

Proof We use the representation

$$\mathbb{1} - T(\tau) = \mathbb{1} - \mathrm{e}^{-\tau A} + \mathrm{e}^{-\tau A}(\mathbb{1} - \mathrm{e}^{-\tau B}). \tag{9.3.36}$$

Since $\|\mathrm{e}^{-\tau A}\| \leq \mathrm{e}^{-\tau} < 1$, the operator $\mathbb{1} - \mathrm{e}^{-\tau A}$ is boundedly invertible provided $\tau > 0$. Hence, (9.3.36) implies the representation: $\mathbb{1} - T(\tau) = (\mathbb{1} - \mathrm{e}^{-\tau A})\,(\mathbb{1} + Y(\tau))$, where we have used the notation (9.3.17). Now by (9.3.22) and $a_* < 1$ we obtain that $\mathbb{1} - T(\tau)$ is also invertible. Moreover, we have the estimate:

$$\|(\mathbb{1} - T(\tau))^{-1}u\| \leq \frac{1}{1-a_*}\left\|\left(\mathbb{1} - \mathrm{e}^{-\tau A}\right)^{-1}u\right\|, \quad u \in \mathfrak{H}\,. \tag{9.3.37}$$

Now for an upper bound of the right-hand side of (9.3.37) we use the spectral representation of the self-adjoint operator A and obtain inequality

$$(\mathbb{1} - \mathrm{e}^{-\tau A}) \geq \tau\, A\,(\mathbb{1} + \tau\, A)^{-1}\,, \quad \tau > 0\,, \tag{9.3.38}$$

between two positive self-adjoint operators. Given that $A \geq \mathbb{1}$, the operator in the right-hand side of (9.3.38) is invertible. Together with invertibility of the left-hand side this observation yields for $\tau > 0$ inequality: $(\mathbb{1} - \mathrm{e}^{-\tau A})^{-1} \leq \mathbb{1} + (\tau\, A)^{-1}$, and as a cosequence the upper bound for the right-hand side of (9.3.37):

$$\left\|\left(\mathbb{1} - \mathrm{e}^{-\tau A}\right)^{-1}u\right\| \leq \|u\| + \frac{1}{\tau}\,\|A^{-1}u\|\,, \quad u \in \mathfrak{H}\,,\ \tau > 0\,. \tag{9.3.39}$$

Then the estimates (9.3.37), (9.3.39) lead to the desired inequality (9.3.35). □

Lemma 9.3.7 *Let boundedly invertible A and operator B be generators of bounded semigroups, with B satisfying the conditions* H2*(i),(ii) for some (not necessarily bounded by one) a and a_*. Assume that operator $H := (A + B)$ is a boundedly invertible generator of some (bounded) semigroup.*

Then there exists a constant L_1 such that for all $\tau > 0$:

$$\left\|A^{-1}\left(e^{-\tau A}e^{-\tau B} - e^{-\tau(A+B)}\right)\right\| \leq L_1\tau\,, \tag{9.3.40}$$

$$\left\|\left(e^{-\tau A}e^{-\tau B} - e^{-\tau(A+B)}\right)A^{-1}\right\| \leq L_1\tau\,. \tag{9.3.41}$$

Proof See Lemma 10.2.3. □

Lemma 9.3.8 *Let A, B and $H = A + B$ be the same as in Lemma 9.3.7. Then, there exists a constant L_2 such that for all $\tau \geq 0$:*

$$\left\| A^{-1} \left(e^{-\tau A} e^{-\tau B} - e^{-\tau(A+B)} \right) A^{-1} \right\| \leq L_2 \tau^2 . \tag{9.3.42}$$

Proof See Lemma 10.2.4. □

Now we are in position to prove the following operator-norm estimate.

Theorem 9.3.9 *Let A be a self-adjoint operator and $A \geq \mathbb{1}$. Let B satisfy conditions* H2*(i)–(iv). Then there is a constant $L > 0$ such that the estimate*

$$\left\| \left(e^{-tA/n} e^{-tB/n} \right)^n - e^{-tH} \right\| \leq L \, \frac{\ln(n)}{n}, \qquad n = 3, 4, \ldots, \tag{9.3.43}$$

holds uniformly in $t \geq 0$, where operator $H := A + B$ on dom $(H) =$ dom (A).

Proof Let $\tau = t/n$ and $U(\tau) := \mathrm{e}^{-\tau H}$, which is by Theorem 9.3.1 a holomorphic contraction semigroup. We start with *telescopic* representation for operators $T(\tau)$ (9.3.13) and $U(\tau)$:

$$\begin{aligned} & T(\tau)^n - U(\tau)^n \\ & = T(\tau)^{n-1}(T(\tau) - U(\tau)) + \sum_{m=1}^{n-1} T(\tau)^{n-m-1}(T(\tau) - U(\tau))U(\tau)^m . \end{aligned} \tag{9.3.44}$$

Since by Lemma 9.3.6 the operator $(\mathbb{1} - T(\tau))^{-1}$ exists for any $\tau > 0$, we find

$$\begin{aligned} T(\tau)^n - U(\tau)^n &= T(\tau)^{n-1}(T(\tau) - U(\tau)) \\ &+ \sum_{m=1}^{n-1} T(\tau)^{n-m-1}(\mathbb{1} - T(\tau))(\mathbb{1} - T(\tau))^{-1}(T(\tau) - U(\tau))U(\tau)^m . \end{aligned}$$

This leads to the estimate

$$\begin{aligned} & \|T(\tau)^n - U(\tau)^n\| \leq \|T(\tau)^{n-1}(T(\tau) - U(\tau))\| \\ & + \sum_{m=1}^{n-1} \|T(\tau)^{n-m-1}(\mathbb{1} - T(\tau))\| \, \|(\mathbb{1} - T(\tau))^{-1}(T(\tau) - U(\tau))U(\tau)^m\| . \end{aligned} \tag{9.3.45}$$

Then, by (9.3.35), we obtain

$$\begin{aligned} & \|(\mathbb{1} - T(\tau))^{-1}(T(\tau) - U(\tau))U(\tau)^m\| \\ & \leq \frac{1}{1 - a_*} \left\{ \|(T(\tau) - U(\tau))U(\tau)^m\| + \frac{1}{\tau} \left\| A^{-1}(T(\tau) - U(\tau))U(\tau)^m \right\| \right\} . \end{aligned} \tag{9.3.46}$$

By identity $(T(\tau) - U(\tau))U(\tau)^m = (T(\tau) - U(\tau))A^{-1}AH^{-1}HU(\tau)^m$, one gets

$$\|(T(\tau) - U(\tau))U(\tau)^m\| \leq \left\|(T(\tau) - U(\tau))A^{-1}\right\| \|AH^{-1}\| \|HU(\tau)^m\| . \tag{9.3.47}$$

Applying the estimate (9.3.34) from Proposition 9.3.5 to the last factor in (9.3.47), we infer the bound

$$\|(T(\tau) - U(\tau))U(\tau)^m\| \leq \left\|(T(\tau) - U(\tau))A^{-1}\right\| \|AH^{-1}\| \frac{C_H}{\tau} \frac{1}{m} .$$

Note that by $AH^{-1} = (\mathbb{1} + BA^{-1})^{-1}$ and $\|BA^{-1}\| \leq a < 1$ we obtain $\|AH^{-1}\| \leq 1/(1-a)$. Hence,

$$\|(T(\tau) - U(\tau))U(\tau)^m\| \leq \left\|(T(\tau) - U(\tau))A^{-1}\right\| \frac{C_H}{(1-a)\tau} \frac{1}{m} . \tag{9.3.48}$$

Similarly, one gets

$$\|A^{-1}(T(\tau) - U(\tau))U(\tau)^m\| \leq \left\|A^{-1}(T(\tau) - U(\tau))A^{-1}\right\| \frac{C_H}{(1-a)\tau} \frac{1}{m} . \tag{9.3.49}$$

Inserting (9.3.48) and (9.3.49) into (9.3.46) we find

$$\begin{aligned} &\|(\mathbb{1} - T(\tau))^{-1}(T(\tau) - U(\tau))U(\tau)^m\| \\ &\leq \left\{\frac{1}{\tau} \|(T(\tau) - U(\tau))A^{-1}\| + \frac{1}{\tau^2} \left\|A^{-1}(T(\tau) - U(\tau))A^{-1}\right\|\right\} \\ &\quad \times \frac{C_H}{(1-a_*)(1-a)} \frac{1}{m}. \end{aligned} \tag{9.3.50}$$

Now, applying to the estimate (9.3.50) the results of Lemmas 9.3.7 and 9.3.8 we deduce

$$\|(\mathbb{1}-T(\tau))^{-1}(T(\tau)-U(\tau))U(\tau)^m\| \leq C_H \frac{L_1 + L_2}{(1-a_*)(1-a)} \frac{1}{m} = \frac{L_3}{m}, \tag{9.3.51}$$

where $L_3 := C_H(L_1 + L_2)/(1 - a_*)(1 - a)$.

To proceed, we recall that Theorem 9.3.4 yields

$$\|T(\tau)^{n-m-1}(\mathbb{1} - T(\tau))\| \leq \frac{C}{n-m} , \quad n = 2, 3, \ldots , \quad m = 0, 1, 2, \ldots, n-1 . \tag{9.3.52}$$

Inserting (9.3.51) and (9.3.52) into (9.3.45) one gets

$$\|T(\tau)^n - U(\tau)^n\| \leq \|T(\tau)^{n-1}(T(\tau) - U(\tau))\| + CL_3 \sum_{m=1}^{n-1} \frac{1}{(n-m)\, m}\,. \tag{9.3.53}$$

Seeing that

$$\sum_{m=1}^{n-1} \frac{1}{(n-m)\, m} = \frac{2}{n} \sum_{m=1}^{n-1} \frac{1}{m} \leq$$
$$\frac{2}{n}\,(1 + \ln(n-1)) \leq 4\,\frac{\ln(n)}{n}\,, \qquad n = 2, 3, \ldots,$$

we obtain from (9.3.53) inequality

$$\|T(\tau)^n - U(\tau)^n\| \leq \|T(\tau)^{n-1}(F(\tau) - U(\tau))\| + 4CL_3\,\frac{\ln(n)}{n}\,. \tag{9.3.54}$$

Now it remains to estimate $\|T(\tau)^{n-1}(T(\tau) - U(\tau))\|$. Again by invertibility of $\mathbb{1} - T(\tau)$ (see Lemma 9.3.6) we find

$$\begin{aligned} &\|T(\tau)^{n-1}(T(\tau) - U(\tau))\| \\ &\leq \|T(\tau)^{n-1}(\mathbb{1} - T(\tau))\|\, \|(\mathbb{1} - T(\tau))^{-1}(T(\tau) - U(\tau))\|\,. \end{aligned} \tag{9.3.55}$$

Note that by (9.3.35), one obtains

$$\begin{aligned} &\|(\mathbb{1} - T(\tau))^{-1}(T(\tau) - U(\tau))\| \\ &\leq \frac{1}{1-a_*}\left\{\|T(\tau) - U(\tau)\| + \frac{1}{\tau}\left\|A^{-1}(T(\tau) - U(\tau))\right\|\right\} \\ &\leq \frac{1}{1-a_*}\left\{2 + \frac{1}{\tau}\left\|A^{-1}(T(\tau) - U(\tau))\right\|\right\}. \end{aligned} \tag{9.3.56}$$

Then applying to the right-hand side of (9.3.56) the estimate (9.3.40) from Lemma 9.3.7, we get

$$\|(\mathbb{1} - T(\tau))^{-1}(T(\tau) - U(\tau))\| \leq \frac{2 + L_1}{1 - a_*}. \tag{9.3.57}$$

Insertion of the estimate (9.3.57) into (9.3.55) yields

$$\|T(\tau)^{n-1}(T(\tau) - U(\tau))\| \leq \frac{2 + L_1}{1 - a_*}\|T(\tau)^{n-1}(\mathbb{1} - F(\tau)\|\,. \tag{9.3.58}$$

Then, applying Theorem 14.2.7 to the right-hand side of (9.3.58), we obtain

$$\|T(\tau)^{n-1}(T(\tau) - U(\tau))\| \leq \frac{2 + L_1}{1 - a_*}\frac{C}{n}, \quad n = 1, 2, \dots . \tag{9.3.59}$$

By insertion of (9.3.59) into (9.3.54), one finds the estimate (9.3.43):

$$\|T(\tau)^n - U(\tau)^n\| \leq L\,\frac{\ln(n)}{n}, \qquad n = 3, 4, \dots ,$$

for $T(\tau) = \mathrm{e}^{-\tau A}\mathrm{e}^{-\tau B}$, (9.3.13), and $\tau = t/n$. Here $L := C(2 + L_1)(1 - a_*)^{-1} + 4CL_3$, and we used that $1/n < \ln(n)/n$ to estimate (9.3.59), for $n = 3, 4, \dots$. □

Corollary 9.3.10 *Let A and B satisfy the conditions of Theorem* 9.3.9. *Then there are constants L', L'' such that estimates of the convergence rates of two forms of the Trotter product formulæ, non-symmetrised and symmetrised*:

$$\left\|\left(e^{-tB/n}e^{-tA/n}\right)^n - e^{-tH}\right\| \leq L'\,\frac{\ln(n)}{n}, \qquad n = 3, 4, \dots , \tag{9.3.60}$$

$$\left\|\left(e^{-tA/2n}e^{-tB/n}e^{-tA/2n}\right)^n - e^{-tH}\right\| \leq L''\,\frac{\ln(n)}{n}, \qquad n = 3, 4, \dots . \tag{9.3.61}$$

hold uniformly in $t \geq 0$.

Remark 9.3.11 Since B^* satisfies conditions of Theorem 9.3.9, we find also the rate of convergence for the Trotter product formula in the form:

$$\left\|\left(\mathrm{e}^{-tA/n}\mathrm{e}^{-tB^*/n}\right)^n - \mathrm{e}^{-tH^*}\right\| \leq L_*\,\frac{\ln(n)}{n}. \tag{9.3.62}$$

Notice that operator $H^* = A + B^*$ is well-defined and m-accretive. By taking the adjoint sequence of factors in (9.3.62), we find (9.3.60). Although adjoint of (9.3.43) gives (9.3.60) for B^*. In order to obtain for B^* the symmetrised formula

$$\left\|\left(\mathrm{e}^{-tA/2n}\mathrm{e}^{-tB^*/n}\mathrm{e}^{-tA/2n}\right)^n - \mathrm{e}^{-tH^*}\right\| \leq L''_*\,\frac{\ln(n)}{n}, \tag{9.3.63}$$

note that the line of reasoning in the proof of Theorem 9.3.9 follows *verbatim* for the both symmetrised formulae, with B or with B^*.

9.4 Convergence Rate: Fractional Condition on Accretive Perturbations

9.4.1 Preliminaries

In Sect. 9.3, our analysis of the operator-norm convergence rate for the Trotter product formula, when one of the involved semigroups has a non-self-adjoint m-accretive generator B, was based on the condition of the relative *Kato-smallness* of B. In particular we demand that the adjoint operator B^* satisfies conditions:

$$\mathrm{dom}(A) \subseteq \mathrm{dom}(B^*)\,, \tag{9.4.1}$$

and

$$\|B^* f\| \le a_* \|Af\|, \quad f \in \mathrm{dom}(A), \quad \text{for some} \quad a_* \in (0, 1)\,, \tag{9.4.2}$$

see H2*(ii)* and H2*(iv)*.

Then, the Trotter product formula operator-norm bound for convergence rate is of the order $O(\ln(n)/n)$, see Theorem 9.3.9 and Corollary 9.3.10. It turns out that the proofs crucially depend on the smallness conditions (9.4.1) and (9.4.2) and, consequently, on the domain $\mathrm{dom}(B^*)$.

In Sect. 9.4, we shall treat the case when conditions of *smallness*: (9.4.1) and (9.4.2), on operator B^* are not satisfied. To this aim, we first construct in Sect. 9.4.2 an example of a pair of self-adjoint operator A and m-accretive operator B such that the conditions H2*(i)* and H2*(iv)* are satisfied:

$$\mathrm{dom}(A) \subseteq \mathrm{dom}(B)\,, \tag{9.4.3}$$

and

$$\|Bf\| \le a\|Af\|, \quad f \in \mathrm{dom}(A), \quad \text{for some} \quad a \in (0, 1)\,, \tag{9.4.4}$$

but there is only *null-vector* in intersection of domains:

$$\mathrm{dom}\,(A) \cap \mathrm{dom}\,(B^*) = \{0\}\,. \tag{9.4.5}$$

This demonstrates that condition (9.4.1) is not a consequence of (9.4.3) and (9.4.4). Note that under the *minimal* conditions (9.4.3) and (9.4.4) $H = A + B$ and $H^* = (A + B)^*$ are well-defined m-accretive operators, but $H^* \neq A + B^*$ for the case indicated in (9.4.5). This is the reason why the techniques used in Sect. 9.3 should be refined if one would like to relax the conditions (9.4.1), (9.4.2) and to establish the operator-norm error bound estimates for the Trotter product formula.

For that reason, in Sect. 9.4.3, we shall look for condition, which is weaker than (9.4.1), but which is still enough to provide error-bound estimates for the Trotter product formula. An *intermediate* condition of this type is:

$$\mathrm{dom}((H^*)^\alpha) \subseteq \mathrm{dom}(A^\alpha) \cap \mathrm{dom}((B^*)^\alpha) \neq \{0\}, \quad \text{for some} \quad \alpha \in (0, 1]. \tag{9.4.6}$$

Notice that conditions (9.4.1) and (9.4.2) imply (9.4.6) for $\alpha = 1$. Assuming (9.4.3), (9.4.4) and (9.4.6) for a *strictly* m-accretive operator B (Definition 9.4.2) we prove in Sect. 9.4.3 that the estimates

$$\left\| \left(e^{-tB/n}e^{-tA/n}\right)^n - e^{-tH} \right\| \leq L_\alpha \frac{\ln(n)}{n^\alpha}, \qquad n = 3, 4, \ldots, \tag{9.4.7}$$

and

$$\left\| \left(e^{-tA/n}e^{-tB/n}\right)^n - e^{-tH} \right\| \leq L_\alpha' \frac{\ln(n)}{n^\alpha}, \qquad n = 3, 4, \ldots, \tag{9.4.8}$$

hold uniformly in $t \geq 0$ for a corresponding $\alpha \in (0, 1)$. Similarly, these estimates are valid for adjoint semigroup generated by H^*, as well as, for *symmetric* (exponential) Trotter product formulæ, cf. (9.3.61) and (9.3.63).

In Sect. 9.4.4, we present some applications for the case of m-accretive and m-sectorial operators B. They are related to the Kato *square-root problem* and to the *Miyazaki condition*, which has its own interest.

First, we prove that our *minimal* conditions (9.4.3), (9.4.4) for the m-accretive B imply the *intermediate* condition (9.4.6) for any $\alpha \in (0, 1/2)$. As a consequence, one gets error-bound estimates (9.4.7), (9.4.8) for those α's with appropriate L_α, L_α'.

Second, we show that the Miyazaki condition ensures (9.4.6) for $\alpha = 1/2$, if B is m-sectorial and satisfying (9.4.3), (9.4.4).

In Sect. 9.4.5, we show that to prove (9.4.7), (9.4.8) for $\alpha = 1/2$ one can relax the above conditions to the following *three* requirements:

(i) the *minimal* conditions (9.4.3), (9.4.4);
(ii) the operator B is m-sectorial;
(iii) the inclusion: $\mathrm{dom}(a) \subseteq \mathrm{dom}(b)$, for sesquilinear forms a and b associated to operators A and B.

9.4.2 *Basic Example*

In this section, we construct a pair of operators A and B, where A is a positive self-adjoint obeying $A \geq \mathbb{1}$ and B is *m-accretive* such that in spite of conditions (9.4.3), (9.4.4), it also satisfies (9.4.5).

Recall that a closed operator K in a Hilbert space $\mathfrak{H}$ is m-accretive, if

$$\mathrm{Re}\,(Kf, f) \geq 0, \quad f \in \mathrm{dom}(K),$$

and $\mathrm{ran}(\lambda\mathbb{1} + K) = \mathfrak{H}$ for $\mathrm{Re}\,(\lambda) > 0$, that is, the operator $\mathbb{1} + K$ is boundedly invertible (Definition 1.8.7). The *Cayley transform* of K is defined by

$$T := (\mathbb{1} - K)\,(\mathbb{1} + K)^{-1}.$$

If K is m-accretive, then operator T is a contraction with $\mathrm{dom}\,(T) = \mathfrak{H}$. Moreover, one has the following claim.

Proposition 9.4.1 *A contraction T is the Cayley transform of an m-accretive operator K if and only if*

$$ker\,(\mathbb{1} + T) = \{0\}. \tag{9.4.9}$$

Moreover, the m-accretive operator K corresponding to contraction T is defined by the properties :

$$\mathrm{dom}(K) = \mathrm{ran}(\mathbb{1} + T)\,, \tag{9.4.10}$$

and

$$Kf = (\mathbb{1} - T)\,(\mathbb{1} + T)^{-1}f, \quad f \in \mathrm{ran}(\mathbb{1} + T)\,. \tag{9.4.11}$$

The map $T \to K$ is bijective.

For the proof see Notes to Chap. 9 (Notes to Sects. 9.4, 9.4.2).

Next, let R and S be two non-negative self-adjoint operators such that

$$\mathrm{dom}(R) \cap \mathrm{dom}(S) = \{0\}\,. \tag{9.4.12}$$

The latter is possible on account of the *von Neumann* theorem, see comments in Notes to Chap. 9 (Notes to Sects. 9.4, 9.4.2). If now we set

$$T := (\mathbb{1} + R)^{-1}(\mathbb{1} + S)^{-1}\,, \tag{9.4.13}$$

then, operator T is a contraction. To this end, we verify for it the condition (9.4.9). If we choose $f = (\mathbb{1} + S)g$, $g \in \mathrm{dom}(S)$, then the condition (9.4.9) transforms into

$$(\mathbb{1} + S)g + (\mathbb{1} + R)^{-1}g = 0.$$

Since $S \geq 0$ and $(\mathbb{1} + R)^{-1} \geq 0$, we get $g = 0$, which yields $f = 0$. Consequently, by Proposition 9.4.1 the contraction T defined by (9.4.13) is the Cayley transform of an m-accretive operator K, which is given by (9.4.10), (9.4.11). Note that one also obtains equations:

$$\begin{aligned}&\mathrm{dom}\,(K) = \mathrm{ran}\,(\mathbb{1} + T) = \mathrm{ran}((\mathbb{1} + R)^{-1}(\mathbb{1} + R + (\mathbb{1} + S)^{-1})) = \\ &\mathrm{ran}((\mathbb{1} + R)^{-1}) = \mathrm{dom}(R)\ .\end{aligned} \tag{9.4.14}$$

Since the Cayley transform of the m-accretive operator K^* is T^*, applying again Proposition 9.4.1 we obtain that

$$\begin{aligned}&\mathrm{dom}\,(K^*) = \mathrm{ran}\,(\mathbb{1} + T^*) = \mathrm{ran}((\mathbb{1} + S)^{-1}(\mathbb{1} + S + (\mathbb{1} + R)^{-1})) = \\ &\mathrm{ran}((\mathbb{1} + S)^{-1}) = \mathrm{dom}(S)\ .\end{aligned} \tag{9.4.15}$$

As a result, we have constructed an m-accretive operator K such that (9.4.14), (9.4.15) and condition (9.4.12) yield

$$\mathrm{dom}(K) \cap \mathrm{dom}(K^*) = \{0\}\ . \tag{9.4.16}$$

Now, we can construct a pair of operators A and B with desirable property (9.4.5). To this aim, we set

$$B := aK\ , \quad \text{for some} \quad a \in (0, 1),$$

and

$$A := \sqrt{\mathbb{1} + K^*K}.$$

By definition, the operator B is m-accretive with $\mathrm{dom}(B) = \mathrm{dom}(K)$, whereas A is a self-adjoint operator obeying $A \geq \mathbb{1}$ and

$$\mathrm{dom}(A) = \mathrm{dom}(\sqrt{K^*K}) = \mathrm{dom}(|K|) = \mathrm{dom}(K). \tag{9.4.17}$$

Moreover, because of $B = aK$, we obtain

$$\|Bf\| = a\|Kf\| \leq a\|\sqrt{\mathbb{1} + K^*K}\,f\| = a\,\|Af\|\,, \quad f \in \mathrm{dom}(A). \tag{9.4.18}$$

Hence, $\mathrm{dom}(A) \subseteq \mathrm{dom}(B)$. Since $B^* = aK^*$, by (9.4.16) and (9.4.17) we finally obtain (9.4.5) since

$$\mathrm{dom}(B^*) \cap \mathrm{dom}(A) = \mathrm{dom}(K^*) \cap \mathrm{dom}(K) = \{0\}. \tag{9.4.19}$$

Let us to illustrate this abstract construction of A and B for intersection (9.4.5) by realisation of unbounded non-negative self-adjoint operators R and S with the property (9.4.12) in a concrete Hilbert space $\mathfrak{H}$.

Let $\mathfrak{H} = L^2([0, 1])$ and let R be one-dimensional *Laplace* operator on the interval $[0, 1]$ with *Dirichlet* boundary conditions:

$$\begin{aligned} \mathrm{dom}(R) &:= \{f \in W^{2,2}([0, 1]) : f(0) = f(1) = 0\}, \\ (Rf)(x) &:= -\,\partial_x^2 f(x), \quad f \in \mathrm{dom}(R)\,. \end{aligned}$$

Then, the operator R is self-adjoint and positive.

Now let $\{r_n\}_{n=1}^{\infty}$ be set of rational numbers of the open interval $(0, 1)$. To define the operator S, we first set

$$q(x) := \sum_{n=1}^{\infty} \frac{c_n}{|x - r_n|^{\alpha}}\,, \quad \alpha \in (1/2, 1), \quad c_n > 0\,, \ \sum_{n=1}^{\infty} c_n < \infty\,.$$

Notice that function $q \in L^1([0, 1])$. Indeed,

$$\int_0^1 dx\, q(x) \le \sum_{n=1}^{\infty} c_n \int_0^1 dx\, \frac{1}{|x - r_n|^{\alpha}} \le$$

$$\sum_{n=1}^{\infty} c_n \int_{-2}^{2} dx\, \frac{1}{|x|^{\alpha}} = \sum_{n=1}^{\infty} 2\, c_n \int_0^2 dx\, \frac{1}{|x|^{\alpha}} = \frac{2^{2-\alpha}}{1-\alpha} \sum_{n=1}^{\infty} c_n < +\infty.$$

Then, we define S as the *multiplication* operator:

$$\begin{aligned} \mathrm{dom}(S) &:= \{f \in L^2([0, 1]) : q(x)f(x) \in L^2([0, 1])\}, \\ (Sf)(x) &:= q(x)f(x), \quad f \in \mathrm{dom}(S)\,, \end{aligned}$$

in Hilbert space $L^2([0, 1])$. By construction, the operator S is unbounded, non-negative and self-adjoint.

Now, to verify (9.4.12) we assume that $\mathrm{dom}(R) \cap \mathrm{dom}(S)$ is *non-trivial*, and

$$f \in \mathrm{dom}(R) \cap \mathrm{dom}(S) \subseteq W^{2,2,}([0, 1]) \cap \mathrm{dom}(S). \tag{9.4.20}$$

Because of (9.4.20), the function f is continuous and such that

$$q(x)f(x) = \sum_{n=1}^{\infty} \frac{c_n}{|x - r_n|^{\alpha}}\, f(x) \in L^2([0, 1])\,. \tag{9.4.21}$$

Then, we obtain

$$\int_0^1 dx\, |q(x)f(x)|^2 \ge c_n^2 \int_0^1 dx\, \frac{1}{|x - r_n|^{2\alpha}} |f(x)|^2\,,$$

for any $n = 1, 2, \ldots$. Since f is continuous, the function $|f|$ is also continuous. Suppose that $f(r_n) \neq 0$. Therefore, by continuity, there is a sufficiently small neighbourhood $(r_n - \epsilon, r_n + \epsilon) \subset (0, 1)$, $\epsilon > 0$, of r_n and a $\delta > 0$, such that $|f(x)| \geq \delta$ for $x \in (r_n - \epsilon, r_n + \epsilon)$. Consequently,

$$\int_0^1 dx\, |q(x)f(x)|^2 \geq c_n^2 \delta^2 \int_{r_n-\epsilon}^{r_n+\epsilon} dx\, \frac{1}{|x - r_n|^{2\alpha}} = +\infty\,,$$

which is impossible since by (9.4.21) one has $q(x)f(x) \in L^2([0, 1])$. Therefore, it must be that $f(r_n) = 0$, $n = 1, 2, \ldots$. Seeing that $x \mapsto f(x)$ is continuous, this yields $f(x) = 0$ for all $x \in [0, 1]$. Hence, $\mathrm{dom}(R) \cap \mathrm{dom}(S) = \{0\}$ as in (9.4.12).

As a result, referring to the abstract setting (9.4.11), (9.4.13), we have constructed in the Hilbert space $\mathfrak{H} = L^2([0, 1])$ the self-adjoint operator $A = \sqrt{\mathbb{1} + K^*K}$ with $\mathrm{dom}(A) = \mathrm{dom}(K) = \mathrm{dom}(R)$, see (9.4.14) and (9.4.17), as well as, operator $B^* = a\,K^*$ with $\mathrm{dom}(B^*) = \mathrm{dom}(K^*) = \mathrm{dom}(S)$, see (9.4.15). Hence, intersections: $\mathrm{dom}\,(A) \cap \mathrm{dom}\,(B^*) = \mathrm{dom}(R) \cap \mathrm{dom}(S) = \{0\}$, give the illustration of the claim (9.4.5) in $\mathfrak{H} = L^2([0, 1])$. More precisely:

$$\mathrm{dom}(A) \subseteq W^{2,2}([0, 1]) \quad \text{and} \quad \mathrm{dom}(B^*) \cap W^{2,2}([0, 1]) = \{0\}\,.$$

This finishes a promised (at the beginning of the subsection) construction of a concrete realisation of desired basic example of the null-vector intersection (9.4.5).

9.4.3 *Intermediate Condition*

The example of the previous subsection shows that conditions of a relative A-smallness of the operator B, see (9.4.3) and (9.4.4), do not guarantee that

$$\mathrm{dom}(H^*) \subseteq \mathrm{dom}(A) \cap \mathrm{dom}(B^*) \neq \{0\}\,, \tag{9.4.22}$$

where H is defined as operator sum:

$$Hf := Af + Bf, \quad f \in \mathrm{dom}(H) = \mathrm{dom}(A)\,. \tag{9.4.23}$$

Remark that intersection in (9.4.22) is non-trivial if we assume (as in Sect. 9.3) that in addition to (9.4.3) and (9.4.4) the conditions (9.4.1) and (9.4.2) of a relative A-smallness of the operator B^* are also satisfied. In this case (9.4.22) takes the form

$$\mathrm{dom}(H^*) = \mathrm{dom}(A) \subseteq \mathrm{dom}(A) \cap \mathrm{dom}(B^*)\,. \tag{9.4.24}$$

Moreover, (9.4.24) guarantee the Trotter product formula error bounds (9.4.7), (9.4.8) with $\alpha = 1$, see Sect. 9.3.

In this section, we show that one can relax conditions (9.4.24) to the *intermediate* condition (9.4.6), which ensures non-trivial operator-norm error-bound estimates (9.4.7), (9.4.8) even when $\mathrm{dom}(A) \cap \mathrm{dom}(B^*) = \{0\}$ (9.4.5).

Definition 9.4.2 Operator B is called *strictly* m-accretive operator if there exists $\epsilon > 0$ such that

$$\mathrm{Re}(Bf, f) \geq \epsilon \|f\|^2, \quad f \in \mathrm{dom}(B) . \tag{9.4.25}$$

As a consequence, the condition (9.4.25) yields that the resolvent set $\rho(B)$ of operator B contains the open half-plane $\mathbb{C}_-(\epsilon) := \{\zeta \in \mathbb{C} : \mathrm{Re}(\zeta) < \epsilon\}$. Then our *minimal* conditions (9.4.3) and (9.4.4) imply that H defined by (9.4.23) is also strictly m-accretive.

Remark 9.4.3 The m-accretivity of B and H implies that B^* and H^* are also m-accretive and that fractional powers of all of these operators: B^α, H^α, $(B^*)^\alpha$, $(H^*)^\alpha$ are well-defined for $\alpha \in (0, 1)$. The *strict* m-accretivity of B guarantees invertibility of these operators.

Now we recall some properties of operator $H = A + B$ (9.4.23), cf. Sect. 1.8).

Proposition 9.4.4 *Let* $A = A^* \geq \mathbb{1}$, *and let* B *be a* m*-accretive operator such that conditions* (9.4.3) *and* (9.4.4) *are satisfied. Then, for* H *defined by* (9.4.23), *one has:*

(i) *The operator* H *is a boundedly invertible generator of contraction holomorphic semigroup* $\{e^{-tH}\}_{t\geq 0}$ *in the sector* $S_\theta = \{t \in \mathbb{C} : |\arg t| < \theta = \arccos(a)\}$, *cf.* (9.4.4).
(ii) *For any* $\alpha > 0$ *and* $t > 0$ *the range* $\mathrm{ran}(e^{-tH}) \subseteq \mathrm{dom}(H^\alpha)$, *and there is* $d_\alpha(H) > 0$ *such that*

$$\|H^\alpha e^{-tH}\| \leq \frac{d_\alpha(H)}{t^\alpha}, \quad t > 0 . \tag{9.4.26}$$

Next we prove the following assertion:

Lemma 9.4.5 *Let* $A \geq \mathbb{1}$ *and let* B *be a strictly* m*-accretive operator obeying conditions* (9.4.3) *and* (9.4.4). *If* H *is given by* (9.4.23), *then there is a constant* $M > 0$ *such that*

$$\left\| \left(e^{-\tau B} e^{-\tau A} - e^{-\tau H} \right) A^{-1} \right\| \leq M\, \tau , \tag{9.4.27}$$

for $\tau \geq 0$. *If, in addition, the intermediate condition* (9.4.6) *is satisfied, then for any* $\alpha \in (0, 1]$ *there is* $N_1(\alpha) > 0$ *such that*

$$\left\| H^{-\alpha} \left(e^{-\tau B} e^{-\tau A} - e^{-\tau H} \right) \right\| \leq N_1(\alpha)\, \tau^\alpha , \quad \tau \geq 0 . \tag{9.4.28}$$

Proof The first estimate (9.4.27) readily follows from the inequality

$$\left\| \left(e^{-\tau B} e^{-\tau A} - e^{-\tau H} \right) A^{-1} \right\|$$
$$\leq \left\| \int_0^\tau ds\, e^{-sB} B A^{-1} e^{-\tau A} \right\| + \left\| \left(\mathbb{1} - e^{-\tau A} \right) A^{-1} \right\| + \left\| \int_0^\tau ds\, e^{-sH} H A^{-1} \right\| ,$$

and from conditions (9.4.3), (9.4.4), which also give $M := \|BA^{-1}\| + 1 + \|HA^{-1}\| \leq 2(a+1)$.

To prove the second estimate (9.4.28), we note that for any strictly m-accretive operator C the fractional powers $C^{-\alpha}$, for $0 < \alpha < 1$, are defined by

$$C^{-\alpha} = \frac{\sin \alpha\pi}{\pi} \int_0^\infty d\lambda\, \lambda^{-\alpha} (C + \lambda\, \mathbb{1})^{-1} ,$$

see Notes to Chap. 9 (Notes to Sects. 9.4, 9.4.3). Hence, $(C^{-\alpha})^* = (C^*)^{-\alpha}$, which allows to pass from (9.4.28) to the norm-estimate of the conjugate operator:

$$\left(H^{-\alpha} \left(e^{-\tau B} e^{-\tau A} - e^{-\tau H} \right) \right)^* = \left(e^{-\tau A} e^{-\tau B^*} - e^{-\tau H^*} \right) (H^*)^{-\alpha}.$$

To this aim, we use identity

$$\left(e^{-\tau A} e^{-\tau B^*} - e^{-\tau H^*} \right) (H^*)^{-\alpha} \tag{9.4.29}$$
$$= (e^{-\tau A} - \mathbb{1})(H^*)^{-\alpha} - e^{-\tau A}(\mathbb{1} - e^{-\tau B^*})(H^*)^{-\alpha} + (\mathbb{1} - e^{-\tau H})(H^*)^{-\alpha}$$
$$= (e^{-\tau A} - \mathbb{1})\, A^{-\alpha} A^{\alpha} (H^*)^{-\alpha} + e^{-\tau A}(e^{-\tau B^*} - \mathbb{1})(B^*)^{-\alpha} (B^*)^{\alpha} (H^*)^{-\alpha}$$
$$-(e^{-\tau H^*} - \mathbb{1})\,(H^*)^{-\alpha} .$$

Since for any strictly m-accretive operator C and $0 < \alpha < 1$, one has representation

$$\left(e^{-tC} - \mathbb{1} \right) C^{-\alpha} = \frac{\sin \alpha\pi}{\pi} \int_0^\infty d\lambda\, \lambda^{-\alpha} (C + \lambda\, \mathbb{1})^{-1} (e^{-tC} - \mathbb{1}) ,$$

we obtain for $\xi > 0$:

$$\left(e^{-tC} - \mathbb{1} \right) C^{-\alpha} = \frac{\sin \alpha\pi}{\pi} \int_0^\xi d\lambda\, \lambda^{-\alpha} \left(C(C + \lambda\, \mathbb{1})^{-1} \right) C^{-1} \left(e^{-tC} - \mathbb{1} \right)$$
$$+ \frac{\sin \alpha\pi}{\pi} \int_\xi^\infty d\lambda\, \lambda^{-\alpha} (C + \lambda\, \mathbb{1})^{-1} \left(e^{-tC} - \mathbb{1} \right) .$$

Because above identity is valid for any $\xi > 0$, by inequalities: $\|(\mathrm{e}^{-tC} - \mathbb{1})C^{-1}\| \leq t$, $\|C(C + \lambda\,\mathbb{1})^{-1}\| \leq 1$, and $\|(C + \lambda\,\mathbb{1})^{-1}\| \leq 1/\lambda$ we obtain estimate

$$\|(\mathrm{e}^{-tC} - \mathbb{1})C^{-\alpha}\| \leq \frac{\sin\alpha\pi}{\pi} \inf_{\xi>0}\left(\frac{t}{1-\alpha}\,\xi^{1-\alpha} + \frac{2}{\alpha}\,\xi^{-\alpha}\right) \tag{9.4.30}$$

$$= \frac{\sin\alpha\pi}{\pi}\,\frac{2^{1-\alpha}}{\alpha(1-\alpha)}\,t^{\alpha}\,, \qquad \alpha \in (0,1)\,,\ t \geq 0\,.$$

Applying the estimate (9.4.30) to generators A, B^* and H^*, by virtue of identity (9.4.29), we find the assertion (9.4.28) with the constant:

$$N_1(\alpha) := \frac{2^{1-\alpha}\sin\alpha\pi}{\pi\alpha\,(1-\alpha)}\left(\|A^{\alpha}(H^*)^{-\alpha}\| + \|(B^*)^{\alpha}(H^*)^{-\alpha}\| + 1\right), \quad \alpha \in (0,1)\,.$$

Here $\|A^{\alpha}(H^*)^{-\alpha}\|$ and $\|(B^*)^{\alpha}(H^*)^{-\alpha}\|$ are bounded because of our *intermediate* condition (9.4.6). The estimate (9.4.28) can be extended to $\alpha = 1$ by setting $N_1(\alpha = 1) := \lim_{\alpha\to 1} N_1(\alpha) = \|A(H^*)^{-1}\| + \|B^*(H^*)^{-1}\| + 1$. □

Lemma 9.4.6 *Let self-adjoint operator* $A \geq \mathbb{1}$, *and let* B *be a strictly m-accretive operator obeying conditions* (9.4.3) *and* (9.4.4). *If the operator* H *defined by* (9.4.23) *satisfies the intermediate condition* (9.4.6), *then for any* $\alpha \in (0,1]$ *there is a constant* $N_2(\alpha) > 0$ *such that*

$$\left\|H^{-\alpha}\left(e^{-\tau B}e^{-\tau A} - e^{-\tau H}\right)A^{-1}\right\| \leq N_2(\alpha)\,\tau^{1+\alpha}\,, \tag{9.4.31}$$

for all $\tau \geq 0$.

Proof We use the following representation

$$H^{-\alpha}\left(\mathrm{e}^{-\tau B}\mathrm{e}^{-\tau A} - \mathrm{e}^{-\tau H}\right)A^{-1} =$$

$$H^{-\alpha}(\mathbb{1} - \mathrm{e}^{-\tau B})(\mathbb{1} - \mathrm{e}^{-\tau A})A^{-1} + \tag{9.4.32}$$

$$H^{-\alpha}(\mathrm{e}^{-\tau B} + \mathrm{e}^{-\tau A} - \mathrm{e}^{-\tau H} - \mathbb{1})A^{-1}. \tag{9.4.33}$$

Let $\alpha \in (0,1)$. By virtue of (9.4.30) one estimates the term (9.4.32) as follows:

$$\|H^{-\alpha}(\mathbb{1} - \mathrm{e}^{-\tau B})(\mathbb{1} - \mathrm{e}^{-\tau A})A^{-1}\| \tag{9.4.34}$$

$$\leq \|(\mathbb{1} - \mathrm{e}^{-\tau B^*})(H^*)^{-\alpha}\|\,\|(\mathbb{1} - \mathrm{e}^{-\tau A})A^{-1}\|$$

$$\leq \|(\mathbb{1} - \mathrm{e}^{-\tau B^*})(B^*)^{-\alpha}\|\,\|(B^*)^{\alpha}(H^*)^{-\alpha}\|\tau$$

$$\leq \frac{2^{1-\alpha}\,\sin\alpha\pi}{\pi\,\alpha\,(1-\alpha)}\,\tau^{1+\alpha}\,\|(B^*)^{\alpha}(H^*)^{-\alpha}\|\,.$$

To estimate the last term (9.4.33), we use the representation

$$\begin{aligned} &\mathrm{e}^{-\tau B} + \mathrm{e}^{-\tau A} - \mathrm{e}^{-\tau H} - \mathbb{1} = \\ &(\mathrm{e}^{-\tau B} - \mathbb{1} + \tau B) + (\mathrm{e}^{-\tau A} - \mathbb{1} + \tau A) - (\mathrm{e}^{-\tau H} - \mathbb{1} + \tau H)\,. \end{aligned} \tag{9.4.35}$$

Seeing that

$$\mathrm{e}^{-\tau C} - \mathbb{1} + \tau C = \int_0^\tau ds\,(\mathbb{1} - \mathrm{e}^{-sC})C\,, \tag{9.4.36}$$

one gets that for a *strictly* m-accretive operator C :

$$\begin{aligned} \|H^{-\alpha}(\mathrm{e}^{-\tau C} - \mathbb{1} + \tau C)A^{-1}\| &\leq \left\| \int_0^\tau ds\,(\mathbb{1} - \mathrm{e}^{-sC^*})(H^*)^{-\alpha} \right\| \|CA^{-1}\| \\ \leq \int_0^\tau ds \|(\mathbb{1} - \mathrm{e}^{-sC^*})(C^*)^{-\alpha}\| &\|(C^*)^{\alpha}(H^*)^{-\alpha}\| \|CA^{-1}\|\,. \end{aligned}$$

Then, by (9.4.30) and (9.4.35), (9.4.36) this yields for (9.4.33) the estimate

$$\begin{aligned} &\|H^{-\alpha}(\mathrm{e}^{-\tau B} + \mathrm{e}^{-\tau A} - \mathrm{e}^{-\tau H} - \mathbb{1})A^{-1}\| \leq \\ &\frac{\tau^{1+\alpha}}{1+\alpha}\,\frac{2^{1-\alpha}\sin\alpha\pi}{\pi\,\alpha\,(1-\alpha)} \left(\|(B^*)^{\alpha}(H^*)^{-\alpha}\|\|BA^{-1}\| + \|A^{\alpha}(H^*)^{-\alpha}\| + \|HA^{-1}\| \right). \end{aligned} \tag{9.4.37}$$

Hence, (9.4.34) and (9.4.37) verify assertion (9.4.31) with the constant

$$\begin{aligned} N_2(\alpha) :=& \frac{2^{1-\alpha}\sin\alpha\pi}{\pi\alpha(1-\alpha)} \\ &\times \left\{ \|(B^*)^{\alpha}(H^*)^{-\alpha}\| + \frac{1}{1+\alpha}\left(a\|(B^*)^{\alpha}(H^*)^{-\alpha}\| + \|A^{\alpha}(H^*)^{-\alpha}\| + 1 + a\right) \right\}. \end{aligned}$$

Setting $N_2(1) := \lim_{\alpha\to 1} N_2(\alpha)$ one easily gets the estimate (9.4.31) for $\alpha = 1$. □

Theorem 9.4.7 *Let self-adjoint operator* $A \geq \mathbb{1}$, *and let* B *be strictly* m*-accretive operators obeying conditions* (9.4.3) *and* (9.4.4). *If the operator* H, *which is defined by* (9.4.23), *satisfies* (9.4.6), *then there is a constant* $L_\alpha > 0$ *such that the estimates*

$$\left\| \left(e^{-tB/n}e^{-tA/n}\right)^n - e^{-tH} \right\| \leq L_\alpha\,\frac{\ln(n)}{n^\alpha}\,, \tag{9.4.38}$$

$$\left\| \left(e^{-tA/n}e^{-tB^*/n}\right)^n - e^{-tH^*} \right\| \leq L_\alpha\,\frac{\ln(n)}{n^\alpha}\,, \tag{9.4.39}$$

hold for $\alpha \in (0, 1]$ *and* $n = 3, 4, \ldots$, *uniformly in* $t \geq 0$.

Proof Remark that (9.4.38) is equivalent to (9.4.39). So, we can start with (9.4.39).

Let $\tau = t/n$ and $F(\tau) := \mathrm{e}^{-\tau B}\mathrm{e}^{-\tau A}$, that is, $F^*(\tau) = \mathrm{e}^{-\tau A}\mathrm{e}^{-\tau B^*}$. Then, we use the operator *telescopic* identity

$$F^*(\tau)^n - \mathrm{e}^{-n\tau H^*} = \sum_{m=0}^{n-1} F^*(\tau)^{(n-m-1)}\left(F^*(\tau) - \mathrm{e}^{-\tau H^*}\right)\mathrm{e}^{-m\tau H^*}, \tag{9.4.40}$$

and the representation

$$\mathbb{1} - F^*(\tau) = (\mathbb{1} - \mathrm{e}^{-\tau A})(\mathbb{1} + X(\tau)),$$

where $X(\tau) = (\mathbb{1} - \mathrm{e}^{-\tau A})^{-1}\mathrm{e}^{-\tau A}(\mathbb{1} - \mathrm{e}^{-\tau B^*})$. By Lemma 9.3.6 the operator $\mathbb{1} + X(\tau)$ is invertible that yields estimate

$$\begin{aligned} &\|(\mathbb{1} - F^*(\tau))^{-1}u\| \leq \\ &\frac{1}{1-a}\,\|(\mathbb{1} - \mathrm{e}^{-\tau A})^{-1}u\| \leq \frac{1}{1-a}\left(\|u\| + \frac{1}{\tau}\left\|A^{-1}u\right\|\right). \end{aligned} \tag{9.4.41}$$

Given that $(H^*)^\alpha$ is invertible, one can use (9.4.41) to estimate (9.4.40) as follows:

$$\begin{aligned} &\|F^*(\tau)^n - \mathrm{e}^{-n\tau H^*}\| \\ &\leq \|F^*(\tau)^{n-1}(\mathbb{1} - F^*(\tau))\|\,\|(\mathbb{1} - F^*(\tau))^{-1}(F^*(\tau) - \mathrm{e}^{-\tau H^*})\| \\ &+ \sum_{m=1}^{n-1} \|F^*(\tau)^{n-m-1}(\mathbb{1} - F^*(\tau))\|\,\|(\mathbb{1} - F^*(\tau))^{-1}(F^*(\tau) - \mathrm{e}^{-\tau H^*}) \\ &(H^*)^{-\alpha}\|\,\|(H^*)^\alpha \mathrm{e}^{-m\tau H^*}\|. \end{aligned} \tag{9.4.42}$$

Note that for the first factor in the right-hand side of (9.4.42) Theorem 9.3.4 yields

$$\|F^*(\tau)^s(\mathbb{1} - F^*(\tau))\| \leq \frac{c}{s+1}, \qquad c > 0 \text{ and } s = 0, 1, 2, \ldots\,. \tag{9.4.43}$$

Then, on account of estimate (9.4.41), we obtain

$$\begin{aligned} &\|(\mathbb{1} - F^*(\tau))^{-1}(F^*(\tau) - \mathrm{e}^{-\tau H^*})\| \leq \\ &\frac{1}{1-a}\left\{\|F^*(\tau) - \mathrm{e}^{-\tau H^*}\| + \frac{1}{\tau}\left\|A^{-1}\left(F^*(\tau) - \mathrm{e}^{-\tau H^*}\right)\right\|\right\}, \end{aligned}$$

and

$$\|(\mathbb{1} - F^*(\tau))^{-1}(F^*(\tau) - \mathrm{e}^{-\tau H^*})(H^*)^{-\alpha}\|$$
$$\le \frac{1}{1-a}\left\{\|(F^*(\tau) - \mathrm{e}^{-\tau H^*})(H^*)^{-\alpha}\| + \frac{1}{\tau}\|A^{-1}\left(F^*(\tau) - \mathrm{e}^{-\tau H^*}\right)(H^*)^{-\alpha}\|\right\}.$$

Therefore, by Lemmata 9.4.5 and 9.4.6 (taking the adjoint), one infers estimates:

$$\|(\mathbb{1} - F^*(\tau))^{-1}(F^*(\tau) - \mathrm{e}^{-\tau H^*})\| \le \frac{1}{1-a}(2+M),$$
$$\|(\mathbb{1} - F^*(\tau))^{-1}(F^*(\tau) - \mathrm{e}^{-\tau H^*})(H^*)^{-\alpha}\| \le \frac{\tau^\alpha}{1-a}(N_1(\alpha) + N_2(\alpha)),$$

for $\alpha \in (0, \tau]$ and $\tau \ge 0$.

Remark that Proposition 9.4.4 yields

$$\|(H^*)^\alpha \,\mathrm{e}^{-m\,\tau\, H^*}\| \le \frac{d_\alpha(H^*)}{(m\,\tau)^\alpha}. \tag{9.4.44}$$

Then, using (9.4.42) and inequalities (9.4.43)–(9.4.44), we deduce that

$$\|F^*(\tau)^n - \mathrm{e}^{-n\tau H^*}\| \tag{9.4.45}$$
$$\le \frac{c}{n(1-a)}(2+M) + \sum_{m=1}^{n-1}\frac{c}{n-m}\,\frac{1}{1-a}\,(N_1(\alpha) + N_2(\alpha))\,\frac{d_\alpha(H^*)}{m^\alpha},$$

for $n = 3, 4, \ldots$. Since for $0 < \alpha < 1$ and $n \ge 3$, one gets

$$\sum_{m=1}^{n-1}\frac{1}{(n-m)m^\alpha} = \frac{1}{n}\sum_{m=1}^{n-1} m^{1-\alpha}\left(\frac{1}{m} + \frac{1}{n-m}\right) \le \frac{2\ln(n)}{n^\alpha}. \tag{9.4.46}$$

Then in virtue of (9.4.45) and (9.4.46) we obtain (9.4.39) where

$$L_\alpha := \frac{2\,c}{1-a}\left\{(2+M)/2 + (N_1(\alpha) + N_2(\alpha))\, d_\alpha(H^*)\right\},$$

for $\alpha \in (0, 1]$ and $\tau \ge 0$. □

Corollary 9.4.8 *Let $A = A^* \ge \mathbb{1}$ and let B be a strictly m-accretive operator obeying* (9.4.3) *and* (9.4.4). *If operator H is defined by* (9.4.23) *and satisfies the intermediate condition* (9.4.6), *then we have the following estimates*

$$\left\|\left(e^{-tA/n}e^{-tB/n}\right)^n - e^{-tH}\right\| \le L_\alpha^{'}\,\frac{\ln(n)}{n^\alpha}, \tag{9.4.47}$$

$$\left\| \left(e^{-tA/2n} e^{-tB/n} e^{-tA/2n} \right)^n - e^{-tH} \right\| \leq L''_\alpha \, \frac{\ln(n)}{n^\alpha} \,, \tag{9.4.48}$$

for $\alpha \in (0, 1]$ *and* $n = 3, 4, \ldots,$ *uniformly in* $t \geq 0$.

Proof Let $T(\tau) := \mathrm{e}^{-\tau A}\mathrm{e}^{-\tau B}$ and $\tau = t/n$. To proceed, we use the identity:

$$T(\tau)^n - \mathrm{e}^{-tH} = \mathrm{e}^{-\tau A}(F(\tau)^{n-1} - \mathrm{e}^{-tH})\mathrm{e}^{-\tau B} + (\mathrm{e}^{-\tau A} - \mathbb{1})\mathrm{e}^{-tH}\mathrm{e}^{-\tau B}$$
$$+(\mathrm{e}^{-tH} H^\alpha)(H^{-\alpha}(\mathrm{e}^{-\tau B} - \mathbb{1})).$$

Since $F(\tau)^{n-1} - \mathrm{e}^{-tH} = F(\tau)^{n-1}(\mathbb{1} - F(\tau)) + (F(\tau)^n - \mathrm{e}^{-tH})$, by Theorem 9.3.4 for quasi-sectorial contraction $F(\tau)$ and by (9.4.38), one gets estimates

$$\|\mathrm{e}^{-\tau A}(F(\tau)^{n-1} - \mathrm{e}^{-tH})\mathrm{e}^{-\tau B}\| \leq \tag{9.4.49}$$
$$\|F(\tau)^{n-1} - \mathrm{e}^{-tH}\| \leq \frac{c}{n} + L_\alpha \, \frac{\ln(n)}{n^\alpha}, \quad n \geq 3 \,.$$

So, following the line of reasoning developed in proofs of Lemmata 9.4.5 and 9.4.6, we infer that

$$\left\| (\mathrm{e}^{-\tau A} - \mathbb{1})\, \mathrm{e}^{-tH} \mathrm{e}^{-\tau B} \right\| \leq \tag{9.4.50}$$
$$\|(\mathrm{e}^{-\tau A} - \mathbb{1})A^{-1}\| \, \|AH^{-1}\| \, \|H\mathrm{e}^{-tH}\| \leq \frac{1}{n} \frac{d_1(H)}{1-a}, \quad n \geq 1,$$

and obtain inequalities

$$\|H^\alpha \mathrm{e}^{-tH} H^{-\alpha}(\mathrm{e}^{-\tau B} - \mathbb{1})\| \leq \tag{9.4.51}$$
$$\|H^\alpha \mathrm{e}^{-tH}\| \, \|(\mathrm{e}^{-\tau B^*} - \mathbb{1})(B^*)^{-\alpha}\| \, \|(B^*)^\alpha (H^*)^{-\alpha}\| \leq$$
$$\frac{1}{n^\alpha} \, d_\alpha(H) \, \frac{2^{1-\alpha} \sin \alpha\pi}{\pi \, \alpha \, (1-\alpha)} \, \|(B^*)^\alpha \, (H^*)^{-\alpha}\| \,.$$

Consequently, estimates (9.4.49)–(9.4.51) yield the first assertion (9.4.47), where

$$L'_\alpha = L_\alpha + c + \frac{d_1(H)}{1-a} + d_\alpha(H) \, \frac{2^{1-\alpha} \sin \alpha\pi}{\pi\alpha \, (1-\alpha)} \, \|(B^*)^\alpha (H^*)^{-\alpha}\| \,,$$

for $\alpha \in (0, 1)$ and $\tau \geq 0$.

If $\alpha = 1$, then one has to replace $\{2^{1-\alpha} \sin \alpha\pi / \pi\alpha \, (1-\alpha)\}$ by 1. The proof of the second assertion (9.4.48) is similar. □

9.4.4 Applications to Accretive and Sectorial Generators

In this subsection, we present *sufficient* conditions, which give the operator-norm error-bound estimates (9.4.7), (9.4.8) for some $0 < \alpha \leq 1$. Because for m-accretive generator B these estimates are guaranteed by the intermediate condition (9.4.6) (see Sect. 9.4.3), we scrutinise here sufficient conditions ensuring (9.4.6) for some $\alpha \in (0, 1)$.

We start by the following simple observation. Let the mapping $J : \mathfrak{H} \to \mathfrak{H}$, on a Hilbert space $\mathfrak{H}$, be *anti-unitary*, that is, an *anti-linear* operator such that

$$(Jf, Jg) = (g, f), \quad f, g \in \mathfrak{H}. \tag{9.4.52}$$

If operator J obeys the proviso:

$$JAJ = A^*(= A) \quad \text{and} \quad JBJ = B^*, \tag{9.4.53}$$

then requirements (9.4.3) and (9.4.4) imply conditions (9.4.1) and (9.4.2), where $a_* = a$. Therefore, the assumptions of Theorem 9.3.9 are satisfied. That yields the operator-norm error bound estimates (9.4.7), (9.4.8) with exponent $\alpha = 1$.

As an illustrating example, one can consider the *Schrödinger* operator. If A is the (minus) *Laplace* operator in $\mathfrak{H} = L^2(\mathbb{R}^n)$ and B is a multiplication operator arising from some complex potential $V(\cdot)$ such that $\operatorname{Re} V \geq 0$ and conditions (9.4.3), (9.4.4) are satisfied, then denoting by operator J the *complex conjugation*, one verifies that conditions (9.4.52) and (9.4.53) are valid.

The problem becomes more difficult if we loose the condition $\operatorname{dom}(A) \subseteq \operatorname{dom}(B^*)$, in particular, if we face up to $\operatorname{dom}(A) \cap \operatorname{dom}(B^*) = \{0\}$. Then it seems to be a delicate problem to satisfy the *intermediate* condition (9.4.6). But it turns out that our *minimal* conditions (9.4.3) and (9.4.4) are sufficient to imply (9.4.6) for any $0 \leq \alpha < 1/2$.

Remark 9.4.9 This comes into view from the following line of reasoning:

(i) Notice that for any m-accretive operators A and B satisfying (9.4.3) one has $\operatorname{dom}(A^\alpha) \subseteq \operatorname{dom}(B^\alpha)$ for $0 \leq \alpha \leq 1$. Since operator H is m-accretive and $\operatorname{dom}(H) \subseteq \operatorname{dom}(A)$ (in fact, $\operatorname{dom}(H) = \operatorname{dom}(A)$, (9.4.23)), we also obtain that $\operatorname{dom}(H^\alpha) \subseteq \operatorname{dom}(A^\alpha)$ for $0 \leq \alpha \leq 1$.

(ii) Since B and H are m-accretive operators, the *Kato fractional power theorem* (see Notes to Chap. 9 (Sect. 9.4.4)) yields: $\operatorname{dom}(B^\alpha) = \operatorname{dom}((B^*)^\alpha)$ and $\operatorname{dom}(H^\alpha) = \operatorname{dom}((H^*)^\alpha)$, for $0 \leq \alpha < 1/2$.

(iii) Summing up the statements in (i) and (ii) one infers that $\operatorname{dom}((H^*)^\alpha) = \operatorname{dom}(H^\alpha) \subseteq \operatorname{dom}(A^\alpha) \subseteq \operatorname{dom}(A^\alpha) \cap \operatorname{dom}(B^\alpha) = \operatorname{dom}(A^\alpha) \cap \operatorname{dom}((B^*)^\alpha)$ for $0 \leq \alpha < 1/2$. Hence, we obtain the *intermediate* condition (9.4.6) for $0 \leq \alpha < 1/2$.

Thus, taking into account Theorem 9.4.7, we obtain the following statement.

Theorem 9.4.10 *Let self-adjoint operator $A \geq \mathbb{1}$ and let B be a strictly m-accretive operator. If the conditions* (9.4.3) *and* (9.4.4) *are satisfied, then estimates* (9.4.7) *and* (9.4.8) *hold for any* $0 < \alpha < 1/2$ *uniformly in* $t \in [0, \infty)$.

Therefore, to cover the case of the error bound for $0 < \alpha < 1/2$ our *minimal* conditions (9.4.3), (9.4.4) are sufficient, in spite of the fact that a trivial intersection: $\mathrm{dom}(A) \cap \mathrm{dom}(B^*) = \{0\}$ may happen, see Example 7.4.12(d).

Now we are going to demonstrate how we can ensure *intermediate* condition (9.4.6) in the limit case $\alpha = 1/2$. Again we assume our *minimal* conditions (9.4.3), (9.4.4), and we suppose that B is *strictly* m-accretive and *sectorial* (that is, m-sectorial) with semi-angle $\theta_B \in [0, \pi/2)$, i.e. with *numerical range* in the sector $S_{\theta_B} := \{t \in \mathbb{C} : |\arg t| < \theta_B\}$. For simplicity, we set in (9.4.25) $\epsilon = 1$. We associate with B a sesquilinear form $\tilde{b}$ defined by

$$\tilde{b}[f, g] := (Bf, g), \quad f, g \in \mathrm{dom}(\tilde{b}) = \mathrm{dom}(B). \tag{9.4.54}$$

Because the form $\tilde{b}$ is closable (Sect. 9.1.1) we consider below its closure b. Definition (9.4.54) implies

$$\mathrm{dom}(B) \subseteq \mathrm{dom}(b) .$$

Recall that one can associate with b its *real* part b_R defined by

$$b_R[f, g] = \frac{1}{2}\{b[f, g] + \overline{b[g, f]}\}, \quad \mathrm{dom}(b_R) = \mathrm{dom}(b) . \tag{9.4.55}$$

The form b_R is symmetric and, by virtue of (9.4.25), it is semibounded from below with lower bound $\epsilon = 1$. Since B is m-accretive, the form b_R is closed. By the *second representation theorem* (see Sect. 9.1.2, Proposition 9.1.4), there is a non-negative self-adjoint operator $B_R \geq \mathbb{1}$ such that

$$\begin{aligned} & b_R[f, g] = (B_R^{1/2} f, B_R^{1/2} g)\,, \quad |b_R[f, g]| \leq \|B_R^{1/2} f\| \, \|B_R^{1/2} g\|\,, \\ & f, g \in \mathrm{dom}(B_R^{1/2}) = \mathrm{dom}(b_R)\,. \end{aligned} \tag{9.4.56}$$

Operator $B_R := \mathrm{Re}B$ is the *real* part of B (9.1.11). Similarly, we introduce the *imaginary* part b_I of b which is defined by

$$b_I[f, g] := \frac{1}{2i}\{b[f, g] - \overline{b[g, f]}\}, \quad f, g \in \mathrm{dom}(b_I) = \mathrm{dom}(b) . \tag{9.4.57}$$

Since B is m-sectorial, the associated form b is also m-sectorial. It is known (Sect. 9.1, (9.1.4)) that in this case, there is a non-negative constant $c := \mathrm{tg}\theta_B$ such that

$$|b_I[f, g]| \leq c\, b_R[f, f]^{1/2} b_R[g, g]^{1/2}, \quad f, g \in \mathrm{dom}(b) = \mathrm{dom}(b_R)\,,$$

which is equivalent to

$$|b_I[f, g]| \le c \, \|B_R^{1/2} f\| \, \|B_R^{1/2} g\| \,, \tag{9.4.58}$$
$$f, g \in \mathrm{dom}(b_I) = \mathrm{dom}(b_R) = \mathrm{dom}(B_R^{1/2}) \,.$$

On account of (9.4.55)–(9.4.58)

$$|b[f, g]| \le (1 + c) \, \|B_R^{1/2} f\| \, \|B_R^{1/2} g\| \,, \tag{9.4.59}$$

for $f, g \in \mathrm{dom}(B_R^{1/2})$.

Now we *assume* that there is a constant M_β such that the inequality (the *Miyazaki condition*, cf. Notes to Chap. 9 (Notes to Sects. 9.4, 9.4.4)):

$$|b_I[f, g]| \le M_\beta \|B_R^{\beta/2} f\| \, \|B_R^{\beta/2} g\|, \quad f, g \in \mathrm{dom}(B_R^{1/2}) \,, \tag{9.4.60}$$

holds for some $\beta \in [0, 1)$. Since $B_R \ge \mathbb{1}$, one has

$$\|B_R^{\beta/2} f\| \le \|B_R^{1/2} f\|, \quad f \in \mathrm{dom}(B_R^{1/2}) \subseteq \mathrm{dom}(B_R^{\beta/2}) \,,$$

which shows that condition (9.4.60) is *stronger* than (9.4.58).

Let a be a closed sesquilinear form associated with a positive self-adjoint operator A:

$$a[f, g] = (A^{1/2} f, A^{1/2} g), \quad \mathrm{dom}(a) = \mathrm{dom}(A^{1/2}) \,.$$

By condition (9.4.3) the domain

$$\mathcal{D} := \mathrm{dom}(a) \cap \mathrm{dom}(b) \,, \tag{9.4.61}$$

is dense in $\mathfrak{H}$. On that account, the sum of closed sectorial forms

$$h[f, g] := a[f, g] + b[f, g], \quad f, g \in \mathrm{dom}(h) = \mathcal{D} \,, \tag{9.4.62}$$

is a well-defined on $\mathcal{D}$ closed sectorial form, Sect. 9.1.1. Then by the *first representation theorem* (see Sect. 9.1.2, Proposition 9.1.3), there exists an m-sectorial operator $H := A \dot{+} B$ (the *form-sum* of A and B) such that one has

$$h[f, g] = (Hf, g), \quad f \in \mathrm{dom}(H) \subseteq \mathrm{dom}(h) \,, \quad g \in \mathrm{dom}(h) \,.$$

Note that because of conditions (9.4.3) and (9.4.4) the *form-sum*: $H = A \dot{+} B$, coincides with operator defined by the *operator* sum: $A + B$, on $\mathrm{dom}(A)$, see (9.4.23).

Theorem 9.4.11 *Let self-adjoint operator $A \geq \mathbb{1}$ and let B be a strictly m-sectorial operator obeying conditions* (9.4.3) *and* (9.4.4). *If in addition the operator B satisfies conditions* (9.4.25), *with $\epsilon = 1$, and* (9.4.60), *then there are constants $L_{\alpha=1/2}$ and $L'_{\alpha=1/2}$ such that the estimates* (9.4.7) *and* (9.4.8) *hold for $\alpha = 1/2$ uniformly in $[0, \infty)$.*

Proof On account of (9.4.25), (9.4.55), (9.4.56) and $\epsilon = 1$, we obtain

$$\mathrm{Re}(b[f, f]) = \|B_R^{1/2} f\| \geq \|f\|^2 , \quad f \in \mathrm{dom}(B_R^{1/2}) ,$$

that is, the sesquilinear form b is *strongly coercive*. The *Miyazaki condition* (9.4.60) implies

$$\mathrm{dom}(B^{1/2}) = \mathrm{dom}((B^*)^{1/2}) = \mathrm{dom}(B_R^{1/2}) = \mathrm{dom}(b) . \tag{9.4.63}$$

Next, the real part h_R of the form h (9.4.62) is equal to

$$h_R[f, g] = a[f, g] + b_R[f, g] , \quad f, g \in \mathrm{dom}(h_R) = \mathrm{dom}(h) . \tag{9.4.64}$$

The corresponding self-adjoint operator we denote by H_R. Then by the same line of reasoning as above for the form b, the sesquilinear form h is also *strongly coercive*. By definition (9.4.64) we have $B_R \leq H_R$. Therefore, taking into account the *Heinz inequality* for positive self-adjoint operators B_R and H_R, we obtain that on domain $\mathrm{dom}\,(H_R^{\beta})$:

$$B_R^{\beta} \leq H_R^{\beta}, \quad \beta \in [0, 1) . \tag{9.4.65}$$

On account of (9.4.62) for the imaginary part of h, we obtain :

$$h_I[f, g] = \frac{1}{2i}\{h[f, g] - \overline{h[g, f]}\} = b_I[f, g] , \quad f, g \in \mathrm{dom}(h_I) = \mathrm{dom}(h) .$$

Taking into account (9.4.60) and (9.4.65), we find that the *Miyazaki condition*:

$$|h_I[f, g]| \leq M_\beta \, \|H_R^{\beta/2} f\| \, \|H_R^{\beta/2} g\|, \quad f, g \in \mathrm{dom}(b_I), \quad \beta \in [0, 1),$$

is valid for the form h. Consequently, one gets

$$\mathrm{dom}(H^{1/2}) = \mathrm{dom}((H^*)^{1/2}) = \mathrm{dom}(H_R^{1/2}) = \mathrm{dom}(h) .$$

By (9.4.61) and (9.4.62), we infer that

$$\mathrm{dom}(h) = \mathrm{dom}(a) \cap \mathrm{dom}(b) ,$$

which together with (9.4.63) yields

$$\mathrm{dom}((H^*)^{1/2}) = \mathrm{dom}(A^{1/2}) \cap \mathrm{dom}((B^*)^{1/2}) \, .$$

Hence, the *intermediate* condition (9.4.6) is satisfied for $\alpha = 1/2$. In virtue of Theorem 9.4.7 and Corollary 9.4.8, this claim implies the estimates (9.4.7) and (9.4.8) for $\alpha = 1/2$. □

9.4.5 More About the Case $\alpha = 1/2$

The aim of this subsection is to relax the *Miyazaki condition* (9.4.60), which is *sufficient* to ensure requirement (9.4.6) and, therefore, to allow application of our main Theorem 9.4.7 for the case $\alpha = 1/2$.

To this end, we propose in this subsection an alternative strategy of the proof, which is based on the following *Kato inequality* (for bounded operators):

Proposition 9.4.12 *Let R, S be bounded accretive operators in Hilbert spaces $\mathfrak{H}$, $\mathfrak{H}'$, respectively, and let T be a bounded operator from $\mathfrak{H}$ to $\mathfrak{H}'$. Then*

$$\|R^\alpha T\, S^\alpha\| \le e^{\pi^2 \alpha(1-\alpha)/2}\, \|T\|^{1-\alpha}\, \|R\, T\, S\|^\alpha \, , \quad 0 \le \alpha \le 1 \, .$$

For comments see Notes to Chap. 9 (Notes to Sects. 9.4, 9.4.5).

Below we shall prove the operator-norm error-bound estimates (9.4.7) and (9.4.8) for the rate of operator-norm convergence the Trotter product approximants in the case $\alpha = 1/2$ under assumptions that avoid *intermediate* condition (9.4.6). In fact, they involve *only* operators A and B without any conditions on the adjoint operator B^*. We begin a validation of this claim by setting $\hat{F}(\tau) := \mathrm{e}^{-\tau A/2}\mathrm{e}^{-\tau B}\mathrm{e}^{-\tau A/2}$ and $F(\tau) := \mathrm{e}^{-\tau B}\mathrm{e}^{-\tau A}$, for $\tau \ge 0$. Then

$$\hat{F}(\tau)^n = \mathrm{e}^{-\tau A/2} F(\tau)^{n-1} \mathrm{e}^{-\tau B} \mathrm{e}^{-\tau A/2} \, . \tag{9.4.66}$$

Lemma 9.4.13 *Let self-adjoint operator $A \ge \mathbb{1}$, and let B be an m-accretive operator. If the minimal conditions* (9.4.3) *and* (9.4.4) *are satisfied, then there is a constant $K > 0$ such that for $n \ge 1$ one has*

$$\left\| \hat{F}(\tau)^n \left(\mathbb{1} - \hat{F}(\tau)\right) \right\| \le \frac{K}{n} \, , \quad \tau > 0 \, . \tag{9.4.67}$$

Proof By (9.4.66), we deduce

$$\begin{aligned} \hat{F}(\tau)^n \left(\mathbb{1} - \hat{F}(\tau)\right) &= \mathrm{e}^{-\tau A/2} F(\tau)^{n-1} \mathrm{e}^{-\tau B} \mathrm{e}^{-\tau A/2} \left(\mathbb{1} - \hat{F}(\tau)\right) \\ &= \mathrm{e}^{-\tau A/2} F(\tau)^{n-1} \left(\mathbb{1} - F(\tau)\right) \mathrm{e}^{-\tau B} \mathrm{e}^{-\tau A/2} \, , \end{aligned}$$

which yields the estimate

$$\left\| \hat{F}(\tau)^n \left(\mathbb{1} - \hat{F}(\tau)\right) \right\| \leq \left\| F(\tau)^{n-1} (\mathbb{1} - F(\tau)) \right\| .$$

Then, by Theorem 9.3.4, we obtain (9.4.67). □

Lemma 9.4.14 *Let self-adjoint operator* $A \geq \mathbb{1}$ *and let* B *be an m-accretive operator. If conditions* (9.4.3) *and* (9.4.4) *are satisfied, then* $(\mathbb{1} - \hat{F}(\tau))$ *is boundedly invertible for any* $\tau > 0$.

Proof We use the representation

$$\mathbb{1} - \hat{F}(\tau) = \mathbb{1} - \mathrm{e}^{-\tau A} + \mathrm{e}^{-\tau A/2} (\mathbb{1} - \mathrm{e}^{-\tau B}) \mathrm{e}^{-\tau A/2} .$$

Setting $Y(\tau) := \mathrm{e}^{-\tau A/2} (\mathbb{1} - \mathrm{e}^{-\tau B}) \mathrm{e}^{-\tau A/2} (\mathbb{1} - \mathrm{e}^{-\tau A})^{-1}$, we infer

$$\mathbb{1} - \hat{F}(\tau) = (\mathbb{1} + Y(\tau)) (\mathbb{1} - \mathrm{e}^{-\tau A}) . \tag{9.4.68}$$

On account of the functional calculus for self-adjoint operators one verifies the estimate: $\|\tau A \mathrm{e}^{-\tau A/2} (\mathbb{1} - \mathrm{e}^{-\tau A})^{-1}\| \leq 1$, $\tau \geq 0$. We also note that conditions H2*(i)*,*(ii)* and the representation (9.3.19) lead to the estimate $\|(\mathbb{1} - \mathrm{e}^{-\tau B}) (\tau A)^{-1}\| \leq a$.

$\mathbb{1} + Y(\tau)$ is boundedly invertible. Since for $\tau > 0$, the operator $(\mathbb{1} - \mathrm{e}^{-\tau A})$ is also boundedly invertible, the same is valid for $(\mathbb{1} - \hat{F}(\tau))$, see (9.4.68). □

Lemma 9.4.15 *Let self-adjoint operator* $A \geq \mathbb{1}$, *and let* B *be an m-accretive operator. If conditions* (9.4.3) *and* (9.4.4) *are satisfied, then there is a constant* $b > 0$ *such that for any* $\tau > 0$ *one has*

$$\left\| (\mathbb{1} - \hat{F}(\tau))^{-1/2} h \right\| \leq b \left(\|h\| + \frac{1}{\sqrt{\tau}} \|A^{-1/2} h\| \right) , \quad h \in \mathfrak{H} . \tag{9.4.69}$$

Proof From (9.4.68), we find the representation

$$\left(\mathbb{1} - \hat{F}(\tau)\right)^{-1} (\mathbb{1} + Y(\tau)) = (\mathbb{1} - \mathrm{e}^{-\tau A})^{-1} , \quad \tau > 0 . \tag{9.4.70}$$

Since $\|\hat{F}(\tau)\| < 1$ for $\tau > 0$, the bounded operators $\mathbb{1} - \hat{F}(\tau)$ and $(\mathbb{1} - \hat{F}(\tau))^{-1}$ are *accretive*. The same holds for operators $\mathbb{1} - \mathrm{e}^{-\tau A}$ and $(\mathbb{1} - \mathrm{e}^{-\tau A})^{-1}$, if $\tau > 0$. Consequently, we can apply the *Kato inequality* (Proposition 9.4.12) for $\alpha = 1/2$ and for operators: $R = \mathbb{1} - \hat{F}(\tau)$, $S = (\mathbb{1} - \mathrm{e}^{-\tau A})^{-1}$, $T = \mathbb{1}$, where $\tau > 0$, Then on account of (9.4.70) one gets the estimate

$$\left\| (\mathbb{1} - \hat{F}(\tau))^{-1/2} (\mathbb{1} + Y(\tau)) f \right\| \leq \mathrm{e}^{\pi^2/8} \|\mathbb{1} + Y(\tau)\|^{1/2} \|(\mathbb{1} - \mathrm{e}^{-\tau A})^{-1/2} f\| , \quad f \in \mathfrak{H} .$$

Because of $\|Y(\tau)\| \leq a < 1$, we obtain

$$\left\|(\mathbb{1} - \hat{F}(\tau))^{-1/2} h\right\| \leq \tag{9.4.71}$$
$$e^{\pi^2/8}\sqrt{1+a}\ \|(\mathbb{1} - e^{-\tau A})^{-1/2}(\mathbb{1} + Y(\tau))^{-1} h\| , \quad h \in \mathfrak{H} ,$$

for $h = (\mathbb{1} + Y(\tau)) f$. Moreover,

$$(\mathbb{1} + Y(\tau))\,(\mathbb{1} - e^{-\tau A})^{1/2} = (\mathbb{1} - e^{-\tau A})^{1/2}(\mathbb{1} + Z(\tau)) , \tag{9.4.72}$$

where $Z(\tau)$ is bounded operator

$$Z(\tau) := \frac{e^{-\tau A/2}}{(\mathbb{1} - e^{-\tau A})^{1/2}}\,(\mathbb{1} - e^{-\tau B})\,\frac{e^{-\tau A/2}}{(\mathbb{1} - e^{-\tau A})^{1/2}} .$$

We comment that conditions H2*(i)*,*(ii)* on B yield that the operator $Z(\tau)$ is m-accretive, that is, $\|(\mathbb{1} + Z(\tau))^{-1}\| \leq 1$. Then by (9.4.71), (9.4.72), and by the *spectral* representation for self-adjoint operator $(\mathbb{1} - e^{-\tau A/2})^{-1/2}$ one infers inequalities

$$\begin{aligned}\left\|(\mathbb{1} - \hat{F}(\tau))^{-1/2} h\right\| &\leq e^{\pi^2/8}\sqrt{1+a}\ \left\|(\mathbb{1} - e^{-\tau A/2})^{-1/2} h\right\| \\ &\leq e^{\pi^2/8}\sqrt{1+a}\ \left(\|h\| + \sqrt{\frac{2}{\tau}}\ \|A^{-1/2} h\|\right) .\end{aligned}$$

Now, by setting $b := \sqrt{2}\, e^{\pi^2/8}\sqrt{1+a}$, we obtain the estimate (9.4.69). □

Lemma 9.4.16 *In a Hilbert space* $\mathfrak{H}$*, for any* m*-sectorial operator* C *with a real part* C_R*, one has the following estimate:*

$$\left\|C_R^{1/2} e^{-tC}\right\| \leq \sqrt{\frac{d_1(C)}{t}} , \quad t > 0 . \tag{9.4.73}$$

Here $d_1(C)$ *is defined as in* (9.4.26).

Proof First, we remark that by definition of the real part C_R (Sect. 9.1.2, (9.1.9) and (9.1.11))

$$\frac{d}{dt}\,\|e^{-tC} u\|^2 = -2\,\mathrm{Re}(e^{-tC} u, C e^{-tC} u) = -2\,\|C_R^{1/2} e^{-tC} u\|^2 \leq 0, \quad h \in \mathfrak{H} . \tag{9.4.74}$$

Then, because of $0 \leq \|e^{-tC}u\| - \|e^{-(t+\tau)C}u\| \leq \|e^{-tC}u - e^{-(t+\tau)C}u\|$, for $t, \tau > 0$, one infers (cf. (9.4.26)) that

$$-\frac{d}{dt}\|e^{-tC}u\| = \lim_{h\to 0} h^{-1}(\|e^{-tC}u\| - \|e^{-(t+h)C}u\|) \leq \tag{9.4.75}$$
$$\lim_{h\to 0} h^{-1}\|e^{-tC}u - e^{-(t+h)C}u\| \leq \left\|\frac{d}{dt}e^{-tC}u\right\| \leq \|Ce^{-tC}u\| \leq \frac{d_1(C)}{t}\|u\| .$$

By (9.4.74) and (9.4.75), we find the inequality

$$2\,\|C_R^{1/2}e^{-tC}u\|^2 = -2\,\|e^{-tC}u\|\,\frac{d}{dt}\|e^{-tC}u\| \leq 2\,\frac{d_1(C)}{t}\,\|u\|^2 ,$$

which yields the estimate (9.4.73). □

Lemma 9.4.17 *Let $A \geq \mathbb{1}$ be a self-adjoint operator and let B be an m-sectorial operator with semi-angle $\theta_B \in [0, \pi/2)$. If* $\mathrm{dom}(A^{1/2}) \subseteq \mathrm{dom}(b)$. *Then, we have estimate*

$$\left\|A^{-1/2}(\mathbb{1} - e^{-\tau B})\right\| \leq (1 + \mathrm{tg}\theta_B)\,\|B_R^{1/2}A^{-1/2}\|\,2\,\tau^{1/2}d_1(B)^{1/2}, \quad \tau \geq 0 . \tag{9.4.76}$$

Proof For any $f, g \in \mathfrak{H}$, we have identity:

$$\left(A^{-1/2}(\mathbb{1} - e^{-\tau B})f, g\right) = \int_0^\tau ds\,\left(Be^{-sB}f,\, A^{-1/2}g\right) .$$

It implies

$$\left(A^{-1/2}(\mathbb{1} - e^{-\tau B})f, g\right) = \int_0^\tau ds\, b\left[e^{-sB}f,\, A^{-1/2}g\right] , \tag{9.4.77}$$

where $b[\cdot,\cdot]$ is a closed sesquilinear sectorial form corresponding to operator B.

By inequality (9.4.59), we find that

$$|b[u, v]| \leq (1 + \mathrm{tg}\theta_B)b_R[u]^{1/2}b_R[v]^{1/2} = (1 + \mathrm{tg}\theta_B)\,\|B_R^{1/2}u\|\,\|B_R^{1/2}v\| .$$

Thus, for the integrand in (9.4.77), we obtain the estimate

$$|b[e^{-sB}f,\, A^{-1/2}g]| \leq (1 + \mathrm{tg}\theta_B)\,\|B_R^{1/2}e^{-sB}f\|\,\|B_R^{1/2}A^{-1/2}g\| .$$

Then, on account of inequality (9.4.73), this gives:

$$
\begin{aligned}
&|\left(A^{-1/2}(\mathbb{1} - \mathrm{e}^{-\tau B})f, g\right)| \leq (1 + \mathrm{tg}\theta_B)\, \|B_R^{1/2} A^{-1/2} g\| \int_0^\tau ds \sqrt{\frac{d_1(B)}{s}}\, \|f\| \\
&\leq (1 + \mathrm{tg}\theta_B)\, \|B_R^{1/2} A^{-1/2}\|\, 2\, d_1(B)^{1/2} \tau^{1/2}\, \|f\|\, \|g\|\,,
\end{aligned}
$$

which implies the announced estimate (9.4.76). Here we used that by condition of lemma, one finds that $\mathrm{dom}(A^{1/2}) \subseteq \mathrm{dom}(b) = \mathrm{dom}(B_R^{1/2})$, cf. (9.4.63), and thus $\|B_R^{1/2} A^{-1/2}\| < \infty$. □

Lemma 9.4.18 *Let self-adjoint operator* $A \geq \mathbb{1}$*, and let* B *be an* m*-sectorial operator obeying conditions* (9.4.3) *and* (9.4.4). *If in addition the condition* $\mathrm{dom}(A^{1/2}) \subseteq \mathrm{dom}(b)$ *is satisfied, then there are constants* $M_{1/2}, M^{'}_{1/2} > 0$ *such that the estimates*

$$\left\| A^{-1/2} \left(\hat{F}(\tau) - e^{-\tau H}\right) \right\| \leq M_{1/2} \tau^{1/2}\,, \tag{9.4.78}$$

and

$$\left\| A^{-1/2} \left(\hat{F}(\tau) - e^{-\tau H}\right) H^{-1} \right\| \leq M'_{1/2} \tau^{3/2}\,, \tag{9.4.79}$$

are valid for $\tau \geq 0$.

Proof To prove (9.4.78), we use the identity

$$
\begin{aligned}
&A^{-1/2}(\mathrm{e}^{-\tau A/2}\mathrm{e}^{-\tau B}\mathrm{e}^{-\tau A/2} - \mathrm{e}^{-\tau H}) = \\
&-\mathrm{e}^{-\tau A/2} A^{-1/2}(\mathbb{1} - \mathrm{e}^{-\tau B})\mathrm{e}^{-\tau A} - A^{-1/2}(\mathbb{1} - \mathrm{e}^{-\tau A}) + A^{-1/2}(\mathbb{1} - \mathrm{e}^{-\tau H})\,.
\end{aligned}
$$

By the functional calculus for the self-adjoint A, one readily gets the estimate $\|A^{-1/2}(\mathbb{1} - \mathrm{e}^{-\tau A})\| \leq (2\tau/e)^{1/2}$. Then taking into account Lemma 9.4.17 we can estimate $A^{-1/2}(\mathbb{1} - \mathrm{e}^{-\tau B})$ and $A^{-1/2}(\mathbb{1} - \mathrm{e}^{-\tau H})$. So, we obtain (9.4.78) for the constant:

$$M_{1/2} := (2/e)^{1/2} + 2\,(1+\mathrm{tg}\theta_B)\,(\,\|B_R^{1/2} A^{-1/2}\|\, d_1(B)^{1/2} + \|H_R^{1/2} A^{-1/2}\|\, d_1(H)^{1/2})\,.$$

To demonstrate (9.4.79), we use the representation

$$A^{-1/2}\,(\hat{F}(\tau) - \mathrm{e}^{-\tau H})\,H^{-1} = A^{-1/2}\,(\mathbb{1} - \mathrm{e}^{-\tau A/2})\,(\mathbb{1} - \mathrm{e}^{-\tau B})\,\mathrm{e}^{-\tau A/2}\,H^{-1} \tag{9.4.80}$$

$$+ A^{-1/2}\,(\mathbb{1} - \mathrm{e}^{-\tau B})\,(\mathbb{1} - \mathrm{e}^{-\tau A/2})\,H^{-1} \tag{9.4.81}$$

$$+ A^{-1/2}\,(\mathrm{e}^{-\tau A} + \mathrm{e}^{-\tau B} - \mathrm{e}^{-\tau H} - \mathbb{1})\,H^{-1}\,. \tag{9.4.82}$$

Since by the functional calculus for the self-adjoint A, one has: $\|A^{-1/2}(\mathbb{1} - e^{-\tau A/2})\| \leq (\tau/e)^{1/2}$ and because $\|(\mathbb{1} - e^{-\tau B})\, A^{-1}\| \leq \tau\, \|BA^{-1}\|$, we find for the term (9.4.80) the following estimate :

$$\| A^{-1/2}(\mathbb{1} - e^{-\tau A/2})\,(\mathbb{1} - e^{-\tau B})\, e^{-\tau A/2} H^{-1}\| \leq \frac{1}{\sqrt{e}}\, \|B\, A^{-1}\|\, \|A\, H^{-1}\|\, \tau^{3/2}\,. \tag{9.4.83}$$

Then, for the second term (9.4.81), we find by Lemma 9.4.17 that

$$\begin{aligned}\|A^{-1/2}(\mathbb{1} - e^{-\tau B})\,(\mathbb{1} - e^{-\tau A/2})\, A^{-1} A H^{-1}\| \leq \\ (1 + \mathrm{tg}\theta_B)\, d_1(B)^{1/2} \|B_R^{1/2} A^{-1/2}\| \tau^{3/2} \|AH^{-1}\|\,.\end{aligned} \tag{9.4.84}$$

Next, taking into account (9.4.35) and (9.4.36), we obtain representation for the third term (9.4.82):

$$\begin{aligned}&A^{-1/2}(e^{-\tau A} + e^{-\tau B} - e^{-\tau H} - \mathbb{1})\, H^{-1} = \\ &\int_0^\tau ds\; A^{-1/2}(\mathbb{1} - e^{-sA})\, AH^{-1} + \int_0^\tau ds\; A^{-1/2}(\mathbb{1} - e^{-sB})\, BA^{-1}AH^{-1} + \\ &\int_0^\tau ds\; A^{-1/2}(e^{-sH} - \mathbb{1})\,.\end{aligned}$$

Since $\|A^{-1/2}(\mathbb{1} - e^{-sA})\| \leq (2s/e)^{1/2}$, then applying Lemma 9.4.17 to $A^{-1/2}(\mathbb{1} - e^{-sB})$ and to $A^{-1/2}(e^{-sH} - \mathbb{1})$, one gets the estimate for the third term (9.4.82):

$$\begin{aligned}&\|A^{-1/2}(e^{-\tau A} + e^{-\tau B} - e^{-\tau H} - \mathbb{1}) H^{-1}\| \leq \\ &\frac{2}{3}\, \tau^{3/2} \sqrt{\frac{2}{e}}\, \|AH^{-1}\| + 2\,(1 + \mathrm{tg}\theta_B)\, \Big(d_1(B)^{1/2} \|B_R^{1/2} A^{-1/2}\|\, \|BA^{-1}\|\, \|AH^{-1}\| \\ &+ d_1(H)^{1/2}\, \|H_R^{1/2} A^{-1/2}\|\Big)\, \tau^{3/2}.\end{aligned} \tag{9.4.85}$$

Together with (9.4.3) and (9.4.4), this gives for (9.4.85) the estimate:

$$\begin{aligned}&\|A^{-1/2}(e^{-\tau A} + e^{-\tau B} - e^{-\tau H} - \mathbb{1})\, H^{-1}\| \leq \frac{2^{3/2}\, \tau^{3/2}}{3\,\sqrt{e}\,(1-a)} + \\ &2\,(1 + \mathrm{tg}\theta_B)\, \Big(d_1(B)^{1/2} \|B_R^{1/2} A^{-1/2}\|\, \frac{a}{1-a} + d_1(H)^{1/2}\, \|H_R^{1/2} A^{-1/2}\|\Big)\, \tau^{3/2}.\end{aligned} \tag{9.4.86}$$

Applying conditions (9.4.3), (9.4.4) to (9.4.82), (9.4.83) one gets similar estimates, which together with (9.4.86) give the proof of lemma for the constant:

$$M'_{1/2} := \frac{a}{\sqrt{e}(1-a)} + \frac{1+\mathrm{tg}\theta_B}{1-a}\|B_R^{1/2}A^{-1/2}\|d_1(B)^{1/2} + \frac{2^{3/2}}{3\sqrt{e}\,(1-a)} \qquad (9.4.87)$$
$$+\,2\,(1+\mathrm{tg}\theta_B)\left(d_1(B)^{1/2}\|B_R^{1/2}A^{-1/2}\|\frac{a}{1-a} + d_1(H)^{1/2}\|H_R^{1/2}A^{-1/2}\|\right),$$

see (9.4.79). □

Theorem 9.4.19 *Let $A \geq \mathbb{1}$ be a self-adjoint operator. Let B be an m-sectorial operator obeying the minimal conditions* (9.4.3) *and* (9.4.4) *with associated closed sectorial form b. If, in addition, one has*

$$\mathrm{dom}(A^{1/2}) \subseteq \mathrm{dom}(b)\,, \qquad (9.4.88)$$

then there exists a constant $\hat{L} > 0$ such that

$$\left\|\left(e^{-tA/2n}e^{-tB/n}e^{-tA/2n}\right)^n - e^{-tH}\right\| \leq \hat{L}\,\frac{\ln(n)}{\sqrt{n}}\,, \qquad (9.4.89)$$

uniformly in $t \in [0,\infty)$.

Proof We begin by *telescopic* identity, which involves approximants (contractions) $\{\hat{F}(\tau = t/n) = \mathrm{e}^{-tA/2n}\mathrm{e}^{-tB/n}\mathrm{e}^{-tA/2n}\}_{n\in\mathbb{N}}$:

$$\hat{F}(\tau)^n - \mathrm{e}^{-\tau nH} = \sum_{m=0}^{n-1}\hat{F}(\tau)^{n-m-1}(\hat{F}(\tau) - \mathrm{e}^{-\tau H})\,\mathrm{e}^{-m\tau H}\,, \quad n \in \mathbb{N}. \qquad (9.4.90)$$

Since operator $(\mathbb{1} - \hat{F}(\tau))^{-1/2}$ exists (Lemma 9.4.14), we obtain representation:

$$\hat{F}(\tau)^n - \mathrm{e}^{-\tau nH}$$
$$= \sum_{m=0}^{n-1}\hat{F}(\tau)^{n-m-1}(\mathbb{1} - \hat{F}(\tau))^{1/2}(\mathbb{1} - \hat{F}(\tau))^{-1/2}(\hat{F}(\tau) - \mathrm{e}^{-\tau H})\,\mathrm{e}^{-m\tau H}\,,$$

which leads to the estimate

$$\|\hat{F}(\tau)^n - \mathrm{e}^{-\tau nH}\| \leq \qquad (9.4.91)$$
$$\sum_{m=0}^{n-1}\|\hat{F}(\tau)^{n-m-1}(\mathbb{1} - \hat{F}(\tau))^{1/2}\|\;\|(\mathbb{1} - \hat{F}(\tau))^{-1/2}(\hat{F}(\tau) - \mathrm{e}^{-\tau H})\,\mathrm{e}^{-m\tau H}\|\,.$$

The bound $\|\hat{F}(\tau)\| \leq 1$ implies that the operator $\mathbb{1} - \hat{F}(\tau)$ is m-accretive, and we can use the *Kato inequality* (Proposition 9.4.12) for $R = \mathbb{1}$, $T = \hat{F}(\tau)^{n-m-1}$ and $S = (\mathbb{1} - \hat{F}(\tau))$. Then together with Lemma 9.4.13, this observation yields:

$$\begin{aligned}
&\|\hat{F}(\tau)^{n-m-1}(\mathbb{1} - \hat{F}(\tau))^{1/2}\| \\
&\leq \mathrm{e}^{\pi^2/8}\|\hat{F}(\tau)^{n-m-1}\|^{1/2}\|\hat{F}(\tau)^{n-m-1}(\mathbb{1} - \hat{F}(\tau))\|^{1/2} \\
&\leq \mathrm{e}^{\pi^2/8}\|\hat{F}(\tau)^{n-m-1}(\mathbb{1} - \hat{F}(\tau))\|^{1/2} \leq \mathrm{e}^{\pi^2/8}\sqrt{\frac{c}{n-m-1}}\,.
\end{aligned}$$

To estimate in (9.4.90), (9.4.91) the term corresponding to $m = 0$, we use again Proposition 9.4.12 as well as Lemmata 9.4.15 and 9.4.18. They provide:

$$\begin{aligned}
&\|\hat{F}(\tau)^{n-1}(\hat{F}(\tau) - \mathrm{e}^{-\tau H})\| \\
&\leq \|\hat{F}(\tau)^{n-1}(\mathbb{1} - \hat{F}(\tau))^{1/2}\|\,\|(\mathbb{1} - \hat{F}(\tau))^{-1/2}(\hat{F}(\tau) - \mathrm{e}^{-\tau H})\| \\
&\leq \mathrm{e}^{\pi^2/8}\sqrt{\frac{c}{n-1}}\,b\left(\|\hat{F}(\tau) - \mathrm{e}^{-\tau H}\| + \frac{1}{\sqrt{\tau}}\|A^{-1/2}(\hat{F}(\tau) - \mathrm{e}^{-\tau H})\|\right) \\
&\leq \mathrm{e}^{\pi^2/8}\sqrt{\frac{c}{n-1}}\,b\,(2 + M_{1/2})\,.
\end{aligned}$$

Now, to estimate the terms with $m > 0$ we use Lemmata 9.4.15 and 9.4.18. Then

$$\begin{aligned}
&\|(\mathbb{1} - \hat{F}(\tau))^{-1/2}(\hat{F}(\tau) - \mathrm{e}^{-\tau H})\,H^{-1}\| \qquad (9.4.92)\\
&\leq b\left(\|(\hat{F}(\tau) - \mathrm{e}^{-\tau H})H^{-1}\| + \frac{1}{\sqrt{\tau}}\|A^{-1/2}(\hat{F}(\tau) - \mathrm{e}^{-\tau H})\,H^{-1}\|\right) \\
&\leq b\,(M'\tau + M'_{1/2}\tau)\,.
\end{aligned}$$

Here we remark that the estimate by $M'\tau$, one obtains along the same line of reasoning as in (9.4.79). Since $\|H\mathrm{e}^{-m\tau H}\| \leq d_1(H)/m\tau$, we obtain from (9.4.92) that

$$\|(\mathbb{1} - \hat{F}(\tau))^{-1/2}(\hat{F}(\tau) - \mathrm{e}^{-\tau H})\,\mathrm{e}^{-m\tau H}\| \leq b\,d_1(H)\,(M' + M'_{1/2})/m\,. \qquad (9.4.93)$$

By (9.4.91)–(9.4.93), one gets the inequalities

$$\begin{aligned}
&\|\hat{F}(\tau)^n - \mathrm{e}^{-\tau n H}\| \leq \\
&\mathrm{e}^{\pi^2/8}\sqrt{\frac{c}{n-1}}\,b\,(2 + M_{1/2}) + \sum_{m=1}^{n-1}\mathrm{e}^{\pi^2/8}\sqrt{\frac{c}{n-m-1}}\,b\,d_1(H)\,(M' + M'_{1/2})\,\frac{1}{m} \\
&\leq \mathrm{e}^{\pi^2/8}\sqrt{\frac{c}{n-1}}\,b\,(2 + M_{1/2}) + \mathrm{e}^{\pi^2/8}\sqrt{c}\,b\,d_1(H)\,(M' + M'_{1/2})\,\frac{2\,\ln(n)}{\sqrt{n}}\,,
\end{aligned}$$

where we used (9.4.46). Therefore, setting

$$\hat{L} := \mathrm{e}^{\pi^2/8}\sqrt{c}\, b\, \{2\,(2 + M_{1/2}) + 4\, d_1(H)\,(M' + M'_{1/2})\}\,,$$

we obtain (9.4.89). □

Corollary 9.4.20 *Let $A \geq \mathbb{1}$ be a self-adjoint operator. Let B be an m-sectorial operator of semi-angle θ_B, obeying the minimal conditions* (9.4.3) *and* (9.4.4)*, with associated closed sectorial form b. If in addition the condition* (9.4.88) *is satisfied, then there are $L > 0$ and $\hat{L}' > 0$ such that*

$$\left\| \left(e^{-tB/n} e^{-tA/n}\right)^n - e^{-tH} \right\| \leq L\, \frac{\ln(n)}{\sqrt{n}}\,, \tag{9.4.94}$$

$$\left\| \left(e^{-tA/n} e^{-tB/n}\right)^n - e^{-tH} \right\| \leq L'\, \frac{\ln(n)}{\sqrt{n}}\,, \tag{9.4.95}$$

uniformly in $t \in [0, \infty)$.

Proof We use the identity (cf. (9.4.66)) for $\tau = t/n$:

$$F(\tau)^n - \mathrm{e}^{-\tau H} = \mathrm{e}^{-\tau B}\mathrm{e}^{-\tau A/2}(\hat{F}(\tau)^{n-1} - \mathrm{e}^{-\tau H})\,\mathrm{e}^{-\tau A/2} + \tag{9.4.96}$$

$$+ (\mathrm{e}^{-\tau B} - \mathbb{1})\,\mathrm{e}^{-\tau A/2}\mathrm{e}^{-\tau H}\mathrm{e}^{-\tau A/2} + \tag{9.4.97}$$

$$+ (\mathrm{e}^{-\tau A/2} - \mathbb{1})\,\mathrm{e}^{-\tau H}\mathrm{e}^{-\tau A/2} + \tag{9.4.98}$$

$$+ \mathrm{e}^{-\tau H}(\mathrm{e}^{-\tau A/2} - \mathbb{1})\,. \tag{9.4.99}$$

Seeing that $\hat{F}(\tau)^{n-1} - \mathrm{e}^{-\tau H} = \hat{F}(\tau)^{n-1}(\mathbb{1} - \hat{F}(\tau)) + (\hat{F}(\tau)^n - \mathrm{e}^{-\tau H})$, by estimates (9.4.67) and (9.4.89) we obtain estimate for the *first* term (9.4.96). For the *second* term (9.4.97) we get

$$\begin{aligned} &\|(\mathrm{e}^{-\tau B} - \mathbb{1})\,\mathrm{e}^{-\tau A/2}\mathrm{e}^{-\tau H}\mathrm{e}^{-\tau A/2}\| \leq \\ &\|(\mathrm{e}^{-\tau B} - \mathbb{1})B^{-1}\|\,\|B\,A^{-1}\|\,\|\mathrm{e}^{-\tau A/2}\|\,\|AH^{-1}\|\,\|H\mathrm{e}^{-\tau H}\| \leq \\ &\frac{\tau\, a}{1-a}\,\frac{d_1(H)}{t} \leq \frac{d_1(H)\, a}{n\,(1-a)}\,. \end{aligned} \tag{9.4.100}$$

For the *third* term (9.4.98), we find

$$\begin{aligned} &\|(\mathrm{e}^{-\tau A/2} - \mathbb{1})\,\mathrm{e}^{-\tau H}\mathrm{e}^{-\tau A/2}\| \leq \\ &\|(\mathrm{e}^{-\tau A/2} - \mathbb{1})A^{-1}\|\,\|AH^{-1}\|\,\|H\mathrm{e}^{-\tau H}\| \leq \frac{\tau\, d_1(H)}{2\,(1-a)\,t} \leq \frac{d_1(H)}{2\,n\,(1-a)}\,. \end{aligned} \tag{9.4.101}$$

To estimate the *last* term (9.4.99) we apply the identity $A^{1/2}A^{-1/2} = \mathbb{1}$. Then

$$\begin{aligned}\|e^{-\tau H}A^{1/2}A^{-1/2}(e^{-\tau A/2}-\mathbb{1})\| &\le \|e^{-\tau H}H_R^{1/2}\|\,\|H_R^{-1/2}A^{1/2}\| \\ &\quad \|A^{-1/2}(e^{-\tau A/2}-\mathbb{1})\| \\ &\le \|H_R^{1/2}e^{-\tau H^*}\|\,\|A^{1/2}H_R^{-1/2}\|\sqrt{\tau/e} \\ &\le \frac{1}{\sqrt{e}}\,d_1(H^*)^{1/2}\,\|A^{1/2}H_R^{-1/2}\|\,n^{-1/2}\,, \end{aligned} \tag{9.4.102}$$

where we used estimate (9.4.73) and $\|A^{-1/2}(e^{-\tau A/2}-\mathbb{1})\| \le (\tau/e)^{1/2}$ for the self-adjoint operator $A \ge \mathbb{1}$. We remind that $\|A^{1/2}H_R^{-1/2}\| < \infty$ because $\mathrm{dom}\,(A^{1/2}) = \mathrm{dom}\,(a) = \mathrm{dom}\,(h) = \mathrm{dom}\,(H_R^{1/2})$.

Taking into account the estimate of (9.4.96) and estimates (9.4.100), (9.4.101), (9.4.102), we deduce for $\tau = t/n$ and uniformly in $t \ge 0$ that

$$\begin{aligned}\left\|F(\tau)^n - e^{-\tau H}\right\| &\le \frac{c}{n-1} + \hat{L}\,\frac{\ln(n)}{\sqrt{n}} + \frac{d_1(H)\,a}{n\,(1-a)} + \frac{d_1(H)}{2\,n\,(1-a)} + \\ &\quad + \frac{1}{\sqrt{e}}\,\|A^{1/2}H_R^{-1/2}\|\,\frac{d_1(H^*)^{1/2}}{\sqrt{n}}\,,\end{aligned}$$

which satisfies the estimate (9.4.94).

In order to obtain a similar estimate for approximants generated by $T(\tau) = e^{-\tau A}e^{-\tau B}$, we use the identity :

$$T(\tau)^n - e^{-\tau H} = e^{-\tau A/2}(\hat{F}(\tau)^{n-1} - e^{-tH})\,e^{-\tau A/2}e^{-\tau B} + \tag{9.4.103}$$
$$+e^{-\tau A/2}e^{-\tau H}\,e^{-\tau A/2}(e^{-\tau B}-\mathbb{1}) + \tag{9.4.104}$$
$$+(e^{-\tau A/2}-\mathbb{1})\,e^{-\tau H}e^{-\tau A/2} + \tag{9.4.105}$$
$$+e^{-\tau H}(e^{-\tau A/2}-\mathbb{1}). \tag{9.4.106}$$

We comment that the *last* two terms (9.4.105) and (9.4.106) are identical to those in (9.4.98) and (9.4.99).

Now, to estimate the *first* term (9.4.103), one has to follow *verbatim* the proof of (9.4.96). For the *second* term (9.4.104), we get the estimate :

$$\begin{aligned}&\|e^{-\tau A/2}e^{-\tau H}e^{-\tau A/2}(e^{-\tau B}-\mathbb{1})\| \le \\ &\|e^{-\tau H}A^{1/2}\|\,\|A^{-1/2}(e^{-\tau B}-\mathbb{1})\| \le \|A^{1/2}e^{-\tau H^*}\|\,\|A^{-1/2}(e^{-\tau B}-\mathbb{1})\| \le \\ &\|A^{1/2}H_R^{-1/2}\|\,\frac{d_1(H^*)^{1/2}}{\sqrt{n}}\,(1+\mathrm{tg}\theta_B)\,\|B_R^{1/2}A^{-1/2}\|\,2\,d_1(B)^{1/2}\,.\end{aligned}$$

To this end, we applied Lemma 9.4.17, the estimate

$$\|e^{-\tau H} A^{1/2}\| \leq \|H_R^{1/2} e^{-\tau H^*}\| \, \|A^{1/2} H_R^{-1/2}\| , \quad \tau = t/n ,$$

that we used in the same way as it has been exploited in (9.4.102), and also the fact that $\mathrm{dom}\,(A^{1/2}) = \mathrm{dom}\,(H_R^{1/2}) \subseteq \mathrm{dom}\,(B_R^{1/2})$. □

For more comments and for references, we invite to consult the Notes to Chap. 9 (Notes to Sect. 9.4).

9.5 Notes to Chap. 9

In conclusion of Chap. 9, few remarks and comments about *self-adjoint* versus *non-self-adjoint* product formulæ are in order:

(i) First, taking into account results of Sects. 7.4 and Theorem 9.4.7, it seems natural to accept that the following *conjecture* is true: Let A be a *self-adjoint* operator such that $A \geq \mathbb{1}$, let B be an m-accretive operator and let H be generator of a holomorphic semigroup such that the condition

$$\mathrm{dom}\,(H^{\alpha'}) \subseteq \mathrm{dom}\,(A^{\alpha'}) \cap \mathrm{dom}\,(B^{\alpha'}) , \tag{9.5.1}$$

holds for some $\alpha' \in (0, 1]$ and the *intermediate* condition (9.4.6) is valid for some $\alpha \in (0, 1]$. If one has $\alpha' + \alpha > 1$, then the error-bound estimate for the Trotter product formula convergence is of the order $O(1/n^{\alpha'+\alpha-1})$.

(ii) Note that if B is a self-adjoint operator, then one has $\alpha' = \alpha$. Hence $\alpha' + \alpha > 1$ yields $\alpha' = \alpha > 1/2$. This assumption is used in Sect. 7.4.

(iii) If $\alpha' = \alpha = 1$ and $B = B^*$, then, as it was proven in Sect. 7.2, the error-bound estimate for the Trotter product formula should be $O(1/n)$ instead of $O(\ln(n)/n)$. Moreover, (cf. Sect. 7.2) this error-bound is *optimal*. On the other hand, in Sect. 7.4 the *optimal* error-bound $O(1/n^{2\alpha-1})$ is proven for $\alpha' = \alpha \in (1/2, 1)$ and $B = B^*$ under *intermediate* condition (9.4.6), cf. condition (7.4.8).

(iv) The conjecture (i) implicitly contains another one: it is that the *smallness* of $B^{\alpha'}$ with respect to $A^{\alpha'}$ is *not* indispensable. As we have shown in Sect. 6.2 in the case $B = B^*$ and $\alpha' = \alpha = 1$ the smallness of B with respect to A is indeed not necessary. But for the case $\alpha' = \alpha \in (1/2, 1)$ this problem remains open, cf. Sect. 6.4 and Notes to Sect. 9.4 below.

(v) Projected to the present *accretive* situation, that is, when B is maximal accretive, the last comment gives a strong indication that *minimal* assumption of smallness: (9.4.3) and (9.4.4), could be replaced by condition (9.5.1). As the first step it would be useful to study the case $\alpha' = \alpha = 1$, and after that to make an extension to the general case $\alpha' + \alpha > 1$.

Notes to Sect. 9.1

These preliminaries are standard and can be acquired in many sources. Here we basically follow a very complete presentation in the book [Kato80, Ch.VI and Ch.VII].

Notes to Sect. 9.2

We follow our short paper [CZ99]. Some of details from this paper were elucidated, developed and completed by comments and proofs. The idea of the *analytic extension* method as a tool for proving the strongly convergent Trotter product formula for m-sectorial generators was formulated by B. Simon. It was presented as a private communication to the author in *Addendum* of the Kato paper [Kato78].

For the *Vitali* convergence theorem, see [HiPh57, Theorem 3.14.1] and a variation in [Si15a, Theorem 6.2.8]. We use here a version that includes the *Montel* compactness principle, see [Mar83, Chapter 6.8].

The proof of the second part of Theorem 9.2.3 needs certain preliminaries. In fact, they reduce to the proof that compactness of $(\mathbb{1} + \mathrm{Re}A)^{-1/2}(\mathbb{1} + \mathrm{Re}B)^{-1/2}$ yields compactness of the family: $(\mathbb{1} + A(x))^{-1/2}(\mathbb{1} + B(x))^{-1/2}$ for $x \in \mathfrak{I}_{A,B}(\delta_1)$, see [CZ99, Proof of Theorem 1]. Then, seeing that now Proposition 9.2.1 implies (9.2.17) the rest of the proof follows along the steps of Remark 9.2.14.

We also comment that since conditions of Proposition 9.2.2 (Theorem H.0.1) are satisfied for the families of self-adjoint operators $\{A(x)\}_{\mathcal{I}_{AB}(\lambda)}$ and $\{B(x)\}_{\mathcal{I}_{AB}(\lambda)}$, the estimate (H.0.1) imposes a uniform in $x \in \mathcal{I}_{AB}(\lambda)$ error bound $O(1/n^{2\alpha-1})$ for the rate of operator-norm convergence in (9.2.23). In contrast to that, the uniform operator-norm convergence (9.2.24) on compact $\mathfrak{D}_{A,B}(\lambda, \delta)$ due to the *Vitali* theorem can not give any operator-norm estimate of the rate of convergence away from the real interval $\mathcal{I}_{AB}(\lambda)$, for example at $z = i$.

In virtue of that Theorem 9.2.4 does not provide the operator-norm estimate for the rate of convergence.

For the *Löwner-Heinz-Kato* theorem (Proposition 9.2.15) see, for example, [BS87, Chapter 10, §4], or [Tan75, Subsection 2.3.3].

Notes to Sect. 9.3

In this section, we essentially follow our paper [CNZ01]. We have shown that the operator-norm error-bound estimates for the rate of convergence of Trotter (i.e., *exponential*) product formula persist if for a couple of generators $\{A, B\}$, where A is self-adjoint, we *pass* from a *self-adjoint* A-small perturbation B (see Chap. 7) to an *m-accretive* A-small perturbation B.

The Theorem 9.3.9 together with Corollary 9.3.10 and Remark 9.3.11 generalise the results of Sect. 7.1 (see Notes to Chap. 7 (Notes to Sect. 7.1)) to the case of a *non-self-adjoint* B under conditions of the *Kato-smallness* of both operators B and B^*. That is, under conditions: $\|Bu\| \leq a\|Au\|$, $u \in \mathrm{dom}\,(A)$ with $a \in (0, 1)$ and $\|B^*u\| \leq a_*\|Au\|$, $u \in \mathrm{dom}\,(A)$ with $a_* \in (0, 1)$.

Now a few comments about these conditions are in order:

C1. To this aim, let us first consider the following special case: $B(z) := zV$, where $z \in \mathbb{C}$ and non-negative V is a self-adjoint operator, which is relatively small with respect to $A \geq \mathbb{1}$, that is, $\mathrm{dom}\,(A) \subseteq \mathrm{dom}\,(V)$ and

$$\|Vf\| \leq a\,\|Af\|, \quad f \in \mathrm{dom}\,(A), \quad \text{for some} \quad a \in (0, 1)\,.$$

Then, the operators $B(z) = zV$ and $B(z)^* = \overline{z}V$ are obviously m-accretive, and they satisfy the conditions H2(i)–(iii) for Re$z \geq 0$. If, in addition, $|z| \leq 1$, then the *relative bounds*: a, a_*, for $B(z)$ remain smaller than 1 (see H2(iv)). Applying Theorem 9.3.9 we find, for instance, that the error estimate

$$\|(\mathrm{e}^{-tA/n}\,\mathrm{e}^{-tzV/n})^n - \mathrm{e}^{-tH(z)}\| \leq L\,\frac{\ln(n)}{n}, \quad n = 2, 3, \ldots, \tag{9.5.2}$$

holds uniformly in $t \geq 0$ and in z such that Re$z \geq 0$, $|z| \leq 1$, where the operator $H(z) = A + zV$ is defined on $\mathrm{dom}\,(H(z)) = \mathrm{dom}\,(A)$. In particular, estimate (9.5.2) is valid for $z = i$. Therefore, it covers the case of perturbation of a self-adjoint contraction semigroup $\{\mathrm{e}^{-tA}\}_{t\geq 0}$ by the strongly continuous *unitary group* $\{\mathrm{e}^{-itB}\}_{t\in\mathbb{R}}$.

C2. Next, it is relevant to ask: why do we need in Theorem 9.3.9 the condition of self-adjointness for operator A?

First, this condition is important for estimation of $\|(z\mathbb{1} - \mathrm{e}^{-\tau A})^{-1}\|$, see (9.3.30), as well as for bound of integral (9.3.32) in the proof of Theorem 9.3.4.

Second, it makes possible to use the spectral calculus of self-adjoint operators to obtain the estimates (9.3.9) and (9.3.35).

Suppose, for example, that instead to be self-adjoint the operator A is m-sectorial. Then, operator-norm continuity of the holomorphic semigroup $U_A(t) = \mathrm{e}^{-tA}$ for $t > 0$, allows to localise the *spectrum* of this semigroup $\sigma(\mathrm{e}^{-tA})$ but *not* the *numerical range* $W(\mathrm{e}^{-tA})$! However, for evaluation the norm of resolvent $(z\mathbb{1} - \mathrm{e}^{-\tau A})^{-1}$, the localisation of the spectrum is not sufficient, except for the *normal* (in particular self-adjoint) operator A, when $\overline{W(\mathrm{e}^{-tA})}$ is a convex hull of $\sigma(\mathrm{e}^{-tA})$. Thus, in this section, we have solved the problem of the Trotter product formula operator-norm convergence rate only for a couple of self-adjoint A and m-accretive B, that is, for the limiting case of semi-angles: $\theta_A = 0$ and $\theta_B = \pi/2$.

C3. Although the method of the next Chap. 10 allows to pass from the self-adjoint to m-sectorial operator A with angle $0 < \theta_A < \pi/2$, and to find a certain operator-norm convergence rate error bound estimate, the price is that B and B^* should be *infinitesimally* A-small. This means the limit case of infinitesimally small *relative bounds*: a and a_*, cf. [CZ01a] and [Zag23].

C4. We conclude by remark that our Theorem 9.3.9 is valid if, instead of exponential e^{-tA}, one considers the operator $f(tA)$. Here $f : [0, +\infty) \to [0, 1]$ is a Borel function from the generic class $\mathcal{K}$ of the Kato functions, Sect. 5.2, but on the subject of *additional* conditions formulated in [NZ98] as boundedness of constants:

$$C_0 := \sup_{x>0}\left\{\frac{x\sqrt{f(x)}}{1-f(x)}\right\}, \; C_1 := \sup_{x>0}\left\{\frac{1-f(x)}{x}\right\},$$
$$C_2 := \sup_{x>0}\left|\left(f(x)-\frac{1}{1+x}\right)\frac{1}{x^2}\right|,$$

and *supplemented* by inequalities $0 < C_0\, a < 1$. Then, we obtain

$$\left\|\left(f(tA/n)\,\mathrm{e}^{-tB/n}\right)^n - \mathrm{e}^{-t(A+B)}\right\| \leq L_0\,\frac{\ln(n)}{n}, \quad n \geq 3, \tag{9.5.3}$$

uniformly in $t \geq 0$. The proof needs only some minor modifications of Lemmata 9.3.6–9.3.8 as it was elucidated in our paper [NZ98].

Notes to Sect. 9.4

Sections 9.4.1 and 9.4.2.

Here we basically follow our paper [CNZ02]. The proof of Proposition 9.4.1 is relatively long and not very much instructive to be present here. Therefore, we refer to [CNZ02, Lemma 2.1].

For comments on the *von Neumann theorem* [vN29] and related *pathology* properties of unbounded operators, see [ArZag15, Section 2.3] and [ArTr20] for more examples.

Section 9.4.3.

We develop here results of [CNZ02, Section 3]. More details about *strictly* accretive operators and about claims in Remark 9.4.3 one can find in [Kato80, Chapter V, §11]. For the proof of Proposition 9.4.4 see, for example, [CZ01a, Prposition 2.16].

The proof of representation for fractional power $C^{-\alpha}$, $0 < \alpha < 1$, of strictly m-accretive operator C one can find , for example, in [Tan79, Chapter 2, Section 2.3], formula (2.33).

Section 9.4.4.

Here we elaborate our results from [CNZ02, Section 4]. For the proof of the *Kato fractional power theorem* (Remark 9.4.9 (ii)), see [Kato61b, Theorem 1.1]. Our considerations are relied besides [CNZ02] on paper by *Miyazaki* [Mi91], which improved some results of *Kato* [Kato61b, Kato62] and of *Shimakura* [Shim68].

Proof of the *Heinz inequality* (9.4.65) one can find, for example, in [BS87, Ch.10, §4, Theorem 3], see also references in paper [Kato61a] about generalisations of this inequality known also as the *Löwner-Heinz-Kato* inequality.

Section 9.4.5.

We were motivated by [CNZ02, Section 5]. Recall that the *Miyazaki* condition (9.4.60), which allow to treat the case $\alpha = 1/2$ was formulated in [Mi91, Theorem 1]. Our alternative approach to this problem is based on Proposition 9.4.12, which is the *Kato inequality* (bounded operators), see [Kato61a, Theorem 3].

We remind that a stronger *Kato inequality* (unbounded operators) [Kato61a, Theorem 1] says that for closed, m-accretive operators A, B in respectively Hilbert spaces: $\mathfrak{H}$, $\mathfrak{H}'$, with a bounded operator T from $\mathfrak{H}$ to $\mathfrak{H}'$, such that $T(\mathrm{dom}\,(A)) \subset \mathrm{dom}\,(B)$ and $\|B\,T\,u\| \leq M\,\|A\,u\|$, for $u \in \mathrm{dom}\,(A)$, one gets $T(\mathrm{dom}\,(A^\alpha)) \subset \mathrm{dom}\,(B^\alpha)$ as well as

$$\|B^\alpha T\,u\| \leq \mathrm{e}^{c\,\alpha(1-\alpha)/2}\,M^\alpha\,\|T\|^{1-\alpha}\,\|A^\alpha\,u\|\,, \quad u \in \mathrm{dom}\,(A^\alpha)\,, \quad 0 \leq \alpha \leq 1\,. \tag{9.5.4}$$

Here $c \geq 0$ is an absolute constant, which in general can be taken as $c = \pi^2$, whereas for self-adjoint non-negative operators A and B one can take $c = 0$. As a consequence, for $T = \mathbb{1}$ and $\mathfrak{H} = \mathfrak{H}'$, the estimate (9.5.4) yields the *Löwner-Heinz-Kato inequality*, cf. (9.4.65), where $M = 1$.

In conclusion of this Hilbert space portion of the Part III (before the *Banach* space Chap. 10), we add a bit more comments on the subject of the *self-adjoint* versus *non-self-adjoint* product formulæ.

Resuming Chap. 9 we conclude that previous observations on *self-adjoint* product formulæ in Part II are in many aspects based on the self-adjoint results from Chap. 6 and Chap. 7. In particular, the self-adjointness allows methodically utilise the *spectral calculus*. We deduce from Chap. 9 that an alternative tool to employ for analysis is owing to properties of the *holomorphic* semigroups.

In [CZ01a] (see Chap. 10), we proved the Trotter product formula for *holomorphic contraction* semigroup on a Banach space generated by operator A and for a quite small perturbation by generator B. To this aim, we used a line of reasoning that dates back to [Zag95]. It allowed to prove the operator-norm convergence with the error-bound estimate of the rate: $O((\ln(n))^2/n)$, cf. [Zag23], but only for those generators B, which are *infinitesimally* small with respect to generator A (Definition 1.9.7).

Chapter 10
Operator-Norm Trotter Product Formula on Banach Spaces

Here we collect results concerning the *operator-norm* convergent *Trotter* product formula for C_0-semigroups $\{e^{-tA}\}_{t\geq 0}$, $\{e^{-tB}\}_{t\geq 0}$ on a Banach space. The proof on a Banach space is demonstrated under hypothesis that one of the involved into product formula contraction C_0-semigroup (e.g. $\{e^{-tA}\}_{t\geq 0}$) is *holomorphic* and that another one is a C_0-semigroup generated by operator B, which satisfies conditions of the relative *infinitesimally* A-smallness. The line of reasoning puts ahead the idea that holomorphy of $\{e^{-tA}\}_{t\geq 0}$ is a principle requirement for the result. This observation allows to revisit the problem of the operator-norm error-bound estimate for the rate of convergent Trotter product formula for non-self-adjoint semigroups.

To scrutinise the question about pertinency of the principle condition of holomorphy, we study example of the *operator-norm* convergent (with error-bound estimate) *Trotter* product formula on a Banach space for C_0-semigroups $\{e^{-tA}\}_{t\geq 0}$ and $\{e^{-tB}\}_{t\geq 0}$, where hypothesis on operator A is (surprisingly) relaxed to generator of *contraction* semigroup, whereas B is a bounded generator.

10.1 Preliminaries

10.1.1 Bounded and Contraction Semigroups on $\mathfrak{X}$

For what follows the properties of holomorphic (contraction) semigroups on a Banach space $\mathfrak{X}$ are essential. Therefore, we start by a suitable for our aim recall of details concerning the bounded, holomorphic semigroups and fractional powers of their generators. We begin by recall, which introduces certain notations adapted in this section for semigroups on a (separable) $\mathfrak{X}$. Hence, this chapter is readable *almost* independently of material we presented in the previous chapters.

V. A. Zagrebnov et al., *Trotter-Kato Product Formulæ*, Operator Theory: Advances and Applications 296, https://doi.org/10.1007/978-3-031-56720-9_10

Definition 10.1.1 Remind that a family $\{U(t)\}_{t\geq 0}$ of bounded linear operators $\mathcal{L}(\mathfrak{X})$ on a Banach space $\mathfrak{X}$ is a C_0-semigroup if it satisfies the conditions:

(i) $U(0) = \mathbb{1}$,
(ii) $U(s+t) = U(s)U(t)$ for all $s, t \geq 0$,
(iii) $\lim_{t\to+0} U(t)x = x$ for all $x \in \mathfrak{X}$.

Here are some essentials for the following sequels of this definition and properties of the *orbit* mappings: $\{t \mapsto U(t)x\}_{t\in\mathbb{R}_0^+}$ for $x \in \mathfrak{X}$ (see Chap. 1 for details):

- The orbit mappings, which are right differentiable at $t = +0$ for $x \in \mathcal{D} \subset \mathfrak{X}$, define a linear operator A with $\mathrm{dom}(A) = \mathcal{D}$ such that $U(t)(\mathrm{dom}(A)) \subseteq \mathrm{dom}(A)$ for all $t \geq 0$.
- For a C_0-semigroup $\{U(t)\}_{t\geq 0}$, the operator A is closed, is densely defined, and determines this semigroup uniquely. It is owing to *generator* defined by $\lim_{t\to+0}(U(t)x - x)/t = -Ax$, for any $x \in \mathrm{dom}(A)$, that conventional notation: $U(t) = \mathrm{e}^{-tA}$ (or $U_A(t) := \mathrm{e}^{-tA}$), makes a sense for $t \geq 0$.
- There exist constants $M_A \geq 1$ and $\beta_A \in \mathbb{R}$, depending on generator A of the semigroup, such that $\|U_A(t)\| \leq M_A \mathrm{e}^{\beta_A t}$ for all $t \geq 0$.
- The resolvent of the generator satisfies the estimate $\|R_A(-\lambda)\| = \|(A + \lambda\mathbb{1})^{-1}\| \leq M_A/(\mathrm{Re}(\lambda) - \beta_A)$ for all λ with $\mathrm{Re}(\lambda) > \beta_A$; thus, the open half-plane with $\mathrm{Re}(z) < -\beta_A$ is contained in the resolvent set of A, which is defined as $\rho(A) = \{z \in \mathbb{C} : \|R_A(z)\| < +\infty\}$.

Recall that if $\beta_A \leq 0$, then $U_A(t)$ is a bounded semigroup; otherwise, $U_A(t)$ is a quasi-bounded semigroup of type $\beta_A > 0$). If $\|U_A(t)\| \leq 1, t \geq 0$, the semigroup is contraction.

In the following we shall use certain observations about generators of *bounded* and *contraction* semigroups. First we recall that the space of linear bounded functionals $\mathfrak{X}^* = \mathcal{L}(\mathfrak{X}, \mathbb{C})$ is a *dual* of the Banach space $\mathfrak{X}$ and that $\mathfrak{X}^*$ is itself a Banach space; see Sect. 1.5. According to Definition 1.9.20, a linear operator A in $\mathfrak{X}$ is *accretive* if for all pairs $(u, \phi) \in \mathrm{dom}(A) \times \mathfrak{X}^*$ with $\|u\|_{\mathfrak{X}} = 1$ and $\|\phi\|_{\mathfrak{X}^*} = 1$, such that functional $\langle u, \phi\rangle = 1$, one has $\mathrm{Re}\,\langle Au, \phi\rangle \geq 0$, cf. (1.9.47). Finally, by Proposition 1.9.21 a densely defined in $\mathfrak{X}$ accretive operator A is a generator of *contraction* semigroup if the range $\mathrm{ran}\,(\lambda\mathbb{1} + A) = \mathfrak{X}$ for $\lambda > 0$,

Now we prove a series of estimates indispensable throughout this Chap. 10.

Lemma 10.1.2 *Let $U_A(t)$ be a bounded C_0-semigroup with boundedly invertible generator A. Then for all $t \geq 0$, and for any $n \in \mathbb{N}$, we have*

$$\left(U_A(t) - \sum_{k=0}^{n} \frac{(-tA)^k}{k!}\right) A^{-n-1} = -\int_0^t d\tau \left(U_A(\tau) - \sum_{k=0}^{n-1} \frac{(-\tau A)^k}{k!}\right) A^{-n}, \tag{10.1.1}$$

$$\left\| \left(U_A(t) - \sum_{k=0}^{n} \frac{(-tA)^k}{k!}\right) A^{-n-1} \right\| \leq M_A \frac{t^{n+1}}{(n+1)!}. \tag{10.1.2}$$

Proof We proceed by induction, and we first prove that

$$(U_A(t) - \mathbb{1})\, x = -\int_0^t d\tau\, U_A(\tau)\, A\, x\,, \quad x \in \operatorname{dom}(A). \tag{10.1.3}$$

Note that for any $\epsilon > 0$, the semigroup properties yield the representation

$$\begin{aligned}\int_0^t ds\, U_A(s)\, \frac{U_A(\epsilon) - \mathbb{1}}{\epsilon} &= \int_0^t ds\, \frac{U_A(s+\epsilon) - U_A(s)}{\epsilon} \\ &= \int_t^{t+\epsilon} ds\, \frac{U_A(s)}{\epsilon} - \int_0^{\epsilon} ds\, \frac{U_A(s)}{\epsilon} \\ &= (U_A(t) - \mathbb{1})\, \frac{1}{\epsilon} \int_0^{\epsilon} ds\, U_A(s)\,.\end{aligned}$$

Then because $\lim_{\epsilon\to 0} \epsilon^{-1} \int_0^{\epsilon} ds\, U_A(s)\, x = x$, for $x \in \mathfrak{X}$, and $\lim_{\epsilon\to 0} \epsilon^{-1}(U_A(\epsilon) - \mathbb{1})\, x = -A\, x$, for $x \in \operatorname{dom}(A)$, we infer (10.1.3). As a consequence, given that A is boundedly invertible, we obtain (10.1.1) for $n = 0$. Furthermore, since $U_A(t)$ is bounded by M_A, we obtain the estimate (10.1.2) for $n = 0$.

Now, suppose that (10.1.1) and (10.1.2) are true for some n; then a straightforward calculation leads to (10.1.1) for $n+1$. Hence, using this representation (10.1.1) for $n + 1$ and estimate (10.1.2), for n, to evaluate the integrand, we deduce (10.1.2) for $n + 1$, which completes the proof by induction. □

Similarly, we obtain a representation for a restricted development of $(\mathbb{1} + A)^{-1}$.

Lemma 10.1.3 *Let A be as in Lemma* 10.1.2. *Then for any* $n \geq 0$

$$(\mathbb{1} + A)^{-1} A^{-n-1} = \left(\sum_{k=0}^{n} (-A)^k\right) A^{-n-1} + (-1)^{n+1} (\mathbb{1} + A)^{-1}. \tag{10.1.4}$$

Proof For $n = 0$, the representation (10.1.4) follows from the resolvent formula:

$$(\mathbb{1} + A)^{-1} - A^{-1} = -(\mathbb{1} + A)^{-1} A^{-1}. \tag{10.1.5}$$

Suppose that (10.1.4) holds for an integer $n > 1$; then

$$(\mathbb{1} + A)^{-1} A^{-n-2} = \left(\sum_{k=0}^{n} (-A)^k\right) A^{-n-2} + (-1)^{n+1} (\mathbb{1} + A)^{-1} A^{-1}. \tag{10.1.6}$$

Applying (10.1.5) to the last term of (10.1.6), we get the representation (10.1.4) for $n + 1$, and thus for any $n \in \mathbb{N} \cup \{0\}$ by induction. □

Lemma 10.1.4 *If $U_A(t)$ is a bounded C_0-semigroup with boundedly invertible generator A, then*

$$\left\| \frac{1}{t^2} \left((\mathbb{1} + tA)^{-1} - U_A(t) \right) A^{-2} \right\| \leq 3\, M_A/2\,, \qquad t > 0\,. \tag{10.1.7}$$

Proof By Lemma 10.1.2 one gets

$$\|(U_A(t) - \mathbb{1} + tA)\, \frac{1}{t^2}\, A^{-2}\| \;\leq\; \frac{M_A}{2}\,. \tag{10.1.8}$$

On the other hand, by Lemma 10.1.3, we have

$$\left\| \left((\mathbb{1} + t\,A)^{-1} - \mathbb{1} + t\,A \right) \frac{1}{t^2}\, A^{-2} \right\| \;=\; \|(\mathbb{1} + t\,A)^{-1}\| \;\leq\; M_A\,. \tag{10.1.9}$$

Here the last estimate follows from $(\mathbb{1} + t\,A)^{-1} = (1/t)\, R_A(-1/t)$ and $\|R_A(-\lambda)\| \leq M_A/(\lambda + \delta)$, $\delta \geq 0$, which is valid for bounded semigroups with boundedly invertible generators. Hence (10.1.7) follows from (10.1.8) and (10.1.9).

□

10.1.2 Holomorphic Contraction Semigroups on $\mathfrak{X}$

Now let mappings $U : \overline{S}_{\theta-\epsilon} \ni z \mapsto U(z)$, be a family of bounded operators. Here $\overline{S}_{\theta-\epsilon}$ is a closed sector of the complex plane, where $\epsilon \in (0, \theta)$ and

$$S_\theta = \{z \in \mathbb{C} : \; z \neq 0 \text{ and } |\arg(z)| < \theta\}\,, \tag{10.1.10}$$

for semi-angle $\theta \in (0, \pi/2)$. Then we elucidate and adapt to our purposes' notations, definitions and assertions from Sect. 1.7.

Definition 10.1.5 Recall that the family of operators $\{U(z)\}_{z \in S_\theta}$ is a bounded holomorphic semigroup of semi-angle $\theta \in (0, \pi/2)$ on a Banach space $\mathfrak{X}$ if it satisfies the following conditions:

(i) If $0 < \epsilon < \theta$, then $\|U(z)\| \leq M_\epsilon$ for all $z \in \overline{S}_{\theta-\epsilon}$ and some $M_\epsilon < \infty$.
(ii) $U(z_1)U(z_2) = U(z_1 + z_2)$ for all $z_1, z_2 \in S_\theta$.
(iii) $U : z \mapsto U(z)$ is analytic function of $z \in S_\theta$ (Remark 1.7.2).
(iv) If $x \in \mathfrak{X}$ and $0 < \epsilon < \theta$, then $\lim_{z \to 0} U(z)\,x = x$ provided $z \in S_{\theta-\epsilon}$.

Let $\sigma(A) = \mathbb{C} \setminus \rho(A)$ denote the spectrum of A; here $\rho(A)$ is its resolvent set. Then $\sigma(A)$ (or $\rho(A)$) can be used for the following characterisation of the holomorphic semigroup generators.

Proposition 10.1.6 *Operator A in a Banach space $\mathfrak{X}$ is the generator of a bounded holomorphic semigroup of semi-angle $\theta \in (0, \pi/2]$ if and only if A is a closed operator with a dense domain* dom(A) *such that*

$$\sigma(A) \subseteq \left\{z \in \mathbb{C},\ |\arg(z)| \leq \frac{\pi}{2} - \theta\right\}, \quad 0 < \theta \leq \frac{\pi}{2},$$

and

$$\|(z\,\mathbb{1} + A)^{-1}\| \leq \frac{N_\epsilon}{|z|} \quad \text{for} \quad N_\epsilon > 0\,, \quad z \in S_{\theta+\pi/2-\epsilon}\ .$$

where $0 < \epsilon < \theta$.

For applications it is also useful an alternative characterisation of these semigroups; cf. Proposition 1.7.7.

Proposition 10.1.7 *If $U_A(z)$ is a bounded holomorphic semigroup of a semi-angle $\theta \in (0, \pi/2)$ with generator A, then for all $z \in S_\theta$ and $n \in \mathbb{N}$, one has $U_A(z)\,\mathfrak{X} \subseteq$* dom$(A^n)$. *Moreover, there are positive constants M'_A, $M_A^{(n)}$, $n > 1$, such that for $t > 0$:*

$$\left\|\frac{dU_A(t)}{dt}\right\| = \|AU_A(t)\| \leq \frac{M'_A}{t} \quad \text{and} \quad \left\|\frac{d^n U_A(t)}{dt^n}\right\| = \left\|A^n U_A(t)\right\| \leq \frac{M_A^{(n)}}{t^n}\ . \tag{10.1.11}$$

For $\theta \in (0, \pi/2)$ the estimates in (10.1.11) *are valid for complex argument $t \in S_\theta$ and constants depending on* $\theta = \arg(t)$.

The converse of this assertion is also true for sector S_θ determined by M'_A.

Remark 10.1.8 Similar to strongly continuous semigroups, a family $\{U(z)\}_{z \in S_\theta}$ is called a *quasi-bounded* holomorphic semigroup of semi-angle θ if there exists a constant $\beta > 0$ such that restriction of $\{\mathrm{e}^{-\beta z}\, U(z)\}_{z \in S_\theta}$ to $\mathbb{R}_0^+$ is a bounded C_0-semigroup. The class of semigroups that we consider in this section is reduced to *holomorphic contraction* semigroups.

To this aim we reformulate definition of this notion below; cf. Definition 1.7.10.

Definition 10.1.9 We say that $\{U_A(z)\}_{z \in S_\theta}$ is a *holomorphic contraction* semigroup with generator $A \in \mathscr{H}_c(\theta, 0)$, if the restriction $\{U_A(t)\}_{t \geq 0}$ on $\mathbb{R}_0^+$ is a contraction C_0-semigroup, that is, $\mathscr{H}_c(\theta, 0) \subseteq \mathscr{H}(\theta, 0) \cap \mathscr{G}(1, 0)$.

Note that this class of semigroups is *not* empty (see Remark 1.7.11) and that the corresponding generators have the following properties:

(i) Let $\{U_A(t)\}_{t \geq 0}$ be a *contraction* C_0-semigroup on a Banach space $\mathfrak{X}$ with generator $A \in \mathscr{G}(1, 0)$. If $M = 1$ and $\|AU(t)\| \leq M_1 t^{-1}$ for $e\,M_1 > 1$ with all $t > 0$, then there exists $\theta = \arcsin{(eM_1)^{-1}}$ such that $\{U_A(t)\}_{t \geq 0}$ may be continued analytically to *contraction holomorphic* semigroup of semi-angle θ; see Proposition 1.7.7.

(ii) Let A be a sectorial operator in a Hilbert space $\mathfrak{H}$, that is, its numerical range $W = \{(Au, u) : u \in \text{dom}(A) \text{ and } \|u\| = 1\} \subset S_{\pi/2-\theta}$ for $0 < \theta \leq \pi/2$. If the operator A is closed, then it is generator of a holomorphic *contraction* semigroup of semi-angle θ. Moreover, $\|U_A(z)\| \leq 1$ in the closed sector $\overline{S}_{\theta-\epsilon}$, for $\epsilon \in (0, \theta]$; see Proposition 1.8.13 in Sect. 1.8.

(iii) Let A be a generator of a bounded holomorphic semigroup on a Banach space $\mathfrak{X}$, that is, $A \in \mathscr{H}(\theta, 0)$ for some $\theta \leq \pi/2$, Definition 1.7.1 and Remark 1.7.3. If operator A is also *accretive*, Definition 1.9.20 and Proposition 1.9.21, then on account of (1.8.4) and (1.9.48), $A \in \mathscr{H}(\theta, 0) \cap \mathscr{G}(1, 0)$ generates a holomorphic *contraction* semigroup in the sense of Definition 10.1.9.

(iv) If A is the self-adjoint generator of a strongly continuous unitary *group* $\{U_A(t) = \mathrm{e}^{-itA}\}_{t\in\mathbb{R}}$ on a Hilbert space $\mathfrak{H}$; then $A_+ := iA$ and $A_- := -iA$ both satisfy Proposition 1.3.4 for $M = 1$ and $\beta = 0$; cf. Example 1.10.3(c) and Remark 13.1.5. Then $A_\pm \in \mathscr{G}(1, 0)$ are generators of contraction C_0-semigroups $\{\mathrm{e}^{-tA_\pm}\}_{t\geq 0}$, whereas the operator $A_+A_- = A^2 \in \mathscr{H}_c(\pi/2, 0)$ is generator of *contraction holomorphic* semigroup $\{\mathrm{e}^{-tA^2}\}_{t\geq 0}$ of semi-angle $\pi/2$ on the Hilbert space $\mathfrak{H}$. See Notes to Chap. 10 (Sect. 10.1.2).

10.1.3 Fractional Powers of Generators

For further preparation we scrutinise in this section some properties of fractional powers of generators for holomorphic semigroups in a Banach space $\mathfrak{X}$. Recall that for $0 < \alpha < 1$ the fractional power A^α of generator of a bounded (i.e. $\{\|U_A(t)\| \leq M_A\}_{t\geq 0}$) holomorphic semigroup $\{U_A(z)\}_{z\in S_\theta}$ can be expressed by the integral *Balakrishnan* representation:

$$A^\alpha x = \frac{1}{\Gamma(-\alpha)} \int_0^\infty d\lambda\, \lambda^{-\alpha-1} (U_A(\lambda) - \mathbb{1})\, x\,, \quad x \in \text{dom}\,(A), \tag{10.1.12}$$

where $\Gamma(\cdot)$ is the Gamma-function and the branch of λ^α is chosen to be positive for $\lambda > 0$. Since for any $x \in \text{dom}(A)$ and $0 < \alpha < 1$ the integral (10.1.12) is convergent, $\text{dom}(A) \subseteq \text{dom}(A^\alpha)$. We set $A^0 := \mathbb{1}$ and define $A^\alpha = A^{\alpha-[\alpha]}A^{[\alpha]}$ for any $\alpha > 0$, where $[\alpha]$ denotes the *integer* part of α,

Proposition 10.1.10 *For each $\alpha \in [0, 1]$, there exists a constant $C_{A,\alpha}$, depending only on M_A, and α, such that for all $\mu > 0$,*

$$\left\| A^\alpha (A + \mu\mathbb{1})^{-1} \right\| \leq \frac{C_{A,\alpha}}{\mu^{1-\alpha}}\,. \tag{10.1.13}$$

Proof For $\alpha = 0$ or $\alpha = 1$, the result follows directly from the estimate of the resolvent. Let $0 < \alpha < 1$ and $x \in \mathfrak{X}$. Note that $\mathrm{ran}(A + \mu\mathbb{1})^{-1} = \mathrm{dom}(A) \subseteq \mathrm{dom}(A^{\alpha})$. As a consequence on account of (10.1.12), one gets

$$A^{\alpha}(A+\mu\mathbb{1})^{-1}\,x = \frac{1}{\Gamma(-\alpha)}\int_0^{\infty} d\lambda\,\lambda^{-\alpha-1}(U_A(\lambda)-\mathbb{1}))\,(A+\mu\mathbb{1})^{-1}\,x\,,\; x\in\mathfrak{X}. \tag{10.1.14}$$

We split integral (10.1.14) into two parts corresponding to domains: $0 < \lambda \leq \mu^{-1}$ and $\lambda > \mu^{-1}$. Then we use (10.1.3) to infer that

$$\begin{aligned} A^{\alpha}(A+\mu\mathbb{1})^{-1}x = &\frac{1}{\Gamma(-\alpha)}\int_0^{\mu^{-1}} d\lambda\,\lambda^{-\alpha-1} \\ &\int_0^{\lambda} dt\,(-U_A(t))\,(\mathbb{1}-\mu(A+\mu\mathbb{1})^{-1})\,x \\ &+\frac{1}{\Gamma(-\alpha)}\int_{\mu^{-1}}^{\infty} d\lambda\,\lambda^{-\alpha-1}(U_A(\lambda)-\mathbb{1})(A+\mu\mathbb{1})^{-1}\,x\,. \end{aligned}$$

Now by estimate of the resolvent: $\|(A+\mu)^{-1}\| \leq M_A/\mu$ for all $\mu > 0$, one obtains

$$\begin{aligned} \|A^{\alpha}(A+\mu\mathbb{1})^{-1}x\| &\leq \frac{M_A(1+M_A)}{|\Gamma(-\alpha)|}\left(\int_0^{\mu^{-1}} d\lambda\,\lambda^{-\alpha} + \frac{1}{\mu}\int_{\mu^{-1}}^{\infty} d\lambda\,\lambda^{-\alpha-1}\right)\|x\| \\ &\leq \frac{M_A(1+M_A)\mu^{\alpha-1}}{\alpha(1-\alpha)\,|\Gamma(-\alpha)|}\,\|x\|\,, \quad x\in\mathfrak{X}. \end{aligned}$$

Setting $C_{A,\alpha} := M_A(1+M_A)/(\alpha(1-\alpha)|\Gamma(-\alpha)|)$ we obtain the estimate (10.1.13). □

Next we recall the following well-known property of the semigroup generator A (see Notes to Chap. 10 (Sect. 10.1.3)):

Lemma 10.1.11 $\mathrm{dom}((A+\delta\,\mathbb{1})^{\alpha}) = \mathrm{dom}(A^{\alpha})$ *for all* $\delta > 0$ *and* $0 < \alpha < 1$.

Theorem 10.1.12 *Let* $U_A(t)$ *be a bounded holomorphic semigroup with generator* A. *Then for any real* $\alpha > 0$, *we have*

$$\sup_{t>0}\left\|t^{\alpha}A^{\alpha}U_A(t)\right\| = M_{\alpha} < \infty. \tag{10.1.15}$$

Proof Let $0 < \alpha < 1$. By $\mathrm{dom}(A) \subseteq \mathrm{dom}(A^{\alpha})$ one gets $\mathrm{dom}(A^{\alpha}U_A(t)) = \mathfrak{X}$ for $t > 0$. Hence by (10.1.12) we infer

$$A^{\alpha}U_A(t) = \frac{1}{\Gamma(-\alpha)}\int_0^{\infty} d\lambda\,\lambda^{-\alpha-1}(U_A(t+\lambda)-U_A(t))\,. \tag{10.1.16}$$

Now we split the integral (10.1.16) in two parts, $0 < \lambda \le t$ and $\lambda > t$, and we use the estimate of the derivative of the holomorphic semigroup (see Proposition 10.1.7) to obtain

$$\|U_A(t+\lambda) - U_A(t)\| \le \lambda \sup_{t \le \tau \le t+\lambda} \|\partial_\tau U_A(\tau)\| \le \lambda \frac{M'_A}{t} . \tag{10.1.17}$$

This leads to the estimate

$$\begin{aligned}\|A^\alpha U_A(t)\| &\le \frac{1}{|\Gamma(-\alpha)|}\left(\int_0^t d\lambda\, \lambda^{-\alpha} \frac{M'_A}{t} + \int_t^\infty d\lambda\, 2\, M_A\, \lambda^{-\alpha-1}\right) \\ &\le \frac{t^{-\alpha}}{|\Gamma(-\alpha)|}\left(\frac{M'_A}{1-\alpha} + \frac{2\, M_A}{\alpha}\right).\end{aligned}$$

Therefore one obtains (10.1.15) for $0 < \alpha < 1$ by setting

$$M_\alpha := |\Gamma(-\alpha)|^{-1}(M'_A/(1-\alpha) + 2\, M_A/\alpha) .$$

For integer powers α, (10.1.15) follows directly from Proposition 10.1.7. Notice that by Proposition 10.1.7 $\operatorname{ran}(U_A(t)) \subseteq \operatorname{dom}(A^n)$ for $t > 0$. Then result (10.1.15) follows for any non-integer $\alpha > 1$, from:

- the observation that $\operatorname{dom}(A^\alpha = A^{\alpha-[\alpha]}A^{[\alpha]}) \supset \operatorname{dom}(A^{[\alpha]+1})$,
- the representation (10.1.16) together with the estimate (10.1.11) for derivatives of order $[\alpha]+1$.

□

10.2 Operator-Norm Trotter Product Formula

10.2.1 Perturbation of Holomorphic Contraction Semigroups

In this subsection we prove the *operator-norm* convergence of the Trotter product formula in Banach space $\mathfrak{X}$. The proof goes by estimating the *rate* of convergence for the case of *small* perturbations of the holomorphic contraction semigroups.

To this aim we require that the operator A generates a holomorphic contraction semigroup and that perturbation B satisfies the following hypothesis:

(H1) B is generator of a contraction semigroup on $\mathfrak{X}$.
(H2) There exists $\alpha \in [0, 1)$ such that $\operatorname{dom}(A^\alpha) \subseteq \operatorname{dom}(B)$ and that for adjoint operator A^* in the dual space $\mathfrak{X}^*$ one has $\operatorname{dom}(A^*) \subseteq \operatorname{dom}(B^*)$.

Notice that we can also suppose that operator A is boundedly invertible. If it is not the case, one considers $A + \eta\, \mathbb{1}$ for some $\eta > 0$. Then by Lemma 10.1.11 we still have $\operatorname{dom}((A+\eta)^\alpha) = \operatorname{dom}(A^\alpha) \subseteq \operatorname{dom}(B)$.

Remark 10.2.1 We note that assumption (H2) implies that B is relatively bounded with respect to A with the relative bound equal to *zero*. Indeed, for $\eta > 0$ by $\mathrm{dom}(A + \eta\mathbb{1}) \subseteq \mathrm{dom}(A^\alpha) \subseteq \mathrm{dom}(B)$ and by Proposition 10.1.10, (10.1.13), one gets (here operator A is supposed to be boundedly invertible):

$$\|B(A + \eta\,\mathbb{1})^{-1}\| \leq \|BA^{-\alpha}\|\,\|A^\alpha(A + \eta\mathbb{1})^{-1}\| \leq \frac{C_\alpha}{\eta^{1-\alpha}}\,\|BA^{-\alpha}\|\,. \qquad (10.2.1)$$

Since the operators A^α and B are closed, the inclusions in (H2) are equivalent to A^α-boundedness of operator B and the A^*-boundedness of operator B^*. In particular, $\|BA^{-\alpha}\| \leq d$ and $\|B^*A^{*-1}\| \leq d'$ for some $d, d' > 0$. Therefore, in virtue of (10.2.1) and $x = (A + \eta\,\mathbb{1})^{-1}\,u$, for $u \in \mathfrak{X}$, we obtain the estimate

$$\|Bx\| \leq \frac{C_\alpha\,d}{\eta^{1-\alpha}}\,\|Ax\| + \eta^\alpha C_\alpha\,d\,\|x\|\,, \quad x \in \mathrm{dom}(A) \subseteq \mathrm{dom}(B)\,. \qquad (10.2.2)$$

Note that the relative bound $C_\alpha\,d/\eta^{1-\alpha}$ in (10.2.2) can be made *infinitesimally* small for a large enough shift $\eta > 0$, that is, unbounded operator $B \in \mathcal{P}_{0^+}(A)$ by Definition 1.9.7.

For this class of (weakly) unbounded perturbations of holomorphic contraction semigroup with generator $A \in \mathscr{H}_c(\theta, 0)$ (Definition 10.1.9), one gets the following result:

Lemma 10.2.2 *Let* $\{\mathrm{e}^{-zA}\}_{z \in S_\theta}$ *be on* $\mathfrak{X}$ *a holomorphic contraction semigroup of semi-angle* $\theta \in (0, \pi/2]$ *and perturbation* $B \in \mathcal{P}_{0^+}(A)$ *satisfy the hypothesis* (H1) *and* (H2). *Then the operator sum:* $A + B$, *defined on* $\mathrm{dom}(A + B) = \mathrm{dom}(A)$ *is also a generator of holomorphic contraction semigroup with the same semi-angle, that is,* $(A + B) \in \mathscr{H}_c(\theta, 0)$.

Proof To this end we verify conditions of Proposition 10.1.6. Let $\epsilon \in (0, \theta)$. Then by (10.2.2) and $x = (A + \eta\,\mathbb{1})^{-1}\,u$, for $u \in \mathfrak{X}$, we obtain inequality

$$\|B(A+z\mathbb{1})^{-1}\| \leq \frac{C_\alpha\,\|B\,A^{-\alpha}\|}{\eta^{1-\alpha}}\|A\,(A+z\mathbb{1})^{-1}\| + \eta^\alpha C_\alpha\,\|B\,A^{-\alpha}\|\,\|(A+z\mathbb{1})^{-1}\|\,,$$

for $|\arg(z)| < \theta + \pi/2 - \epsilon$. Seeing that for generator of holomorphic semigroups one gets by Proposition 10.1.6: $\|(z\,\mathbb{1} + A)^{-1}\| \leq M_\epsilon\,|z|^{-1}$, for $M_\epsilon > 0$ and $z \in \overline{S}_{\theta+\pi/2-\epsilon}$, this inequality leads to

$$\|B\,(A + z\mathbb{1})^{-1}\| \leq \frac{C_\alpha\|B\,A^{-\alpha}\|}{\eta^{1-\alpha}}\,(1 + N_\epsilon) + \eta^\alpha C_\alpha\,\|B\,A^{-\alpha}\|\,\frac{M_\epsilon}{|z|}\,, \qquad (10.2.3)$$

for $z \in \overline{S}_{\theta+\pi/2-\epsilon}$. Therefore, the Neumann series for $(A + B + z\mathbb{1})^{-1}$ converges if the right-hand side of (10.2.3) is smaller than 1. Since we can choose η and $z \in S_{\theta+\pi/2-\epsilon}$ in such a way that the right-hand side of the estimate (10.2.3) becomes smaller than 1, we obtain

$$\|(A+B+z\mathbb{1})^{-1}\| \leq \frac{M}{|z-\gamma|} .$$

Here M and γ are some positive constants. Then by Proposition 10.1.6 we conclude that operator $(A+B) \in \mathscr{H}(\theta,\gamma)$, that is, it generates a *quasi-bounded* holomorphic semigroup $\{U_{A+B}(z)\}_{z\in S_\theta}$.

On the other hand, the conditions of lemma imply that operators A and B are *accretive*; thus operator $A+B$ is also accretive, and by Proposition 1.9.17 we conclude that operator $(A+B) \in \mathscr{G}(1,0)$, that is, generates a contraction semigroup. Therefore, owing to the previous paragraph, this semigroup is also holomorphic, which yields $(A+B) \in \mathscr{H}(\theta,\gamma) \cap \mathscr{G}(1,0)$ and, thus, finishes the proof. □

The proof of the main statements of Sect. 10.2 involves *three* technical lemmata. For the two of them, we need only that operators B (B^*) are $A(A^*)$-bounded in the *Kato sense* (Definition 1.9.7), that is, there are positive constants a and b such that:

$$x \in \mathrm{dom}(A) \subseteq \mathrm{dom}(B), \ \|Bx\| \leq a\|Ax\| + b\|x\|, \tag{10.2.4}$$

$$\phi \in \mathrm{dom}(A^*) \subseteq \mathrm{dom}(B^*), \ \|B^*\phi\| \leq a\|A^*\phi\| + b\|\phi\|. \tag{10.2.5}$$

If A is boundedly invertible, then we are able to substitute a by the relative bound $a + b\|A^{-1}\|$ and to skip the last term with b.

Lemma 10.2.3 *Let boundedly invertible A and operator B be generators of bounded C_0-semigroups. Let B and B^* satisfy* (10.2.4), (10.2.5) *and suppose that operator $H = (A+B)$ on* $\mathrm{dom}(H) = \mathrm{dom}(A)$ *is boundedly invertible generator of a bounded semigroup. Then there exists a constant L_1 such that for all $\tau \geq 0$*

$$\left\|A^{-1}\left(\mathrm{e}^{-\tau B}\mathrm{e}^{-\tau A} - \mathrm{e}^{-\tau(A+B)}\right)\right\| \leq L_1\tau, \tag{10.2.6}$$

$$\left\|\left(\mathrm{e}^{-\tau B}\mathrm{e}^{-\tau A} - \mathrm{e}^{-\tau(A+B)}\right)A^{-1}\right\| \leq L_1\tau. \tag{10.2.7}$$

Proof By virtue of identity

$$A^{-1}\left(\mathrm{e}^{-\tau B}\mathrm{e}^{-\tau A} - \mathrm{e}^{-\tau(A+B)}\right) = A^{-1}\left(\mathrm{e}^{-\tau B} - \mathbb{1}\right)\mathrm{e}^{-\tau A}$$
$$+ A^{-1}\left(\mathrm{e}^{-\tau A} - \mathbb{1}\right) + A^{-1}HH^{-1}\left(\mathbb{1} - \mathrm{e}^{-\tau H}\right),$$

and by Lemma 10.1.2, we infer (10.2.6):

$$\begin{aligned}\left\|A^{-1}\left(e^{-\tau B}e^{-\tau A}-e^{-\tau H}\right)\right\| &\le \left\|\int_0^\tau ds\, A^{-1}B\, e^{-sB}\right\| \\ &+\left\|A^{-1}\left(e^{-\tau A}-\mathbb{1}\right)\right\|+\|A^{-1}H\|\,\left\|H^{-1}\left(\mathbb{1}-e^{-\tau H}\right)\right\| \\ &\le \|A^{-1}B\|\, M_B\tau + M_A\tau + \|A^{-1}H\|\, M_H\tau\,,\end{aligned}$$

where coefficients M_B and M_H are defined similar to M_A in Sect. 10.1.3.

Finally we remark that (10.2.5) implies a boundedness of the closed operator $\overline{A^{-1}B}$ and that $\|A^{-1}H\| \le \|\mathbb{1}+A^{-1}B\| \le 1+a+b\|A^{-1}\|$. To prove (10.2.7) one has to use (10.2.4), and the same line of reasoning as above to put finally $L_1 = M_B\, a' + M_A + M_H\,(1+a')$, where $a' := a + b\,\|A^{-1}\|$. □

Lemma 10.2.4 *Let A, B and $H = A + B$ be the same as in Lemma* 10.2.3. *Then there exists a constant L_2 such that for all $\tau \ge 0$:*

$$\left\|A^{-1}\left(e^{-\tau B}e^{-\tau A}-e^{-\tau(A+B)}\right)A^{-1}\right\| \le L_2\tau^2, \tag{10.2.8}$$

$$\left\|A^{-1}\left(e^{-\tau A}e^{-\tau B}-e^{-\tau(A+B)}\right)A^{-1}\right\| \le L_2\tau^2. \tag{10.2.9}$$

Proof By virtue of equation

$$\begin{aligned}e^{-\tau B}e^{-\tau A}-e^{-\tau H} &= \left(\mathbb{1}-e^{-\tau B}\right)\left(\mathbb{1}-e^{-\tau A}\right)+\left(e^{-\tau A}-(\mathbb{1}+\tau A)^{-1}\right) \\ &+\left(e^{-\tau B}-(\mathbb{1}+\tau B)^{-1}\right)+\left((\mathbb{1}+\tau H)^{-1}-e^{-\tau H}\right) \\ &+\tau H\,(\mathbb{1}+\tau H)^{-1}-\tau A\,(\mathbb{1}+\tau A)^{-1}-\tau B\,(\mathbb{1}+\tau B)^{-1}\,,\end{aligned}$$

and by identity

$$\begin{aligned}&A^{-1}\left(\tau H\,(\mathbb{1}+\tau H)^{-1}-\tau A\,(\mathbb{1}+\tau A)^{-1}-\tau B\,(\mathbb{1}+\tau B)^{-1}\right)A^{-1} \\ &=\tau^2\left((\mathbb{1}+\tau A)^{-1}+A^{-1}B(\mathbb{1}+\tau B)^{-1}BA^{-1}-A^{-1}H(\mathbb{1}+\tau H)^{-1}HA^{-1}\right),\end{aligned}$$

we obtain the representation

$$\begin{aligned}&A^{-1}(e^{-\tau B}e^{-\tau A}-e^{-\tau H})A^{-1} = \\ &\quad A^{-1}(\mathbb{1}-e^{-\tau B})(\mathbb{1}-e^{-\tau A})A^{-1} \\ &\quad+\left(e^{-\tau A}-(\mathbb{1}+\tau A)^{-1}\right)A^{-2}+A^{-1}\left(e^{-\tau B}-(\mathbb{1}+\tau B)^{-1}\right)A^{-1}\end{aligned}$$

$$+A^{-1}H\left((\mathbb{1}+\tau H)^{-1}-\mathrm{e}^{-\tau H}\right)H^{-2}HA^{-1}+\tau^{2}(\mathbb{1}+\tau A)^{-1}$$
$$+\tau^{2}A^{-1}B(\mathbb{1}+\tau B)^{-1}BA^{-1}-\tau^{2}A^{-1}H(\mathbb{1}+\tau H)^{-1}HA^{-1}\,.$$

This presentation yields the following estimate:

$$\begin{aligned}
&\frac{1}{\tau^2}\left\|A^{-1}(\mathrm{e}^{-\tau B}\mathrm{e}^{-\tau A}-\mathrm{e}^{-\tau H})\,A^{-1}\right\| \le \\
&\quad \frac{1}{\tau}\,\|A^{-1}B\|\left\|\int_0^{\tau} ds\,\mathrm{e}^{-sB}\right\|\,\frac{1}{\tau}\,\left\|(\mathbb{1}-\mathrm{e}^{-\tau A})A^{-1}\right\| \\
&\quad +\frac{1}{\tau^2}\left\|\left(\mathrm{e}^{-\tau A}-(\mathbb{1}+\tau A)^{-1}\right)A^{-2}\right\| \\
&\quad +\frac{1}{\tau^2}\|A^{-1}B\|\left\|\int_0^{\tau} ds_1\int_0^{s_1} ds_2\,\mathrm{e}^{-s_2 B}-\tau^2(\mathbb{1}+\tau B)^{-1}\right\|\,\|BA^{-1}\| \\
&\quad +\frac{1}{\tau^2}\,\|A^{-1}H\|\,\left\|\left((\mathbb{1}+\tau H)^{-1}-\mathrm{e}^{-\tau H}\right)H^{-2}\right\|\,\|HA^{-1}\| \\
&\quad +1+\|A^{-1}B\|\,\|BA^{-1}\|+\|A^{-1}H\|\,\|HA^{-1}\|\,.
\end{aligned}$$

Now Lemmata 10.1.2 and 10.1.4 together with (10.2.4), (10.2.5) imply (10.2.8), where we can take $L_2 = a' M_A M_B + 3\,M_A/2 + 3\,M_B\,a'^2/2 + 3\,M_H(1+a')^2/2 + 1 + a'^2 + (1+a')^2$ with $a' = a + b\,\|A^{-1}\|$.

In a similar way, one obtains (10.2.9). □

Note that for the proof of the third lemma (Lemma 10.2.5), we do need *conditions* (H2), as well as requirement that semigroups are *holomorphic* and *contraction*. We remind that by convention $\mathrm{dom}(A^{\alpha=0}) = \mathfrak{X}$. Therefore, if $\alpha = 0$, then operator B in the Lemma 10.2.5 is bounded.

Lemma 10.2.5 *Let A be a boundedly invertible generator of holomorphic contraction semigroup. If B is generator of a contraction semigroup and there exists $\alpha \in [0, 1)$ such that $\mathrm{dom}(A^\alpha) \subseteq \mathrm{dom}(B)$, then for any $k \geq 1$ and $\tau > 0$, one gets the estimates*

$$\left\|\left(\mathrm{e}^{-\tau B}\mathrm{e}^{-\tau A}\right)^k A\right\| \le \frac{L_3}{\tau^\alpha}+\frac{M_A'}{k\tau}\,, \qquad \text{if } \alpha > 0\,, \tag{10.2.10}$$

$$\left\|\left(\mathrm{e}^{-\tau B}\mathrm{e}^{-\tau A}\right)^k A\right\| \le \widetilde{L}_3(1+\ln k)+\frac{M_A'}{k\tau}\,, \quad \text{if } \alpha = 0\,. \tag{10.2.11}$$

Proof We begin with the following chain of estimates:

$$\begin{aligned}\left\|\left(e^{-\tau B}e^{-\tau A}\right)^k A\right\| &\le \left\|\left(\left(e^{-\tau B}e^{-\tau A}\right)^k - e^{-k\tau A}\right)A\right\| + \left\|e^{-k\tau A}A\right\| \\ &\le \left\|\sum_{j=0}^{k-1}\left(e^{-\tau B}e^{-\tau A}\right)^{k-1-j}\left(e^{-\tau B} - \mathbb{1}\right)e^{-\tau A}e^{-j\tau A}A\right\| + \left\|e^{-k\tau A}A\right\| \\ &\le \sum_{j=0}^{k-1}\left\|\int_0^{\tau} ds e^{-sB}BA^{-\alpha}\right\|\left\|A^{\alpha}e^{-(j+1)\tau A}A\right\| + \left\|e^{-k\tau A}A\right\| .\end{aligned}$$

Notice that the second inequality is in particular due to contractions of e^{-tA} and e^{-tB} and to equation (10.1.3) of Lemma 10.1.2. From the hypothesis $\mathrm{dom}(A^{\alpha}) \subseteq \mathrm{dom}(B)$, we deduce that $\|BA^{-\alpha}\| \le d$; see Remark 10.2.1. By Propositions 10.1.7 and Theorem 10.1.12, we obtain, respectively:

$$\left\|e^{-k\tau A}A\right\| \le \frac{M'_A}{k\tau} \quad \text{and} \quad \left\|A^{1+\alpha}e^{-(j+1)\tau A}\right\| \le \frac{M_{\alpha}}{((j+1)\tau)^{1+\alpha}} .$$

Therefore, we conclude that

$$\left\|\left(e^{-\tau B}e^{-\tau A}\right)^k A\right\| \le \frac{M_{\alpha}\, d}{\tau^{\alpha}}\sum_{j=0}^{k-1}\frac{1}{(j+1)^{1+\alpha}} + \frac{M'_A}{k\tau} .$$

If $\alpha > 0$, this gives the announced result (10.2.10) for

$$L_3 = M_{\alpha}\, d \sum_{j=1}^{\infty}(1/j)^{1+\alpha} ,$$

as well as (10.2.11) for $\alpha = 0$, where $\widetilde{L}_3 = \|B\| < \infty$, since by virtue of $\mathrm{dom}(A^0) = \mathfrak{X}$ the operator B is bounded. □

It is not surprising that since $\mathrm{dom}(A^{\alpha}) \subseteq \mathrm{dom}(B)$ implies $\mathrm{dom}(A^{\alpha'}) \subseteq \mathrm{dom}(B)$ for $\alpha' \ge \alpha$, the estimate (10.2.10) is also valid for any $\alpha' \ge \alpha$.

10.2.2 Operator-Norm Convergence Rate

The main assertions of Sect. 10.2 (Theorems 10.2.6, 10.2.7 and Corollary 10.2.8) with *estimates* of the rate of operator-norm convergence of the corresponding approximants in the Trotter product formulæ are collected below.

Theorem 10.2.6 *Let* $\{e^{-zA}\}_{z\in S_\theta}$ *be a holomorphic contraction semigroup, that is,* $A \in \mathscr{H}_c(\theta, 0)$. *Let* B *be generator of a contraction semigroup on* $\mathfrak{X}$. *If there exists* $\alpha \in [0, 1)$ *such that* $\mathrm{dom}(A^\alpha) \subseteq \mathrm{dom}(B)$ *and* $\mathrm{dom}(A^*) \subseteq \mathrm{dom}(B^*)$, *then there are constants* $M_1, M_2, \tilde{M}_2, \eta > 0$, *such that for any* $t \geq 0$ *and* $n > 2$ *one gets estimates*

$$\left\|\left(e^{-tB/n}e^{-tA/n}\right)^n - e^{-t(A+B)}\right\| \leq e^{\eta t}(M_1 + M_2 t^{1-\alpha})\frac{\ln(n)}{n^{1-\alpha}}, \quad \alpha > 0, \tag{10.2.12}$$

$$\left\|\left(e^{-tB/n}e^{-tA/n}\right)^n - e^{-t(A+B)}\right\| \leq e^{\eta t}(M_1 + \tilde{M}_2 t)\frac{2\,(\ln(n))^2}{n}, \quad \alpha = 0. \tag{10.2.13}$$

Proof Because B satisfies hypotheses (H1) and (H2), by Lemma 10.2.2 the operator $H = (A + B)$ is generator of a holomorphic contraction semigroup. If operator A has no bounded inverse, let $\tilde{A} := A + \eta\mathbb{1}$ and $\tilde{H} := \tilde{A} + B$ for some $\eta > 0$ (see Remark 10.2.1). Then both operators are boundedly invertible. As we indicated above (Lemma 10.1.11), these changes of generators do not modify inclusions of domains. If we want to obtain $\|B\tilde{A}^{-1}\| < 1$, then by the estimate (10.2.1) we have to choose a sufficiently large shift parameter $\eta > 0$. This gives the estimate $\|\tilde{A}\tilde{H}^{-1}\| = \|(\mathbb{1} + B\tilde{A}^{-1})^{-1}\| \leq 1/(1 - a)$, where we set $a = \|B\tilde{A}^{-1}\|$.

Now we put $\tau := t/n$, $\tilde{U}(t) := e^{-t\tilde{H}}$, and $\tilde{T}(\tau) := e^{-\tau B}e^{-\tau\tilde{A}}$. To estimate the left-hand side of (10.2.12), we use

$$\left(e^{-tB/n}e^{-tA/n}\right)^n - e^{-t(A+B)} = (\tilde{T}^n(\tau) - \tilde{U}^n(\tau))\,e^{t\eta}, \tag{10.2.14}$$

and the *telescopic* identity:

$$\begin{aligned}\tilde{T}(\tau)^n - \tilde{U}(\tau)^n &= \sum_{m=0}^{n-1} \tilde{T}(\tau)^{n-m-1}(\tilde{T}(\tau) - \tilde{U}(\tau))\tilde{U}(\tau)^m \\ &= \tilde{T}(\tau)^{n-1}\tilde{A}\tilde{A}^{-1}(\tilde{T}(\tau) - \tilde{U}(\tau)) \\ &\quad + (\tilde{T}(\tau) - \tilde{U}(\tau))\tilde{A}^{-1}\tilde{A}\tilde{H}^{-1}\tilde{H}\tilde{U}(\tau)^{n-1} \\ &\quad + \sum_{m=1}^{n-2} \tilde{T}(\tau)^{n-m-1}\tilde{A}\tilde{A}^{-1}(\tilde{T}(\tau) - \tilde{U}(\tau))\tilde{A}^{-1}\tilde{A}\tilde{H}^{-1}\tilde{H}\tilde{U}(\tau)^m,\end{aligned}$$

which implies

$$\begin{aligned}\left\|\tilde{T}(\tau)^n-\tilde{U}(\tau)^n\right\| &\leq \|\tilde{T}(\tau)^{n-1}\tilde{A}\|\,\|\tilde{A}^{-1}(\tilde{T}(\tau)-\tilde{U}(\tau))\|\\ &+\|(\tilde{T}(\tau)-\tilde{U}(\tau))\tilde{A}^{-1}\|\,\|\tilde{A}\tilde{H}^{-1}\|\,\|\tilde{H}\tilde{U}(\tau)^{n-1}\|\\ &+\sum_{m=1}^{n-2}\|\tilde{T}(\tau)^{n-m-1}\tilde{A}\|\,\|\tilde{A}^{-1}(\tilde{T}(\tau)-\tilde{U}(\tau))\tilde{A}^{-1}\|\,\|\tilde{A}\tilde{H}^{-1}\|\\ &\times\|\tilde{H}\tilde{U}(\tau)^m\|\,.\end{aligned}$$

Hence by Lemmata 10.2.3, 10.2.4 and 10.2.5 (it is at this point that we use the hypothesis of *contraction*) and by Proposition 10.1.7 (*holomorphy*), we obtain the estimates

$$\begin{aligned}\|\tilde{T}(\tau)^n-\tilde{U}(\tau)^n\| &\leq \left(\frac{L_3}{\tau^\alpha}+\frac{M'_A}{(n-1)\tau}\right)L_1\tau+\frac{L_1}{1-a}\frac{M'_H}{n-1}\\ &+\sum_{m=1}^{n-2}\left(L_3\tau^{1-\alpha}+\frac{M'_A}{n-m-1}\right)\frac{L_2}{1-a}\frac{M'_H}{m}\\ &\leq L_3L_1\frac{t^{1-\alpha}}{n^{1-\alpha}}+\frac{L_1}{n-1}\left(M'_A+\frac{M'_H}{1-a}\right)+\frac{L_3L_2M'_H}{1-a}\frac{t^{1-\alpha}}{n^{1-\alpha}}\sum_{m=1}^{n-2}\frac{1}{m}\\ &+\frac{L_2M'_HM'_A}{1-a}\sum_{m=1}^{n-2}\frac{1}{n-m-1}\cdot\frac{1}{m}\\ &\leq L_3L_1\frac{t^{1-\alpha}}{n^{1-\alpha}}+\frac{L_1}{n-1}\left(M'_A+\frac{M'_H}{1-a}\right)\\ &+2\,\frac{L_3L_2M'_H}{1-a}\,t^{1-\alpha}\,\frac{\ln(n)}{n^{1-\alpha}}+4\,\frac{L_2M'_HM'_A}{1-a}\,\frac{\ln(n)}{n}\,, \qquad (10.2.15)\end{aligned}$$

due to $\sum_{m=1}^{n-1}1/((n-m)\,m)=(2/n)\sum_{m=1}^{n-1}1/m\leq(2/n)\;(1+\ln(n-1))\leq 4\,(\ln(n))/n$ for $n>2$. Then estimates (10.2.14) and (10.2.15) imply the announced result (10.2.12) for $\alpha>0$, where $M_1:=4\,L_1\left(M'_A+\frac{M'_H}{1-a}\right)+4\,\frac{L_2M'_HM'_A}{1-a}$ and $M_2:=2\,L_3L_1+2\,\frac{L_3L_2M'_H}{1-a}$.

In a similar way, one infers for $\alpha = 0$ and $n > 2$ that

$$\|\tilde{T}(\tau)^n - \tilde{U}(\tau)^n\| \leq \tilde{L}_3(1+\ln(n-1))L_1\frac{t}{n} + \frac{L_1 M'_A}{n-1} + \frac{L_1 M'_H}{1-a}\frac{1}{n-1}$$
$$+\sum_{m=1}^{n-2}\left(\tilde{L}_3\frac{t}{n}(1+\ln(n-m-1)) + \frac{M'_A}{n-m-1}\right)\frac{L_2}{1-a}\frac{M'_H}{m}.$$

This estimate together with (10.2.14) yields (10.2.13), where we put $\tilde{M}_2 := 2\,\tilde{L}_3\,L_1 + 2\,\tilde{L}_3\,L_2\,M'_H/(1-a)$. □

Now we need hypothesis, which is in a sense “adjoint” to hypothesis (H1), (H2).

Theorem 10.2.7 *Let* $\{e^{-zA}\}_{z\in S_\theta}$ *be a holomorphic contraction semigroup, that is,* $A \in \mathscr{H}_c(\theta, 0)$. *Let* B *be generator of a contraction semigroup, and there exists* $\alpha \in [0, 1)$ *such that* $\mathrm{dom}((A^\alpha)^*) \subseteq \mathrm{dom}(B^*)$ *and* $\mathrm{dom}(A) \subseteq \mathrm{dom}(B)$. *If in addition* $\mathrm{dom}(A^*) \subseteq \mathrm{dom}(B^*)$ *(if* $\mathfrak{X}$ *is not reflexive), then there are constants* M_3, M_4, $\tilde{M}_4$, $\eta > 0$, *such that for any* $t \geq 0$ *and* $n > 2$

$$\left\|\left(e^{-tA/n}e^{-tB/n}\right)^n - e^{-t(A+B)}\right\| \leq e^{\eta t}(M_3 + M_4\,t^{1-\alpha})\,\frac{\ln(n)}{n^{1-\alpha}}\,,\ \alpha > 0, \tag{10.2.16}$$

$$\left\|\left(e^{-tA/n}e^{-tB/n}\right)^n - e^{-t(A+B)}\right\| \leq e^{\eta t}(M_3 + \tilde{M}_4\,t)\,\frac{2\,(\ln(n))^2}{n}\,,\ \alpha = 0, \tag{10.2.17}$$

for any $t \geq 0$ *and* $n > 2$.

Proof Let $\tilde{F}(\tau) := e^{-\tau\tilde{A}}e^{-\tau B}$. Then by the same arguments as in the proof of Theorem 10.2.6, one obtains

$$\tilde{U}(\tau)^n - \tilde{F}(\tau)^n = \sum_{m=0}^{n-1}\tilde{U}(\tau)^{n-m-1}(\tilde{U}(\tau) - \tilde{F}(\tau))\tilde{F}(\tau)^m$$
$$= \tilde{U}(\tau)^{n-1}\tilde{H}\,\tilde{H}^{-1}\tilde{A}\,\tilde{A}^{-1}(\tilde{U}(\tau) - \tilde{F}(\tau))$$
$$+(\tilde{U}(\tau) - \tilde{F}(\tau))\tilde{A}^{-1}\tilde{A}\,\tilde{F}(\tau)^{n-1}$$
$$+\sum_{m=1}^{n-2}\tilde{U}(\tau)^{n-m-1}\tilde{H}\,\tilde{H}^{-1}\tilde{A}\,\tilde{A}^{-1}(\tilde{U}(\tau) - \tilde{F}(\tau))\tilde{A}^{-1}\tilde{A}\tilde{F}(\tau)^m.$$

Notice that Lemmata 10.2.3 and 10.2.4 hold for $\tilde{F}(\tau)$. By a simple modification of Lemma 10.2.5, where one uses $\|\tilde{A}^{-\alpha}B\| = \|B^*(\tilde{A}^{-\alpha})^*\| < \infty$, we find that

$$\left\|\tilde{A}\left(e^{-\tau\tilde{A}}e^{-\tau B}\right)^k\right\| \leq \frac{L_4}{\tau^\alpha} + \frac{M'_A}{k\tau}, \qquad \alpha > 0,$$

$$\left\|\tilde{A}\left(e^{-\tau\tilde{A}}e^{-\tau B}\right)^k\right\| \leq \tilde{L}_4(1+\ln k) + \frac{M'_A}{k\tau}, \quad \alpha = 0.$$

These ingredients (cf. (10.2.10) and (10.2.11)) ensure the estimates (10.2.16) and (10.2.17). □

Corollary 10.2.8 *Under conditions of Theorem* 10.2.6 *and Theorem* 10.2.7, *we have the operator-norm convergence of the symmetric Trotter formula. That is, there exists* M_5, M_6, $\tilde{M}_6$, $\eta > 0$, *such that for any* $t \geq 0$ *and* $n > 2$:

$$\left\|\left(e^{-tA/2n}e^{-tB/n}e^{-tA/2n}\right)^n - e^{-t(A+B)}\right\| \leq e^{\eta t}(M_5 + M_6 t^{1-\alpha})\frac{\ln(n)}{n^{1-\alpha}}, \quad \text{for } 0 < \alpha < 1, \tag{10.2.18}$$

$$\left\|\left(e^{-tA/2n}e^{-tB/n}e^{-tA/2n}\right)^n - e^{-t(A+B)}\right\| \leq e^{\eta t}(M_5 + \tilde{M}_6 t)\frac{2(\ln(n))^2}{n}, \quad \text{for } \alpha = 0. \tag{10.2.19}$$

Proof Because Lemmata 10.2.3, 10.2.4 and 10.2.5 can be extended to the symmetric approximants $\{e^{-\tau A/2n}e^{-\tau B/n}e^{-\tau A/2n}\}_{n\geq 1}$, the proof of Theorem 10.2.6 and Theorem 10.2.7 carries through *verbatim* to yield (10.2.18) and (10.2.19). □

Remark 10.2.9 (a) Seeing that in Theorems 10.2.6 and 10.2.7 and in Corollary 10.2.8 the perturbation B of a *dominating* generator A is either *infinitesimally* A-small ($B \in \mathcal{P}_{0^+}(A)$, Definition 1.9.7), or simply a *bounded* operator, the corresponding results in Banach space $\mathfrak{X}$ sound (in that sense) as much *weaker* than those in a Hilbert space $\mathfrak{H}$; see Chap. 9. In Sects. 9.3 and 9.4, the perturbation B in $\mathfrak{H}$ is *Kato-small* with the relative bound less than *one*, that is, $B \in \mathcal{P}_1(A)$ (Definition 1.9.7).
(b) On the other hand, it is even more important that in Theorems 10.2.6 and 10.2.7 and in Corollary 10.2.8 instead of the *self-adjointness* of semigroup $\{e^{-tA}\}_{t\geq 0}$ (see Sects. 9.3 and 9.4) we used for establishing the operator-norm estimate for the rate of convergence the Trotter product formula a *weaker* hypothesis of *analyticity*.
For that reason the holomorphic (contraction) semigroups are a natural generalisation of the self-adjoint semigroups in order to prove the operator-norm convergence of the Trotter product formula with estimate for the rate of convergence.

We shall discuss the impact of Remark 10.2.9 on revision of results in Chap. 9 for non-self-adjoint semigroups in Sect. 10.2.3.

10.2.3 Non-Self-Adjoint Semigroups: Revisited

Notice that results of Sect. 10.2 are evidently valid on a Hilbert space $\mathfrak{H}$, that is, for $\mathfrak{X}$, which is substituted by $\mathfrak{H}$. This permits to compare them with our conclusions concerning the operator-norm convergence of the Trotter product formula for non-self-adjoint semigroups in Chap. 9.

First of all we conclude that Theorems 10.2.6 and 10.2.7 and Corollary 10.2.8 hold *mutatis mutandis* for a Hilbert space $\mathfrak{H}$ (instead of $\mathfrak{X}$) with the rates of the operator-norm convergence of the Trotter product formula defined by (10.2.12) and (10.2.13), (10.2.16) and (10.2.17) or (10.2.18) and (10.2.19). These *weaker* (with respect to the bounds in (9.3.43), (9.3.60) and (9.3.61), or (9.4.7) and (9.4.8)) rates of the operator-norm convergence are because of *weaker* conditions on operator A. Now it is generator of the contraction holomorphic semigroup with (a bit) *stronger* than in Sect. 9.3.1, (H2)(i)–(iv), conditions on generator B; see (H1), (H2) in Sect. 10.2.1.

For example, Theorem 10.2.6 in a Hilbert space $\mathfrak{H}$ gets the form:

Theorem 10.2.10 *Let* $\{e^{-zA}\}_{z \in S_\theta}$ *be a holomorphic contraction semigroup, that is,* $A \in \mathscr{H}_c(\theta, 0)$. *Let* B *be generator of a contraction semigroup on* $\mathfrak{H}$. *If there exists* $\alpha \in [0, 1)$ *such that* $\mathrm{dom}(A^\alpha) \subseteq \mathrm{dom}(B)$ *and* $\mathrm{dom}(A^*) \subseteq \mathrm{dom}(B^*)$, *then there are constants* $M_1, M_2, \tilde{M}_2, \eta > 0$, *such that for any* $t \geq 0$ *and* $n > 2$ *one gets estimates*

$$\left\| \left(e^{-tB/n} e^{-tA/n} \right)^n - e^{-t(A+B)} \right\| \leq e^{\eta t} (M_1 + M_2 t^{1-\alpha}) \frac{\ln(n)}{n^{1-\alpha}}, \quad \alpha > 0, \tag{10.2.20}$$

$$\left\| \left(e^{-tB/n} e^{-tA/n} \right)^n - e^{-t(A+B)} \right\| \leq e^{\eta t} (M_1 + \tilde{M}_2 t) \frac{2\,(\ln(n))^2}{n}, \quad \alpha = 0. \tag{10.2.21}$$

This result shows that hypothesis of self-adjointness in the case of a Hilbert space has only a technical importance. It allows to improve the rate the operator-norm convergence of the Trotter product formula for non-self-adjoint semigroups. In fact, the evident advantage is *bona fide*: one can use the spectral theorem, which is widely exploited in Chap. 9 (Sects. 9.3 and 9.4).

The method developed in Sects. 10.2.1 and 10.2.2 for the operator-norm convergence of the Trotter product formula on Banach spaces distinguishes analyticity of dominating semigroup $\{e^{-tA}\}_{t \geq 0}$ as a sufficient property for:

- providing the operator-norm convergence *and*
- supplying the estimate for the rate of convergence.

This allows to revise the problem (Theorem 9.2.4) of operator-norm convergence of the Trotter product formula for non-self-adjoint semigroups generated by sectorial operators. Note that it is solved by the analytic extension method in Sect. 10.2.2 *without* estimate of the rate of convergence.

Application of the method presented in Sects. 10.2.1 and 10.2.2 yields the following statement.

Theorem 10.2.11 *Let A and B be m-sectorial operators with semi-angles $\theta_A < \pi/2$ and $\theta_B < \pi/2$ and with domains in a Hilbert space $\mathfrak{H}$. If they satisfy assumptions (i)–(v) of Sect.* 9.2.1 *and there exists $\alpha \in [0, 1)$ such that by conditions* (H1), (H2):$\mathrm{dom}(A^\alpha) \subseteq \mathrm{dom}(B)$ *and* $\mathrm{dom}(A^*) \subseteq \mathrm{dom}(B^*)$, *then there are constants $M_1, M_2, \tilde{M}_2$, such that*

- *the Trotter product approximants $\{(e^{-tA/n}e^{-tB/n})^n\}_{n\geq 1}$ converge in the operator-norm topology to e^{-tH}, locally-uniformly in $t \in S_{\theta^*}$, and*
- *for any $t \geq 0$ and $n > 2$ one gets estimates*

$$\left\| \left(e^{-tB/n}e^{-tA/n}\right)^n - e^{-tH} \right\| \leq (M_1 + M_2 t^{1-\alpha}) \frac{\ln(n)}{n^{1-\alpha}}, \quad \alpha > 0, \qquad (10.2.22)$$

$$\left\| \left(e^{-tB/n}e^{-tA/n}\right)^n - e^{-tH} \right\| \leq (M_1 + \tilde{M}_2 t) \frac{2\,(\ln(n))^2}{n}, \quad \alpha = 0. \qquad (10.2.23)$$

Here operator $H = A \dotplus B$ is the form-sum of A and B, and $\theta^* < \theta = \pi/2 - \max\{\theta_A, \theta_B\}$ is semi-angle of the sector S_{θ^*} in the complex plane.

Proof The arguments mimic the line of reasoning of Theorem 9.2.4 and Theorem 10.2.6 (or Theorem 10.2.10) for $\mathfrak{H}$.

First we note that by condition (i) operators A and B are invertible. Thus, the shift parameter $\eta = 0$ in (10.2.22) and (10.2.23), cf. (10.2.20) and (10.2.21).

Next, by conditions (H1), (H2) operator B is infinitesimally A-small, and owing to (ii)–(iv) (valid for $\alpha \in [0, 1)$ because of (H1), (H2)) the operator $H = A \dotplus B$, as a form-sum of A and B, is well-defined and sectorial with semi-angle θ_A.

Finally, since operator A is m-sectorial, the semigroup $\{e^{-tA}\}_{t\geq 0}$ is contraction and holomorphic in open sector with semi-angle $\theta = \pi/2 - \theta_A$.

As a consequence, operators A and B satisfy conditions of Theorem 10.2.10 that provides the proof of (10.2.22) and (10.2.23) for $\eta = 0$. □

10.3 Example

Resuming Remark 10.2.9, the question arises: Whether the Trotter product formula converges in the *operator-norm* topology if the condition on a *dominating* generator $A \in \mathscr{H}_c(\theta, 0)$ is relaxed, for example, to a much weaker hypothesis that $A \in \mathscr{G}(1, 0)$? More precisely, let A be generator of a *contraction* (but *not* holomorphic

(!); cf. Definition 1.7.10) C_0-semigroup on Banach space $\mathfrak{X}$, whereas $B \in \mathcal{L}(\mathfrak{X})$ is, for simplicity, a *bounded* generator.

The aim of this section is to provide a *positive* answer to this question using example from a particular class of generators and semigroups. It turns out that appropriate for this purpose is the class of generators of the *evolution* semigroups introduced in Chap. 2 on Banach space $L^p(\mathcal{I}, \mathfrak{X})$ for evolution equations in the L^p-setting.

10.3.1 Evolution Semigroups and Trotter Product Formula

To proceed further we need some key notions from the *evolution* semigroups theory and in particular the notion of *solution operator*; cf. Chap. 2 (Sect. 2.2). We shall return to a detailed analysis of this theory in Chap. 11.

A strongly continuous mapping $U(\cdot,\cdot) : \Delta \longrightarrow \mathcal{L}(\mathfrak{X})$, where domain $\Delta := \{(t,s) : 0 < s \le t \le T\}$ and $\mathcal{L}(\mathfrak{X})$ is the set of bounded operators on a *separable* Banach space $\mathfrak{X}$, is called a *solution operator* if the conditions

$$\text{(i)} \quad \sup_{(t,s)\in\Delta} \|U(t,s)\|_{\mathcal{L}(\mathfrak{X})} < \infty ,$$

$$\text{(ii)} \quad U(t,s) = U(t,r)U(r,s), \quad 0 < s \le r \le t \le T ,$$

are satisfied.

Let us consider the Banach space $L^p(\mathcal{I}, \mathfrak{X})$ for $\mathcal{I} := [0,T]$ and $p \in [1,\infty)$. The operator $\mathcal{K}$ is an *evolution generator* of the evolution semigroup $\{\mathcal{U}(\tau) := e^{-\tau\mathcal{K}}\}_{\tau\ge 0}$ if there is a solution operator such that the *Howland-Evans-Neidhardt* (HEN) representation (2.2.3) (or Appendix F.2):

$$(e^{-\tau\mathcal{K}} f)(t) = U(t, t-\tau)\chi_{\mathcal{I}}(t-\tau) f(t-\tau), \quad f \in L^p(\mathcal{I}, \mathfrak{X}) , \tag{10.3.1}$$

holds for almost all (a.a.) $t \in \mathcal{I}$ and $\tau \ge 0$. Seeing that on account of (10.3.1) the semigroup $\{e^{-\tau\mathcal{K}}\}_{\tau\ge 0}$ is *nilpotent*: $e^{-\tau\mathcal{K}} f = 0$ for $\tau \ge T$, the evolution generator $\mathcal{K}$ can *never* be generator of a holomorphic semigroup.

A simple example of evolution generator is the operator of differentiation:

$$\begin{aligned} (D_0 f)(t) &:= \partial_t f(t) , \quad t \in \mathcal{I} , \\ f \in \operatorname{dom}(D_0) &:= \{ f \in W^{1,p}(\mathcal{I}, \mathfrak{X}) : f(0) = 0 \} , \end{aligned} \tag{10.3.2}$$

where $W^{1,p}(\mathcal{I}, \mathfrak{X})$ is the *Sobolev* space of order $(1, p)$ of the *Bochner* p-integrable vector-functions. Then by (10.3.2) one gets the contraction shift semigroup:

$$(e^{-\tau D_0} f)(t) = \chi_{\mathcal{I}}(t-\tau) f(t-\tau), \quad f \in L^p(\mathcal{I}, \mathfrak{X}), \tag{10.3.3}$$

for a.a. $t \in \mathcal{I}$ and $\tau \geq 0$. Hence, (10.3.1) and (10.3.3) imply that the corresponding solution operator for the *non-holomorphic* evolution semigroup $\{e^{-\tau D_0}\}_{\tau \geq 0}$ is given by identity: $U_{D_0}(t, s) = \mathbb{1}$, for all $(t, s) \in \Delta$.

Note that in Chap. 11 we consider the operator $\mathcal{K}_0 := \overline{D_0 + \mathcal{A}}$, where $\mathcal{A}$ is a *multiplication* operator induced by the generator A of a *holomorphic contraction* semigroup on $\mathfrak{X}$. More precisely, for $t \in \mathcal{I}$:

$$(\mathcal{A}f)(t) := Af(t), \text{ and } (e^{-\tau \mathcal{A}} f)(t) = e^{-\tau A} f(t) ,$$
$$f \in \text{dom}(\mathcal{A}) := \{f \in L^p(\mathcal{I}, \mathfrak{X}) : Af(\cdot) \in L^p(\mathcal{I}, \mathfrak{X})\} .$$

Then the perturbation of the shift semigroup (10.3.3) by $\mathcal{A}$ corresponds to the semigroup with generator $\mathcal{K}_0$. We check that $\mathcal{K}_0$ is the evolution generator of a contraction semigroup on $L^p(\mathcal{I}, \mathfrak{X})$, which is *never* holomorphic. Indeed, since the generators D_0 and $\mathcal{A}$ commute, the representation (10.3.1) for evolution semigroup $\{e^{-\tau \mathcal{K}_0}\}_{\tau \geq 0}$ takes the form:

$$(e^{-\tau \mathcal{K}_0} f)(t) = e^{-\tau A} \chi_{\mathcal{I}}(t - \tau) f(t - \tau), \quad f \in L^p(\mathcal{I}, \mathfrak{X}) ,$$

for a.a. $t \in \mathcal{I}$ and $\tau \geq 0$, with solution operator $U_0(t, s) = e^{-(t-s)A}$. Therefore, again $e^{-\tau \mathcal{K}_0} f = 0$ for $\tau \geq T$, that is, semigroup $\{e^{-\tau \mathcal{K}_0}\}_{\tau \geq 0}$ is *nilpotent*.

Furthermore, if $\{B(t)\}_{t \in \mathcal{I}}$ is a *strongly measurable* family of generators of *contraction* semigroups on $\mathfrak{X}$, i.e., $B(\cdot) : \mathcal{I} \longrightarrow \mathcal{G}(1, 0)$, then the induced *multiplication* operator $\mathcal{B}$:

$$(\mathcal{B}f)(t) := B(t) f(t) , \tag{10.3.4}$$
$$f \in \text{dom}(\mathcal{B}) := \left\{ f \in L^p(\mathcal{I}, \mathfrak{X}) : \begin{array}{c} f(t) \in \text{dom}(B(t)) \text{ for a.a. } t \in \mathcal{I} \\ B(t) f(t) \in L^p(\mathcal{I}, \mathfrak{X}) \end{array} \right\} ,$$

is a generator of *contraction* semigroup on $L^p(\mathcal{I}, \mathfrak{X})$.

In Sect. 10.3.2 we consider perturbation of generator $\mathcal{K}_0$ by multiplication operator $\mathcal{B}$ (10.3.4). Thereupon we construct by means of the *Trotter* product formula approach a corresponding perturbed semigroup.

Remark 10.3.1 We conclude this subsection by remarks concerning *notations* used below throughout Sect. 10.3.

1. For characterisation of the rate of convergence, we use so-called Landau symbols:

$$g(n) = O(f(n)) \Longleftrightarrow \limsup_{n \to \infty} \left| \frac{g(n)}{f(n)} \right| < \infty ,$$

$$g(n) = o(f(n)) \Longleftrightarrow \limsup_{n \to \infty} \left| \frac{g(n)}{f(n)} \right| = 0 ,$$

$$g(n) = \Theta(f(n)) \Longleftrightarrow 0 < \liminf_{n\to\infty} \left|\frac{g(n)}{f(n)}\right| \leq \limsup_{n\to\infty} \left|\frac{g(n)}{f(n)}\right| < \infty\,,$$

$$g(n) = \omega(f(n)) \Longleftrightarrow \limsup_{n\to\infty} \left|\frac{g(n)}{f(n)}\right| = \infty\,.$$

2. We also use the notation $C^{0,\beta}(\mathcal{I}) := \{f : \mathcal{I} \to \mathbb{C} :$ there is $k > 0$ such that $|f(x) - f(y)| \leq k\,|x - y|^{\,\beta}$, for any $x, y \in \mathcal{I}$ and $\beta \in (0, 1]\}$.
3. Below we consider the Banach space $L^p(\mathcal{I}, \mathfrak{X})$ for $\mathcal{I} := [0, T]$ and $p \in [1, \infty)$.

10.3.2 Trotter Product Formula

We reminisce that semigroup $\{\mathcal{U}(\tau)\}_{\tau\geq 0}$, on the Banach space $L^p(\mathcal{I}, \mathfrak{X})$, is called the evolution semigroup if there is a solution operator: $\{U(t, s)\}_{(t,s)\in\Delta}$, such that representation (10.3.1) holds.

Let $\mathcal{K}_0$ be generator of an evolution semigroup $\{\mathcal{U}_0(\tau)\}_{\tau\geq 0}$, and let $\mathcal{B}$ be a multiplication operator induced by a measurable family $\{B(t)\}_{t\in\mathcal{I}}$ of generators of contraction semigroups. Note that in this case the multiplication operator $\mathcal{B}$ (10.3.4) is a generator of a contraction semigroup $(\mathrm{e}^{-\tau\,\mathcal{B}} f)(t) = \mathrm{e}^{-\tau\, B(t)} f(t)$, on the Banach space $L^p(\mathcal{I}, \mathfrak{X})$. Since $\{\mathcal{U}_0(\tau)\}_{\tau\geq 0}$ is an evolution semigroup, then by (10.3.1) there exists a solution operator $\{U_0(t, s)\}_{(t,s)\in\Delta}$ such that the HEN representation

$$(\mathcal{U}_0(\tau) f)(t) = U_0(t, t - \tau)\, \chi_{\mathcal{I}}(t - \tau) f(t - \tau), \quad f \in L^p(\mathcal{I}, \mathfrak{X}),$$

holds for a.a. $t \in \mathcal{I}$ and $\tau \geq 0$. Then we define $\tau_n := (t - s)/n$, for $n \in \mathbb{N}$, and

$$G_j(t, s; n) := U_0(s + j\,\tau_n\,,\; s + (j - 1)\,\tau_n)\,\mathrm{e}^{-\,\tau_n\, B\left(s+(j-1)\,\tau_n\right)}\,, \quad (t, s) \in \Delta\,,$$

where $j \in \{1, 2, \ldots, n\}$, $n \in \mathbb{N}$, $(t, s) \in \Delta$, and we set

$$V_n(t, s) := \prod_{j=n}^{1} G_j(t, s; n), \quad n \in \mathbb{N}, \quad (t, s) \in \Delta\,.$$

That is, the product is increasingly ordered in j from the *right* to the *left*. Then a straightforward computation shows that the representation

$$\left(\left(\mathrm{e}^{-\tau\mathcal{K}_0/n}\mathrm{e}^{-\tau\mathcal{B}/n}\right)^n f\right)(t) = V_n(t, t - \tau)\, \chi_{\mathcal{I}}(t - \tau) f(t - \tau)\,, \tag{10.3.5}$$

$f \in L^p(\mathcal{I}, \mathfrak{X})$, is valid for each $\tau \geq 0$ and a.a. $t \in \mathcal{I}$.

Theorem 10.3.2 *Let $\mathcal{K}$ and $\mathcal{K}_0$ be generators of evolution semigroups on the Banach space $L^p(\mathcal{I}, \mathfrak{X})$ for some $p \in [1, \infty)$. Further, let $\{B(t) \in \mathcal{G}(1, 0)\}_{t \in \mathcal{I}}$ be a strongly measurable family of generators of contraction semigroups on $\mathfrak{X}$. Then*

$$\sup_{\tau \geq 0} \left\| e^{-\tau\mathcal{K}} - \left(e^{-\tau\mathcal{K}_0/n} e^{-\tau\mathcal{B}/n}\right)^n \right\|_{\mathcal{L}(L^p(\mathcal{I},\mathfrak{X}))} \tag{10.3.6}$$
$$= \operatorname*{ess\,sup}_{(t,s)\in\Delta} \|U(t,s) - V_n(t,s)\|_{\mathcal{L}(\mathfrak{X})}, \quad n \in \mathbb{N}.$$

Proof Let $\{L(\tau)\}_{\tau \geq 0}$ be the *left-shift* semigroup on the Banach space $L^p(\mathcal{I}, \mathfrak{X})$:

$$(L(\tau)f)(t) := \chi_{\mathcal{I}}(t+\tau) f(t+\tau), \quad f \in L^p(\mathcal{I}, \mathfrak{X}).$$

Owing to this definition, we obtain

$$\left(L(\tau)\left(e^{-\tau\mathcal{K}} - \left(e^{-\tau/n\mathcal{K}_0} e^{-\tau\mathcal{B}/n}\right)^n\right) f\right)(t)$$
$$= \{U(t+\tau, t) - V_n(t+\tau, t)\}\, \chi_{\mathcal{I}}(t+\tau) f(t),$$

for $\tau \geq 0$, $n \in \mathbb{N}$, and a.a. $t \in \mathcal{I}$.

So, it turns out that for each $n \in \mathbb{N}$ the operator $L(\tau)\Big(e^{-\tau\mathcal{K}} - \left(e^{-\tau/n\mathcal{K}_0} e^{-\tau\mathcal{B}/n}\right)^n\Big)$ is a multiplication operator induced by $\{(U(t+\tau, t) - V_n(t+\tau, t))\, \chi_{\mathcal{I}}(t+\tau)\}_{t \in \mathcal{I}}$. As a consequence (Proposition 2.2.16), for each $\tau \geq 0$ we have

$$\left\| L(\tau)\left(e^{-\tau\mathcal{K}} - \left(e^{-\tau\mathcal{K}_0/n} e^{-\tau\mathcal{B}/n}\right)^n\right)\right\|_{\mathcal{L}(L^p(\mathcal{I},\mathfrak{X}))}$$
$$= \operatorname*{ess\,sup}_{t \in \mathcal{I}} \left(\|U(t+\tau, t) - V_n(t+\tau, t)\|_{\mathcal{L}(\mathfrak{X})}\, \chi_{\mathcal{I}}(t+\tau)\right).$$

Note that for strongly continuous mappings: $\tau \mapsto L(\tau)\Big(e^{-\tau\mathcal{K}} - \big(e^{-\tau\mathcal{K}_0/n}\, e^{-\tau\mathcal{B}/n}\big)^n\Big)$, where $\tau \geq 0$ and $n \in \mathbb{N}$:

$$\sup_{\tau \geq 0} \left\| L(\tau)\left(e^{-\tau\mathcal{K}} - \left(e^{-\tau\mathcal{K}_0/n} e^{-\tau\mathcal{B}/n}\right)^n\right)\right\|_{\mathcal{L}(L^p(\mathcal{I},\mathfrak{X}))}$$
$$= \operatorname*{ess\,sup}_{\tau \geq 0} \left\| L(\tau)\left(e^{-\tau\mathcal{K}} - \left(e^{-\tau\mathcal{K}_0/n} e^{-\tau\mathcal{B}/n}\right)^n\right)\right\|_{\mathcal{L}(L^p(\mathcal{I},\mathfrak{X}))}.$$

As a result, we find

$$\sup_{\tau \geq 0} \left\| L(\tau)\left(e^{-\tau\mathcal{K}} - \left(e^{-\tau\mathcal{K}_0/n} e^{-\tau\mathcal{B}/n}\right)^n\right)\right\|_{\mathcal{L}(L^p(\mathcal{I},\mathfrak{X}))}$$
$$= \operatorname*{ess\,sup}_{\tau \geq 0}\ \operatorname*{ess\,sup}_{t \in \mathcal{I}} \left(\|U(t+\tau, t) - V_n(t+\tau, t))\|_{\mathcal{L}(\mathfrak{X})}\, \chi_{\mathcal{I}}(t+\tau)\right).$$

Further, if $\Phi(\cdot,\cdot) : \mathbb{R}^+ \times \mathcal{I} \to \mathcal{L}(\mathfrak{X})$ is a strongly measurable function, then

$$\operatorname*{ess\,sup}_{(\tau,t)\in\mathbb{R}^+\times\mathcal{I}} \|\Phi(\tau,t)\|_{\mathcal{L}(\mathfrak{X})} = \operatorname*{ess\,sup}_{\tau\geq 0}\ \operatorname*{ess\,sup}_{t\in\mathcal{I}} \|\Phi(\tau,t)\|_{\mathcal{L}(\mathfrak{X})}.$$

Then, taking into account two last equalities, one infers

$$\begin{aligned}
&\sup_{\tau\geq 0}\left\|L(\tau)\left(\mathrm{e}^{-\tau\mathcal{K}} - \left(\mathrm{e}^{-\tau\mathcal{K}_0/n}\mathrm{e}^{-\tau\mathcal{B}/n}\right)^n\right)\right\|_{\mathcal{L}(L^p(\mathcal{I},\mathfrak{X}))}\\
&\quad= \operatorname*{ess\,sup}_{(\tau,t)\in\mathbb{R}^+\times\mathcal{I}} \|U(t+\tau,t) - V_n(t+\tau,t)\|_{\mathcal{L}(\mathfrak{X})}\,\chi_{\mathcal{I}}(t+\tau) =\\
&\quad= \operatorname*{ess\,sup}_{(t,s)\in\Delta} \|U(t,s) - V_n(t,s)\|_{\mathcal{L}(\mathfrak{X})}\,,
\end{aligned}$$

that proves (10.3.6). □

For details in the last step of the proof, we invite to consult the argument proving (11.5.77) and (11.5.78) in Theorem 11.5.18.

Next we study bounded perturbations of the evolution generator $\mathcal{K}_0 := D_0$ (10.3.2). To this aim we consider $\mathcal{I} = [0,1]$, $\mathfrak{X} = \mathbb{C}$, and we denote by $L^p(\mathcal{I})$ the Banach space $L^p(\mathcal{I},\mathbb{C})$.

For $t \in \mathcal{I}$, let $q : t \mapsto q(t) \in L^\infty(\mathcal{I})$. Then, function q induces a *bounded* multiplication operator Q on the Banach space $L^p(\mathcal{I})$:

$$(Qf)(t) = q(t)f(t), \quad f \in L^p(\mathcal{I}).$$

For simplicity we assume that function $q \geq 0$. Then Q generates on $L^p(\mathcal{I})$ a contraction semigroup $\{\mathrm{e}^{-\tau Q}\}_{\tau\geq 0}$. Since generator Q is bounded, the closed operator $\mathcal{A} := D_0 + Q$, on domain $\mathrm{dom}\,(\mathcal{A}) = \mathrm{dom}\,(D_0)$, is generator of a C_0-semigroup on $L^p(\mathcal{I})$. Then by the *Trotter* product formula in the *strong* operator topology, it follows immediately that

$$\lim_{n\to\infty}\left(\mathrm{e}^{-\tau D_0/n}\mathrm{e}^{-\tau Q/n}\right)^n f = \mathrm{e}^{-\tau(D_0+Q)}f, \quad f \in L^p(\mathcal{I}), \tag{10.3.7}$$

uniformly in $\tau \in [0,T]$ on the bounded time intervals; cf. Sect. 1.9.6.

That being so, we define on Banach space $\mathfrak{X} = \mathbb{C}$ a family of bounded operators $\{V(t)\}_{t\in\mathcal{I}}$ by

$$V(t) := \mathrm{e}^{-\int_0^t ds\,q(s)}\,, \quad t \in \mathcal{I}\,.$$

Note that for almost every (a.e.) $t \in \mathcal{I}$, these operators are positive. Then $V^{-1}(t)$ exists, and it has the form

$$V^{-1}(t) = \mathrm{e}^{\int_0^t ds\,q(s)}\,, \quad t \in \mathcal{I}\,.$$

The operator families $\{V(t)\}_{t\in\mathcal{I}}$ and $\{V^{-1}(t)\}_{t\in\mathcal{I}}$ induce two bounded multiplication operators $\mathcal{V}$ and $\mathcal{V}^{-1}$ on $L^p(\mathcal{I})$, respectively. Then invertibility implies that $\mathcal{V}\,\mathcal{V}^{-1} = \mathcal{V}^{-1}\,\mathcal{V} = \mathbb{1}\big|_{L^p(\mathcal{I})}$. Using the operator $\mathcal{V}$, one easily verifies that $D_0 + Q$ is similar to D_0, that is, one has

$$\mathcal{V}^{-1}(D_0 + Q)\mathcal{V} = D_0 \text{ or } D_0 + Q = \mathcal{V}\, D_0\, \mathcal{V}^{-1} .$$

Hence, the semigroup generated on $L^p(\mathcal{I})$ by $D_0 + Q$ gets the explicit form:

$$\left(\mathrm{e}^{-\tau(D_0+Q)} f\right)(t) = \left(\mathcal{V}\,\mathrm{e}^{-\tau D_0}\,\mathcal{V}^{-1} f\right)(t) = \tag{10.3.8}$$
$$= \mathrm{e}^{-\int_{t-\tau}^{t} dy\, q(y)}\, f(t-\tau)\, \chi_{\mathcal{I}}(t-\tau) .$$

Since by (10.3.1) the solution operator $U(t,s)$ that corresponds to evolution semigroup (10.3.8) is defined by equation

$$\left(\mathrm{e}^{-\tau(D_0+Q)}\right) f(t) = U(t, t-\tau) f(t-\tau)\, \chi_{\mathcal{I}}(t-\tau) ,$$

we deduce that it is equal to $U(t,s) = \mathrm{e}^{-\int_s^t dy\, q(y)}$.

Now we study the corresponding Trotter product formula. For a fixed $\tau \geq 0$ and $n \in \mathbb{N}$, we define *approximates* $\{V_n\}_{n\geq 1}$ by

$$\left(\left(\mathrm{e}^{-\tau D_0/n}\mathrm{e}^{-\tau Q/n}\right)^n f\right)(t) =: V_n(t, t-\tau)\, \chi_{\mathcal{I}}(t-\tau) f(t-\tau) .$$

Then by straightforward calculations, which are similar to (10.3.5), one finds that approximants have the following explicit form:

$$V_n(t,s) = \mathrm{e}^{-\tau_n \sum_{k=0}^{n-1} q\,(s+k\,\tau_n)} , \quad (t,s) \in \Delta , \quad \tau_n = (t-s)/n , \quad n \in \mathbb{N} .$$

Theorem 10.3.3 *Let measurable function $q \in L^\infty(\mathcal{I})$ be non-negative. Then*

$$\sup_{\tau\geq 0} \left\| \mathrm{e}^{-\tau(D_0+Q)} - \left(\mathrm{e}^{-\tau D_0/n}\mathrm{e}^{-\tau Q/n}\right)^n \right\|_{\mathcal{L}(L^p(\mathcal{I}))} =$$
$$\Theta\left(\operatorname*{ess\,sup}_{(t,s)\in\Delta} \left| \int_s^t dy\, q(y) - \tau_n \sum_{k=0}^{n-1} q\,(s+k\,\tau_n) \right| \right) , \quad n \in \mathbb{N} ,$$

as $n \to \infty$, where Θ is the Landau symbol defined in Sect. 10.3.1.

Proof First, by Theorem 10.3.2 and by $U(t,s) = \mathrm{e}^{-\int_s^t dy\, q(y)}$ we obtain that

$$\sup_{\tau \geq 0} \left\| \mathrm{e}^{-\tau(D_0+Q)} - \left(\mathrm{e}^{-\tau D_0/n}\mathrm{e}^{-\tau Q/n}\right)^n \right\|_{\mathcal{L}(L^p(\mathcal{I}))} \tag{10.3.9}$$
$$= \operatorname*{ess\,sup}_{(t,s)\in\Delta} \left| \mathrm{e}^{-\int_s^t dy\, q(y)} - \mathrm{e}^{-\tau_n \sum_{k=0}^{n-1} q(s+k\,\tau_n)} \right| .$$

Then, using inequality

$$\mathrm{e}^{-\max\{x,y\}}\,|x-y| \leq |\mathrm{e}^{-x} - \mathrm{e}^{-y}| \leq |x-y|, \quad 0 \leq x, y ,$$

for $0 \leq s < t \leq 1$, we obtain the estimates

$$\mathrm{e}^{-\|q\|_{L^\infty}} R_n(t,s;q) \leq \left| \mathrm{e}^{-\int_s^t dy\, q(y)} - \mathrm{e}^{-\tau_n \sum_{k=0}^{n-1} q(s+k\,\tau_n)} \right| \leq R_n(t,s;q) ,$$

where

$$R_n(t,s,q) := \left| \int_s^t dy\, q(y) - \tau_n \sum_{k=0}^{n-1} q(s+k\,\tau_n) \right|, \quad (t,s) \in \Delta , \quad n \in \mathbb{N} . \tag{10.3.10}$$

Hence, for the left-hand side of (10.3.9), one gets the estimates

$$\mathrm{e}^{-\|q\|_{L^\infty}} R_n(q) \leq \sup_{\tau \geq 0} \left\| \mathrm{e}^{-\tau(D_0+Q)} - \left(\mathrm{e}^{-\tau D_0/n}\mathrm{e}^{-\tau Q/n}\right)^n \right\|_{\mathcal{L}(L^p(\mathcal{I}))} \leq R_n(q) ,$$

where $R_n(q) := \operatorname{ess\,sup}_{(t,s)\in\Delta} R_n(t,s;q)$, $n \in \mathbb{N}$. These estimates together with definition of Θ prove the assertion. □

Note that by virtue of (10.3.10) and Theorem 10.3.3, the operator-norm convergence rate of the Trotter product formula for pair $\{D_0, Q\}$ coincides with the convergence rate of the *Darboux-Riemann* integral sum approximation for the *Lebesgue* integral.

10.3.3 Rate of the Operator-Norm Convergence

First we consider the case of a real *Hölder-continuous* function $q \in C^{0,\beta}(\mathcal{I})$.

Theorem 10.3.4 *If* $q \in C^{0,\beta}(\mathcal{I})$ *is non-negative, then*

$$\sup_{\tau \geq 0} \left\| \mathrm{e}^{-\tau\,(D_0+Q)} - \left(\mathrm{e}^{-\tau\, D_0/n}\,\mathrm{e}^{-\tau\, Q/n}\right)^n \right\|_{\mathcal{L}(L^p(\mathcal{I}))} = O(1/n^\beta) ,$$

as $n \to \infty$.

Proof One gets for $\tau_n = (t-s)/n$ and $n \in \mathbb{N}$:

$$\int_s^t dy\, q(y) - \tau_n \sum_{k=0}^{n-1} q\,(s + k\,\tau_n)$$
$$= \sum_{k=0}^{n-1} \int_{k\,\tau_n}^{(k+1)\tau_n} dy\ (q(s+y) - q(s + k\,\tau_n))\ ,$$

which yields the estimate

$$\left| \int_s^t dy\, q(y) - \tau_n \sum_{k=0}^{n-1} q\,(s + k\tau_n) \right|$$
$$\leq \sum_{k=0}^{n-1} \int_{k\tau_n}^{(k+1)\tau_n} dy\ |q(s+y) - q(s + k\,\tau_n)|\ .$$

Since $q \in C^{0,\beta}(\mathcal{I})$, there is a constant $L_\beta > 0$ such that for $y \in [k\tau_n, (k+1)\tau_n]$ one has

$$|q(s+y) - q(s + k\tau_n)| \leq L_\beta\, |y - k\tau_n|^\beta \leq L_\beta\, \frac{(t-s)^\beta}{n^\beta}\ .$$

Hence, we find

$$\left| \int_s^t dy\, q(y) - \tau_n \sum_{k=0}^{n-1} q(s + k\tau_n) \right| \leq L_\beta \frac{(t-s)^{1+\beta}}{n^\beta} \leq L_\beta \frac{1}{n^\beta}\ ,$$

which proves

$$\operatorname*{ess\,sup}_{(t,s)\in\Delta} \left| \int_s^t dy\, q(y) - \tau_n \sum_{k=0}^{n-1} q(s + k\,\tau_n) \right| = O\left(\frac{1}{n^\beta}\right)\ .$$

Applying now Theorem 10.3.3, one completes the proof. □

The next natural question is: What happens, when q is only *continuous* ?

Theorem 10.3.5 *If $q : \mathcal{I} \to \mathbb{C}$ is continuous and non-negative, then*

$$\left\| \mathrm{e}^{-\tau(D_0+Q)} - \left(\mathrm{e}^{-\tau D_0/n}\mathrm{e}^{-\tau Q/n}\right)^n \right\|_{\mathcal{L}(L^p(\mathcal{I}))} = o(1)\ , \qquad (10.3.11)$$

as $n \to \infty$.

Proof Seeing that q is continuous, for any $\varepsilon > 0$ there is $\delta > 0$ such that for $y, x \in \mathcal{I}$ and $|y - x| < \delta$ we have $|q(y) - q(x)| < \varepsilon$. Therefore, if $1/n < \delta$, then for $y \in (k\tau_n, (k+1)\tau_n)$ we have

$$|q(s+y) - q(s + k\,\tau_n)| < \varepsilon, \quad (t,s) \in \Delta\,.$$

Hence,

$$\left| \int_s^t dy\, q(y) - \tau_n \sum_{k=0}^{n-1} q(s + k\,\tau_n) \right| \le \varepsilon(t-s) \le \varepsilon\,,$$

which yields

$$\operatorname*{ess\,sup}_{(t,s)\in\Delta} \left| \int_s^t dy\, q(y) - \tau_n \sum_{k=0}^{n-1} q(s + k\,\tau_n) \right| = o(1)\,. \tag{10.3.12}$$

Now it remains only to apply Theorem 10.3.3 to continuous function q and asymptotics (10.3.12). □

It is worth mentioning here that for a *general* continuous function q, one can say nothing about the convergence rate. Indeed, it can be shown that in (10.3.11) the rate of the operator-norm convergence to zero may be *arbitrary slow* (see (10.3.13)) for a *bounded* perturbation Q. Note that this is drastically different to the case, when dominating generator corresponds to a holomorphic semigroup and perturbation operator is bounded, cf. (10.2.17), (10.2.19) for $\alpha = 0$, or to the case of unbounded perturbation, when $0 < \alpha < 1$; see (10.2.16), (10.2.18).

Theorem 10.3.6 *Let $\delta_n > 0$ be a sequence with $\delta_n \to 0$ as $n \to \infty$. Then there exists a continuous function $q : \mathcal{I} = [0,1] \to \mathbb{R}$, such that*

$$\sup_{\tau\ge 0} \left\| \mathrm{e}^{-\tau(D_0+Q)} - \left(\mathrm{e}^{-\tau D_0/n} e^{-\tau Q/n}\right)^n \right\|_{\mathcal{L}(L^p(\mathcal{I}))} = \omega\,(\delta_n)\,, \tag{10.3.13}$$

as $n \to \infty$, where ω is the Landau symbol defined in Sect. 10.3.1. *Moreover, the convergence of the product formula approximants in* (10.3.13) *is arbitrary slow.*

Proof Taking into account the *Walsh-Sewell* theorem, we find that for any sequence $\{\delta_n\}_{n\in\mathbb{N}}$, $\delta_n > 0$ satisfying $\lim_{n\to\infty} \delta_n = 0$, there exists a continuous function $f : [0, 2\pi] \longrightarrow \mathbb{R}$, such that

$$\left| \int_0^{2\pi} dx\, f(x) - \frac{2\pi}{n} \sum_{k=1}^{n} f(2k\pi/n) \right| = \omega\,(\delta_n)\,,$$

as $n \to \infty$. Setting $q(y) := f(2\pi(1-y))$ for $y \in [0,1]$, we get a continuous function $q : [0,1] \longrightarrow \mathbb{R}$, such that

$$\left|\int_0^1 dy\, q(y) - \frac{1}{n}\sum_{k=0}^{n-1} q(k/n)\right| = \omega(\delta_n)\,.$$

Given that function q is continuous, we find

$$\begin{aligned}\operatorname*{ess\,sup}_{(t,s)\in\Delta}\left|\int_s^t dy\, q(y) - \tau_n \sum_{k=0}^{n-1} q(s+k\,\tau_n)\right| \\ \geq \left|\int_0^1 dy\, q(y) - \frac{1}{n}\sum_{k=0}^{n-1} q(k/n)\right|,\end{aligned}$$

which yields

$$\operatorname*{ess\,sup}_{(t,s)\in\Delta}\left|\int_s^t dy\, q(y) - \tau_n \sum_{k=0}^{n-1} q(s+k\,\tau_n)\right| = \omega(\delta_n)\,. \tag{10.3.14}$$

Applying now Theorem 10.3.3 to the continuous function q, which is constructed above, we obtain by (10.3.14) the asymptotic (10.3.13).

Finally, since by Theorem 10.3.5, (10.3.11), the left-hand side of (10.3.13) tends to zero, we infer by the right-hand side of (10.3.13) that it converges to zero *arbitrary slow*. □

Our final comment concerns the case, when q is only *measurable*. Then it can happen that the Trotter product formula for that pair $\{D_0, Q\}$ does *not* converge in the *operator-norm* topology.

Theorem 10.3.7 *There is a non-negative function* $q \in L^\infty([0,1])$ *such that*

$$\limsup_{n\to\infty}\sup_{\tau\geq 0}\left\| \mathrm{e}^{-\tau(D_0+Q)} - \left(\mathrm{e}^{-\tau D_0/n}\mathrm{e}^{-\tau Q/n}\right)^n\right\|_{\mathcal{L}(L^p(\mathcal{I}))} > 0\,. \tag{10.3.15}$$

Proof Let us introduce open intervals:

$$\begin{aligned}\Delta_{0,n} &:= (0, \tfrac{1}{2^{2n+2}}),\\ \Delta_{k,n} &:= (t_{k,n} - \tfrac{1}{2^{2n+2}},\ t_{k,n} + \tfrac{1}{2^{2n+2}}), \quad k = 1, 2, \ldots, 2^n - 1,\\ \Delta_{2^n,n} &:= (1 - \tfrac{1}{2^{2n+2}},\ 1),\end{aligned}$$

$n \in \mathbb{N}$, where $t_{k,n} = k/2^n$, $\quad k = 0, \ldots, n$. Notice that $t_{0,n} = 0$ and $t_{2^n,n} = 1$. One can check that the intervals $\Delta_{k,n}$, $k = 0, \ldots, 2^n$, are mutually disjoint.

Further, we introduce open sets:

$$\mathcal{O}_n = \bigcup_{k=0}^{2^n} \Delta_{k,n} \subseteq \mathcal{I}, \quad n \in \mathbb{N}.$$

and

$$\mathcal{O} = \bigcup_{n \in \mathbb{N}} \mathcal{O}_n \subseteq \mathcal{I}.$$

Then it is clear that $|\,\mathcal{O}_n\,| = 1/2^{n+1}$, for $n \in \mathbb{N}$ and $|\,\mathcal{O}\,| \le 1/2$. Therefore, the *Lebesgue* measure of the closed set $\mathcal{C} := \mathcal{I} \setminus \mathcal{O} \subseteq \mathcal{I}$ can be estimated by $|\,\mathcal{C}\,| \ge 1/2$. Using the characteristic function $\chi_{\mathcal{C}}(\cdot)$ of the set $\mathcal{C}$, we define

$$q(t) := \chi_{\mathcal{C}}(t), \quad t \in \mathcal{I}\,.$$

The function q is measurable, and it satisfies condition: $0 \le q(t) \le 1$, $t \in \mathcal{I}$.

Let $\varepsilon \in (0, 1)$. We choose $s \in (0, \varepsilon)$ and $t \in (1 - \varepsilon, 1)$, and we set

$$\xi_{k,n}(t, s) := s + k\,\frac{t - s}{2^n}, \quad k = 0, \ldots, 2^n - 1, \quad n \in \mathbb{N}, \quad (t, s) \in \Delta.$$

Note that $\xi_{k,n}(t, s) \in (0, 1)$, $k = 0, \ldots, 2^n - 1$, $n \in \mathbb{N}$. Moreover, we have

$$t_{k,n} - \xi_{k,n}(t, s) = k\,\frac{1}{2^n} - s - k\,\frac{t - s}{2^n} = k\,\frac{1 - t + s}{2^n} - s\,,$$

which leads to estimate:

$$|\,t_{k,n} - \xi_{k,n}(t, s)\,| \le \varepsilon\;(1 + k/2^{n-1}), \quad k = 0, \ldots, 2^n - 1, \quad n \in \mathbb{N}\,.$$

Therefore,

$$|\,t_{k,n} - \xi_{k,n}(t, s)\,| \le 3\varepsilon, \quad k = 0, \ldots, 2^n - 1, \quad n \in \mathbb{N}.$$

Let $\varepsilon_n := 1/(3 \cdot 2^{2n+2})$ for $n \in \mathbb{N}$. Then we get that $\xi_{k,n}(t, s) \in \Delta_{k,n}$ for $k = 0, \ldots, 2^n - 1$, $n \in \mathbb{N}$, $s \in (0, \varepsilon_n)$ and for $t \in (1 - \varepsilon_n, 1)$.

Now let

$$S_n(t, s; q) := \tau_n \sum_{k=0}^{n-1} q\,(s + k\,\tau_n)\,, \quad (t, s) \in \Delta\,, \quad \tau_n = (t - s)/n\,, \quad n \in \mathbb{N}\,. \tag{10.3.16}$$

We consider

$$S_{2^n}(t,s;q) = \frac{t-s}{2^n}\sum_{k=0}^{2^n-1} q\left(s + k\,(t-s)/2^n\right) = \frac{t-s}{2^n}\sum_{k=0}^{2^n-1} q\left(\xi_{k,n}(t,s)\right), \tag{10.3.17}$$

$n \in \mathbb{N}$, $(t,s) \in \Delta$. If $s \in (0, \varepsilon_n)$ and $t \in (1-\varepsilon_n, 1)$, then $S_{2^n}(t,s;q) = 0$, $n \in \mathbb{N}$, that is,

$$\left|\int_s^t dy\, q(y) - S_{2^n}(t,s;q)\right| = \int_s^t dy\, q(y)\,, \qquad n \in \mathbb{N}\,,$$

for $s \in (0, \varepsilon_n)$ and $t \in (1-\varepsilon_n, 1)$. In particular, this yields

$$\begin{aligned}\operatorname*{ess\,sup}_{(t,s)\in\Delta}\left|\int_s^t dy\, q(y) - S_{2^n}(t,s;q)\right| &\geq \operatorname*{ess\,sup}_{(t,s)\in\Delta}\int_s^t dy\, q(y)\\ \geq \int_{\mathcal{I}} dy\, \chi_C(y) \geq \frac{1}{2}\,.\end{aligned}$$

As a result, we obtain

$$\limsup_{n\to\infty}\ \operatorname*{ess\,sup}_{(t,s)\in\Delta}\left|\int_s^t dy\, q(y) - S_{2^n}(t,s;q)\right| \geq \frac{1}{2}, \tag{10.3.18}$$

for *Darboux-Riemann* sums (10.3.16) for a measurable function q defined in (10.3.17). Now, applying Theorem 10.3.3 for (10.3.18), one infers the estimate (10.3.15). □

Remark 10.3.8 We note that Theorem 10.3.7 does not exclude the convergence of the *Trotter* product formula for the pair $\{D_0, Q\}$ in the *strong* operator topology. In fact it is true. The pair $\{D_0, Q\}$ as generators of contraction C_0-semigroups on Banach space $L^p(\mathcal{I})$ is *Trotter-stable* (Definition 1.9.37) and satisfies conditions of Proposition 1.9.35 (Lie-Trotter product formula).

We would remind that the same kind of *dichotomy* (*convergence/non-convergence* in two topologies) is known for the Trotter product formula on a Hilbert space; see the *Hiroshi Tamura* Example in Notes to Chap. 7 (Notes to Sect. 7.4) and Appendix H, Example H.0.5. By virtue of (10.3.7) and (10.3.15), Theorem 10.3.7 yields an example of this kind of dichotomy in Banach space $L^p(\mathcal{I})$.

We stress that there is a drastic difference between the origin of these phenomena in a Hilbert space for *unbounded* perturbation of a *holomorphic* semigroup (see Notes to Sect. 7.4) and in a Banach space for *bounded* perturbation of a (non-holomorphic) *contraction* semigroup; see Theorem 10.3.7 and Notes to Sect. 10.3.

10.4 Notes to Chap. 10

In Chap. 10 we have collected results concerning the *operator-norm* convergent *Trotter* product formula for two semigroups $\{e^{-tA}\}_{t\geq 0}$, $\{e^{-tB}\}_{t\geq 0}$, with densely defined generators A and B in a *Banach* space. Although the *strong* convergence in Banach space for contraction semigroups is known since the seminal paper by Trotter [Trot59], which after more than three decades was extended to convergence in the *operator-norm* topology in *Hilbert* spaces by Rogava [Rog93], a similar result about the operator-norm convergence in a *Banach* space was proved only in [CZ01a].

For the first time, this result was established under hypothesis that one of the involved into the product formula contraction semigroups, e.g. $\{e^{-tA}\}_{t\geq 0}$, is *holomorphic* together with additional conditions of *infinitesimal smallness* on generators B and B^* with respect to generators A and A^*, cf. Definition 1.9.7. Note that in spite of quite strong assumptions on operators A and B the proof of the operator-norm convergent Trotter product formula and the estimate of its rate of convergence on a Banach space rests (unexpectedly) involved and technical, even for bounded operator B; see Sect. 10.2.

The observation that holomorphy is a principal condition for the proof allows to revisit the problem of the error-bound estimate for the rate of the operator-norm convergent Trotter product formula for *non-self-adjoint* semigroups.

A new important (and surprising) result was obtained in paper [NSZ18a]. We develop it in Sect. 10.3. This result elucidates the question about conditions on A and B in assertions concerning the operator-norm convergence of the Trotter and Trotter-Kato product formilæ on Hilbert and Banach spaces. As above, a usual condition is *holomorphy* of the one of semigroups involved in the product formula.

We show in Sect. 10.3 an Example of the *operator-norm* convergent Trotter product formula for two C_0-semigroups $\{e^{-tA}\}_{t\geq 0}$ and $\{e^{-tB}\}_{t\geq 0}$ on a Banach space, where hypothesis on the operator A is relaxed to condition that A is a generator of *contraction* semigroup and B is a bounded operator.

Notes to Sect. 10.1

Sections 10.1.1 and 10.1.2

Besides some standard matter, we follow here [CZ01a, Section 2]. Examples (i)–(iv) of the *holomorphic contraction* semigroups at the end of Sect. 10.1.2 result from discussions in the previous sections.
We would like also to remark that demand of contraction is not as superfluous as one could suppose. For demonstration we address the reader to an instructive example by Trotter [Trot59], where he called it: "the norm condition". Otherwise, one has to use the *equivalent* norm; see Remark 1.3.3.

The last example (iv) is motivated by Example 1.10.3(c) and a general case of A, which is due to R.Nagel [Nag86, A-II, Theorem 1.15]; see also [EN00, 4.9 Corollary, Chapter II]. An immediate example of generator A is the self-adjoint generator $D := ((1/i)\,\partial_x)_{s-a}$ in Hilbert space $\mathfrak{H} = L^2(\mathbb{R})$. Then non-negative self-adjoint operator $D^2 := (-\partial_x^2)_{s-a}$ is generator of *contraction holomorphic* semigroup $\{e^{-t D^2}\}_{t \geq 0}$ of semi-angle $\pi/2$ on Hilbert space $L^2(\mathbb{R})$. See also Remark 13.1.5.

Section 10.1.3

For more details about the fractional powers of closed operators, one can find, for example, in [Yos80], Chapter IX, Section 11. For the proof of Lemma 10.1.11, see, e.g. [Tan75], Lemma 2.3.5. For related points we refer also to Sect. 7.4.

A proof of the *V. Balakrishnan* representation (10.1.12), one can find, for example, in [Yos80, Chapter IX, Section 11].

Notes to Sect. 10.2

Sections 10.2.1 and 10.2.2

Section 10.2 contains the main result of the paper [CZ01a, Section 3]. There the *operator-norm* convergence of the Trotter product formula on a Banach space $\mathfrak{X}$ with the error-bound estimate of the rate of convergence has been proven (up to our knowledge) for the first time.

We note that idea of the proof in [CZ01a, Theorem 3.6] retraces back to arguments in [Zag95] for a Hilbert space. Then one proves Theorem 10.2.6 and Theorem 10.2.7 for $\alpha = 0$ (a bounded operator as perturbation) with the rate $O(\ln(n)^2/n)$ for error bound of the operator-norm convergence, which is the same as for the case of a Hilbert space; cf. Notes to Chap. 7 (Notes to Sect. 7.1). The proof of Theorem 10.2.6 for $\alpha \in (0, 1)$ needs more refined arguments, which are due to results in [CZ01a, Section 3] and [Zag23, §3].

Another direction in the theory of C_0-semigroups on a Banach space $\mathfrak{X}$ concerns a more straightforward (less arduous) problem of approximations of $\{e^{-tA}\}_{t \geq 0}$. We mention here an approach, [GoTo14], [GoKoTo19], to approximation formulæ with estimates, which are *restricted* to domains of powers of the semigroup generators. For example, if semigroup is bounded holomorphic, then for *Euler approximants* this approach gives in [GoTo14, Corollary 1.6] the *point-wise* estimate:

$$\left\| \left(e^{-tA} - (\mathbb{1} + tA/n)^{-n}\right) x \right\| \leq C\, t^{\alpha}\, \|A^{\alpha} x\|/n\,, \quad x \in \mathrm{dom}\,(A^{\alpha})\,, \tag{10.4.1}$$

for $t \in \mathbb{R}^+$, $C > 0$, and every $\alpha \in [0, 1]$, cf. with the *uniform* (operator-norm) estimate (3.2.39) for the Euler formula on a Hilbert space $\mathfrak{H}$.

In the recent paper, [Rog22] Dzh. Rogava used *linear-fractional* operator approximants to obtain for analogue of inequality (10.4.1), when $\alpha = 1$, the estimate $O\left(t\, \|A\, x\|\, \ln(n)/n^2\right)$, where x is restricted to domain $\mathrm{dom}\,(A)$.

Similar results of the *point-wise* estimates, but in the general set-up of the Chernoff semigroup approximations are due to [GaRe22].

Section 10.2.3

This subsection is dedicated to application of the *holomorphic-based* approach developed in Sects. 10.2.1 and 10.2.2 for estimation of the rate of the operator-norm convergence of the Trotter product formula on Banach space to non-self-adjoint (quasi-sectorial contraction) semigroups on a Hilbert space. It permits to establish in Theorem 10.2.11 the estimate missed in Theorem 9.2.4.

Note that in contrast to Theorem 9.2.4, the proof of Theorem 10.2.11 allows only infinitesimally A-small generator $B \in \mathcal{P}_{0^+}(A)$.

Notes to Sect. 10.3

In contrast to Sect. 10.2, where operator-norm convergence holds true if dominating operator $A \in \mathscr{H}_c(\theta, 0)$ generates a *holomorphic* contraction semigroup and $B \in \mathcal{P}_{0^+}(A)$ is an infinitesimally A-small generator of a *contraction* semigroup (in particular, if B is a bounded operator), we present here an example that it is also possible even if conditions on generator A are considerably *relaxed.*

The conditions are (cf. Sect. 10.3.2):

(1) Operator $A = \mathcal{K}_0$ generates a *contraction* (not holomorphic!) semigroup.
(2) $B = \mathcal{B}$ is a *bounded* operator.

In this section we follow essentially papers [NSZ18a] and [Zag23]. There it is also demonstrated that

- the operator-norm convergence generally may *fail* even for a *bounded* operator $\mathcal{B}$ if the *unbounded* $\mathcal{K}_0$ is *not* a *holomorphic* generator;
- the rate of the operator-norm convergence of the Trotter product formula may be *arbitrary slow*; cf. Sect. 10.3.3.

This case is again very different to holomorphic contraction case: $A \in \mathscr{H}_c(\theta, 0)$ (Sect. 10.2.2), where the rate of operator-norm convergence is of the order $O((\ln(n))^2/n)$ for *any* bounded perturbation B ($\alpha = 0$); see Theorem 10.2.6.

For the *Walsh-Sewell* theorem, we refer to [WaSe37].

We mention here a recent result [GaRe22] with another example of the *arbitrary slow* convergence to C_0-semigroup of the Chernoff approximants. In contrast to [NSZ18a], this is convergence in the strong operator topology, and approximants can be "tuned" to the opposite regime of the fast convergence.

To estimate a variety of applications of the Chernoff approximation method, we refer to [But20] and especially to [MMRS23].

Part IV
Time-Dependent Product Formulæ

Chapter 11
Time-Dependent Product formulæ: Banach Space

The aim of this chapter is to summarise the results about development of the product formula approach to construction of solutions of the abstract non-Autonomous Cauchy Problem (nACP) in the normed vector spaces, Chap. 2.

Here we construct the product formula approximation (*approximants*) for the *solution operator* of nACP in a Banach space, and we postpone the case of the Hilbert space to the next chapter.

In this chapter, we present three principle results that concern convergence of the product formula approximations of the nACP solution operator:

- The first one (Sect. 11.1) proves that product formula approximants converge to solution operator of nACP in the *strong* operator topology. This result includes the case when *domain* of the nACP generator is *time-dependent*.
- The second result advocates the *evolution semigroups* method for analysis of nACP, Sects. 11.2 and 11.3. Owing to this approach, we establish convergence of the product formula approximants to solution operator in a certain *integral topology*, Sect. 11.4.
- The main observation of the chapter is *uplifting* (improvement) of the strong operator convergence of the product formula approximants (Sect. 11.1) to convergence in the *operator-norm* topology, Sect. 11.5. The evolution semigroups approach allows also to establish the estimate for the *rate* of this operator-norm convergence as a function of smoothness of the family of nACP generators.

11.1 Product Approximation of Solution Operator

We consider here construction of solution operator $\{U(t,s)\}_{(t,s)\in\Delta}$ (Definition 2.1.5) for abstract non-Autonomous Cauchy Problem (nACP) (2.1.4) generated by the family of linear operators $\{A(t)\}_{t\in[0,T]}$ in a Banach space $\mathfrak{X}$ by means of convergent *product approximation*. Similar to the Trotter–Kato product formulæ the

V. A. Zagrebnov et al., *Trotter-Kato Product Formulæ*, Operator Theory: Advances and Applications 296, https://doi.org/10.1007/978-3-031-56720-9_11

foremost interest of this approach is to use it for treatment of perturbations of linear evolution equations when the nACP is generated by a family of non-stationary generators: $\{A(t) + B(t)\}_{t\in[0,T]}$.

Under appropriate hypotheses regarding operators $A(t)$ and $B(t)$, we study below the nACP corresponding to linear evolution equation:

$$\begin{aligned} \partial_t u(t) &= -\,(A(t) + B(t))u(t)\,, \quad \text{for} \quad t \in [s, T] \subset \mathbb{R}_0^+\,, \\ u(s) &= x \in \mathfrak{X}\,, \quad s \in (0, T), \end{aligned} \tag{11.1.1}$$

from the point of view which we have just outlined for the case when domain $\mathrm{dom}\,(A(t)) \cap \mathrm{dom}\,(B(t))$ may be *time-dependent.*

To benefit of full power of the semigroup theory, we use for solution operator $\{U_{A+B}(t,s)\}_{(t,s)\in\Delta}$ the semigroup *product approximants* $\{V_{A,B}^{(n)}(t,s)\}_{(t,s)\in\Delta}$ of the form

$$\begin{aligned} V_{A,B}^{(n)}(t,s) := \prod_{q=n-1}^{0} &\exp\left[-\frac{t-s}{n}\,A\left(s + q\,\frac{t-s}{n}\right)\right] \\ &\times \exp\left[-\frac{t-s}{n}\,B\left(s + q\,\frac{t-s}{n}\right)\right]. \end{aligned} \tag{11.1.2}$$

The main assertion of this chapter concerns the proof of convergence of these *product approximants*:

$$U_{A+B}(t,s) = \tau - \lim_{n\to\infty} V_{A,B}^{(n)}(t,s)\,, \quad (t,s) \in \Delta\,. \tag{11.1.3}$$

Here τ indicates the appropriate operator topology of convergence on $\mathcal{L}(\mathfrak{X})$. In this chapter, we pay particular attention to the *strong* operator topology: $\tau = s$, and the *uniform* operator (*operator-norm*) topology: $\tau = \|\cdot\|$.

In this section, we establish (11.1.3) for $\tau = s$. To this end, we assume that the following *hypotheses* are valid:

(A1) For each $t \in [0, T]$ linear operator $A(t)$ is the infinitesimal generator of a *holomorphic* semigroup $\{\exp(-s\,A(t))\}_{s\geq 0}$ on $\mathfrak{X}$, and we suppose that $0 \in \rho(A(t))$ for every such t, where $\rho(A(t))$ denotes the resolvent set of operator $A(t)$.

(A2) The function $t \mapsto A^{-1}(t)$ is continuously differentiable in the operator-norm topology of $\mathcal{L}(\mathfrak{X})$ and there exist constants $a_2 \in \mathbb{R}^+, \widehat{a}_2 \in (0, 1]$ such that the Hölder continuity estimate

$$\left\|\partial_t A^{-1}(t) - \partial_s A^{-1}(s)\right\| \leq a_2\,|t-s|^{\widehat{a}_2}\,, \tag{11.1.4}$$

is valid for every $s, t \in (0, T)$.

Note that hypothesis (A1) implies the existence of constants $\theta \in (0, \pi/2)$, $\epsilon > 0$ and $c_* \in \mathbb{R}^+$ such that $\{A(t) \in \mathscr{H}(\theta - \epsilon, \beta)\}_{t\in[0,T]}$, sector $\widehat{\mathcal{S}}_\theta \subset \bigcap_{t\in[0,T]} \rho(A(t))$ and inequality

$$\left\| R_{A(t)}(\lambda) \right\| \leq c_* \, (1 + |\lambda|)^{-1} \,, \tag{11.1.5}$$

holds for every $t \in [0, T]$ and $\lambda \in \widehat{\mathcal{S}}_\theta$, where

$$R_{A(t)}(\lambda) = (A(t) - \lambda \, \mathbb{1})^{-1} \,,$$

and the closed sector

$$\widehat{\mathcal{S}}_\theta := \{\lambda \in \mathbb{C} : |\arg \lambda| \geq \theta\} \cup \{0\} \,. \tag{11.1.6}$$

Furthermore, hypotheses (A1) and (A2) also imply the operator-norm differentiability of the function $t \mapsto R_{A(t)}(\lambda)$ in $[0, T]$, whose derivative we require to satisfy the following hypothesis:

(A3) There exist constants $a_3 \in \mathbb{R}^+$, $\widehat{a}_3 \in (0, 1]$ such that the inequality

$$\left\| \partial_t R_{A(t)}(\lambda) \right\| \leq a_3 \, |\lambda|^{-\widehat{a}_3} \,, \tag{11.1.7}$$

holds for every $t \in (0, T)$ and every $\lambda \in \widehat{\mathcal{S}}_\theta \setminus \{0\}$.

Remark 11.1.1 Since for every $t \in [0, T]$ the operator $A(t)$ is generator of a *holomorphic* semigroup, the hypotheses (A1)–(A3) are the basis of the existence theorem for non-autonomous linear *parabolic* equations when dom $(A(t))$ *varies* with time, cf. Proposition 2.1.13. We note that for $B(t) = 0$ the product approximation (11.1.3) is an *alternative* to the approach for construction of solution operator in the case of nACP (2.1.4), or (2.1.12).

As long as we have in mind a generalisation of (2.1.4) to (11.1.1), it is natural to ask whether the product approximation by (11.1.2) yields the solution operator for a suitable class of perturbations of $A(t)$. We shall prove that this is indeed the case provided we impose the following conditions:

(B1) For every $t \in [0, T]$ the linear operator $B(t)$ is closed in $\mathfrak{X}$ ($B(t) \in \mathcal{C}(\mathfrak{X})$) and dom $(B(t)) \supseteq$ dom $(A(t))$. Moreover, there exist constants $a \in [0, 1)$ and $b \in [0, (1 - a)/c_*)$, where c_* is the constant in (11.1.5), such that the inequality

$$\|B(t)u\| \leq a \, \|(A(t) - \lambda)u\| + b \, \|u\| \,, \tag{11.1.8}$$

holds for every $u \in$ dom $(A(t))$, every $t \in [0, T]$ and every $\lambda \in \widehat{\mathcal{S}}_\theta$.

(B2) The function $[0, 1] \ni t \mapsto B(t)A^{-1}(t)$ is continuously differentiable in $[0, T]$ with respect to the operator-norm topology and there exist constants $b_2 \in \mathbb{R}^+$,

$\widehat{b_2} \in (0, 1]$ such that the *Hölder* continuity estimate

$$\left\| \partial_t (B(t)A^{-1}(t)) - \partial_s (B(s)A^{-1}(s)) \right\| \leq b_2 \, |t-s|^{\widehat{b_2}} \, ,$$

is valid for every $s, t \in (0, T)$.

(B3) The function $t \mapsto B(t)R_{A(t)}(\lambda)$ is continuously differentiable in $[0, T]$ with respect to the operator-norm topology for every $\lambda \in \widehat{\mathcal{S}}_\theta$ and there exists a constant $c_{**} \in \mathbb{R}^+$ such that the inequality

$$\left\| \partial_t (B(t)R_{A(t)}(\lambda) \right\| \leq c_{**} \, ,$$

holds for every $t \in (0, T)$ and every $\lambda \in \widehat{\mathcal{S}}_\theta$.

Remark 11.1.2 Note that the Kato relative boundedness condition (B1) includes also a smallness condition on the value of b. Therefore, even bounded perturbations $B(t)$ will be admissible only if they satisfy the operator-norm bound (11.1.8) for appropriate b.

We show that the preceding hypotheses guarantee the existence of solution operator $\{U_{A+B}(t, s)\}_{(t,s)\in\Delta}$ (Proposition 11.1.9), but they are not sufficient to allow the proof of the product approximation (11.1.3) in the strong operator topology $\tau = s$. To this end, we need the following additional conditions.

(A4) Operator $A(t) \in \mathscr{H}(\theta - \epsilon, 0)$ is the infinitesimal generator of holomorphic semigroup $\{\exp(-s\, A(t))\}_{s\geq 0}$ of *contractions* on $\mathfrak{X}$ for every $t \in [0, T]$.

(B4) Operator $B(t) \in \mathscr{H}(\theta^* - \epsilon, \beta)$ is the infinitesimal generator of *holomorphic* semigroup $\{\exp(-s\, B(t))\}_{s\geq 0}$ on $\mathfrak{X}$ for every $t \in [0, T]$. In addition, the function $t \mapsto R_{B(t)}(\lambda)$ is continuous on $[0, T]$ uniformly in $\lambda \in \widehat{\mathcal{S}}_{\theta^*}$ in the strong operator topology. Here the closed sector $\widehat{\mathcal{S}}_{\theta^*} \subset \mathbb{C}$, defined as in (11.1.6), is such that $\widehat{\mathcal{S}}_{\theta^*} \subset \rho(B(t))$ for some $\theta^* \in (0, \pi/2)$, and every $t \in [0, T]$.

(D) There exists a dense set $\mathcal{D} \subset \mathfrak{X}$ satisfying

$$\mathcal{D} \subset \bigcap_{t\in[0,T]} \operatorname{dom}(A(t) + B(t)) \, , \tag{11.1.9}$$

such that for each $u \in \mathcal{D}$

$$\sup_{t\in[0,T]} \|A(t)u\| < +\infty \, , \tag{11.1.10}$$

and

$$\sup_{t\in[0,T]} \|B(t)u\| < +\infty \, . \tag{11.1.11}$$

Note that similar to (11.1.5), the hypothesis (B.4) implies

$$\left\| R_{B(t)}(\lambda) \right\| \leq c'_* \,(1 + |\lambda|)^{-1} \,, \tag{11.1.12}$$

which holds for every $t \in [0, T]$ and $\lambda \in \widehat{S}_{\theta*}$. Moreover, for time-independent case $B(t) = B$ the hypothesis (B.4) can be relaxed to condition that C_0-semigroup $\{\exp(-s\,B)\}_{s\geq 0}$ is a contraction on $\mathfrak{X}$.

Theorem 11.1.3 *Assume that hypotheses* (A1)–(A3) *and* (B1)–(B3) *hold. Then, there exists a solution operator* $U_{A+B}(t, s)_{0\leq s\leq t\leq T}$ *for nACP* (11.1.1) *such that for all* s, t, *with* $0 \leq s < t \leq T$, *the following properties are valid:*

(1) *The range of* $U_{A+B}(t, s)$ *satisfies*

$$\operatorname{ran}(U(t, s)) \subseteq \operatorname{dom}(A(t) + B(t)) = \operatorname{dom}(A(t)) \,.$$

(2) *The mapping* $t \mapsto U_{A+B}(t, s)$ *is operator-norm continuously differentiable function for* $t > s$, *and we have*

$$\partial_t U_{A+B}(t, s) = -(A(t) + B(t))U_{A+B}(t, s) \in \mathcal{L}(\mathfrak{X}),$$

along with

$$\|\partial_t U_{A+B}(t, s)\| \leq c\,(t - s)^{-1} \,,$$

for some $c \in \mathbb{R}^+$.

(3) *The mapping* $s \mapsto U(t, s)$ *is operator-norm differentiable function for* $s < t$ *and we have*

$$\partial_s U_{A+B}(t, s) \in \mathcal{L}(\mathfrak{X}) \,,$$

which is equal to bounded extension of operator $U(t, s)(A(t) + B(t))$, *with estimate*

$$\|\partial_s U_{A+B}(t, s)\| \leq c'\,(t - s)^{-1} \,,$$

for some $c' \in \mathbb{R}^+$.

(4) *If, in addition, the hypotheses* (A4), (B4) *and* (D) *hold, then for solution operator* $\{U_A(t, s)\}_{(t,s)\in\Delta}$ *we obtain approximation by the product formula for approximants* (11.1.2):

$$U_{A+B}(t, s) = \text{s-}\lim_{n\to\infty} V^{(n)}_{A,B}(t, s) \,, \tag{11.1.13}$$

for $0 \leq s \leq t \leq T$ *in the strong operator topology* .

Remark 11.1.4

(a) Since the formula (11.1.13) is valid for operators $A(t)$ and $B(t)$ shifted by any number, the condition $0 \in (\rho(A(t)) \cap \rho(B(t)))$ (in (A1) and (B4)) is imposed only for convenience.
(b) We note also that for $B(t) = 0$ the approximation product formula (11.1.13):

$$U_A(t, s) := \text{s-}\lim_{n\to\infty} \prod_{q=n-1}^{0} \exp\left[-\frac{t-s}{n} A\left(s + q\,\frac{t-s}{n}\right)\right], \qquad (11.1.14)$$

yields solution operator for nACP (2.1.4). Representation (11.1.14) is *alternative* to a standard construction of solution operator $\{U_A(t, s)\}_{(t,s)\in\Delta}$ for *parabolic* nACP by strongly convergent product approximants.

We begin the proof of Theorem 11.1.3 by series of preliminary lemmata and the following remark.

Remark 11.1.5 Note that for any $n \geq 3$ and two families of bounded operators $\{U_k\}_{k\in\{1,\ldots,n\}}$ and $\{V_k\}_{k\in\{1,\ldots,n\}}$, one has identity

$$\prod_{k=n}^{1} U_k - \prod_{k=n}^{1} V_k = \prod_{l=n}^{2} U_l \times (U_1 - V_1) +$$

$$+ \sum_{k=2}^{n-1} \prod_{l=n}^{k+1} U_l \times (U_k - V_k) \times \prod_{l=k-1}^{1} V_l + (U_n - V_n) \times \prod_{l=n-1}^{1} V_l\,. \qquad (11.1.15)$$

Similarly, for natural $n \in \mathbb{N}$ and for any pair of operators $U, V \in \mathcal{L}(\mathfrak{X})$ we obtain the *telescopic* identity:

$$U^n - V^n = \sum_{k=1}^{n} U^{n-k}(U - V)V^{k-1}\,. \qquad (11.1.16)$$

Lemma 11.1.6 *Assume that hypotheses* (A1) *and* (B1) *hold. Then, operator* $C(t) := A(t) + B(t)$ *is the infinitesimal generator of a holomorphic semigroup* $\{e^{-s\,C(t)}\}_{s\geq 0}$ *on* $\mathfrak{X}$ *for each* $t \in [0, T]$.

Proof It is sufficient to prove that sector $\widehat{\mathcal{S}}_\theta \subset \rho(C(t))$ and that

$$\left\| R_{C(t)}(\lambda)\right\| \leq \widehat{c}_*(1 + |\lambda|)^{-1}\,, \qquad (11.1.17)$$

for any $t \in [0, T]$ and $\lambda \in \mathcal{S}_\theta$, for some $\widehat{c}_* \in \mathbb{R}^+$, where $\widehat{\mathcal{S}}_\theta$ is given by (11.1.6). Let $\lambda \in \widehat{\mathcal{S}}_\theta$ and let $u = R_{A(t)}(\lambda)\, v$, for arbitrary $v \in \mathfrak{X}$, be inserted into (11.1.8). Then, by virtue of (11.1.5) and due to choice of a, b as in hypothesis (B1), we obtain

$$\left\| B(t)R_{A(t)}(\lambda)v \right\| \leq (a + bc_*)\, \|v\| < \|v\| , \tag{11.1.18}$$

so that $(\mathbb{1} + B(t)R_{A(t)}(\lambda))^{-1} \in \mathcal{L}(\mathfrak{X})$. As a result, we get identity

$$\begin{aligned} R_{C(t)}(\lambda) &= R_{A(t)}(\lambda)(\mathbb{1} + B(t)R_{A(t)}(\lambda))^{-1} \\ &= R_{A(t)}(\lambda) \sum_{m=0}^{+\infty} (-1)^m \left(B(t)R_{A(t)}(\lambda)\right)^m \in \mathcal{L}(\mathcal{B}) , \end{aligned} \tag{11.1.19}$$

that yields (11.1.17) on account of (11.1.5) and (11.1.18). □

Note that for $\lambda = 0$ and $t \in [0, T]$, we obtain from (11.1.19) the representation:

$$C^{-1}(t) = A^{-1}(t)\mathsf{D}(t) , \tag{11.1.20}$$

where

$$\mathsf{D}(t) := (\mathbb{1} + B(t)A^{-1}(t))^{-1} = \sum_{m=0}^{\infty} (-1)^m \left(B(t)A^{-1}(t)\right)^m . \tag{11.1.21}$$

Next we show that the *Hölder continuity* of derivatives (cf. (11.1.4)) persists for perturbed operator $C(t) = A(t) + B(t)$.

Lemma 11.1.7 *Assume that hypotheses* (A1), (A2) *and* (B1), (B2) *hold. Then, the function* $t \mapsto C^{-1}(t)$ *is continuously differentiable with respect to the operator-norm topology and there exist constants* $c_2 \in \mathbb{R}^+$, $\widehat{c_2} \in (0, 1]$ *such that the Hölder continuity estimate*

$$\left\| \partial_t\, C^{-1}(t) - \partial_s\, C^{-1}(s) \right\| \leq c_2\, |t - s|^{\widehat{c_2}} ,$$

is valid for $s, t \in [0, T]$.

Proof According to hypothesis (A2) and (11.1.20), it is sufficient to prove that the function $t \mapsto \mathsf{D}(t)$ is continuously differentiable in the operator-norm topology and that its derivative $t \mapsto \partial_t\, \mathsf{D}(t)$ is Hölder continuous.

To this aim, we define by

$$S_N(t) = \sum_{m=0}^{N} (-1)^m \left(B(t)A^{-1}(t)\right)^m , \quad t \in [0, T] , \tag{11.1.22}$$

the sequence $\{S_N\}_{N\in\mathbb{N}} \subset \mathcal{C}([0,T], \mathcal{L}(\mathfrak{X}))$. Here $\mathcal{C}([0,T], \mathcal{L}(\mathfrak{X}))$ is the Banach space of all continuous $\mathcal{L}(\mathfrak{X})$-valued functions defined on $[0, T]$, endowed with the *uniform* topology, which is defined by the norm $\|\cdot\|_{\mathcal{C}([0,T],\mathcal{L}(\mathfrak{X}))}$.

Then, by virtue of (11.1.21), (11.1.22), together with estimate (11.1.18) for $\lambda = 0$, one gets that

$$\lim_{N\to\infty} \|S_N - \mathsf{D}\|_{\mathcal{C}([0,T],\mathcal{L}(\mathfrak{X}))} = 0. \tag{11.1.23}$$

Since for $M > N$

$$\begin{aligned}&\partial_t S_M(t) - \partial_t S_N(t)\\ &= \sum_{m=N+1}^{M} (-1)^m \sum_{k=0}^{m-1} \left(B(t)A^{-1}(t)\right)^k \partial_t (B(t)A^{-1}(t))dt \left(B(t)A^{-1}(t)\right)^{m-k-1},\end{aligned}$$

by setting $r := a + bc^*$ and by invoking again (11.1.18) for $\lambda = 0$ together with boundedness of function: $t \mapsto \partial_t(B(t)A^{-1}(t))$ (B2), we obtain

$$\begin{aligned}&\sup_{t\in[0,T]} \|\partial_t S_M(t) - \partial_t S_N(t)\|\\ &\leq c \sup_{t\in[0,T]} \sum_{m=N+1}^{M} m \left\|B(t)A^{-1}(t)\right\|^{m-1} \leq c \sum_{m=N+1}^{M} m r^{m-1} \to 0,\end{aligned}$$

as $M, N \to \infty$ because $r \in [0, 1)$. Consequently, $\{\partial_t S_N\}_{N\in\mathbb{N}}$ is a *Cauchy* sequence in Banach space $\mathcal{C}([0,T], \mathcal{L}(\mathcal{B}))$, which, together with (11.1.23), indeed entails the operator-norm continuous differentiability of D and representation:

$$\partial_t \mathsf{D}(t) = \sum_{m=1}^{\infty} (-1)^m \sum_{k=0}^{m-1} \left(B(t)A^{-1}(t)\right)^k \partial_t (B(t)A^{-1}(t)) \left(B(t)A^{-1}(t)\right)^{m-k-1}, \tag{11.1.24}$$

on $[0, T]$.

In order to prove the *Hölder* continuity of (11.1.24), we use the representation:

$$\begin{aligned}&\partial_t \mathsf{D}(t) - \partial_s \mathsf{D}(s)\\ &= \sum_{m=1}^{\infty} (-1)^m \sum_{k=0}^{m-1} \left(\left(B(t)A^{-1}(t)\right)^k \partial_t (B(t)A^{-1}(t)) \left(B(t)A^{-1}(t)\right)^{m-k-1}\right.\\ &\left. - \left(B(s)A^{-1}(s)\right)^k \partial_s (B(s)A^{-1}(s)) \left(B(s)A^{-1}(s)\right)^{m-k-1}\right).\end{aligned}$$

For the operator-norm estimation of the difference in the preceding expression we use (11.1.15) for $n = 3$. This leads to inequality

$$\begin{aligned}
&\|\partial_t \mathsf{D}(t) - \partial_s \mathsf{D}(s)\| \\
&\le c \sum_{m=2}^{\infty} \sum_{k=0}^{m-2} r^k \left\| \left(B(t)A^{-1}(t)\right)^{m-k-1} - \left(B(s)A^{-1}(s)\right)^{m-k-1} \right\| \\
&+ \left(\sum_{m=1}^{\infty} m r^{m-1}\right) \left\| \partial_t (B(t)A^{-1}(t)) - \partial_s (B(s)A^{-1}(s)) \right\| \\
&+ c \sum_{m=2}^{\infty} \sum_{k=1}^{m-1} r^{m-k-1} \left\| \left(B(t)A^{-1}(t)\right)^{k} - \left(B(s)A^{-1}(s)\right)^{k} \right\| . \qquad (11.1.25)
\end{aligned}$$

Now we write the operator differences in the first and third terms on the right-hand side of (11.1.25) by using the *telescopic* identity (11.1.16). This gives

$$\begin{aligned}
&\left(B(t)A^{-1}(t)\right)^{m-k-1} - \left(B(s)A^{-1}(s)\right)^{m-k-1} \\
&= \sum_{l=1}^{m-k-1} \left(B(t)A^{-1}(t)\right)^{m-k-1-l} (B(t)A^{-1}(t) - B(s)A^{-1}(s)) \left(B(s)A^{-1}(s)\right)^{l-1} ,
\end{aligned} \qquad (11.1.26)$$

for $k \in \{0, ..., m-2\}$ and

$$\begin{aligned}
&\left(B(t)A^{-1}(t)\right)^{k} - \left(B(s)A^{-1}(s)\right)^{k} \\
&= \sum_{l=1}^{k} \left(B(t)A^{-1}(t)\right)^{k-l} (B(t)A^{-1}(t) - B(s)A^{-1}(s)) \left(B(s)A^{-1}(s)\right)^{l-1} ,
\end{aligned} \qquad (11.1.27)$$

for $k \in \{1, ..., m-1\}$, respectively. Then for the norms of (11.1.26) and (11.1.27) we obtain

$$\begin{aligned}
&\left\| \left(B(t)A^{-1}(t)\right)^{m-k-1} - \left(B(s)A^{-1}(s)\right)^{m-k-1} \right\| \\
&\le (m-k-1)\, r^{m-k-2} \left\| B(t)A^{-1}(t) - B(s)A^{-1}(s) \right\| , \qquad (11.1.28)
\end{aligned}$$

and

$$\left\| \left(B(t)A^{-1}(t)\right)^k - \left(B(s)A^{-1}(s)\right)^k \right\|$$
$$\leq k\, r^{k-1} \left\| B(t)A^{-1}(t) - B(s)A^{-1}(s) \right\| . \tag{11.1.29}$$

Substitution of estimates (11.1.28) and (11.1.29) into (11.1.25) yield

$$\begin{aligned}
&\|\partial_t \mathsf{D}(t) - \partial_s \mathsf{D}(s)\| \\
&\leq c \left(\sum_{m=2}^{\infty} \sum_{k=0}^{m-2} (m-k-1)\, r^{m-2} \right) \left\| B(t)A^{-1}(t) - B(s)A^{-1}(s) \right\| \\
&+ \left(\sum_{m=1}^{\infty} m\, r^{m-1} \right) \left\| \partial_t (B(t)A^{-1}(t)) - \partial_s (B(s)A^{-1}(s)) ds \right\| \\
&+ c \left(\sum_{m=2}^{\infty} \sum_{k=1}^{m-1} k\, r^{m-2} \right) \left\| B(t)A^{-1}(t) - B(s)A^{-1}(s) \right\| \\
&\leq c \left(\left\| B(t)A^{-1}(t) - B(s)A^{-1}(s) \right\| + \left\| \partial_t (B(t)A^{-1}(t)) - \partial_s (B(s)A^{-1}(s)) \right\| \right),
\end{aligned}$$

since all involved in these inequalities infinite series are convergent. The desired *Hölder* continuity follows then from hypothesis (B2). □

In the following lemma, we show that the estimate (11.1.7) persists also for the perturbed operator $C(t) = A(t) + B(t)$.

Lemma 11.1.8 *Assume that hypotheses* (A1), (A3), (B1) *and* (B3) *hold. Then there exist constants* $c_3 \in \mathbb{R}^+$, $\widehat{c_3} \in (0, 1]$ *such that the inequality*

$$\left\| \frac{\partial}{\partial t} R_{C(t)}(\lambda) \right\|_{\infty} \leq c_3\, |\lambda|^{-\widehat{c_3}} \tag{11.1.30}$$

holds for any $t \in [0, T]$ *and* $\lambda \in S_\theta \setminus \{0\}$.

Proof Our starting point is (11.1.19), which we rewrite as

$$R_{C(t)}(\lambda) = R_{A(t)}(\lambda)\mathsf{D}(t, \lambda).$$

Here

$$\mathsf{D}(t, \lambda) := \sum_{m=0}^{\infty} (-1)^m \left(B(t) R_{A(t)}(\lambda)\right)^m .$$

Then, by virtue of (11.1.18), the function $t \mapsto \mathsf{D}(t, \lambda)$ is bounded in the operator-norm topology on the set $[0, T]$ uniformly in $\lambda \in S_\theta$. On the other hand, by repeating *verbatim* the arguments from the first part of the proof of Lemma 11.1.7, after having replaced $B(t)A^{-1}(t)$ by $B(t)R_{A(t)}(\lambda)$ therein, we conclude that the function $t \mapsto \mathsf{D}(t, \lambda)$ is continuously differentiable in the operator-norm topology on the set $[0, T]$ and that representation

$$\begin{aligned}&\partial_t \mathsf{D}(t, \lambda)\\&= \sum_{m=1}^{\infty}(-1)^m \sum_{k=0}^{m-1} \left(B(t)R_{A(t)}(\lambda)\right)^k \partial_t (B(t)R_{A(t)}(\lambda) \left(B(t)R_{A(t)}(\lambda)\right)^{m-k-1},\end{aligned}$$

holds as a convergent series in the $\|\cdot\|_{C([0,T],\mathcal{L}(\mathfrak{X}))}$-norm.

From this fact together with estimate (11.1.18)and hypothesis (B3), we then infer that the estimate

$$\sup_{t\in[0,T]} \|\partial_t \mathsf{D}(t, \lambda)\| \le \left(\sum_{m=1}^{\infty} m\, r^{m-1}\right) \left\|\partial_t (B(t)R_{A(t)}(\lambda))\right\| \le c < \infty,$$

is valid uniformly in $\lambda \in S_\theta$. Then combining this estimate with identity

$$\partial_t R_{C(t)}(\lambda) = R_{A(t)}(\lambda)\, \partial_t \mathsf{D}(t, \lambda) + (\partial_t R_{A(t)}(\lambda))\, \mathsf{D}(t, \lambda),$$

by virtue of (11.1.5), (11.1.7) and the fact that $\widehat{a}_3 \in (0, 1]$, we obtain

$$\begin{aligned}\left\|\partial_t R_{C(t)}(\lambda)\right\| &\le c \left(\left\|R_{A(t)}(\lambda)\right\| + \left\|\partial_t R_{A(t)}(\lambda))\right\|\right)\\&\le c \left(c_* (1 + |\lambda|)^{-1} + a_3\, |\lambda|^{-\widehat{a}_3}\right) \le c_3\, |\lambda|^{-\widehat{a}_3},\end{aligned}$$

for some $c_3 \in \mathbb{R}^+$ and for any $\lambda \in \widehat{S}_\theta$. Note that we may thus choose $\widehat{c}_3 = \widehat{a}_3$. □

The preceding results exhibit a stability of the basic properties of the *parabolic* nACP (11.1.1) with generator $\{A(t)\}_{t\in[0,T]}$ (see Sect. 2.1.1) for a class of *non-autonomous* perturbations $\{B(t)\}_{t\in[0,T]}$ and thereby lead to the following assertion.

Proposition 11.1.9 *Assume that hypotheses* (A1)–(A3) *and* (B1)–(B3) *hold. Then, there exists solution operator* $U_{A+B}(t, s)_{0\le s\le t\le T}$ *for nACP* (11.1.1) *such that the operator range condition claimed by Theorem* 11.1.3 (1):

$$\operatorname{ran}(U_{A+B}(t, s)) \subseteq \operatorname{dom}(C(t)) = \operatorname{dom}(A(t)), \tag{11.1.31}$$

holds for every s, t *with* $0 \le s < t \le T$. *Moreover, under the very same conditions the statements* (2) *and* (3) *of Theorem* 11.1.3 *are valid.*

Proof The proof is a consequence of Lemmata 11.1.6–11.1.8 together with Proposition 2.1.13, regarding dom $(A(t))$ (11.1.31) as a *regularity* set for parabolic nACP (2.1.12), see Sect. 2.1.1. □

Proof of Theorem 11.1.3(4) For $n \in \mathbb{N}$ we set $h = (t-s)/n$ and define sequence of bounded operators $\{\Delta_n(t,s)\}_{n\geq 1} \subset \mathcal{L}(\mathcal{B})$ by

$$\Delta_n(t,s) = U_{A+B}(t,s) - \prod_{k=n}^{1} \exp\left[-hA\left(s+(k-1)h\right)\right] \exp\left[-hB\left(s+(k-1)h\right)\right].$$

So, in order to prove product formula(11.1.13) for approximants (11.1.2) it is sufficient to show that $\Delta_n(t,s)u \to 0$ as $n \to \infty$ for every $u \in \mathcal{D}$ from the dense (by hypothesis (D)) set $\mathcal{D}$. □

To this aim, we use *telescopic* identities (11.1.16) for two families $\{U_k\}_{k\in\{1,\ldots,n\}}$ and $\{V_k\}_{k\in\{1,\ldots,n\}}$ where

$$\begin{aligned} U_k &:= U(s+kh, s+(k-1)h), \\ V_k &:= \exp\left[-hA\left(s+(k-1)h\right)\right] \exp\left[-hB\left(s+(k-1)h\right)\right]. \end{aligned}$$

Owing to the basic Chapman–Kolmogorov composition law: $U(t,\tau)U(\tau,s) = U(t,s)$, for solution operator $\{U(t,s)\}_{0\leq s\leq t\leq T}$ (Definition 2.1.4) and by (11.1.15), we obtain after some rearrangements:

$$\begin{aligned} \Delta_n(t,s) = & \sum_{k=1}^{n-1} \prod_{l=n}^{k+1} V_l \left(U_k - V_k\right) U_{A+B}(s+(k-1)h, s) + \\ & + (U_n - V_n) U_{A+B}(s+(n-1)h, s). \end{aligned}$$

Since $\| V_k \| \leq 1$, the last formula gives for $x \in \mathfrak{X}$ the estimate

$$\begin{aligned} \|\Delta_n(t,s)x\| &\leq \sum_{k=1}^{n} \|(U_k - V_k) U_{A+B}(s+(k-1)h, s)x\| \qquad (11.1.32) \\ &\leq n \sup_{r\in[s,s+(n-1)h]} \|U_{A+B}(r+h,s)x - \exp[-hA(r)] \exp[-hB(r)]\, U_{A+B}(r,s)x\| \\ &\leq n \sup_{r\in[s,t]} \|\{U_{A+B}(r+h,r) - \exp[-hA(r)] \exp[-hB(r)]\} U_{A+B}(r,s)x\|. \end{aligned}$$

The remaining lemmata are preparatory statements for the proof of the product formula (11.1.13) for approximants (11.1.2), see part (4) of Theorem 11.1.3.

From now on we may assume that $0 \leq s < t \leq T$ since (11.1.13) trivially holds for $t = s$. We begin by the following *uniformity* result adapted for our purposes.

Lemma 11.1.10 *Let* $0 \leq s < t \leq T$ *and* $\{R(h,r)\}_{(h,r)\in\mathbb{R}^+\times[s,t]}$ *be a family of bounded operators satisfying*

$$\sup_{(h,r)\in\mathbb{R}^+\times[s,t]} \|R(h,r)\| \leq C_{s,t} < \infty\,. \tag{11.1.33}$$

If we assume that for a compact $\mathcal{K} \subset \mathfrak{X}$ *the limit*

$$\lim_{h\to+0} R(h,r)w = 0, \tag{11.1.34}$$

exists for each $w \in \mathcal{K}$ *uniformly in* $r \in [s,t]$, *then* (11.1.34) *also holds uniformly in* $w \in \mathcal{K}$:

$$\lim_{h\to+0} \sup_{r\in[s,t]} \sup_{w\in\mathcal{K}} R(h,r)w = 0\,. \tag{11.1.35}$$

Proof If not, then for any sequence $h_n \downarrow 0$, there exist $\varepsilon > 0$ and two sequences $\{w_{n,\varepsilon}\}_{n\geq1} \subset \mathcal{K}$ and $\{r_{n,\varepsilon}\}_{n\geq1} \subset [s,t]$ such that

$$\left\|R(h_n, r_{n,\varepsilon})w_{n,\varepsilon}\right\| \geq \epsilon\,. \tag{11.1.36}$$

But by compactness, the sequence $\{w_{n,\varepsilon}\}_{n\geq1}$ has at least one accumulation point $w^* \in \mathcal{K}$. Therefore, by virtue of (11.1.33) and (11.1.34) the estimate

$$\begin{aligned}
&\left\|R(h_n, r_{n,\varepsilon})w_{n,\varepsilon}\right\| \leq \left\|R(h_n, r_{n,\varepsilon})(w_{n,\varepsilon} - w^*)\right\| + \left\|R(h_n, r_{n,\varepsilon})w^*\right\| \\
&\leq C_{s,t}\left\|w_{n,\varepsilon} - w^*\right\| + \left\|R(h_n, r_{n,\varepsilon})v^*\right\| \leq \epsilon\,,
\end{aligned}$$

is valid for all $n \geq N(\epsilon, w^*)$ and the corresponding $N(\epsilon, w^*)$, thereby contradicting (11.1.36). □

Now we introduce three families of bounded linear operators defined on $\mathfrak{X}$ by

$$\begin{aligned}
E(h,r) &= h^{-1}\left(\mathbb{1} - \exp\left[-hA(r)\right]\right) - A(s),\\
F(h,r) &= h^{-1}\exp\left[-hA(r)\right]\left(\mathbb{1} - \exp\left[-hB(r)\right]\right) - B(r),\\
G(h,r) &= h^{-1}\left(\mathbb{1} - U_{A+B}(r+h,r)\right) - C(r)\,,
\end{aligned}$$

for $(h,r) \in \mathbb{R}^+ \times [s,t]$. These relations and definition of generator $C(r)$ yield

$$\begin{aligned}
&(E(h,r) + F(h,r) - G(h,r))u\\
&= h^{-1}\{U_{A+B}(r+h,r) - \exp\left[-hA(r)\right]\exp\left[-hB(r)\right]\}u,
\end{aligned} \tag{11.1.37}$$

for any $u \in \mathcal{D}$.

With help of Lemma 11.1.10, we elucidate behaviour of (11.1.37) when $h \to +0$.

Lemma 11.1.11 *Assume that the same hypotheses as in Proposition* 11.1.9 *hold. If in addition we assume the hypotheses* (A4),(B4) *and* (D)*, then*

$$\lim_{h\to+0} \sup_{r\in[s,t]} \|E(h,r)U_{A+B}(r,s)u\| = \lim_{h\to+0} \sup_{r\in[s,t]} \|F(h,r)U_{A+B}(r,s)u\|$$

$$= \lim_{h\to+0} \sup_{r\in[s,t]} \|G(h,r)U_{A+B}(r,s)u\| = 0, \tag{11.1.38}$$

for any $t \in (0,T)$ *and* $u \in \mathcal{D}$.

Proof Since by (11.1.9), (11.1.10) and (11.1.31) one has $U(r,s)u \in \mathcal{D}(A(r))$ for any $r \in [s,t]$ if $u \in \mathcal{D}$, definition of $E(h,r)$ and properties of the $\mathcal{C}_0$-semigroup $\{e^{-\tau A(r)}\}_{\tau\geq 0}$ yield

$$E(h,r)U(r,s)u = h^{-1}\int_0^h d\tau\,(\exp[-\tau A(r)] - \mathbb{1})\,A(r)U_{A+B}(r,s)u\,.$$

Note that for each $u \in \mathcal{D}$ the integrand is norm-continuous vector-valued function on $[0,h]$. Then by the *integral mean value* theorem we obtain estimate

$$\|E(h,r)U_{A+B}(r,s)u\| \leq \sup_{\tau\in[0,h]} \|(\exp[-\tau A(r)] - \mathbb{1})\;A(r)U_{A+B}(r,s)u\| =$$

$$\|(\exp[-\widehat{\tau}(h)A(r)] - \mathbb{1})\;A(r)U_{A+B}(r,s)u\|\,,\quad \widehat{\tau}(h) \in [0,h]\,. \tag{11.1.39}$$

Note that in (11.1.39) $\lim_{h\to+0}\widehat{\tau}(h) = 0$. Hence, to prove the first assertion in (11.1.38) it is sufficient to show

$$\lim_{h\to+0} \sup_{r\in[s,t]} \|(\exp[-\widehat{\tau}(h)A(r)] - \mathbb{1})\;A(r)U_{A+B}(r,s)u\| = 0\,, \tag{11.1.40}$$

for any $u \in \mathcal{D}$.

To this end, we define operator

$$R(h,r) := \exp[-hA(r)] - \mathbb{1}\,, \tag{11.1.41}$$

and set

$$\mathcal{K}_{[s,t]} := \{w \in \mathfrak{X} : w = A(\eta)\,U_{A+B}(\eta,s)u\,,\quad u \in \mathcal{D},\ \eta \in [s,t]\}\,. \tag{11.1.42}$$

Since by (A4) the C_0-semigroup in (11.1.41) is contraction , $R(h, r)$ satisfies condition (11.1.33). Moreover, from condition (D) (11.1.10) we obtain from (11.1.41) for every $u \in \mathcal{D}$ the estimate

$$\|R(h,r)u\| \leq \int_0^h d\tau \; \|\exp[-\tau A(r)]\, A(r)u\| \leq h \sup_{r\in[0,T]} \|A(r)u\| \;,$$

uniformly in $r \in [0, T]$. Since by hypothesis (D) the invariant $\mathcal{D} \subset \mathfrak{X}$ is dense, this estimate implies

$$\lim_{h\to+0} \sup_{r\in[s,t]} R(\widehat{\tau}(h), r)\, x = 0\,, \quad x \in \mathfrak{X}. \tag{11.1.43}$$

By definition of the set $\mathcal{K}_{[s,t]}$ for any $u \in \mathcal{D}$ one gets

$$\sup_{r\in[s,t]} \|(\exp[-\widehat{\tau}(h)A(r)] - \mathbb{1})\; A(r)U_{A+B}(r,s)u\| \leq \tag{11.1.44}$$
$$\leq \sup_{r\in[s,t]} \sup_{w\in\mathcal{K}_{[s,t]}} \|(\exp[-\widehat{\tau}(h)A(r)] - \mathbb{1})\; w\| \;.$$

Therefore (11.1.40) would follow from (11.1.44) if

$$\lim_{h\to+0} \sup_{r\in[s,t]} \sup_{w\in\mathcal{K}_{[s,t]}} \|R(\widehat{\tau}(h), r)\; w\| = 0\,. \tag{11.1.45}$$

Since $R(\widehat{\tau}(h), r)$ satisfies conditions (11.1.33) and (11.1.43), Lemma 11.1.10 (limit (11.1.35)) would yield (11.1.45) if $\mathcal{K}_{[s,t]}$ is a compact.

To check that the set (11.1.42) is a compact, one has to verify that the mapping of compact $[s, t]$: $r \mapsto A(r)U_{A+B}(r, s)u$ is strongly continuous for any $u \in \mathcal{D}$. To this aim, we use for $u \in \mathcal{D}$ the representation

$$A(r)U_{A+B}(r,s)u = (\mathbb{1} - B(r)A(r)^{-1}\mathrm{D}(r))(A(r) + B(r))U_{A+B}(r,s)u\,. \tag{11.1.46}$$

Then, continuity of the mapping follows from (B2), (11.1.20), Theorem 11.1.3(1) and condition (D), for $r = s$. This completes the proof of (11.1.40) and consequently the first limit in (11.1.38).

To prove the second limit in (11.1.38), we proceed by starting with integral representation

$$F(h,r)U_{A+B}(r,s)u = h^{-1}\int_0^h d\tau\; (\exp[-hA(r)]\exp[-\tau B(r)] - \mathbb{1})$$
$$\times\; B(r)U_{A+B}(r,s)u$$

which for $u \in \mathcal{D}$ is valid on compact: $r \in [s,t]$. It yields the estimate

$$\|F(h,r)U_{A+B}(r,s)u\| \leq \sup_{\tau\in[0,h]} \|(\exp[-hA(r)]\exp[-\tau B(r)] - \mathbb{1})$$
$$B(r)U_{A+B}(r,s)u\|$$
$$= \|(\exp[-hA(r)]\exp[-\widehat{\tau}(h)B(r)] - \mathbb{1})\, B(r)U_{A+B}(r,s)u\| \,, \qquad (11.1.47)$$

where $\widehat{\tau}(h) \in [0,h]$.

This time we define $\widetilde{R}(h,r)$ and the set $\widetilde{\mathcal{K}}_{[s,t]}$ by

$$\widetilde{R}(h,\tau,r) := \exp[-hA(r)]\exp[-\tau B(r)] - \mathbb{1} \qquad (11.1.48)$$

and

$$\widetilde{\mathcal{K}}_{[s,t]} := \{w \in \mathfrak{X} : w = B(\eta)U_{A+B}(\eta,s)u, \quad u \in \mathcal{D},\ \eta \in [s,t]\}\,. \qquad (11.1.49)$$

From contraction properties of the involved in (11.1.48) C_0-semigroups, it follows that $\widetilde{R}(h,\tau,r)$ satisfies (11.1.33) and that

$$\lim_{h\to+0} \sup_{r\in[s,t]} \widetilde{R}(h,\widehat{\tau}(h),r)\,x = 0\,, \quad x \in \mathfrak{X}. \qquad (11.1.50)$$

Following the same line of reasoning as in (11.1.44), we obtain from definitions (11.1.48) and (11.1.49) for any $u \in \mathcal{D}$ the estimate

$$\sup_{r\in[s,t]} \|(\exp[-hA(r)]\exp[-\widehat{\tau}(h)B(r)] - \mathbb{1})\, B(r)U_{A+B}(r,s)u\| \leq$$
$$\leq \sup_{r\in[s,t]} \sup_{w\in\widetilde{\mathcal{K}}_{[s,t]}} \|\widetilde{R}(h,\widehat{\tau}(h),r)\,w\|. \qquad (11.1.51)$$

By virtue of (11.1.47) and (11.1.51), the second assertion in (11.1.38) would follow from the limit

$$\lim_{h\to+0} \sup_{r\in[s,t]} \sup_{w\in\widetilde{\mathcal{K}}_{[s,t]}} \|\widetilde{R}(h,\widehat{\tau}(h),r)\,w\| = 0\,. \qquad (11.1.52)$$

Since $\widetilde{R}(h,\widehat{\tau}(h),r)$ satisfies conditions (11.1.33) and (11.1.50), Lemma 11.1.10 (limit (11.1.35)) would yield (11.1.52) if the set $\widetilde{\mathcal{K}}_{[s,t]}$ is a compact in $\mathfrak{X}$. For compactness it is sufficient to note that function

$$[s,t] \ni r \mapsto B(r)U_{A+B}(r,s)\,u = (C(r) - A(r))U_{A+B}(r,s)\,u\,,$$

is continuous for any $u \in \mathcal{D}$. This follows from Theorem 11.1.3(2) and from continuity of function (11.1.46). Accordingly, this accomplishes the proof of the second limit in (11.1.38).

Since by Theorem 11.1.3(2), the function

$$\tau \mapsto (C(\tau)U_{A+B}(\tau, s) - C(r)U_{A+B}(r, s)), \tag{11.1.53}$$

on the compact $[r, r+h]$ is strongly continuous, we can use for the proof of the third limit in (11.1.38) the representation

$$G(h, r)U_{A+B}(r, s)x = h^{-1} \int_r^{r+h} d\tau\, (C(\tau)U_{A+B}(\tau, s) - C(r)U_{A+B}(r, s))\, x, \tag{11.1.54}$$

where $x \in \mathfrak{X}$. By continuity of the integrand in (11.1.54) the integral mean value theorem yields

$$G(h, r)U_{A+B}(r, s)x = (C(\widehat{\tau}(h))U_{A+B}(\widehat{\tau}(h), s) - C(r)U_{A+B}(r, s))\, x, \tag{11.1.55}$$

where $\widehat{\tau}(h) \in [r, r+h]$. Therefore, by Theorem 11.1.3(1)–(2), conditions (D), continuity (11.1.53) and equation (11.1.55)

$$\lim_{h \to +0} \sup_{r \in [s,t]} (C(\widehat{\tau}(h))U_{A+B}(\widehat{\tau}(h), s) - C(r)U_{A+B}(r, s))\, u = 0, \tag{11.1.56}$$

for $u \in \mathcal{D}$. This finishes the proof of (11.1.38). □

Finally, to proceed further with the proof of Theorem 11.1.3(4), we note that (11.1.32) and (11.1.37) yield

$$\|\Delta_n(t, s)u\| \leq (t-s)\|(E(h, r) + F(h, r) - G(h, r))U_{A+B}(r, s)u\|, \tag{11.1.57}$$

for any $u \in \mathcal{D}$. Now the proof of the strong limit (11.1.13) follows from Lemma 11.1.11 and (11.1.57), since $\|\Delta_n(t, s)\| \leq 2$ and domain $\mathcal{D}$ is dense in $\mathfrak{X}$. □

11.2 The Evolution Semigroups Method in L^p-Setting

We present in this section yet another approach to analysis of product approximation the solution operator $\{U_{A+B}(t, s)\}_{0 \leq s \leq t \leq T}$, which yields a stronger convergence than in the *strong* operator topology of Theorem 11.1.3(4). It turns out that assuming some reinforced hypotheses, see Assumptions 11.2.12, one can strengthen topology, (and therefore our results) to the *operator-norm* convergence, including the operator-norm estimate of *error bound* for the convergence rate.

This approach is based on the *evolution semigroup* method of solution the abstract non-Autonomous Cauchy Problem (nACP) that we outlined in Sect. 2.1 and in Sect. 2.2. As it is indicated in Sect. 2.2.1, the C-setting (Definition 2.1.16) seems to be too restrictive. For that reason, we shall essentially focus below on the outlined in Sect. 2.2 the L^p-setting in the Banach space $\mathfrak{X}_T^p := L^p([0,T],\mathfrak{X})$ for some $p \in [1,\infty)$ endowed with the norm

$$\|f\|_{\mathfrak{X}_T^p} := \left(\int_0^T dt\, \|f(t)\|_{\mathfrak{X}}^p\right)^{1/p}, \quad \text{for } f \in \mathfrak{X}_T^p \text{ and } f(t) \in \mathfrak{X}. \tag{11.2.1}$$

We denote by $\|\cdot\|_{\mathcal{L}(\mathfrak{X}_T^p)}$ the operator norm

$$\|C\|_{\mathcal{L}(\mathfrak{X}_T^p)} := \sup_{\{f\in\mathfrak{X}_T^p:\, \|f\|_{\mathfrak{X}_T^p}=1\}} \|Cf\|_{\mathfrak{X}_T^p}, \quad \text{for } C \in \mathcal{L}(\mathfrak{X}_T^p), \tag{11.2.2}$$

of bounded on $\mathfrak{X}_T^p$ set of operators $\mathcal{L}(\mathfrak{X}_T^p)$. cf. Sect. 2.2 (Sect. 2.2.1) and Appendix E.3 (Sect. E.3.2).

To proceed, we first need a certain number of preliminaries.

Definition 11.2.1 Let $\{C(t)\}_{t\in[0,T]}$ be a family of linear operators in the Banach space $\mathfrak{X}$. We note that domains $\{\mathrm{dom}\,(C(t))\}_{t\in[0,T]}$ may *depend* on the time-parameter t. The corresponding multiplication operator $\mathcal{C}$ in the Banach space $\mathfrak{X}_T^p = L^p([0,T],\mathfrak{X})$, which is induced by the family $\{C(t)\}_{t\in[0,T]}$, is defined by

$$(\mathcal{C}f)(t) := C(t)f(t), \quad \text{for a.a. } t \in [0,T], \quad \text{with domain}$$
$$\mathrm{dom}\,(\mathcal{C}) := \left\{ f \in \mathfrak{X}_T : \begin{array}{c} f(t) \in \mathrm{dom}\,(C(t)) \text{ for a.a. } t \in [0,T] \\ [0,T] \ni t \mapsto C(t)f(t) \in L^p([0,T],\mathfrak{X}) \end{array} \right\}. \tag{11.2.3}$$

We note that *a priori* the operator (11.2.3) is *not* dense in $\mathfrak{X}_T^p$, cf. Definition 2.1.21 and Definition 2.2.14. To proceed, we recall that for a family of densely defined closed operators $\{C(t)\}_{t\in[0,T]}$ in $\mathfrak{X}$, we have the following *three* key assertions, that are proved in Sect. 2.2, see Lemma 2.2.15, Theorems 2.2.17 and 2.2.18, respectively. For the reader's convenience, we explicitly reformulate them below in a suitable for our purposes form.

Proposition 11.2.2 *If $\{C(t)\}_{t\in[0,T]}$ is a family of closed linear operators in $\mathfrak{X}$, then multiplication operator $\mathcal{C}$ induced according to* (11.2.3) *in $\mathfrak{X}_T^p$ is also closed.*

Proposition 11.2.3 *Let $\{C(t)\}_{t\in[0,T]}$ be a family of generators in $\mathfrak{X}$ such that for almost all $t \in [0,T]$ it holds that $C(t) \in \mathcal{G}(M,\beta)$ for some $M \geq 1$ and $\beta \in \mathbb{R}$. If the function $[0,T] \ni t \mapsto (C(t) + \xi\,\mathbb{1})^{-1}x \in \mathfrak{X}$ is strongly measurable for $\xi > \beta$, $x \in \mathfrak{X}$, then induced multiplication operator $\mathcal{C}$ is a densely defined in the space $\mathfrak{X}_T^p$,*

$p \in [1, \infty)$, generator and the corresponding semigroup $\{e^{-\tau \mathcal{C}}\}_{\tau \geq 0}$ is given by

$$(e^{-\tau \mathcal{C}} f)(t) = e^{-\tau C(t)} f(t) , \quad f \in \mathfrak{X}^p_T, \text{ for a.a. } t \in [0, T].$$

Moreover, the generator $\mathcal{C} \in \mathcal{G}(M, \beta)$ and equality of norms :

$$\| e^{-\tau \mathcal{C}} \|_{\mathcal{L}(\mathfrak{X}^p_T)} = \operatorname*{ess\,sup}_{t \in [0,T]} \| e^{-\tau C(t)} \|_{\mathcal{L}(\mathfrak{X})} , \tag{11.2.4}$$

holds for $\tau \in \mathbb{R}^+_0$.

Proposition 11.2.4 *Let $\{C(t)\}_{t \in [0,T]}$ be a family of generators of holomorphic on $\mathfrak{X}$ semigroups such that $C(t) \in \mathscr{H}(\theta, 0)$ for some $\theta \in (0, \pi/2]$ and for all $t \in [0, T]$. If the function $[0, T] \ni t \mapsto (C(t) + \xi \mathbb{1})^{-1} x$ is measurable for each $x \in \mathfrak{X}$ and $\xi > 0$, then induced in $\mathfrak{X}^p_T$ multiplication operator $\mathcal{C}$ is generator of a holomorphic in the open sector $S_\theta \subset \mathbb{C}_+$ semigroup $\{\mathcal{U}_{\mathcal{C}}(\tau) := e^{-\tau \mathcal{C}}\}_{\tau \in S_\theta}$ on $\mathfrak{X}^p_T$, that is, $\mathcal{C} \in \mathscr{H}(\theta, 0)$.*

For comments on these assertions, see Appendix E (Sect. E.3). The equality of norms (11.2.4) is due to Theorem E.3.12.

Remark 11.2.5 We also note that domain of generator $\mathcal{C}$ does *not* necessarily have a dense *cross section* in $\mathfrak{X}$ since elements of dom $(\mathcal{C})$ may be not continuous, see Definition 2.2.5. On the other hand, the *accretivity* of operators $\{C(t)\}_{t \in [0,T]}$ in Proposition 11.2.3, cf. $C(t) \in \mathcal{G}(M, \beta)$, and similarly $C(t) \in \mathscr{H}(\theta, 0)$ in Proposition 11.2.4, ensure that operator $\mathcal{C}$ is *densely* defined in $\mathfrak{X}^p_T$.

If $\{C(t) = A_0\}_{t \in [0,T]}$, that is, the family does not depend on t, then according to (11.2.3) and Proposition 11.2.2 a densely defined in $\mathfrak{X}$ closed operator A_0 trivially induces in $\mathfrak{X}^p_T$ a closed multiplication operator $\mathcal{A}_0$ given by

$$(\mathcal{A}_0 f)(t) := A_0 f(t) \text{ for a.a. } t \in [0, T] , \tag{11.2.5}$$

with domain:

$$\text{dom}\,(\mathcal{A}_0) := \left\{ f \in L^p([0, T], \mathfrak{X}) : \begin{array}{c} f(t) \in \text{dom}\,(A_0) \text{ for a.a. } t \in [0, T] \\ [0, T] \ni t \mapsto A_0 f(t) \in L^p([0, T], \mathfrak{X}) \end{array} \right\} .$$

Then, Proposition 11.2.3, or Proposition 11.2.4, yields the following result, which also makes sure that dom $(\mathcal{A}_0)$ is dense in $\mathfrak{X}^p_T$.

Corollary 11.2.6 *Let operator A_0 be generator of quasi-bounded, or holomorphic C_0-semigroup on $\mathfrak{X}$. Then, the induced multiplication operator $\mathcal{A}_0$ is densely defined generator of a C_0-semigroup in $\mathfrak{X}^p_T$, that is, $\mathcal{A}_0 \in \{\mathcal{G}(M, \beta)$ or $\mathscr{H}(\theta, 0)\}$, given by*

$$(e^{-\tau \mathcal{A}_0} f)(t) = e^{-\tau A_0} f(t) , \text{ a.a. } t \in [0, T] , \quad f \in \mathfrak{X}^p_T , \tag{11.2.6}$$

seeing that function : $[0, T] \ni t \mapsto (A_0 + \xi\, \mathbb{1})^{-1}x \in \mathfrak{X}$ *is strongly measurable for any* $x \in \mathfrak{X}$ *when* $\xi > \beta$.

In order to fix the notations let A_0 be generator of a *quasi-bounded* C_0-semigroup defined on the Banach space $\mathfrak{X}$, that is, $A_0 \in \mathcal{G}(M_0, \beta_0)$. According to (2.2.21) (or (11.2.3)) the operator A_0 induces a multiplication operator $\mathcal{A}_0$ in the Banach space $\mathfrak{X}^p_T$. Since $A_0 \in \mathcal{G}(M_0, \beta_0)$, the operator $\mathcal{A}_0$ generates on $\mathfrak{X}^p_T$ a quasi-bounded C_0-semigroup (11.2.6) such that in turn, the generator $\mathcal{A}_0 \in \mathcal{G}(M_0, \beta_0)$, Proposition 11.2.3. Recall that in Sect. 2.2.2 we also introduced in the Banach space $\mathfrak{X}^p_T$ generator $D_{0p} \in \mathcal{G}(1, 0)$ (2.2.19) of the *nilpotent* right-shift *contraction* semigroup (2.2.20): $\{e^{-\tau D_{0p}} = \mathcal{S}_R(\tau)\}_{\tau \geq 0}$.

Remark 11.2.7 Since in the following we shall consider *only* the L^p-setting, below it will be no confusion with the C-setting. For that reason, we denote the above generator by $D_0 := D_{0p}$. Similarly, we shall drop *subindex* p attributed to all other operators introduced in $\mathfrak{X}^p_T$ (Sect. 2.2.2) which we are going to use in this chapter.

Remark 11.2.8 Let operator $\widetilde{\mathcal{K}}_0 := D_0 + \mathcal{A}_0$ on $\mathrm{dom}\,(\widetilde{\mathcal{K}}_0) = \mathrm{dom}\,(D_0) \cap \mathrm{dom}\,(\mathcal{A}_0)$. Since A_0 is *time-independent*, generators $\mathcal{A}_0$ and D_0 *commute*. Consequently, the product of two commuting semigroups:

$$\mathcal{T}_0(\tau) := e^{-\tau D_0} e^{-\tau \mathcal{A}_0}, \quad \tau \geq 0, \tag{11.2.7}$$

defines on $\mathfrak{X}^p_T$ a unique nilpotent C_0-semigroup $\{\mathcal{T}_0(\tau) := e^{-\tau \mathcal{K}_0}\}_{\tau \geq 0}$. We denote by $\mathcal{K}_0$ its generator, which *commutes* with operators $\mathcal{A}_0$ and D_0.

Lemma 11.2.9 *Let operator* $A_0 \in \mathcal{G}(M_0, \beta_0)$ *be generator of a quasi-bounded* C_0-*semigroup on* $\mathfrak{X}$. *Then the following holds* :

(i) *The operator* $\widetilde{\mathcal{K}}_0$ *is a densely defined evolution pre-generator, Definition* 2.2.8.
(ii) *The closure*: $\mathcal{K}_0 := \overline{\widetilde{\mathcal{K}}_0}$, *of this pre-generator, is generator of (nilpotent on* $\mathfrak{X}^p_T$) *evolution semigroup* $\{\mathcal{T}_0(\tau)\}_{\tau \geq 0}$. *The corresponding solution operator* $\{U_0(t, s)\}_{(t,s) \in \Delta}$ *is given by*

$$U_0(t, s) := e^{-(t-s)A_0}, \quad (t, s) \in \Delta. \tag{11.2.8}$$

(iii) *Generator* $\mathcal{K}_0 \in \mathcal{G}(M_0, \beta_0)$ *if and only if* $A_0 \in \mathcal{G}(M_0, \beta_0)$.
(iv) *The operator* $\mathcal{K}_0$ *is invertible and the resolvent set* $\rho(\mathcal{K}_0) = \mathbb{C}$.

Proof

(i) Recall that $\mathrm{dom}\,(D_0) = \{f \in W^{1,p}([0, T], \mathfrak{X}) : f(0) = 0\}$ (2.2.19) and by definition of $\widetilde{\mathcal{K}}_0$: $\mathrm{dom}\,(\widetilde{\mathcal{K}}_0) \subseteq \mathrm{dom}\,(D_0)$. Thus, $\mathrm{dom}\,(\widetilde{\mathcal{K}}_0) \subseteq \mathfrak{X}^p_T \cap \mathfrak{X}^C_T$. Since for $\phi \in W^{1,p}([0, T])$ and for any $u \in \mathrm{dom}\,(A_0)$, the element $\phi_u(t) := \phi(t) \otimes u$, $t \in [0, T]$, belongs to $\mathrm{dom}\,(\widetilde{\mathcal{K}}_0)$, one gets that by definition (2.1.44), we obtain for cross section at $t \in [0, T]$:

$$[\mathrm{dom}\,(\widetilde{\mathcal{K}}_0)]_t \supseteq \mathrm{dom}\,(A_0)\,.$$

Seeing that dom (A_0) is dense in $\mathfrak{X}$, the cross section of domain dom $(\widetilde{\mathcal{K}}_0) \subseteq \mathfrak{X}_T^p$:

$$[\mathrm{dom}\,(\widetilde{\mathcal{K}}_0)]_{\mathfrak{X}_T^p} = \bigcap_{t\in[0,T]} [\mathrm{dom}\,(\widetilde{\mathcal{K}}_0)]_t \,,$$

is dense in $\mathfrak{X}$ by Definition 2.2.5. Now, given that for any $\varphi \in W^{1,p}([0,T])$, one has for scalar multiplication operator: $\mathcal{M}_p(\varphi)\mathrm{dom}\,(\widetilde{\mathcal{K}}_0) \subseteq \mathrm{dom}\,(\widetilde{\mathcal{K}}_0)$, by Lemma 2.2.19 we infer that operator $\widetilde{\mathcal{K}}_0 = D_0 + \mathcal{A}_0$ is a densely defined evolution pre-generator.

(ii) Let $f \in \mathrm{dom}\,(\widetilde{\mathcal{K}}_0) := \mathrm{dom}\,(D_0) \cap \mathrm{dom}\,(\mathcal{A}_0)$. Since by definition of the C_0-semigroup $\{\mathcal{T}_0(\tau)_{\tau\geq 0}$ there exists generator $\mathcal{K}_0$ such that $\{\mathcal{T}_0(\tau) = \mathrm{e}^{-\tau\,\mathcal{K}_0}\}_{\tau\geq 0}$, the representation (11.2.7) yields

$$\mathcal{K}_0 f = \lim_{\tau\to+0} \frac{\mathcal{I} - \mathcal{T}_0(\tau)}{\tau} f = D_0 f + \mathcal{A}_0 f\,, \quad f \in \mathrm{dom}\,(\widetilde{\mathcal{K}}_0)\,. \tag{11.2.9}$$

This means that restriction $\mathcal{K}_0 \restriction \mathrm{dom}\,(\widetilde{\mathcal{K}}_0) = \widetilde{\mathcal{K}}_0$.

Note that by definition of semigroups $\{\mathcal{S}_R(\tau)\}_{\tau\geq 0}$ and $\{\mathrm{e}^{-\tau\,\mathcal{A}_0}\}_{\tau\geq 0}$ together with representation (11.2.7) one gets a *nilpotent* for any $f \in \mathrm{dom}\,(\widetilde{\mathcal{K}}_0) \subset \mathfrak{X}_T^p$ mapping:

$$f(t) \mapsto (\mathcal{T}_0(\tau)f)(t) = \mathrm{e}^{-\tau A_0}\chi_{[0,T]}(t-\tau)f(t-\tau)\,. \tag{11.2.10}$$

Because the right-hand side of (11.2.10) again belongs to dom $(\widetilde{\mathcal{K}}_0)$, we obtain

$$\mathcal{T}_0(\tau)\,\mathrm{dom}\,(\widetilde{\mathcal{K}}_0) \subset \mathrm{dom}\,(\widetilde{\mathcal{K}}_0), \quad \tau \geq 0\,. \tag{11.2.11}$$

Consequently, dom $(\widetilde{\mathcal{K}}_0)$ is a core for generator $\mathcal{K}_0$. That is, $\widetilde{\mathcal{K}}_0$ has a unique closed extension:

$$\mathcal{K}_0 = \overline{\mathcal{K}_0 \restriction \mathrm{dom}\,(\widetilde{\mathcal{K}}_0)} = \overline{\widetilde{\mathcal{K}}_0}\,. \tag{11.2.12}$$

To show that $\mathcal{K}_0$ is an evolution generator, we note that (11.2.10) and definitions (11.2.7), (11.2.8) yield the representation

$$\begin{aligned} (\mathcal{T}_0(\tau)f)(t) &= \mathrm{e}^{-\tau A_0}\chi_{[0,T]}(t-\tau)f(t-\tau) \\ &= U_0(t,t-\tau)\chi_{[0,T]}(t-\tau)f(t-\tau) \end{aligned}\,, \quad t \in [0,T], \tag{11.2.13}$$

for $\tau \geq 0$ and $f \in \mathfrak{X}_T^p$. Given that $\{U_0(t,s)\}_{(t,s)\in\Delta}$ (11.2.8) is the solution operator of ACP (1.1.23), the fundamental one-to-one correspondence (2.2.3), Definition 2.2.2 (HEN formula) and representation (11.2.13) yield that semigroup $\{\mathcal{T}_0(\tau)\}_{\tau\geq 0}$ is an evolution semigroup. On the other hand, by Theorem 2.2.11, Corollary 2.2.12, and taking into account the one-to-one correspondence (2.2.12), we infer that generator $\mathcal{K}_0$ of semigroup $\{\mathcal{T}_0(\tau)\}_{\tau\geq 0}$ is the evolution generator, cf. also Theorem 2.2.10.

(iii) If $A_0 \in \mathcal{G}(M_0, \beta_0)$, then $\mathcal{A}_0 \in \mathcal{G}(M_0, \beta_0)$ by (11.2.6). On account of (11.2.7), (11.2.10) and $D_0 \in \mathcal{G}(1, 0)$ we obtain for $\tau \geq 0$ estimates:

$$\begin{aligned}&\|\mathcal{T}_0(\tau)\|_{\mathcal{L}(\mathfrak{X}^p_T)} \leq \|e^{-\tau D_0}\|_{\mathcal{L}(\mathfrak{X}^p_T)} \, \|e^{-\tau \mathcal{A}_0}\|_{\mathcal{L}(\mathfrak{X}^p_T)} \leq \\ &\|e^{-\tau A_0}\|_{\mathcal{L}(\mathfrak{X})} \leq M_0 \, e^{\beta_0 \tau} \leq M_0 \, e^{\beta_0 T} =: M_{0,T} \,, \quad \tau \geq 0 \,,\end{aligned} \tag{11.2.14}$$

that is, $\mathcal{K}_0 \in \mathcal{G}(M_0, \beta_0)$. Compare with arguments (2.1.57) and (2.1.58) for the case of accretive generator, that is, for $M_0 = 1$, $\beta_0 = 0$.

Conversely, let $\mathcal{K}_0 \in \mathcal{G}(M_0, \beta_0)$. Then using the right-hand side of (11.2.10) and $\|\mathcal{T}_0(\tau)\|_{\mathcal{L}(\mathfrak{X}^p_T)} \leq M_0 \, e^{\beta_0 \tau}$ we deduce that $A_0 \in \mathcal{G}(M_0, \beta_0)$.

(iv) Because by (11.2.7) and (11.2.13), the semigroup $\{e^{-\tau \mathcal{K}_0}\}_{\tau\geq 0}$ is nilpotent, its representation (11.2.10) yields for $f \in \mathfrak{X}^p_T$ and for any $t \in [0, T]$, $\xi \in \mathbb{C}$:

$$((\mathcal{K}_0 + \xi\, \mathcal{I})^{-1} f)(t) = \int_0^T d\tau \, e^{-\xi \tau} \, e^{-\tau A_0} \, \chi_{[0,T]}(t-\tau) f(t-\tau) \,, \tag{11.2.15}$$

cf. Corollary 2.2.12. Then owing to (11.2.14), the resolvent of generator $\mathcal{K}_0$ is bounded on $\mathfrak{X}^p_T$ for all $\xi \in \mathbb{C}$ given that (11.2.15) yields

$$\sup_{\{f\in\mathfrak{X}^p_T :\, \|f\|_{\mathfrak{X}^p_T}=1\}} \|(\mathcal{K}_0 + \xi\, \mathcal{I})^{-1} f\|_{\mathfrak{X}^p_T} \leq T\, M_0 \, e^{(|\xi|+\beta_0)\, T} \,. \tag{11.2.16}$$

Therefore, the operator $\mathcal{K}_0$ is invertible and the resolvent set $\rho(\mathcal{K}_0) = \mathbb{C}$.

□

Remark 11.2.10 In the most cases, the operator $\widetilde{\mathcal{K}}_0$ is not closed and domain of $\mathcal{K}_0$ is larger than dom $(D_0) \cap$ dom $(\mathcal{A}_0)$. However, if operator A_0 is maximal L^p-regular, see Definition 2.2.23, then $\widetilde{\mathcal{K}}_0 = \mathcal{K}_0$.

Below we scrutinise evolution equation (2.1.4) for *parabolic* nACP (which for a *well-posed* nACP is equivalent to (2.1.12), see Proposition 2.1.9), where the family $\{C_0(t)\}_{t\in[0,T]}$ has a special form:

$$C_0(t) := A_0 + B(t), \quad \text{dom}\,(C_0(t)) = \text{dom}\,(A_0) \cap \text{dom}\,(B(t)) \,, \tag{11.2.17}$$

of the *time-dependent* perturbation. Here A_0 is generator of a C_0-semigroup on the Banach space $\mathfrak{X}$ and perturbation $\{B(t)\}_{t\in[0,T]}$ is a family of, in general, unbounded closed, densely defined linear operators in $\mathfrak{X}$. The *parabolic* property of the nACP is elucidated by Lemma 11.2.15, cf. Definition 2.1.11.

So, for the evolution semigroup generator, one has to consider instead of $\mathcal{K}_0 = (D_0 + \mathcal{A}_0)_{ext} = \overline{D_0 + \mathcal{A}_0}$, (11.2.12), a perturbed by $\mathcal{B}$ generator $\mathcal{K} := (D_0 + \mathcal{A}_0 + \mathcal{B})_{ext}$. Here $(-)_{ext}$ denotes a *closed extension* of the operator sums and $\mathcal{B}$ is a multiplication operator in $\mathfrak{X}_T^p$ generated by family $\{B(t)\}_{t\in[0,T]}$ according to definition (11.2.3).

Remark 11.2.11 We scrutinise two *modus operandi* for construction of the evolution semigroup generator $\mathcal{K}$ from operator sum $D_0 + \mathcal{A}_0 + \mathcal{B}$. The first one corresponds to perturbation of generator D_0 by the closed operator $\mathcal{C}_0 := (\mathcal{A}_0 + \mathcal{B})_{ext}$, whereas the second one considers perturbation of $\mathcal{K}_0 = \overline{D_0 + \mathcal{A}_0}$, (11.2.12), by operator $\mathcal{B}$. As a consequence, in the next Sect. 11.3, we present two versions of *splitting* the evolution semigroup generator $\mathcal{K} = (D_0 + \mathcal{A}_0 + \mathcal{B})_{ext}$:

$$\mathcal{K} = (D_0 + \mathcal{C}_0)_{ext} \quad \text{and} \quad \mathcal{K} = (\mathcal{K}_0 + \mathcal{B})_{ext} , \tag{11.2.18}$$

see Theorem 11.3.1, Theorems 11.3.2 and 11.3.5.

To proceed with application of the *evolution semigroup* method to time-dependent nACP of the form (11.2.17), we suppose that following assumptions are valid.

Assumptions 11.2.12 Let $\alpha \in (0, 1)$ and $\mathfrak{X}$ be a (*separable*) Banach space.

(C1) The operator A_0 is generator of a *bounded* holomorphic semigroup on $\mathfrak{X}$, (i.e., it is from the class $\mathcal{G}(M_0, 0) \cap \mathscr{H}(\theta, 0)$ for some $M_0 \geq 1$ and $\theta \in (0, \pi/2]$), which is such that $0 \in \varrho(A_0)$.

(C2) The operators $\{B(t)\}_{t\in[0,T]}$ are densely defined and closed for $t \in [0, T]$. In addition $\mathrm{dom}\,(A_0) \subset \mathrm{dom}\,(B(t))$ for a.a. $t \in [0, T]$ and for any $x \in \mathrm{dom}\,(A_0) \subset \mathfrak{X}$ the function $t \mapsto B(t)x$ is *strongly* measurable.

(C3) $\mathrm{dom}\,(A_0^\alpha) \subset \mathrm{dom}\,(B(t))$ holds for a.a. $t \in [0, T]$ and

$$C_\alpha := \text{ess sup}_{t\in[0,T]} \|B(t)A_0^{-\alpha}\|_{\mathcal{L}(\mathfrak{X})} < \infty\,, \quad \alpha \in (0, 1)\,,$$

in the operator norm on $\mathcal{L}(\mathfrak{X})$.

(C4) Let $\{B(t)\}_{t\in[0,T]}$ be a family of generators in $\mathfrak{X}$ such that for all $t \in [0, T]$ they belong to the class $\mathcal{G}(M_B, \beta_B)$ and let function $[0, T] \ni t \mapsto (B(t) + \xi\mathbb{1})^{-1}x \in \mathfrak{X}$ be strongly measurable for any $x \in \mathfrak{X}$ and any $\xi > \beta_B$.

(C5) We assume that $\mathrm{dom}\,(A_0^*) \subset \mathrm{dom}\,(B(t)^*)$ and

$$C_1^* := \operatorname*{ess\,sup}_{t\in[0,T]} \|B(t)^*(A_0^*)^{-1}\|_{\mathcal{L}(\mathfrak{X}^*)} < \infty\,.$$

Here A_0^* and $B(t)^*$ denote operators adjoint to, respectively, A_0 and $B(t)$.

(C6) There exists $\beta \in (\alpha, 1)$ and a constant $L_\beta > 0$ such that for a.a. $t, s \in [0, T]$ the estimate

$$\|A_0^{-1}(B(t) - B(s))\, A_0^{-\alpha}\|_{\mathcal{L}(\mathfrak{X})} \leq L_\beta\, |t - s|^\beta\,,$$

is satisfied.

(C7) We assume that $\mathrm{dom}\,((A_0^\alpha)^*) \subset \mathrm{dom}\,(B(t)^*)$ and

$$C_\alpha^* := \operatorname*{ess\,sup}_{t\in[0,T]} \|B(t)^*(A_0^{-\alpha})^*\|_{\mathcal{L}(\mathfrak{X}^*)} < \infty\,.$$

Here $(A_0^\alpha)^*$ and $B(t)^*$ denote operators adjoint to, respectively, A_0^α and $B(t)$.

Note that $((A_0^\alpha)^*)^{-1} = (A_0^{-\alpha})^*$ and that (C7) implies (C5) since (C7) is stronger than (C5).

Remark 11.2.13 Similar to Chap. 10, our ultimate aim in Chap. 11 is the proof of the *operator-norm* product approximations in Banach space for evolution semigroup and solution operator.

In contrast to the strong operator approximations (Sect. 11.1), there is more similarity in problematic between autonomous and non-autonomous cases in Banach space for the operator-norm product approximations. For that reason certain of Assumptions 11.2.12 are motivated by analysis of the Trotter product formula operator-norm convergence in Banach space, see Sect. 10.2.

A key common feature is expressed by conditions (H1), (H2) (Sect. 10.2) and assumptions (C1), (C3) (Assumptions 11.2.12). The (H1) and (C1) guarantee that initial unperturbed semigroup is holomorphic, whereas (H2) and (C3) that perturbation is *infinitesimally-small*, that is, so "gentle" that a relative bound of perturbation is equal to $+0$. (Note that a similar restricted result is presented in Chap. 12 for time-dependent product formulæ in a Hilbert space.) Similar to (H1), (H2), the assumptions (C1), (C3) together with *smoothness* of the perturbation family condition (C6) are essential in definition of the *rate* of the operator-norm convergence of time-dependent product formulae, which is defined by exponents α and β.

A specific for this section condition (C4) serves to ensure existence of the evolution semigroup *generators* in the extended Banach space $\mathfrak{X}_T^p := L^p([0, T], \mathfrak{X})$. The assumptions (C5) and (C7) provide technical tools in dual space $\mathfrak{X}^*$ for estimate the operator-norm *rate* of convergence to *solution operator* of the product approximants generated by the Trotter product formula for *evolution* semigroup.

To decode the assumption (C2), we note that by Definition 11.2.1 multiplication operator $\mathcal{B}$ induced by the family $\{B(t)\}_{t\in[0,T]}$ in the Banach space $\mathfrak{X}_T^p$ is such that for each $f \in \mathrm{dom}\,(\mathcal{B})$ the function:

$$[0, T] \ni t \mapsto B(t)f(t) \in L^p([0, T], \mathfrak{X}) \quad f \in \mathrm{dom}\,(\mathcal{B})\,,$$

is strongly measurable. This measurability implements the construction of multiplication operators in $\mathfrak{X}_T^p$, for example, definition of domain of operator (11.2.5), as well as, a restriction, or extension, of multiplication operators, see Lemmata 11.2.16, 11.2.18 and Theorem 11.3.1.

Finally, about the assumption (C2), we have the following elucidating assertion.

Theorem 11.2.14 *Let $\{B(t)\}_{t\in[0,T]}$ be a family operators in Banach space $\mathfrak{X}$. If the assumptions* (C1), (C3) *and* (C4) *are satisfied, then* (C2) *holds.*

Proof From (C3) it follows that $\operatorname{dom}(A_0) \subset \operatorname{dom}(A_0^\alpha) \subset \operatorname{dom}(B(t))$ for a.a. $t \in [0, T]$. Let $x \in \mathfrak{X}$ and $f_n(t)\,x := \xi\, B(t)(B(t)/n + \xi\, \mathbb{1})^{-1}\, x$, for $n \in \mathbb{N}$, $\xi > \beta_B$, $t \in [0, T]$. Then, on account of

$$f_n(t)\,x = n\,\xi\,x - n^2\,\xi^2 (B(t) + n\,\xi\,\mathbb{1})^{-1}\,x\,,$$

the assumption (C4) implies that function $t \mapsto f_n(t)\,x$ is measurable for each $n \in \mathbb{N}$ and $x \in \mathfrak{X}$.

If $x \in \operatorname{dom}(A_0)$, then by (C3) and (C4) we obtain representation:

$$f_n(t)\,x = \xi\,(B(t)/n + \xi\,\mathbb{1})^{-1}\,B(t)\,x\,, \quad t \in [0, T]\,.$$

Hence, $\lim_{n\to\infty} f_n(t)\,x = B(t)\,x$ for a.a. $t \in [0, T]$. As a consequence the function $t \mapsto B(t)\,x$ is (strongly) measurable for any $x \in \operatorname{dom}(A_0)$. □

Lemma 11.2.15 *Let assumptions* (C1), (C2) *and* (C3) *be satisfied. Then, operator family $\{C_0(t)\}_{t\in[0,T]}$, which is defined by*

$$C_0(t)\,x = A_0\,x + B(t)\,x, \quad x \in \operatorname{dom}(C_0(t))\,, \tag{11.2.19}$$

has domain $\operatorname{dom}(A_0) = \operatorname{dom}(C_0(t))$ *and it is a family of generators of holomorphic semigroups on $\mathfrak{X}$. Moreover, for a.a. $t \in [0, T]$, the operator $C_0(t) \in \mathcal{G}(M, \beta) \cap \mathscr{H}(\theta, \beta)$ for some $M \geq 1$, $\beta \geq 0$.*

Proof The assumptions (C1), (C2), (C3) imply that for a.a. $t \in [0, T]$ the operator $B(t)$ is *infinitesimally* small with respect to A_0. Indeed, fix $t \in [0, T]$. If $\eta > 0$, then by Proposition 10.1.10 and (10.1.13) we obtain for $\alpha \in (0, 1)$

$$\operatorname{dom}(A_0 + \eta\mathbb{1}) = \operatorname{dom}(A_0) \subset \operatorname{dom}(A_0^\alpha) \subset \operatorname{dom}(B(t)),$$

and also the existence of a constant $C := C(M_0, \alpha) > 0$ such that for $t \in [0, T]$

$$\|B(t)(A_0 + \eta\mathbb{1})^{-1}\|_{\mathcal{L}(\mathfrak{X})} \leq \|B(t)A_0^{-\alpha}\|_{\mathcal{L}(\mathfrak{X})} \cdot \|A_0^\alpha (A_0 + \eta\mathbb{1})^{-1}\|_{\mathcal{L}(\mathfrak{X})} \leq \frac{C}{\eta^{1-\alpha}}\,.$$

Therefore, we get for any $x \in \mathrm{dom}\,(A_0) \subset \mathrm{dom}\,(B(t))$ estimate

$$\|B(t)x\| \leq \frac{C}{\eta^{1-\alpha}} \cdot \|(A_0 + \eta\mathbb{1})x\| \leq \frac{C}{\eta^{1-\alpha}}\|A_0 x\| + C\eta^{\alpha}\|x\| . \tag{11.2.20}$$

Since for sufficiently large η the relative bound $C/\eta^{1-\alpha}$ in (11.2.20) can be arbitrarily small and $A_0 \in \mathcal{G}(M_0, 0) \cap \mathscr{H}(\theta, 0)$, Theorem 1.9.9 yields that operator $C_0(t) = A_0 + B(t)$ with $\mathrm{dom}\,(C_0(t)) = \mathrm{dom}\,(A_0)$ is generator of a quasi-bounded holomorphic semigroup for a.a. $t \in [0, T]$.

For holomorphic semigroup with generator A_0 and perturbation with infinitesimally small relative bound (11.2.20), we infer that operator $C_0(t) \in \mathcal{G}(M, \beta) \cap \mathscr{H}(\theta, \beta)$. Here uniform in $t \in [0, T]$ estimates for parameters: $M \geq 1$, $\beta \geq 0$, follow from inequalities (1.9.21), (1.9.22) (see Theorem 1.9.9), whereas the value of M_0 is due to a suitable choice of a large η, cf. Theorem 11.3.1. Note that the semi-angle θ rests intact. □

The next two lemmata evoke how domains of induced multiplication operators in $\mathfrak{X}^p_T$, cf. (11.2.3), can be described by domains of the corresponding operators in the space $\mathfrak{X}$ under two types of conditions.

Lemma 11.2.16 *Let assumptions* (C1) *and* (C2) *be satisfied. Then operators*: $\mathcal{A}_0$ *on* $\mathrm{dom}\,(\mathcal{A}_0)$ *and* $\mathcal{B}$ *on* $\mathrm{dom}\,(\mathcal{B})$, *are closed. If*

$$\operatorname*{ess\,sup}_{t\in[0,T]} \|B(t)x\|_{\mathfrak{X}} \leq C_0\|A_0 x\|_{\mathfrak{X}} < \infty , \quad \text{for } x \in \mathrm{dom}\,(A_0) , \tag{11.2.21}$$

is valid, then $\mathrm{dom}\,(\mathcal{A}_0) \subset \mathrm{dom}\,(\mathcal{B})$.

Proof The closeness of induced in $\mathfrak{X}^p_T$ multiplication operators: $\mathcal{A}_0 \in \mathcal{C}(\mathfrak{X}^p_T)$, $\mathcal{B} \in \mathcal{C}(\mathfrak{X}^p_T)$, is inheritance of this property stems from the corresponding operator families in the space $\mathfrak{X}$, see (C1), (C2) and Proposition 11.2.2.

Now let $f \in \mathrm{dom}\,(\mathcal{A}_0)$. Then, by definition of $\mathrm{dom}\,(\mathcal{A}_0)$, one gets $f(t) \in \mathrm{dom}\,(A_0)$ for a.a. $t \in [0, T]$ and by (C2): $f(t) \in \mathrm{dom}\,(B(t))$ for a.a. $t \in [0, T]$. Note that (C1) and (11.2.21) yield

$$\operatorname*{ess\,sup}_{t\in[0,T]} \|B(t)A_0^{-1}\|_{\mathcal{L}(\mathfrak{X})} \leq C_0 .$$

Hence, for $f \in \mathrm{dom}\,(\mathcal{A}_0)$, we obtain the estimate

$$\|B(t)f(t)\|_{\mathfrak{X}} = \|B(t)A_0^{-1}A_0 f(t)\|_{\mathfrak{X}} \leq C_0\|A_0 f(t)\|_{\mathfrak{X}} , \quad t \in [0, T] ,$$

which implies that function $t \mapsto B(t)f(t)$ belongs to the space $\mathfrak{X}^p_T = L^p([0, T], \mathfrak{X})$. Thus, $f \in \mathrm{dom}\,(\mathcal{B})$, that is, $\mathrm{dom}\,(\mathcal{A}_0) \subset \mathrm{dom}\,(\mathcal{B})$. □

Corollary 11.2.17 *On account of assumption* (C1) *and Corollary* 11.2.6 *the operator* $\mathcal{A}_0$ *is densely defined. Seeing that* $\mathrm{dom}\,(\mathcal{A}_0) \subset \mathrm{dom}\,(\mathcal{B})$, *the operator* $\mathcal{B}$ *is also densely defined.*

Lemma 11.2.18 *Let assumptions* (C1), (C2) *and* (C3) *be satisfied. Then induced operators*: $\mathcal{A}_0$ *on* $\mathrm{dom}\,(\mathcal{A}_0)$ *and* $\mathcal{B}$ *on* $\mathrm{dom}\,(\mathcal{B})$, *are closed, densely defined and* $\mathrm{dom}\,(\mathcal{A}_0) \subset \mathrm{dom}\,(\mathcal{B})$.

Proof Owing to Lemma 11.2.16 and Corollary 11.2.17, it rests only to show that there is a constant $C_0 > 0$ such that for any function $t \mapsto f(t) \in \mathrm{dom}\,(A_0)$ on $[0, T]$:

$$\|B(t)f(t)\|_{\mathfrak{X}} \leq C_0 \|A_0 f(t)\|_{\mathfrak{X}} \quad \text{for a.a.} \quad t \in [0, T]\,. \tag{11.2.22}$$

To this aim, we note that (C1), (C2) and (C3) yield estimate:

$$\begin{aligned}\|B(t)f(t)\|_{\mathfrak{X}} &= \|B(t)A_0^{-\alpha} A_0^{-(1-\alpha)} A_0 f(t)\|_{\mathfrak{X}} \\ &\leq \|B(t)A_0^{-\alpha}\|_{\mathcal{L}(\mathfrak{X})}\, \|A_0^{-(1-\alpha)}\|_{\mathcal{L}(\mathfrak{X})}\, \|A_0 f(t)\|_{\mathfrak{X}}\,,\end{aligned}$$

for a.a. $t \in [0, T]$. Using (C3) and setting $C_0 := C_\alpha \|A_0^{-(1-\alpha)}\|_{\mathcal{L}(\mathfrak{X})}$ we prove (11.2.22).

Let $f \in \mathrm{dom}\,(\mathcal{A}_0)$. Then, by definition of $\mathrm{dom}\,(\mathcal{A}_0)$ of multiplication operator in $\mathfrak{X}_T^p$ one gets $f(t) \in \mathrm{dom}\,(A_0)$ for a.a. $t \in [0, T]$ and by (C2) $f(t) \in \mathrm{dom}\,(B(t))$ for a.a. $t \in [0, T]$. On account of (11.2.22) and (C2), we obtain that the map $t \mapsto B(t)f(t) \in \mathfrak{X}$ is strongly measurable and belongs to $L^p([0, T], \mathfrak{X})$. Therefore, $f \in \mathrm{dom}\,(\mathcal{B})$, that is, $\mathrm{dom}\,(\mathcal{A}_0) \subset \mathrm{dom}\,(\mathcal{B})$. □

The next statement translates the *fractional* condition in (C3) into relation for corresponding multiplication operators in $\mathfrak{X}_T^p$.

Lemma 11.2.19 *Let the assumptions* (C1), (C2) *and* (C3) *be satisfied. Then estimate*: $\|\mathcal{B}\mathcal{A}_0^{-\alpha}\|_{\mathcal{L}(\mathfrak{X}_T^p)} \leq C_\alpha$, *holds for* $\alpha \in (0, 1)$.

Proof Let $f \in \mathfrak{X}_T^p$. Then for multiplication operators $\mathcal{B}$ and $\mathcal{A}_0$ one gets by (C3)

$$(\mathcal{B}\,\mathcal{A}_0^{-\alpha}\, f)(t) = B(t)\, A_0^{-\alpha}\, f(t), \quad \alpha \in (0, 1), \quad \text{for a.a. } t \in [0, T]\,,$$

that is, $\mathrm{dom}\,(\mathcal{A}_0^{\alpha}) \subset \mathrm{dom}\,(\mathcal{B})$. Now again owing to (C3), we obtain for any $f \in \mathfrak{X}_T^p$

$$\|(\mathcal{B}\mathcal{A}_0^{-\alpha} f)(t)\|_{\mathfrak{X}} = \|B(t)A_0^{-\alpha} f(t)\|_{\mathfrak{X}} \leq C_\alpha \|f(t)\|_{\mathfrak{X}}\,, \quad \text{for a.a. } t \in [0, T],$$

that by definition (2.2.1) yields $\|\mathcal{B}\mathcal{A}_0^{-\alpha} f\|_{\mathfrak{X}_T^p} \leq C_\alpha \|f\|_{\mathfrak{X}_T^p}$, and, consequently, the operator-norm estimate $\|\mathcal{B}\mathcal{A}_0^{-\alpha}\|_{\mathcal{L}(\mathfrak{X}_T^p)} \leq C_\alpha$ for $\alpha \in (0, 1)$. □

11.3 Evolution Semigroups for Time-Dependent Perturbations

Here we study the evolution equation (2.1.4) (for a well-posed nACP, it is equivalent to (2.1.12)) where family of operators $\{C_0(t) = A_0+B(t)\}_{t\in[0,T]}$ given by (11.2.19), cf. Chap. 2. So, the corresponding (*parabolic*) nACP takes the form

$$\begin{aligned} \partial_t u(t) &= -(A_0 + B(t))u(t), \quad t \in [s, T], \\ u(s) &= x, \quad s \in (0, T). \end{aligned} \tag{11.3.1}$$

With nACP (11.3.1), we associate in the space $\mathfrak{X}^p_T = L^p([0, T], \mathfrak{X})$ the evolution *pre-generator* (cf. Definition 2.2.8 and Remark 11.2.11, the first *splitting* in (11.2.18))

$$\widetilde{\mathcal{K}} := D_0 + \mathcal{C}_0\,, \quad \mathrm{dom}\,(\widetilde{\mathcal{K}}) = \mathrm{dom}\,(D_0) \cap \mathrm{dom}\,(\mathcal{C}_0)\,, \tag{11.3.2}$$

where $\mathcal{C}_0$ is multiplication operator defined by the family $\{C_0(t) = A_0 + B(t)\}_{t\in[0,T]}$, as it is affirmed in Definition 11.2.1.

As we note in Remark 11.2.11, according to Definition 11.2.1, there are *two ways* for construction of the multiplication operator $\mathcal{C}_0$ in the Banach space $\mathfrak{X}^p_T$:

(I) The first one is to apply Definition 11.2.1 directly to the family $\{C_0(t)\}_{t\in[0,T]}$, which is defined by (11.2.17).
(II) The second one consists of construction the multiplication operators $\mathcal{A}_0$ and $\mathcal{B}$ and then a closed extension $(\mathcal{A}_0 + \mathcal{B})_{ext}$, cf. (11.2.18).

Theorem 11.3.1 *If assumptions* (C1), (C2) *and* (C3) *are satisfied, then closed multiplication operator* $\mathcal{C}_0$ *induced by the family* $\{C_0(t) := A_0 + B(t)\}_{t\in[0,T]}$ *is generator* $\mathcal{C}_0 \in \{\mathcal{G}(M, \beta)$ and $\mathscr{H}(\theta, \beta)\}$ *of a* C_0*-semigroup on the space* $\mathfrak{X}^p_T$.

Proof By virtue of Lemma 11.2.15 operator $C_0(t) = A_0+B(t)$ with $\mathrm{dom}\,(C_0(t)) = \mathrm{dom}\,(A_0)$ is generator of a quasi-bounded holomorphic semigroup for a.a. $t \in [0, T]$. That is, $C_0(t) \in \{\mathcal{G}(M, \beta)$ and $\mathscr{H}(\theta, \beta)\}$ for some $M \geq 1$, $\beta \geq 0$ and $\theta \in (0, \pi/2)$. Consequently, according to Definition 11.2.1 and Proposition 11.2.2, the family $\{C_0(t)\}_{t\in[0,T]}$ defines a closed multiplication operator $\mathcal{C}_0$ on $\mathrm{dom}\,(\mathcal{C}_0) = \mathrm{dom}\,(\mathcal{A}_0)$.

To check that $\mathcal{C}_0$ is densely defined generator of a C_0-semigroup on the space $\mathfrak{X}^p_T$ we have to verify that for each $x \in \mathfrak{X}$ the function

$$t \mapsto (C_0(t) + \xi\,\mathbb{1})^{-1}x \in \mathfrak{X}\,, \quad t \in [0, T]\,, \tag{11.3.3}$$

is strongly measurable if $\xi > \beta$, see Proposition 11.2.3 and Proposition 11.2.4.

To this aim, we use representation

$$(C_0(t) + \xi\,\mathbb{1})^{-1} = (A_0 + \xi\,\mathbb{1})^{-1}(\mathbb{1} + B(t)/(A_0 + \xi\,\mathbb{1}))^{-1}\,, \quad \xi > \xi_0\,, \tag{11.3.4}$$

which is justified by condition (C1) and by estimate (11.2.20)

$$\|B(t)/(A_0 + \xi\, \mathbb{1})\|_{\mathcal{L}(\mathfrak{X})} \leq \frac{C}{\xi^{1-\alpha}} < 1\,, \quad \text{for } \xi > \xi_0\,, \tag{11.3.5}$$

where $C/\xi_0^{1-\alpha} = 1$. Note that by assumption (C2) the function

$$t \mapsto (\mathbb{1} + B(t)/(A_0 + \xi\, \mathbb{1}))\, x\,, \;\; t \in [0, T]\,, \quad \text{for } \xi > 0\,, \tag{11.3.6}$$

is strongly measurable for any $x \in \mathfrak{X}$. Given that by (11.3.5) operators in (11.3.6) are invertible, the function

$$t \mapsto (\mathbb{1} + B(t)/(A_0 + \xi\, \mathbb{1}))^{-1}\, x\,, \quad t \in [0, T] \text{ for } \;\; \xi > \xi_0\,, \tag{11.3.7}$$

is also strongly measurable for any $x \in \mathfrak{X}$. As a consequence, by (11.3.4) and (11.3.7) the function (11.3.3) is strongly measurable for $\xi > \xi_0$ and each $x \in \mathfrak{X}$.

Therefore, on account of Proposition 11.2.3 and Proposition 11.2.4, the corresponding closed multiplication operator $\mathcal{C}_0$ is generator of a quasi-bounded holomorphic C_0-semgroup on $\mathfrak{X}^p_T$. That is, $\mathcal{C}_0 \in \{\mathcal{G}(M, \beta)$ and $\mathscr{H}(\theta, \beta)\}$, for $\beta = \xi_0$. □

This yields construction of multiplication operator $\mathcal{C}_0$ along the first line of reasoning (I). Now we examine the second way (II).

Theorem 11.3.2 *If assumptions* (C1), (C2) *and* (C3) *are satisfied, then closed densely defined multiplication operator $\mathcal{C}_0$ induced by the family $\{C_0(t) := A_0 + B(t)\}_{t\in[0,T]}$, see Theorem* 11.3.1, *has representation*

$$\mathcal{C}_0 = \mathcal{A}_0 + \mathcal{B}\,, \quad \operatorname{dom}(\mathcal{C}_0) = \operatorname{dom}(\mathcal{A}_0)\,. \tag{11.3.8}$$

Here $\mathcal{A}_0$ and $\mathcal{B}$ are multiplication operators in $\mathfrak{X}^p_T$ generated respectively by family $\{A_0\}_{t\in[0,T]}$ and family $\{B(t)\}_{t\in[0,T]}$ according to Definition 11.2.1.

Proof By virtue of assumption (C1) and by Corollary 11.2.6, the closed and densely defined multiplication operator $\mathcal{A}_0 \in \mathscr{H}(\theta, 0)$ is generator of a bounded holomorphic semigroup on $\mathfrak{X}^p_T$.

On account of (C2) and Proposition 11.2.2, the multiplication operator $\mathcal{B}$ is closed. Moreover, owing to Lemma 11.2.18 operator $\mathcal{B}$ is also densely defined since $\operatorname{dom}(\mathcal{A}_0) \subset \operatorname{dom}(\mathcal{B})$. The latter allows to introduce densely defined operator

$$\widetilde{\mathcal{C}_0} := \mathcal{A}_0 + \mathcal{B}\,, \quad \operatorname{dom}(\widetilde{\mathcal{C}_0}) = \operatorname{dom}(\mathcal{A}_0)\,. \tag{11.3.9}$$

Note that assumption (C3) and Lemma 11.2.15 provide the relative smallness (11.2.20) of any $B(t)$, $t \in [0, T]$, with respect to A_0. Then for induced in $\mathfrak{X}^p_T$ multiplication operators $\mathcal{B}$ and $\mathcal{A}_0$ by Lemmata 11.2.18 and 11.2.19, this smallness yields (for appropriate η) inequality

$$\|\mathcal{B} f\|_{\mathfrak{X}_T^p} \leq \varepsilon \,\|\mathcal{A}_0 f\|_{\mathfrak{X}_T^p} + \delta \,\|f\|_{\mathfrak{X}_T^p} , \quad f \in \mathrm{dom}\,(\mathcal{A}_0),\ 0 < \varepsilon < 1\,,\ \delta \geq 0\,. \tag{11.3.10}$$

As a result, (11.3.10) gives for the sum of operators in (11.3.9) estimates

$$\begin{aligned}(1-\varepsilon)\,\|\mathcal{A}_0 f\|_{\mathfrak{X}_T^p} - \delta\,\|f\|_{\mathfrak{X}_T^p} &\leq \|(\mathcal{A}_0 + \mathcal{B})\, f\|_{\mathfrak{X}_T^p} \leq \\ \leq (1+\varepsilon)\,\|\mathcal{A}_0 f\|_{\mathfrak{X}_T^p} + \delta\,\|f\|_{\mathfrak{X}_T^p} , &\quad f \in \mathrm{dom}\,(\mathcal{A}_0)\,.\end{aligned} \tag{11.3.11}$$

Since operator $\mathcal{A}_0$ is closed, the estimates (11.3.11) ensure the *closeness* of operator (11.3.9). For that reason, the operator $\widetilde{\mathcal{C}}_0$ is closed:

$$\overline{\widetilde{\mathcal{C}}_0} = \widetilde{\mathcal{C}}_0\,, \quad \mathrm{dom}\,(\widetilde{\mathcal{C}}_0) = \mathrm{dom}\,(\mathcal{A}_0)\,. \tag{11.3.12}$$

Given that $\mathcal{A}_0 \in \mathscr{H}(\theta, 0)$, by *infinitesimally* small (for large η) relative bound (11.3.10) and perturbation theory for holomorphic semigroups (see Chap. 2, Theorem 1.9.9 for $b = 0$) we obtain that $(\mathcal{A}_0 + \mathcal{B}) \in \mathcal{G}(M, \beta') \cap \mathscr{H}(\theta, \beta')$, for $\beta' \geq 0$.

By construction of the closed multiplication operator (11.3.12), we obtain

$$((\mathcal{A}_0 + \mathcal{B}) f)(t) = A_0\, f(t) + B(t) f(t)\,, \quad \text{for } f \in \mathrm{dom}\,(\mathcal{A}_0)\,,\ t \in [0, T]\,, \tag{11.3.13}$$

where by assumption (C2) the function

$$t \mapsto A_0\, f(t) + B(t) f(t)\,, \quad f \in \mathrm{dom}\,(\mathcal{A}_0)\,,\ t \in [0, T]\,, \tag{11.3.14}$$

is strongly measurable and belongs to $L^p([0, T], \mathfrak{X}) = \mathfrak{X}_T^p$. Seeing that closed multiplication operator $\mathcal{C}_0$ is constructed in Theorem 11.3.1 from the same family $\{C_0(t) = A_0 + B(t)\}_{t\in[0,T]}$, that is:

$$(\mathcal{C}_0 f)(t) = A_0\, f(t) + B(t) f(t)\,, \quad \text{for } f \in \mathrm{dom}\,(\mathcal{A}_0)\,,\ t \in [0, T]\,, \tag{11.3.15}$$

by assumption (C2), the function (11.3.15) possesses the same property as (11.3.14). Hence, Definition 11.2.1 yields identity (11.3.8) and consequently, $\beta' = \beta$. □

Corollary 11.3.3 *We note that due to estimates* (11.3.11) *the graph-norms* $\|\cdot\|_{\mathcal{A}_0}$ *and* $\|\cdot\|_{\mathcal{C}_0}$ *on* $\mathfrak{X}_T^p$ *are equivalent. Therefore, if set* $\mathcal{D} = \mathrm{core}\,(\mathcal{A}_0)$, *then* $\mathcal{D}$ *is also a core for the operator* $\mathcal{C}_0$ *and conversely.*

To proceed with analysis of splitting (11.2.18), we use the following useful fact.

Remark 11.3.4 Note that by virtue of (C1) and Theorem 10.1.12 for holomorphic semigroup $\{U_{A_0}(t)\}_{t\geq 0}$ generated by A_0, one has inequality

$$\|A_0^\alpha\, U_{A_0}(t)\|_{\mathcal{L}(\mathfrak{X})} \leq \frac{M_{\alpha,A_0}}{t^\alpha}\,, \quad t>0\,, \quad \alpha>0\,. \tag{11.3.16}$$

Definition of multiplication operator $\mathcal{A}_0$ (11.2.5) and the corresponding semigroup (11.2.6) yield

$$\begin{aligned}&\|\mathcal{A}_0^\alpha\, \mathrm{e}^{-\tau\mathcal{A}_0}\, f\|^p_{\mathfrak{X}^p_T} = \int_0^T dt\, \|A_0^\alpha\, \mathrm{e}^{-\tau\, A_0}\, f(t)\|^p_{\mathfrak{X}} \leq \\ &(M_{\alpha,A_0}/\tau^\alpha)^p \int_0^T dt\, \|f(t)\|^p_{\mathfrak{X}} = (M_{\alpha,A_0}/\tau^\alpha)^p\, \|f\|^p_{\mathfrak{X}^p_T}\,, \quad \tau>0\,,\ f\in\mathfrak{X}^p_T\,,\end{aligned} \tag{11.3.17}$$

cf. definition (2.2.1). Then, (11.3.17) uplifts the operator-norm estimate (11.3.16) in the space $\mathfrak{X}$ up to the operator-norm estimate

$$\|\mathcal{A}_0^\alpha\, \mathrm{e}^{-\tau\mathcal{A}_0}\|_{\mathcal{L}(\mathfrak{X}^p_T)} \leq \frac{M_{\alpha,A_0}}{\tau^\alpha}\,, \quad \tau>0\,, \quad \alpha>0\,, \tag{11.3.18}$$

in the Banach space $\mathfrak{X}^p_T$.

Recall that operator $\widetilde{\mathcal{K}}_0 = D_0 + \mathcal{A}_0$ is densely defined on $\mathrm{dom}\,(\widetilde{\mathcal{K}}_0) = \mathrm{dom}\,(D_0)\cap\mathrm{dom}\,(\mathcal{A}_0)$ and that its closure $\mathcal{K}_0$ is generator of (nilpotent) semigroup, see Lemma 11.2.9. Then owing to definition (11.3.2) and Theorem 11.3.2, we obtain

$$\mathrm{dom}\,(\widetilde{\mathcal{K}}) = \mathrm{dom}\,(D_0)\cap\mathrm{dom}\,(\mathcal{C}_0) = \mathrm{dom}\,(D_0)\cap\mathrm{dom}\,(\mathcal{A}_0) = \mathrm{dom}\,(\widetilde{\mathcal{K}}_0)\,.$$

For that reason $\mathrm{dom}\,(\widetilde{\mathcal{K}})$ has dense cross-section in $\mathfrak{X}$. Indeed, the following assertion about perturbation of generator $\mathcal{K}_0 := \overline{\widetilde{\mathcal{K}}_0} = \overline{D_0+\mathcal{A}_0}$ by operator $\mathcal{B}$ is consequent.

Theorem 11.3.5 *Let assumptions* (C1), (C2) *and* (C3) *be satisfied. Then, closure* $\mathcal{K} := \overline{D_0+\mathcal{A}_0+\mathcal{B}} = \overline{D_0+\mathcal{C}_0}$ *is generator of semigroup* $\{\mathrm{e}^{-\tau\mathcal{K}}\}_{\tau\geq 0}$ *on* $\mathfrak{X}^p_T$. *In particular,* $\mathcal{K} = \mathcal{K}_0 + \mathcal{B}$ *and* $\mathrm{dom}\,(\mathcal{K}) = \mathrm{dom}\,(\mathcal{K}_0)$.

Proof Let $\mathfrak{D} := \mathrm{dom}\,(D_0)\cap\mathrm{dom}\,(\mathcal{A}_0)\cap\mathrm{dom}\,(\mathcal{B})$. Since $\mathrm{dom}\,(\mathcal{A}_0)\subset\mathrm{dom}\,(\mathcal{B})$, see Lemma 11.2.18, and given that by Theorem 11.3.2, one has $\mathrm{dom}\,(\mathcal{C}_0) = \mathrm{dom}\,(\mathcal{A}_0)$, we figure out that by definition of operators: $\widetilde{\mathcal{K}}_0$, in Remark 11.2.8 and $\widetilde{\mathcal{K}}$, in (11.3.2), one gets

$$\begin{aligned}\mathfrak{D} &= \mathrm{dom}\,(D_0)\cap\mathrm{dom}\,(\mathcal{A}_0) = \mathrm{dom}\,(\widetilde{\mathcal{K}}_0) = \\ &= \mathrm{dom}\,(D_0)\cap\mathrm{dom}\,(\mathcal{C}_0) = \mathrm{dom}\,(\widetilde{\mathcal{K}})\,.\end{aligned} \tag{11.3.19}$$

On that account, the restriction:

$$(D_0 + \mathcal{C}_0) \restriction \mathfrak{D} = D_0 + \mathcal{A}_0 + \mathcal{B} = \widetilde{\mathcal{K}}_0 + \mathcal{B} = (\mathcal{K}_0 + \mathcal{B}) \restriction \mathfrak{D}\,. \tag{11.3.20}$$

Seeing that by Lemma 11.2.9 operator $\widetilde{\mathcal{K}}_0$ is densely defined in $\mathfrak{X}_T^p$, domain $\mathfrak{D} \subset \mathrm{dom}\,(\mathcal{K}_0)$ is also dense in $\mathfrak{X}_T^p$ because of (11.3.19).

Owing to the fact that operator $\mathcal{A}_0$ is generated in $\mathfrak{X}_T^p$ by the time-independent family (11.2.5), one gets

$$\mathrm{e}^{-\tau \mathcal{A}_0}\,\mathrm{dom}\,(D_0) \subseteq \mathrm{dom}\,(D_0) \quad \text{and} \quad \mathrm{e}^{-\tau D_0}\,\mathrm{dom}\,(\mathcal{A}_0) \subseteq \mathrm{dom}\,(\mathcal{A}_0)\,,$$

for $\tau \geq 0$. Then by commutativity of representation (11.2.7) we obtain:

$$\mathcal{T}_0(\tau)\,\mathfrak{D} = \mathrm{e}^{-\tau D_0}\mathrm{e}^{-\tau \mathcal{A}_0}\,\mathfrak{D} \subseteq \mathrm{e}^{-\tau D_0}\,\mathrm{dom}\,(\mathcal{A}_0) \subseteq \mathrm{dom}\,(\mathcal{A}_0)\,,$$

and

$$\mathcal{T}_0(\tau)\,\mathfrak{D} = \mathrm{e}^{-\tau \mathcal{A}_0}\mathrm{e}^{-\tau D_0}\,\mathfrak{D} \subseteq \mathrm{e}^{-\tau \mathcal{A}_0}\,\mathrm{dom}\,(D_0) \subseteq \mathrm{dom}\,(D_0)\,.$$

As a consequence, $\mathcal{T}_0(\tau)\,\mathfrak{D} \subseteq \mathrm{dom}\,(D_0) \cap \mathrm{dom}\,(\mathcal{A}_0) = \mathfrak{D}$ for $\tau \geq 0$. For that reason by (11.3.19)

$$\mathfrak{D} = \mathrm{core}\,(\mathcal{K}_0) = \mathrm{dom}\,(\widetilde{\mathcal{K}}_0) = \mathrm{dom}\,(\widetilde{\mathcal{K}})\,, \tag{11.3.21}$$

compare with arguments (11.2.10), (11.2.11).

To prove that for any $f \in \mathfrak{D}$, the map $\tau \mapsto \mathcal{B}\,\mathcal{T}_0(\tau)\,f \in \mathfrak{X}_T^p$ is continuous, we note that $\mathfrak{D} \subseteq \mathrm{dom}\,(\mathcal{A}_0) \subseteq \mathrm{dom}\,(\mathcal{A}_0^\alpha)$ implies the representation

$$\mathcal{B}\,\mathcal{T}_0(\tau)\,f = \mathcal{B}\mathcal{A}_0^{-\alpha}\,\mathcal{T}_0(\tau)\,\mathcal{A}_0^\alpha\,f, \quad \tau \geq 0, \quad f \in \mathfrak{D}\,.$$

Then by (C3) and Lemma 11.2.19 the operator $\mathcal{B}\,\mathcal{A}_0^{-\alpha}$ is bounded: $\|\mathcal{B}\mathcal{A}_0^{-\alpha}\|_{\mathcal{L}(\mathfrak{X}_T^p)} \leq C_\alpha$ for $\alpha \in (0,1)$. Consequently, the mapping $\tau \mapsto \mathcal{B}\,\mathcal{T}_0(\tau)\,f \in \mathfrak{X}_T^p$ is continuous for $\tau \geq 0$ and any $f \in \mathfrak{D}$. This observation, together with inclusions $\mathfrak{D} \subset \mathrm{dom}\,(\mathcal{K}_0) \cap \mathrm{dom}\,(\mathcal{B})$, see (11.3.20), and $\mathcal{T}_0(\tau)\,\mathfrak{D} \subseteq \mathfrak{D}$, proves *assumption* (i) of Proposition 1.9.27.

To check the assumption (ii) of Proposition 1.9.27, we use representation

$$\mathcal{B}\,\mathcal{T}_0(\tau)\,f = \mathcal{B}\,\mathcal{A}_0^{-\alpha}\,\mathcal{A}_0^\alpha\,\mathrm{e}^{-\tau \mathcal{A}_0}\mathrm{e}^{-\tau D_0}\,f\,. \quad f \in \mathfrak{D},$$

Because the shift semigroup $\{\mathrm{e}^{-\tau D_0}\}_{\tau \geq 0}$ is contraction, we obtain that

$$\|\mathcal{B}\,\mathcal{T}_0(\tau) f\|_{\mathfrak{X}_T^p} \leq \|\mathcal{B}\,\mathcal{A}_0^{-\alpha}\|_{\mathcal{L}(\mathfrak{X}_T^p)} \|\mathcal{A}_0^\alpha\,\mathrm{e}^{-\tau \mathcal{A}_0}\|_{\mathcal{L}(\mathfrak{X}_T^p)} \|f\|_{\mathfrak{X}_T^p}.$$

Therefore, on account of (C3), Lemma 11.2.19 and estimate (11.3.18)

$$\|\mathcal{B}\mathcal{T}_0(\tau)f\| \leq \frac{C_\alpha M_{\alpha,\mathcal{A}_0}}{\tau^\alpha}\|f\|, \quad \tau > 0, \quad \alpha \in (0,1), \quad f \in \mathfrak{D}.$$

As a consequence, for $a > 0$, $\tau > 0$ and any $f \in \mathfrak{D}$ we obtain estimate:

$$\int_0^a d\tau\, \|\mathcal{B}e^{-\tau\mathcal{K}_0}f\| \leq C_\alpha M_{\alpha,\mathcal{A}_0}\|f\| \int_0^a d\tau\, \frac{1}{\tau^\alpha} = \frac{C_\alpha M_{\alpha,\mathcal{A}_0}}{(1-\alpha)}\, a^{1-\alpha}\|f\|. \tag{11.3.22}$$

If we choose

$$a < [(1-\alpha)/C_\alpha M_{\alpha,\mathcal{A}_0}]^{1/(1-\alpha)},$$

then *assumption* (ii) of Proposition 1.9.27 is also satisfied, cf. (1.9.56) and (11.3.22).

Consequently, by Proposition 1.9.27 (c), there is a unique semigroup with generator, which is the closure $\overline{(\mathcal{K}_0 + \mathcal{B}) \restriction \mathfrak{D}}$ with domain dom $(\mathcal{K}_0)$. Thus, on account of (11.3.19), (11.3.20) and definition of operators: $\widetilde{\mathcal{K}} = D_0 + \mathcal{C}_0$, $\mathcal{K} = \overline{D_0 + \mathcal{A}_0 + \mathcal{B}}$, one gets:

$$\begin{aligned} \overline{(\mathcal{K}_0 + \mathcal{B}) \restriction \mathfrak{D}} &= \overline{D_0 + \mathcal{A}_0 + \mathcal{B}} = \mathcal{K}, \\ \overline{(D_0 + \mathcal{C}_0) \restriction \mathfrak{D}} &= \overline{D_0 + \mathcal{C}_0} = \overline{\widetilde{\mathcal{K}}} = \mathcal{K}. \end{aligned} \tag{11.3.23}$$

Note that operator $\mathcal{B}$ is closed, whereas operator $\widetilde{\mathcal{K}}_0 = D_0 + \mathcal{A}_0$ is closable and such that dom $(\widetilde{\mathcal{K}}_0) \subset$ dom $(\mathcal{B})$, see (C2). Then by (11.3.20) and (11.3.23) the closure $\overline{\widetilde{\mathcal{K}}_0 + \mathcal{B}} = \mathcal{K}_0 + \mathcal{B}$. So, operator $\mathcal{K} = \mathcal{K}_0 + \mathcal{B}$, on domain dom $(\mathcal{K})$ = dom $(\mathcal{K}_0)$, is generator of the *evolution* semigroup $\{e^{-\tau\mathcal{K}}\}_{\tau\geq 0}$. □

Note that arguments in Theorem 11.3.5 are independent of parameter T in $[0, T]$. As a consequence, this theorem defines in the space $\mathfrak{X}^p_{T=\infty}$ a quasi-bounded C_0-semigroup with generator $\mathcal{K} \in \mathcal{G}(M_\mathcal{K}, \beta_\mathcal{K})$. Then, Proposition 1.9.27, formulæ (1.9.58), and Lemma 11.2.9 (iii) provide the values of $M_\mathcal{K}$ and $\beta_\mathcal{K}$ for generator $\mathcal{K} \in \mathcal{G}(M_\mathcal{K}, \beta_\mathcal{K})$.

Remark 11.3.6 By Remark 2.2.13, (2.2.20), the right-shift semigroup $\{e^{-\tau\, D_0}\}_{\tau\geq 0}$ on the Banach space $\mathfrak{X}^p_T$ is nilpotent. That is, $e^{-\tau\, D_0} = 0$ for $t > T$. Then, by definition (11.2.7), the semigroup $\{\mathcal{T}_0(\tau) = e^{-\tau\,\mathcal{K}_0}\}_{\tau\geq 0}$ on $\mathfrak{X}^p_T$ inherits this property. As a consequence, the semigroup $\{e^{-\tau\mathcal{K}}\}_{\tau\geq 0}$ on $\mathfrak{X}^p_T$ constructed in Theorem 11.3.5 by perturbation of $\{e^{-\tau\mathcal{K}_0}\}_{\tau\geq 0}$ is also nilpotent. On that account $e^{-\tau\,\mathcal{K}} = 0$ for $t > T$, and thus *bounded*, although by Theorem 11.3.5 it is quasi-bounded C_0-semigroup with $\mathcal{K} \in \mathcal{G}(M_\mathcal{K}, \beta_\mathcal{K})$ on the space $\mathfrak{X}^p_{T=\infty}$.

To elucidate this point, we remind that for nACP (11.3.1), there exists a one-to-one correspondence between the set of all evolution semigroups $\{e^{-\tau\mathcal{K}}\}_{\tau\geq 0}$ on $\mathfrak{X}^p_T$ and the corresponding set of *quasi-bounded* solution operators $\{U(t,s)\}_{(t,s)\in\Delta}$ on

$\mathfrak{X}$ due to the *Howland-Evans-Neidhardt* formula (see Theorem 2.2.11):

$$(e^{-\tau\mathcal{K}} f)(t) = U(t, t-\tau)\chi_{[0,T]}(t-\tau) f(t-\tau), \quad f \in \mathfrak{X}_T^p, \text{ a.a. } t \in [0, T]. \tag{11.3.24}$$

Here $\|U(t,s)\| \le M\,e^{\beta(t-s)}$, $\beta \ge 0$, see Remark 2.1.14, (2.1.21) and (2.2.12). As a consequence, for generator $\mathcal{K} \in \mathscr{G}(M_{\mathcal{K}}, \beta_{\mathcal{K}})$ one gets $M_{\mathcal{K}} \le M$ and $\beta_{\mathcal{K}} \le \beta$. But besides that, formula (11.3.24) yields estimate

$$\|e^{-\tau\mathcal{K}} f\|_{\mathfrak{X}_T^p} \le M\,e^{\beta\,\tau}\|f\|_{\mathfrak{X}_T^p}, \quad f \in \mathfrak{X}_T^p.$$

This means that nilpotent on the Banach space $\mathfrak{X}_T^p$ evolution semigroup $\{e^{-\tau\mathcal{K}}\}_{\tau\ge 0}$ is *bounded*:

$$\|e^{-\tau\mathcal{K}}\|_{\mathcal{L}(\mathfrak{X}_T^p)} \le M_T, \quad \tau \ge 0, \tag{11.3.25}$$

where $M_T := M\,e^{\beta\,T}$, cf. (2.2.13).

11.4 Strong and Integrated Convergence of Product Approximants

We evoke that the aim of Chap. 11 is analysis of time-dependent product formulæ on Banach spaces. For non-Autonomous Cauchy Problem (11.1.1), it is the question of construction and proof of convergence the *product approximants* to solution operator $\{U(t,s)\}_{(t,s)\in\Delta}$, see Definition 2.1.5 in Chap. 2. To this end, we advanced in Sect. 11.1 a *direct* method to prove the convergence of approximants in the *strong* operator topology (11.1.13) on the space $\mathcal{L}(\mathfrak{X})$.

Now we advocate in this section the *evolution semigroup* method developed in Chap. 2 and in preceding sections of this chapter. The *first* step in application of this method is the proof of the *strong* operator convergence of the Trotter product formula (Proposition 1.9.41) to *evolution* semigroup $\{e^{-\tau\mathcal{K}}\}_{\tau\ge 0}$ on the Banach space $\mathfrak{X}_T^p$ when generator $\mathcal{K}$ is splitted into a *Trotter-stable* (Definition 1.9.37) pair of generators in approximants. This allows to proceed with the *second* step, that is, to establish a convergence to solution operator $\{U(t,s)\}_{(t,s)\in\Delta}$ of the corresponding *product approximants* $\{U_n(t,s)\}_{n\ge 1}$ in the *integral* topology (11.4.3), which we shall define in this section below.

Note that convergence of the product approximants $\{U_n(t,s)\}_{n\ge 1}$ in the *integral* topology is *weaker* than convergence in the *strong* operator topology that we established in Sect. 11.1. However, in Sect. 11.5, we shall improve topologies for both of these steps. We establish the convergence of $\{U_n(t,s)\}_{n\ge 1}$ in the *operator-norm* topology on $\mathcal{L}(\mathfrak{X})$. The key point for this improvement will be the proof of assertion about *uplifting* (strengthening) the convergence of the Trotter product formula for the *evolution* semigroup $\{e^{-\tau\mathcal{K}}\}_{\tau\geq 0}$ from the *strong* to the *operator-norm* topology on $\mathcal{L}(\mathfrak{X}^p_T)$.

Following Remark 11.2.11, we consider two kinds of the *product approximants* for evolution semigroup $\{e^{-\tau\mathcal{K}}\}_{\tau\geq 0}$, which correspond to two ways in splittig the generator $\mathcal{K}$:

$$\mathcal{K} = \overline{\mathcal{D}_0 + \mathcal{C}_0} \quad \text{and} \quad \mathcal{K} = \mathcal{K}_0 + \mathcal{B}\,,$$

see (11.2.18) and Theorem 11.3.5.

As in preceding Sect. 11.3, we consider evolution equation (11.3.1) for nACP, where the family $\{C_0(t)\}_{t\in[0,T]}$ has a special form of *time-dependent* perturbation:

$$C_0(t) := A_0 + B(t), \quad \operatorname{dom}(C_0(t)) = \operatorname{dom}(A_0) \cap \operatorname{dom}(B(t))\,. \tag{11.4.1}$$

For assumptions (C1)–(C3) the Theorem 11.3.2 ensures that family $\{C_0(t)\}_{t\in[0,T]}$ defines in $\mathfrak{X}^p_T$ a multiplication operator $\mathcal{C}_0$, which is generator of a quasi-bounded holomorphic semigroup. On that account, we consider first the splitting $\mathcal{K} = \overline{\mathcal{D}_0 + \mathcal{C}_0}$, cf. Theorem 11.3.5.

Theorem 11.4.1 *Let assumptions* (C1), (C2) *and* (C3) *be satisfied. Let, in addition, the stability condition*

$$M' := \sup_{n\in\mathbb{N}} \operatorname*{ess\,sup}_{(t,s)\in\Delta} \left\| \prod_{k=n-1}^{0} e^{-\frac{t-s}{n} C_0\left(t+k\frac{t-s}{n}\right)} \right\|_{\mathcal{L}(\mathfrak{X})} < \infty\,, \tag{11.4.2}$$

be satisfied. Then, approximation of solution operator $\{U(t,s)\}_{(t,s)\in\Delta}$ *in the* integral topology:

$$\lim_{n\to\infty} \sup_{\tau\in[0,T]} \int_0^{T-\tau} ds\, \|(U_n(s+\tau,s) - U(s+\tau,s))\,x\|^p_{\mathfrak{X}} = 0\,, \quad x \in \mathfrak{X}\,, \tag{11.4.3}$$

holds for product approximants :

$$U_n(t,s) := \prod_{k=n-1}^{0} e^{-\frac{(t-s)}{n} C_0\left(s+k\frac{t-s}{n}\right)}, \quad (t,s) \in \Delta, \quad n \in \mathbb{N}. \tag{11.4.4}$$

Proof First, we prove an appropriate version of the Trotter product formula (Chap. 2, Proposition 1.9.41) for evolution semigroup $\{e^{-\tau\mathcal{K}}\}_{\tau\geq 0}$ on the Banach space $\mathfrak{X}^p_T$.

Note that for assumptions (C1)–(C3), the Theorem 11.3.5 ensures splitting of the evolution semigroup generator $\mathcal{K} = \overline{\mathcal{D}_0 + \mathcal{C}_0}$ for the pair of generators $\{\mathcal{D}_0, \mathcal{C}_0\}$. To check for this pair, the Trotter stability condition (1.9.113):

$$\|(e^{-\tau D_0/n}e^{-\tau C_0/n})^n\|_{\mathcal{L}(\mathfrak{X}^p_T)} \leq M e^{\beta\tau}, \quad M \geq 1, \beta \in \mathbb{R}, \quad n \in \mathbb{N}, \tau \in \mathbb{R}^+_0, \tag{11.4.5}$$

we refer to Lemma 2.2.28 that claims equivalence of stability conditions (11.4.2) and (11.4.5). Given that all conditions of Proposition 1.9.41 are satisfied, we obtain the Trotter product formula

$$\lim_{n\to\infty} \sup_{\tau\in[0,T]} \|(e^{-\tau D_0/n}e^{-\tau C_0/n})^n f - e^{-\tau\mathcal{K}} f\|_{\mathfrak{X}^p_T} = 0, \quad f \in \mathfrak{X}^p_T, \tag{11.4.6}$$

in the strong operator topology on $\mathcal{L}(\mathfrak{X}^p_T)$.

Now, for proving (11.4.3), we note that by (11.4.4), together with definition of multiplication and shift operators in $\mathfrak{X}^p_T$, one gets representation:

$$\left(\left(e^{-\tau D_0/n}e^{-\tau C_0/n}\right)^n f\right)(t) = U_n(t, t-\tau)\, \chi_{[0,T]}(t-\tau)\, f(t-\tau), \quad n \in \mathbb{N}, \tag{11.4.7}$$

for $\tau \in \mathbb{R}^+_0$ and a.a. $t \in [0, T]$, cf. (2.1.60), (2.1.61). Then, (11.4.7) and the HEN *representation* (11.3.24) for evolution semigroup $\{e^{-\tau\mathcal{K}}\}_{\tau\geq 0}$ yield for $f \in \mathfrak{X}^p_T$:

$$\left\|(e^{-\tau D_0/n}e^{-\tau C_0/n})^n f - e^{-\tau\mathcal{K}} f\right\|^p_{\mathfrak{X}^p_T} =$$

$$= \int_0^{T-\tau} ds\, \left\|(U_n(s+\tau, s) - U(s+\tau, s)) f(s)\right\|^p_{\mathfrak{X}}.$$

As a consequence, the Trotter product formula (11.4.6) on vector $f(t) = (\mathbb{1}_T \otimes x)(t)$, $x \in \mathfrak{X}$, from the subspace $\mathfrak{X}_{T\otimes} \subset \mathfrak{X}^p_T$ of *constant* functions (see Remark 2.2.3) yields formula (11.4.3) for product approximants of solution operator. □

Remark 11.4.2

(a) For approximants (11.4.4) formula (11.4.3) does *not* require that operators $\{B(t)\}_{[0,T]}$ are generators. This requirement concerns $\{C_0(t)\}_{[0,T]}$, (11.4.1).
(b) The assumption $0 \in \varrho(A_0)$ exists just for simplicity. Otherwise, the generator A_0 can be shifted by a constant $\eta > 0$. The domain of the fractional power of A_0 does not change in this case, see Lemma 10.1.11.

Next, we examine the convergence of product approximants in the *integral topology* (11.4.3) when operators $\{B(t)\}_{[0,T]}$ are assumed to be generators, that is, when the assumption (C4) holds. Note that owing to Theorem 11.2.14, the assertions in Sects. 11.2 and 11.3 are valid if in Assumptions 11.2.12, one retains only conditions (C1), (C3) and (C4).

We also note that on account of (C4) and Proposition 11.2.3 the multiplication operator $\mathcal{B}$, induced in $\mathfrak{X}_T^p$ by the family $\{B(t)\}_{t\in[0,T]}$, is generator of a quasi-bounded semigroup, $\mathcal{B} \in \mathcal{G}(M_B, \beta_B)$. Owing to (C1) and Lemma 11.2.9 (iii), the operator $\mathcal{K}_0 \in \mathcal{G}(M_0, 0)$ is also generator of a quasi-bounded semigroup. Then, pair of generators $\{\mathcal{B}, \mathcal{K}_0\}$ is *Trotter-stable* if they satisfy (cf. Definition 1.9.37) the condition:

$$\sup_{n\in\mathbb{N}} \sup_{\tau\in\mathbb{R}_0^+} \|(\mathrm{e}^{-\tau\mathcal{B}/n}\mathrm{e}^{-\tau\mathcal{K}_0/n})^n\|_{\mathcal{L}(\mathfrak{X}_T^p)} < \infty \,. \tag{11.4.8}$$

Hence, by Proposition 1.9.41 and Theorem 11.3.5, the Trotter formula for product approximants $\{(\mathrm{e}^{-\tau\mathcal{B}/n}\mathrm{e}^{-\tau\mathcal{K}_0/n})^n\}_{n\geq 1}$ converges on the space $\mathfrak{X}_T^p$:

$$\mathrm{e}^{-\tau\mathcal{K}} = \text{s-}\lim_{n\to\infty} (\mathrm{e}^{-\tau\mathcal{B}/n}\mathrm{e}^{-\tau\mathcal{K}_0/n})^n \,,$$

strongly and uniformly in τ on compact intervals $[a, b] \subset \mathbb{R}_0^+$ to quasi-bounded semigroup with generator $\mathcal{K} = \overline{(\mathcal{K}_0 + \mathcal{B})}$.

Lemma 11.4.3 *Let assumptions* (C1), (C3) *and* (C4) *be satisfied. Then, the following assertions hold:*

(i) *The pair $\{\mathcal{B}, \mathcal{K}_0\}$ is Trotter-stable if and only if the family $\{B(t)\}_{t\in[0,T]}$ is right A_0-stable, that is,*

$$M_{A_0}^G := \sup_{n\in\mathbb{N}} \operatorname*{ess\,sup}_{(t,s)\in\Delta} \left\| \prod_{k=n}^{1} G_n^{A_0}(t, s; k) \right\|_{\mathcal{L}(\mathfrak{X})} < \infty \,, \tag{11.4.9}$$

where

$$G_n^{A_0}(t, s; k) := \mathrm{e}^{-\frac{t-s}{n} B(s+k\frac{t-s}{n})}\mathrm{e}^{-\frac{t-s}{n} A_0}, \quad (t, s) \in \Delta, \quad n \in \mathbb{N}, \tag{11.4.10}$$

for $k = 0, 1, 2, \ldots, n$.

(ii) *The pair $\{\mathcal{K}_0, \mathcal{B}\}$ is Trotter-stable if and only if the family $\{B(t)\}_{t\in[0,T]}$ is left A_0-stable, that is,*

$$M_{A_0}^F := \sup_{n\in\mathbb{N}} \operatorname*{ess\,sup}_{(t,s)\in\Delta} \left\| \prod_{k=n-1}^{0} F_n^{A_0}(t, s; k) \right\|_{\mathcal{L}(\mathfrak{X})} < \infty \,, \tag{11.4.11}$$

where

$$F_n^{A_0}(t,s;k) := \mathrm{e}^{-\frac{t-s}{n}A_0}\mathrm{e}^{-\frac{t-s}{n}B(s+k\frac{t-s}{n})}, \quad (t,s)\in\Delta, \quad n\in\mathbb{N}, \tag{11.4.12}$$

for $k = 0, 1, 2, \ldots, n$.

(iii) *The family* $\{B(t)\}_{t\in[0,T]}$ *is right* A_0*-stable if and only if it is left* A_0*-stable.*

Proof

(i) First, we note that by definitions of induced in $\mathfrak{X}_T^p$ multiplication operators, the semigroup $\{\mathcal{T}_0(\tau) := \mathrm{e}^{-\tau\mathcal{K}_0}\}_{\tau\geq 0}$, (11.2.7), and the right-shift semigroup $\{\mathcal{S}_R(\tau)\}_{\tau\geq 0}$, see (2.2.20), one gets

$$\begin{aligned}(\mathrm{e}^{-\tau/n\,\mathcal{B}}\mathrm{e}^{-\tau/n\,\mathcal{K}_0}f)(t) &= (\mathrm{e}^{-\tau/n\,\mathcal{B}}\mathrm{e}^{-\tau/n\,\mathcal{A}_0}\mathcal{S}_R(\tau/n)f)(t)\\ &= \mathrm{e}^{-\tau/n\,B(t)}\mathrm{e}^{-\tau/n\,A_0}\chi_{[0,T]}(t-\tau/n)f(t-\tau/n), \quad f\in\mathfrak{X}_T^p,\end{aligned} \tag{11.4.13}$$

The n-times iteration of (11.4.13) yields for $f \in \mathfrak{X}_T^p$ and for a.a. $t \in [0, T]$:

$$((\mathrm{e}^{-\tau/n\,\mathcal{B}}\mathrm{e}^{-\tau/n\,\mathcal{K}_0})^n f)(t) = \prod_{j=0}^{n-1}\mathrm{e}^{-\tau/n\,B(t-j\,\tau/n)}\,\mathrm{e}^{-\tau/n\,A_0}$$

$$\chi_{[0,T]}(t-\tau)\,f(t-\tau)\,. \tag{11.4.14}$$

Let $\{\mathcal{S}_L(\tau)\}_{\tau\geq 0}$ be the left-shift (contraction) semigroup on $\mathfrak{X}_T^p$ defined by (2.2.39). Then, on account of (11.4.14)

$$\begin{aligned}&(\mathcal{S}_L(\tau)(\mathrm{e}^{-\tau/n\,\mathcal{B}}\mathrm{e}^{-\tau/n\,\mathcal{K}_0})^n f)(t)\\ &= \prod_{j=0}^{n-1}\mathrm{e}^{-\tau/n\,B(t+(n-j)\,\tau/n)}\,\mathrm{e}^{-\tau/n\,A_0}\,\chi_{[0,T]}(t+\tau)\,f(t)\,,\end{aligned} \tag{11.4.15}$$

for a.a. $t \in [0, T]$ and $f \in \mathfrak{X}_T^p$. Since (for any $\tau \geq 0$) $\mathcal{S}_L(\tau)\left(\mathrm{e}^{-\tau\mathcal{B}/n}\mathrm{e}^{-\tau\mathcal{K}_0/n}\right)^n$ is a bounded multiplication operator on $\mathfrak{X}_T^p$, we obtain (cf. Proposition 2.2.16 in Chap. 2, as well as Sect. E.3.3)

$$\begin{aligned}&\left\|\mathcal{S}_L(\tau)\left(\mathrm{e}^{-\tau\mathcal{B}/n}\mathrm{e}^{-\tau\mathcal{K}_0/n}\right)^n\right\|_{\mathcal{L}(\mathfrak{X}_T^p)}\\ &\qquad= \operatorname*{ess\,sup}_{t\in[0,T-\tau]}\left\|\prod_{k=n}^{1}\left(\mathrm{e}^{-\frac{\tau}{n}B(t+k\,\frac{\tau}{n})}\,\mathrm{e}^{-\frac{\tau}{n}A_0}\right)\right\|_{\mathcal{L}(\mathfrak{X})},\end{aligned}$$

which yields for $n \in \mathbb{N}$

$$\begin{aligned}
&\operatorname*{ess\,sup}_{\tau\in[0,T]} \left\| \mathcal{S}_L(\tau)\left(\mathrm{e}^{-\tau\mathcal{B}/n}\mathrm{e}^{-\tau\mathcal{K}_0/n}\right)^n \right\|_{\mathcal{L}(\mathfrak{X}^p_T)} = \\
&\operatorname*{ess\,sup}_{\tau\in[0,T]}\ \operatorname*{ess\,sup}_{t\in[0,T-\tau]} \left\| \prod_{k=n-1}^{0} \left(\mathrm{e}^{-\frac{\tau}{n}B(t+k\frac{\tau}{n})}\mathrm{e}^{-\frac{\tau}{n}A_0}\right)\right\|_{\mathcal{L}(\mathfrak{X})} = \\
&\operatorname*{ess\,sup}_{(t,s)\in\Delta} \left\| \prod_{k=n-1}^{0} \left(\mathrm{e}^{-\frac{t-s}{n}B(t+k\frac{t-s}{n})}\mathrm{e}^{-\frac{t-s}{n}A_0}\right)\right\|_{\mathcal{L}(\mathfrak{X})} = \\
&\operatorname*{ess\,sup}_{(t,s)\in\Delta} \left\| \prod_{k=n-1}^{0} G_n^{A_0}(t,s;k)\right\|_{\mathcal{L}(\mathfrak{X})} .
\end{aligned} \tag{11.4.16}$$

Now, if the pair $\{\mathcal{B}, \mathcal{K}_0\}$ is Trotter-stable, then by $\|\mathcal{S}_L(\tau)\|_{\mathcal{L}(\mathfrak{X}^p_T)} \leq 1$ and (11.4.8)

$$\sup_{n\in\mathbb{N}}\ \operatorname*{ess\,sup}_{\tau\in[0,T]} \left\| \mathcal{S}_L(\tau)\left(\mathrm{e}^{-\tau\mathcal{B}/n}\,\mathrm{e}^{-\tau\mathcal{K}_0/n}\right)^n \right\|_{\mathcal{L}(\mathfrak{X}^p_T)} < \infty .$$

As a consequence, owing to (11.4.16) the condition (11.4.9) is satisfied, which means that the family $\{B(t)\}_{t\in[0,T]}$ is right A_0-stable.

Conversely, if the family $\{B(t)\}_{t\in[0,T]}$ is right A_0-stable, then by (11.4.9) and (11.4.16)

$$\sup_{n\in\mathbb{N}}\ \operatorname*{ess\,sup}_{\tau\in[0,T]} \left\| \mathcal{S}_L(\tau)\left(\mathrm{e}^{-\tau\mathcal{B}/n}\mathrm{e}^{-\tau\mathcal{K}_0/n}\right)^n \right\|_{\mathcal{L}(\mathfrak{X}^p_T)} < \infty .$$

Since semigroups $\{\mathcal{S}_L(\tau)\}_{\tau\geq 0}$, $\{\mathrm{e}^{-\tau\mathcal{K}_0}\}_{\tau\geq 0}$ and $\{\mathrm{e}^{-\tau\mathcal{B}}\}_{\tau\geq 0}$ are C_0-continuous, we also have inequality

$$\sup_{n\in\mathbb{N}}\ \sup_{\tau\in[0,T]} \left\| \mathcal{S}_L(\tau)\left(\mathrm{e}^{-\tau\mathcal{B}/n}\mathrm{e}^{-\tau\mathcal{K}_0/n}\right)^n \right\|_{\mathcal{L}(\mathfrak{X}^p_T)} < \infty .$$

Seeing that the right-shift semigroup $\{\mathcal{S}_R(\tau)\}_{\tau\geq 0}$ is a contraction (2.2.20), this inequality yields

$$\sup_{n\in\mathbb{N}}\ \sup_{\tau\in[0,T]} \left\| \mathcal{S}_R(\tau)\mathcal{S}_L(\tau)\left(\mathrm{e}^{-\tau\mathcal{B}/n}\mathrm{e}^{-\tau\mathcal{K}_0/n}\right)^n \right\|_{\mathcal{L}(\mathfrak{X}^p_T))} < \infty . \tag{11.4.17}$$

Let $h_\tau(t) := (\mathcal{S}_L(\tau)(\mathrm{e}^{-\tau/n\,\mathcal{B}}\mathrm{e}^{-\tau/n\,\mathcal{K}_0})^n f)(t)$. Then on account of (11.4.15), definition $\mathcal{S}_R(\tau)$, see (2.2.20), and (11.4.14) we obtain

$$\begin{aligned}(\mathcal{S}_R(\tau)h_\tau)(t) = \chi_{[0,T]}(t-\tau)\prod_{j=0}^{n-1} \mathrm{e}^{-\tau/n\,B(t-\tau+(n-j)\,\tau/n)}\,\mathrm{e}^{-\tau/n\,A_0} \\ \chi_{[0,T]}(t)\,f(t-\tau) \\ = ((\mathrm{e}^{-\tau/n\,\mathcal{B}}\mathrm{e}^{-\tau/n\,\mathcal{K}_0})^n f)(t)\,, \quad t\in[0,T]\,, \quad f\in\mathfrak{X}^p_T)\,.\end{aligned} \tag{11.4.18}$$

As a consequence, (11.4.18) yields for $n \in \mathbb{N}$

$$\mathcal{S}_R(\tau)\mathcal{S}_L(\tau)\left(\mathrm{e}^{-\tau\mathcal{B}/n}\mathrm{e}^{-\tau\mathcal{K}_0/n}\right)^n = \left(\mathrm{e}^{-\tau\mathcal{B}/n}\mathrm{e}^{-\tau\mathcal{K}_0/n}\right)^n, \quad \tau\geq 0\,. \tag{11.4.19}$$

Then (11.4.17) and (11.4.19) provide the Trotter stability (11.4.8).

(ii) The proof of this assertion is *verbatim* of that in (i).

(iii) By Lemma 1.9.38 (Chap. 1) one gets that $\{\mathcal{B},\mathcal{K}_0\}$ is Trotter-stable if and only if $\{\mathcal{K}_0,\mathcal{B}\}$ is Trotter-stable. Hence, the left and the right A_0-stability of $\{B(t)\}_{t\in[0,T]}$ are equivalent.

□

It is worth noting here that $\mathcal{S}_R(\tau)\mathcal{S}_L(\tau) \neq \mathcal{I}$. Indeed, by definitions of $\mathcal{S}_R(\tau)$ (2.2.20) and $\mathcal{S}_L(\tau)$ (2.2.39) we obtain that on the space $\mathfrak{X}^p_T$

$$\begin{aligned}(\mathcal{S}_R(\tau)\mathcal{S}_L(\tau)\,f)(t) = \chi_{[0,T]}(t-\tau)\,f(t) \text{ and } (\mathcal{S}_L(\tau)\mathcal{S}_R(\tau)\,f)(t) \\ = \chi_{[0,T]}(t+\tau)\,f(t)\,,\end{aligned} \tag{11.4.20}$$

for any $f \in \mathfrak{X}^p_T$ and $t \in [0,T]$. We note also that operators in (11.4.20) are *idempotent*: $(\mathcal{S}_R(\tau)\mathcal{S}_L(\tau))^2 = \mathcal{S}_R(\tau)\mathcal{S}_L(\tau)$ and $(\mathcal{S}_L(\tau)\mathcal{S}_R(\tau))^2 = \mathcal{S}_L(\tau)\mathcal{S}_R(\tau)$. For that reason the equality in formula (11.4.19) is due to the product structure of (11.4.14), which is identical to the statement:

$$\mathrm{ran}\,((\mathrm{e}^{-\tau/n\,\mathcal{B}}\mathrm{e}^{-\tau/n\,\mathcal{K}_0})^n) \subseteq \mathrm{ran}\,(\mathcal{S}_R(\tau)\mathcal{S}_L(\tau))\,, \quad \tau\geq 0\,, \quad n\in\mathbb{N}\,.$$

Remark 11.4.4 Since by Lemma 11.4.3 the family $\{B(t)\}_{t\in[0,T]}$ is simultaneously left and right A_0-stable, we say in the following simply that this family is A_0-stable.

Theorem 11.4.5 *Let assumptions* (C1), (C3) *and* (C4) *be satisfied. If the operator family* $\{B(t)\}_{t\in[0,T]}$ *is* A_0*-stable, then one has the Trotter product formulæ on* $\mathfrak{X}^p_T$:

$$e^{-\tau\mathcal{K}} = \text{s-}\lim_{n\to\infty} (e^{-\tau\mathcal{B}/n} e^{-\tau\mathcal{K}_0/n})^n = \text{s-}\lim_{n\to\infty} (e^{-\tau\mathcal{K}_0/n} e^{-\tau\mathcal{B}/n})^n\,, \tag{11.4.21}$$

with uniform convergence in τ *on compact intervals* $[a, b] \subset \mathbb{R}^+_0$.

Proof From Lemma 11.4.3, we get that the pair $\{\mathcal{K}_0, \mathcal{B}\}$ is Trotter-stable. Then, proof follows thanks to Proposition 1.9.41 and Theorem 11.3.5. □

Theorem 11.4.6 *Let assumptions* (C1), (C3) *and* (C4) *be satisfied. If the family* $\{B(t)\}_{t\in[0,T]}$ *is* A_0*-stable, then*

$$\lim_{n\to\infty} \sup_{\tau\in[0,T]} \int_0^{T-\tau} ds\, \|\{U_n^{A_0}(s+\tau, s) - U(s+\tau, s)\}x\|^p_{\mathfrak{X}} = 0, \quad x \in \mathfrak{X}, \tag{11.4.22}$$

and

$$\lim_{n\to\infty} \sup_{\tau\in[0,T]} \int_0^{T-\tau} ds\, \|\{V_n^{A_0}(s+\tau, s) - U(s+\tau, s)\}x\|^p_{\mathfrak{X}} = 0, \quad x \in \mathfrak{X}, \tag{11.4.23}$$

where (see (11.4.10) *and* (11.4.12)*)*

$$U_n^{A_0}(t, s) := \prod_{k=n-1}^{0} G_n^{A_0}(t, s; k), \quad (t, s) \in \Delta, \quad n \in \mathbb{N}, \tag{11.4.24}$$

and

$$V_n^{A_0}(t, s) := \prod_{k=n}^{1} F_n^{A_0}(t, s; k), \quad (t, s) \in \Delta, \quad n \in \mathbb{N}. \tag{11.4.25}$$

Proof Similarly to (11.4.7), we note that by definition of multiplication and shift operators in $\mathfrak{X}^p_T$ and on account of (11.4.10), (11.4.24) the representation

$$((e^{-\tau\mathcal{B}/n}\, e^{-\tau\mathcal{K}_0/n})^n f)(t) = U_n^{A_0}(t, t-\tau)\chi_{[0,T]}(t-\tau) f(t-\tau)\,, \tag{11.4.26}$$

holds for a.a. $t \in [0, T]$ and $f \in \mathfrak{X}^p_T$. Next, due to Theorem 11.4.5 we obtain

$$\lim_{n\to\infty} \sup_{\tau\in[0,T]} \|(e^{-\tau\mathcal{B}/n} e^{-\tau\mathcal{K}_0/n})^n f - e^{-\tau\mathcal{K}} f\|^p_{\mathfrak{X}^p_T} = 0\,. \tag{11.4.27}$$

Therefore, by (11.4.26) and by the HEN *representation* (2.2.12) (Chap. 2) for evolution semigroup $\{e^{-\tau\mathcal{K}}\}_{\tau\geq 0}$ one gets

$$\left\| \left((e^{-\tau\mathcal{B}/n} e^{-\tau\mathcal{K}_0/n})^n - e^{-\tau K} \right) f \right\|^p_{\mathfrak{X}^p_T}$$
$$= \int_0^T dt \, \|\{U_n^{A_0}(t, t-\tau) - U(t, t-\tau)\}\chi_{[0,T]}(t-\tau) f(t-\tau)\|^p_{\mathfrak{X}}$$
$$= \int_0^{T-\tau} ds \, \|\{U_n^{A_0}(s+\tau, s) - U(s+\tau, s)\} f(s)\|^p_{\mathfrak{X}} \, .$$

Then, on the set of vectors $\{f \in \mathfrak{X}_{T\otimes} : f(t) = (\mathbb{1}_T \otimes x)(t), \ x \in \mathfrak{X}, \ t \in [0, T]\}$, from the subspace $\mathfrak{X}_{T\otimes} \subset \mathfrak{X}^p_T$ of *constant* functions (see Remark 2.2.3), the Trotter product formula (11.4.27) gives the limit:

$$\lim_{n\to\infty} \sup_{\tau\in[0,T]} \left\| \left((e^{-\tau\mathcal{B}/n} e^{-\tau\mathcal{K}_0/n})^n - e^{-\tau K} \right) (\mathbb{1}_T \otimes x) \right\|^p_{\mathfrak{X}^p_T}$$
$$= \lim_{n\to\infty} \sup_{\tau\in[0,T]} \int_0^{T-\tau} ds \, \|(U_n^{A_0}(s+\tau, s) - U(s+\tau, s))\, x\|^p_{\mathfrak{X}} = 0 \, ,$$

which proves (11.4.22), that is convergence in the *integral* topology (11.4.3).

The proof of (11.4.23) is similar. To this end, we note that by definition of multiplication and shift operators in $\mathfrak{X}^p_T$ owing to (11.4.12), (11.4.25) the representation

$$((e^{-\tau\mathcal{K}_0/n}\, e^{-\tau\mathcal{B}/n})^n\, f)(t) = V_n^{A_0}(t, t-\tau)\chi_{[0,T]}(t-\tau) f(t-\tau), \qquad (11.4.28)$$

holds for a.a. $t \in [0, T]$ and $f \in \mathfrak{X}^p_T$. On account of Theorem 11.4.5 we obtain

$$\lim_{n\to\infty} \sup_{\tau\in[0,T]} \|(e^{-\tau\mathcal{K}_0/n}\, e^{-\tau\mathcal{B}/n})^n\, f - e^{-\tau\mathcal{K}}\, f\|^p_{\mathfrak{X}^p_T} = 0 \, . \qquad (11.4.29)$$

As a consequence, by (11.4.28) and by the HEN *representation* (2.2.12) for evolution semigroup $\{e^{-\tau\mathcal{K}}\}_{\tau\geq 0}$, it follows that

$$\left\| \left((e^{-\tau\mathcal{K}_0/n}\, e^{-\tau\mathcal{B}/n})^n - e^{-\tau K} \right) f \right\|^p_{\mathfrak{X}^p_T}$$
$$= \int_0^T dt \, \|\{V_n^{A_0}(t, t-\tau) - U(t, t-\tau)\}\chi_{[0,T]}(t-\tau) f(t-\tau)\|^p_{\mathfrak{X}}$$
$$= \int_0^{T-\tau} ds \, \|\{V_n^{A_0}(s+\tau, s) - U(s+\tau, s)\} f(s)\|^p_{\mathfrak{X}} \, .$$

The rest of the proof is *verbatim* of the arguments which are used for approximants (11.4.24). □

11.5 Operator-Norm Convergence of Product Approximants

In the previous section, we proved the convergence of the product *approximants*: $\{U_n(t,s)\}_{n\geq 1}$, (11.4.4), or $\{U_n^{A_0}(t,s)\}_{n\geq 1}$, (11.4.24) and $\{V_n^{A_0}(t,s)\}_{n\geq 1}$, (11.4.25), to solution operator $\{U(t,s)\}_{(t,s)\in\Delta}$, Definition 2.1.5 (Chap. 2) in the *integral* topology (11.4.3).

By inspection and comparison of (11.4.24), (11.4.25) with the strong operator limit (11.1.13), one infers that convergence of the product approximants $\{U_n(t,s)\}_{n\geq 1}$ in the *integral topology* is weaker than convergence in the *strong* operator topology that we established in Sect. 11.1 within a more general setting where domains of generators are *time-dependent* .

In this section, we prove this convergence in the *operator-norm* topology for *time-independent* domains of generators. The uplifting to this stronger topology has not only an abstract interest. Convergence in the operator-norm topology is more useful in applications, especially if the *rate* of convergence can be estimated, which is our ultimate aim.

We estimate the rate of convergence in the operator-norm topology on Banach space $\mathfrak{X}$ for approximants $\{U_n^{A_0}(t,s)\}_{n\geq 1}$:

$$\sup_{(t,s)\in\Delta} \|U_n^{A_0}(t,s) - U(t,s)\|_{\mathcal{L}(\mathfrak{X})} \to 0 \quad \text{as} \quad n\to\infty\,, \tag{11.5.1}$$

Similar results and proofs for approximants $\{U_n(t,s)\}_{n\geq 1}$ and $\{V_n^{A_0}(t,s)\}_{n\geq 1}$ can be reproduced *verbatim et literatim* following the case of $\{U_n^{A_0}(t,s)\}_{n\geq 1}$.

It is worthy to note that the operator-norm convergence in (11.5.1) results from *improvement* to the operator-norm convergence on $\mathfrak{X}_T^p = L^p([0,T],\mathfrak{X})$ (cf. Sect. E.3.3) of the Trotter product formula for the *evolution* semigroup:

$$\lim_{n\to\infty}\sup_{\tau\geq 0}\|(\mathrm{e}^{-\tau\mathcal{B}/n}\mathrm{e}^{-\tau\mathcal{K}_0/n})^n - \mathrm{e}^{-\tau\mathcal{K}}\|_{\mathcal{L}(\mathfrak{X}_T^p)} = 0\,. \quad \text{as} \quad n\to\infty\,. \tag{11.5.2}$$

Moreover, the rate of convergence in (11.5.1) is uniquely determined by the rate of convergence of the Trotter product formula (11.5.2).

11.5.1 Technical Lemmata

Here we present and prove technical lemmata indispensable for demonstration of convergence and for estimate of the error bound for the Trotter product formula approximations (11.5.2) in the operator-norm topology on the set of bounded operators $\mathcal{L}(\mathfrak{X}_T^p)$, see Sect. E.3.2. They are formulated in the framework of Assumptions 11.2.12.

Lemma 11.5.1 *Assume* (C1), (C2) *and* (C7) *for some* $\alpha \in (0, 1]$. *Then, the operator* $\overline{\mathcal{A}_0^{-\alpha}\mathcal{B}}$ *is bounded on* $\mathfrak{X}_T^p$ *and the norm of this closure:* $\left\|\overline{\mathcal{A}_0^{-\alpha}\mathcal{B}}\right\|_{\mathcal{L}(\mathfrak{X}_T^p)} \leq C_\alpha^*$.

Proof Let $x \in \operatorname{dom}(A_0) \subset \operatorname{dom}(B(t))$. Then by (C7) the estimate

$$|(A_0^{-\alpha}B(t)x, \phi)| = |(x, B(t)^*(A_0^{-\alpha})^*\phi)| \leq C_\alpha^* \|x\|_{\mathfrak{X}} \|\phi\|_{\mathfrak{X}^*}, \quad \text{a.a. } t \in [0, T],$$

holds for any $\phi \in \mathfrak{X}^*$. Since by assumption (C1) dom (A_0) is dense in $\mathfrak{X}$, we infer that for the norm of the closure of product one has: $\operatorname{ess\,sup}_{t\in[0,T]} \|\overline{A_0^{-\alpha}B(t)}\|_{\mathcal{L}(\mathfrak{X})} \leq C_\alpha^*$.

Let $\Gamma \in (\mathfrak{X}_T^p)^*$. By Proposition E.2.7 (iii) there exists a w^*-measurable function $\Psi_\Gamma : [0, T] \to \mathfrak{X}^*$, such that

$$\Gamma(\mathcal{A}_0^{-\alpha}\mathcal{B}h) = \int_{[0,T]} dt\, (A_0^{-\alpha}B(t)h(t), \Psi_\Gamma(t)), \quad h \in \operatorname{dom}(\mathcal{B}).$$

Consequently, the estimate $\operatorname{ess\,sup}_{t\in[0,T]} \|\overline{A_0^{-\alpha}B(t)}\|_{\mathcal{L}(\mathfrak{X})} \leq C_\alpha^*$ implies

$$|\Gamma(\mathcal{A}_0^{-\alpha}\mathcal{B}h)| \leq C_\alpha^* \|h\|_{\mathfrak{X}_T^p} \left(\int_{[0,T]} dt\, \|\Psi_\Gamma(t)\|^{p'}\right)^{1/p'}, \quad 1/p + 1/p' = 1,$$

for all $h \in \operatorname{dom}(\mathcal{B})$ and any $\Gamma \in (\mathfrak{X}_T^p)^* = \mathfrak{X}_T^{p'}$, which in turn yields:

$$|\Gamma(\mathcal{A}_0^{-\alpha}\mathcal{B}h)| \leq C_\alpha^* \|h\|_{\mathfrak{X}_T^p} \|\Gamma\|_{(\mathfrak{X}_T^p)^*}.$$

As a consequence, we obtain estimate $\|\mathcal{A}_0^{-\alpha}\mathcal{B}h\|_{\mathfrak{X}_T^p} \leq C_\alpha^* \|h\|_{\mathfrak{X}_T^p}$, $h \in \operatorname{dom}(\mathcal{B})$. Given that by Corollary 11.2.17 domain dom $(\mathcal{B})$ is dense in $\mathfrak{X}_T^p$, this estimate proves the claim. □

Remark 11.5.2

(a) By assumption $(C1)$ the operator $A_0 \in \mathcal{G}(M_0, 0) \cap \mathscr{H}(\theta, 0)$ generates on the space $\mathfrak{X}_T^p$ a holomorphic semigroup $\{e^{-\tau\mathcal{A}_0}\}_{\tau\geq 0}$ with generator $\mathcal{A}_0 \in \mathcal{G}(M_0, 0) \cap \mathscr{H}(\theta, 0)$, see Theorem 2.2.18. On the contrary, the operator $\mathcal{K}_0 = \overline{\widetilde{\mathcal{K}}_0}$, where $\widetilde{\mathcal{K}}_0 = D_0 + \mathcal{A}_0$, is not a generator of holomorphic semigroup. Indeed, by (11.2.7) (Remark 11.2.8)

$$(e^{-\tau\mathcal{K}_0} f)(t) = (e^{-\tau D_0} e^{-\tau\mathcal{A}_0} f)(t) = e^{-\tau A_0} f(t - \tau)\chi_{[0,T]}(t - \tau), \qquad (11.5.3)$$

is zero for $\tau \geq T$, $f \in \mathfrak{X}_T^p$. So, the bounded by M_0 semigroup $\{e^{-\tau\mathcal{K}_0}\}_{\tau\geq 0}$ is *nilpotent* and, consequently, has *no* analytic extension $\tau \in \mathbb{C}$ to the complex plane.

(b) By the very same reason, integrating (11.5.3) over $\tau \geq 0$ we deduce estimate:

$$\sup_{\{f\in\mathfrak{X}_T^p:\ \|f\|_{\mathfrak{X}_T^p}=1\}} \|(\mathcal{K}_0 + \xi\,\mathcal{I})^{-1} f\|_{\mathfrak{X}_T^p} \leq T\, M_0\, e^{|\xi|\,T}\,, \tag{11.5.4}$$

for any $\xi \in \mathbb{C}$, cf. Lemma 11.2.9 (iv). So, the resolvent of $\mathcal{K}_0$ is bounded on $\mathfrak{X}_T^p$ for all $\xi \in \mathbb{C}$, that is, generator $\mathcal{K}_0 \in \mathscr{G}(M_0, 0)$ is invertible and the resolvent set $\rho(\mathcal{K}_0) = \mathbb{C}$.

We also comment that although $\mathrm{dom}\,(\widetilde{\mathcal{K}}_0) \subset \mathrm{dom}\,(\mathcal{A}_0)$, for the closure of $\mathcal{K}_0$ of this operator the inclusion $\mathrm{dom}\,(\mathcal{K}_0) \subseteq \mathrm{dom}\,(\mathcal{A}_0)$, in general, does not hold (see Remarks 11.2.8 and 11.2.10), but one has the following assertion:

Lemma 11.5.3 *Let* (C1) *be satisfied. Then for any* $\alpha \in [0, 1)$ *one gets*

$$\mathrm{dom}\,(\mathcal{K}_0) \subset \mathrm{dom}\,(\mathcal{A}_0^\alpha)\,. \tag{11.5.5}$$

Proof Since semigroup $\{e^{-\tau\mathcal{K}_0}\}_{\tau\geq 0}$ is nilpotent, its generator has empty spectrum, $\sigma(\mathcal{K}_0) = \emptyset$. Therefore, it is invertible and owing to Remark 11.2.8, we obtain:

$$\mathcal{A}_0^\alpha \mathcal{K}_0^{-1} f = \mathcal{A}_0^\alpha \int_0^\infty d\tau\, e^{-\tau\mathcal{K}_0} f = \int_0^T d\tau\, e^{-\tau D_0} \mathcal{A}_0^\alpha e^{-\tau\mathcal{A}_0} f\,, \quad f \in \mathfrak{X}_T^p.$$

By (C1) and Theorem 2.2.18 the semigroup $\{e^{-\tau\mathcal{A}_0}\}_{\tau\geq 0}$ is bounded and holomorphic in sector S_θ. Then from Theorem 10.1.12 we obtain, for $\alpha \in [0, 1)$ and $\tau > 0$, the estimate $\|\mathcal{A}_0^\alpha e^{-\tau\mathcal{A}_0}\|_{\mathcal{L}(\mathfrak{X}_T^p)} \leq M_{\alpha,\mathcal{A}_0}/\tau^\alpha$ (11.3.18). On account of this estimate and of $D_0 \in \mathscr{G}(1, 0)$, the integrand $e^{-\tau D_0} \mathcal{A}_0^\alpha e^{-\tau\mathcal{A}_0} f$ is integrable on $[0, \infty)$ and

$$\|\mathcal{A}_0^\alpha \mathcal{K}_0^{-1}\|_{\mathcal{L}(\mathfrak{X}_T^p)} \leq \frac{T^{1-\alpha}\, M_{\alpha,\mathcal{A}_0}}{1-\alpha}\,,$$

that implies the claim for $\alpha \in [0, 1)$. □

In what follows, we shall use the *Gronwall-type* inequality. It can be proved by iteration of the corresponding *Volterra* integral equation, for details see Appendix G, Lemma G.2.1.

Lemma 11.5.4 (Gronwall-Type Inequality) *Let* $F(\cdot) : \mathbb{R}_0^+ \to \mathbb{R}_0^+$ *be a real continuous function satisfying the integral inequality*

$$F(t) \leq c_1\, t^{-\alpha} + c_2 \int_0^t ds\, (t-s)^{-\alpha}\, F(s)\,, \quad t > s \geq 0\,,$$

for positive coefficients $c_1, c_2 > 0$ *and* $\alpha \in (0, 1)$. *Then there exists a constant* $C := 2\,c_1$ *and* $t_0 := t_0(\alpha, c_2) > 0$ *such that* $F(t) \leq C/t^{\alpha}$ *for* $t \in (0, t_0]$.

With the help of Lemma 11.5.4, one can prove the following assertion.

Lemma 11.5.5 *Let* (C1), (C2) *and* (C3) *be satisfied. Then, there is a positive constant* $\Lambda_\alpha > 0$ *such that operator norm*

$$\|\mathcal{A}_0^{\alpha} e^{-\tau\mathcal{K}}\|_{\mathcal{L}(\mathfrak{X}_T^p)} \leq \frac{\Lambda_\alpha}{\tau^{\alpha}}, \quad \tau > 0. \tag{11.5.6}$$

Proof Note that the following equation holds

$$\begin{aligned}\partial_\sigma e^{-(\tau-\sigma)\mathcal{K}_0} e^{-\sigma\mathcal{K}} f &= e^{-(\tau-\sigma)\mathcal{K}_0}\mathcal{K}_0 e^{-\sigma\mathcal{K}} f + e^{-(\tau-\sigma)\mathcal{K}_0}(-\mathcal{K}) e^{-\sigma\mathcal{K}} f \\ &= -e^{-(\tau-\sigma)\mathcal{K}_0}\mathcal{B} e^{-\sigma\mathcal{K}} f,\end{aligned}$$

for $f \in \mathrm{dom}\,(\mathcal{K} = \mathcal{K}_0 + \mathcal{B}) = \mathrm{dom}\,(\mathcal{K}_0)$ (Theorem 11.3.5). Then we obtain

$$\int_0^{\tau} d\sigma\, e^{-(\tau-\sigma)\mathcal{K}_0}\mathcal{B} e^{-\sigma\mathcal{K}} f = e^{-\tau\mathcal{K}_0} f - e^{-\tau\mathcal{K}} f, \tag{11.5.7}$$

and, since $\mathrm{dom}\,(\mathcal{K}) = \mathrm{dom}\,(\mathcal{K}_0) \subset \mathrm{dom}\,(\mathcal{A}_0^{\alpha})$ (see Lemma 11.5.3), also

$$\mathcal{A}_0^{\alpha} e^{-\tau\mathcal{K}} f = \mathcal{A}_0^{\alpha} e^{-\tau\mathcal{K}_0} f - \mathcal{A}_0^{\alpha}\int_0^{\tau} d\sigma\, e^{-(\tau-\sigma)\mathcal{K}_0}\mathcal{B} e^{-\sigma\mathcal{K}} f. \tag{11.5.8}$$

Now, we estimate two terms on the right-hand side of (11.5.8). First, by commutativity of generators D_0, $\mathcal{A}_0$ (Remark 11.2.8) and by analyticity of semigroup $\{e^{-\tau\mathcal{A}_0}\}_{\tau\geq 0}$ (Theorem 2.2.18), or (11.3.18), we get estimate

$$\|\mathcal{A}_0^{\alpha} e^{-\tau\mathcal{K}_0} f\|_{\mathfrak{X}_T^p} \leq \|\mathcal{A}_0^{\alpha} e^{-\tau\mathcal{A}_0} f\|_{\mathfrak{X}_T^p} \leq \frac{M_{\alpha,\mathcal{A}_0}}{\tau^{\alpha}}\|f\|_{\mathfrak{X}_T^p}, \quad f \in \mathfrak{X}_T^p, \tag{11.5.9}$$

for $\tau > 0$. Recall that $\mathrm{dom}\,(\mathcal{K}) = \mathrm{dom}\,(\mathcal{K}_0) \subset \mathrm{dom}\,(\mathcal{A}_0^{\alpha}) \subset \mathrm{dom}\,(\mathcal{B})$. Then semigroup properties yield for the last term in (11.5.8) where $f \in \mathrm{dom}\,(\mathcal{K})$:

$$\mathcal{A}_0^{\alpha}\int_0^{\tau} d\sigma\, e^{-(\tau-\sigma)\mathcal{K}_0}\mathcal{B} e^{-\sigma\mathcal{K}} f = \int_0^{\tau} d\sigma\, \mathcal{A}_0^{\alpha} e^{-(\tau-\sigma)\mathcal{K}_0}\mathcal{B}\mathcal{A}_0^{-\alpha}\mathcal{A}_0^{\alpha} e^{-\sigma\mathcal{K}} f. \tag{11.5.10}$$

Given that $\|\mathcal{B}\mathcal{A}^{-\alpha_0}\|_{\mathcal{L}(\mathfrak{X}_T^p)} \leq C_\alpha$ (Lemma 11.2.19), by (11.5.9) and (11.5.10) we find the estimate for the norm in the left-hand side of (11.5.8):

$$\begin{aligned}
&\|\mathcal{A}_0^\alpha e^{-\tau\mathcal{K}} f\|_{\mathfrak{X}_T^p} \leq \frac{M_{\alpha,\mathcal{A}_0}}{\tau^\alpha}\|f\|_{\mathfrak{X}_T^p} + \\
&+ \int_0^\tau d\sigma\, \|\mathcal{A}_0^\alpha e^{-(\tau-\sigma)\mathcal{K}_0}\|_{\mathcal{L}(\mathfrak{X}_T^p)} \|\mathcal{B}\mathcal{A}_0^{-\alpha}\|_{\mathcal{L}(\mathfrak{X}_T^p)} \|\mathcal{A}_0^\alpha e^{-\sigma\mathcal{K}} f\|_{\mathfrak{X}_T^p} \\
&\leq \frac{M_{\alpha,\mathcal{A}_0}}{\tau^\alpha}\|f\|_{\mathfrak{X}_T^p} + C_\alpha \int_0^\tau d\sigma\, \frac{M_{\alpha,\mathcal{A}_0}}{(\tau-\sigma)^\alpha}\|\mathcal{A}_0^\alpha e^{-\sigma\mathcal{K}} f\|_{\mathfrak{X}_T^p}, \quad f \in \mathrm{dom}\,(\mathcal{K})\,.
\end{aligned} \tag{11.5.11}$$

Let $\mathfrak{B} := \{f \in \mathrm{dom}\,(\mathcal{K}) : \|f\|_{\mathfrak{X}_T^p} \leq 1\}$. For each $f \in \mathfrak{B}$ we introduce the function $\tau \mapsto F_f(\tau) := \|\mathcal{A}_0^\alpha e^{-\tau\mathcal{K}} f\|_{\mathfrak{X}_T^p}$, $\tau \geq 0$. Then (11.5.11) yields inequality

$$F_f(\tau) \leq c_1 \tau^{-\alpha} + c_2 \int_0^\tau d\sigma\, (\tau-\sigma)^{-\alpha}\, F_f(\sigma)\,,$$

where $c_1 := M_{\alpha,\mathcal{A}_0}$ and $c_2 := C_\alpha M_{\alpha,\mathcal{A}_0}$. By Lemma 11.5.4 a solution of this inequality satisfies the estimate $F_f(\tau)\,\tau^\alpha \leq 2c_1$ for $\tau \in (0, \tau_0]$, where parameter $\tau_0 = \tau_0(\alpha, c_2) > 0$. Hence, we obtain for $\tau \in (0, \tau_0]$

$$\|\mathcal{A}_0^\alpha\, e^{-\tau\mathcal{K}}\, f\|_{\mathfrak{X}_T^p} \leq \frac{2\, M_{\alpha,\mathcal{A}_0}}{\tau^\alpha}, \quad f \in \mathfrak{B}\,.$$

So, operator $\mathcal{A}_0^\alpha\, e^{-\tau\mathcal{K}}$ is bounded on domain $\mathrm{dom}\,(\mathcal{K})$, which is dense in $\mathfrak{X}_T^p$. Then the *principle of extension by continuity* (Appendix A.1.1) provides a unique, preserving boundedness, extension of this operator to domain $\mathfrak{X}_T^p$. We again denote it by $\mathcal{A}_0^\alpha\, e^{-\tau\mathcal{K}}$ and infer that

$$\|\mathcal{A}_0^\alpha\, e^{-\tau\mathcal{K}}\|_{\mathcal{L}(\mathfrak{X}_T^p)} \leq \frac{2\, M_{\alpha,\mathcal{A}_0}}{\tau^\alpha}, \tag{11.5.12}$$

for $\tau \in (0, \tau_0]$.

Since $e^{-\tau\mathcal{K}} f = 0$ for $f \in \mathfrak{X}_T^p$ and $\tau > T$ (Remark 11.3.6), it remains to consider the case $\tau \in (\tau_0, T]$. By (11.5.12), one gets

$$\|\overline{\mathcal{A}_0^\alpha e^{-\tau\mathcal{K}}}\|_{\mathcal{L}(\mathfrak{X}_T^p)} \leq \|\overline{\mathcal{A}_0^\alpha e^{-\tau_0\mathcal{K}} e^{-(\tau-\tau_0)\mathcal{K}}}\|_{\mathcal{L}(\mathfrak{X}_T^p)} \leq \frac{2\, M_{\alpha,\mathcal{A}_0}\, M_T}{\tau_0^\alpha},$$

where $\|e^{-\tau\mathcal{K}}\|_{\mathcal{L}(\mathfrak{X}^p_T)} \le M_T$ for $\tau \ge 0$. Here we keep in mind that under conditions (C1)–(C3) the evolution semigroup on the space $\mathfrak{X}^p_T$ is bounded, see Remark 11.3.6. As a consequence, we obtain for $\tau \in (\tau_0, T]$:

$$\|\overline{\mathcal{A}_0^\alpha e^{-\tau\mathcal{K}}}\|_{\mathcal{L}(\mathfrak{X}^p_T)} \le \frac{\tau^\alpha}{\tau_0^\alpha}\frac{2\,M_{\alpha,\mathcal{A}_0}\,M_T}{\tau^\alpha} \le \frac{T^\alpha}{\tau_0^\alpha}\frac{2\,M_{\alpha,\mathcal{A}_0}\,M_T}{\tau^\alpha}\,.$$

By setting $\Lambda_\alpha := \max\{2\,M_{\alpha,\mathcal{A}_0}\,, (2\,M_{\alpha,\mathcal{A}_0}\,M_T T^\alpha)/\tau_0^\alpha\}$, we complete the proof. □

In the following, we shall use the notation

$$\mathcal{T}(\tau) := e^{-\tau\mathcal{B}}e^{-\tau\mathcal{K}_0}, \quad \tau \in \mathbb{R}_0^+\,, \tag{11.5.13}$$

cf. $\mathcal{T}_0(\tau) = e^{-\tau D_0}e^{-\tau\mathcal{A}_0} = e^{-\tau\mathcal{K}_0}$, $\tau \in \mathbb{R}_0^+$, (11.2.7) in Remark 11.2.8.

Lemma 11.5.6 *Let* (C1), (C3) *and* (C4) *be satisfied. If the family of generators* $\{B(t)\}_{t\in[0,T]}$ *is* A_0*-stable, then there exist constants* $c_1 > 0$ *and* $c_2 > 0$ *such that*

$$\|\overline{\mathcal{T}(\tau)^k\mathcal{A}_0}\|_{\mathcal{L}(\mathfrak{X}^p_T)} \le \frac{c_1}{\tau^\alpha} + \frac{c_2}{k\tau}, \quad \tau > 0\,, \quad k \in \mathbb{N}. \tag{11.5.14}$$

Proof For $k\tau > T$ we have $\mathcal{T}(\tau)^k = 0$. Hence, one has to prove the estimate (11.5.14) only for $k\tau \le T$. Since the family of generators $\{B(t)\}_{t\in[0,T]}$ is A_0-stable, Lemma 11.4.3 (iii) ensures that pairs $\{\mathcal{K}_0, \mathcal{B}\}$ and $\{\mathcal{B}, \mathcal{K}_0\}$ are Trotter-stable. Then Lemma 1.9.38, (1.9.114), yields for the pair $\{\mathcal{B}, \mathcal{K}_0\}$ the estimate $\|\mathcal{T}(\tau)^k\|_{\mathcal{L}(\mathfrak{X}^p_T)} \le M$, $\tau \ge 0$, where $k \in \mathbb{N}$ and a positive constant $M \ge 1$.

Let $f \in \operatorname{dom}(\mathcal{A}_0)$. Then

$$\begin{aligned}
&\|\mathcal{T}(\tau)^k\mathcal{A}_0 f\| \\
&\le \|(\mathcal{T}(\tau)^k - e^{-k\tau\mathcal{K}_0})\mathcal{A}_0 f\| + \|e^{-k\tau\mathcal{K}_0}\mathcal{A}_0 f\| \\
&\le \|\sum_{j=0}^{k-1}\mathcal{T}(\tau)^{k-(j+1)}(e^{-\tau\mathcal{B}} - \mathbb{1})e^{-(j+1)\tau\mathcal{K}_0}\mathcal{A}_0 f\| + \|e^{-k\tau\mathcal{K}_0}\mathcal{A}_0 f\| \\
&\le M\sum_{j=0}^{k-1}\int_0^\tau d\sigma\,\|e^{-\sigma\mathcal{B}}\mathcal{B}\mathcal{A}_0^{-\alpha}\|\,\|\mathcal{A}_0^{1+\alpha}e^{-(j+1)\tau\mathcal{K}_0}f\| + \|e^{-k\tau\mathcal{K}_0}\mathcal{A}_0 f\|\,.
\end{aligned}$$

Here we used that $\mathbb{1} - e^{-\tau\mathcal{B}} = \int_0^\tau d\sigma\;\mathcal{B}e^{-\sigma\mathcal{B}}$, and assumption (C3). Moreover, due to (C4) and Proposition 11.2.3 generator $\mathcal{B} \in \mathscr{G}(M_B, \beta_B)$ where $M_B \ge 1$, $\beta_B \ge 0$. Then, we have

$$\|e^{-\sigma\mathcal{B}}\|_{\mathcal{L}(\mathfrak{X}^p_T)} \le M_B\,e^{\beta_B\,\sigma} \le M_B\,e^{\beta_B\,T} =: M_T^{\mathcal{B}}\,, \quad \sigma \in [0, T]\,. \tag{11.5.15}$$

By (C1) and Theorem 10.1.12 on account of (11.2.7) and $D_0 \in \mathscr{G}(1,0)$ we deduce:

$$\begin{aligned} \|\mathcal{A}_0^{1+\alpha} e^{-(j+1)\tau\mathcal{K}_0} f\| &\leq \|\mathcal{A}_0^{1+\alpha} e^{-(j+1)\tau\mathcal{A}_0} f\| \\ &\leq \frac{M_{1+\alpha,\mathcal{A}_0}}{((j+1)\tau)^{1+\alpha}} \|f\|, \end{aligned} \tag{11.5.16}$$

as well as

$$\|\mathcal{A}_0 e^{-k\tau\mathcal{K}_0} f\| \leq \|\mathcal{A}_0 e^{-k\tau\mathcal{A}_0} f\| \leq \frac{M_{1,\mathcal{A}_0}}{k\tau} \|f\|.$$

For that reason, one obtains for any $f \in \mathrm{dom}(\mathcal{A}_0)$ the estimate:

$$\begin{aligned} \|\mathcal{T}(\tau)^k \mathcal{A}_0 f\| &\leq \frac{M M_T^{\mathcal{B}} M_{1+\alpha,\mathcal{A}_0} C_\alpha \tau}{\tau^{\alpha+1}} \sum_{j=0}^{k-1} \frac{1}{(j+1)^{\alpha+1}} \|f\| + \frac{M_{1,\mathcal{A}_0}}{k\tau} \|f\| \\ &\leq \frac{M M_T^{\mathcal{B}} M_{1+\alpha,\mathcal{A}_0} C_\alpha \, \zeta(\alpha+1)}{\tau^\alpha} \|f\| + \frac{M_{1,\mathcal{A}_0}}{k\tau} \|f\|, \quad \tau k \leq T, \end{aligned}$$

where $\zeta(\alpha+1) := \sum_{n=1}^\infty 1/n^{\alpha+1}$ is the *Riemann* ζ-function. Since $\mathcal{T}(\tau)^k = 0$ for $\tau k > T$, we have estimate

$$\|\mathcal{T}(\tau)^k \mathcal{A}_0 f\| \leq \frac{M M_T^{\mathcal{B}} M_{1+\alpha,\mathcal{A}_0} C_\alpha \, \zeta(\alpha+1)}{\tau^\alpha} \|f\| + \frac{M_1^{\mathcal{A}_0}}{k\tau} \|f\|, \tag{11.5.17}$$

for $f \in \mathrm{dom}(\mathcal{A}_0)$ and any $k \in \mathbb{N}$.

Given that $\mathrm{dom}(\mathcal{A}_0)$ is dense in the space $\mathfrak{X}_T^p$, the *principle of extension by continuity* (Appendix A.1.1) provides a unique and preserving the bound extension $\overline{\mathcal{T}(\tau)^k \mathcal{A}_0} \supset \mathcal{T}(\tau)^k \mathcal{A}_0$ to domain $\mathfrak{X}_T^p$ of the operator in (11.5.17). As a consequence, the estimate (11.5.14) follows from (11.5.17). □

Lemma 11.5.7 *Let* (C1), (C3), (C4) *and* (C5) *be satisfied. Then, there is a constant* $c > 0$ *such that*

$$\|(\mathcal{T}(\tau) - e^{-\tau\mathcal{K}})\mathcal{A}_0^{-\alpha}\|_{\mathcal{L}(\mathfrak{X}_T^p)} \leq c\,\tau, \tag{11.5.18}$$

and

$$\|\mathcal{A}_0^{-1}(\mathcal{T}(\tau) - e^{-\tau\mathcal{K}})\|_{\mathcal{L}(\mathfrak{X}_T^p)} \leq c\,\tau, \tag{11.5.19}$$

for $\tau \geq 0$.

Proof To prove (11.5.18), we use definition (11.5.13) and identity

$$\mathcal{T}(\tau)g - e^{-\tau\mathcal{K}}g = (e^{-\tau\mathcal{B}} - \mathbb{1})e^{-\tau\mathcal{K}_0}g + e^{-\tau\mathcal{K}_0}f - e^{-\tau\mathcal{K}}g, \quad g \in \mathfrak{X}_T^p,$$

for $\tau \geq 0$. Let $f \in \mathfrak{X}_T^p$ and $g := \mathcal{A}_0^{-\alpha} f \in \mathrm{dom}\,(\mathcal{A}_0^{\alpha})$. Then this identity yields

$$\|(\mathcal{T}(\tau) - \mathrm{e}^{-\tau\mathcal{K}})\mathcal{A}_0^{-\alpha} f\|_{\mathfrak{X}_T^p}$$
$$\leq \|(\mathrm{e}^{-\tau\mathcal{B}} - \mathbb{1})\mathcal{A}_0^{-\alpha}\,\mathrm{e}^{-\tau\mathcal{K}_0} f\|_{\mathfrak{X}_T^p} + \|(\mathrm{e}^{-\tau\mathcal{K}_0} - \mathrm{e}^{-\tau\mathcal{K}})\mathcal{A}_0^{-\alpha} f\|_{\mathfrak{X}_T^p}, \quad f \in \mathfrak{X}_T^p . \tag{11.5.20}$$

From representation (cf. assumption (C4))

$$(\mathbb{1} - \mathrm{e}^{-\tau\mathcal{B}})\, h = \int_0^\tau ds\, \mathcal{B}\,\mathrm{e}^{-s\,\mathcal{B}}\, h, \quad h \in \mathrm{dom}\,(\mathcal{B}),$$

by estimate $\|\mathrm{e}^{-s\,\mathcal{B}}\|_{\mathcal{L}(\mathfrak{X}_T^p)} \leq M_{\mathcal{B}}\,\mathrm{e}^{\beta_{\mathcal{B}}\,s}$ (Proposition 11.2.3) and assumption (C3), i.e., $\mathrm{dom}\,(\mathcal{B}) \supset \mathrm{dom}\,(\mathcal{A}_0^{\alpha})$ (Lemma 11.2.19), we get for $\tau \geq 0$

$$\|(\mathbb{1} - \mathrm{e}^{-\tau\mathcal{B}})\mathcal{A}_0^{-\alpha}\,\mathrm{e}^{-\tau\mathcal{K}_0} f\|_{\mathfrak{X}_T^p}$$
$$\leq \tau\, M_{\mathcal{B}}\,\mathrm{e}^{\beta_{\mathcal{B}}\tau}\, \|\mathcal{B}\mathcal{A}_0^{-\alpha}\|_{\mathcal{L}(\mathfrak{X}_T^p)}\, \|\mathrm{e}^{-\tau\mathcal{K}_0} f\|_{\mathfrak{X}_T^p}, \quad f \in \mathfrak{X}_T^p .$$

Then, using the semigroup quasi-boundedness: $\|\mathrm{e}^{-\tau\mathcal{K}_0} = \mathrm{e}^{-\tau D_0}\mathrm{e}^{-\tau\mathcal{A}_0}\| \leq M_0$, $\tau \geq 0$, see (C1) and Lemma 11.2.9 (iii), together with $\|\mathcal{B}\mathcal{A}_0^{-\alpha}\|_{\mathcal{L}(\mathfrak{X}_T^p)} \leq C_\alpha$ (Lemma 11.2.19), one gets for the first term on the right-hand side of (11.5.20) the estimate

$$\|(\mathbb{1} - \mathrm{e}^{-\tau\mathcal{B}})\mathcal{A}_0^{-\alpha}\mathrm{e}^{-\tau\mathcal{K}_0} f\|_{\mathfrak{X}_T^p} \leq T\, M_{\mathcal{B}}\,\mathrm{e}^{\beta_{\mathcal{B}}T}\, C_\alpha\, M_0\, \|f\|_{\mathfrak{X}_T^p}, \quad f \in \mathfrak{X}_T^p, \quad \tau \geq 0 . \tag{11.5.21}$$

Here we used that on the space $\mathfrak{X}_T^p$ semigroup $\{\mathrm{e}^{-\tau\mathcal{K}_0}\}_{\tau\geq 0}$ is nilpotent, that is, $\mathrm{e}^{-\tau\mathcal{K}_0} = 0$ for $\tau > T$. Note that by Lemma 11.5.3 and by (C3): $\mathrm{dom}\,(\mathcal{K}_0) \subset \mathrm{dom}\,(\mathcal{A}_0^{\alpha}) \subset \mathrm{dom}\,(\mathcal{B})$. Then, by semigroup property: $\mathrm{e}^{-\sigma\mathcal{K}}(\mathrm{dom}\,(\mathcal{K}_0)) \subseteq \mathrm{dom}\,(\mathcal{K}_0)$, we obtain for $g \in \mathrm{dom}\,(\mathcal{K}_0)$

$$\mathrm{e}^{-\tau\mathcal{K}} g - \mathrm{e}^{-\tau\mathcal{K}_0} g = \int_0^\tau d\sigma\, \frac{d}{d\sigma}\, \{\mathrm{e}^{-\sigma\mathcal{K}}\mathrm{e}^{-(\tau-\sigma)\mathcal{K}_0}\} g$$
$$= -\int_0^\tau d\sigma\, \mathrm{e}^{-\sigma\mathcal{K}}\mathcal{B}\mathrm{e}^{-(\tau-\sigma)\mathcal{K}_0} g .$$

Let $g := \mathcal{A}_0^{-\alpha}\, y$, where $y \in \mathcal{A}_0^{\alpha}(\mathrm{dom}\,(\mathcal{K}_0))$. Then,

$$(\mathrm{e}^{-\tau\mathcal{K}} - \mathrm{e}^{-\tau\mathcal{K}_0})\mathcal{A}_0^{-\alpha}\, y = -\int_0^\tau d\sigma\, \mathrm{e}^{-\sigma\mathcal{K}}\mathcal{B}\mathrm{e}^{-(\tau-\sigma)\mathcal{K}_0}\mathcal{A}_0^{-\alpha}\, y ,$$

which by commutativity (Remark 11.2.8) and by (11.3.25) together with (C3) yield for $\tau \geq 0$ the estimate

$$\|(\mathrm{e}^{-\tau\mathcal{K}} - \mathrm{e}^{-\tau\mathcal{K}_0})\mathcal{A}_0^{-\alpha}\, y\|_{\mathfrak{X}_T^p} \leq \tau\, M_T\, C_\alpha\, M_0\|y\|_{\mathfrak{X}_T^p}\,, \quad y \in \mathcal{A}_0^\alpha(\mathrm{dom}\,(\mathcal{K}_0))\,. \tag{11.5.22}$$

Because generator $\mathcal{K}_0$ of nilpotent semigroup is invertible the set in (11.5.22) has representation

$$\mathcal{A}_0^\alpha\,(\mathrm{dom}\,(\mathcal{K}_0)) = \mathrm{ran}\,(\mathcal{A}_0^\alpha\, \mathcal{K}_0^{-1})\,. \tag{11.5.23}$$

Seeing that by assumption (C1) and Lemma 11.5.3, the operator $\mathcal{A}_0^\alpha \mathcal{K}_0^{-1}$ is bounded and invertible, i.e.: $(\mathcal{A}_0^\alpha\, \mathcal{K}_0^{-1})^{-1} = \mathcal{K}_0\, \mathcal{A}_0^{-\alpha}$, we obtain

$$\mathrm{ran}\,(\mathcal{A}_0^\alpha\, \mathcal{K}_0^{-1}) = \mathrm{dom}\,(\mathcal{K}_0\, \mathcal{A}_0^{-\alpha})\,. \tag{11.5.24}$$

By $\mathrm{dom}\,(\mathcal{K}_0) \subset \mathrm{dom}\,(\mathcal{A}_0^\alpha)$, (11.5.5), domain $\mathrm{dom}\,(\mathcal{K}_0\, \mathcal{A}_0^{-\alpha})$ is dense in $\mathfrak{X}_T^p$. Then (11.5.23) and (11.5.24) imply that the set $\mathcal{A}_0^\alpha$ $(\mathrm{dom}\,(\mathcal{K}_0))$ is also dense in $\mathfrak{X}_T^p$.

As a consequence, the *principle of extension by continuity* (Appendix A.1.1) supplies a unique and preserving the bound extension of the operator in (11.5.22) to domain $\mathfrak{X}_T^p$. That is, one gets the operator-norm bound:

$$\|(\mathrm{e}^{-\tau\mathcal{K}} - \mathrm{e}^{-\tau\mathcal{K}_0})\, \mathcal{A}_0^{-\alpha}\|_{\mathcal{L}(\mathfrak{X}_T^p)} \leq \tau\, M_T\, C_\alpha\, M_0\,. \tag{11.5.25}$$

Taking into account (11.5.20), (11.5.21), and (11.5.25) we establish (11.5.18).

To prove (11.5.19), we note that

$$\begin{aligned} &\|\mathcal{A}_0^{-1}(\mathrm{e}^{-\tau\mathcal{B}}\mathrm{e}^{-\tau\mathcal{K}_0} - \mathrm{e}^{-\tau\mathcal{K}})\, f\|_{\mathfrak{X}_T^p} \\ &\leq \|\mathcal{A}_0^{-1}(\mathrm{e}^{-\tau\mathcal{B}} - \mathbb{1})\, \mathrm{e}^{-\tau\mathcal{K}_0} f\|_{\mathfrak{X}_T^p} + \|\mathcal{A}_0^{-1}(\mathrm{e}^{-\tau\mathcal{K}_0} - \mathrm{e}^{-\tau\mathcal{K}})\, f\|_{\mathfrak{X}_T^p}\,, \end{aligned} \tag{11.5.26}$$

for $f \in \mathrm{dom}\,(\mathcal{K}_0) = \mathrm{dom}\,(\mathcal{K})$. By virtue of (C1), (C3) and (C4) equation

$$\mathcal{A}_0^{-1}(\mathbb{1} - \mathrm{e}^{-\tau\mathcal{B}})\, \mathrm{e}^{-\tau\mathcal{K}_0} f = \int_0^\tau d\sigma\, \mathcal{A}_0^{-1}\mathcal{B}\, \mathrm{e}^{-\sigma\mathcal{B}}\, \mathrm{e}^{-\tau\mathcal{K}_0}\, f\,,$$

yields estimate for the first term in the right-hand side of (11.5.26):

$$\|\mathcal{A}_0^{-1}(\mathbb{1} - \mathrm{e}^{-\tau\mathcal{B}})\, \mathrm{e}^{-\tau\mathcal{K}_0} f\|_{\mathfrak{X}_T^p} \leq \tau\, C_1^* M_0\, M_T^{\mathcal{B}}\, \|f\|_{\mathfrak{X}_T^p}, \quad \tau \geq 0\,. \tag{11.5.27}$$

Here again we used that semigroup $\{\mathrm{e}^{-\tau\mathcal{K}_0}\}_{\tau\geq 0}$ on $\mathfrak{X}_T^p$ is nilpotent and the estimate of the norm $\|\mathrm{e}^{-\sigma\mathcal{B}}\|_{\mathcal{L}(\mathfrak{X}_T^p)}$ due to (11.5.15).

For the second term on the right-hand side of (11.5.26), we use identity

$$
\begin{aligned}
e^{-\tau\mathcal{K}_0} f - e^{-\tau\mathcal{K}} f &= \int_0^\tau d\sigma \, \frac{d}{d\sigma}\{e^{-\sigma\mathcal{K}} e^{-(\tau-\sigma)\mathcal{K}}\} \, f \\
&= \int_0^\tau d\sigma \, e^{-\sigma\mathcal{K}_0} \, \mathcal{B} \, e^{-(\tau-\sigma)\mathcal{K}} f \, ,
\end{aligned}
$$

$f \in \mathrm{dom}\,(\mathcal{K})$, which leads to

$$
\mathcal{A}_0^{-1}(e^{-\tau\mathcal{K}_0} - e^{-\tau\mathcal{K}}) \, f = \int_0^\tau d\sigma \, \mathcal{A}_0^{-1} \, \mathcal{B} \, e^{-\sigma\mathcal{K}_0} \, e^{-(\tau-\sigma)\mathcal{K}} f \, ,
$$

for any $f \in \mathrm{dom}\,(\mathcal{K}_0) = \mathrm{dom}\,(\mathcal{K})$. Hence, by (C5) one gets estimate

$$
\|\mathcal{A}_0^{-1}(e^{-\tau\mathcal{K}_0} - e^{-\tau\mathcal{K}}) \, f\|_{\mathfrak{X}_T^p} \leq \tau \, C_1^* \, M_0 \, M_T \, \|f\|_{\mathfrak{X}_T^p}, \quad \tau \geq 0 \, . \tag{11.5.28}
$$

Summarising (11.5.27) and (11.5.28) we obtain estimate (11.5.26) on a dense in the space $\mathfrak{X}_T^p$ domain $\mathrm{dom}\,(\mathcal{K}_0)$. Therefore, the *principle of extension by continuity* (Appendix A.1.1) allows to uplift (11.5.26) to the operator-norm estimate, which gives (11.5.19). □

Lemma 11.5.8 *Let* (C1), (C3), (C4), (C5), *and* (C6) *be satisfied. If* $\beta \in (\alpha, 1)$, *then there exists a constant* $Z(\beta) > 0$ *such that*

$$
\|\mathcal{A}_0^{-1}(\mathcal{T}(\tau) - e^{-\tau\mathcal{K}}) \, \mathcal{A}_0^{-\beta}\|_{\mathcal{L}(\mathfrak{X}_T^p)} \leq Z(\beta) \, \tau^{1+\beta} \, , \quad \tau \geq 0 \, . \tag{11.5.29}
$$

Proof Let $f \in \mathrm{dom}\,(\mathcal{K}_0) = \mathrm{dom}\,(\mathcal{K})$. Then, definition (11.5.13) and identity

$$
\begin{aligned}
\frac{d}{d\sigma}\mathcal{T}(\sigma)e^{-(\tau-\sigma)\mathcal{K}} f &= \frac{d}{d\sigma} e^{-\sigma\mathcal{B}} e^{-\sigma\mathcal{K}_0} \, e^{-(\tau-\sigma)\mathcal{K}} f \\
&= - \, e^{-\sigma\mathcal{B}} \mathcal{B} e^{-\sigma\mathcal{K}_0} \, e^{-(\tau-\sigma)\mathcal{K}} f - e^{-\sigma\mathcal{B}} e^{-\sigma\mathcal{K}_0} \mathcal{K}_0 e^{-(\tau-\sigma)\mathcal{K}} f + \\
&\quad e^{-\sigma\mathcal{B}} e^{-\sigma\mathcal{K}_0} \mathcal{K} e^{-(\tau-\sigma)\mathcal{K}} f \\
&= - \, e^{-\sigma\mathcal{B}} \mathcal{B} e^{-\sigma\mathcal{K}_0} e^{-(\tau-\sigma)\mathcal{K}} f + e^{-\sigma\mathcal{B}} e^{-\sigma\mathcal{K}_0} \mathcal{B} e^{-(\tau-\sigma)\mathcal{K}} f \\
&= e^{-\sigma\mathcal{B}}\{e^{-\sigma\mathcal{K}_0} \mathcal{B} f - \mathcal{B} e^{-\sigma\mathcal{K}_0}\} e^{-(\tau-\sigma)\mathcal{K}} f,
\end{aligned}
$$

provides relation

$$
\begin{aligned}
\mathcal{T}(\tau) f - e^{-\tau\mathcal{K}} f &= \int_0^\tau d\sigma \, \frac{d}{d\sigma}\mathcal{T}(\sigma) \, e^{-(\tau-\sigma)\mathcal{K}} f \\
&= \int_0^\tau d\sigma \, e^{-\sigma\mathcal{B}}\{e^{-\sigma\mathcal{K}_0} \mathcal{B} - \mathcal{B} \, e^{-\sigma\mathcal{K}_0}\} \, e^{-(\tau-\sigma)\mathcal{K}} f \, .
\end{aligned} \tag{11.5.30}
$$

On the other hand, we have the following identity:

$$
\begin{aligned}
& e^{-\sigma\mathcal{B}}\left(e^{-\sigma\mathcal{K}_0}\mathcal{B}-\mathcal{B}e^{-\sigma\mathcal{K}_0}\right)e^{-(\tau-\sigma)\mathcal{K}}f \\
&=(e^{-\sigma\mathcal{B}}-\mathbb{1})\left\{e^{-\sigma\mathcal{K}_0}\mathcal{B}-\mathcal{B}e^{-\sigma\mathcal{K}_0}\right\}\left(e^{-(\tau-\sigma)\mathcal{K}}-e^{-(\tau-\sigma)\mathcal{K}_0}\right)f+ \\
&\quad (e^{-\sigma\mathcal{B}}-\mathbb{1})\left\{e^{-\sigma\mathcal{K}_0}\mathcal{B}-\mathcal{B}e^{-\sigma\mathcal{K}_0}\right\}e^{-(\tau-\sigma)\mathcal{K}_0}f+ \\
&\quad +\left\{e^{-\sigma\mathcal{K}_0}\mathcal{B}-\mathcal{B}e^{-\sigma\mathcal{K}_0}\right\}\left(e^{-(\tau-\sigma)\mathcal{K}}-e^{-(\tau-\sigma)\mathcal{K}_0}\right)f+ \\
&\quad +\{e^{-\sigma\mathcal{K}_0}\mathcal{B}-\mathcal{B}e^{-\sigma\mathcal{K}_0}\}e^{-(\tau-\sigma)\mathcal{K}_0}f\,,
\end{aligned}
$$

which after substitution $f =: \mathcal{A}_0^{-\beta}g$ yields:

$$
\begin{aligned}
& \mathcal{A}_{0}{}_0^{-1}e^{-\sigma\mathcal{B}}\left(e^{-\sigma\mathcal{K}_0}\mathcal{B}-\mathcal{B}e^{-\sigma\mathcal{K}_0}\right)e^{-(\tau-\sigma)\mathcal{K}}\mathcal{A}_0^{-\beta}g= \\
&=\mathcal{A}_0^{-1}(e^{-\sigma\mathcal{B}}-\mathbb{1})\{e^{-\sigma\mathcal{K}_0}\mathcal{B}-\mathcal{B}e^{-\sigma\mathcal{K}_0}\}(e^{-(\tau-\sigma)\mathcal{K}}-e^{-(\tau-\sigma)\mathcal{K}_0})\mathcal{A}_0^{-\beta}g \\
&\quad +\mathcal{A}_0^{-1}(e^{-\sigma\mathcal{B}}-\mathbb{1})\,\{e^{-\sigma\mathcal{K}_0}\mathcal{B}-\mathcal{B}e^{-\sigma\mathcal{K}_0}\}\mathcal{A}_0^{-\beta}e^{-(\tau-\sigma)\mathcal{K}_0}g \\
&\quad +\mathcal{A}_0^{-1}\{e^{-\sigma\mathcal{K}_0}\mathcal{B}-\mathcal{B}e^{-\sigma\mathcal{K}_0}\}(e^{-(\tau-\sigma)\mathcal{K}}-e^{-(\tau-\sigma)\mathcal{K}_0})\mathcal{A}_0^{-\beta}g \\
&\quad +\mathcal{A}_0^{-1}\{(e^{-\sigma\mathcal{K}_0}-e^{-\sigma D_0})\mathcal{B}-\mathcal{L}(e^{-\sigma\mathcal{K}_0}-e^{-\sigma D_0})\}e^{-(\tau-\sigma)\mathcal{K}_0}\mathcal{A}_0^{-\beta}g \\
&\quad +\mathcal{A}_0^{-1}(e^{-\sigma D_0}\mathcal{B}-\mathcal{B}e^{-\sigma D_0})\mathcal{A}_0^{-\beta}e^{-(\tau-\sigma)\mathcal{K}_0}g\,.
\end{aligned}
\tag{11.5.31}
$$

Here (cf. (11.5.23) and (11.5.24))

$$
g\in \operatorname{ran}(\mathcal{A}_0^{\beta}\,\mathcal{K}_0^{-1})\,,\quad \alpha<\beta<1\,. \tag{11.5.32}
$$

Below we estimate separately *five* terms on the right-hand side of identity (11.5.31).

To this aim, we remind that $\mathcal{A}_0$ and $\mathcal{K}_0$ commute. This gives

$$
(e^{-(\tau-\sigma)\mathcal{K}}-e^{-(\tau-\sigma)\mathcal{K}_0})\,\mathcal{A}_0^{-\beta}\,g=\int_0^{\tau-\sigma}dr\,e^{-(\tau-\sigma-r)\mathcal{K}}\,\mathcal{B}\,\mathcal{A}_0^{-\beta}\,e^{-r\mathcal{K}_0}\,g\,.
$$

As a consequence, for the *first* term, we obtain

$$
\begin{aligned}
& \mathcal{A}_0^{-1}(e^{-\sigma\mathcal{B}}-\mathbb{1})\{e^{-\sigma\mathcal{K}_0}\mathcal{B}-\mathcal{B}e^{-\sigma\mathcal{K}_0}\}(e^{-(\tau-\sigma)\mathcal{K}}-e^{-(\tau-\sigma)\mathcal{K}_0})\mathcal{A}_0^{-\beta}g \\
&\quad =-\int_0^{\sigma}dr\,\mathcal{A}_0^{-1}\mathcal{B}e^{-r\mathcal{B}}\,[e^{-\sigma\mathcal{K}_0},\mathcal{B}]\,\mathcal{A}_0^{-\beta} \\
&\qquad \times\sigma\int_0^{\tau-\sigma}dr\,\mathcal{A}_0^{\beta}e^{-(\tau-\sigma-r)\mathcal{K}}\mathcal{B}\mathcal{A}_0^{-\beta}e^{-r\mathcal{K}_0}g\,,
\end{aligned}
$$

where commutator:

$$[e^{-\sigma\mathcal{K}_0}, \mathcal{B}]\, f = \{e^{-\sigma\mathcal{K}_0}\mathcal{B} - \mathcal{B}e^{-\sigma\mathcal{K}_0}\} f, \quad f \in \operatorname{dom}(\mathcal{K}_0), \quad \tau \geq 0.$$

Then using Lemma 11.5.5, we obtain the estimate

$$\begin{aligned}
&\|\mathcal{A}_0^{-1}(e^{-\sigma\mathcal{B}} - \mathbb{1})\{e^{-\sigma\mathcal{K}_0}\mathcal{B} - \mathcal{B}e^{-\sigma\mathcal{K}_0}\}(e^{-(\tau-\sigma)K} - e^{-(\tau-\sigma)\mathcal{K}_0})\mathcal{A}_0^{-\beta} g\|_{\mathfrak{X}_T^p} \\
&\quad \leq \sigma\; 2C_1^* C_\beta^2 \Lambda_\beta\, M_0^2\, M_B\, e^{\beta_B T} \int_0^{\tau-\sigma} dr\, \frac{1}{(\tau-\sigma-r)^\beta}\, \|g\|_{\mathfrak{X}_T^p} \\
&\quad \leq \sigma(\tau-\sigma)^{1-\beta}\, \frac{2\, C_1^*\, C_\beta^2\, \Lambda_\beta\, M_0^2\, M_B\, e^{\beta_B T}}{1-\beta}\, \|g\|_{\mathfrak{X}_T^p}\,,
\end{aligned} \tag{11.5.33}$$

for $g \in \operatorname{ran}(\mathcal{A}_0^\beta\, \mathcal{K}_0^{-1})$, $\sigma \in [0, \tau]$ and $\tau \geq 0$.

For the *second* term, one can readily establish the estimate

$$\begin{aligned}
&\|\mathcal{A}^{-1}(e^{-\sigma\mathcal{B}} - \mathbb{1})\{e^{-\sigma\mathcal{K}_0}\mathcal{B} - \mathcal{B}\, e^{-\sigma\mathcal{K}_0}\}\mathcal{A}_0^{-\beta}\, e^{-(\tau-\sigma)\mathcal{K}_0}\, g\|_{\mathfrak{X}_T^p} \\
&\quad \leq \sigma\; 2\, C_1^*\, C_\beta\, M_0^2\, M_B\, e^{\beta_B T}\, \|g\|_{\mathfrak{X}_T^p}\,,
\end{aligned} \tag{11.5.34}$$

again for $g \in \operatorname{ran}(\mathcal{A}_0^\beta\, \mathcal{K}_0^{-1})$, $\sigma \in [0, \tau]$ and $\tau \geq 0$.

Now note that by virtue of relation

$$e^{-(\tau-\sigma)\mathcal{K}} f - e^{-(\tau-\sigma)\mathcal{K}_0} f = \int_0^{\tau-\sigma} dr\, e^{-(\tau-r-\sigma)\mathcal{K}}\, \mathcal{B}\, e^{-r\mathcal{K}_0} f\,, \quad f \in \operatorname{dom}(\mathcal{K}_0)\,,$$

we obtain for the *third* term the estimate

$$\begin{aligned}
&\|\mathcal{A}_0^{-1}\{e^{-\sigma\mathcal{K}_0}\mathcal{B} - \mathcal{B}\, e^{-\sigma\mathcal{K}_0}\}(e^{-(\tau-\sigma)K} - e^{-(\tau-\sigma)\mathcal{K}_0})\, \mathcal{A}_0^{-\beta} g\|_{\mathfrak{X}_T^p} \\
&\quad \leq (\tau-\sigma)\, 2\, C_1^* C_\beta\, M_0^2\, M_T\, \|g\|_{\mathfrak{X}_T^p}\,,
\end{aligned} \tag{11.5.35}$$

for $g \in \operatorname{ran}(\mathcal{A}_0^\beta\, \mathcal{K}_0^{-1})$, $\sigma \in [0, \tau]$ and $\tau \geq 0$.

On account of equality

$$e^{-\sigma\mathcal{K}_0} f - e^{-\sigma D_0} f = -\int_0^\sigma dr\, e^{-r\mathcal{K}_0}\, \mathcal{A}_0\, e^{-(\sigma-r)D_0} f\,, \quad f \in \operatorname{dom}(\mathcal{K}_0)\,,$$

we get for the *fourth* term:

$$\mathcal{A}_0^{-1}\{(\mathrm{e}^{-\sigma\mathcal{K}_0} - \mathrm{e}^{-\sigma D_0})\mathcal{B} - \mathcal{B}(\mathrm{e}^{-\sigma\mathcal{K}_0} - \mathrm{e}^{-\sigma D_0})\}\mathrm{e}^{-(\tau-\sigma)\mathcal{K}_0}\mathcal{A}_0^{-\beta} g =$$
$$\Big(-\int_0^\sigma dr\, \mathrm{e}^{-r\mathcal{K}_0}\mathrm{e}^{-(\sigma-r)D_0}\mathcal{B}\mathcal{A}_0^{-\beta}\mathcal{A}_0^{-1}\mathcal{B}$$
$$\int_0^\sigma dr\, \mathrm{e}^{-r\mathcal{K}_0}\mathcal{A}_0^{1-\beta}\mathrm{e}^{-(\sigma-r)D_0}\Big)\mathrm{e}^{-(\tau-\sigma)\mathcal{K}_0}\, g\,,$$

which (cf. inequality (11.5.16)) yields the estimate

$$\begin{aligned}
&\|\mathcal{A}_0^{-1}\,\{\,(\,\mathrm{e}^{-\sigma\mathcal{K}_0} - \mathrm{e}^{-\sigma D_0})\,\mathcal{B} - \mathcal{B}\,(\mathrm{e}^{-\sigma\mathcal{K}_0} - \mathrm{e}^{-\sigma D_0}\,)\,\}\,\mathrm{e}^{-(\tau-\sigma)\mathcal{K}_0}\mathcal{A}_0^{-\beta}\, g\|_{\mathfrak{X}_T^p}\\
&\le \sigma\, C_\alpha\, M_0\,\|g\| + C_1^*\, M_0\, M_{1-\beta,\mathcal{A}_0}\int_0^\sigma dr\,\frac{1}{r^{1-\beta}}\,\|g\|_{\mathfrak{X}_T^p}\\
&= \left(\sigma\, C_\beta\, M_0 + \sigma^\beta\,\frac{C_1^*\, M_0\, M_{1-\beta,\mathcal{A}_0}}{\beta}\right)\,\|g\|_{\mathfrak{X}_T^p}\,,
\end{aligned} \tag{11.5.36}$$

for $g \in \operatorname{ran}(\mathcal{A}_0^\beta\,\mathcal{K}_0^{-1})$, $\sigma \in [0, \tau]$ and $\tau \ge 0$.

To estimate the *fifth* term, note that

$$\begin{aligned}
((\mathrm{e}^{-\sigma D_0}\mathcal{B} - \mathcal{B}\mathrm{e}^{-\sigma D_0})\, f)(\tau) &= \mathrm{e}^{-\sigma D_0}B(t)f(\tau) - B\chi_{\mathcal{I}}(\tau-\sigma)\, f(\tau-\sigma) =\\
&= \chi_{\mathcal{I}}(\tau-\sigma)B(\tau-\sigma)f(\tau-\sigma) - B(\tau)\chi_{\mathcal{I}}(\tau-\sigma)\, f(\tau-\sigma) =\\
&= \chi_{\mathcal{I}}(\tau-\sigma)\{B(\tau-\sigma) - B(\tau)\}f(\tau-\sigma)\,, \quad 0 \le \sigma \le \tau \le T\,.
\end{aligned}$$

Then, as a consequence, one gets for $g \in \operatorname{ran}(\mathcal{A}_0^\beta\,\mathcal{K}_0^{-1})$ the estimate

$$\begin{aligned}
&\|\mathcal{A}_0^{-1}(\mathrm{e}^{-\sigma D_0}\,\mathcal{B} - \mathcal{B}\,\mathrm{e}^{-\sigma D_0})\,\mathcal{A}_0^{-\beta}\mathrm{e}^{-(\tau-\sigma)\mathcal{K}_0} g\|_{\mathfrak{X}_T^p}\\
&\qquad = \|\mathcal{A}_0^{-1}\{\mathrm{e}^{-\sigma D_0}\,\mathcal{B} - \mathcal{B}\,\mathrm{e}^{-\sigma D_0}\}\,\mathcal{A}_0^{-\beta} g\|_{\mathfrak{X}_T^p}\\
&\qquad \le \operatorname*{ess\,sup}_{0\le\sigma\le\tau\le T}\,\|\,A_0^{-1}\{B(\tau-\sigma) - B(\tau)\}\,A_0^{-\beta}\|_{\mathcal{L}(\mathfrak{X})}\,\|g\|\\
&\qquad \le L_\beta\sigma^\beta\|g\|_{\mathfrak{X}_T^p}\,.
\end{aligned} \tag{11.5.37}$$

Identity (11.5.31) yields for $g \in \operatorname{ran}(\mathcal{A}_0^\beta \mathcal{K}_0^{-1})$ inequality

$$\begin{aligned}
&\|\mathcal{A}_0^{-1} e^{-\sigma\mathcal{B}}\left(e^{-\sigma\mathcal{K}_0}\mathcal{B} - \mathcal{B}e^{-\sigma\mathcal{K}_0}\right) e^{-(\tau-\sigma)\mathcal{K}} \mathcal{A}_0^{-\beta} g\| \\
&\quad \le \|\mathcal{A}_0^{-1}(e^{-\sigma\mathcal{B}} - \mathbb{1})\{e^{-\sigma\mathcal{K}_0}\mathcal{B} - \mathcal{B}e^{-\sigma\mathcal{K}_0}\}(e^{-(\tau-\sigma)\mathcal{K}} - e^{-(\tau-\sigma)\mathcal{K}_0})\mathcal{A}_0^{-\beta} g\| \\
&\quad + \|\mathcal{A}_0^{-1}(e^{-\sigma\mathcal{B}} - \mathbb{1})\{e^{-\sigma\mathcal{K}_0}\mathcal{B} - \mathcal{B}\, e^{-\sigma\mathcal{K}_0}\}\mathcal{A}_0^{-\beta} e^{-(\tau-\sigma)\mathcal{K}_0} g\| \\
&\quad + \|\mathcal{A}_0^{-1}\{e^{-\sigma\mathcal{K}_0}\mathcal{B} - \mathcal{B}e^{-\sigma\mathcal{K}_0}\}(e^{-(\tau-\sigma)\mathcal{K}} - e^{-(\tau-\sigma)\mathcal{K}_0})\mathcal{A}_0^{-\beta} g\| \\
&\quad + \|\mathcal{A}_0^{-1}\{(e^{-\sigma\mathcal{K}_0} - e^{-\sigma D_0})\mathcal{B} - \mathcal{B}(e^{-\sigma\mathcal{K}_0} - e^{-\sigma D_0})\}e^{-(\tau-\sigma)\mathcal{K}_0} \mathcal{A}_0^{-\beta} g\| \\
&\quad + \|\mathcal{A}_0^{-1}(e^{-\sigma D_0}\mathcal{B} - \mathcal{B}e^{-\sigma D_0})\mathcal{A}_0^{-\beta}\, e^{-(\tau-\sigma)\mathcal{K}_0} g\|\,,
\end{aligned}$$

where $\sigma \in [0,\tau]$ and $\tau \ge 0$. Now taking into account (11.5.33) - (11.5.37) and the fact that semigroup $\{e^{-\tau\mathcal{K}_0}\}_{\tau\ge 0}$ on $\mathfrak{X}_T^p$ is nilpotent we obtain estimate

$$\begin{aligned}
&\|\mathcal{A}_0^{-1} e^{-\sigma\mathcal{B}}\left(e^{-\sigma\mathcal{K}_0}\mathcal{B} - \mathcal{B}e^{-\sigma\mathcal{K}_0}\right) e^{-(\tau-\sigma)\mathcal{K}} \mathcal{A}_0^{-\alpha} g\|_{\mathfrak{X}_T^p} \le \\
&\quad \le \Big\{\sigma(\tau-\sigma)^{1-\beta}\, \frac{2\, C_1^*\, C_\beta^2\, \Lambda_\beta\, M_\mathcal{B}\, M_0^2\, e^{\beta_\mathcal{B} T}}{1-\beta} + \sigma\, 2\, C_1^*\, C_\beta\, M_\mathcal{B}\, M_0^2\, e^{\beta_\mathcal{B} T} + \\
&\quad (\tau-\sigma)\, 2\, C_1^*\, C_\beta\, M_0^2\, M_T + \sigma\, C_\beta\, M_0 \\
&\qquad + \sigma^\beta\, \frac{C_1^*\, M_0\, M_{1-\beta,\mathcal{A}_0}}{\beta} + \sigma^\beta\, L_\beta\Big\}\, \|g\|_{\mathfrak{X}_T^p}\,,
\end{aligned}$$

for $g \in \operatorname{ran}(\mathcal{A}_0^\beta \mathcal{K}_0^{-1})$, $\sigma \in [0,\tau]$ and $\tau \ge 0$. Therefore, setting

$$Z_1 := \frac{2C_1^*\, C_\beta^2\, \Lambda_\beta\, M_\mathcal{B}\, M_0^2\, e^{\beta_\mathcal{B} T}}{1-\beta}\,, \qquad Z_2 := 2\, C_1^*\, C_\beta M_\mathcal{B}\, M_0^2\, e^{\beta_\mathcal{B} T} + C_\beta\, M_0\,,$$

$$Z_3 := 2C_1^*\, C_\beta\, M_0^2\, M_T\,, \quad Z_4 := \frac{C_1^*\, M_0\, M_{1-\beta,\mathcal{A}_0}}{\beta} + L_\beta\,,$$

we deduce inequality

$$\begin{aligned}
&\|\mathcal{A}_0^{-1} e^{-\sigma\mathcal{B}}\left(e^{-\sigma\mathcal{K}_0}\mathcal{B} - \mathcal{B}e^{-\sigma\mathcal{K}_0}\right) e^{-(\tau-\sigma)\mathcal{K}} \mathcal{A}_0^{-\beta} g\|_{\mathfrak{X}_T^p} \\
&\quad \le \left\{Z_1\, \sigma(\tau-\sigma)^{1-\beta} + Z_2\, \sigma + Z_3\, (\tau-\sigma) + Z_4\, \sigma^\beta\right\} \|g\|_{\mathfrak{X}_T^p}\,.
\end{aligned} \tag{11.5.38}$$

To advance, we remark that (11.5.30) provides representation:

$$\mathcal{A}_0^{-1}(\mathcal{T}(\tau) - \mathrm{e}^{-\tau\mathcal{K}})\mathcal{A}_0^{-\beta} g$$
$$= \int_0^\tau d\sigma\, \mathcal{A}_0^{-1}\mathrm{e}^{-\sigma\mathcal{B}}\{\mathrm{e}^{-\sigma\mathcal{K}_0}\mathcal{B} - \mathcal{B}\mathrm{e}^{-\sigma\mathcal{K}_0}\}\mathrm{e}^{-(\tau-\sigma)\mathcal{K}}\mathcal{A}_0^{-\beta} g\,,$$

which yields estimate

$$\|\mathcal{A}_0^{-1}(\mathcal{T}(\tau) - \mathrm{e}^{-\tau\mathcal{K}})\mathcal{A}_0^{-\beta} g\|_{\mathfrak{X}_T^p}$$
$$\leq \int_0^\tau d\sigma\, \|\mathcal{A}_0^{-1}\mathrm{e}^{-\sigma\mathcal{B}}\{\mathrm{e}^{-\sigma\mathcal{K}_0}\mathcal{B} - \mathcal{B}\mathrm{e}^{-\sigma\mathcal{K}_0}\}\mathrm{e}^{-(\tau-\sigma)\mathcal{K}}\mathcal{A}_0^{-\beta} g\|_{\mathfrak{X}_T^p}\,.$$

Inserting into this estimate inequality (11.5.38) and using identity

$$\int_0^\tau d\sigma\, \sigma\, (\tau - \sigma)^{1-\beta} = \tau^{3-\beta}\int_0^1 dx\, x(1-x)^{1-\beta} = \tau^{3-\beta}B(2, 2-\beta)\,,$$

(here $B(\cdot,\cdot)$ denotes the *Euler* Beta-function), we infer for $0 \leq \tau < T$, inequality

$$\|\mathcal{A}_0^{-1}(\mathcal{T}(\tau) - \mathrm{e}^{-\tau\mathcal{K}})\mathcal{A}_0^{-\beta}\, g\|_{\mathfrak{X}_T^p} \leq$$
$$\left(Z_1\, B(2, 2-\beta)\, \tau^{2-2\beta} + \frac{Z_2 + Z_3}{2}\, \tau^{1-\beta} + Z_4\right)\, \tau^{1+\beta}\|g\|_{\mathfrak{X}_T^p}\,.$$

Because semigroups $\{\mathrm{e}^{-\tau\mathcal{K}_0}\}_{\tau\geq 0}$ and $\{\mathrm{e}^{-\tau\mathcal{K}}\}_{\tau\geq 0}$ on the space $\mathfrak{X}_T^p$ are nilpotent, that is, $\mathcal{T}(\tau) = \mathrm{e}^{-\tau\mathcal{B}}\mathrm{e}^{-\tau\mathcal{K}_0} = 0$ and $\mathrm{e}^{-\tau\mathcal{K}} = 0$ for $\tau \geq T$, we obtain by virtu of the last inequality for *any* $\tau \geq 0$ that

$$\|\mathcal{A}_0^{-1}(\mathcal{T}(\tau) - \mathrm{e}^{-\tau\mathcal{K}})\mathcal{A}_0^{-\beta}\, g\|_{\mathfrak{X}_T^p} \leq Z(\beta)\, \tau^{1+\beta}\, \|g\|_{\mathfrak{X}_T^p}\,, \quad g \in \mathrm{ran}\,(\mathcal{A}_0^{\beta}\, \mathcal{K}_0^{-1})\,, \tag{11.5.39}$$

where

$$Z(\beta) := \left(Z_1\, B(2, 2-\beta)\, T^{2-2\beta} + \frac{Z_2 + Z_3}{2}\, T^{1-\beta} + Z_4\right)\,,$$

Seeing that arguments in (11.5.23) and (11.5.24) cover the case (11.5.32), the set $\mathrm{ran}\,(\mathcal{A}_0^{\beta}\, \mathcal{K}_0^{-1})$, for $\alpha < \beta < 1$, is dense in $\mathfrak{X}_T^p$. Consequently, the *principle of extension by continuity* (Appendix A.1.1) provides extension of the estimate (11.5.39) to $g \in \mathfrak{X}_T^p$. That is, we establish the operator-norm bound (11.5.29), and thus, we prove assertion of the lemma. □

11.5.2 Operator-Norm Product Formulæ for Evolution Semigroup

To establish the operator-norm convergence of product approximants for evolution semigroup (see Theorem 11.5.10), we need two technical estimates.

Lemma 11.5.9 *Let $\beta \in [0, 1)$. Then the estimates*

$$(a) \sum_{m=1}^{n-1} \frac{1}{m^{\beta}} \leq \frac{n^{1-\beta}}{1-\beta} \quad \text{and} \quad (b) \sum_{m=1}^{n-1} \frac{1}{(n-m)m^{\beta}} \leq \frac{2}{1-\beta}\frac{1}{n^{\beta}} + \frac{\ln(n)}{n^{\beta}}, \tag{11.5.40}$$

hold for $n = 2, 3, \dots$.

Proof The function $f(x) = x^{-\beta}$, $x > 0$, is decreasing. Hence

$$\sum_{m=1}^{n-1} \frac{1}{m^{\beta}} \leq \int_0^{n-1} dx\, \frac{1}{x^{\beta}} \leq \frac{(n-1)^{1-\beta}}{1-\beta} \leq \frac{n^{1-\beta}}{1-\beta}, \quad \beta < 1, \tag{11.5.41}$$

for $n = 2, 3, \dots$, that proves (11.5.40)(a).

Further, we have

$$\sum_{m=1}^{n-1} \frac{1}{(n-m)m^{\beta}} = \frac{1}{n}\sum_{m=1}^{n-1} \frac{n}{(n-m)m^{\beta}} = \frac{1}{n}\sum_{m=1}^{n-1} \frac{1}{m^{\beta}} + \frac{1}{n}\sum_{m=1}^{n-1} \frac{m^{1-\beta}}{n-m}.$$

On account of (11.5.41), we infer the estimate

$$\sum_{m=1}^{n-1} \frac{1}{(n-m)m^{\beta}} \leq \frac{1}{1-\beta}\frac{1}{n^{\beta}} + \frac{1}{n}\sum_{m=1}^{n-1} \frac{m^{1-\beta}}{n-m}. \tag{11.5.42}$$

Note that the function $g(x) := x^{1-\beta}/(n-x)$, $x \in [0, n)$, is increasing. Hence,

$$\sum_{m=1}^{n-1} \frac{m^{1-\beta}}{n-m} \leq \int_1^{n-1} dx\, \frac{x^{1-\beta}}{n-x} \leq n^{1-\beta} \int_{1/n}^{1} ds\, \frac{(1-s)^{1-\beta}}{s}$$

$$\leq n^{1-\beta} \int_{1/n}^{1} ds\, \frac{1}{s} = n^{1-\beta} \ln(n), \quad n = 3, 4, \dots .$$

Consequently, we obtain the estimate

$$\frac{1}{n}\sum_{m=1}^{n-1}\frac{m^{1-\beta}}{n-m}=\frac{1}{n}\sum_{m=1}^{n-2}\frac{m^{1-\beta}}{n-m}+\frac{1}{n}(n-1)^{1-\beta}\le\frac{\ln(n)}{n^{\beta}}+\frac{1}{1-\beta}\frac{1}{n^{\beta}},\quad n=2,3,\ldots,$$

which together with (11.5.42) proves (11.5.40)(b). □

Theorem 11.5.10 *Let* (C1), (C3), (C4), (C5) *and* (C6) *be satisfied. If the family of generators* $\{B(t)\}_{t\in\mathcal{I}}$ *is* A_0*-stable and* $\beta\in(\alpha,1)$*, then there exists a constant* $C_{\alpha,\beta}>0$ *such that*

$$\|(\mathrm{e}^{-\tau\mathcal{B}/n}\mathrm{e}^{-\tau\mathcal{K}_0/n})^n-\mathrm{e}^{-\tau\mathcal{K}}\|_{\mathcal{L}(\mathfrak{X}^p_T)}\le\frac{C_{\alpha,\beta}}{n^{\beta-\alpha}},\tag{11.5.43}$$

for $\tau\ge 0$ *and* $n=2,3,\ldots$*, gives the operator-norm estimate for the rate of convergence of the Trotter product formula approximants* $\{(\mathrm{e}^{-\tau\mathcal{B}/n}\mathrm{e}^{-\tau\mathcal{K}_0/n})^n\}_{n\ge 2}$.

Proof Let $\mathcal{T}(\sigma):=\mathrm{e}^{-\sigma\mathcal{B}}\mathrm{e}^{-\sigma\mathcal{K}_0}$ (11.5.13) and $\mathcal{U}(\sigma):=e^{-\sigma\mathcal{K}}$, $\sigma\ge 0$. Then the following identity holds:

$$\begin{aligned}\mathcal{T}(\sigma)^n-\mathcal{U}(\sigma)^n&=\sum_{m=0}^{n-1}\mathcal{T}(\sigma)^{n-m-1}(\mathcal{T}(\sigma)-\mathcal{U}(\sigma))\,\mathcal{U}(\sigma)^m\\&=\mathcal{T}(\sigma)^{n-1}(\mathcal{T}(\sigma)-\mathcal{U}(\sigma))+(\mathcal{T}(\sigma)-\mathcal{U}(\sigma))\,\mathcal{U}(\sigma)^{n-1}+\\&\quad+\sum_{m=1}^{n-2}\mathcal{T}(\sigma)^{n-m-1}(\mathcal{T}(\sigma)-\mathcal{U}(\sigma))\,\mathcal{U}(\sigma)^m\\&=\mathcal{T}(\sigma)^{n-1}\mathcal{A}\mathcal{A}_0^{-1}(\mathcal{T}(\sigma)-\mathcal{U}(\sigma))+(\mathcal{T}(\sigma)-\mathcal{U}(\sigma))\,\mathcal{A}_0^{-\alpha}\mathcal{A}_0^{\alpha}\,\mathcal{U}(\sigma)^{n-1}+\\&\quad+\sum_{m=1}^{n-2}\mathcal{T}(\sigma)^{n-m-1}\mathcal{A}_0\,\mathcal{A}_0^{-1}(\mathcal{T}(\sigma)-\mathcal{U}(\sigma))\,\mathcal{A}_0^{-\beta}\mathcal{A}_0^{\beta}\,\mathcal{U}(\sigma)^m\,.\end{aligned}$$

It readily yields inequality

$$\begin{aligned}&\|\mathcal{T}(\sigma)^n-\mathcal{U}(\sigma)^n\|_{\mathcal{L}(\mathfrak{X}^p_T)}\le\\&\|\mathcal{T}(\sigma)^{n-1}(\mathcal{T}(\sigma)-\mathcal{U}(\sigma))\|_{\mathcal{L}(\mathfrak{X}^p_T)}+\|(\mathcal{T}(\sigma)-\mathcal{U}(\sigma))\,\mathcal{U}(\sigma)^{n-1}\|_{\mathcal{L}(\mathfrak{X}^p_T)}\\&+\sum_{m=1}^{n-2}\|\mathcal{T}(\sigma)^{n-m-1}(\mathcal{T}(\sigma)-\mathcal{U}(\sigma))\,\mathcal{U}(\sigma)^m\|_{\mathcal{L}(\mathfrak{X}^p_T)}\le\\&\|\overline{\mathcal{T}(\sigma)^{n-1}\mathcal{A}_0}\|_{\mathcal{L}(\mathfrak{X}^p_T)}\,\|\mathcal{A}_0^{-1}(\mathcal{T}(\sigma)-\mathcal{U}(\sigma))\|_{\mathcal{L}(\mathfrak{X}^p_T)}\end{aligned}$$

$$+ \|(\mathcal{T}(\sigma) - \mathcal{U}(\sigma))\mathcal{A}_0^{-\alpha}\| \, \|\mathcal{A}_0^{\alpha}\, \mathcal{U}(\sigma)^{n-1}\|_{\mathcal{L}(\mathfrak{X}_T^p)}$$

$$+ \sum_{m=1}^{n-2} \|\overline{\mathcal{T}(\sigma)^{n-m-1}\mathcal{A}_0}\| \, \|\mathcal{A}_0^{-1}(\mathcal{T}(\sigma) - \mathcal{U}(\sigma))\, \mathcal{A}_0^{-\beta}\|_{\mathcal{L}(\mathfrak{X}_T^p)} \, \|\mathcal{A}_0^{\beta}\, \mathcal{U}(\sigma)^m\|_{\mathcal{L}(\mathfrak{X}_T^p)} \,.$$

By virtue of Lemma 11.5.6 we get for $\sigma \in (0, \tau]$ estimates

$$\|\overline{\mathcal{T}(\sigma)^{n-1}\mathcal{A}_0}\|_{\mathcal{L}(\mathfrak{X}_T^p)} \le \frac{c_1}{\sigma^\alpha} + \frac{c_2}{(n-1)\,\sigma}\,, \qquad n \ge 2\,.$$

Now, due to Lemma 11.5.7, we find that

$$\|\mathcal{A}_0^{-1}(\mathcal{T}(\sigma) - \mathcal{U}(\sigma))\|_{\mathcal{L}(\mathfrak{X}_T^p)} \le c\,\sigma \quad \text{and} \quad \|(\mathcal{T}(\sigma) - \mathcal{U}(\sigma))\, \mathcal{A}_0^{-\alpha}\|_{\mathcal{L}(\mathfrak{X}_T^p)} \le c\,\sigma\,,$$
$$n \ge 2\,.$$

This entails that $\|\overline{\mathcal{T}(\sigma)^{n-1}\mathcal{A}_0}\|_{\mathcal{L}(\mathfrak{X}_T^p)} \, \|\mathcal{A}_0^{-1}(\mathcal{T}(\sigma) - \mathcal{U}(\sigma))\|_{\mathcal{L}(\mathfrak{X}_T^p)} \le c_1\, c\, \sigma^{1-\alpha} + c_2\, c/(n-1)$ and $\|(\mathcal{T}(\sigma) - \mathcal{U}(\sigma))\ \mathcal{A}_0^{-\alpha}\|_{\mathcal{L}(\mathfrak{X}_T^p)} \ \|\mathcal{A}_0^{\alpha}\ \mathcal{U}(\sigma)^{n-1}\|_{\mathcal{L}(\mathfrak{X}_T^p)} \le c\,\Lambda_\alpha\, \sigma^{1-\alpha}/(n-1)^\alpha$, where we also used Lemma 11.5.5, (11.5.6). Since by Lemma 11.5.8

$$\|\mathcal{A}_0^{-1}(\mathcal{T}(\sigma) - \mathrm{e}^{-\sigma\mathcal{K}})\, \mathcal{A}_0^{-\beta}\|_{\mathcal{L}(\mathfrak{X}_T^p)} \le Z(\beta)\, \sigma^{1+\beta}\,, \qquad \sigma \in [0, \tau)\,,$$

one gets for $m \ge 1$ and $n > m$:

$$\|\overline{\mathcal{T}(\sigma)^{n-m-1}\mathcal{A}_0}\|_{\mathcal{L}(\mathfrak{X}_T^p)} \, \|\mathcal{A}_0^{-1}(\mathcal{T}(\sigma) - \mathcal{U}(\sigma))\, \mathcal{A}_0^{-\beta}\|_{\mathcal{L}(\mathfrak{X}_T^p)} \, \|\mathcal{A}_0^{\beta}\, \mathcal{U}(\sigma)^m\|_{\mathcal{L}(\mathfrak{X}_T^p)}$$
$$\le c_1\, Z(\beta)\, \Lambda_\beta\, \frac{\sigma^{1-\alpha}}{m^\beta} + c_2\, Z(\beta)\, \Lambda_\beta\, \frac{1}{(n-1-m)\, m^\beta}\,.$$

Owing to Lemma 11.5.9, we obtain inequalities

$$\sum_{m=1}^{n-2} \|\overline{\mathcal{T}(\sigma)^{n-m-1}\mathcal{A}_0}\|_{\mathcal{L}(\mathfrak{X}_T^p)} \, \|\mathcal{A}_0^{-1}(\mathcal{T}(\sigma) - \mathcal{U}(\sigma))\, \mathcal{A}_0^{-\beta}\|_{\mathcal{L}(\mathfrak{X}_T^p)} \, \|\mathcal{A}_0^{\beta}\, \mathcal{U}(\sigma)^m\|_{\mathcal{L}(\mathfrak{X}_T^p)}$$
$$\le c_1\, Z(\beta)\, \Lambda_\beta\, \sigma^{1-\alpha} \sum_{m=1}^{n-2} \frac{1}{m^\beta} + c_2\, Z(\beta)\, \Lambda_\beta \sum_{m=1}^{n-2} \frac{1}{(n-1-m)\, m^\beta}$$
$$\le \frac{c_1\, Z(\beta)\, \Lambda_\beta}{1-\beta} (n-1)^{1-\beta} \sigma^{1-\alpha} + \frac{2c_2\, Z(\beta)\, \Lambda_\beta}{1-\beta}\, \frac{1}{(n-1)^\beta}$$
$$+ c_2\, Z(\beta)\, \Lambda_\beta\, \frac{\ln(n-1)}{(n-1)^\beta}\,.$$

Summarising now all these ingredients, we figure out for $n \geq 2$ the estimate

$$
\begin{aligned}
&\|\mathcal{T}(\sigma)^n - U(\sigma)^n\|_{\mathcal{L}(\mathfrak{X}_T^p)} \\
&\leq c_1 c\, \sigma^{1-\alpha} + \frac{c_2\, c}{n-1} + \frac{c\, \Lambda_\alpha}{(n-1)^\alpha}\, \sigma^{1-\alpha} \\
&\quad + \frac{c_1\, Z(\beta)\, \Lambda_\beta}{1-\beta} (n-1)^{1-\beta} \sigma^{1-\alpha} + \frac{2c_2\, Z(\beta)\, \Lambda_\beta}{1-\beta} \frac{1}{(n-1)^\beta} \\
&\quad + c_2\, Z(\beta)\, \Lambda_\beta\, \frac{\ln(n-1)}{(n-1)^\beta}\,.
\end{aligned}
$$

Note that because of *nilpotentness*, one has: $\mathcal{T}(\sigma) = 0$ and $\mathcal{U}(\sigma) = 0$ for $\sigma > T$. For that reason, if we set $\sigma := \tau/n$, then

$$
\begin{aligned}
&\|\mathcal{T}(\tau/n)^n - \mathcal{U}(\tau/n)^n\|_{\mathcal{L}(\mathfrak{X}_T^p)} \\
&\leq \frac{c_1 c\, T^{1-\alpha}}{(n-1)^{1-\alpha}} + \frac{c_2\, c}{n-1} + \frac{c\, \Lambda_\alpha\, T^{1-\alpha}}{(n-1)} \\
&\quad + \frac{c_1\, Z(\beta)\, \Lambda_\beta\, T^{1-\alpha}}{1-\beta} \frac{1}{(n-1)^{\beta-\alpha}} + \frac{2c_2\, Z(\beta)\, \Lambda_\beta}{1-\beta} \frac{1}{(n-1)^\beta} \\
&\quad + c_2\, Z(\beta)\, \Lambda_\beta \frac{\ln(n-1)}{(n-1)^\beta}.
\end{aligned}
$$

for any $\tau \geq 0$ and $n = 2, 3, \ldots$. As a result, there exists a constant $C_{\alpha,\beta} > 0$ such that (11.5.43) holds for $\beta \in (\alpha, 1)$. □

Corollary 11.5.11 *If in condition* (C6) *we put* $\beta = 1$, *then there exists* $L_\gamma > 0$ *such that for each* $\gamma \in (\alpha, 1)$ *one has* $L_{\beta=1}\, |t-s| \leq L_\gamma\, |t-s|^\gamma$ *for* $t, s \in [0, T]$. *As a consequence, there is* $C_{\alpha,\gamma} > 0$ *so that*

$$
\|(\mathrm{e}^{-\tau B/n}\mathrm{e}^{-\tau \mathcal{K}_0/n})^n - \mathrm{e}^{-\tau \mathcal{K}}\|_{\mathcal{L}(\mathfrak{X}_T^p)} \leq \frac{C_{\alpha,\gamma}}{n^{\gamma-\alpha}}, \quad \gamma \in (\alpha, 1), \tag{11.5.44}
$$

for $\tau \geq 0$ *and* $n = 1, 2, 3, \ldots$, *gives the operator-norm estimate of the rate of convergence for the Trotter product formula approximants* $\{(\mathrm{e}^{-\tau \mathcal{B}/n}\mathrm{e}^{-\tau \mathcal{K}_0/n})^n\}_{n\geq 1}$.

Remark 11.5.12

(a) We infer that in Theorem 11.5.10 and in assertion (11.5.44) *neither* of operators $\mathcal{K}_0$ and $\mathcal{B}$ generates a *holomorphic* semigroup.
(b) If the family of generators $\{B(t)\}_{t\in\mathcal{I}}$ is time-independent: $B(t) = B$, then condition (C6) is automatically satisfied for any $\beta > 0$. In particular, we can set $\beta = 1$. Seeing that now both $\mathcal{A}_0$ and $\mathcal{B}$ commute with generator D_0, we obtain for approximants representation:

$$
\mathcal{T}(\tau/n)^n = (\mathrm{e}^{-\tau \mathcal{B}/n}\mathrm{e}^{-\tau \mathcal{A}_0/n})^n \mathrm{e}^{-\tau D_0}, \quad \tau \geq 0, \quad n \in \mathbb{N}. \tag{11.5.45}
$$

(c) It deserves to mention here that if one of the operators: A or B, is generator of a *holomorphic contraction* semigroup and another one is generator of a *contraction* semigroup on $\mathfrak{X}$, then Theorem 10.2.6 provides the existence of constants $M_1 > 0$, $M_2 > 0$ and $\eta > 0$ such that estimate

$$\|(e^{-\tau B/n}e^{-\tau A/n})^n - e^{-\tau C}\|_{\mathcal{L}(\mathfrak{X})} \leq e^{\tau\eta}(M_1 + M_2\tau^{1-\alpha})\,\frac{\ln(n)}{n^{1-\alpha}}\,, \tag{11.5.46}$$

holds for $\tau \geq 0$ and $n \in \mathbb{N}$. Applying (11.5.46) to approximants (11.5.45), we deduce the existence of a constant $R > 0$ such that

$$\|(e^{-\tau\mathcal{B}/n}e^{-\tau\mathcal{K}_0/n})^n - e^{-\tau\mathcal{K}}\|_{\mathcal{L}(\mathfrak{X}_T^p)} \leq R\,\frac{\ln(n)}{n^{1-\alpha}}\,,$$

holds for $\tau \geq 0$ and $n \in \mathbb{N}$. Note that this estimate is *sharper* than the estimate established for convergence rate in (11.5.44).

We comment that the question about possibility to interchange the *order* of factors in the Trotter product formula (11.5.43), still remains open. We shall elucidate this question below. As the first step we prove the analogue of Lemma 11.5.5, cf. (11.5.6).

Lemma 11.5.13 *Let* (C1), (C2) *and* (C7) *be satisfied. Then there exits a constant* $\Lambda'_\alpha > 0$ *such that the norm of the closure* :

$$\|\,\overline{e^{-\tau\mathcal{K}}\mathcal{A}_0^\alpha}\,\|_{\mathcal{L}(\mathfrak{X}_T^p)} \leq \frac{\Lambda'_\alpha}{\tau^\alpha}\,, \qquad \tau > 0\,. \tag{11.5.47}$$

Proof If $\mathcal{A}_0^\alpha f \in \mathrm{dom}\,(\mathcal{K})$, which by invertibility of operator A_0 (C1), and thence of $\mathcal{A}_0$, is equivalent to $f \in \mathcal{A}_0^{-\alpha}(\mathrm{dom}\,(\mathcal{K}))$, then from (11.5.7) we get representation

$$e^{-\tau\mathcal{K}}\mathcal{A}_0^\alpha f \leq e^{-\tau\mathcal{K}_0}\mathcal{A}_0^\alpha f - \int_0^\tau d\sigma\, e^{-(\tau-\sigma)\mathcal{K}_0}\,\mathcal{B}\,e^{-\sigma\mathcal{K}}\,\mathcal{A}_0^\alpha f\,. \tag{11.5.48}$$

Seeing that $\mathcal{K}_0 = \overline{\mathcal{D}_0 + \mathcal{A}_0}$ and on account of (11.2.7), (11.5.9) we obtain estimate

$$\|\,e^{-\tau\mathcal{K}_0}\mathcal{A}_0^\alpha f\,\|_{\mathfrak{X}_T^p} = \|\mathcal{A}_0^\alpha e^{-\tau\mathcal{A}_0} f\,\|_{\mathfrak{X}_T^p} \leq \frac{M_{\alpha,\mathcal{A}_0}}{\tau^\alpha}\,\|\,f\,\|_{\mathfrak{X}_T^p}\,, \qquad \tau > 0\,. \tag{11.5.49}$$

Here we used that in the space $\mathfrak{X}_T^p$ generator $\mathcal{D}_0 \in \mathscr{G}(1,0)$ of nilpotent contraction semigroup commutes with generator $\mathcal{A}_0$ of holomorphic semigroup, see Remark 11.2.8 and (C1).

On account of (11.5.48) and (11.5.49) for any $h \in \mathrm{ran}\,(\mathcal{A}_0^\alpha) \cap \mathrm{dom}\,(\mathcal{K})$, that is, $h := \mathcal{A}_0^\alpha f \in \mathrm{dom}\,(\mathcal{K})$, representation

$$e^{-\tau\mathcal{K}}\,h = e^{-\tau\mathcal{K}_0}\,h - \int_0^\tau d\sigma\, e^{-(\tau-\sigma)\mathcal{K}_0}\,\mathcal{A}_0^\alpha\,\mathcal{A}_0^{-\alpha}\,\mathcal{B}\,e^{-\sigma\mathcal{K}}\,h\,, \tag{11.5.50}$$

is well-defined because $e^{-\sigma\mathcal{K}}h \in \text{dom}(\mathcal{B})$, for $\sigma > 0$, since $\mathcal{K} = \mathcal{K}_0 + \mathcal{B}$. Note that Lemma 11.2.19 yields $\|\overline{\mathcal{A}_0^{-\alpha}\mathcal{B}}\|_{\mathcal{L}(\mathfrak{X}_T^p)} \leq C_\alpha^*$. Given that by Lemma 11.5.3

$$\text{dom}(\mathcal{K}) = \text{dom}(\mathcal{K}_0) \subset \text{dom}(\mathcal{A}_0^\alpha) \subset \text{dom}(\mathcal{B}),$$

and that operator $\mathcal{A}_0^\alpha$ commutes with $\mathcal{K}_0$, we obtain

$$\int_0^\tau d\sigma\, e^{-(\tau-\sigma)\mathcal{K}_0}\,\mathcal{B}e^{-\sigma\mathcal{K}}h = \int_0^\tau d\sigma\, \mathcal{A}_0^\alpha\, e^{-(\tau-\sigma)\mathcal{K}_0}\,\overline{\mathcal{A}_0^{-\alpha}\mathcal{B}}\,e^{-\sigma\mathcal{K}}h\,. \tag{11.5.51}$$

Next, by Lemma 11.2.19, there is a constant $C_\alpha > 0$, such that $\|\mathcal{B}\mathcal{A}_0^{-\alpha}\|_{\mathcal{L}(\mathfrak{X}_T^p)} \leq C_\alpha$. Then collecting (11.5.49) - (11.5.51) we deduce the estimate:

$$\begin{aligned}
\|e^{-\tau\mathcal{K}}\mathcal{A}_0^\alpha f\|_{\mathfrak{X}_T^p} &\leq \frac{M_{\alpha,\mathcal{A}_0}}{\tau^\alpha}\|f\|_{\mathfrak{X}_T^p} + \\
&+ \int_0^\tau d\sigma\, \|\mathcal{A}_0^\alpha e^{-(\tau-\sigma)\mathcal{K}_0}\|_{\mathcal{L}(\mathfrak{X}_T^p)}\, \|\overline{\mathcal{A}_0^{-\alpha}\mathcal{B}}\|_{\mathcal{L}(\mathfrak{X}_T^p)}\, \|e^{-\sigma\mathcal{K}}\mathcal{A}_0^\alpha f\|_{\mathfrak{X}_T^p} \\
&\leq \frac{M_{\alpha,\mathcal{A}_0}}{\tau^\alpha}\|f\|_{\mathfrak{X}_T^p} + C_\alpha^* \int_0^\tau d\sigma\, \frac{M_{\alpha,\mathcal{A}_0}}{(\tau-\sigma)^\alpha}\|e^{-\sigma\mathcal{K}}\mathcal{A}_0^\alpha f\|_{\mathfrak{X}_T^p}\,.
\end{aligned} \tag{11.5.52}$$

To proceed, we define the subset

$$\mathfrak{B}_0 := \{f \in \mathfrak{X}_T^p : f \in \mathcal{A}_0^{-\alpha}(\text{dom}(\mathcal{K})),\ \|f\|_{\mathfrak{X}_T^p} \leq 1\}. \tag{11.5.53}$$

If for each $f \in \mathfrak{B}_0$, we introduce function $\tau \mapsto F_f(\tau) := \|e^{-\tau\mathcal{K}}\mathcal{A}_0^\alpha f\|_{\mathfrak{X}_T^p}$, $\tau \geq 0$, then on account of (11.5.52) the inequality

$$F_f(\tau) \leq c_1\tau^{-\alpha} + c_2\int_0^\tau d\sigma\,(\tau-\sigma)^{-\alpha}F_f(\sigma),$$

holds for $c_1 = M_{\alpha,\mathcal{A}_0}$ and $c_2 = C_\alpha^* M_{\alpha,\mathcal{A}_0}$. As a consequence, Lemma 11.5.4 states that the estimate $F_f(\tau)\tau^\alpha \leq 2c_1$ is valid for $\tau \in [0, \tau_0]$, where $\tau_0 > 0$ depends on α and c_2. Hence,

$$\|e^{-\tau\mathcal{K}}\mathcal{A}_0^\alpha f\|_{\mathfrak{X}_T^p} \leq \frac{2\,M_{\alpha,\mathcal{A}_0}}{\tau^\alpha}, \quad f \in \mathfrak{B}_0, \quad \tau \in (0, \tau_0]. \tag{11.5.54}$$

Seeing that on $\mathfrak{X}_T^p$ the semigroup $\{e^{-\tau\mathcal{K}}\}_{\tau\geq 0}$ is nilpotent (Remark 11.3.6), it remains to consider only the case $\tau \in (\tau_0, T]$. Then by (11.5.49) and (11.3.25)

$$\|e^{-\tau\mathcal{K}}\mathcal{A}_0^\alpha f\|_{\mathfrak{X}_T^p} \leq \|e^{-(\tau-\tau_0)\mathcal{K}}e^{-\tau_0\mathcal{K}}\mathcal{A}_0^\alpha f\|_{\mathfrak{X}_T^p} \leq \frac{2\,M_{\alpha,\mathcal{A}_0}\,M_T}{\tau_0^\alpha}, \quad \tau > \tau_0,$$

where $\|e^{-t\mathcal{K}}\|_{\mathcal{L}(\mathfrak{X}^p_T)} \le M_T$ for $t \ge 0$. So, on the interval $\tau \in (\tau_0, T]$ for $f \in \mathfrak{B}$ we obtain

$$\|e^{-\tau\mathcal{K}} \mathcal{A}_0^\alpha f\|_{\mathfrak{X}^p_T} \le \frac{\tau^\alpha}{\tau_0^\alpha} \frac{2\, M_{\alpha,\mathcal{A}_0}\, M_T}{\tau^\alpha} \le \frac{T^\alpha}{\tau_0^\alpha} \frac{2\, M_{\alpha,\mathcal{A}_0}\, M_T}{\tau^\alpha} . \tag{11.5.55}$$

Introducing constant

$$\Lambda'_\alpha := \max\left\{2\, M_{\alpha,\mathcal{A}_0},\ (2\, M_{\alpha,\mathcal{A}_0}\, M_T T^\alpha)/\tau_0^\alpha\right\} ,$$

we deduce from (11.5.54) and (11.5.55) the estimate for any $\tau > 0$:

$$\|e^{-\tau\mathcal{K}}\, \mathcal{A}_0^\alpha\, f\|_{\mathfrak{X}^p_T} \le \frac{\Lambda'_\alpha}{\tau^\alpha}, \quad f \in \mathfrak{B}_0 . \tag{11.5.56}$$

For extension of (11.5.56) from $\mathfrak{B}_0$ to the *unit* ball $\mathfrak{B}_1 := \{f \in \mathfrak{X}^p_T : \|f\|_{\mathfrak{X}^p_T} \le 1\}$ we note that generator $\mathcal{K}_0$ is invertible, see Remark 11.5.2, and $\mathcal{A}_0^{-\alpha}$ is invertible by assumption (C1). Then,

$$\mathcal{A}_0^{-\alpha}(\mathrm{dom}\,(\mathcal{K})) = \mathcal{A}_0^{-\alpha}(\mathrm{dom}\,(\mathcal{K}_0)) = \mathrm{ran}\,(\mathcal{A}_0^{-\alpha}\, \mathcal{K}_0^{-1}) . \tag{11.5.57}$$

We claim that the range in (11.5.57) is dense in the space $\mathfrak{X}^p_T$. By *reductio ad absurdum*: if not, then the closure $\overline{\mathrm{ran}\,(\mathcal{A}_0^{-\alpha}\, \mathcal{K}_0^{-1})} \subset \mathfrak{X}^p_T$ is a proper subspace in $\mathfrak{X}^p_T$. As a consequence, the inverse operator $((\mathcal{A}_0^{-\alpha}\, \mathcal{K}_0^{-1}))^{-1} : \mathrm{ran}\,(\mathcal{A}_0^{-\alpha}\, \mathcal{K}_0^{-1}) \to \mathfrak{X}^p_T$ (if it exists) should not be densely defined. But operator $\mathcal{K}_0\, \mathcal{A}_0^\alpha \subset ((\mathcal{A}_0^{-\alpha}\, \mathcal{K}_0^{-1}))^{-1}$ is well-defined on a dense domain $\mathrm{dom}\,(\mathcal{A}_0^\alpha)$. Indeed, $\mathcal{A}_0^\alpha : \mathrm{dom}\,(\mathcal{A}_0^\alpha) \to \mathrm{ran}\,(\mathcal{A}_0^\alpha) = \mathrm{dom}\,(\mathcal{A}_0^{-\alpha})$, where $\mathrm{dom}\,(\mathcal{A}_0^{-\alpha}) = \mathfrak{X}^p_T$ by (C1). Therefore, $\mathrm{ran}\,(\mathcal{A}_0^\alpha) \supset \mathrm{dom}\,(\mathcal{K}_0)$ and consequently the product $\mathcal{K}_0\, \mathcal{A}_0^\alpha$ is densely defined on $\mathrm{dom}\,(\mathcal{A}_0^\alpha)$. This proves the claim.

On that account, the *principle of extension by continuity* (Appendix A.1.1) supplies a unique and preserving the bound extension of the operator in (11.5.56) to operator $\overline{e^{-\tau\mathcal{K}}\mathcal{A}_0^\alpha}$ on $\mathfrak{B}_1$. Then, taking supremum over unit ball $\mathfrak{B}_1$ we complete the proof of (11.5.47). □

Lemma 11.5.14 *Let* (C1), (C2) *and* (C3) *be satisfied. Then, there is a constant* $C_{\mathcal{K}_0} > 0$ *such that*

$$\|(e^{-\tau\mathcal{K}_0} - e^{-\tau\mathcal{K}})\, e^{-\sigma\mathcal{K}}\|_{\mathcal{L}(\mathfrak{X}^p_T)} \le C_{\mathcal{K}_0}\, \frac{\tau}{\sigma^\alpha} , \tag{11.5.58}$$

for $\tau \ge 0$ *and* $\sigma > 0$.

Proof Because $\mathrm{dom}\,(\mathcal{K}) = \mathrm{dom}\,(\mathcal{K}_0) \subset \mathrm{dom}\,(\mathcal{A}_0^\alpha)$ (see Theorem 11.3.5 and Lemma 11.5.3) the representation

$$(\mathrm{e}^{-\tau\mathcal{K}_0} - \mathrm{e}^{-\tau\mathcal{K}})\mathrm{e}^{-\sigma\mathcal{K}} f = (\mathrm{e}^{-\tau\mathcal{K}_0} - \mathrm{e}^{-\tau\mathcal{K}})\, \mathcal{A}_0^{-\alpha}\, \mathcal{A}_0^{\alpha}\, \mathrm{e}^{-\sigma\mathcal{K}}\, f\,,$$

$f \in \mathrm{dom}\,(\mathcal{K})$, and semigroup property yield inequality

$$\|(\mathrm{e}^{-\tau\mathcal{K}_0} - \mathrm{e}^{-\tau\mathcal{K}})\,\mathrm{e}^{-\sigma\mathcal{K}}\, f\|_{\mathfrak{X}_T^p} \le \|(\mathrm{e}^{-\tau\mathcal{K}_0} - \mathrm{e}^{-\tau\mathcal{K}})\, \mathcal{A}_0^{-\alpha}\|_{\mathcal{L}(\mathfrak{X}_T^p)}\, \|\mathcal{A}_0^{\alpha}\, \mathrm{e}^{-\sigma\mathcal{K}}\, f\|_{\mathfrak{X}_T^p}\,.$$

Then, taking into account (11.5.22) and (11.5.6), we obtain the estimate

$$\|(\mathrm{e}^{-\tau\mathcal{K}_0} - \mathrm{e}^{-\tau\mathcal{K}})\,\mathrm{e}^{-\sigma\mathcal{K}} f\|_{\mathfrak{X}_T^p} \le \tau\, C_\alpha\, M_T\, M_0\, \frac{\Lambda_\alpha}{\sigma^\alpha}\, \|f\|_{\mathfrak{X}_T^p}\,,$$

for $f \in \mathrm{dom}\,(\mathcal{K})$ and $\tau \ge 0$. Since $\overline{\mathrm{dom}\,(\mathcal{K})} = \mathfrak{X}_T^p$, the *principle of extension by continuity* provides this estimate also for $f \in \mathfrak{X}_T^p$. Then setting $C_{\mathcal{K}_0} := C_\alpha M_T\, M_0 \Lambda_\alpha$ we complete the proof of (11.5.58). □

Lemma 11.5.15 *Let assumptions* (C1), (C2), (C3), (C4) *and* (C7) *be satisfied. Then there is a constant* $C_{\mathcal{B}_0} > 0$ *such that*

$$\|\mathrm{e}^{-\sigma\mathcal{K}}(\mathcal{I} - \mathrm{e}^{-\tau\mathcal{B}})\|_{\mathcal{L}(\mathfrak{X}_T^p)} \le C_{\mathcal{B}}\, \frac{\tau}{\sigma^\alpha}\,, \tag{11.5.59}$$

for $\tau \ge 0$ *and* $\sigma > 0$.

Proof On account of (C1) and (C4), we use representation

$$\mathrm{e}^{-\sigma\mathcal{K}}(\mathcal{I} - \mathrm{e}^{-\tau\mathcal{B}}) f = \mathrm{e}^{-\sigma\mathcal{K}} \mathcal{A}_0^{\alpha} \mathcal{A}_0^{-\alpha} (\mathcal{I} - \mathrm{e}^{-\tau\mathcal{B}})\, f\,,$$

for $f \in \mathrm{dom}\,(\mathcal{B})$, to deduce estimate

$$\|\,\mathrm{e}^{-\sigma\mathcal{K}}(\mathcal{I} - \mathrm{e}^{-\tau\mathcal{B}}) f\,\|_{\mathfrak{X}_T^p} \le \|\overline{\mathrm{e}^{-\sigma\mathcal{K}} \mathcal{A}_0^{\alpha}}\|_{\mathcal{L}(\mathfrak{X}_T^p)}\, \|\mathcal{A}_0^{-\alpha} (\mathcal{I} - \mathrm{e}^{-\tau\mathcal{B}})\, f\,\|_{\mathfrak{X}_T^p}\,.$$

Then, using (11.5.47), we obtain

$$\|\mathrm{e}^{-\sigma\mathcal{K}}(\mathcal{I} - \mathrm{e}^{-\tau\mathcal{B}})\, f\,\|_{\mathfrak{X}_T^p} \le \frac{\Lambda'_\alpha}{\sigma^\alpha}\, \|\mathcal{A}_0^{-\alpha}\, (\mathcal{I} - \mathrm{e}^{-\tau\mathcal{B}})\, f\,\|_{\mathfrak{X}_T^p}\,. \tag{11.5.60}$$

Seeing that for $f \in \mathrm{dom}\,(\mathcal{B})$

$$\mathcal{A}_0^{-\alpha}(\mathcal{I} - \mathrm{e}^{-\tau\mathcal{B}}) f = \int_0^\tau d\sigma\, \mathcal{A}_0^{-\alpha}\, \mathrm{e}^{-\sigma\mathcal{B}}\, \mathcal{B}\, f\,,$$

one gets by Lemma 11.5.1

$$\mathcal{A}_0^{-\alpha}(\mathcal{I} - \mathrm{e}^{-\tau\mathcal{B}})\, f = \int_0^{\tau} d\sigma\, \overline{\mathcal{A}_0^{-\alpha}\, \mathcal{B}}\, \mathrm{e}^{-\sigma\mathcal{B}}\, f\,,$$

and by (C7) the estimate

$$\|\mathcal{A}_0^{-\alpha}\, (\mathcal{I} - \mathrm{e}^{-\tau\mathcal{B}})\, f\|_{\mathfrak{X}_T^p} \leq \tau\, C_\alpha^*\, M_B\, \|f\|_{\mathfrak{X}_T^p}\,, \quad \tau \geq 0\,.$$

As a consequence, inequality (11.5.60) implies the estimate

$$\|\mathrm{e}^{-\sigma\mathcal{K}}(\mathcal{I} - \mathrm{e}^{-\tau\mathcal{B}})f\|_{\mathfrak{X}_T^p} \leq \frac{\tau}{\sigma^\alpha}\, \Lambda'_\alpha\, C_\alpha^*\, M_B\, \|f\|_{\mathfrak{X}_T^p}\,, \quad f \in \mathrm{dom}\,(\mathcal{B})\,,$$

for $\tau \geq 0$ and $\sigma > 0$. Since the closure $\overline{\mathrm{dom}\,(\mathcal{B})} = \mathfrak{X}_T^p$, this estimate and the *principle of extension by continuity* provide also the same for any $f \in \mathfrak{X}_T^p$. Then setting $C_{\mathcal{B}} := \Lambda'_\alpha\, C_\alpha^*\, M_B$, we get the operator-norm estimate (11.5.59). □

Theorem 11.5.16 *Let assumptions* (C1), (C3), (C4), (C5), (C6) *and* (C7) *be satisfied. If the family of generators* $\{B(t)\}_{t\in[0,T]}$ *is* A_0*-stable and if* $\beta \in (\alpha, 1)$, *then there exists a constant* $C'_{\alpha,\beta} > 0$ *such that, cf.* (11.5.43),

$$\|(\mathrm{e}^{-\tau\mathcal{K}_0/n}\mathrm{e}^{-\tau\mathcal{B}/n})^n - \mathrm{e}^{-\tau\mathcal{K}}\|_{\mathcal{L}(\mathfrak{X}_T^p)} \leq \frac{C'_{\alpha,\beta}}{n^{\beta-\alpha}}\,, \tag{11.5.61}$$

for $\tau \geq 0$ *and* $n \in \mathbb{N}$.

Proof Let us replace in (11.5.43) τ by $\tau\, n/(n+1)$. Then

$$\|(\mathrm{e}^{-\tau\mathcal{B}/(n+1)}\mathrm{e}^{-\tau\mathcal{K}_0/(n+1)})^n - \mathrm{e}^{-n\tau\mathcal{K}/(n+1)}\|_{\mathcal{L}(\mathfrak{X}_T^p)} \leq \frac{C_{\alpha,\beta}}{n^{\beta-\alpha}}\,, \tag{11.5.62}$$

for $\tau \geq 0$ and $n = 2, 3, \ldots$.

To proceed, we consider equation

$$\begin{aligned}
&(\mathrm{e}^{-\tau\mathcal{K}_0/(n+1)}\mathrm{e}^{-\tau\mathcal{B}/(n+1)})^{n+1} - \mathrm{e}^{-\tau\mathcal{K}} \\
&= \mathrm{e}^{-\tau\mathcal{K}_0/(n+1)}(\mathrm{e}^{-\tau\mathcal{B}/(n+1)}\mathrm{e}^{-\tau\mathcal{K}_0/(n+1)})^n\mathrm{e}^{-\tau\mathcal{B}/(n+1)} - \mathrm{e}^{-\tau\mathcal{K}} \\
&= \mathrm{e}^{-\tau\mathcal{K}_0/(n+1)}\left\{(\mathrm{e}^{-\tau\mathcal{B}/(n+1)}\mathrm{e}^{-\tau\mathcal{K}_0/(n+1)})^n - \mathrm{e}^{-n\tau\mathcal{K}/(n+1)}\right\}\mathrm{e}^{-\tau\mathcal{B}/(n+1)} \\
&\quad + (\mathrm{e}^{-\tau\mathcal{K}_0/(n+1)} - \mathrm{e}^{-\tau\mathcal{K}/n+1})\mathrm{e}^{-n\tau\mathcal{K}/(n+1)}\mathrm{e}^{-\tau\mathcal{B}/(n+1)} \\
&\quad + \mathrm{e}^{-\tau\mathcal{K}}(\mathrm{e}^{-\tau\mathcal{B}/(n+1)} - \mathcal{I})\,,
\end{aligned}$$

which implies by (11.2.14) and (11.5.15) inequality

$$
\begin{aligned}
&\|(e^{-\tau\mathcal{K}_0/(n+1)}e^{-\tau\mathcal{B}/(n+1)})^{n+1} - e^{-\tau\mathcal{K}}\|_{\mathcal{L}(\mathfrak{X}^p_T)} \\
&\le M_{0,T}\, M^{\mathcal{B}}_T\, \|(e^{-\tau\mathcal{B}/(n+1)}e^{-\tau\mathcal{K}_0/(n+1)})^n - e^{-n\tau\mathcal{K}/(n+1)}\|_{\mathcal{L}(\mathfrak{X}^p_T)} \\
&\quad + M^{\mathcal{B}}_T\, \|(e^{-\tau\mathcal{K}_0/(n+1)} - e^{-\tau\mathcal{K}/n+1})e^{-n\tau\mathcal{K}/(n+1)}\|_{\mathcal{L}(\mathfrak{X}^p_T)} \\
&\quad + \|e^{-\tau\mathcal{K}}(e^{-\tau\mathcal{B}/(n+1)} - \mathcal{I})\|_{\mathcal{L}(\mathfrak{X}^p_T)}\,.
\end{aligned}
\tag{11.5.63}
$$

For estimation of the right-hand side of (11.5.63), we note that on account of Lemma 11.5.14, (11.5.58),

$$
\left\|(e^{-\tau\mathcal{K}_0/(n+1)} - e^{-\tau\mathcal{K}/n+1})e^{-n\tau\mathcal{K}/(n+1)}\right\|_{\mathcal{L}(\mathfrak{X}^p_T)} \le C_{\mathcal{K}_0}\, \frac{\tau^{1-\alpha}}{n^\alpha (n+1)^{1-\alpha}}\,,
$$

which by *nilpotentness* of semigroups $\{e^{-\tau\mathcal{K}_0}\}_{\tau\ge 0}$ and $\{e^{-\tau\mathcal{K}}\}_{\tau\ge 0}$ on $\mathfrak{X}^p_T$ yields for $\tau \ge 0$ and $n \in \mathbb{N}$

$$
\begin{aligned}
\left\|(e^{-\tau\mathcal{K}_0/(n+1)} - e^{-\tau\mathcal{K}/n+1})e^{-n\tau\mathcal{K}/(n+1)}\right\|_{\mathcal{L}(\mathfrak{X}^p_T)} &\le 2^\alpha\, C_{\mathcal{K}_0}\, \frac{\tau^{1-\alpha}}{n+1} \le \\
&\le 2^\alpha\, C_{\mathcal{K}_0}\, T^{1-\alpha}\, \frac{1}{n+1}\,.
\end{aligned}
\tag{11.5.64}
$$

Then, owing to Lemma 11.5.15 and again by the *nilpotentness* of semigroups on $\mathfrak{X}^p_T$ we obtain inequalities

$$
\left\|e^{-\tau\mathcal{K}}(e^{-\tau\mathcal{B}/(n+1)} - \mathcal{I})\right\|_{\mathcal{L}(\mathfrak{X}^p_T)} \le C_{\mathcal{B}} \frac{\tau^{1-\alpha}}{n+1} \le C_{\mathcal{B}} T^{1-\alpha} \frac{1}{n+1}\,. \tag{11.5.65}
$$

Inserting the estimates (11.5.62), (11.5.64) and (11.5.65) into (11.5.63) we find

$$
\begin{aligned}
&\left\|(e^{-\tau\mathcal{K}_0/(n+1)}e^{-\tau\mathcal{B}/(n+1)})^{n+1} - e^{-\tau\mathcal{K}}\right\|_{\mathcal{L}(\mathfrak{X}^p_T)} \le \\
&\le\; 2^{\beta-\alpha} M_{0,T}\, M^{\mathcal{B}}_T\, C_{\alpha,\beta} \frac{1}{(n+1)^{\beta-\alpha}} + 2^\alpha M^{\mathcal{B}}_T\, C_{\mathcal{K}_0} T^{1-\alpha} \frac{1}{n+1} \\
&\quad + 2^\alpha C_{\mathcal{K}_0} T^{1-\alpha} \frac{1}{n+1}\,, \quad \tau \ge 0\,, \quad n \in \mathbb{N}\,,
\end{aligned}
$$

which yields the estimate (11.5.61) together with a proper value of $C'_{\alpha,\beta}$. □

Similarly, one infers the proof of the following corollary about convergence of the Trotter product formulæ for symmetric approximants.

Corollary 11.5.17 *Let* (C1), (C3), (C4), (C5), (C6) *and* (C7) *be satisfied. If the family of generators* $\{B(t)\}_{t\in[0,T]}$ *is* A_0*-stable and* $\beta \in (\alpha, 1)$, *then there is a constant* $\widetilde{C}_{\alpha,\beta} > 0$ *such that*

$$\left\| (e^{-\tau\mathcal{K}_0/2n} e^{-\tau\mathcal{B}/n} e^{-\tau\mathcal{K}_0/2n})^n - e^{-\tau\mathcal{K}} \right\|_{\mathcal{L}(\mathfrak{X}_T^p)} \leq \frac{\widetilde{C}_{\alpha,\beta}}{n^{\beta-\alpha}}, \tag{11.5.66}$$

as well as

$$\left\| (e^{-\tau\mathcal{B}/2n} e^{-\tau\mathcal{K}_0/n} e^{-\tau\mathcal{B}/n})^n - e^{-\tau\mathcal{K}} \right\|_{\mathcal{L}(\mathfrak{X}_T^p)} \leq \frac{\widetilde{C}_{\alpha,\beta}}{n^{\beta-\alpha}}, \tag{11.5.67}$$

for $\tau \geq 0$ *and* $n \in \mathbb{N}$.

11.5.3 Operator-Norm Product Formulæ for Solution Operator

The ultimate aim of Chap. 11 is to prove the *operator-norm* product formulæ approximations for *solution operator* of the non-autonomous Cauchy problem in a Banach space $\mathfrak{X}$.

In contrast to a direct approach developed in Sect. 11.1, which proves the convergence of approximants in the *strong* operator topology, a method based on the *evolution semigroup* approach and the *Howland-Evans–Neidhardt* formula (see Theorem 2.2.11) allows to solve this problem of convergence in the *operator-norm* topology on $\mathcal{L}(\mathfrak{X})$. The key point (instrument) of this method is the proof in Sect. 11.5.2 (Theorems 11.5.10, 11.5.16 and Corollary 11.5.17) about *operator-norm* convergence of symmetric and non-symmetric Trotter product formulæ to evolution semigroup $\{e^{-\tau\mathcal{K}}\}_{\tau\geq 0}$ on the Banach space $\mathfrak{X}_T^p$.

So, after this important preparatory point on the evolution semigroup Trotter product formulae, we collected a sufficient number of argument to establish the operator-norm product approximations of solution operator $\{U(t,s)\}_{(t,s)\in\Delta}$ for nACP (11.3.1) with a time-dependent perturbation of generator A_0. They ensure, for example, the operator-norm convergence to $\{U(t,s)\}_{(t,s)\in\Delta}$ of approximants $\{U_n^{A_0}(t,s)\}_{(t,s)\in\Delta}$, for $n \in \mathbb{N}$, given by (11.4.10) and (11.4.24).

Note that concept of A_0-stability (Lemma 11.4.3 and Remark 11.4.4) is important for control of *acceptable* perturbations in the assertion below.

Theorem 11.5.18 *Let assumptions* (C1), (C3), (C4), (C5) *and* (C6) *be satisfied. If the family* $\{B(t)\}_{t\in[0,T]}$ *is* A_0*-stable and* $\beta \in (\alpha, 1)$, *then there exists constant* $C_{\alpha,\beta} > 0$ *such that*

$$\operatorname*{ess\,sup}_{(t,s)\in\Delta} \|U_n^{A_0}(t,s) - U(t,s)\|_{\mathcal{L}(\mathfrak{X})} \leq \frac{C_{\alpha,\beta}}{n^{\beta-\alpha}}, \quad n \in \mathbb{N}. \tag{11.5.68}$$

The constant $C_{\alpha,\beta}$ *coincides with that in the estimate* (11.5.43) *from Theorem* 11.5.10.

Proof We use approximants $\{U_n^{A_0}(t,s)\}_{(t,s)\in\Delta}$, (11.4.10), (11.4.24) to set

$$S_n(t,s) := U_n^{A_0}(t,s) - U(t,s), \quad (t,s) \in \Delta, \quad n \in \mathbb{N}. \tag{11.5.69}$$

Remind that due to the *Howland–Evans–Neidhardt* formula (2.2.12) for evolution semigroup $\{e^{-\tau\mathcal{K}}\}_{\tau\geq 0}$ and the corresponding solution operator $\{U(t,s)\}_{(t,s)\in\Delta}$ we have representation:

$$(e^{-\tau\mathcal{K}} f)(t) = U(t, t-\tau)\chi_{[0,T]}(t-\tau)f(t-\tau), \quad f \in \mathfrak{X}_T^p, \text{ a.a. } t \in [0,T], \tag{11.5.70}$$

as well as, see (11.4.26), representation

$$((e^{-\tau\mathcal{B}/n}e^{-\tau\mathcal{K}_0/n})^n f)(t) = U_n^{A_0}(t, t-\tau)\chi_{[0,T]}(t-\tau)f(t-\tau), \tag{11.5.71}$$

for $\tau \geq 0$, a.a. $t \in [0,T]$, and for any $f \in \mathfrak{X}_T^p$.

Using the left-shift contraction semigroup $\{\mathcal{S}_L(\tau)\}_{\tau\geq 0}$ (2.2.39) and representations (11.5.70), (11.5.71), we obtain

$$(\mathcal{S}_L(\tau)e^{-\tau\mathcal{K}}) f)(t) = \chi_{[0,T]}(t+\tau)\, U(t+\tau, t)\chi_{[0,T]}(t)f(t), \quad \tau \geq 0, \tag{11.5.72}$$

and

$$(\mathcal{S}_L(\tau)(e^{-\tau/n\,\mathcal{B}}e^{-\tau/n\,\mathcal{K}_0})^n f)(t) = \chi_{[0,T]}(t+\tau)U_n^{A_0}(t+\tau, t)\,\chi_{[0,T]}(t)\,f(t). \tag{11.5.73}$$

Then on account of definition (11.5.69) and (11.5.72), (11.5.73) we introduce on the space $\mathfrak{X}_T^p$, for $n \in \mathbb{N}$, operator

$$\mathfrak{S}_n(\tau) := \mathcal{S}_L(\tau)\{(e^{-\tau\mathcal{B}/n}e^{-\tau\mathcal{K}_0/n})^n - e^{-\tau\mathcal{K}}\}, \quad \tau \geq 0, \tag{11.5.74}$$

which has representation:

$$(\mathfrak{S}_n(\tau)f)(t) = S_n(t+\tau, t)\chi_{[0,T]}(t+\tau)f(t), \quad f \in \mathfrak{X}_T^p, \quad n \in \mathbb{N}. \tag{11.5.75}$$

Thus, for each $n \in \mathbb{N}$ and $\tau \in [0,T]$ the operator (11.5.74) is a multiplication operator on $\mathfrak{X}_T^p$ induced by the family $\{S_n(\cdot+\tau,\cdot)\chi_{[0,T]}(\cdot+\tau)\}_{n\in\mathbb{N}}$ of bounded operators.

Since C_0-semigroups $\{e^{-\tau\mathcal{K}_0}\}_{\tau\geq 0}$ and $\{e^{-\tau\mathcal{K}}\}_{\tau\geq 0}$ are nilpotent on $\mathfrak{X}^p_T$, we consider only $\tau \in [0, T]$. Then on account of the left-shift operator $\{\mathcal{S}_L(\tau)\}_{\tau\geq 0}$ we have

$$\sup_{\tau\in[0,T]} \|(e^{-\tau\mathcal{B}/n} e^{-\tau\mathcal{K}_0/n})^n - e^{-\tau\mathcal{K}}\|_{\mathcal{L}(\mathfrak{X}^p_T)} =$$
$$= \sup_{\tau\in[0,T]} \|\mathcal{S}_L(\tau)\{(e^{-\tau\mathcal{B}/n} e^{-\tau\mathcal{K}_0/n})^n - e^{-\tau\mathcal{K}}\}\|_{\mathcal{L}(\mathfrak{X}^p_T)} . \tag{11.5.76}$$

As a consequence, (11.5.75) and (11.5.76) yield

$$\sup_{\tau\in[0,T]} \|(e^{-\tau\mathcal{B}/n} e^{-\tau\mathcal{K}_0/n})^n - e^{-\tau\mathcal{K}}\|_{\mathcal{L}(\mathfrak{X}^p_T)} = \sup_{\tau\in[0,T]} \|\mathfrak{S}_n(\tau)\|_{\mathcal{L}(\mathfrak{X}^p_T)} . \tag{11.5.77}$$

Then owing to relation (2.2.22) for operator norm of multiplication operators in Banach spaces $\mathfrak{X}^p_T$ and $\mathfrak{X}$ (Proposition 2.2.16, or Theorem E.3.12 in Sect. E.3.3), together with (11.5.69), (11.5.75), we obtain

$$\sup_{\tau\in[0,T]} \|\mathfrak{S}_n(\tau)\|_{\mathcal{L}(\mathfrak{X}^p_T)} = \sup_{\tau\in[0,T]} \operatorname*{ess\,sup}_{t\in[0,T]} \|S_n(t+\tau, t)\chi_{[0,T]}(t+\tau)\|_{\mathcal{L}(\mathfrak{X})}$$
$$= \sup_{\tau\in[0,T]} \operatorname*{ess\,sup}_{t\in[0,T]} \|\{U_n^{A_0}(t+\tau, t) - U(t+\tau, t)\}\chi_{[0,T]}(t+\tau)\|_{\mathcal{L}(\mathfrak{X})}$$
$$= \operatorname*{ess\,sup}_{t\in[0,T-\tau]} \|U_n^{A_0}(t+\tau, t) - U(t+\tau, t)\|_{\mathcal{L}(\mathfrak{X})} . \tag{11.5.78}$$

For that reason by equations (11.5.76)–(11.5.78) and by Theorem 11.5.10 we deduce the estimate:

$$\operatorname*{ess\,sup}_{t\in(0,T-\tau]} \|U_n^{A_0}(t+\tau, t) - U(t+\tau, t)\|_{\mathcal{L}(\mathfrak{X})} \leq \frac{C_{\alpha,\beta}}{n^{\beta-\alpha}}, \quad \tau \in [0, T], \quad n = 2, 3, \ldots .$$

It yields the operator-norm convergence rate (11.5.68) for product approximants $\{U_n^{A_0}(t, s)\}_{(t,s)\in\Delta}$ of solution operator $\{U(t, s)\}_{(t,s)\in\Delta}$. □

Theorem 11.5.19 *Let assumptions* (C1), (C3), (C4), (C5), (C6) *and* (C7) *be satisfied. If the family* $\{B(t)\}_{t\in[0,T]}$ *is* A_0*-stable and* $\beta \in (\alpha, 1)$*, then there exists a constant* $C'_{\alpha,\beta} > 0$ *such that*

$$\operatorname*{ess\,sup}_{(t,s)\in\Delta} \|V_n^{A_0}(t, s) - U(t, s)\|_{\mathcal{L}(\mathfrak{X})} \leq \frac{C'_{\alpha,\beta}}{n^{\beta-\alpha}}, \quad n \in \mathbb{N} . \tag{11.5.79}$$

The constant $C'_{\alpha,\beta}$ *coincides with that in the estimate* (11.5.61) *from Theorem* 11.5.16.

Proof Note that by (11.4.12) and (11.4.25), the product approximants $\{V_n^{A_0}(t,s)\}_{(t,s)\in\Delta}$ take the form:

$$V_n^{A_0}(t,s) = \prod_{k=n}^{1} \mathrm{e}^{-\frac{t-s}{n}A_0}\mathrm{e}^{-\frac{t-s}{n}B(s+k\frac{t-s}{n})}, \quad n \in \mathbb{N}. \tag{11.5.80}$$

Then, by definition of semigroup generators $\mathcal{K}_0$, $\mathcal{B}$ and owing to (11.5.80) the representation (cf. (11.4.28))

$$((\mathrm{e}^{-\tau\mathcal{K}_0/n}\,\mathrm{e}^{-\tau\mathcal{B}/n})^n f)(t) = V_n^{A_0}(t, t-\tau)\chi_{[0,T]}(t-\tau)f(t-\tau), \tag{11.5.81}$$

holds for a.a. $t \in [0,T]$, $\tau \geq 0$ and $f \in \mathfrak{X}_T^p$.

If we introduce $S_n'(t,s) := V_n^{A_0}(t,s) - U(t,s)$, $(t,s) \in \Delta$, $n \in \mathbb{N}$ and operators

$$\mathfrak{S}_n'(\tau) := \mathcal{S}_L(\tau)\{(\mathrm{e}^{-\tau\mathcal{K}_0/n}\,\mathrm{e}^{-\tau\mathcal{B}/n})^n - \mathrm{e}^{-\tau\mathcal{K}}\}, \quad \tau \geq 0, \quad n \in \mathbb{N},$$

then $(\mathfrak{S}_n'(\tau)f)(t) = S_n'(t+\tau,t)\chi_{[0,T]}(t+\tau)f(t)$, for any $f \in \mathfrak{X}_T^p$, is multiplication operator.

Now following *verbatim* the line of reasoning in the proof of Theorem 11.5.18 we obtain (cf. (11.5.78))

$$\sup_{\tau\in[0,T]} \|\mathfrak{S}_n'(\tau)\|_{\mathcal{L}(\mathfrak{X}_T^p)} = \sup_{\tau\in[0,T]}\ \operatorname*{ess\,sup}_{t\in[0,T-\tau]} \|V_n^{A_0}(t+\tau,t) - U(t+\tau,t)\|_{\mathcal{L}(\mathfrak{X})}\,. \tag{11.5.82}$$

Then, by Theorem 11.5.16, we infer the estimate (11.5.79). □

Remark 11.5.20 Let approximants $\{W_n^{A_0}(t,s)\}_{n\geq 1}$ have symmetric form:

$$W_n^{A_0}(t,s) := \prod_{k=n-1}^{0} \mathrm{e}^{-\frac{t-s}{2n}A_0}\,\mathrm{e}^{-\frac{t-s}{n}B(s+(k+1/2)\frac{t-s}{n})}\,\mathrm{e}^{-\frac{t-s}{2n}A_0}, \quad n \in \mathbb{N}. \tag{11.5.83}$$

Then, on account of Theorem 11.5.19, the estimate (11.5.66) yields

$$\operatorname*{ess\,sup}_{(t,s)\in\Delta} \|W_n^{A_0}(t,s) - U(t,s)\|_{\mathcal{L}(\mathfrak{X})} \leq \frac{\widetilde{C}_{\alpha,\beta}}{n^{\beta-\alpha}}, \quad n \in \mathbb{N}, \tag{11.5.84}$$

for $\beta \in (\alpha, 1)$. The proof of (11.5.84) follows *verbatim* the line of reasoning of Theorems 11.5.18 and 11.5.19.

11.6 Notes to Chap. 11

Notes to Sects. 11.1

In this section, we essentially follow [VWZ08] and [VWZ09]. There it was proven that approximants in the product formula converge to solution operator of the *parabolic*-type nACP in the *strong operator* topology, including the case, when domain of the nACP generator in a separable Banach space is *time-dependent*. We remark here that a *time-independent* case for domain of the nACP generator was carried out in a similar setup for the product formula approximation of solution operator in [Far67b].

We note that aside from hypotheses ***(C1)–(C3), there exist several other sufficient conditions that would have allowed the construction of solution operator when $B(t) = 0$. We refer to [AcqTer87] for a general and thorough investigation of such conditions and for the relations between them. In particular, we could have used hypotheses of that paper in the somewhat stronger form introduced in [RRSV00] and [Schn00] to get such a result. There also exist various sufficient conditions which could have led to the existence of perturbed solution operator $\{U_{A+B}(t,s)\}_{0\leq s<t\leq T}$ of nACP for suitable classes of perturbation $\{B(t)\}_{0\leq t\leq T}$, for example those put forward in [RRSV00] and [Schn00]. However, a basic difficulty emerges there when one tries to prove for them a product formula such as (11.1.3).

Thus, while the $U_{A+B}(t,s)$ of Theorem 9.19 in [Schn00] are lacking good differentiability properties, those of Theorem 4.2 in [RRSV00] are only weakly locally differentiable relative to the time variable and satisfy an equation such as (11.1.1) only almost everywhere. A direct consequence of this is that such evolution systems are not directly amenable to the way we prove product formula (11.1.2)–(11.1.3), which requires the $U_{A+B}(t,s)$ to be continuously differentiable in t for the strong operator topology on $\mathcal{L}(\mathfrak{X})$.

Furthermore, this strong smoothness property of solution operator in [RRSV00] and [Schn00] does not readily follow from our hypotheses regarding $\{B(t)\}_{0\leq t\leq T}$ unless we assume more regularity properties on this perturbation. On that account, it is by far the general framework of [Tan79] that has allowed us to prove Theorem 11.1.3. Of course, whether one can prove for parabolic case a *strongly* convergent product formula such as (11.1.13) under sole conditions of [RRSV00, Schn00] or under more general conditions, remains to our knowledge an open problem.

The product formula approximation as a method of construction of solution operator (Theorem 11.1.3 (4)) offers an alternative approach to the Sobolevskiĭ–Kato theory of non-autonomous *parabolic* evolution equations on Banach space, which was developed at the end of fifties, see [Sob61, Kato61]. Since that time this theory has been successfully applied to numerous problems, particularly when the domains of the operators involved into nACP are *time-independent*. However, when those domains become *time-dependent* the Kato theory ([Kato61, Tan79], Section 5.3) imposes rather strong *invariance* conditions, which are as a rule difficult to check in practice, see, e.g., [VWZ09, Section 1]. This remark applies, for instance, to the verification of the first product formula in [Nick00], which, incidentally, does bear some formal resemblance with approximants (11.1.14). On the contrary, our result [VWZ09], which is based on the product formula approximation method, does not require any such invariance conditions.

Notes to Sects. 11.2

In contrast with direct approach to construction of the nACP solution operator by means of product formula approximation developed in Sect. 11.1, which proves the convergence of approximants in the strong operator topology, a method based on the Howland–Evans–Neidhardt *evolution semigroup* approach [How74, Ev76, Nei81], allows to prove convergence of these product approximants in the *operator-norm* topology.

In this section, we start a rather *round-about way* to justify this alternative method for the operator-norm product formula approximation of the nACP solution operator in the L^p-setting. To this aim we rectify some results of Sect. 2.2 and prove certain preparatory assertions. We also introduce in this section assumptions (C1)–(C7), plainly formulated for the first time in [NSZ20]. They provide technical tools, which are indispensable in the proceeding sections of the Chap. 11. Note that on account of assumptions (C1)–(C7) domain of the nACP generator in a separable Banach space is *time-independent*, cf. Sect. 11.1, which covers the case of the time-dependent domain.

In presentation of this section, we follow ideas and results that appeared in [Nei81, Nei81a, NSZ17, NSZ20].

Notes to Sects. 11.3

We studied the nACP for time-dependent perturbation in a Banach space in [NSZ17, NSZ20]. In this section we rectify and adapt results of Sect. 2.2 to a special form (11.3.1) of the nACP generator, where time-independent part is generator of a holomorphic semigroup, cf. (C1).

Notes to Sects. 11.4

In this section, we vindicate the *evolution semigroup* method for proving the *strong* operator convergence of the Trotter product formula (Proposition 1.9.41) for evolution semigroup $\{e^{-\tau\mathcal{K}}\}_{\tau\geq 0}$ on Banach space $\mathfrak{X}^p_T$ when generator is splitted into Trotter stable pairs of generators. This result allows to proceed with the second step and to advance with convergence to solution operator $\{U(t,s)\}_{(t,s)\in\Delta}$ of the product approximants $\{U_n(t,s)\}_{n\geq 1}$ in the *integral* topology, which was introduced in [NSZ20], Theorem 1.3.

This was an intermediate step before uplifting the strong topology for convergence of $\{U_n(t,s)\}_{n\geq 1}$ to the operator-norm topology as it is guessed in Sect. 11.2.

Notes to Sects. 11.5

In this section, we prove convergence the product approximants $\{U_n^{A_0}(t,s)\}_{n\geq 1}$ (and $\{V_n^{A_0}(t,s)\}_{n\geq 1}$) in the *operator-norm* topology. This result is due to [NSZ20]. The uplifting from the *strong* to the *operator-norm* topology with estimate of the convergence *rate* is the ultimate aim of Chap. 11.

Although presentation in this section is complete and detailed, see Theorems 11.5.18 and 11.5.19, as well as, Theorems 11.5.10 and 11.5.16, the reader can find original formulations of the main assertions in [NSZ20, Theorem 7.11, Remark 7.12] and in review [NSZ18b, Theorem 3.9].

Section 11.5.1

Here we present and prove technical lemmata indispensable for demonstration of convergence and for estimate of the error bound for the Trotter product formula approximations (11.5.2) in the operator-norm topology on the set of bounded operators $\mathcal{L}(\mathfrak{X}_T^p)$. They are formulated in the framework of Assumptions 11.2.12.

As the main instrument, we use a generalised *Gronwall-type* lemma. It is proved by iteration of the corresponding *Volterra* integral equation, see Appendix G, Lemma G.2.1. Our generalisation is motivated by [HuNa01, Theorem 2.25]. This instrument is bolstered by the standard *principle of extension by continuity* for bounded operators, see Appendix A, Sect. A.1.1.

Section 11.5.2

1. We prove in this subsection the operator-norm Trotter product formulæ for evolution semigroup, see Theorems 11.5.10 and 11.5.16. These assertions are instructive to compare with Theorems 10.2.6 and 10.2.7 in Sect. 10.2. There the perturbation B (Remark 10.2.1) in the space $\mathfrak{X}$ is *time-independent*, whereas the perturbation $\mathcal{B}$ in the space $\mathfrak{X}_T^p$ is a τ*-independent* multiplication operator generated by the family $\{B(t)\}_{t\in[0,T]}$ that satisfies Assumptions 11.2.12.

   This fundamental difference between *time-independent* setup for the Trotter product formula on $\mathfrak{X}$ and the τ*-independent* setup for the Trotter product formula on $\mathfrak{X}_T^p$ reflects in differences of techniques for the proofs of convergence and for values of the operator-norm rates of convergence. Compare, for example, (10.2.13), (10.2.17) and (11.5.43), (11.5.61). In the latter case the rate of convergence depends on the parameter $\beta > 0$, which is the Hölder exponent that controls the *smoothness* of the family $\{B(t)\}_{t\in[0,T]}$, see Assumptions 11.2.12 (C6).
2. Proof of the estimate (11.5.44) in Corollary 11.5.11 for $\beta = 1$ follows *verbatim et literatim* the proof of Theorem 11.5.10, [NSZ20, §7.2]. Similarly one gets for $\beta = 1$ the estimate (11.5.44) (instead of (11.5.61)) in Theorem 11.5.16.

   Note that under a bit stronger condition than in Corollary 11.5.11 for $\beta = 1$, we obtain the following result, cf. [NSZ17, Theorem 5.4].

Proposition 11.6.1 *Let assumptions* (C1), (C3), (C4), (C5) *be satisfied and let instead of* (C6) *the assumption*:
(C6*) *For* $\beta = 1$ *there exists a constant* $L_{\alpha,1} > 0$ *such that estimate*

$$\|A_0^{-\alpha}(B(t) - B(s))\, A_0^{-\alpha}\|_{\mathcal{L}(\mathfrak{X})} \leq L_{\alpha,1}\, |t - s|\,,$$

be satisfied for almost all $t, s \in [0, T]$.

If $\alpha \in (1/2, 1)$ and the family of generators $\{B(t)\}_{t\in[0,T]}$ is A_0-stable, then there exists a depending on $\alpha > 0$ and on T coefficient $C_{\alpha,1} > 0$ such that

$$\|(\mathrm{e}^{-\tau\mathcal{B}/n}\,\mathrm{e}^{-\tau\mathcal{K}_0/n})^n - \mathrm{e}^{-\tau\mathcal{K}}\|_{\mathcal{L}(\mathfrak{X}_T^p)} \le \frac{C_{\alpha,1}}{n^{1-\alpha}}\,, \tag{11.6.1}$$

for $\tau \ge 0$ and $n \in \mathbb{N}$.

The result (11.6.1) is *stronger* than (11.5.43) and (11.5.44). since assumption (C6*) is stronger than (C6). Indeed, by virtue of (C6*) one gets for $\beta \le 1$ and for appropriate $L_\beta > 0$ the estimate

$$\|A_0^{-1}(B(t)-B(s))\,A_0^{-\alpha}\|_{\mathcal{L}(\mathfrak{X})} \le \|\overline{(A_0^{-1}A_0^{\alpha})}\,A_0^{-\alpha}(B(t)-B(s))\,A_0^{-\alpha}\|_{\mathcal{L}(\mathfrak{X})} \le$$
$$\|A_0^{-\alpha}(B(t)-B(s))\,A_0^{-\alpha}\|_{\mathcal{L}(\mathfrak{X})} \le L_{\alpha,1}\,|t-s| \le L_\beta\,|t-s|^\beta\,, \quad t,s\in[0,T]\,,$$

which coincides with assumption (C6).

We return to the assumption (C6*) and the proof of Proposition 11.6.1 in Chap. 12. It turns out that they coincide *mutatis mutandis* with assumptions and arguments that we used for scrutinising the time-dependent product formulæ in a Hilbert space.

3. Until the papers [NSZ17, NSZ18a, NSZ20], the error bound estimates in the operator-norm topology for the Trotter product formulæ on *Banach spaces* have been proven only under *assumption* that at least one of the involved operators is generator of a *holomorphic* contraction semigroup, see [CZ01a].

   Note that evolution semigroups in Theorems 11.5.10 and 11.5.16 are *nilpotent* and thus can be never holomorphic. Therefore, although motivated by [CZ01a] the operator-norm Trotter product formulæ in Theorems 11.5.10 and 11.5.16, together with Example 10.3 (Chap. 10), are the first results, where assumption of analyticity is *dropped*, see review [NSZ18b, Sections 3 and 4].

   As it is shown in Chap. 10 (Sect. 10.3) the known examples of the *operator-norm* convergent Trotter product formulæ *without* hypothesis of analyticity follows a rather *round-about way* that involves Banach spaces for *evolution* semigroups. This concerns, for example, the "solvable" example in [NSZ18b, Subsection 4.1] as well as Theorem 10.3.3 in Chap. 10.
4. If the family of generators is independent of $t \in [0,T]$, i.e. $B(t) = B$, then condition (C6) is automatically satisfied for any $\beta \ge 0$. In particular, we can set $\beta = 1$. Since $\mathcal{A}$ and $\mathcal{B}$ commute with D_0, we get, cf. (11.5.45):

$$(\mathrm{e}^{-\tau\mathcal{B}/n}\mathrm{e}^{-\tau\mathcal{K}_0/n})^n = (\mathrm{e}^{-\tau\mathcal{B}/n}\mathrm{e}^{-\tau\mathcal{A}/n})^n\,\mathrm{e}^{-\tau D_0}, \quad \tau \ge 0\,, \quad n\in\mathbb{N}. \tag{11.6.2}$$

5. Now we comment that if one of the operators: A or B, is generator of a *holomorphic* contraction semigroup and another one (*infinitesimally* relative small) of a contraction semigroup on a Banach space $\mathfrak{X}$, then from [CZ01a, Theorem 3.6] we get the existence of constants $b_1 > 0$, $b_2 > 0$ and $\eta > 0$

such that the estimate

$$\|(\mathrm{e}^{-\tau B/n}\mathrm{e}^{-\tau A/n})^n - \mathrm{e}^{-\tau C}\|_{\mathcal{L}(\mathfrak{X})} \leq (b_1 + b_2\tau^{1-\alpha})\, \mathrm{e}^{\tau\eta}\, \frac{\ln(n)}{n^{1-\alpha}}, \quad \alpha \in (0,1),$$

holds for $\tau \geq 0$ and $n \in \mathbb{N}$. Applying this result to (11.6.2), we immediately obtain the existence of a constant $R > 0$ such that

$$\|(\mathrm{e}^{-\tau\mathcal{B}/n}\mathrm{e}^{-\tau\mathcal{K}_0/n})^n - \mathrm{e}^{-\tau\mathcal{K}}\|_{\mathcal{L}(\mathfrak{X}^p_T)} \leq R\, \frac{\ln(n)}{n^{1-\alpha}}, \quad \alpha \in (0,1),$$

is valid for $\tau \geq 0$ and $n \in \mathbb{N}$. Note that this estimate is *sharper* than the estimate in (11.5.44).

Section 11.5.3

Operator-norm product formulæ for *solution operator*, see Theorems 11.5.18 and 11.5.19, follow from Theorems 11.5.10 and 11.5.16 about the evolution semigroup Trotter product formulæ as consequence of the *Howland–Evans–Neidhardt* formula, see equations (11.5.70), (11.5.71). In particular, they ensure the operator-norm convergence of approximants $\{U_n^{A_0}(t,s)\}_{(t,s)\in\Delta}$, $n \in \mathbb{N}$, given by (11.4.10), (11.4.24), to solution operator $\{U(t,s)\}_{(t,s)\in\Delta}$ of nACP (11.3.1) for a time-dependent perturbation of generator A_0.

Similarly to Theorem 11.5.18 and with necessary modifications, one infers from Proposition 11.6.1 and (11.6.1) the corresponding results for the case $\beta = 1$:

$$\operatorname*{ess\,sup}_{(t,s)\in\Delta} \|U_n^{A_0}(t,s) - U(t,s)\|_{\mathcal{L}(\mathfrak{X})} \leq \frac{C^*_{\alpha,1}}{n^{1-\alpha}}, \quad \alpha \in (1/2,1), \quad n \in \mathbb{N}. \tag{11.6.3}$$

We note that within assumptions (C1)–(C7) and (C6*) this programme was formulated and developed in [NSZ17, NSZ20].

Chapter 12
Time-Dependent Product Formulæ: Hilbert Space

In this chapter we present the product-formula approximations of *solution operator* for non-autonomous Cauchy problem (nACP) in a Hilbert space. Under a similar to Chap. 11 the *infinitesimal smallness* and *time-continuity* conditions on perturbation in the nACP, the product approximation can be established within such approach in the *operator-norm* topology, but with (in a certain sense) better estimate for the bound of the rate of convergence. The latter is essentially due to a supplementary condition of the *self-adjointness*.

We compare this approach with the evolution semigroup method in the L^p-setting for Banach spaces developed for the nACP in Chap. 11. To this aim we first present the Hilbert space analogue of the *evolution semigroups* method in L^2-setting. This (indirect) method yields the operator-norm product formula approximations for *solution operator* of non-autonomous Cauchy problem in Hilbert spaces. Similar to the case of the Banach L^p-setting, our main instrument will be the operator-norm Trotter product formula for the evolution semigroup corresponding to nACP, together with the *Howland-Evans-Neidhardt* (HEN) formula.

In the second part of this chapter, we present a direct *Ichinose-Tamura* approach to the Trotter-product formula approximations for solution operator of the nACP in a Hilbert space. It gives quite accurate estimates for the operator-norm rate of convergence for the product of the nACP solution operator approximants.

12.1 The Evolution Semigroups Method in L^2-Setting

12.1.1 The L^2-Setting in the Hilbert Space Framework

The aim of the subsection is to introduce the *evolution semigroup* formalism for non-autonomous Cauchy problem within the L^2-setting in the Hilbert space framework.

V. A. Zagrebnov et al., *Trotter-Kato Product Formulæ*, Operator Theory: Advances and Applications 296, https://doi.org/10.1007/978-3-031-56720-9_12

After that, similar to the Banach L^p-setting (Chap. 11), we shall establish the product formula approximations for the corresponding *solution operator*.

To this end we mimic the arguments in Sect. 11.3 and consider in *separable* Hilbert space $\mathfrak{H}$ the evolution equation with time-dependent perturbation $\{B(t)\}_{t\in[0,T]}$ of the type

$$\begin{aligned} &\partial_t u(t) = -\, C(t)u(t), \quad u(s) = u\,, \quad s \in [0, T) \subset \mathbb{R}_0^+\,, \\ &C(t) := A + B(t)\,, \end{aligned} \qquad t \in [0, T], \tag{12.1.1}$$

where $\mathbb{R}_0^+ = \{0\} \cup \mathbb{R}^+$ and the initial condition $u \in \mathfrak{H}$. We consider this *parabolic* non-autonomous Cauchy problem (nACP) under following assumptions; cf. Assumptions 11.2.12 in a Banach space:

Assumptions 12.1.1 Let $\mathfrak{H}$ be a separable Hilbert space and $\alpha \in [0, 1)$.

(D1) The operator $A \geq \mathbb{1}$ is self-adjoint in $\mathfrak{H}$ and $\{B(t)\}_{t\in[0,T]}$ is a family of non-negative self-adjoint operators in $\mathfrak{H}$ so that function $(\mathbb{1} + B(\cdot))^{-1} : [0, T] \longrightarrow \mathcal{L}(\mathfrak{H})$ is strongly measurable.

(D2) There exists an $\alpha \in [0, 1)$ such that for *almost all* (a.a.) $t \in [0, T]$ it holds the inclusion $\operatorname{dom}(A^\alpha) \subseteq \operatorname{dom}(B(t))$. In addition, the function $B(\cdot)A^{-\alpha}$: $[0, T] \longrightarrow \mathcal{L}(\mathfrak{H})$ is strongly measurable and essentially bounded, that is,

$$C_\alpha := \operatorname*{ess\,sup}_{t\in[0,T]} \|B(t)A^{-\alpha}\|_{\mathcal{L}(\mathfrak{H})} < \infty\,. \tag{12.1.2}$$

(D3) The mapping $A^{-\alpha}B(\cdot)A^{-\alpha} : [0, T] \longrightarrow \mathcal{L}(\mathfrak{H})$ is *Hölder*-continuous, that is, for some $\beta \in (0, 1)$ there is a constant $L_{\alpha,\beta} > 0$ such that one has operator-norm estimate

$$\|A^{-\alpha}(B(t) - B(s))A^{-\alpha}\|_{\mathcal{L}(\mathfrak{H})} \leq L_{\alpha,\,\beta}\,|t - s|^\beta, \quad (t, s) \in [0, T] \times [0, T]\,, \tag{12.1.3}$$

cf. Assumptions 11.2.12 (C6) and Proposition 11.6.1 (C6*).

Assumptions (D1) and (D2) provide that for any $t \in [0, T]$ the positive self-adjoint operator $C(t)$ in (12.1.1) is generator of a contraction holomorphic semigroup, namely, $C(t) \in \mathscr{H}_c(\pi/2, 0)$, Theorem 9.3.1. Therefore, by Definition 2.1.11 the nACP (12.1.1) is of the *parabolic* type.

Definition 12.1.2 We consider the Hilbert space $\mathfrak{H}_T := L^2([0, T], \mathfrak{H})$ consisting of measurable functions: $f(\cdot) : [0, T] \longrightarrow \mathfrak{H},\ g(\cdot) : [0, T] \longrightarrow \mathfrak{H}$, with scalar product

$$(f, g)_{\mathfrak{H}_T} := \int_0^T dt\,(f(t), g(t))_{\mathfrak{H}}\,, \qquad f, g \in \mathfrak{H}_T\,. \tag{12.1.4}$$

The space $\mathfrak{H}_T$ is endowed with the norm

$$\|f\|_{\mathfrak{H}_T} := \left(\int_0^T dt\, \|f(t)\|^2_{\mathfrak{H}}\right)^{1/2}, \quad \text{for} \quad f \in \mathfrak{H}_T, \tag{12.1.5}$$

cf. (2.2.1) for L^p-setting in Sect. 2.2.1. Below for (12.1.5) we shall use $\|f\|$, where $f \in \mathfrak{H}_T$, if it will not produce a confusion. By $\|\cdot\|_{\mathcal{L}(\mathfrak{H}_T)}$ we denote the operator norm on the Hilbert space $\mathfrak{H}_T$. We shall also skip the operator-norm subindex $\mathcal{L}(\mathfrak{H}_T)$, when it does not produce any confusion.

Remark 12.1.3 On account of Definition 12.1.2, the rest of notations, definitions and statements concerning the L^2-setting in the Hilbert space framework follows *mutatis mutandis* from the L^p-setting of Sect. 2.2. Below we make explicit only points where these two settings are different. If not, we use appropriate results of the L^p-setting directly.

Let D_0 be generator of the right-shift semigroup on $\mathfrak{H}_T$, that is, for any $t \in [0, T]$:

$$(e^{-\tau D_0} f)(t) = \chi_{[0,T]}(t-\tau) f(t-\tau), \quad \tau \in \mathbb{R}_0^+, \quad f \in \mathfrak{H}_T. \tag{12.1.6}$$

Seeing that, the right-shift semigroup$\{e^{-\tau D_0}\}_{\tau \geq 0}$ on the space $\mathfrak{H}_T$ is *nilpotent* since $e^{-\tau D_0} = 0$ for $\tau \geq T$. The generator D_0 in (12.1.6) is given by

$$(D_0 f)(t) := \partial_t f(t),$$
$$f \in \mathrm{dom}(D_0) := W_0^{1,2}([0, T], \mathfrak{H}) = \{f \in W^{1,2}([0, T], \mathfrak{H}) : f(0) = 0\}.$$

We remark that operator D_0 is *not* self-adjoint, but it is a closure of *maximal symmetric* operator, and that semigroup $\{e^{-\tau D_0}\}_{\tau \geq 0}$ is a contraction: $\|e^{-\tau D_0}\|_{\mathcal{L}(\mathfrak{H}_T)} \leq 1$.

Next we define in the Hilbert space $\mathfrak{H}_T$ a *time-independent* (unbounded) multiplication operator $\mathcal{A}$.

$$(\mathcal{A} f)(t) := A f(t),$$
$$f \in \mathrm{dom}\,(\mathcal{A}) := \left\{ f \in \mathfrak{H}_T : \begin{array}{c} f(t) \in \mathrm{dom}\,(A) \quad \text{for a.a. } t \in [0, T] \\ A f(t) \in \mathfrak{H}_T \end{array} \right\}.$$

If (D1) is satisfied, then by definition of scalar product (12.1.4) the operator $\mathcal{A}$ is self-adjoint and $\mathcal{A} \geq \mathcal{I}$ for time-independent *unit* operator $(\mathcal{I})(t) = \mathbb{1}$, $t \in [0, T]$. For resolvent, one has for $f \in \mathfrak{H}_T$ the representation

$$((\mathcal{A} - z\,\mathcal{I})^{-1} f)(t) = (A - z\,\mathbb{1})^{-1} f(t), \quad t \in [0, T]. \quad z \in \rho(\mathcal{A}) = \rho(A),$$

The corresponding semigroup $\{e^{-\tau\mathcal{A}}\}_{\tau\in\mathbb{R}_0^+}$ is contraction: $\|e^{-\tau\mathcal{A}}\|_{\mathcal{L}(\mathfrak{H}_T)} \le 1$, which is defined by

$$(e^{-\tau\mathcal{A}}f)(t) = e^{-\tau A}f(t), \quad f \in \mathfrak{H}_T, \quad \tau \in \mathbb{R}_0^+. \tag{12.1.7}$$

Note that for the time-independent $\mathcal{A}$, semigroups (12.1.6) and (12.1.7) commute. Then, similar to (11.2.7) on the space $\mathfrak{X}_T^p = L^p([0,T],\mathfrak{H})$, one defines a *contraction* and *nilpotent* semigroup on the space $\mathfrak{H}_T = L^2([0,T],\mathfrak{H})$:

$$\mathcal{U}_0(\tau) := e^{-\tau D_0}e^{-\tau\mathcal{A}}, \quad \tau \in \mathbb{R}_0^+. \tag{12.1.8}$$

On account of arguments from Lemma 11.2.9 (i),(ii), and Theorem 11.3.5 (11.3.19), (11.3.21), which are adapted for the Hilbert space L^2-setting, we infer that *generator* of the (*non-self-adjoint*) contraction C_0-semigroup (12.1.8) is

$$\mathcal{K}_0 := D_0 + \mathcal{A}, \quad \mathrm{dom}\,(\mathcal{K}_0) := \mathrm{dom}\,(D_0) \cap \mathrm{dom}\,(\mathcal{A}), \tag{12.1.9}$$

that is, $\{\mathcal{U}_0(\tau) = e^{-\tau\mathcal{K}_0}\}_{\tau\ge 0}$ with $\|\mathcal{U}_0(\tau)\|_{\mathcal{L}(\mathfrak{H}_T)} \le 1$ and $\mathcal{U}_0(\tau) = 0$ for $\tau \ge T$.

Further we consider the multiplication operator $\mathcal{B}$, defined in the space $\mathfrak{H}_T$ as

$$(\mathcal{B}f)(t) := B(t)f(t)$$

$$f \in \mathrm{dom}\,(\mathcal{B}) := \left\{ f \in \mathfrak{H}_T : \begin{array}{c} f(t) \in \mathrm{dom}\,(B(t)) \quad \text{for a.a. } t \in [0,T] \\ B(t)f(t) \in \mathfrak{H}_T \end{array} \right\}.$$

If (D1) is satisfied, then multiplication operator $\mathcal{B}$ on domain $\mathrm{dom}\,(\mathcal{B}) \subset \mathfrak{H}_T$ is self-adjoint and non-negative. Consequently, it is a generator of *contraction* C_0-semigroup $\{e^{-\tau\mathcal{B}}\}_{\tau\in\mathbb{R}_0^+}$, $\|e^{-\tau\mathcal{B}}\| \le 1$, on $\mathfrak{H}_T$. For the resolvent of $\mathcal{B}$, we have

$$((\mathcal{B} - z\mathcal{I})^{-1}f)(t) = (B(t) - z\mathbb{1})^{-1}f(t), \quad f \in \mathfrak{H}_T, \quad z \notin \mathbb{R}_0^+,$$

and semigroup $\{e^{-\tau\mathcal{B}}\}_{\tau\in\mathbb{R}_0^+}$ admits for a.a. $t \in [0,T]$ the representation

$$(e^{-\tau\mathcal{B}}f)(t) = e^{-\tau B(t)}f(t), \quad f \in \mathfrak{H}_T. \tag{12.1.10}$$

In construction of *evolution semigroup* for nACP (12.1.1), we consider $\mathcal{B}$ as a perturbation of generator $\mathcal{K}_0$ (12.1.9). To this aim we use the *Miyadera-Voigt* perturbation theory (Proposition 1.9.27) and mimic the arguments of Theorem 11.3.5.

To proceed we first check *assumption* (i) of Proposition 1.9.27. To prove that for any $f \in \mathrm{dom}\,(\mathcal{K}_0)$ the map $\tau \mapsto \mathcal{B}\,\mathcal{U}_0(\tau)f \in \mathfrak{H}_T$ is continuous, we note that $\mathrm{dom}\,(\mathcal{K}_0) \subseteq \mathrm{dom}\,(\mathcal{A}) \subseteq \mathrm{dom}\,(\mathcal{A}^\alpha)$ and commutativity of (12.1.8) imply for $\alpha \in (0,1]$ the representation

$$\mathcal{B}\,\mathcal{U}_0(\tau)f = \mathcal{B}\,\mathcal{A}^{-\alpha}\,\mathcal{U}_0(\tau)\,\mathcal{A}^\alpha f, \quad \tau \ge 0, \quad f \in \mathrm{dom}\,(\mathcal{K}_0).$$

Then by (D1), (D2) and by Lemma 11.2.19, adapted for $\mathfrak{H}_T$, the operator $\mathcal{B}\mathcal{A}^{-\alpha}$ is bounded:

$$\|\mathcal{B}\mathcal{A}^{-\alpha}\|_{\mathcal{L}(\mathfrak{H}_T)} \leq C_\alpha \quad \alpha \in (0, 1]. \tag{12.1.11}$$

As a consequence, the mapping $\tau \mapsto \mathcal{B}\,\mathcal{U}_0(\tau)\, f \in \mathfrak{H}_T$ is continuous for $\tau \geq 0$ and any $f \in \mathrm{dom}\,(\mathcal{K}_0)$. Since by (D2) and (12.1.9) one has $\mathrm{dom}\,(\mathcal{K}_0) = \mathrm{dom}\,(\mathcal{K}_0) \cap \mathrm{dom}\,(\mathcal{B})$ and seeing that $\mathcal{U}_0(\tau)\,\mathrm{dom}\,(\mathcal{K}_0) \subseteq \mathrm{dom}\,(\mathcal{K}_0)$, the established continuity ensures *assumption* (i) of Proposition 1.9.27 for $\mathcal{D} = \mathrm{dom}\,(\mathcal{K}_0)$.

To check the *assumption* (ii) of Proposition 1.9.27, we use representation

$$\mathcal{B}\,\mathcal{U}_0(\tau)\, f = \mathcal{B}\,\mathcal{A}^{-\alpha}\,\mathcal{A}^{\alpha}\,\mathrm{e}^{-\tau\mathcal{A}}\mathrm{e}^{-\tau D_0}\, f\,. \quad f \in \mathcal{D} = \mathrm{dom}\,(\mathcal{K}_0)\,,$$

Because the right-shift semigroup $\{\mathrm{e}^{-\tau D_0}\}_{\tau\geq 0}$ is a contraction, we obtain that

$$\|\mathcal{B}\,\mathcal{U}_0(\tau) f\| \leq \|\mathcal{B}\,\mathcal{A}^{-\alpha}\|_{\mathcal{L}(\mathfrak{H}_T)}\|\mathcal{A}^{\alpha}\,\mathrm{e}^{-\tau\mathcal{A}}\|_{\mathcal{L}(\mathfrak{H}_T)}\|f\|. \tag{12.1.12}$$

Given that operator $\mathcal{A}$ in the space $\mathfrak{H}_T$ is *self-adjoint* and $\mathcal{A} \geq \mathcal{I}$, by the spectral calculus one gets for some $M_\alpha^{\mathcal{A}} > 0$ the estimate

$$\|\mathcal{A}^{\alpha}\,\mathrm{e}^{-\tau\mathcal{A}}\|_{\mathcal{L}(\mathfrak{H}_T)} \leq \frac{M_\alpha^{\mathcal{A}}}{\tau^\alpha}\,, \quad \tau > 0\,, \quad \alpha > 0\,, \tag{12.1.13}$$

Then estimates (12.1.11)–(12.1.13) yield for $a > 0$, $\tau > 0$ and $f \in \mathrm{dom}\,(\mathcal{K}_0)$:

$$\int_0^a d\tau\, \|\mathcal{B}\mathrm{e}^{-\tau\mathcal{K}_0} f\| \leq C_\alpha M_\alpha^{\mathcal{A}}\,\|f\| \int_0^a d\tau\, \frac{1}{\tau^\alpha} = \frac{C_\alpha M_\alpha^{\mathcal{A}}}{(1-\alpha)}\, a^{1-\alpha}\,\|f\|\,. \tag{12.1.14}$$

If we choose

$$a < [(1-\alpha)/C_\alpha M_\alpha^{\mathcal{A}}]^{1/(1-\alpha)}\,,$$

then *assumption* (ii) of Proposition 1.9.27 is also satisfied; cf. (1.9.56) and (12.1.14).

Consequently, by Proposition 1.9.27 (c) there is a *unique* semigroup with generator $\mathcal{K}$, which is the closure $\overline{(\mathcal{K}_0 + \mathcal{B}) \restriction \mathrm{dom}\,(\mathcal{K}_0)}$ with domain $\mathrm{dom}\,(\mathcal{K}_0)$. Then owing to the fact that by (D2) and (12.1.9) one has $\mathrm{dom}\,(\mathcal{K}_0) = \mathrm{dom}\,(\mathcal{K}_0) \cap \mathrm{dom}\,(\mathcal{B})$, we conclude that

$$\mathcal{K} = \mathcal{K}_0 + \mathcal{B}\,, \quad \mathrm{dom}\,(\mathcal{K}) = \mathrm{dom}\,(\mathcal{K}_0) \cap \mathrm{dom}\,(\mathcal{B})\,. \tag{12.1.15}$$

Therefore, operator (12.1.15) is generator of *evolution* semigroup $\{\mathcal{U}(\tau) = \mathrm{e}^{-\tau\mathcal{K}}\}_{\tau\in\mathbb{R}_0^+}$ on $\mathfrak{H}_T$ for nACP (12.1.1) in the L^2-setting.

Remark 12.1.4 Seeing that by (12.1.8), (12.1.9) operator $\mathcal{K}_0 \in \mathscr{G}(1,0)$ is generator of *contraction* semigroup: $\|e^{-\tau\mathcal{K}_0}\|_{\mathcal{L}(\mathfrak{H}_T)} \leq 1$, and that self-adjoint non-negative perturbation $\mathcal{B} \geq 0$ is also generator of *contraction* semigroup (12.1.10): $\| e^{-\tau\mathcal{B}}\|_{\mathcal{L}(\mathfrak{H}_T)} \leq 1$, by Proposition 1.9.20 and (D2) we infer that operator $\mathcal{K} \in \mathscr{G}(1,0)$. Accordingly, the evolution semigroup $\{\mathcal{U}(\tau)\}_{\tau\in\mathbb{R}_0^+}$ is a *contraction*, that is, $\| \mathcal{U}(\tau)\|_{\mathcal{L}(\mathfrak{H}_T)} \leq 1$.

Similar to the L^p-setting, the contraction semigroup $\{\mathcal{U}(\tau)\}_{\tau\in\mathbb{R}_0^+}$ is related to *solution operator* of the non-autonomous Cauchy problem (12.1.1) by the *Howland-Evans-Neidhardt* (HEN) formula (Appendix F.2).

Proposition 12.1.5 *If* $\{\mathcal{U}(\tau)\}_{\tau\in\mathbb{R}_0^+}$ *is evolution semigroup for nACP* (12.1.1), *then there exists a strongly continuous operator family* $\{U(t,s)\}_{(t,s)\in\Delta_0}$, $\Delta_0 := \{(t,s) \in \mathcal{I}_0 \times \mathcal{I}_0 : s \leq t\}$, $\mathcal{I}_0 := (0,T]$, *on* $\mathfrak{H}$, *such that the evolution semigroup admits the representation*

$$(\mathcal{U}(\tau)f)(t) = U(t, t-\tau)\chi_{[0,T]}(t-\tau)f(t-\tau), \qquad f \in \mathfrak{H}_T\,, \tag{12.1.16}$$

which entails that evolution semigroup is nilpotent: $\mathcal{U}(\tau) = 0$ *for* $\tau \geq T$. *Here* $\{U(t,s)\}_{(t,s)\in\Delta_0}$ *is solution operator of the nACP* (12.1.1).

Hence, there is a *one-to-one* correspondence between the set of evolution semigroups on $\mathfrak{H}_T$ and the set of solution operators of nACP on $\mathfrak{H}$. The simplest examples of the HEN formula (12.1.16) give relation (12.1.6) and representation

$$(\mathcal{U}_0(\tau)f)(t) = e^{-\tau A}\,\chi_{[0,T]}(t-\tau)f(t-\tau), \qquad f \in \mathfrak{H}_T\,, \tag{12.1.17}$$

where we used (12.1.6) and (12.1.7), (12.1.8). Representation confirms that semigroup $\{\mathcal{U}_0(\tau)\}_{\tau\in\mathbb{R}_0^+}$ is nilpotent.

Now, to prepare in the L^2-setting the product formulæ approximations of evolution semigroup and then of solution operator for the nACP (12.1.1), we define the *product approximants*; cf. Sect. 11.4.

Definition 12.1.6 Using contraction semigroups (12.1.8) and (12.1.10) (Remark 12.1.4), we introduce two contraction families of operators:

$$T(\tau) := e^{-\tau\mathcal{K}_0}\,e^{-\tau\mathcal{B}} \text{ and } \widetilde{T}(\tau) := e^{-\tau\mathcal{B}}\,e^{-\tau\mathcal{K}_0}\,, \quad \tau \in \mathbb{R}_0^+\,. \tag{12.1.18}$$

Note that $\mathcal{B} = \mathcal{B}^*$, whereas $\mathcal{K}_0 \neq \mathcal{K}_0^*$, and that by *nilpotentness* of semigroup $\{\mathcal{U}_0(\tau) = e^{-\tau\mathcal{K}_0}\}_{\tau\in\mathbb{R}_0^+}$ (12.1.17) the families (12.1.18) are also nilpotent on the space $\mathfrak{H}_T$: $T(\tau) = \widetilde{T}(\tau) = 0$ for $\tau \geq T$. On account of (12.1.18), we define the corresponding families of the Trotter *product approximants*:

$$\{(T(\tau/n))^n\}_{n\geq 1} \text{ and } \{(\widetilde{T}(\tau/n))^n\}_{n\geq 1}\,, \tag{12.1.19}$$

for $\tau \geq 0$.

Remark 12.1.7 By virtue of definitions (12.1.10) and (12.1.17), we obtain that $(T(\tau)f)(t) = \mathrm{e}^{-\tau A}\,\mathrm{e}^{-\tau B(t-\tau)}\chi_{[0,T]}(t-\tau)f(t-\tau)$ for $f \in \mathfrak{H}_T$. As a consequence:

$$((T(\tau/n))^n f)(t) = U_n(t, t-\tau)\,\chi_{[0,T]}(t-\tau)f(t-\tau)\,, \quad \tau \geq 0\,, \tag{12.1.20}$$

where, for $\Delta = \{(t, s) \in [0, T] \times [0, T] : s \leq t\}$,

$$U_n(t, s) := \prod_{k=n}^{1} \mathrm{e}^{-\frac{t-s}{n}A_0}\,\mathrm{e}^{-\frac{t-s}{n}B(s+k\frac{t-s}{n})}, \quad (t, s) \in \Delta, \quad n \in \mathbb{N}\,, \tag{12.1.21}$$

Similarly, one gets

$$((\widetilde{T}(\tau/n))^n f)(t) = \widetilde{U}_n(t, t-\tau)\,\chi_{[0,T]}(t-\tau)f(t-\tau)\,, \quad \tau \geq 0\,, \tag{12.1.22}$$

for

$$\widetilde{U}_n(t, s) := \prod_{k=n-1}^{0} \mathrm{e}^{-\frac{t-s}{n}B(s+k\frac{t-s}{n})}\,\mathrm{e}^{-\frac{t-s}{n}A_0}\,, \quad (t, s) \in \Delta, \quad n \in \mathbb{N}\,, \tag{12.1.23}$$

From (12.1.20) and (12.1.22), it follows that the Trotter product approximants (12.1.19) are *nilpotent*, that is, $(T(\tau/n))^n = (\widetilde{T}(\tau/n))^n = 0$, for $\tau \geq T$. Seeing that semigroups involved in definition of approximants (12.1.19) are contractions (Remark 12.1.4). they are also *Trotter-stable*; see Definition 1.9.37.

12.1.2 Operator-Norm Product Approximations: Preliminaries

We start by a few comments and observations concerning conditions (D1)–(D3).

Remark 12.1.8

(a) If the conditions (D1)–(D3) are satisfied for some $\alpha \in (0, 1]$ and $\beta \in (0, 1)$, then they also hold for each $\gamma \in [\alpha, 1]$ in the following form:

- The condition (D1) rests unaffected.
- To show (D2) we note that $\mathrm{dom}\,(A^\gamma) \subseteq \mathrm{dom}\,(A^\alpha) \subseteq \mathrm{dom}\,(B(t))$ for a.a. $t \in \mathcal{I}$, where $\mathcal{I} := [0, T]$. Using the representation

$$B(t)A^{-\gamma} = B(t)A^{-\alpha}A^{-(\gamma-\alpha)}\,, \quad \gamma \geq \alpha\,, \tag{12.1.24}$$

for a.a. $t \in \mathcal{I}$ we get that the map $B(\cdot)A^{-\gamma} : \mathcal{I} \longrightarrow \mathcal{L}(\mathfrak{H})$ is strongly measurable and

$$C_\gamma := \operatorname*{ess\,sup}_{t\in\mathcal{I}} \|B(t)A^{-\gamma}\| \leq \operatorname*{ess\,sup}_{t\in\mathcal{I}} \|B(t)A^{-\alpha}\| = C_\alpha < \infty.$$

– Condition (D3) and representation (12.1.24) yield

$$\|A^{-\gamma}(B(t) - B(s))A^{-\gamma}\| \leq \|A^{-\alpha}(B(t) - B(s))A^{-\alpha}\| \leq L_{\alpha,\beta}|t - s|^{\beta},$$

$(t, s) \in \mathcal{I} \times \mathcal{I}$, which shows that there is a constant $L_{\gamma,\beta} \leq L_{\alpha,\beta}$ such that

$$\|A^{-\gamma}(B(t) - B(s))A^{-\gamma}\| \leq L_{\gamma,\beta}|t - s|^{\beta}, \quad (t, s) \in \mathcal{I} \times \mathcal{I},$$

holds for $\gamma \geq \alpha$.

(b) Seeing that operator A is self-adjoint and non-negative (D1), we obtain by the spectral calculus for A the operator-norm estimates: $\|A^\gamma\, \mathrm{e}^{-\tau A}\| = \|\overline{\mathrm{e}^{-\tau A}\, A^\gamma}\| \leq 1/\tau^\gamma$, for $\tau > 0$ and $\gamma > 0$, on the space $\mathfrak{H}$. The second norm is of the closure of the densely defined product, which is a bounded operator on $\mathfrak{H}$. Then on account of (12.1.7) and of (12.1.8), (12.1.17), one deduces for $\tau > 0$ the operator-norm estimates:

$$\begin{aligned} \|\mathcal{A}^\gamma\, \mathrm{e}^{-\tau\mathcal{A}}\|_{\mathcal{L}(\mathfrak{H}_T)} &= \|\overline{\mathrm{e}^{-\tau\mathcal{A}}\,\mathcal{A}^\gamma}\|_{\mathcal{L}(\mathfrak{H}_T)} \leq \frac{1}{\tau^\gamma}, \\ \|\mathcal{A}^\gamma\, \mathrm{e}^{-\tau\mathcal{K}_0}\|_{\mathcal{L}(\mathfrak{H}_T)} &= \|\overline{\mathrm{e}^{-\tau\mathcal{K}_0}\,\mathcal{A}^\gamma}\|_{\mathcal{L}(\mathfrak{H}_T)} \leq \frac{1}{\tau^\gamma}, \end{aligned} \tag{12.1.25}$$

on the Hilbert space $\mathfrak{H}_T$, where generator $\mathcal{K}_0$ is closed symmetric operator (12.1.9).

We continue by a series of estimates, which are necessary to prove the Trotter product formulæ for evolution semigroup $\{\mathcal{U}(\tau) = \mathrm{e}^{-\tau\mathcal{K}}\}_{\tau\geq 0}$ with generator (12.1.15) and the Trotter product approximants: $\{T(\tau/n)^n\}_{n\geq 1}$, $\{\widetilde{T}(\tau/n)^n\}_{n\geq 1}$, for $\tau \in \mathbb{R}_0^+$.

Lemma 12.1.9 *Let the assumptions* (D1) *and* (D2) (*Remark* 12.1.8) *be satisfied. Then for every* $\gamma \in [\alpha, 1)$ *there is a constant* $\Lambda_\gamma > 0$ *such that operator norms :*

$$\|\mathcal{A}^\gamma \mathrm{e}^{-\tau\mathcal{K}}\|_{\mathcal{L}(\mathfrak{H}_T)} \leq \frac{\Lambda_\gamma}{\tau^\gamma} \quad \textit{and} \quad \|\overline{\mathrm{e}^{-\tau\mathcal{K}}\mathcal{A}^\gamma}\|_{\mathcal{L}(\mathfrak{H}_T)} \leq \frac{\Lambda_\gamma}{\tau^\gamma}, \qquad \tau > 0, \tag{12.1.26}$$

hold on $\mathfrak{H}_T$.

Proof The proof of the first estimate follows from Lemma 11.5.5 and Remark 12.1.8. The second estimate can be proved similarly by mimicking *mutatis mutandis* the arguments of Lemma 11.5.13 and then applying again Remark 12.1.8. □

Lemma 12.1.10 *Let the assumptions* (D1) *and* (D2) *be satisfied. Then for every* $\gamma \in [\alpha, 1)$ *one has estimates*

$$\|\mathcal{A}^{-\gamma}(T(\tau) - \mathcal{U}(\tau))\| \le 2\,C_\gamma\,\tau \quad \textit{and} \quad \|(T(\tau) - \mathcal{U}(\tau))\mathcal{A}^{-\gamma}\| \le 2\,C_\gamma\,\tau\,, \tag{12.1.27}$$

for $\tau \ge 0$*, where* $C_\gamma := \operatorname{ess\,sup}_{t\in\mathcal{I}} \|B(t)A^{-\gamma}\|$.

Proof The proof of the first estimate follows from Lemma 11.5.7 (i) and Remark 12.1.8. The specific value of the constant: $2\,C_\gamma$, is owing to inspection of the proof of Lemma 11.5.7. Similarly, the second estimate is a *verbatim et literatim* consequence of the proof of Lemma 11.5.7 (ii) and Remark 12.1.8. □

The next assertion is more accurate than Lemma 11.5.8 in the Banach space $\mathfrak{X}_T^p$.

Lemma 12.1.11 *Let the assumptions* (D1)–(D3) (*Remark* 12.1.8) *be satisfied. Then for any* $\gamma \in [\alpha, 1)$ *and* $\beta \in (0, 1)$ *there is a constant* $Z_{\gamma,\beta} > 0$ *such that*

$$\|\mathcal{A}^{-\gamma}(T(\tau) - \mathcal{U}(\tau))\mathcal{A}^{-\gamma}\| \le Z_{\gamma,\beta}\,\tau^{1+\varkappa}, \quad \tau \in \mathbb{R}_0^+, \tag{12.1.28}$$

holds, where $\varkappa := \min\{\gamma, \beta\}$.

Proof We start by representation:

$$\begin{aligned}
&\frac{d}{d\sigma}\mathrm{e}^{-(\tau-\sigma)\mathcal{K}}\mathrm{e}^{-\sigma\mathcal{K}_0}\mathrm{e}^{-\sigma\mathcal{B}} = \\
&\mathrm{e}^{-(\tau-\sigma)\mathcal{K}}\left\{\mathcal{K}\mathrm{e}^{-\sigma\mathcal{K}_0} - \mathrm{e}^{-\sigma\mathcal{K}_0}\mathcal{K}_0 - \mathrm{e}^{-\sigma\mathcal{K}_0}\mathcal{B}\right\}\mathrm{e}^{-\sigma\mathcal{B}} = \\
&\mathrm{e}^{-(\tau-\sigma)\mathcal{K}}\left\{\mathcal{B}\mathrm{e}^{-\sigma\mathcal{K}_0} - \mathrm{e}^{-\sigma\mathcal{K}_0}\mathcal{B}\right\}\mathrm{e}^{-\sigma\mathcal{B}},
\end{aligned} \tag{12.1.29}$$

where $\tau \ge \sigma > 0$.

For the right-hand side of (12.1.29), we have identity

$$\begin{aligned}
&\mathrm{e}^{-(\tau-\sigma)\mathcal{K}}\left\{\mathcal{B}\mathrm{e}^{-\sigma\mathcal{K}_0} - \mathrm{e}^{-\sigma\mathcal{K}_0}\mathcal{B}\right\}\mathrm{e}^{-\sigma\mathcal{B}} \\
&\quad=(\mathrm{e}^{-(\tau-\sigma)\mathcal{K}} - \mathrm{e}^{-(\tau-\sigma)\mathcal{K}_0})\left\{\mathcal{B}\mathrm{e}^{-\sigma\mathcal{K}_0} - \mathrm{e}^{-\sigma\mathcal{K}_0}\mathcal{B}\right\}(\mathrm{e}^{-\sigma\mathcal{B}} - \mathcal{I})+ \\
&\qquad\mathrm{e}^{-(\tau-\sigma)\mathcal{K}_0}\left\{\mathcal{B}\mathrm{e}^{-\sigma\mathcal{K}_0} - \mathrm{e}^{-\sigma\mathcal{K}_0}\mathcal{B}\right\}(\mathrm{e}^{-\sigma\mathcal{B}} - \mathcal{I})+
\end{aligned}$$

$$(\mathrm{e}^{-(\tau-\sigma)\mathcal{K}} - \mathrm{e}^{-(\tau-\sigma)\mathcal{K}_0})\Big\{\mathcal{B}\mathrm{e}^{-\sigma\mathcal{K}_0} - \mathrm{e}^{-\sigma\mathcal{K}_0}\mathcal{B}\Big\} +$$
$$\mathrm{e}^{-(\tau-\sigma)\mathcal{K}_0}\Big\{\mathcal{B}\mathrm{e}^{-\sigma\mathcal{K}_0} - \mathrm{e}^{-\sigma\mathcal{K}_0}\mathcal{B}\Big\},$$

and as a consequence:

$$\begin{aligned}
&\mathcal{A}^{-\gamma}\mathrm{e}^{-(\tau-\sigma)\mathcal{K}}\Big\{\mathcal{B}\mathrm{e}^{-\sigma\mathcal{K}_0} - \mathrm{e}^{-\sigma\mathcal{K}_0}\mathcal{B}\Big\}\mathrm{e}^{-\sigma\mathcal{B}}\mathcal{A}^{-\gamma}\\
&=\mathcal{A}^{-\gamma}(\mathrm{e}^{-(\tau-\sigma)\mathcal{K}} - \mathrm{e}^{-(\tau-\sigma)\mathcal{K}_0})\Big\{\mathcal{B}\mathrm{e}^{-\sigma\mathcal{K}_0} - \mathrm{e}^{-\sigma\mathcal{K}_0}\mathcal{B}\Big\}(\mathrm{e}^{-\sigma\mathcal{B}} - \mathcal{I})\mathcal{A}^{-\gamma} +\\
&\mathrm{e}^{-(\tau-\sigma)\mathcal{K}_0}\mathcal{A}^{-\gamma}\Big\{\mathcal{B}\mathrm{e}^{-\sigma\mathcal{K}_0} - \mathrm{e}^{-\sigma\mathcal{K}_0}\mathcal{B}\Big\}(\mathrm{e}^{-\sigma\mathcal{B}} - \mathcal{I})\mathcal{A}^{-\gamma} +\\
&\mathcal{A}^{-\gamma}(\mathrm{e}^{-(\tau-\sigma)\mathcal{K}} - \mathrm{e}^{-(\tau-\sigma)\mathcal{K}_0})\Big\{\mathcal{B}\mathrm{e}^{-\sigma\mathcal{K}_0} - \mathrm{e}^{-\sigma\mathcal{K}_0}\mathcal{B}\Big\}\mathcal{A}^{-\gamma} +\\
&\mathrm{e}^{-(\tau-\sigma)\mathcal{K}_0}\mathcal{A}^{-\gamma}\Big\{\mathcal{B}\mathrm{e}^{-\sigma\mathcal{K}_0} - \mathrm{e}^{-\sigma\mathcal{K}_0}\mathcal{B}\Big\}\mathcal{A}^{-\gamma},
\end{aligned}$$

which leads to the estimate

$$\Big\|\mathcal{A}^{-\gamma}\mathrm{e}^{-(\tau-\sigma)\mathcal{K}}\Big\{\mathcal{B}\mathrm{e}^{-\sigma\mathcal{K}_0} - \mathrm{e}^{-\sigma\mathcal{K}_0}\mathcal{B}\Big\}\mathrm{e}^{-\sigma\mathcal{B}}\mathcal{A}^{-\gamma}\Big\| \leq$$
$$\Big\|\mathcal{A}^{-\gamma}(\mathrm{e}^{-(\tau-\sigma)\mathcal{K}} - \mathrm{e}^{-(\tau-\sigma)\mathcal{K}_0})\Big\{\mathcal{B}\mathrm{e}^{-\sigma\mathcal{K}_0} - \mathrm{e}^{-\sigma\mathcal{K}_0}\mathcal{B}\Big\}(\mathrm{e}^{-\sigma\mathcal{B}} - \mathcal{I})\mathcal{A}^{-\gamma}\Big\| + \tag{12.1.30}$$
$$\Big\|\mathrm{e}^{-(\tau-\sigma)\mathcal{K}_0}\mathcal{A}^{-\gamma}\Big\{\mathcal{B}\mathrm{e}^{-\sigma\mathcal{K}_0} - \mathrm{e}^{-\sigma\mathcal{K}_0}\mathcal{B}\Big\}(\mathrm{e}^{-\sigma\mathcal{B}} - \mathcal{I})\mathcal{A}^{-\gamma}\Big\| + \tag{12.1.31}$$
$$\Big\|\mathcal{A}^{-\gamma}(\mathrm{e}^{-(\tau-\sigma)\mathcal{K}} - \mathrm{e}^{-(\tau-\sigma)\mathcal{K}_0})\Big\{\mathcal{B}\mathrm{e}^{-\sigma\mathcal{K}_0} - \mathrm{e}^{-\sigma\mathcal{K}_0}\mathcal{B}\Big\}\mathcal{A}^{-\gamma}\Big\| + \tag{12.1.32}$$
$$\Big\|\mathrm{e}^{-(\tau-\sigma)\mathcal{K}_0}\mathcal{A}^{-\gamma}\Big\{\mathcal{B}\mathrm{e}^{-\sigma\mathcal{K}_0} - \mathrm{e}^{-\sigma\mathcal{K}_0}\mathcal{B}\Big\}\mathcal{A}^{-\gamma}\Big\|. \tag{12.1.33}$$

Notice that on account of (12.1.25) and Lemma 12.1.10, we obtain

$$\begin{aligned}
\|\mathrm{e}^{-\sigma\mathcal{K}_0}\mathcal{B}\| &= \|\overline{\mathrm{e}^{-\sigma\mathcal{K}_0}\mathcal{A}^{\gamma}}\;\overline{\mathcal{A}^{-\gamma}\mathcal{B}}\| \leq \sigma^{-\gamma}\|\overline{\mathcal{A}^{-\gamma}\mathcal{B}}\| = C_\gamma\sigma^{-\gamma},\\
\|\mathcal{B}\mathrm{e}^{-\sigma\mathcal{K}_0}\| &= \|\mathcal{B}\mathcal{A}^{-\gamma}\;\mathcal{A}^{\gamma}\mathrm{e}^{-\sigma\mathcal{K}_0}\| \leq \sigma^{-\gamma}\|\mathcal{B}\mathcal{A}^{-\gamma}\| = C_\gamma\sigma^{-\gamma},
\end{aligned} \tag{12.1.34}$$

for $\sigma > 0$. Then for estimation of (12.1.30), we notice that by (12.1.34) one gets

$$\begin{aligned}
&\Big\|\mathcal{A}^{-\gamma}(\mathrm{e}^{-(\tau-\sigma)\mathcal{K}} - \mathrm{e}^{-(\tau-\sigma)\mathcal{K}_0})\Big\{\mathcal{B}\mathrm{e}^{-\sigma\mathcal{K}_0} - \mathrm{e}^{-\sigma\mathcal{K}_0}\mathcal{B}\Big\}(\mathrm{e}^{-\sigma\mathcal{B}} - \mathcal{I})\mathcal{A}^{-\gamma}\Big\|\\
&\quad\leq 2\,C_\gamma\sigma^{-\gamma}\Big\|\mathcal{A}^{-\gamma}(\mathrm{e}^{-(\tau-\sigma)\mathcal{K}} - \mathrm{e}^{-(\tau-\sigma)\mathcal{K}_0})\Big\|\;\Big\|(\mathrm{e}^{-\sigma\mathcal{B}} - \mathcal{I})\mathcal{A}^{-\gamma}\Big\|.
\end{aligned}$$

Owing to the fundamental property of semigroup $\{e^{-\tau \mathcal{B}}\}_{\tau \geq 0}$, one gets

$$\mathcal{I} - e^{-\sigma \mathcal{B}} = \int_0^\sigma d\tau \, \mathcal{B} e^{-\tau \mathcal{B}} , \tag{12.1.35}$$

and due to $C_\gamma := \text{ess sup}_{t \in \mathcal{I}} \|B(t)A^{-\gamma}\|$ (see Lemma 12.1.10), we then infer that

$$\left\|(e^{-\sigma \mathcal{B}} - \mathcal{I})\mathcal{A}^{-\gamma}\right\| \leq \|\mathcal{B}\mathcal{A}^{-\gamma}\| \, \sigma \leq C_\gamma \, \sigma, \quad \sigma \in \mathbb{R}_0^+ , \tag{12.1.36}$$

and for $\tau \geq \sigma$

$$\left\|\mathcal{A}^{-\gamma}(e^{-(\tau-\sigma)\mathcal{K}} - e^{-(\tau-\sigma)\mathcal{K}_0})\right\| \leq C_\gamma \, (\tau - \sigma), \quad \sigma \in \mathbb{R}_0^+ .$$

Therefore, thanks to (12.1.34) we deduce for (12.1.30) the estimate

$$\begin{aligned} &\left\|\mathcal{A}^{-\gamma}(e^{-(\tau-\sigma)\mathcal{K}} - e^{-(\tau-\sigma)\mathcal{K}_0})\left\{\mathcal{B}e^{-\sigma\mathcal{K}_0} - e^{-\sigma\mathcal{K}_0}\mathcal{B}\right\}(e^{-\sigma\mathcal{B}} - \mathcal{I})\mathcal{A}^{-\gamma}\right\| \\ &\leq 2\, C_\gamma^3 \, \sigma^{1-\gamma} \, (\tau - \sigma), \quad 0 \leq \sigma \leq \tau . \end{aligned} \tag{12.1.37}$$

Now, to estimate (12.1.31) we recall that operators $\mathcal{K}_0$ and $\mathcal{A}$ commute. Then $C_\gamma := \text{ess sup}_{t \in \mathcal{I}} \|B(t)A^{-\gamma}\|$ and (12.1.36) yield

$$\begin{aligned} &\left\|e^{-(\tau-\sigma)\mathcal{K}_0}\mathcal{A}^{-\gamma}\left\{\mathcal{B}e^{-\sigma\mathcal{K}_0} - e^{-\sigma\mathcal{K}_0}\mathcal{B}\right\}(e^{-\sigma\mathcal{B}} - \mathcal{I})\mathcal{A}^{-\gamma}\right\| \\ &\leq 2\, C_\gamma \, \|(e^{-\sigma\mathcal{B}} - \mathcal{I})\mathcal{A}^{-\gamma}\| \leq 2\, C_\gamma^2 \, \sigma, \quad 0 \leq \sigma \leq \tau . \end{aligned} \tag{12.1.38}$$

On account of previous arguments, the estimate of (12.1.32) is straightforward:

$$\begin{aligned} &\left\|\mathcal{A}^{-\gamma}(e^{-(\tau-\sigma)\mathcal{K}} - e^{-(\tau-\sigma)\mathcal{K}_0})\left\{\mathcal{B}e^{-\sigma\mathcal{K}_0} - e^{-\sigma\mathcal{K}_0}\mathcal{B}\right\}\mathcal{A}^{-\gamma}\right\| \\ &\leq 2\, C_\gamma \, \|\mathcal{A}^{-\gamma}(e^{-(\tau-\sigma)\mathcal{K}} - e^{-(\tau-\sigma)\mathcal{K}_0})\| \leq 2\, C_\gamma^2 \, (\tau - \sigma) , \end{aligned} \tag{12.1.39}$$

here $0 \leq \sigma \leq \tau$.

Finally, to estimate (12.1.33) we use the representation

$$\begin{aligned} &\mathcal{A}^{-\gamma}\left\{\mathcal{B}e^{-\sigma\mathcal{K}_0} - e^{-\sigma\mathcal{K}_0}\mathcal{B}\right\}\mathcal{A}^{-\gamma} \\ &= \mathcal{A}^{-\gamma}\mathcal{B}(e^{-\sigma\mathcal{A}} - \mathcal{I})\mathcal{A}^{-\gamma}e^{-\sigma\mathcal{D}_0} - e^{-\sigma\mathcal{D}_0}\mathcal{A}^{-\gamma}(e^{-\sigma\mathcal{A}} - \mathcal{I})\mathcal{B}\mathcal{A}^{-\gamma} + \\ &\quad \mathcal{A}^{-\gamma}\left\{\mathcal{B}e^{-\sigma\mathcal{D}_0} - e^{-\sigma\mathcal{D}_0}\mathcal{B}\right\}\mathcal{A}^{-\gamma} , \end{aligned}$$

which yields for operator-norm inequality

$$
\begin{aligned}
&\left\| \mathcal{A}^{-\gamma} \left\{ \mathcal{B} \mathrm{e}^{-\sigma \mathcal{K}_0} - \mathrm{e}^{-\sigma \mathcal{K}_0} \mathcal{B} \right\} \mathcal{A}^{-\gamma} \right\| \\
&\quad \leq \left\| \mathcal{A}^{-\gamma} \mathcal{B} (\mathrm{e}^{-\sigma \mathcal{A}} - \mathcal{I}) \mathcal{A}^{-\gamma} \right\| + \left\| \mathcal{A}^{-\gamma} (\mathrm{e}^{-\sigma \mathcal{A}} - \mathcal{I}) \mathcal{B} \mathcal{A}^{-\gamma} \right\| + \\
&\quad \left\| \mathcal{A}^{-\gamma} \left\{ \mathcal{B} \mathrm{e}^{-\sigma \mathcal{D}_0} - \mathrm{e}^{-\sigma \mathcal{D}_0} \mathcal{B} \right\} \mathcal{A}^{-\gamma} \right\| .
\end{aligned}
\tag{12.1.40}
$$

Then by $C_\gamma := \operatorname{ess\,sup}_{t \in \mathcal{I}} \| B(t) A^{-\gamma} \|$ and by the semigroup fundamental property (12.1.35), we obtain estimates:

$$
\left\| \mathcal{A}^{-\gamma} \mathcal{B} (\mathrm{e}^{-\sigma \mathcal{A}} - \mathcal{I}) \mathcal{A}^{-\gamma} \right\| \leq \frac{C_\gamma}{\gamma} \sigma^\gamma , \quad \sigma \in \mathbb{R}_0^+ ,
$$

and

$$
\left\| \mathcal{A}^{-\gamma} (\mathrm{e}^{-\sigma \mathcal{A}} - \mathcal{I}) \mathcal{B} \mathcal{A}^{-\gamma} \right\| \leq \frac{C_\gamma}{\gamma} \sigma^\gamma , \quad \sigma \in \mathbb{R}_0^+ ,
$$

for the first two terms in the right-hand side of inequality (12.1.40). To estimate the last term in the right-hand side of inequality (12.1.40), we refer to assumption (D3) for α substituted by γ and to definition of operators induced in the space $\mathfrak{H}_T$; cf. Sect. 12.1.1. Then

$$
\left\| \mathcal{A}^{-\gamma} \left\{ \mathcal{B} \mathrm{e}^{-\sigma D_0} - \mathrm{e}^{-\sigma \mathcal{D}_0} \mathcal{B} \right\} \mathcal{A}^{-\gamma} \right\| \leq L_{\gamma,\beta} \, \sigma^\beta , \quad \sigma \in \mathbb{R}_0^+ .
$$

Summing up these estimates for (12.1.40), we find for (12.1.33)

$$
\left\| \mathcal{A}^{-\gamma} \left\{ \mathcal{B} \mathrm{e}^{-\sigma \mathcal{K}_0} - \mathrm{e}^{-\sigma \mathcal{K}_0} \mathcal{B} \right\} \mathcal{A}^{-\gamma} \right\| \leq \frac{2\, C_\gamma}{\gamma} \sigma^\gamma + L_{\gamma,\beta} \, \sigma^\beta , \quad \sigma \in \mathbb{R}_0^+ .
\tag{12.1.41}
$$

Using the estimates (12.1.37), (12.1.38), (12.1.39) and (12.1.41), we get

$$
\begin{aligned}
&\left\| \mathcal{A}^{-\gamma} \mathrm{e}^{-(\tau-\sigma)\mathcal{K}} \left\{ \mathcal{B} \mathrm{e}^{-\sigma \mathcal{K}_0} - \mathrm{e}^{-\sigma \mathcal{K}_0} \mathcal{B} \right\} \mathrm{e}^{-\sigma \mathcal{B}} \mathcal{A}^{-\gamma} \right\| \\
&\quad \leq 2\, C_\gamma^3 \, \sigma^{1-\gamma} \, (\tau - \sigma) + 2\, C_\gamma^2 \, \sigma + 2\, C_\gamma^2 \, (\tau - \sigma) + \frac{2\, C_\gamma}{\gamma} \sigma^\gamma + L_{\gamma,\beta} \, \sigma^\beta \\
&\quad = 2\, C_\gamma^3 \, \sigma^{1-\gamma} \, (\tau - \sigma) + 2\, C_\gamma^2 \, \tau + \frac{2\, C_\gamma}{\gamma} \sigma^\gamma + L_{\gamma,\beta} \, \sigma^\beta .
\end{aligned}
$$

Then on account of representation (12.1.29), this estimate yields

$$\begin{aligned}&\left\|\mathcal{A}^{-\gamma}\frac{d}{d\sigma}\,\mathrm{e}^{-(\tau-\sigma)\mathcal{K}}\mathrm{e}^{-\sigma\mathcal{K}_0}\mathrm{e}^{-\sigma\mathcal{B}}\mathcal{A}^{-\gamma}\right\|\\ &\leq 2\,C_\gamma^3\sigma^{1-\gamma}(\tau-\sigma)+2\,C_\gamma^2\,\tau+\frac{2\,C_\gamma}{\gamma}\sigma^\gamma+L_{\gamma,\beta}\,\sigma^\beta\,,\end{aligned}\tag{12.1.42}$$

for $0\leq\sigma\leq\tau$.

Owing to the fact that

$$\mathcal{A}^{-\gamma}(\mathrm{e}^{-\tau\mathcal{K}_0}\mathrm{e}^{-\tau\mathcal{B}}-\mathrm{e}^{-\tau\mathcal{K}})\mathcal{A}^{-\gamma}=\int_0^\tau d\sigma\,\mathcal{A}^{-\gamma}\frac{d}{d\sigma}\mathrm{e}^{-(\tau-\sigma)\mathcal{K}}\mathcal{A}^{-\gamma}\mathrm{e}^{-\sigma\mathcal{K}_0}\mathrm{e}^{-\sigma\mathcal{B}}\,,$$

we find inequality

$$\begin{aligned}&\left\|\mathcal{A}^{-\gamma}(\mathrm{e}^{-\tau\mathcal{K}_0}\mathrm{e}^{-\tau\mathcal{B}}-\mathrm{e}^{-\tau\mathcal{K}})\mathcal{A}^{-\gamma}\right\|\\ &\leq\int_0^\tau d\sigma\left\|\mathcal{A}^{-\gamma}\frac{d}{d\sigma}\mathrm{e}^{-(\tau-\sigma)\mathcal{K}}\mathrm{e}^{-\sigma\mathcal{K}_0}\mathrm{e}^{-\sigma\mathcal{B}}\mathcal{A}^{-\gamma}\right\|\end{aligned}$$

which by virtue of (12.1.42) yields the estimate

$$\begin{aligned}&\left\|\mathcal{A}^{-\gamma}(\mathrm{e}^{-\tau\mathcal{K}_0}\mathrm{e}^{-\tau\mathcal{B}}-\mathrm{e}^{-\tau\mathcal{K}})\mathcal{A}^{-\gamma}\right\|\\ &\leq\frac{2\,C_\gamma^3}{(2-\gamma)(3-\gamma)}\,\tau^{3-\gamma}+2\,C_\gamma^2\,\tau^2+\frac{2\,C_\gamma}{(1+\gamma)\,\gamma}\,\tau^{1+\gamma}+\frac{L_{\gamma,\beta}}{1+\beta}\,\tau^{1+\beta}\,,\end{aligned}\tag{12.1.43}$$

for $\tau\in\mathbb{R}_0^+$.

Note that if $\gamma\in[\alpha,1)$ and $\gamma\leq\beta<1$, then (12.1.43) gives

$$\begin{aligned}&\left\|\mathcal{A}^{-\gamma}(\mathrm{e}^{-\tau\mathcal{K}_0}\mathrm{e}^{-\tau\mathcal{B}}-\mathrm{e}^{-\tau\mathcal{K}})\mathcal{A}^{-\gamma}\right\|\\ &\leq\Big(\frac{2\,C_\gamma^3}{(2-\gamma)(3-\gamma)}\tau^{2-2\gamma}+2\,C_\gamma^2\tau^{1-\gamma}+\frac{2\,C_\gamma}{(1+\gamma)\,\gamma}+\frac{L_{\gamma,\beta}}{1+\beta}\tau^{\beta-\gamma}\Big)\,\tau^{1+\gamma}\,,\end{aligned}\tag{12.1.44}$$

$\tau\in\mathbb{R}_0^+$, which immediately yields (12.1.28).

However, if $\gamma \in [\alpha, 1)$ and $0 < \beta < \gamma$, then (12.1.43) predicates

$$\left\| \mathcal{A}^{-\gamma} (\mathrm{e}^{-\tau \mathcal{K}_0} \mathrm{e}^{-\tau \mathcal{B}} - \mathrm{e}^{-\tau \mathcal{K}}) \mathcal{A}^{-\gamma} \right\| \tag{12.1.45}$$
$$\leq \left(\frac{2\, C_\gamma^3}{(2-\gamma)(3-\gamma)} \tau^{2-\gamma-\beta} + 2\, C_\gamma^2 \tau^{1-\beta} + \frac{2\, C_\gamma}{(1+\gamma)\,\gamma} \tau^{\gamma-\beta} + \frac{L_{\gamma,\beta}}{1+\beta} \right) \tau^{1+\beta},$$

$\tau \in \mathbb{R}_0^+$, which shows (12.1.28) for this choice of β and γ.

Finally, since $\{T(\tau)\}_{\tau \geq 0}$ and $\{\mathcal{U}(\tau)\}_{\tau \geq 0}$ on the space $\mathfrak{H}_T$ are *nilpotent*, for $\gamma \in [\alpha, 1)$ and $\gamma \leq \beta < 1$, we find from (12.1.44) that

$$Z_{\gamma,\beta} := \frac{2C_\gamma^3}{(2-\gamma)(3-\gamma)} T^{2-2\gamma} + 2C_\gamma^2 T^{1-\gamma} + \frac{2C_\gamma}{(1+\gamma)\,\gamma} + \frac{L_{\gamma,\beta}}{1+\beta} T^{\beta-\gamma}\,. \tag{12.1.46}$$

On the other hand, for $\gamma \in [\alpha, 1)$ and $0 < \beta < \gamma$, we get from (12.1.45) that

$$Z_{\gamma,\beta} := \frac{2C_\gamma^3}{(2-\gamma)(3-\gamma)} T^{2-\gamma-\beta} + 2C_\gamma^2 T^{1-\beta} + \frac{2C_\gamma}{(1+\gamma)\,\gamma} T^{\gamma-\beta} + \frac{L_{\gamma,\beta}}{1+\beta}\,.$$

Recall that $C_\gamma = \operatorname{ess\,sup}_{t \in \mathcal{I}} \|\mathcal{B}\mathcal{A}^{-\gamma}\|$ and $L_{\gamma,\beta}$ is the Hölder constant in (12.1.3) of assumption (D3). □

Lemma 12.1.12 *Let the assumptions* (D1) *and* (D2) *be satisfied. Then*

$$\|\mathcal{A}^\gamma (\mathcal{U}(\tau) - T(\tau)) \mathcal{A}^{-\gamma}\| \leq \left(\frac{\Lambda_\gamma}{1-\gamma} + 1 \right) C_\gamma\, \tau^{1-\gamma}, \quad \tau \in \mathbb{R}_0^+, \tag{12.1.47}$$

for $\gamma \in [\alpha, 1)$.

Proof We use the representation

$$\mathcal{A}^\gamma (\mathcal{U}(\tau) - T(\tau)) \mathcal{A}^{-\gamma} = \mathcal{A}^\gamma (\mathrm{e}^{-\tau \mathcal{K}} - \mathrm{e}^{-\tau \mathcal{K}_0}) \mathcal{A}^{-\gamma}$$
$$+ \mathcal{A}^\gamma \mathrm{e}^{-\tau \mathcal{K}_0} (\mathcal{I} - \mathrm{e}^{-\tau \mathcal{B}})\, \mathcal{A}^{-\gamma}.$$

Since $\mathcal{K} = \mathcal{K}_0 + \mathcal{B}$ (12.1.15), for the first term in the right-hand side of this representation, we get

$$\mathcal{A}^\gamma (\mathrm{e}^{-\tau \mathcal{K}} - \mathrm{e}^{-\tau \mathcal{K}_0}) \mathcal{A}^{-\gamma} = -\int_0^\tau dx\, \mathcal{A}^\gamma \mathrm{e}^{-(\tau - x)\mathcal{K}}\, \mathcal{B}\, \mathcal{A}^{-\gamma} \mathrm{e}^{-x\mathcal{K}_0}\,.$$

Consequently, by Lemma 12.1.9 and $C_\gamma = \operatorname{ess\,sup}_{t\in\mathcal{I}} \|\mathcal{B}\mathcal{A}^{-\gamma}\|$, we get estimate

$$\begin{aligned} &\|\mathcal{A}^\gamma(\mathrm{e}^{-\tau\mathcal{K}} - \mathrm{e}^{-\tau\mathcal{K}_0})\mathcal{A}^{-\gamma}\| \le \int_0^\tau dx\, \|\mathcal{A}^\gamma \mathrm{e}^{-(\tau-x)\mathcal{K}}\|\, \|\mathcal{B}\mathcal{A}^{-\gamma}\| \\ &\le \Lambda_\gamma C_\gamma \int_0^\tau dx\, \frac{1}{(\tau-x)^\gamma} = \frac{\Lambda_\gamma}{1-\gamma} C_\gamma\, \tau^{1-\gamma}. \end{aligned} \tag{12.1.48}$$

To estimate the second term in the right-hand side of the representation, we use the operator-norm inequality

$$\|\mathcal{A}^\gamma \mathrm{e}^{-\tau\mathcal{K}_0}(\mathcal{I} - \mathrm{e}^{-\tau\mathcal{B}})\,\mathcal{A}^{-\gamma}\| \le \|\mathcal{A}^\gamma \mathrm{e}^{-\tau\mathcal{K}_0}\|\, \|(\mathcal{I} - \mathrm{e}^{-\tau\mathcal{B}})\,\mathcal{A}^{-\gamma}\|.$$

Then by (12.1.25), $C_\gamma = \operatorname{ess\,sup}_{t\in\mathcal{I}} \|\mathcal{B}\mathcal{A}^{-\gamma}\|$ and the semigroup fundamental property (12.1.35), we obtain

$$\|\mathcal{A}^\gamma \mathrm{e}^{-\tau\mathcal{K}_0}(\mathcal{I} - \mathrm{e}^{-\tau\mathcal{B}})\,\mathcal{A}^{-\gamma}\| \le C_\gamma\, \tau^{1-\gamma}. \tag{12.1.49}$$

As a consequence, estimates (12.1.48) and (12.1.49) yield (12.1.47). □

Lemma 12.1.13 *Let assumption* (D1) (*see comments in Remark* 12.1.8) *be satisfied. If for each* $\gamma \in [\alpha, 1)$, *there is a constant* $M_\gamma > 0$ *such that*

$$\|\mathcal{A}^\gamma T(\tau)^m\| \le \frac{M_\gamma}{(m\tau)^\gamma}, \quad m \in \mathbb{N}, \quad \tau \in \mathbb{R}^+, \tag{12.1.50}$$

is satisfied, then estimates

$$\|\mathcal{A}^\sigma T(\tau)^m\| \le \frac{M_\gamma^\delta}{(m\tau)^\sigma}, \quad m \in \mathbb{N}, \quad \tau \in \mathbb{R}^+, \tag{12.1.51}$$

hold for $\sigma \in (0, \gamma)$ *and* $\delta := \sigma/\gamma$.

Proof If (12.1.50) is satisfied, then

$$\|\overline{(T(\tau)^*)^m \mathcal{A}^\gamma}\| \le \frac{M_\gamma}{(m\tau)^\gamma}, \quad m \in \mathbb{N},$$

holds for $\tau > 0$, which is equivalent to $\mathcal{A}^\gamma T(\tau)^m \overline{(T(\tau)^*)^m \mathcal{A}^\gamma} \le M_\gamma^2/(m\tau)^{2\gamma}$, or (seeing that the self-adjoint operator $\mathcal{A}$ is invertible) to

$$T(\tau)^m (T(\tau)^*)^m \le \frac{M_\gamma^2}{(m\tau)^{2\gamma}} \mathcal{A}^{-2\gamma}, \quad m \in \mathbb{N}.$$

Note that $T(\tau)^m (T(\tau)^*)^m$ and $\mathcal{A}^{-2\gamma}$ are bounded self-adjoint operators and that $\delta \in (0,1)$. Therefore, we can use the *Heinz inequality* to infer estimates

$$\Big(T(\tau)^m (T(\tau)^*)^m\Big)^{\delta} \le \frac{M_\gamma^{2\delta}}{(m\tau)^{2\delta\gamma}} \mathcal{A}^{-2\delta\gamma}, \quad m \in \mathbb{N}, \quad \tau > 0.$$

Moreover, since $T(\tau)^m (T(\tau)^*)^m$ is a self-adjoint contraction and $\delta \in (0,1)$, we have

$$T(\tau)^m (T(\tau)^*)^m \le \Big(T(\tau)^m (T(\tau)^*)^m\Big)^{\delta}, \quad m \in \mathbb{N},$$

which yields $T(\tau)^m (T(\tau)^*)^m \le \mathcal{A}^{-2\delta\gamma} M_\gamma^{2\delta}/(m\tau)^{2\delta\gamma}$, as well as inequality

$$\mathcal{A}^{\delta\gamma}\, T(\tau)^m\, \overline{(T(\tau)^*)^m\, \mathcal{A}^{\delta\gamma}} \le \frac{M_\gamma^{2\delta}}{(m\tau)^{2\delta\gamma}}, \quad m \in \mathbb{N}.$$

Given that $(\mathcal{A}^{\delta\gamma}\, T(\tau)^m)^* = \overline{(T(\tau)^*)^m\, \mathcal{A}^{\delta\gamma}}$, this inequality provides two estimates:

$$\left\|\overline{(T(\tau)^*)^m\, \mathcal{A}^{\delta\gamma}}\right\| \le \frac{M_\gamma^{\delta}}{(m\tau)^{\delta\gamma}} \quad \text{and} \quad \left\|\mathcal{A}^{\delta\gamma}\, T(\tau)^m\right\| \le \frac{M_\gamma^{\delta}}{(m\tau)^{\delta\gamma}}, \quad m \in \mathbb{N},\ \tau > 0.$$

Then setting $\delta = \sigma/\gamma$ we prove estimates (12.1.51) for $\tau > 0$. □

In the next assertion, we check the validity of condition (12.1.50) requested by Lemma 12.1.13.

Lemma 12.1.14 *Let assumptions* (D1) *and* (D2) (*Remark* 12.1.8) *be satisfied, and let* $\gamma \in (\alpha, 1)$. *Then there is a constant* $M_\gamma > 0$ *such that*

$$\|\mathcal{A}^{\gamma}\, T(\tau)^m\| \le \frac{M_\gamma}{(m\tau)^{\gamma}}, \quad m = 1, 2, \ldots n, \tag{12.1.52}$$

holds for $\tau \in (0, T/n)$, *where* $n \ge n_0$. *Here* $n_0 := [(2\,(\Lambda_\gamma/(1-\gamma) + 1)\, C_\gamma)^{1/(1-\gamma)}\, T] + 1$ *and* $[x]$ *denotes the integer part of* $x \in \mathbb{R}_0^+$.

Proof Let $M_\gamma > 0$ be a constant which satisfies the inequality

$$5\,\Lambda_\gamma + 2\,\frac{\Lambda_\gamma C_\gamma T^{1-\gamma}}{1-\gamma}\,\frac{M_\gamma}{n^{1-\gamma}} + 2\,\Lambda_\gamma C_\gamma M_\gamma^{\alpha/\gamma}\, B(1-\alpha, 1-\gamma)\, T^{1-\alpha} \le M_\gamma\,, \tag{12.1.53}$$

for $n \ge n_0$ where the constants Λ_γ and C_γ are defined by Lemma 12.1.9 and Lemma 12.1.10, respectively, but $B(\cdot,\cdot)$ denotes the *Euler* Beta-function. (Note that such constant $M_\gamma > 0$ always exists; see Remark 12.1.15 .)

Let $m = 1$. Then by definition of operator $T(\tau) = \mathrm{e}^{-\tau\mathcal{K}_0}\mathrm{e}^{-\tau\mathcal{B}}$, by (12.1.26) and by assumption (12.1.53), we obtain

$$\|\mathcal{A}^\gamma T(\tau)\| \leq \|\mathcal{A}^\gamma \mathrm{e}^{-\tau\mathcal{K}_0}\| \leq \frac{\Lambda_\gamma}{\tau^\gamma} \leq \frac{M_\gamma}{\tau^\gamma},$$

for $\tau > 0$ and, in particular, for $\tau \in (0, T/n)$. Hence (12.1.52) holds for $m = 1$.

Now let us assume that (12.1.52) holds for $l = 1, 2, \ldots, m-1$ and $m \leq n$, that is,

$$\|\mathcal{A}^\gamma T(\tau)^l\| \leq \frac{M_\gamma}{(l\tau)^\gamma}, \tag{12.1.54}$$

for $\tau \in (0, T/n)$. We are going to show that (12.1.54) also holds for $l = m$.

To this aim we use representation

$$\mathcal{U}(\tau)^m - T(\tau)^m = \sum_{k=0}^{m-1} \mathcal{U}(\tau)^{m-1-k}(\mathcal{U}(\tau) - T(\tau))\, T(\tau)^k, \quad m = 2, 3, \ldots,$$

which implies

$$\begin{aligned}\mathcal{A}^\gamma T(\tau)^m &= \mathcal{A}^\gamma \mathcal{U}(\tau)^m - \mathcal{A}^\gamma \mathcal{U}(\tau)^{m-1}(\mathcal{U}(\tau) - T(\tau)) \\ &- \mathcal{A}^\gamma(\mathcal{U}(\tau) - T(\tau))T(\tau)^{m-1} - \sum_{k=1}^{m-2} \mathcal{A}^\gamma \mathcal{U}(\tau)^{m-1-k}(\mathcal{U}(\tau) - T(\tau))T(\tau)^k.\end{aligned}$$

for $m = 3, 4, \ldots$. This expression yields the estimate

$$\begin{aligned}\|\mathcal{A}^\gamma\, T(\tau)^m\| &\leq \|\mathcal{A}^\gamma \mathcal{U}(\tau)^m\| + \|\mathcal{A}^\gamma \mathcal{U}(\tau)^{m-1}(\mathcal{U}(\tau) - T(\tau))\| + \\ &\|\mathcal{A}^\gamma(\mathcal{U}(\tau) - T(\tau))T(\tau)^{m-1}\| + \sum_{k=1}^{m-2} \|\mathcal{A}^\gamma \mathcal{U}(\tau)^{m-1-k}(\mathcal{U}(\tau) - T(\tau))\, T(\tau)^k\|,\end{aligned} \tag{12.1.55}$$

for $m = 3, 4, \ldots$.

From Lemma 12.1.9 we get the estimates $\|\mathcal{A}^\gamma\, \mathcal{U}(\tau)^m\| \leq \Lambda_\gamma/(m\tau)^\gamma$. Then owing to $\|\mathcal{U}(\tau)\|_{\mathcal{L}(\mathfrak{H}_T)} \leq 1$ and $\|T(\tau)\|_{\mathcal{L}(\mathfrak{H}_T)} \leq 1$, see Remark 12.1.4, it follows that

$$\|\mathcal{A}^\gamma \mathcal{U}(\tau)^{m-1}(\mathcal{U}(\tau) - T(\tau))\| \leq \frac{2\Lambda_\gamma}{((m-1)\tau)^\gamma} \leq \frac{4\Lambda_\gamma}{(m\tau)^\gamma}, \quad m = 2, 3, \ldots.$$

As a consequence we find for the first two terms in the right-hand side of (12.1.55) the estimate

$$\|\mathcal{A}^\gamma \mathcal{U}(\tau)^m\| + \|\mathcal{A}^\gamma \mathcal{U}(\tau)^{m-1}(\mathcal{U}(\tau) - T(\tau))\| \leq \frac{5\Lambda_\gamma}{(m\tau)^\gamma}, \quad m = 2, 3, \dots . \tag{12.1.56}$$

Next we estimate the third term:

$$\|\mathcal{A}^\gamma (\mathcal{U}(\tau) - T(\tau))\, T(\tau)^{m-1}\| \leq \|\mathcal{A}^\gamma (\mathcal{U}(\tau) - T(\tau))\mathcal{A}^{-\gamma}\|\, \|\mathcal{A}^\gamma\, T(\tau)^{m-1}\|,$$

$m = 2, 3, \dots$. Then using Lemma 12.1.12, we find

$$\|\mathcal{A}^\gamma (\mathcal{U}(\tau) - T(\tau))\, T(\tau)^{m-1}\| \leq \frac{\Lambda_\gamma C_\gamma}{1-\gamma}\, \tau^{1-\gamma}\, \|\mathcal{A}^\gamma\, T(\tau)^{m-1}\|, \quad m = 2, 3, \dots ,$$

and by assumption (12.1.54) also

$$\|\mathcal{A}^\gamma (\mathcal{U}(\tau) - T(\tau))\, T(\tau)^{m-1}\| \leq \frac{\Lambda_\gamma M_\gamma C_\gamma}{1-\gamma}\, \frac{1}{((m-1)\tau)^\gamma}\, \tau^{1-\gamma}, \quad \tau \in (0, T/n) .$$

This leads to estimate

$$\|\mathcal{A}^\gamma (\mathcal{U}(\tau) - T(\tau))\, T(\tau)^{m-1}\| \leq \frac{\Lambda_\gamma M_\gamma C_\gamma}{1-\gamma}\, \frac{2}{(m\tau)^\gamma}\, \tau^{1-\gamma}, \quad m = 2, 3, \dots , \tag{12.1.57}$$

for $\tau \in (0, T/n)$.

Finally, for the sum in the right-hand side of (12.1.55), we have estimate

$$\sum_{k=1}^{m-2} \|\mathcal{A}^\gamma\, \mathcal{U}(\tau)^{m-1-k}\, (\mathcal{U}(\tau) - T(\tau))\, T(\tau)^k\|$$
$$\leq \sum_{k=1}^{m-2} \|\mathcal{A}^\gamma\, \mathcal{U}(\tau)^{m-1-k}\|\, \|(\mathcal{U}(\tau) - T(\tau))\, \mathcal{A}^{-\alpha}\|\, \|\mathcal{A}^\alpha\, T(\tau)^k\| , \; m = 2, 3, \dots .$$

Then on account of Lemmas 12.1.9 and 12.1.10, we obtain for $\tau > 0$

$$\sum_{k=1}^{m-2} \|\mathcal{A}^\gamma\, \mathcal{U}(\tau)^{m-1-k}\, (\mathcal{U}(\tau) - T(\tau))\, T(\tau)^k\|$$
$$\leq \sum_{k=1}^{m-2} \frac{2\, C_\alpha \tau\, \Lambda_\gamma}{((m-1-k)\tau)^\gamma}\, \|\mathcal{A}^\alpha\, T(\tau)^k\|, \quad m = 2, 3, \dots .$$

Now, using assumption (12.1.54) and Lemma 12.1.13 we deduce, for $\tau \in (0, T/n)$ and $\delta = \alpha/\gamma$, the estimate

$$\sum_{k=1}^{m-2} \|\mathcal{A}^\gamma \mathcal{U}(\tau)^{m-1-k}(\mathcal{U}(\tau) - T(\tau))\, T(\tau)^k\|$$
$$\leq 2\Lambda_\gamma C_\gamma M_\gamma^{\alpha/\gamma} \left(\sum_{k=1}^{m-2} \frac{1}{(m-1-k)^\gamma}\,\frac{1}{k^\alpha}\right) \tau^{1-\gamma-\alpha}, \quad m = 2, 3, \dots .$$

Thanks to Remark 12.1.16 (see below):

$$\sum_{k=1}^{m-2} \frac{1}{(m-1-k)^\gamma}\,\frac{1}{k^\alpha} \leq B(1-\alpha, 1-\gamma)\,(m-1)^{1-\gamma-\alpha}, \quad m = 2, 3, \dots , \tag{12.1.58}$$

where $B(\cdot, \cdot)$ is the *Euler* Beta-function. Hence

$$\sum_{k=1}^{m-2} \|\mathcal{A}^\gamma \mathcal{U}(\tau)^{m-1-k}(\mathcal{U}(\tau) - T(\tau))\, T(\tau)^k\|$$
$$\leq 2\,\Lambda_\gamma\, C_\gamma\, M_\gamma^{\alpha/\gamma}\; B(1-\alpha, 1-\gamma)\tau^{1-\gamma-\alpha}\,(m-1)^{1-\gamma-\alpha}, \quad m = 2, 3, \dots ,$$

which in turn leads, for $m = 2, 3, \dots$ and $\tau \in (0, T/n)$, to

$$\sum_{k=1}^{m-2} \|\mathcal{A}^\gamma\, \mathcal{U}(\tau)^{m-1-k}\, (\mathcal{U}(\tau) - T(\tau))\, T(\tau)^k\| \tag{12.1.59}$$
$$\leq \frac{2\,\Lambda_\gamma\, C_\gamma\, M_\gamma^{\alpha/\gamma}\; B(1-\alpha, 1-\gamma)}{(m\tau)^\gamma}\,\tau^{1-\alpha}\, m^{1-\alpha} .$$

As a result, (12.1.55)–(12.1.57) and (12.1.59) yield, for $m = 2, 3, \dots$ and for $\tau \in (0, T/n)$, the estimate

$$\|\mathcal{A}^\gamma\, T(\tau)^m\| \leq \tag{12.1.60}$$
$$\left\{5\,\Lambda_\gamma + 2\,\frac{\Lambda_\gamma C_\gamma\, M_\gamma\, T^{1-\gamma}}{1-\gamma}\,\frac{1}{n^{1-\gamma}}\right.$$
$$\left. +2\,\Lambda_\gamma C_\gamma\, M_\gamma^{\alpha/\gamma}\; B(1-\alpha, 1-\gamma)\, T^{1-\alpha}\right\} \frac{1}{(m\tau)^\gamma},$$

which concludes the induction (12.1.54) for $l = m$.

Seeing that by assumption (12.1.53) we have for $n > n_0$:

$$5\,\Lambda_\gamma + 2\,\frac{\Lambda_\gamma\,C_\gamma\,M_\gamma\,T^{1-\gamma}}{1-\gamma}\,\frac{1}{n^{1-\gamma}} + 2\,\Lambda_\gamma\,C_\gamma\,M_\gamma^{\alpha/\gamma}\,B(1-\alpha,1-\gamma)\,T^{1-\alpha} \leq M_\gamma,$$

this inequality together with estimate (12.1.60) yields (12.1.54) for $l = 1, 2, \ldots, n$. Because (12.1.54) for $l = m \leq n$ and $n > n_0$ implies (12.1.52), we conclude the proof of lemma up to Remarks 12.1.15 and 12.1.16. □

Remark 12.1.15 To check that condition (12.1.53) is always satisfied for sufficiently *large* $M = M_\gamma > 0$ and natural $n \in \mathbb{N}$ such that

$$n > n_0 = [(2\,(\Lambda_\gamma/(1-\gamma)+1)\,C_\gamma)^{1/(1-\gamma)}\,T] + 1\,, \tag{12.1.61}$$

we introduce for $\alpha \in (0,1)$ and $\gamma \in (\alpha, 1)$

$$c_0 := 5\,\Lambda_\gamma\,,\;\; c_1 := 2\left(\frac{\Lambda_\gamma}{1-\gamma}+1\right)C_\gamma\,T^{1-\gamma}\,,$$
$$c_2 := 4\,\Lambda_\gamma\,C_\gamma\,B(1-\alpha,1-\gamma)\,T^{1-\alpha}.$$

Then condition (12.1.53) gets the form: $c_0 + c_1\,M/n^{1-\gamma} + c_2\,M^{\alpha/\gamma} \leq M$, which yields

$$\frac{c_0}{M} + \frac{c_2}{M^{1-\alpha/\gamma}} \leq 1 - \frac{c_1}{n^{1-\gamma}}\,. \tag{12.1.62}$$

Since $n > n_0$, inequality (12.1.61) yields $1 - c_1/n^{1-\gamma} > 0$, whereas the left-hand side of (12.1.62) tends to zero if $M \to \infty$. On that account, choosing M sufficiently large we can guarantee existence of $M = M_\gamma$ such that condition (12.1.53) is satisfied for any $n > n_0$, (12.1.61).

Remark 12.1.16 Let $\alpha \in (0,1)$ and $\gamma \in [\alpha, 1)$. Then estimate (12.1.58):

$$\sum_{k=1}^{n-1}\frac{1}{(n-k)^\gamma}\,\frac{1}{k^\alpha} \leq B(1-\alpha,1-\gamma)\,n^{1-\gamma-\alpha},\quad n \in 2, 3, \ldots\,,$$

holds. Here $B(\cdot,\cdot)$ is the *Euler* Beta-function

$$B(1-\alpha,1-\gamma) := \int_0^1 dx\,\frac{1}{x^\alpha\,(1-x)^\gamma}\,.$$

To verify this claim, we note that for $\alpha \in [0, 1)$, $\gamma \in [\alpha, 1)$ and $x \in (k-1, k]$ one has

$$\frac{1}{(n-k)^\gamma\, k^\alpha} \le \frac{1}{(n-1-x)^\gamma\, x^\alpha}, \qquad k = 1, 2, \ldots, n-1\,.$$

Therefore, we get estimate

$$\begin{aligned}\frac{1}{(n-k)^\gamma\, k^\alpha} &= \int_{k-1}^{k} dx\, \frac{1}{(n-k)^\gamma\, k^\alpha} \\ &\le \int_{k-1}^{k} dx\, \frac{1}{x^\alpha\, (n-1-x)^\gamma}, \qquad x \in (k-1, k],\end{aligned}$$

and as a consequence the inequality (12.1.58):

$$\sum_{k=1}^{n-1} \frac{1}{(n-k)^\gamma k^\alpha} \le \sum_{k=1}^{n-1} \int_{k-1}^{k} dx\, \frac{1}{x^\alpha\, (n-1-x)^\gamma} =$$
$$\int_0^{n-1} dx\, \frac{1}{x^\alpha\, (n-1-x)^\gamma} = B(1-\alpha, 1-\gamma)\, n^{1-\alpha-\gamma}\,.$$

12.1.3 Operator-Norm Product Approximations: Proofs

In this section we collect the proofs of results based on the evolution semigroup approach to the product formula approximations to the nACP solution operator in the L^2-setting. They are based on preliminaries established in the *operator-norm* topology on the Hilbert space $\mathfrak{H}_T$; see Sect. 12.1.2.

Theorem 12.1.17 *Let the assumptions* (D1)–(D3) *be satisfied, and let* $\beta > 2\alpha - 1$. *Then there is a constant* $R_\beta > 0$ *such that*

$$\sup_{\tau \in \mathbb{R}_0^+} \|\mathcal{U}(\tau) - T(\tau/n)^n\| \le \frac{R_\beta}{n^\beta}, \tag{12.1.63}$$

holds for $n \in \mathbb{N}$ *and* $\tau \in \mathbb{R}_0^+$.

Proof Taking into account the representation

$$\mathcal{U}(\tau/n)^n - T(\tau/n)^n = \sum_{m=0}^{n-1} \mathcal{U}(\tau/n)^{n-m-1}(\mathcal{U}(\tau/n) - T(\tau/n))\, T(\tau/n)^m, \quad n \in \mathbb{N},$$

or, identically, expression

$$
\begin{aligned}
&\mathcal{U}(\tau/n)^n - T(\tau/n)^n \\
&\qquad = \mathcal{U}(\tau/n)^{n-1}(\mathcal{U}(\tau/n) - T(\tau/n)) + (\mathcal{U}(\tau/n) - T(\tau/n))\, T(\tau/n)^{n-1} \\
&\qquad + \sum_{m=1}^{n-2} \mathcal{U}(\tau/n)^{n-m-1}(\mathcal{U}(\tau/n) - T(\tau/n))\, T(\tau/n)^m, \quad n = 3, 4, \ldots,
\end{aligned}
$$

we infer inequality

$$
\begin{aligned}
&\| \mathcal{U}(\tau/n)^n - T(\tau/n)^n \| \\
&\leq \| \mathcal{U}(\tau/n)^{n-1} \mathcal{A}^\gamma \| \, \| \mathcal{A}^{-\gamma} (\mathcal{U}(\tau/n) - T(\tau/n)) \| \\
&+ \| (\mathcal{U}(\tau/n) - T(\tau/n))\, \mathcal{A}^{-\gamma} \| \, \| \mathcal{A}^\gamma \; T(\tau/n)^{n-1} \| \qquad (12.1.64) \\
&+ \sum_{m=1}^{n-2} \| \mathcal{U}(\tau/n)^{n-m-1} \mathcal{A}^\gamma \| \, \| \mathcal{A}^{-\gamma} (\mathcal{U}(\tau/n) - T(\tau/n)) \mathcal{A}^{-\gamma} \| \, \| \mathcal{A}^\gamma \; T(\tau/n)^m \| ,
\end{aligned}
$$

for $\gamma \in (\alpha, 1)$, $\alpha \in (0, 1)$, and for $n = 3, 4, \ldots$.

Using Lemmas 12.1.9 and 12.1.10, one gets for the first term in the right-hand side of (12.1.64) estimate

$$
\|\mathcal{U}(\tau/n)^{n-1}\, \mathcal{A}^\gamma \| \, \| \mathcal{A}^{-\gamma} (\mathcal{U}(\tau/n) - T(\tau/n)) \| \leq 2 \, \frac{\Lambda_\gamma \, C_\gamma}{(\tau(n-1)/n)^\gamma} \, \frac{\tau}{n} ,
$$

which yields

$$
\|\mathcal{U}(\tau/n)^{n-1} \mathcal{A}^\gamma \| \, \| \mathcal{A}^{-\gamma} (\mathcal{U}(\tau/n) - T(\tau/n)) \| \leq 2^{1+\gamma} \, \Lambda_\gamma \, C_\gamma \, T^{1-\gamma} \, \frac{1}{n} , \tag{12.1.65}
$$

for $n = 3, 4, \ldots$ and for any $\tau \in [0, T]$.

Next, owing to Lemma 12.1.10 and on account of Lemma 12.1.14, for $m = n-1$ and $\tau \in [0, T]$, we find for the second term in the right-hand side of (12.1.64) estimate

$$
\| (\mathcal{U}(\tau/n) - T(\tau/n)) \mathcal{A}^{-\gamma} \| \, \| \mathcal{A}^\gamma \; T(\tau/n)^{n-1} \| \leq 2 \, C_\gamma \, \frac{\tau}{n} \, \frac{M_\gamma}{(\tau(n-1)/n)^\gamma} ,
$$

for $n \geq n_0$, where n_0 is defined in Lemma 12.1.14. Hence,

$$
\| (\mathcal{U}(\tau/n) - T(\tau/n))\, \mathcal{A}^{-\gamma} \| \, \| \mathcal{A}^\gamma \; T(\tau/n)^{n-1} \| \leq 2^{1+\gamma} \, C_\gamma \, M_\gamma \, T^{1-\gamma} \, \frac{1}{n} . \tag{12.1.66}
$$

Promptly, taking into account Lemmas 12.1.9, 12.1.11 (for $\varkappa = \min\{\gamma, \beta\}$) and Lemma 12.1.14, we obtain for the sum in the right-hand side of (12.1.64) estimate:

$$\sum_{m=1}^{n-2} \|\mathcal{U}(\tau/n)^{n-m-1}\mathcal{A}^{\gamma}\| \, \|\mathcal{A}^{-\gamma}(\mathcal{U}(\tau/n) - T(\tau/n))\mathcal{A}^{-\gamma}\| \, \|\mathcal{A}^{\gamma}\, T(\tau/n)^m\| \leq$$

$$\sum_{m=1}^{n-2} \frac{\Lambda_\gamma\, Z_{\gamma,\beta}}{((n-m-1)\,\tau/n)^{\gamma}} \left(\frac{\tau}{n}\right)^{1+\varkappa} \frac{M_\gamma}{(m\,\tau/n)^{\gamma}}$$

$$= \frac{\Lambda_\gamma Z_{\gamma,\beta} M_\gamma \tau^{1+\varkappa-2\gamma}}{n^{1+\varkappa-2\gamma}} \sum_{m=1}^{n-2} \frac{1}{(n-m-1)^{\gamma}} \frac{1}{m^{\gamma}},$$

for $n > \max\{2, n_0\}$ and $\tau \in [0, T]$. Then by (12.1.58) one gets

$$\sum_{m=1}^{n-2} \|\mathcal{U}(\tau/n)^{n-m-1}\mathcal{A}^{\gamma}\| \, \|\mathcal{A}^{-\gamma}(\mathcal{U}(\tau/n) - T(\tau/n))\mathcal{A}^{-\gamma}\| \, \|\mathcal{A}^{\gamma}\, T(\tau/n)^m\|$$

$$\leq \frac{\Lambda_\gamma Z_{\gamma,\beta} M_\gamma \tau^{1+\varkappa-2\gamma}}{n^{1+\varkappa-2\gamma}} \, B(1-\gamma, 1-\gamma)\, n^{1-2\gamma},$$

and, because of $\tau \in [0, T]$, also

$$\sum_{m=1}^{n-2} \|\mathcal{U}(\tau/n)^{n-m-1}\mathcal{A}^{\gamma}\| \, \|\mathcal{A}^{-\gamma}(\mathcal{U}(\tau/n) - T(\tau/n))\mathcal{A}^{-\gamma}\| \, \|\mathcal{A}^{\gamma}\, T(\tau/n)^m\|$$
$$\leq \Lambda_\gamma\, Z_{\gamma,\beta}\, M_\gamma\, B(1-\gamma, 1-\gamma)\, T^{1+\varkappa-2\gamma}\, \frac{1}{n^{\varkappa}}. \tag{12.1.67}$$

As a consequence, on account of (12.1.64), (12.1.65), (12.1.66) and (12.1.67), we infer, for $n > \max\{2, n_0\}$ and $\tau \in [0, T]$, the estimate

$$\begin{aligned}
\|\mathcal{U}(\tau)^n - T(\tau/n)^n\| &= \|\mathcal{U}(\tau/n)^n - T(\tau/n)^n\| \\
&\leq 2^{1+\gamma}\Lambda_\gamma C_\gamma T^{1-\gamma}\, \frac{1}{n} + 2^{1+\gamma} C_\gamma M_\gamma T^{1-\gamma}\, \frac{1}{n} \\
&\quad + \Lambda_\gamma Z_{\gamma,\beta} M_\gamma\, B(1-\gamma, 1-\gamma) T^{1+\varkappa-2\gamma}\, \frac{1}{n^{\varkappa}} \\
&\leq \Big\{2^{1+\gamma}\Lambda_\gamma C_\gamma T^{1-\gamma} + 2^{1+\gamma} C_\gamma M_\gamma T^{1-\gamma} \\
&\quad + \Lambda_\gamma Z_{\gamma,\beta} M_\gamma\, B(1-\gamma, 1-\gamma)\, T^{1+\varkappa-2\gamma}\Big\}\, \frac{1}{n^{\varkappa}}.
\end{aligned}$$

If $\alpha < \beta < 1$, then we choose $\gamma = \beta$, that is, $\varkappa = \min\{\gamma, \beta\} = \beta$ and $1 + \varkappa - 2\gamma = 1 - \beta \geq 0$. Setting

$$R'_\beta := 2^{1+\beta} \Lambda_\beta C_\beta \, T^{1-\beta} + 2^{1+\beta} C_\beta M_\beta \, T^{1-\beta} + \Lambda_\beta Z_{\beta,\beta} M_\beta \; B(1-\beta, 1-\beta) \, T^{1-\beta} \,,$$

one obtains the estimate

$$\|\mathcal{U}(\tau)^n - T(\tau/n)^n\| \leq \frac{R'_\beta}{n^\beta} \,, \tag{12.1.68}$$

for $n > \max\{2, n_0\}$ and $\tau \in [0, T]$.

Now let $0 < \beta \leq \alpha$. Since by condition of theorem: $1 + \beta - 2\alpha > 0$, there exists $\gamma \in (\alpha, 1)$ such that $1 + \beta - 2\gamma > 0$. Therefore, $\varkappa = \min\{\gamma, \beta\} = \beta$. Setting

$$R''_\beta := 2^{1+\gamma} \Lambda_\gamma C_\gamma \; T^{1-\gamma} + 2^{1+\gamma} C_\gamma M_\gamma \; T^{1-\gamma} + \Lambda_\gamma Z_{\gamma,\beta} M_\gamma \; B(1-\gamma, 1-\gamma) \, T^{1+\beta-2\gamma} \,,$$

we obtain (12.1.68) for $n > \max\{2, n_0\}$.

Both results imply that there is a constant R_β such that (12.1.63) holds for $\tau \in [0, T]$ and $n \in \mathbb{N}$. Finally, using the nilpotence: $\mathcal{U}(\tau > T) = 0$ (Proposition 12.1.5) and $T(\tau/n)^n = 0$ (Remark 12.1.7) for $\tau \geq T$, we obtain (12.1.63) for any $\tau \in \mathbb{R}_0^+$.

□

Corollary 12.1.18 *Let the assumptions* (D1)–(D3) *be satisfied and* $\beta > 2\alpha - 1$. *Then there exists* $\widetilde{R}_\beta > 0$ *such that estimate*

$$\sup_{\tau \in \mathbb{R}_0^+} \|\mathcal{U}(\tau) - \widetilde{T}(\tau/n)^n\| \leq \frac{\widetilde{R}_\beta}{n^\beta} \,, \tag{12.1.69}$$

holds for $n \in \mathbb{N}$ *and* $\tau \in \mathbb{R}_0^+$.

Proof Since $\widetilde{T}(\tau/n)^{n+1} = \mathrm{e}^{-\tau\mathcal{B}/n}\, T(\tau/n)^n\, \mathrm{e}^{-\tau\mathcal{K}_0/n}$, (12.1.18), for $\tau \in \mathbb{R}_0^+$ and $n \in \mathbb{N}$,

$$\begin{aligned}
&\mathcal{U}((n+1)\tau/n) - \widetilde{T}(\tau/n)^{n+1} = \mathrm{e}^{-(n+1)\tau\mathcal{K}/n} - \mathrm{e}^{-\tau\mathcal{B}/n} T(\tau/n)^n \mathrm{e}^{-\tau\mathcal{K}_0/n} = \\
&\mathrm{e}^{-(n+1)\tau\mathcal{K}/n} - \mathrm{e}^{-\tau\mathcal{B}/n}\mathrm{e}^{-\tau\mathcal{K}}\mathrm{e}^{-\tau\mathcal{K}_0/n} + \mathrm{e}^{-\tau\mathcal{B}/n}(\mathcal{U}(\tau) - T(\tau/n)^n)\mathrm{e}^{-\tau\mathcal{K}_0/n} = \\
&(\mathcal{I} - \mathrm{e}^{-\tau\mathcal{B}/n})\mathrm{e}^{-\tau\mathcal{K}}\mathrm{e}^{-\tau\mathcal{K}_0/n} + \mathrm{e}^{-\tau\mathcal{K}}(\mathrm{e}^{-\tau\mathcal{K}/n} - \mathrm{e}^{-\tau\mathcal{K}_0/n}) + \\
&\mathrm{e}^{-\tau\mathcal{B}/n}(\mathcal{U}(\tau) - T(\tau/n)^n)\mathrm{e}^{-\tau\mathcal{K}_0/n}, \quad \tau \in \mathbb{R}_0^+, \quad n \in \mathbb{N}\,.
\end{aligned}$$

On account of Remark 12.1.4, this representation yields estimate

$$\begin{aligned} &\| \mathcal{U}((n+1)\tau/n) - \widetilde{T}(\tau/n)^{n+1} \| \\ &\qquad \leq \| (\mathcal{I} - \mathrm{e}^{-\tau\mathcal{B}/n}) \mathrm{e}^{-\tau\mathcal{K}} \| + \| \mathrm{e}^{-\tau\mathcal{K}} (\mathrm{e}^{-\tau\mathcal{K}/n} - \mathrm{e}^{-\tau\mathcal{K}_0/n}) \| \\ &\qquad + \| \mathcal{U}(\tau) - T(\tau/n)^n \|, \quad \tau \in \mathbb{R}_0^+, \quad n \in \mathbb{N}. \end{aligned} \tag{12.1.70}$$

Note that $\|(\mathcal{I} - \mathrm{e}^{-\tau\mathcal{B}/n}) \mathrm{e}^{-\tau\mathcal{K}}\| \leq \|(\mathcal{I} - \mathrm{e}^{-\tau\mathcal{B}/n}) \mathcal{A}^{-\alpha}\| \, \|\mathcal{A}^{\alpha} \mathrm{e}^{-\tau\mathcal{K}}\|$, where $\tau \in \mathbb{R}_0^+$, $n \in \mathbb{N}$. Now using

$$(\mathcal{I} - \mathrm{e}^{-\tau\mathcal{B}/n}) \mathcal{A}^{-\alpha} = \int_0^{\tau/n} d\sigma \, \mathrm{e}^{-\sigma\mathcal{B}} \mathcal{B} \mathcal{A}^{-\alpha}, \quad \tau \in \mathbb{R}_0^+, \quad n \in \mathbb{N},$$

one gets

$$\|(\mathcal{I} - \mathrm{e}^{-\tau\mathcal{B}/n}) \mathcal{A}^{-\alpha}\| \leq C_\alpha \, \frac{\tau}{n}, \quad \tau \in \mathbb{R}_0^+, \quad n \in \mathbb{N}.$$

Then taking into account assumption (D2) and Lemma 12.1.9, we obtain estimate for the first term in the right-hand side of (12.1.70):

$$\|(\mathcal{I} - \mathrm{e}^{-\tau\mathcal{B}/n}) \mathrm{e}^{-\tau\mathcal{K}}\| \leq C_\alpha \, \Lambda_\alpha \, \frac{\tau^{1-\alpha}}{n} \leq C_\alpha \, \Lambda_\alpha \, T^{1-\alpha} \, \frac{1}{n}. \tag{12.1.71}$$

Here $\tau \in \mathbb{R}_0^+$, $n \in \mathbb{N}$ and we used that semigroup $\{\mathrm{e}^{-\tau\mathcal{K}}\}_{\tau \geq 0}$ is nilpotent because of Proposition 12.1.5.

Further, to estimate the second term in the right-hand side of (12.1.70), we use that $\|\mathrm{e}^{-\tau\mathcal{K}} (\mathrm{e}^{-\tau\mathcal{K}/n} - \mathrm{e}^{-\tau\mathcal{K}_0/n})\| \leq \|\mathrm{e}^{-\tau\mathcal{K}} \mathcal{A}^{\alpha}\| \, \|\mathcal{A}^{-\alpha} (\mathrm{e}^{-\tau\mathcal{K}/n} - \mathrm{e}^{-\tau\mathcal{K}_0/n})\|$. Then thanks to representation

$$\mathcal{A}^{-\alpha} (\mathrm{e}^{-\tau\mathcal{K}/n} - \mathrm{e}^{-\tau\mathcal{K}_0/n}) = - \int_0^{\tau/n} d\sigma \, \mathrm{e}^{-\sigma\mathcal{K}_0} \, \overline{\mathcal{A}^{-\alpha} \mathcal{B}} \mathrm{e}^{-(\tau-\sigma)\mathcal{K}},$$

where $\tau \in \mathbb{R}_0^+$, $n \in \mathbb{N}$, we infer the estimate

$$\|\mathcal{A}^{-\alpha} (\mathrm{e}^{-\tau\mathcal{K}/n} - \mathrm{e}^{-\tau\mathcal{K}_0/n})\| \leq C_\alpha \, \frac{\tau}{n}, \quad \tau \in \mathbb{R}_0^+.$$

Applying again Lemma 12.1.9 and nilpotentness of semigroup $\{\mathrm{e}^{-\tau\mathcal{K}}\}_{\tau \geq 0}$, one gets

$$\|\mathrm{e}^{-\tau\mathcal{K}} (\mathrm{e}^{-\tau\mathcal{K}/n} - \mathrm{e}^{-\tau\mathcal{K}_0/n})\| \leq C_\alpha \Lambda_\alpha T^{1-\alpha} \, \frac{1}{n}, \quad \tau \in \mathbb{R}_0^+, \quad n \in \mathbb{N}. \tag{12.1.72}$$

Thus, insertion of (12.1.71) and (12.1.72) into (12.1.70) yields

$$\| \mathcal{U}((n+1)\tau/n) - \widetilde{T}(\tau/n)^{n+1} \| \leq 2\, C_\alpha\, \Lambda_\alpha\, \frac{T^{1-\alpha}}{n} + \| \mathcal{U}(\tau) - T(\tau/n)^n \| \,.$$

Then by Theorem 12.1.17 we obtain

$$\| \mathcal{U}((n+1)\tau/n) - \widetilde{T}(\tau/n)^{n+1} \| \leq 2\, C_\alpha \Lambda_\alpha\, \frac{T^{1-\alpha}}{n} + \frac{R_\beta}{n^\beta}, \quad \tau \in \mathbb{R}_0^+, \quad n \in \mathbb{N}\,,$$

that is, $\| \mathcal{U}((n+1)\tau/n) - \widetilde{T}(\tau/n)^{n+1} \| \leq \widetilde{R}'_\beta / n^\beta$, where $\widetilde{R}'_\beta := 2\, C_\alpha\, \Lambda_\alpha\, T^{1-\alpha} + R_\beta$, seeing that $\beta \leq 1$. This yields

$$\sup_{\tau \in \mathbb{R}_0^+} \| \mathcal{U}((n+1)\tau/n) - \widetilde{T}(\tau/n)^{n+1} \| \leq \frac{\widetilde{R}'_\beta}{n^\beta}, \quad n \in \mathbb{N}\,.$$

Let $\tau = \tau' n/(n+1)$ for $\tau' \in \mathbb{R}_0^+$. Then

$$\sup_{\tau \in \mathbb{R}_0^+} \| \mathcal{U}((n+1)\tau/n) - \widetilde{T}(\tau/n)^{n+1} \| = \sup_{\tau' \in \mathbb{R}_0^+} \| \mathcal{U}(\tau') - \widetilde{T}(\tau'/(n+1))^{n+1} \| \leq \frac{\widetilde{R}'_\beta}{n^\beta}\,,$$

or, for $n \in \mathbb{N}$,

$$\sup_{\tau' \in \mathbb{R}_0^+} \| \mathcal{U}(\tau') - \widetilde{T}(\tau'/(n+1))^{n+1} \| \leq 2^\beta\, \frac{\widetilde{R}'_\beta}{(n+1)^\beta}\,.$$

Setting $\widetilde{R}_\beta := \max\{2, 2^\beta\, \widetilde{R}'_\beta\}$ we prove (12.1.69). □

Using the HEN formula (Proposition 12.1.5), the approximations by the Trotter product formulæ (12.1.63) and (12.1.69) for *evolution semigroup* $\{\mathcal{U}(\tau)\}_{\tau \geq 0}$ on $\mathfrak{H}_T = L^2([0,T], \mathfrak{H})$ can be extended to the operator-norm product formulæ approximations of *solution operator* $\{U(t,s)\}_{(t,s)\in\Delta_0}$ on Hilbert space $\mathfrak{H}$.

Theorem 12.1.19 *Let the assumptions* (D1)–(D3) *be satisfied. Let* $\{U(t,s)\}_{(t,s)\in\Delta_0}$ *be solution operator of the nACP corresponding to evolution generator* $\mathcal{K}$*, and let the product approximants* $\{U_n(t,s)\}_{(t,s)\in\Delta_0}$ *and* $\{\widetilde{U}_n(t,s)\}_{(t,s)\in\Delta_0}$ *be defined* *by* (12.1.21) *and* (12.1.23), *respectively. If* $\beta > 2\alpha - 1$, *then the estimates*

$$\operatorname*{ess\,sup}_{(t,s)\in\Delta_0} \|U(t,s) - U_n(t,s)\| \le \frac{R_\beta}{n^\beta}, \tag{12.1.73}$$

$$\operatorname*{ess\,sup}_{(t,s)\in\Delta_0} \|U(t,s) - \widetilde{U}_n(t,s)\| \le \frac{\widetilde{R}_\beta}{n^\beta}, \tag{12.1.74}$$

hold for $n \in \mathbb{N}$, *where the constants* R_β *and* $\widetilde{R}_\beta$ *are those of Theorem* 12.1.17 *and Corollary* 12.1.18.

Proof To this aim we use the left-shift contraction semigroup $\{\mathcal{S}_L(\tau)\}_{\tau\ge 0}$ (2.2.39) and HEN formulæ (12.1.16), (12.1.20) to define on the space $\mathfrak{H}_T$ operators:

$$(\mathcal{S}_L(\tau)\,\mathrm{e}^{-\tau\mathcal{K}})\,f)(t) = \chi_{[0,T]}(t+\tau)\,U(t+\tau,t)\chi_{[0,T]}(t)f(t)\,, \quad \tau \ge 0\,, \tag{12.1.75}$$

and, for each $n \in \mathbb{N}$,

$$(\mathcal{S}_L(\tau)(\mathrm{e}^{-\tau/n\,\mathcal{K}_0}\mathrm{e}^{-\tau/n\,\mathcal{B}})^n\, f)(t) = \chi_{[0,T]}(t+\tau)U_n(t+\tau,t)\,\chi_{[0,T]}(t)\,f(t)\,. \tag{12.1.76}$$

Then on account of (12.1.75) and (12.1.76), we introduce on $\mathfrak{H}_T$ (nilpotent) operators

$$\mathfrak{R}_n(\tau) := \mathcal{S}_L(\tau)\{\mathrm{e}^{-\tau\mathcal{K}} - (\mathrm{e}^{-\tau\mathcal{K}_0/n}\,\mathrm{e}^{-\tau\mathcal{B}/n})^n\}\,, \quad \tau \ge 0\,, \quad n \in \mathbb{N}\,, \tag{12.1.77}$$

which have representations:

$$(\mathfrak{R}_n(\tau)f)(t) = (U(t+\tau,t) - U_n(t+\tau,t))\chi_{[0,T]}(t+\tau)f(t)\,, \quad f \in \mathfrak{H}_T\,. \tag{12.1.78}$$

Consequently, for each $n \in \mathbb{N}$ and $t \in [0,T]$ the operator (12.1.77) is a *multiplication* operator on $\mathfrak{H}_T$ induced by the family $\{(U(\cdot+\tau,\cdot) - U_n(\cdot+\tau,\cdot))\chi_{[0,T]}(\cdot+\tau)\}_{n\in\mathbb{N}}$ of bounded operators.

Since operators (12.1.77) are nilpotent on $\mathfrak{H}_T$, we consider only $\tau \in [0,T]$. Then on account of the left-shift operator $\{\mathcal{S}_L(\tau)\}_{\tau\ge 0}$, one gets equality

$$\sup_{\tau\in[0,T]} \|\mathrm{e}^{-\tau\mathcal{K}} - (\mathrm{e}^{-\tau\mathcal{K}_0/n}\,\mathrm{e}^{-\tau\mathcal{B}/n})^n\|_{\mathcal{L}(\mathfrak{H}_T)} =$$
$$= \sup_{\tau\in[0,T]} \|\mathcal{S}_L(\tau)\{\mathrm{e}^{-\tau\mathcal{K}} - (\mathrm{e}^{-\tau\mathcal{K}_0/n}\,\mathrm{e}^{-\tau\mathcal{B}/n})^n\}\|_{\mathcal{L}(\mathfrak{H}_T)}\,. \tag{12.1.79}$$

For that reason, Eqs. (12.1.77) and (12.1.79) yield

$$\sup_{\tau\in[0,T]} \|\mathrm{e}^{-\tau\mathcal{K}} - (\mathrm{e}^{-\tau\mathcal{K}_0/n}\,\mathrm{e}^{-\tau\mathcal{B}/n})^n\|_{\mathcal{L}(\mathfrak{H}_T)} = \sup_{\tau\in[0,T]} \|\mathfrak{R}_n(\tau)\|_{\mathcal{L}(\mathfrak{H}_T)}\,. \tag{12.1.80}$$

Then by virtue of relation (2.2.22) for operator norms of multiplication operators in $\mathfrak{H}_T$, see Proposition 2.2.16 for a general case of Banach spaces, and by (12.1.78) it follows that

$$\begin{aligned}
&\sup_{\tau\in[0,T]} \|\mathfrak{R}_n(\tau)\|_{\mathcal{L}(\mathfrak{H}_T} = \\
&= \sup_{\tau\in[0,T]}\ \operatorname*{ess\,sup}_{t\in[0,T]} \|\{U(t+\tau,t) - U_n(t+\tau,t)\}\chi_{[0,T]}(t+\tau)\|_{\mathcal{L}(\mathfrak{H})} \qquad (12.1.81)\\
&= \sup_{\tau\in[0,T]}\ \operatorname*{ess\,sup}_{t\in[0,T-\tau]} \|U(t+\tau,t) - U_n(t+\tau,t)\|_{\mathcal{L}(\mathfrak{H})}\,.
\end{aligned}$$

Therefore, on account of Eqs. (11.5.76)–(11.5.78) and Theorem 12.1.17, we deduce the estimate:

$$\operatorname*{ess\,sup}_{t\in(0,T-\tau]} \|U(t+\tau,t) - U_n(t+\tau,t)\|_{\mathcal{L}(\mathfrak{H})} \leq \frac{R_\beta}{n^\beta}, \quad \tau\in[0,T], \quad n = 2,3,\ldots\,.$$

It yields the operator-norm convergence rate (12.1.73) for product approximants $\{U_n(t,s)\}_{(t,s)\in\Delta}$ of solution operator $\{U(t,s)\}_{(t,s)\in\Delta}$.

To prove (12.1.74) for the product approximants $\{\widetilde{U}_n(t,s)\}_{(t,s)\in\Delta}$, we introduce on the space $\mathfrak{H}_T$ operators

$$\widetilde{\mathfrak{R}}_n(\tau) := \mathcal{S}_L(\tau)\{\mathrm{e}^{-\tau\mathcal{K}} - (\mathrm{e}^{-\tau\mathcal{B}/n}\,\mathrm{e}^{-\tau\mathcal{K}_0/n})^n\}\,, \quad \tau\geq 0\,, \quad n\in\mathbb{N}\,. \tag{12.1.82}$$

Owing to the fact that for each $n\in\mathbb{N}$:

$$(\widetilde{\mathfrak{R}}_n(\tau)f)(t) = (U(t+\tau,t) - \widetilde{U}_n(t+\tau,t))\chi_{[0,T]}(t+\tau)f(t)\,, \quad f\in\mathfrak{H}_T\,, \tag{12.1.83}$$

$\{\widetilde{\mathfrak{R}}_n(\tau)\}_{n\geq 1}$ is the set of τ-dependent multiplication operators on $\mathfrak{H}_T$, that is, we can use Proposition 2.2.16 for operator-norm estimate in $\mathfrak{H}_T$ (L^2-setting).

Now following *verbatim* the line of reasoning in the proof of Theorem 12.1.19, we obtain (cf. (12.1.81) and (12.1.83))

$$\sup_{\tau\in[0,T]} \|\widetilde{\mathfrak{R}}_n(\tau)\|_{\mathcal{L}(\mathfrak{H}_T)} = \sup_{\tau\in[0,T]}\ \operatorname*{ess\,sup}_{t\in[0,T-\tau]} \|U(t+\tau,t) - \widetilde{U}_n(t+\tau,t)\|_{\mathcal{L}(\mathfrak{H})}\,. \tag{12.1.84}$$

Then by (12.1.82) and (12.1.84), the estimate (12.1.74) is stemming from (12.1.69), Corollary 12.1.18. □

It is instructive at this point to compare the results for estimates of the rate of the operator-norm convergence of the product approximants for the nACP solution operators in a Banach space: (11.5.68), (11.5.79), for $\alpha < \beta < 1$ (Theorems 11.5.18 and 11.5.16), and in a Hilbert space: (12.1.73), (12.1.74), for $2\alpha - 1 < \beta < 1$, (Theorem 12.1.19).

See Notes to Chap. 12 (Notes to Sects. 12.1 and 12.1.3).

12.2 The Ichinose-Tamura Approach

In this section we re-consider the problem of construction of the product formula approximation of solution operator for the *parabolic type* nACP (12.1.1), Sect. 12.1. The essential difference now is that, instead of a rather *round-about* way via the Trotter product formula for auxiliary *evolution semigroup* on the extended space $\mathfrak{H}_T = L^2([0, T], \mathfrak{H})$ and the *Howland-Evans-Neidhardt* representation formula, we develop here in Hilbert space $\mathfrak{H}$ a direct *Ichinose-Tamura* approach to this problem.

It deserves to mention that for the *Lipschitz* condition: $\beta = 1$, (12.1.3), which is *not* covered by (D3), this approach due to the *self-adjointness* yields for convergence rates in (12.1.73) and (12.1.74), the best of the actually known estimate of the order $O(\ln(n)/n)$, for $n \to \infty$; see (12.2.68) and (12.2.71).

12.2.1 Preliminaries: Assumptions and Theorem

On the Hilbert space level, the set-up in the present section is the same as that we formulated at the beginning of Sect. 12.1.

Let $\mathfrak{H}$ be a separable Hilbert space and A be a strictly positive self-adjoint operator with a dense dom $(A) \subset \mathfrak{H}$. Let $\{B(t)\}_{t\in[0,T]}$ be a family of non-negative self-adjoint operators such that their domains satisfy for some $\alpha : 0 \leq \alpha < 1$ the inclusion:

$$\text{dom}\,(A^{\alpha}) \subset \text{dom}\,(B(t))\,, \quad t \in [0, T]\,. \tag{12.2.1}$$

Since by Remark 10.2.1 the operators $B(t)$ for every $t \in [0, T]$ are A-bounded with *infinitesimally* small relative bound, the sum $C(t) := A + B(t)$ is a well-defined positive self-adjoint operator with dom $(C(t)) =$ dom (A) independent of $t \in [0, T]$. Hence for each $t \in [0, T]$, operator $C(t)$ is generator of holomorphic semigroup $\mathfrak{H}$, which makes the corresponding nACP parabolic. If, in addition, the family $\{B(t)\}_{t\in[0,T]}$ satisfies a suitable *continuity* condition, then $C(t)$ generates a solution operator $\{U(t,s)\}_{(t,s)\in\Delta}$ (also known as fundamental solution, evolution family or solution operator) for evolution equation:

$$\begin{aligned} \partial_t U(t,s) + C(t)\, U(t,s) &= 0\,, \quad (t,s) \in \Delta_> := \{(t,s) \in \Delta : t > s\}\,,\\ \lim_{t\to s} U(t,s) &= \mathbb{1}\,, \end{aligned} \tag{12.2.2}$$

in the *strong* operator sense on the space $\mathcal{L}(\mathfrak{H})$, or equivalently,

$$\begin{aligned} \partial_s U(t,s)u - U(t,s)\, C(s)\, u &= 0\,, \quad u \in \mathrm{dom}\,(A)\,, \ (t,s) \in \Delta_>\,,\\ \lim_{s\to t} U(t,s) &= \mathbb{1}\,. \end{aligned} \tag{12.2.3}$$

cf. Eqs. (2.1.13) and (2.1.15) in Sect. 2.1.1, as well as the nACP (11.1.1), (12.1.1).

Owing to (12.2.2) and positivity of self-adjoint operators $\{C(t)\}_{t\in[0,T]}$, we infer the contraction property of solution operator $U(t,s) : \mathfrak{H} \to \mathfrak{H}$ for $(t,s) \in \Delta$. Indeed,

$$\begin{aligned} \partial_t \|U(t,s)u\|^2 &= (\partial_t U(t,s)u, U(t,s)u) + (U(t,s)u, \partial_t U(t,s)u)\\ &= -(C(t)U(t,s)u, U(t,s)u) - (U(t,s)u, C(t)U(t,s)u)\\ &= -2\,(U(t,s)u, C(t)U(t,s)u) \le 0, \qquad u \in \mathfrak{H}\,, \end{aligned} \tag{12.2.4}$$

and, as a consequence, we obtain for $U(s,s) := \lim_{s\to t} U(t,s)$:

$$\|U(t,s)u\|^2 = \|U(s,s)u\|^2 + \int_0^t d\tau\; \partial_\tau \|U(\tau,s)u\|^2 \le \|U(s,s)u\|^2 = \|u\|^2\,, \tag{12.2.5}$$

wherefore $\|U(t,s)\| \le 1$ for $(t,s) \in \Delta$.

Now we define a family of *product approximats* for solution operator $\{U(t,0)\}_{t\in[0,T]}$. To this aim we let

$$0 = t_0 < t_1 < \ldots < t_{n-1} < t_n = t, \quad t_j = j\tau, \quad \tau := t/n\,, \quad n \in \mathbb{N}\,,$$

for $1 \le j \le n$ and for some $0 < t \le T$.

Definition 12.2.1 To take an advantage of the full power of self-adjointness in the proof of convergence, we introduce a *symmetrised* version of two product approximants $\{F_j(\tau)\}_{1\leq j\leq n}$ and $\{\widetilde{F}_{n-j}(\tau)\}_{1\leq j\leq n}$, which are generated by the self-adjoint factors $\{K_i(\tau)\}_{1\leq i\leq n}$:

$$K_i(\tau) := \exp(-\tau A/2)\exp(-\tau B(t_{i-1}))\exp(-\tau A/2)\,, \quad 1 \leq i \leq n\,,$$

$$F_j(\tau) := \prod_{i=j}^{1} K_i(\tau) \quad \text{and} \quad \widetilde{F}_{n-j}(\tau) := \prod_{i=n}^{j+1} K_i(\tau)\,. \tag{12.2.6}$$

Note that operators in (12.2.6) are contractions.

Assumptions 12.2.2 Let $\mathfrak{H}$ be a separable Hilbert space and $\alpha \in [0, 1]$.

(E1) The operator $A \geq c\,\mathbb{1}$ is self-adjoint in $\mathfrak{H}$ and $\{B(t) \geq c\,\mathbb{1}\}_{t\in[0,T]}$ is a family of positive self-adjoint operators in $\mathfrak{H}$, for some $c > 0$.

(E2) There exists an $\alpha \in [0, 1)$ such that for a.a. $t \in [0, T]$ it holds the inclusion $\text{dom}\,(A^\alpha) \subseteq \text{dom}\,(B(t))$. In addition, the function $B(\cdot)A^{-\alpha} : [0, T] \longrightarrow \mathcal{L}(\mathfrak{H})$ is strongly measurable and essentially bounded, that is,

$$C_\alpha := \operatorname*{ess\,sup}_{t\in[0,T]} \|B(t)A^{-\alpha}\|_{\mathcal{L}(\mathfrak{H})} < \infty\,. \tag{12.2.7}$$

(E3) The mapping $A^{-\alpha}B(\cdot)A^{-\alpha} : [0, T] \longrightarrow \mathcal{L}(\mathfrak{H})$ is Lipschitz continuous, that is, there is a constant $L_\alpha > 0$ such that one has estimate

$$\|A^{-\alpha}(B(t) - B(s))A^{-\alpha}\|_{\mathcal{L}(\mathfrak{H})} \leq L_\alpha\,|t - s|, \quad (t, s) \in [0, T] \times [0, T]\,, \tag{12.2.8}$$

for $\alpha \in [0, 1)$.

Remark 12.2.3 A few comments are now in order.

(a) It is known that Assumptions 12.2.2 are sufficient for justification of the product formula

$$U(t, 0) = \operatorname*{s\text{-}lim}_{n\to\infty} F_n(t/n)\,, \quad t \in [0, T]\,, \tag{12.2.9}$$

in the *strong* operator topology on $\mathcal{L}(\mathfrak{H})$. The proof follows *mutatis mutandis* the line of reasoning of the proof of Theorem 11.1.3.

Moreover, in this case the condition (E2) concerning the *infinitesimal* smallness of perturbation $\{B(t)\}_{t\in[0,T]}$ is superfluous. It can be relaxed to the *Kato-smallness* of this perturbation; see (B1) (11.1.8).

(b) Note that assumption (A2) of Theorem 11.1.3 allows to extend the proof of the product formula (12.2.9) to the case of a family of *non-stationary* generators: $\{A(t)\}_{t\in[0,T]}$. By inspection of this proof, we find that the equivalent to (A2) additional condition, which is sufficient for the strong convergence (12.2.9), has the form

$$\text{(E3)}^* \qquad \|A^{-\alpha}(A(t) - A(s))A^{-\alpha}\|_{\mathcal{L}(\mathfrak{H})} \leq L_\alpha^* \, |t - s| \,, \quad A := A(0) \,. \tag{12.2.10}$$

Here $\alpha \in [0, 1)$, $L_\alpha^* > 0$, $(t, s) \in [0, T] \times [0, T]$ and $\mathrm{dom}\,(A(t)) = \mathrm{dom}\,(A(0))$.

(c) For infinitesimally small perturbation $\{B(t)\}_{t \in [0,T]}$, the result mentioned in (b) can be *uplifted* to convergence of the product formula (12.2.9) in the operator-norm topology on $\mathcal{L}(\mathfrak{H})$.

We return to a detailed discussion of the claims (a)–(c) and to generalisations of the product formula (12.2.9) in Sect. 12.2.4, but first of all we prove (c). To apply a full power of *self-adjointness* in the proof and for estimate of the rate of convergence, we start by the case when involved into the product formula (12.2.9) factors $\{K_i(\tau)\}_{1 \leq i \leq n}$ (12.2.6) are self-adjoint.

Theorem 12.2.4 *Let positively semibounded from below self-adjoint operators A and $\{B(t)\}_{t \in [0,T]}$ satisfy Assumptions* 12.2.2 : (E1)–(E3). *Then*

$$\|U(t, 0) - F_n(t/n)\| = O(\ln(n)/n) \,, \quad n \to \infty \,, \tag{12.2.11}$$

uniformly on any interval $\mathcal{I} = [0, T] \subset \mathbb{R}_0^+$.

We consider in Sect. 12.2.4 extensions of the product formula (12.2.11) to the case of *non-self-adjoint* factors in approximants; cf. (12.2.6) and (12.2.68), (12.2.69).

12.2.2 Evolution Equation and Solution Operator

We summarise here some essential properties of solution operator $\{U(t, s)\}_{(t,s) \in \Delta}$ (12.2.2) that will be indispensable for the proofs in the proceeding subsections.

Lemma 12.2.5 *Let the assumptions* (E1)–(E3) *be satisfied. Then we infer that* :

(a) *For any $t \in [0, T]$ operator $C(t) := A + B(t) \geq c\,\mathbb{1}$, for $c > 0$, that is, $0 \in \rho(C(t))$. This operator is self-adjoint on the independent of $t \in [0, T]$ domain,* $\mathrm{dom}\,(C(t)) = \mathrm{dom}\,(A)$, *and the estimate*:

$$\|(z\,\mathbb{1} + C(t))^{-1}\| \leq \frac{M_\varepsilon}{|z|}, \quad z \in \overline{S}_{\pi/2-\varepsilon} \,, \quad t \in [0, T], \tag{12.2.12}$$

holds for $\varepsilon \in (0, \pi/2)$ and the corresponding $M_\varepsilon > 0$, that is, operators $\{C(t)\}_{t \in [0,T]}$ are generators of self-adjoint holomorphic contraction semi-groups.

(b) *For any* $(t, s) \in [0, T] \times [0, T]$ *one has inequality*

$$\|C(t)^{\rho}(C(t)^{-1} - C(s)^{-1})\, C(s)^{\rho}\| \leq \widehat{L}_{\rho}\, |t - s|\,, \tag{12.2.13}$$

for $\rho := 1 - \alpha > 0$ *and some* $\widehat{L}_{\rho} > 0$.

Proof

(a) Owing to Remark 10.2.1, the assumption (E2) entails that operators $B(t)$ are A-bounded with *infinitesimally* small relative bound for every $t \in [0, T]$. Then due to stability of self-adjointness and lower bound of operator A the sum $C(t) := A + B(t)$ is positive self-adjoint operator with domain $\mathrm{dom}\,(C(t)) = \mathrm{dom}\,(A)$ independent of $t \in [0, T]$ and with $0 \in \rho(C(t))$.

By assumptions (E1), (E2), operator $A \in \mathscr{H}_c(\pi/2, 0)$ is generator of a holomorphic contraction semigroup and $\{B(t)\}_{t\in[0,T]}$ is family of the A-bounded non-negative self-adjoint operators. As a result, by Lemma 10.2.2 in a Hilbert space $\mathfrak{H}$, the operators $C(t) \in \mathscr{H}_c(\pi/2, 0)$, for $t \in [0, T]$, are in turn generators of self-adjoint holomorphic contraction semigroups. This proves the estimate (12.2.12): cf. Sects. 1.8.2 and 1.9.2.

(b) Using for $\alpha \in (0, 1)$ invertibility of operator $A^{\alpha} \geq c^{\alpha}\mathbb{1}$, (E1), one can rewrite the left-hand side of (12.2.13) as

$$\|C(t)^{\rho}(C(t)^{-1} - C(s)^{-1})C(s)^{\rho}\| =$$
$$\|C(t)^{-\alpha} A^{\alpha}[A^{-\alpha}(B(s) - B(t))A^{-\alpha}]A^{\alpha}C(s)^{-\alpha}\|\,. \tag{12.2.14}$$

for $(t, s) \in [0, T] \times [0, T]$. Note that on $\mathrm{dom}\,(A) = \mathrm{dom}\,(C(t))$ the corresponding self-adjoint operators satisfy for $t \in [0, T]$ inequality $A \leq C(t)$. As a consequence, one gets (the *Löwner-Heinz* inequality, Notes to Chap. 12 (Sect. 12.2.2)) that

$$\|A^{\alpha}\, u\| \leq \|C(t)^{\alpha}\, u\|\,, \quad u \in \mathrm{dom}\,(A) = \mathrm{dom}\,(C(t))\,, \quad \alpha \in (0, 1)\,, \tag{12.2.15}$$

for $t \in [0, T]$. Then (12.2.15) and invertibility of operator $C(t)^{\alpha} \geq c^{\alpha}\mathbb{1}$ give for operator norms' estimates: $\|A^{\alpha}C(s)^{-\alpha}\| \leq 1$ and $\|C(t)^{-\alpha}A^{\alpha}\| \leq 1$, for $t, s \in [0, T]$. Here $C(t)^{-\alpha}A^{\alpha} := (A^{\alpha}C(t)^{-\alpha})^{*}$. These estimates together with (12.2.14) yield (12.2.13) for $\widehat{L}_{\rho} = L_{\alpha}$, see (E3).

□

As a corollary by Lemma 12.2.5 (a) and by Definition 2.1.11, the nACP for $\{C(t)\}_{t\in[0,T]}$ is of a parabolic type.

Proposition 12.2.6 *Let the assumptions* (E1)–(E3) *be satisfied. Then for parabolic* nACP (12.1.1), *there exists a unique solution operator* $\{U(t,s)\}_{(t,s)\in\Delta}$ *(Definition* 2.1.5) *with the following properties:*

(a) *The operator-valued function:* $(t,s) \mapsto U(t,s) \in \mathcal{L}(\mathfrak{H})$, *is strongly continuous on* $\Delta = \{(t,s) \in [0,T] \times [0,T] : 0 \le s \le t \le T\}$.

(b) *For* $(t,s) \in \Delta_> := \{(t,s) \in \Delta : t > s\}$, *the range*: $\mathrm{ran}(U(t,s)) \subset \mathrm{dom}(A)$, *and* $\|C(t)\,U(t,s)\| \le M(t-s)^{-1}$, $\|C(t)\,U(t,s)\,C(s)^{-1}\| \le M$ *for some* $M > 0$.

(c) *The function* $(t,s) \mapsto U(t,s)$ *is strongly continuously differentiable in* t *for* $(t,s) \in \Delta_>$, *with* $\partial_t U(t,s) \in \mathcal{L}(\mathfrak{H})$, *and satisfies*

$$\begin{aligned} &\partial_t U(t,s) = -\,(A + B(t))\,U(t,s)\,, \qquad (t,s) \in \Delta_>\,, \\ &U(s,s) := \lim_{t\to s} U(t,s) = \mathbb{1}\,, \end{aligned} \tag{12.2.16}$$

in the strong operator sense.

(d) *Function* $t \mapsto u(t) = U(t,s)\,u_s$ *is a unique classical solution of* nACP (12.1.1) *for initial condition* $u(s) = u_s$ *from* $\mathfrak{H}_s = \mathrm{dom}(A)$ *(regularity set), that is, function* $u \in C([s,T], \mathrm{dom}\,(A)) \cap C^1([s,T],\mathfrak{H})$, *and satisfies* (12.1.1).

(e) *For any* $u \in \mathfrak{H}$ *and* $t > s$ *the function* $t \mapsto u_s(t) := U(t,s)u \in \mathrm{dom}(A)$ *and* $u_s \in C([s,T),\mathfrak{H}) \cap C^1((s,T),\mathfrak{H})$. *Then on account of Eq.* (12.2.16), *the function* u_s *satisfies* (12.1.1) *in* $\Delta_>$. *It is a* strict *solution of* nACP (12.1.1) *for initial condition* $u = \lim_{t\to s} u_s(t)$.

Proof

(a) On account of Lemma 12.2.5, Definition 2.1.11 and Propositions 2.1.9, 2.1.12, the nACP (12.1.1) is a well-posed parabolic non-autonomous Cauchy problem (2.1.12) with unique solution operator $\{U(t,s)\}_{(t,s)\in\Delta}$, which satisfies (2.1.13) and (2.1.15).

(b)–(c) Since by (a) the nACP (12.1.1) is a well-posed and parabolic, these properties follow from Proposition 2.1.12 (a)–(c) together with Eqs. (2.1.13) and (2.1.15) for solution operator, which are equivalent to (12.2.2) and (12.2.3). As a consequence, we also infer (12.2.16).

(d) Given that dom(A) is a *regularity* set (Definition 2.1.3), this property is a straightforward consequence of (b) and Definition 2.1.2 of the *classical* solution of nACP (2.1.12).

(e) This property is stemming from (b), definition of the *strict* solution (see Remark 2.1.1(b)) and strong continuity of unique solution operator $\{U(t,s)\}_{(t,s)\in\Delta}$ which satisfies (2.1.13) and (2.1.15).

□

So, due to Propositions 2.1.9, 2.1.12 and Assumptions 12.2.2, (E1)–(E3), we obtain that Lemma 12.2.5 and Proposition 12.2.11 guarantee the existence and general properties of a *unique* solution operator $\{U(t,s)\}_{(t,s)\in\Delta}$ of nACP (2.1.12). To scrutinise further the properties of solution operator, we consider more carefully the Assumption 12.2.2(E2) about infinitesimal smallness of perturbation $\{B(t)\}_{t\in[0,T]}$.

For construction of the solution operator $\{U(t,s)\}_{(t,s)\in\Delta}$, we consider the equivalent to system (12.2.16) the *integral equation*;cf. Remark 2.1.1 (c). Seeing that by assumption (E1) $\{U_A(t) := \mathrm{e}^{-tA}\}_{t\geq 0}$ is a holomorphic (contraction) semigroup, by virtue of (12.2.2) and Proposition 12.2.11 (c) one gets for solution of inhomogeneous equation (12.2.16) the identity:

$$\partial_\tau(U_A(t-\tau)U(\tau,s)) = -\,U_A(t-\tau)\,B(\tau)\,U(\tau,s)\,,\quad 0\leq s<\tau<t\leq T\,. \tag{12.2.17}$$

By Proposition 12.2.11 (b) and assumption (E3), function: $\tau \mapsto U_A(t-\tau)\,B(\tau)\,U(\tau,s)$, is a strongly measurable on interval $[s,t]$. Then integrating (12.2.17) over $[s,t]$, we obtain that solution operator $\{U(t,s)\}_{(t,s)\in\Delta}$ satisfies in the strong operator sense the integral equation:

$$U(t,s) = U_A(t-s) - \int_s^t d\tau\; U_A(t-\tau)\,B(\tau)\,U(\tau,s)\,,\quad U(s,s)=\mathbb{1}\,, \tag{12.2.18}$$

for $(t,s)\in\Delta$ and for the *Bochner* integral in (12.2.18).

Similarly, by virtue of (12.2.3) and $\mathrm{ran}(U_A(t))\subset\mathrm{dom}(A)$, for $t>0$, one infers measurability of function: $\tau\mapsto U(t,\tau)\,B(\tau)\,U_A(\tau-s)$ on Δ and deduces on this interval the integral equation:

$$U(t,s) = U_A(t-s) - \int_s^t d\tau\; U(t,\tau)\,B(\tau)\,U_A(\tau-s)\,,\quad U(s,s)=\mathbb{1}\,, \tag{12.2.19}$$

that holds in the strong operator sense on $\mathcal{L}(\mathfrak{H})$.

Theorem 12.2.7 *Let the assumptions* (E1)–(E3) *be satisfied. Then integral equation* (12.2.18) *has a unique strongly continuous solution* $\{U(t,s)\}_{(t,s)\in\Delta}$.

Proof On account of assumptions (E1)–(E2), one gets for the self-adjoint holomorphic contraction semigroup $\{U_A(t)\}_{t\geq 0}$ and for the family $\{B(\tau)\}_{\tau\in[0,T]}$:

$$\|A^\alpha U_A(t-s)\| \leq \frac{M_\alpha}{(t-s)^\alpha}\quad\text{and}\quad \|B(\tau)\,A^{-\alpha}\|\leq C_\alpha\,,\quad \alpha\in(0,1)\,, \tag{12.2.20}$$

where $0\leq s<t\leq T,\ M_\alpha\leq 1$ and the function: $\tau\mapsto B(\tau)\,A^{-\alpha}$, is operator-norm continuous. Then the function: $\Delta_>\ni(t,s)\mapsto B(t)\,U_A(t-s)$, is also operator-norm continuous, and (12.2.20) yields the estimate

$$\|B(t)\,U_A(t-s)\|\leq\frac{C_\alpha\,M_\alpha}{(t-s)^\alpha}\,,\quad \alpha\in(0,1),\quad (t,s)\in\Delta_>\,. \tag{12.2.21}$$

Since singularity in (12.2.21) is integrable, one can construct solution operator $\{U(t,s)\}_{(t,s)\in\Delta}$ of nACP (12.2.16) as a *Dyson-Phillips* series $\sum_{n=0}^{\infty} S_n(t,s)$, which converges in the operator-norm topology uniformly for $(t,s) \in \Delta$, by iteration of Eq. (12.2.18).

To this aim we define for $0 \leq s \leq t \leq T$ the recurrence relations

$$\begin{aligned} S_0(t,s) &= U_A(t-s), \\ S_n(t,s) &= \int_s^t d\tau\, U_A(t-\tau)\,(-B(\tau))\,S_{n-1}(\tau,s), \quad n \in \mathbb{N}. \end{aligned} \tag{12.2.22}$$

On account of (E1)–(E3), the bounded strongly continuous on Δ and strongly differentiable in $\Delta_> = \{(t,s) \in \Delta : t > s\}$ operator-valued functions $\{S_n(t,s)\}_{n\geq 1}$, defined by (12.2.22), are the n-fold strongly convergent *Bochner* integrals with conventions that $\tau_0 = s$ and $\tau_{n+1} = t$:

$$\begin{aligned} S_n(t,s) = \int_s^t d\tau_n \int_s^{\tau_n} d\tau_{n-1} \ldots \int_s^{\tau_2} d\tau_1\, U_A(t-\tau_n) \times \\ \times (-B(\tau_n))\,U_A(\tau_n-\tau_{n-1}) \ldots U_A(\tau_2-\tau_1)\,(-B(\tau_1))\,U_A(\tau_1-s)\,. \end{aligned} \tag{12.2.23}$$

Since by (E1) the holomorphic semigroup $\{U_A(t)\}_{t\geq 0}$ is a contraction, (12.2.21) and (12.2.23) provide estimates

$$\|S_n(t,s)\| \leq (C_\alpha\, M_\alpha)^n\, \mathcal{I}_n(t,s)\,, \quad \alpha \in (0,1)\,, \quad n \in \mathbb{N}\,, \tag{12.2.24}$$

for integral $\mathcal{I}_n(t,s) =: \kappa_n(\alpha)\,(t-s)^{n\,(1-\alpha)}$, cf. (I.0.5). Here $(t,s) \in \Delta$ and

$$\begin{aligned} \mathcal{I}_n(t,s) := \int_s^t d\tau_n \int_s^{\tau_n} d\tau_{n-1} \ldots \int_s^{\tau_2} d\tau_1\, (\tau_n-\tau_{n-1})^{-\alpha} \cdots \\ \cdots (\tau_2-\tau_1)^{-\alpha}(\tau_1-s)^{-\alpha} = \frac{\Gamma(1-\alpha)^n}{\Gamma(n(1-\alpha)+1)}\,(t-s)^{n(1-\alpha)}\,. \end{aligned} \tag{12.2.25}$$

Thanks to estimates in Appendix I, (I.0.5) and (I.0.8), we obtain for $\alpha \in (0,1)$

$$\kappa_n(\alpha) \leq \frac{1}{(1-\alpha)\sqrt{2\pi\, n}} \left(\frac{\sqrt{2\pi}\;\Gamma(1-\alpha)}{\Gamma(z_*)\, n^{(1-\alpha)}} \right)^n, \quad n \geq 1\,. \tag{12.2.26}$$

Here $\Gamma(z_*) := \min_{\{1-\alpha\,\leq\, z\,\leq\, 2-\alpha\}} \Gamma(z)$ and $\Gamma(\cdot)$ denotes the *Euler* Gamma-function.

For $n \to \infty$ the upper bound (12.2.26) is of order $O(n^{-n(1-\alpha)})$. Then inequality (12.2.24) yields the *uniform* in $(t, s) \in \Delta$ operator-norm converges of the *Dyson-Phillips* series $\sum_{n=0}^{\infty} S_n(t, s)$ to a bounded operator-valued function

$$U(t, s) := \sum_{n=0}^{\infty} S_n(t, s), \quad (t, s) \in \Delta, \tag{12.2.27}$$

with estimate $\|U(t, s)\| \leq 1 + \sum_{n=1}^{\infty} (C_\alpha M_\alpha)^n \kappa_n(\alpha) T^{n(1-\alpha)} =: C_T(\alpha)$.

For analysis of continuity of function (12.2.27), we consider $0 \leq s \leq t \leq t' \leq T$. Then by Assumptions 12.2.2 (E1) and by (12.2.27), one gets, for $s < t$ and for any $u \in \mathfrak{H}$, the estimate:

$$\|(S_0(t', s) - S_0(t, s))\, u\| \leq \left(\frac{t - s}{t' - s}\right)^{(t-s)/(t'-t)} (t' - t)\, \|u\| \leq (t' - t)\, \|u\|, \tag{12.2.28}$$

whereas for $s = t$ we infer by the strong continuity of semigroup $\{U_A(\tau)\}_{\tau \geq 0}$ that $\lim_{t' \to s} \|(\mathbb{1} - S_0(t', s))\, u\| = 0$. Further, for the terms $\{S_n(t, s)\}_{n \geq 1}$, we use inequality

$$\|(S_n(t', s) - S_n(t, s))\| \leq \left\| \int_t^{t'} d\tau_n\, U_A(t' - \tau_n)\, L_n(\tau_n, s) \right\| + \tag{12.2.29}$$
$$\int_s^t d\tau_n\, \|(U_A(t' - \tau_n) - U_A(t - \tau_n))\, L_n(\tau_n, s)\|,$$

where we introduced family $\{L_n(\tau_n, s)\}_{n \geq 1}$, cf. (12.2.23), defined by equation

$$S_n(t, s) = \int_s^t d\tau_n\, U_A(t - \tau_n)\, L_n(\tau_n, s). \tag{12.2.30}$$

Since semigroup $\{U_A(\tau)\}_{\tau \geq 0}$ is contraction, by (12.2.24), by (12.2.25) and by virtue of (12.2.30), we obtain for $n \geq 1$

$$\left\| \int_t^{t'} d\tau_n\, U_A(t' - \tau_n)\, L_n(\tau_n, s) \right\| \leq (C_\alpha M_\alpha)^n \kappa_n(\alpha)\, (t' - t)^{n(1-\alpha)}. \tag{12.2.31}$$

To estimate the second term in the right-hand side of (12.2.29), we note that by (12.2.24) and (12.2.30)

$$\int_s^t \mathrm{d}\tau_n \, \|(U_A(t' - \tau_n) - U_A(t - \tau_n))\, L_n(\tau_n, s)\| \leq$$

$$\int_s^t \mathrm{d}\tau_n \, \|U_A(t' - \tau_n) - U_A(t - \tau_n)\| \, \|L_n(\tau_n, s)\| \leq (C_\alpha\, M_\alpha)^n\, \kappa_n(\alpha) \times$$

$$\times\ (t' - t) \int_s^t \mathrm{d}\tau_n \left(\frac{t - \tau_n}{t' - \tau_n}\right)^{(t-\tau_n)/(t'-t)} n\,(1-\alpha)\,(\tau_n - s)^{n\,(1-\alpha)-1}\,, \quad n \geq 1. \tag{12.2.32}$$

As a result, summarising (12.2.28)–(12.2.32) we infer that all terms in (12.2.27) are strongly continuous in $(t, s) \in \Delta$. Given that series (12.2.27) converges in operator-norm topology uniformly on Δ, we conclude that $\{U(t, s)\}_{(t,s)\in\Delta}$ is a strongly continuous operator-valued function on Δ.

Inserting the strongly continuous function (12.2.27) into (12.2.18) and taking into account representations (12.2.23), we obtain an identity. As a consequence, the defined by (12.2.27) function $\{U(t, s)\}_{(t,s)\in\Delta}$ is a strongly continuous solution of the integral equation (12.2.18) on Δ.

To prove the uniqueness of this solution, we use (12.2.18). If there are two solutions $\{U_1(t, s)\}_{(t,s)\in\Delta}$ and $\{U_2(t, s)\}_{(t,s)\in\Delta}$, then $\delta U(t, s) := U_1(t, s) - U_2(t, s)$ is a solution of the homogeneous equation

$$\delta U(t, s) = \int_s^t d\tau\, U_A(t - \tau)\,(-B(\tau))\,\delta U(\tau, s)\,, \quad \delta U(s, s) = 0\,. \tag{12.2.33}$$

First we note that strong continuity of solutions $\{U_1(t, s)\}_{(t,s)\in\Delta}$ and $\{U_2(t, s)\}_{(t,s)\in\Delta}$ yields continuity of the function: $(t, s) \mapsto \|\delta U(t, s)\, u\|$ on Δ, for any $u \in \mathfrak{H}$. Next, by Assumptions 12.2.2 (E1), (E2), the closure $\overline{U_A(t - \tau)\, B(\tau)} = (B(\tau)\, U_A(t - \tau))^*$ is a bounded operator with operator-norm estimate (cf. (12.2.21))

$$\|\overline{U_A(t - \tau)\, B(\tau)}\| \leq \frac{C_\alpha\, M_\alpha}{(t - \tau)^\alpha}\,, \quad \alpha \in (0, 1), \quad s \leq \tau < t\,, \tag{12.2.34}$$

and also the function: $\tau \mapsto \overline{U_A(t - \tau)\,(-B(\tau))}$ is strongly measurable. Then by (12.2.33), (12.2.34) and properties of the *Bochner* integral, we obtain inequality:

$$\|\delta U(t, s)\, u\| \leq \int_s^t d\tau\, \frac{C_\alpha\, M_\alpha}{(t - \tau)^\alpha}\, \|\delta U(\tau, s)\, u\|\,, \quad \alpha \in (0, 1)\,, \quad u \in \mathfrak{H}\,. \tag{12.2.35}$$

After n iterations of the integral inequality (12.2.35) (cf. the *Gronwall* inequality in Appendix G.1, Corollary G.1.2 (c)), one gets for $\alpha \in (0, 1)$ the estimate

$$\|\delta U(t,s)\,u\| \leq \kappa_n(\alpha)\,(t-s)^{n\,(1-\alpha)} \sup_{(\tau,s)\in\Delta} \|\delta U(\tau,s)\|\,\|u\|\,, \quad u \in \mathfrak{H}\,. \tag{12.2.36}$$

On account of (12.2.26) and $\|U(t,s)\| \leq C_T(\alpha)$, the estimate (12.2.36) ensures that inequality (12.2.35) has for any $u \in \mathfrak{H}$ only trivial solution, which yields the unicity: $U_1(t,s) = U_2(t,s)$ for $(t,s) \in \Delta$. □

Corollary 12.2.8 *Let the assumptions* (E1)–(E3) *be satisfied. Then integral equation* (12.2.19) *has a unique strongly continuous solution* $\{\widetilde{U}(t,s)\}_{(t,s)\in\Delta}$.

Proof It is *mutatis mutandis* of the proof of Theorem 12.2.7. To this aim we define for $0 \leq s \leq t \leq T$ the recurrence relation (cf. (12.2.22))

$$\begin{aligned} &\widetilde{S}_0(t,s) = U_A(t-s)\,, \\ &\widetilde{S}_n(t,s) = \int_s^t d\tau\, \widetilde{S}_{n-1}(t,\tau)\,(-B(\tau))\,U_A(\tau - s)\,, \quad n \in \mathbb{N}\,. \end{aligned} \tag{12.2.37}$$

Given that the operator-valued functions $\{\widetilde{S}_n(t,s)\}_{n\geq 1}$, defined by (12.2.37), are the n-fold strongly convergent *Bochner* integrals:

$$\begin{aligned} &\widetilde{S}_n(t,s) = \int_s^t d\tau_1 \int_{\tau_1}^t d\tau_2 \ldots \int_{\tau_{n-1}}^t d\tau_n\, U_A(t-\tau_n)\,\cdot \\ &(-B(\tau_n))\,U_A(\tau_n - \tau_{n-1}) \cdots U_A(\tau_2 - \tau_1)\,(-B(\tau_1))\,U_A(\tau_1 - s)\,. \end{aligned} \tag{12.2.38}$$

Then estimates similar to (12.2.24)–(12.2.26) ensure the *uniform* in $(t,s) \in \Delta$ *operator-norm* converges of the *Dyson-Phillips* series $\sum_{n=0}^{\infty} \widetilde{S}_n(t,s)$ to a bounded operator-valued function

$$\widetilde{U}(t,s) := \sum_{n=0}^{\infty} \widetilde{S}_n(t,s)\,, \quad (t,s) \in \Delta\,, \tag{12.2.39}$$

with estimate $\sup_{(t,s)\in\Delta} \|\widetilde{U}(t,s)\| \leq \widetilde{C}_T(\alpha)$.

The rest of the proof follows *verbatim* the arguments of Theorem 12.2.7. As a consequence, the function $\{\widetilde{U}(t,s)\}_{(t,s)\in\Delta}$ (12.2.39) is a unique strongly continuous on Δ solution of the integral equation (12.2.19). □

Corollary 12.2.9 *Since function* $\{U(t,s)\}_{(t,s)\in\Delta}$ (12.2.27) *is strongly continuous solution of integral equation* (12.2.18) *on* Δ, *by assumptions* (E1)–(E3) *the integrand:* $\tau \to U_A(t-\tau)\,B(\tau)\,U(\tau,s)$ *in* (12.2.18), *is a strongly continuous function on* Δ. *Moreover, by* (E1) *the semigroup* $\{U_A(\tau)\}_{\tau\geq 0}$ *is strongly continuously differentiable for* $\tau > 0$. *Then for solution* (12.2.27), *the right-hand side of identity* (12.2.18) *is strongly continuously differentiable for* $(t,s) \in \Delta_>$, *and one* *infers that*

$$\partial_t U(t,s) = -\,(A + B(t))\,U(t,s)\,, \quad (t,s) \in \Delta_>\,, \tag{12.2.40}$$

in the strong operator sense on the space $\mathcal{L}(\mathfrak{H})$. *That is, function* $\{U(t,s)\}_{(t,s)\in\Delta}$ (12.2.27), *with* $\mathrm{ran}\,(U(t,s)) \subset \mathrm{dom}\,(A)$ *for* $(t,s) \in \Delta_>$, *is the solution operator of nACP* (12.2.16).

Owing to (12.2.40) *and in virtue of* (12.2.4), (12.2.5) *the solution operator* $\{U(t,s)\}_{(t,s)\in\Delta}$ (12.2.27) *is a contraction.*

Corollary 12.2.10 *Since function* $\{\widetilde{U}(t,s)\}_{(t,s)\in\Delta}$ (12.2.39) *is strongly continuous solution of integral equation* (12.2.19) *on* Δ, *the assertion, which is similar to Corollary* 12.2.9, *holds for* (12.2.39).

To this end we note that now by assumptions (E1)–(E3) *and Corollary* 12.2.8, *the integrand:* $\tau \to \widetilde{U}(t,\tau)\,B(\tau)\,U_A(\tau - s)$ *in* (12.2.19), *is a strongly continuous function on* Δ *and derivative of* (12.2.19) *with respect to* s *in* $\Delta_>$ *can be calculated explicitly on* $\mathrm{dom}\,(A)$. *This provides that family* (12.2.39) *is a solution of equation*

$$\partial_s \widetilde{U}(t,s)\,u = \widetilde{U}(t,s)\,(A{+}B(s))\,u\,, \quad u \in \mathrm{dom}\,(A)\,, \quad (t,s) \in \Delta_>\,. \tag{12.2.41}$$

As a consequence, function $\{\widetilde{U}(t,s)\}_{(t,s)\in\Delta}$ (12.2.39) *is a solution operator for nACP* (12.2.3). *On account of* (12.2.41) *and* (12.2.4), (12.2.5), *the solution operator* $\{\widetilde{U}(t,s)\}_{(t,s)\in\Delta}$ (12.2.27) *is also a contraction.*

Finally, due to (12.2.40) *and* (12.2.41), *one gets* $\partial_\tau(\widetilde{U}(t,\tau)\,U(\tau,s)\,u) = 0$, *for any* $u \in \mathrm{dom}\,(A)$ *and* $s < \tau < t$. *Seeing that* $\mathrm{dom}\,(A)$ *is dense in* $\mathfrak{H}$, *the integration of this equality on interval* $[s,t]$ *yields identity* :

$$U(t,s) = \widetilde{U}(t,s)\,, \quad (t,s) \in \Delta\,, \tag{12.2.42}$$

of two solution operators (12.2.27) *and* (12.2.39).

We now mention properties of the solution operator $\{U(t,s)\}_{(t,s)\in\Delta}$ (Theorem 12.2.7) which are required for proving the main Theorem 12.2.4.

Lemma 12.2.11 *Let* $0 \le \gamma < 1$. *Then the solution operator* $\{U(t,s)\}_{(t,s)\in\Delta}$ (12.2.27) *has the following properties* :

(1) $\|A^\gamma\,U(t,s)\| = O((t-s)^{-\gamma})$, $\qquad \|\overline{U(t,s)\,A^\gamma}\| = O((t-s)^{-\gamma})$.

(2) $\|A^\gamma U(t,s) A^{-\gamma}\| = O(1)$, $\qquad \|\overline{A^{-\gamma} U(t,s) A^\gamma}\| = O(1)$.

(3) $\|A\,U(t,s)\,A^{-\gamma}\| = O((t-s)^{\gamma-1})$, $\qquad \|\overline{A^{-\gamma}\,U(t,s)\,A}\| = O((t-s)^{\gamma-1})$.

Proof

(1) On account of integral equation (12.2.18) and Corollary 12.2.9, we obtain, for $(t, s) \in \Delta_>$, the representation:

$$A^\gamma U(t,s) = A^\gamma U_A(t-s) - \int_s^t d\tau\, A^\gamma U_A(t-\tau)\, B(\tau)\, A^{-\gamma} A^\gamma U(\tau,s)\,. \tag{12.2.43}$$

Then by Corollary 12.2.9, $\operatorname{ran}(U(t,s)) \subset \operatorname{dom}(A)$, for $(t,s) \in \Delta_>$, i.e. the family $\{A^\gamma U(t,s)\}_{(t,s)\in\Delta_>}$ is a bounded operator-valued function. Because of that and of assumptions E(1)–E(2), cf. (12.2.20)–(12.2.21), the representation (12.2.43) yields inequality

$$\|A^\gamma U(t,s)\| \le \frac{M_\gamma}{(t-s)^\gamma} + \int_s^t d\tau\, \frac{M_\gamma\, C_\gamma}{(t-\tau)^\gamma}\, \|A^\gamma U(\tau,s)\|\,, \quad (t,s) \in \Delta_>\,. \tag{12.2.44}$$

By iteration of (12.2.44), one gets, following (G.3.4)–(G.3.8) in Theorem G.3.2 (see Appendix G.3), the first part of assertion (1).

To prove the second part of assertion (1), we note owing to (12.2.18) one also has for $(t,s) \in \Delta_>$ and any $u \in \operatorname{dom}(A)$ the representation

$$U(t,s)\, A^\gamma u = U_A(t-s)\, A^\gamma u - \int_s^t d\tau\, U_A(t-\tau)\, B(\tau)\, U(\tau,s)\, A^\gamma u\,. \tag{12.2.45}$$

Seeing that similar to (12.2.34) the norm

$$\|\overline{U_A(t-\tau)\, B(\tau)}\| = \|(B(\tau)\, U_A(t-\tau))^*\| \le \frac{M_\alpha\, C_\alpha}{(t-\tau)^\alpha}\,, \quad \alpha \in (0,1)\,, \quad t > \tau\,,$$

representation (12.2.45) provides inequality

$$\|U(t,s)\, A^\gamma u\| \le \frac{M_\gamma}{(t-s)^\gamma}\, \|u\| + \int_s^t d\tau\, \frac{M_\alpha\, C_\alpha}{(t-\tau)^\alpha}\, \|U(\tau,s)\, A^\gamma u\|\,, \tag{12.2.46}$$

for any $u \in \operatorname{dom}(A)$ and $(t,s) \in \Delta_>$.

By iteration of (12.2.46), we infer (according to (G.3.4)–(G.3.8) in Theorem G.3.2, see Appendix G.3) the estimate

$$\|U(t,s)\,A^{\gamma}\,u\| \leq \frac{M_\gamma}{(t-s)^\gamma}\left\{1+\sum_{k=1}^{\infty}\frac{\Gamma(1-\gamma)\,\{M_\alpha\,C_\alpha\,\Gamma(1-\alpha)\,(t-s)^{1-\alpha}\}^k}{\Gamma(1-\gamma+k(1-\alpha))}\right\}\|u\|\,. \tag{12.2.47}$$

Given that dom (A) is dense in $\mathfrak{H}$, we infer by estimate (12.2.47) that for each $(t,s)\in\Delta_>$ the closure $\overline{U(t,s)\,A^\gamma}$ is a bounded operator on the space $\mathfrak{H}$. Asymptotic of the operator norm: $\|\overline{U(t,s)\,A^\gamma}\| = O((t-s)^{-\gamma})$ for small $(t-s)>0$, follows from (12.2.45) and (12.2.47). This proves the second part of assertion (1).

(2) On account of integral equation (12.2.18) and Corollary 12.2.9, we obtain, for $(t,s)\in\Delta_>$, the representation:

$$A^\gamma\,U(t,s)A^{-\gamma} = A^\gamma\,U_A(t-s)A^{-\gamma} - \int_s^t d\tau\;A^\gamma\,U_A(t-\tau)\,B(\tau)\,A^{-\gamma}\,A^\gamma\,U(\tau,s)\,A^{-\gamma}\,,$$

which yields for the family of bounded operators $\{A^\gamma\,U(t,s)\,A^{-\gamma}\}_{(t,s)\in\Delta_>}$ inequality

$$\|A^\gamma\,U(t,s)\,A^{-\gamma}\| \leq 1+\int_s^t d\tau\;\frac{M_\gamma\,C_\gamma}{(t-\tau)^\gamma}\,\|A^\gamma\,U(\tau,s)\,A^{-\gamma}\|\,,\quad (t,s)\in\Delta_>\,. \tag{12.2.48}$$

By iteration of (12.2.48), one gets, following (G.3.4)–(G.3.8) in Theorem G.3.2 (see Appendix G.3), the first part of assertion (2).

To prove the second part of assertion (2), one has to consider the family of vectors $\{A^{-\gamma}\,U(t,s)\,A^\gamma\,u\}_{(t,s)\in\Delta_>}$ for $u\in$ dom (A). Then (12.2.18) provides representation:

$$A^{-\gamma}\,U(t,s)\,A^\gamma\,u = U_A(t-s)\,u - \int_s^t d\tau\;A^{-\gamma}\,U_A(t-\tau)\,B(\tau)\,U(\tau,s)\,A^\gamma\,u\,.$$

Owing to $\|U_A(t-\tau)\,A^{-\gamma}\;B(\tau)\| \leq C_\gamma$ and (12.2.47), we obtain estimate

$$\|A^{-\gamma}\,U(t,s)\,A^\gamma\| \leq 1+\int_s^t d\tau\;C_\gamma\,\|\overline{U(\tau,s)\,A^\gamma}\| = O(1)\,, \tag{12.2.49}$$

which proves the second part of (2).

(3) By virtue of Proposition 12.2.6 (b) for $C(t) = A + B(t)$, we have

$$\|C(t)U(t,s)\| \leq \frac{C}{(t-s)}, \quad \|C(t)U(t,s)C(s)^{-1}\| \leq C, \quad (t,s) \in \Delta_> . \tag{12.2.50}$$

Then taking into account that perturbations $\{B(t)\}_{t\geq 0}$ is infinitesimally small (see Definition 1.9.7), estimates (12.2.50) yield

$$\|A\,U(t,s)\| = O((t-s)^{-1}), \quad \|A\,U(t,s)\,A^{-1}\| = O(1), \quad (t,s) \in \Delta_> . \tag{12.2.51}$$

The rest of the proof results from combinations of (1) and (2).

□

Remark 12.2.12 Below we shall systematically skip the *sign* of the *closure* for boundedly closed products of operators. For example, instead of the explicit closure $\overline{U(\tau,s)\,A^\gamma}$, we shall simply write $U(\tau,s)\,A^\gamma$, in the case when it does not produce a confusion.

12.2.3 Strategy of the Proof: Three Key Lemmata

For simplicity and briefness, we prove Theorem 12.2.4 for the *case* $t = 1$. To this aim we evaluate the norm of difference $U(1,0) - F_n(\tau)$ with $\tau = 1/n$.

Let $\varphi_0 \in C_0^\infty([0,\infty),[0,1])$ be a smooth cut-off function such that $0 \leq \varphi_0 \leq 1$ and

$$\varphi_0(\lambda) = 1 \text{ for } 0 \leq \lambda \leq 1, \quad \varphi_0(\lambda) = 0 \text{ for } \lambda \geq 2. \tag{12.2.52}$$

We introduce the auxiliary operators

$$C_\epsilon(t) := A + B_\epsilon(t), \qquad B_\epsilon(t) := \varphi_0(\epsilon A)\,B(t)\,\varphi_0(\epsilon A)$$

for $0 < \epsilon < 1$, *small enough*, and we denote by $\{U_\epsilon(t,s)\}_{(t,s)\in\Delta_1}$ the solution operator generated by $C_\epsilon(t)$, where $\Delta_1 := \{0 \leq s \leq t \leq 1\}$. Besides the nACP equations:

$$\begin{aligned} \partial_t U_\epsilon(t,s) &= -\,C_\epsilon(t)\,U_\epsilon(t,s)\,, \\ \partial_s U_\epsilon(t,s) &= \,U_\epsilon(t,s)\,C_\epsilon(s)\,, \end{aligned} \tag{12.2.53}$$

for $(t, s) \in \Delta_>$, it also satisfies integral formulae:

$$U_\epsilon(t, s) = U_A(t - s) - \int_s^t d\tau\, U_A(t - \tau)\, B_\epsilon(\tau)\, U_\epsilon(\tau, s)\,,$$

$$U_\epsilon(t, s) = U_A(t - s) - \int_s^t d\tau\, U_\epsilon(t, \tau)\, B_\epsilon(\tau)\, U_A(\tau - s)\,, \tag{12.2.54}$$

$$U(t, s) = U_\epsilon(t, s) + \int_s^t d\tau\, U_\epsilon(t, \tau)\,(B_\epsilon(\tau) - B(\tau))\, U(\tau, s)\,, \quad .$$

in the strong operator sense for $(t, s) \in \Delta$ and $U_\epsilon(s, s) = \mathbb{1}$; cf. (12.2.18), (12.2.19).

Remark 12.2.13

(a) As a consequence, the solution operator $\{U_\epsilon(t, s)\}_{(t,s)\in\Delta}$ has the same properties as in Lemma 12.2.11.
(b) We further use the notations $K_{\epsilon,j}(\tau)$, $F_{\epsilon,j}(\tau)$ and $\widetilde{F}_{\epsilon,n-j}(\tau)$, which are defined in a similar way as in (12.2.6) with $B(t)$ replaced by $B_\epsilon(t)$.

With these notations, the difference in question is decomposed into the sum of three operators

$$U(1, 0) - F_n(\tau) = I_{\epsilon,1} + I_{\epsilon,2}(\tau) + I_{\epsilon,3}(\tau)\,,$$

where $I_{\epsilon,1} := U(1, 0) - U_\epsilon(1, 0)$ and

$$I_{\epsilon,2}(\tau) := U_\epsilon(1, 0) - F_{\epsilon,n}(\tau), \qquad I_{\epsilon,3}(\tau) := F_{\epsilon,n}(\tau) - F_n(\tau)\,.$$

We are going to show that these three differences of operators obey the bounds

$$\|I_{\epsilon,1}\| = O(\epsilon)\,, \qquad \|I_{\epsilon,3}(\tau)\| = O(\epsilon)\,,$$

uniformly in τ, and

$$\|I_{\epsilon,2}(\tau)\| = \epsilon^{-1} O((\ln(n)/n)^2) + O(n^{-1}).$$

Seeing that we shall choose for ϵ the value: $\ln(n)/n$. Then

$$\|U(1, 0) - F_n(\tau)\| = O(\ln(n)/n)\,,$$

which gives the one claimed in Theorem 12.2.4 error bound (12.2.11).

Throughout the discussion below, τ and ϵ are fixed as follows:

$$\tau := 1/n, \qquad \epsilon := \ln(n)/n. \tag{12.2.55}$$

As a consequence the proof of Theorem 12.2.4 is reduced to demonstration of the *three key* estimates that we split into following lemmata.

Lemma 12.2.14 *Let positively semibounded from below self-adjoint operators* A *and* $\{B(t)\}_{t\in[0,T]}$ *satisfy Assumptions* 12.2.2 : (E1)–(E2). *Then*

$$\|I_{\epsilon,1}\| = O(\ln(n)/n)\,, \quad n \to \infty\,.$$

Lemma 12.2.15 *Let positively semibounded from below self-adjoint operators* A *and* $\{B(t)\}_{t\in[0,T]}$ *satisfy Assumptions* 12.2.2 : (E1)–(E3). *Then one gets for* $n \to \infty$ *the same asymptotics* :

$$\|I_{\epsilon,2}(\tau)\| = O(\ln(n)/n)\,,$$

as in Lemma 12.2.14.

Lemma 12.2.16 *Under the same conditions as in Lemma* 12.2.15, *we infer that*

$$\|I_{\epsilon,3}(\tau)\| = O(\ln(n)/n)\,, \quad n \to \infty\,,$$

which gives the same asymptotics as for two limits above.

Proof of Lemma 12.2.14

Remark 12.2.17 For notational brevity, we write below: φ_0 and φ_∞, for $\varphi_0(\epsilon A)$ and $\varphi_\infty(\epsilon A)$, respectively, where $\varphi_\infty(\lambda) = 1 - \varphi_0(\lambda)$.

Lemma 12.2.18 *Let positive self-adjoint operators* A *and* $\{B(t)\}_{t\in[0,1]}$ *satisfy Assumptions* 12.2.2 : (E1)–(E2). *Then*

$$\begin{aligned}
&\|(U(t,s) - U_A(t-s))\,\varphi_\infty\| = (t-s)^{-\alpha}\, O(\epsilon),\\
&\|\varphi_\infty\,(U(t,s) - U_A(t-s))\| = (t-s)^{-\alpha}\, O(\epsilon),\\
&\|(U_\epsilon(t,s) - U_A(t-s))\,\varphi_\infty\| = (t-s)^{-\alpha}\, O(\epsilon),\\
&\|\varphi_\infty\,(U_\epsilon(t,s) - U_A(t-s))\| = (t-s)^{-\alpha}\, O(\epsilon)\,.
\end{aligned}$$

Here $U_A(t-s) = \exp(-(t-s)A)$ *and all the order estimates for* $\alpha \in (0,1)$ *are uniform in* $(t,s) \in \Delta_1$.

Proof We prove only the first relation. The same argument, together with (12.2.54), may be applied for the proving other relations.

For short, we prove this assertion only for the case $s = 0$. Then the difference under consideration is written in the integral form (12.2.19)

$$(U(t,0) - U_A(t))\,\varphi_\infty = -\int_0^t d\eta\, U(t,\eta)\, B(\eta)\, U_A(\eta)\, \varphi_\infty\,. \tag{12.2.56}$$

Note that by definition of operators $U_A(\eta)$ and φ_∞ we have

$$\| \exp(-\eta A)\, (\mathbb{1} - \varphi_0(\epsilon A)) \| \le e^{-\eta/\epsilon}\,, \tag{12.2.57}$$

and that by Lemma 12.2.11 (1), for $\alpha \in (0, 1)$, we obtain

$$\|U(t, \eta)\, B(\eta)\| \le \|U(t, \eta)\, A^\alpha\|\, \|A^{-\alpha}\, B(\eta)\| = O((t - \eta)^{-\alpha})\,.$$

As a result, the norm of the integrand in (12.2.56) is bounded by $O((t-\eta)^{-\alpha})\, e^{-\eta/\epsilon}$. On account of integration (12.2.56), this observation yields for the first relation of lemma a desired bound $t^{-\alpha} O(\epsilon)$ for $s = 0$. □

Proof of Lemma 12.2.14 We use for $I_{\epsilon,1}$ the integral form (12.2.54):

$$I_{\epsilon,1} = U(1, 0) - U_\epsilon(1, 0) = \int_0^1 ds\, U_\epsilon(1, s)\, (B_\epsilon(s) - B(s))\, U(s, 0)\,.$$

The difference $B(s) - B_\epsilon(s)$ in the integrand can be represented as

$$B(s) - B_\epsilon(s) = \varphi_\infty B(s) + \varphi_0 B(s) \varphi_\infty\,.$$

Note that by Lemma 12.2.18, (12.2.57) and due to Remark 12.2.17

$$\|U_\epsilon(1, s)\, \varphi_\infty\| \le e^{-(1-s)/\epsilon} + (1-s)^{-\alpha} O(\epsilon)\,, \quad \|\varphi_\infty\, U(s, 0)\| \le e^{-s/\epsilon} + s^{-\alpha} O(\epsilon)\,.$$

On the other hand, by Lemma 12.2.11 (1) one gets for $\alpha \in (0, 1)$

$$\|U_\epsilon(1, s)\, \varphi_0\, B(s)\| = O((1 - s)^{-\alpha})\,, \quad \|B(s)\, U(s, 0)\| = O(s^{-\alpha})\,.$$

If now one takes into account these estimates and agreement (12.2.55), then the desired bound $O(\epsilon) = O(\ln(n)/n)$ follows. □

Proof of Lemma 12.2.15 The proof of Lemma 12.2.15 is stemming (in turn) from a series of lemmata. □

Lemma 12.2.19 *Let* $0 \le \gamma \le 1$ *and let* $r \in [0, 1]$. *Then for assumptions* (E1)–(E2), *one infers that*

$$\|A^\gamma\, \exp(-s\, B_\epsilon(r))\, A^{-\gamma}\| = O(1)\,, \quad 0 \le s \le \tau\,,$$

holds uniformly in arguments.

Proof By the *interpolation* of the norm between $\gamma = 0$ and $\gamma = 1$ (see Notes to Chap. 12 (Notes to Sects. 12.2 and 12.2.3)), it suffices to prove the lemma for the case $\gamma = 1$, since owing to (E1) the order $O(1)$ holds for $\gamma = 0$. For conciseness, we consider only the case $r = 0$. Set $B = B(0)$ and $B_\epsilon = \varphi_0\, B(0)\, \varphi_0$. Given that by assumptions (E1)–(E2) the operator $B\,A^{-1}$ is bounded, on account of (12.2.52)) and agreement (12.2.55), we obtain the asymptotic estimate

$$\|A\,sB_\epsilon\,A^{-1}\| \le \tau\,\|A\varphi_0\|\,\|B\,A^{-1}\| \le \tau\,\epsilon^{-1}\,\|B\,A^{-1}\|$$
$$= O(1/\ln(n)) = o(1)\,, \quad n \to \infty\,.$$

In turn this yields the asymptotic estimate

$$\|A\,\mathrm{e}^{-s\,B_\epsilon(r)}\,A^{-1}\| \le \sum_{k=0}^{\infty}(k!)^{-1}\,s^k\,\|A\,B_\epsilon\,A^{-1}\|^k = O(1)\,,$$

and the proof is complete. □

Now, taking into account the notation $K_{\epsilon,j}(\tau)$, see Remark 12.2.13, we define

$$W_{\epsilon,j}(\tau) := U_\epsilon(t_j, t_{j-1}) - K_{\epsilon,j}(\tau), \quad 1 \le j \le n. \tag{12.2.58}$$

Lemma 12.2.20 *Let $W_{\epsilon,j}(\tau)$ be as above and let $\alpha \le \sigma < 1$. Then :*

(1) $\|A^{-\alpha}\,W_{\epsilon,j}(\tau)\| = O(\tau)$, $\|W_{\epsilon,j}(\tau)\,A^{-\alpha}\| = O(\tau)$.
(2) $\|A^{-\sigma}\,W_{\epsilon,j}(\tau)\,A^{\sigma}\| = O(\tau^{1-\sigma})$, $\|A^{\sigma}\,W_{\epsilon,j}(\tau)\,A^{-\sigma}\| = O(\tau^{1-\sigma})$.
(3) $\|A^{-\alpha}\,W_{\epsilon,j}(\tau)\,A\| = O(1)$, $\|A\,W_{\epsilon,j}(\tau)\,A^{-\alpha}\| = O(1)$,

hold for agreement (12.2.55) *and* $1 \le j \le n$.

Proof We prove only the *first* of the couple of relations in statements (1), (2) and (3), seeing that a similar argument applies to each of the second relation. We also consider only the case $j = 1$ and use again the notations: $B = B(0)$ and $B_\epsilon = B_\epsilon(0)$.

(1) We write $W_{\epsilon,1}(\tau)$ as the sum $W_{\epsilon,1}(\tau) = V_{\epsilon,1}(\tau) + V_{\epsilon,2}(\tau)$, where

$$V_{\epsilon,1}(\tau) = U_\epsilon(\tau, 0) - \exp(-\tau A), \quad V_{\epsilon,2}(\tau) = \exp(-\tau A) - K_{\epsilon,1}(\tau).$$

These two operators can be further rewritten as

$$\begin{aligned} V_{\epsilon,1}(\tau) &= -\int_0^\tau ds\;U_\epsilon(\tau, s)\,B_\epsilon(s)\,\exp(-sA)\,, \\ V_{\epsilon,2}(\tau) &= \exp(-\tau A/2)\,(\mathbb{1} - \exp(-\tau B_\epsilon))\,\exp(-\tau A/2). \end{aligned} \tag{12.2.59}$$

Then (12.2.59) yields for $\alpha \in (0, 1)$ formulae:

$$A^{-\alpha} V_{\epsilon,1}(\tau) = -\int_0^\tau ds\, A^{-\alpha} U_\epsilon(\tau, s) A^\alpha A^{-\alpha} B_\epsilon(s) \exp(-sA),$$
$$A^{-\alpha} V_{\epsilon,2}(\tau) = \exp(-\tau A/2) \int_0^\tau ds\, A^{-\alpha} B_\epsilon \exp(-s B_\epsilon) \exp(-\tau A/2). \tag{12.2.60}$$

As a result, by assumptions (E1)–(E2) and by Lemma 12.2.11 (2), together with Remark 12.2.13, it follows from (12.2.60) that

$$\|A^{-\alpha} V_{\epsilon,1}(\tau)\| + \|A^{-\alpha} V_{\epsilon,2}(\tau)\| = O(\tau),$$

and hence the first relation in (1) is proved.

(2) Again, by virtue of assumptions (E1)–(E2) and by Lemma 12.2.11 (2), together with Remark 12.2.13, one infers that similar to arguments in (12.2.60):

$$\|A^{-\sigma} V_{\epsilon,1}(\tau) A^\sigma\| = O(1) \int_0^\tau ds\, s^{-\sigma} = O(\tau^{1-\sigma}), \quad \sigma \in (0, 1),$$

and by seeing that $\|A^{-\sigma}(\mathbb{1} - \exp(-\tau B_\epsilon))\| = O(\tau)$, $\|\exp(-\tau A/2)\, A^\sigma\| = O(\tau^{-\sigma})$:

$$\|A^{-\sigma} V_{\epsilon,2}(\tau) A^\sigma\| = O(\tau)\, \|\exp(-\tau A/2)\, A^\sigma\| = O(\tau^{1-\sigma}).$$

Thus the first relation in (2) is also proved.

(3) First we note that by assumptions (E1)–(E2) one gets estimates for the norms $\|\exp(-\tau A/2)\, A\| \le 2\,(e\,\tau)^{-1}$ and $\|A^{-\alpha} B_\epsilon \exp(-s B_\epsilon)\| \le C$. Then owing to representation (12.2.60), we obtain the bound $\|A^{-\alpha} V_{\epsilon,2}(\tau)\, A\| = O(1)$.

Next we consider the operator $V_{\epsilon,1}(\tau)$. We further decompose this operator as

$$A^{-\alpha} V_{\epsilon,1}(\tau)\, A = Q_{\epsilon,1}(\tau) + Q_{\epsilon,2}(\tau),$$

where

$$Q_{\epsilon,1}(\tau) = A^{-\alpha} \int_0^\tau ds\, U_\epsilon(\tau, s)\, B_\epsilon\, \partial_s(\exp(-s A)),$$
$$Q_{\epsilon,2}(\tau) = A^{-\alpha} \int_0^\tau ds\, U_\epsilon(\tau, s)\, (B_\epsilon - B_\epsilon(s))\, A \exp(-s A).$$

On account of integration by parts and thanks to (12.2.53), one gets

$$Q_{\epsilon,1}(\tau) = A^{-\alpha}\, B_\epsilon\, \exp(-\tau\, A) - A^{-\alpha}\, U_\epsilon(\tau, 0)\, A^\alpha\, A^{-\alpha} B_\epsilon -$$
$$\int_0^\tau ds\; A^{-\alpha}\, U_\epsilon(\tau, s)\, (A + B_\epsilon)\, B_\epsilon\, A^{-\alpha}\, A^\alpha\, \exp(-s\, A)\,.$$

Then by assumptions (E1)–(E2) and by Lemma 12.2.11 (2), (3), together with Remark 12.2.13 and $\|A^\alpha \exp(-s\, A)\| = O(s^{-\alpha})$, it follows that

$$\|Q_{\epsilon,1}(\tau)\| = O(1) + \int_0^\tau ds\; O((\tau - s)^{\alpha-1})\, \|A^\alpha\, \exp(-s\, A)\| = O(1)\,.$$

On the other hand,

$$Q_{\epsilon,2}(\tau) = A^{-\alpha} \int_0^\tau ds\; U_\epsilon(\tau, s)\, A^\alpha\, A^{-\alpha}\, (B_\epsilon - B_\epsilon(s))\, A^{-\alpha}\, A^{\alpha+1}\, \exp(-s\, A)\,.$$

Then by assumption (E3) and Lemma 12.2.11 (2) with Remark 12.2.13, the norm of $Q_{\epsilon,2}(\tau)$ can be evaluated as

$$\|Q_{\epsilon,2}(\tau)\| = O(1) \int_0^\tau ds\; s\, \|A^{1+\alpha} \exp(-sA)\| = O(1) \int_0^\tau ds\; s^{-\alpha} = O(\tau^{1-\alpha})\,.$$

This completes the proof of the first relation in (3). □

Now we decompose $W_{\epsilon,j}(\tau)$ into the sum: $W_{\epsilon,j1}(\tau) + W_{\epsilon,j2}(\tau)$, where

$$W_{\epsilon,j1}(\tau) = U_\epsilon(t_j, t_{j-1}) - \exp(-\tau C_\epsilon(t_{j-1})),$$
$$W_{\epsilon,j2}(\tau) = \exp(-\tau C_\epsilon(t_{j-1})) - K_{\epsilon,j}(\tau)$$

with $C_\epsilon(t_{j-1}) = A + B_\epsilon(t_{j-1})$ and $1 \le j \le n$, $\tau = 1/n$.

Lemma 12.2.21 *Let $W_{\epsilon,j1}(\tau)$ and $W_{\epsilon,j2}(\tau)$ be defined as above. Then :*

(1) $\|A^{-\alpha}\, W_{\epsilon,j1}(\tau)\, A^{-\alpha}\| = O(\tau^2)$.
(2) $\|A^{-1}\, W_{\epsilon,j2}(\tau)\, A^{-1}\| = \epsilon^{-1}\, O(\tau^3)$.

Proof We prove the lemma only for the case $j = 1$.

(1) Let $C_\epsilon := A + B_\epsilon$ with $B_\epsilon := B_\epsilon(0)$. Then

$$W_{\epsilon,11}(\tau) = U_\epsilon(\tau, 0) - \exp(-\tau\, C_\epsilon).$$

This can be written in the integral form

$$W_{\epsilon,11}(\tau) = -\int_0^\tau ds\; U_\epsilon(\tau, s)\, A^\alpha\, A^{-\alpha}\, (B_\epsilon(s) - B_\epsilon)\, A^{-\alpha}\, A^\alpha\, \exp(-s\, C_\epsilon)\,.$$

Thus (1) follows from assumption (E3), Lemma 12.2.11 (2) with Remark 12.2.13, and (12.2.15).

(2) The proof of this assertion repeatedly uses (without further references) the following *commutation* relation:

$$\big[\exp(-t\,Y),\, Z\big] = \int_0^t ds\, \exp(-s\,Y)\,[Z,\, Y]\,\exp(-(t-s)\,Y)\,, \quad t \geq 0\,.$$

It is stemming from integration of derivative $\partial_s(\exp(-s\,Y)\,Z\,\exp(-(t-s))$.

Let

$$K_\epsilon(t) = \exp(-t\,A/2)\,\exp(-t\,B_\epsilon)\,\exp(-t\,A/2)\,.$$

Then it is solution of inhomogeneous equation

$$\partial_t K_\epsilon(t) = -\,C_\epsilon\,K_\epsilon(t) + R_\epsilon(t)\,, \tag{12.2.61}$$

where $R_\epsilon(t) = R_{\epsilon,1}(t) + R_{\epsilon,2}(t)$ and

$$R_{\epsilon,1}(t) = [B_\epsilon, \exp(-tA/2)]\exp(-tB_\epsilon)\exp(-tA/2),$$
$$R_{\epsilon,2}(t) = \exp(-tA/2)[A/2, \exp(-tB_\epsilon)]\exp(-tA/2).$$

We shall evaluate the norm of these two remainder operators.

To this aim we calculate the commutator appeared in operator $R_{\epsilon,1}(t)$ as

$$[B_\epsilon, \exp(-tA/2)] = \int_0^t ds\, \exp(-s\,A/2)\,[A/2,\, B_\epsilon]\,\exp(-(t-s)\,A/2)$$
$$= t\,[A/2,\, B_\epsilon]\,\exp(-tA/2) + R_{\epsilon,3}(t),$$

where

$$R_{\epsilon,3}(t) := \int_0^t ds\, [\exp(-s\,A/2), [A/2,\, B_\epsilon]]\,\exp(-(t-s)\,A/2)\,.$$

Since the double commutator satisfies

$$\|A^{-1}\,[A,\,[A,\,B_\epsilon]]\,A^{-1}\| = O(\epsilon^{-1})\,,$$

we see by Lemma 12.2.19 that $R_{\epsilon,3}(t)$ obeys the bound

$$\|A^{-1}\,R_{\epsilon,3}(t)\,A^{-1}\| = \epsilon^{-1}O(t^2).$$

Thus $R_{\epsilon,1}(t)$ takes the form

$$R_{\epsilon,1}(t) = t\,[A/2,\ B_\epsilon]\,K_\epsilon(t) + \epsilon^{-1} A\, O_p(t^2)\, A\,.$$

Here we remark that $O_p(t^\nu)$ denotes the *class* of *bounded* operators with bound $O(t^\nu)$ as $t \to 0$, cf. Sect. 7.2.

A similar argument applies to the other remainder operator $R_{\epsilon,2}(t)$. On account of estimate

$$\|A^{-1}\,[B_\epsilon,\ [A,\ B_\epsilon]]\,A^{-1}\| = O(\epsilon^{-1})\,,$$

we obtain representation

$$R_{\epsilon,2}(t) = t\,[B_\epsilon,\ A/2]\,K_\epsilon(t) + \epsilon^{-1}\,A\,O_p(t^2)\,A$$

in the same way as above.

Therefore, by virtue of (12.2.61) we infer that

$$\partial_t K_\epsilon(t) = -\,C_\epsilon\ K_\epsilon(t) + \epsilon^{-1}\ A\ O_p(t^2)\ A\,. \tag{12.2.62}$$

By the *Duhamel principle*, see Notes to Chap. 12 (Notes to Sects. 12.2 and 12.2.2) for integration of inhomogeneous equation (12.2.61)

$$K_\epsilon(t) = \exp(-t\,C_\epsilon) + \int_0^t ds\ \exp(-(t-s)\,C_\epsilon)\ R_\epsilon(s)\,.$$

As a consequence, owing to (12.2.62) together with Lemma 12.2.19 for $K_\epsilon(t) = K_{\epsilon,1}(t)$ (see (12.2.6) and Remark 12.2.13) (b)), we obtain asymptotics

$$\|A^{-1}\ W_{\epsilon,12}(\tau)\ A^{-1}\| = \|A^{-1}\ (\exp(-\tau C_\epsilon) - K_{\epsilon,1}(\tau))\ A^{-1}\| = \epsilon^{-1}\ O(\tau^3)\,.$$

The proof of (2) is now complete. □

Lemma 12.2.22 *Let* $0 \le \sigma < 1$. *Then there exists* $M := M_\sigma > 0$ *such that*

$$\|A^\sigma\ F_{\epsilon,k}(\tau)\| \le M\,(k\tau)^{-\sigma}\,, \qquad \|\widetilde{F}_{\epsilon,n-k}(\tau)\ A^\sigma\,\| \le M\,((n-k)\tau)^{-\sigma}\,,$$

for $1 \le k \le n-1$ *and* $n > 2$.

Proof We prove only the *first* inequality, since a similar argument can be applied to the second one. By assumption (E2) and since it holds for $\sigma = 0$, due to *interpolation*, cf. Lemma 12.2.19, it suffices to prove this inequality, when $k = 1$, for the case $\sigma > \alpha$. Then it can be verified by *induction* on $k \in \mathbb{N}$.

On account of definition (12.2.6), the case $k = 1$ is apparent. Assume now that

$$\|A^\sigma\ F_{\epsilon,k}(\tau)\| \le M\,(k\tau)^{-\sigma}\,,$$

for the case $1 \leq k \leq m-1$, where $m < n$. Then for $\alpha < \sigma$ we obtain by the *interpolation* that:

$$\|A^{\alpha}\, F_{\epsilon,k}(\tau)\| \leq M^{\delta}\, (k\tau)^{-\alpha}\,, \quad 1 \leq k \leq m-1\,, \tag{12.2.63}$$

where $\delta = \alpha/\sigma < 1$.

Now we shall prove the case $k = m \leq n-1$. To this aim, we consider the (*telescopic*) difference:

$$U_{\epsilon}(t_m, 0) - F_{\epsilon,m}(\tau) = \sum_{j=1}^{m} X_{\epsilon,jm}(\tau)\,,$$

where because of (12.2.58) and Remark 12.2.13

$$X_{\epsilon,jm}(\tau) := U_{\epsilon}(t_m, t_j)\, W_{\epsilon,j}(\tau)\, F_{\epsilon,j-1}(\tau)\,, \tag{12.2.64}$$

with $F_{\epsilon,0}(\tau) = \mathbb{1}$. Since by Remark 12.2.13 (a): $\|A^{\sigma} U_{\epsilon}(t_m, 0)\| \leq c\,(m\tau)^{-\sigma}$ for some $c > 0$, the telescopic difference yields

$$\|A^{\sigma}\, F_{\epsilon,m}(\tau)\| \leq c\,(m\tau)^{-\sigma} + \sum_{j=1}^{m} \|A^{\sigma}\, X_{\epsilon,jm}(\tau)\|\,. \tag{12.2.65}$$

Owing to definition (12.2.64) and Lemma 12.2.11, with Remark 12.2.13 (a), one gets estimate for the first term of the sum in (12.2.65) as:

$$\|A^{\sigma}\, X_{\epsilon,1m}(\tau)\| \leq \|A^{\sigma}\, U_{\epsilon}(t_m, t_1)\, W_{\epsilon,1}(\tau)\| \leq c'\,((m-1)\,\tau)^{-\sigma}\,,$$

for some $c' > 0$. Next, it follows from (12.2.64), Lemma 12.2.20 (2) and, due to induction, from (12.2.63) that the last term of the sum in (12.2.65) has estimate

$$\begin{aligned}\|A^{\sigma}\, X_{\epsilon,mm}(\tau)\| &= \|A^{\sigma}\, W_{\epsilon,m}(\tau)\, A^{-\sigma}\, A^{\sigma}\, F_{\epsilon,m-1}(\tau)\| = \\ &\leq O(\tau^{1-\sigma})\, \|A^{\sigma}\, F_{\epsilon,m-1}(\tau)\| = M\, o(1)\, ((m-1)\,\tau)^{-\sigma}\,, \quad \tau \to 0\,.\end{aligned}$$

Now seeing that $m\tau \leq 1$, by Remark 12.2.13 (a), Lemma 12.2.20 (1) and (12.2.63), we obtain estimate of the rest the sum in (12.2.65):

$$\sum_{j=2}^{m-1} \|A^{\sigma}\, X_{\epsilon,jm}(\tau)\| = \sum_{j=2}^{m-1} \|A^{\sigma}\, U_{\epsilon}(t_m, t_j)\, W_{\epsilon,j}(\tau)\, A^{-\sigma}\, A^{\sigma}\, F_{\epsilon,j-1}(\tau)\| =$$

$$O(\tau) \sum_{j=2}^{m-1} \|A^{\sigma} U_{\epsilon}(t_m, t_j)\|\; \|A^{\alpha} F_{\epsilon,j-1}(\tau)\|$$

$$= O(\tau^{1-\alpha-\sigma})\, M^{\delta} \sum_{j=2}^{m-1} (m-j)^{-\sigma} (j-1)^{-\alpha}$$

$$\leq M^{\delta}\, O(\tau^{1-\alpha-\sigma})\, \Gamma_{\alpha,\sigma}\, m^{1-\alpha-\sigma} \leq c''\, M^{\delta}\, (m\tau)^{-\sigma}\,, \quad \tau \to 0\,,$$

for some $c'' > 0$. Finally, summing up we infer for (12.2.65) inequality

$$\|A^{\sigma}\, F_{\epsilon,m}(\tau)\| \leq \tag{12.2.66}$$
$$c\,(m\tau)^{-\sigma} + c'\,((m-1)\,\tau)^{-\sigma} + M\, o(1)\,((m-1)\,\tau)^{-\sigma} + c''\, M^{\delta}\,(m\tau)^{-\sigma}\,.$$

Since $\delta = \alpha/\sigma < 1$, one can choose $M > 1$ in (12.2.66) so large that

$$c\,(m\tau)^{-\sigma} + c'\,((m-1)\,\tau)^{-\sigma} + M\, o(1)\,((m-1)\,\tau)^{-\sigma} + c''\, M^{\delta}\,(m\tau)^{-\sigma} \leq M\,(m\tau)^{-\sigma}\,.$$

Consequently, (12.2.66) provides a desired estimate for any $k = m \leq n-1$, and thus the proof of the *first* inequality. □

Lemma 12.2.23 *There exists $M > 0$ such that*

$$\|A\, F_{\epsilon,k}(\tau)\| \leq M\,(k\tau)^{-1}\, \ln(n)\,, \quad \|\widetilde{F}_{\epsilon,n-k}(\tau)\, A\| \leq M\,((n-k)\tau)^{-1}\, \ln(n)\,,$$

for $1 \leq k \leq n-1$ and $n > 2$.

Proof We again verify only the first inequality. A similar argument applies to the second one. It is enough to prove this for $k > 1$ large enough.

Let $X_{\epsilon,jk}(\tau)$, $1 \leq j \leq k$, be as in the proof of Lemma 12.2.22. Then by the *telescopic* representation, we have

$$A\, F_{\epsilon,k}(\tau) = A\, U_{\epsilon}(t_k, 0) - \sum_{j=1}^{k} A\, X_{\epsilon,jk}(\tau)\,.$$

By virtue of (12.2.51) and Remark 12.2.13 (a), we obtain for $j = 1$:

$$\|A\, U_{\epsilon}(t_k, 0)\| + \|A\, X_{\epsilon,1k}(\tau)\| = O((k\tau)^{-1})\,,$$

whereas for the term, $j = k$, Lemma 12.2.20 (3) and Lemma 12.2.22 yield estimate

$$\|A\, X_{\epsilon,kk}(\tau)\| = O(1)\, \|A^{\alpha}\, F_{\epsilon,k-1}(\tau)\| = O(((k-1)\tau)^{-\alpha}) \leq O(((k-1)\tau)^{-1})\,.$$

Now by (12.2.51) with Remark 12.2.13 (a), definition (12.2.64) and Lemma 12.2.20 (1), Lemma 12.2.22, the rest of the sum is evaluated as follows:

$$\sum_{j=2}^{k-1} \|A\, X_{\epsilon,jk}(\tau)\| = \sum_{j=2}^{k-1} \|A\, U_\epsilon(t_k, t_j)\, W_{\epsilon,j}(\tau)\, A^{-\sigma}\ A^{\sigma}\ F_{\epsilon,j-1}(\tau)\| =$$

$$O(\tau) \sum_{j=2}^{k-1} \|A\, U_\epsilon(t_k, t_j)\|\ \|A^{\alpha}\ F_{\epsilon,j-1}(\tau)\| = O(\tau^{-\alpha}) \sum_{j=2}^{k-1} (k-j)^{-1} (j-1)^{-\alpha}$$

$$\leq O(((k-1)\,\tau)^{-\alpha})\ \frac{2}{1-\alpha}\ \ln(k-1)\ \leq\ M\ (k\,\tau)^{-1}\ \ln(n)\,,\qquad \tau \to 0\,.$$

for $1 \leq k \leq n-1$ and some $M > 0$. This proves the lemma. □

We are now in position to prove the second key lemma in question.

Proof of Lemma 12.2.15 Recall that $t_n = 1$ and $\tau = 1/n$. For $X_{\epsilon,jn}(\tau)$, $1 \leq j \leq n$, see definition (12.2.64), we can write the difference operator $I_{\epsilon,2}(\tau)$ as

$$I_{\epsilon,2}(\tau) = U_\epsilon(t_n, 0) - F_{\epsilon,n}(\tau) = \sum_{j=1}^{n} X_{\epsilon,jn}(\tau).$$

By Lemmas 12.2.20 and 12.2.22, we have for two terms of the sum estimates:

$$\|X_{\epsilon,1n}(\tau)\| \leq \|U_\epsilon(t_n, t_1)\, A^{\alpha}\|\ \|A^{-\alpha}\ W_{\epsilon,1}(\tau)\| = O(\tau) = O(n^{-1})\,,$$

$$\|X_{\epsilon,nn}(\tau)\| \leq \|W_{\epsilon,n}(\tau)\, A^{-\alpha}\|\ \|A^{\alpha}\ F_{\epsilon,n-1}(\tau)\| = O(\tau) = O(n^{-1})\,.$$

To evaluate the other terms, we write $X_{\epsilon,jn}(\tau)$, $2 \leq j \leq n-1$, as

$$X_{\epsilon,jn}(\tau) = Y_{\epsilon,jn}(\tau) + Z_{\epsilon,jn}(\tau)\,,$$

where

$$Y_{\epsilon,jn}(\tau) := U_\epsilon(t_n, t_j)\, W_{\epsilon,j1}(\tau)\, F_{\epsilon,j-1}(\tau),$$

$$Z_{\epsilon,jn}(\tau) := U_\epsilon(t_n, t_j)\, W_{\epsilon,j2}(\tau)\, F_{\epsilon,j-1}(\tau).$$

By Lemmas 12.2.11, 12.2.21 and 12.2.22, one gets

$$\sum_{j=2}^{n-1} \|Y_{\epsilon,jn}(\tau)\| = O(\tau^{2-2\alpha}) \sum_{j=2}^{n-1} (n-j)^{-\alpha}\, (j-1)^{-\alpha} = O(n^{-1})\,.$$

On the other hand, by Lemmas 12.2.11, 12.2.21 and 12.2.23,

$$\sum_{j=2}^{n-1} \|Z_{\epsilon,jn}(\tau)\| = \epsilon^{-1}\, O(\tau)\, \ln(n) \sum_{j=2}^{n-1} (n-j)^{-1}\,(j-1)^{-1} = \epsilon^{-1}\, O(n^{-2}(\ln(n))^2).$$

Thus the proof of the lemma is now complete. □

Proof of Lemma 12.2.16 Finally we prove the last of key lemmata, Lemma 12.2.16. In turn, it is proved by means of several lemmata. □

Lemma 12.2.24 *Let $r \in [0, 1]$. Then one has for $\tau \to 0$ the asymptotics:*

$$\|(\exp(-\tau B(r)) - \exp(-\tau B_\epsilon(r)))\, A^{-\alpha}\| = O(\tau)\,,$$

$$\|A^{-\alpha}\,(\exp(-\tau B(r)) - \exp(-\tau B_\epsilon(r)))\| = O(\tau)\,,$$

and hence

$$\|(K_j(\tau) - K_{\epsilon,j}(\tau))\, A^{-\alpha}\| + \|A^{-\alpha}\,(K_j(\tau) - K_{\epsilon,j}(\tau))\| = O(\tau)\,.$$

Proof We shall prove the first relation. To his aim, we rewrite the difference at issue in integral form

$$\int_0^\tau ds\, \exp(-s\, B(r))\,(B_\epsilon(r) - B(r))\, \exp(-(\tau - s)B_\epsilon(r))\, A^{-\alpha}\,.$$

Since operator $(B_\epsilon(r) - B(r))\, A^{-\alpha}$, is uniformly bounded by assumptions (E1)–(E2), the desired bound follows from Lemma 12.2.19.

The second relation can be proved in a similar way, and the third one is a consequence of definitions. □

Lemma 12.2.25 *The difference $K_{\epsilon,j}(\tau) - K_j(\tau)$, $1 \le j \le n$, has the following asymptotic form :*

$$K_{\epsilon,j}(\tau) - K_j(\tau) = \varphi_\infty\, O_p(\tau)\, A^\alpha + A^\alpha\, O_p(\tau)\,\varphi_\infty + A^\alpha\, O_p(\tau^2)\, A^\alpha\,, \quad \tau > 0\,.$$

Remark that $O_p(\tau^\nu)$ denotes here a class of bounded operators with bound $O(\tau^\nu)$, as $\tau \to 0$; cf. proof of Lemma 12.2.15.

Proof For brevity, we prove the lemma only for the case $j = 1$, and we write again B and B_ϵ for $B(0)$ and $B_\epsilon(0)$, respectively. Then the difference under consideration is represented as

$$K_{\epsilon,1}(\tau) - K_1(\tau) = \exp(-\tau A/2)\,(\exp(-\tau B_\epsilon) - \exp(-\tau B))\, \exp(-\tau A/2)\,.$$

This can be further rewritten in the form

$$K_{\epsilon,1}(\tau) - K_1(\tau) = \sum_{j=1}^{3} \exp(-\tau A/2)\, \Gamma_{\epsilon,j}(\tau)\, \exp(-\tau A/2)\,,$$

where $\{\Gamma_{\epsilon,j}(\tau)\}_{j=1}^{3}$ are integrals:

$$\Gamma_{\epsilon,1}(\tau) = \int_0^\tau ds\, \exp(-sB_\epsilon)\, \varphi_\infty\, B\, \varphi_0\, \exp(-(\tau - s)B)\,,$$

$$\Gamma_{\epsilon,2}(\tau) = \int_0^\tau ds\, \exp(-sB_\epsilon)\, \varphi_0\, B\, \varphi_\infty\, \exp(-(\tau - s)B)\,,$$

$$\Gamma_{\epsilon,3}(\tau) = \int_0^\tau ds\, \exp(-sB_\epsilon)\, \varphi_\infty\, B\, \varphi_\infty\, \exp(-(\tau - s)B)\,.$$

To proceed we decompose operator $\exp(-s\,B)$ as

$$\exp(-sB) = \exp(-sB_\epsilon) + (\exp(-sB) - \exp(-sB_\epsilon))\,.$$

Then, owing to $\epsilon^{-\alpha}s \leq 1$ for $0 \leq s \leq \tau = 1/n$, Lemmas 12.2.19 and 12.2.24 allow to infer representation

$$B\, \varphi_0\, \exp(-sB) = O_p(s^0)\, A^\alpha + \epsilon^{-\alpha}\, O_p(s)\, A^\alpha = O_p(s^0)\, A^\alpha\,.$$

Here $\nu = 0$ cf. remark about operators of the *class*: $O_p(\tau^\nu)$, at the end Lemma 12.2.25.

Since operator $A^{-\alpha}[B_\epsilon,\, \varphi_0]$ is uniformly bounded in $\epsilon \in (0, 1)$, it follows from Lemma 12.2.19 that

$$[\varphi_\infty,\, \exp(-sB_\epsilon)] = [\exp(-sB_\epsilon),\, \varphi_0] =$$

$$\int_0^s d\sigma\, \exp(-\sigma B_\epsilon)\, [B_\epsilon,\, \varphi_0]\, \exp(-(s - \sigma)B_\epsilon) = A^\alpha\, O_p(s)\,.$$

A similar argument shows that $[\exp(-sB_\epsilon),\, \varphi_0] = O_p(s)\, A^\alpha$. Thus we have

$$\exp(-s\, B_\epsilon)\, \varphi_\infty = \varphi_\infty\, O_p(s^0) + A^\alpha\, O_p(s)\,,$$

and hence

$$\Gamma_{\epsilon,1}(\tau) = \varphi_\infty\, O_p(\tau)\, A^\alpha + A^\alpha\, O_p(\tau^2)\, A^\alpha.$$

Note that on account of Lemma 12.2.19, we obtain:

$$\exp(-sB_\epsilon)\,\varphi_0\,B = A^\alpha\,O_p(s^0)\,, \qquad \exp(-sB_\epsilon)\,\varphi_\infty\,B = A^\alpha\,O_p(s^0)\,,$$

and by Lemma 12.2.24 we also get

$$\big[\exp(-sB) - \exp(-sB_\epsilon),\ \varphi_0\big] = O_p(s)\,A^\alpha\,.$$

This implies that $[\exp(-sB),\ \varphi_0] = O_p(s)\,A^\alpha$ and consequently

$$\varphi_\infty\,\exp(-sB) = \exp(-sB)\,\varphi_\infty + [\exp(-sB),\ \varphi_0] = O_p(s^0)\,\varphi_\infty + O_p(s)A^\alpha\,.$$

As a result

$$\Gamma_{\epsilon,2}(\tau) = A^\alpha\,O_p(\tau)\,\varphi_\infty + A^\alpha\,O_p(\tau^2)\,A^\alpha\,.$$

A similar argument applies *mutatis mutandis* to calculation of $\Gamma_{\epsilon,3}(\tau)$. Accordingly, this operator has the same structure as $\Gamma_{\epsilon,2}(\tau)$. Thus, the proof of the lemma is now concluded. □

Lemma 12.2.26 *When $\tau \to 0$ the asymptotics:*

$$\|\varphi_\infty F_{\epsilon,k}(\tau)\| = O(\mathrm{e}^{-k\tau/\epsilon}) + \epsilon\,O((k\tau)^{-\alpha})\,,$$
$$\|\widetilde{F}_{\epsilon,n-k}(\tau)\varphi_\infty\| = O(\mathrm{e}^{-(n-k)\tau/\epsilon}) + \epsilon\,O(((n-k)\tau)^{-\alpha})\,,$$

hold for $2 \le k \le n-2$, where $n > 3$ and $\epsilon \in (0,1)$, $\alpha \in [0,1)$.

Proof We prove only the first relation since the second one follows *verbatim.* Seeing that from the *telescopic* representation (cf. Lemma 12.2.22), one gets

$$\|\varphi_\infty\,(U_\epsilon(t_k,0) - F_{\epsilon,k}(\tau))\| \le \sum_{j=1}^{k} \|\varphi_\infty\,X_{\epsilon,jk}(\tau)\|\,,$$

we obtain from definition (12.2.20) and Lemma 12.2.20 (1) that the left-hand side obeys for $\tau \to 0$ asymptotics

$$\begin{aligned} &\|\varphi_\infty\,(U_\epsilon(t_k,0) - F_{\epsilon,k}(\tau))\| = \\ &\quad O(\tau)\sum_{j=2}^{k-1}\|\varphi_\infty\,U_\epsilon(t_k,t_j)\|\;\|A^\alpha\,F_{\epsilon,j-1}(\tau)\| + \\ &\quad O(\tau)\,(\|U_\epsilon(t_k,t_1)\,A^\alpha\| + \|A^\alpha\,F_{\epsilon,k-1}(\tau)\|)\,. \end{aligned} \tag{12.2.67}$$

As it follows from the proof of Lemma 12.2.18:

$$\|\varphi_\infty\, U_\epsilon(t_k, t_j)\| = O(\mathrm{e}^{-(k-j)\tau/\epsilon}) + \epsilon\, O(((k-j)\tau)^{-\alpha})\,.$$

and consequently we infer by Lemma 12.2.22 that the sum in the right-hand side of (12.2.67) is estimated as

$$\begin{aligned} O(\tau^{1-\alpha}) &\sum_{j=2}^{k-1} (j-1)^{-\alpha} (\mathrm{e}^{-(k-j)\tau/\epsilon} + \epsilon\, ((k-j)\tau)^{-\alpha}) \\ &= O(\tau^{1-\alpha}) \left\{ \sum_{j=2}^{r} + \sum_{j=r+1}^{k-1} \right\} (j-1)^{-\alpha} \mathrm{e}^{-(k-j)\tau/\epsilon} + \epsilon\, O((k\tau)^{1-2\alpha}) \\ &= O(\tau^{1-\alpha})\, (k^{1-\alpha} \mathrm{e}^{-k\tau/2\epsilon} + k^{-\alpha}(1-\mathrm{e}^{-\tau/\epsilon})^{-1}) + \epsilon\, O((k\tau)^{1-2\alpha})\,. \end{aligned}$$

Here $r = [(k-1)/2]$, where $[x]$ is the integer part of $x \geq 0$. This shows that the sum obeys the bound $\epsilon\, O((k\tau)^{-\alpha})$.

Note that the second and third terms in the right-hand side of (12.2.67) also obey the same bound as above. Indeed, the estimate of the second term is due to Lemma 12.2.11 (1) and Remark 12.2.13 (1), whereas the bound on the third term follows again from Lemma 12.2.22.

Thus the proof is complete. □

With help of definitions (12.2.6), (12.2.55) and Remark 12.2.17, we can formulate the next two lemmata, cf. Lemmas 12.2.22 and 12.2.26.

Lemma 12.2.27 *Let $0 \leq \sigma < 1$. Then there exists $M = M_\sigma > 0$ such that*

$$\|A^\sigma\, F_k(\tau)\| \leq M\, (k\tau)^{-\sigma}, \quad \|\widetilde{F}_{n-k}(\tau)\, A^\sigma\,\| \leq M\, ((n-k)\tau)^{-\sigma}\,,$$

for $1 \leq k \leq n-1$ and $n > 2$.

Lemma 12.2.28 *Let $0 \leq \alpha < 1$ and $\epsilon \in (0, 1)$. Then*

$$\begin{aligned} \|\varphi_\infty F_k(\tau)\| &= O(\mathrm{e}^{-k\tau/\epsilon}) + \epsilon O((k\tau)^{-\alpha}), \\ \|\widetilde{F}_{n-k}(\tau)\varphi_\infty\| &= O(\mathrm{e}^{-(n-k)\tau/\epsilon}) + \epsilon O(((n-k)\tau)^{-\alpha})\,, \end{aligned}$$

for $2 \leq k \leq n-2$ and $n > 3$.

To this end we observe that if one defines the operator $W_j(\tau)$ similarly to (12.2.58):

$$W_j(\tau) := U(t_j, t_{j-1}) - K_j(\tau)\,, \quad 1 \leq j \leq n\,,$$

then it can be shown in exactly the same way as in the proof of Lemma 12.2.20 that this operator has the same properties as $W_{\epsilon,j}(\tau)$ (12.2.58). This enables us to prove these two lemmata in almost the same way as in the proof of Lemmas 12.2.22 and 12.2.26.

We are now in position for proving the third key lemma.

Proof of Lemma 12.2.16 We rewrite the difference $I_{\epsilon,3}(\tau)$ as a *telescopic* sum:

$$I_{\epsilon,3}(\tau) = F_{\epsilon,n}(\tau) - F_n(\tau) = \sum_{j=1}^{n} G_{\epsilon,jn}(\tau),$$

where $\epsilon \in (0,1)$, $\tau = 1/n$ and

$$G_{\epsilon,jn}(\tau) = \widetilde{F}_{\epsilon,n-j}(\tau)\,(K_{\epsilon,j}(\tau) - K_j(\tau))\,F_{j-1}(\tau).$$

By virtue of Lemmas 12.2.22, 12.2.24 and 12.2.27, we deduce:

$$\|G_{\epsilon,1n}(\tau)\| + \|G_{\epsilon,2n}(\tau)\| = O(\tau) = O(n^{-1}),$$
$$\|G_{\epsilon,nn}(\tau)\| + \|G_{\epsilon,(n-1)n}(\tau)\| = O(n^{-1}).$$

A a consequence, by Lemmas 12.2.22 and 12.2.25–12.2.28, we obtain representation

$$\|I_{\epsilon,3}(\tau)\| = O(\tau^{1-\alpha})\sum_{j=3}^{n-2}(j-1)^{-\alpha}\mathrm{e}^{-(n-j)\tau/\epsilon} +$$
$$\epsilon\, O(\tau^{1-2\alpha})\sum_{j=3}^{n-2}(j-1)^{-\alpha}\,(n-j)^{-\alpha}$$
$$+\, O(\tau^{2-2\alpha})\sum_{j=3}^{n-2} j^{-\alpha}\,(n-j)^{-\alpha} + O(n^{-1}).$$

Now we estimate these three sums in the right-hand side of the representation. The first sum was calculated in the proof of Lemma 12.2.26. Then together with factor $O(\tau^{1-\alpha})$ one gets that the first term in the right-hand side of the representation has asymptotics $O(\epsilon^{1-\alpha})$.

The values of the second and the third sums are provided by arguments in the proof of Lemma 12.2.22, or by explicit formula for $\alpha = \gamma \in (0,1)$ in Remark 12.1.16. Then the second term in the right-hand side of the representation obeys the bound $O(\epsilon)$, whereas the third one obeys the bound $O(\tau)$.

All together these arguments complete the proof of Lemma 12.2.16. □

12.2.4 Generalisations

Here we discuss generalisations of the operator-norm product formula approximation for the nACP solution operator (12.2.11) and present *two* comments on the main Theorem 12.2.4.

First we note that the *same* error bound as in (12.2.11) remains to be true for *non-symmetric* approximants; cf. Definition 12.2.1. Namely, for Assumptions 12.2.2, (E1)–(E3), we obtain a generalisation:

$$\left\| U(t,0) - \prod_{j=n}^{1} G_j(\tau) \right\| = O(\ln(n)/n)\,, \quad n \geq 2\,, \tag{12.2.68}$$

uniformly in $0 \leq t \leq T$, where $\tau = t/n$ and the *non-symmetric approximants*

$$G_j(\tau) := \exp(-\tau A)\, \exp(-\tau B(t_{j-1}))\,, \quad t_j = j\,\tau\,,\; j = 1, 2, \ldots, n\,, \tag{12.2.69}$$

see the proof which follows below in Theorem 12.2.29.

The second comment concerns the case when operator $\{A(t)\}_{t\in[0,T]}$ is also time dependent in the sum $C(t) = A(t) + B(t)$, which defines the nACP (12.2.3). In Sect. 11.1 this problem has been scrutinised in a full generality, including the case when domain dom $(A(t))$ is *time-dependent*. There the product formula approximation for the nACP solution operator generated by $C(t)$ was established in a Banach space but only in the *strong* operator topology; see Notes to Chap. 12 (Notes to Sect. 12.2).

The *Ichinose-Tamura* approach allows to generalise this result to the *operator-norm* topology in a Hilbert space when domain dom $(A(t))$ is *time-independent* and family $\{A(t)\}_{t\in[0,T]}$ satisfies a continuity condition similar to Assumptions 12.2.2, (E1)–(E3), without any essential modification of the proof. To this end see Remark 12.2.30.

Theorem 12.2.29 *Let positively semibounded from below self-adjoint operators A and* $\{B(t)\}_{t\in[0,T]}$ *satisfy Assumptions* 12.2.2: (E1)–(E3), *for each* $t \in [0,T]$. *Then asymptotic estimate* (12.2.68) *holds uniformly on any interval* $\mathcal{I} = [0,T] \subset \mathbb{R}_0^+$.

Proof For brevity, we again prove (12.2.68) for $t = 1$. Set

$$E_n(\tau) := \prod_{j=n}^{1} G_j(\tau), \quad \tau = 1/n\,.$$

Let $K_j(\tau)$ be as in (12.2.6). Then we use Theorem 12.2.4, (12.2.6), to obtain that

$$\begin{aligned} E_n(\tau) &= \exp(-\tau A/2) \prod_{j=n}^{2} K_j(\tau)\ \exp(-\tau A/2)\exp(-\tau B) \\ &= \exp(-\tau A/2)U(1,t_1)\exp(-\tau A/2)\exp(-\tau B) + O_p(\ln(n)/n)\,, \end{aligned}$$

with $B := B(t_0) = B(0)$ according to (12.2.69).

Next we calculate the commutator

$$\begin{aligned} &\big[\exp(-\tau A/2),\ U(1,t_1)\big] \\ &= \int_0^{\tau} ds\ \exp(-s\,A/2)[U(1,t_1),\ A/2]\exp(-(\tau-s)\,A/2)\,. \end{aligned}$$

By Lemma 12.2.11 commutator $[U(1,t_1),\ A]$ is uniformly bounded, so that

$$\big\|[\exp(-\tau A/2), U(1,t_1)]\big\| = O(\tau) = O(n^{-1}).$$

This implies that

$$E_n(\tau) = U(1,t_1)\ \exp(-\tau A)\exp(-\tau B) + O_p(\ln(n)/n). \tag{12.2.70}$$

On account of representation

$$\exp(-\tau A)\exp(-\tau B) - \exp(-\tau A) = \tau\, A^{\alpha}\ \exp(-\tau A)\ A^{-\alpha}\,(\exp(-\tau B) - \mathbb{1})\,\tau^{-1}\,,$$

and assumptions (E1)–(E2), one gets

$$\exp(-\tau A)\exp(-\tau B) = \exp(-\tau A) + A^{\alpha}O_p(\tau) = K_1(\tau) + A^{\alpha}O_p(\tau)\,.$$

As a result, the representation (12.2.70) with Lemma 12.2.11 and Theorem 12.2.4 for product (12.2.6) yield

$$E_n(\tau) = U(1,0) + O_p(\ln(n)/n)\,.$$

Thus, by definition of the *class* of bounded operators: $O_p(\cdot)$, the desired error bound is established.

Similarly one shows that

$$\left\| U(t,0) - \prod_{j=n}^{1} \tilde{G}_j(\tau) \right\| = O(\ln(n)/n)\,, \tag{12.2.71}$$

where $\tilde{G}_j(\tau) := \exp(-\tau B(t_{j-1}))\exp(-\tau A)$.

This completes the proof. □

Remark 12.2.30 The main Theorem 12.2.4 can be also extended to the case in which $C(t)$ takes the form: $C(t) = A(t)+B(t)$, for a time-dependent family of self-adjoint operators $\{A(t) \geq c > 0\}_{t\in[0,T]}$ with domain $\text{dom}\,(A(t)) = \text{dom}\,(A(0))$, which is independent of $t \in [0, T]$.

Suppose that $B(t)$ fulfills Assumptions 12.2.2, (E1)–(E3) for $A = A(0)$. In other words, $B(t)$ is assumed to satisfy

$$\left\| A^{-\alpha}\,(B(t) - B(s))\,A^{-\alpha} \right\| = O(|t-s|)\,, \quad (t,s) \in [0,T] \times [0,T]\,,$$

for some α, such that $0 \leq \alpha < 1$. If, in addition (cf. assumption (E3)* in (12.2.10)) we assume that $(t, s) \mapsto A(t)A(s)^{-1}$ is uniformly bounded and

$$\left\| A^{-\alpha}\,(A(t) - A(s))\,A^{-\alpha} \right\| = O(|t-s|)\,, \quad (t,s) \in [0,T] \times [0,T]\,,$$

for the same α as above, then it follows that

$$\left\| U(t,0) - \prod_{j=n}^{1} P_j(\tau) \right\| = O(\ln(n)/n)\,,$$

where $P_j(\tau) := \exp(-\tau A(t_{j-1}))\,\exp(-\tau B(t_{j-1}))$.

To this end, it is efficacious to use

$$C_\epsilon(t) = A(t) + B_\epsilon(t)\,, \quad B_\epsilon(t) = \varphi_0(\epsilon A(t))\,B(t)\,\varphi_0(\epsilon A(t))\,, \quad \epsilon \in (0,1)\,,$$

as the auxiliary operators. Then the proof follows *mutatis mutandis* the arguments of Theorem 12.2.29.

12.3 Notes to Chap. 12

Notes to Sect. 12.1

In this section we follow the paper [NSZ19]. It is devoted to the L^2-setting for evolution semigroups in the framework of a Hilbert space $\mathfrak{H}$. Similar to Chap. 11, this is a method for construction of the product approximation for solution operator of a parabolic type nACP in the operator-norm topology on $\mathcal{L}(\mathfrak{H})$. The method essentially follows the L^p-setting for Banach spaces developed in Chap. 11 but with (in a certain sense) better operator-norm estimates of the rate of convergence. This improvement is due to the self-adjointness, which yields some number of straightforward estimates available in the Hilbert space L^2-setting.

Section 12.1.1

Assumptions 12.1.1 (D1)–(D3) are motivated by the Ichinose-Tamura conditions formulated by the first time in [IT98a], cf. Assumptions 12.2.2, (E1)–(E3). For both the *evolution semigroup* method and the *Ichinose-Tamura* approach, these conditions ensure the *operator-norm* product approximation for solution operator of a parabolic type nACP generated by family $\{C(t) = A + B(t)\}_{t\in[0,T]}$. Note that the proofs are heavily based on condition of the *infinitesimal* relative A-smallness of perturbation $\{B(t)\}_{t\in[0,T]}$, see (D2), (E2). We remained that one does not need this *severe* condition for proving convergence of the product approximants in the *strong* operator topology; see Sect. 11.1.

Sections 12.1.2 and 12.1.3

These subsections contain an extended and rather detailed presentation of results of the paper [NSZ19].

The results in Hilbert space are stronger than in Banach space, see comments in [NSZ19, Section 1].

Notes to Sect. 12.2

This section is an extended version of the paper [IT98a]. It was the first result about a direct construction of the operator-norm convergent product formula approximation of solution operator for the *parabolic type* nACP (12.1.1) under very similar to Sect. 12.1 conditions: compare (D1)–(D3) and (E1)–(E3).

The essential difference between Sects. 12.1 and 12.2 is that, instead of a rather *round-about* way via the Trotter product formula for evolution semigroups on extended space $L^2([0, T], \mathfrak{H})$, we develop in Sect. 12.2 the direct *Ichinose-Tamura* approach to this problem in a Hilbert space $\mathfrak{H}$.

Section 12.2.1

The set-up and Assumptions 12.2.2, (E1)–(E3), (E3)* are due to [IT98a]. It is instructive to compare them with softer hypothesis and conditions in Sect. 11.1 for the *strong* convergence of the product formula approximants to solution operator for a *parabolic type* nACP scrutinised in [Far67b], [VWZ08] and [VWZ09].

Section 12.2.2

In discussion of the parabolic-type nACP, we partially follow [Sob61, Theorem 1], and [Tan75, Theorem 5.2.1]. For construction and properties of solution operator, we use some ideas from [SKre67, Chapter II, §4] and [Zag21, Section 2].

For the proof of the *Heinz* inequality, see, for example, [BS87, Chapter 10, §4], or [Tan75, Subsection 2.3.3] for a more general the *Löwner-Heinz-Kato* inequality.

The method of solution of the operator-valued inhomogeneous equations of the type (12.2.16), comes from the theory of the ordinary differential equations. The integration of (12.2.16) via identity (12.2.17) and representation (12.2.18) is known as the *Variation of Constant* formula, or the *Duhamel principle*; see, e.g. [Yos80, Chapter XIV.5], or [Haa06, Chapter 9, Subsection 9.3.1].

Section 12.2.3

The strategy of the proof, which is divided into three lemmata, comes from [IT98a, Section 3]. To this end we use for establishing certain operator-norm estimates the method of *interpolation*. For the reader convenience, we remained below the basic idea, which is behind it, and we refer to [RS75, Appendix to IX.4], for details.

Proposition 12.3.1 (Hadamard's Three Lines Theorem) *Let $\phi(z)$ be a complex-valued function, bounded and continuous on the closed strip*

$$\mathfrak{S} := \{z \in \mathbb{C} : 0 \le \operatorname{Re} z \le 1\},$$

which is analytic in the interior of $\mathfrak{S}$ and satisfying

$$|\phi(z)| \le M_0, \quad \operatorname{Re} z = 0,$$

$$|\phi(z)| \le M_1, \quad \operatorname{Re} z = 1,$$

with positive constants M_0, M_1. Then

$$|\phi(z)| \le M_0^{1-\operatorname{Re} z} M_1^{\operatorname{Re} z}, \quad z \in \mathfrak{S}. \tag{12.3.1}$$

Corollary 12.3.2 *Seeing that complex plane $\mathbb{C}$ is a particular case of Banach space with absolute value $|\cdot|$ as a norm, the Hadamard three lines theorem holds for any Banach space-valued analytic functions and, consequently, for the Banach space of bounded operators $\mathcal{L}(\mathfrak{H})$.*

Section 12.2.4

The first note about *non-symmetric* extensions (Theorem 12.2.29) for the product *approximants* sounds as a natural routine, whereas the second comment concerns a delicate problem of a *more time-dependence*.

The second comment concerns a time-dependent family of generators $\{A(t)\}_{t\in[0,T]}$. The *Ichinose-Tamura* approach allows a generalisation of Theorem 12.2.4 (as well as Theorem 12.2.29) for time-dependent self-adjoint family of operators $\{A(t)\}_{t\in[0,T]}$ satisfying a continuity condition similar to Assumptions 12.2.2: (E1)–(E3), if domain $\operatorname{dom}(A(t))$ is *time-independent*.

A further possible generalisation concerns the time-dependent domains in the spirit of results in Sect. 11.1, where domain $\operatorname{dom}(A(t))$ of the time-dependent generator $\{A(t)\}_{t\in[0,T]}$ is also *time-dependent*.

Part V
Unitary and Zeno Product Formulæ

Chapter 13
Unitary Product Formulæ

In this chapter, we first recollect the main results about convergence of the unitary *Trotter* product formula in the *strong* operator topology, Sect. 13.1.

A particular case of the unitary *Trotter* product formula, which converges in the *operator-norm* topology, is due to the *Ichinose–Tamura* commutator conditions. It is a subject of Sect. 13.2, There we also discuss some instructive examples, as well as counterexamples.

In Sect. 13.3, we focus on sufficient conditions ensuring convergence of the unitary *Trotter–Kato* product formulæ in a more general set-up, which involves a new, L^2*-topology* of convergence. In contrast to the *operator-norm* convergent unitary *Trotter* product formula (Sect. 13.2), we scrutinise in Sect. 13.3 a convergence of the unitary *Trotter-Kato* product formulæ for a *holomorphic* Kato functions in a *palliative* L^2-topology of the *Fréchet* space.

On the other hand, for a class of *admissible* Kato functions, we prove convergence of the unitary *Trotter-Kato* product formulæ in the *strong* operator topology.

13.1 Unitary Trotter Product Formula: Strong Operator Topology

13.1.1 Unitary Groups and Nelson's Result

Owing to our knowledge of already introduced m-accretive generators (Definition 1.8.7), there is an abstract way to introduce the *groups* of unitary operators, see Notes to Chap. 13 (Notes to Sects. 13.1, 13.1.1). However, below we prefer to do this traditionally, involving the *Stone* theorem since it will be useful later for discussion of the *unitary* product formulae.

Definition 13.1.1 A family $\{U(t)\}_{t\in\mathbb{R}}$ of *unitary* operators on a Hilbert space $\mathfrak{H}$ is called a one-parameter *strongly continuous unitary group* if:

V. A. Zagrebnov et al., *Trotter-Kato Product Formulæ*, Operator Theory: Advances and Applications 296, https://doi.org/10.1007/978-3-031-56720-9_13

(a) The mapping $\mathbb{R} \ni t \mapsto U(t)$ is continuous in the *strong* operator topology on $\mathcal{L}(\mathfrak{H})$.
(b) The *group law*: $U(t+s) = U(t)\,U(s)$, holds for all $t, s \in \mathbb{R}$.

The group law implies $U(0) = \mathbb{1}$ and also $U(-t) = U(t)^{-1}$, which by the *unitarity* of operators yields: $U^*(t) = U(-t)$ as well as $\|U(t)\| = 1$ for $t \in \mathbb{R}$.

Remark 13.1.2

(1) By the group law (b) and unitarity, it follows that similarly to the C_0-semigroup (cf. Proposition 1.2.3) the strong continuity only at $t = 0$ is equivalent to (a). Indeed,

$$\lim_{\delta\to 0} \|(U(t+\delta) - U(t))\,u\| \leq \lim_{\delta\to 0} \|(U(\delta) - \mathbb{1})\,u\| = 0\,, \quad u \in \mathfrak{H}\,,$$

for any $t \in \mathbb{R}$. Moreover, the *group law* (b) and the *weak continuity* of the family $\{U(t)\}_{t\in\mathbb{R}}$ at $t = 0$ is equivalent to (a). In fact, the limits

$$\lim_{t\to 0} \|U(t)\,u - u\|^2 = \lim_{t\to 0} \Big(\|U(t)\,u\|^2 + \|u\|^2 - 2\,\mathrm{Re}\,(U(t)\,u, u)\Big) \leq$$
$$\lim_{t\to 0} \Big(2\,\|u\|^2 - 2\,\mathrm{Re}\,(U(t)\,u, u)\Big) = 0\,, \quad u \in \mathfrak{H}\,,$$

provide the strong continuity at $t = 0$ and, hence, for any $t \in \mathbb{R}$.
(2) In this connection, we also note that similar to C_0-semigroups (cf. Notes to Sect. 1.2) the *weak* measurability of unitary group $\{U(t)\}_{t\in\mathbb{R}}$ on a *separable* Hilbert space $\mathfrak{H}$ ensures (by the *von Neumann* theorem) the *strong* continuity of this group. See Notes to Chap. 13 (Notes to Sects. 13.1, 13.1.1).
(3) Using the (spectral) *functional calculus*, one can associate any self-adjoint operator A on domain $\mathrm{dom}\,(A) \subseteq \mathfrak{H}$ with unitary group $\{U_A(t) := \mathrm{e}^{-itA}\}_{t\in\mathbb{R}}$, cf. (13.1.4).

Proposition 13.1.3 *Let A with domain* $\mathrm{dom}\,(A)$ *be a self-adjoint operator in $\mathfrak{H}$ and family $\{U_A(t) := \mathrm{e}^{-itA}\}_{t\in\mathbb{R}}$ be given by the functional calculus. Then,*

(a) *$\{U_A(t)\}_{t\in\mathbb{R}}$ is a one-parameter strongly continuous unitary group and $U_A(t)$:* $\mathrm{dom}\,(A) \to \mathrm{dom}\,(A)$, *for all $t \in \mathbb{R}$.*
(b) *For any $v \in$* $\mathrm{dom}\,(A)$

$$\lim_{t\to 0} \frac{1}{t}\,(U_A(t) - \mathbb{1})\,v = -\,i\,A\,v\,. \tag{13.1.1}$$

(c) *Conversely, if* $\lim_{t\to 0} t^{-1}\,(U_A(t) - \mathbb{1})\,v$ *exists, then $v \in$* $\mathrm{dom}\,(A)$.

Proof

(a) Let $f \in \mathcal{B}(\mathbb{R})$, be (in general) a complex-valued *Borel* function on $\mathbb{R}$. On account of the *Borel functional calculus*: $f \mapsto f(A)$, for self-adjoint operator

A, there exists a unique family of orthogonal projections: $\{P_A(-\infty, \lambda) =: E_A(\lambda)\}_{\lambda\in\mathbb{R}}$, which is a *resolution* of the identity: $E_A(+\infty) = \mathbb{1}$, $E_A(-\infty) = 0 \cdot \mathbb{1}$, and $E_A(\lambda)\, E_A(\mu) = E_A(\lambda \wedge \mu)$. For more details, see Notes to Sect. 13.1 (Sect. 13.1.1).

As a consequence, for any $u \in \mathfrak{H}$ the *projection-valued* measure $\{dE_A(\lambda)\}_{\lambda\in\mathbb{R}}$ corresponds to a well-defined on $\mathbb{R}$ *Borel* measure $\{d(E_A(\lambda)u, u)\}_{\lambda\in\mathbb{R}}$, such that in accordance with the functional calculus:

$$(f(A)v,\, v) = \int_{\mathbb{R}} d(E_A(\lambda)v,\, v)\, f(\lambda)\,, \quad v \in \mathrm{dom}\,(f(A))\,. \tag{13.1.2}$$

Here the complex measure $\{d(E_A(\lambda)u_1, u_2)\}_{\lambda\in\mathbb{R}}$ for $u_1, u_2 \in \mathfrak{H}$ is defined by the *polarisation* identity and domain in (13.1.2) by condition $\|f(A)v\| < \infty$. Given that $\{E_A(\lambda)\}_{\lambda\in\mathbb{R}}$ is a resolution of the identity, one gets

$$\begin{aligned} &\|f(A)\,v\|^2 = \int_{\mathbb{R}} d(E_A(\lambda)v,\; f(A)v)\, f(\lambda) = \int_{\mathbb{R}} d\overline{(f(A)v,\; E_A(\lambda)\,v)}\, f(\lambda) = \\ &\int_{\mathbb{R}} d \int_{\mathbb{R}} d_\mu \overline{(E_A(\mu\wedge\lambda)v,\; v)\, f(\mu)}\, f(\lambda) \\ &= \int_{\mathbb{R}} d \int_{-\infty}^{\lambda} d_\mu \overline{(E_A(\mu)v,\; v)}\, \overline{f(\mu)} f(\lambda) = \int_{\mathbb{R}} d(E_A(\lambda)v,\; v)\, |f(\lambda)|^2 < \infty\,. \end{aligned} \tag{13.1.3}$$

If now we apply the Borel functional calculus (13.1.2), (13.1.3) for the complex-valued Borel function $f_t : \lambda \mapsto \mathrm{e}^{-it\lambda}$, $t \in \mathbb{R}$ and $\lambda \in \mathbb{R}$, then operator $f_t(A) = \mathrm{e}^{-itA}$ is also customary denoted as a *spectral* representation by the *Stiltjes* integral:

$$U_A(t) = \int_{\mathbb{R}} dE_A(\lambda)\, \mathrm{e}^{-it\lambda}\,, \quad \mathrm{dom}\,(U_A(t)) = \mathfrak{H}\,, \tag{13.1.4}$$

where due to the fact that $\{dE_A(\lambda)\}_{\lambda\in\mathbb{R}}$ is a *resolution* of the identity one has $U_A(t = 0) = \mathbb{1}$. As a consequence, by (13.1.3) and (13.1.4), one obtains unitarity: $U_A^*(t) = U_A^{-1}(t) = U_A(-t)$. By unitarity and by arguments similar to (13.1.3), we infer for any $t, t' \in \mathbb{R}$ the *group law* $U_A(t)\, U_A(t') = U_A(t+t')$. To this end we note that

$$\begin{aligned} &(U_A(t)\, U_A(t')\, u,\; u) = \int_{\mathbb{R}} d \int_{\mathbb{R}} d_\mu \overline{(E_A(\mu\wedge\lambda)u,\; u)\, \mathrm{e}^{it\mu}}\, \mathrm{e}^{-it'\lambda} = \\ &\int_{\mathbb{R}} d \int_{-\infty}^{\lambda} d_\mu \overline{(E_A(\mu)u,\; u)}\, \mathrm{e}^{-it\mu}\, \mathrm{e}^{-it'\lambda} = \int_{\mathbb{R}} d(E_A(\lambda)u,\; u)\, \mathrm{e}^{-i\,(t+t')\,\lambda} \\ &= (U_A(t+t')\, u,\; u)\,. \end{aligned} \tag{13.1.5}$$

Again, by virtue of (13.1.3), one concludes that $\|A\,v\|^2 = \int_{\mathbb{R}} d(E_A(\lambda)v,\ v)\ |\lambda|^2 < \infty$, for $v \in \mathrm{dom}\,(A)$, implies

$$\|A\,U_A(t)\,v\|^2 = \int_{\mathbb{R}} d(E_A(\lambda)v,\ v)\ |\lambda|^2 < \infty\,. \tag{13.1.6}$$

This proves that $U_A(t) : \mathrm{dom}\,(A) \to \mathrm{dom}\,(A)$, for all $t \in \mathbb{R}$.

To establish the strong continuity of the group (13.1.4) at $t = 0$, we note that owing to (13.1.3)

$$\|(U_A(t) - \mathbb{1})\,u\|^2 = \int_{\mathbb{R}} d(E_A(\lambda)u,\ u)\,|\,\mathrm{e}^{-it\lambda} - 1|^2\,, \quad u \in \mathfrak{H}\,.$$

Then, estimate $|\,\mathrm{e}^{-it\lambda} - 1|^2 \leq 2$ and the pointwise convergence $\lim_{t\to 0}|\,\mathrm{e}^{-it\lambda} - 1|^2 = 0$ yield

$$\lim_{t\to 0}\|(U_A(t) - \mathbb{1})\,u\|^2 = 0\,, \quad u \in \mathfrak{H}\,, \tag{13.1.7}$$

by virtue of the *Lebesgue* dominated convergence theorem since function $f(\lambda) = 2$ is $d(E_A(\lambda)u,\ u)$-integrable on $\mathbb{R}$. By the group law (13.1.5), the convergence (13.1.7) can be extended to any $t \in \mathbb{R}$.

(b) Now we use the *Borel* functional calculus (13.1.2), (13.1.3) for complex-valued Borel function $g_t : \lambda \mapsto (\mathrm{e}^{-it\lambda} - 1)/t$, for $t \in \mathbb{R}\setminus\{0\}$ and $\lambda \in \mathbb{R}$. Then,

$$\frac{U_A(t) - \mathbb{1}}{t} = \int_{\mathbb{R}} dE_A(\lambda)\,\frac{\mathrm{e}^{-it\lambda} - 1}{t}\,, \quad t \in \mathbb{R}\setminus\{0\}\,, \tag{13.1.8}$$

is a family of bounded operators. For proving (13.1.1), we use (13.1.3) to infer the representation

$$\left\|\frac{U_A(t) - \mathbb{1}}{t}\,v + i\,A\,v\right\|^2 = \int_{\mathbb{R}} d(E_A(\lambda)\,v,\ v)\,\left|\frac{1}{t}(\mathrm{e}^{-it\lambda} - 1) + i\,\lambda\right|^2, \tag{13.1.9}$$

where $t \neq 0$ and $v \in \mathrm{dom}\,(A)$. Seeing that singularity at $t \neq 0$ is removable, one gets for $t \in \mathbb{R}$, $\lambda \in \mathbb{R}$ and some $C > 0$ the estimate: $|(\mathrm{e}^{-it\lambda} - 1)/t + i\,\lambda|^2 \leq C\,t^2\lambda^2$. Then, on account of (13.1.6) and (13.1.9), we obtain the limit (13.1.1).

(c) To prove the converse, let us define domain $D_{\tilde{A}}$ and corresponding operator $\tilde{A}$ as

$$\begin{aligned} D_{\tilde{A}} &:= \{w \in \mathfrak{H} : \lim_{t\to 0}\frac{1}{t}\,(U_A(t) - \mathbb{1})\,w \ \text{ exists}\}\,,\\ \tilde{A}\,w &:= \lim_{t\to 0}\frac{i}{t}\,(U_A(t) - \mathbb{1})\,w\,, \quad w \in D_{\tilde{A}}\,. \end{aligned} \tag{13.1.10}$$

Since $\{U_A(t)\}_{t\in\mathbb{R}}$ is a unitary group, operator $\tilde{A}$ in (13.1.10) is symmetric: $\tilde{A} \subset \tilde{A}^*$. By definition of $\{U_A(t)\}_{t\in\mathbb{R}}$, together with (13.1.1) and (13.1.10), one also infers that $\tilde{A} \supseteq A$. Then, $A^* \supseteq \tilde{A}^* \supseteq \tilde{A} \supseteq A$ and $A^* = A$ yield $\tilde{A} = A$, thus $D_{\tilde{A}} = \mathrm{dom}\,(A)$.

□

Note that (13.1.10) suggests a definition of generator of a unitary group. The part (c) of Proposition 13.1.3 shows a pertinence of this definition in the case when unitary group is defined via functional calculus for a self-adjoint operator. The following proposition the (*Stone* theorem) states that *every* strongly continuous unitary group arises as the exponential of a self-adjoint operator. Similar to Proposition 13.1.3 (a) and (b), the key step is a differentiation of unitary group $\{U(t)\}_{t\in\mathbb{R}}$ at $t = 0$.

Proposition 13.1.4 (Stone Theorem) *Let $\{U(t)\}_{t\in\mathbb{R}}$ be a one-parameter strongly continuous unitary group on a Hilbert space $\mathfrak{H}$. Then there exists a self-adjoint operator A, with dense domain* $\mathrm{dom}\,(A) \subset \mathfrak{H}$, *such that $U(t) = \mathrm{e}^{-itA}$, where for any $t \in \mathbb{R}$ the right-hand side denotes the Stiltjes integral* (13.1.4).

For the proof, details, and references concerning the *Stone* theorem, see Proposition 13.4.1 and Notes to Chap. 13 (Notes to Sects. 13.1, 13.1.1).

The next observation creates a link between one-parameter strongly continuous unitary groups on a Hilbert space $\mathfrak{H}$ and contraction C_0-semigroups. This relation serves for translation of some properties established for C_0-semigroups to similar properties for unitary groups.

Remark 13.1.5

(a) Let $\{U(t) = \mathrm{e}^{-itA}\}_{t\in\mathbb{R}}$ be strongly continuous unitary group on $\mathfrak{H}$ generated by self-adjoint operator A with dom (A), which is dense in $\mathfrak{H}$. We have:

$$\|(A - \zeta\,\mathbb{1})^{-1}\| \leq \frac{1}{|\mathrm{Im}\,\zeta|}\,, \quad \zeta \in \mathbb{C}\,. \tag{13.1.11}$$

Now we introduce operators $A_\pm := \pm i\,A$. Then, owing to (13.1.11), one gets

$$(-\infty, 0) \subset \rho(A_\pm) \quad \text{and} \quad \|(A_\pm + \lambda\,\mathbb{1})^{-1}\| \leq \lambda^{-1}\,, \quad \lambda > 0\,. \tag{13.1.12}$$

Therefore, $A_\pm$ are m-accretive operators (Definition 1.8.7). Then on account of the *Lumes-Phillips* theorem (Proposition 1.1.1, or Proposition 1.2.7) conditions (13.1.12) yield that operators $A_\pm \in \mathscr{G}(1, 0)$. That is, they are generators of contraction C_0-semigroups: $\{\mathrm{e}^{-tA_+}\}_{t\geq 0}$ and $\{\mathrm{e}^{-tA_-}\}_{t\geq 0}$, see Definition (1.3.1).

(b) A link between unitary group $\{U(t) = e^{-itA}\}_{t\in\mathbb{R}}$ and contraction C_0-semigroups is established by equation :

$$e^{-itA} = \begin{cases} e^{-tA_+} & , \ t > 0 , \\ \mathbb{1} & , \ t = 0 , \\ e^{tA_-} & , \ t < 0 , \end{cases} \tag{13.1.13}$$

where generators $A_\pm = \pm i\, A \in \mathscr{G}(1, 0)$.

(c) Formula (13.1.13) provides a hint for checking whether a given C_0-*semigroup* $\{U(t)\}_{t\geq 0}$ on $\mathfrak{H}$ can be *embedded* into a C_0-*group* $\{\widehat{U}(t)\}_{t\in\mathbb{R}}$ on $\mathfrak{H}$.

To this end, we note that if $\{U(t)\}_{t\geq 0}$ is *invertible* for some $t_0 > 0$, then by the semigroup law one gets: $U(t_0) = U(t_0 - t)\, U(t) = U(t)\, U(t_0 - t)$ for $0 \leq t \leq t_0$. So, $U(t_0 - t)$ is also invertible and

$$U^{-1}(t_0 - t) = U^{-1}(t_0)\, U(t) = U(t)\, U^{-1}(t_0) \,, \quad t \in [0, t_0] \,,$$

To extend this observation to any $t \geq t_0$ note that $t = s + k\, t_0$ for $s \in [0, t_0)$ and $k \in \mathbb{N}$. Consequently, $U(t) = U(s)\, U(k\, t_0)$ is invertible. As a result, the semigroup $\{U(t)\}_{t\geq 0}$ is invertible and thus can be extended to the C_0-group $\{\widehat{U}(t)\}_{t\in\mathbb{R}}$ by definition:

$$\widehat{U}(t) = \begin{cases} U(t) & , \ t > 0 , \\ \mathbb{1} & , \ t = 0 , \\ U^{-1}(-t) & , \ t < 0 . \end{cases} \tag{13.1.14}$$

In (b), one finds a realisation (13.1.13) of extension (13.1.14).

After these preliminaries, we return to the problem of Trotter product formula approximations for unitary groups. As we learned from Parts I–III the contraction (Definition 1.2.5) and, in particular, quasi-sectorial contraction (Definition 3.1.2) properties of semigroups are helpful for *uplifting* the strong operator convergence of the Trotter–Kato product formulæ to convergence in the operator-norm topology. This advantage of contractions becomes evident also for the case of product formulæ corresponding to the unitary groups of operators.

On that account, the Nelson proof (see Proposition 13.1.6) of the *strong* operator convergence of the unitary Trotter product formula is quite instructive and deserves that we recall it first. Note that for Nelson's result the assumption of *self-adjointness* of $A + B$ for two self-adjoint generators A and B sounds as rather restrictive. We comment that it coincides with assumption in the ITTZ Theorem 7.1.9 (cf. Corollary 7.1.13 for the Trotter product formula), Sect. 7.1.

The self-adjointness of $A + B$ comes from a special tuning of operators A and B, which *a priori* are not related by any *subordination*. In the case of subordination, for example, by the *Kato-smallness* of the one operator with respect to another one

(Definition 1.9.6), the self-adjointness is stable by the *Kato–Rellich* theorem, see Corollary 1.9.15. As a consequence, the Nelson proof is obviously valid for such couple of subordinated generators A and B as well.

Proposition 13.1.6 (Nelson's Result) *Let A and B be self-adjoint generators of unitary groups on $\mathfrak{H}$. Suppose that operator sum*

$$C v := A v + B v \,, \quad v \in \mathrm{dom}\,(A) \cap \mathrm{dom}\,(B) \,,$$

is, in turn, a self-adjoint operator C on domain $D :=$ $\mathrm{dom}\,(A) \cap \mathrm{dom}\,(B)$. Then, the unitary group $\{U_C(t)\}_{t\,\in\mathbb{R}}$ can be approximated on $\mathfrak{H}$ in the strong operator topology by the Trotter product formula, that is:

$$\mathrm{e}^{-itC}\, u = \lim_{n\to\infty} (\mathrm{e}^{-itA/n}\, \mathrm{e}^{-itB/n})^n\, u \,, \tag{13.1.15}$$

for all $u \in \mathfrak{H}$ locally-uniformly for t in $\mathbb{R}$.

Proof First, we define the operator family $\{K(\tau) := \tau^{-1}(\mathrm{e}^{-i\tau A}\mathrm{e}^{-i\tau B} - \mathrm{e}^{-i\tau C})\}_{\tau\,\in\mathbb{R}}$. Then, by definition of the operator C, one gets $\lim_{\tau\to 0} K(\tau)\, v = 0$ for each vector $v \in D =: \mathrm{dom}\,(C)$.

Since C is self-adjoint and thus closed on D, we can consider domain D as a Banach $\mathcal{B}$ space endowed with the norm $\|v\|_C := \|C\, v\| + \|v\|$, $v \in D$. Then each of maps $K(\tau) : D \to \mathfrak{H}$ for $\tau \in \mathbb{R}$ is bounded, and one has $\lim_{\tau\to\{0\,\vee\,\infty\}} K(\tau)\, v = 0$ for each $v \in D$, and thus by the *uniform boundedness principle* in $\mathcal{B}$, there exists a constant M such that for all $v \in D$ and $\tau \in \mathbb{R}$ we obtain the estimate $\|K(\tau)\, v\| \le M\, \|v\|_C$. Therefore, on any $\|\cdot\|_C$-compact subset $\mathcal{C}$ of domain D by a standard argument we obtain $K(\tau)\, v \to 0$, when $\tau \to 0$, uniformly in $v \in \mathcal{C}$.

Given that C is self-adjoint generator of the group $\{U_C(t)\}_{t\,\in\mathbb{R}}$ and that $\mathrm{dom}\,(C) = D$, one gets $v \mapsto \mathrm{e}^{-i\tau C}\, v \in D$, if $v \in D$. Moreover, by definition of the space $\mathcal{B}$, one gets that $\tau \mapsto \mathrm{e}^{-i\tau C}\, v$ is a continuous mapping of $\mathbb{R}$ into $\mathcal{B}$. Thus, for each fixed vector $v \in D$, the set $\{\mathrm{e}^{-i\tau C}\, v\}_{\tau\,\in[-1,1]}$ is a $\|\cdot\|_C$-compact in D.

Summarising these two observations, we conclude that for any $v \in D$

$$\lim_{s\to 0} \frac{1}{s}\, \{\mathrm{e}^{-isA}\, \mathrm{e}^{-isB} - \mathrm{e}^{-isC}\}\, \mathrm{e}^{-i\tau C}\, v = 0 \,, \tag{13.1.16}$$

uniformly in $\tau \in [-1, 1]$. Then, we can use the *telescopic* representation:

$$\{(\mathrm{e}^{-itA/n}\, \mathrm{e}^{-itB/n})^n - \mathrm{e}^{-itC}\}\, v =$$

$$\sum_{k=0}^{n-1} (\mathrm{e}^{-itA/n}\, \mathrm{e}^{-itB/n})^k\, \{\mathrm{e}^{-itA/n}\, \mathrm{e}^{-itB/n} - \mathrm{e}^{-itC/n}\}\, \mathrm{e}^{-i\,(n-k-1)\,t\,C/n}\, v \,,$$

to prove the proposition. Indeed, for any fixed $t \in \mathbb{R}$, the unitarity yields the estimate of the norm on the right-hand side by

$$|t| \max_{|\tau|<t} \|(t/n)^{-1} \{e^{-itA/n} e^{-itB/n} - e^{-itC/n}\} e^{-i\tau C} v\| .$$

Together with (13.1.16), this prove the claim (13.1.15) for $v \in D$ uniformly for t in any compact subset of $\mathbb{R}$. Then, by virtue of density of domain D and by the boundedness due to unitarity of families $\{e^{-itA/n}\}_{n\geq 1}$ and $\{e^{-itB/n}\}_{n\geq 1}$, for $t \in \mathbb{R}$, one can extend the above conclusion to all $u \in \mathfrak{H}$. □

Remark 13.1.7 Seeing that *approximants* $\{U_n(t) := (e^{-itA/n} e^{-itB/n})^n\}_{n\geq 1}$, in the Trotter product formula (13.1.15) are unitary, by the arguments similar to those in Remark 13.1.2.(1), we infer that a *weakly* convergent unitary product formula converges also in the *strong* operator topology.

13.1.2 Strongly Convergent Product Approximants

We comment that *self-adjointness* of the sum $A + B$ in the previous subsection is decisive for construction of the *axillary* space $\mathcal{B}$ and for invariance of domain D under the group $\{U_C(t)\}_{t\in\mathbb{R}}$. For that reason, a *tiny* deviation from the condition of self-ajointness makes the proof of the Trotter product approximation for unitary groups nontrivial even in the strong operator topology. For example, if the sum $A + B$ in the previous subsection is not self-adjoint, but still *essentially* self-adjoint, the proof of convergence of the unitary Trotter product formula is quite different to the *Nelson* arguments in Proposition 13.1.6.

In this subsection, our line of reasoning is based on application of the *Chernoff approach* to product formulæ approximations that we utilised in Sect. 1.9.6 (Theorem 1.9.34) and in Sect. 3.1.2 (Lemma 1.9.33) for C_0-semigroups on Banach spaces. Note that in Sect. 4.1 (Theorem 4.1.3), we applied this approach to self-adjoint *semigroups* on a Hilbert space $\mathfrak{H}$. The aim of this subsection is to use the Chernoff product formula (4.1.7) for justification the Trotter product formula for *unitary* groups in the *strong* operator topology on $\mathcal{L}(\mathfrak{H})$.

Proposition 13.1.8 *Let $t \in \mathbb{R}$ and $\Psi : t \mapsto \Psi(t)$ be a strongly continuous function to unitary operators on a separable Hilbert space $\mathfrak{H}$. If for any $t \in \mathbb{R}$, there exists limit*

$$U(t)u = \lim_{n\to\pm\infty} (\Psi(t/n))^n u , \quad u \in \mathfrak{H} , \tag{13.1.17}$$

then family of bounded operators $\{U(t)\}_{t\in\mathbb{R}}$ is a one-parameter strongly continuous unitary group on $\mathfrak{H}$.

Proof

(a) Since the product *approximants*: $\{(\Psi(t/n))^n\}_{n\geq 1}$, are unitary operators, one gets, for any $u \in \mathfrak{H}$ and $t \in \mathbb{R}$, the estimate:

$$| \, \|U(t)\,u\|^2 - \|u\|^2| = |(U(t)\,u,\, U(t)\,u) - ((\Psi(t/n))^n\,u,\, (\Psi(t/n))^n\,u)| \leq$$
$$\|(U(t) - (\Psi(t/n))^n)\,u\| \, \|U(t)\,u\| + \|(\Psi(t/n))^n\,u\| \, \|(U(t) - (\Psi(t/n))^n)\,u\| \,. \tag{13.1.18}$$

Because of the unitarity, which yields $\|(\Psi(t))\| = 1$, and of the fact that for any $t \in \mathbb{R}$ the strong limit: $\text{s-lim}_{n\to+\infty}(\Psi(t/n))^n = U(t) \in \mathcal{L}(\mathfrak{H})$, we deduce from (13.1.17) and (13.1.18) the *isometry*: $\|U(t)\,u\| = \|u\|$, as well as, (by the *polarisation identity*):

$$(U(t)\,u, U(t)\,v) = (u, v)\,, \quad u, v \in \mathfrak{H}\,. \tag{13.1.19}$$

As a consequence of isometry, the mapping $U(t) : \mathfrak{H} \to \mathfrak{H}$ is one-one (*injective*), that is: $U(t)\,u_1 = U(t)\,u_2$ implies $u_1 = u_2$ for any $u_{1,2} \in \mathfrak{H}$. Hence, there exists inverse operator $U^{-1}(t)$ on the ran $(U(t))$. Thus if ran $(U(t)) \subset \mathfrak{H}$, then $U^*(t)$ is only *partially* isometric since $U^*(t) = U^{-1}(t)$ on the ran $(U(t))$, but $U^*(t)\,v = 0$ for $v \in$ ran $(U(t))^{\perp}$.

(b) Seeing that isometry $U(t)$ is *unitary* if and only if the mapping $U(t) : \mathfrak{H} \to \mathfrak{H}$, is one-one onto (*bijective*), we need to check that ran $(U(t)) = \mathfrak{H}$, that is, there exists a bounded $U^{-1}(t) = U^*(t)$. To this aim, we use the strong limits to $\pm\infty$ in (13.1.17):

$$\begin{aligned} U(-t) &= \text{s-}\lim_{n\to+\infty}(\Psi(-t/n))^n = \text{s-}\lim_{n\to-\infty}(\Psi(-t/n))^n \\ &= \text{s-}\lim_{m\to+\infty}(\Psi(t/m))^{-m}. \end{aligned} \tag{13.1.20}$$

On account of the strong continuity of multiplication of strongly convergent *sequences* of bounded operators (see Remark 1.6.11 and (1.6.16)) the limits (13.1.17) and (13.1.20) yield

$$U(t)\,U(-t) = \text{s-}\lim_{n\to+\infty}(\Psi(t/n))^n(\Psi(t/n))^{-n} = U(-t)\,U(t) = \mathbb{1}\,, \quad t \in \mathbb{R}\,. \tag{13.1.21}$$

So, a (two-sided) bounded inverse operator exists: $U^{-1}(t) = U(-t)$, and, consequently, $\{U(t)\}_{t\in\mathbb{R}}$ is a *family* of *unitary* operators and by (13.1.19), (13.1.21) the adjoint operator $U^*(t) = U(-t)$.

As a consequence, on account of unitarity: $\Psi^*(t) = \Psi^{-1}(t)$, and of the limits (13.1.17), (13.1.20):

$$U^*(t) \equiv (\text{s-}\lim_{n\to+\infty}(\Psi(t/n))^n)^* = \text{s-}\lim_{n\to+\infty}(\Psi^*(t/n))^n = \text{s-}\lim_{n\to+\infty}(\Psi(t/n))^{-n} = U(-t)\,. \tag{13.1.22}$$

Hence, although *a priori* the conjugation mapping: $(\Psi(t/n))^n \mapsto ((\Psi(t/n))^n)^*$, is *not* continuous in the strong operator topology, the limits (13.1.22) show that it is continuous because of condition (13.1.17).

(c) Again, because of the strong continuity of multiplication of strongly convergent *sequences* of bounded operators (see Remark 1.6.11 and (1.6.16)) and owing to (13.1.17), we infer the following composition property for any $k \in \mathbb{Z}$:

$$U(k\,t) = \text{s-}\lim_{n\to\infty}(\Psi(k\,t/n))^n = \text{s-}\lim_{m\to\infty}(\Psi(t/m))^{k\,m} = [\,\text{s-}\lim_{m\to\infty}(\Psi(t/m))^m\,]^k = U^k(t)\,, \qquad t \in \mathbb{R}\,. \tag{13.1.23}$$

Note that being the strong limit (13.1.17) of strongly continuous functions, the mapping: $t \mapsto U(t)$ is locally *strongly* Borel measurable on $\mathbb{R}$. Then, by virtue of the property (13.1.23), there exists a *strongly* continuous one-parameter *unitary* group $\{\widetilde{U}(t)\}_{t\in\mathbb{R}}$ on a *separable* Hilbert space $\mathfrak{H}$ such that $\widetilde{U}(t) = U(t)$ *almost everywhere* (a.e.) on $\mathbb{R}$.

(d) On account of Remark 13.1.2 (2), one infers that equality: $U(t) = \widetilde{U}(t)$, holds actually for all $t \in \mathbb{R}$. So, the unitary group (13.1.17) is *strongly* continuous.

□

For more details concerning the proof of Proposition 13.1.8, including the points (c) and (d), we suggest Notes to Chap. 13 (Notes to Sects. 13.1, 13.1.2).

By the *Stone* theorem (Proposition 13.1.4), the strongly continuous one-parameter unitary group (13.1.17) is generated on a separable Hilbert space $\mathfrak{H}$ by a self-adjoint operator C. We denote this unitary group by $\{U_C(t)\}_{t\in\mathbb{R}}$.

Now we are in position to elucidate the *unitary Trotter* product formula (13.1.15) in the strong operator topology for the case where, in contrast to the *Nelson* result, Proposition 13.1.6, the sum $A + B$ is only *essentially* self-adjoint. To this end, we apply the Chernoff approximation formula developed in Sect. 1.9.6 of Sect. 1.9 (cf. Theorem 1.9.34) in a Banach space, and Theorem 4.1.3, for the case of self-adjoint C_0-semigroups.

Proposition 13.1.9 *Let* $t \in \mathbb{R}$ *and* $\Psi : t \mapsto \Psi(t)$ *be a strongly continuous mapping to unitary operators on a Hilbert space* $\mathfrak{H}$ *such that* $\Psi(0) = \mathbb{1}$.

Let $\{U_C(t) = \mathrm{e}^{-itC}\}_{t\in\mathbb{R}}$ *be a unitary* C_0*-group, and let* $D \subset \mathrm{dom}(C)$ *be a core for the generator* C. *Let the unitary family* $\{\Psi(t)\}_{t\in\mathbb{R}}$ *be strongly (right) differentiable at* $t = 0$, *that is,* $\Psi'(0)\,v$ *exists for any* $v \in \mathrm{dom}(\Psi'(0))$ *and*

$$\Psi'(0)\,v := \lim_{t\to 0}\frac{1}{t}\,(\Psi(t) - \mathbb{1})\,v = -i\,C\,v\,, \tag{13.1.24}$$

for all $v \in D \subseteq \mathrm{dom}(\Psi'(0))$.

Then, the strong operator limit

$$\text{s-}\lim_{n\to\infty}\{\Psi(t/n)\}^n = U_C(t)\,, \tag{13.1.25}$$

holds for all $t \in \mathbb{R}$.

Proof On account of Proposition 13.1.8 and conditions of Proposition 13.1.9, the limit (13.1.25) (if exists) yields on the Hilbert space $\mathfrak{H}$ a one-parameter strongly continuous unitary group. By the *Stone* theorem (Proposition 13.1.4) it is generated by a densely defined self-adjoint operator $\widetilde{C}$, that is, $\{U_{\widetilde{C}}(t) = \mathrm{e}^{-it\widetilde{C}}\}_{t\in\mathbb{R}}$.

The rest of the proof is devoted to the identification $\widetilde{C}$ with self-adjoint operator C steaming from the properties of the family $\{\Psi(t)\}_{t\in\mathbb{R}}$. Note that by virtue of Remark 13.1.5 (a), the operators $C_\pm := \pm i\,C$ are m-accretive and thus they are generators of contraction semigroups: $C_\pm \in \mathscr{G}(1, 0)$,

First, we consider C_+. Then, for $t \geq 0$ the condition (13.1.24) coincides with condition (1.9.100) in Theorem 1.9.34. As a consequence, the Chernoff product formula (1.9.101), or (3.1.18), provides

$$\text{s-}\lim_{n\to\infty}\{\Psi(t/n)\}^n = \mathrm{e}^{-t\,C_+} \quad \text{for} \quad t \geq 0\,. \tag{13.1.26}$$

Similarly, we deduce for generator C_- the product formula

$$\text{s-}\lim_{n\to\infty}\{\Psi(t/n)\}^n = \mathrm{e}^{\,t\,C_-} \quad \text{for} \quad t \leq 0\,. \tag{13.1.27}$$

Therefore, by (13.1.26), (13.1.27) and Remark 13.1.5 (b), the strong operator limit (13.1.25) holds for all $t \in \mathbb{R}$ and $\{U_C(t) = U_{\widetilde{C}}(t)\}_{t\in\mathbb{R}}$ by uniqueness of the limit. □

Proposition 13.1.10 (Unitary Trotter Product Formula I) *Let* $\{U_1(t)\}_{t\in\mathbb{R}}$ *and also* $\{U_2(t)\}_{t\in\mathbb{R}}$ *be strongly continuous unitary groups on a separable Hilbert space* $\mathfrak{H}$. *Suppose that the strong limit:*

$$\text{s-}\lim_{n\to\infty}(U_1(t/n)\,U_2(t/n))^n = U(t)\,, \tag{13.1.28}$$

exists for all $t \in \mathbb{R}$. *Then* $\{U(t)\}_{t\in\mathbb{R}}$ *is, in turn, a strongly continuous unitary group on the Hilbert space* $\mathfrak{H}$.

Proof Note that approximants $\{U^{(n)}(t) := (U_1(t/n)\,U_2(t/n))^n\}_{n\geq 1}$ are unitary families for $t \in \mathbb{R}$. Then on account of (13.1.28) the limit $\{U(t)\}_{t\in\mathbb{R}}$ is an isometry.

On the other hand, the strong limit (13.1.28) also provides convergence in the weak operator topology for adjoint approximants:

$$\text{w-}\lim_{n\to\infty}(U^{(n)}(t))^* = U^*(t)\,, \quad t \in \mathbb{R}\,. \tag{13.1.29}$$

Seeing that adjoint approximants can be represented in the form:

$$(U^{(n)}(t))^* = U_1(t/n)\,(U_1(-t/n)\,U_2(-t/n))^n U_1(t/n)\,, \quad t \in \mathbb{R}\,, \tag{13.1.30}$$

we obtain by the strong continuity of $\{U_{1,2}(t)\}_{t\in\mathbb{R}}$ together with the strong limit (13.1.28) and the strong continuity of the product in (13.1.30) (Remark 1.6.11) that:

$$\text{s-}\lim_{n\to\infty}(U^{(n)}(t))^* = \text{s-}\lim_{n\to\infty} U_1(t/n)\,(U_1(-t/n)\,U_2(-t/n))^n U_1(t/n) = U(-t)\,. \tag{13.1.31}$$

Since in turn the approximants $\{(U^{(n)}(t))^* := (U_2(-t/n)\,U_1(-t/n))^n\}_{n\geq 1}$ are unitary families for $t \in \mathbb{R}$, the limits (13.1.30) and (13.1.31) provide that $(U^*(t)) = U(-t)$ and that $\{U^*(t)\}_{t\in\mathbb{R}}$ is also an isometry. As a consequence,

$$(U(t)U^*(t)\,v, U(t)U^*(t)\,v) = (U^*(t)U(t)\,v, U^*(t)U(t)\,v) = (v, v)\,, \quad v \in \mathfrak{H}\,,$$

and hence, $\{U(t)\}_{t\in\mathbb{R}}$ is a one-parameter *unitary* family of operators.

Given that $\{U_1(t)\}_{t\in\mathbb{R}}$ and $\{U_2(t)\}_{t\in\mathbb{R}}$ are strongly continuous, the strong limit $\{U(t)\}_{t\in\mathbb{R}}$ (13.1.28) is also *strongly* continuous and, thus, provides on $\mathbb{R}$ a locally strongly *measurable* unitary family. Now we take into account that by Remark 1.6.11 one has for any $k \in \mathbb{Z}$:

$$U(k\,t) = \text{s-}\lim_{n\to\infty}(U_1(k\,t/n)\,U_2(k\,t/n))^n = \text{s-}\lim_{m\to\infty}(U_1(t/m)\,U_2(t/m))^{k\,m} =$$

$$[\,\text{s-}\lim_{m\to\infty}(U_1(t/m)\,U_2(t/m))^m\,]^k = U^k(t)\,, \qquad t \in \mathbb{R}\,.$$

Then following *mutatis mutandis* the line of reasoning in the Proof (c),(d) of Proposition 13.1.8, we deduce that the limit (13.1.28) yields on $\mathfrak{H}$ strongly continuous unitary group $\{U(t)\}_{t\in\mathbb{R}}$. □

Proposition 13.1.11 (Unitary Trotter Product Formula II) *Let A, B and C be self-adjoint generators of unitary groups on $\mathfrak{H}$. Suppose that sum*

$$C\,v = A\,v + B\,v\,, \quad v \in D\,, \tag{13.1.32}$$

holds for $D \subset (\text{dom}\,(A) \cap \text{dom}\,(B))$, where $D = \text{core}\,(C)$ is a core of operator C. Then:

(a) $\{\Psi_{AB}(t) := e^{-itA}e^{-itB}\}_{t\in\mathbb{R}}$ *is a unitary family that satisfies condition* (13.1.24) *of Proposition* 13.1.9*:* $\Psi'(0)\,v = -i\,(A+B)\,v$, *for* $v \in D$.

(b) *The unitary group* $\{U_C(t) = e^{-itC}\}_{t\in\mathbb{R}}$ *can be approximated in the strong operator topology on* $\mathcal{L}(\mathfrak{H})$ *by the Trotter product formula:*

$$e^{-itC}\,u = \lim_{n\to\infty}(e^{-itA/n}e^{-itB/n})^n\,u\,, \quad u \in \mathfrak{H}\,, \tag{13.1.33}$$

for all $t \in \mathbb{R}$ *and the closed generator* $C := \overline{(A+B)\restriction D}$.

Proof

(a) Seeing that families $\{e^{-itA}\}_{t\in\mathbb{R}}$ and $\{e^{-itB}\}_{t\in\mathbb{R}}$ are separately unitary groups, the assertion steams from formulæ (13.1.24) and (13.1.32).

(b) Proposition 13.1.10 provides that the limit (13.1.33) is a strongly continuous unitary group on $\mathfrak{H}$. To identify this limit, we use Proposition 13.1.9. Owing to the fact that by virtue of (13.1.32), the family $\{\Psi_{AB}(t)\}_{t\in\mathbb{R}}$ satisfies condition (13.1.24) the rest of the proof follows from (13.1.25).

□

Remark 13.1.12 For the unitary Trotter product formula (13.1.33), the strong convergence is equivalent to convergence in the *weak* operator topology:

$$(e^{-itC}\,u, v) = \lim_{n\to\infty}((e^{-itA/n}e^{-itB/n})^n\,u, v)\,, \quad u, v \in \mathfrak{H}\,. \tag{13.1.34}$$

Indeed, given that approximants $\{U_n(t) := (e^{-itA/n}e^{-itB/n})^n\}_{n\geq 1}$ and family $\{U_C(t) = e^{-itC}\}_{t\in\mathbb{R}}$ are *unitary* operators,

$$\|U_n(t)\,u - U_C(t)\,u\|^2 = 2\,\|u\|^2 - 2\,\mathrm{Re}\,(U_n(t)\,u, U_C(t)\,u)\,, \quad u \in \mathfrak{H}\,. \tag{13.1.35}$$

Then, (13.1.34) and (13.1.35) yield (13.1.33).

We warn that these arguments are *not* apt when applied to the Trotter product formula for C_0-semigroups, see, e.g., Proposition 1.9.35. For instance, a weak convergence of Trotter product formula for *contraction* C_0-semigroups is *not* equivalent to the strong convergence since one gets only:

$$\|(e^{-t\,A/n}e^{-t\,B/n})^n\,u - e^{-t\,C}\,u\|^2 = \|(e^{-t\,A/n}e^{-t\,B/n})^n\,u\|^2 + \|\,e^{-t\,C}\,u\|^2 -$$
$$2\,\mathrm{Re}\,((e^{-t\,A/n}e^{-t\,B/n})^n\,u, e^{-t\,C}\,u) \leq \|u\|^2 - \|\,e^{-t\,C}\,u\|^2\,, \quad u \in \mathfrak{H}\,. \tag{13.1.36}$$

Therefore, *a fortiori*, the observation (13.1.36) alludes that proviso in (13.1.34), (13.1.35), as well as that $U_C(t) = e^{-itC}$, for $t \in \mathbb{R}$, are unitary operators, can not be dropped, or relaxed to *isometry*.

For more comments about possible relaxation of conditions discussed in Remark 13.1.12, we refer to Notes to Chap. 13 (Notes to Sects. 13.1, 13.1.2).

13.2 Unitary Trotter Product Formula: Operator-Norm Topology

13.2.1 Examples

Note that conditions for the *strong* operator Trotter product approximations of contraction semigroup (Proposition 1.9.35) and for convergence of the *unitary* product formula (Proposition 13.1.11) are identical. On the other hand, similarly to the case of C_0-semigroups, under these conditions the convergence of the unitary Trotter product formula (13.1.33) does not generally hold in the *operator-norm* topology.

In this subsection, we present examples showing that conditions of Proposition 13.1.11 are neither *sufficient* (Example (a)), nor *necessary* (Example (b)) for convergence of the unitary product formula in the operator-norm topology. To this end, we study self-adjoint operators A and B such that operator $A + B$ on $\operatorname{dom}(A) \cap \operatorname{dom}(B)$ is *not* self-adjoint, but *essentially* self-adjoint, cf. conditions of Proposition 13.1.11. For the first pair of these operators (Example (a)), the product formula (13.1.33) does not hold in the *operator-norm* topology, but for the second pair under the *very same* conditions (Example (b)), the formula (13.1.33) holds in the *operator-norm* topology.

Separately, we consider a particular case (Example (c)), when operator $A + B$ is *self-adjoint* on $\operatorname{dom}(A) \cap \operatorname{dom}(B)$, cf. conditions of the ITTZ-Theorem 7.1.9 in Chap. 7 (Sects. 7.1 and 7.2).

Let $(A\,f)(x) := -i\partial_x f(x)$ and $(Bf)(x) := V(x)f(x)$, $x \in \mathbb{R}$, be *self-adjoint* operators in Hilbert space $\mathfrak{H} = L^2(\mathbb{R})$, which are defined by differentiation with respect to $x \in \mathbb{R}$ and by multiplication of vectors $f \in L^2(\mathbb{R})$ by a real-valued C^∞-function V. As a consequence, $\operatorname{dom}(A) = W^{1,2}(\mathbb{R})$ and $\operatorname{dom}(B) = \{f \in L^2(\mathbb{R}) : Vf \in L^2(\mathbb{R})\}$.

We remaind that the operator sum: $(A + B)\,u(x) = (-i\,\partial_x + V(x))\,u(x)$, $x \in \mathbb{R}$, is essentially self-adjoint on $C_0^\infty(\mathbb{R})$ and hence on $\operatorname{dom}(A) \cap \operatorname{dom}(B) = \{f \in W^{1,2}(\mathbb{R}) : Vf \in L^2(\mathbb{R})\}$. Therefore, with $C = \overline{(A + B)}$ being the closure of $A+B$, the unitary Trotter product formula (13.1.33) holds in the *strong* operator topology. But in general, $\{(e^{-itA/n}e^{-itB/n})^n\}_{n\geq 1}$ does *not* converge to e^{-itC}, $t \in \mathbb{R}$, in the *operator-norm* topology when $n \to \infty$,

Example (a) Let $x \mapsto V(x) := x^k$, for $k \in \mathbb{N}$ and $x \in \mathbb{R}$. First, we note that as above the operator $A + B$ is essentially self-adjoint on $C_0^\infty(\mathbb{R})$ and *a fortiori* on a *larger* set $D := \operatorname{dom}(A) \cap \operatorname{dom}(B) = \{f \in W^{1,2}(\mathbb{R}) : x^k f \in L^2(\mathbb{R})\}$. Thereby, the closure $C := \overline{A + B}$ is a unique self-adjoint extension of operator $A + B$.

Indeed, this closure is nontrivial since the sum $A + B$ on D is *not* a self-adjoint operator. To show this let $u(x) := \varphi(x)\exp[-ix^{k+1}/(k + 1)]$ with $\varphi(x) = (1 +$

$x^2)^{-1/2}$. Then $u \in L^2(\mathbb{R})$ and $\partial_x u(x) = (\varphi'(x) - i x^k \varphi(x)) \exp[-i x^{k+1}/(k+1)]$, so that

$$[-i\partial_x + x^k]u(x) = -i\varphi'(x)\exp[-ix^{k+1}/(k+1)] \in L^2(\mathbb{R}) .$$

On the other hand, vector u does *not* belong to the *Sobolev* space $W^{1,2}(\mathbb{R})$ of order $(1, 2)$ and, hence, to D. So, the operator $(A + B) \restriction D$ is not self-adjoint for any $k \in \mathbb{N}$.

Next, we are going to show that for $n \to \infty$ one gets:

$$\|(e^{-itA/n}\,\mathrm{e}^{-itB/n})^n - \mathrm{e}^{-itC}\| = \begin{cases} 2, & k \geq 2\,, \\ O(n^{-1})\,, & k = 1. \end{cases} \tag{13.2.1}$$

This means that *strongly* convergent unitary Trotter product formula also holds in the *operator-norm* topology for $k = 1$, but *not* for $k \geq 2$.

To this end, let $f \in C_0^\infty(\mathbb{R})$. Then for $x \in \mathbb{R}$

$$\big(\mathrm{e}^{-itA} f\big)(x) = \big(\mathrm{e}^{-t\,\partial_x} f\big)(x) = f(x-t)\,, \quad t \in \mathbb{R}\,.$$

As a consequence, one gets for $n \in \mathbb{N}$

$$\begin{aligned}
\big([\mathrm{e}^{-itA}\mathrm{e}^{-itB}]\, f\big)(x) &= \mathrm{e}^{-t\,\partial_x}\big(\mathrm{e}^{-itx^k} f\big)(x) = \mathrm{e}^{-it\,(x-t)^k} f(x-t)\,, \\
\big([\mathrm{e}^{-itA}\mathrm{e}^{-itB}]^2 f\big)(x) &= \mathrm{e}^{-t\,\partial_x}\big(\mathrm{e}^{-it[x^k+(x-t)^k]} f\big)(x) \\
&= \mathrm{e}^{-it[(x-t)^k+(x-2t)^k]} f(x-2t)\,, \\
&\ \ \ldots\,, \\
\big([\mathrm{e}^{-itA}\mathrm{e}^{-itB}]^n f\big)(x) &= \mathrm{e}^{-it\sum_{j=1}^n (x-jt)^k} f(x-nt)\,.
\end{aligned} \tag{13.2.2}$$

Therefore, we obtain the representation

$$\big([\mathrm{e}^{-itA/n}\mathrm{e}^{-itB/n}]^n f\big)(x) = \mathrm{e}^{-i\,(t/n)\sum_{j=1}^n (x-jt/n)^k} f(x-t)\,, \quad n \in \mathbb{N}\,. \tag{13.2.3}$$

Because function $x \mapsto V(x) = x^k$, for $k \in \mathbb{N}$, is locally integrable, by letting $n \to \infty$ on the right-hand side, we define the unitary group

$$\big(\mathrm{e}^{-itC} f\big)(x) := \exp\Big[-i\int_0^t ds\; V(x-s)\Big] f(x-t)\,, \quad t \in \mathbb{R}\,, \tag{13.2.4}$$

with generator $C = \overline{A + B}$. Here

$$((A+B)g)(x) = \lim_{t\to 0} \frac{i}{t}\left(\exp\Big[-i\int_0^t ds\; V(x-s)\Big]g(x-t) - g(x)\right), \quad g \in D.$$

Representation (13.2.3) and definition (13.2.4) yield

$$\begin{aligned}
&\| [e^{-itA/n}e^{-itB/n}]^n - e^{-itC} \|^2 \\
&= \sup_{\|f\|=1} \int_{\mathbb{R}} dx \left|\exp\Big[-i\frac{t}{n}\sum_{j=1}^{n} V(x-jt/n)\Big] - \exp\Big[-i\int_0^t ds\; V(x-s)\Big]\right|^2 \\
&\qquad\qquad\qquad\qquad \times | f(x-t) |^2 \\
&= \sup_{\|f\|=1} \int_{\mathbb{R}} dx\, 2\left(1 - \cos\Big[\frac{t}{n}\sum_{j=1}^{n} V(x-jt/n) - \int_0^t ds\; V(x-s)\Big]\right) |f(x-t)|^2 \\
&= \sup_{\|f\|=1} 4\int_{\mathbb{R}} dx\, \sin^2\frac{1}{2}\Big[\frac{t}{n}\sum_{j=1}^{n} V(x-jt/n) - \int_0^t ds\; V(x-s)\Big] |f(x-t)|^2 \\
&= \sup_{\|f\|=1} 4\int_{\mathbb{R}} dx\, \sin^2\frac{1}{2}\Big[\frac{t}{n}\sum_{j=1}^{n} V(x+t-jt/n) - \int_0^t ds\; V(x+t-s)\Big] |f(x)|^2 .
\end{aligned} \tag{13.2.5}$$

Let $k \geq 2$. Then, we obtain that

$$\begin{aligned}
&\|[e^{-itA/n}e^{-itB/n}]^n - e^{-itC}\| \qquad (13.2.6)\\
&= 2\sup_{x\in\mathbb{R}} \left|\sin\frac{1}{2}\Big[\frac{t}{n}\sum_{j=1}^{n} V(x+t-jt/n) - \int_0^t ds\; V(x+t-s)\Big]\right| \\
&= 2\sup_{x\in\mathbb{R}} \left|\sin\frac{1}{2}\Big[\frac{t}{n}\sum_{j=1}^{n} V(x+t-jt/n) - \frac{1}{k+1}[(x+t)^{k+1} - x^{k+1}]\Big]\right| = 2 .
\end{aligned}$$

Consequently, the *operator-norm* convergence of the unitary Trotter product formula for this pair of A and B does not hold. Although, it *does* in the *strong* operator topology since by calculations in (13.2.5) for $f \in C_0^\infty(\mathbb{R}) \subset L^2(\mathbb{R})$ one infers that

$$\lim_{n\to\infty} \| ([e^{-itA/n}e^{-itB/n}]^n - e^{-itC})\, f\|^2 = 0,$$

which implies convergence for any $f \in L^2(\mathbb{R})$ by the uniform boundedness principle.

Example (b) Let $k = 1$, that is, $V(x) = x$. Then, by direct calculations, one infers that

$$\frac{t}{n}\sum_{j=1}^{n} V(x+t-jt/n) - \frac{1}{k+1}[(x+t)^{k+1} - x^{k+1}]\Big|_{k=1} = \frac{t^2}{2n}\,.$$

As a result, by virtue of (13.2.6), we obtain for $k = 1$

$$\left\| [\mathrm{e}^{-itA/n}\,\mathrm{e}^{-itB/n}]^n - \mathrm{e}^{-itC} \right\| = 2\left|\sin\frac{t^2}{4n}\right|, \tag{13.2.7}$$

on the interval $[-T, T]$ for each $T > 0$. Then expression (13.2.7) yields for $n \to \infty$ the operator-norm convergence with asymptotic (13.2.1).

As a consequence, the unitary Trotter product formula holds in the operator-norm topology although similarly to the *first* pair A, B the operator $A + B$ is *not* self-adjoint. Note also that the rate of the operator-norm convergence is $O(n^{-1})$. This error bound is the same as for a couple of noncommuting bounded operators A and B in the Lie–Trotter product formula, cf. Proposition 1.10.10. Consequently, the error bound $O(n^{-1})$ is *optimal* by explicit calculation (13.2.7), cf. Chap. 7 (Sect. 7.2).

Recall that for contraction semigroups on a Hilbert space $\mathfrak{H}$, we established the *optimality* of the operator-norm error bound $O(n^{-1})$ versus the *self-adjointness* of operator sum $A + B$ in Chap. 7. Here we scrutinise in Example (c) the condition of self-adjointness for unitary Trotter product formula.

Example (c) Let $x \mapsto V(x)$ be a real-valued C^1-function with bounded derivative, that is, $\|V'\|_\infty = \sup_{x\in\mathbb{R}} |\partial_x V(x)| < \infty$. Then $A + B$ is *essentially* self-adjoint on $\mathrm{dom}\,(A) \cap \mathrm{dom}\,(B) = \{f \in W^{1,2}(\mathbb{R}) : Vf \in L^2(\mathbb{R})\}$. If in addition, $V \in C^1(\mathbb{R})$ itself is a bounded function, then the corresponding operator B is bounded. Hence, operator $A+B$ is *self-adjoint* on domain $\mathrm{dom}\,(A) \cap \mathrm{dom}\,(B) = \mathrm{dom}\,(A) = W^{1,2}(\mathbb{R})$. Given that, we define $C := A + B$ with $\mathrm{dom}\,(C) = \mathrm{dom}\,(A)$.

By the same line of reasoning as in Example (a) (13.2.3), we find for $f \in C_0^\infty(\mathbb{R})$,

$$\big([\mathrm{e}^{-itA/n}\,\mathrm{e}^{-itB/n}]^n f\big)(x) = \exp[-i\,(t/n)\sum_{j=1}^{n} V(x - jt/n)]\; f(x-t),$$

as well as by letting n tend to infinity, we find the representation (13.2.4) for unitary group $\{e^{-itC}\}_{t\in\mathbb{R}}$. Then

$$\begin{aligned}
&\left\| [e^{-itA/n} e^{-itB/n}]^n - e^{-itC} \right\| \\
&= 2 \sup_{x\in\mathbb{R}} \left| \sin \frac{1}{2}\Big[\frac{t}{n} \sum_{j=1}^{n} V(x+t-jt/n) - \int_0^t ds\, V(x+t-s)\Big] \right| \qquad (13.2.8) \\
&= 2 \sup_{x\in\mathbb{R}} \left| \sin \frac{1}{2}\Big[\int_0^t ds \Big(V(x+t-[ns/t]\,t/n) - V(x+t-s)\Big)\Big] \right| .
\end{aligned}$$

Here $[ns/t]$ denotes *integer* part of number $ns/t \geq 0$, so as $[ns/t] \leq ns/t < [ns/t] + 1$. Then, with proviso that $[ns/t]\,t/n = 0$ for $t = 0$ we obtain

$$0 \leq s - [ns/t]\,t/n < t/n\,, \quad n \in \mathbb{N}\,, \quad t, s \in \mathbb{R}\,. \qquad (13.2.9)$$

On account of $\|V'\|_\infty < \infty$, the inequalities (13.2.9) yield estimate:

$$\begin{aligned}
&\left| \int_0^t ds \Big(V(x+t-[ns/t]\,t/n) - V(x+t-s)\Big)\Big] \right| \leq \\
&\|V'\|_\infty \int_0^t ds\, (s - [ns/t]\,t/n) \leq \|V'\|_\infty \frac{t^2}{n}, \quad n \in \mathbb{N}, \quad t \in \mathbb{R}\,.
\end{aligned} \qquad (13.2.10)$$

Since $|\sin(X)| \leq |X|$, for $X \in \mathbb{R}$, by virtue of (13.2.8) and (13.2.10), we obtain estimate

$$\left\| [e^{-itA/n} e^{-itB/n}]^n - e^{-itC} \right\| \leq \|V'\|_\infty \frac{t^2}{n}, \quad n \in \mathbb{N}, \quad t \in \mathbb{R}, \qquad (13.2.11)$$

uniformly on $[-T, T]$ for each $T > 0$. So, the unitary Trotter product formula (13.2.11) holds under conditions of Example (c) in the operator-norm topology with error bound $O(n^{-1})$ for $C = A + B$.

Remark 13.2.1

(1) Example (a) and Example (b) show that condition of *essential* self-adjointness of the sum $A + B$ is neither *sufficient*, nor *necessary* for convergence of the unitary product formula in the operator-norm topology, but *sufficient* for the strong operator convergence, cf. Proposition 13.1.11 and comments in the Notes to Chap. 13 (Notes to Sects. 13.1, 13.1.2).

(2) Similarly, Example (b) and Example (c) demonstrate that *self-adjointness* of the sum $A + B$ is neither *sufficient*, nor *necessary* for the *optimal* rate of convergence $O(n^{-1})$ in the *operator-norm* topology, This condition of *self-adjointness* for operator-norm convergence of the product formula with error bound $O(n^{-1})$ we studied in a general set-up for self-adjoint *semigroups* in

Chap. 7 (Sect. 7.2). There we also provide an example of *essential* self-adjoint sum $A + B$ (cf. Sect. 7.4, Example in Sect. 7.4.4), which in a similar vein as Example (a), breaks down the operator-norm convergence of product formula by reducing it to convergence only in the strong operator topology.

(3) Summarising, it seems plausible that *a priori* for the operator-norm convergence of product formulæ a proviso about self-adjointness of the sum $A + B$ cannot be dropped without additional conditions on generators A and B. We return to this observation in Sect. 13.3.

In the next subsection, we shall continue analysis of the operator-norm convergent Trotter product formulæ for unitary groups, which is motivated by certain *commutator* conditions imposed on the self-adjoint operators A and B.

13.2.2 Commutator Conditions

In this subsection, we show that a systematic investigation of convergence of the unitary Trotter product formulæ in the operator-norm topology can be based on a certain *commutator* conditions.

Let A and B be self-adjoint operators in a Hilbert space $\mathfrak{H}$ with domains dom (A) and dom (B) such that the operator sum $A + B$ is *essentially* self-adjoint on dom $(A) \cap$ dom (B). Both A and B may *not* be *semibounded* neither from below, nor from above. We denote a unique self-adjoint extension, that is, the closure of the sum $A + B$, by $C := \overline{(A + B)}$ and by $C := A + B$ if the sum is itself already self-adjoint. Then, it is proved in Sect. 13.1.1 that the unitary product formulae:

$$\text{s-}\lim_{n\to\infty} \left(e^{-itA/n} e^{-itB/n}\right)^n = e^{-itC} , \tag{13.2.12}$$

and the corresponding *symmetric* version

$$\text{s-}\lim_{n\to\infty} \left(e^{-itB/2n} e^{-itA/n} e^{-itB/2n}\right)^n = e^{-itC} , \tag{13.2.13}$$

hold in the strong operator topology, uniformly on each t-bounded interval (i.e., *locally-uniformly*) in $\mathbb{R}$.

The aim of this subsection is to prove the operator-norm convergence of the unitary Trotter product formulæ (13.2.12) and (13.2.13) with (if possible) the optimal operator-norm error-bound estimate, under certain *commutator* conditions on self-adjoint operators A and B.

To this end, we assume that in $\mathfrak{H}$ there exists a dense set $\mathcal{D} \subset$ dom $(A) \cap$ dom (B) with the following properties:

($d1$) e^{-itA} (and e^{-itB}) : $\mathcal{D} \to \mathcal{D}$ for $t \in \mathbb{R}$.

($d2$) There exists on $\mathcal{D}$ a symmetric operator E and one has

$$i\,\big((Bu, Av) - (Au, Bv)\big) = (Eu, v)\,,$$

for every $u,\ v \in \mathcal{D}$.

($d3$) On $\mathcal{D}$ there are symmetric operators E_A and E_B such that

$$i\,\big((Eu, Av) - (Au, Ev)\big) = (E_A u, v)\,,$$
$$i\,\big((Eu, Bv) - (Bu, Ev)\big) = (E_B u, v)\,,$$

for every $u,\ v \in \mathcal{D}$, where $E :\ \mathrm{dom}\,(E) \subset \mathfrak{H} \to \mathfrak{H}$ is the operator defined in ($d2$).

($d4$) It holds that $\mathrm{e}^{-itC} : \mathcal{D} \to \mathcal{D}$ for $t \in \mathbb{R}$.

Remark 13.2.2 The above operators $E,\ E_A$ and E_B are formally represented in the form of commutators: $E = i[A, B] = i(AB - BA)$, and

$$E_A = i[A, E] = -[A, [A, B]], \quad E_B = i[B, E] = -[B, [A, B]]\,.$$

Theorem 13.2.3 *Let A and B be self-adjoint operators with domains* dom (A) *and* dom (B) *respectively. Suppose that operator* $A + B$ *is essentially self-adjoint on* dom $(A) \cap$ dom (B) *and denote by* C *its unique self-adjoint extension. If* ($d1$) *and* ($d2$) *are fulfilled and if* E *in* ($d2$) *admits a bounded extension on* $\mathfrak{H}$, *then*

$$\|\big(\mathrm{e}^{-i(t/n)A}\mathrm{e}^{-i(t/n)B}\big)^n - \mathrm{e}^{-itC}\| = O(n^{-1}), \tag{13.2.14}$$
$$\|\big(\mathrm{e}^{-i(t/n)B/2}\mathrm{e}^{-i(t/n)A}\mathrm{e}^{-i(t/n)B/2}\big)^n - \mathrm{e}^{-itC}\| = O(n^{-1})\,, \tag{13.2.15}$$

as $n \to \infty$, *locally-uniformly in* $t \in \mathbb{R}$.

Theorem 13.2.4 *Let operator* C *be as in Theorem* 13.2.3. *If conditions* ($d1$)–($d3$) *hold and if* E_A *and* E_B *admit bounded extensions on* $\mathfrak{H}$, *then*

$$\|\big(\mathrm{e}^{-i(t/n)B/2}\mathrm{e}^{-i(t/n)A}\mathrm{e}^{-i(t/n)B/2}\big)^n - \mathrm{e}^{-itC}\| = O(n^{-2}), \tag{13.2.16}$$

as $n \to \infty$, *locally-uniformly in* $t \in \mathbb{R}$.

It is worth noting here that in contrast to Theorem 13.2.3 (13.2.15), the convergence rate $O(n^{-2})$ in (13.2.16) for symmetrised product formula becomes *optimal* (cf. the *Lie* product formulæ for matrices (or bounded operators) in Proposition 1.10.10(b)), due to the supplementary condition (d3).

Theorem 13.2.5 *Let C be again as in* 13.2.3. *If conditions* (d1)–(d4) *hold and if E, as well as, E_A and E_B, respectively, admit bounded extensions on* $\mathfrak{H}$, *then*

$$\|(e^{-i(t/n)A}e^{-i(t/n)B})^n - e^{-itC} - i(t/2n)[e^{-itC}, A]\| = O(n^{-2}), \tag{13.2.17}$$

$$\|(e^{-i(t/n)B}e^{-i(t/n)A})^n - e^{-itC} - i(t/2n)[e^{-itC}, B]\| = O(n^{-2}), \tag{13.2.18}$$

and

$$\left\|\frac{1}{2}\left\{(e^{-i(t/n)A}e^{-i(t/n)B})^n + (e^{-i(t/n)B}e^{-i(t/n)A})^n\right\} - e^{-itC}\right\| = O(n^{-2}), \tag{13.2.19}$$

as $n \to \infty$, locally-uniformly in $\mathbb{R}$.

If conditions of these theorems are not satisfied, the operator-norm convergence of the Trotter product formulæ for strongly continuous unitary groups does not *a priory* hold. In Sect. 13.2.1, we collected examples where a unitary product formula holds in the operator-norm topology as well as when it does not hold. The question of optimality of the error bound is related to our discussion in Sect. 7.2 for operator-norm convergence of the Trotter product formula for self-adjoint strongly continuous *semigroups*.

We also comment that (13.2.17) and (13.2.18) in Theorem 13.2.5 show how large is a deviation of symmetric (13.2.16) from nonsymmetric approximants, which is expressed by terms with commutators. The asymptotic (13.2.19) manifests just another type of symmetrisation.

13.2.3 Proof of Theorems

For proving theorems of Sect. 13.2.2, we need some preparatory lemmata.

To this aim, we make use of the (formal) commutator relation

$$[e^{-itX}, Y] = i\int_0^t ds\, e^{-isX}[Y, X]\, e^{-i(t-s)X}. \tag{13.2.20}$$

and also define for $t \in \mathbb{R}$ one-parameter families

$$G(t) := e^{-itA}\, e^{-itB}, \tag{13.2.21}$$

$$K(t) := e^{-itB/2}\, e^{-itA}\, e^{-itB/2}. \tag{13.2.22}$$

Lemma 13.2.6 *Assume that conditions* (d1)–(d3) *hold. Let* $u,\ v\ \in\mathcal{D}$. *Then for* $t\in\mathbb{R}$:

(1) $\big((A\mathrm{e}^{-itB}-\mathrm{e}^{-itB}A)\,u,v\big)=-\int_0^t ds\,(E\,\mathrm{e}^{-i(t-s)B}u,\mathrm{e}^{isB}v)$,

(2) $\big((B\mathrm{e}^{-itA}-\mathrm{e}^{-itA}B)\,u,v\big)=\int_0^t ds\,(E\,\mathrm{e}^{-i(t-s)A}u,\mathrm{e}^{isA}v)$,

(3) $\big((E\mathrm{e}^{-itB}-\mathrm{e}^{-itB}E)\,u,v\big)=-\int_0^t ds\,(E_B\mathrm{e}^{-i(t-s)B}u,\mathrm{e}^{isB}v)$,

(4) $\big((E\mathrm{e}^{-itA}-\mathrm{e}^{-itA}E)\,u,v\big)=-\int_0^t ds\,(E_A\mathrm{e}^{-i(t-s)A}u,\mathrm{e}^{isA}v)$.

Proof We show only the relation (1). The others can be established similarly.
Put $f(t):=\big((A\mathrm{e}^{-itB}-\mathrm{e}^{-itB}A)\,u,v\big)$. Then,

$$f(t)=-(\mathrm{e}^{-isB}A\mathrm{e}^{-i(t-s)B}u,v)\Big|_{s=0}^{t}=-\int_0^t ds\,\partial_s(\mathrm{e}^{-isB}A\mathrm{e}^{-i(t-s)B}u,v)$$
$$=-\int_0^t ds\,i\,\Big[(B\,\mathrm{e}^{-i(t-s)B}u,\ A\,\mathrm{e}^{isB}v)-(A\,\mathrm{e}^{-i(t-s)B}u,\ B\,\mathrm{e}^{isB}v)\Big].$$

Since by $(d1)$, both $\mathrm{e}^{-i(t-s)B}u$ and $\mathrm{e}^{isB}v$ are in $\mathcal{D}$, we obtain

$$f(t)=-\int_0^t ds\,(E\,\mathrm{e}^{-i(t-s)B}u,\mathrm{e}^{isB}v),\quad u,\ v\ \in\mathcal{D},$$

which, by $(d2)$, showing the desired relation. □

Lemma 13.2.7 *On account of the same hypothesis as in Theorem* 13.2.4, *one has*

$$\|\mathrm{e}^{-itC}-G(t)\|=O(t^2),\quad t\to 0. \tag{13.2.23}$$

Proof Put $g(t)=(G(t)u,v)$ for arbitrary $u,\ v\ \in\mathcal{D}$, see (13.2.21). Then, on account of the *Stone* theorem (Proposition 13.1.4), we obtain for derivative $\partial_t g(t)$:

$$\begin{aligned}(\partial_t G(t)u,v)&=(-i\,AG(t)u,v)+(\mathrm{e}^{-itA}(-i\,B)\mathrm{e}^{-itB}u,v)\\&=(-i\,CG(t)u,v)+(-i\,[\mathrm{e}^{-itA},B]\mathrm{e}^{-itB}u,v).\end{aligned}\tag{13.2.24}$$

Since $u,\ v\ \in\mathcal{D}$, on account of Lemma 13.2.6 (2) and condition that operator E admits a bounded extension on $\mathfrak{H}$, the second term in the last member of (13.2.24) can be estimated from above by $|t|\ \|E\|\ \|u\|\ \|v\|$. Then integrating (with help of the *Duhamel principle*) inhomogeneous equation (13.2.24), we figure out that by the density of $\mathcal{D}\subset\mathfrak{H}$ the relation:

$$\mathrm{e}^{-itC}-G(t)=i\int_0^t ds\,\mathrm{e}^{-i\,(t-s)C}[\mathrm{e}^{-isA},B]\,\mathrm{e}^{-isB},\quad t\in\mathbb{R},$$

holds on the space $\mathfrak{H}$ and yields a desired asymptotic (13.2.23) for the operator-norm bound of the left-hand side. □

Lemma 13.2.8 *Owing to the same hypothesis as in Theorem* 13.2.4, *one gets asymptotic*

$$\|e^{-itC} - K(t)\| = O(|t|^3), \quad t \to 0. \tag{13.2.25}$$

Proof Put $k(t) = (K(t)\,u, v)$ for arbitrary $u, v \in \mathcal{D}$, see (13.2.22), and calculate $\partial_t k(t)$. Because of (13.2.22) one gets decomposition $\partial_t k(t) = k_0(t) + k_1(t) + k_2(t)$, where for $t \in \mathbb{R}$ we deduce by the *Stone* theorem the identities:

$$\begin{aligned}
k_0(t) &= \big(-i\,(B/2)\,K(t)u, v\big) = \big(-i\,e^{-itA}\,e^{-it(B/2)}u, e^{it(B/2)}\,(B/2)v\big),\\
k_1(t) &= \big(e^{-it(B/2)}(-i\,A)\,e^{-itA}\,e^{-it(B/2)}u, v\big)\\
&= \big(-i\,Ae^{-itA}\,e^{-it(B/2)}u, e^{it(B/2)}v\big),\\
k_2(t) &= \big(e^{-it(B/2)}\,e^{-itA}(-i\,B/2)\,e^{-it(B/2)}u, v\big)\\
&= \big(-i\,(B/2)\,e^{-it(B/2)}u, e^{itA}\,e^{it(B/2)}v\big).
\end{aligned}$$

Now applying Lemma 13.2.6 (1) to

$$k_1(t) = -i\,\big(e^{-it(B/2)}A\,e^{-itA}e^{-it(B/2)}u, v\big)\,,$$

we obtain $k_1(t) = \big(-i\,A\,K(t)u, v\big) + r_1(t)$, where

$$\begin{aligned}
r_1(t) &= -\frac{i}{2}\int_0^t ds\,\big(e^{-is(B/2)}\,E\,e^{-i(t-s)(B/2)}(e^{-itA}\,e^{-it(B/2)}u),\ v\big)\\
&= -\frac{i}{2}\int_0^t ds\,\big(E\,e^{-is(B/2)}(e^{-itA}e^{-it(B/2)}u),\ e^{i(t-s)(B/2)}v\big)\,,
\end{aligned}$$

and the last line by making a change of variable.

Similarly we apply Lemma 13.2.6 (2) to

$$k_2(t) = -\frac{i}{2}\,\big(e^{-itA}\,B\,(e^{-it(B/2)}u), e^{it(B/2)}v\big)\,.$$

This yields $k_2(t) = \big(-i\,(B/2)\,K(t)u, v\big) + r_2(t)$, where

$$r_2(t) = \frac{i}{2}\int_0^t ds\,\big(e^{-isA}\,E\,(e^{-i(t-s)A}\,e^{-it(B/2)}u), e^{it(B/2)}v\big).$$

Next we apply Lemma 13.2.6 (3), (4) to study $r_1(t)$ and $r_2(t)$. Then,

$$r_1(t) = -r_0(t) + r_{10}(t), \quad r_2(t) = r_0(t) + r_{20}(t),$$

where $r_0(t) = (i/2)\, t\, \big(E\, \mathrm{e}^{-itA}\, \mathrm{e}^{-it(B/2)}u, \mathrm{e}^{it(B/2)}v\big)$ and

$$r_{10}(t) = (R_{10}(t)u, v), \quad r_{20}(t) = (R_{20}(t)u, v)\,,$$

for $t \in \mathbb{R}$

$$R_{10}(t) = -\frac{i}{4}\int_0^t ds \int_0^s d\sigma\, \mathrm{e}^{-i\,(t-\sigma)(B/2)}\, E_B\, \mathrm{e}^{-i\,\sigma(B/2)}\, \mathrm{e}^{-itA}\, \mathrm{e}^{-it(B/2)},$$

$$R_{20}(t) = -\frac{i}{2}\int_0^t ds \int_0^s d\sigma\, \mathrm{e}^{-it(B/2)}\, \mathrm{e}^{-i\sigma A}\, E_A\, \mathrm{e}^{-i\,(t-\sigma)A}\, \mathrm{e}^{-it(B/2)}\,.$$

By assumptions of lemma (cf. Theorem 13.2.4) operators E_A and E_B admit bounded extensions on $\mathfrak{H}$. As a consequence, operators $R_{10}(t)$ and $R_{20}(t)$ are bounded, and operator norms have asymptotic

$$\|R_{10}(t)\| + \|R_{20}(t)\| = O(t^2), \quad t \to 0. \tag{13.2.26}$$

Now let $u(t) = K(t)\,u$ for $t \in \mathbb{R}$ and $u \in \mathcal{D}$. Then, by assumption (d1), the function: $t \mapsto u(t) \in \mathcal{D}$, satisfies equation

$$(\partial_t u(t), v) = (-i\, C\, u(t), v) + \big((R_{10}(t) + R_{20}(t))\, u, v\big)\,, \quad u, v \in \mathcal{D}\,. \tag{13.2.27}$$

Note that u and v belong to a dense set $\mathcal{D} \subset \mathfrak{H}$ and $R_{10}(t) + R_{20}(t)$ is a family of bounded operators such that there is asymptotic (13.2.26). Then by the *Duhamel principle* for inhomogeneous equation (13.2.27) (cf. Notes to Chap. 12 (Sect. 12.2.2)) we obtain representation

$$\mathrm{e}^{-itH} - K(t) = -\int_0^t ds\, \mathrm{e}^{-i\,(t-s)C}\,(R_{10}(s) + R_{20}(s))\,, \quad t \in \mathbb{R}\,,$$

which holds on the space $\mathfrak{H}$ and yields desired asymptotic (13.2.25) for the operator-norm bound of the left-hand side. □

The proof of Theorem 13.2.5 needs the following lemma.

Lemma 13.2.9 *Assume that conditions* (d1) *and* (d2) *are fulfilled. If operator* E *admits a bounded extension on* $\mathfrak{H}$*, then*

$$\|G(t) - K(t)\| = O(t^2)\,, \quad t \to 0. \tag{13.2.28}$$

Proof By definitions (13.2.21) and (13.2.22), we obtain for $t \in \mathbb{R}$

$$G(t) = K(t) + [\mathrm{e}^{-itA},\, \mathrm{e}^{-it(B/2)}]\, \mathrm{e}^{-it(B/2)}\,. \tag{13.2.29}$$

Then, by formula (13.2.20) and by Lemma 13.2.6 (1), one gets

$$\begin{aligned}[\mathrm{e}^{-itA},\, \mathrm{e}^{-it(B/2)}] &= \mathrm{e}^{-isA}\, \mathrm{e}^{-it(B/2)}\, \mathrm{e}^{-i(t-s)A}\Big|_{s=0}^{t} \\ &= i \int_0^t ds\, \mathrm{e}^{-isA} [\mathrm{e}^{-it(B/2)},\, A]\, \mathrm{e}^{-i(t-s)A}\,,\end{aligned} \tag{13.2.30}$$

and for $t \in \mathbb{R}$

$$[\mathrm{e}^{-it(B/2)},\, A] = \frac{1}{2} \int_0^t ds\, \mathrm{e}^{-is(B/2)}\, E\, \mathrm{e}^{-i(t-s)(B/2)}\,.$$

By assumption, operator E admits a bounded extension, and hence, it follows that

$$\left\| [\mathrm{e}^{-it(B/2)},\, A] \right\| = O(|t|)\,, \quad t \to 0\,.$$

This asymptotic together with (13.2.29) and (13.2.30) proves (13.2.28). □

Now we are in position for proving the promised theorems, cf. Sect. 13.2.2.

***Proof** (Theorem 13.2.3)* We show (13.2.14). The proof of (13.2.15) follows, *mutatis mutandis*, the arguments for the asymptotic (13.2.14).

Let $\tau = t/n$ and $D_1(\tau) := \mathrm{e}^{-i\tau C} - G(\tau)$ for $t \in \mathbb{R}$, $n \in \mathbb{N}$. Then we get the *telescopic* representation

$$\mathrm{e}^{-itC} - G(\tau)^n = \sum_{k=1}^{n} X_k(\tau)\,, \tag{13.2.31}$$

where

$$X_k(\tau) := \mathrm{e}^{-i(k-1)\tau C} D_1(\tau)\, G(\tau)^{n-k}, \quad 1 \le k \le n\,.$$

Given that by Lemma 13.2.7): $\|X_k(\tau)\| \le \|D_1(\tau)\| \le O(\tau^2)$, for every $k \ge 1$, with help of representation (13.2.31) we obtain $\|\mathrm{e}^{-itC} - G(\tau)^n\| = O(|\tau|)$, showing the asymptotic (13.2.14). One can show (13.2.15) similarly, noting that Lemma 13.2.7) and Lemma 13.2.9) yield: $\|\mathrm{e}^{-i\tau C} - K(\tau)\| = O(\tau^2)$, and then using the *telescopic* representation for $\mathrm{e}^{-itC} - K(\tau)^n$. This proves the theorem. □

Proof (Theorem 13.2.4) Let $\tau = t/n$ and $D_2(\tau) := \mathrm{e}^{-i\tau C} - K(\tau)$ for $t \in \mathbb{R}$, $n \in \mathbb{N}$. Then, we use the *telescopic* representation

$$\mathrm{e}^{-itC} - K(\tau)^n = \sum_{k=1}^{n} Y_k(\tau),$$

where

$$Y_k(\tau) := \mathrm{e}^{-i(k-1)\tau C} D_2(\tau)\, K(\tau)^{n-k}, \quad 1 \le k \le n.$$

Since by Lemma 13.2.8): $\|Y_k(\tau)\| \le \|D_2(\tau)\| \le O(|\tau|^3)$, for every $k \ge 1$, with help of the telescopic representation we obtain $\|\mathrm{e}^{-itC} - K(\tau)^n\| = O(\tau^2)$, showing (13.2.16). This proves the theorem. □

Proof (Theorem 13.2.5) Let $\tau = t/n$ for $t \in \mathbb{R}$, $n \in \mathbb{N}$. Recall that for $\rho > 0$ we denote by $O_p(n^{-\rho})$ any operator from the *class* of equivalence of *bounded* operators $\mathcal{L}(\mathfrak{H})$ whose norms obey the asymptotic bound $O(n^{-\rho})$ as $n \to \infty$, cf. Sect. 7.2.

By virtue of assumptions (d1) and (d4), there exists on $\mathcal{D}$ derivative $\partial_s(\mathrm{e}^{-isC} B\, \mathrm{e}^{isC})$, which by integration over interval $[0, t]$ yields:

$$[B, \mathrm{e}^{-itC}] = i\int_0^t ds\, \mathrm{e}^{-isC}\,[C, B]\,\mathrm{e}^{-i(t-s)C} = i\int_0^t ds\, \mathrm{e}^{-isC} E\, \mathrm{e}^{-i(t-s)C}, \tag{13.2.32}$$

here we used (d2). Since operator E admits a bounded extension on $\mathfrak{H}$, we conclude that commutator (13.2.32) is a bounded operator with the norm of order $O(t)$.

Similarly, by integration over interval $[0, \tau = t/n]$ derivative $\partial_\sigma(\mathrm{e}^{-i\sigma B}\, \mathrm{e}^{-itC}\, \mathrm{e}^{i\sigma B})$ and taking into account (13.2.32) we also estimate commutator:

$$[\mathrm{e}^{-itC}, \mathrm{e}^{-i\tau B}] = i\int_0^\tau d\sigma\, \mathrm{e}^{-i\sigma B}\,[B, \mathrm{e}^{-itC}]\,\mathrm{e}^{-i(\tau-\sigma)B} = O_p(n^{-1}). \tag{13.2.33}$$

Now we use Theorem 13.2.4 and Lemma 13.2.7 to deduce representation:

$$\begin{aligned} G(t/n)^n &= \mathrm{e}^{-i\tau A}\, \mathrm{e}^{-i\tau(B/2)} K(\tau)^{n-1} \mathrm{e}^{-i\tau(B/2)} \\ &= \mathrm{e}^{-i\tau A} \mathrm{e}^{-i\tau(B/2)} \mathrm{e}^{-i(n-1)\tau C} \mathrm{e}^{-i\tau(B/2)} + O_p(n^{-2}) \\ &= \mathrm{e}^{-itC} + \mathrm{e}^{-i\tau A} \mathrm{e}^{-i\tau(B/2)}\,[\mathrm{e}^{-i(n-1)\tau C}, \mathrm{e}^{-i\tau(B/2)}] + O_p(n^{-2}). \end{aligned}$$

To calculate commutator $F := [\mathrm{e}^{-i\,(n-1)\tau C},\ \mathrm{e}^{-i\,\tau(B/2)}]$ in the last member of this representation we apply (13.2.33). Then,

$$F = \frac{i}{2}\int_0^\tau d\sigma\ \mathrm{e}^{-i\,\sigma(B/2)}[B, \mathrm{e}^{-i\,(n-1)\tau C}]\,\mathrm{e}^{-i\,(\tau-\sigma)\,(B/2)}$$
$$= \frac{i\tau}{2}\,\mathrm{e}^{-i\,\tau(B/2)}\,[B, \mathrm{e}^{-i\,(n-1)\tau C}] + O_p(n^{-2})\,.$$

As a consequence, on account of Lemmata 13.2.8, 13.2.9 and because of estimate $\|B,\ [\mathrm{e}^{-i\,\tau H}]\,\| = O(n^{-1})$ (13.2.32) one gets

$$\begin{aligned}
G(t/n)^n &= \mathrm{e}^{-i\,t\,C} + (i\tau/2)\mathrm{e}^{-i\,\tau A}\mathrm{e}^{-i\,\tau B}[B,\ \mathrm{e}^{-i\,(n-1)\tau H}] + O_p(n^{-2})\\
&= \mathrm{e}^{-i\,t\,C} + (i\tau/2)K(\tau)[B,\ \mathrm{e}^{-i\,(n-1)\tau H}] + O_p(n^{-2})\\
&= \mathrm{e}^{-i\,t\,C} + (i\tau/2)\mathrm{e}^{-i\,\tau H}[B,\ \mathrm{e}^{-i\,(n-1)\tau H}] + O_p(n^{-2})\\
&= \mathrm{e}^{-i\,t\,C} + (i\tau/2)\mathrm{e}^{-i\,\tau H}\big(B\mathrm{e}^{-i\,(n-1)\tau H} - \mathrm{e}^{-i\,(n-1)\tau H}B\big) + O_p(n^{-2})\\
&= \mathrm{e}^{-i\,t\,C} + (i\tau/2)\big(\mathrm{e}^{-i\,\tau H}B\mathrm{e}^{-i\,(n-1)\tau H} - \mathrm{e}^{-i\,t\,C}B\big) + O_p(n^{-2})\\
&= \mathrm{e}^{-i\,t\,C} + (i\tau/2)[B,\ \mathrm{e}^{-i\,t\,C}] + O_p(n^{-2})\\
&= \mathrm{e}^{-i\,t\,C} + (i\tau/2)[\mathrm{e}^{-i\,t\,C},\ A] + O_p(n^{-2})\,.
\end{aligned}$$

We remind that here $O_p(n^{-\rho})$, $\rho > 0$, denotes the *class* of equivalence of bounded operators $\mathcal{L}(\mathfrak{H})$ whose norms obey the asymptotic bound $O(n^{-\rho})$ as $n \to \infty$, cf. Sect. 7.2. So, this yields a desired relation (13.2.17).

To check (13.2.18), we note that

$$\big(\mathrm{e}^{-i\,(t/n)B}\mathrm{e}^{-i\,(t/n)A}\big)^n = (G(-t/n)^n)^*\,, \quad t \in \mathbb{R}\,.$$

Then, on account of (13.2.17), we deduce

$$\|\big(\mathrm{e}^{-i\,(t/n)B}\mathrm{e}^{-i\,(t/n)A}\big)^n - \mathrm{e}^{-i\,tC} - i\,(t/2n)[A, \mathrm{e}^{-i\,tC}]\| = O(n^{-2})\,,$$

and from equation: $[\mathrm{e}^{-i\,tC},\ A] + [\mathrm{e}^{-i\,tC},\ B] = 0$, the relation (13.2.18).

Finally, thanks to the last equation for commutators, the relation (13.2.19) follows from (13.2.17), (13.2.18) for $t \in \mathbb{R}$:

$$\begin{aligned}
&\Big\|\big(\mathrm{e}^{-i\,(t/n)A}\,\mathrm{e}^{-i\,(t/n)B}\big)^n + \big(\mathrm{e}^{-i(t/n)B}\,\mathrm{e}^{-i(t/n)A}\big)^n - 2\,\mathrm{e}^{-i\,tC}\Big\| \le\\
&\Big\|\big(\mathrm{e}^{-i\,(t/n)A}\mathrm{e}^{-i\,(t/n)B}\big)^n - \mathrm{e}^{-i\,tC} - i\,(t/2n)[\mathrm{e}^{-i\,tC},\ A]\Big\| +\\
&\Big\|\big(\mathrm{e}^{-i\,(t/n)B}\mathrm{e}^{-i\,(t/n)A}\big)^n - \mathrm{e}^{-i\,tC} - i\,(t/2n)[\mathrm{e}^{-i\,tC},\ B]\Big\| = O(n^{-2})\,.
\end{aligned} \tag{13.2.34}$$

That finishes the proof of the theorem. □

As a concluding remark, we want to comment that these theorems actually hold if there exists a *dense* subspace $\mathcal{D}$ for which all equations in the proofs above have sense. Note that assumptions: $(d1)$–$(d3)$ and/or $(d4)$, together with condition that operators E, E_A, E_B, defined on $\mathcal{D}$, admit a *bounded extensions* on a Hilbert space $\mathfrak{H}$, are in fact sufficient for this.

13.2.4 Examples and Counterexamples

Similarly to Sect. 13.2.1, we discern here via new examples that in general the unitary Trotter product formula in the operator-norm topology does not hold. In this subsection, we extend the list of examples to some nontrivial special cases of the Dirac operator and the relativistic Schrödinger operator. To this aim, we apply below our Theorems 13.2.3–13.2.5.

We start by example when unitary Trotter product formula holds in the operator-norm topology, for the case of a Dirac operator. Let A be a self-adjoint $k \times k$-system of differential operators of the first order in the Hilbert space $L^2(\mathbb{R}^d)^k$ and B be a multiplication operator defined by a real-valued C^1 (or C^2 smooth) function $V(x)$, which has bounded derivatives $\partial_x^\alpha V(x)$ for $|\alpha| = 1$ (or $|\alpha| = 2$). In particular, this set-up holds for the Dirac operator $A = D_0 := -i\sum_{j=1}^3 \alpha_j \partial_{x_j} + m\beta$, with mass $m \geq 0$ for $k = 4$, in $L^2(\mathbb{R}^3)^4$, and $B = V(x)$ a real-valued C^1 or C^2 smooth function. Here $\{\alpha_j\}_{j=1}^3$ and β are Dirac's the 4×4-matrices. By C we denote a *unique* self-adjoint extension of $A + B = D_0 + V(x)$.

Proposition 13.2.10 (Dirac Operator in $\mathfrak{H} = L^2(\mathbb{R}^3)^4$)

(i) *Assume that $V(x)$ is a C^1-smooth function with bounded derivatives $\partial_x V(x)$. Then, as $n \to \infty$, one has for $t \in \mathbb{R}$*

$$\|(\mathrm{e}^{-itV/n}\mathrm{e}^{-itD_0/n})^n - \mathrm{e}^{-itC}\| = O(n^{-1}), \tag{13.2.35}$$

$$\|(\mathrm{e}^{-itD_0/n}\mathrm{e}^{-itV/n})^n - \mathrm{e}^{-itC}\| = O(n^{-1}). \tag{13.2.36}$$

(ii) *Assume that $V(x)$ is a real-valued C^2-smooth function with bounded derivatives $\partial_x^\alpha V(x)$ for $|\alpha| = 2$. Then, as $n \to \infty$, one gets for $t \in \mathbb{R}$*

$$\|(\mathrm{e}^{-itV/2n}\mathrm{e}^{-itD_0/n}\mathrm{e}^{-itV/2n})^n - \mathrm{e}^{-itC}\| = O(n^{-2}). \tag{13.2.37}$$

(iii) *Assume that $V(x)$ is a real-valued C^2-smooth function with bounded derivatives $\partial_x^\alpha V(x)$ for $1 \leq |\alpha| \leq 2$. Then, as $n \to \infty$, one has for $t \in \mathbb{R}$*

$$\|\big(\mathrm{e}^{-i(t/n)D_0}\mathrm{e}^{-i(t/n)V}\big)^n - \mathrm{e}^{-itC} - i(t/2n)[\mathrm{e}^{-itC}, D_0]\| = O(n^{-2}), \tag{13.2.38}$$

$$\|\big(\mathrm{e}^{-i(t/n)V}\mathrm{e}^{-i(t/n)D_0}\big)^n - \mathrm{e}^{-itC} - i(t/2n)[\mathrm{e}^{-itC}, V]\| = O(n^{-2}), \tag{13.2.39}$$

$$\|\frac{1}{2}[\big(\mathrm{e}^{-i(t/n)D_0}\mathrm{e}^{-i(t/n)V}\big)^n + \big(\mathrm{e}^{-i(t/n)V}\mathrm{e}^{-i(t/n)D_0}\big)^n] - \mathrm{e}^{-itC}\| = O(n^{-2}). \tag{13.2.40}$$

In formulæ (13.2.35)–(13.2.40) *the convergence is locally uniform in* $\mathbb{R}$.

Proof Let $\mathcal{D} := \{u \in L^2(\mathbb{R}^3)^4;\ \langle x\rangle^\gamma u \in W^{1,2}(\mathbb{R}^3)^4$ for every $\gamma \geq 0\}$, where $\langle x\rangle := (1+x^2)^{1/2}$. This yields assumption $(d1)$ and ensures that operators E, E_A and E_B act on $\mathcal{D}$ as commutators $i\,[D_0, V]$, $-[D_0, [D_0, V]]$ and $-[V, [D_0, V]]$, cf. Remark 13.2.2. It is clear because of commutators $[D_0, V]$ are basically derivatives $\partial_x V(x)$ and double-commutators $[D_0, [D_0, V]]$ are essentially the derivatives $\partial_x^\alpha V(x)$ for $|\alpha| = 2$, whereas $[V, [D_0, V]] = 0$. Then we also see that $(d4)$ is satisfied. As a consequence, Proposition 13.2.10 follows from Theorems 13.2.3-13.2.5. □

We note that besides Example (b) (Sect. 13.2.1) with optimal rate of convergence the product formula, there is a peculiar case of the *Stark* Hamiltonian $H = A + B := -\Delta + x_1$ in $L^2(\mathbb{R}^d)$ with $x = (x_1, \ldots, x_d) \in \mathbb{R}^d$, where Theorem 13.2.4 holds because the double commutators $[A, [A, B]]$ and $[B, [A, B]]$ are bounded in this case. On the other hand, the unitary Trotter product formula in operator norm will *not* in general hold for the Schrödinger operator $-\Delta + V(x)$ with nontrivial potential $V(x)$, since it does not satisfy the conditions of the theorems. However, *it does hold* for the relativistic Schrödinger operator $H = A + B := H_0 + V \equiv \sqrt{-\Delta + m^2} + V(x)$ in $L^2(\mathbb{R}^d)$ with some real-valued potentials $V(x)$. It is known that H is essentially self-adjoint on $C_0^\infty(\mathbb{R}^d)$ if the potential $V(x)$ is a function in $L^2_{loc}(\mathbb{R}^d)$ satisfying $V(x) \geq -c_1\mathrm{e}^{c_2|x|}$ with c_1, c_2 being non-negative constants, in particular, under the assumption of the following proposition. Denote the unique self-adjoint extension of $H_0 + V$ by the same H.

Proposition 13.2.11 (Massive (Respectively Massless) Relativistic Schrödinger Operator in $\mathfrak{H} = L^2(\mathbb{R}^d)$)

(i) *Assume in case $m > 0$ (respectively $m = 0$) that $V(x)$ is a real-valued C^2- smooth function with bounded derivatives $\partial_x^\alpha V(x)$ for $1 \leq |\alpha| \leq 2$ (respectively $0 \leq |\alpha| \leq 2$). Then, as $n \to \infty$, one has for $t \in \mathbb{R}$*

$$\|(\mathrm{e}^{-itC_0/n}\mathrm{e}^{-itV/n})^n - \mathrm{e}^{-itH}\| = O(n^{-1}), \tag{13.2.41}$$

$$\|(\mathrm{e}^{-itV/n}\mathrm{e}^{-itC_0/n})^n - \mathrm{e}^{-itH}\| = O(n^{-1}). \tag{13.2.42}$$

(ii) *Assume in case* $m > 0$ *(resp.* $m = 0$*) and that* $V(x)$ *is a real-valued* C^4*-smooth function with bounded derivatives* $\partial_x^\alpha V(x)$ *for* $1 \le |\alpha| \le 4$ *(resp.* $0 \le |\alpha| \le 4$*). Then, as* $n \to \infty$*, one gets*

$$\|(e^{-itV/2n}e^{-itH_0/n}e^{-itV/2n})^n - e^{-itH}\| = O(n^{-2}), \tag{13.2.43}$$

$$\|\big(e^{-i(t/n)H_0}e^{-i(t/n)V}\big)^n - e^{-itH} - i(t/2n)[e^{-itH}, H_0]\| = O(n^{-2}), \tag{13.2.44}$$

$$\|\big(e^{-i(t/n)V}e^{-i(t/n)H_0}\big)^n - e^{-itH} - i(t/2n)[e^{-itH}, V]\| = O(n^{-2}), \tag{13.2.45}$$

$$\|\frac{1}{2}\big[\big(e^{-i(t/n)H_0}e^{-i(t/n)V}\big)^n + \big(e^{-i(t/n)V}e^{-i(t/n)H_0}\big)^n\big] - e^{-itH}\| = O(n^{-2}). \tag{13.2.46}$$

In formulæ (13.2.43)–(13.2.46)*, the convergence is locally uniform in* $t \in \mathbb{R}$*.*

The number of assumptions that we impose on the potential $V(x)$ in Proposition 13.2.11 is as minimal as possible. For the proof of this proposition, we shall not consider $H_0 = \sqrt{-\Delta + m^2}$ as the pseudo-differential operator, since its symbol $\sqrt{\xi^2 + m^2}$ is not smooth in the massless case ($m = 0$), so that standard pseudo-differential calculus does not allow us to treat the issue well even if $V(x)$ is C^∞-smooth. Instead, we use for H_0 the following expression as an integral operator with kernel $n(\cdot)$:

$$\begin{aligned}(H_0 u)(x) :&= m\,u(x) - \int_{|y|>0} dy\,[u(x+y) - u(x) - I_{\{|y|<1\}}\, y \cdot \partial_x u(x)]\, n(y) \\ &= m\,u(x) - \lim_{\varepsilon \to +0} \int_{|y| \ge \varepsilon} dy\,[u(x+y) - u(x)]\, n(y)\,, \end{aligned} \tag{13.2.47}$$

for $u \in \mathcal{S}(\mathbb{R}^d)$. Here $I_{\{|y|<1\}}$ is the indicator function of the set $\{|y| < 1\}$, i.e. $I_{\{|y|<1\}}(z) = 1$, if $|z| < 1$, and $= 0$, if $|z| \ge 1$. The integral kernel $n(y)$ is a positive function dependent on $m \ge 0$ and given by

$$n(y) = 2\,(2\pi)^{-(d+1)/2}\, m^{(d+1)/2}\, |y|^{-(d+1)/2}\, K_{(d+1)/2}(m|y|), \quad m > 0,$$

$$n(y) = \pi^{-(d+1)}\, \Gamma\Big(\frac{d+1}{2}\Big)\, |y|^{-(d+1)}, \quad m = 0,$$

where $K_\nu(z)$ is the modified Bessel function of the third kind of order ν and $\Gamma(z)$ the gamma function. It holds that for $m \ge 0$,

$$n_{0,\kappa} := \int_{0<|y|<1} dy\, |y|^\kappa\, n(y) < \infty, \quad 1 < \kappa \le 2, \tag{13.2.48}$$

and for $m > 0$ and $m = 0$, respectively,

$$n_{+;\infty,\kappa} := \int_{|y|\geq 1} dy\, |y|^{\kappa}\, n(y) < \infty, \quad \kappa \geq 0, \quad m > 0,$$

$$n_{0;\infty,0} := \int_{|y|\geq 1} dy\, n(y) < \infty\,, \qquad m = 0. \tag{13.2.49}$$

Further, $\{y_j n(y)\}_{j=1}^{d}$ is a *Calderón–Zygmund* kernel. So, if we define a linear operator K_j by the singular integral

$$(K_j g)(x) = \lim_{\varepsilon\to+0} \int_{\varepsilon\leq|y|<1} dy\, g(x+y)\, y_j\, n(y)\,, \tag{13.2.50}$$

for $g \in \mathcal{S}(\mathbb{R}^d)$, then by the *Calderón–Zygmund* theorem, the limit of the integral on the right of (3.17) exists, and K_j becomes a bounded operator on $L^2(\mathbb{R}^d)$, i.e.,

$$\|K_j g\| \leq C\|g\|, \quad g \in \mathcal{S}(\mathbb{R}^d), \quad 1 \leq j \leq d, \tag{13.2.51}$$

with a constant $C > 0$, where $\|\cdot\|$ on both sides is the L^2 norm. When $d = 1$, we simply write K for K_1.

Therefore and thereby, H_0 (13.2.47) becomes a self-adjoint operator in $L^2(\mathbb{R}^d)$ with domain $W^{1,2}(\mathbb{R}^d)$.

Proof of Proposition 13.2.11 We take $\mathcal{D} = \{u \in L^2(\mathbb{R}^d);\ \langle x\rangle^2 u \in W^{2,2}(\mathbb{R}^d)\}$ for massive case, and $\mathcal{D} = W^{2,2}(\mathbb{R}^d)$ for massless case. Then, assumption $(d1)$ is satisfied, and operators E, E_A and E_B act on $\mathcal{D}$ as commutators, cf. Remark 13.2.2. One can prove (cf. Lemma 13.2.12 below) that under assumptions of the proposition operators: $E = i[H_0, V]$, $E_A = -[V, [H_0, V]]$ and $E_B = -[H_0, [H_0, V]]$, are bounded, and also establish that condition $(d4)$ is satisfied. As a consequence, Proposition 13.2.11 is steaming from Theorems 13.2.3, 13.2.4 and 13.2.5. □

Before going to the lemma, let us notice a difference in assumptions on the potential $V(x)$ in Proposition 13.2.11 between the *massive* and *massless* cases. In fact, in the former case, the function $V(x)$ itself is assumed to be bounded, but not in the latter. (For this, see also the bounds obtained in Lemma 13.2.12 below for three commutators.) To this end, note that in the proof the dense subspace $\mathcal{D}$ in the *massive* case is taken smaller than in the *massless* case.

Lemma 13.2.12 *Let for the both massive and massless cases $V(x)$ satisfy the same assumption as in Proposition* 13.2.11. *Then, commutators:*

$$[H_0, V],\ [V, [H_0, V]],\ [H_0, [H_0, V]] : \mathcal{S}(\mathbb{R}^d) \to L^2(\mathbb{R}^d)\,,$$

admit bounded extensions on $L^2(\mathbb{R}^d)$.

For comments on this lemma, we refer Notes to Chap. 13 (Notes to Sects. 13.2, 13.2.4).

13.3 Unitary Trotter–Kato Product formulæ in L^2-Topology

13.3.1 Motivations and Preliminaries

Seeing the difficulties with the proof of the Trotter product formula (Sect. 13.2), which is *operator-norm* convergent to a unitary group, we scrutinise in this section a different set-up, that implies a *new*, the L^2-topology of convergence.

In particular, it allows to solve the problem of substitution of the essential self-adjointness of $A + B$ in Proposition 13.1.11 by a less restrictive assumption that $A \dot{+} B$ is a *form-sum* of non-negative self-adjoint operators. We shall prove in Sect. 13.3.2 the unitary limit of the Trotter (i.e., exponential) product formula in a *weaker* than the strong operator (13.1.33), the L^2-topology. Then, owing to a *complex deformation* of exponential products, one can extend the convergence in the L^2-topology to a certain class of *non-exponential holomorphic* Kato functions, Sect. 13.3.3. This complex extension allows also to consider a concept of the *admissible* Kato functions and to generalise it for the *Ichinose theory* that establishes unitary limits of non-exponential product formulæ in the *strong* operator topology, see Sect. 13.3.4.

For motivation, we note that it is a longstanding *open* question, whether the unitary Trotter product formula

$$\text{s-}\lim_{n\to\infty}\left(\mathrm{e}^{-itA/n}\mathrm{e}^{-itB/n}\right)^n = \mathrm{e}^{-itC}\,, \quad t \in \mathbb{R}\,, \tag{13.3.1}$$

remains valid for the same assumptions on generators A, B and C as that which justify the Kato product formula for C_0-semigroups in Theorem 5.1.5, see Notes to Chap. 13 (Notes to Sects. 13.3, 13.3.1). For example, let A and B be two non-negative self-adjoint operators in the Hilbert space $\mathfrak{H}$. If domain $dom(A^{1/2}) \cap \mathrm{dom}\,(B^{1/2})$ is *dense* in $\mathfrak{H}$, then for C_0-semigroups the Kato product formula (5.1.15) (cf. (13.3.1) after substitution: $it \to t \geq 0$) converges in the strong operator topology locally-uniformly in $t \in \mathbb{R}_0^+$ for self-adjoint operator $C := A \dot{+} B$, which is the *form-sum* of A and B. Hence, the question, which follows is whether this might be true for (13.3.1)?

Remark 13.3.1

(a) We remind that if A and B are non-negative self-adjoint operators in $\mathfrak{H}$ and *if* the limit in the left-hand side of (13.3.1) *exists* for all $t \in \mathbb{R}$, then $\mathrm{dom}\,(A^{1/2}) \cap \mathrm{dom}\,(B^{1/2})$ is dense in $\mathfrak{H}$ and thus the form-sum $C = A \dot{+} B$ is a well-defined self-adjoint operator, see Notes to Chap. 13 (Notes to Sects. 13.3, 13.3.1).

(b) For that reason, in contrast to the Kato Theorem 5.1.5 and Remark 5.2.6(e), we assume in the following that $\mathrm{dom}\,(A^{1/2}) \cap \mathrm{dom}\,(B^{1/2})$ is *dense* in $\mathfrak{H}$. Furthermore, applying Trotter's result (Proposition 13.1.11) one infers that formula (13.3.1) is valid *if* the operator $C := \overline{A+B}$ is self-adjoint. However, if $A+B$ is *not* essentially self-adjoint, then all attempts to establish the unitary Trotter product formula (13.3.1) have *failed* so far.

A somewhat weaker result (we call it the *Lapidus* product formula) says that

$$\text{s-}\lim_{n\to\infty} \int_{\mathbb{R}} dt\, \varphi(t) \left(\mathrm{e}^{-itA/n}\, \mathrm{e}^{-itB/n}\right)^n = \int_{\mathbb{R}} dt\, \varphi(t)\, \mathrm{e}^{-itC}, \quad C := A \dot{+} B, \tag{13.3.2}$$

holds for all $\varphi \in L^1(\mathbb{R})$, see Sect. 13.3.2.

Another approach was proposed by *T. Ichinose*, see Notes to Chap. 13 (Notes to Sects. 13.3, 13.3.1). In this approach, he conserves convergence in the strong operator topology, but *modifies* the *Trotter* product formula in such a way:

$$\begin{aligned} \mathrm{e}^{-itC} = \text{s-}\lim_{n\to\infty} \Big\{ & \left(\mathrm{e}^{-itA/n} (E_A([0, n\delta/t]) + \mathrm{e}^{-ita/n} E_A((n\delta/t, \infty))\right) \times \\ & \left(\mathrm{e}^{-itB/n} (E_B([0, n\delta/t]) + \mathrm{e}^{-itb/n} E_B((n\delta/t, \infty))\right) \Big\}^n, \quad t \geq 0, \end{aligned} \tag{13.3.3}$$

where $E_A(\cdot)$ and $E_B(\cdot)$ denote, respectively, the spectral measures of the operators A and B, and $a \geq 0$, $b \geq 0$, $0 < \delta < \pi/2$. If one introduces the functions

$$f(\lambda) := \mathrm{e}^{-i\lambda} \chi_{[0,\delta]}(\lambda) + \chi_{(\delta,\infty)}(\lambda), \quad \lambda \geq 0, \tag{13.3.4}$$

then the result of the *Ichinose theory* for $a = b = 0$ acquires a form of the product formula:

$$\text{s-}\lim_{n\to\infty} (f(tA/n) f(tB/n))^n = \mathrm{e}^{-itC}, \quad t \geq 0. \tag{13.3.5}$$

Notice that the above function $f(\lambda)$ is an *admissible* Kato function, that is,

$$|f(x)| \leq 1, \quad x \in [0, \infty), \quad f(0) = 1, \quad \text{and} \quad f'(+0) = -i,$$

and satisfies in addition the conditions $\mathrm{Re}(f(x)) \geq 0$ and $\mathrm{Im}(f(x)) \leq 0$, $x \in \mathbb{R}^+$. See Definition 13.3.13 in Sect. 13.3.4, where this result is generalised.

In Sect. 13.3.2, we shall show that the Trotter product formula makes sense in the L^2-topology, defined by the limits:

$$\lim_{n\to\infty} \int_{-T}^{T} dt\, \left\| \left(\mathrm{e}^{-itA/n}\, \mathrm{e}^{-itB/n}\right)^n h - \mathrm{e}^{-itC} h \right\|^2 = 0, \tag{13.3.6}$$

for $h \in \mathfrak{H}$ and any $T > 0$, see Theorem 13.3.5. Note that it holds without any additional assumptions in comparing to that in Remark 13.3.1(b). In fact, one has about (13.3.6) the following statement with a reassuring conclusion conserning a pointwise convegence on $\mathfrak{H}$.

Proposition 13.3.2 *The locally convex vector space* $L^2_{loc}(\mathbb{R}, \mathfrak{H})$ *endowed with topology generated by seminorms, which correspond to the family of integrals in* (13.3.6) *innumerate by* $T > 0$*, is a complete, metrisable Fréchet space.*

Remark 13.3.3 Of course, the limits (13.3.6) do not solve the problem of the strong operator convergence in (13.3.1). Nevertheless, we note that convergence (13.3.6) in the weaker L^2-topology of the *Fréchet* space $L^2_{loc}(\mathbb{R}, \mathfrak{H})$ implies the existence of *subsequences* $\{n_k\}_{k \in \mathbb{N}}$ (see Corollary 13.3.6) such that one has *pointwise* (that is, *strong* operator) convergence along it for *almost all* (a.a.) $t \in [-T, T]$. From the point of view of applications in mathematical/theoretical physics, this observation seems to be quite satisfactory.

For remark about the proof of Proposition 13.3.2 and for comments on Remark 13.3.3, we refer to Notes to Chap. 13 (Notes to Sects. 13.3, 13.3.1).

Next, using the concept of the *holomorphic* Kato functions we prove in Sect. 13.3.3 the *Trotter-Kato* product formula in the L^2-topology, that is,

$$\lim_{n\to\infty} \int_{-T}^{T} dt \, \left\| \left(\widetilde{f}(itA/n)\widetilde{g}(itB/n)\right)^n h - \mathrm{e}^{-itC} h \right\|^2 = 0 , \quad T > 0 . \tag{13.3.7}$$

Here $h \in \mathfrak{H}$ and $\widetilde{f}, \widetilde{g}$ are the Borel measurable extensions of holomorphic Kato functions f and g on the imaginary axis, see Lemma 13.3.10 and Theorem 13.3.11. Moreover, we propose a characterisation of the class of *holomorphic* Kato functions.

Finally, in Sect. 13.3.4, we present a generalisation of the result (13.3.3) to the class of *admissible* Kato functions defined above. More precisely, we show that

$$\text{s-}\lim_{n\to\infty} (\phi(tA/n)\psi(tB/n))^n = \mathrm{e}^{-itC} , \tag{13.3.8}$$

where ϕ and ψ are admissible functions such that $\mathrm{Re}(\phi(y)) \geq 0$, $\mathrm{Im}(\phi(y) \leq 0$ and $\mathrm{Im}(\psi(y)) \leq 0$ for $y \in \mathbb{R}^+$, cf. Theorem 13.3.19. Choosing $\phi(y) = \psi(y) = (1 + iy)^{-1}$, $y \in \mathbb{R}$, one recovers *Lapidus'* product formula (13.4.8), see Notes to Chap. 13. Moreover, it turns out that admissible functions can be always *slightly* modified so that the Trotter–Kato product formula is valid in the strong operator topology, see Corollary 13.3.21. In particular, it follows from Corollary 13.3.21 that the Trotter product formula modified *à la Ichinose*:

$$\text{s-}\lim_{n\to\infty} \left(\mathrm{e}^{-itA/n} E_A([0, \pi n/2t]) \, \mathrm{e}^{-itB/n} E_B([0, \pi n/2t])\right)^n = \mathrm{e}^{-itC} , \tag{13.3.9}$$

holds uniformly in $t \in [0, T]$, $T > 0$, cf. (13.3.4) and (13.3.5). Notice that (13.3.9) is similar to (13.3.3).

13.3.2 Lapidus' Product Formula and L^2-Convergence

We start this subsection by proving the following technical lemma.

Lemma 13.3.4 *Let $\{F_n(\cdot)\}_{n\in\mathbb{N}}$ be a family of measurable operator-valued functions $F_n(\cdot) : i\,\mathbb{R} \longrightarrow \mathcal{L}(\mathfrak{H})$ such that $\|F_n(i\,t)\| \leq 1$ holds for a.a. $t \in \mathbb{R}$. Furthermore, let C be a densely defined self-adjoint operator. Then the following assertions are equivalent:*

(i) *For each $\varphi \in L^1(\mathbb{R})$ one has the weak operator limit in $\mathfrak{H}$:*

$$\text{w-}\lim_{n\to\infty} \int_{\mathbb{R}} dt\, \varphi(t)\, F_n(i\,t) = \int_{\mathbb{R}} dt\, \varphi(t)\, \mathrm{e}^{-i\,tC} . \tag{13.3.10}$$

(ii) *For each $h \in \mathfrak{H}$ and any $T > 0$ it holds (for seminorms)*

$$\lim_{n\to\infty} \int_{-T}^{T} dt\, \|F_n(it)\,h - \mathrm{e}^{-i\,tC}\,h\|^2 = 0 . \tag{13.3.11}$$

(iii) *For each $\varphi \in L^1(\mathbb{R})$ one has the strong operator limit:*

$$\text{s-}\lim_{n\to\infty} \int_{\mathbb{R}} dt\, \varphi(t)\, F_n(it) = \int_{\mathbb{R}} dt\, \varphi(t)\, \mathrm{e}^{-i\,tC} . \tag{13.3.12}$$

Proof (i) $\Longrightarrow$ (ii) Since

$$\|F_n(it)\,h - \mathrm{e}^{-i\,tC}\,h\|^2 \leq 2\,\|h\|^2 - 2\,\mathrm{Re}(F_n(it)\,h,\, \mathrm{e}^{-i\,tC}\,h), \quad t \in \mathbb{R},$$

and also

$$\mathrm{e}^{-i\,tC}\,h = \sum_{k=0}^{\infty} \frac{(-i\,t)^k}{k!} C^k\,h , \quad t \in \mathbb{R} ,$$

for $h \in E_C([a,b])\mathfrak{H}$ and $-\infty < a < b < \infty$, we find

$$\|F_n(it)\,h - \mathrm{e}^{-i\,tC}\,h\|^2 \leq 2\,\|h\|^2 - 2\,\mathrm{Re}\left(\sum_{k=0}^{\infty} \frac{i^k\,t^k}{k!}(F_n(it)\,h,\, C^k\,h)\right) ,$$

for a.a. $t \in \mathbb{R}$. This inequality leads to

$$\int_{-T}^{T} dt\, \|F_n(it)\,h - \mathrm{e}^{-i\,tC}\,h\|^2 \leq$$

$$4\,T\,\|h\|^2 - 2\,\mathrm{Re}\left(\sum_{k=0}^{\infty} \frac{i^k}{k!} \int_{-T}^{T} dt\, t^k (F_n(it)^n\,h,\, C^k\,h)\right) ,$$

or, by the *Fubini* theorem,

$$\int_{-T}^{T} dt \; \|F_n(it)\,h - \mathrm{e}^{-itC}\,h\|^2 \leq$$

$$4\,T\;\|h\|^2 - 2\,\mathrm{Re}\left(\sum_{k=0}^{\infty} \frac{i^k}{k!}\left(\int_{-T}^{T} dt\; t^k\; F_n(it)\,h,\; C^k\,h\right)\right),$$

for $t \geq 0$. Then on account to (13.3.10) we obtain

$$\lim_{n\to\infty}\left(\int_{-T}^{T} dt\; t^k\; F_n(it)\,h,\; C^k\,h\right) = \left(\int_{-T}^{T} dt\; t^k \mathrm{e}^{-itC}\,h,\; C^k\,h\right).$$

As a consequence,

$$\int_{-T}^{T} dt \; \|F_n(it)\,h - \mathrm{e}^{-itC}\,h\|^2 \leq$$

$$4\,T\;\|g\|^2 - 2\,\mathrm{Re}\left(\sum_{k=0}^{\infty} \frac{i^k}{k!}\left(\int_{-T}^{T} dt\; t^k \mathrm{e}^{-itC}\,h,\; C^k\,h\right)\right).$$

Therefore, one infers for the limit:

$$\limsup_{n\to\infty} \int_{-T}^{T} dt \; \|F_n(it)\,h - \mathrm{e}^{-itC}\,h\|^2 \leq$$

$$4\,T\;\|h\|^2 - 2\,\mathrm{Re}\int_{-T}^{T} dt\; (\mathrm{e}^{-itC}\,h,\; \mathrm{e}^{-itC}\,h) = 0,$$

which proves (13.3.11).

(ii) $\Longrightarrow$ (iii) Note that the following estimate holds:

$$\left\|\int_{-T}^{T} dt\; \varphi(t)\, \left(F_n(it) - \mathrm{e}^{-itC}\right) h\right\| \leq \int_{-T}^{T} dt\; |\varphi(t)|\, \left\|F_n(it)\,h - \mathrm{e}^{-itC}\,h\right\|.$$

From (13.3.11), we obtain the convergence in (the *Lebesgue*) measure: $\Delta \mapsto \mu\,|\Delta|$, $\Delta \subseteq [-T, T]$, that is, for each $\varepsilon > 0$ one has

$$\lim_{n\to\infty} \mu\left|\{t \in [-T, T] : \|F_n(it)\,h - \mathrm{e}^{-itC}\,h\| \geq \varepsilon\}\right| = 0.$$

Setting $\Delta_{\varepsilon,n} := \{t \in [-T, T] : \|F_n(it)\,h - \mathrm{e}^{-itC}\,h\| \geq \varepsilon\}$ we find the estimate

$$\left\| \int_{-T}^{T} dt\, \varphi(t) \left(F_n(it) - \mathrm{e}^{-itC}\right) h \right\|$$
$$\leq \varepsilon \int_{[-T,T]\setminus\Delta_{\varepsilon,n}} dt\, |\varphi(t)| \; + 2 \int_{\Delta_{\varepsilon,n}} dt\, |\varphi(t)|\,, \quad n \in \mathbb{N}\,.$$

In view of (13.3.11) we obtain in the limit $n \to \infty$ the inequality

$$\limsup_{n\to\infty} \left\| \int_{-T}^{T} dt\, \varphi(t) \left(F_n(it) - \mathrm{e}^{-itC}\right) h \right\| \leq \varepsilon \int_{\mathbb{R}} dt\, |\varphi(t)|\,,$$

for any $\varepsilon > 0$. Seeing that, for ε small enough one gets

$$\limsup_{n\to\infty} \left\| \int_{\mathbb{R}} dt\, \varphi(t) \left(F_n(it) - \mathrm{e}^{-itC}\right) h \right\| \leq \varepsilon \int_{\mathbb{R}} dt\, |\varphi(t)| + 2 \int_{\mathbb{R}\setminus[-T,T]} dt\, |\varphi(t)|\,.$$

Since T can be chosen sufficiently large and ε can be arbitrary small, we obtain

$$\limsup_{n\to\infty} \left\| \int_{\mathbb{R}} dt\, \varphi(t) \left(F_n(it) - \mathrm{e}^{-itC}\right) h \right\| = 0\,,$$

which yields

$$\text{s-}\lim_{n\to\infty} \int_{\mathbb{R}} dt\, \varphi(t)\, F_n(it)\, h = \int_{\mathbb{R}} dt\, \varphi(t)\, \mathrm{e}^{-itC}\,, \quad h \in \mathfrak{H}\,.$$

(iii) $\Longrightarrow$ (i) The strong limit (13.3.12) obviously implies (13.3.10). □

We note that Lemma 13.3.4 allows us to reformulate the *Lapidus* result about product formula (13.3.2) in the following form.

Theorem 13.3.5 *Let A and B two non-negative self-adjoint operators in a Hilbert space $\mathfrak{H}$. If the form sum $C := A \dotplus B$ is densely defined, then product formula* (13.3.6) *holds in the L^2-topology for any $h \in \mathfrak{H}$ and $T > 0$.*

Proof We set

$$F_n(it) := \left(\mathrm{e}^{-itA/n}\mathrm{e}^{-itB/n}\right)^n, \quad n \in \mathbb{N}, \quad t \in \mathbb{R}\,.$$

From (13.3.2), we infer (13.3.12), which yields (13.3.11). As a consequence, applying Lemma 13.3.4 we obtain convergence (13.3.6) in topology of the locally convex (*Fréchet*) space $L^2_{loc}(\mathbb{R}, \mathfrak{H})$, Proposition 13.3.2. □

Note that Theorem 13.3.5 *partially* solves the open question posed in (13.3.1) by a (slight) modification of topology. Indeed, from Theorem 13.3.5 we get that limit

(13.3.1) holds in *measure*, that is, for any real number $\eta > 0$, $h \in \mathfrak{H}$ and $T > 0$:

$$\lim_{n\to\infty} \mu \left| \left\{ t \in [-T, T] : \left\| \left(\mathrm{e}^{-itA/n} \mathrm{e}^{-itB/n} \right)^n h - \mathrm{e}^{-itC} h \right\| \geq \eta \right\} \right| = 0 , \tag{13.3.13}$$

here $\mu|\cdot|$ denotes the *Lebesgue* measure of the set $|\cdot|$. On the contrary, (13.3.1) requires a *uniform* convergence in $t \in [-T, T]$, that is, for any $\eta > 0$, $h \in \mathfrak{H}$ and $T > 0$

$$\lim_{n\to\infty} \sup_{t\in[-T,T]} \left\| \left(\mathrm{e}^{-tA/n} \mathrm{e}^{-tB/n} \right)^n h - \mathrm{e}^{-tC} h \right\| = 0.$$

On the other hand, in addition to (13.3.13) we bring the following observation, which bolsters a conclusion that L^2-convergence in Theorem 13.3.5 is a *partial* solution of the problem posed in (13.3.1), cf. Remark 13.3.3.

Corollary 13.3.6 *The convergence in measure* (13.3.13) *takes place if and only if any subsequence of* $\left\{ \left(\mathrm{e}^{-itA/n} \, \mathrm{e}^{-itB/n} \right)^n \right\}_{n\in\mathbb{N}}$ *in* (13.3.1) *contains a subsequence* $\left\{ \left(\mathrm{e}^{-itA/n_k} \, \mathrm{e}^{-itB/n_k} \right)^{n_k} \right\}_{k\in\mathbb{N}}$*, for* $\{n_k \in \mathbb{N}\}_{k\in\mathbb{N}}$*, which converges strongly to* e^{-itC} *for a.a.* $t \in [-T, T]$*:*

$$\text{s-}\lim_{k\to\infty} \left(\mathrm{e}^{-itA/n_k} \mathrm{e}^{-itB/n_k} \right)^{n_k} = \mathrm{e}^{-itC} . \tag{13.3.14}$$

As a result, the convergence in the L^2*-topology* (13.3.6) *yields* (13.3.1) *for a.a.* $t \in [-T, T]$ *by subsequences* (13.3.14).

Remark 13.3.7

(a) From the viewpoint of applications in physics, the formula (13.3.6) allows us to extend the Trotter-type definition of *Feynman* integrals for *Schrödinger* operators to some sufficiently wide class of potentials.
(b) We indicate that the *Feynman* integral $\mathcal{F}^t_{\mathrm{TP}}(V)$ associated with the potential V can be defined as the strong operator limit

$$\mathcal{F}^t_{\mathrm{TP}}(V) := \text{s-}\lim_{n\to\infty} \left(\mathrm{e}^{-itH_0/n} \mathrm{e}^{-itV/n} \right)^n , \quad t \in \mathbb{R} ,$$

where $H_0 := -\frac{1}{2}\Delta$. Here $-\Delta$ is the *Laplacian* operator in $L^2(\mathbb{R}^d)$ defined in the usual way. Then, one infers that the *Feynman* integral exists if $V : \mathbb{R}^d \longrightarrow \mathbb{R}$ is *Lebesgue* measurable, non-negative, and satisfies condition: $V \in L^2_{\mathrm{loc}}(\mathbb{R}^d)$. See Notes to Chap. 13 (Notes to Sects. 13.3, 13.3.2) for more comments and details.

(c) With Theorem 13.3.5 in mind, it is possible to extend Trotter-type definition of the *Feynman* integrals if one replaces the $L^2(\mathbb{R}^d)$-topology by the $L^2([-T,T]\times \mathbb{R}^d)$-topology. Indeed, let us define the generalised *Feynman* integral $\mathcal{F}^t_{\mathrm{gTP}}(V)$ by

$$\lim_{n\to\infty}\int_{-T}^{T} dt\, \left\|\left(\mathrm{e}^{-it\,H_0/n}\mathrm{e}^{-itV/n}\right)^n h - \mathcal{F}^t_{\mathrm{gTP}}(V)\,h\right\|^2 = 0\,,$$

for $h \in L^2(\mathbb{R}^d)$ and $T > 0$. Obviously, the existence of $\mathcal{F}^t_{\mathrm{TP}}(V)$ yields the existence of $\mathcal{F}^t_{\mathrm{gTP}}(V)$ while the converse is in general not true. By Theorem 13.3.5 one can immediately conclude that the generalised *Feynman* integral exists if $V : \mathbb{R}^d \longrightarrow \mathbb{R}$ is Lebesgue measurable, non-negative, and satisfies $V \in L^1_{\mathrm{loc}}(\mathbb{R}^d)$. This substantially extends the class of *acceptable* potentials. The same class of potentials is covered by the so-called *modified* Feynman integral $\mathcal{F}^t_M(V)$ defined by the *Trotter–Kato* product formula

$$\mathcal{F}^t_M(V) := \text{s-}\lim_{n\to\infty}\left([\mathbb{1} + i(t/n)H_0]^{-1}[\mathbb{1} + i(t/n)V]^{-1}\right)^n .$$

However, in this case, the exponents are replaced by the *resolvents*, which leads to loss of the typical structure of *Feynman* integrals and the related physical insights.

13.3.3 Trotter–Kato Product Formulæ in L^2-Topology

The original *Lapidus* result (13.3.2) (cf. Theorem 13.3.5) relies on, so-called, the *Vitali classical* and the *Vitali extended* theorems. We reformulate them below adapting to our particular set-up.

Remark 13.3.8

(a) Let $\Phi_n(z)$, $n \in \mathbb{N}$, be a sequence of uniformly bounded holomorphic functions in the right half-plane $\mathbb{C}_+$ which for $\mathrm{Re}\, z = x \in \mathbb{R}^+$ converges to function $\Phi(x)$, that is, $\lim_{n\to\infty}\Phi_n(x) = \Phi(x)$ for $x \in \mathbb{R}^+$. Then, $\Phi(x)$ admits a holomorphic continuation $\Phi(z)$ to $\mathbb{C}_+$ such that $\Phi(z) = \lim_{n\to\infty}\Phi_n(z)$, for $z \in \mathbb{C}_+$, with the same bound (*Vitali classical* theorem).

(b) Since $\Phi_n(z)$ and $\Phi(z)$ are uniformly bounded holomorphic functions in $\mathbb{C}_+$, the limits $\Phi_n(iy) := \lim_{\epsilon\to+0}\Phi_n(\epsilon+iy)$, $n \in \mathbb{N}$, and $\Phi(iy) := \lim_{\epsilon\to+0}\Phi(\epsilon+iy)$ exist for a.a. $y \in \mathbb{R}$. The *Vitali extended* theorem yields

$$\lim_{n\to\infty}\int_{\mathbb{R}} dy\, \varphi(y)\, \Phi_n(iy) = \int_{\mathbb{R}} dy\, \varphi(y)\, \Phi(iy)\,, \quad \varphi \in L^1(\mathbb{R})\,. \qquad (13.3.15)$$

(c) It is worth noting that conclusion (13.3.15) cannot be deduced from arguments based on the *Lapidus* result (13.3.2) since it requires that functions $\Phi_n(z)$ and $\Phi(z)$ must admit a *continuous* extension to $\mathbb{C}_+ \cup i\,\mathbb{R}$. However, applying a generalisation of this result due to the *Cachia* lemma , see Notes to Chap. 13 (Notes to Sects. 13.3, 13.3.3), one infers that this conclusion holds.

Now we make precise the notion of the *holomorphic* Kato functions, that we mentioned in Sect. 13.3.1.

Definition 13.3.9 Let $f : \mathbb{R}_0^+ \longrightarrow \mathbb{R}_0^+$ be a generic Kato function of the class $\mathcal{K}$, Definition 5.2.1. Then it is called a *holomorphic* Kato function if $\mathbb{R}^+ \ni x \mapsto f(x)$ admits a holomorphic continuation into the right half-plane $\mathbb{C}_+$ such that

$$|f(z)| \leq 1, \quad z \in \mathbb{C}_+ .$$

The *standard* holomorphic Kato functions are: $f_k(x) = (1 + x/k)^{-k}$, $x \in \mathbb{R}_0^+$, $k \in \mathbb{N}$, and, of course, $f(x) = \mathrm{e}^{-x}$, $x \in \mathbb{R}_0^+$. At the end of this subsection, we give a characterisation of the *holomorphic* Kato functions and indicate some non-standard examples of them. It turns out that for this standard holomorphic Kato functions the limit to the imaginary axis $i\,\mathbb{R}$ exists *everywhere*. For $z = \epsilon + it \in \mathbb{C}_+$ this yields $f(i\,tA) = \text{s-lim}_{\epsilon\to+0} f((\epsilon + it)A)$ for any $t \in \mathbb{R}$.

However, if f is a general (including non-standard) holomorphic Kato function, then for a self-adjoint operator A, the relation $f(itA) = \text{s-lim}_{\epsilon\to+0} f((\epsilon + it)A)$ cannot be expected. Indeed, this is because of the fact that for $z = x + iy \in \mathbb{C}_+$ the limit $f(iy) = \lim_{x\to+0} f(x + iy)$ may exist only for *almost all* (a.a.) $y \in \mathbb{R}$. Hence, the limit function $f(iy)$ is not in general a Borel measurable, that makes it impossible to apply in this case the *functional calculus* for self-adjoint operators. We note that still, the limit function $f(iy)$, defined for all those $y \in \mathbb{R}$ for which the limit $f(iy)$ exists, admits an extension to the whole axis $i\,\mathbb{R}$, and this extension *is* a Borel measurable function.

Lemma 13.3.10 *Let $f : \mathbb{R}_0^+ \longrightarrow \mathbb{R}_0^+$ be a holomorphic Kato function. Then, there is a Borel measurable function $\widetilde{f} : i\,\mathbb{R} \longrightarrow \mathbb{C}$ satisfying $|\widetilde{f}(iy)| \leq 1$, $y \in \mathbb{R}$, such that $\widetilde{f}(iy) = \lim_{\epsilon\to+0} f(\epsilon + iy)$ for a.a. $y \in \mathbb{R}$.*

Proof We set $f_R(z) := \mathrm{Re}(f(z))$ and $f_I(z) = \mathrm{Im}(f(z))$, $z \in \mathbb{C}_+$. Since by Definition 13.3.9, one has $|f(z)| \leq 1$,

$$|f_R(z)|^2 + |f_I(z)|^2 \leq 1, \quad z \in \mathbb{C}_+ .$$

Further, let $f_R^{\pm}(z) := \max\{0, \pm f_R(z)\} \geq 0$ and $f_I^{\pm}(z) := \max\{0, \pm f_I(z)\} \geq 0$, $z \in \mathbb{C}_+$. Since the function f is holomorphic in $\mathbb{C}_+$ the functions $f_R^{\pm}$ and $f_I^{\pm}$ are Borel measurable. Note that

$$f(z) = f_R^+(z) - f_R^-(z) + i(f_I^+(z) - f_I^-(z)), \quad z \in \mathbb{C}_+ , \tag{13.3.16}$$

and

$$|f_R^+(z)|^2 + |f_R^-(z)|^2 + |f_I^+(z)|^2 + |f_I^-(z)|^2 \le 1, \quad z \in \mathbb{C}_+ . \tag{13.3.17}$$

We set $\widetilde{f_R^{\pm}}(iy) := \liminf_{\epsilon\to+0} f_R^{\pm}(\epsilon+iy)$ and $\widetilde{f_I^{\pm}}(iy) := \liminf_{\epsilon\to+0} f_I^{\pm}(\epsilon+iy)$, $y \in \mathbb{R}$. Since $f_R^{\pm}(z)$ and $f_I^{\pm}(z)$, $z \in \mathbb{C}_+$, are Borel measurable functions, the functions $\widetilde{f_R^{\pm}}(iy)$ and $\widetilde{f_I^{\pm}}(iy)$ are also Borel measurable. From

$$\inf_{0<\eta\le\epsilon} f_R^{\pm}(\eta+iy) \le f_R^{\pm}(\epsilon+iy) \quad \text{and} \quad \inf_{0<\eta\le\epsilon} f_I^{\pm}(\eta+iy) \le f_I^{\pm}(\epsilon+iy), \quad y \in \mathbb{R},$$

we find

$$\begin{aligned} &\sum_{j=\pm}\left(\left|\inf_{0<\eta\le\epsilon} f_R^j(\eta+iy)\right|^2 + \left|\inf_{0<\eta\le\epsilon} f_I^j(\eta+iy)\right|^2\right) \\ &\quad \le |f_R^+(\epsilon+iy)|^2 + |f_R^-(\epsilon+iy)|^2 + |f_I^+(\epsilon+iy)|^2 + |f_I^-(\epsilon+iy)|^2 \le 1, \end{aligned}$$

$y \in \mathbb{R}$, which yields

$$|\widetilde{f_R^+}(iy)|^2 + |\widetilde{f_R^-}(iy)|^2 + |\widetilde{f_I^+}(iy)|^2 + |\widetilde{f_I^-}(iy)|^2 \le 1, \quad y \in \mathbb{R}.$$

Now setting, cf. (13.3.16):

$$\widetilde{f}(iy) := \widetilde{f_R^+}(iy) - \widetilde{f_R^-}(iy) + i(\widetilde{f_I^+}(iy) - \widetilde{f_I^-}(iy)), \quad y \in \mathbb{R},$$

we define a Borel measurable function. Then, (13.3.17) yields

$$\begin{aligned} |\widetilde{f}(iy)|^2 &= |\widetilde{f_R^+}(iy) - \widetilde{f_R^-}(iy)|^2 + |\widetilde{f_I^+}(iy) - \widetilde{f_I^-}(iy)|^2 \\ &\le |\widetilde{f_R^+}(iy)|^2 + |\widetilde{f_R^-}(iy)|^2 + |\widetilde{f_I^+}(iy)|^2 + |\widetilde{f_I^-}(iy)|^2 \le 1, \quad y \in \mathbb{R}. \end{aligned}$$

Whence follows relation $\widetilde{f}(iy) = \lim_{\epsilon\to+0} f(\epsilon+iy)$ for a.a. $y \in \mathbb{R}$. □

Since $|\widetilde{f}(iy)| \le 1$, $y \in \mathbb{R}$, the operator-valued function $\widetilde{f}(i\tau A)$, $\tau \in \mathbb{R}$, is well-defined for any self-adjoint operator A.

Setting $F(z) := f(zA)\, g(zB)$, for $z \in \mathbb{C}_+$, it turns out that the operator-valued family $\{F(z)\}_{z\in\mathbb{C}_+}$ is contraction and holomorphic, if f and g are holomorphic Kato functions.

Theorem 13.3.11 *Let A and B two non-negative self-adjoint operators in the Hilbert space $\mathfrak{H}$. Assume that form-sum operator $C := A \dotplus B$ is densely defined. If f and g are holomorphic Kato functions, then limit* (13.3.7) *holds for any $h \in \mathfrak{H}$ and $T > 0$.*

Proof We set $F_n(z) := F(z/n)^n$ for $z \in \mathbb{C}_+ \cup i\,\mathbb{R}$ and $n \in \mathbb{N}$ and note that if $z = t \in \mathbb{R}_0^+$, then

$$F_n(t) := (f(tA/n)g(tB/n))^n = \left(\mathrm{e}^{-tA/n}\mathrm{e}^{-tB/n}\right)^n, \quad t \in \mathbb{R}_0^+, \quad n \in \mathbb{N}\,.$$

From Theorem 5.2.5 we find $\text{s-lim}_{n\to\infty} F_n(t) = \mathrm{e}^{-tC}$, which yields

$$\lim_{n\to\infty} (F_n(t)h, g) = (\mathrm{e}^{-tC}h, g)\,, \quad t \in \mathbb{R}_0^+\,,$$

for $h, g \in \mathfrak{H}$. Setting $\Phi_n(t) := (F_n(t)h, g)$, $n \in \mathbb{N}$, and $\Phi(t) = (\mathrm{e}^{-tC}h, g)$, $t \in \mathbb{R}_0^+$, by Vitali's *classical* theorem (Remark 13.3.8(a)) we find $\lim_{n\to\infty} \Phi_n(z) = \Phi(z)$, $z \in \mathbb{C}_+$. Moreover, taking into account the *extended* Vitali theorem (Remark 13.3.8(b) and (c)), we obtain

$$\lim_{n\to\infty} \int_{\mathbb{R}} dt\, \varphi(t)\, \Phi_n(it) = \int_{\mathbb{R}} dt\, \varphi(t)\, \Phi(it)\,,$$

for all $\varphi \in L^1(\mathbb{R})$. The last relation yields

$$\text{w-}\lim_{n\to\infty} \int_{\mathbb{R}} dt\, \varphi(t)\, F_n(it) = \int_{\mathbb{R}} dt\, \varphi(t)\, \mathrm{e}^{-itC}\,,$$

for all $\varphi \in L^1(\mathbb{R})$, where $F_n(it) := \text{s-lim}_{\epsilon\to+0} F_n(\epsilon + it)$, for a.a. $t \in \mathbb{R}$, by the operator-valued *Fatou–Privalov* theorem, see Notes to Chap. 13 (Notes to Sects. 13.3, 13.3.3). Then applying Lemma 13.3.4 we obtain

$$\lim_{n\to\infty} \int_{-T}^{T} dt\, \|F_n(it)\, h - \mathrm{e}^{-itC}\, h\|^2 = 0\,, \quad h \in \mathfrak{H}\,, \quad T > 0\,.$$

It remains to show that $F_n(it)$ can be replaced by $\left(\widetilde{f}(itA/n)\widetilde{g}(itB/n)\right)^n$ for each $n \in \mathbb{N}$. We find

$$\begin{aligned}
&\int_{-T}^{T} dt\, \left\|f((\epsilon + it)A/n)h - \widetilde{f}(itA/n)h\right\|^2 \\
&= \int_{-T}^{T} dt \int_{[0,\infty)} d(E_A(\lambda)h, h)\, |f((\epsilon + it)\lambda/n) - \widetilde{f}(it\lambda/n)|^2 \\
&= \int_{[0,\infty)} d(E_A(\lambda)h, h) \int_{-T}^{T} dt\, |f((\epsilon + it)\lambda/n) - \widetilde{f}(it\lambda/n)|^2
\end{aligned}$$

for each $n \in \mathbb{N}$ and $h \in \mathfrak{H}$. For any $\lambda \in [0, \infty)$ we have

$$\lim_{\epsilon \to +0} \int_{-T}^{T} dt \, |f((\epsilon + it)\lambda/n) - \widetilde{f}(it\lambda/n)|^2 = 0$$

which yields

$$\lim_{\epsilon \to +0} \int_{-T}^{T} dt \, \left\| f((\epsilon + it)A/n)h - \widetilde{f}(itA/n)h \right\|^2 = 0$$

for each $n \in \mathbb{N}$ and $h \in \mathfrak{H}$. Since also

$$\lim_{\epsilon \to +0} \int_{-T}^{T} dt \, \|g((\epsilon + it)B/n)h - \widetilde{g}(itB/n)h\|^2 = 0$$

holds for each $n \in \mathbb{N}$ and $h \in \mathfrak{H}$ we immediately find that

$$\lim_{\epsilon \to +0} \int_{-T}^{T} dt \, \Big\| (f((\epsilon + it)A/n)g((\epsilon + it)B/n))^n \, h - \\ \left(\widetilde{f}(itA/n)\widetilde{g}(itB/n)\right)^n h \Big\|^2 = 0$$

for each $n \in \mathbb{N}$ and $h \in \mathfrak{H}$. Hence

$$\begin{aligned} 0 = & \lim_{\epsilon \to +0} \int_{-T}^{T} dt \, \left\| F_n(\epsilon + it)h - \left(\widetilde{f}(itA/n)\widetilde{g}(itB/n)\right)^n h \right\|^2 \\ = & \int_{-T}^{T} dt \, \left\| F_n(it)h - \left(\widetilde{f}(itA/n)\widetilde{g}(itB/n)\right)^n h \right\|^2 \end{aligned}$$

for each $n \in \mathbb{N}$ and $h \in \mathfrak{H}$ which yields $F_n(it) = \left(\widetilde{f}(itA/n)\widetilde{g}(itB/n)\right)^n$ for a.a. $t \in \mathbb{R}$ and $n \in \mathbb{N}$. □

Using the standard Kato function: $f_k(x) := (1 + x/k)^{-k}$, $x \in \mathbb{R}_0^+$, $k \in \mathbb{N}$, one infers that Theorem 13.3.11 provides

$$\lim_{n \to \infty} \int_{-T}^{T} dt \, \left\| \left((\mathbb{1} + itA/kn)^{-k}(\mathbb{1} + itB/kn)^{-k}\right)^n h - e^{-itC}h \right\|^2 = 0,$$

for any $h \in \mathfrak{H}$ and $T > 0$, in the L^2-topology. On the other hand, we remind that for particular case: $k = 1$, there is a stronger result (13.4.8). Namely,

$$e^{-itC} = \text{s-}\lim_{n \to \infty} \left((\mathbb{1} + itA/n)^{-1}(\mathbb{1} + itB/n)^{-1}\right)^n, \tag{13.3.18}$$

holds uniformly in $t \in [0,T]$, $T > 0$, see Notes to Chap. 13 (Notes to Chap. 13, Sect. 13.3.1). This discrepancy will be elucidated in Corollary 13.3.20 of the next Sect. 13.3.4.

Note that the set of *holomorphic* Kato functions has a complete characterisation. We return to this observation below.

Theorem 13.3.12 *If f is a holomorphic Kato function* (*Definition* 13.3.9), *then:*

(i) *There is an at most countable set of complex numbers $\{\xi_k\}_{k\geq 1}$, $\xi_k \in \mathbb{C}_+$ with $\mathrm{Im}(\xi_k) \geq 0$, $k \in \mathbb{N}$, satisfying the condition*

$$\varkappa := 4 \sum_{k\geq 1} \frac{\mathrm{Re}(\xi_k)}{|\xi_k|^2} \leq 1\,. \tag{13.3.19}$$

(ii) *There is a Borel measure ν defined on $\mathbb{R}_0^+ = [0,\infty)$ obeying $\nu(\{0\}) = 0$ and*

$$\int_{\mathbb{R}^+} \frac{d\nu(t)}{1+t^2}\, d\nu(t) < \infty\,,$$

such that the limit $\beta := \lim_{x\to+0} \frac{2}{\pi}\int_{\mathbb{R}^+} d\nu(t)(x^2+t^2)^{-1}$ exists and satisfies the condition $\beta \leq 1-\varkappa$.

(iii) *The Kato function f admits the representation*

$$f(x) = D(x)\, \exp\left\{-\frac{2x}{\pi}\int_{\mathbb{R}^+} \frac{d\nu(t)}{x^2+t^2}\right\} \mathrm{e}^{-\alpha x}\,, \quad x \in \mathbb{R}^+\,, \tag{13.3.20}$$

where $\alpha := 1-\varkappa-\beta$ and $D(x)$ is a Blaschke-type product given by

$$D(x) := \prod_{k\geq 1} \frac{x^2 - 2x\,\mathrm{Re}(\xi_k) + |\xi_k|^2}{x^2 + 2x\,\mathrm{Re}(\xi_k) + |\xi_k|^2}\,, \quad x \in \mathbb{R}^+\,. \tag{13.3.21}$$

The factor $D(x)$ is absent if the set $\{\xi_k\}_{k\geq 1}$ is empty. In that case, we set $\varkappa := 0$.

Conversely, if a real function f admits the representation (13.3.20) *such that the assumptions (i) and (ii) are satisfied and the condition $\alpha+\varkappa+\beta = 1$ holds, then f is a holomorphic Kato function and its holomorphic extension to $\mathbb{C}_+$ is given by*

$$f(z) = D(z)\, \exp\left\{-\frac{2z}{\pi}\int_{\mathbb{R}^+} \frac{d\nu(t)}{z^2+t^2}\right\} e^{-\alpha z}, \quad z \in \mathbb{C}_+\,.$$

Above, we have already indicated several *standard* holomorphic Kato functions such as $f(z) = \mathrm{e}^{-z}$, $f_k(z) = (1+iz/k)^{-k}$, $k \in \mathbb{N}$, $z \in \mathbb{C}_+$. The last theorem allows us to give examples of some *non-standard* holomorphic Kato functions.

1. If a holomorphic Kato function f has *no* zeros in $\mathbb{C}_+$ and $\nu \equiv 0$, then $f(z) = \mathrm{e}^{-z}$, $z \in \mathbb{C}_+$, where $\alpha = 1$ follows from condition $\alpha = 1 - \varkappa - \beta$, where $\varkappa = \beta = 0$. Then one obviously gets for restriction on the imaginary axis: $\widetilde{f}(iy) = \lim_{\epsilon \to +0} f(\epsilon + iy) = \mathrm{e}^{-iy}$ for $y \in \mathbb{R}$.
2. If a holomorphic Kato function f has zeros and the measure $\nu \equiv 0$, then f is of the form $f(z) = D(z)\,\mathrm{e}^{-\alpha z}$, where the Blaschke-type product $D(z)$ is given by (13.3.21). If, in particular, the function f has a *single* zero at ξ, i.e. $k = 1$, we find the representation

$$f(z) = \frac{z^2 - 2z\mathrm{Re}(\xi) + |\xi|^2}{z^2 + 2z\mathrm{Re}(\xi) + |\xi|^2}\,\mathrm{e}^{-\alpha z}, \quad z \in \mathbb{C}_+ ,$$

where $\xi \in \mathbb{C}_+$ is such that

$$\alpha + 4\,\frac{\mathrm{Re}(\xi)}{|\xi|^2} = 1 .$$

This gives the representation

$$f(z) = \frac{z^2 - 2\eta\,(z - 2/(1-\alpha))}{z^2 + 2\eta\,(z + 2/(1-\alpha))}\,\mathrm{e}^{-\alpha z}, \quad z \in \mathbb{C}_+ ,$$

$0 < \eta \leq 4/(1-\alpha)$, $0 \leq \alpha \leq 1$, where we have denoted $\xi = \eta + i\tau$, $\eta > 0$, and $\tau = \sqrt{4/(1-\alpha)^2 - (\eta - 2/(1-\alpha))^2}$. This yields

$$\widetilde{f}(iy) = \lim_{\epsilon \to +0} f(\epsilon + iy) = \frac{y^2 + 4\,\eta\,(1-\alpha)^{-1} + 2\,i\,\eta\,y}{y^2 - 4\,\eta\,(1-\alpha)^{-1} + 2\,i\,\eta\,y}\,\mathrm{e}^{-i\alpha y}, \quad y \in \mathbb{R} .$$

3. If a holomorphic Kato function $f(z)$ has no zeros and the measure ν is *atomic*, then $f(z)$ admits the representation

$$f(z) = \exp\left\{ -\frac{2z}{\pi} \sum_{l \geq 1} \frac{1}{z^2 + s_l^2}\,\nu(\{s_l\}) \right\} \mathrm{e}^{-\alpha z}, \quad z \in \mathbb{C}_+ ,$$

where $\{s_l\}_{l \geq 1}$ are the points where $\nu(\{s_l\}) \neq 0$. In a particular case of *one* atom, when $d\nu(t) = c\,\delta(t-s)\,dt$, $s > 0$, we have

$$f(z) = \exp\left\{ -\frac{2z\,c}{\pi}\,\frac{1}{z^2 + s^2} \right\} \mathrm{e}^{-\alpha z} ,$$

and $\alpha + 2c/(\pi s^2) = 1$, which yields $c = \frac{1}{2}(1-\alpha)\pi s^2$ and

$$f(z) := \exp\left\{-z\,(1-\alpha)\,\frac{s^2}{z^2+s^2}\right\}\,\mathrm{e}^{-\alpha z}\,, \qquad z \in \mathbb{C}_+\,.$$

As a consequence, one gets

$$\widetilde{f}(iy) = \begin{cases} \lim_{\epsilon\to+0} f(\epsilon+iy) \\ \quad = \exp\left\{iy\,(1-\alpha)\,s^2/(y^2-s^2)\right\}\,\mathrm{e}^{-i\alpha y} & , \; y \neq \pm s\,, \\ 0 & , \; y = \pm s\,, \end{cases}$$

where $y \in \mathbb{R}$.

4. If a holomorphic Kato function $f(z)$ has no zeros and the measure ν is *absolutely continuous*, that is, $d\nu(t) := h(t)\,dt$, $h(t)\,(1+t^2)^{-1} \in L^1(\mathbb{R}_0^+)$, then $f(z)$ admits the representation

$$f(z) = \exp\left\{-\frac{2z}{\pi}\int_0^\infty dt\,\frac{h(t)}{z^2+t^2}\right\}\,\mathrm{e}^{-\alpha z}\,, \qquad z \in \mathbb{C}_+\,,$$

such that

$$\alpha + \lim_{x\to+0}\frac{2}{\pi}\int_0^\infty dt\,\frac{h(t)}{x^2+t^2} = 1\,. \tag{13.3.22}$$

If h is Hölder continuous, then $\lim_{\epsilon\to+0} 2(\epsilon+iy)(\pi)^{-1}\int_0^\infty dt\,h(t)[(\epsilon+iy)^2 + t^2]^{-1}$ exists for each $y \in \mathbb{R}$ and one gets

$$\widetilde{f}(iy) = \exp\left\{-\lim_{\epsilon\to+0}\frac{2(\epsilon+iy)}{\pi}\int_0^\infty dt\,\frac{h(t)}{(\epsilon+iy)^2+t^2}\right\}\,\mathrm{e}^{-iy}\,, \qquad y \in \mathbb{R}\,.$$

For example, if $f(x) = (1+x/k)^{-k}$ for $x \in \mathbb{R}_0^+$, $k \in \mathbb{N}$, then for

$$h_k(t) := \frac{k}{2}\,\ln\left(1+\frac{t^2}{k^2}\right),$$

we obtain

$$\lim_{x\to+0}\frac{k}{\pi}\int_{\mathbb{R}_0^+} dt\,\frac{1}{x^2+t^2}\,\ln\left(1+\frac{t^2}{k^2}\right) = 1\,,$$

which by (13.3.22) yields $\alpha = 0$ and consequently

$$f(z) = \exp\left\{-\frac{kz}{\pi}\int_{\mathbb{R}_0^+} dt\, \frac{1}{z^2+t^2}\, \ln\left(1+\frac{t^2}{k^2}\right)\right\},$$

for $z \in \mathbb{C}_+$ and $\widetilde{f}(iy) = (1+iy/k)^{-k}$, where $y \in \mathbb{R}$, $k \in \mathbb{N}$.

13.3.4 Admissible Kato Functions and the Ichinose Theory

In this subsection, we present a generalisation of the *Ichinose* theory (cf. (13.3.3), (13.3.9)) for the class of so-called *admissible* Kato functions.

Definition 13.3.13 A Borel measurable function $\phi : \mathbb{R}_0^+ \longrightarrow \mathbb{C}$, is called *admissible* Kato function if the conditions

$$|\phi(y)| \le 1, \quad y \in [0,\infty), \quad \phi(0) = 1, \quad \phi'(0) = -i\,,$$

are satisfied.

We set $\phi_{\mathcal{R}}(y) := \mathrm{Re}(\phi(y))$ and $\phi_{\mathcal{I}}(y) := \mathrm{Im}(\phi(y))$, $y \in \mathbb{R}_0^+$. Then we obviously get

$$|\phi_{\mathcal{R}}(y)| \le 1, \quad y \in \mathbb{R}_0^+, \quad \phi_{\mathcal{R}}(0) = 1 \quad \text{and} \quad \phi_{\mathcal{R}}'(0) = 0\,,$$

as well as

$$|\phi_{\mathcal{I}}(y)| \le 1, \quad y \in \mathbb{R}_0^+, \quad \phi_{\mathcal{I}}(0) = 0 \quad \text{and} \quad \phi_{\mathcal{I}}'(0) = -1\,.$$

Let $\Sigma := \{y \in \mathbb{R}_0^+ : \phi(y) = 0\}$ and $\Omega := \mathbb{R}_0^+ \setminus \Sigma$. We set

$$\varphi(y) := \begin{cases} 1/\phi(y)\,, & y \in \Omega\,,\\ 1\,, & y \in \Sigma\,. \end{cases} \tag{13.3.23}$$

Notice that $\chi_\Omega(y) := \varphi(y)\,\phi(y)$, $y \in \mathbb{R}_0^+$, where χ_Ω is a characteristic function of the set Ω. The function φ obeys

$$|\varphi(y)| \ge 1, \quad y \in \mathbb{R}_0^+, \quad \varphi(0) = 1, \quad \varphi'(0) = i\,.$$

Moreover, we find

$$\varphi_{\mathcal{R}}(y) := \mathrm{Re}(\varphi(y)) = \begin{cases} \phi_{\mathcal{R}}(y)/|\phi(y)|^2, & y \in \Omega\\ 1, & y \in \Sigma \end{cases}$$

and

$$\varphi_{\mathcal{I}}(y) := \operatorname{Im}(\varphi(y)) = \begin{cases} -\phi_{\mathcal{I}}(y)/|\phi(y)|^2, & y \in \Omega \\ 0, & y \in \Sigma. \end{cases}$$

as well as

$$\varphi_{\mathcal{R}}(0) = 1, \quad \varphi'_{\mathcal{R}}(0) = 0 \quad \text{and} \quad \varphi_{\mathcal{I}}(0) = 0, \quad \varphi'_{\mathcal{I}}(0) = 1\,.$$

Let $E_\tau := \chi_\Omega(\tau A)$, $\tau > 0$ for a self-adjoint operator A. Then $\{E_\tau\}_{\tau>0}$ is a family of orthogonal projections corresponding to the set Ω generated by the spectral measure of A. To proceed, we consider the operator-valued function

$$K(\tau) := \frac{\varphi(\tau A) - \mathbb{1}}{\tau} + E_\tau \frac{\mathbb{1} - \psi(\tau B)}{\tau} E_\tau\,, \quad \tau > 0\,, \tag{13.3.24}$$

where $\psi : \mathbb{R}_0^+ \longrightarrow \mathbb{C}$, is also an *admissible* Kato function. We note that

$$K_{\mathcal{R}}(\tau) := \operatorname{Re}(K(\tau)) = \frac{\varphi_{\mathcal{R}}(\tau A) - \mathbb{1}}{\tau} + E_\tau \frac{\mathbb{1} - \psi_{\mathcal{R}}(\tau B)}{\tau} E_\tau\,, \tag{13.3.25}$$

and

$$K_{\mathcal{I}}(\tau) := \operatorname{Im}(K(\tau)) = \frac{\varphi_{\mathcal{I}}(\tau A) - E_\tau \psi_{\mathcal{I}}(\tau B) E_\tau}{\tau}\,. \tag{13.3.26}$$

If imaginary part of admissible functions are non-positive: $\phi_{\mathcal{I}}(y) \leq 0$ and $\psi_{\mathcal{I}}(y) \leq 0$, $y \in \mathbb{R}_0^+$, then $K_{\mathcal{I}}(\tau) \geq 0$. Furthermore, we set

$$L_\gamma(\tau) := \gamma K_{\mathcal{R}}(\tau) + K_{\mathcal{I}}(\tau), \quad \gamma \in [0, 1]\,.$$

Let us introduce the functions

$$f_\gamma(y) := \gamma(\varphi_{\mathcal{R}}(y) - 1) + \varphi_{\mathcal{I}}(y), \quad y \in \mathbb{R}_0^+, \tag{13.3.27}$$

and

$$g_\gamma(y) := \gamma(1 - \psi_{\mathcal{R}}(y)) - \psi_{\mathcal{I}}(y), \quad y \in \mathbb{R}_0^+, \tag{13.3.28}$$

for $\gamma \in [0, 1]$. If $\phi(\cdot)$ and $\psi(\cdot)$ are admissible functions, then

$$f_\gamma(0) = 0, \quad f'_\gamma(0) = 1 \quad \text{and} \quad g_\gamma(0) = 0, \quad g'_\gamma(0) = 1. \tag{13.3.29}$$

Using the functions $f_\gamma(\cdot)$ and $g_\gamma(\cdot)$ one gets the representation

$$L_\gamma(\tau) = \frac{f_\gamma(\tau A)}{\tau} + E_\tau \frac{g_\gamma(\tau B)}{\tau} E_\tau . \tag{13.3.30}$$

If $\phi_\mathcal{R}(y) \geq 0$ and $\phi_\mathcal{I}(y) \leq 0$ for $y \in \mathbb{R}_0^+$, then

$$f_\gamma(y) = \begin{cases} \gamma\phi_\mathcal{R}(y)(\mathbb{1} - \phi_\mathcal{R}(y)) - \phi_\mathcal{I}(y)(1 + \gamma\phi_\mathcal{I}(y))/|\phi(y)|^2, & y \in \Omega \\ 0, & y \in \Sigma \end{cases}$$

yields $f_\gamma(y) \geq 0$ for $y \in \mathbb{R}_0^+$ and $\gamma \in [0, 1]$. Hence $f_\gamma(\tau A) \geq 0$ for $\tau > 0$ and $\gamma \in [0, 1]$. Similarly, if $\psi_\mathcal{I}(y) \leq 0$ for $y \in \mathbb{R}_0^+$, then $g_\gamma(y) \geq 0$ for $y \in \mathbb{R}_0^+$ which implies $g_\gamma(\tau B) \geq 0$ for $\tau > 0$ and $\gamma \in [0, 1]$. Hence, one has $L_\gamma(\tau) \geq 0$ and $\tau > 0$ and $\gamma \in [0, 1]$, which shows that $(\mu\mathbb{1} + L_\gamma(\tau))^{-1}$ exists and is bounded for $\mu > 0$, $\tau > 0$ and $\gamma \in [0, 1]$.

Lemma 13.3.14 *Let A and B be non-negative self-adjoint operators such that the intersection* dom $(A^{1/2}) \cap dom(B^{1/2})$ *is dense in $\mathfrak{H}$. If $f(\cdot) : \mathbb{R}_0^+ \longrightarrow \mathbb{R}$ and $g(\cdot) : \mathbb{R}_0^+ \longrightarrow \mathbb{R}$ are finite-valued non-negative Borel measurable functions satisfying*

$$f(0) = 0, \quad f'(0) = 1 \quad \textit{and} \quad g(0) = 0, \quad g'(0) = 1 , \tag{13.3.31}$$

and

$$0 \leq g(y) \leq 1, \quad y \in \mathbb{R}_0^+ , \tag{13.3.32}$$

then

$$\text{s-}\lim_{\tau \to +0} (\mu\mathbb{1} + L(\tau))^{-1} = (\mu\mathbb{1} + C)^{-1} \tag{13.3.33}$$

for $\mu > 0$, where the form-sum: $C = A \dot{+} B$, and

$$L(\tau) := \frac{1}{\tau} f(\tau A) + \frac{1}{\tau} E_\tau\, g(\tau B)\, E_\tau , \quad \tau > 0 , \tag{13.3.34}$$

where $E_\tau = \chi_\Omega(\tau A)$, $\Omega \supseteq$ supp$(f) \cup \{0\}$, supp$(f) := \{y \in \mathbb{R}_0^+ : f(y) > 0\}$.

Proof Since $f(\cdot)$ takes only finite values the operator $f(\tau A)$, $\tau > 0$, is densely defined. Moreover, the operator $g(\tau B)$ is bounded. Hence the operator $L(\tau)$ is well-defined. We set

$$p(y) := \frac{1}{1 + f(y)} \quad \text{and} \quad q(y) := 1 - g(y), \quad y \in \mathbb{R}_0^+ . \tag{13.3.35}$$

We note that

$$0 \leq p(y) \leq 1, \quad y \in \mathbb{R}_0^+, \quad p(0) = 1, \quad \text{and} \quad p'(0) = -1\,,$$

as well as

$$0 \leq q(y) \leq 1, \quad y \in \mathbb{R}_0^+, \quad q(0) = 1, \quad \text{and} \quad q'(0) = -1\,.$$

Hence, $p(\cdot)$ and $q(\cdot)$ are generic Kato functions from the class $\mathcal{K}$. We have

$$L(\tau) = \\ p(\tau A)^{-1/2}\left(\frac{\mathbb{1} - p(\tau A)}{\tau} + \sqrt{p(\tau A)} E_\tau \frac{\mathbb{1} - q(\tau B)}{\tau} E_\tau \sqrt{p(\tau A)}\right) p(\tau A)^{-1/2}.$$

Let

$$\widehat{F}(\tau) := p(\tau A)^{1/2} E_\tau q(\tau B) E_\tau p(\tau A)^{1/2}, \quad \tau \geq 0\,.$$

Since $p(y) = 1$ for $y \in \Sigma := \mathbb{R} \setminus \Omega \subseteq \ker(f)$, $\ker(f) := \{y \in \mathbb{R}_0^+ : f(y) = 0\}$, we find

$$L(\tau) = p(\tau A)^{-1/2} E_\tau \widehat{S}_\tau E_\tau p(\tau A)^{-1/2}\,, \tag{13.3.36}$$

where

$$\widehat{S}_\tau := \frac{\mathbb{1} - \widehat{F}(\tau)}{\tau}, \quad \tau > 0\,.$$

Hence, the representation

$$(\mu \mathbb{1} + L(\tau))^{-1} = \sqrt{p(\tau A)}\left(\mu p(\tau A) + E_\tau \widehat{S}_\tau E_\tau\right)^{-1} \sqrt{p(\tau A)}\,,$$

holds. Since $0 \leq p(\tau A) \leq \mathbb{1}$ we obtain

$$\sqrt{p(\tau A)}\left(\mu \mathbb{1} + E_\tau \widehat{S}_\tau E_\tau\right)^{-1} \sqrt{p(\tau A)} \leq (\mu \mathbb{1} + L(\tau))^{-1}\,, \tag{13.3.37}$$

for $\mu > 0$. By the formula

$$(\mu \mathbb{1} + E_\tau \widehat{S}_\tau E_\tau)^{-1} = \frac{1}{\mu} E_\tau^\perp + E_\tau (\mu E_\tau + E_\tau \widehat{S}_\tau E_\tau)^{-1} E_\tau\,, \tag{13.3.38}$$

where $E_\tau^\perp = \mathbb{1} - E_\tau$, we get

$$\sqrt{p(\tau A)} E_\tau \left(\mu E_\tau + E_\tau \widehat{S}_\tau E_\tau\right)^{-1} E_\tau \sqrt{p(\tau A)} \leq (\mu \mathbb{1} + L(\tau))^{-1}\,,$$

for $\tau > 0$. Setting $\widehat{p}(y) := p(y)\chi_\Omega(y)$, $y \in \mathbb{R}_0^+$, we find the representation $\widehat{F}(\tau) = \sqrt{\widehat{p}(\tau A)}q(\tau B)\sqrt{\widehat{p}(\tau A)}$, $\tau > 0$. Since $\widehat{p}(\cdot)$ and $q(\cdot)$ are Kato functions we deduce

$$\text{s-}\lim_{\tau\to+0}(\mu\mathbb{1} + \widehat{S}_\tau(\tau))^{-1} = (\mu\mathbb{1} + C)^{-1},$$

for $\mu > 0$. Then taking into account formula (13.3.38) and $\text{s-lim}_{\tau\to+0} E_\tau = \mathbb{1}$, we find

$$\text{s-}\lim_{\tau\to+0} E_\tau(\mu E_\tau + E_\tau \widehat{S}_\tau E_\tau)^{-1}E_\tau = (\mu\mathbb{1} + C)^{-1}, \tag{13.3.39}$$

for $\mu > 0$. From (13.3.37) and (13.3.39) we finally get

$$((\mu\mathbb{1} + C)^{-1}h, h) \le \liminf_{\tau\to+0}\left((\mu\mathbb{1} + L(\tau))^{-1}h, h\right)^{-1},$$

for $h \in \mathfrak{H}$, $\mu > 0$. Moreover, from (13.3.30) we find

$$L(\tau) \ge \frac{f(\tau A)}{\tau}E_A([0,a)) + E_\tau\frac{g(\tau B)}{\tau}E_B([0,b))E_\tau, \quad a, b \in (0,\infty),$$

which gives the estimate

$$(\mu\mathbb{1} + L(\tau))^{-1} \le \left(\mu\mathbb{1} + \frac{f(\tau A)}{\tau}E_A([0,a)) + E_\tau\frac{g(\tau B)}{\tau}E_B([0,b))E_\tau\right)^{-1},$$

for $\mu > 0$ and $a, b \in (0,\infty)$. Using that $\text{s-lim}_{\tau\to+0} E_\tau = \mathbb{1}$ we obtain

$$\limsup_{\tau\to+0}\left((\mu\mathbb{1} + L(\tau))^{-1}h, h\right) \le \left((\mu\mathbb{1} + AE_A([0,a)) + BE_B([0,b)))^{-1}h, h\right),$$

for $h \in \mathfrak{H}$, $\mu > 0$ and $a, b \in (0,\infty)$. Since $a, b \in (0,\infty)$ are arbitrary we infer

$$\limsup_{\tau\to+0}\left((\mu\mathbb{1} + L(\tau))^{-1}h, h\right) \le \left((\mu\mathbb{1} + C)^{-1}h, h\right),$$

for $h \in \mathfrak{H}$, $\mu > 0$. Hence

$$\text{w-}\lim_{\tau\to+0}(\mu\mathbb{1} + L(\tau))^{-1} = (\mu\mathbb{1} + C)^{-1}, \tag{13.3.40}$$

for $\mu > 0$, and consequently,

$$\text{w-}\lim_{\tau\to+0}(\mu\mathbb{1} + L(\tau))^{-1/2} = (\mu\mathbb{1} + C)^{-1/2}, \tag{13.3.41}$$

for $\mu > 0$. From (13.3.40) and (13.3.41) we get (13.3.33). □

Lemma 13.3.15 *Let A and B be non-negative self-adjoint operators such that the intersection* dom $(A^{1/2}) \cap dom(B^{1/2})$ *is dense in* $\mathfrak{H}$. *If* ϕ *and* ψ *are admissible Kato functions such that*

$$\phi_{\mathcal{R}}(y) \geq 0, \quad \phi_{\mathcal{I}}(y) \leq 0 \quad \text{and} \quad \psi_{\mathcal{R}}(y) \geq 0, \quad \psi_{\mathcal{I}}(y) \leq 0, \quad y \in \mathbb{R}_0^+, \tag{13.3.42}$$

then the self-adjoint operators $L_\gamma(\tau)$, $\tau > 0$ *are well-defined and non-negative, and it holds*

$$\text{s-}\lim_{\tau \to +0} (\mu \mathbb{1} + L_\gamma(\tau))^{-1} = (\mu \mathbb{1} + C)^{-1}, \tag{13.3.43}$$

for the form-sum: $C = A \dotplus B$, *and* $\mu > 0$, $\gamma \in [0, 1]$.

Proof One can verify that the functions $f_\gamma(\cdot)$ and $g_\gamma(\cdot)$ defined by (13.3.27) and (13.3.28) satisfy the assumptions (13.3.31) and (13.3.32) for each $\gamma \in [0, 1]$. Setting $\Omega := \text{supp}(\phi) := \{y \in \mathbb{R}_0^+ : \phi(y) \neq 0\}$ we find $\Omega \supseteq \text{supp}(f_\gamma) \cup \{0\}$ for $\gamma = [0, 1]$. Moreover, the definition of $L_\gamma(\tau)$ given by (13.3.30) coincides with that one of $L(\tau)$ for each $\tau > 0$ and $\gamma \in [0, 1]$, see (13.3.34). Applying Lemma 13.3.14 we arrive at the sought conclusion (13.3.43). □

For purposes of the next statement, we introduce the operators

$$M_\gamma(\tau) := L_\gamma(\tau) + (1 + \gamma)\, \varphi_{\mathcal{R}}(\tau A) + (1 - \gamma)\, \varphi_{\mathcal{I}}(\tau A), \tag{13.3.44}$$

with $\tau > 0$ and $\gamma \in [0, 1]$. Since $L_\gamma(\tau) \geq 0$, $\varphi_{\mathcal{R}}(\tau A) \geq 0$ and $\varphi_{\mathcal{I}}(\tau A) \geq 0$ we get $M_\gamma(\tau) \geq 0$ for $\gamma \in [0, 1]$.

Lemma 13.3.16 *Let A and B be non-negative self-adjoint operators such that the intersection* dom $(A^{1/2}) \cap dom(B^{1/2})$ *is dense in* $\mathfrak{H}$. *If* ϕ *and* ψ *are admissible Kato functions such that the conditions* (13.3.42) *are satisfied, then* $M_\gamma(\tau) \geq 0$ *and*

$$\text{s-}\lim_{\tau \to +0} (\mu \mathbb{1} + M_\gamma(\tau))^{-1} = ((1 + \mu + \gamma)\mathbb{1} + C)^{-1}, \tag{13.3.45}$$

holds for $C = A \dotplus B$ *and* $\mu > 0$, $\gamma \in [0, 1]$.

Proof We note that $M_\gamma(\tau) \geq L_\gamma(\tau) + (1+\gamma)\varphi_{\mathcal{R}}(\tau A) \geq 0$ for $\tau > 0$ and $\gamma \in [0, 1]$. Let $\Omega_{\mathcal{R}}^{(n)} := \{y \in \mathbb{R} : \varphi_{\mathcal{R}}(y) \leq n\}$. We set

$$\varphi_{\mathcal{R}}^{(n)}(y) := \begin{cases} \varphi_{\mathcal{R}}(y), & y \in \Omega_{\mathcal{R}}^{(n)} \\ n, & y \in \mathbb{R}_0^+ \setminus \Omega_{\mathcal{R}}^{(n)} \end{cases}$$

for any $n \in \mathbb{N}$. Obviously we have $0 \le \varphi_{\mathcal{R}}^{(n)}(y) \le n$, $y \in \mathbb{R}_0^+$, and $0 \le \varphi_{\mathcal{R}}^{(n)}(y) \le \varphi_{\mathcal{R}}(y)$, $y \in \mathbb{R}_0^+$. Therefore one obtains

$$M_\gamma(\tau) \ge L_\gamma(\tau) + (1+\gamma)\varphi_{\mathcal{R}}^{(n)}(\tau A) \ge 0\,,$$

for $\tau > 0$ and $\gamma \in [0, 1]$, which yields

$$(\mu\mathbb{1} + M_\gamma(\tau))^{-1} \le (\mu\mathbb{1} + L_\gamma(\tau) + (1+\gamma)\varphi_{\mathcal{R}}^{(n)}(\tau A))^{-1},$$

for $\mu > 0$ and $\gamma \in [0, 1]$. Since $\text{s-lim}_{\tau\to+0}\,\varphi_{\mathcal{R}}^{(n)}(\tau A) = \mathbb{1}$, we obtain from Lemma 13.3.15 that

$$\text{s-}\lim_{\tau\to+0}(\mu\mathbb{1} + L_\gamma(\tau) + (1+\gamma)\varphi_{\mathcal{R}}^{(n)}(\tau A))^{-1} = ((1+\mu+\gamma)\mathbb{1} + C)^{-1}.$$

Hence,

$$\limsup_{\tau\to+0}\left((\mu\mathbb{1} + M_\gamma(\tau))^{-1}h, h\right) \le \left(((1+\mu+\gamma)\mathbb{1} + C)^{-1}h, h\right)\,, \qquad (13.3.46)$$

for $\mu > 0$ and $\gamma \in [0, 1]$. Furthermore, we note that

$$M_\gamma(\tau) \le (1+\gamma)\mathbb{1} + L_\gamma(\tau) + (1+\gamma)\rho(\tau A)\,,$$

where $\rho(y) := \varphi_{\mathcal{R}}(y) + \varphi_{\mathcal{I}}(y) - 1$, $y \in \mathbb{R}_0^+$. One has $\rho(0) = 0$ and $\rho'(0) = 1$. Hence, we find

$$M_\gamma(\tau) \le (1+\gamma)\mathbb{1} + L_\gamma(\tau) + \tau_0(1+\gamma)\frac{\rho(\tau A)}{\tau}\,,$$

for $0 < \tau \le \tau_0$. By

$$1 + \rho(y) = \frac{\phi_{\mathcal{R}}(y) - \phi_{\mathcal{I}}(y)}{|\phi(y)|^2} \ge \frac{\phi_{\mathcal{R}}^2(y) + \phi_{\mathcal{I}}^2(y)}{|\phi(y)|^2} \ge 1\,, \quad y \in \mathbb{R}_0^+,$$

we find $\rho(y) \ge 0$, $y \in \mathbb{R}_0^+$. We set

$$\widetilde{f}_\gamma(y) := \frac{1}{(1+\tau_0+\tau_0\gamma)}\left(f_\gamma(y) + \tau_0(1+\gamma)\rho(y)\right), \quad y \in \mathbb{R}_0^+, \quad \tau_0 > 0\,,$$

where $f_\gamma(y)$ is given by (13.3.27). It holds $\widetilde{f}_\gamma(0) = 0$ and $\widetilde{f}'_\gamma(0) = 1$ as well as $\widetilde{f}_\gamma(y) = 0$ for $y \in \Sigma$. One gets

$$L_\gamma(\tau) + \tau_0(1+\gamma)\frac{\rho(\tau A)}{\tau} \le (1+\tau_0+\gamma\tau_0)\left(\frac{\widetilde{f}_\gamma(\tau A)}{\tau} + E_\tau \frac{g_\gamma(\tau B)}{\tau} E_\tau\right),$$

for $\gamma \in [0,1]$ and $\tau \in (0, \tau_0]$, where $g_\gamma(y)$ is given by (13.3.28). Setting

$$\widetilde{L}_\gamma(\tau) := \frac{\widetilde{f}_\gamma(\tau A)}{\tau} + E_\tau \frac{g_\gamma(\tau B)}{\tau} E_\tau, \quad \tau > 0, \quad \gamma \in [0,1],$$

we obtain

$$M_\gamma(\tau) \le (1+\gamma)\,\mathbb{1} + (1+\tau_0+\gamma\tau_0)\,\widetilde{L}_\gamma(\tau), \quad \tau > 0, \quad \gamma \in [0,1],$$

which yields

$$\begin{aligned}&((1+\mu+\gamma)\,\mathbb{1} + (1+\tau_0+\gamma\tau_0)\widetilde{L}_\gamma(\tau))^{-1}\\ &\quad\le (\mu\mathbb{1} + M_\gamma(\tau))^{-1}, \quad \mu > 0, \quad 0 < \tau \le \tau_0,\end{aligned}$$

and $\gamma \in [0,1]$. Let $\lambda := (1+\mu+\gamma)/(1+\tau_0+\gamma\tau_0)$, we find

$$(\lambda\mathbb{1} + \widetilde{L}_\gamma(\tau))^{-1} \le (1+\tau_0+\gamma\tau_0)\,(\mu\mathbb{1} + M_\gamma(\tau))^{-1},$$

for $\mu > 0$, $0 < \tau \le \tau_0$ and $\gamma \in [0,1]$. Applying Lemma 13.3.14, we get

$$((\lambda\mathbb{1} + C)^{-1}h, h) \le (1+\tau_0+\gamma\tau_0)\,\liminf_{\tau\to+0}((\mu + M_\gamma(\tau))^{-1}h, h),$$

for $\mu > 0$, $\tau_0 > 0$, $\gamma \in [0,1]$ and $h \in \mathfrak{H}$. Since $\tau_0 > 0$ is arbitrary, we finally obtain

$$(((1+\mu+\gamma)\,\mathbb{1} + C)^{-1}h, h) \le \liminf_{\tau\to+0}((\mu\mathbb{1} + M_\gamma(\tau))^{-1}h, h), \tag{13.3.47}$$

for $\mu > 0$, $\gamma \in [0,1]$ and $h \in \mathfrak{H}$. From (13.3.46) and (13.3.47) we deduce that

$$\text{w-}\lim_{\tau\to+0}(\mu\mathbb{1} + M_\alpha(\tau))^{-1} = ((1+\mu+\gamma)\mathbb{1} + C)^{-1}, \tag{13.3.48}$$

holds for $\mu > 0$ and $\gamma \in [0,1]$. Since the relation (13.3.48) is valid for every $\mu > 0$, we get

$$\text{w-}\lim_{\tau\to+0}(\mu\mathbb{1} + M_\alpha(\tau))^{-1/2} = ((1+\mu+\gamma)\,\mathbb{1} + C)^{-1/2},$$

for $\mu > 0$, which yields

$$\text{s-}\lim_{\tau\to+0}(\mu\mathbb{1} + M_\alpha(\tau))^{-1/2} = ((1+\mu+\gamma)\,\mathbb{1} + C)^{-1/2},$$

for $\mu > 0$. The last relation proves (13.3.45). □

Let us introduce the operator-valued function

$$T(\tau) := \frac{1}{\sqrt{\mathbb{1} + M_0(\tau)}}\,(K_{\mathcal{R}}(\tau) + \varphi_{\mathcal{R}}(\tau A) - \varphi_{\mathcal{I}}(\tau A))\,\frac{1}{\sqrt{\mathbb{1} + M_0(\tau)}}, \quad \tau > 0, \tag{13.3.49}$$

where $M_0(\tau) = K_{\mathcal{I}}(\tau) + \varphi_{\mathcal{R}}(\tau A) + \varphi_{\mathcal{I}}(\tau A) \geq 0$, see (13.3.26), (13.3.30) and (13.3.44).

Lemma 13.3.17 *Let A and B be non-negative self-adjoint operators in a separable Hilbert space $\mathfrak{H}$ such that* $\operatorname{dom}(A^{1/2}) \cap \operatorname{dom}(B^{1/2})$ *is dense in $\mathfrak{H}$. If ϕ and ψ are admissible Kato functions such that the conditions* (13.3.42) *are satisfied, then $T(\tau) \geq -\mathbb{1}$, $\tau > 0$, and*

$$\text{s-}\lim_{\tau\to+0}(i\,\mathbb{1} + T(\tau))^{-1} = (i\,\mathbb{1} + (2\mathbb{1} + C)^{-1})^{-1}, \tag{13.3.50}$$

where $T(\tau)$ is defined by (13.3.49).

Proof Since $M_\gamma(\tau) \geq 0$ for $\gamma \in [0, 1]$ and $\tau > 0$, we find $\mathbb{1} + T(\tau) \geq 0$, which yields $T(\tau) \geq -\mathbb{1}$. Hence $\gamma T(\tau) \geq -\gamma\,\mathbb{1}$ holds for $\gamma \in [0, 1]$, and therefore the operator $\mathbb{1} + \gamma T(\tau)$ is boundedly invertible for $\gamma \in [0, 1)$ and we have the representation

$$(\mathbb{1} + M_\gamma(\tau))^{-1} = \frac{1}{\sqrt{\mathbb{1} + M_0(\tau)}}\,(\mathbb{1} + \gamma T(\tau))^{-1}\,\frac{1}{\sqrt{\mathbb{1} + M_0(\tau)}},$$

for $\gamma \in [0, 1)$. Setting $\gamma = 0$ we find from Lemma 13.3.16 that

$$\text{s-}\lim_{\tau\to+0}\frac{1}{\sqrt{\mathbb{1} + M_0(\tau)}} = (2\mathbb{1} + C)^{-1/2}.$$

Since $\text{s-}\lim_{\tau\to+0}(\mathbb{1} + M_\gamma(\tau))^{-1} = ((2+\gamma)\mathbb{1} + C)^{-1}$ for $\gamma \in [0, 1]$, by Lemma 13.3.16 we get that $\text{w-}\lim_{\tau\to+0}(\mathbb{1} + \gamma\,T(\tau))^{-1}$ exists for $\gamma \in [0, 1)$ and is given by

$$\text{w-}\lim_{\tau\to+0}(\mathbb{1} + \gamma T(\tau))^{-1} = \frac{2\mathbb{1} + C}{(2+\gamma)\mathbb{1} + C}, \quad \gamma \in [0, 1).$$

Hence, we get

$$\text{w-}\lim_{\tau\to+0}(\nu\mathbb{1}+T(\tau))^{-1}=\frac{2\mathbb{1}+C}{\mathbb{1}+\nu(2\mathbb{1}+C)}=(\nu\mathbb{1}+(2\mathbb{1}+C)^{-1})^{-1},\quad \nu\in(1,\infty).$$

However, this relation yields

$$\text{s-}\lim_{\tau\to+0}(\nu\mathbb{1}+T(\tau))^{-1}=(\nu\mathbb{1}+(2\mathbb{1}+C)^{-1})^{-1},\quad \nu\in(1,\infty)\,,$$

which immediately implies (13.3.50). □

Finally, for technical reasons, we need the following lemma. First, we recall that a bounded operator X is called *accretive*, if $\mathrm{Re}\,(X\,h,h)\geq 0$ for any $h\in\mathfrak{H}$.

Lemma 13.3.18 *Let $\{X(\tau)\}_{\tau>0}$ be a sequence of bounded accretive operators in $\mathfrak{H}$. If there is self-adjoint operator Y such that*

$$\text{w-}\lim_{\tau\to+0}(X(\tau)-\xi\,\mathbb{1})^{-1}=(i\,Y-\xi\,\mathbb{1})^{-1}\,,$$

for some $\mathrm{Re}(\xi)\leq 0$*, then*

$$\text{s-}\lim_{\tau\to+0}(X(\tau)-\xi\,\mathbb{1})^{-1}=(i\,Y-\xi\,\mathbb{1})^{-1}\,.$$

Proof We set

$$W(\tau):=(X(\tau)+\overline{\xi}\,\mathbb{1})\,(X(\tau)-\xi\,\mathbb{1})^{-1}\quad\text{and}\quad W:=(i\,Y+\overline{\xi}\,\mathbb{1})(i\,Y-\xi\,\mathbb{1})^{-1},$$

$\tau>0$. Note that by definitions and by condition of lemma one infers that $\{W(\tau)\}_{\tau>0}$ is a family of contractions and that $\text{w-}\lim_{\tau\to+0}W(\tau)=W$. Then in virtue of $\|Wh\|=\|h\|$ and

$$\|W(\tau)h-Wh\|^2=\|W(\tau)h\|^2+\|h\|^2-2\,\mathrm{Re}((W(\tau)h,Wh)),\quad \tau>0\,,$$

we find

$$\limsup_{\tau\to+0}\|W(\tau)h-Wh\|^2\leq 2\|h\|^2-2\lim_{\tau\to+0}(W(\tau)h,Wh)=0\,,$$

which completes the proof. □

Theorem 13.3.19 *Let A and B be non-negative self-adjoint operators such that the intersection* $\mathrm{dom}\,(A^{1/2})\cap dom(B^{1/2})$ *is dense in $\mathfrak{H}$. If ϕ and ψ admissible Kato functions such that the conditions* (13.3.42) *are satisfied, then the strong limit:*

$$\text{s-}\lim_{n\to\infty}(\phi(tA/n)\psi(tB/n))^n=\mathrm{e}^{-itC}\,,\tag{13.3.51}$$

holds for generator $C = A \dot{+} B$ *uniformly in* $t \in [-T, T]$, *where* $T > 0$, *cf.* (13.3.8).

Proof Taking into account the representation

$$\frac{1}{\sqrt{\mathbb{1} + M_0(\tau)}} (i\,\mathbb{1} + T(\tau))^{-1} \frac{1}{\sqrt{\mathbb{1} + M_0(\tau)}} = (i\,\mathbb{1} + (1+i)\,\varphi(\tau A) + K(\tau))^{-1},$$

we find

$$\frac{1}{\sqrt{\mathbb{1} + M_0(\tau)}} (i\,\mathbb{1} + T(\tau))^{-1} \frac{1}{\sqrt{\mathbb{1} + M_0(\tau)}} = \sqrt{\widetilde{\phi}(\tau A)}\,(Z(\tau) - \xi_0\,\mathbb{1})^{-1} \sqrt{\widetilde{\phi}(\tau A)}\,, \tag{13.3.52}$$

for $\tau > 0$. Here $\widetilde{\phi}(y) := 1/\varphi(y)$ for $y \in \mathbb{R}_0^+$, see (13.3.23), and $\xi_0 = -(1+i)$, where

$$Z(\tau) := i\,\widetilde{\phi}(\tau A) + \widetilde{S}_\tau\,, \quad \tau > 0\,,$$

and

$$\widetilde{S}_\tau := \sqrt{\widetilde{\phi}(\tau A)}\; K(\tau) \sqrt{\widetilde{\phi}(\tau A)}\,, \quad \tau > 0\,.$$

Hence, by definition of $K(\tau)$ (13.3.24) together with $\widetilde{\phi}(y) = 1/\varphi(y)$ and $E_\tau = \chi_\Omega(\tau A)$, we obtain that for $\tau > 0$

$$\widetilde{S}_\tau = E_\tau \left(\frac{\mathbb{1} - \phi(\tau A)}{\tau} + \sqrt{\phi(\tau A)}\, \frac{\mathbb{1} - \psi(\tau B)}{\tau} \sqrt{\phi(\tau A)} \right) E_\tau = E_\tau S_\tau E_\tau\,, \tag{13.3.53}$$

where

$$S_\tau := \frac{\mathbb{1} - F(\tau)}{\tau}\,, \quad \tau > 0\,,$$

and $F(\tau) := \sqrt{\phi(\tau A)}\; \psi(\tau B) \sqrt{\phi(\tau A)}$. Since for each $\tau > 0$ the operators S_τ are accretive and $\mathrm{Re}(\widetilde{\phi}(\tau A)) \geq 0$, $\tau \geq 0$, the operator $Z(\tau)$ is accretive and the inverse operator $(Z(\tau) - \xi_0\,\mathbb{1})^{-1}$ exists and its norm is bounded by one for $\tau > 0$. From the representation (13.3.52), Lemma 13.3.16 and Lemma 13.3.17, we get

$$\text{w-}\lim_{\tau \to +0} (Z(\tau) - \xi_0\,\mathbb{1})^{-1} = (i\,C - \xi\,\mathbb{1})^{-1}, \quad \xi = -1 - 2i\,, \quad C = A \dot{+} B\,, \tag{13.3.54}$$

where we have used that s-$\lim_{\tau\to+0}\sqrt{\phi(\tau A)} = \mathbb{1}$. Since $\widetilde{S}_\tau$ is accretive, we infer that

$$\begin{aligned}&(i\,\mathbb{1} + \widetilde{S}_\tau - \xi_0\,\mathbb{1})^{-1} - (Z(\tau) - \xi_0\,\mathbb{1})^{-1}\\&\quad = i\,(i\,\mathbb{1} + \widetilde{S}_\tau - \xi_0\,\mathbb{1})^{-1}(\widetilde{\phi}(\tau A) - \mathbb{1})\,(Z(\tau) - \xi_0\,\mathbb{1})^{-1}\,,\end{aligned} \tag{13.3.55}$$

for $\tau > 0$. Then from (13.3.54), (13.3.55) and s-$\lim_{\tau\to+0}\widetilde{\phi}(\tau A) = \mathbb{1}$, we deduce

$$\text{w-}\lim_{\tau\to+0}(\widetilde{S}_\tau - \xi\,\mathbb{1})^{-1} = (i\,C - \xi\,\mathbb{1})^{-1}.$$

Therefore, applying Lemma 13.3.18, we obtain

$$\text{s-}\lim_{\tau\to+0}(\widetilde{S}_\tau - \xi\,\mathbb{1})^{-1} = (i\,C - \xi\,\mathbb{1})^{-1}\,,\quad \xi = -1 - 2i\,,\quad C = A \dotplus B\,,$$

which also yields

$$\text{s-}\lim_{\tau\to+0}(\mu\mathbb{1} + \widetilde{S}_\tau)^{-1} = (\mu\mathbb{1} + i\,C)^{-1},\quad \mu > 0\,. \tag{13.3.56}$$

Next, we use (13.3.56) for representation

$$\begin{aligned}&(\mu\mathbb{1} + S_\tau)^{-1} - (\mu\mathbb{1} + \widetilde{S}_\tau)^{-1} =\\&(\mu\mathbb{1} + S_\tau)^{-1}(\widetilde{S}_\tau - S_\tau)\,[(\mu\mathbb{1} + \widetilde{S}_\tau)^{-1} - (\mu\mathbb{1} + i\,C)^{-1}] +\\&(\mu\mathbb{1} + S_\tau)^{-1}(\widetilde{S}_\tau - S_\tau)\,(\mu\mathbb{1} + i\,C)^{-1}\,,\quad \tau > 0\,,\quad \mu > 0\,.\end{aligned} \tag{13.3.57}$$

On account of (13.3.53), one get $\|(\mu\mathbb{1} + S_\tau)^{-1}(\widetilde{S}_\tau - S_\tau)\| \le M_\mu$ for $\mu > 0$, uniformly in $\tau > 0$. Then owing to (13.3.56) the s-$\lim_{\tau\to+0}$ of the second line in representation (13.3.57) is equal to zero.

For the third line, let $\{E_A(\Delta)u\}_{\Delta\subset\mathbb{R}_0^+}$ for $u \in \mathfrak{H}$, be a dense in $\mathfrak{H}$ set of vectors from the core (C), which is generated by the spectral projectors of operator A. Consider the operator of the third line on the dense set $\{(\mu\mathbb{1} + i\,C)\,E_A(\Delta)u\}_{\Delta\subset\mathbb{R}_0^+}$ for $u \in \mathfrak{H}$. Consequently, because of (13.3.53) and s-$\lim_{\tau\to+0} E_\tau = \mathbb{1}$ one obtains that

$$(\mu\mathbb{1} + S_\tau)^{-1}(E_\tau S_\tau E_\tau E_A(\Delta) - S_\tau E_A(\Delta))\,u = (\mu\mathbb{1} + S_\tau)^{-1}(E_\tau - \mathbb{1})S_\tau E_A(\Delta)\,u\,,$$

for sufficiently small $\tau > 0$, as well as that the right-hand side tends to zero for $\tau \to +0$ on the dense set: $\{E_A(\Delta)u\}_{\Delta\subset\mathbb{R}_0^+}$, $u \in \mathfrak{H}$. As a result, we infer from (13.3.57) that

$$\text{s-}\lim_{\tau\to+0}(\mu\mathbb{1} + S_\tau)^{-1} = (\mu\mathbb{1} + i\,C)^{-1},\quad \mu > 0\,. \tag{13.3.58}$$

On account of Proposition 1.9.30 (*Trotter–Kato–Neveu* theorem) and Proposition 13.1.9 (or by another way around, *Chernoff* product Theorem 4.1.3 (Sect. 4.1)) the strong limit (13.3.58) yields product formula: $\text{s-lim}_{n\to\infty} F(t/n)^n = \mathrm{e}^{-itC}$, for $t \in \mathbb{R}$, where $\{F(\tau) = \sqrt{\phi(\tau A)}\, \psi(\tau B) \sqrt{\phi(\tau A)}\}_{\tau \in \mathbb{R}}$. Finally, since a product of bounded operators is continuous in the strong operator topology, the extension of the limit $\text{s-lim}_{n\to\infty} F(t/n)^n$ to the nonsymmetric case (13.3.51) is straightforward, that completes the proof. □

Corollary 13.3.20 *Let A and B be non-negative self-adjoint operators such that* dom $(A^{1/2}) \cap dom(B^{1/2})$ *is dense in $\mathfrak{H}$. If $\phi(y) = \psi(y) = (1 + iy)^{-1}$, $y \in \mathbb{R}_0^+$, then for the form-sum: $C = A \dot{+} B$,*

$$\text{s-}\lim_{n\to\infty} \left((\mathbb{1} + itA/n)^{-1}(\mathbb{1} + itB/n)^{-1}\right)^n = \mathrm{e}^{-itC}, \tag{13.3.59}$$

uniformly in $t \in [0, T]$ for $T > 0$.

Proof One easily verifies that $\phi(\cdot)$ and $\psi(\cdot)$ are admissible functions. Moreover, one has

$$\phi_{\mathcal{R}}(y) = \psi_{\mathcal{R}}(y) = \frac{1}{1+y^2} \geq 0 \quad \text{and} \quad \phi_{\mathcal{I}}(y) = \psi_{\mathcal{I}}(y) = -\frac{y}{1+y^2} \leq 0, \tag{13.3.60}$$

which shows that the assumptions (13.3.42) are satisfied. Applying Theorem 13.3.19 we arrive at the sought conclusion. □

As a result, for admissible Kato functions (13.3.60) one obtains the *Lapidus* product formula (13.3.59) for *imaginary resolvents*. See also (13.4.8) and comments in Notes to Chap. 13 (Notes to Sects. 13.3, 13.3.1).

Another choice of admissible functions leads to the *Ichinose* product formula (13.3.61). Note that function $\phi(y) = \mathrm{e}^{-iy}$, $y \in \mathbb{R}_0^+$, does not satisfy the conditions $\mathrm{Re}(\phi(y)) \geq 0$ and $\mathrm{Im}(\phi(y)) \leq 0$. However, its modification $\phi(y) := \mathrm{e}^{-iy}\chi_{[0,\pi/2]}(y)$, $y \in \mathbb{R}_0^+$, obeys $\mathrm{Re}(\phi(y)) \geq 0$ and $\mathrm{Im}(\phi(y)) \leq 0$, $y \in \mathbb{R}_0^+$. In particular, the function (13.3.4) satisfies the conditions $\mathrm{Re}(f(y)) \geq 0$ and $\mathrm{Im}(f(y)) \leq 0$. The last observation leads to the following result.

Corollary 13.3.21 *Let A and B be non-negative self-adjoint operators in a separable Hilbert space $\mathfrak{H}$ such that* dom $(A^{1/2}) \cap$ dom $(B^{1/2})$ *is dense in $\mathfrak{H}$. Let ϕ and ψ be admissible functions. Then there are real numbers $\delta_\phi > 0$ and $\delta_\psi > 0$ such that*

$$\text{s-}\lim_{n\to\infty} \left(\phi(tA/n)E_A(([0, n\delta_\phi/t])\psi(tB/n)E_B(([0, n\delta_\psi/t])\right)^n = \mathrm{e}^{-itC}, \tag{13.3.61}$$

holds uniformly in $t \in [0, T]$ for $T > 0$.

Proof If the function ϕ is admissible, then $\phi(0) := \lim_{y\to+0}\phi(y) = 1$ and $\phi'(0) = \lim_{y\to+0}(\phi(y)-1)/y = -i$. In particular, this yields $\mathrm{Re}(\phi(0)) = \lim_{y\to+0}\phi_{\mathcal{R}}(y) = 1$ and $\mathrm{Im}(\phi(0)) = \lim_{y\to+0}\phi_{\mathcal{I}}(y) = 0$, as well as, $\mathrm{Re}(\phi'(0)) = \lim_{y\to+0}(\phi_{\mathcal{R}}(y)-1)/y = 0$ and $\mathrm{Im}(\phi'(0)) = \lim_{y\to+0}(\phi_{\mathcal{I}}(y))/y = -1$, where $\phi_{\mathcal{R}} := \mathrm{Re}(\phi(y))$ and $\phi_{\mathcal{I}} := \mathrm{Im}(\phi(y))$, $y \in \mathbb{R}_0^+$. Hence there is a $\delta_\phi > 0$ such that $\phi_{\mathcal{R}}(y) \geq 0$ and $\phi_{\mathcal{I}}(y) \leq 0$ for $y \in [0, \delta_\phi]$, and consequently, the function $\phi(y)\chi_{[0,\delta_\phi]}(y)$ satisfies the assumptions of Theorem 13.3.19. Similar considerations are valid for ψ. □

Corollary 13.3.21 yields the product formula *à la* Ichinose, (13.3.9), mentioned in Sect. 13.3.1. See also Notes to Chap. 13 (Notes to Sects. 13.3, 13.3.1 and 13.3.4).

13.4 Notes to Chapter 13

Notes to Sect. 13.1

The main part of this section is standard. Few remarks together with references serve for illumination a content of certain points of presentation of the Sect. 13.1.

Section 13.1.1

Definition 13.1.1 of unitary groups on a Hilbert space $\mathfrak{H}$ is standard. However, below we shall discuss another version of this definition motivated by the C_0-semigroups and m-accretive generators.

The statement in Remark 13.1.2 (2) is the *von Neumann* theorem:

If one-parameter unitary group on a separable Hilbert space $\mathfrak{H}$ is weakly measurable, then it is strongly continuous. See, for example, [RS80, Theorem VIII.9].

We remind a similar result for semigroups on a Banach space $\mathfrak{X}$, that is, allude to [HiPh57], Theorem 3.5.3, saying that a vector-valued *weakly* measurable function with a *separable* image is *strongly* measurable. Then, a one-sided weak continuity implies strong measurability. As a consequence, on account of [HiPh57], Theorem 10.2.3, one concludes that the *weak* one-sided continuity of semigroup $\{U(t)\}_{t>0}$ yields the *strong* continuity of this semigroup away from $t = 0$. (cf. Notes to Sect. 1.2.)

We also remark that there is yet another, than the *functional calculus* approach (cf. Definition 13.1.1, Remark 13.1.2 (3) and Proposition 13.1.3, (13.1.4)), method for construction and analysis of unitary C_0-groups. It is based on the *Lumer–Phillips* theorem (Proposition 1.8.10) about C_0-semigroups for m-accretive generators and *embedding* of C_0-semigroup in a unitary C_0-group. The main idea of the embedding a contraction C_0-semigroup in a unitary C_0-group is presented in Remark 13.1.5 (a)–(d). For more details we would mention, e.g., [EN00, Chapter II, Section 3.11] and [Haa06, Theorem A.8.9].

We use this method below for the proof of the *Stone* theorem (cf. Proposition 13.1.4), which essentially (see, e.g., [Haa06, Theorem C.7.4] and [Amr09, Proposition 5.1]) is the following claim.

Proposition 13.4.1 *A densely defined in a Hilbert space $\mathfrak{H}$ operator A generates a C_0-group $\{U_A(t) = \mathrm{e}^{-i\,t\,A}\}_{t\in\mathbb{R}}$ of unitary operators if and only if the operator A is self-adjoint in $\mathfrak{H}$.*

Proof If A with dense $\mathrm{dom}\,(A) \subset \mathfrak{H}$ is self-adjoint in $\mathfrak{H}$, then operators $A_+ := i\,A$ and $A_- := -i\,A$ are both m-accretive, see Definition 1.8.7 and Remark 13.1.5. On account of the *Lumer-Phillips* theorem (Proposition 1.8.10) both operators generate C_0-contraction semigroups. Then on account of Remark 13.1.5 the operator A generates a unitary C_0-group by formula (13.1.13).

To prove the converse assertion we suppose that $\{U(t)\}_{t\in\mathbb{R}}$ is a C_0-group of unitary operators. Then for $z \in \mathbb{C}$, such that $\mathrm{Im}\,z > 0$, one can define operator

$$R(z) := i\int_0^\infty dt\;\mathrm{e}^{\,i\,z\,t}\,U(t)\,,\qquad z\in\mathbb{C}^+\,, \tag{13.4.1}$$

which is bounded: $R(z) \in \mathcal{L}(\mathfrak{H})$, and $R(z)\,U(t) = U(t)\,R(z)$. Accordingly, by (13.4.1) we infer that closure of the range: $\overline{\mathrm{ran}\,(R(z))} = \mathfrak{H}$, and also obtain for $u \in \mathfrak{H}$ the representation:

$$\frac{i}{t}\,(U(t)-\mathbb{1})\,R(z)\,u = \frac{i}{t}\,(\mathrm{e}^{\,-i\,z\,t}-1)\,R(z)\,u + \frac{1}{t}\,\mathrm{e}^{\,-i\,z\,t}\int_0^t ds\;\mathrm{e}^{\,i\,z\,s}\,U(s)\,u\,. \tag{13.4.2}$$

As $t \to 0$ the right-hand side of (13.4.2) converges (because of the strong continuity of $\{U(t)\}_{t\in\mathbb{R}}$) to $z\,R(z)\,u+u$ for any $u \in \mathfrak{H}$. On account of (13.4.2) this limit defines on domain $\mathrm{dom}\,(A) = \mathrm{ran}\,(R(z))$ operator A by

$$\lim_{t\to 0}\frac{i}{t}\,(U(t)-\mathbb{1})\,R(z)\,u =: A\,R(z)\,u = z\,R(z)\,u + u\,,\qquad u\in\mathfrak{H}\,. \tag{13.4.3}$$

By virtue of $\overline{\mathrm{ran}\,(R(z))} = \mathfrak{H}$ the operator A is densely defined in $\mathfrak{H}$ and

$$(A - z\,\mathbb{1})\,R(z)\,u = u\,,\qquad u\in\mathfrak{H}\,,\qquad \mathrm{Im}\,z > 0\,. \tag{13.4.4}$$

Since $\mathrm{dom}\,(A) = \mathrm{ran}\,(R(z))$, by taking $z = i$ in (13.4.4) we infer that $\mathrm{ran}\,(A - i\,\mathbb{1}) = \mathfrak{H}$.

Now one can *extend* the operator (13.4.1) to $z \in \mathbb{C}$, such that $\mathrm{Im}\,z < 0$, by setting for this case

$$R(z) := -i\int_{-\infty}^0 dt\;\mathrm{e}^{\,i\,z\,t}\,U(t)\,,\qquad z\in\mathbb{C}^-\,, \tag{13.4.5}$$

and by repeating the arguments above. Then (13.4.4) holds for $\mathrm{Im}\,z < 0$ and in particular for $z = -\,i$ we obtain that $\mathrm{ran}\,(A + i\,\mathbb{1}) = \mathfrak{H}$.

Finally, the unitarity of $\{U(t)\}_{t\in\mathbb{R}}$ and definition (13.4.2) of operator A yield that for any $u, v \in \mathrm{dom}\,(A)$

$$\begin{aligned}(A\,u, v) &= \lim_{t\to 0}\,(i/t\,(U(t) - \mathbb{1})\,u, v) = \lim_{t\to 0}\,(u,\, i/t\,(\mathbb{1} - U(-t))\,v)\\ &= \lim_{s\to 0}\,(u,\, i/s\,(U(s) - \mathbb{1})\,v) = (u, A\,v)\,.\end{aligned} \tag{13.4.6}$$

Therefore, on account of $\mathrm{ran}\,(A \mp i\,\mathbb{1}) = \mathfrak{H}$ and (13.4.6) the operators A on domain $\mathrm{dom}\,(A)$ is a self-adjoint generator of the unitary C_0-group $\{U(t) = \mathrm{e}^{-i\,t\,A}\}_{t\in\mathbb{R}}$, cf. representation by formula (13.1.13). □

The Proposition 13.1.6 is Nelson's result [Nel64] for a Hilbert space $\mathfrak{H}$. Note that it is a special case of the Trotter theory, see [Trot58] and [Trot59], when operator $A+B$ is *self-adjoint*. Our proof of Proposition 13.1.6 is an adjustment of arguments for a Hilbert space $\mathfrak{H}$ from [Nel64, Theorem 9], cf. [RS80, Theorem VIII.30].

Section 13.1.2

Here we basically follow the *Chernoff approach* to product formulæ approximations, see [Cher74, §2].

The Proposition 13.1.8 is motivated by [Cher74, Corollary 2.5.4]. Here it is supposed that the strong limit $\{U(t)\}_{t\in\mathbb{R}}$ of the product approximants exists. Existence of a *strongly* continuous *unitary* group $\{\widetilde{U}(t)\}_{t\in\mathbb{R}}$ on a *separable* Hilbert space $\mathfrak{H}$ such that $\widetilde{U}(t) = U(t)$ *almost everywhere* (a.e.) on $\mathbb{R}$ in the proof of (c) results from [Cher74, Corollary 2.5.2]. The proof of (d) follows from the von Neumann theorem, see Remark 13.1.2 (2) and notes to Sect. 13.1.1.

Propositions 13.1.9 and 13.1.11 show that in contrast to the case of C_0-semigroups (cf. Theorem 1.9.34, or Theorem 3.1.16) the application of the *Chernoff approach* to the product formula approximation of unitary groups is quite restricted. These assertions for *self-adjoint* and *essentially* self-adjoint sums of unbounded self-adjoint operators: $A + B$, are *conditioned*. The latter is equivalent to demand of validity the *Trotter-Kato-Neveu* theorem, as, e.g., in (1.9.104). Note that these demands are largely *excessive* in the case of C_0-semigroups.

We shall return to this problem in the Notes to Sect. 13.3 (see below) for the case of positive self-adjoint operators $A > 0$ and $B > 0$. These conditions allow to advance in construction of the strong unitary product approximations, but in a rather *round-about way*.

On that account it appears difficult to generalise a well-known unitary Trotter product theorem [Trot59] for *essentially* self-adjoint operator $A + B$ to the case of arbitrary pairs of self-adjoint generators A and B.

Notes to Sect. 13.2

Besides indications of references, we complete here some missing from the text of Sect. 13.2 proofs and comments.

Section 13.2.1

For essential self-adjointness of the operator sum $(A + B)\,u(x) = (-i\,\partial_x + V(x))\,u(x)$ on $C_0^\infty(\mathbb{R})$ we refer to [Cher73, Theorems 1.3 and 2.2], cf. [Cher77, p. 362].

The rest of this subsection is motivated by [Cher74, §5] and essentially based on results from [Ich03, Ch.4].

Section 13.2.2

We follow here the paper [IT04a]. Note that *commutator conditions* and their relevance to the unitary Trotter product formulæ were demonstrated for the first time in [IT04, Sect.3].

In Sect. 13.2.1, we have observed that there are examples with $A = -i\,\partial_x$ and $B = V(\cdot)$ in $L^2(\mathbb{R})$ for which the unitary product formula in operator norm holds *as well as* does not hold. This point is scrutinised in Sect. 13.2.3.

Section 13.2.3

To prove theorems, we use the method of *commutators* suggested in paper [IT04] for optimal error-bound estimates in the case of self-adjoint Trotter product formulæ for semigroups generated by Schrödinger operators.

The proofs are essentially taken from [IT04a, Section 2].

Section 13.2.4

Here again we basically follow the paper [IT04a, Section 3. Applications].

For some recent example with applications of results from Sect. 13.2.4 in mathematical physics, we refer to the paper [RBW22].

About the *Calderón–Zygmund* theorem, see, for example, [St70, Ch.II, Section 3, Theorem 2].

For the proof of Lemma 13.2.12, we suggest to consult [IT04a, Lemma 3.1].

Notes to Sect. 13.3

Note that apart from a pure mathematical interest, the *unitary* Trotter product formula is tightly related to the Quantum Mechanics. In particular, it provides a natural way to define the *Feynman path* integral, see [Dal60, Nel64] and books [Ex85, JL00]. The fundamental problem is a conflict between demands formulated on the pair of operators A, B in (mathematical) physics and rather strict conditions on them indispensable for mathematical proof of the *unitary* Trotter product formula even in the strong operator topology. Seeing that for the unitary Trotter product formulæ the strong and the weak operator topologies coincide (Remark 13.1.7), one is looking for *new* (or *alternative*) topologies of convergence.

Section 13.3.1

First, we remind a *partially solved* problem of extension to *unitary* groups of the strongly convergent semigroup *Kato* product formula in Theorem 5.1.5, or by another way around, a partial generalisation of the strongly convergent *unitary* Trotter product formula to the *Kato* conditions on generators A, B when $C =$

$A \dot{+} B$ is self-adjoint *form-sum*, cf. Remark 13.3.1(a). A solution is presented in Sect. 13.3.4 for *admissible* Kato functions satisfying supplementary conditions that *exclude* exponential and, consequently, the conclusion of Remark 13.3.1(b) is still valid.

The fact that we retreated in [EI05] and [ENZ11] into that corner of the product formula convergence in topology of the locally convex *Fréchet* space $L^2_{loc}(\mathbb{R}, \mathfrak{H})$, which for bounded operators $\mathcal{L}(\mathfrak{H})$ is weaker than the strong operator topology, indicates how far we are from the ultimate aim to *uplift* this convergence to the operator-norm topology.

The comment about density of the form-sum domain $\mathrm{dom}\,(A^{1/2}) \cap \mathrm{dom}\,(B^{1/2})$ (Remark 13.3.1(a)) is due to [JL00, Problem 11.3.9, Proposition 11.7.3].

For introduction to locally convex topological vector spaces see, e.g., [KF89, Chapter III, §5] and [RS80, Section V.2] for *Fréchet* spaces. About the proof of Proposition 13.3.2 one can consult, for example, [BEH94, 1.4.4 Theorem]. More comments on Remark 13.3.3 that convergence (13.3.6) in the L^2-topology of the *Fréchet* space implies the existence of *subsequences* such that they converge in the *strong* operator topology for a.a. $t \in [-T, T]$ one finds in [FP08, Section 11].

In order to put our message into a proper context, we recall first some known results relevant for our presentation. In [Ich80, §3], this observation was generalised to functions $\zeta(t, \lambda)$ from the class denoted as $\mathcal{F}_{\nu,\,\mu}(\tau, \gamma, \varepsilon)$, $0 < \tau \le \infty$, $0 < \mu < \nu \le 1$, $\gamma \in \mathbb{R}$ and $\varepsilon = \pm 1$, defined in a slightly *cumbersome* way.

Consider a particular case. Let f be an admissible (Kato) function. Choosing $\gamma = 0$ and $\varepsilon = -1$ one can verify that $\zeta(t, \lambda) := f(t\lambda) \in \mathcal{F}_{\nu,\mu}(\tau, 0, -1)$ if and only if $f(\cdot)$ is continuous with :

$$\mathrm{Im}(f(x)) \le 0 \quad \text{and} \quad |1 - f(x)| \le \min\left\{2\mu, \frac{1}{\nu}|\mathrm{Im}(f(x))|\right\}, \qquad x \in \mathbb{R}^+_0 . \tag{13.4.7}$$

In particular, there exists a $\delta > 0$ such that the conditions (13.4.7) are satisfied for

$$f(x) = \frac{1}{1+ix}\chi_{[0,\delta]}(x) + \chi_{(\delta,\infty)}(x)\,, \quad x \ge 0\,,$$

which yields

$$\mathrm{e}^{-itC} = \text{s-}\lim_{n\to\infty} (f(tA/n)f(tB/n))^n\,, \quad t \ge 0\,.$$

In [Lap82, Theorem], see also [Lap85, Theorem 1] or [JL00, Corollary 11.3.5], *Lapidus* showed a slightly stronger result, namely that

$$\mathrm{e}^{-itC} = \text{s-}\lim_{n\to\infty} \left((\mathbb{1} + itA/n)^{-1}(\mathbb{1} + itB/n)^{-1}\right)^n, \tag{13.4.8}$$

holds uniformly in $t \in [0, T]$, $T > 0$. There we prove (13.4.8) within the *Ichinose* theory of the *admissible* Kato functions, see (13.3.59) in Corollary 13.3.20.

Some averaging formulæ were proposed then in [Lap82] for *real* parameters in cases of linear and non-linear semigroups. But it was *Cachia* [Ca05] who for the first time linked the *imaginary* parameter averaging formulæ to the L^2-convergence. In fact, inspired by the paper [EI05], he proved in [Ca05, Theorem 1.1] that the limit

$$\lim_{n\to\infty} \int_0^T dt \left\| \left(\frac{e^{-2itA/n} + e^{-2itB/n}}{2} \right)^n h - e^{-itC} h \right\|^2 = 0,$$

holds for any $h \in \mathfrak{H}$ and $T > 0$, that is, on a complete locally convex metrisable *Fréchet* space $L^2_{loc}(\mathbb{R}, \mathfrak{H})$, [KF89, Chapter III,§5.3], or [RS80, Chapter V.2]. In [Ca05] he also introduced the notion of *holomorphic* Kato functions.

We recall that Kato function $f(\cdot)$ is called *holomorphic*, if it admits a holomorphic extension to the right complex half-plane, $\mathbb{C}_+ := \{z \in \mathbb{C} : \mathrm{Re}(z) > 0\}$, such that $|f(z)| \leq 1$ for $z \in \mathbb{C}_+$. For holomorphic Kato functions, the limit $f(iy) := \lim_{\epsilon\to+0} f(\epsilon + iy)$ exists for a.a. $y \in \mathbb{R}$. In the following we are going to show that there is a Borel measurable function $\widetilde{f}(\cdot) : i\,\mathbb{R} \longrightarrow \mathbb{C}$ satisfying $|\widetilde{f}(iy)| \leq 1$, $y \in \mathbb{R}$, such that $f(iy) = \widetilde{f}(iy)$ for a.a. $y \in \mathbb{R}$, cf. Lemma 13.3.10. Since the $\widetilde{f}(\cdot)$ is Borel measurable the expression $\widetilde{f}(isA)$ is well defined by the functional calculus for self-adjoint A and any $s \in \mathbb{R}$. Moreover, one has $\|\widetilde{f}(isA)\| \leq 1$ for $s \in \mathbb{R}$. It was shown in [Ca05] that if f and g are *holomorphic* Kato functions, then

$$\lim_{n\to\infty} \int_0^T dt \left\| \left(\frac{\widetilde{f}(2itA/n) + \widetilde{g}(2itB/n)}{2} \right)^n h - e^{-itC} h \right\|^2 = 0,$$

holds for any $h \in \mathfrak{H}$ and $T > 0$. A natural question is:
To what extend these observations, including our result (13.3.6), are satisfactory from the point of view of applications in *theoretical* physics?
This was an important point of discussion in [FP08, Section 11].

Section 13.3.2

Here we basically follow [ENZ11, Section 2].

Theorem 13.3.5 establishes equivalence between the *Lapidus* weak convergence (13.3.2) [Lap82, Proposition 3.2] and convergence in the L^2-topology (13.3.6). The latter implies the strong operator convergence by subsequences (13.3.14), for a.a. $t \in [-T, T]$, which is important for interpretations and applications.

Following [Nel64] and [JL00, Definition 11.2.21], we indicate in Remark 13.3.7 a link between *Feynman* integral for *Schrödinger* operator and its *unitary* Trotter-type definition, $\mathcal{F}^t_{\mathrm{TP}}(V)$, cf. [Kato72] and [JL00, Corollary 11.2.22]. The *modified* (resolvent) Feynman integral, $\mathcal{F}^t_{\mathrm{M}}(V)$, for Schrödinger operator is due to [JL00, Definition 11.4.4] and [JL00, Corollary 11.4.5].

Section 13.3.3

In this subsection, we follow [ENZ11, Section 3].

The *Lapidus* result (13.3.2) for the (unitary) *Trotter* product formula relies on the vector-valued *classical* ([ABHN01, Appendix A], [Si15a, Chapter 6.2]) and *extended* ([JL00, Theorem 11.7.1], [Lap82, Appendix]) *Vitali* theorems.

The aim of Sect. 13.3.3 is a generalisation of Lapidus' result (13.3.2) to the *Fréchet* L^2-topology for the *Trotter–Kato* product formulæ that involve *holomorphic* Kato functions, see Definition 13.3.9.

It is worth noting that this generalisation cannot be deduced from Lapidus' version of the Vitali *extended* theorem, cf. [JL00, Theorem 11.7.1] and [Lap82, Appendix], since it is required that the functions $\Phi_n(z)$ and $\Phi(z)$ in (13.3.15) must admit a *continuous* extension on the closure $\overline{\mathbb{C}}_+$. By the other way around, one can use an improvement of the *Vitali extended* theorem due to the *Cachia* lemma [Ca05, Lemma 2.1] and the operator-valued version of the *Fatou–Privalov* theorem [Pri56], developed in [SzNF70, Section V.2]. This allows to figure out Theorem 13.3.11 saying that desirable conclusion holds in topology of the *Fréchet* space $L^2_{loc}(\mathbb{R}, \mathfrak{H})$.

Characterisation of the *holomorphic* Kato functions was given in [ENZ11]. The proof of Theorem 13.3.12 one can find there, see [ENZ11, Theorem 3.4].

Section 13.3.4

We follow here [ENZ11, Section 4], where results presented by *Ichinose* in [Ich80] are revised and generalised.

A notion of the *admissible* Kato functions (Definition 13.3.13), which is pertinent for analysis of convergence in Theorem 13.3.19 and Corollary 13.3.21, was introduced in [EINZ07, Definition 1].

Theorem 13.3.19 shows that not only the problem for *unitary* product formula evoked in Remark 13.3.1(b), but also restriction to the case of *admissible* Kato function are far from a satisfactory solution. In the later case the most natural exponential admissible Kato functions: $\phi(y) = \psi(y) = \mathrm{e}^{-i\,y}$ for $y \in \mathbb{R}$ do not satisfy supplementary conditions demanded in Theorem 13.3.19, whereas functions: $\phi(y) = \psi(y) = (1 + i\,y)^{-1}$ do, cf. Corollary 13.3.20.

Chapter 14
Zeno Product Formulæ

This chapter is devoted to study a link between *Trotter*, or *Trotter-Kato* product formulæ, and product formulæ including *projections*. Formally, the *Zeno* product formulæ can be seen as the *Trotter-Kato* product formulæ that involve, together with unitary groups, a constant *degenerate* (semi)groups of projections.

We conclude the Notes to Chap. 14 by Addendum with a short *historical* survey accompanied by comments on *Zeno*- and *anti-Zeno* effects.

14.1 Zeno Product Formulae in Fréchet Space

14.1.1 Product Formulæ for Projections

This subsection is dedicated to recall a link between the Trotter product formula and product formulæ for *projections*. We defer introduction of a formal definition of the *Zeno* product formula and preliminary results to the next subsections. Instead we concentrate here on preliminaries that elucidate the rôle of projections in product formulæ. Note that formally the *Zeno* product formula can be seeing as a *Trotter-Kato* product formula that involves, together with a unitary group, a *degenerate constant* semigroup of (orthogonal) *projections* $\{P\}_{t\in\mathbb{R}}$; see Remark 1.6.4.

For illustration and motivation, we first consider the case, when self-adjoint generator C of a unitary group in the product formula (1.6.3) is a bounded operator $C \in \mathcal{L}(\mathfrak{H})$. Then the Zeno product formula is a *degenerate* (*unitary*) limit

$$U_Z(t) := \tau - \lim_{n\to\infty} \left(P\, \mathrm{e}^{-i\,t\,C/n}\, P\right)^n , \quad t \in \mathbb{R}, \quad \|C\| < \infty , \qquad (14.1.1)$$

where τ indicates an *appropriate* topology of convergence on $\mathfrak{H}$, or on $\mathcal{L}(\mathfrak{H})$. The family of bounded operators $\{U_Z(t)\}_{t\in\mathbb{R}}$ is called the *Zero* dynamics, and we denote

V. A. Zagrebnov et al., *Trotter-Kato Product Formulæ*, Operator Theory: Advances and Applications 296, https://doi.org/10.1007/978-3-031-56720-9_14

by $\{U_n(t) := (P\, e^{-itC/n}\, P)^n\}_{n\geq 1}$ the family of *Zeno approximants* corresponding to this dynamics.

For the case (14.1.1),where $C \in \mathcal{L}(\mathfrak{H})$, one can use representation $e^{-itC} = \mathbb{1} - itC + \eta(t)$, for a family of bounded operators $\{\eta(t)\}_{t\in\mathbb{R}}$ such that $\lim_{t\to 0} \|\eta(t)\|/t = 0$. As a consequence

$$U_n(t) = P\left(\mathbb{1} - i\, P\, C\, P\,\frac{t}{n} + P\,\eta(t/n)\,P\right)^n, \quad t\in\mathbb{R}, \quad n\in\mathbb{N}. \qquad (14.1.2)$$

On account of (14.1.2), the limit (14.1.1) exists (*Euler* formula) in the operator-norm topology: $\tau = \|\cdot\|$, locally-uniformly in $t \in \mathbb{R}$, and yields

$$U_Z(t) = \|\cdot\| - \lim_{n\to\infty} U_n(t) = P\, e^{-itC_Z} \quad \text{for} \quad C_Z := P\,C\,P\,, \quad t\in\mathbb{R}. \qquad (14.1.3)$$

Seeing that C_Z commutes with P, the family $\{U_Z(t)\}_{t\in\mathbb{R}}$ is a *degenerate unitary* group on $\mathfrak{H}$: $U_Z(t_1)\,U_Z(t_2) = U_Z(t_1+t_2)$ for $t_1, t_2 \in \mathbb{R}$ with $U_Z(t=0) = P(\neq \mathbb{1})$.

Remark 14.1.1 We observe that in the limit (14.1.3) the quantum *Zeno effect* (or *Zeno paradox*) takes place:

- If the initial wave-function $v \in P\,\mathfrak{H}$, then it *forever* rests in the *subspace of observation*, $\mathfrak{H}_P := P\,\mathfrak{H}$, evolving according to (degenerate) unitary Zero dynamics (14.1.3), that is, $U_Z(t) : v \mapsto U_Z(t)\,v \in \mathfrak{H}_P$ for $t\in\mathbb{R}$.
- Whereas, for $u \in \mathfrak{H}$ and any $t \in \mathbb{R}$, one gets $u \mapsto U_Z(t)\,u = U_Z(t)\,v \in \mathfrak{H}_P$, that is, the *immediate* Zeno effect.

It is known that set-up and the proof of the Zeno product formula are drastically changed if generator C in (14.1.1) is *unbounded* (self-adjoint) operator in $\mathfrak{H}$. In this case operator C_Z in (14.1.3) differs, in *general*, from the operator PCP, which may be *not* self-adjoint in $\mathfrak{H}_P$, *nor* even closed, seeing that PC is *not* necessarily closed, though the operator CP is. For illustration we consider an example showing that for unbounded C the Zeno product formula (14.1.1) is, in general, not valid. For example, the limit (14.1.1) may exist *without* selecting a (degenerate) *unitary* group for the *Zeno* dynamics $\{U_Z(t)\}_{t\in\mathbb{R}}$.

Let $\mathfrak{H} = L^2(\mathbb{R})$ and let $C := -i\,\partial_x$ be differential operator, which is self-adjoint on domain: $\mathrm{dom}\,(C) = W^{1,2}(\mathbb{R}) \subset \mathfrak{H}$. Then C is generator of the unitary group $\{e^{-itC}\}_{t\in\mathbb{R}}$ of *bilateral* shifts on $\mathfrak{H}$. Let operator $P := P_{[a,b]}$ be the orthogonal projection of $\mathfrak{H}$ onto subspace $\mathfrak{H}_P = P\,\mathfrak{H} = L^2([a,b])$. Hence, P is a multiplication operator: $(P_{[a,b]}\,u)(x) = \chi_{[a,b]}(x)\,u(x)$ on $\mathfrak{H}$, for the characteristic function: $\chi_{[a,b]}$, of interval $[a, b]$ and any $u \in \mathfrak{H}$. As a consequence we infer that

$$(P\, e^{-itC}\, P\, u)(x) = \begin{cases} \chi_{[a,b]}(x)\,\chi_{[a+t,\,b+t]}(x)\,u(x-t)\,, & |t| \leq b-a\,, \\ 0\,, & |t| > b-a\,, \end{cases} \quad x\in\mathbb{R}. \qquad (14.1.4)$$

Then a calculation similar to (14.1.4) yields for $s, t \in \mathbb{R}$

$$((P\,\mathrm{e}^{-isC}P)\,(P\,\mathrm{e}^{-itC}P)\,u)(x) = \chi_{[a+t+s,\,b]}(x)\,u(x-t-s)\,, \quad x \in \mathbb{R}\,,$$

which on account of representation (14.1.4) provides the *composition* law:

$$(P\,\mathrm{e}^{-isC}P)\,(P\,\mathrm{e}^{-itC}P) = P\,\mathrm{e}^{-i\,(t+s)\,C}P\,. \tag{14.1.5}$$

Given (14.1.5), we obtain that the limit in the Zeno product formula (14.1.1) exists in the strong operator topology ($\tau = s$) on the subspace $\mathfrak{H}_P$:

$$U_Z(t) = \text{s-}\lim_{n\to\infty}\left(P\,\mathrm{e}^{-itC/n}P\right)^n = P\,\mathrm{e}^{-itC} \upharpoonright \mathfrak{H}_P\,, \quad t \in \mathbb{R}\,. \tag{14.1.6}$$

Note that restriction $C \upharpoonright \mathfrak{H}_P$ offers a *non-self-adjoint* operator in subspace $\mathfrak{H}_P$. Consequently, the Zeno dynamics $\{U_Z(t)\}_{t\in\mathbb{R}}$ (14.1.6) on $\mathfrak{H}_P$ satisfying the composition law (14.1.5) is strongly continuous since (e.g. for $t \geq 0$):

$$\begin{aligned}\|U_Z(t)\,u - U_Z(0)\,u\|^2_{\mathfrak{H}_P} &= \int_a^{b-t} dx\ |u(x)|^2 + \int_a^b dx\ |u(x)|^2 - \\ 2\,\mathrm{Re}\left(\int_{a+t}^b dx\ u(x-t)\,\overline{u(x)}\right) &\leq \int_a^{a+t} dx\ |u(x)|^2 \leq t\ \|u\|^2_{\mathfrak{H}_P}\,,\end{aligned} \tag{14.1.7}$$

but it is *not* unitary on $\mathfrak{H}_P$. Indeed, on account of (14.1.4) and (14.1.6), one gets

$$\|U_Z(t)\,u\|^2_{\mathfrak{H}_P} = \int_{a+t}^b dx\ |u(x-t)|^2 \leq \|u\|^2_{\mathfrak{H}_P}\,, \quad t \geq 0\,, \tag{14.1.8}$$

that is, $\lim_{t\to\infty}\|U_Z(t)\,u\|_{\mathfrak{H}_P} = 0$ (in fact, for $t \geq (b-a)$), for any $u \in \mathfrak{H}_P$. This means that Zeno dynamics (14.1.6) is a so-called *completely non-unitary* contraction C_0-semigroup for $t \geq 0$ on the subspace $\mathfrak{H}_P$.

Moreover, formula (14.1.8) indicates also a complete "leakage out" of the wave-functions from the subspace of *observation* $\mathfrak{H}_P$, since the family $\{U_Z(t)\}_{t\in\mathbb{R}}$ is *nilpotent*: $U_Z(t)\big|_{t\,>\,b-a} = 0$. As a consequence, according to Remark 14.1.1, there is *no* Zeno effect in this situation. That is, instead of a unitary, the Zeno dynamics is rather a contraction semigroup which describes a "leakage".

Remark 14.1.2 Note that this example of violation of the Zeno effect is covered by the following abstract observation based of a link between unitary groups and contraction C_0-semigroups; see Remark 13.1.5.

Let $\{U_K(t)\}_{t\geq 0}$ be a contraction C_0-semigroup with m-accretive generator $K \in \mathscr{G}(1,0)$ in a Hilbert space $\mathfrak{H}'$, Definition 1.8.7. Then there exists a unitary group

$\{U_C(t)\}_{t\in\mathbb{R}}$ on Hilbert space $\mathfrak{H} \supset \mathfrak{H}'$ such that if $P : \mathfrak{H} \to \mathfrak{H}_P$ is the *orthogonal* projection onto $\mathfrak{H}_P = \mathfrak{H}'$, then one gets representation

$$U_K(t) = P\, U_C(t) \restriction \mathfrak{H}_P\,, \quad t \geq 0\,. \tag{14.1.9}$$

The representation (14.1.9) is unique up to isomorphism, if the closure $\{U_C(t)\mathfrak{H}_P : t \in \mathbb{R}\} = \mathfrak{H}$. We note that in this case the unitary group $\{U_C(t)\}_{t\in\mathbb{R}}$ is called the *minimal unitary dilation* of contraction C_0-semigroup $\{U_K(t)\}_{t\geq 0}$.

To apply this concept to the Zeno problem (14.1.4), let C be a minimal self-adjoint *dilation* in $\mathfrak{H}$ corresponding to m-accretive operator K defined in the subspace $\mathfrak{H}_P \subset \mathfrak{H}$. Since by representation (14.1.9)

$$\mathrm{e}^{-tK} = P\, \mathrm{e}^{-itC} \restriction \mathfrak{H}_P = P\, \mathrm{e}^{-itC} P\,, \quad t \geq 0\,,$$

we find a contraction Zeno dynamics, which violates the Zeno effect:

$$U_Z(t) := \text{s-}\lim_{n\to\infty} \left(P\, \mathrm{e}^{-itC/n} P\right)^n = \mathrm{e}^{-tK}\,, \quad t \geq 0\,,$$

Here *strong* convergence of the product formula is considered on $\mathfrak{H}_P$, cf. (14.1.6).

See Notes to Chap. 14 (Notes to Sects. 14.1 and 14.1.1) for additional comments and references.

Other examples (see Notes to Chap. 14 (Notes to Sect. 14.4)) also show that violation of the *Zeno effect* is related to the fact that generator C is not *semibounded*, similar to (14.1.4). This observation suggests a limitation of the *Zeno product formula* analysis by *self-adjoint* generators C bounded from *below*.

On the other hand, it has to be stressed that, *a priori*, a restriction to self-adjoint generators bounded from below, for example, $C \geq 0$, does not ensure the existence of the limit (14.1.1) in the strong operator topology $\tau = s$ on the subspace $\mathfrak{H}_P$. Indeed, if the set $\mathrm{dom}\,(\sqrt{C}) \cap \mathfrak{H}_P$ is not *dense* in $\mathfrak{H}_P$, then it may happen that the Zeno product formula does *not* converge.

To elucidate this concern, we first consider on a Hilbert space $\mathfrak{H}$ the simplest example of the *one-dimensional* orthogonal projection on unit vector: $P_e\, u := (u, e)\, e$, for $u \in \mathfrak{H}$ and a unitary group $\{\mathrm{e}^{-itC}\}_{t\in\mathbb{R}}$ generated by a non-negative self-adjoint operator C. For this set-up we obtain

$$P_e\, \mathrm{e}^{-itC} P_e\, u = (\mathrm{e}^{-itC} e, e)\, P_e\, u\,, \quad u \in \mathfrak{H}\,, \quad t \in \mathbb{R}\,, \tag{14.1.10}$$

and consequently the Zeno approximants are

$$U_n^{P_e}(t) = (P_e\, \mathrm{e}^{-itC/n} P_e)^n = (\mathrm{e}^{-itC/n} e, e)^n\, P_e\,, \quad n \in \mathbb{N}\,, \quad t \in \mathbb{R}\,. \tag{14.1.11}$$

Now we *suppose* that $e \in \mathrm{dom}\,(\sqrt{C})$. Then

$$(\mathrm{e}^{-it C/n} e, e) = \mathbb{1} - i\,\frac{t}{n}\,(\sqrt{C}\, e,\, \sqrt{C}\, e) + O((t/n)^2)\,.$$

As a result the product formula for approximants (14.1.11) converges to

$$U_Z(t) = \text{s-}\lim_{n\to\infty} U_n^{P_e}(t) = P_e\, \mathrm{e}^{-it\,(\sqrt{C}\, e,\, \sqrt{C}\, e)\, P_e} \restriction \mathfrak{H}_e\,, \tag{14.1.12}$$

where $\mathfrak{H}_e := P_e\, \mathfrak{H}$ and thus the strong operator topology of convergence coincides with the operator-norm topology $\|\cdot\|$. On account of (14.1.12) and identity

$$\begin{aligned}(\sqrt{C}\, e,\, \sqrt{C}\, e)\, P_e\, u &= (C\, e,\, e)\,(u,\, e)\, e = P_e\, C\,(u,\, e)\, e\\ &= P_e\, C\, P_e\, u = (\sqrt{C}\, P_e)^*(\sqrt{C}\, P_e)\, u\,,\end{aligned}$$

where $u \in \mathfrak{H}$, we obtain for the Zeno product formula the representation:

$$U_Z(t) = P_e\, \mathrm{e}^{-it\, C_Z} \quad \text{for} \quad C_Z := (\sqrt{C}\, P_e)^*(\sqrt{C}\, P_e)\,, \quad t \in \mathbb{R}\,, \tag{14.1.13}$$

that is, the *Zeno effect* on the one-dimensional subspace $\mathfrak{H}_e$, cf. (14.1.3).

Remark 14.1.3 On account of arguments leading to the Zeno product formula (14.1.13) its extension to a *finite-dimensional* ($\dim \mathrm{ran}\,(P^{(r)}) \leq r$) orthogonal projection operator such that:

$$P^{(r)}\, u := \sum_{k=1}^{r} (u, e_k)\, e_k\,, \quad u \in \mathfrak{H}\,,\ e_k \in \mathrm{dom}\,(\sqrt{C})\,,\ \|e_k\| = 1\,,\ 1 \leq k \leq r\,, \tag{14.1.14}$$

is straightforward. Indeed, by definition and conditions in (14.1.14),

$$\sqrt{C}\, P^{(r)}\, u := \sum_{k=1}^{r} (u, e_k)\, f_k\,, \quad f_k := \sqrt{C}\, e_k\,, \quad u \in \mathfrak{H}\,. \tag{14.1.15}$$

Then $\sqrt{C}\, P^{(r)}$, as well as $(\sqrt{C}\, P^{(r)})^*$, are also bounded finite-dimensional operators with *range* spanned by vectors $\{f_1, f_2, \ldots, f_k\}$ and with *kernel*, which is orthogonal to the subspace spanned by, not necessarily orthogonal, vectors $\{e_1, e_2, \ldots, e_k\}$. Therefore, (14.1.15) is the *canonical form* of the *final-rank* operators.

As a consequence $P^{(r)} C\, P^{(r)}$ is a *finite-rank* operator, and on account of that, we use for any $u \in \mathfrak{H}$ and $t \in \mathbb{R}$ representation (cf. (14.1.10)) :

$$P^{(k)}\, \mathrm{e}^{-it\, C}\, P^{(k)}\, u = P^{(k)}\,(\mathbb{1} - it\, C)\, P^{(k)}\, u + P^{(k)}\,(\mathrm{e}^{-it\, C} + it\, C - \mathbb{1})\, P^{(k)}\, u\,, \tag{14.1.16}$$

where

$$P^{(k)}\,(\mathbb{1} - i\,t\,C)\,P^{(k)} = (\mathbb{1} - i\,t\,P^{(k)}\,C\,P^{(k)})\,P^{(k)}\,. \tag{14.1.17}$$

Let

$$D^{(k)}(t) := [(\mathrm{e}^{-i\,t\,C} - \mathbb{1})\,t^{-1} + i\,C]\,P^{(k)}\,. \tag{14.1.18}$$

Then because of (14.1.16)–(14.1.18) and orthogonality of projection $P^{(k)}$,

$$P^{(k)}\,\mathrm{e}^{-i\,t\,C}\,P^{(k)} = [\mathbb{1} - i\,t\,P^{(k)}\,C\,P^{(k)} + t\,P^{(k)}\,D^{(k)}(t)\,]\,P^{(k)}\,,$$

that provides for $t \in \mathbb{R}$ and $n \in \mathbb{N}$ the Zeno product approximants; cf. (14.1.11):

$$\begin{aligned} U_n^{P^{(k)}}(t) &:= (P^{(k)}\,\mathrm{e}^{-i\,t\,C/n}\,P^{(k)})^n \\ &= P^{(k)}\,[\mathbb{1} - i\,t/n\,P^{(k)}\,C\,P^{(k)} + t/n\,P^{(k)}\,D^{(k)}(t/n)\,]^n\,P^{(k)}\,. \end{aligned} \tag{14.1.19}$$

Owing to definition (14.1.14), the operator $P^{(k)}\,C\,P^{(k)} \in \mathcal{L}(\mathfrak{H})$. Moreover, by virtue of the strong continuity of the unitary group $\{\mathrm{e}^{-i\,t\,C}\}_{t\in\mathbb{R}}$, one infers from (14.1.18) that

$$\lim_{t\to 0}\|\,P^{(k)}\,D^{(k)}(t)\,\| = 0\,.$$

Then thanks to representation (14.1.19) for approximants, the Zeno product formula holds locally-uniformly for $t \in \mathbb{R}$ in the *operator-norm* topology :

$$U_Z(t) = \|\cdot\| - \lim_{n\to\infty} U_n^{P^{(k)}}(t) = P^{(k)}\,\mathrm{e}^{-i\,t\,C_Z}\,, \quad C_Z := (\sqrt{C}\,P^{(k)})^*(\sqrt{C}\,P^{(k)})\,, \tag{14.1.20}$$

as a limit of the *Euler* formula (in the operator-norm topology) on the subspace $\mathfrak{H}_{P^{(k)}} := P^{(k)}\,\mathfrak{H}$; compare to (14.1.12) and (14.1.13). Since on subspace $\mathfrak{H}_{P^{(k)}}$ the Zeno dynamics (14.1.20) is unitary, one observes on this subspace the *Zeno effect.*

Note that the choice of the orthogonal projection $P^{(k)}$ in (14.1.14) is quite special. The proof of the Zeno product formula for *arbitrary* finite-dimensional orthogonal projection is much less trivial and the corresponding result holds *only* in the *strong* operator topology; see Notes to Chap. 14 (Sects. 14.1 and 14.1.1).

In the next Sect. 14.1.2, we start scrutinising the Zeno product formula for the case of arbitrary *infinite-dimensional* orthogonal projection P. In this case one needs for analysis and for proofs a full power the Trotter-Kato product formulæ technique even for some partial (*non-ultimate*) results.

First of all it concerns the problem of *existence* of the Zeno dynamics $\{U_Z(t)\}_{t\in\mathbb{R}}$. This includes elucidation of *topology* τ of convergence and existence the limit (14.1.1). Then it needs, in a sequence, to study the *unitarity* of $\{U_Z(t)\}_{t\in\mathbb{R}}$ on the subspace of *observation*, that is, the proper Zeno effect.

14.1.2 Zeno Dynamics: Fréchet Space

Throughout this (sub)section, H will be a non-negative (or positive) self-adjoint operator in a *separable* Hilbert space $\mathfrak{H}$, and P will be an orthogonal projection. The non-negativity assumption is imposed only for convenience, since our main result extends easily to any self-adjoint operator H bounded from below as well as one bounded from above, that is, to each semi-bounded self-adjoint operator in $\mathfrak{H}$.

Consider the quadratic form: $u \mapsto \|H^{1/2}Pu\|^2$, with form-domain $\mathrm{dom}\,(H^{1/2}P)$. Note that $H^{1/2}P$ involved here is a closed operator and HP has the same property. Let $H_P := (H^{1/2}P)^*(H^{1/2}P)$ be the self-adjoint operator associated with this quadratic form. In general, H_P may *not* be densely defined in which case it is a self-adjoint operator in a closed subspace of $\mathfrak{H}$. More specifically, it is obviously defined and acts nontrivially in a closed subspace $\mathrm{ran}\,(P)$ of the closure of the form domain $\mathrm{dom}\,(H^{1/2}P)$, while in the orthogonal complement $(\mathrm{ran}\,(P))^{\perp}$ it acts as *zero*.

Remark 14.1.4 The quadratic form: $u \mapsto \|H^{1/2}Pu\|^2$, defined on $\mathrm{dom}\,(H^{1/2}P)$ is a *closed extension* of the form: $u \mapsto (Pu, HPu)$, defined on $\mathrm{dom}\,(HP)$. We notice that the former is *not*, in general, a simple closure of the latter. Indeed, if operator H is unbounded, $\mathrm{dom}\,(H)$ is a proper subspace of $\mathrm{dom}\,(H^{1/2})$.

Indeed, take $u_0 \in \mathrm{dom}\,(H^{1/2})\backslash\mathrm{dom}\,(H)$ such that the vector $H^{1/2}u_0$ is non-zero. Let P be orthogonal projection onto the one-dimensional subspace spanned by u_0. Taking into account that $\mathrm{dom}\,(HP) = \{u \in \mathfrak{H} : \ Pu \in \mathrm{dom}\,(H)\}$, to which $u_0 = Pu_0$ does not belong, we find $HPu_0 = 0$ for $u_0 \in \mathrm{dom}\,(HP)$, while $H^{1/2}Pu_0 = H^{1/2}u_0 \neq 0$ by assumption.

To describe the results of this subsection, we introduce a topology of convergence $\tau = \mathfrak{f}$ and a locally convex vector space $\mathfrak{F}$ endowed with this topology, cf. Sect. 13.3 (Proposition 13.3.2).

Definition 14.1.5 By $\mathfrak{F} := L^2_{\mathrm{loc}}(\mathbb{R}, \mathfrak{H}) \simeq L^2_{\mathrm{loc}}(\mathbb{R}) \otimes \mathfrak{H}$ we denote the topological complete metrisable *Fréchet* vector space of the $\mathfrak{H}$-valued strongly measurable functions: $t \mapsto f(t)$, on $\mathbb{R}$ such that they are locally square integrable there, with the topology $\mathfrak{f}$ induced by a countable family of *seminorms*:

$$f \longmapsto \left(\int_{-m_1}^{m_2} dt\, \|f(t)\|^2\right)^{1/2}, \quad m_1, m_2 \in \mathbb{N}. \tag{14.1.21}$$

In a similar way, one defines the *Fréchet* space $\mathfrak{F}_0^+ := L^2_{\mathrm{loc}}(\mathbb{R}_0^+, \mathfrak{H}) \simeq L^2_{\mathrm{loc}}(\mathbb{R}_0^+)\otimes\mathfrak{H}$.

The first result that we are going to discuss in this subsection is the following assertion about the *Zeno* product formula and the *Zeno dynamics* for a family $\{P(t)\}_{t\in\mathbb{R}}$ of *orthogonal* projections on $\mathfrak{H}$.

Theorem 14.1.6 *Let H be a non-negative self-adjoint operator on a separable Hilbert space $\mathfrak{H}$ and P an orthogonal projection. Let: $t \mapsto P(t)$, be a strongly continuous function whose values are orthogonal projections in $\mathfrak{H}$, defined in some neighbourhood of zero, with $P(0) =: P$. Moreover, suppose that* $\mathrm{dom}\,(H^{1/2}P(t)) \supset \mathrm{dom}\,(H^{1/2}P)$ *and* $\lim_{t\to 0} \|H^{1/2}P(t)g\| = \|H^{1/2}Pg\|$ *holds for* $g \in \mathrm{dom}\,(H^{1/2}P)$. *If the operator $H_P := (H^{1/2}P)^*(H^{1/2}P)$ is densely defined in the whole Hilbert space $\mathfrak{H}$, then for every $f \in \mathfrak{H}$ and $\varepsilon = \pm 1$ the limits :*

$$[P(1/n)\exp(-i\varepsilon t H/n)P(1/n)]^n f \longrightarrow P\,\exp(-i\varepsilon t H_P)\,Pf\,, \tag{14.1.22}$$

$$[P(1/n)\,\exp(-i\varepsilon t H/n)]^n\, f \longrightarrow P\,\exp(-i\varepsilon t H_P)\,Pf\,, \tag{14.1.23}$$

$$[\exp(-i\varepsilon t H/n)\,P(1/n)]^n\, f \longrightarrow P\,\exp(-i\varepsilon t H_P)\,Pf\,, \tag{14.1.24}$$

hold in topology $\mathfrak{f}$ of the Fréchet space $\mathfrak{F}$ as $n \to \infty$.

Note that in general the generator H_P differs from the operator PHP. It may be non-self-adjoint in $\mathfrak{H}$ and even not closed, because operator PH is not necessarily closed, though HP is and operator H_P is a self-adjoint extension of PHP. Remark that requirement of the theorem that H_P is densely defined in $\mathfrak{H}$ means nothing else but that the domain $\mathrm{dom}\,(H^{1/2}P)$ of the quadratic form in question is dense in $\mathfrak{H}$.

We also remark that for $\varepsilon = 1$, the theorem concerns a non-negative self-adjoint operator $\varepsilon H = H$, while for $\varepsilon = -1$, we get product formulae for the non-positive self-adjoint operator $\varepsilon H = -H$. Moreover, the result is preserved when H is replaced by a shifted operator $H + c\,\mathbb{1}$, that is, for any semi-bounded self-adjoint operator in a separable Hilbert space $\mathfrak{H}$.

An important particular case, which is the most often encountered in applications, concerns the situation when the projection-valued function is constant.

Corollary 14.1.7 *Let H be a self-adjoint operator bounded from below in a separable Hilbert space $\mathfrak{H}$ and P an orthogonal projection. If the operator H_P specified above is densely defined, then for every $f \in \mathfrak{H}$ and $\varepsilon = \pm 1$ we have in topology $\mathfrak{f}$ of the Fréchet space $\mathfrak{F}$ the limit*

$$\mathfrak{f} - \lim_{n\to\infty} [P\exp(-i\,\varepsilon t\; H/n)P]^n f = P\,\exp(-i\,\varepsilon t\; H_P)\,Pf\,, \tag{14.1.25}$$

as well as its nonsymmetric counterparts obtained by setting $P(1/n) = P$ in (14.1.23) *and* (14.1.24).

From the viewpoint of the *quantum* Zeno effect described in Sect. 14.1.1, the optimal result would be a *strong* convergence on $\mathfrak{H}$ for a fixed value of the time variable, moreover locally-uniformly in $t \in \mathbb{R}$. Our Theorem 14.1.6 implies the following weaker result about a pointwise convergence.

Corollary 14.1.8 *Under the same hypotheses, as in Theorem* 14.1.6*, there exists a set $M \subset \mathbb{R}$ of Lebesgue measure zero and a strictly increasing sequence $\{n'\}$ of positive integers along which we have*

$$\text{s-}\lim_{n'\to\infty} [P(1/n') \exp(-i\varepsilon t H/n') P(1/n')]^{n'} = P \exp(-i\varepsilon t H_P)\, P\,, \quad (14.1.26)$$

$$\text{s-}\lim_{n'\to\infty} [P(1/n') \exp(-i\varepsilon t H/n'))]^{n'} = P \exp(-i\varepsilon t H_P)\, P\,, \quad (14.1.27)$$

$$\text{s-}\lim_{n'\to\infty} [P \exp(-i\varepsilon t H/n')\, P(1/n')]^{n'} = P \exp(-i\varepsilon t H_P)\, P\,, \quad (14.1.28)$$

strongly on $\mathfrak{H}$ for all $t \in \mathbb{R} \setminus M$ and $\varepsilon = \pm 1$.

For the proof and more consequences owing to Theorem 14.1.6, see Sect. 14.1.3.

As we have indicated in Sect. 14.1.1, one does not need to resort to subsequences in a particular case when the projections involved in product formulae are *finite-dimensional*. Moreover, under conditions (14.1.14) the convergence to a *unitary* Zeno dynamics holds in the *operator-norm* topology (14.1.20); cf. Remark 14.1.3. In a more general setting, we shall scrutinise the case of finite-dimensional projections in Sect. 14.1.4.

Remark 14.1.9 Although a necessity to select a subsequence makes the *pointwise* convergence result weaker than desired, notice that from the physical point of view the convergence in the Fréchet space $\mathfrak{F} = L^2_{\text{loc}}(\mathbb{R}, \mathfrak{H})$ can be regarded as satisfactory. In fact any actual measurement, in particular that of time, is burdened with errors. Suppose thus we perform the Zeno experiment on numerous copies of the system. The time value in the results will be characterised by a *probability distribution* density $\phi : \mathbb{R}^+ \to \mathbb{R}^+$, which is typically a bounded, compactly supported function—in a precisely posed experiment it is sharply peaked, of course. As a consequence Corollary 14.1.7 yields

$$\int_{\mathbb{R}^+} dt\, \phi(t)\, \left\| [P \exp(-i\,\varepsilon t\, H/n) P]^n f - \exp(-i\,\varepsilon t\, H_P)\, Pf \right\|^2 \to 0\,, \quad (14.1.29)$$

as $n \to \infty$. In other words, the Zeno dynamics limit is valid after averaging over experimental errors, however small they are.

Remark 14.1.10 While the proof of strong convergence in $\mathfrak{H}$ in Theorem 14.1.6 and Corollaries 14.1.7 and 14.1.8 remains *elusive* without the finite-dimension assumption, such a claim can be easily established in the orthogonal complement of the subspace $P\,\mathfrak{H}$. Indeed, taking $f \in Q\,\mathfrak{H}$, where $Q := \mathbb{1} - P$, we have

$$(P(1/n)\, \mathrm{e}^{-i\,\varepsilon t\, H/n}\, P(1/n))^n f = (P(1/n)\, \mathrm{e}^{-i\,\varepsilon t\, H/n}\, P(1/n))^n P(1/n)\, Q\, f\,,$$
$$(e^{-i\,\varepsilon t\, H/n}\, P(1/n))^n f = (e^{-i\,\varepsilon t\, H/n}\, P(1/n))^n P(1/n)\, Q\, f\,,$$

which converge to *zero*, locally-uniformly in $t \in \mathbb{R}$, as $n \to \infty$, because $P(\tau) \xrightarrow{s} P$ in the strong operator topology as $\tau \to 0$. This gives the result for (14.1.26) and (14.1.28), while for (14.1.27), one has to employ in addition the relation

$$\left(P(1/n)\, e^{-it\,H/n}\right)^n g - \left(P(1/n)\, e^{-it\,H/n} P(1/n)\right)^n g$$
$$= - \left(P(1/n)\, \mathrm{e}^{-it\,H/n}\, P(1/n)\right)^{n-1} [\, e^{-it\,H/n},\, P(1/n)\,]\, g\,,$$

for any $g \in \mathfrak{H}$.

Remark 14.1.11 The *fact* that existence of product formulae requires the operator H_P to be densely defined is *nontrivial*. Recall the example [Ex85, Remark 2.4.9] (see also Notes to Chap. 14 (Notes to Sects. 14.1 and 14.1.2)), in which H is the multiplication operator, $(H\psi)(x) = x\psi(x)$ on $L^2(\mathbb{R}^+)$, and P is the one-dimensional projection onto the subspace spanned by the vector $\psi_0 : \psi_0(x) = [(\pi/2)(1+x^2)]^{-1/2}$. In this case $H_P = (H^{1/2}P)^*(H^{1/2}P)$ is obviously a *zero* operator on $\mathrm{dom}\,(H_P) = \{\psi_0\}^{\perp}$. On the other hand, operator $P\,\mathrm{e}^{-itH}\,P$ acts on the $\mathrm{ran}\,(P)$ as a multiplication by the function

$$v(t) := \mathrm{e}^{-t} - \frac{i}{\pi}\left[\mathrm{e}^{-t}\,\overline{\mathrm{Ei}}(t) - \mathrm{e}^{t}\,\mathrm{Ei}(-t)\right] = 1 + \frac{2i}{\pi}\,t\,\ln t + O(t),$$

where $\mathrm{Ei}(-t)$ and $\overline{\mathrm{Ei}}(t)$ are special functions (the *exponential integral*). Because of rapid oscillations of the imaginary part of this expression for $t \downarrow 0$, a pointwise limit of $v(t/n)^n$ for $n \to \infty$ does not exist. Notice also that *different* limits may be obtained in this example along suitably chosen subsequences $\{n_k\}_{k\geq 1}$.

For more comments and references, we invite to consult Notes to Chap. 14 (Notes to Sects. 14.1 and 14.1.2).

14.1.3 The Main Result

In this subsection we describe (*depict* the key steps of) the proof of Theorem 14.1.6. We leave details for comments and references collected in Notes to Chap. 14 (Notes to Sects. 14.1 and 14.1.3).

Below we present the argument for $\varepsilon = 1$ since the case $\varepsilon = -1$ can be treated similarly. We start in (a) by proving the $\mathfrak{f}$-convergence of formula (14.1.22), and then formulae (14.1.23), (14.1.24) in (b).

(a) We start with the *symmetric* product case and outline the proof of the formula (14.1.22) with $\varepsilon = 1$. We shall check the locally-uniform convergence in (14.1.22) on an arbitrary compact t-interval in the right half-line $\mathbb{R}_0^+ = [0, \infty)$. The proof for t-intervals in the left half-line $\mathbb{R}_0^-$ is analogous, and in addition, it can be included in the case $\varepsilon = -1$ with the convergence in (14.1.22) on compact t-intervals of the closed right half-line $[0, \infty)$.

Put $Q(t) := \mathbb{1} - P(t)$ and $Q := Q(0) = \mathbb{1} - P(0) = \mathbb{1} - P$, where $\mathbb{1}$ is the identity operator on $\mathfrak{H}$. Since by assumption, the operator H is non-negative and self-adjoint, there exists a spectral measure $E(d\lambda)$ on the non-negative real line such that $H = \int_{0-}^{\infty} E(d\lambda)\,\lambda$. For $\zeta \in \mathbb{C}$ with $\mathrm{Re}\,\zeta \geq 0$ and $\tau > 0$, we put

$$F(\zeta, \tau) := P(\tau)\, e^{-\zeta\,\tau\, H}\, P(\tau)\,, \tag{14.1.30}$$

which is a contraction, and

$$S(\zeta, \tau) := \tau^{-1}\,[\mathbb{1} - F(\zeta, \tau)] = \tau^{-1}\,[\mathbb{1} - P(\tau)\, e^{-\zeta\,\tau\, H}\, P(\tau)]\,, \tag{14.1.31}$$

which exists as a bounded operator on $\mathfrak{H}$ with $\mathrm{Re}\,(S(\zeta, \tau) f,\ f) \geq 0$ for every $f \in \mathfrak{H}$. For a non-zero $\zeta \in \mathbb{C}$ with $\mathrm{Re}\,\zeta \geq 0$, we put also

$$H(\zeta) := \zeta^{-1}[\mathbb{1} - \mathrm{e}^{-\zeta\, H}]\,. \tag{14.1.32}$$

We note that each element $v(\cdot)$ in the *Fréchet* space $\mathfrak{F}_0^+ = L^2_{\mathrm{loc}}(\mathbb{R}_0^+, \mathfrak{H})$ is an *equivalence class* such that any two representatives of it are equal a.e. on $\mathbb{R}_0^+$. However, at some places we shall not avoid an abuse of notation using for a particular representative of such an element the same symbol $v(\cdot)$. At the same time, in the following the convergence of a family of vectors $\{v(\cdot, \tau)\}_{\tau>0}$ to $v(\cdot)$ in the topology of the *Fréchet* space $\mathfrak{F}_0^+$ as τ tends to zero will be often written as $v(t, \tau) \to v(t)$; this will be the case when writing $v(\cdot, \tau) \to v(\cdot)$ would require to introduce a separate symbol for this $v(t, \tau)$ the meaning of which is clear from the context.

The key ingredient of the proof is the following lemma.

Lemma 14.1.12 *Operators* $(\mathbb{1} + S(it, \tau))^{-1}$, $\tau > 0$, *converge to* $(\mathbb{1} + it\, H_P)^{-1} P$ *as* $\tau \to 0$ *strongly on* $\mathfrak{F}_0^+$, *in other words, for all* $f \in \mathfrak{H}$ *and every finite* $T > 0$ *one has*

$$\lim_{\tau \to 0} \int_0^T dt\, \|(\mathbb{1} + S(it, \tau))^{-1} f - (\mathbb{1} + it\, H_P)^{-1}\, P\, f\|^2 = 0\,. \tag{14.1.33}$$

The proof of Lemma 14.1.12 is based on the *Vitali* theorem for holomorphic vector-valued functions and arguments analogous to those used in the *Kato* paper about a strong convergence of the self-adjoint *Trotter* product formula where the semigroup generator is a *form-sum* of pair of non-negative self-adjoint operators.

We postpone a discussion of Lemma 14.1.12 to Notes to Chap. 14 (Sects. 14.1 and 14.1.3). For the moment we are going to *accept* its claim and use it for demonstration that (14.1.33) implies the proof of symmetric product formula (14.1.22) in Theorem 14.1.6.

First we scrutinise the strong limit (as $\tau \to 0$) of resolvents $\{(\mathbb{1}+S(i\,t,\tau))^{-1}\}_{\tau>0}$ on $\mathfrak{H}$, which are involved in the integrand of (14.1.33). To this end we carry through and implement later the following steps:

Step 1. Let $\{m_n\}$ be a strictly increasing sequence of positive integers, i.e. a subsequence of the sequence of all positive integers. We have only to show that there exists a subsequence $\{n'\}$ in any such sequence $\{m_n\}$ along which (14.1.22) holds. Then by a standard argument, we can conclude that (14.1.22) actually holds along the sequence of all positive integers n. For if this were not the case, there would exist a subsequence $\{n'\}$ of strictly increasing positive integers along which (14.1.22) does not converge. However, we see that there is a subsequence $\{n''\}$ of $\{n'\}$ along which the convergence takes place to the same limit, which is a contradiction.

Step 2. Fix $\{m_n\}$ and $f \in \mathfrak{H}$. Given that Lemma 14.1.12 holds, in particular, along the sequence $\{\tau_n\}$ with $\tau_n := 1/m_n$, and since L^2 convergence implies pointwise convergence a.e. along a subsequence, there exist a subset M_f of Lebesgue measure zero of the variable t in $\mathbb{R}_0^+ = [0,\infty)$ and a subsequence $\{\tau_{f,n}\}$ of $\{\tau_n\}$, both dependent on f, such that

$$(\mathbb{1} + S(it, \tau_{f,n}))^{-1} f \to (\mathbb{1} + itH_P)^{-1} Pf\,,$$

holds strongly on $\mathfrak{H}$ for $t \in \mathbb{R}_0^+ \setminus M_f$. Since $\mathfrak{H}$ is separable by assumption, we can choose a countable dense subset $\mathcal{D} = \{f_\ell\}_{\ell=1}^\infty$ in $\mathfrak{H}$. Then we infer that for $f_1 \in \mathcal{D}$ there exist a set $M_1 := M_{f_1}$ of *Lebesgue* measure zero and a subsequence $\{\tau_{1,n}\}$ of $\{\tau_n\}$ along which $(\mathbb{1} + S(it,\tau_{1,n}))^{-1} f$ converges to $(\mathbb{1} + itH_P)^{-1} Pf$ for every $t \notin M_1$. Next, for $f_2 \in \mathcal{D}$ there exist a set $M_2 := M_{f_2}$ of *Lebesgue* measure zero and a subsequence $\{\tau_{2,n}\}$ of $\{\tau_{1,n}\}$ along which $(\mathbb{1} + S(it,\tau_{2,n}))^{-1} f$ converges to $(\mathbb{1} + itH_P)^{-1} Pf$ for every $t \notin M_2$. Proceeding in this way, we associate in the ℓ-th step with $f_\ell \in \mathcal{D}$ a set $M_\ell := M_{f_\ell}$ of *Lebesgue* measure zero and a subsequence $\{\tau_{\ell,n}\}$ of $\{\tau_{\ell-1,n}\}$ along which $\{(\mathbb{1}+S(it,\tau_{\ell,n}))^{-1} f\}_{n\ge1}$ converges to $(\mathbb{1}+ itH_P)^{-1} Pf$ for every $t \notin M_\ell$.

Step 3. Now we put $\tau_n' := \tau_{n,n}$ and $n' := [1/\tau_n']$, so that $\{n'\}$ is a subsequence of the strictly increasing sequence $\{m_n\}$ of positive integers from which we have started. Then it follows that for every $t \in \mathbb{R}_0^+ \setminus \cup_{\ell=1}^\infty M_\ell$, the sequence $\{(\mathbb{1}+S(it,\tau_n'))^{-1} f\}_{n\ge1}$ converges to $(\mathbb{1}+itH_P)^{-1} Pf$ strongly in $\mathfrak{H}$ as $\tau_n' \to 0$ for every $f \in \mathcal{D}$, and therefore also in $\mathfrak{H}$, because both $\{(\mathbb{1} + S(it,\tau_{\ell,n}))^{-1} f\}_{n\ge1}$ and $(\mathbb{1} + itH_P)^{-1} P$ are bounded operators on $\mathfrak{H}$ with the norms not exceeding one. We denote $M := \cup_{\ell=1}^\infty M_\ell$, which is, of course, again a set of *Lebesgue* measure zero. In this way we have found a subsequence $\{\tau'_n\}$ of $\{\tau_n = 1/m_n\}$ and an *exceptional* subset M of $\mathbb{R}_0^+$ such that

$$(\mathbb{1} + S(i\,t,\, \tau'_n))^{-1} f = (\mathbb{1} + S(i\,t,\, 1/n'))^{-1} f$$
$$\to (\mathbb{1} + i\,tH_P)^{-1} P f\,, \quad f \in \mathfrak{H}\,, \qquad (14.1.34)$$

i.e. strongly on $\mathfrak{H}$ as $\tau'_n \to 0$, or the corresponding $n' \to \infty$, for every $f \in \mathfrak{H}$ and for each fixed $t \notin M$. It is important that M is *independent* of f.

Lemma 14.1.13 *For the sequence* $\{n'\}$ *specified in* (14.1.34), *we have*

$$\lim_{n'\to\infty} [P(1/n') \exp(-itH/n') P(1/n')]^{n'} f = \mathrm{e}^{-itH_P} P f\,, \quad f \in \mathfrak{H}\,, \qquad (14.1.35)$$

provided $t \notin M$.

We note that this claim is in fact the "symmetric" product formula (14.1.26) for $\varepsilon = +1$; see Corollary 14.1.8.

Proof of Lemma 14.1.13 Below, our arguments are essentially based on the *Chernoff* $\sqrt{n}$-Lemma (Lemma 1.9.33) and the *Trotter-Kato-Neveu* convergence theorem (Proposition 1.9.30) for self-adjoint operator $H \geq 0$.

We divide the proof into two steps referring to f, which first belongs to $P\,\mathfrak{H}$ and in the second step to its orthogonal complement $Q\,\mathfrak{H} = (\mathbb{1} - P)\,\mathfrak{H}$.

Suppose first that $f \in P\,\mathfrak{H}$. Then for $t \notin M$ and a fixed $\tau > 0$, operator $S(i\,t,\, \tau)$ (14.1.31) generates a strongly continuous semigroup $\{e^{-\theta\, S(i\,t,\,\tau)}\}_{\theta\geq 0}$ on $\mathfrak{H}$. As a consequence, the convergence of resolvents (14.1.34) implies by the *Trotter-Kato-Neveu* theorem the strong operator convergence of the corresponding semigroups. So, we obtain

$$\text{s-}\lim_{n'\to\infty} \mathrm{e}^{-\theta\, S(i\,t,1/n')} = \mathrm{e}^{-i\,\theta\, tH_P}\,,$$

for $t \notin M$, locally-uniformly in $\theta \in \mathbb{R}_0^+$. In particular, choosing $\theta = 1$, we get for each $t \in \mathbb{R}_0^+ \setminus M$

$$\text{s-}\lim_{n'\to\infty} \mathrm{e}^{-S(i\,t,1/n')} = \mathrm{e}^{-i\,t\,H_P}\,. \qquad (14.1.36)$$

Then due to Proposition 1.9.30, the same equivalence yields for any $\lambda \geq 0$ and $t \in \mathbb{R}_0^+ \setminus M$:

$$\text{s-}\lim_{n'\to\infty} (\mathbb{1} + \lambda\, S(i\,t,\, 1/n'))^{-1} = (\mathbb{1} + i\,\lambda t\, H_P)^{-1} P\,.$$

In particular, using a *diagonal trick*: $\lambda = 1/\sqrt{n'}$, we obtain

$$\left(\mathbb{1} + \frac{1}{\sqrt{n'}} S(i\,t, 1/n')\right)^{-1} f \xrightarrow{s} P f \quad \text{as } n' \to \infty, \tag{14.1.37}$$

for every $t \in \mathbb{R}_0^+ \setminus M$ and any $f \in \mathfrak{H}$.

Next, we apply to contractions (14.1.30) the *Chernoff* $\sqrt{n}$-Lemma (Lemma 1.9.33), which gives for any $g \in \mathfrak{H}$ estimate

$$\left\| F(i\,t,\, 1/n')^{n'} g - \mathrm{e}^{-n'[\mathbb{1} - F(i\,t,\,1/n')]} g \right\| \le \sqrt{n'}\; \| \left(\mathbb{1} - F(i\,t, 1/n')\right) g \| \,.$$

Then choosing

$$g := \left(\mathbb{1} + \frac{1}{\sqrt{n'}} S(i\,t, 1/n')\right)^{-1} f\,,$$

we infer on account of definitions (14.1.30) and (14.1.31) that

$$\begin{aligned} &\left\| \left(F(i\,t, 1/n')^{n'} - \mathrm{e}^{-S(i\,t,1/n')}\right) \left(\mathbb{1} + \frac{1}{\sqrt{n'}} S(i\,t, 1/n')\right)^{-1} f \right\| \\ &\le \left\| f - \left(\mathbb{1} + \frac{1}{\sqrt{n'}} S(i\,t, 1/n')\right)^{-1} f \right\| . \end{aligned} \tag{14.1.38}$$

Since we consider the case $f \in P\,\mathfrak{H}$, by virtue of (14.1.37) the right-hand side of (14.1.38) tends to zero as $n' \to \infty$.

Now, owing to $\|F(i\,t, 1/n')^{n'}\| \le 1$ and (14.1.38) we obtain estimate

$$\begin{aligned} &\left\| F(i\,t, 1/n')^{n'} f - \mathrm{e}^{-S(i\,t,1/n')} f \right\| \le \\ &\left\| F(i\,t, 1/n')^{n'} - \mathrm{e}^{-S(i\,t,1/n')} \right\| \left\| f - \left(\mathbb{1} + \frac{1}{\sqrt{n'}} S(i\,t, 1/n')\right)^{-1} f \right\| + \\ &\left\| (F(i\,t, 1/n')^{n'} - \mathrm{e}^{-S(i\,t,1/n')}) \left(\mathbb{1} + \frac{1}{\sqrt{n'}} S(i\,t, 1/n')\right)^{-1} f \right\| \le \\ &3 \left\| f - \left(\mathbb{1} + \frac{1}{\sqrt{n'}} S(i\,t, 1/n')\right)^{-1} f \right\| . \end{aligned} \tag{14.1.39}$$

Using (14.1.37) once again we deduce from (14.1.39)

$$\left\| F(i\,t, 1/n')^{n'} f - \mathrm{e}^{-S(i\,t,1/n')} f \right\| \to 0\,, \tag{14.1.40}$$

for $f \in P\,\mathfrak{H}$, as $n' \to \infty$.

The sought relation (14.1.35) immediately follows from (14.1.36) and (14.1.40), since by (14.1.30) we have $F(it, 1/n')^{n'} = [P(1/n') \exp(-itH/n')P(1/n')]^{n'}$.

Now let $f \in Q\,\mathfrak{H}$. This case is easier being independent of the arguments in (*Steps* 1.–3.) preceding Lemma (14.1.13). We have, along the sequence of *all* $n \in \mathbb{N}$:

$$[P(1/n) \exp(-itH/n)P(1/n)]^n f \to 0\,, \quad f \in Q\,\mathfrak{H}\,, \tag{14.1.41}$$

for each $t \in \mathbb{R}_0^+$, since $P(1/n)f = P(1/n)Qf$ converges by assumption $f \in Q\,\mathfrak{H}$ to $P\,Q\,f = 0$ as $n \to \infty$, while $\exp(-itH_P)\,P\,f = 0$. □

(a) Hence, this lemma yields desired result (14.1.22) because

$$\{[P(1/n') \exp(-itH/n')P(1/n')]^{n'}\}_{n'>1}\,,$$

is a bounded sequence for any $t \geq 0$ and by the *Lebesgue* dominated convergence theorem it tends to the expected limit in topology of the *Fréchet* space $L^2_{\text{loc}}([0, \infty); \mathfrak{H})$. Next, using the standard *subsequence* trick mentioned above in *Steps* 1.–3., we have thus shown that Lemma 14.1.13 implies the *symmetric* product formula (14.1.22) of Theorem 14.1.6.

(b) Let us turn to cases of *nonsymmetric* product formulae, namely, to prove that (14.1.22) implies (14.1.23) and (14.1.24).

Proof of (14.1.23) We use the standard notation, $[U, V] = UV - VU$, for the commutator of bounded operators U and V and observe the following fact. □

Lemma 14.1.14 *Under conditions of Theorem* 14.1.6 *the limit* :

$$\lim_{\tau\to 0}\,[\,e^{-it\tau H},\ P(\tau)\,]\,u = 0 \tag{14.1.42}$$

holds locally-uniformly in variable $t \in \mathbb{R}$ *for each* $w \in \mathfrak{H}$.

Proof By virtue of definition (14.1.32), we obtain for $\zeta = i\,t\,\tau$:

$$[\,e^{-it\tau H},\ P(\tau)\,] = i\,\Big(\,P(\tau)\,t\,\tau\,H(i\,t\,\tau) - t\,\tau\,H(i\,t\,\tau)\,P(\tau)\,\Big)\,,$$

and, hence, for any $w \in \mathfrak{H}$ we have estimate

$$\big\|[\,e^{-it\tau H},\ P(\tau)\,]\,w\big\| \leq \|\,t\,\tau\,H(i\,t\,\tau)\,w\| + \|\,t\,\tau H(i\,t\,\tau)\,P(\tau)\,w\|\,.$$

We rewrite (14.1.32) for $\zeta = i\,t\,\tau$ as

$$i\,H(it\,\tau) = \frac{\mathbb{1} - \cos(t\,\tau H)}{t\,\tau} + i\,\frac{\sin(t\,\tau H)}{t\,\tau} =: B(t\,\tau) + i\,A(t\,\tau)\,, \tag{14.1.43}$$

recall that we assume $t \neq 0$. Here $B(t\,\tau)$ and $A(t\,\tau)$ are bounded self-adjoint operators on $\mathfrak{H}$, and, in addition, $B(t\,\tau)$ is non-negative. Then we get for $w \in \mathfrak{H}$

$$\begin{aligned}\|t\,\tau\, H(it\,\tau)w\|^2 &= \|\,[t\,\tau\, B(t\,\tau) + it\,\tau\, A(t\,\tau)]\, w\|^2\\ &= \|\,[(\mathbb{1} - \cos(t\,\tau H)) + i\sin(t\,\tau H)]\, w\|^2\\ &= (\,[(\mathbb{1} - \cos(t\,\tau H))^2 + \sin^2(t\,\tau H)]\, w,\, w)\\ &= 4\,\|\sin(t\,\tau H/2)\, w\|^2 \to 0\,,\end{aligned}$$

as $\tau \to 0$, locally-uniformly for variable $t \in \mathbb{R}\setminus\{0\}$. Since $\text{s-lim}_{\tau\to 0} P(\tau) = P$, in this way we obtain the proof of the claim (14.1.42) locally-uniformly in $t \in \mathbb{R}\setminus\{0\}$ for each $w \in \mathfrak{H}$. □

Now we apply for $w \in \mathfrak{H}$ the identity

$$\begin{aligned}&\big(P(1/n)\, e^{-itH/n}\big)^n w - \big(P(1/n)\, e^{-itH/n} P(1/n)\big)^n w = \\ &-\big(P(1/n)\,\mathrm{e}^{-itH/n} P(1/n)\big)^{n-1} [\,e^{-itH/n},\; P(1/n)\,]\, w\,.\end{aligned} \tag{14.1.44}$$

Since $\{\big(P(1/n)\, e^{-itH/n} P(1/n)\big)^{n-1}\}_{n\geq 1}$ are contractions, by Lemma 14.1.14 the right-hand side of (14.1.44) converges to zero for all $w \in \mathfrak{H}$ and $t \neq 0$. Then this also holds for the left-hand side of (14.1.44) and by the *Lebesgue* dominated convergence theorem it tends to zero in topology of the *Fréchet* space $\mathfrak{F}_0^+ = L^2_{\text{loc}}([0,\infty);\mathfrak{H})$. Therefore, symmetric formula (14.1.22) and (14.1.44) yield formula (14.1.23).

Proof of (14.1.24) In view of the already proven formula (14.1.22), we have for every $f \in \mathfrak{H}$ and $T > 0$ the following chain of relations,

$$\begin{aligned}T\,\|Pf\|^2 &\geq \limsup \int_0^T dt\, \|(e^{-itH/n} P(1/n))^n f\|^2\\ &= \limsup \Big[\int_0^T dt\, \|P(1/n)(e^{-itH/n} P(1/n))^n f\|^2\\ &\quad + \int_0^T dt\, \|Q(1/n)(e^{-itH/n} P(1/n))^n f\|^2\Big]\\ &\geq \limsup \int_0^T dt\, \|(P(1/n)e^{-itH/n} P(1/n))^n f\|^2\\ &= \int_0^T dt\, \|e^{-itH_P} Pf\|^2\, dt = T\,\|P\,f\|^2\,,\end{aligned}$$

with the lim sup taken along $n \to \infty$, because $\mathbb{1} = P(1/n) + Q(1/n)$ and $P(\tau) \xrightarrow{s} P$ in the strong operator topology as $\tau \to 0$. As a consequence

$$\lim_{n\to\infty} \int_0^T dt \, \| Q(1/n)(e^{-itH/n} P)^n f \|^2 = 0 .$$

Then for any $v \in \mathfrak{F}_0^+ = L^2_{\text{loc}}(\mathbb{R}_0^+, \mathfrak{H})$ and $f \in \mathfrak{H}$ due to *symmetric* product formula (14.1.22), we conclude that for every $T > 0$:

$$\begin{aligned}
&\int_0^T dt \, ((e^{-itH/n} P(1/n))^n f, \; v(t)) = \\
&\int_0^T dt \, ((P(1/n)\, \mathrm{e}^{-itH/n} P(1/n))^n f, \; P(1/n)\, v(t)) \\
&\qquad + \int_0^T dt \, (Q(1/n)(e^{-itH/n} P(1/n))^n f, \; Q(1/n)\, v(t)) \\
&\longrightarrow \int_0^T dt \, (e^{-itH_P} Pf, \; v(t)) ,
\end{aligned}$$

as $n \to \infty$. It means that the sequence of vectors $\{(e^{-itH/n} P(1/n))^n f\}_{n\geq 1}$ converges to $e^{-itH_P} Pf$ *weakly* in $\mathfrak{F}_0^+ = L^2_{\text{loc}}(\mathbb{R}_0^+, \mathfrak{H})$ together with *all* seminorms. As a consequence, this convergence is, in fact, *strong* in the *Fréchet* space $\mathfrak{F}_0^+$. This yields the formula (14.1.24). □

Consequently, it remains only to prove Lemma 14.1.12 on which the above arguments were based. To this end, see Notes to Chap. 14 (Sects. 14.1 and 14.1.3).

14.1.4 Finite-Dimensional Projections

As we have indicated in Sect. 14.1.1 (Remark 14.1.3), one does not need to resort to *subsequences* in particular case when the projections involved in product formulae are finite-dimensional.

On the other hand, in contrast to (14.1.20) the question about: whether in this case the Zeno formula converges to a *unitary* Zeno dynamics in the operator-norm topology *without* supplementary condition (14.1.14) on the *image* of projection, leading to (14.1.15), nonetheless rests open.

Theorem 14.1.15 *Let, in addition to the hypotheses of Theorem* 14.1.6, *the orthogonal projection P as well as $P(t)$ be of finite dimension (i.e. finite-rank) operators. Then*:

(i) *The formulae* (14.1.22)–(14.1.24) *hold, as $n \to \infty$, in the strong operator topology on $\mathcal{L}(\mathfrak{H})$, locally-uniformly in $t \in \mathbb{R} \setminus \{0\}$.*

(ii) *It also holds for* $\varepsilon = \pm 1$ *that as* $n \to \infty$

$$\text{s-}\lim_{n\to\infty} [P(t/n)\exp(-i\,\varepsilon t\, H/n)P(t/n)]^n\, u = P\,\exp(-i\varepsilon t H_P)\,P\,u\,,$$

$$\text{s-}\lim_{n\to\infty} [P(t/n)\,\exp(-i\varepsilon t H/n)]^n\, u = P\,\exp(-i\,\varepsilon t\, H_P)\,P\,u\,,$$

$$\text{s-}\lim_{n\to\infty} [\,\exp(-i\varepsilon t H/n)\,P(t/n)]^n\, u = P\,\exp(-i\,\varepsilon t\, H_P)\,P\,u\,,$$

locally-uniformly in $t \in \mathbb{R}$ *for any* $u \in \mathfrak{H}$.

Before begin the proof of this statement we need some preparatory assertions.

Note that since the closed operator $H^{1/2}P$ is supposed to be densely defined, domain $\mathrm{dom}\,(H^{1/2}P) = \mathfrak{H}$. For restriction $H^{1/2}P \upharpoonright P\,\mathfrak{H}$ of the operator $H^{1/2}P$ to the *finite-dimensional* subspace $P\,\mathfrak{H}$ is densely defined, so its domain must coincide with $P\,\mathfrak{H}$, and it acts as zero on $Q\,\mathfrak{H}$, where $Q = \mathbb{1} - P$.

The same is valid for $H^{1/2}P(t)$ when $P(t)$ is of a finite dimension. As a result, $H^{1/2}P$ and $H_P = (H^{1/2}P)^*(H^{1/2}P)$ as well as $H^{1/2}P(t)$ are bounded operators on $\mathfrak{H}$ by the *closed-graph* theorem. By the assumptions common with Theorem 14.1.6, for each fixed $f \in \mathrm{dom}\,(H^{1/2}P) = \mathfrak{H}$ the family $\{H^{1/2}P(t)f\}_{t\in\mathbb{R}}$ converges to $H^{1/2}Pf$ as $t \to 0$, and hence is uniformly bounded with respect to t near to zero, say, for $t \in [-1, 1]$. Then by the *uniform boundedness principle* we conclude that $\sup_{|t|\leq 1} \|H^{1/2}P(t)\| < \infty$.

To prove the assertions (i) and (ii) of Theorem 14.1.15 simultaneously, take a fixed parameter $a \in \mathbb{R}^+$ and consider instead of $F(\zeta, \tau)$, $S(\zeta, \tau)$ defined by (14.1.30) and (14.1.31), respectively, the following operators

$$F_a(\zeta,\tau) := P(a\tau)\,\exp(-\zeta\,\tau\,H)\,P(a\tau)\,, \quad S_a(\zeta,\tau) := \tau^{-1}[\mathbb{1} - F_a(\zeta,\tau)]\,. \tag{14.1.45}$$

In fact, we shall employ $F_a(it, \tau)$, $S_a(it, \tau)$ instead of $F(it, \tau)$, $S(it, \tau)$ in the proof of Lemmas 14.1.12 and 14.1.13. Similarly $u(t, \tau)$ used above will be replaced by $u_a(t, \tau) = (\mathbb{1} + S_a(it, \tau))^{-1} f$ corresponding to a given $f \in \mathfrak{H}$.

Lemma 14.1.16 *For any* $t, t' \geq 0$ *and* $0 < \tau \leq 1$ *we have*

$$\|u_a(t,\tau) - u_a(t',\tau)\| \leq C(a)\,|t - t'|\,\|f\|\,, \quad f \in \mathfrak{H}\,, \tag{14.1.46}$$

with $C(a) > 0$ *independent of* t, t', *which is uniformly bounded as a function of* $a > 0$ *on each compact interval of* $\mathbb{R}^+$.

Proof By the resolvent equation we have for $f \in \mathfrak{H}$

$$
\begin{aligned}
& u_a(t,\tau) - u_a(t',\tau) = \\
& (\mathbb{1} + S_a(i\,t,\tau))^{-1} f - (\mathbb{1} + S_a(i\,t',\tau))^{-1} f = \\
& (\mathbb{1} + S_a(i\,t,\tau))^{-1} P(a\tau)\,\tau^{-1}[e^{-i\,t\,\tau\,H} - \mathrm{e}^{-i\,t'\,\tau\,H}]\,P(a\tau)\,(\mathbb{1} + S_a(i\,t',\tau))^{-1} f = \\
& (\mathbb{1} + S_a(i\,t,\tau))^{-1} P(a\tau)\,\frac{1}{\tau}\int_{t'}^{t} ds\,\frac{d}{ds}\,e^{-i\,s\,\tau\,H}\;P(a\tau)\,(\mathbb{1} + S_a(i\,t',\tau))^{-1} f = \\
& = -i\;(\mathbb{1} + S_a(i\,t,\tau))^{-1} P(a\tau)\int_{t'}^{t} ds\;H\,\mathrm{e}^{-i\,s\,\tau\,H}\;P(a\tau)(\mathbb{1} + S_a(i\,t',\tau))^{-1} f = \\
& = -i\;(\mathbb{1} + S_a(i\,t,\tau))^{-1}(H^{1/2}\,P(a\tau))^* \;\times \\
& \times \int_{t'}^{t} ds\;e^{-is\,\tau\,H}\;(H^{1/2}\,P(a\tau))\,(\mathbb{1} + S_a(i\,t',\tau))^{-1} f\,.
\end{aligned}
$$

In the beginning of this section, we have argued that the operators $H^{1/2}P(a\tau)$ are uniformly bounded on $\mathfrak{H}$ for $0 < a\tau \leq 1$. It follows that

$$
\|u_a(t,\tau) - u_a(t',\tau)\| \leq C(a)\,|t - t'|\;\|f\|\,, \quad f \in \mathfrak{H}\,,
$$

with $C(a) := \sup_{0<a\tau\leq 1} \|H^{1/2}P(a\tau)\,\|^2$. By the arguments preceding the lemma the function: $a \mapsto C(a)$, is uniformly bounded on each compact a-interval in $\mathbb{R}^+$. This yields the claim. □

Proof of Theorem 14.1.15 From (14.1.46) one gets that continuous vector-valued family $\{u_a(t,\tau)\}_{t\in\mathbb{R}^+}$ in $\mathfrak{H}$, is uniformly bounded and equicontinuous. Hence, we infer by the *Arzelà - Ascoli* theorem that there exists a sequence $\{\tau_n\}_{n\geq 1}$ such that for each t the corresponding sequence $\{u_a(t,\tau_n)\}_{n\geq 1}$ converges strongly to $u(t)$ uniformly in $t \in \mathbb{R}_0^+$. As a consequence the function: $t \mapsto u(t)$, is strongly continuous in $t \in \mathbb{R}_0^+$ and coincides with $(\mathbb{1} + i\,tH_P)^{-1} f$ for all $t \geq 0$. Thus for any $a \in \mathbb{R}^+$, we obtain instead of Lemma 14.1.12 the following claim:

$$
\lim_{\tau\to 0}\,(\mathbb{1} + S_a(i\,t,\tau))^{-1}\,f = (\mathbb{1} + i\,tH_P)^{-1} P\,f\,, \tag{14.1.47}
$$

locally-uniformly in variable $t \in \mathbb{R}_0^+$ for each $f \in \mathfrak{H}$.

Next we shall modify the reasoning in the proof of Lemma 14.1.13 with the aim to show the symmetric product case for *contractions* $\{F_a(i,\,t/n)\}_{n\geq 1}$ (cf. (14.1.45)) with parameter $a > 0$:

$$
\text{s-}\lim_{n\to\infty} F_a(i,\,t/n)^n = \exp(-i\,tH_P)\,P\,. \tag{14.1.48}
$$

To this end we note that the resolvent convergence (14.1.47) is equivalent (for $t = 1$) the convergence of the corresponding semigroups. As a result

$$\lim_{\tau \to 0} \mathrm{e}^{-\theta S_a(i,\tau)} f = \mathrm{e}^{-i\theta H_P} f, \quad f \in \mathfrak{H}, \tag{14.1.49}$$

locally-uniformly in variable $\theta \geq 0$. Applying this equivalence again, but now to (14.1.49), we obtain for any $\lambda \geq 0$ the strong limit

$$\text{s-}\lim_{\tau \to 0} (\mathbb{1} + \lambda S_a(i,\tau))^{-1} = (\mathbb{1} + i\lambda H_P)^{-1} P.$$

In particular, taking $\tau = \theta/n$ and using the *diagonal trick* we infer that

$$\text{s-}\lim_{n \to \infty} (\mathbb{1} + \frac{\theta}{\sqrt{n}} S_a(i,\theta/n))^{-1} = P, \tag{14.1.50}$$

holds locally-uniformly in $\theta \in \mathbb{R}_0^+$.

Now similar to the line of reasoning in the proof of Lemma 14.1.13, we apply to contractions $\{F_a(i, t/n)\}_{n\geq 1}$ (14.1.45)) the *Chernoff* $\sqrt{n}$-Lemma (Lemma 1.9.33), which yields for any $g \in \mathfrak{H}$ estimates

$$\left\| F_a(i,\theta/n)^n g - \mathrm{e}^{-n(\mathbb{1} - F_a(i,\theta/n))} g \right\| \leq \sqrt{n} \, \| (\mathbb{1} - F_a(i,\theta/n))\, g \|, \quad n \in \mathbb{N}.$$

Let $f \in P\mathfrak{H}$. Then choosing $g := \left(\mathbb{1} + \theta\, S_a(i,\theta/n)/\sqrt{n}\right)^{-1} f$, we obtain inequality

$$\begin{aligned} &\left\| \left(F_a(i,\theta/n)^n - \mathrm{e}^{-\theta S_a(i,\theta/n)}\right) \left(\mathbb{1} + \frac{\theta}{\sqrt{n}} S_a(i,\theta/n)\right)^{-1} f \right\| \\ &\leq \left\| f - \left(\mathbb{1} + \frac{\theta}{\sqrt{n}} S_a(i,\theta/n)\right)^{-1} f \right\|. \end{aligned} \tag{14.1.51}$$

Note that since $f \in P\mathfrak{H}$ the right-hand side of (14.1.51) tends to zero as $n \to \infty$ by (14.1.50).

Given that by (14.1.45) $\{F_a(i, \theta/n)^n\}_{n\geq 1}$ and $\{e^{-\theta S_a(i,\theta/n)}\}_{n\geq 1}$ are contractions one can mimic arguments from (14.1.39) to show that (14.1.51) yields

$$\lim_{n \to \infty} \| F_a(i,\theta/n)^n f - e^{-\theta S_a(i,\theta/n)} f \| = 0, \quad f \in P\mathfrak{H}, \tag{14.1.52}$$

locally-uniformly on each compact θ-interval in $\mathbb{R}_0^+$.

Choosing now $\theta = t$ we see that the validity of (14.1.48) on $P\mathfrak{H}$ follows from (14.1.49) and (14.1.52). Consequently, on the subspace $P\mathfrak{H}$ the assertion (i) of Theorem 14.1.15 is obtained by taking $a = 1/t$ for any t belonging to a compact interval in $\mathbb{R} \setminus \{0\}$ whereas the assertion (ii) of Theorem 14.1.15 by choosing $a = 1$.

The case $f \in Q\,\mathfrak{H}$ can be treated for symmetric product formula similarly to the proof of this case in Lemma 14.1.13. The arguments follow *verbatim* the line of reasoning that leads to (14.1.41). As a result, on the vector $f \in Q\,\mathfrak{H}$ the limit (14.1.48) is *zero* for $a = 1/t$ as well as for $a = 1$, that proves Theorem 14.1.15 (i) and (ii) on the subspace $Q\,\mathfrak{H}$.

The proof of the *non-symmetric* product formulae is *mutatis mutandis* the arguments developed in Sect. 14.1.3(b) for non-symmetric formulae (14.1.23) and (14.1.24), cf. Lemma 14.1.14.

This concludes the proof of Theorem 14.1.15. □

We summarise by remark that discussion about problem concerning the Zeno product formula (cf. Remark 14.1.9) is not exhausted. The Zeno product formulae in the *Fréchet* space does not sound as an *optimal* result. For that reason we shall revise setup of the problem in the next sections.

For references, comments and more details see Notes to Chap. 14 (Notes to Sects. 14.1 and 14.1.4).

14.2 Non-exponential Zeno Product Formulæ

14.2.1 Admissible Functions

Similarly to Sect. 14.1.2 we assume in this section that H is a non-negative self-adjoint operator in a Hilbert space $\mathfrak{H}$ and orthogonal projection P such that $\mathrm{dom}\,(\sqrt{H}) \cap \mathfrak{H}_P$ is dense in subspace $P\,\mathfrak{H}$. Under these assumptions the limit of the Zeno product formula (14.1.25) in topology of the larger (*Fréchet*) space (14.1.21) is the unitary group $\{\mathrm{e}^{-itK}\}_{t\in\mathbb{R}}$, on $\mathfrak{H}_P$, generated by the non-negative self-adjoint operator $K := PHP$ associated with the closed *sesquilinear* form $\mathfrak{k}$,

$$\mathfrak{k}(f, g) := (\sqrt{H}f, \sqrt{H}g), \quad f, g \in \mathrm{dom}\,(\mathfrak{k}) = \mathrm{dom}\,(\sqrt{H}) \cap \mathfrak{H}_P \subseteq \mathfrak{H}_P. \quad (14.2.1)$$

This claim is based upon Theorem 14.1.6 and Corollary 14.1.7. There, it is shown that convergence of the Zeno product formula in the *Fréchet* space explicitly means that for each $f \in \mathfrak{H}_P$ one has

$$\lim_{n\to\infty} \int_{-T_1}^{T_2} dt\, \left\| \left(P\mathrm{e}^{-itH/n}P\right)^n f - P\,\mathrm{e}^{-itK} f \right\|^2 = 0, \quad \text{for any} \quad T_1, T_2 > 0\,. \tag{14.2.2}$$

This result also yields existence of the limit of the Zeno product formula for *almost all* (a.a.) $t \in \mathbb{R}$ in the *strong* operator topology (Corollary 14.1.8), along infinite *subsequences* $\{n'\}_{n'\in\mathbb{N}}$.

Remark 14.2.1 The reason behind of this claim, which is weaker than a natural conjecture, is caused by the following problem. For $n \to \infty$, the *exponential* function involved into the *interlaced* product in (14.2.2) manifests oscillations, which are not easy to control. See comments in Notes to Chap. 14 (Sects. 14.2 and 14.2.1). One of the main point of the present section is observation that we can avoid the mentioned problem when exponent $\phi(x) = \mathrm{e}^{-ix}$ in the Zeno product approximants is replaced by a function, whose *imaginary* part has a constant sign.

To this aim, in analogy with functions of the *Kato* class (Sect. 5.2), it is useful to consider a class of *admissible* Kato functions with the *constant sign* for imaginary part, cf. Definition 13.3.13 in Sect. 13.3.4.

Definition 14.2.2 We call a Borel measurable function $\phi : \mathbb{R}_0^+ \to \mathbb{C}$ *admissible* with the *constant sign* for imaginary part if the conditions

$$|\phi(x)| \leq 1, \quad x \in \mathbb{R}_0^+, \qquad \phi(0) = 1, \qquad \text{and} \quad \phi'(+0) = -i\,, \tag{14.2.3}$$

are satisfied and which, in addition, obeys the property

$$\mathrm{Im}(\phi(x)) \leq 0\,, \quad x \in \mathbb{R}_0^+\,, \tag{14.2.4}$$

that is, of *nonpositivity* of imaginary part.

Recall that examples of such kind of *admissible* functions are:

$$\phi(x) = (1 + ix)^{-1} \quad \text{and} \quad \phi(x) = (1 + ix/2)^{-2}, \quad x \in \mathbb{R}_0^+\,.$$

Unfortunately, not all functions with properties (14.2.3) do satisfy the condition of constant sign for the imaginary part (14.2.4). For example, the functions $\phi(x) = (1 + ix/3)^{-3}$ and $\phi(x) = \mathrm{e}^{-ix}$, $x \in \mathbb{R}_0^+$ satisfy (14.2.3), but do not obey (14.2.4).

The basic thought of this Sect. 14.2 is observation about two kinds of the Zeno product formulae:

(a) The *strongly* convergent Zeno product formulae for *non-exponential* admissible functions from Definition 14.2.2, see Sect. 14.2.2.
(b) The *strongly* and *operator-norm* convergent Zeno product formulae for a *generalised* admissible functions when proviso (14.2.4) in Definition 14.2.2 is *dropped*, see Sect. 14.2.3.

We first formulate and comment below the assertion from the point (a), but we postpone its proof to Sect. 14.2.2.

Theorem 14.2.3 *Let H be a non-negative self-adjoint operator in $\mathfrak{H}$, and let $\mathfrak{H}_P$ be a closed subspace of $\mathfrak{H}$ such that $P : \mathfrak{H} \to \mathfrak{H}_P$ is the orthogonal projection from $\mathfrak{H}$ onto $\mathfrak{H}_P$. If* dom $(\sqrt{H}) \cap \mathfrak{H}_P$ *is dense in $\mathfrak{H}_P$ and ϕ is admissible function which obeys*

$$\mathrm{Im}(\phi(x)) \leq 0, \quad x \in \mathbb{R}_0^+,$$

then for any $t_0 > 0$ *one has*

$$\text{s-}\lim_{n\to\infty} (P\phi(tH/n)P)^n = P\,\mathrm{e}^{-itK}\,, \tag{14.2.5}$$

uniformly in $t \in [0, t_0]$, *where the generator* $K := PHP$ *is defined by* (14.2.1) *and the strong convergence is meant on* $\mathfrak{H}_P$.

We note that function: $\phi(x) = \mathrm{e}^{-ix}$, $x \in \mathbb{R}_0^+$, does not satisfy conditions of Theorem 14.2.3, namely (14.2.4). In particular, this yields that the convergence problem for the *original* (exponential) Zeno product formula (14.1.1) is not solved by Theorem 14.2.3 and the question about exponential product formula (14.2.5) remains *open*.

However, Theorem 14.2.3 suggests the following *regularisation* procedure. To this end we set

$$\Delta_\phi := \{x \in \mathbb{R}_0^+ : \mathrm{Im}(\phi(x)) \le 0\}.$$

By (14.2.3) the set Δ_ϕ contains a neighborhood of zero. If the subset $\Delta \subseteq \Delta_\phi$ also contains a neighborhood of zero, then

$$\phi_\Delta(x) := \phi(x)\chi_\Delta(x), \quad x \in \mathbb{R}_0^+,$$

defines an admissible function obeying $\mathrm{Im}(\phi_\Delta(x)) \le 0$, $x \in \mathbb{R}_0^+$. By Theorem 14.2.3 we obtain that for any $t_0 > 0$ one has the strong limit:

$$\text{s-}\lim_{n\to\infty} (P\phi_\Delta(tH/n)P)^n = P\,\mathrm{e}^{-itK}\,,$$

uniformly in $t \in [0, t_0]$. Applying this procedure to $\phi(x) = \mathrm{e}^{-ix}$, $x \in \mathbb{R}_0^+$, one has to choose a subset $\Delta \subseteq \Delta_\phi := \cup_{m=0}^{\infty}[2m\pi, (2m+1)\pi]$ containing a neighborhood of zero. From $\phi(x) = \mathrm{e}^{-ix}$ one can construct a "cutoff" admissible function $\phi_\Delta(x) := \mathrm{e}^{-ix}\chi_\Delta(x)$, $x \in \mathbb{R}_0^+$, obeying (14.2.4). In particular, for $\Delta = [0, \pi)$, the function

$$\phi_\Delta(x) = \mathrm{e}^{-ix}\chi_\Delta(x), \quad x \in \mathbb{R}_0^+\,,$$

is admissible and obeys (14.2.4). This immediately leads to the following corollary.

Corollary 14.2.4 *Since the assumptions of Theorem* 14.2.3 *are satisfied for admissible functions* $\phi(\cdot)$ *below, for any* $t_0 > 0$ *one has the strong limits:*

$$\text{s-}\lim_{n\to\infty} \left(P(\mathbb{1} + itH/n)^{-1}P\right)^n = \text{s-}\lim_{n\to\infty} \left(P(\mathbb{1} + itH/2n)^{-2}P\right)^n = \mathrm{e}^{-itK}\,, \tag{14.2.6}$$

and

$$\text{s-}\lim_{n\to\infty}\left(P E_H([0,\pi n/t))\mathrm{e}^{-i\,t\,H/n}\,P\right)^n = \mathrm{e}^{-i\,tK}\,, \tag{14.2.7}$$

uniformly in $t \in [0, t_0]$, *where* $E_H(\cdot)$ *is the spectral measure of self-adjoint operator* H, *that is* , $H = \int_{\mathbb{R}_0^+} dE_H(\lambda)\,\lambda$.

14.2.2 Strong Operator Convergence

Section 14.2.2 is entirely devoted to the proof of Theorem 14.2.3.

To this aim we first start by setting:

$$F(\tau) := P\,\phi(\tau H)\,P : \mathfrak{H}_P \to \mathfrak{H}_P\,, \quad \tau \geq 0\,, \tag{14.2.8}$$

and

$$S(\tau) := \frac{\mathbb{1}_{\mathfrak{H}_P} - F(\tau)}{\tau} : \mathfrak{H}_P \to \mathfrak{H}_P\,, \quad \tau > 0\,, \tag{14.2.9}$$

where $\mathbb{1}_{\mathfrak{H}_P} := \mathbb{1} \restriction \mathfrak{H}_P$ denotes the *unit* operator on $\mathfrak{H}_P$. If operator X is defied on domain in the space $\mathfrak{X}$, we use the notation PXP for the operator such that $PXP \restriction \mathfrak{H}_P := PX \restriction \mathfrak{H}_P : \mathfrak{H}_P \to \mathfrak{H}_P$, whereas $PXP \restriction \mathfrak{H}_P^{\perp} := 0 \cdot P^{\perp}$, where $\mathbb{1} = P + P^{\perp}$.

Let us assume that

$$\mathrm{dom}\,(T) := \mathrm{dom}\,(\sqrt{H}) \cap \mathfrak{H}_P\,, \tag{14.2.10}$$

is dense in $\mathfrak{H}_P$. We define a linear operator $T : \mathfrak{H}_P \to \mathfrak{H}$, by

$$Tf := \sqrt{H}f, \quad f \in \mathrm{dom}\,(T)\,. \tag{14.2.11}$$

Since $\sqrt{H}$ is closed and $\mathrm{dom}\,(\sqrt{H}) \cap \mathfrak{H}_P$ is dense, the operator T is closed. Then $T^*T : \mathfrak{H}_P \to \mathfrak{H}_P$ is a self-adjoint operator, which is identical to K defined by (14.2.1), that is,

$$K := T^*T : \mathfrak{H}_P \to \mathfrak{H}_P \tag{14.2.12}$$

defines a non-negative self-adjoint operator in $\mathfrak{H}_P$.

Further, let us represent the function ϕ as

$$\phi(x) = \psi(x) - i\omega(x), \quad x \in \mathbb{R}_0^+\,,$$

where $\psi, \omega : \mathbb{R}_0^+ \to \mathbb{R}$ are real-valued, Borel measurable functions obeying

$$|\psi(x)| \leq 1, \quad \psi(0) = 1, \quad \psi'(+0) = 0\,, \tag{14.2.13}$$

and

$$0 \leq \omega(x) \leq 1, \quad \omega(0) = 0, \quad \omega'(+0) = 1\,.$$

Now setting

$$\varphi(x) := 1 - \omega(x), \quad x \in \mathbb{R}_0^+\,,$$

one gets

$$0 \leq \varphi(x) \leq 1, \quad \varphi(0) = 1, \quad \varphi'(+0) = -1\,, \tag{14.2.14}$$

which shows that $\varphi \in \mathcal{K}$ is a *Kato function* (Definition 5.2.1 in Sect. 5.2). In terms of ψ, φ, the function ϕ admits the representation

$$\phi(x) = \psi(x) - i(1 - \varphi(x)), \quad x \in \mathbb{R}_0^+. \tag{14.2.15}$$

We set

$$\begin{aligned} p_-(x) &:= \begin{cases} 1, & x = 0, \\ \inf_{s\in(0,x]}(1-\varphi(s))/s, & x > 0, \end{cases} \\ p_+(x) &:= \begin{cases} 1, & x = 0, \\ \sup_{s\in(0,x]}(1-\varphi(s))/s, & x > 0. \end{cases} \end{aligned} \tag{14.2.16}$$

Both functions $p_\mp$ are bounded on $\mathbb{R}_0^+$ and obey

$$0 \leq p_-(x) \leq 1 \leq p_+(x) < \infty, \quad x \in \mathbb{R}_0^+. \tag{14.2.17}$$

The function p_- is decreasing: $p_-(x) \geq p_-(y)\,,\ 0 \leq x \leq y$, and p_+ is increasing: $p_+(x) \leq p_+(y)\,,\ 0 \leq x \leq y$.

We define now sesquilinear forms:

$$\begin{aligned} &\mathfrak{t}_\tau^-(f, g) := (p_-(\tau H)\sqrt{H}f, \sqrt{H}g), \\ &\qquad f, g \in \mathrm{dom}\,(\mathfrak{t}_\tau^-) := \mathrm{dom}\,(\sqrt{H}) \cap \mathfrak{H}_P, \quad \tau \geq 0\,, \end{aligned}$$

and

$$\begin{aligned} &\mathfrak{t}_\tau^+(f, g) := (p_+(\tau H)\sqrt{H}f, \sqrt{H}g), \\ &\qquad f, g \in \mathrm{dom}\,(\mathfrak{t}_\tau^+) := \mathrm{dom}\,(\sqrt{H}) \cap \mathfrak{H}_P, \quad \tau \geq 0\,. \end{aligned}$$

Notice that for $\tau = 0$ one has $\mathfrak{t}_0^- = \mathfrak{t}_0^+ = \mathfrak{t}$, where the sesquilinear form $\mathfrak{t}$ is defined by (14.2.1). Obviously, both forms $\mathfrak{t}_\tau^\pm$ are non-negative for each $\tau \geq 0$. Moreover, the form $\mathfrak{t}_\tau^-$ is closable for each $\tau > 0$ and its closure is a bounded form on $\mathfrak{H}_P$ while the form $\mathfrak{t}_\tau^+$ is already closed for each $\tau \geq 0$.

By $\{K_\tau^\pm\}_{\tau \geq 0}$ we denote the associated non-negative self-adjoint operators in $\mathfrak{H}_P$. We note that $K_0^\pm = K$, cf. (14.2.11) and (14.2.12). Then by (14.2.17) we get

$$\mathfrak{t}_\tau^-(f, f) \leq \mathfrak{t}(f, f) \leq \mathfrak{t}_\tau^+(f, f), \quad f \in \operatorname{dom}(\mathfrak{t}_\tau^-) = \operatorname{dom}(\mathfrak{t}) = \operatorname{dom}(\mathfrak{t}_\tau^+), \quad \tau \geq 0,$$

which yields

$$K_\tau^- \leq K \leq K_\tau^+, \quad \tau \geq 0.$$

Since p_- is decreasing, the family $\{K_\tau^-\}_{\tau \geq 0}$ is increasing as $\tau \to +0$. Further, from (14.2.16) one gets that $\text{s-lim}_{\tau \to +0} \, p_-(\tau H) = \mathbb{1}$. Since $\mathfrak{t}_\tau^- \leq \mathfrak{t}$ and

$$\lim_{\tau \to +0} \mathfrak{t}_\tau^-(f, g) = \lim_{\tau \to +0} (p_-(\tau H)\sqrt{H} f, \sqrt{H} g) = \mathfrak{t}(f, g),$$
$$f, g \in \operatorname{dom}(\mathfrak{t}_\tau^-) = \operatorname{dom}(\mathfrak{t}),$$

we obtain by virtue of monotonicity that

$$\text{s-}\lim_{\tau \to +0} (\mathbb{1}_{\mathfrak{H}_P} + K_\tau^-)^{-1} = (\mathbb{1}_{\mathfrak{H}_P} + K)^{-1}, \tag{14.2.18}$$

where $\mathbb{1}_{\mathfrak{H}_P} = \mathbb{1} \restriction \mathfrak{H}_P$ is the unit operator on $\mathfrak{H}_P$.

Further, since p_+ is increasing the family $\{K_\tau^+\}_{\tau \geq 0}$ is decreasing as $\tau \to +0$. By $\text{s-lim}_{\tau \to +0} \, p_+(\tau H) = \mathbb{1}$, we find

$$\lim_{\tau \to +0} \mathfrak{t}_\tau^+(f, g) = \lim_{\tau \to +0} (p_+(\tau H)\sqrt{H} f, \sqrt{H} g) = \mathfrak{t}(f, g),$$
$$f, g \in \operatorname{dom}(\mathfrak{t}_\tau^+) = \operatorname{dom}(\mathfrak{t}).$$

Since $\mathfrak{t}$ is closed, then again on account of monotonicity, we obtain

$$\text{s-}\lim_{\tau \to +0} (\mathbb{1}_{\mathfrak{H}_P} + K_\tau^+)^{-1} = (\mathbb{1}_{\mathfrak{H}_P} + K)^{-1}. \tag{14.2.19}$$

Lemma 14.2.5 *Let $\{X(\tau)\}_{\tau>0}$, $\{Y(\tau)\}_{\tau>0}$, and $\{A(\tau)\}_{\tau>0}$ be families of bounded non-negative self-adjoint operators in $\mathfrak{H}_P$ such that the condition*

$$0 \leq X(\tau) \leq A(\tau) \leq Y(\tau), \quad \tau > 0,$$

is satisfied. If $\text{s-lim}_{\tau \to 0} X(\tau) = \text{s-lim}_{\tau \to 0} Y(\tau) = A$, where A is a bounded self-adjoint operator in $\mathfrak{H}_P$, then $\text{s-lim}_{\tau \to 0} A(\tau) = A$.

Proof Since for each $f \in \mathfrak{H}_P$ we have

$$(X(\tau)f, f) \leq (A(\tau)f, f) \leq (Y(\tau)f, f), \quad \tau > 0,$$

we get $\lim_{\tau\to+0}(A(\tau)f, f) = (Af, f)$, $f \in \mathfrak{H}_P$, or $\text{w-lim}_{\tau\to+0} A(\tau) = A$. Hence,

$$\text{w-}\lim_{\tau\to+0} (Y(\tau) - A(\tau)) = 0.$$

Since $Y(\tau) - A(\tau) \geq 0$, $\tau > 0$, we find

$$\text{s-}\lim_{\tau\to+0} (Y(\tau) - A(\tau))^{1/2} = 0,$$

which yields $\text{s-lim}_{\tau\to+0}(Y(\tau) - A(\tau)) = 0$. Hence, $\text{s-lim}_{\tau\to+0} A(\tau) = A$. □

From (14.2.9) and (14.2.15), we obtain

$$S(\tau) = \frac{1}{\tau} P\,(\mathbb{1} - \psi(\tau H))\,P + i\,\frac{1}{\tau} P\,(\mathbb{1} - \varphi(\tau H))\,P, \quad \tau > 0. \tag{14.2.20}$$

Let

$$L_0(\tau) := \frac{1}{\tau}\, P\,(\mathbb{1} - \varphi(\tau H))\,P : \mathfrak{H}_P \to \mathfrak{H}_P, \quad \tau > 0. \tag{14.2.21}$$

Lemma 14.2.6 *Let H be a non-negative self-adjoint operator in $\mathfrak{H}$, and let $\mathfrak{H}_P = P\,\mathfrak{H}$ be a closed subspace of $\mathfrak{H}$. If* $\text{dom}\,(\sqrt{H}) \cap \mathfrak{H}_P$ *is dense in $\mathfrak{H}_P$ and $\varphi \in \mathcal{K}$ is a Kato function, then we have*

$$\text{s-}\lim_{\tau\to+0} \left(\mathbb{1}_{\mathfrak{H}_P} + L_0(\tau)\right)^{-1} = (\mathbb{1}_{\mathfrak{H}_P} + K)^{-1}, \tag{14.2.22}$$

where $\mathbb{1}_{\mathfrak{H}_P} = \mathbb{1} \upharpoonright \mathfrak{H}_P$.

Proof Since

$$\mathfrak{t}_\tau^-(f, f) \leq \left(\frac{\mathbb{1} - \varphi(\tau H)}{\tau} f, f\right) \leq \mathfrak{t}_\tau^+(f, f), \quad f \in \text{dom}\,(\sqrt{H}) \cap \mathfrak{H}_P,$$

we find

$$K_\tau^- \leq P\,\frac{\mathbb{1} - \varphi(\tau H)}{\tau}\,P \leq K_\tau^+, \quad \tau > 0.$$

Hence

$$\begin{aligned} X(\tau) &:= (\mathbb{1}_{\mathfrak{H}_P} + K_\tau^+)^{-1} \leq \left(\mathbb{1}_{\mathfrak{H}_P} + P\,\frac{\mathbb{1} - \varphi(\tau H)}{\tau}\,P\right)^{-1} \\ &\leq (\mathbb{1}_{\mathfrak{H}_P} + K_\tau^-)^{-1} =: Y(\tau), \end{aligned}$$

$\tau > 0$. Taking into account (14.2.18),(14.2.19) and applying Lemma 14.2.5, we prove (14.2.22). □

Now we set

$$L(\tau) := \frac{1}{\tau} P(\mathbb{1} - \psi(\tau H)) P + \frac{1}{\tau} P(\mathbb{1} - \varphi(\tau H)) P : \mathfrak{H}_P \to \mathfrak{H}_P, \quad \tau > 0.$$

Lemma 14.2.7 *Let H be a non-negative self-adjoint operator in $\mathfrak{H}$, and let $\mathfrak{H}_P$ be a closed subspace of $\mathfrak{H}$. If* dom $(\sqrt{H}) \cap \mathfrak{H}_P$ *is dense in $\mathfrak{H}_P$, the real-valued Borel measurable function ψ obeys* (14.2.13) *and $\varphi \in \mathcal{K}$ is a Kato function, then*

$$\text{s-}\lim_{\tau \to +0} (\mathbb{1}_{\mathfrak{H}_P} + L(\tau))^{-1} = (\mathbb{1}_{\mathfrak{H}_P} + K)^{-1}. \tag{14.2.23}$$

Proof Let

$$\zeta(x) := \frac{\psi(2x) + \varphi(2x)}{2}, \quad x \in \mathbb{R}_0^+.$$

Notice that ζ is a Kato function. Setting

$$\widetilde{L}_0(\tau) := \frac{1}{\tau} P(\mathbb{1} - \zeta(\tau H)) P, \quad \tau > 0,$$

we obtain from Lemma 14.2.6 that $\text{s-}\lim_{\tau \to +0} = (\mathbb{1}_{\mathfrak{H}_P} + \widetilde{L}_0(\tau))^{-1} = (\mathbb{1}_{\mathfrak{H}_P} + K)^{-1}$. By $L(2\tau) = \widetilde{L}_0(\tau)$ we prove (14.2.23). □

Next we define

$$M(\tau) := (\mathbb{1}_{\mathfrak{H}_P} + L_0(\tau))^{-1/2} P \frac{\mathbb{1} - \psi(\tau H)}{\tau} P (\mathbb{1}_{\mathfrak{H}_P} + L_0(\tau))^{-1/2}, \quad \tau > 0.$$

Lemma 14.2.8 *Let H be a non-negative self-adjoint operator in $\mathfrak{H}$ and let $\mathfrak{H}_P$ be a closed subspace of $\mathfrak{H}$. If* dom $(\sqrt{H}) \cap \mathfrak{H}_P$ *is dense in $\mathfrak{H}_P$, the real-valued, Borel measurable functions ψ obeys* (14.2.13) *and $\varphi \in \mathcal{K}$ is a Kato function, then*

$$\text{s-}\lim_{\tau \to +0} (\mathbb{1} + M(\tau))^{-1} = \mathbb{1}. \tag{14.2.24}$$

Proof A straightforward computation provides representation

$$(\mathbb{1}_{\mathfrak{H}_P} + L(\tau))^{-1} = (\mathbb{1}_{\mathfrak{H}_P} + L_0(\tau))^{-1/2} \left(\mathbb{1}_{\mathfrak{H}_P} + M(\tau)\right)^{-1} \times (\mathbb{1}_{\mathfrak{H}_P} + L_0(\tau))^{-1/2}, \quad \tau > 0.$$

By (14.2.22) and (14.2.23), we get

$$\text{w-}\lim_{\tau\to+0}\left(\mathbb{1}_{\mathfrak{H}_P}+M(\tau)\right)^{-1}=\mathbb{1}_{\mathfrak{H}_P}\,,$$

which yields

$$\text{s-}\lim_{\tau\to+0}\left(\mathbb{1}_{\mathfrak{H}_P}-\left(\mathbb{1}_{\mathfrak{H}_P}+M(\tau)\right)^{-1}\right)^{1/2}=0\,.$$

Hence

$$\text{s-}\lim_{\tau\to+0}\left(\mathbb{1}_{\mathfrak{H}_P}-\left(\mathbb{1}_{\mathfrak{H}_P}+M(\tau)\right)^{-1}\right)=0\,,$$

which proves (14.2.24). □

From (14.2.24) one gets

$$\text{s-}\lim_{\tau\to+0}\left(i\,\mathbb{1}_{\mathfrak{H}_P}+M(\tau)\right)^{-1}=(-i)\,\mathbb{1}_{\mathfrak{H}_P}\,.$$

Hence

$$\text{s-}\lim_{\tau\to+0}(\mathbb{1}_{\mathfrak{H}_P}+L_0(\tau))^{-1/2}\left(i\,\mathbb{1}_{\mathfrak{H}_P}+M(\tau)\right)^{-1}(\mathbb{1}_{\mathfrak{H}_P}+L_0(\tau))^{-1/2}$$
$$=-i\,(\mathbb{1}_{\mathfrak{H}_P}+K)^{-1}\,,$$

or

$$\text{s-}\lim_{\tau\to+0}\left(i\,\mathbb{1}_{\mathfrak{H}_P}+\frac{1}{\tau}\,P\,(\mathbb{1}_{\mathfrak{H}_P}-\psi(\tau H))\,P+i\,L_0(\tau)\right)^{-1}=(i\,\mathbb{1}_{\mathfrak{H}_P}+i\,K)^{-1}.$$

Using (14.2.20) and(14.2.21), we obtain

$$\text{s-}\lim_{\tau\to+0}(i\,\mathbb{1}_{\mathfrak{H}_P}+S(\tau))^{-1}=(i\,\mathbb{1}_{\mathfrak{H}_P}+i\,K)^{-1}\,,$$

which yields for non-negative self-adjoint operator K (14.2.12) the strong limit:

$$\text{s-}\lim_{\tau\to+0}(\mathbb{1}_{\mathfrak{H}_P}+S(\tau))^{-1}=(\mathbb{1}_{\mathfrak{H}_P}+iK)^{-1}\,. \qquad (14.2.25)$$

Finally, on account of the limit (14.2.25), the proof of Theorem 14.2.3 follows from the *Chernoff* product Theorem 4.1.3 (Sect. 4.1).

14.2.3 Operator-Norm Convergence

In the present section, we *relax* condition on *admissible* functions under a *stronger* assumption on domain, namely: $\mathfrak{H}_P \subseteq \mathrm{dom}\,(\sqrt{H})$. We show that under this assumption a modified Zeno product formula converges to e^{-itK} for admissible Kato functions not necessary satisfying the additional *constant sign* for imaginary part condition (14.2.4). In particular, one has

$$\text{s-}\lim_{n\to\infty}\left(P\mathrm{e}^{-it\,H/n}P\right)^n = \mathrm{e}^{-itK}\,, \tag{14.2.26}$$

uniformly in $t \in [0, t_0]$ for any $t_0 > 0$ and vector in $\mathfrak{H}$. Moreover, we shall demonstrate here that under some *stronger* conditions (unfortunately too restrictive from the viewpoint of physical applications) even the *operator-norm* convergence can be obtained.

From Sect. 14.2.2 one infers that Theorem 14.2.3 needs the additional assumption (14.2.4), and it is unclear whether this assumption can be dropped; cf. Sect. 14.3. In this subsection we are going to show that under stronger assumptions on domain of operator $\sqrt{H}$, the condition (14.2.4) is indeed *not* necessary.

Theorem 14.2.9 *Let H be a non-negative self-adjoint operator on $\mathfrak{H}$, and let $\mathfrak{H}_P$ be a closed subspace of $\mathfrak{H}$ such that $P : \mathfrak{H} \to \mathfrak{H}_P$, is the orthogonal projection of $\mathfrak{H}$ onto $\mathfrak{H}_P$. If $\mathfrak{H}_P \subseteq$* dom $(\sqrt{H})$ *and ϕ satisfies* (14.2.3), *then*

$$\text{s-}\lim_{n\to\infty}\,(P\,\phi(t\,H/n)\,P)^n = \mathrm{e}^{-itK}\,, \tag{14.2.27}$$

uniformly in $t \in [0, t_0]$ for any $t_0 > 0$ and vector in $\mathfrak{H}$, where generator K is defined by (14.2.1), (14.2.12).

Proof We note that $\mathfrak{H}_P \subseteq \mathrm{dom}\,(\sqrt{H})$ implies that $T = \sqrt{H}\,P$ is a bounded operator, and consequently, operator $K = T^*T$ is also bounded. Now we may use representation

$$\left(\frac{\mathbb{1}-\phi(\tau H)}{\tau}\,f,\,g\right) = \left(p(\tau H)\sqrt{H}f,\,\sqrt{H}g\right), \quad \tau > 0\,, \tag{14.2.28}$$

for $f \in \mathrm{dom}\,(H)$ and $g \in \mathrm{dom}\,(\sqrt{H})$, where

$$p(x) := \begin{cases} i\,, & x = 0 \\ (1-\phi(x))/x\,, & x > 0 \end{cases}.$$

Since $C_p := \sup_{x\in\mathbb{R}_0^+} |p(x)| < \infty$, by (14.2.3) one gets $\|p(\tau H)\|_{\mathcal{L}(\mathfrak{H})} \leq C_p$, $\tau > 0$. Hence, the equality (14.2.28) extends to $f, g \in \mathrm{dom}\,(\sqrt{H})$, in particular to $f, g \in \mathfrak{H}_P$. This leads to the representation

$$(\mathbb{1}_{\mathfrak{H}_P} - F(\tau))\,f = T^*p(\tau H)\,Tf\,, \quad \tau > 0\,, \quad f \in \mathfrak{H}_P\,, \tag{14.2.29}$$

or

$$S(\tau)f - iKf = T^*(p(\tau H) - i\,\mathbb{1})\,Tf\,, \quad \tau > 0\,, \quad f \in \mathfrak{H}_P\,. \tag{14.2.30}$$

By assumption (14.2.3) we find $\text{s-lim}_{\tau\to 0}\, p(\tau H) = i\,\mathbb{1}$, which yields $\text{s-lim}_{\tau\to 0}\, S(\tau) = i\,K$. In this way we obtain the limit; cf. (14.2.25),

$$\operatorname*{s-lim}_{\tau\to 0}(\mathbb{1}_{\mathfrak{H}_P} + S(\tau))^{-1} = (\mathbb{1}_{\mathfrak{H}_P} + i\,K)^{-1}, \tag{14.2.31}$$

and using the *Chernoff* product Theorem 4.1.3 (Sect. 4.1) we prove (14.2.27). □

It turns out that convergence (14.2.27) can be improved to the operator-norm convergence under some stronger assumptions.

Corollary 14.2.10 *Let the assumptions of Theorem* 14.2.9 *be satisfied. Then one infers the operator-norm limit*

$$\lim_{n\to\infty}\left\|(P\phi(t\,H/n)P)^n - \mathrm{e}^{-itK}\right\|_{\mathcal{L}(\mathfrak{H}_P)} = 0\,, \tag{14.2.32}$$

uniformly in $t \in [0, t_0]$ *for any* $t_0 > 0$ *if in addition* :

(i) *the operator* $T := \sqrt{H}\,P$ *is compact, or*
(ii) *there is* $\alpha > 0$ *such that* $\mathfrak{H}_P \subseteq \operatorname{dom}(\sqrt{H^{1+\alpha}})$ *and* $C_\alpha := \sup_{x\in(0,\infty)}|p_\alpha(x)| < \infty$, *where*

$$p_\alpha(x) := \begin{cases} 0, & x = 0 \\ (p(x) - i)/x^\alpha, & x > 0 \end{cases}.$$

Proof Note that from (14.2.30), (14.2.31) and compactness of T, we infer

$$\lim_{\tau\to 0}\|S(\tau) - iK\|_{\mathcal{L}(\mathfrak{H}_P)} = 0. \tag{14.2.33}$$

On the other hand, if $\mathfrak{H}_P \subseteq \operatorname{dom}(\sqrt{H^{1+\alpha}})$ for some $\alpha > 0$, then we set $T_\alpha := \sqrt{H^{1+\alpha}}\,P$ and $K_\alpha := T_\alpha^* T_\alpha$. Notice that T_α is a bounded operator. From (14.2.30) we obtain the representation

$$S(\tau) - iK = \tau^\alpha\, T_\alpha^*\, p_\alpha(\tau H)\, T_\alpha\,, \quad \tau > 0\,.$$

Hence, we find estimate

$$\|S(\tau) - iK\|_{\mathcal{L}(\mathfrak{H}_P)} \le \tau^\alpha C_\alpha \|K_\alpha\|_{\mathcal{L}(\mathfrak{H}_P)}\,, \quad \tau > 0\,,$$

which again yields (14.2.33).

Now, using the representation:

$$\mathrm{e}^{-itK} - \mathrm{e}^{-t\,S(t/n)} = \int_0^t ds\, \mathrm{e}^{-(t-s)\,S(t/n)}\,(S(t/n) - iK)\,\mathrm{e}^{-isK}\,,$$

we get estimate

$$\left\|\mathrm{e}^{-itK} - \mathrm{e}^{-t\,S(t/n)}\right\|_{\mathcal{L}(\mathfrak{H}_P)} \le t\,\|S(t/n) - iK\|_{\mathcal{L}(\mathfrak{H}_P)}, \quad t \ge 0.$$

So, by virtue of (14.2.33)

$$\lim_{n\to\infty}\left\|\mathrm{e}^{-t\,S(t/n)} - \mathrm{e}^{-itK}\right\|_{\mathcal{L}(\mathfrak{H}_P)} = 0\,, \tag{14.2.34}$$

holds for any $t > 0$, uniformly in $t \in [0, t_0]$. We shall combine this limit with the *telescopic* estimate

$$\begin{aligned}\left\|F(t/n)^n - \mathrm{e}^{-itK}\right\|_{\mathcal{L}(\mathfrak{H}_P)} &\le \\ \le \left\|F(t/n)^n - \mathrm{e}^{-t\,S(t/n)}\right\|_{\mathcal{L}(\mathfrak{H}_P)} &+ \left\|\mathrm{e}^{-t\,S(t/n)} - \mathrm{e}^{-itK}\right\|_{\mathcal{L}(\mathfrak{H}_P)}\,,\end{aligned} \tag{14.2.35}$$

where the first term in the right-hand side of (14.2.35) can be estimated by the *Chernoff* $\sqrt{n}$-Lemma 1.9.33

$$\left\|\left(F(t/n)^n - \mathrm{e}^{-t\,S(t/n)}\right) f\right\| \le \sqrt{n}\,\left\|\left(F(t/n) - \mathbb{1}_{\mathfrak{H}_P}\right) f\right\|\,, \quad f \in \mathfrak{H}_P\,. \tag{14.2.36}$$

In turn, owing to representation (14.2.29) for $\tau = t/n$, we can estimate the right-hand side of (14.2.36) by

$$\left\|\left(F(t/n) - \mathbb{1}_{\mathfrak{H}_P}\right) f\right\| \le \frac{t}{n}\,\left\|T^* p(t\,H/n)\,Tf\right\|\,, \quad f \in \mathfrak{H}_P\,, \quad t > 0\,.$$

Since $\|p(\tau H)\|_{\mathcal{L}(\mathfrak{H})} \le C_p\,, \tau > 0$, we find

$$\left\|\left(F(t/n) - \mathbb{1}_{\mathfrak{H}_P}\right) f\right\| \le C_p\,\frac{t}{n}\,\|K\|_{\mathcal{L}(\mathfrak{H}_P)}\|f\|\,, \quad f \in \mathfrak{H}_P\,.$$

Inserting this estimate into (14.2.36), we obtain

$$\left\|F(t/n)^n - \mathrm{e}^{-t\,S(t/n)}\right\|_{\mathcal{L}(\mathfrak{H}_P)} \le C_p\,\frac{t}{\sqrt{n}}\,\|K\|_{\mathcal{L}(\mathfrak{H}_P)}\,,$$

which yields

$$\lim_{n\to\infty} \left\| F(t/n)^n - \mathrm{e}^{-tS(t/n)} \right\|_{\mathcal{L}(\mathfrak{H}_P)} = 0\,, \tag{14.2.37}$$

for any $t > 0$, uniformly in $t \in [0, t_0]$. Taking into account (14.2.34), (14.2.35) and (14.2.37), we arrive at the sought limit (14.2.32). □

Now few remarks are in order.

Remark 14.2.11 Since $\phi(x) = \mathrm{e}^{-ix}$, $x \in \mathbb{R}_0^+$, satisfies conditions (14.2.3), we get from Theorem 14.2.9 that under the assumptions $\mathfrak{H}_P \subseteq \mathrm{dom}\,(\sqrt{H})$ the original Zeno product formula (14.2.26) holds and that under *additional* assumptions: either that $\sqrt{H}P$ is *compact* or that $\mathfrak{H}_P \subseteq \mathrm{dom}\,(\sqrt{H^{1+\alpha}})$, for some $\alpha > 0$, the original Zeno product formula (14.2.26) converges even in the *operator-norm* topology.

Remark 14.2.12 Obviously, the conclusion (14.2.32) is valid if $\mathfrak{H}_P \subseteq \mathrm{dom}\,(\sqrt{H})$ and $\mathfrak{H}_P$ is a finite-dimensional subspace. Indeed, in this case the operator $T = \sqrt{H}\,P$ is finite dimensional and therefore compact. This gives an alternative proof of some known results; see Notes to Chap. 14 (Notes to Sects. 14.2 and 14.2.3).

14.3 The Exner-Ichinose Result

In this section we elucidate and improve the theory developed in Sect. 14.1 in topology of the *Fréchet* space. Here we present the proof of the key assertions and develop a theory devoted to the *exponential* Zeno product formula in the *strong* operator topology; cf. Sect. 14.2.2. We call this result the *Exner-Ichinose* product formula.

14.3.1 Introduction and Set-Up

The aim of subsection is to present and examine two main theorems, which are basic statements of the *Exner-Ichinose* approach.

Theorem 14.3.1 *Let H be a non-negative self-adjoint operator acting on a separable Hilbert space $\mathfrak{H}$ and P an orthogonal projection onto a closed subspace of $\mathfrak{H}$. Suppose that $H^{1/2}P$ is densely defined in $\mathfrak{H}$, so that $H_P := (H^{1/2}P)^*(H^{1/2}P)$ is a self-adjoint operator. Then for any $f \in \mathfrak{H}$ and $\varepsilon = \pm 1$, the following limits hold,*

$$\lim_{n\to\infty} (P\,\mathrm{e}^{-\varepsilon itH/n}P)^n f = \mathrm{e}^{-\varepsilon itH_P}Pf\,, \tag{14.3.1}$$

$$\lim_{n\to\infty} (\mathrm{e}^{-\varepsilon itH/n}P)^n f = \mathrm{e}^{-\varepsilon itH_P}Pf\,, \tag{14.3.2}$$

$$\lim_{n\to\infty} (P\, \mathrm{e}^{-\varepsilon i t H/n})^n f = \mathrm{e}^{-\varepsilon i t H_P} P f\,, \tag{14.3.3}$$

and convergence is locally uniform in $t \in \mathbb{R}$.

Needless to say, that this claim is nontrivial only if the operators: H and P, are not commuting.

In fact, we shall prove in Sect. 14.3.3 a more general claim with a fixed P replaced by a projection-valued function: $t \mapsto P(t)$, satisfying certain regularity assumptions; cf. Theorem 14.1.6 for the Zeno product formula in the *Fréchet* space $\mathfrak{F} = L^2_{\mathrm{loc}}(\mathbb{R}, \mathfrak{H})$.

Theorem 14.3.2 *Let $\mathfrak{H}$ be a separable Hilbert space, and H, P, as well as H_P, be the same as in Theorem 14.3.1. Let further : $\tau \mapsto P(\tau)$, be a strongly continuous function whose values are orthogonal projections on $\mathfrak{H}$ defined in a right neighbourhood of $\tau = 0$, satisfying $P(0) = P$. Moreover, suppose that*

$$\lim_{\tau\to+0} [\tau^{-1}(\mathbb{1} - \mathrm{e}^{-it\tau H})]^{1/2} P(\tau)\, v = \mathrm{e}^{\pi i/4}(t\, H)^{1/2} P\, v\,, \tag{14.3.4}$$

for every $v \in \mathrm{dom}\,(H^{1/2}P)$. Then for any $f \in \mathfrak{H}$ and $\varepsilon = \pm 1$, we have

$$\lim_{n\to\infty} (P(1/n)\, \mathrm{e}^{-\varepsilon i t H/n} P(1/n))^n f = \mathrm{e}^{-\varepsilon i t H_P} P f\,, \tag{14.3.5}$$

$$\lim_{n\to\infty} (\mathrm{e}^{-\varepsilon i t H/n} P(1/n))^n f = \mathrm{e}^{-\varepsilon i t H_P} P f\,, \tag{14.3.6}$$

$$\lim_{n\to\infty} (P(1/n)\, \mathrm{e}^{-\varepsilon i t H/n})^n f = \mathrm{e}^{-\varepsilon i t H_P} P f\,, \tag{14.3.7}$$

where convergence in the Hilbert space norm is locally uniform in $t \in \mathbb{R}$.

Remark 14.3.3 We comment that the hypothesis in Theorem 14.3.2 about convergence of $[\tau^{-1}(\mathbb{1} - \mathrm{e}^{-\tau H})]^{1/2} P(\tau)\, v$, (14.3.4), is slightly weaker in comparison with Theorem 14.1.6, where we assumed that $\mathrm{dom}\,(H^{1/2}P(\tau)) \supset \mathrm{dom}\,(H^{1/2}P)$ as well as that $\lim_{\tau\to+0} \|H^{1/2}P(\tau)v\| = \|H^{1/2}P\, v\|$ hold for every $v \in \mathrm{dom}\,(H^{1/2}P)$.

Note that the assumption of *positivity* of operator H in Theorems 14.3.1 and 14.3.2 is presumed only for convenience. It is obvious that the result remains to be valid for any self-adjoint operator bounded from below. On the other hand, the hypothesis about *density* of domain $\mathrm{dom}\,(H^{1/2}P)$ in $\mathfrak{H}$ is crucial.

In Remark 14.1.11 we present an example showing that without this hypothesis the sequence $\{(\mathrm{e}^{-it\,H/n}P)^n\}_{n\geq 1}$ may not converge, for $n \to \infty$, in any sense. It may also happen that it converges, but not in the strong operator sense. See Notes to Sect. 14.3 (Sect. 14.3.1).

We now briefly describe the strategy for proving Theorem 14.3.2. The main tool is the *Chernoff* product approximation theorem; see Theorem 1.9.34 and, for self-adjoint case, Theorem 4.1.3. For reader's convenience we shall reproduce them in a form suitable for our purpose in Sect. 14.3.3.

This theorem would yield a desired result if we show that the τ-family

$$\{\widetilde{S}(\varepsilon\, it, \tau) := [\mathbb{1} + \tau^{-1}(\mathbb{1} - P(\tau)\, \mathrm{e}^{-\varepsilon\, it\, \tau H} P(\tau))]^{-1}\}_{\tau>0}\,, \quad \varepsilon = \pm 1\,,$$

converges, as $\tau \to +0$, to $(\mathbb{1} + \varepsilon\, it\, H_P)^{-1}$ in the *strong* operator topology on $\mathcal{L}(\mathfrak{H})$. The proof might basically follow the argument used in the *Kato* theorem (Theorem 5.1.5) to establish the celebrated self-adjoint *Trotter-Kato* product formula (Sect. 5.2) for the form-sum of two non-negative self-adjoint operators, since it appears that these two problems have notable similarities. However, a straightforward analogy with Kato's argument is not sufficient because of a difficulty which one encounters. It will be specified in Sect. 14.3.3, in *Step* IV, of the proof. The point is that this argument *can* be applied *only* to a certain *class* of functions: $\mathbb{R} \ni x \mapsto \phi(x)$, which contains besides the real exponentials (*Kato* functions $\mathcal{K}$; see Sect. 5.2), also *admissible* Kato functions (e.g. $\phi(x) = (1 + \varepsilon i x)^{-1}$, Sect. 14.2.1, Definition 14.2.2) with the *constant sign* for imaginary part. But unfortunately these classes *fail* to include the function: $\phi(x) = \mathrm{e}^{-\varepsilon\, i x}$ for $t \in \mathbb{R}$, which corresponds to the unitary group $\{\mathrm{e}^{-itH}\}_{t\in\mathbb{R}}$.

We comment that the *Exner-Ichinose* way to overcome this obstacle is to start with a weaker result in the *Fréchet* space, Sect. 14.1. What we did in Sect. 14.1.3 was an improvement of modified *Kato's* argument by the *Vitali* theorem from the complex function theory. In that way we obtain (cf. Lemma 14.1.12 and Notes to Sects. 14.1 and 14.1.3) that for each $f \in \mathfrak{H}$ the τ-family $\{\widetilde{S}(\varepsilon\, it, \tau)\, f\}_{\tau>0}$ converges:

$$\mathfrak{f} - \lim_{\tau\to+0} \widetilde{S}(\varepsilon\, it, \tau)\, f = (\mathbb{1} + \varepsilon\, it\, H_P)^{-1} f\,,$$

in topology $\mathfrak{f}$ of the Fréchet space $\mathfrak{F} = L^2_{\mathrm{loc}}(\mathbb{R}; \mathfrak{H})$.

This conclusion serves as a *starting* point, although it only implies that for some set $M_f \subset \mathbb{R}^+_0$ of Lebesgue measure *zero*, the τ-family $\{S(\varepsilon\, it, \tau) f\}_{\tau>0}$ converges in the space $\mathfrak{H}$ strongly for every $t \in \mathbb{R}^+_0 \setminus M_f$, not excluding the possibility that the convergence *does* hold at some points of M_f; see *Step* 1–*Step* 3 in Sect. 14.1.3. Furthermore, using the separability of $\mathfrak{H}$, we can choose a countable dense subset $\mathcal{D} := \{f_l\}_{l=1}^{\infty} \subset \mathfrak{H}$. Putting then

$$M = M_{\mathcal{D}} := \bigcup_{l=1}^{\infty} M_{f_l}\,,$$

which is also a set of Lebesgue measure *zero*, we may say that the above indicated τ-family converges for all $t \in \mathbb{R} \setminus M$ and for every $f \in \mathcal{D}$, and therefore, in view of the density of $\mathcal{D}$, also for every $f \in \mathfrak{H}$.

To pass from the "almost all t" to the "all t" claims, one has to demonstrate that the exceptional set M is, in fact, empty. This task may seem like a small step, but in reality, it is a profound and highly non-trivial question. One way to deal with this problem is to establish the *equicontinuity* of the above τ-family (see Lemma 14.3.11), which would allow us to achieve our goal by means of the *Arzelà-Ascoli* theorem.

14.3.2 The Limiting Operator H_P

Throughout the subsection H will be a non-negative self-adjoint operator in a separable Hilbert space $\mathfrak{H}$, and P will be an orthogonal projection. As we have indicated above, the non-negativity assumption is supposed for *convenience*. Our main result extends easily to any self-adjoint operator H bounded from below as well as, by sign change, to one bounded from above, that is, to each semi-bounded self-adjoint operator in $\mathfrak{H}$.

Our aim here is to elucidate what is the Zeno generator H_P appearing in Theorems 14.3.1 and 14.3.2 and to indicate some of its properties, which we will need in the sequel. Consider the quadratic form $u \mapsto \|H^{1/2}Pu\|^2$ with *form domain* $\mathrm{dom}\,(H^{1/2}P)$, being the domain of the *operator* $H^{1/2}P$. While the domains of operators HP and $H^{1/2}P$ are nontrivial only as subsets in $P\mathfrak{H}$, we also consider these operators as acting in the whole Hilbert space $\mathfrak{H}$. To this aim we write, if necessary, that operator $HP = (HP\restriction P\mathfrak{H}) \oplus (0\restriction(\mathbb{1}-P)\mathfrak{H})$ has domain

$$\mathrm{dom}\,(HP) = \mathrm{dom}\,(HP\restriction P\mathfrak{H}) \oplus (\mathbb{1}-P)\mathfrak{H}, \quad \mathrm{dom}\,(HP\restriction P\mathfrak{H}) = \mathrm{dom}\,(H) \cap P\mathfrak{H}, \tag{14.3.8}$$

and a similar representation for operator $H^{1/2}P$. Note that the latter is a *closed* linear operator, because $H^{1/2}$ is closed. By the same reason, the operator HP is also closed. We have, in general, $(H^{1/2}P)^* \supset PH^{1/2}$. On the other hand, the operator $PH^{1/2}$ may not be closed. This would be the case if and only if there were $c_0,\ c_1 > 0$ such that $\|H^{1/2}u\| \le c_0\|PH^{1/2}u\| + c_1\|u\|$ holds for all u in the domain $\mathrm{dom}\,(H^{1/2})$ of $PH^{1/2}$; however, the necessary part of this condition may not be satisfied. The same is true for PH.

By H_P we denote the unique self-adjoint operator associated with the above mentioned quadratic form: $u \mapsto \|H^{1/2}Pu\|^2$. Its domain $\mathrm{dom}\,(H_P)$ is a subspace of $\mathrm{dom}\,(H^{1/2}P)$ consisting of all $u \in \mathrm{dom}\,(H^{1/2}P)$, which satisfies $|(H^{1/2}P\,u,\ H^{1/2}P\,v)| \le C\|v\|$ for all $v \in \mathrm{dom}\,(H^{1/2}P)$ with a constant $C \ge 0$. So, by the *first representation theorem* (Sect. 9.1.2, Proposition 9.1.3), there exists operator H_P such that

$$H_P := (H^{1/2}P)^*(H^{1/2}P)\,. \tag{14.3.9}$$

We remind that the form-domain of H_P is the (operator) domain of $H_P^{1/2}$, for which we have by *polar decomposition*

$$H_P^{1/2} = |H^{1/2}P| = ((H^{1/2}P)^*(H^{1/2}P))^{1/2},$$
$$\mathrm{dom}\,(H_P^{1/2}) = \mathrm{dom}\,(H^{1/2}P) = (\mathrm{dom}\,(H^{1/2}) \cap P\mathfrak{H}) \oplus (\mathbb{1} - P)\mathfrak{H}. \tag{14.3.10}$$

Then one obtains the following claim:

Proposition 14.3.4 *The operator H_P in* (14.3.9) *decomposes as*

$$H_P = (H_P \restriction P\mathfrak{H}) \oplus (0 \restriction (\mathbb{1} - P)\mathfrak{H}), \tag{14.3.11}$$

having the dense domain

$$\mathrm{dom}\,(H_P) := \mathrm{dom}\,(H_P \restriction P\mathfrak{H}) \oplus (\mathbb{1} - P)\mathfrak{H} \subset P\mathfrak{H} \oplus (\mathbb{1} - P)\mathfrak{H} = \mathfrak{H}.$$

Here $H_P \restriction P\mathfrak{H}$ is the $P\mathfrak{H}$-component of H_P, which is the self-adjoint operator in the subspace $P\mathfrak{H} \subset \mathfrak{H}$ associated with the quadratic form in $P\mathfrak{H}$ with the form domain $\mathrm{dom}\,({H_P}^{1/2}) \cap P\mathfrak{H}$, *such that*

$$\mathrm{dom}\,({H_P}^{1/2}) \cap P\mathfrak{H} = \mathrm{dom}\,(H^{1/2}P) \cap P\mathfrak{H} \ni w \mapsto \|H^{1/2}Pw\|^2 = \|H_P^{1/2}w\|^2, \tag{14.3.12}$$

and the zero operator $0\restriction(\mathbb{1} - P)\mathfrak{H}$ is its $(\mathbb{1}-P)\mathfrak{H}$-component, trivially bounded and self-adjoint on the subspace $(\mathbb{1}-P)\mathfrak{H}$ orthogonal to $P\mathfrak{H}$.

From Proposition 14.3.4, we can see that H_P is in general *not* a restriction of H, and furthermore, the inclusion $\mathrm{dom}\,(H_P) \subset \mathrm{dom}\,(H)$ does not hold either. This is obvious, since H is the unique self-adjoint operator in $\mathfrak{H}$ associated with the quadratic form

$$\mathrm{dom}\,(H^{1/2}) \ni w \mapsto \|H^{1/2}w\|_{\mathfrak{H}}^2, \tag{14.3.13}$$

so that $(\mathbb{1} - P)\mathfrak{H}$ is not a subset of $\mathrm{dom}\,(H)$ as long as H is *unbounded*.

We can make, however, a weaker claim described in the following proposition. We note in passing that it is well illustrated by the inclusion $\mathrm{dom}\,((-\Delta)_\Omega) \subset \mathrm{dom}\,(-\Delta)$ from the example treated in Sect. 14.4 (Sect. 14.3.3).

Proposition 14.3.5 *Let H be a non-negative unbounded self-adjoint operator acting on a Hilbert space $\mathfrak{H}$ and P an orthogonal projection which may not commute with H. Assume that $H^{1/2}P$ is densely defined, then the $P\mathfrak{H}$-component $H_P \restriction P\mathfrak{H}$ of H_P in* (14.3.11) *satisfies* $\mathrm{dom}\,(H_P \restriction P\mathfrak{H}) \subset \mathrm{dom}\,(H)$.

Proof The form (14.3.13) restricted to $\mathrm{dom}\,(H^{1/2}P)$ becomes the form (14.3.12), so that $\|H^{1/2}w\|^2 = \|H^{1/2}Pw\|^2 = \|H_P^{1/2}w\|^2$ holds if $w \in \mathrm{dom}\,(H_P^{1/2})$ and proving thus the claim. □

Remark 14.3.6 One can modify H_P restricting it to the self-adjoint operator H_P^{mod} such that $H_P^{\mathrm{mod}} \restriction P\mathfrak{H} = H_P \restriction P\mathfrak{H}$ *and, at the same time,* $\mathrm{dom}\,(H_P^{\mathrm{mod}}) \subset \mathrm{dom}\,(H)$. To this end, it is enough to replace the zero operator in (14.3.11) on $(\mathbb{1} - P)\mathfrak{H}$ by its restriction to $(\mathbb{1} - P)\mathrm{dom}\,(H)$, so that

$$H_P^{\mathrm{mod}} := H_P \restriction P\mathfrak{H} \oplus 0 \restriction (\mathbb{1} - P)\,\mathrm{dom}\,(H)\,, \tag{14.3.14}$$

$$\mathrm{dom}\,(H_P^{\mathrm{mod}}) := \mathrm{dom}\,(H_P \restriction P\mathfrak{H}) \oplus (\mathbb{1} - P)\,\mathrm{dom}\,(H)\,.$$

Let us stress that in general H is not an extension of H_P^{mod} either.

Unless H is bounded, the operator H_P is, in general, *different* from PHP. When HP is densely defined, the symmetric operator PHP is not necessarily essentially self-adjoint on $\mathrm{dom}\,(HP)$, and consequently, H_P may not be the closure of PHP either.

On the other hand, the quadratic form $u \mapsto \|H^{1/2}Pu\|^2$ defined on $\mathrm{dom}\,(H^{1/2}P)$ is a closed extension of the form $u \mapsto (Pu, HPu)$ defined on $\mathrm{dom}\,(HP)$, but in general the former is not the closure of the latter, because $\mathrm{dom}\,(HP)$ is not necessarily dense in $\mathrm{dom}\,(H^{1/2}P)$. Indeed, if H is unbounded, $\mathrm{dom}\,(H)$ is a proper subspace of $\mathrm{dom}\,(H^{1/2})$. Take $u_0 \in \mathrm{dom}\,(H^{1/2}) \setminus \mathrm{dom}\,(H)$ such that the vector $H^{1/2}u_0$ is nonzero, and set P to be the orthogonal projection onto the one-dimensional subspace spanned by u_0. Taking into account that $\mathrm{dom}\,(HP) = \{u \in \mathfrak{H};\; Pu \in \mathrm{dom}\,(H)\}$ which $u_0 = Pu_0$ does not belong to, we find $HPu = 0$ for $u \in \mathrm{dom}\,(HP)$, while $H^{1/2}Pu_0 = H^{1/2}u_0 \neq 0$ by assumption.

14.3.3 *Principal Results*

Note that the most parts of arguments in this subsection carry on for a general Hilbert space $\mathfrak{H}$; however, there is a place, where we have to assume $\mathfrak{H}$ to be *separable*. We also remark that it is clearly sufficient to prove formula (14.3.5) in Theorem 14.3.2 because (14.3.6) and (14.3.7) follow from it, and furthermore, it is enough to consider only $\varepsilon = 1$ and $t \geq 0$.

The non-negative self-adjoint operator H we deal with can be conventionally represented through its spectral family $\{E(\lambda)\}_{\lambda \geq 0}$,

$$H = \int_0^\infty E(\mathrm{d}\lambda)\,\lambda\,. \tag{14.3.15}$$

Given $\kappa > 0$, we introduce

$$K(\kappa) := \frac{1}{\kappa}[\mathbb{1} - e^{-i\kappa H}] = \frac{\mathbb{1} - \cos \kappa H}{\kappa} + i\,\frac{\sin \kappa H}{\kappa}$$

$$=: G(\kappa) + i\,H(\kappa)\,, \tag{14.3.16}$$

where $G(\kappa)$ and $H(\kappa)$ are bounded self-adjoint operators, and $G(\kappa)$ is non-negative. Let $H^{\pm}(\kappa)$ be non-negative bounded self-adjoint operators which are the non-negative/negative parts of $H(\kappa)$, respectively. Then owing to the *polar decomposition* of bounded operators, we have

$$|K(\kappa)| = [G(\kappa)^2 + H(\kappa)^2]^{1/2} = \frac{|\sin(\kappa H)/2|}{\kappa/2}\,,$$

$$|\mathbb{1} + K(\kappa)| = \left(\mathbb{1} + (1+\kappa)\left(\frac{\sin(\kappa H)/2}{\kappa/2}\right)^2\right)^{1/2}, \tag{14.3.17}$$

$$H(\kappa) = H^+(\kappa) - H^-(\kappa)\,, \quad |H(\kappa)| = H^+(\kappa) + H^-(\kappa)\,.$$

Using the spectral family $\{E(\lambda)\}_{\lambda\geq 0}$ from (14.3.15), we denote by $E_H^+(\kappa)$ the orthogonal projection

$$E_H^+(\kappa) : \mathfrak{H} \to \bigoplus_{m=0}^{\infty} E\left([2m\pi/\kappa, (2m+1)\pi/\kappa)\right)\mathfrak{H}$$

$$\equiv E\left(\bigcup_{m=0}^{\infty} [2m\pi/\kappa, (2m+1)\pi/\kappa)\right)\mathfrak{H}\,,$$

and we put $E_H^-(\kappa) := \mathbb{1} - E_H^+(\kappa)$. Note that $E_H^+(\kappa) \xrightarrow{s} \mathbb{1}$ and $E_H^-(\kappa) \xrightarrow{s} 0$ holds as $\kappa \to +0$. Of course, both the $E_H^{\pm}(\kappa)$ commute with $H(\kappa)$. As $E_H^+(\kappa)$ and $E_H^-(\kappa)$ are nothing but the projections onto the closed subspaces of $\mathfrak{H}$ where $H(\kappa)$ becomes, respectively, a non-negative and negative self-adjoint operator, we have

$$H^+(\kappa) = H(\kappa)E_H^+(\kappa)\,, \quad H^-(\kappa) = -H(\kappa)E_H^-(\kappa)\,. \tag{14.3.18}$$

We put $\kappa = |t|\,\tau$ with $\tau > 0$, so that $\kappa = \pm t\,\tau$ holds for $\pm t \geq 0$.

We shall use in the following notations (14.1.30)–(14.1.32) for operator-valued τ-families: $\{F(\zeta;\tau) = P(\tau)\,e^{-\zeta\tau H}P(\tau)\}_{\tau>0}$ and $\{S(\zeta;\tau) = \tau^{-1}[\mathbb{1} - F(\zeta;\tau)]\}_{\tau>0}$ with $\mathrm{Im}\,\zeta \geq 0$. These families are uniformly bounded and holomorphic for $\operatorname{Im}\zeta > 0$, which are the properties we used there. However, we need only the value $\zeta = it$, that is, only

$$F(it, \tau) = P(\tau)\,\mathrm{e}^{-it\,\tau H}\,P(\tau)\,, \tag{14.3.19}$$

$$S(it, \tau) = \tau^{-1}[\mathbb{1} - F(it, \tau)] = \tau^{-1}[\mathbb{1} - P(\tau)\,\mathrm{e}^{-it\,\tau H}\,P(\tau)]\,. \tag{14.3.20}$$

It is easy to see that $F(it, \tau)$ in (14.3.19) is a contraction and $S(it, \tau)$ in (14.3.20) satisfies

$$\begin{aligned}\operatorname{Re}(S(it, \tau)f,\, f) &= \tau^{-1}\big[(f, f) - (P(\tau)\,\mathrm{e}^{-it\,\tau H}\,P(\tau)f,\, f)\big]\\ &\geq \tau^{-1}\big[\|f\|^2 - \|f\|\,\|P(\tau)\,\mathrm{e}^{-it\,\tau H}\,P(\tau)f\|\big]\\ &\geq \tau^{-1}\big[\|f\|^2 - \|f\|^2\big] = 0\,.\end{aligned}$$

for all $f \in \mathfrak{H}$, and consequently, $S(it, \tau)$ is an m-accretive operator. This means that $\mathbb{1} + S(it, \tau)$ has a bounded inverse and that $(\mathbb{1} + S(it, \tau))^{-1}$ is also a contraction. The crucial observation is that to prove Theorem 14.3.2 it is sufficient to refer to the *Chernoff* product Theorem 4.1.3 (cf. Theorem 1.9.34) and to verify that

$$(\mathbb{1} + S(it, \tau))^{-1} \xrightarrow{s} (\mathbb{1} + itH_P)^{-1}P\,, \quad \text{as } \tau \to +0\,, \tag{14.3.21}$$

holds for $t \in \mathbb{R}$. Given that function $F(it, \tau)$ in (14.3.19) differs from the canonical *Chernoff* set up by projections $\{P(\tau)\}_{\tau\geq 0}$, we mimic the proof of Lemma 14.1.13, which concerns the similar case.

Let us stress that our proof requires to demonstrate the convergence in (14.3.21) *pointwise for any fixed* t, since this will be sufficient to establish the convergence in the product formulæ (14.3.5)–(14.3.7) as *locally uniform* in $t \in \mathbb{R}$. In fact, we have only to deal with the limit (14.3.5); see remark at the beginning of the present Sect. 14.3.3.

To proceed with the argument, we rewrite relation (14.3.20) as follows:

$$\begin{aligned}S(it, \tau) &= \tau^{-1}\big[\mathbb{1} - P(\tau)\,(\cos t\,\tau H - i\,\sin t\,\tau H)\,P(\tau)\big]\\ &= \tau^{-1}(\mathbb{1} - P(\tau)) + P(\tau)\,t\,G(t\,\tau)\,P(\tau) + i\,P(\tau)\,t\,H(t\,\tau)\,P(\tau)\,,\end{aligned}$$

where $t\,G(t\,\tau) := \tau^{-1}(\mathbb{1} - \cos t\,\tau H)$ and $t\,H(t\,\tau) := \tau^{-1}\,\sin t\,\tau H$. As a consequence,

$$\begin{aligned}&\mathbb{1} + S(it, \tau) \\ &= \mathbb{1} + \tau^{-1}(\mathbb{1} - P(\tau)) + P(\tau)\,tG(t\,\tau)\,P(\tau) + iP(\tau)\,tH(t\,\tau)\,P(\tau)\\ &= (1 + \tau^{-1})\,(\mathbb{1} - P(\tau)) \oplus P(\tau)\,(\mathbb{1} + tG(t\,\tau) + i\,tH(t\,\tau))\,P(\tau)\,.\end{aligned} \tag{14.3.22}$$

Here $\oplus$ denotes the direct sum corresponding to the decomposition of the Hilbert space $\mathfrak{H}$ into $P(\tau)\mathfrak{H}$ and its orthogonal complement $(P(\tau)\mathfrak{H})^\perp$.

To prove the desired relation (14.3.21), we need first a couple of lemmata.

Lemma 14.3.7 *The inverse of* $\mathbb{1} + S(it, \tau)$ *in* (14.3.22) *is given by*

$$(\mathbb{1} + S(it, \tau))^{-1} = \qquad (14.3.23)$$
$$(1 + \tau^{-1})^{-1}(\mathbb{1} - P(\tau)) \oplus \left[P(\tau)(\mathbb{1} + tG(t\,\tau) + itH(t\,\tau))\,P(\tau)\right]^{-1}.$$

Proof The above expression and (14.3.22) clearly multiply to identity. □

Lemma 14.3.8 *Let* $u \in \mathrm{dom}\,(H^{1/2})$, *then in the limit* $\kappa \to +0$ *we have*

(i) $G(\kappa)^{1/2}u \longrightarrow 0$,

(ii) $H^+(\kappa)^{1/2}u \longrightarrow H^{1/2}u\,,\; H^-(\kappa)^{1/2}u \longrightarrow 0$, and $|H(\kappa)|^{1/2}u \longrightarrow H^{1/2}u$

Proof Using the spectral representation together with the dominated convergence theorem, we infer that for $u \in \mathrm{dom}\,(H^{1/2})$:

$$\lim_{\kappa\to+0} G(\kappa)^{1/2}u = \lim_{\kappa\to+0}\int_0^\infty E(\mathrm{d}\lambda)\,\lambda^{1/2}\left|\frac{\sin(\kappa\lambda/2)}{(\kappa\lambda/2)}\right|^{1/2} |\sin(\kappa\lambda/2)|^{1/2}\,u = 0\,,$$

$$\begin{aligned}\lim_{\kappa\to+0}|H(\kappa)|^{1/2}u &= \lim_{\kappa\to+0}\int_0^\infty E(\mathrm{d}\lambda)\left|\frac{\sin\kappa\lambda}{\kappa}\right|^{1/2} u \\ &= \int_0^\infty E(\mathrm{d}\lambda)\,\lambda^{1/2}u = H^{1/2}u\,,\end{aligned}$$

and as $\kappa \to +0$ one gets

$$\begin{aligned}H^+(\kappa)^{1/2}u &= |H(\kappa)|^{1/2}\,E_H^+(\kappa)\,u \\ &= \int_0^\infty E(\mathrm{d}\lambda)\left|\frac{\sin\kappa\lambda}{\kappa}\right|^{1/2} \chi_{\bigcup_{m=0}^\infty[2m\pi/\kappa,(2m+1)\pi/\kappa)}(\lambda)\;u \\ &\longrightarrow \int_0^\infty E(\mathrm{d}\lambda)\,\lambda^{1/2}\,u = H^{1/2}u\,.\end{aligned}$$

Then by (14.3.17) the limit $\lim_{\kappa\to+0} H^-(\kappa)^{1/2}u = 0$ for $u \in \mathrm{dom}\,(H^{1/2})$. □

After these preliminaries, we are going to start the proof of (14.3.21). As we mentioned in Sect. 14.3.1, a core for our reasoning is the argument used in the *Kato* theorem (Theorem 5.1.5). In the same vein, with (14.3.20) in mind, we put for $\tau > 0$ and $t \in \mathbb{R}$

$$u_{\tau,f}(t) := (\mathbb{1} + S(it, \tau))^{-1} f\,, \qquad (14.3.24)$$

for an arbitrary but *fixed* $f \in \mathfrak{H}$. We see from the expression (14.3.23) of $(\mathbb{1} + S(it, \tau))^{-1}$ that $u_{\tau,f}(t)$ is strongly continuous in t. At the same time, the τ-family $\{u_{\tau,f}(t)\}_{\tau>0} \subset \mathfrak{H}$ with $u_{\tau,f}(t)$ defined by (14.3.24) and $t \in \mathbb{R}$ fixed is uniformly bounded by $\|f\|$, because $(\mathbb{1} + S(it, \tau))^{-1}$ is a contraction.

Our next aim is to show that for each fixed $t \in \mathbb{R}$, the family $\{u_{\tau,f}(t)\}_{\tau>0}$ converges in the Hilbert space norm to some $u_f(t) \in \mathfrak{H}$ as $\tau \to +0$ and that

$$u_f(t) := (\mathbb{1} + itH_P)^{-1} Pf\,, \quad f \in \mathfrak{H}\,. \tag{14.3.25}$$

To achieve this goal, we have to analyse, in particular a new *uniform* property of the τ-family $\{u_{\tau,f}(t)\}_{\tau>0}$ at the final stage of the argument.

Using the identity (14.3.22), we can invert relation (14.3.24) explicitly as

$$\begin{aligned} f &= (\mathbb{1} + S(it, \tau))\, u_{\tau,f}(t) \\ &= u_{\tau,f}(t) + \tau^{-1}(\mathbb{1} - P(\tau))u_{\tau,f}(t) + P(\tau)\, tG(t\,\tau)\, P(\tau)u_{\tau,f}(t) \\ &\qquad\qquad + i\, P(\tau)\, tH(t\,\tau)\, P(\tau)u_{\tau,f}(t) \\ &= (1 + \tau^{-1})\, (\mathbb{1} - P(\tau))\, u_{\tau,f}(t) \\ &\oplus\; P(\tau)\, [\mathbb{1} + tG(t\,\tau) + itH(t\,\tau)]\, P(\tau)\, u_{\tau,f}(t)\,. \end{aligned} \tag{14.3.26}$$

Taking the scalar product of f with $(\mathbb{1} - P(\tau))u_{\tau,f}(t)$ and $P(\tau)u_{\tau,f}$, we get

$$(f,\, (\mathbb{1} - P(\tau))u_{\tau,f}(t)) = (1 + \tau^{-1})\|(\mathbb{1} - P(\tau))u_{\tau,f}(t)\|^2\,, \tag{14.3.27}$$

and for $\tau > 0$

$$\begin{aligned} (f,\, P(\tau)u_{\tau,f}(t)) &= (u_{\tau,f}(t),\, P(\tau)u_{\tau,f}(t)) + (tG(t\,\tau)\, P(\tau)u_{\tau,f}(t),\, P(\tau)u_{\tau,f}(t)) \\ &\quad + i\left(tH(t\,\tau)P(\tau)\, u_{\tau,f}(t),\, P(\tau)u_{\tau,f}(t)\right) \qquad (14.3.28) \\ &= \|P(\tau)\, u_{\tau,f}(t)\|^2 + \|(|t|\, G(t\,\tau))^{1/2} P(\tau)\, u_{\tau,f}(t)\|^2 \\ &\quad + i\left(\|(|t|\, H^+(t\,\tau))^{1/2} P(\tau)\, u_{\tau,f}(t)\|^2\right. \\ &\quad \left. - \|(|t|\, H^-(t\,\tau))^{1/2} P(\tau)\, u_{\tau,f}(t)\|^2\right) \end{aligned}$$

Recall that $H(\kappa) = H^+(\kappa) - H^-(\kappa)$.

From the *real part* of (14.3.28) together with Eq. (14.3.27), we infer by *Cauchy-Bouniakovski-Schwarz* inequality, that the three τ-families: $\{P(\tau)\, u_{\tau,f}(t)\}_{\tau>0}$, $\{(\mathbb{1} - P(\tau))\, u_{\tau,f}(t)\}_{\tau>0}$, and $\{\tau^{-1}(\mathbb{1} - P(\tau))\, u_{\tau,f}(t)\}_{\tau>0}$, are uniformly bounded by $\|f\|$. The last claim naturally extends to $\{\tau^{-1/2}(\mathbb{1} - P(\tau))\, u_{\tau,f}(t)\}$ as long as $\tau \leq 1$. Moreover, using the real part of (14.3.28) again, we can conclude that the same is also true for the family $\{(|t|\, G(\tau))^{1/2} P(\tau)\, u_{\tau,f}(t)\}_{\tau>0}$.

Now let $f \in \mathfrak{H}$ and $t \in \mathbb{R}$ be arbitrary, but fixed. Then it follows from the acquired uniform boundedness that for each $t \in \mathbb{R}$, there exist a (sub)sequence $\{\tau'\}_{0<\tau'\leq 1}$ of the family $\{\tau\}_{0<\tau\leq 1}$ with $\tau' \to +0$ and *three vectors*: $u_f(t)$, $u_0(t)$ and $g(t)$ in $\mathfrak{H}$ such that the sequences $\{u_{\tau',f}(t)\}_{0<\tau'\leq 1}$, $\{(\tau')^{-1/2}(\mathbb{1} - P(\tau'))\, u_{\tau',f}(t)\}_{0<\tau'\leq 1}$ and $\{t^{1/2}G(|t|\tau')^{1/2}\, u_{\tau',f}(t)\}_{0<\tau'\leq 1}$ converge, as $\tau' \to +0$, *weakly* in the Hilbert space $\mathfrak{H}$ to $u_f(t)$, $u_0(t)$ and $g(t)$, respectively. Thus $\{P(\tau')\, u_{\tau',f}(t)\}_{0<\tau'\leq 1}$ converges as $\tau' \to +0$ weakly to $P\, u_f(t)$. We keep in mind that, up to our knowledge, the limit $u_f(t)$ may *depend* on the chosen sequence $\{u_{\tau',f}(t)\}_{0<\tau'\leq 1}$.

Lemma 14.3.9 *One infers that three vectors mentioned above are* :

$$u_f(t) = P\, u_f(t)\,, \quad u_0(t) = 0\,, \quad g(t) = 0\,, \tag{14.3.29}$$

and therefore, as $\tau' \to +0$, *we obtain*

$$\begin{aligned} &u_{\tau',f}(t) \xrightarrow{w} u_f(t)\,, \quad (\tau')^{-1/2}(\mathbb{1} - P(\tau'))\, u_{\tau',f}(t) \xrightarrow{w} 0\,, \\ &P(\tau')\, u_{\tau',f}(t) \xrightarrow{w} P\, u_f(t)\,, \quad (|t|\, G(t\,\tau'))^{1/2} P(\tau')\, u_{\tau',f}(t) \xrightarrow{w} 0\,. \end{aligned} \tag{14.3.30}$$

Proof To begin with, the second limit in (14.3.30) implies, in particular, that $(\mathbb{1} - P(\tau'))u_{\tau',f}(t) \to 0$; hence we have $(\mathbb{1} - P)\, u_f(t) = 0$ or $u_f(t) = P\, u_f(t)$. Indeed, the uniform boundedness of $\|(\tau')^{-1/2}(\mathbb{1}-P(\tau'))u_{\tau',f}(t)\|$ means that $\|(\mathbb{1}-P(\tau'))u_{\tau',f}(t)\| \to 0$ which together with the weak convergence implies $u_f(t) = P\, u_f(t)$. Moreover, the same uniform boundedness following from (14.3.27) means that $\tau^{-1}(\mathbb{1} - P(\tau))u_{\tau,f}(t)$ also converges weakly along $\{\tau'\}_{0<\tau'\leq 1}$, which implies $u_0(t) = 0$.

Next, for any fixed $w \in \mathrm{dom}\,(H^{1/2})$ and $t \in \mathbb{R}$, we have the relation

$$\begin{aligned} &(g(t),\ w) = \lim_{\tau'\to+0} ((|t|\, G(t\,\tau'))^{1/2} P(\tau')u_{\tau',f}(t),\ w) = \\ &\lim_{\tau'\to+0} (P(\tau')u_{\tau',f}(t),\ (|t|\, G(t\,\tau'))^{1/2} w) = (Pu_f(t),\ 0) = 0\,, \end{aligned}$$

because by Lemma 14.3.8 we have $(|t|\, G(t\,\tau'))^{1/2}w \to 0$ as $\tau' \to +0$. This means that $g(t) = 0$, because $\mathrm{dom}\,(H^{1/2})$ is dense by assumption. □

Afterwards, we want to find out what one can learn about the properties of operators $H^{\pm}(t\,\tau)$ introduced in (14.3.17), (14.3.18) from the *imaginary part* of (14.3.28).

To this aim we have to employ a topology *different* from those used up till now. Given a dense subspace $\mathcal{K} \subset \mathfrak{H}$, the symbol $\sigma(\mathfrak{H}, \mathcal{K})$ denotes the *weak topology* on $\mathfrak{H}$ defined by the *dual pairing* $\langle \mathfrak{H}, \mathcal{K}\rangle$ or the *$\mathcal{K}$-weak topology* on $\mathfrak{H}$. In general, it is weaker (*coarser*) than the usual *weak* topology on $\mathfrak{H}$, the latter being nothing else than $\sigma(\mathfrak{H}, \mathfrak{H})$ in the just introduced notation.

We see from the imaginary part of (14.3.28) that the difference between the respective elements of the τ-families:

$$\{ \| (|t|\, H^{+}(t\,\tau))^{1/2} P(\tau)\, u_{\tau,f}(t) \|^2 \}_{\tau>0} \quad \text{and}$$
$$\{ \| (|t|\, H^{-}(t\,\tau))^{1/2} P(\tau)\, u_{\tau,f}(t) \|^2 \}_{\tau>0} \,,$$

is uniformly bounded. This is because on account of the Cauchy-Bouniakovski-Schwarz inequality the modulus of the left-hand side in

$$\operatorname{Im}(f,\ P u_f(t)) =$$
$$\lim_{\tau\to+0} \left(\|(|t|H^{+}(t\,\tau))^{1/2} P(\tau) u_{\tau,f}(t)\|^2 - \|(|t|H^{-}(t\,\tau))^{1/2} P(\tau) u_{\tau,f}(t)\|^2 \right),$$

does not exceed $\|f\|^2$. This fact, unfortunately, does not tell us whether the two τ-families of vectors, $\{(|t|H^{\pm}(t\,\tau))^{1/2} P(\tau) u_{\tau,f}(t)\}_{\tau>0}$, are separately (uniformly) bounded, that is, whether each of them is (uniformly) weakly bounded. We have, however, at least the following result.

Proposition 14.3.10 *Let $\{u_{\tau',f}(t)\}_{0<\tau'\leq 1}$ be the subsequence from Lemma* 14.3.9, *weakly convergent to $u_f(t)$. Then the τ'-families $\{(|t|H^{\pm}(t\,\tau'))^{1/2} P(\tau') u_{\tau',f}(t)\}_{0<\tau'\leq 1}$ are Cauchy sequences in the $\sigma(\mathfrak{H}, \operatorname{dom}(H^{1/2}))$-weak topology. As a result they are $\sigma(\mathfrak{H}, \operatorname{dom}(H^{1/2}))$-weakly bounded and the family $\{(|t|\, H^{-}(t\,\tau'))^{1/2} P(\tau')\, u_{\tau',f}(t)\}_{0<\tau'\leq 1}$ converges to zero in this topology as $\tau' \to +0$.*

Proof Take an arbitrary $\phi \in \operatorname{dom}(H^{1/2})$. We use Lemma 14.3.8 which states, in particular that $(|t|\, H^{+}(t\,\tau))^{1/2}\phi \to (|t|H)^{1/2}\phi$ holds when $\tau \to +0$. In combination with (14.3.30), this yields

$$\begin{aligned} &\lim_{\tau'\to+0} ((|t|\, H^{+}(t\,\tau'))^{1/2} P(\tau') u_{\tau',f}(t),\ \phi) \\ &= \lim_{\tau'\to+0} (P(\tau') u_{\tau',f}(t),\ (|t|\, H^{+}(t\,\tau'))^{1/2}\phi) \\ &= (P\, u_f(t),\ (|t|\, H)^{1/2}\phi)\,. \end{aligned}$$

On the other hand, Lemma 14.3.8 says that $\lim_{\tau\to+0}(|t|\, H^{-}(t\,\tau))^{1/2}\,\phi = 0$. As a consequence

$$\lim_{\tau'\to+0} \left((|t|\, H^{-}(t\,\tau'))^{1/2} P(\tau')\, u_{\tau',f}(t),\ \phi \right) = 0\,, \tag{14.3.31}$$

because the self-adjointness of operator $(|t|\, H^{-}(t\,\tau'))^{1/2}$ in combination with the Cauchy-Bouniakovski-Schwarz inequality gives estimate

$$\begin{aligned} |(P(\tau') u_{\tau',f}(t),\ (|t|\, H^{-}(t\,\tau'))^{1/2}\phi)| &\leq \|P(\tau') u_{\tau',f}(t)\|\ \|(|t|\, H^{-}(t\,\tau'))^{1/2}\phi\| \\ &\leq \|f\|\ \|(|t|\, H^{-}(t\,\tau'))^{1/2}\phi\|\,. \end{aligned}$$

This yields the stated assertion, including the fact that by virtue of limit (14.3.31) the family $\{(|t|\, H^{-}(t\,\tau'))^{1/2} P(\tau')\, u_{\tau',f}(t)\}_{0<\tau'\leq 1}$ is $\sigma(\mathfrak{H}, \mathrm{dom}\,(H^{1/2}))$-weakly convergent to zero. □

On the other hand, it is not clear whether $Pu_f(t)$ belongs to $\mathrm{dom}\,(H^{1/2})$, that is, whether the plus-sign family: $\{(|t| H^{+}(t\,\tau))^{1/2} P(\tau) u_{\tau,f}(t)\}_{\tau>0}$, converges (as $\tau \to +0$) to $H^{1/2} Pu_f(t)$ in the $\sigma(\mathfrak{H}, \mathrm{dom}\,(H^{1/2}))$ topology. Consequently, the information that we were able to deduce in this way about the convergence of the families $\{(|t|\, H^{\pm}(t\,\tau'))^{1/2} P(\tau')\, u_{\tau',f}(t)\}_{\tau'>0}$ is too limited. It does not seem possible to apply the same procedure as we used in Lemma 14.3.9 for $\{(|t|\, G(t\,\tau'))^{1/2}\, P(\tau')\, u_{\tau',f}(t)\}_{\tau'>0}$.

This forces us to seek a different strategy for the proof of Theorem 14.3.2 that would allow us to identify the vector $u_f(t)$ with $(\mathbb{1} + itH_P)^{-1} Pf$, in other words, to demonstrate relation (14.3.21) claiming that for every fixed $t \in \mathbb{R}$, the family $\{(\mathbb{1} + S(it, \tau))^{-1} f\}_{\tau>0}$, or otherwise $\{u_{\tau,f}(t)\}_{\tau>0}$ in accordance with (14.3.24), converges to $(\mathbb{1} + itH_P)^{-1} Pf$ in the Hilbert space norm as $\tau \to +0$.

The argument is somewhat subtle and relies on Sects. 14.1 and 14.2 about the Zeno product formulæ related to Theorems 14.3.1 and 14.3.2. In these sections we demonstrated that the family $\{u_{\tau,f}(t)\}_{\tau>0}$ has a unique limit, namely, $(\mathbb{1} + itH_P)^{-1} Pf$, as $\tau \to +0$. As we mentioned in the introduction, however, the obtained convergence referred neither to the norm of $\mathfrak{H}$ nor even to the weak topology.

Precisely speaking, it is shown in Sect. 14.1 that (14.3.1)–(14.3.3) as well as (14.3.5)–(14.3.7) hold in the topology of the *Fréchet* space $L^2_{\mathrm{loc}}(\mathbb{R}, \mathfrak{H}) = L^2_{\mathrm{loc}}(\mathbb{R}) \otimes \mathfrak{H}$:

$$\lim_{\tau\to+0} \int_a^b \mathrm{d}t\ \|u_{\tau,f}(t) - (\mathbb{1} + itH_P)^{-1} Pf\|^2 = 0\,, \tag{14.3.32}$$

for every bounded closed interval $[a, b] \subset \mathbb{R}$.

Note also that Lemma 14.1.12 and *Step* 1–*Step* 3 imply that for every $f \in \mathfrak{H}$, there exist a set $M_f \subset \mathbb{R}$ of Lebesgue measure zero, possibly dependent on f, and a (sub)sequence $\{\tau'\}_{0<\tau'\leq 1}$ of $\{\tau\}_{0<\tau\leq 1}$ along which it holds that for all $s \in \mathbb{R}\setminus M_f$,

$$u_{\tau',f}(s) \longrightarrow (\mathbb{1} + isH_P)^{-1} Pf \quad \text{in the norm of } \mathfrak{H}\,, \tag{14.3.33}$$

as $\tau' \to +0$. With the coming argument in mind, it is useful to note that the set $\mathbb{R} \setminus M_f$ at which the convergence takes place is *dense* in $\mathbb{R}$. Furthermore, since $\mathfrak{H}$ is separable by assumption, we can choose a countable dense subset $\mathcal{D} := \{f_l\}_{l=1}^{\infty}$ in $\mathfrak{H}$. Putting $M = M_{\mathcal{D}} := \cup_{l=1}^{\infty} M_{f_l}$, which is also a set of Lebesgue measure zero, we may then say that (14.3.33) holds for all $s \in \mathbb{R} \setminus M$ and for every $f \in \mathcal{D}$, and hence, in view of the density, also for every $f \in \mathfrak{H}$.

Moreover, we note that $s = 0$ does not belong to M_f for any $f \in \mathfrak{H}$, and therefore it neither belongs to M. Indeed, using Lemma 14.3.7 and (14.3.23) in combination with the continuity of $\tau \mapsto P(\tau)$, it is easy to see that

$$u_{\tau,f}(0) = (\mathbb{1} + S(0;\tau))^{-1} f = (\mathbb{1} + \tau^{-1})^{-1}(\mathbb{1} - P(\tau)) \oplus P(\tau) f \longrightarrow Pf, \tag{14.3.34}$$

as $\tau \to +0$, which means that $0 \notin M_f$.

Now let us first briefly outline our plan of *how to complete the proof of* Theorem 14.3.2. We recall that one has to verify relation (14.3.21) showing that for every $t \in \mathbb{R}$, $\{u_{\tau,f}(t) = (\mathbb{1} + S(it,\tau))^{-1} f\}$ converges to $(\mathbb{1} + itH_P)^{-1} Pf$ in the Hilbert space norm for all t as $\tau \to +0$. Let us remind here that, as already mentioned in the text just following (14.3.21), the use of the Chernoff product theorem (e.g. Theorem 4.1.3) requires to establish in (14.3.21) only the *pointwise* convergence for all $t \in \mathbb{R}$. Here and in the following we keep in mind that $\{u_{\tau,f}(t)\}_{0<\tau\leq 1}$ is uniformly bounded in both t and $0 < \tau \leq 1$, that is,

$$\sup_{0<\tau\leq 1} \sup_{t\in\mathbb{R}} \|u_{\tau,f}(t)\| \leq \|f\|, \tag{14.3.35}$$

since $(\mathbb{1} + S(it,\tau))^{-1}$ is a contraction. We need not strive to show its local uniformity; however, the following reasoning will establish this property with respect to $t \in \mathbb{R} \setminus \{0\}$. We shall proceed in *four* steps demonstrating validity of the following claims:

I. The τ-family $\{u_{\tau,f}(t)\}_{0<\tau\leq 1}$ of vectors $u_{\tau,f} : \mathbb{R} \ni t \mapsto u_{\tau,f}(t) \in \mathfrak{H}$ is *equicontinuous* in topology of the Hilbert space $\mathfrak{H}$ for $t \in \mathbb{R} \setminus \{0\}$.

II. The family $\{u_{\tau,f}(t)\}_{0<\tau\leq 1}$ converges as $\tau \to +0$ for each fixed $t \in \mathbb{R}$ to *some* $u_f(t) \in \mathfrak{H}$ in the weak topology of $\mathfrak{H}$, and furthermore, the convergence is locally uniform for $t \in \mathbb{R} \setminus \{0\}$. The limit function $t \mapsto u_f(t)$ turns out to be continuous in the weak topology of the space $\mathfrak{H}$ for $t \in \mathbb{R}$.

III. The limit satisfies equation (14.3.25), that is,

$$u_f(t) = (\mathbb{1} + itH_P)^{-1} Pf\,, \quad \text{for all } t \in \mathbb{R} \text{ and } f \in \mathfrak{H}\,. \tag{14.3.36}$$

IV. Finally, the family $\{u_{\tau,f}(t)\}_{0<\tau\leq 1}$ converges as $\tau \to +0$ for any fixed $t \in \mathbb{R}$ to $u_f(t) \equiv (\mathbb{1} + itH_P)^{-1} Pf$ in topology of the space $\mathfrak{H}$, and moreover, the convergence is locally uniform for $t \in \mathbb{R} \setminus \{0\}$.

Step I. The claim is expressed in the following lemma about the local equicontinuity of the τ-family $\{u_{\tau,f}(t)\}_{0<\tau\leq 1}$ in a neighbourhood of $s \in \mathbb{R} \setminus \{0\}$. It plays a crucial role in concluding the proof of Theorem 14.3.2.

Lemma 14.3.11 *Let $f \in \mathfrak{H}$. Then the τ-family $\{u_{\tau,f}(t)\}_{0<\tau\leq 1}$ of vector functions $u_{\tau,f} : t \mapsto u_{\tau,f}(t) \in \mathfrak{H}$ on $\mathbb{R} \setminus \{0\}$ are locally equicontinuous for variable t in topology on $\mathfrak{H}$. More explicitly, for every $\varepsilon > 0$ and for every $s \in \mathbb{R} \setminus \{0\}$ there exists* *an s-dependent constant* $\delta = \delta(f, \varepsilon, s) > 0$ *such that if* $t, s > 0$, *or* $t, s < 0$, *with* $|t - s| < \delta$, *then* $\| u_{\tau,f}(t) - u_{\tau,f}(s) \| < \varepsilon$ *holds for all* $0 < \tau \le 1$.

We postpone (for the moment) the proof of Lemma 14.3.11, returning to it in Sect. 14.3.4, and accept its claim, to finish first the proof of Theorem 14.3.2.
Step II. Without loss of generality, we may suppose $f \neq 0$. If $t = 0$, the family $\{u_{\tau,f}(0)\}_{0<\tau\le 1}$ converges strongly to Pf as $\tau \to +0$ as mentioned above; cf. (14.3.34). Consider thus a nonzero $t \in \mathbb{R} \setminus \{0\}$ and take *arbitrary* (sub)sequence $\{\tau'\}_{0<\tau'\le 1}$ of $\{\tau\}_{0<\tau\le 1}$ with $\tau' \to +0$. By Lemma 14.3.11, to be yet proven, we see the τ'-family $\{u_{\tau',f}(t)\}_{0<\tau'\le 1}$ is *equicontinuous* in topology of $\mathfrak{H}$ and therefore in the *weak* topology, because the “full” τ-family $\{u_{\tau,f}(t)\}_{0<\tau\le 1}$ is; cf. *Step* I.

We observed in (14.3.35) that the vectors $u_{\tau,f}(t)$ with any $t \in \mathbb{R} \setminus \{0\}$ and $\tau \in (0, 1]$ lie in the closed ball $\bar{B}(0; \|f\|) \subset \mathfrak{H}$ with the center at the origin and radius $\|f\|$, which is *weakly compact*. To proceed, note that $\bar{B}(0; \|f\|)$ is *metrisable* in the weak topology, since the space $\mathfrak{H}$ is by assumption *separable*. Thus the equicontinuity holds with respect to the metric on the space $\bar{B}(0; \|f\|)$ which is equivalent to the weak topology on it. Then, by virtue of the *Ascoli–Arzelà* theorem, there exists a (sub)sequence $\{\tau''\}_{0<\tau''\le 1}$ of $\{\tau'\}_{0<\tau'\le 1}$ with $\tau'' \to +0$, along which $\{u_{\tau,f}(t)\}$ converges *weakly* to some limit $u_f(t)$, and moreover, the convergence is *(locally) uniform* in $t \in \mathbb{R} \setminus \{0\}$. This means that the numerical family $\{(u_{\tau''}(t),\ \psi)\}_{0<\tau''\le 1}$ converges for every fixed $\psi \in \mathfrak{H}$ as $\tau'' \to +0$, (locally) uniformly in $t \in \mathbb{R} \setminus \{0\}$.

Putting this together with the case $t = 0$ mentioned above (14.3.34), we see that this limit, which is equal to $(u_f(t),\ \psi)$, is a t-continuous function on $\mathbb{R}$, in other words, that: $t \mapsto u_f(t)$, is t-continuous everywhere on $\mathbb{R}$ in the weak topology of $\mathfrak{H}$.
Step III. We have established above two ways of convergence of the original τ-family $\{u_{\tau,f}(t)\}_{0<\tau\le 1}$ of $\mathfrak{H}$-valued t-continuous functions as $\tau \to +0$, or more specifically, convergence with respect to *two* different topologies. One is the convergence to $u_f(t)$ in the *weak* topology of $\mathfrak{H}$, (locally) uniformly for $t \in \mathbb{R} \setminus \{0\}$.

The other one is the convergence of this τ-family $\{u_{\tau,f}(t)\}_{0<\tau\le 1}$ to $(\mathbb{1} + itH_P)^{-1}Pf$ in the *Fréchet* topology on the space $L^2_{\text{loc}}(\mathbb{R}, \mathfrak{H})$; cf. (14.3.32).

The latter would allow us to conclude that these two limit vectors coincide for all t, in other words, to establish the equality (14.3.36) if we deduce from (14.3.32) convergence in the weak topology of $\mathfrak{H}$. To this end we note that for any $\psi \in \mathfrak{H}$ and an arbitrary bounded interval (a, b) with $a < b$ there is the following elementary estimate:

$$\begin{aligned}&\int_a^b \mathrm{d}t\ |\langle \psi, u_f(t) - (\mathbb{1} + itH_P)^{-1}Pf \rangle| \\ &\le \int_a^b \mathrm{d}t\ |\langle \psi, u_f(t) - u_{\tau,f}(t) \rangle| + \int_a^b \mathrm{d}t\ |\langle \psi, u_{\tau,f}(t) - (\mathbb{1} + itH_P)^{-1}Pf \rangle|,\end{aligned} \tag{14.3.37}$$

the last term of which tends to zero as $\tau \to +0$ by (14.3.32) and therefore, of course, as $\tau'' \to +0$. At the same time, the first term on the right-hand side of (14.3.37) also tends to zero in view of the weak convergence and the *dominated convergence* theorem. On that account, the integrand in the left-hand side of (14.3.37) vanishes identically:

$$(u_f(t) - (\mathbb{1} + itH_P)^{-1} Pf,\ \psi) = 0\,,$$

for all t and for all $\psi \in \mathfrak{H}$, which yields the desired claim (14.3.36).

As a consequence, in this way we obtain relations

$$\begin{aligned} u_f(t) &= Pu_f(t) \in \operatorname{dom}(H_P) = \operatorname{dom}((H^{1/2}P)^*(H^{1/2}P))\,, \\ Pf &= P(\mathbb{1} + itH_P)u_f(t) = P[\mathbb{1} + it(H^{1/2}P)^*(H^{1/2}P)]u_f(t)\,. \end{aligned} \tag{14.3.38}$$

This shows that the weak limit $u_f(t)$ (14.3.25) does not depend on the chosen subsequence $\{u_{\tau'',f}(t)\}_{0<\tau''\leq 1}$. For that reason, by a standard argument, the limit: $\lim_{\tau'\to+0} u_{\tau',f}(t) = u_f(t)$, is independent of the choice of subsequence. As a consequence, $\{u_{\tau,f}(t)\}_{\tau>0}$ converges to $u_f(t) = (\mathbb{1} + itH_P)^{-1}Pf$ (14.3.36) in the weak topology of $\mathfrak{H}$ along $\tau' \to +0$, and therefore also along $\tau \to +0$.

Step IV. It remains to demonstrate that $\{u_{\tau,f}(t)\}_{\tau>0}$ converges to $u_f(t) = (\mathbb{1} + itH_P)^{-1}Pf$ also *in the Hilbert space norm* for all $t \in \mathbb{R}$. For the sake of completeness, we shall show that the same claim can also be made about the τ-family $\{(tG(\tau))^{1/2}P(\tau)u_{\tau,f}(t)\}_{\tau>0}$ in (14.3.28). Note that the analogous question concerning $\{(|t|H^{\pm}(t\,\tau))^{1/2}\,P(\tau)\,u_{\tau,f}(t)\}_{\tau>0}$ also appearing in (14.3.28) is more complicated and we are able to provide only a partial answer to it; cf. Proposition 14.3.10.

Since we have already established for all t the weak convergence of the τ-family $\{u_{\tau,f}(t)\}_{\tau>0}$, we need only to show that the τ-families of the norms of these vectors converge. To this end, we observe again the real part of (14.3.28), however, replacing now f by $P(\tau)f$ in the inner product. We write also the imaginary part for purpose of a further discussion. Writing the left-hand side of (14.3.28) as $(P(\tau)f,\ u_{\tau,f}(t))$, we obtain for its real and imaginary part

$$\operatorname{Re}(P(\tau)f,\ u_{\tau,f}(t)) = \|P(\tau)u_{\tau,f}(t)\|^2 + \|(|t|\,G(t\,\tau))^{1/2}\,P(\tau)\,u_{\tau,f}(t)\|^2\,, \tag{14.3.39}$$

$$\begin{aligned} \operatorname{Im}(P(\tau)f,\ u_{\tau,f}(t)) = \|(|t|\,H^{+}(t\,\tau))^{1/2}\,P(\tau)\,u_{\tau,f}(t)\|^2 \\ - \|(|t|\,H^{-}(t\,\tau))^{1/2}\,P(\tau)\,u_{\tau,f}(t)\|^2\,. \end{aligned} \tag{14.3.40}$$

In view of (14.3.38), the continuity of the projection family $\{P(\tau)\}_{\tau>0}$, and the weak convergence of $\{u_{\tau,f}(t)\}_{\tau>0}$ as $\tau \to +0$, which we have already established, the left-hand sides of the last two relations converge, leading to the following limits:

$$\mathrm{Re}\,(P(\tau)f,\, u_{\tau,f}(t)) \longrightarrow (P(\tau)f,\, u_f(t)) = \|Pu_f(t)\|^2\,, \tag{14.3.41}$$

$$\mathrm{Im}\,(P(\tau)f,\, u_{\tau,f}(t)) \longrightarrow (tH_P u_f(t),\, u_f(t)) = \|(|t|\,H)^{1/2}\,P\,u_f(t)\|^2\,, \tag{14.3.42}$$

valid for *all* t. Comparing now the right-hand sides of relations (14.3.39) and (14.3.41), we get

$$\begin{aligned}
\|Pu_f(t)\|^2 &= \lim_{\tau\to+0} \left[\|P(\tau)u_{\tau,f}(t)\|^2 + \|(|t|G(t\,\tau))^{1/2}P(\tau)u_{\tau,f}(t)\|^2\right] \\
&= \liminf_{\tau\to+0} \left[\|P(\tau)u_{\tau,f}(t)\|^2 + \|(|t|G(t\,\tau))^{1/2}P(\tau)u_{\tau,f}(t)\|^2\right] \\
&\geq \liminf_{\tau\to+0} \|P(\tau)u_{\tau,f}(t)\|^2 + \liminf_{\tau\to+0} \|(|t|G(\tau))^{1/2}P(\tau)u_{\tau,f}(t)\|^2 \\
&\geq \|Pu_f(t)\|^2 + \|0\|^2 = \|Pu_f(t)\|^2,
\end{aligned}$$

which means that the two terms on the right-hand side of (14.3.39) converge to $\|Pu_f(t)\|^2$ and to zero, respectively, yielding thus the sought convergence in the Hilbert space norm along the sequence $\tau \to +0$,

$$P(\tau)u_{\tau,f}(t) \longrightarrow Pu_f(t)\,, \quad (|t|G(t\,\tau))^{1/2}P(\tau)u_{\tau,f}(t) \longrightarrow 0\,. \tag{14.3.43}$$

The first convergence in (14.3.43) also implies $u_{\tau,f}(t) \to u_f(t) = P\,u_f(t)$, i.e. convergence of τ-family $\{u_{\tau,f}(t)\}_{\tau>0}$ to $u_f(t) = P\,u_f(t)$ in the Hilbert space norm, because we know already that $(\mathbb{1} - P(\tau'))\,u_{\tau',f}(t) \to 0$ as $\tau' \to +0$.

This justifies our statement about convergence (14.3.21) for the family $\{u_{\tau,f}(t)\}_{\tau>0}$ which, with the reference to the Chernoff product formula, concludes the proof of Theorem 14.3.2. □

The proof of Lemma 14.3.11 will be given in the next section.

14.3.4 Auxiliary Statements

First recall that the Hilbert space $\mathfrak{H}$ we are handling is assumed to be separable which is the property we needed to construct the exceptional set M; cf. the text following (14.3.33). Recall also the notation we introduced in (14.3.24) for the vector obtained by application of the operator $(\mathbb{1} + S(it, \tau))^{-1}$ to an arbitrary $f \in \mathfrak{H}$,

$$\begin{aligned} u_{\tau,f}(t) &= (\mathbb{1} + S(it, \tau))^{-1} f \\ &= \left\{(1 + \tau^{-1})^{-1}(\mathbb{1} - P(\tau)) \oplus [P(\tau)(\mathbb{1} + tG(t\,\tau) + itH(t\,\tau))P(\tau)]^{-1}\right\} f \\ &= \left\{(1 + \tau^{-1})^{-1}(\mathbb{1} - P(\tau)) \oplus T^P(t, \tau)\right\} f\,. \end{aligned} \tag{14.3.44}$$

Here operator

$$T^P(t, \tau) := [P(\tau)(\mathbb{1} + tG(t\,\tau) + itH(t\,\tau))P(\tau)]^{-1}\,, \tag{14.3.45}$$

may be also considered on the whole Hilbert space $\mathfrak{H}$, although in the proper sense it is an operator with the domain and range included in the closed subspace $P(\tau)\mathfrak{H}$. So, one may regard it as vanishing on the orthogonal complement $(\mathbb{1} - P(\tau))\mathfrak{H}$.

Next, for fixed $t, s > 0$ or $t, s < 0$, we put

$$\begin{aligned} D(t, s, \tau)f :&= u_{\tau,f}(t) - u_{\tau,f}(s) \\ &= [(\mathbb{1} + S(it, \tau))^{-1} - (\mathbb{1} + S(is, \tau))^{-1}]f \\ &= (\mathbb{1} + \tau^{-1})^{-1}(\mathbb{1} - P(\tau)) \oplus \left[T^P(t, \tau) - T^P(s, \tau)\right] f\,. \end{aligned} \tag{14.3.46}$$

With the above direct sum decomposition in mind, one may for a fixed τ deal with operator (14.3.46) on the subspace $P(\tau)\mathfrak{H}$,

$$T^P(t, s, \tau) := D(t, s, \tau) \restriction_{P(\tau)\mathfrak{H}} = T^P(t, \tau) - T^P(s, \tau)\,, \tag{14.3.47}$$

however, since τ is varying, we have to consider it on the whole space $\mathfrak{H}$. Note that the τ-family $\{D(t, s, \tau)\}_{0<\tau\leq 1}$ is strongly continuous in $0 < \tau \leq 1$, and uniformly bounded, i.e. $\|D(t, s, \tau)\| \leq 2$, because both $(\mathbb{1} + S(it, \tau))^{-1}$ and $(\mathbb{1} + S(is, \tau))^{-1}$ are contractions. Note also that we have $D(t, s, \tau) = P(\tau)D(t, s, \tau)$.

To verify the assertion of Lemma 14.3.11, we need to show that for any fixed $f \in \mathfrak{H}$, the difference $D(t, s, \tau)f$ in (14.3.46) with $t, s \in \mathbb{R} \setminus \{0\}$ converges to zero in the Hilbert space norm as $|t - s| \to 0$, and that the convergence is uniform with respect to $\tau \in (0, 1]$.

To this end we use a small trick showing first that it is sufficient to establish the claim of Lemma 14.3.11 under the *additional assumption* that one of the t and s, say the latter, belongs to $\mathbb{R} \setminus (M \cup \{0\})$. Indeed, if this is the case, i.e. if for any fixed $s \in \mathbb{R} \setminus (M \cup \{0\})$ and arbitrary $f \in \mathfrak{H}$, $\varepsilon > 0$, there is $\delta = \delta(f; \varepsilon, s) > 0$ such that

$$\|u_{\tau,f}(t) - u_{\tau,f}(s)\| < \varepsilon/2 \ \text{ holds for all } \ t \in (s - \delta, s + \delta) \ \text{ and } \ \tau \in (0, 1],$$

the lemma is valid in the general case as well. To see that, we take any two points $t_1, t_2 \in \mathbb{R} \setminus \{0\}$ in the vicinity of the chosen s satisfying $|t_i - s| < \frac{\delta}{2}$ for $i = 1, 2$, then

$$|t_1 - t_2| \le |t_1 - s| + |s - t_2| < \delta/2 + \delta/2 = \delta,$$

$$\|u_{\tau,f}(t_1) - u_{\tau,f}(t_2)\| \le \|u_{\tau,f}(t_1) - u_{\tau,f}(s)\| + \|u_{\tau,f}(s) - u_{\tau,f}(t_2)\| < \varepsilon/2 + \varepsilon/2 = \varepsilon\,,$$

holds independently of τ. This yields the complete claim of Lemma 14.3.11 in view of the fact that the set $\mathbb{R} \setminus (M \cup \{0\})$ from which the number s is chosen is dense in $\mathbb{R} \setminus \{0\}$.

Let us thus turn to a *nontrivial* part of the proof which consists of establishing the assertion of Lemma 14.3.11 for $s \in \mathbb{R} \setminus (M \cup \{0\})$ and $t \in \mathbb{R} \setminus \{0\}$; the aim is to complete the proof of our main result in the forthcoming Lemma 14.3.16(ii). We will work out the argument assuming that $t, s > 0$, the case with the opposite signs is completely analogous.

To begin with, we use functional calculus with (14.3.16), (14.3.17) to rewrite the differences containing $H(t\,\tau)$ and $G(t\,\tau)$ in (14.3.46) as

$$tH(t\,\tau) - sH(s\tau) = (\sin t\,\tau H - \sin s\tau H)/\tau = 2\,\tau^{-1}\cos(\tfrac{t+s}{2}\tau H)\,\sin(\tfrac{t-s}{2}\tau H), \tag{14.3.48}$$

$$tG(t\,\tau) - sG(s\tau) = (-\cos t\,\tau H + \cos s\tau H)/\tau = 2\,\tau^{-1}\sin(\tfrac{t+s}{2}\tau H)\,\sin(\tfrac{t-s}{2}\tau H). \tag{14.3.49}$$

To simplify expressions in the discussion to follow, we recall the quantity $K(\kappa)$ introduced in (14.3.16) which for $\kappa = t\,\tau$ allows us to write

$$\mathbb{1} + tK(t\,\tau) = \mathbb{1} + tG(t\,\tau) + itH(s\tau) = \mathbb{1} + (\mathbb{1} - \cos t\,\tau H)/\tau + i\,(\sin t\,\tau H)/\tau, \tag{14.3.50}$$

and similarly for $k = s\tau$. Furthermore, for a given $\tau \in (0, 1]$, we introduce the self-adjoint operator

$$H_\tau := H(\mathbb{1} + \tau H)^{-1}\,, \tag{14.3.51}$$

which is positive and bounded, and note that $\mathbb{1} + |s|H_\tau$ has a bounded inverse for any $s \in \mathbb{R}$. The difference $D(t, s, \tau)$ in (14.3.46) which can be identified with its nontrivial part $T^P(t,s,\tau)$ in (14.3.47) can be then rewritten as

$$\begin{aligned}
D(t,s,\tau) &= T^P(t,s,\tau) \\
&= [P(\tau)(\mathbb{1} + tK(t\,\tau))P(\tau)]^{-1} - [P(\tau)(\mathbb{1} + sK(s\tau))P(\tau)]^{-1} \\
&= [P(\tau)(\mathbb{1} + tK(t\,\tau))P(\tau)]^{-1} \\
&\quad \cdot \big\{P(\tau)(\mathbb{1} + sK(s\tau))P(\tau) - P(\tau)(\mathbb{1} + tK(t\,\tau))P(\tau)\big\} \\
&\qquad \times [P(\tau)(\mathbb{1} + sK(s\tau))P(\tau)]^{-1} \\
&= [P(\tau)(\mathbb{1} + tK(t\,\tau))P(\tau)]^{-1}[P(\tau)(sK(s\tau) - tK(t\,\tau))P(\tau)] \\
&\quad \times [P(\tau)(\mathbb{1} + sK(s\tau))P(\tau)]^{-1} \\
&= [P(\tau)(\mathbb{1} + tK(t\,\tau))P(\tau)]^{-1} \\
&\quad \times [(sK(s\tau) - tK(t\,\tau))(\mathbb{1} + |s|H_\tau)^{-1}] \\
&\qquad \times (\mathbb{1} + |s|H_\tau)P(\tau)[P(\tau)(\mathbb{1} + sK(s\tau))P(\tau)]^{-1} \\
&=: T_1(t,\tau)\, T_2(t,s,\tau)\, T_3(s,\tau)\,, \qquad (14.3.52)
\end{aligned}$$

where we have introduced

$$T_1(t,\tau) = T^P(t,\tau) = [P(\tau)(\mathbb{1} + tK(t\,\tau))P(\tau)]^{-1}, \qquad (14.3.53)$$

$$T_2(t,s,\tau) = (sK(s\tau) - tK(t\,\tau))(\mathbb{1} + |s|H_\tau)^{-1}, \qquad (14.3.54)$$

$$T_3(s,\tau) = (\mathbb{1} + |s|H_\tau)P(\tau)[P(\tau)(\mathbb{1} + sK(s,\tau))P(\tau)]^{-1}, \qquad (14.3.55)$$

which are all bounded operators on $P(\tau)\mathfrak{H}$ as well as on $\mathfrak{H}$; in (14.3.54) we are able to drop the projections $P(\tau)$ appearing in the last formula since the operator (14.3.45) maps the subspace $P(\tau)\mathcal{H}$ onto itself.

Next we consider the following two operator-valued functions, which may be thought of as the τ-limits of the families $\{T_1(t,\tau)\}_{0<\tau\le 1}$ in (14.3.53), and $\{T_3(s,\tau)\}_{0<\tau\le 1}$ in (14.3.55), namely,

$$T_1(t) := [(\mathbb{1} + itH_P)^{-1}P], \qquad (14.3.56)$$

$$T_3(s) := (\mathbb{1} + |s|H)P[(\mathbb{1} + isH_P)^{-1}P] = (\mathbb{1} + |s|H)[(\mathbb{1} + isH_P)^{-1}P], \qquad (14.3.57)$$

where we can remove again one P from the second expression of (14.3.57) since $(\mathbb{1} + isH_P)^{-1}$ maps $P\mathfrak{H}$ to itself. We already know from (14.3.43) that $T_1(t)$ is the strong limit as $\tau \to 0+$ of the family $\{T_1(t,\tau)\}$ of contractions, for the moment at least as long as $t \in \mathbb{R} \setminus (M \cup \{0\})$.

Upcoming we are going to show that $T_3(s)$ can be extended to a bounded operator on $\mathfrak{H}$. We begin with a crucial observation.

Lemma 14.3.12 *Let H be our non-negative self-adjoint operator acting in $\mathfrak{H}$ and H_P the self-adjoint operator referring to the orthogonal projection P. Consider the operator*

$$T_0 := (\mathbb{1} + H)[(\mathbb{1} + H_P)^{-1}P], \tag{14.3.58}$$

in $\mathfrak{H}$ for which we have:

(i) *the domain and range of T_0 are*

$$\begin{aligned} \mathrm{dom}\,(T_0) &= P(\mathbb{1} + H_P)\mathrm{dom}\,(HP) = P(\mathbb{1} + H_P)(\mathrm{dom}\,(H) \cap P\mathfrak{H}) \oplus (P\mathfrak{H})^{\perp}, \\ \mathrm{ran}\,[T_0] &= (\mathbb{1} + H)\mathrm{dom}\,(HP) = (\mathbb{1} + H)(\mathrm{dom}\,(H) \cap P\mathfrak{H}), \end{aligned} \tag{14.3.59}$$

as T_0 is the direct sum, $T_0 = T_0 \restriction P\mathfrak{H} \oplus 0$, in accordance with (14.3.8),

(ii) *in addition,*

$$T_0\, g = g\,, \quad g \in \mathrm{dom}\,(T_0 \restriction P\mathfrak{H}), \tag{14.3.60}$$

thus T_0 can be extended to a bounded operator $\tilde{T}_0$ on $\mathfrak{H}$ such that

$$\begin{aligned} &\textit{the closure of } T_0 \restriction_{P\mathfrak{H}} \textit{ is the identity operator } I_{P\mathfrak{H}} \textit{ on } P\mathfrak{H}, \\ &\textit{and } \tilde{T}_0 = I_{P\mathfrak{H}} \oplus 0 \ \text{ on } \mathfrak{H} = P\mathfrak{H} \oplus (P\mathfrak{H})^{\perp}. \end{aligned} \tag{14.3.61}$$

In particular, $\tilde{T}_0$ is a contraction.

Proof Note that for domain of product of two linear operators A and B in $\mathfrak{H}$ one obtains: $\mathrm{dom}\,(AB) = B^{-1}\mathrm{dom}\,(A) := \{\, g \in \mathrm{dom}\,(B) : \ Bg \in \mathrm{dom}\,(A)\,\}$. Since P commutes with H_P, we can rewrite (14.3.58) as $T_0 := [(\mathbb{1} + H)P][(\mathbb{1} + H_P)^{-1}P]$ with the domain

$$\begin{aligned} \mathrm{dom}\,(T_0) &= [P(\mathbb{1} + H_P)^{-1}]^{-1}\mathrm{dom}\,((\mathbb{1} + H)P) \\ &= \{\, g \in \mathrm{dom}\,((\mathbb{1} + H_P)^{-1}P) : \ (\mathbb{1} + H_P)^{-1}Pg \in \mathrm{dom}\,((\mathbb{1} + H)P)\} \\ &= \{\, g \in \mathrm{dom}\,(H_P) : \ (\mathbb{1} + H_P)^{-1}Pg \in \mathrm{dom}\,(HP)\} \\ &= P(\mathbb{1} + H_P)\mathrm{dom}\,(HP) = P(\mathbb{1} + H_P)(\mathrm{dom}\,(H) \cap P\mathfrak{H}) \oplus (P\mathfrak{H})^{\perp}\,, \end{aligned}$$

and the range

$$\begin{aligned}\operatorname{ran}(T_0) &= T_0 \operatorname{dom}(T_0)\\ &= [(\mathbb{1}+H)P][(\mathbb{1}+H_P)^{-1}P]P(\mathbb{1}+H_P)\operatorname{dom}(HP)\\ &= [(\mathbb{1}+H)P]\operatorname{dom}(HP) = (\mathbb{1}+H)(\operatorname{dom}(H)\cap P\mathfrak{H}).\end{aligned}$$

Let us now turn to the claim (ii). Just as H_P is the self-adjoint operator in $\mathfrak{H}$ associated with the quadratic form $u \mapsto \|H^{1/2}Pu\|^2$ defined on $\operatorname{dom}(H^{1/2}P) = (\operatorname{dom}(H^{1/2}) \cap P\mathfrak{H}) \oplus (P\mathfrak{H})^{\perp}$, the self-adjoint operator $P(\mathbb{1}+H_P)$ is associated with the form: $u \mapsto \|(\mathbb{1}+H)^{1/2}Pu\|^2$ well-defined in view of the inequalities

$$\frac{1}{\sqrt{2}}(\mathbb{1}+H^{1/2}) \le (\mathbb{1}+H)^{1/2} \le \mathbb{1}+H^{1/2},$$

on the same domain $\operatorname{dom}((\mathbb{1}+H)^{1/2}P) = \operatorname{dom}(H^{1/2}P)$. Consequently,

$$P(\mathbb{1}+H_P) = P(\mathbb{1}+H)_P = ((\mathbb{1}+H)^{1/2}P)^*(\mathbb{1}+H)^{1/2}P\,. \tag{14.3.62}$$

We are going to use it to show (14.3.60). The adjoint $((\mathbb{1}+H)^{1/2}P)^*$ to $(\mathbb{1}+H)^{1/2}P$ is a closed extension of the closable (in general, non-closed) operator $P(\mathbb{1}+H)^{1/2}$, in other words, $((\mathbb{1}+H)^{1/2}P)^* \supset P(\mathbb{1}+H)^{1/2}$, which yields

$$((\mathbb{1}+H)^{1/2}P)^*(\mathbb{1}+H)^{1/2}P \supset P(\mathbb{1}+H)^{1/2}(\mathbb{1}+H)^{1/2}P = P(\mathbb{1}+H)P,$$

i.e. the operator on the left-hand side is an extension of the operator on the right. As both sides are invertible, the analogous inclusion holds for their inverses,

$$[(\mathbb{1}+H_P)^{-1}P] = [((\mathbb{1}+H)^{1/2}P)^*(\mathbb{1}+H)^{1/2}P]^{-1} \supset [P(\mathbb{1}+H)P]^{-1}.$$

It follows that for any $g \in \operatorname{dom}(T_0)$ specified in (14.3.61), we have

$$T_0\, g = (\mathbb{1}+H)[(\mathbb{1}+H_P)^{-1}P]g = (\mathbb{1}+H)[P(\mathbb{1}+H)P]^{-1}g = Pg,$$

which yields the desired claim (14.3.60), since T_0 is reduced by the projection P and $\operatorname{dom}(T_0 \upharpoonright P\mathfrak{H}) \subseteq P\mathfrak{H}$. Thus we see that the closure of $T_0 \upharpoonright P\mathfrak{H}$ is the identity operator $\mathbb{1}_{P\mathfrak{H}}$ on $P\mathfrak{H}$, and $\tilde{T_0}$ as the closed extension of T_0 to the whole $\mathfrak{H}$ has the norm not exceeding one. □

Lemma 14.3.13 *The operator $T_3(s)$ in* (14.3.57), *acting in $\mathfrak{H}$ with the domain $P(\mathbb{1}+isH_P)\operatorname{dom}(HP) = (\mathbb{1}+isH_P)(\operatorname{dom}(H)\cap P\mathfrak{H})$, can be extended to a bounded operator on the whole space $\mathfrak{H}$ with the norm satisfying $\|T_3(s)\| \le \sqrt{2}$.*

Proof Note that the claim of Lemma 14.3.12 remains valid when we replace H_P, H by $|s|H_P$, $|s|H$, respectively, for any $s \neq 0$. We rewrite the operator $T_3(s)$ in (14.3.57) as

$$T_3(s) = \big\{(\mathbb{1} + |s|H)[(\mathbb{1} + |s|H_P)]^{-1}P]\big\} \\ \cdot \big\{[P(\mathbb{1} + |s|H_P)][(\mathbb{1} + isH_P)^{-1}P]\big\}.$$

By the *spectral theorem*, the norms of the first and second factors on the right-hand side are one and $\sqrt{2}$, respectively, independently of s, and Lemma 14.3.12 (ii) in combination with the above observation implies that the first factor is a contraction, which gives $\|T_3(s)\| \leq \sqrt{2}$. □

In the next step, we turn to investigation of the second and third factors, (14.3.54) and (14.3.55), of the operator $T^P(t, s, \tau)$ in (14.3.52). We begin with proving a crucial property of the τ-family $\{T_3(s, \tau)\}_{0<\tau\leq 1}$ with $s \in \mathbb{R} \setminus (M \cup \{0\})$, where the set $M \subseteq \mathbb{R}$ of Lebesgue measure zero was introduced in the text that follows the mapping (14.3.33). Recall that at this point we adopt the separability assumption. Afterwards, we shall focus on the τ-family $\{T_2(t, s, \tau)\}_{0<\tau\leq 1}$ with $t,\ s \in \mathbb{R} \setminus \{0\}$.

Lemma 14.3.14 *Let $\mathfrak{H}$ be a separable Hilbert space. Let number $s \in \mathbb{R} \setminus (M \cup \{0\})$ be such that family $\{[P(\tau)(\mathbb{1} + sK(s\tau))P(\tau)]^{-1}f\}_{\tau>0}$ strongly converges to $[(\mathbb{1} + isH_P)^{-1}P]f$, as $\tau \to +0$, for any vector $f \in \mathfrak{H}$. Then the following claims are valid*:

(i) *For fixed s, the operator family $\{T_3(s, \tau)\}_{0<\tau\leq 1}$ defined by* (14.3.55) *is uniformly bounded on $\mathfrak{H}$ and converges strongly to $T_3(s)$ in* (14.3.57) *as $\tau \to +0$.*
(ii) *To be specific, for a fixed $s \in \mathbb{R} \setminus (M \cup \{0\})$ there is a constant $C_{T_3}(s) \geq \sqrt{2}$ such that*

$$\|T_3(s, \tau)\| \leq C_{T_3}(s) \quad \text{for all } \tau \in (0, 1]\,. \tag{14.3.63}$$

Proof (i) Our aim is to verify that $T_3(s, \tau)g$ converges for $g \in \mathfrak{H}$ to $T_3(s)g$ as $\tau \to +0$ in the Hilbert space norm and to use this fact to establish the uniform boundedness of $\{T_3(s, \tau)g\}_{0<\tau\leq 1}$.

To this end, let us recall the notion of operator *convergence* in the *strong graph* sense. Given a sequence $\{A_n\}_{n=1}^{\infty}$ of operators in a Hilbert space $\mathfrak{H}$, we say that $(\psi, \varphi) \in \mathfrak{H} \times \mathfrak{H}$ belongs to the strong graph *limit* of operators $\{A_n\}_{n=1}^{\infty}$ if one can find $\psi_n \in \mathrm{dom}\,(A_n)$, $n = 1, 2, \ldots$, such that

$$\psi_n \longrightarrow \psi, \quad A_n\psi_n \longrightarrow \varphi \quad \text{as } n \to \infty\,.$$

We denote the set of such pairs (ψ, φ) by Γ^s_∞. If Γ^s_∞ is the graph of an operator A, that is, $\Gamma^s_\infty = \{(\psi, A\psi) \in \mathfrak{H} \times \mathfrak{H} : \psi \in \mathrm{dom}\,(A)\}$, then A is also said to be the *strong graph limit* of $\{A_n\}_{n=1}^{\infty}$.

Recall that there is the following result.

Proposition 14.3.15 *Let $\{A_n\}_{n=1}^{\infty}$ and A be self-adjoint operators in $\mathfrak{H}$. Then sequence $\{A_n\}_{n\in\mathbb{N}}$ converges to A in the strong resolvent sense if and only if A is the strong graph limit of $\{A_n\}_{n\in\mathbb{N}}$.*

To make use of this result, let us consider the family $\{A_\tau\}_{0<\tau\le 1}$ of the following bounded, self-adjoint operators

$$A_\tau := \mathbb{1} + |s|\, H_\tau\,, \quad 0 < \tau \le 1\,, \quad s \in \mathbb{R}\,.$$

By the spectral theorem ,elements of this family converge in the strong operator topology (as $\tau \to +0$) to the self-adjoint operator $\mathbb{1} + |s|H$, and consequently operator $\mathbb{1} + |s|\,H$ by Proposition 14.3.15 is the strong graph limit of $\{A_\tau\}_{0<\tau\le 1}$. If $s \in \mathbb{R} \setminus (M \cup \{0\})$ holds, then we also obtain

$$\psi_\tau := [P(\tau)(\mathbb{1} + sK(s\tau))P(\tau)]^{-1} g \longrightarrow \psi := [(\mathbb{1} + is\, H_P)^{-1} P]\, g\,,$$

as $\tau \to +0$ for $g \in \mathfrak{H}$, for

$$[(\mathbb{1} + isH_P)^{-1} P]\, g \in \operatorname{dom}(H_P) \subset \operatorname{dom}((\mathbb{1} + |s|\, H)P) \subset \operatorname{dom}(\mathbb{1} + |s|\, H)\,.$$

In this way, the pair $(\psi, (\mathbb{1}+|s|H)\psi)$ belongs to the set Γ_∞^s resulting from the strong graph limit of the family $\{A_\tau\}_{\tau>0}$, and consequently, $A_\tau\psi_\tau \to (\mathbb{1} + |s|\, H)\psi$, that is,

$$\lim_{\tau\to+0} A_\tau [P(\tau)(\mathbb{1} + sK(s\tau))P(\tau)]^{-1} g = (\mathbb{1} + |s|\, H)[(\mathbb{1} + is\, H_P)^{-1} P]\, g\,,$$

or, in other words, the sought convergence, $\lim_{\tau\to+0} T_3(s,\tau)g = T_3(s)g$. This implies, by the *uniform boundedness* principle (*Banach-Steinhaus* theorem, Appendix A.1.2), that family $\{T_3(s,\tau)\}_{0\le\tau<1}$ is uniformly bounded in the operator norm.

(ii) Consider again arbitrary $g \in \mathfrak{H}$. In view of the convergence established above in combination with the bound on the norm of $T_3(s)$ from Lemma 14.3.13(ii), we see that for a fixed $s \in \mathbb{R} \setminus (M \cup \{0\})$ there is a positive, g-dependent number $\tau_s = \tau_s(g) \le 1$ such that

$$\|T_3(s,\tau)g\| \le \|T_3(s)g\| + \frac{1}{2} \le \|T_3(s)\|\, \|g\| + \frac{1}{2} \le \sqrt{2}\, \|g\| + \frac{1}{2}\,, \qquad \tau \in (0, \tau_s]\,.$$

This helps us to complement our knowledge of the uniform boundedness using the estimate

$$\sup_{0<\tau\le 1} \|T_3(s,\tau)g\| \le \max\left\{\sqrt{2}\, \|g\| + \frac{1}{2}\,,\ \sup_{\tau_s<\tau\le 1} \|T_3(s,\tau)g\|\right\}.$$

We note that the function $\tau \mapsto \|T_3(s,\tau)g\|$ is continuous on the compact set $[\tau_s, 1] \subset \mathbb{R}$. Hence the right-hand side is bounded for any g and, applying again the uniform boundedness principle, we conclude that the τ-family $\{T_3(s,\tau)\}_{0<\tau\le 1}$ is uniformly bounded by a constant $C_{T_3}(s) \ge \sqrt{2}$ depending on s. This completes the proof of Lemma 14.3.14. □

It remains to study the middle factor $T_2(t,s,\tau)$ in (14.3.54). We rewrite it as

$$\begin{aligned} T_2(t,s,\tau) &= i(sH(s\tau) - tH(t\,\tau))\,(\mathbb{1} + |s|H_\tau)^{-1} \\ &\quad + (sG(s\tau) - tG(t\,\tau))\,(\mathbb{1} + |s|H_\tau)^{-1} \\ &=: iT_{2,H}(t,s,\tau) + T_{2,G}(t,s,\tau)\,. \end{aligned} \tag{14.3.64}$$

For a fixed $\tau \in (0,1]$, the operators $T_{2,H}(t,s,\tau)$, $T_{2,G}(t,s,\tau)$, as well as their linear combination, are bounded operators which are in the functional calculus sense obtained from self-adjoint H as its appropriate bounded and continuous functions. By virtue of (14.3.48) and (14.3.49), we have

$$\begin{aligned} T_{2,H}(t,s,\tau) &= \phi_H(t,s;\tau,H) := (tH(t\,\tau) - sH(s\tau))\,(\mathbb{1} + |s|\,H_\tau)^{-1} \\ &= \frac{2\,\tau^{-1}\,\sin((t-s)\,\tau\,\lambda/2)\,\cos((s+t)\,\tau\,\lambda/2)}{\mathbb{1} + |s|\,\lambda\,(\mathbb{1} + \tau\,\lambda)^{-1}}\,, \\ T_{2,G}(t,s,\tau) &= \phi_G(t,s;\tau,H) := (tG(t\,\tau) - sG(s\tau))\,(\mathbb{1} + |s|H_\tau)^{-1} \\ &= \frac{2\,\tau^{-1}\,\sin((t-s)\,\tau\,\lambda/2)\,\sin((s+t)\,\tau\,\lambda/2)}{\mathbb{1} + |s|\,\lambda\,(\mathbb{1} + \tau\,\lambda)^{-1}}\,, \end{aligned}$$

and

$$T_2(t,s,\tau) = \phi(t,s;\tau,H) := i\phi_H(t,s;\tau,H) + \phi_G(t,s;\tau,H)\,.$$

Then it is straightforward to check that

$$|\phi(t,s;\tau,\lambda)|^2 = \left|\frac{2\,\tau^{-1}\,\sin((t-s)\,\tau\,\lambda/2)}{\mathbb{1} + |s|\lambda(\mathbb{1} + \tau\lambda)^{-1}}\right|^2, \tag{14.3.65}$$

which means that for any $g \in \mathfrak{H}$ we obtain

$$\begin{aligned} \|T_2(t,s,\tau)g\|^2 &= \|T_{2,H}(t,s,\tau)g\|^2 + \|T_{2,G}(t,s,\tau)g\|^2 \\ &= \int_{0-}^{\infty} \|E(\mathrm{d}\lambda)\,g\|^2 \left|\frac{2\,\tau^{-1}\,\sin((t-s)\,\tau\,\lambda/2)}{\mathbb{1} + |s|\lambda(\mathbb{1} + \tau\lambda)^{-1}}\right|^2. \end{aligned} \tag{14.3.66}$$

The following lemma represents the second crucial element in establishing Lemma 14.3.11. In this way, it proves the ultimate of auxiliary statements indispensable for our principal results in Sect. 14.3.3.

Lemma 14.3.16 *Consider $t, s \in \mathbb{R} \setminus \{0\}$ such that either $t, s > 0$ or $t, s < 0$. Then for $T_2(t, s, \tau)$ given by* (14.3.54) *and for $D(t, s, \tau)$ given by* (14.3.46) *or* (14.3.52) *the following holds*:

(i) *The τ-family $\{T_2(t, s; \tau\}$ of bounded operators in* (14.3.64) *is bounded locally-uniformly for $t, s \in \mathbb{R} \setminus \{0\}$. Moreover, for every $f \in \mathfrak{H}$ and any $\varepsilon > 0$, there is an s-dependent number $\delta = \delta(f; \varepsilon, s) > 0$ such that*

$$t \in \mathbb{R} \setminus \{0\} \text{ with } |t - s| < \delta \quad \Longrightarrow \quad \|T_2(t, s, \tau) f\| < \varepsilon\,, \tag{14.3.67}$$

uniformly with respect to $\tau \in (0, 1]$.

(ii) *Let further $s \in \mathbb{R} \setminus (M \cup \{0\})$, then for every $f \in \mathfrak{H}$ and any $\varepsilon > 0$, there is a number $\delta = \delta(f; \varepsilon, s) > 0$ such that*

$$t \in \mathbb{R} \setminus \{0\} \;\; with \;\; |t - s| < \delta \quad \Longrightarrow \quad \|D(t, s, \tau) f\| < \varepsilon\,, \tag{14.3.68}$$

uniformly with respect to $\tau \in (0, 1]$.

Proof

(i) The τ-families of bounded operators appearing in (14.3.64) converge in the strong operator topology for any fixed pair $t, s \in \mathbb{R} \setminus \{0\}$, then by the functional calculus, we find that

$$T_{2,H}(t, s, \tau) \xrightarrow{s} T_{2,H}(t, s) := \big((t - s) H\big)/\big(\mathbb{1} + |s|\, H\big)\,,$$

$$T_{2,G}(t, s, \tau) \xrightarrow{s} T_{2,G}(t, s) := 0\,,$$

$$T_2(t, s, \tau) \xrightarrow{\;\;s\;\;} T_2(t, s) := i\, T_{2,H}(t, s) + T_{2,G}(t, s)\,,$$

holds as $\tau \to +0$, and therefore

$$|T_2(t, s, \tau)|^2 \xrightarrow{\;\;s\;\;} |T_2(t, s)|^2 = T_{2,H}(t, s)^2 + T_{2,G}(t, s)^2$$
$$= \Big(((t - s) H)/(\mathbb{1} + |s|\, H)\Big)^2.$$

Our way to prove the claim is to apply the *uniform boundedness principle* to these operator-families. From (14.3.66) we obtain

$$\|T_2(t, s, \tau)\| = \sup_{\|g\|=1} \|T_2(t, s, \tau) g\| = \sup_{\lambda \geq 0} |\phi(t, s; \tau, \lambda)|\,,$$

and for $T_2(t,s) = \phi(t,s;H)$, with $\phi(t,s;\lambda) := ((t-s)\,\lambda)/(1+|s|\,\lambda)$, we have similarly

$$\|T_2(t,s)\| = \sup_{\lambda \ge 0} |\phi(t,s;\lambda)| = |\,t-s|/|s|\,. \tag{14.3.69}$$

We note that $\phi(t,s;\tau,\lambda)$ is continuous with respect to all the variables, $\tau \in (0,1]$, $\lambda \in [0,\infty)$, and $t,s \in \mathbb{R}\setminus\{0\}$. Our intention is now to regard the family $\{T_2(t,s,\tau)\}_{\tau>0}$ as depending on a *broader* set of parameters. To this aim we fix a non-zero s and consider τ in $(0,1]$, together with t in a neighborhood of s, or more precisely:

$$\tau \in (0,1], \quad t \in D(s) := \{t \in \mathbb{R}\setminus\{0\} : |t-s| \le |s|/2\}\,. \tag{14.3.70}$$

For each fixed $f \in \mathfrak{H}$ and $s \in \mathbb{R}\setminus\{0\}$, we put

$$\Phi_{f,s}(t,\tau) := \begin{cases} \|T_2(t,s,\tau)f\| & \ldots \quad 0<\tau\le 1\,, \\ \|T_2(t,s)f\| & \ldots \quad \tau = 0\,, \end{cases} \tag{14.3.71}$$

which is a bounded and continuous function on the compact set $D(s)\times[0,1] \subset \mathbb{R}^2$. Then $C_{f,s} := \max_{t,\tau\in D(s)\times[0,1]} \Phi_{f,s}(t,\tau)$ exists, depending on f and s, but being independent of $t \in D(s)$ and $\tau \in (0,1]$, and from the strong convergence of $\{T_2(t,s,\tau)\}$ to $T_2(t,s)$ noted above, we infer that

$$\|T_2(t,s,\tau)f\| \le C_{f,s}\,,$$

holds for given $f \in \mathfrak{H}$, $s \neq 0$, and all $(t,\tau) \in D(s)\times(0,1]$. Applying now the *uniform boundedness principle* to this operator family with the extended set of parameters (14.3.70), we establish the existence of a constant $C_{T_2}(s) > 0$, depending on s, but independent of $t \in D(s)$ and $\tau \in (0,1]$, such that

$$\|T_2(t,s,\tau)\| \le C_{T_2}(s)\,. \tag{14.3.72}$$

In view of the strong convergence, it cannot be smaller than the norm of $T_2(t,s)$ in (14.3.69), that is, $C_{T_2}(s) \ge |\,t-s|/|s|\,$.

Since $E([\lambda,\infty)) \xrightarrow{s} 0$ holds as $\lambda \to \infty$, one can find for a fixed s, any $f \in \mathfrak{H}$, and each $\varepsilon > 0$ a number $\lambda_0 = \lambda_0(f;\varepsilon,s) > 0$ such that

$$\|E([\lambda_0,\infty))f\|^2 < C_{T_2}(s)^{-2}\,\frac{\varepsilon^2}{2}\,. \tag{14.3.73}$$

In this case using (14.3.65) and (14.3.73), we obtain

$$\begin{aligned}\|T_2(t,s,\tau)f\|^2 &= \|\phi(t,s;\tau,H)f\|^2\\ &= \left(\int_{0-}^{\lambda_0} + \int_{\lambda_0}^{\infty}\right) \|E(d\lambda)f\|^2\, |\phi(t,s;\tau,\lambda)|^2\\ &\le \int_{0-}^{\lambda_0} \|E(d\lambda)f\|^2 \left|\frac{2\tau^{-1}\sin((t-s)\tau\lambda/2)}{1+s\lambda(1+\tau\lambda)^{-1}}\right|^2\\ &\quad + C_{T_2}(s)^2\, \|E([\lambda_0,\infty))f\|^2\\ &\le (|t-s|\lambda_0)^2\, \|E([0,\lambda_0))f\|^2 + \frac{\varepsilon^2}{2}\,. \end{aligned} \tag{14.3.74}$$

Here the last inequality comes from the estimate of the numerator of the squared modulus factor in the above integral,

$$\left|2\tau^{-1}\sin((t-s)\tau\lambda/2)\right| \le |t-s|\,\lambda_0\,, \quad 0\le\lambda\le\lambda_0\,.$$

Since the set $\{(\tau,\lambda);\ 0<\tau\le 1,\ 0\le\lambda\le\lambda_0\}$ is bounded, and therefore its closure is compact, we infer that for a fixed $s\in\mathbb{R}\setminus\{0\}$ there is a number $\delta=\delta(f;\varepsilon,s)>0$ such that

$$t\in\mathbb{R}\setminus\{0\}\ \text{with}\ |t-s|<\delta \quad\Longrightarrow\quad (|t-s|\lambda_0)^2\,(1+\|f\|^2) < \frac{\varepsilon^2}{2}\,,$$

uniformly for $0<\tau\le 1$. In fact, we can find it explicitly. It is enough to choose

$$\delta(f;\varepsilon,s) := \min\left\{\frac{\varepsilon}{\sqrt{2}\,\lambda_0\,(1+\|f\|^2)^{1/2}}\,,\ \frac{|s|}{2}\right\}.$$

As a result, we deduce on account of (14.3.74) that

$$\|T_2(t,s,\tau)f\|^2 \le (1+\|f\|^2)^{-1}\,\frac{\varepsilon^2}{2}\,\|E([0,\lambda_0))f\|^2 + \frac{\varepsilon^2}{2} < \varepsilon^2\,,$$

in other words,

$$\|T_2(t,s;\tau,H)f\| < \varepsilon\,,$$

which yields the implication (14.3.67).

(ii) To prove the remaining claim of the lemma, we first note that $T_1(t,\tau)\equiv T^P(t,\tau)$ is a contraction, $\|T_1(t,\tau)\|\le 1$, and therefore

$$\begin{aligned}\|D(t,s,\tau)f\| = \|T^P(t,s,\tau)f\| &= \|T_1(t,\tau)T_2(t,s,\tau)T_3(s,\tau)f\|\\ &\le \|T_2(t,s,\tau)T_3(s,\tau)f\|\,.\end{aligned}$$

To scrutinise the last norm, we repeat the argument from the part (i) replacing f by $T_3(s,\tau)f$, and deducing for any $\lambda_0 > 0$, the estimate:

$$
\begin{aligned}
&\|T_2(t,s,\tau)T_3(s,\tau)f\|^2 \\
&= \left(\int_{0-}^{\lambda_0} + \int_{\lambda_0}^{\infty}\right) \|E(\mathrm{d}\lambda)T_3(s,\tau)f\|^2 \, |\phi(t,s;\tau,\lambda)|^2 \\
&\le \Big[\int_{0-}^{\lambda_0} \|E(\mathrm{d}\lambda)\, T_3(s,\tau)f\|^2 \left|\frac{2\,\tau^{-1}\,\sin((t-s)\,\tau\,\lambda/2)}{1+s\,\lambda\,(1+\tau\lambda)^{-1}}\right|^2 \\
&\qquad\qquad + C_{T_2}(s)^2\, \|E([\lambda_0,\infty))\, T_3(s,\tau)f\|^2\Big] \qquad (14.3.75) \\
&\le (|t-s|\lambda_0)^2\, \|E([0,\lambda_0))\, T_3(s,\tau)f\|^2 + C_{T_2}(s)^2\, \|E([\lambda_0,\infty))\, T_3(s,\tau)f\|^2 ,
\end{aligned}
$$

with the constant $C_{T_2}(s)$ from (14.3.72). The second term of the last expression converges to zero as $\lambda_0 \to \infty$ for any *fixed* $\tau \in (0,1]$.

In any case, to use this estimate, we need to find a λ_0 independent of τ in order to obtain convergence *uniform* with respect to $0 < \tau \le 1$. This can be achieved, however, because for fixed $f \in \mathfrak{H}$ and $s \in \mathbb{R} \setminus (M \cup \{0\})$, the set

$$V_f := \{T_3(s,\tau)f;\ 0 < \tau \le 1\} \subset \mathfrak{H},$$

is bounded by (14.3.63) and has a compact closure, as a consequence of the strong continuity of $T_3(s,\tau)f$ with respect to the variable $\tau > 0$. Then for every $f \in \mathfrak{H}$ and every $\varepsilon > 0$, one can find an s-dependent number $\lambda_{00} = \lambda_{00}(f;\varepsilon,s) > 0$ such that

$$\sup_{0<\tau\le 1} \|E([\lambda_{00},\infty))T_3(s,\tau)f\|^2 < C_{T_2}(s)^{-2}\,\frac{\varepsilon^2}{2}, \qquad (14.3.76)$$

with the constant $C_{T_2}(s)$ from (14.3.72).

To elucidate the last claim, we mention the following. Since V_f is totally bounded, there is a finite family of vectors $\{y_j\}_{j=1}^N \subset V_f$, $y_j = T_3(s;\tau_j)f$ for some $\tau_j \in (0,1]$, and an open ball $B(0;\varepsilon/\sqrt{8}C_{T_2}(s))$ with the center at the origin and radius $\varepsilon/\sqrt{8}C_{T_2}(s)$ for which we have

$$V_f \subset \bigcup_{j=1}^{N} \left(y_j + B(0;\varepsilon/\sqrt{8}C_{T_2}(s))\right).$$

Using again the fact that $E([\lambda,\infty)) \xrightarrow{s} 0$ holds as $\lambda \to \infty$, we infer that there is a family $\{\lambda_j\}_{j=1}^N$ of large positive numbers such that $C_{T_2}(s)\,\|E([\lambda_j,\infty))y_j\| \le \varepsilon/\sqrt{8}$ holds for $j = 1,2,\ldots,N$ and we put $\lambda_{00} := \max_{j=1,2,\ldots,N} \lambda_j$. The finite union coverage means that any $\phi \in V_f$ satisfies $\phi \in y_{j_\phi} + B(0; \varepsilon/\sqrt{8}C_{T_2}(s))$ for some $1 \le j_\phi \le N$ with $0 < \tau_{j_\phi} \le 1$. Noting that $\|E([\lambda_{00}, \infty))\| \le 1$, we infer

$$\begin{aligned}
&C_{T_2}(s)^2 \, \|E([\lambda_{00}, \infty))\phi\|^2 \\
&\le C_{T_2}(s)^2 \left(\max_{j=1,2,\dots,N} \|E([\lambda_{00}, \infty))y_j\| + \varepsilon/\sqrt{8}C_{T_2}(s)\right)^2 \\
&\le 2C_{T_2}(s)^2 \left[\max_{j=1,2,\dots,N} \|E([\lambda_{00}, \infty))y_j\|^2 + \left(\varepsilon/\sqrt{8}C_{T_2}(s)\right)^2\right] \\
&< \frac{\varepsilon^2}{4} + \frac{\varepsilon^2}{4} = \frac{\varepsilon^2}{2}.
\end{aligned}$$

Returning to the first term on the right-hand side of (14.3.75) where taking $\lambda_0 = \lambda_{00}$, and noting that $\|T_3(s,\tau)f\| \le C_{T_3}(s)\|f\|$ holds in view of (14.3.63), we see that there exists a number $\delta = \delta(f; \varepsilon, s) > 0$ such that for a fixed s we obtain

$$t \in \mathbb{R} \setminus \{0\} \text{ with } |t-s| < \delta \implies (|t-s|\lambda_{00})^2 (C_{T_3}(s)\|f\|)^2 < \frac{\varepsilon^2}{2},$$

uniformly for all $\tau \in (0, 1]$. Here one can be again explicit and choose

$$\delta(f; \varepsilon, s) := \min\left\{ \frac{\varepsilon}{\sqrt{2}\lambda_{00}\,(1 + (C_{T_3}(s)\|f\|)^2)^{1/2}}, \frac{|s|}{2} \right\}.$$

The estimate (14.3.75) then gives

$$\|T_2(t,s,\tau)T_3(s,\tau)f\|^2 < \frac{\varepsilon^2}{2} + \frac{\varepsilon^2}{2} = \varepsilon^2,$$

or in other words

$$\|D(t,s,\tau)f\| \le \|T_2(t,s,\tau)T_3(s,\tau)f\| < \varepsilon,$$

which yields (14.3.68). This concludes the proof of the present Lemma 14.3.16, and in this way also of Lemma 14.3.11, as we have outlined in the opening of this Sect. 14.3.4. □

14.4 Notes to Chapt. 14

Notes to Sect. 14.1
The main part of this section is standard.

Section 14.1.1

We start the notes for this subsection by a brief reminder of essentials and heuristics related to what is called the quantum *Zeno effect* (or *Zeno paradox*).

For simplicity let a quantum system, which is defined by the self-adjoint Hamiltonian H in a (complex) Hilbert space $\mathfrak{H}$, be *prepared* at the time $t = 0$ in the (*normalised*) vector state $\psi_0 \in \mathfrak{H}$. Then according to *postulates* of the Quantum Mechanics this state evolves under the action of unitary group $\{e^{-itH}\}_{t\in\mathbb{R}}$, and the probability to observe this system for $t > 0$ *still* in the state ψ_0 is equal to

$$p(t) := |(e^{-itH}\,\psi_0, \psi_0)|^2 . \tag{14.4.1}$$

It is known as the *survival probability* with evident value $p(0) = 1$ at $t = 0$. Since the *Zeno paradox* comes from interlaced *short-time* measurements we develop (formally), the evolution in (14.4.1) for small times ($t \ll \tau_Z$):

$$p(t) = 1 - \frac{t^2}{\tau_Z^2} + \ldots \text{ for } \tau_Z^{-2} := (H^2\,\psi_0, \psi_0) - (H\,\psi_0, \psi_0)^2, \quad \psi_0 \in \mathrm{dom}\,(H) . \tag{14.4.2}$$

In fact, what matters is the vanishing derivative at zero, which is the case if ψ_0 belong to the *form-domain* of H; see [Ex13, Proposition 6.1].

Now we remind that according to the Quantum Mechanics *protocol* of measurements in order to check whether the system after a shot time t/n is *still* in its initial state ψ_0, one uses the survival probability (14.4.1). If it is positive *and* the system is still in the initial state, the evolution starts *again* from ψ_0. As a consequence of (supposed) independence of measurements, the survival probability $p_n(t)$ after n acts is the product:

$$p_n(t) := p(t/n)^n = |(e^{-itH/n}\,\psi_0, \psi_0)|^{2n} . \tag{14.4.3}$$

If the *Zeno time* τ_Z is positive and bounded, by (14.4.1)–(14.4.3), these heuristic arguments yield in the limit of "continuous observation", $n \to \infty$, the value for the limiting survival probability:

$$p_\infty(t) := \lim_{n\to\infty} p_n(t) = \lim_{n\to\infty}\left(1 - \frac{t^2}{n^2\,\tau_Z^2} + \ldots\right)^n = \lim_{n\to\infty} e^{-t^2/(n\,\tau_Z^2)} = 1 . \tag{14.4.4}$$

The observation (14.4.4) is, according to the *Misra-Sudarshan* terminology [MS77], the quantum Zeno paradox:

The infinitely frequent measurements halt the quantum mechanical evolution and freeze the system in its initial state.

We note that is was *von Neumann* [vN32] who for the first time understood the unusual properties of the short-time quantum evolution. Then [BN67] discussed the "continuous observation" in the context of the freezing of decay of unstable particles. In paper [Yo68] this phenomenon was called the *Turing paradox* before the allusion with the *classical* Zeno paradox was invented in [MS77].

For more details about theoretical and experimental aspects and references concerning the quantum Zeno effect (paradox), we suggest the book [Ex85] and a very complete review article [FP08].

The observation in Remark 14.1.2 is due to an important link (*Sz.- Nagy* theorem (1953)) between contraction C_0-semigroups and their unitary *dilations* on a Hilbert space. For comments with proof, see [SzNF70, Chapter I, Theorem 8.1] as well as a short comprehensive presentation in [Dav80, Chapter 6, Section 3].

In Sects. 13.1 and 13.1.1 we consider a similar problem of *embedding* a given contraction C_0-semigroup in a (unitary) C_0-group. See Remark 13.1.5 (a)–(d) and Notes to Chap. 13 (Sects. 13.1 and 13.1.1) for comments and references.

Let us mention another family of related results. Note that the paper [Ca05] was inspired, in fact, by results obtained by P.Exner and T.Ichinose in [EI05]. This article was devoted to the so-called Zeno product formula, which can be regarded as a kind of *degenerated* Lie-Trotter product formula. In this formula one replaces the unitary factor e^{-itA} by an orthogonal projection onto some closed subspace $\mathfrak{H}_P \subseteq \mathfrak{H}$ and defines C as the self-adjoint operator associated with the quadratic form $\mathfrak{t}(h, g) := \left(\sqrt{B}h, \sqrt{B}g\right)$, $h, g \in \mathrm{dom}\,(\mathfrak{g}) := \mathrm{dom}\,(\sqrt{B}) \cap \mathfrak{H}_P$, where it is assumed that $\mathrm{dom}\,(\mathfrak{g})$ is dense in $\mathfrak{H}_P$. It was proved in [EI05] that the limit:

$$\lim_{n\to\infty} \int_0^T dt \, \left\| \left(P\mathrm{e}^{-itB/n}P\right)^n h - \mathrm{e}^{-itC}h \right\| = 0\,,$$

holds for any $h \in \mathfrak{H}_P$ and $T > 0$, where P is the orthogonal projection from $\mathfrak{H}$ onto $\mathfrak{H}_P$. Subsequently, an attempt was made in [EINZ07] to replace the strong L^2-topology of [EI05] by the usual strong topology of $\mathfrak{H}$. For admissible functions ϕ satisfying $\mathrm{Im}(\phi(x)) \leq 0$, $x \in \mathbb{R}_0^+$, it was shown in [EINZ07] that

$$\mathrm{e}^{-itC} = \text{s-}\lim_{n\to\infty} \left(P\phi(tB/n)P\right)^n \,,$$

holds uniformly in $t \in [0, T]$ for any $T > 0$.

On the other hand, we would like to stress that the function $\phi(x) = \mathrm{e}^{-ix}$, $x \in \mathbb{R}_0^+$, is admissible, but it does *not* satisfy the condition: $\mathrm{Im}(\mathrm{e}^{-ix}) \leq 0$ for $x \in \mathbb{R}_0^+$. Thus in this case the question about convergence of the Zeno product formula in the strong topology of $\mathfrak{H}$ is solved in [EI05].

Section 14.1.2

Remark In their study of Trotter-type formulae involving projections *Matolcsi* and *Shvidkoy* [MSh03], have presented two examples in which expressions of the type $[\exp(-iH/n)P]^n$ do not converge in the *strong* operator topology on $\mathcal{L}(\mathfrak{H})$. However, this result does *not* answer the question, whether the product formulae considered in Theorem 14.1.6 converge in the strong operator topology or not, because our assumptions are *not satisfied* there. In the first example of [MSh03], the analogue of the operator H_P is *not* densely defined; in the second one, H is *not* semibounded.

Section 14.1.3

The proof of Lemma 14.1.12 in [EI05] is based on the *Vitali* theorem for holomorphic vector-valued functions and arguments analogous to those used in the *Kato* paper [Kato78] about a strong convergence of the self-adjoint *Trotter* product formula where the semigroup generator is a *form-sum* of pair of non-negative self-adjoint operators. We invite the readers to consult [EI05, Section 4] for detailed presentation of the proof and corresponding line of reasoning.

Example ([EI05, Section 6]) This example serves for illustration of the main result of Sect. 14.1.3.

To this aim, we consider an open domain $\Omega \subset \mathbb{R}^d$ with a smooth boundary, and denote by P the orthogonal projection on Hilbert space $L^2(\mathbb{R}^d)$ defined as the multiplication operator by the *indicator* function χ_Ω of the set Ω. Consider further the free quantum Hamiltonian $H := -\Delta$, that is, the Laplacian in $\mathbb{R}^d$ which is a non-negative self-adjoint operator in $L^2(\mathbb{R}^d)$, and the *Dirichlet* Laplacian $-\Delta_\Omega$ in $L^2(\Omega)$ defined in the standard way [Kato80, Chapter VI, §2.3] as the *Friedrichs* extension of the appropriate quadratic form.

We consider a Zeno dynamics in the subspace $L^2(\Omega)$ corresponding to a permanent reduction of the wavefunction to the region Ω, which may be identified with the volume of *detector*. In the sense of the *Fréchet* topology ($\mathfrak{f}$ − lim), which in the framework of *theoretical physics* seems to be a quite plausible (see Remark 14.1.9), we claim that a generator of the Zeno dynamics in $L^2(\Omega)$ is just an appropriate *Dirichlet* Laplacian:

$$\mathfrak{f} - \lim_{n\to\infty} (Pe^{-it(-\Delta/n)}P)^n = P\,\mathrm{e}^{-it(-\Delta_\Omega)}\,,$$

cf. Definition 14.1.5. Otherwise, it can be stated as follows.

Proposition 14.4.1 *The self-adjoint operator*

$$-\Delta_P := ((-\Delta)^{1/2}P)^*((-\Delta)^{1/2}P)\,,$$

is densely defined in $L^2(\mathbb{R}^d)$*, and its restriction to the subspace* $L^2(\Omega)$ *is nothing but the Dirichlet Laplacian* $-\Delta_\Omega$ *of the region* Ω*, with the domain* $\mathrm{dom}\,(-\Delta_\Omega) = W_0^1(\Omega) \cap W^2(\Omega)$.

Proof Let $u \in \mathrm{dom}\,(-\Delta_P)$, so that u and $-\Delta_P u$ belong to $L^2(\mathbb{R}^d)$. We have

$$(-\Delta_P u, \varphi) = (u, -\Delta\varphi) = (-\Delta u, \varphi)\,,$$

for any $\varphi \in C_0^\infty(\Omega)$ because φ has a compact support in Ω. Thus $-\Delta_P u = -\Delta u$ holds in Ω in the sense of distributions, which means that $\Delta u|_\Omega \in L^2(\Omega)$. On the other hand, since $(-\Delta)^{1/2} P u \in L^2(\mathbb{R}^d)$, we have $\chi_\Omega u \in W^1(\mathbb{R}^d)$. Since we have

$$\nabla(\chi_\Omega u) = \nabla((\chi_\Omega)^2 u) = (\nabla \chi_\Omega)\chi_\Omega u(x) + \chi_\Omega \nabla(\chi_\Omega u),$$

in order to belong to $L^2(\mathbb{R}^d)$, the function $\nabla(\chi_\Omega u)$ must not contain the δ-type singular term, which requires $u(\cdot) = 0$ on the boundary of Ω. This combined with the fact that $u|_\Omega$, $\Delta u|_\Omega \in L^2(\Omega)$—see, e.g. [LM, Theorem 5.4]—implies that $u|_\Omega$ belongs to $W^2(\Omega)$ and $W_0^1(\Omega)$.

Thus we have shown that $u|_\Omega \in \operatorname{dom}(-\Delta_\Omega)$ and $(-\Delta_P u)|_\Omega = -\Delta_\Omega(u|_\Omega)$, or $-\Delta_\Omega \supset -\Delta_P|_{L^2(\Omega)}$, but both operators are self-adjoint, so they coincide. □

In this sense therefore our result given in Theorem 14.1.6 provides a rigorous version of the formal result by Facchi et al. [FPSS01].

Section 14.1.4

We follow here the proof in [EI05, Section 5]. We note that in contrast to (14.1.20) the question about whether within Theorem 14.1.15 the Zeno product formula may converge to a *unitary* Zeno dynamics in the *operator-norm* topology *without* additional condition (14.1.14) on domain of projection rests open.

We conclude the Notes to Sect. 14.1 by mentioning a recent progress in understanding of the quantum Zeno effect in *open* quantum systems; see [BDS21], which is, alas, out of the scope of this book. This paper suggests a Zeno-type product formula for open systems whose evolution is governed by the quantum *dynamical semigroups* and proves the existence of a quantum Zeno limit for open systems in the case of unbounded generators.

Notes to Sect. 14.2

Section 14.2.1

Ideas of replacing the unitary group $\{e^{-itH}\}$ by a resolvent, cf. (14.2.6), or to employ a spectral cut-off together with e^{-itC}, cf. (14.2.7), are not new: they were used to derive a modification of the unitary Lie-Trotter formula in [Ich80] and [Lap85], respectively, both for the form sum of two non-negative self-adjoint operators. See also [Ca05].

More comments and details concerning Remark 14.2.1, one can find in a still contemporary and very complete book [Ex85, Sect. 2.4].

Finally, let us note that formula (14.2.7) admits a *physical interpretation* in the context of the Zeno effect. To this end we note that the combination of the energy filtering and non-decay measurement following immediately one after another, see (14.2.7), can be regarded as *a single generalised measurement.* In fact, a product of two, in general non-commuting,[1] projections represents the simplest non-trivial example of *generalised observables*[2] introduced by *E.B.Davies*. They are implemented as positive maps on the respective spaces of *density matrices* [Dav76, Sec. 2.1]. Thus formula (14.2.7) corresponds to a modified Zeno situation with such generalised measurements, which depend on n and tend to a standard non-decay *yes-no* experiment when $n \to \infty$.

Section 14.2.2

Here we follow [EINZ07] and [Ich17].

Section 14.2.3

Remark 14.4.2 In connection with Remark 14.2.12, let us mention that in the finite-dimensional case, there is one more way to prove the claim suggested by G.M. Graf and A. Guekos for the special case $\phi(x) = \mathrm{e}^{-ix}$. The argument is based on the observation that

$$\lim_{t\to 0} t^{-1} \left\| P\mathrm{e}^{-itC} P - P\mathrm{e}^{-itK} P \right\|_{\mathcal{L}(\mathfrak{H}_P)} = 0\,, \tag{14.4.5}$$

implies $\|(P\mathrm{e}^{-itC/n}P)^n - \mathrm{e}^{-itK}\|_{\mathcal{L}(\mathfrak{H}_P)} = n\,o(t/n)$ as $n \to \infty$ by means of a natural telescopic estimate. To establish (14.4.5) one first proves that

$$t^{-1}\Big[(f, P\mathrm{e}^{-itC}Pg) - (f,g) - i\,t(\sqrt{H}Pf, \sqrt{H}Pg)\Big] \to 0\,,$$

as $t \to 0$ for all f, g from $\mathrm{dom}\,(\sqrt{H}P)$ which coincides in this case with $\mathfrak{H}_P$ by assumption. The last expression is equal to

$$\left(\sqrt{H}Pf, \left[\frac{\mathrm{e}^{-itC} - \mathbb{1}}{tC} - i\,\mathbb{1}\right]\sqrt{H}Pg\right),$$

and the square bracket tends to zero strongly by the (self-adjoint) functional calculus, which yields the sought conclusion. We note that the operator in the square brackets is well-defined by the functional calculus even if H is not invertible. In the same way, we find that

$$t^{-1}\Big[(f, P\mathrm{e}^{-itK}Pg) - (f,g) - i\,t(\sqrt{K}f, \sqrt{K}g)\Big] \to 0\,, \quad f, g \in \mathfrak{H}_P\,,$$

[1] We are primarily interested, of course, in the nontrivial case when the P does not commute with H, and thus also with the spectral projections $E_H([0, \pi n/t))$.

[2] Since involved spectral projections commute with the corresponding *evolution operator*, one can also replace the product $PE_H([0, \pi t/n))$ in our formulæ by $E_H([0, \pi t/n))PE_H([0, \pi t/n))$. Such generalised observables represented by symmetrised projection products have been studied as corresponding to the *almost sharp quantum effects* [ArGu04].

holds as $t \to 0$. Next we note that $(\sqrt{K} f, \sqrt{K} g) = (\sqrt{H} P f, \sqrt{H} P g)$, and consequently, the expression contained in (14.4.5) tends to zero weakly as $t \to 0$. However, in a finite dimensional $\mathfrak{H}_P$, the weak and operator-norm topologies are equivalent.

Conjecture 14.4.3 *Comparing the results of the present section with Sect.* 14.1, *we conjecture that if we drop the assumption* (14.2.4), *then at least the convergence*

$$\lim_{n\to\infty} \int_0^T \left\| (\phi(t\,C/n))^n f - \mathrm{e}^{-itK} f \right\|^2 dt = 0\,, \tag{14.4.6}$$

holds for for each $f \in \mathfrak{H}_P$, $T > 0$ *and arbitrary admissible functions* ϕ. *The proofs of* [EI05] *rely heavily on the analytic properties of the exponential function* $\phi(x) = \mathrm{e}^{-ix}$. *For admissible functions analytic properties are not required which yields a necessity to look for a different idea of the proof.*

Notes to Sect. 14.3

Here we tightly follow the paper [EI21], which exposes the *Exner-Ichinose* approach and results concerning the *strong* convergence of the exponential *Zeno* product formula; cf Sect. 14.2.2.

Section 14.3.1

For the *Arzelà-Ascoli* theorem, see, e.g. [KF89, Chapter II, §7.4], or a detailed and comprehensive presentation in the volumes [Si15, Theorem 2.3.14] and [Si15b, Theorem 1.5.3].

Sections 14.3.2–14.3.4

Here we present, respectively, Sections 2–4 of [EI21].
About σ-strong and σ-weak topologies on $\mathcal{L}(\mathfrak{H})$ (in Proposition 14.3.10), see, e.g. [BEH94, Section 6.3].

Addendum: Remarks on Zeno and Anti-Zeno Effects

We conclude this Notes to Chap. 14 by a short (*historical*) survey about *Zeno* and *anti-Zeno* effects, as well as a related remark on behaviour of unstable systems with *internal degrees* of freedom. They are motivated by introductions to papers [EI05], [EINZ07], [EI21], and by the book [Ex85].

(1) *Alan Turing* was the first who noticed in [Ho04] that if we *ascertain repeatedly* whether a quantum system is in a given state, then in the limit of *infinite* measurement frequency, it becomes *impossible* for the system to leave this state. The idea was rediscovered in the context of unstable system decays—see, e.g. [BN67, Fr72]—but it attracted a wide attention only after Misra and Sudarshan [MS77] invented a catchy name calling such a behaviour: *quantum Zeno effect*, in an allusion to the classical *Zeno aporia* about a flying arrow. More about the early history can be found in [Ex85].

The second breaking moment came in 1990 when Itano et al. [IHBW90] demonstrated the existence of this effect experimentally. Since then it became object of an extensive examination, both from the theoretical and experimental points of view, and it led even to various practical applications; a partial summary can be found in the review paper [FP08].

It is natural to expect that such a "Zeno dynamics" will be governed by the part of the original Hamiltonian H acting in the subspace $P\mathfrak{H}$ of the Hilbert space $\mathfrak{H}$, and it was shown in [Ex85, Sect. 2.4] that the generator is indeed associated with appropriate quadratic form constructed from the operator H and the projection P. It was *not* easy, however, to establish the existence of the Zeno dynamics beyond the situations when the dimension of P is finite or the operator H is bounded; needless to say that this is often not the case with actual physical systems. A prime example is a permanent ascertaining whether a free quantum particle dwells within a prescribed region Ω of the configuration space discussed in [FPSS01]; see also [FP08], with the conclusion that the Zeno generator is (the multiple of) the corresponding Dirichlet Laplacian. The argument made use of the stationary phase method, but the existence of the limit was not actually established by the authors.

Motivated by the said papers, P. Exner and T. Ichinose addressed in [EI05] the question of existence of the Zeno dynamics. There they have managed to establish the existence of the limits of the expressions appearing on the left-hand sides of (14.3.1)–(14.3.3) in topology of (a larger than Hilbert space $\mathfrak{H}$, namely) the *Fréchet* space $L^2_{\mathrm{loc}}(\mathbb{R}, \mathfrak{H}) = L^2_{\mathrm{loc}}(\mathbb{R}) \otimes \mathfrak{H}$, provided that H is semibounded (from below) and the operator H_P is densely defined. The validity of the formulæ is preserved if the exponential in (14.3.1)–(14.3.3) is replaced by functions of a wider class, in particular by the resolvent $(\mathbb{1} + itH)^{-1}$ [EINZ07, Ich17]. We argued in [EI05] that such a result can be regarded as sufficient from the viewpoint of (theoretical) physics because of the fact that every measurement, in particular that of time, is burdened with errors, and any actual experiment typically involves *averaging* over a large number of system copies.

It is desirable, though, to answer the question without such an underpinning by demonstrating the result with the convergence in a stronger sense, namely, that of the *strong* operator topology of $\mathfrak{H}$; cf. a result in [EI21]. In addition to the described physical motivation, the obtained relation is of independent mathematical interest belonging to the genre of the product formulæ of Trotter and Trotter-Kato [Kato78, Trot59]; see also [RS80].

Before we close this survey, let us mention a related to the *Zeno* effect (introduced in Sect. 14.1.1), the *anti-Zeno* effect [ChSM77], which in turn merits separate comments as an *addendum* to survey; see Remark 14.4.4. Recall that the ideal *anti-Zeno* effect means that a perpetual observation leads to an *immediate* disappearance of the unstable system. In *addendum* we follow the paper [Ex05]. There it is shown that the *gap* between *Zeno* and *anti-Zeno* effects is in fact very *narrow*.

(2) The next remark is the *addendum* about the nature of the anti-Zeno effect.

Remark 14.4.4 (The Anti-Zeno Effect) For illustration we shall consider example of *pure* states only. If quantum system is prepared at the initial instant $t = 0$ in a pure state $\psi \in P\,\mathfrak{H}$, its *decay law* unperturbed by measurements, or *non-decay* probability at a later time $t > 0$, is

$$p(t) = \|P\, e^{-iHt}\psi\|^2 . \tag{14.4.7}$$

In particular, for the case $\dim P = 1$, one gets $p_\psi(t) := |(e^{-iHt}\psi, \psi)|^2$; see (14.4.1). If now we perform *non-decay* measurements at times: $t/n,\ 2t/n\ \ldots,\ t$, all with the *positive* outcome, the resulting non-decay probability is

$$p_n(t) = p_\psi(t/n)p_{\psi_1}(t/n)\cdots p_{\psi_{n-1}}(t/n)\,, \tag{14.4.8}$$

where ψ_{j+1} is the normalized projection of $e^{-iHt/n}\psi_j$, $j = 0, 1, \ldots n-2$, on $P\,\mathfrak{H}$ and $\psi_0 := \psi$, in particular

$$p_n(t) = (p_\psi(t/n))^n\,, \tag{14.4.9}$$

if $\dim P = 1$. Combining (14.4.9) with the fact that $\lim_{n\to\infty}(f(t/n))^n = \exp(\dot f(+0)\, t)$, whenever $f(0) = 1$ and the one-sided derivative $\dot f(+0)$ exists, we see that the *Zeno effect*, that is, $p(t) := \lim_{n\to\infty} p_n(t) = 1$ for all $t > 0$, and its *anti-Zeno* counterpart, that is. $p(t) = 0$ for any $t > 0$, require that derivative $\dot p_\psi\,(+0)$ *must* be, respectively, *zero* (cf. (14.4.2)) and *infinitely negative*.

Note that a *simply negative* leads to the *exponential* decay. The same is true if $\dim P > 1$ provided the derivative $\dot p_\psi\,(+0)$ has such a property for *any* $\psi \in P\,\mathfrak{H}$.

It is thus crucial to estimate the quantity $1 - p_\psi(t)$, or more explicitly $(P\,\psi, \psi) - (e^{iHt}Pe^{-iHt}\,\psi, \psi)$, to find its behaviour for small values of $t > 0$. First, we can rewrite it in the form

$$1 - p_\psi(t) = 2\,\mathrm{Re}\,(P(\mathbb{1} - e^{-iHt})\,\psi,\,\psi) - \|P(\mathbb{1} - e^{-iHt})\,\psi\|^2 . \tag{14.4.10}$$

By the Cauchy-Bouniakovski-Schwarz inequality the right-hand side of (14.4.10) is non-negative. Our aim is to find tighter *upper* and *lower* bounds for it.

To this aim we use the spectral representation of e^{-iHt} in terms of the spectral measure E_H of the Hamiltonian H (generated by a non-decreasing projection-valued function $\lambda \mapsto E^H_\lambda := E_H((-\infty, \lambda]))$ and define $d\omega_\psi(\lambda) := d(E^H_\lambda\psi, \psi)$. Then for the case $\dim P = 1$ using (due to $\|\psi\| = 1$) the spectral-measure normalisation, $\int_{-\infty}^{\infty} d\omega_\psi(\lambda) = 1$, we can rewrite the difference (14.4.10) as

$$1 - p_\psi(t) = 2\int_{-\infty}^{\infty}\int_{-\infty}^{\infty} d\omega_\psi(\lambda)\, d\omega_\psi(\mu)\, \sin^2\frac{(\lambda - \mu)\,t}{2}\,. \tag{14.4.11}$$

Take $\alpha \in (0, 2]$, then we infer from (14.4.11) the upper bound for the ration:

$$\frac{1 - p_\psi(t)}{t^\alpha} \le 2 \int_{-\infty}^{\infty} \int_{-\infty}^{\infty} d\omega_\psi(\lambda)\, d\omega_\psi(\mu)\, (|\lambda|^\alpha + |\mu|^\alpha)$$
$$\le 4 \int_{-\infty}^{\infty} d(E_\lambda^H \psi, \psi)\, |\lambda|^\alpha .$$

This means that $1 - p_\psi(t) = \mathcal{O}(t^\alpha)$ if $\psi \in \mathrm{dom}\,(|H|^{\alpha/2})$. For that reason, if this is true for some $\alpha > 1$, then we obtain $\dot{p}_\psi\ (+0) = 0$, and consequently the *Zeno effect*.

Note that this requirement is slightly stronger than the other known *sufficient* conditions. Recall that derivative $\dot{p}_\psi\ (+0) = 0$ holds whenever the *expectation*

$$\langle |H| \rangle_\psi := \int_{-\infty}^{\infty} d(E_\lambda^H \psi, \psi)\, |\lambda| , \qquad (14.4.12)$$

is *finite* as has been known for a longtime, see [Ex85, Theorem 1.3.1], as well as references there.

On the other hand, by negation we infer that $\psi \notin \mathrm{dom}\,(|H|^{1/2})$ is a *necessary condition* for the (one-dimensional) *anti-Zeno effect*. Some may believe that by this the problem is closed—it is a common mistake that only states from the domain of the Hamiltonian have physical sense. In reality one can never test experimentally that a given state does *not* belong to $\mathrm{dom}\,(|H|^{1/2})$, see [Ex85, Section I.6].

To find a *sufficient* condition for the *anti-Zeno effect* (case $\dim P = 1$) let us observe that for $\lambda, \mu \in [-1/t, 1/t]$ there is a positive C independent of $t > 0$ such that by (14.4.11) and (14.4.12) we obtain the *lower* bound:

$$\frac{1 - p_\psi(t)}{t} \ge 4\, C^2\, t\, \left\{ \langle H_N^2 \rangle_\psi \, \langle I_N \rangle_\psi - \langle H_N \rangle_\psi^2 \right\} , \quad N := 1/t . \qquad (14.4.13)$$

Here we introduced operators $H_N^\gamma := H^\gamma\, E_H(\Delta_N)$ for $\gamma > 0$ with the spectral cut-off to the interval $\Delta_N := (-N, N)$; in particular, we denote $I_N := E_H(-N, N)$.

This lower bound yields that the *anti-Zeno effect* occurs if the right-hand side of (14.4.13) *diverges* as $t \to +0$ and $\dot{p}_\psi\ (+0) = -\infty$, which is true if

$$\langle H_N^2 \rangle_\psi \langle I_N \rangle_\psi - \langle H_N \rangle_\psi^2 \ge c\, N^\alpha , \qquad (14.4.14)$$

holds for any $N = 1/t > 0$ and for some $c > 0,\ \alpha > 1$.

Note that in most decay models, ψ belongs to the *absolutely continuous* spectral subspace of H in which case one has representation

$$d\omega_\psi(\lambda) = d(E_\lambda^H \psi, \psi) = \omega_\psi(\lambda)\, d\lambda , \qquad (14.4.15)$$

for some $\omega_\psi \in L^1$. As an *example*, consider a Hamiltonian H bounded from below and ψ from its absolutely continuous spectral subspace such that the corresponding distribution function (14.4.15) behaves as $\omega_\psi(\lambda) \approx c\lambda^{-\beta}$, while $\lambda \to +\infty$, for some $r > 0$ and $\beta \in (1, 2)$. Then by definition of expectations in (14.4.14) and distribution of function (14.4.15), the asymptotic behaviour of the left-hand side of (14.4.14) as $N \to +\infty$ contains two terms:

$$c_1\, N^{3-\beta} - c_2\, N^{4-2\beta}\,, \quad c_{1,2} > 0\,,$$

where for $\beta \in (1, 2)$ the first term is dominating and $(\alpha =)\ 3 - \beta > 1$. By (14.4.13) this gives for derivative: $\dot{p}_\psi\ (+0) = -\infty$, so the *anti-Zeno effect* occurs.

The above argument shows that in the *same* situation the value $\beta > 2$ leads to the Zeno effect. This demonstrates that the exponential decay law (14.4.7) indeed walks a thin rope between the *Scylla* of æternal *preservation* and *Charybda* of immediate *destruction.*

(3) Next we note how the situation looks like when the unstable system is allowed to have *internal degrees* of freedom: $\dim P > 1$. One might expect that the sufficient condition is given by (14.4.14) again, but this guess is *wrong*; see [Ex05].

Remark 14.4.5 (Internal Degrees of Freedom) To comment the answer, we denote by $\{e_j\}_{j\geq 1}$ an orthonormal basis in $P\mathfrak{H}$ and introduce $d\omega_{jk}(\lambda) := d(E_\lambda^H\, e_j,\, e_k)$, which is naturally a real-valued measure symmetric with respect to interchange of the indices $j, k \in \mathbb{N}$. Then with help of measure

$$d\|E_\lambda^H\, \psi\|^2 = \sum_{jk} c_j\, \bar{c}_k\, d\omega_{jk}(\lambda)\,, \quad \psi = \sum_j c_j\, e_j\,, \quad \|\psi\| = 1\,,$$

we can cast the quantity of interest: $(1 - p(t))$, into the form (14.4.11). As a consequence, one deduces a *generalisation* of (14.4.14) to the sufficient condition for the anti-Zeno effect ($\dim P > 1$) by mimicking the arguments for $\dim P = 1$. We refer to [Ex05] for details.

To elucidate a drastic difference between the case $\dim P = 1$ and $\dim P > 1$, we observe that in the case $\dim P > 1$ a system subject to a perpetual observation can exhibit a more complicated behaviour. A simple example is a *combination* of Zeno and anti-Zeno effect for $\dim P = 2$.

To this aim take *two* copies of Hilbert spaces, one-dimensional projectors on unit vectors and Hamiltonians in these spaces, that is, $\{\mathfrak{H}_j,\, P_{e_j},\, H_j\}$, $j = 1, 2$, with $\dim P_{e_j} = 1$. Let derivatives: $\dot{P}_{e_1}\ (+0) = 0$ and $\dot{P}_{e_1}\ (+0) = -\infty$. Consider the combined system, which is described by the triple

$$\{\mathfrak{H}_1 \oplus \mathfrak{H}_2\,,\ P_{e_1} \oplus P_{e_2}\,,\ H_1 \oplus H_2\}\,.$$

If the initial state ψ of such an unstable system is represented by a non-trivial linear combination $\psi = c_1\psi_1 + c_2\psi_2$ and we monitor it continuously, the second component of ψ disappears *immediately*, while the first one will survive *forever*. Of course, one can imagine various more complicated combinations, especially in the case when dim P is *infinite*. For more comments we refer to [Ex05].

Appendix A
Linear Operators in Normed Spaces

A.1 Continuous Operators

A.1.1 The Principle of Extension by Continuity

Recall that a linear operator A with domain $\mathrm{dom}\,(A) \subseteq \mathfrak{X}$ and range $\mathrm{ran}\,(A) \subseteq \mathfrak{X}$ in a Banach space $\mathfrak{X}$ is *continuous* at $u \in \mathrm{dom}\,(A)$ if for any sequence $\{u_n\}_{n\geq 1} \subset \mathrm{dom}\,(A)$ such that $\|u_n - u\|_{\mathfrak{X}} \to 0$ one gets $\|A\,u_n - A\,u\|_{\mathfrak{X}} \to 0$.

Note that A is continuous on $\mathrm{dom}\,(A)$ if and only if A is *bounded* on $\mathrm{dom}\,(A)$, i.e. $\|A\,u\| \leq M\|u\|$, $u \in \mathrm{dom}\,(A)$. The smallest number M is called the *bound* of A and is denoted by $\|A\|$.

Proposition A.1.1 (Principle of Extension by Continuity) *A bounded linear operator A defined on domain* $\mathrm{dom}\,(A) \subset \mathfrak{X}$*, which is dense in a complete normed linear space $\mathfrak{X}$, can be extended to a unique linear operator $\overline{A}$ defined on the entire space $\mathfrak{X}$ without increase of the bound $\|A\|$, which then coincides with*

$$\|\overline{A}\| := \sup_{x\in\mathfrak{X}} \frac{\|\overline{A}\,x\|_{\mathfrak{X}}}{\|x\|_{\mathfrak{X}}}\,, \tag{A.1.1}$$

called the norm of the extension $\overline{A}$.

Proof Given that $\mathrm{dom}\,(A)$ is dense in $\mathfrak{X}$, for any $x \in \mathfrak{X}\setminus\mathrm{dom}\,(A)$ there is a sequence $\{u_n\}_{n\geq 1} \subset \mathrm{dom}\,(A)$ such that $\|u_n - x\|_{\mathfrak{X}} \to 0$. Since A is bounded on $\mathrm{dom}\,(A)$ and $\{u_n\}_{n\geq 1}$ is the Cauchy sequence,

$$\|A\,u_n - A\,u_{n+m}\|_{\mathfrak{X}} \leq \|A\|\,\|u_n - u_{n+m}\|_{\mathfrak{X}} \to 0\,,\ \text{for},\quad n \to \infty\,,\quad m \in \mathbb{N}\,.$$

V. A. Zagrebnov et al., *Trotter-Kato Product Formulæ*, Operator Theory: Advances and Applications 296, https://doi.org/10.1007/978-3-031-56720-9

Thus $\{A\,u_n\}_{n\geq 1}$ is in turn a Cauchy sequence, which by completeness of the space $\mathfrak{X}$ converges in the norm $\|\cdot\|_{\mathfrak{X}}$ to some limit denoted as

$$\overline{A}\,x := \lim_{n\to\infty} A\,u_n\,. \tag{A.1.2}$$

If there exists another sequence $\{w_n\}_{n\geq 1} \subset \mathrm{dom}\,(A)$ such that $x = \lim_{n\to\infty} w_n$, then $\lim_{n\to\infty} \|u_n - w_n\|_{\mathfrak{X}} = 0$, wherefore $\lim_{n\to\infty} \|A\,u_n - A\,w_n\|_{\mathfrak{X}} = 0$. As a consequence $\overline{A}\,x := \lim_{n\to\infty} A\,w_n$. Note that for $x \in \mathrm{dom}\,(A)$ and $\{u_n\}_{n\geq 1} \subset \mathrm{dom}\,(A)$ such that $\|u_n - x\|_{\mathfrak{X}} \to 0$, one gets $\overline{A}\,x = A\,x$. On that account the operator $\overline{A}$ (A.1.2) is uniquely defined on the entire space $\mathfrak{X}$ as extension of A.

By virtue of (A.1.2) the operator $\overline{A}$ inherits linearity of A. Moreover, taking the limit (A.1.2) in inequality $\|A\,u_n\|_{\mathfrak{X}} \leq \|A\|\,\|u_n\|_{\mathfrak{X}}$, we obtain $\|A\,x\|_{\mathfrak{X}} \leq \|A\|\,\|x\|_{\mathfrak{X}}$, and thus $\|\overline{A}\| \leq \|A\|$ by (A.1.1). Owing to the fact that under extension of operator the norm cannot decrease, we infer that $\|\overline{A}\| = \|A\|$. □

The *principle of extension by continuity* is (evidently) valid for any linear bounded operator with the *range* in a *complete* normed linear space. This is the case of operators in a Banach space $\mathfrak{X}$, but also the case of linear functionals since the range of the mapping (the complex numbers) is a Banach space.

In addition, we remind an assertion, which is similar to Proposition A.1.1 and is known as the *bounded linear transformation* (B.L.T.) theorem; see [RS80, Chapter I.2, Theorem 1.7].

Proposition A.1.2 (B.L.T. Theorem) *Suppose T is a bounded linear transformation from a normed linear space $(V_1, \|\cdot\|_1)$ to a complete normed linear space $(V_2, \|\cdot\|_2)$. Then T can be uniquely extended to a bounded linear transformation (with the same bound) $\tilde{T}$, from the completion $\tilde{V_1}$, of the space V_1, to $(V_2, \|\cdot\|_2)$.*

Note that the proof of Proposition A.1.1 follows basically the same line of reasoning as the proof of Proposition A.1.2; cf. [RS80, Chapter I.2, Theorem 1.7.]

A.1.2 The Principle of Uniform Boundedness

This result, which is among the basic theorems in the Banach space theory, is also known as the *Banach-Steinhaus* theorem, or the *uniform boundedness principle*; see, e.g. [Dav07, Section 1.3]. For our purpose we formulate it as follows.

Proposition A.1.3 *Let $\{T_\lambda\}_{\lambda\in\Lambda} \subset \mathcal{L}(\mathfrak{X})$, $\Lambda \subset \mathbb{R}$, be a family of bounded operators on Banach space $\mathfrak{X}$. Then the following assertions are equivalent:*

(i) $\sup_{\lambda\in\Lambda} \|T_\lambda\| = M < \infty$.
(ii) $\sup_{\lambda\in\Lambda} \|T_\lambda\,x\| = M_x < \infty$ *for every* $x \in \mathfrak{X}$.
(iii) $\sup_{\lambda\in\Lambda} \langle T_\lambda\,x, \phi\rangle = M_{x,\phi} < \infty$ *for every* $x \in \mathfrak{X}$ *and* $\phi \in \mathfrak{X}^*$.

Proof Clearly (i)⇒(ii)⇒(iii) for corresponding M, M_x, and $M_{x,\phi}$.

Now suppose that (ii) holds but (i) does not. Then we construct sequences $\{x_k\}_{k\geq 1} \subset \mathfrak{X}$ and $\{\lambda_k\}_{k\geq 1} \subset \Lambda$ as follows:

(a) Given vectors $x_1, x_2, \ldots, x_{n-1}$ satisfying $\|x_k\| = 1/4^k$ for $k \in \{1, 2, \ldots, n-1\}$, let

$$c_{n-1} := \sup_{\lambda\in\Lambda} \|T_\lambda(x_1 + x_2 + \ldots + x_{n-1})\| , \quad n > 1 . \tag{A.1.3}$$

(b) Since (i) does not hold, there exists $\lambda_n \in \Lambda$ such that operator norm $\|T_{\lambda_n}\|$ is large:

$$\|T_{\lambda_n}\| \geq 4^{n+1} (n + c_{n-1}) . \tag{A.1.4}$$

By the same reason one can also find vector $x_n \in \mathfrak{X}$, with $\|x_n\| = 1/4^n$, and choose a lower bound that we tune as

$$\|T_{\lambda_n} x_n\| \geq \frac{2}{3} \|T_{\lambda_n}\| \|x_n\| . \tag{A.1.5}$$

Consider vector $x \in \mathfrak{X}$ defined by convergent series: $x = \sum_{n\geq 1} x_n$. Then

$$\|T_{\lambda_n} x\| \geq \|T_{\lambda_n} x_n\| - \|T_{\lambda_n} (x - x_n)\| . \tag{A.1.6}$$

Seeing that

$$\|T_{\lambda_n} (x - x_n)\| \leq \|T_{\lambda_n} (x_1 + x_2 + \ldots + x_{n-1})\| + \|T_{\lambda_n}\| \sum_{k=n+1}^{\infty} \|x_k\| ,$$

by virtue of definition (A.1.3) and inequalities (A.1.4)–(A.1.6), we obtain estimate

$$\|T_{\lambda_n} x\| \geq \frac{2}{3} \|T_{\lambda_n}\| 4^{-n} - c_{n-1} - \frac{1}{3} \|T_{\lambda_n}\| 4^{-n} \geq n , \quad n > 1 ,$$

which contradicts to (ii) and thus implies (i) for some finite $M > 0$.

The proof of (iii)⇒(ii) uses (ii)⇒(i) and follows a similar line of reasoning. □

The next assertion scrutinises a relation between bound M and the set pointwise bounds $\{M_x\}_{x\in\mathfrak{X}}$, which are declared in Proposition A.1.3(ii).

Proposition A.1.4 *Let $\{T_n\}_{n\in\mathbb{N}} \subset \mathcal{L}(\mathfrak{X})$ be bounded operators on a Banach space $\mathfrak{X}$, such that $\{T_n x\}_{n\in\mathbb{N}}$ is a Cauchy sequence at every point $x \in \mathfrak{X}$. Then operator norms: $\{\|T_n\|\}_{n\in\mathbb{N}}$, are uniformly bounded*

$$\sup_{n\in\mathbb{N}} \|T_n\| = M < \infty . \tag{A.1.7}$$

Proof For the reasoning *ad absurdum* we suppose that (A.1.7) is *not* true. Then first of all one can conclude that the set $\{\|T_n\, x\|\}_{n\in\mathbb{N}}$ is *not* bounded on *any* closed ball $\overline{B}(x_0, r) = \{x \in \mathfrak{X} : \|x - x_0\| \le r\}$, where $x_0 \in \mathfrak{X}$ and $r > 0$.

To bolster this conclusion let us assume the *contrary*, i.e. that

$$\|T_n\, x\| \le C(z_0, \rho) < \infty\,, \qquad n \in \mathbb{N}\,, \tag{A.1.8}$$

for any x in *some* ball $\overline{B}(z_0, \rho)$. Since any element

$$x_\xi := \frac{\rho}{\|\xi\|}\, \xi + z_0 \in \overline{S}(z_0, \rho)\,, \qquad \xi \in \mathfrak{X}\,,$$

on account of (A.1.8) we obtain

$$\|T_n\, x_\xi\| \le C(z_0, \rho) < \infty\,, \qquad n \in \mathbb{N}\,,$$

and for that reason this yields inequalities:

$$\frac{\rho}{\|\xi\|}\, \|T_n\, \xi\| - \|T_n\, z_0\| \le \|\rho\, \|\xi\|^{-1}\, T_n\, \xi + T_n\, z_0\| \le C(z_0, \rho)\,.$$

As a consequence, we obtain estimate

$$\|T_n\, \xi\| \le \frac{C(z_0, \rho) + \|T_n\, z_0\|}{\rho}\, \|\xi\|\,, \qquad n \in \mathbb{N}\,. \tag{A.1.9}$$

Seeing that $\{T_n\, z_0\}_{n\in\mathbb{N}}$ for a given $z_0 \in \mathfrak{X}$ is the Cauchy sequence in $\mathfrak{X}$, the limit: $\lim_{n\to\infty} T_n\, z_0 = z \in \mathfrak{X}$. On that account $\sup_{n\in\mathbb{N}} \|T_n\, z_0\| =: M_{z_0} < \infty$. Then (A.1.9) yields for any $\xi \in \mathfrak{X}$:

$$\|T_n\, \xi\| \le M(z_0, \rho)\, \|\xi\|\,, \quad M(z_0, \rho) := \frac{C(z_0, \rho) + M_{z_0}}{\rho}\,, \qquad n \in \mathbb{N}\,. \tag{A.1.10}$$

The estimate (A.1.10) proves the conclusion that negation of (A.1.7) implies that the set $\{\|T_n\, x\|\}_{n\in\mathbb{N}}$ is *not* bounded on *any* closed ball in $\mathfrak{X}$.

Now let $\overline{B}_0(x_0, r_0) \subset \mathfrak{X}$ be a closed ball. Since the sequence $\{\|T_n\, x\|\}_{n\in\mathbb{N}}$ is not bounded on it, there are $n_1 \in \mathbb{N}$ and $x_1 \in \overline{B}_0(x_0, r_0)$ such that $\|T_{n_1}\, x_1\| > 1$. Note that by continuity of operator T_{n_1} this estimate from below also holds in some closed ball $\overline{B}_1(x_1, r_1) \subset \overline{B}_0(x_0, r_0)$. Given that again the sequence $\{\|T_n\, x\|\}_{n\in\mathbb{N}}$ is not bounded on $\overline{B}_1$, there are $n_2(> n_1)$ and vector $x_2 \in \overline{B}_1(x_1, r_1)$ such that $\|T_{n_2}\, x_2\| > 2$, and so forth, for $r_k \to 0$ as $k \to \infty$, where $k \in \mathbb{N}$. As a consequence, in Banach space $\mathfrak{X}$ there exists vector

$$x^* \in \bigcap_{k\ge 0} \overline{B}_k(x_k, r_k)\,,$$

such that $\|T_{n_k}\, x^*\| \ge k$ for $n_k \in \mathbb{N}$.

The latter estimate (from below) implies that *negation* of (A.1.7) also provides that $\{T_n x^*\}_{n\in\mathbb{N}}$ is *not* the Cauchy sequence in $\mathfrak{X}$. This contradicts the condition of theorem and completes the proof of (A.1.7) *ad absurdum* for $M := M(z_0, \rho)$. □

Corollary A.1.5 *Since* $\{T_n x\}_{n\in\mathbb{N}}$ *for every* $x \in \mathfrak{X}$ *is the Cauchy sequence in* $\mathfrak{X}$, *there exists a limit* $y = \lim_{n\to\infty} T_n x$, *which defines the mapping:* $x \mapsto y := Tx$. *By construction of* T *it is a linear operator with* $\mathrm{dom}\,(T) = \mathfrak{X}$.

On account of the principle of uniform boundedness (A.1.10) *one gets*

$$\|T x\| \leq M \,\|x\|\,, \quad x \in \mathfrak{X}\,, \quad M = \sup_{n\in\mathbb{N}} \|T_n\| = M(z_0, \rho)\,, \tag{A.1.11}$$

that is, the operator T *is bounded.*

Note that by virtue of (A.1.8) *we have* $M_x = C(z_0, \rho)$*; cf. Proposition* A.1.3(ii).

In the framework of implementation of the *Banach-Steinhaus* theorem, we recall the following statements, which are often helpful (see, e.g. Proposition 1.9.30 and Theorem 1.9.34) for justifying and uplifting the *strong* operator convergence on a Banach space $\mathfrak{X}$ from convergence on dense or compact subsets.

Proposition A.1.6 *Let* $\mathcal{L}(\mathfrak{X})$ *be set of bounded operators. Then on any bounded subset of* $\mathcal{L}(\mathfrak{X})$ *the following topologies coincide*:

(a) *The strong operator topology on* $\mathcal{L}(\mathfrak{X})$
(b) *The topology of pointwise convergence on any dense subset* $\mathcal{D}$ *of* $\mathfrak{X}$

See, for example, [Kato80, Lemma 3.5, Chapter III], [Sch80, Section III.4.5] or [EN00, A.3 Proposition].

Note that one can demonstrate a stronger claim if instead of hypothesis (b) in Proposition A.1.6 the subset $\mathcal{D}$ of a Banach space $\mathfrak{X}$ is supposed to be *compact.*

Proposition A.1.7 *Let sequence of bounded operators* $\{L_n\}_{n\geq 1} \subset \mathcal{L}(\mathfrak{X})$ *converge strongly to L, that is,* $\lim_{n\to\infty} L_n\, x = L\, x$ *for* $x \in \mathfrak{X}$. *If* $\mathfrak{X}_c \subset \mathfrak{X}$ *is a compact subset of* $\mathfrak{X}$, *then* $\{L_n\}_{n\geq 1}$ *converges on* $\mathfrak{X}_c$ *uniformly, that is,*

$$\lim_{n\to\infty} \sup_{x\in\mathfrak{X}_c} \|\, L_n\, x - L\, x\,\| = 0\,. \tag{A.1.12}$$

See, for example, [Kato80, Lemma 3.7, Chapter III].

We use Proposition A.1.6 for the proof of the *Trotter-Kato-Neveu* theorem; see Proposition 1.9.30. Another application of Proposition A.1.6 (see proof of the strong convergence of the Euler formula in Proposition 1.1.1) is combined with a preliminary abstract result, which is Proposition A.1.8; see, e.g. [Kato80, Chapter III, Problem 2.9] or [Bour60, Theorem 2.3.1].

Proposition A.1.8 *Let bounded operator* $T \in \mathcal{L}(\mathfrak{X})$ *be such that its range* $\mathrm{ran}\,(T)$ *is dense in* $\mathfrak{X}$. *Then* $T : \mathcal{D} \to T(\mathcal{D})$ *is mapping of a dense subset* $\mathcal{D} \subset \mathfrak{X}$ *onto* $T(\mathcal{D})$, *which is also dense in* $\mathfrak{X}$.

Corollary A.1.9

(a) *If A is a closed operator with dense domain* $\mathrm{dom}\,(A) \subset \mathfrak{X}$ *such that* $\|(A + \lambda\,\mathbb{1})^{-1}\| \le \lambda^{-1}$ *for* $\lambda > 0$, *then* $T_\lambda := (A + \lambda\,\mathbb{1})^{-1}$ *is the bounded operator with dense range* $\mathrm{ran}\,(T_\lambda) = \mathrm{dom}\,(A)$. *Therefore, owing to Proposition* A.1.8 *the subset* $T_\lambda\,(\mathrm{dom}\,(A)) = \mathrm{dom}\,(A^2) = \{x \in \mathrm{dom}\,(A) : Ax \in \mathrm{dom}\,(A)\}$ *is also dense in* $\mathfrak{X}$.

(b) *On account of Proposition* A.1.6 *we infer that a sequence* $\{L_n\}_{n\in\mathbb{N}}$ *of uniformly bounded operators:* $\|L_n\| \le M,\ n \ge 1$, *which is pointwise convergent on* $\mathrm{dom}\,(A^2) \subset \mathfrak{X}$, *converges on the Banach space* $\mathfrak{X}$ *strongly.*

Appendix B
Defect Number, Spectrum and Resolvent of Closed Operators

B.1 Spectrum and Resolvents

B.1.1 Sets $\sigma(A)$ and $\widehat{\sigma}(A)$

1. Let A be a densely defined *closed* linear operator in the Hilbert space $\mathfrak{H}$ and denote $A_\zeta := A - \zeta\mathbb{1}$, with the *spectral* parameter $\zeta \in \mathbb{C}$. We denote by $\mathrm{def}(A_\zeta) := \dim((\mathrm{ran}\,(A_\zeta))^\perp)$ the *deficiency* (or the *defect* number) of the operator A_ζ.
2. The set of ζ such that A_ζ has *continuous* inverse *only* on the range $\mathrm{ran}\,(A_\zeta)$ is called the *quasi-regular set* $\widehat{\rho}(A)$ of the operator A. Note that the set $\widehat{\rho}(A)$ is open and the integer-valued function $\mathrm{def}(A_\zeta)$ is constant on each connected component of $\widehat{\rho}(A)$. We denote by $\rho(A) := \{\zeta \in \widehat{\rho}(A) : \mathrm{def}(A_\zeta) = 0\}$ the set of *regular points* of A. Since operators $A_\zeta^{-1} \in \mathcal{L}(\mathfrak{H})$ for $\zeta \in \rho(A)$, the set $\rho(A)$ is traditionally called the *resolvent set* of A. This set is obviously also open since it consists of components $\widehat{\rho}(A)$, where defect $\mathrm{def}(A_\zeta) = 0$.
3. The set $\sigma(A) := \mathbb{C} \setminus \rho(A)$ is known as the *spectrum* of the closed operator A, whereas the complement to the quasi-regular set: $\widehat{\sigma}(A) := \mathbb{C} \setminus \widehat{\rho}(A)$, is called the *core* of $\sigma(A)$. It is clear that $\widehat{\sigma}(A) \subseteq \sigma(A)$ and that the both of these sets are open.
4. Note that assumption of *closeness* makes the spectrum of the linear operator A *nontrivial*. Indeed, let A be *not* closed. Suppose that for some $\zeta \in \mathbb{C}$ we have that A_ζ^{-1} is everywhere defined on $\mathfrak{H}$ bounded operator, that is, $\zeta \in \rho(A)$. Then operator A_ζ^{-1} is closed, which implies that operator A_ζ is also closed and so is A. This contradiction yields: $\rho(A) = \emptyset$. As a consequence, the spectral problem for a non-closed A is *trivial* since in this case $\sigma(A) = \mathbb{C}$.
5. If operator A is closed *and* $\sigma(A) = \mathbb{C}$, then it must be unbounded, since if A is a bounded operator then the resolvent set $\rho(A)$ cannot be non-empty and on that

V. A. Zagrebnov et al., *Trotter-Kato Product Formulæ*, Operator Theory: Advances and Applications 296, https://doi.org/10.1007/978-3-031-56720-9

account $\sigma(A) = \mathbb{C} \setminus \rho(A)$. In particular, the closed unbounded operators provide examples of two extreme cases: $\sigma(A) = \mathbb{C}$ and $\sigma(A) = \emptyset$.

B.1.2 Subsets of the Spectrum $\sigma(A)$

6. By the *point spectrum* of A we call the subset

$$\sigma_{\mathrm{p}}(A) := \{\zeta \in \mathbb{C} : \ker(A_\zeta) \neq \{0\}\} .$$

This subset consists of eigenvalues of A and nontrivial $\ker(A_\zeta)$ are corresponding eigenspaces. In the point spectrum $\sigma_{\mathrm{p}}(A)$ we distinguish a subset of eigenvalues with *infinite multiplicity*:

$$\sigma_{\mathrm{p}}^{\infty}(A) := \{\zeta \in \mathbb{C} : \dim(\ker(A_\zeta)) = \infty\} .$$

7. By the *continuous spectrum* of A we call the subset

$$\sigma_{\mathrm{cont}}(A) := \{\zeta \in \mathbb{C} : \mathrm{ran}\,(A_\zeta) \neq \overline{\mathrm{ran}\,(A_\zeta)} = \mathfrak{H}\} .$$

First we note that the intersection $\sigma_{\mathrm{p}}(A) \cap \sigma_{\mathrm{cont}}(A)$ can be non-empty. Next, we remark that $\sigma_{\mathrm{p}}(A) \subset \widehat{\sigma}(A)$ and $\sigma_{\mathrm{cont}}(A) \subset \widehat{\sigma}(A)$. Finally, one note that $\sigma_{\mathrm{p}}(A) \cup \sigma_{\mathrm{cont}}(A) = \widehat{\sigma}(A)$

8. By the *residual spectrum* of A we call the subset

$$\sigma_{\mathrm{res}}(A) := \sigma(A) \setminus \widehat{\sigma}(A) = \sigma(A) \cap \widehat{\rho}(A) .$$

In other words, it is an *open* set, which consists of components of $\widehat{\rho}(A)$ with nontrivial defects: $\mathrm{def}(A_\zeta) \neq 0$. If $\zeta \in \sigma_{\mathrm{res}}(A)$, then $\mathrm{ran}\,(A_\zeta) \neq \mathfrak{H}$ and the operator $A_\zeta^{-1} \in \mathcal{L}(\mathrm{ran}\,(A_\zeta), \mathfrak{H})$.

9. In virtue of definition in 8. and by representation of the *core* of the spectrum due to 7. one infers that

$$\sigma(A) = \sigma_{\mathrm{p}}(A) \cup \sigma_{\mathrm{cont}}(A) \cup \sigma_{\mathrm{res}}(A). \tag{B.1.1}$$

10. By the *essential spectrum* of A we call the subset

$$\sigma_{\mathrm{ess}}(A) := \sigma_{\mathrm{p}}^{\infty}(A) \cup \sigma_{\mathrm{cont}}(A) = \{\zeta \in \mathbb{C} : \dim(\ker(A_\zeta)) = \infty \vee \mathrm{ran}\,(A_\zeta) \neq \overline{\mathrm{ran}\,(A_\zeta)}\} .$$

We note that the essential spectrum is *closed*: $\sigma_{\mathrm{ess}}(A) = \overline{\sigma_{\mathrm{ess}}(A)}$, and that

$$\sigma_{\mathrm{res}}(A) \cap \sigma_{\mathrm{ess}}(A) = \emptyset .$$

Furthermore, we recall that essential spectrum does not change under compact perturbations. That is, if $K \in \mathcal{C}_\infty(\mathcal{H})$ is a compact operator, the $\sigma_{\text{ess}}(A) = \sigma_{\text{ess}}(A+K)$. Moreover, if A and A' two closed operators such that $(A-\zeta)^{-1} - (A'-\zeta)^{-1}$ is compact for some $\zeta \in \rho(A) \cap \rho(A')$, then $\sigma_{\text{ess}}(A) = \sigma_{\text{ess}}(A')$.

11. By the *discrete spectrum* of A we call the subset

$$\sigma_{\text{d}}(A) := \sigma_{\text{p}}(A) \setminus \sigma_{\text{ess}}(A) . \tag{B.1.2}$$

By definition the essential and discrete spectra are disjoint, that is, $\sigma_{\text{ess}}(A) \cap \sigma_{\text{d}}(A) = \emptyset$. From (B.1.2) we get

$$\sigma_{\text{d}}(A) = \{\zeta \in \sigma_{\text{p}}(A) : \dim(\ker(A_\zeta)) < \infty \ \wedge \ \text{ran}(A_\zeta) = \overline{\text{ran}(A_\zeta)}\}. \tag{B.1.3}$$

It is also useful to introduce the complement of a discrete spectrum in the point spectrum $\sigma_{\text{p}}^*(A) := \sigma_{\text{p}}(A) \setminus \sigma_{\text{d}}(A)$.

12. We remind that a *compact* operator $A \in \mathcal{C}_\infty(\mathcal{H})$ is not invertible. As a consequence, the spectrum $\sigma(A)$ is not empty and $\{0\} \subseteq \sigma(A)$.

On the other hand, by the *Fredholm alternative* the *non-zero* part of the spectrum: $\sigma(A) \setminus \{0\} = \sigma_d(A)$, is a *discrete* spectrum with only accumulation point at zero. On that account the spectrum of a compact operator A has the canonical structure: $\sigma(A) = \sigma_{\text{d}}(A) \cup \{0\}$.

Classification of the *zero* spectral point: $\lambda = 0$, is a subtle problem. On account of (B.1.2) it may belong to *different* parts of essential spectrum; see Sect. B.2.

13. Let A be a closed operator with $\text{dom}(A) \subset \mathfrak{H}$. We say that a complex number λ belongs to the *approximate* spectrum $\sigma_{\text{app}}(A)$ of A if there exists a sequence $\{u_n\}_{n\geq 1} \subset \text{dom}(A)$ such that $\|u_n\| = 1$ and $\lim_{n\to\infty}(A - \lambda \mathbb{1})\, u_n = 0$. We note that $\sigma_{\text{p}}(A) \subseteq \sigma_{\text{app}}(A)$ and $\sigma_{\text{cont}}(A) \subseteq \sigma_{\text{app}}(A)$, together with

$$\sigma(A) = \sigma_{\text{app}}(A) \cup \sigma_{\text{res}}(A) , \tag{B.1.4}$$

although $\sigma_{\text{res}}(A)$ need not to belong to the approximate spectrum.

B.1.3 Resolvent and Pseudo-Resolvent

14. The operator-valued function $R_A(\zeta) = (A_\zeta)^{-1} : \rho(A) \to \mathcal{L}(\mathfrak{H})$ is called the *resolvent* of the closed operator A.

Note that $\zeta_0 \in \widehat{\rho}(A)$ if and only if there exists $\delta > 0$ such that $\|A_\zeta\, u\| \geq \delta_0 \|u\|$ for any $u \in \text{dom}(A)$. Then $\|R_A(\zeta_0)u\| \leq \delta_0^{-1}$ and for regular points $\zeta \in \rho(A)$ one obtains the *resolvent identity*

$$R_A(\zeta) - R_A(\zeta_0) = (\zeta - \zeta_0)\, R_A(\zeta)\, R_A(\zeta_0) . \tag{B.1.5}$$

We comment that (B.1.5) can be regarded as an equation for unknown operator-valued function $R_A(\zeta)$ and that (B.1.5) is soluble in the open disc $D_{\zeta_0}(\delta_0) := \{\zeta \in \mathbb{C} : |\zeta - \zeta_0| < \delta_0\}$. This solution is a uniformly operator norm convergent in the disc $D_{\zeta_0}(\delta_0)$ power series:

$$R_A(\zeta) = \sum_{n=0}^{\infty} (\zeta - \zeta_0)^n \, R_A(\zeta_0)^{n+1} \, . \tag{B.1.6}$$

Therefore, the resolvent is piecewise holomorphic function $R_A : \zeta \mapsto R_A(\zeta)$ on any connected component of the resolvent set $\rho(A)$.

15. It is often useful to consider a *pseudo-resolvent* $\hat{R}(\zeta)$. To this end we note that the strong operator limit of resolvents $\{R_k(\zeta)\}_{k\geq 1}$

$$\hat{R}(\zeta) = \text{s-}\lim_{k\to\infty} R_k(\zeta) \, , \quad \zeta \in D_s \, , \tag{B.1.7}$$

not need be the resolvent of an operator. Here D_s denotes in the regular set for $\{R_k(\zeta)\}_{k\geq 1}$ the region of the *strong convergence*. However, the limit (B.1.7) satisfies the resolvent equation (B.1.5) for $\zeta, \zeta_0 \in D_s$.

Recall that the *pseudo-resolvent* $\hat{R}(\zeta)$ is the resolvent (of some closed operator A) if and only if $\ker(\hat{R}(\zeta)) = \{0\}$ for any $\zeta \in D_s$.

Let D_s be non-empty. Then there is an alternative: either $\hat{R}(\zeta)$ is not invertible for any $\zeta \in D_s$, or $\hat{R}(\zeta) = R_A(\zeta)$ is the resolvent of a unique closed operator A and we have that $D_s = \rho(A) \cap D_b$.

Here D_b is the region of the *norm boundedness* for the sequence $\{R_k(\zeta)\}_{k\geq 1}$ and one gets $D_u \subset D_s \subset D_b$, where D_u is the region of the *norm convergence* for the sequence $\{R_k(\zeta)\}_{k\geq 1}$.

B.2 Example: The Volterra Operator

We present this example because the *Volterra equation/operator* appears in different sections of the book.

(*a*) The Volterra operator $V : \mathfrak{H} \to \mathfrak{H}$ on the Hilbert space $\mathfrak{H} = L^2([0,1])$ is the integral operator

$$(Vu)(t) := \int_0^t ds \, u(s) \, , \quad u \in \text{dom}(V) = L^2([0,1]) \, .$$

To estimate the norm $\|V\|$ of the Volterra operator we note that

$$\|Vu\|^2 = \int_0^1 dt \left| \int_0^t ds\, \sqrt{\cos(\pi s/2)} \frac{u(s)}{\sqrt{\cos(\pi s/2)}} \right|^2 \leq$$

$$\int_0^1 dt \left\{ \int_0^t ds_1 \cos(\pi s_1/2) \int_0^t ds_2 |u(s_2)|^2 / \cos(\pi s_2/2) \right\} =$$

$$= \frac{2}{\pi} \int_0^1 ds \int_s^1 dt\, \sin(\pi t/2) |u(s)|^2 / \cos(\pi s/2) = \left(\frac{2}{\pi}\right)^2 \|u\|^2 .$$

The maximal value: $\|V\| = 2/\pi$, corresponds to the vector $u^*(s) := \cos(\pi s/2)$.

(*b*) Recall that the linear operator $A \in \mathcal{L}(\mathfrak{H})$ belongs to the *Hilbert-Schmidt* class if there exists an orthonormal basis $\{e_n\}_{n\in\mathbb{N}}$ such that

$$\|A\|_2^2 := \sum_{n=1}^{\infty} \|Ae_n\|^2 < \infty .$$

Here $\|A\|_2$ is the Hilbert-Schmidt *norm* of A, which (if finite) does not depend on the choice of the basis. Note that the set of Hilbert-Schmidt operators $\mathcal{C}_2(\mathfrak{H})$ is a $*$-ideal in $\mathcal{L}(\mathfrak{H})$ and that it is also a subset of compact operators: $\mathcal{C}_2(\mathfrak{H}) \subset \mathcal{C}_\infty(\mathfrak{H})$. Moreover, if operators $A_1, A_2 \in \mathcal{C}_2(\mathfrak{H})$, then the product $A_1 A_2$ belongs to the *trace-class* $*$-ideal $\mathcal{C}_1(\mathfrak{H}) \subset \mathcal{C}_2(\mathfrak{H}) \subset \mathcal{C}_\infty(\mathfrak{H})$.

Applying this definition to the Volterra operator on $L^2([0,1])$ for the orthonormal basis $\{e_n(x) = \sqrt{2}\sin(\pi n x)\}_{n\in\mathbb{Z}}$ one finds that $\|Ve_n\|^2 = 3/(n^2\pi^2)$. Therefore, the Volterra operators, together with its adjoint V^*, are Hilbert-Schmidt operators, and consequently they are compact. Note that V is not normal and that by explicit calculations the self-adjoint operator, $V^*V \in \mathcal{C}_1(L^2([0,1]))$, with a *discrete* spectrum of eigenvalues: $\sigma_{\mathrm{d}}(V^*V) = \{\lambda_k(V^*V) = 4/[(2k+1)^2\pi^2]\}_{k\geq 0}$ and $\sigma(V^*V) = \sigma_{\mathrm{d}}(V^*V) \cup \{0\}$. Since $\|V^*V\| = \|V\|^2$ and since for self-adjoint operators the spectral radius $r(V^*V) = \|V^*V\| = \max_{k\geq 0} \lambda_k(V^*V)$, we obtain $\|V\| = 2/\pi$, as in (*a*).

(*c*) The spectrum of compact V has the canonical structure: $\sigma(V) = \sigma_{\mathrm{d}}(V) \cup \{0\}$ (Sect. B.1.2 12.). To study the discrete part $\sigma_{\mathrm{d}}(V)$ one notes that

$$|V^n u|(t) \leq \int_0^t ds\, |u(s)| \frac{(t-s)^{n-1}}{(n-1)!} \leq \|u\| \frac{1}{(n-1)!} ,$$

which implies $\|V^n\| \leq 1/(n-1)!$. Then the spectral radius $r(V) = \lim_{n\to\infty}(1/(n-1)!)^{1/n} = 0$. Hence, V is a *quasi-nilpotent* compact operator with $\sigma_{\mathrm{d}}(V) = \emptyset$.

Recall that $r(A) = 0$ for a self-adjoint operator implies that $A = 0$, cf. with (*b*) for $A = V^*V$, but it is not the case of the Volterra operator. Although it is quasi-nilpotent and its spectrum contains only zero point, $\sigma(V) = \{0\}$, one has

$\|V\| \neq 0$. Consequently, by the *Lidskiĭ theorem* about spectral trace one gets: $\mathrm{Tr}\, V = \Lambda(V) = 0$, for a nontrivial set of matrix elements $\{(e_k, Ve_k)\}_{k\geq 1}$ and for any choice of the orthonormal basis $\{e_k\}_{k\geq 1}$ in $L^2([0,1])$. Here the *spectral trace* is $\Lambda(V) := \sum_{\lambda\in\sigma(V)} \lambda$.

(*d*) To study the *zero* spectral point of (non-self-adjoint) V we remark that

$$\{0\} \subseteq \sigma_{\mathrm{p}}^*(V) \cup \sigma_{\mathrm{cont}}(V) \cup \sigma_{\mathrm{res}}(V)\,; \tag{B.2.1}$$

see Sect. B.1.2 11. Note that according to definition in (*a*) the range $\mathrm{ran}\,(V) = AC_0([0,1])$ is the set of *absolutely continuous* functions, which are zero at $t = 0$.

Now suppose that the *zero* spectral point, $\lambda = 0$, belongs to $\sigma_{\mathrm{p}}^*(V)$ for a nontrivial eigenvector u_0, i.e. one has $Vu_0 = 0$. Then since $Vu_0 \in AC_0([0,1])$, this implies that vector $u_0 = 0$ in $L^2([0,1])$. Hence, $\sigma_{\mathrm{p}}^*(V) = \emptyset$, that is, $\lambda = 0$ does not belong to the point spectrum.

Next we note that closure of $\mathrm{ran}\,(V)$ in $L^2([0,1])$ gives: $\overline{AC_0([0,1])} = L^2([0,1])$. So, the range of the Volterra operator V is dense in the Hilbert space $\mathcal{H}$. Therefore, $\sigma_{\mathrm{res}}(V) = \emptyset$, that is, $\lambda = 0$ does not belong to the residual spectrum.

Thus, owing to (B.2.1) we conclude that $\{0\} = \sigma_{\mathrm{cont}}(V)$, i.e. $\lambda = 0$ belongs to the continuous spectrum.

(*e*) To demonstrate the subtlety of classification the *zero* spectral point for compact operators, see Sect. B.1.2 12., we scrutinise the Volterra operator on a Banach space.

The Volterra operator $V : \mathfrak{X} \to \mathfrak{X}$, on the Banach space $\mathfrak{X} = C([0,1])$ with sup-norm $\|\cdot\|_\infty$, is defined by the same integral operator as in (*a*). Then for $f \in C([0,1])$ and *identity* operator $\{\mathfrak{I} : t \mapsto 1,\ \ t \in [0,1]\}$ we obtain the estimates:

$$|(V^n f)(t)| \leq \frac{1}{n!}\|f\|_\infty \ \text{ and } \ \|V^n\| \geq \|V^n\mathfrak{I}\|_\infty = \frac{1}{n!}\,.$$

Hence, the operator norm $\|V^n\| = 1/n!$ and, consequently, the spectral radius $r(V) = 0$. Therefore, the spectrum of this bounded operator with norm $\|V\| = 1$ is $\sigma(V) = \{0\}$.

Note that the range of this operator, $\mathrm{ran}\,(V) = AC_0([0,1])$, is the set of absolutely continuous functions such that $(Vf)(t = 0) = 0$. Seeing that $|(Vf)(t)| \leq \|f\|_\infty\, t^{1/2}$ and $|(Vf)(t) - (Vf)(s)| \leq \|f\|_\infty\, |t-s|^{1/2}$, by the *Arzela-Ascoli* theorem the image $V(B_1)$ of the unit ball $B_1 := \{f \in \mathfrak{X} : \|f\|_\infty \leq 1\}$ is a compact.

So, V is a *compact* operator on $\mathfrak{X} = C([0,1])$ and $\sigma(V) = \{0\}$ implies by the *Fredholm alternative* that $\sigma_{\mathrm{d}}(V) = \emptyset$ and $\sigma_{\mathrm{p}}^*(V) = \sigma_{\mathrm{p}}(V)$. Now, we have to decide between three *disjoint* possibilities: $\{0\} = \sigma_{\mathrm{p}}(V) \cup \sigma_{\mathrm{cont}}(V) \cup \sigma_{\mathrm{res}}(V)$.

First we claim that $\{0\} \neq \sigma_{\mathrm{p}}(V)$ follows along the same line of reasoning as in (*d*). The second, we note that ran $(V) = AC_0([0, 1])$ is *not* dense in $C([0, 1])$ in topology of the $\|\cdot\|_\infty$-norm. Therefore, $\overline{\operatorname{ran}(V)} \neq \mathfrak{X}$, and consequently $\{0\} = \sigma_{\mathrm{res}}(V)$.

For references to Appendix B with proofs and details, see, e.g. [BS87, Chapter 3 and Chapter 9], [Dav07] and [Kato80, Chapter VIII, §1].

Appendix C
Spectral and Numerical Range Mapping Theorems

C.1 Spectral Mapping Theorems for Semigroups

We collect here some facts concerning relations between spectra of generator and the corresponding semigroup; see, e.g. [EN00, Chapter IV], [Dav07, Section 8.2] and [Zag19, Chapter 4].

(*a*) Let A be a closed unbounded operator in a Banach space $\mathfrak{X}$, with dense domain $\operatorname{dom} A \subset \mathfrak{X}$. Recall that if $\zeta \notin \sigma(A)$, then one has the following *resolvent spectral mapping* theorem:

$$\sigma(R_A(\zeta)) = \{0\} \cup \{\lambda - \zeta)^{-1} : \lambda \in \sigma(A)\}\,.$$

(*b*) Let A be generator of strongly continuous semigroup $\{U_A(t)\}_{t\geq 0}$ on $\mathfrak{X}$. Recall that the relationship between the spectra $\sigma(A)$ and $\sigma(U_A(t))$ is not as simple as in the *resolvent* spectral mapping theorem. In general, one can prove only the following form of this theorem for *strongly* continuous semigroups:

$$\sigma(U_A(t)) \supseteq \{e^{-tz} : z \in \sigma(A)\}\,, \quad t > 0\,. \tag{C.1.1}$$

We denoted this class of continuous semigroups as C_0-semigroups. One can establish a converse statement under some further conditions on the *continuity* of C_0-semigroups.

(*c*) To this end we remind the following version of the spectral mapping theorem. Let C_0-semigroup $\{U_A(t)\}_{t\geq 0}$ generated by unbounded A be *quasi-bounded* of the *type* ω_0: $\|U_A(t)\| \leq M e^{\omega_0 t}$, $M \geq 1$. Let $\widehat{U}_{A,\alpha}(t) := U_A(t)\, R_A(\alpha)$, where $\alpha \notin \sigma(A)$. Then for $t > 0$ and $\operatorname{Re}\alpha < -\omega_0$, we obtain

$$\sigma(\widehat{U}_{A,\alpha}(t)) = \{0\} \cup \{e^{-tz}(z-\alpha)^{-1} :\; z \in \sigma(A)\}\,, \quad t > 0\,.$$

V. A. Zagrebnov et al., *Trotter-Kato Product Formulæ*, Operator Theory: Advances and Applications 296, https://doi.org/10.1007/978-3-031-56720-9

(*d*) Now it is a time of the first result, which is due to improved *continuity* conditions. Let C_0-semigroup $\{U_A(t)\}_{t\geq 0}$ be *eventually norm-continuous* (Sect. 1.6), that is, there exists $t_0 > 0$ such that $t \mapsto U_A(t)$ is an operator-norm continuous function for $t > t_0$. Then inclusion (C.1.1) transforms into $\sigma(U_A(t)) = \{e^{-tz} : z \in \sigma(A)\}$, for $t > t_0$.

(*e*) If one assumes that C_0-semigroup $\{U_A(t)\}_{t\geq 0}$ is *eventually compact* (Sect. 1.6), that is, there exists $t_0 > 0$ such that $t \mapsto U_A(t) \in \mathcal{C}_\infty(\mathfrak{X})$ for $t > t_0$, then (C.1.1) transforms into

$$\sigma(U_A(t)) = \{0\} \cup \{e^{-t\sigma(A)}\}, \quad t > t_0 .$$

For the proof in a Hilbert space $\mathfrak{H}$, see Proposition 1.6.6. By inspection the arguments there hold *verbatim et literatim* also for the proof in the Banach space $\mathfrak{X}$.

(*f*) Counterexamples show that inclusion which is converse to (C.1.1) does not hold in general, but there is a partial result in this direction; cf. [EN00, Chapter IV, Section 3.7].

Let $\{U_A(t)\}_{t\geq 0}$ be quasi-bounded C_0-semigroup on a Banach space $\mathfrak{X}$ generated by unbounded operator A. Then converse to inclusion (C.1.1) concerns only the *point* and the *residual* spectra (Sect. B.1.2 6. and 8.) and it yields:

$$\sigma_{\mathrm{p}}(U_A(t)) = \{0\} \cup \{e^{-tz} : z \in \sigma_{\mathrm{p}}(A)\} ,$$

$$\sigma_{\mathrm{res}}(U_A(t)) = \{0\} \cup \{e^{-tz} : z \in \sigma_{\mathrm{res}}(A)\} .$$

(*g*) For proofs and more details about the spectral mapping theorems for semigroups, we refer to [Zag19, Section 4.1 and 4.2].

C.2 Kato Mapping Theorems for Operator Numerical Range

The *Kato mapping theorems* for the numerical range of a linear operator are analogous to the *spectral mapping* theorems in Sect. C.1. But since the operator numerical range, Definition 1.8.1 (Sect. 1.8.1), is neither open nor closed and since by the *Töplitz-Hausdorff* theorem is always convex [Dav07, Section 9.3], a validity of the corresponding statements is rather restricted compared to the spectral mapping theorems.

For that reason, following *Kato* [Kato65] we remind here *two* versions of the mapping theorem for the numerical range of a bounded operator on a Hilbert space. The first one (global) is when the behaviour of the mapping on the whole complex plane $\mathbb{C}$ is essential. The second version (*local*) is restricted to the functions defined on special subsets of the complex plane $\mathbb{C}$.

1. Let A be a bounded linear operator in a Hilbert space $\mathfrak{H}$. Recall that its spectrum $\sigma(A) \subset \overline{W(A)}$, where $\overline{W(A)}$ is the closure of its numerical range

(Definition 1.8.1). Note that for a normal operator A the set $\overline{W(A)}$ is the closed convex hull of $\sigma(A)$.

We recall that K is a *kernel* of a non-empty set $E \subset \mathbb{C}$ if K is the set of points $z \in \mathbb{C}$ such that E is *star-shaped* relative to each $z \in K$. That is,

$$K := \{z \in E : \text{ segment } [z, \zeta] \subset E \text{ for any } \zeta \in E\}. \tag{C.2.1}$$

Therefore, K may be empty, but if not, then it is known (see references in [Kato65]) that *kernel* K is a *convex* subset of E, which is compact if E is a compact set. For the kernel K to be non-empty it is evident necessary that E be connected. It is also evident that if E is a convex set, then $K = E$. But in the opposite case the kernel $K \subset E$, or even it may be *empty*: $K = \emptyset$.

Now we formulate the *global* Kato mapping theorem [Kato65, Theorem 1], when the corresponding mapping function f has domain that belongs to whole $\mathbb{C}$.

Theorem C.2.1 *Let $f : z \mapsto f(z)$ be a rational complex function such that $f(\infty) = \infty$ and let E' be a compact convex set in $\mathbb{C}$. Let $E := f^{-1}(E')$ and let $K \subset E$ be the convex kernel of E. If A is a bounded operator with numerical range $W(A) \subset K$, then for numerical range of the operator $f(A)$ one gets $W(f(A)) \subset E'$.*

2. There are few remarks in order:

- First, if kernel $K = \emptyset$, then the numerical range mapping Theorem C.2.1 (cf. Proposition 3.1.5) is of no use.
- Second, we note that since E' is assumed to be compact, sets E and K are also compact and f has no poles in E. Hence, since $\sigma(A) \subset \overline{W(A)} \subset K \subset E$, the operator $f(A)$ is well-defined by the *Riesz-Dunford* functional calculus for bounded operators (1.7.1).
- Third, for $K \neq \emptyset$ it is necessary that the set E be connected and contain all *critical* points of f, so that E' contains all *branch* points of the inverse function f^{-1}.

3. The following statement [Kato65, Corollary 3] is in fact owing to the numerical range mapping Theorem C.2.1.

Corollary C.2.2 *Let $W(A) \subseteq \mathbb{D}$ be a subset of the closed unit disc $\mathbb{D}$. Since function $f_n(z) = z^n$ for $z \in \mathbb{C}$ and $n \in \mathbb{N}$ satisfies conditions of Theorem* C.2.1, *the same will hold for operators $\{A^n\}_{n\geq 1}$, that is, $W(A^n) \subseteq \mathbb{D}$, for $n \in \mathbb{N}$.*

On account of Lemma 3.1.6 (cf. Lemma 3.2.13) one obtains a similar result: $W(A^n) \subseteq D_\alpha$, for $n \in \mathbb{N}$ and $D_\alpha \subset \mathbb{D}$. Here subset D_α contains the numerical range of a *quasi-sectorial* contraction A, that is, $W(A) \subseteq D_\alpha$. We point out that subset $D_{\alpha\leq\pi/2}$ is defined by (3.1.1) and $D_{\alpha=\pi/2} = \mathbb{D}$.

4. The *local* mapping theorems [Kato65, Theorems 5–7] are proved only for holomorphic functions f defined on *discs* and on *half-planes*. We refer to an instructive discussion in [Kato65, Remark 8] about conditions of validity, as well as about the similarity and difference of these theorems, comparing them in particular with [Kato65, Corollary 3].

   For references to Appendix C (Sect. C.2) see [Dav07, Section 9.3] and [Kato65].

Appendix D
Remarks About Condition $\omega[f] < \infty$

Condition $\omega[f] < \infty$ (7.1.3) is necessary for the proof of the ITTZ-Theorem 7.1.9, Sect. 7.1. It controls the behaviour of the "tail" of function $f \in \mathcal{K}$ "away from zero", namely: for $x \geq \delta$ and arbitrary $\delta > 0$. This condition selects a proper *subclass* of the Kato functions $\mathcal{K}$, [Zag19, Appendix C].

In this appendix we scrutinise the condition $\omega[f] < \infty$ for generic Kato function $f \in \mathcal{K}$, Definition 5.2.1. To this aim it is useful to introduce a generalisation of the functional (7.1.2):

$$[f]_s := \sup_{x>0} \frac{|1 - f(x) - x|}{x^s}, \quad 1 < s \leq 2; \tag{D.0.1}$$

cf. the Kato functions of the class $\widehat{\mathcal{K}}_s$ in [Zag19, Definition C.3].

We note that $[f]_2 = \varkappa[f]$ (7.1.2), and if $[f]_s < \infty$, then $[f]_{s'} < \infty$ for $1 < s' \leq s$.

Lemma D.0.1 *Let $f \in \mathcal{K}$ be a Kato function. Then condition*

$$\omega[f] := \sup_{x>0} \frac{1}{\sqrt{1 - f(x)}} \left| \frac{1 - f(x)}{x} - 1 \right| < \infty \tag{D.0.2}$$

is satisfied if and only if $[f]_{s=3/2} < \infty$, (D.0.1), *and for each $\delta > 0$ there exists $\varepsilon(\delta) > 0$ such that $f(x) \leq 1 - \varepsilon(\delta)$ for $x \geq \delta$.*

Proof Since f is a generic Kato function: $\gamma[f] < \infty$, (7.1.1), condition $[f]_{s=3/2} < \infty$ for $|(1 - f(x))/x - 1|/x^{3/2}$ ensures the existence of $\delta > 0$ such that

$$\left| \frac{1 - f(x)}{x} - 1 \right| \leq \frac{1}{2}, \quad x \in (0, \delta). \tag{D.0.3}$$

V. A. Zagrebnov et al., *Trotter-Kato Product Formulæ*, Operator Theory: Advances and Applications 296, https://doi.org/10.1007/978-3-031-56720-9

This yields $\sqrt{x}/\sqrt{1-f(x)} \le \sqrt{2}$, $x \in (0,\delta)$. As a consequence, we obtain

$$\sup_{x\in(0,\delta)} \frac{1}{\sqrt{1-f(x)}}\left|\frac{1-f(x)}{x}-1\right| \le \sqrt{2}\left|\frac{1-f(x)-x}{x^{3/2}}\right| \le \sqrt{2}\,[f]_{s=3/2}\,. \tag{D.0.4}$$

Moreover, by assumption of the lemma for $\delta > 0$ we can find an $\varepsilon(\delta) > 0$ such that $f(x) \le 1-\varepsilon(\delta)$ for $x \ge \delta$. Hence, $\sqrt{\varepsilon(\delta)} \le \sqrt{1-f(x)}$ for $x \ge \delta$ and

$$\begin{aligned}\sup_{x>\delta} \frac{1}{\sqrt{1-f(x)}}\left|\frac{1-f(x)}{x}-1\right| &\le \frac{1}{\sqrt{\varepsilon(\delta)}}\sup_{x>\delta}\left|\frac{1-f(x)-x}{x}\right| \\ &\le \frac{1}{\sqrt{\varepsilon(\delta)}}\,[f]_{s=1}\,. \end{aligned} \tag{D.0.5}$$

Now, seeing that by (D.0.1) the inequality $[f]_{3/2} < \infty$ yields $[f]_1 < \infty$, on account of (D.0.4), (D.0.5) and definition (7.1.3) we obtain $\omega[f] < \infty$.

Conversely, if $\omega := \omega[f] < \infty$, (D.0.2), then

$$-\,\omega \le \frac{x-1+f(x)}{x\sqrt{1-f(x)}} \le \omega, \qquad x > 0.$$

Let $h(x) := 1-f(x) \ge 0$. Then

$$-\,x\omega\sqrt{h(x)} \le x-h(x) \le x\omega\sqrt{h(x)}, \qquad x > 0. \tag{D.0.6}$$

Hence, $x \le h(x) + x\omega\sqrt{h(x)}$, or $x + \frac{1}{4}x^2\omega^2 \le (\sqrt{h(x)} + \frac{1}{2}x\omega)^2$. Therefore, we obtain estimate: $\sqrt{x+\frac{1}{4}x^2\omega^2} - \frac{1}{2}x\omega \le \sqrt{h(x)}$, which yields

$$\frac{x}{\sqrt{x+\frac{1}{4}x^2\omega^2}+\frac{1}{2}x\omega} = \left(\sqrt{\tfrac{1}{x}+\tfrac{1}{4}\omega^2}+\tfrac{1}{2}\omega\right)^{-1} =: p(x) \le \sqrt{h(x)}, \quad x > 0. \tag{D.0.7}$$

Now let $\delta > 0$. Then by (D.0.7) the exists $\varepsilon(\delta) := p^2(\delta)$ such that $\varepsilon(\delta) \le h(x) = 1-f(x)$, for $x \ge \delta$. This proves $f(x) \le 1-\varepsilon(\delta)$ for $x \ge \delta$.

Moreover, since f is a Kato function, the hypothesis $\omega := \omega[f] < \infty$, (D.0.2) for $|(1-f(x))/x - 1|/(x\sqrt{1-f(x)})$ ensures the existence of $\delta > 0$ (defined in (D.0.3)) such that $(1-f(x))/x \le 3/2$ for $x \in (0,\delta)$. Then

$$\sup_{x\in(0,\delta)}\left|\frac{1-f(x)-x}{x^{3/2}}\right| \le \sqrt{\frac{3}{2}}\sup_{x\in(0,\delta)}\left|\frac{1-f(x)-x}{x\sqrt{1-f(x)}}\right| \le \sqrt{\frac{3}{2}}\,\omega[f] < \infty\,. \tag{D.0.8}$$

On the other hand, since $f \in \mathcal{K}$, one gets for $x \geq \delta > 0$ that $(1 - f(x))/x \leq 1/\delta$. Then

$$\sup_{x \geq \delta} \left| \frac{1 - f(x) - x}{x^{3/2}} \right| \leq \sqrt{\frac{1}{\delta}} \sup_{x \geq \delta} \left| \frac{1 - f(x) - x}{x\sqrt{1 - f(x)}} \right| \leq \sqrt{\frac{1}{\delta}}\, \omega[f] < \infty . \qquad \text{(D.0.9)}$$

Owing to (D.0.8) and (D.0.9) we obtain that $[f]_{3/2} < \infty$. □

Note that by definition (D.0.2) for any Kato function $f \in \mathcal{K}$ one always has that functional $\omega[f] \geq \sup_{x>0} |(1 - f(x))/x - 1| = 1$. We also note that from (D.0.6) one also gets that $h(x) - x\omega\sqrt{h(x)} \leq x$, $x > 0$, which yields

$$h(x) \leq \left(\tfrac{1}{2}\omega x + \sqrt{x + \tfrac{1}{4}\omega^2 x^2} \right)^2 =: L_\omega(x) , \qquad x > 0.$$

Then in virtue of (D.0.7), we obtain estimates: $1 - L_\omega(x) \leq f(x) \leq 1 - x^2/L_\omega(x)$, for $x > 0$.

The following remark serves for illustration that condition $\omega[f] < \infty$ is *necessary* for proving the ITTZ-Theorem 7.1.9.

Remark D.0.2 Assume that f and g are generic Kato functions from $\mathcal{K}$. Let H be a non-negative self-adjoint operator such that its spectrum σ_H is a collection of eigenvalues accumulating at infinity. Let $\sigma(H) = \{\lambda_j\}_{j=1}^{\infty}$ be the eigenvalues and let $\{\psi_j\}_{j=1}^{\infty}$ be the corresponding normalised eigenvectors of operator H.

1. Suppose that Kato functions f and g be such that $f(x) = g(x) = 1$ for $x > 1$. Let

$$A = B := \tfrac{1}{2}\, H .$$

Fix $n \in \mathbb{N}$ sufficiently large, and take $N(n) \in \mathbb{N}$ so large that $\lambda_{N(n)} > 2n$. Then

$$[f(A/n)g(B/n)]^n \psi_{N(n)} = [f(H/2n)g(H/2n)]^n \psi_{N(n)} =$$
$$[f(\lambda_{N(n)}/2n)g(\lambda_{N(n)}/2n)]^n \psi_{N(n)} = \psi_{N(n)} .$$

On the other hand, we have

$$\lim_{N\to\infty} \mathrm{e}^{-H} \psi_{N(n)} = \lim_{n\to\infty} \mathrm{e}^{-\lambda_{N(n)}} \psi_{N(n)} = 0,$$

For demonstration the operator-norm convergence of the product formula, we use the limit:

$$\lim_{n\to\infty} \|(f(A/n)g(B/n))^n - e^{-H}\|$$
$$= \lim_{n\to\infty} \sup_{u\in\mathfrak{H}:\|u\|=1} \|((f(A/n)g(B/n))^n - e^{-H})\, u\| \geq$$
$$\lim_{n\to\infty} \|((f(A/n)g(B/n))^n - e^{-H})\, \psi_{N(n)}\| = 1\,.$$

This means that approximants $\{(f(A/n)g(B/n))^n\}_{n\geq 1}$ never converge to e^{-H} in the operator norm.

2. Let f and g be Kato functions such that $f(x = 1) = 1$. Let H be a self-adjoint operator as above but with eigenvalues $\{\lambda_j = j\}_{j=1}^{\infty}$. Take $A = H$ and $B = 0 \cdot \mathbb{1}$, that is, $H = A + B$. Then

$$[f(A/n)g(B/n)]^n \psi_n = f(H/n)^n \psi_n = f(1)^n \psi_n = \psi_n\,,$$

which preserves the vector norm. However, one gets that

$$\lim_{n\to\infty} e^{-H} \psi_n = e^{-\lambda_n} \psi_n = 0\,.$$

Then by the same arguments as above we deduce that $\{(f(A/n)g(B/n))^n\}_{n\geq 1}$ does not converge to e^{-H} in the operator-norm topology.

Appendix E
Banach Space of Vector-Valued Functions

E.1 Integration in Banach Space of Vector-Valued Functions

In the next sections we summarise some indispensable basic facts from the integration theory for Banach space valued functions. Here, we follow [ABHN01, Chapter A1], [KF89, Chapter X], [HiPh57, Chapter III] and [CeMe97], [Rao04].

Let $[0, T] \subset \mathbb{R}$ be a compact (time) interval where we fix the Lebesgue measure λ_0 on $[0, T]$. Let $\mathcal{B}$ be the σ-algebra of all *Borel* sets of the interval $[0, T]$. Let $\mathfrak{X}$ be a (not necessarily *separable*) Banach space and $\mathfrak{X}^*$ its dual. In order to define an integral of $f : [0, T] \to \mathfrak{X}$, we have to define some *measurability* of the function f.

Definition E.1.1 For $n \in \mathbb{N}$ we call a function $s_n : [0, T] \to \mathfrak{X}$ *simple* (or *finitely valued*) if it is a finite sum of the form

$$s_n(t) = \sum_{i=1}^{n} \chi_{\Delta_i}(t)\, x_i\,, \tag{E.1.1}$$

where Δ_i are disjoint Borel sets of $[0, T]$ and $x_i \in \mathfrak{X}$.

Since $[0, T] \subset \mathbb{R}$ is compact, the Δ_i are finite and the integral of a simple function s_n is defined as

$$\int_{[0,T]} dt\, s_n(t) = \sum_{i=1}^{n} |\Delta_i|\, x_i\,.$$

Because the Banach space $\mathfrak{X}$ can be endowed with non-equivalent topologies (e.g. with the *strong* or with the *weak* topology), different concepts of measurability are of importance.

V. A. Zagrebnov et al., *Trotter-Kato Product Formulæ*, Operator Theory: Advances and Applications 296, https://doi.org/10.1007/978-3-031-56720-9

Definition E.1.2

(i) A function $f : [0, T] \to \mathfrak{X}$ is called *strongly* measurable if there exists a sequence of simple functions $\{s_n\}_{n\in\mathbb{N}}$ such that

$$\|\cdot\|_{\mathfrak{X}} - \lim_{n\to\infty} s_n(t) = f(t), \tag{E.1.2}$$

in topology of $\mathfrak{X}$ for almost all (*a.a.*) $t \in [0, T]$, or equivalently, pointwise almost everywhere (*a.e.*) on the interval $[0, T]$.

(ii) A function $f : [0, T] \to \mathfrak{X}$ is called *weakly* measurable if for all $x^* \in \mathfrak{X}^*$ the scalar function $x^* \circ f : [0, T] \to \mathbb{R}$ is *Lebesgue* measurable.

Note that if $f : [0, T] \to \mathfrak{X}$ is *countably* valued, that is.

$$f = \sum_{n\in\mathbb{N}} x_n\, \chi_{\Delta_n}, \quad x_n \in \mathfrak{X}, \quad \text{where } \{\Delta_n\} \text{ are disjoint } \textit{Borel} \text{ sets},$$

then Definition E.1.2 implies that the function f is strongly and weakly measurable. To verify measurability of a general function f, it is often useful to use the characterisation which appertains to *A. Pettis*; see [ABHN01, Theorem 1.1.1].

Theorem E.1.3 *The function $f : [0, T] \to \mathfrak{X}$ is strongly measurable if and only if it is weakly measurable and there is a set $\mathcal{N} \subset [0, T]$ of measure zero such that $f([0, T]\backslash\mathcal{N})$ is a separable subset of $\mathfrak{X}$.*

Corollary E.1.4 *Let $f : [0, T] \to \mathfrak{X}$. Then the following assertions hold*:

(i) *If the Banach space $\mathfrak{X}$ is separable, then f is strongly measurable if and only if it is weakly measurable.*

(ii) *If $f : [0, T] \to \mathfrak{X}$ is continuous, then it is strongly measurable.*

(iii) *If $f_n : [0, T] \to \mathfrak{X}$, $n \in \mathbb{N}$, are strongly measurable functions and $\lim_{n\to\infty} f_n(t) = f(t)$ pointwise a.e. on $[0, T]$, then f is strongly measurable.*

We denote the space of $\mathfrak{X}$-valued *continuous* functions on $[0, T]$ by

$$C([0, T], \mathfrak{X}) = \{\{f : [0, T] \to X\} : f \text{ (strongly) continuous}\}. \tag{E.1.3}$$

Since the (*strongly*) continuous image of a separable space (in this case $[0, T]$) is *separable*, the functions (E.1.3) are strongly measurable and the *sup-norm* of the function: $t \mapsto \|f(t)\|_{\mathfrak{X}}$ defines a norm on $C([0, T], \mathfrak{X})$:

$$\sup_{t\in[0,T]} \|f(t)\|_{\mathfrak{X}} =: \|f\|_{C([0,T],\mathfrak{X})}.$$

Given that $[0, T]$ is compact, the space of $\mathfrak{X}$-valued *continuous* functions $C([0, T], \mathfrak{X})$ is a Banach space with respect to the norm $\|\cdot\|_{C([0,T],\mathfrak{X})}$.

Definition E.1.5 A function $f : [0, T] \to \mathfrak{X}$ is called *Bochner* integrable if there exists a sequence of simple functions $\{s_n : [0, T] \to \mathfrak{X}\}_{n\in\mathbb{N}}$ such that

(i) $\|s_n(t) - f(t)\|_{\mathfrak{X}} \to 0$, as $n \to \infty$, pointwise for *a.a.* $t \in [0, T]$, cf. (E.1.2).
(ii) It holds that

$$\int_{[0,T]} dt \, \|f(t) - s_n(t)\|_{\mathfrak{X}} \to 0, \quad n \to \infty .$$

Then the *Bochner integral* of function $f : [0, T] \to \mathfrak{X}$, is

$$\int_{[0,T]} dt \, f(t) = \|\cdot\|_{\mathfrak{X}} - \lim_{n\to\infty} \int_{[0,T]} dt \, s_n(t) . \tag{E.1.4}$$

We note that the *Bochner* integral is well-defined, that is, independent of the choice of the sequence $\{s_n\}_{n\in\mathbb{N}}$. Moreover, the following comprehensive and useful characterisation of the *Bochner* integrable functions holds.

Theorem E.1.6 ([HiPh57, Theorem 3.7.4], [ABHN01, Theorem 1.1.4]) *A function $f : [0, T] \to \mathfrak{X}$, is Bochner integrable if and only if it is (strongly) measurable* (E.1.2) *and the function* : $t \mapsto \|f(t)\|_{\mathfrak{X}}$, *is (Lebesgue) integrable.*

Proof Suppose that function: $t \mapsto f(t) \in \mathfrak{X}$, is *Bochner* integrable on $[0, T]$. Then f is (strongly) measurable, that is, on account of Definition E.1.5 (i), there exists on $[0, T]$ a corresponding to f sequence of simple functions $\{s_n\}_{n\in\mathbb{N}}$, (E.1.2). Therefore, by (E.1.2) and by estimate $| \, \|f(t)\|_{\mathfrak{X}} - \|s_n(t)\|_{\mathfrak{X}} \, | \leq \|f(t) - s_n(t)\|_{\mathfrak{X}}$ the function $\|f(t)\|_{\mathfrak{X}}$ is also measurable. Its integrability follows from estimate

$$\int_{[0,T]} dt \, \|f(t)\|_{\mathfrak{X}} \leq \int_{[0,T]} dt \, \|s_n(t)\|_{\mathfrak{X}} + \int_{[0,T]} dt \, \|f(t) - s_n(t)\|_{\mathfrak{X}} . \tag{E.1.5}$$

and Definition E.1.5 of the *Bochner* integrable function f on $[0, T]$.

Conversely, let f be (strongly) measurable (E.1.2) and the function : $t \mapsto \|f(t)\|_{\mathfrak{X}}$ be (*Lebesgue*) integrable. Then let $\{s_n\}_{n\in\mathbb{N}}$ be a sequence of simple functions approximating f according to Definition E.1.2. Now for *a.a.* $t \in [0, T]$ and $n \in \mathbb{N}$ we define a new simple functions

$$\widehat{s_n}(t) := \begin{cases} s_n(t) , & \|s_n(t)\|_{\mathfrak{X}} \leq \|f(t)\|_{\mathfrak{X}}(1 + 1/n) \\ 0 , & \|s_n(t)\|_{\mathfrak{X}} > \|f(t)\|_{\mathfrak{X}}(1 + 1/n) \end{cases} .$$

As a consequence, $\|\widehat{s_n}(t)\|_{\mathfrak{X}} \leq \|f(t)\|_{\mathfrak{X}}(1 + 1/n)$ for $n \in \mathbb{N}$ and

$$\lim_{n\to\infty} \|f(t) - \widehat{s_n}(t)\|_{\mathfrak{X}} = 0 , \text{ for } a.a. \; t \in [0, T] . \tag{E.1.6}$$

Therefore, $\|\widehat{s_n}(t)\|_{\mathfrak{X}} \leq 2 \, \|f(t)\|_{\mathfrak{X}}$ and $\{\widehat{s_n}\}_{n\in\mathbb{N}}$ are again approximating simple functions for f; see Definition E.1.2.

Since by conditions function $\|f(t)\|_{\mathfrak{X}}$ and norms $\{\|\widehat{s}_n(t)\|_{\mathfrak{X}}\}_{n\in\mathbb{N}}$ of simple functions are integrable for any $n \in \mathbb{N}$, the functions $\{\|\widehat{s}_n(t) - f(t)\|_{\mathfrak{X}}\}_{n\in\mathbb{N}}$ are also integrable on $[0, T]$. Hence, because of the bound $\|\widehat{s}_n(t) - f(t)\|_{\mathfrak{X}} \leq 3\,\|f(t)\|_{\mathfrak{X}}$ and convergence (E.1.6) we can apply the *Lebesgue dominated convergence* theorem to infer that

$$\lim_{n\to\infty} \int_{[0,T]} dt\ \|f(t) - \widehat{s}_n(t)\|_{\mathfrak{X}} = 0\,. \tag{E.1.7}$$

Then limits (E.1.6) and (E.1.7) yield that function $t \mapsto f(t) \in \mathfrak{X}$ is Bochner integrable on $[0, T]$; see Definition E.1.5 (i) and (ii). □

Theorem E.1.7 ([HiPh57, Theorem 3.7.6], [ABHN01, Theorem 1.1.4]) *If function $f : [0, T] \to \mathfrak{X}$ is Bochner integrable, then*

$$\left\| \int_{[0,T]} dt\ f(t) \right\|_{\mathfrak{X}} \leq \int_{[0,T]} dt\ \|f(t)\|_{\mathfrak{X}}\,. \tag{E.1.8}$$

Proof To this aim one can use for $F, S \in \mathfrak{X}$ in a normed space $\mathfrak{X}$ the *triangle inequality* in the form: $|\,\|F\|_{\mathfrak{X}} - \|S\|_{\mathfrak{X}}\,| \leq \|F - S\|_{\mathfrak{X}}$. Then owing to (E.1.4) and (E.1.5) we obtain

$$\begin{aligned}\left\| \int_{[0,T]} dt\ f(t) \right\|_{\mathfrak{X}} &= \lim_{n\to\infty} \left\| \int_{[0,T]} dt\ s_n(t) \right\|_{\mathfrak{X}} \leq \lim_{n\to\infty} \int_{[0,T]} dt\ \|s_n(t)\|_{\mathfrak{X}} \\ &= \int_{[0,T]} dt\ \|f(t)\|_{\mathfrak{X}}\,,\end{aligned}$$

since for *simple* functions the inequality (E.1.8) holds by construction (E.1.1). □

Using Theorems E.1.6 and E.1.7, we can define L^p-spaces.

Definition E.1.8 Let $\mathfrak{X}$ be a Banach space. For $p \in [1, \infty]$, we define the space of p-integrable functions with values in the Banach space $\mathfrak{X}$. For $1 \leq p < \infty$, we set

$$\mathcal{L}^p([0, T], \mathfrak{X}) = \{f : [0, T] \to \mathfrak{X} : f(\cdot)\|f(\cdot)\|_{\mathfrak{X}}^{p-1} \text{ is Bochner integrable}\},$$

$$\|f\|_{\mathcal{L}^p([0,T],\mathfrak{X})} := \left(\int_{[0,T]} dt\ \|f(t)\|_{\mathfrak{X}}^p \right)^{1/p},$$

and for $p = \infty$, we set

$$\mathcal{L}^\infty([0, T], \mathfrak{X}) = \{f : [0, T] \to \mathfrak{X} : \text{strongly measurable and essentially bounded}\},$$

$$\|f\|_{\mathcal{L}^\infty([0,T],\mathfrak{X})} := \operatorname*{ess\,sup}_{t\in[0,T]} \|f(t)\|_{\mathfrak{X}}\,.$$

It turns out that the space $(\mathcal{L}^p([0,T],\mathfrak{X}), \|\cdot\|_{\mathcal{L}^p})$ becomes a complete metric vector space endowed with a semi-norm for any $p \in [1,\infty]$.

Definition E.1.9 For a Banach space $\mathfrak{X}$ we define for $p \in [1,\infty]$,

$$L^p([0,T],\mathfrak{X}) := \mathcal{L}^p([0,T],\mathfrak{X})/\sim\ ,$$

with the *equivalence* relation

$$f \sim g \Leftrightarrow f(t) = g(t) \quad \text{for a.a. } t \in [0,T],$$

endowed with the norm

$$\|f\|_{L^p([0,T],\mathfrak{X})} := \|f\|_{\mathcal{L}^p([0,T],\mathfrak{X})}\,.$$

It turns out that $L^p([0,T],\mathfrak{X})$ is a linear space which becomes a Banach space endowed with the norm $\|\cdot\|_{L^p([0,T],\mathfrak{X})}$. In the following, we just write $f \in L^p([0,T],\mathfrak{X})$ for a function $f : [0,T] \to \mathfrak{X}$ and neglect the equivalence class.

Remark E.1.10 We note that the Banach space of continuous functions $C([0,T],\mathfrak{X})$ can be regarded as *embedded* into $L^p([0,T];\mathfrak{X})$, $p \in [1,\infty]$.

Let $\iota_p : C([0,T],\mathfrak{X}) \to L^p([0,T],\mathfrak{X})$, $p \in [1,\infty]$, be the embedding operator of $C([0,T],\mathfrak{X})$ into $L^P([0,T],\mathfrak{X})$. We point out that $\ker(\iota_p) = \{0\}$ for $p \in [1,\infty)$.

E.2 The Dual Spaces of $C(\mathfrak{I},\mathfrak{X})$ and of $L^p(\mathfrak{I},\mathfrak{X})$

The aim of this section is to characterise the *dual* spaces of $C(\mathfrak{I},\mathfrak{X})$ and $L^p(\mathfrak{I},\mathfrak{X})$ for any bounded (closed) interval $\mathfrak{I} \subset \mathbb{R}_0^+$. To this aim we define the *variation* of a function. For $A \subset \mathfrak{I}$, let $\mathcal{B}(A) := \{A \cap B : B \in \mathcal{B}\}$, where $\mathcal{B}$ denotes the set of all Borel subsets in $\mathbb{R}_0^+$. By $\mathfrak{X}^*$ we denote the dual of a Banach space $\mathfrak{X}$.

Definition E.2.1 Let mapping $\nu : \mathcal{B} \to \mathfrak{X}^*$ be σ-additive. We call $\|\nu\|(\cdot)$ the *variation* of ν defined via

$$\|\nu\|(A) = \sup\left\{\sum_{i=1}^{n} \|\nu(A_i)\|_{\mathfrak{X}^*} : A_i \in \mathcal{B}(A),\ A_i \cap A_j = \emptyset \text{ for } i \neq j\right\},$$

and we say ν has finite variation if $\|\nu\|(\mathfrak{I}) < \infty$.

Note that $\|\nu\|(\cdot)$ is also σ-additive, but in general it can be infinite.

Definition E.2.2 A *scalar* measure $\mu : \mathcal{B} \to \mathbb{R}_0^+$ is called *regular* if it is inner-regular and outer-regular, namely, if

(i) $\mu(B) = \sup\{\mu(A) : A \subset B,\ A \ \text{compact}\}$ and
(ii) $\mu(B) = \inf\{\mu(A) : A \supset B,\ A \ \text{open}\}$,

hold for all $B \in \mathcal{B}$.

We say that a σ-additive measure $\nu : \mathcal{B} \to \mathfrak{X}^*$ is *regular* if its variation $\|\nu\| : \mathcal{B} \to \mathbb{R}_0^+$ is regular in the sense of Definition E.2.2. Moreover, we call a scalar measure $\lambda : \mathcal{B} \to \mathbb{R}_0^+$ *complete* if any subset of a *null* set is λ-measurable and, hence, is also a null set; see, e.g. [KF89, Definition 4, Chapter V, §3].

After these preparations, we introduce the spaces $\mathsf{cab}\, v(\mathfrak{I}, \mathfrak{X}^*)$ and $\mathsf{rcab}\, v(\mathfrak{I}, \mathfrak{X}^*)$.

Definition E.2.3 We denote by $\mathsf{cab}\, v(\mathfrak{I}, \mathfrak{X}^*)$ the space of all σ-additive measures of bounded variation with values in $\mathfrak{X}^*$ endowed with the variation norm $\|\cdot\|$ defined in Definition E.2.1. By $\mathsf{rcab}\, \mathsf{v}(\mathfrak{I}, \mathfrak{X}^*)$ we denote the subspace of $\mathsf{cab}\, v(\mathfrak{I}, \mathfrak{X}^*)$ of regular σ-additive measures of bounded variation.

It turns out that $\mathsf{cab}\, v(\mathfrak{I}, \mathfrak{X}^*)$ is a Banach space and $\mathsf{rcab}\, \mathsf{v}(\mathfrak{I}, \mathfrak{X}^*)$ is a closed subspace and, thus, also a Banach space.

To characterise the dual of $C(\mathfrak{I}, \mathfrak{X})$ we recall that for a finite σ-additive measure $\nu : \mathcal{B} \to \mathfrak{X}^*$ of bounded variation and for each $\mathfrak{X}$-valued simple function $f = \sum_{i=1}^n x_i\, \chi_{A_i}$ the integral can be defined by

$$\int_{\mathfrak{I}} d\nu\, f := \sum_{i=1}^{n} \nu(A_i)(x_i) = \sum_{i=1}^{n} \langle x_i, \nu(A_i) \rangle.$$

The integral is well-defined and since ν has a bounded variation, it can be extended to all $\mathfrak{X}$-valued functions which are uniform limits of simple functions. In particular, the integral makes sense for all functions belonging to $C(\mathfrak{I}, \mathfrak{X})$. So ν defines a continuous linear functional on $C(\mathfrak{I}, \mathfrak{X})$, since it holds (cf. [Rao04, Theorem 7.4.4])

$$\left\| \int_{\mathfrak{I}} d\nu\, f \right\|_{\mathfrak{X}} \leq \|f\|_{\mathcal{L}^\infty(\mathfrak{I}, \mathfrak{X})} \|\nu\|(\mathfrak{I}).$$

Moreover, we have the following *Riesz-Singer* representation.

Proposition E.2.4 ([CeMe97, Theorem 1.7.1]) *Let $\mathfrak{X}$ be a Banach space. Then, the map*

$$\tau : \mathsf{rcab}\, \mathsf{v}(\mathfrak{I}, \mathfrak{X}^*) \to C(\mathfrak{I}, \mathfrak{X})^*,$$

defined by

$$\tau(\nu)(f) := \int_{\mathfrak{I}} \langle f, d\nu \rangle, \quad f \in C(\mathfrak{I}, \mathfrak{X}), \quad \nu \in \mathsf{rcab\,v}(\mathfrak{I}, \mathfrak{X}^*),$$

is an isometric isomorphism onto $C(\mathfrak{I}, \mathfrak{X})^*$.

Another useful tool we are going to develop now. The notation is adopted from [CeMe97]. Let λ be a scalar measure on $\mathcal{B}$ which is finite and complete. We say that a function $\phi : \mathfrak{I} \to \mathfrak{X}^*$ is λ-*essentially* bounded if $\lambda\text{-ess}\sup_{t \in \mathfrak{I}} \|\phi(t)\|_{\mathfrak{X}^*} < \infty$ and that it is w^*-*measurable* if for any $x \in \mathfrak{X}$ the function $\mathfrak{I} \ni t \to \langle x, \phi(t) \rangle$ is measurable. The vector space of all these functions is defined by

$$\mathcal{L}^\infty_{w^*}(\lambda, \mathfrak{X}^*) = \left\{\phi : \mathfrak{I} \to \mathfrak{X}^* : \lambda\text{-essentially bounded } \textit{and } w^*\text{-measurable}\right\}.$$

For a given regular measure $m \in \mathsf{rcab\,v}(\mathfrak{I}, \mathfrak{X}^*)$, its variation $\|m\|$ defines a finite non-negative measure on $\mathfrak{I}$, which can be completed in the usual way. Our aim is to represent a vector measure m by a density function $\phi \in \mathcal{L}^\infty_{w^*}(\|m\|, \mathfrak{X}^*)$, namely, as

$$m(A)(x) = \int_A d\|m\|(t)\, \langle x, \phi(t) \rangle.$$

This can be carried out using the standard *Radon-Nikodym* theorem.

Proposition E.2.5 *Let* $m \in \mathsf{rcab\,v}(\mathfrak{I}, \mathfrak{X}^*)$. *Then there is a function* $\phi \in \mathcal{L}^\infty_{w^*}(\|m\|, \mathfrak{X}^*)$ *such that*:

(i) *For each* $A \in \mathcal{B}$ *and all* $x \in \mathfrak{X}$

$$m(A)(x) = \int_A d\|m\|(t)\, \langle x, \phi(t) \rangle.$$

(ii) *The function* $t \mapsto \|\phi(t)\|$ *is measurable and for each* $A \in \mathcal{B}$ *it holds*

$$\|m(A)\| = \int_A d\|m\|(t)\, \|\phi(t)\|.$$

Proof The assertion is a consequence of [CeMe97, Theorem 1.5.2]. □

We remark that for any given scalar measure λ on $\mathfrak{I}$ each $\phi \in \mathcal{L}^\infty_{w^*}(\lambda, \mathfrak{X}^*)$ defines a measure $\nu \in \mathsf{cab}\, v(\mathcal{B}, \mathfrak{X}^*)$ by

$$\nu(A)(x) = \int_A d\lambda(t)\, \langle x, \phi(t) \rangle, \quad A \in \mathcal{B}, \quad x \in \mathfrak{X}.$$

Lemma E.2.6 *For each $\mu \in C(\mathfrak{I}, \mathfrak{X})^*$ there exists a function $\phi : \mathfrak{I} \to \mathfrak{X}^*$ such that*

(i) *ϕ is bounded and w^*-measurable.*
(ii) *$\mu(f) = \int_{\mathfrak{I}} dm\, f = \int_{\mathfrak{I}} d\|m\|(t)\,\langle f(t), \phi(t)\rangle$ for all $f \in C(\mathfrak{I}, \mathfrak{X})$, where $\mu = \tau(m)$ as in Theorem* E.2.4.

Proof According to Theorem E.2.4, there is a regular measure $m \in \mathsf{rcab}\, v(\mathfrak{I}, \mathfrak{X}^*)$ with its variation $\|m\|(\mathfrak{I}) = \|\nu\|_{C(\mathfrak{I},\mathfrak{X})^*}$. Then, according to Proposition E.2.5, there is $\phi : \mathfrak{I} \to \mathfrak{X}^*$ bounded and w^*-measurable such that for all Borel sets $A \subset \mathfrak{I}$ and $x \in \mathfrak{X}$ it holds that

$$m(A)(x) = \int_A d\|m\|(t)\,\langle x, \phi(t)\rangle.$$

So, for any simple function $h : \mathfrak{I} \to X$, it follows that

$$\int_{\mathfrak{I}} dm\, h = \int_{\mathfrak{I}} d\|m\|(t)\,\langle h(t), \phi(t)\rangle\,.$$

Now, take any $f \in C(\mathfrak{I}, \mathfrak{X})$. Then, there exists a sequence of simple functions $\{f_n\}_{n\in\mathbb{N}}$ which are uniformly bounded and such that $f_n \to f$ pointwise (or, equivalently, they converge in the $\|\cdot\|_{\mathcal{L}^\infty(\mathfrak{I},\mathfrak{X})}$-norm). Since ϕ is bounded, it follows that

$$\langle f_n(t), \phi(t)\rangle \to \langle f(t), \phi(t)\rangle, \quad t \in \mathfrak{I}.$$

According to the definition of the integral of $f \in C(\mathfrak{I}, \mathfrak{X})$ and using the Lebesgue *dominated convergence* theorem , we obtain

$$\int_{\mathfrak{I}} dm\, f = \lim_{n\to\infty} \int_{\mathfrak{I}} dm\, f_n = \lim_{n\to\infty} \int_{\mathfrak{I}} d\|m\|(t)\,\langle f_n(t), \phi(t)\rangle =$$
$$= \int_{\mathfrak{I}} d\|m\|(t)\, \lim_{n\to\infty} \langle f_n(t), \phi(t)\rangle = \int_{\mathfrak{I}} d\|m\|(t)\,\langle f(t), \phi(t)\rangle.$$

So, the claim follows. □

For the L^p-setting, we get the following characterisation of the space dual of $L^p(\mathfrak{I}, \mathfrak{X})$, for $p \in [1, \infty)$.

Proposition E.2.7 ([CeMe97, Theorem 1.5.4]) *Let $1 \le p < \infty$, and let $1 < p' \le \infty$ such that $1/p + 1/p' = 1$. Then for each $\Gamma \in L^p(\mathfrak{I}, \mathfrak{X})^*$ there is a function $\Psi : \mathfrak{I} \to \mathfrak{X}^*$ such that*

(i) *Ψ is w^*-measurable,*
(ii) *the function: $t \mapsto \|\Psi(t)\|_{\mathfrak{X}^*}$ is measurable and belongs to $L^{p'}(\mathfrak{I})$,*

(iii) $\Gamma(f) = \int_{\mathfrak{I}} dt \, \langle f(t), \Psi(t) \rangle$ *for all* $f \in L^p(\mathfrak{I}, \mathfrak{X})$,
(iv) $\|\Gamma\| = \| \, \|\Psi(\cdot)\| \, \|_{L^{p'}}$,
for any bounded (closed) interval $\mathfrak{I} \subset \mathbb{R}_0^+$.

Conversely, each w^**-measurable function* $\Psi : \mathfrak{I} \to \mathfrak{X}^*$, *for which there is* $g \in L^{p'}(\mathfrak{I})$ *such that* $\|\Psi(t)\|_{\mathfrak{X}^*} \leq g(t)$ *for a.a.* $t \in \mathfrak{I}$, *induces by (iii) a continuous linear functional* Γ *on* $L^p(\mathfrak{I}, \mathfrak{X})$, *whose norm is less than or equal to* $\|g\|_{L^{p'}}$.

E.3 Multiplication Operators and Semigroup Generators

In this section we essentially follow [Nei79, NSZ19] and [NSZ20]. Further we use the corresponding notations and assertions in the texts of Chap. 2 and of Part IV.

E.3.1 Multiplication Operators and Generators in $C_0([0, T], \mathfrak{X})$

We start the preliminaries of this section by the following key assertion.

Proposition E.3.1 *For every bounded multiplication operator* $\mathcal{C}$ *on* $\mathfrak{X}_T^C$, *there exists a unique family* $\{C(t)\}_{t\in[0,T]}$ *of bounded linear operators* $t \mapsto C(t) \in \mathcal{L}(\mathfrak{X})$, *such that:*

(i) $(\mathcal{C} f)(t) = C(t) f(t)$, *for any* $f \in \mathfrak{X}_T^C$ *and* $t \in [0, T]$.
(ii) *Uniformly bounded:* $\sup_{t\in[0,T]} \|C(t)\|_{\mathcal{L}(\mathfrak{X})} < \infty$.
(iii) *The operator norm of bounded multiplication operator* $\mathcal{C}$ *is given*

$$\|\mathcal{C}\|_{\mathcal{L}(\mathfrak{X}_T^C)} = \sup_{t\in[0,T]} \|C(t)\|_{\mathcal{L}(\mathfrak{X})} \, . \tag{E.3.1}$$

(iv) *Function* $t \mapsto C(t)$ *is strongly continuous on* $\mathfrak{X}$ *for* $t \in [0, T]$.

Proof of Proposition 2.1.24 By Definition 2.1.23, for scalar multiplication operator in identity (2.1.37) we choose for $\delta > 0$ a bounded non-negative function $\phi_\delta \in C([0, T])$ with compact support $[t - \delta, t + \delta] \subset [0, T]$ such that $\phi_\delta(t) = 1$ is its maximal value. Given that multiplication operator $\mathcal{C}$ is bounded: $\|\mathcal{C}\|_{\mathcal{L}(\mathfrak{X}_T^C)} < \infty$, on account of identity (2.1.37) and definition of scalar multiplication operator $\mathcal{M}(\phi_\delta)$ we obtain for $f \in \mathfrak{X}_T^C$

$$\begin{aligned} &\|\mathcal{M}(\phi_\delta)\, \mathcal{C} f\|_{\mathfrak{X}_T^C} = \|\phi_\delta \, \mathcal{C} f\|_{\mathfrak{X}_T^C} = \|\mathcal{C}\, \mathcal{M}(\phi_\delta)\, f\|_{\mathfrak{X}} \leq \\ &\|\mathcal{C}\|_{\mathcal{L}(\mathfrak{X}_T^C)} \|\mathcal{M}(\phi_\delta)\, f\|_{\mathfrak{X}_T^C} = \|\mathcal{C}\|_{\mathcal{L}(\mathfrak{X}_T^C)} \|\phi_\delta \, f\|_{\mathfrak{X}_T^C} \, . \end{aligned} \tag{E.3.2}$$

By definition of norms (2.1.22) and owing to the choice: $\phi_\delta \in \{C([0,T]) \wedge \phi_\delta(t) = 1\}$, inequality (E.3.2) yields

$$\|(\mathcal{C} f)(t)\|_{\mathfrak{X}} \leq \|\mathcal{C}\|_{\mathcal{L}(\mathfrak{X}_T^C)} \sup_{s\in[t-\delta,t+\delta]} \|f(s)\|_{\mathfrak{X}}, \tag{E.3.3}$$

for any $\delta > 0$ and every $f \in \mathfrak{X}_T^C$. Then (E.3.3) implies that $(\mathcal{C} f)(t) = 0$, if $f(t) = 0$, and consequently the map

$$C(t) : f(t) \mapsto (\mathcal{C} f)(t) \ \ (= C(t)\, f(t)), \tag{E.3.4}$$

for some $f \in \mathfrak{X}_T^C$ such that $f(t) = x \in \mathfrak{X}$ is well-defined.

Because of identity (2.1.37) the map (E.3.4) is linear. Further, on account of (E.3.3) for every $f \in \mathfrak{X}_T^C$ we obtain estimate

$$\sup_{t\in[0,T]} \|(\mathcal{C} f)(t)\|_{\mathfrak{X}} = \sup_{t\in[0,T]} \|C(t)\, f(t)\|_{\mathfrak{X}} \leq \|\mathcal{C}\|_{\mathcal{L}(\mathfrak{X}_T^C)} \sup_{t\in[0,T]} \|f(t)\|_{\mathfrak{X}},$$

which by definition of (2.1.22) and of operator norms provides

$$\sup_{t\in[0,T]} \|C(t)\|_{\mathcal{L}(\mathfrak{X})} = \|\mathcal{C}\|_{\mathcal{L}(\mathfrak{X}_T^C)}, \tag{E.3.5}$$

that is, $C(t) \in \mathcal{L}(\mathfrak{X})$ for $t \in [0, T]$.

Finally, let $f \in \mathfrak{X}_T^C$ and $x = f(t) \in \mathfrak{X}$. Then by (E.3.4) and (E.3.5) we obtain for $t + \varepsilon \in [t - \delta, t + \delta]$

$$\begin{aligned} &\|(C(t+\varepsilon) - C(t))\, x\|_{\mathfrak{X}} \leq \|(\mathcal{C} f)(t+\varepsilon) - (\mathcal{C} f)(t)\|_{\mathfrak{X}} + \\ &\|C(t+\varepsilon)(f(t+\varepsilon) - f(t))\|_{\mathfrak{X}} \leq \|(\mathcal{C} f)(t+\varepsilon) - (\mathcal{C} f)(t)\|_{\mathfrak{X}} + \\ &\|\mathcal{C}\|_{\mathcal{L}(\mathfrak{X}_T^C)} \|(f(t+\varepsilon) - f(t))\|_{\mathfrak{X}}\,. \end{aligned} \tag{E.3.6}$$

Since $f \in \mathfrak{X}_T^C$ and $\mathcal{C} : \mathfrak{X}_T^C \to \mathfrak{X}_T^C$, the estimate (E.3.6) implies the strong continuity of function: $t \mapsto C(t) \in \mathcal{L}(\mathfrak{X})$. □

Recall that in Sect. 2.1 we introduced the subspace $C_0([0,T],\mathfrak{X}) \subset C([0,T],\mathfrak{X})$ defined by

$$C_0([0,T],\mathfrak{X}) = \{f \in C([0,T],\mathfrak{X}) : f(0) = 0\}.$$

Each continuous function $\phi \in C([0,T])$ defines on $C_0([0,T],\mathfrak{X})$ a *scalar* multiplication operator $\mathcal{M}(\phi)$ given by

$$(\mathcal{M}(\phi) f)(t) := \phi(t) f(t), \quad \mathrm{dom}\,(\mathcal{M}(\phi)) = C_0([0,T],\mathfrak{X}).$$

If $\{A(t)\}_{t\in[0,T]}$ is a family of operators in $\mathfrak{X}$, with domains, which in general depend on time, then family $\{A(t)\}_{t\in[0,T]}$ induces on $C_0([0,T],\mathfrak{X})$ a multiplication operator $\mathcal{A}$ such that

$$(\mathcal{A}f)(t) := A(t)f(t),$$

with

$$\operatorname{dom}(\mathcal{A}) := \{f \in C_0([0,T],\mathfrak{X}) : f(t) \in \operatorname{dom}(A(t)), \quad A(t)f(t) \in C_0([0,T],\mathfrak{X})\}.$$

Then for $\phi \in C([0,T])$ these multiplication operators commute:

$$\mathcal{A}\mathcal{M}(\phi)f = \mathcal{M}(\phi)\mathcal{A}f, \quad f \in \operatorname{dom}(\mathcal{A}), \quad \phi \in C([0,T]).$$

Moreover, they inherit in $C_0([0,T],\mathfrak{X})$ the properties to be closed, densely defined and to have similar norm estimates of operators of generating families in $\mathfrak{X}$; see Sect. 2.1.

Furthermore, we additionally have the following lemma.

Lemma E.3.2 *If $\{A(t)\}_{t\in[0,T]}$ is a family of closed operators, then the induced multiplication operator $\mathcal{A}$ is also closed.*

Proof Let $f_n \to f$ be a sequence in $C_0([0,T],\mathfrak{X})$ with $f_n \in \operatorname{dom}(\mathcal{A})$ and $\mathcal{A}f_n \to g \in C_0([0,T],\mathfrak{X})$. Then, it holds pointwise for any $t \in [0,T]$ that

$$f_n(t) \in \operatorname{dom}(A(t)), \quad f_n(t) \to f(t), \quad A(t)f_n(t) \to g(t).$$

Since the operators $A(t)$ are closed, we conclude for $t \in [0,T]$ that $f(t) \in \operatorname{dom}(A(t))$ and $A(t)f(t) = g(t)$. So, $f \in \operatorname{dom}(\mathcal{A})$ and $(\mathcal{A}f)(t) = A(t)f(t) = g(t)$. □

Whether or not the induced operator $\mathcal{A}$ is *densely* defined strongly depends on another assumptions on the operator family $\{A(t)\}_{t\in[0,T]}$. The following example illustrates this fact.

Example E.3.3 On $[0,T] = [0,1]$ we define the scalar-valued function

$$A(t) := \begin{cases} 1, & t \in \left[0, \frac{1}{2}\right) \\ 2, & t \in \left[\frac{1}{2}, 1\right] \end{cases},$$

defined on the Banach space $\mathfrak{X} = \mathbb{C}$. Hence, for any $t \in [0,T]$, the operator $A(t)$ is trivially densely defined. The induced operator $\mathcal{A}$ is not densely defined on $C_0([0,T])$, because $t \mapsto A(t)f(t) \in C_0([0,T])$ only if $f(1/2) = 0$.

This observation leads us to the following definition.

Definition E.3.4 We say the set $\mathfrak{D} \subset C_0([0,T],\mathfrak{X})$ has a dense *cross-section* in $\mathfrak{X}$ if for every $t \in (0,T]$ the set

$$[\mathfrak{D}]_t := \{x \in \mathfrak{X} : \exists f \in \mathfrak{D} : f(t) = x\}, \tag{E.3.7}$$

that is, the cross-section of $\mathfrak{D}$ at t is dense in $\mathfrak{X}$.

The following lemma provides a criteria for the induced multiplication operator being densely defined.

Lemma E.3.5 *Let $\mathfrak{D} \subset C_0([0,T],\mathfrak{X})$ be a linear subset such that for every $\varphi \in C^1([0,T])$ one has $\mathcal{M}(\varphi)\mathfrak{D} \subseteq \mathfrak{D}$. The subset $\mathfrak{D}$ is dense in $C_0([0,T],\mathfrak{X})$ if and only if $\mathfrak{D}$ has dense cross-section in $\mathfrak{X}$.*

Proof Let us assume that $\mathfrak{D}$ is not dense. Then there is a functional $\mu \in C_0([0,T],\mathfrak{X})^*$ such that $\mu(f) = 0$ for $f \in \mathfrak{D}$. By the *Hahn-Banach* theorem there is an extension $\tilde{\mu} \in C([0,T],\mathfrak{X})^*$ of μ such that $\tilde{\mu} \restriction C_0([0,T],\mathfrak{X}) = \mu$ and $\|\tilde{\mu}\| = \|\mu\|$ such that $\tilde{\mu}(f) = 0$ for $f \in \mathfrak{D}$. By Theorem E.2.4 there is a measure $m \in \mathsf{rcab}\,v([0,T],\mathfrak{X}^*)$ such that

$$\tau(m)(f) = \int_{[0,T]} \langle f, dm\rangle, \quad f \in C([0,T],\mathfrak{X}).$$

By Lemma E.2.6 there is a w^*-measurable function $\phi : [0,T] \to \mathfrak{X}^*$

$$0 = \mu(f) = \int_{[0,T]} \langle f, dm\rangle = \int_{[0,T]} \langle f(t), \phi(t)\rangle d\|m\|(t), \quad f \in \mathfrak{D}.$$

Since $\mathcal{M}(\varphi)\mathfrak{D} \subset \mathfrak{D}$ for $\varphi \in C^1([0,T])$, it follows that

$$0 = \int_{[0,T]} \langle \varphi(t) f(t), \phi(t)\rangle d\|m\|(t) = \int_{[0,T]} \varphi(t)\langle f(t), \phi(t)\rangle d\|m\|(t), \quad f \in \mathfrak{D}.$$

The function $t \mapsto \langle f(t), \phi(t)\rangle$ is in $L^1(\|m\|)$. Since $\varphi \in C^1([0,T])$ is arbitrary, we conclude that

$$0 = \langle f(t), \phi(t)\rangle$$

for *a.a.* $t \in [0,T]$ with respect to $\|m\|$. Since $\mathfrak{D}$ has dense cross-section, we have $\phi(t) = 0$ for a.a. $t \in [0,T]$ with respect to $\|m\|$. Hence, $\mu = 0$ which yields that $\mathfrak{D}$ is dense in $C_0([0,T],\mathfrak{X})$.

The converse is plain to see. □

Remark that if $\mathfrak{D} \subset C_0([0,T],\mathfrak{X})$ is dense, then it has dense cross-section in $\mathfrak{X}$. Moreover, for a multiplication operator $\mathcal{A}$ we have the relation

$$[\operatorname{dom}(\mathcal{A})]_t \subset \operatorname{dom}(A(t)) \quad \text{for } t \in (0,T].$$

For generators of C_0-semigroups on $C_0([0,T],\mathfrak{X})$ we have the following theorem.

Theorem E.3.6 *Let $\{A(t)\}_{t\in[0,T]}$ be a family of generators on $\mathfrak{X}$ such that for all $t \in (0,T]$ it holds that $A(t) \in \mathcal{G}(M,\beta)$ for some $M \geq 1$ and $\beta \in \mathbb{R}$. If the function $[0,T] \ni t \mapsto (A(t)+\xi)^{-1}x$ is continuous for each $x \in \mathfrak{X}$ and $\xi > \beta$, then the induced multiplication operator $\mathcal{A}$ is a generator in $C_0([0,T],\mathfrak{X})$ and the corresponding semigroup is given by*

$$(\mathrm{e}^{-\tau\mathcal{A}}f)(t) = \mathrm{e}^{-\tau A(t)}f(t), \quad f \in C_0([0,T],\mathfrak{X}), \quad t \in [0,T].$$

In particular, it holds $\|\mathrm{e}^{-\tau\mathcal{A}}\|_{\mathcal{L}(C_0([0,T],\mathfrak{X}))} = \sup_{t\in(0,T]}\|\mathrm{e}^{-\tau A(t)}\|_{\mathcal{L}(\mathfrak{X})}$, i.e. $\mathcal{A} \in \mathcal{G}(M,\beta)$.

Proof First we show that $\mathcal{A}$ is closed and densely defined. According to Lemma 2.1.22, the induced multiplication operator $\mathcal{A}$ is closed. Notice that $\mathcal{M}(\phi)\operatorname{dom}(\mathcal{A}) \subseteq \operatorname{dom}(\mathcal{A})$ for any $\phi \in C^1([0,T])$. By Lemma E.3.5, it remains to show that $\operatorname{dom}(\mathcal{A})$ has dense cross-section in $\mathfrak{X}$. Let $t_0 \in (0,T]$, $0 < \alpha < \frac{1}{|\beta|}$ and $x \in \mathfrak{X}$. Let $\varphi \in C^1([0,T])$ such that $\varphi(0) = 0$ and $\varphi(t_0) = 1$. We set

$$f_\xi(t) := \varphi(t)\xi(A(t)+\xi)^{-1}x, \quad \xi > \beta.$$

Then, by assumption $f_\xi \in C_0([0,T],\mathfrak{X})$, $f_\xi(t) \in \operatorname{dom}(A(t))$,

$$A(t)f_\xi(t) = \xi\varphi(t)x - \xi^2\varphi(t)(A(t)+\xi)^{-1}x \in C_0([0,T],\mathfrak{X}).$$

Hence, $f_\xi \in \operatorname{dom}(\mathcal{A})$. Since $A(t_0)$ is a generator, we have

$$f_\xi(t_0) = \xi(A(t_0)+\xi)^{-1}x \to x, \quad \text{for } \xi \to \infty \text{ and } x \in \mathfrak{X}.$$

Therefore, for any $\varepsilon > 0$ there is a $\xi > \beta$ such that $\|f_\xi(t_0) - x\|_{\mathfrak{X}} < \varepsilon$. Hence, $\operatorname{dom}(\mathcal{A})$ has dense cross-section in $\mathfrak{X}$.

For any $t \in [0,T]$ the operators $A(t)$ belong to the same class $\mathcal{G}(M,\beta)$. Thus, we have for any $k \in \mathbb{N}$

$$\|(A(t)+\lambda)^{-k}\|_{\mathcal{L}(\mathfrak{X})} \leq \frac{M}{|\lambda-\beta|^k}, \quad \lambda > \beta, \quad k \in \mathbb{N}.$$

Hence, for any $t \in [0, T]$ and $f \in C_0([0, T], \mathfrak{X})$, we get

$$\|((\mathcal{A}+\lambda)^{-k} f)(t)\|_{\mathfrak{X}} = \|(A(t)+\lambda)^{-k}(f(t))\|_{\mathfrak{X}} \leq \frac{M}{|\lambda-\beta|^k}\|f(t)\|_{\mathfrak{X}}, \quad \lambda > \beta.$$

Taking the *supremum* on $[0, T]$, we obtain

$$\|(\mathcal{A}+\lambda)^{-k} f\|_{C_0([0,T],x)} \leq \frac{M}{|\lambda-\beta|^k}\|f\|_{C_0([0,T],x)}, \quad \lambda > \beta, \quad k \in \mathbb{N},$$

and hence it follows that $\mathcal{A}$ is a generator on $C_0([0, T], \mathfrak{X})$. Its semigroup is given by the limit

$$\mathrm{e}^{-\tau\mathcal{A}} f = \lim_{n\to\infty}\left(\mathbb{1}+\frac{\tau}{n}\mathcal{A}\right)^{-n} f, \quad f \in C_0([0, T], \mathfrak{X}).$$

For any $n \in \mathbb{N}$ we have

$$\left(\left(\mathbb{1}+\frac{\tau}{n}\mathcal{A}\right)^{-n} f\right)(t) = \left(\mathbb{1}+\frac{\tau}{n}A(t)\right)^{-n} f(t),$$

which yields

$$\begin{aligned}(\mathrm{e}^{-\tau\mathcal{A}} f)(t) &= \lim_{n\to\infty}\left(\left(\mathbb{1}+\frac{\tau}{n}\mathcal{A}\right)^{-n} f\right)(t) = \lim_{n\to\infty}\left(\mathbb{1}+\frac{\tau}{n}A(t)\right)^{-n} f(t)\\ &= \mathrm{e}^{-\tau A(t)} f(t)\end{aligned}$$

for $f \in C_0([0, T], \mathfrak{X})$.

The last statement is owing to Lemma E.3.7. □

Lemma E.3.7 *Let $\{C(t)\}_{t\in[0,T]}$ be a family of bounded linear operators on $\mathfrak{X}$ such that such that $t \ni t \mapsto C(t)x$ is continuous for each $x \in \mathfrak{X}$. If $\sup_{t\in[0,T]}\|C(t)\|_{\mathcal{L}(\mathfrak{X})} < \infty$, then the induced multiplication operator $\mathcal{C}$ on $C_0([0, T], \mathfrak{X})$ is bounded and*

$$\|\mathcal{C}\|_{\mathcal{L}(C_0([0,T],\mathfrak{X}))} = \sup_{t\in[0,T]}\|C(t)\|_{\mathcal{L}(\mathfrak{X})}.$$

For the proof of this lemma we refer to Theorem E.3.12 in Sect. E.3.3.

E.3.2 Multiplication Operators and Generators in $L^p([0, T], \mathfrak{X})$

In Sect. 2.2 we introduced the Banach spaces $L^p([0, T], \mathfrak{X})$ for $p \in [1, \infty)$. Definition of the both kinds of multiplication operators $\mathcal{M}(\phi)$ and $\mathcal{A}_p$ can be naturally extended to $L^p([0, T], \mathfrak{X})$.

Let $\phi \in L^\infty([0, T])$, $p \in [1, \infty]$. We set

$$(\mathcal{M}_p(\phi) f)(t) := \phi(t) f(t), \quad \mathrm{dom}\,(\mathcal{M}_p(\phi)) = L^p([0, T], \mathfrak{X}),$$

and, similarly to the C-setting, we define operator

$$(\mathcal{A}_p f)(t) := A(t) f(t),$$

with domain

$$\mathrm{dom}\,(\mathcal{A}_p) := \{f \in L^p([0, T], \mathfrak{X}) : f(t) \in \mathrm{dom}\,(A(t)),\\ A(t) f(t) \in L^p([0, T], \mathfrak{X})\}$$

induced by a family of densely defined closed operators $\{A(t)\}_{t\in[0,T]}$.

Obviously, we have $\mathcal{M}_p(\phi)\mathrm{dom}\,(\mathcal{A}_p) \subseteq \mathrm{dom}\,(\mathcal{A}_p)$, $\phi \in L^\infty([0, T])$, and

$$\mathcal{A}_p \mathcal{M}_p(\phi) f = \mathcal{M}_p(\phi) \mathcal{A}_p f, \quad f \in \mathrm{dom}\,(\mathcal{A}_p), \quad \phi \in L^\infty([0, T]),$$

for any $p \in \mathbb{R}_0^+$.

The L^p-setting admits a natural extension of the concept of *cross-section* from $C_0([0, T], \mathfrak{X})$ to the space $L^p([0, T], \mathfrak{X})$, as well as *inheritance* of operator's domain and norm properties, already observed for induced operators in the case of the C-setting.

Definition E.3.8 We say that set $\mathfrak{D} \subset L^p([0, T], \mathfrak{X})$, $p \in [1, \infty]$, has a dense cross-section in $\mathfrak{X}$ if

$$\mathfrak{D} \subseteq L^p([0, T], \mathfrak{X}) \cap C_0([0, T], \mathfrak{X}),$$

and the set $\mathfrak{D}$ regarded as a subset of $C_0([0, T], \mathfrak{X})$ has dense cross-section in $\mathfrak{X}$. Similarly to C-setting the cross-section of $\mathfrak{D}$ at $t \in [0, T]$ is denoted by $[\mathfrak{D}]_{t\in[0,T]}$ and defined as

$$[\mathfrak{D}]_t := \{x \in \mathfrak{X} : \exists f \in \mathfrak{D} \text{ such that } f(t) = x\};$$

cf. Definition 2.1.29.

Lemma E.3.9 *Let $\mathfrak{D} \subseteq L^p([0,T],\mathfrak{X}) \cap C_0([0,T],\mathfrak{X})$, $p \in [1,\infty)$, be a linear set. If $\mathfrak{D}$ has dense cross-section in $\mathfrak{X}$ and it holds $\mathcal{M}_p(\phi)\mathfrak{D} \subseteq \mathfrak{D}$ for every $\phi \in W^{1,\infty}([0,T])$, then $\mathfrak{D}$ is dense in $L^p([0,T],\mathfrak{X})$.*

Proof Let $\mu \in L^p([0,T],\mathfrak{X})^*$ be any functional on $L^p([0,T],\mathfrak{X})$ such that

$$\langle f, \mu \rangle = 0, \quad f \in \mathfrak{D}.$$

We use the characterisation of the dual space $L^p([0,T],\mathfrak{X})^*$ from Proposition E.2.7. Hence, there is a w^*-measurable function $\Psi : [0,T] \to \mathfrak{X}^*$ such that

$$0 = \langle f, \mu \rangle = \int_{[0,T]} \langle f(t), \Psi(t) \rangle \, dt, \text{ for } f \in \mathfrak{D}.$$

Since $M_p(\phi)\mathfrak{D} \subset \mathfrak{D}$ for $\phi \in W^{1,\infty}([0,T])$, it follows that

$$0 = \int_{[0,T]} \langle \phi(t) f(t), \Psi(t) \rangle \, dt = \int_{[0,T]} \phi(t) \langle f(t), \Psi(t) \rangle \, dt.$$

The function $t \mapsto \langle f(t), \phi(t) \rangle$ is in $L^1([0,T])$. Since $\phi \in W^{1,\infty}([0,T])$ is arbitrary, we conclude that $0 = \langle f(t), \Psi(t) \rangle$, for a.a. $t \in [0,T]$. Since $\mathfrak{D}$ has dense cross-section, we have $\Psi(t) = 0$ for a.a. $t \in [0,T]$ and it follows that $\mu = 0$. □

We remark that the *converse* to this statement is in general *not* true. To put it differently, let $\mathfrak{D} \subseteq L^p([0,T],\mathfrak{X}) \cap C_0([0,T],\mathfrak{X})$ be a linear subset such that $\mathcal{M}_p(\phi)\mathfrak{D} \subseteq \mathfrak{D}$ for each $\phi \in W^{1,\infty}([0,T])$. If $\mathfrak{D}$ is dense in $L^p([0,T],\mathfrak{X})$, then *in general* $\mathfrak{D}$ does not have a dense cross-section.

Now, let $\{A(t)\}_{t \in [0,T]}$ be a family of operators on $\mathfrak{X}$ with domains, which might be *time-dependent*. The family $\{A(t)\}_{t \in [0,T]}$ induces a multiplication operator $\mathcal{A}_p$ on $L^p([0,T],\mathfrak{X})$, $p \in [1,\infty]$, given by

$$(\mathcal{A}_p f)(t) := A(t) f(t), \quad t \in [0,T], \quad p \in [1,\infty], \tag{E.3.8}$$

with domain

$$\begin{aligned} \operatorname{dom}(\mathcal{A}_p) := \{ & f \in L^p([0,T],\mathfrak{X}) : f(t) \in \operatorname{dom}(A(t)), \\ & A(t) f(t) \in L^p([0,T],\mathfrak{X}) \}. \end{aligned}$$

To consider multiplication operators (E.3.8) has the advantage that in contrast to the multiplication operator $\mathcal{A}$ in $C_0([0,T],\mathfrak{X})$, the operator $\mathcal{A}_p$ may be densely defined, whereas $\mathcal{A}$ not. For instance, let

$$A(t) := \begin{cases} A, & t \in [0, T/2), \\ A + B, & t \in [T/2, T], \end{cases}$$

where A is a generator of a semigroup in $\mathfrak{X}$ and B is a bounded operator such that $\ker(B) = \{0\}$. One can check that $f \in \mathrm{dom}(\mathcal{A}) \subset C_0([0,T],\mathfrak{X})$ yields $f(T/2) = 0$, which shows that $\mathcal{A}$ is *not* densely defined in $C_0([0,T],\mathfrak{X})$. However, the multiplication operator $\mathcal{A}_p$, is densely defined in $L^p([0,T],\mathfrak{X})$ for any $p \in [1,\infty)$.

We also comment that $\mathcal{A}_p\, \iota_p = \iota_p\, \mathcal{A}$, $p \in [0,\infty]$. Here ι_p is the *embedding* operator for $C_0([0,T],\mathfrak{X}) \overset{d}{\hookrightarrow} L^P([0,T],\mathfrak{X})$, with $\ker(\iota_p) = \{0\}$; cf. (2.2.8) and (2.2.9).

Lemma E.3.10 *If $\{A(t)\}_{t\in[0,T]}$ is a family of closed linear operators on $\mathfrak{X}$, then the induced operator $\mathcal{A}_p$, $p \in [1,\infty]$, is also closed.*

Proof Given a sequence $f_n \to f$ with $f_n \in \mathcal{A}$, $f \in L^p([0,T],\mathfrak{X})$ and $\mathcal{A}f_n \to g$ in $L^p([0,T],\mathfrak{X})$, we conclude that $A(t)f_n(t) \to g(t)$ for a.a. $t \in [0,T]$ for a subsequence. Moreover, $f_n(t) \to f(t)$ for a.a. $t \in [0,T]$ for another subsequence. Choosing a diagonal sequence we can denote the subsequences in the same way as the original sequence. Since $A(t)$ is a closed operator on $\mathfrak{X}$, we conclude $f(t) \in \mathrm{dom}(A(t))$ and $A(t)f(t) = g(t)$ almost everywhere. Since $g \in L^p([0,T],\mathfrak{X})$, we conclude $f \in \mathrm{dom}(\mathcal{A})$ and $\mathcal{A}f = g$ almost everywhere on $[0,T]$. □

For a family of generators, we have the following theorem.

Theorem E.3.11 *Let $\{A(t)\}_{t\in[0,T]}$ be a family of generators in $\mathfrak{X}$ such that $A(t) \in \mathcal{G}(M,\beta)$, $M \geq 1$, $\beta \in \mathbb{R}$ for a.a. $t \in [0,T]$. If the function $[0,T] \ni t \mapsto (A(t) + \xi)^{-1}x \in \mathfrak{X}$ is strongly measurable for $\xi > \beta$, $x \in \mathfrak{X}$, then for each $p \in \mathbb{R}_0^+$ the induced multiplication operator $\mathcal{A}_p$ is a generator in $L^p([0,T],\mathfrak{X})$ of a C_0-semigroup such that*

$$(\mathrm{e}^{-\tau\mathcal{A}_p} f)(t) = \mathrm{e}^{-\tau A(t)} f(t) \quad \textit{for a.a. } t \in [0,T].$$

In particular, $\|\mathrm{e}^{-\tau\mathcal{A}_p}\|_{\mathcal{L}(L^p([0,T],\mathfrak{X}))} = \|\mathrm{e}^{-\tau A(t)}\|_{\mathcal{L}(\mathfrak{X})}$ for $\tau \in \mathbb{R}_+^0$ and $\mathcal{A}_p \in \mathcal{G}(M,\beta)$.

Proof First we show that $\mathcal{A}_p$ is densely defined on $L^p([0,T],\mathfrak{X})$. Let $\mathcal{I} \subset [0,T]$ be some Borel set, $\xi > \beta$ and $x \in \mathfrak{X}$. Let

$$f_{\mathcal{I},\xi}(t) := \xi\,(A(t) + \xi\,\mathbb{1})^{-1}\,\chi_{[0,T]}(t)x, \quad t \in [0,T].$$

Then, by assumption $f_{\mathcal{I},\xi}$ is in $L^p([0,T],\mathfrak{X})$, $f_{\mathcal{I},\xi}(t) \in \mathrm{dom}(A(t))$ for a.a. $t \in [0,T]$, and

$$A(t)f_{\mathcal{I},\xi}(t) = \xi\,\chi_{\mathcal{I}}(t) - \xi^2(A(t) + \xi\,\mathbb{1})^{-1}\chi_{\mathcal{I}}(t)x, \quad t \in [0,T],$$

is in $L^p([0,T],\mathfrak{X})$ for $p \in [1,\infty)$. Hence $f_{[0,T],\xi} \in \mathrm{dom}\,(\mathcal{A}_p)$. Note that

$$f_{\mathcal{I},\xi} \xrightarrow{L^p} \chi_{[}0,T]x, \quad \text{as } \xi \to \infty, \text{ for } x \in \mathfrak{X} \text{ and a.a. } t \in [0,T].$$

Therefore, $\mathrm{dom}\,(\mathcal{A}_p)$ is dense in $L^p([0,T],\mathfrak{X})$.

The rest of the proof can be completed using similar arguments as for the proof of Theorem E.3.6.

The last statement of Theorem E.3.11 is again due to Lemma E.3.7. □

We conclude by the comment that $\iota_p(\mathrm{dom}\,(\mathcal{A})) \subseteq \mathrm{dom}\,(\mathcal{A}_p)$ for any $p \in [1,\infty)$, and

$$\iota_p \mathcal{A} f = \mathcal{A}_p \iota_p f\,, \quad f \in \mathrm{dom}\,(\mathcal{A}), \quad p \in [1,\infty)\,.$$

Here ι_p is the *embedding* operator: $C_0([0,T],\mathfrak{X}) \overset{d}{\hookrightarrow} L^P([0,T],\mathfrak{X})$ for $p \in [1,\infty)$; see definition (2.2.8) and identity (2.2.9).

E.3.3 Norm of the Multiplication Operator

Theorem E.3.12 ([NSZ20, Proposition 2.11]) *Let $\{C(t)\}_{t\in[0,T]}$ be a strongly measurable family of bounded linear operators on (separable) Banach space $\mathfrak{X}$.*

(i) *If* $\mathrm{ess\,sup}_{t\in[0,T]} \|C(t)\|_{\mathcal{L}(\mathfrak{X})} < \infty$, *then the induced multiplication operator $\mathcal{C}$ on the Banach space $\mathfrak{X}_T^C = C_0([0,T],\mathfrak{X})$ is bounded and*

$$\|\mathcal{C}\|_{\mathcal{L}(\mathfrak{X}_T^C)} = \operatorname*{ess\,sup}_{t\in[0,T]} \|C(t)\|_{\mathcal{L}(\mathfrak{X})}\,,$$

for its operator norm on the Banach space $\mathcal{L}(\mathfrak{X}_T^C)$.

(ii) *If* $\mathrm{ess\,sup}_{t\in[0,T]} \|C(t)\|_{\mathcal{L}(\mathfrak{X})} < \infty$, *then the induced multiplication operator $\mathcal{C}$ on the Banach space $\mathfrak{X}_T^p = L^p([0,T],\mathfrak{X})$, $p \in [1,\infty)$, is bounded and*

$$\|\mathcal{C}\|_{\mathcal{L}(\mathfrak{X}_T^p)} = \operatorname*{ess\,sup}_{t\in[0,T]} \|C(t)\|_{\mathcal{L}(\mathfrak{X})}\,, \quad p \in [1,\infty)\,, \tag{E.3.9}$$

for its operator norm on the Banach space $\mathcal{L}(\mathfrak{X}_T^p)$.

Proof

(i) First, for the C-setting, we fix $t_0 \in (0, T]$ and $x \in \mathfrak{X}$ with $\|x\|_{\mathfrak{X}} = 1$. Then there is a function $f \in \mathfrak{X}_T^C := C_0([0, T], \mathfrak{X})$ such that $\|f\|_C = 1$ and $f(t_0) = x$. As a consequence we obtain for f that

$$\|\mathcal{C}\|_{\mathcal{L}(\mathfrak{X}_T^C)} \geq \|\mathcal{C} f\|_{\mathfrak{X}_T^C} = \sup_{t\in[0,T]} \|C(t) f(t)\|_{\mathfrak{X}} \geq \|C(t_0)x\|_{\mathfrak{X}}\,.$$

Since $t_0 \in (0, T]$ is arbitrary, we have

$$\|\mathcal{C}\|_{\mathcal{L}(C_0([0,T],\mathfrak{X}))} \geq \sup_{t\in[0,T]} \|C(t)x\|\,,$$

for any $x \in \mathfrak{X}$, and for that reason

$$\|\mathcal{C}\|_{\mathcal{L}(\mathfrak{X}_T^C)} \geq \sup_{t\in(0,T]} \|C(t)\|_{\mathcal{L}(\mathfrak{X})}\,.$$

Conversely, let $\beta := \sup_{t\in(0,T]} \|C(t)\|_{\mathcal{L}(\mathfrak{X})}$. Then, we infer that

$$\begin{aligned}
\|\mathcal{C}\|_{\mathcal{L}(\mathfrak{X}_T^C)} &= \sup_{\|f\|_{\mathfrak{X}_T^C}=1} \sup_{t\in[0,T]} \|C(t) f(t)\|_{\mathfrak{X}} \\
&= \sup_{\|f\|_{\mathfrak{X}_T^C}=1} \sup_{t\in(0,T]} \|C(t)\| \cdot \|f(t)\|_{\mathfrak{X}} \leq \\
&\leq \beta \sup_{\|f\|_{\mathfrak{X}_T^C}=1} \sup_{t\in[0,T]} \|f(t)\|_{\mathfrak{X}} = \beta\,.
\end{aligned}$$

Consequently, we have $\|\mathcal{C}\|_{\mathcal{L}(\mathfrak{X}_T^C)} = \sup_{t\in(0,T]} \|C(t)\|_{\mathcal{L}(\mathfrak{X})}$.

(ii) Given that $\mathfrak{X}_T^p := L^p([0, T], \mathfrak{X})$, for the L^p-setting we obtain the estimates:

$$\begin{aligned}
\|\mathcal{C}\|^p_{\mathcal{L}(\mathfrak{X}_T^p)} &= \sup_{\|f\|_{\mathfrak{X}_T^p}=1} \|\mathcal{C} f\|^p_{\mathfrak{X}_T^p} = \sup_{\|f\|_{\mathfrak{X}_T^p}=1} \int_{[0,T]} ds\, \|C(s) f(s)\|^p_{\mathfrak{X}} \leq \\
&\leq \sup_{\|f\|_{\mathfrak{X}_T^p}=1} \int_{[0,T]} ds\, \|C(s)\|^p_{\mathcal{L}(\mathfrak{X})} \|f(s)\|^p_{\mathfrak{X}} \leq \\
&\leq \operatorname{ess\,sup}_{t\in[0,T]} \|C(t)\|^p_{\mathcal{L}(\mathfrak{X})} \cdot \sup_{\|f\|_{\mathfrak{X}_T^p}=1} \left(\int_{[0,T]} ds\, \|f(s)\|^p_{\mathfrak{X}} \right) = \\
&= \operatorname{ess\,sup}_{t\in[0,T]} \|C(t)\|^p_{\mathcal{L}(\mathfrak{X})}\,,
\end{aligned} \tag{E.3.10}$$

which prove the assertion in one direction.

Next, we show the converse estimate. For $\epsilon > 0$, let $\Delta_\epsilon = [t, t+\epsilon]$ for a fixed $t \in [0, T)$, whereas for $t = T$ we put $\Delta_\epsilon = [T-\epsilon, T]$. Fix $x \in \mathfrak{X}\backslash\{0\}$. Then vector $\chi_{\Delta_\epsilon} x \in L^p([0, T], \mathfrak{X})$, and by the strong measurability it follows that

$$\frac{1}{\epsilon}\int_{\Delta_\epsilon} ds\, \|C(s)x\|^p_{\mathfrak{X}} = \frac{1}{\epsilon}\int_{[0,T]} ds\, \|C(s)\chi_{\Delta_\epsilon}(s)x\|^p_{\mathfrak{X}} \leq$$
$$\leq \frac{1}{\epsilon}\, \|\mathcal{C}\|^p_{\mathcal{L}(\mathfrak{X}^p_T)} \int_{[0,T]} ds\, \|\chi_{\Delta_\epsilon}(s)x\|^p_X\,.$$

Now let $\epsilon \to +0$. Then again by the strong measurability one infers that

$$\|C(t)x\|^p_{\mathfrak{X}} \leq \|\mathcal{C}\|^p_{\mathcal{L}(\mathfrak{X}^p_T)}\|x\|^p_{\mathfrak{X}} \Rightarrow \frac{\|C(t)x\|_{\mathfrak{X}}}{\|x\|_{\mathfrak{X}}} \leq \|\mathcal{C}\|_{\mathcal{L}(\mathfrak{X}^p_T)}\,, \text{ for a.a. } t \in [0, T]\,.$$

Since $x \in \mathfrak{X} \setminus \{0\}$ is arbitrary, mapping $t \mapsto C(t)$, is bounded and strongly measurable, we conclude that

$$\|C(t)\|_{\mathcal{L}(\mathfrak{X})} \leq \|\mathcal{C}\|_{\mathcal{L}(\mathfrak{X}^p_T)}\,, \text{ for a.a. } t \in [0, T]\,. \tag{E.3.11}$$

Taking in (E.3.11) the *essential supremum* on $[0, T]$ we prove the estimate, which is converse to (E.3.10) and, as a result, the assertion.

□

Remark E.3.13 We note that for a *separable* Banach space $\mathfrak{X}$ and a *bounded* linear map $\mathcal{C}$ on the Banach space $\mathfrak{X}^p_T = L^p([0, T], \mathfrak{X})$, $p \in [1, \infty)$, which is in turn a separable space, one can also obtain equality (E.3.9). It holds for a family $\{C(t)\}_{t\in[0,T]}$ of bounded strongly measurable operators on $\mathfrak{X}$ that can be *reconstructed* from the operator $\mathcal{C}$ under complementary condition (see [Ev76, Theorem 5.7]):

$$\mathcal{C}\, \mathcal{M}_p(\phi) = \mathcal{M}_p(\phi)\, \mathcal{C} \quad \text{for any} \quad \phi \in L^\infty([0, T])\,.$$

Here $\mathcal{M}_p(\cdot)$ is the multiplication operator on the Banach space $\mathfrak{X}^p_T$ defined by

$$(\mathcal{M}_p(\phi) f)(t) := \phi(t) f(t), \quad f \in \mathfrak{X}^p_T\,.$$

Cf. Appendix E.3.2.

Appendix F
Comments on Evolution Semigroups

F.1 The Howland-Evans-Neidhardt Formula: C-Setting

The *Howland-Evans-Neidhardt* (HEN) formula (or representation) (2.1.24) for evolution *semigroups* on a general (not necessarily *separable*) Banach space $\mathfrak{X}$ in the C-setting is due to [Nei79] and [Nei81, Theorem 3.8, Theorem 3.10]. See a preliminary note along the same vein in [Paq79, Theorem 1.8] and the original important advance in consolidation the idea of HEN by Howland [How74] and Evans [Ev76] in a *separable* Banach space.

We make a point that approach developed in [Nei79, Nei81, Nei81a, Nei82] has two aspects: the C-setting and the L^p-setting. The more restricted C-setting basically serves as an auxiliary construction, albeit certain assertions of this section provide, *mutatis mutandis*, results for the L^p-setting.

Section F.1.1

The C-setting means that we consider evolution semigroup $\{\mathcal{U}(\tau)\}_{\tau\geq 0}$ on the Banach space $\mathfrak{X}_T^C := C_0([0,T],\mathfrak{X})$, Sect. 2.1.2. Recall that Banach space

$$C_0([0,T],\mathfrak{X}) := \{f \in C([0,T],\mathfrak{X}) : f(0) = 0\} \tag{F.1.1}$$

is endowed with the sup-norm topology:

$$\|f\|_{\mathfrak{X}_T^C} := \sup_{t\in[0,T]} \|f(t)\|_{\mathfrak{X}} . \tag{F.1.2}$$

For a given solution operator $\{U(t,s)\}_{(t,s)\in\Delta}$ of the nACP (Definition 2.1.4), one can construct by the *Howland-Evans-Neidhardt* (HEN) formula:

$$(\mathcal{U}(\tau)f)(t) = U(t,t-\tau)\ \chi_{[0,T]}(t-\tau)\ f(t-\tau), \quad 0 \leq \tau \leq t \leq T\,, \tag{F.1.3}$$

V. A. Zagrebnov et al., *Trotter-Kato Product Formulæ*, Operator Theory: Advances and Applications 296, https://doi.org/10.1007/978-3-031-56720-9

where $f \in \mathfrak{X}_T^C$ and $\chi_{[0,T]}(\cdot)$ is the characteristic function of interval $[0, T] \subset \mathbb{R}_0^+$, one-parameter family of operators $\{\mathcal{U}(\tau)\}_{\tau \geq 0}$ on $\mathfrak{X}_T^C$.

Theorem F.1.1 *The one-parameter family of operators $\{\mathcal{U}(\tau)\}_{\tau \geq 0}$ defined on Banach space $\mathfrak{X}_T^C$ by the Howland-Evans-Neidhardt formula* (F.1.3) *is a strongly continuous nilpotent semigroup.*

Proof We check these properties of the family $\{\mathcal{U}(\tau)\}_{\tau \geq 0}$ in three steps.

(a) Indeed, the *semigroup* composition law is inheritance of the *Chapman-Kolmogorov* composition law (2.1.6) since by (F.1.3)

$$\begin{aligned}
&(\mathcal{U}(\tau_1 + \tau_2) f)(t) = U(t, t - \tau_1 - \tau_2)\, \chi_{[0,T]}(t - \tau_1 - \tau_2)\, f(t - \tau_1 - \tau_2) = \\
&U(t, t - \tau_1)\, U(t - \tau_1, t - \tau_1 - \tau_2)\, \chi_{[0,T]}(t - \tau_1 - \tau_2)\, f(t - \tau_1 - \tau_2) = \\
&(\mathcal{U}(\tau_1)\, [\mathcal{U}(\tau_2) f\,])\,(t) = (\mathcal{U}(\tau_1)\, \mathcal{U}(\tau_2) f)\,(t)\,, \quad t \in [0, T]\,,
\end{aligned} \tag{F.1.4}$$

for $\tau_1 \geq 0$ and $\tau_2 \geq 0$.

(b) Since (F.1.3) implies $\mathcal{U}(\tau) f = 0$, for $\tau > T$, the semigroup $\{\mathcal{U}(\tau)\}_{\tau \geq 0}$ on $\mathfrak{X}_T^C$ is *nilpotent*; cf. (2.1.23).

(c) To check that $\{\mathcal{U}(\tau)\}_{\tau \geq 0}$ is a *strongly continuous* semigroup on $\mathfrak{X}_T^C$, let us fix $\tau_\varepsilon \in (0, T)$ and consider $0 \leq \tau < \tau_\varepsilon$. Then

$$\begin{aligned}
&\sup_{t \in [0,T]} \|(\mathcal{U}(\tau) - \mathcal{I})\, f)(t)\|_{\mathfrak{X}} \\
&= \sup_{t \in [0,T]} \|U(t, t - \tau)\chi_{[0,T]}(t - \tau) f(t - \tau) - f(t)\|_{\mathfrak{X}} \leq \\
&\sup_{t \in [0,\tau_\varepsilon]} \|(U(t, t - \tau)\chi_{[0,T]}(t - \tau) - \mathbb{1})\, f(t)\|_{\mathfrak{X}} \\
&+ \sup_{t \in [\tau_\varepsilon,T]} \|(U(t, t - \tau) - \mathbb{1})\, f(t)\|_{\mathfrak{X}} + \\
&\sup_{t \in [0,\tau_\varepsilon]} \|U(t, t - \tau)\chi_{[0,T]}(t - \tau)(f(t - \tau) - f(t))\|_{\mathfrak{X}} + \\
&\sup_{t \in [\tau_\varepsilon,T]} \|U(t, t - \tau)(f(t - \tau) - f(t))\|_{\mathfrak{X}}\,,
\end{aligned} \tag{F.1.5}$$

for any fixed $f \in \mathfrak{X}_T^C$.

On account of Definition 2.1.4 (c) we have $\sup_{t \in [0,\tau_\varepsilon]} \|(U(t, t - \tau)\chi_{[0,T]}(t - \tau)\|_{\mathcal{L}(\mathfrak{X})} < M_\varepsilon$. Then, owing to $f(0) = 0$ (F.1.1), for any (small) $\varepsilon > 0$ there exists $\tau_\varepsilon > 0$ such that $\sup_{t \in [0,\tau_\varepsilon]} \|f(t)\|_{\mathfrak{X}} < \varepsilon$ and as a consequence:

$$\begin{aligned}
&\sup_{t \in [0,\tau_\varepsilon]} \|(U(t, t - \tau)\chi_{[0,T]}(t - \tau) - \mathbb{1})\, f(t)\|_{\mathfrak{X}} < (M_\varepsilon + 1)\, \varepsilon\,, \\
&\sup_{t \in [0,\tau_\varepsilon]} \|U(t, t - \tau)\chi_{[0,T]}(t - \tau)(f(t - \tau) - f(t))\|_{\mathfrak{X}} < 2\, M_\varepsilon\, \varepsilon\,.
\end{aligned} \tag{F.1.6}$$

Note that by Definition 2.1.4 (a) and $f \in C_0([0,T],\mathfrak{X})$ the $\mathfrak{X}$-valued function

$$\Delta_\varepsilon \ni (t,s) \mapsto U(t,s)f(t)\,, \quad \Delta_\varepsilon = \{(t,s) \in [\tau_\varepsilon, T] \times [\tau_\varepsilon, T] : \tau_\varepsilon < s \le t \le T\}\,,$$

is continuous on the compact Δ_ε. As a result, for $\tau \in [0, \tau_\varepsilon)$

$$\Phi_\tau : t \mapsto \|(U(t, t-\tau) - \mathbb{1})\, f(t)\|_{\mathfrak{X}}\,, \quad t \in [\tau_\varepsilon, T]\,,$$

is the family of non-negative continuous functions such that $\lim_{\tau\to 0} \Phi_\tau(t) = 0$ for each $t \in [\tau_\varepsilon, T]$. Then (by *Dini* theorem) this convergence is uniform, that is,

$$\lim_{\tau\to 0} \sup_{t\in[\tau_\varepsilon,T]} \|(U(t,t-\tau) - \mathbb{1})\, f(t)\|_{\mathfrak{X}} = 0\,. \tag{F.1.7}$$

Next, owing to Definition 2.1.4 (c) we obtain $\sup_{t\in[\tau_\varepsilon,T]} \|(U(t,t-\tau)\|_{\mathcal{L}(\mathfrak{X})} < M'_\varepsilon$ and thus $\|U(t,t-\tau)(f(t-\tau) - f(t))\|_{\mathfrak{X}} < M'_\varepsilon \|(f(t-\tau) - f(t))\|_{\mathfrak{X}}$. Since a continuous on compact $[\tau_\varepsilon, T]$ function $f \in C_0([0,T],\mathfrak{X})$ is uniformly continuous, there exists $\tau'_\varepsilon > 0$ such that

$$\sup_{t\in[\tau_\varepsilon,T]} \|f(t-\tau) - f(t)\|_{\mathfrak{X}} < \varepsilon\,, \quad \tau < \tau'_\varepsilon\,. \tag{F.1.8}$$

Consequently, on account of (F.1.5)–(F.1.8) and (F.1.2) we infer that

$$\lim_{\tau\to+0} \|(\mathcal{U}(\tau) - \mathcal{I})\, f)\|_{\mathfrak{X}^C_T} = \lim_{\tau\to+0} \sup_{t\in[0,T]} \|(\mathcal{U}(\tau) - \mathcal{I})\, f)(t)\|_{\mathfrak{X}} = 0\,, \tag{F.1.9}$$

for any $f \in \mathfrak{X}^C_T$, that is, the strong continuity of semigroup $\{\mathcal{U}(\tau)\}_{\tau\ge 0}$. □

Corollary F.1.2 *The preceding arguments provide mutatis mutandis the same result for strongly continuous evolution semigroup* $\{\mathcal{U}_p(\tau)\}_{\tau\ge 0}$ *defined by the Howland-Evans-Neidhardt formula* (2.2.3) *on the Banach space* $\mathfrak{X}^p_T := L^p([0,T],\mathfrak{X})$. *Similar to* (F.1.3) *this formula is uniquely determined by a strongly continuous on* $\mathfrak{X}$ *solution operator* $\{U(t,s)\}_{(t,s)\in\Delta}$.

For more comments see Sect. F.2. Thus, we immediately exploit Corollary F.1.2 in the next Sect. F.1.2.

Section F.1.2

Here we turn to another approach when evolution semigroups are defined *without* explicit implication of *solution* operator for some nACP. This approach was advocated by *Hagen Neidhardt* within a scheme, where semigroup $\{\mathcal{U}(\tau)\}_{\tau\ge 0}$ on $\mathfrak{X}^C_T$ is *self-defined* on account of certain conditions (*equation*). To keep a contact with evolution semigroups this *equation* must be *identity* for semigroup defined by the HEN formula (F.1.3).

For the reason that Banach space $\mathfrak{X}^C_T = C_0([0,T],\mathfrak{X})$ is too restrictive for derivation of the requested equation, we start developing necessary arguments first in a larger space $\mathfrak{X}_T^p = L^p([0,T],\mathfrak{X})$. Then taking into account the right-shift $\{\mathcal{S}_R(\tau)\}_{\tau\geq 0}$ (2.1.35) and the left-shift $\{\mathcal{S}_L(\tau)\}_{\tau\geq 0}$ (2.1.50) strongly continuous on $\mathfrak{X}_T^p$ semigroups, this allows to define for two families of projections:

$$(P_{LR}(\tau)\, f)(t) := (S_L(\tau)\, S_R(\tau)\, f)(t) = \chi_{[0,T]}(t+\tau)\, f(t)\,,$$
$$(P_{RL}(\tau)\, f)(t) := (S_R(\tau)\, S_L(\tau)\, f)(t) = \chi_{[0,T]}(t-\tau)\, f(t)\,, \tag{F.1.10}$$

for $\tau \geq 0$ and any $f \in \mathfrak{X}_T^p$. As a consequence we infer from the HEN formula (F.1.3) on $\mathfrak{X}_T^p$ and (F.1.10) the *first* fundamental property of the evolution semigroup:

$$\mathcal{U}_p(\tau)\, P_{LR}(\tau) = \mathcal{U}_p(\tau)\,, \quad \tau \geq 0\,. \tag{F.1.11}$$

Indeed,

$$\begin{aligned}(\mathcal{U}_p(\tau)\, P_{LR}(\tau)\, f)(t) &= U(t,t-\tau)\chi_{[0,T]}(t-\tau)(P_{LR}(\tau)\, f)(t-\tau)\\ &= (\mathcal{U}_p(\tau)\, f)(t)\,, \qquad f \in \mathfrak{X}_T^p\,.\end{aligned}$$

The *second* fundamental property of the evolution semigroup is stemming from commutation of $\{\mathcal{U}_p(\tau)\,\mathcal{S}_L(\tau)\}_{\tau\geq 0}$ with any scalar multiplication operator $\mathcal{M}(\phi)$ for $\phi \in L^\infty([0,T])$:

$$\begin{aligned}&(\mathcal{U}_p(\tau)\,\mathcal{S}_L(\tau)\,\mathcal{M}(\phi)\, f)(t) = U(t,t-\tau)\,\chi_{[0,T]}(t-\tau)\\ &(\mathcal{S}_L(\tau)\,\mathcal{M}(\phi)\, f)(t-\tau) =\\ &U(t,t-\tau)\,\chi_{[0,T]}(t-\tau)\,(\mathcal{M}(\phi)\, f)(t) = (\mathcal{M}(\phi))(t)\\ &U(t,t-\tau)\,\chi_{[0,T]}(t-\tau)\, f(t) =\\ &(\mathcal{M}(\phi)\,\mathcal{U}_p(\tau)\,\mathcal{S}_L(\tau)\, f)(t)\,, \quad f \in \mathfrak{X}_T^p\,.\end{aligned} \tag{F.1.12}$$

Then (see Definition 2.1.23) $\{\mathcal{U}_p(\tau)\,\mathcal{S}_L(\tau)\}_{\tau\geq 0}$ is in turn a family of bounded multiplication operators. That is, by Proposition 2.1.24 (and [Nei81], Proposition 3.2) the commutation in (F.1.12) is equivalent for $f \in \mathfrak{X}_T^C$ to

$$(\mathcal{U}_p(\tau)\,\mathcal{S}_L(\tau) f)(t) = (\mathcal{U}_p(\tau)\,\mathcal{S}_L(\tau))(t)\; f(t)\,, \tag{F.1.13}$$

and $\|\mathcal{U}_p(\tau)\,\mathcal{S}_L(\tau)\|_{\mathcal{L}(\mathfrak{X}_T^C)} = \sup_{t\in[0,T]} \|(\mathcal{U}_p(\tau)\,\mathcal{S}_L(\tau))(t)\|_{\mathcal{L}(\mathfrak{X})}$.

If $f \in \mathfrak{X}_T^p$, then one gets (see Lemma 8.2.1, (F.2.12))

$$\|(\mathcal{U}_p(\tau)\,\mathcal{S}_L(\tau) f)(t)\|_{\mathfrak{X}} \leq \|\mathcal{U}_p(\tau)\,\mathcal{S}_L(\tau)\|_{\mathcal{L}(\mathfrak{X}_T^p)}\; \|f(t)\|_{\mathfrak{X}}\,, \tag{F.1.14}$$

for a.a. $t \in [0,T]$.

Lemma F.1.3 *A C_0-semigroup $\{\mathcal{U}_p(\tau)\}_{\tau\geq 0}$ on Banach space $\mathfrak{X}_T^p$ has fundamental properties* (F.1.11) *and* (F.1.12) *if and only if it satisfies equation*

$$\mathcal{U}_p(\tau)\,\mathcal{M}(\mathcal{S}_L(\tau)\phi) = \mathcal{M}(\phi)\,\mathcal{U}_p(\tau)\,, \quad \tau \geq 0\,, \tag{F.1.15}$$

for every $\phi \in L^\infty([0,T])$.

Proof On account of fundamental properties (F.1.11) and (F.1.12),

$$\begin{aligned}&\mathcal{U}_p(\tau)\,\mathcal{S}_L(\tau)\,\mathcal{M}(\phi)\mathcal{S}_R(\tau) = \mathcal{M}(\phi)\mathcal{U}_p(\tau)\,\mathcal{S}_L(\tau)\,\mathcal{S}_R(\tau) = \\ &\mathcal{M}(\phi)\mathcal{U}_p(\tau)\,P_{LR}(\tau) = \mathcal{M}(\phi)\,\mathcal{U}_p(\tau)\,, \quad \tau \geq 0\,.\end{aligned} \tag{F.1.16}$$

Seeing that

$$\mathcal{S}_L(\tau)\,\mathcal{M}(\phi)\mathcal{S}_R(\tau) = \mathcal{M}(\mathcal{S}_L(\tau)\,\phi)\,, \quad \tau \geq 0\,, \tag{F.1.17}$$

(F.1.16) and (F.1.17) yield (F.1.15).

Conversely, suppose that (F.1.15) holds for any $\phi \in L^\infty([0,T])$. Then by (F.1.17)

$$\begin{aligned}&\mathcal{M}(\phi)\,\mathcal{U}_p(\tau)\,\mathcal{S}_L(\tau) = \mathcal{U}_p(\tau)\,\mathcal{M}(\mathcal{S}_L(\tau)\phi)\mathcal{S}_L(\tau) = \\ &\mathcal{U}_p(\tau)\,\mathcal{S}_L(\tau)\,\mathcal{M}(\phi)\,\mathcal{S}_R(\tau)\,\mathcal{S}_L(\tau) = \mathcal{U}_p(\tau)\,\mathcal{S}_L(\tau)\,\mathcal{M}(\phi)\,P_{RL}(\tau)\,, \quad \tau \geq 0\,.\end{aligned} \tag{F.1.18}$$

Owing to the fact that by (F.1.10)

$$\mathcal{S}_L(\tau)\,\mathcal{M}(\phi)\,P_{RL}(\tau) = \mathcal{S}_L(\tau)\,\mathcal{M}(\phi)\,, \quad \phi \in L^\infty([0,T])\,, \quad \tau \geq 0\,, \tag{F.1.19}$$

(F.1.18) and (F.1.19) yield for any $\phi \in L^\infty([0,T])$ the commutation

$$\mathcal{U}_p(\tau)\,\mathcal{S}_L(\tau)\,\mathcal{M}(\phi) = \mathcal{M}(\phi)\,\mathcal{U}_p(\tau)\,\mathcal{S}_L(\tau)\,, \quad \tau \geq 0\,. \tag{F.1.20}$$

This proves the *second* fundamental property (F.1.12) and hence (F.1.13).

Now, let function $\widehat{\phi}_\tau \in L^\infty([0,T])$ be such that $\widehat{\phi}(t) := \chi_{[0,\tau]}(t)$. Then by definition (2.1.25) of a scalar multiplication operator and (F.1.10)

$$\mathcal{M}(\widehat{\phi}_\tau) = (\mathcal{I} - P_{RL}(\tau))\,, \quad \tau \geq 0\,. \tag{F.1.21}$$

Since $(\mathcal{S}_L(\tau)\widehat{\phi}_\tau)(t) = \chi_{[0,T]}(t+\tau)\,\chi_{[0,\tau]}(t+\tau) = 0$, by virtue of (F.1.15) and (F.1.21) one gets

$$(\mathcal{I} - P_{RL}(\tau))\mathcal{U}_p(\tau) = 0\,, \quad \tau \geq 0\,. \tag{F.1.22}$$

If we insert in formula (F.1.20) the scalar multiplication operator $\mathcal{M}(\widehat{\phi_\tau})$ and use definitions (F.1.10), we obtain

$$\mathcal{U}_p(\tau)\, P_{LR}(\tau)) = P_{RL}(\tau))\,\mathcal{U}_p(\tau)\,, \quad \tau \geq 0\,. \tag{F.1.23}$$

Then (F.1.22) and (F.1.23) imply the *first* fundamental property (F.1.11). □

Section F.1.3

Now we reformulate for Banach space $\mathfrak{X}_T^C$ the concept developed in Sect. F.1.2 for evolution semigroups on Banach space $\mathfrak{X}_T^p$. Within this setup we show that equation (F.1.15) for evolution semigroup allows to extend Lemma F.1.3, that is, to *establish* the HEN formula that relates a given strongly continuous evolution semigroup with corresponding solution operator.

Definition F.1.4 We call a C_0-semigroup $\{\mathcal{U}(\tau)\}_{\tau\geq 0}$ on the space $\mathfrak{X}_T^C = C_0([0,T],\mathfrak{X})$ *evolution semigroup* if it satisfies equation

$$\mathcal{U}(\tau)\,\mathcal{M}(\mathcal{S}_L(\tau)\phi) = \mathcal{M}(\phi)\,\mathcal{U}(\tau)\,, \quad \tau \geq 0\,, \tag{F.1.24}$$

for every $\phi \in C_T([0,T]) := \{\phi \in C([0,T]) : \phi(T) = 0\}$ or $\phi \in C_T^1([0,T])$; cf. (F.1.15).

Owing to (F.1.15) and Definition F.1.4 we prove the assertion, which is converse to Theorem F.1.1.

Theorem F.1.5 *For every evolution semigroup* $\{\mathcal{U}(\tau)\}_{\tau\geq 0}$, *Definition* F.1.4, *exists a unique strongly continuous solution operator* $\{U(t,s)\}_{(t,s)\in\Delta}$ *such that the HEN formula*:

$$(\mathcal{U}(\tau)f)(t) = U(t,t-\tau)\,\chi_{[0,T]}(t-\tau)\,f(t-\tau), \quad 0 \leq \tau \leq t \leq T\,, \tag{F.1.25}$$

holds for any $f \in \mathfrak{X}_T^C = C_0([0,T],\mathfrak{X})$.

Proof On account of Lemma F.1.3, (F.1.11) and (F.1.23), we obtain for evolution semigroup representation

$$\mathcal{U}(\tau) = \mathcal{U}(\tau)\, P_{LR}(\tau) = \mathcal{U}(\tau)\,\mathcal{S}_L(\tau)\,\mathcal{S}_R(\tau)\,, \quad \tau \geq 0\,. \tag{F.1.26}$$

Note that by (F.1.13) $\{\mathcal{U}(\tau)\,\mathcal{S}_L(\tau)\}_{\tau\geq 0}$ are bounded multiplication operators, that is, we define

$$(\mathcal{U}(\tau)\,\mathcal{S}_L(\tau)f)(t) =: W(\tau,t)\,f(t)\,, \quad \tau \geq 0\,, \tag{F.1.27}$$

for any $f \in \operatorname{ran}(\mathcal{S}_R(\tau)) \subset L^p([0,T],\mathfrak{X})$. Given that right-shift $(\mathcal{S}_R(\tau)f)(t) = \chi_{[0,T]}(t-\tau)\,f(t-\tau)$ is a bounded operator from $L^p([0,T],\mathfrak{X})$ into the Banach space $L^p([\tau,T],\mathfrak{X})$ we infer

$$\mathcal{U}(\tau)\,P_{LR}(\tau)(C_0([0,T-\tau],\mathfrak{X})) \subseteq C_\tau([\tau,T],\mathfrak{X})\,, \quad \tau \geq 0\,, \tag{F.1.28}$$

where $C_\tau([\tau,T],\mathfrak{X}) := \{f \in C([\tau,T],\mathfrak{X}) : f(\tau) = 0\}$. For that reason the multiplication operator (F.1.27) is well-defined as a bounded operator on $C_\tau([\tau,T],\mathfrak{X})$ and consequently

$$(\mathcal{U}(\tau)\,\mathcal{S}_L(\tau))(t)\,(\mathcal{S}_R(\tau)f)(t) = W(\tau,t)\,\chi_{[0,T]}(t-\tau)\,f(t-\tau)\,, \quad \tau \geq 0\,, \tag{F.1.29}$$

for $f \in C_0([0,T],\mathfrak{X})$. If we define $U(t,t-\tau) := W(\tau,t)$ and use (F.1.26), (F.1.29), we obtain the HEN formula (F.1.25).

Following arguments in (F.1.5)–(F.1.8) one infers that semigroup property and C_0-continuity of semigroup $\{\mathcal{U}(\tau)\}_{\tau\geq 0}$ imply the *Chapman-Kolmogorov* composition law and strong continuity of the family $\{U(t,s)\}_{(t,s)\in\Delta}$ on $\mathfrak{X}$.

Finally, the (exponential) boundedness of solution operator $\{U(t,s)\}_{(t,s)\in\Delta}$ is stemming from formula (F.1.25) and respective property of evolution semigroup $\{\mathcal{U}(\tau)\}_{\tau\geq 0}$. □

Summarising Theorems F.1.1 and F.1.5 we get the following statement.

Theorem F.1.6 *Between the set of all evolution semigroups $\{\mathcal{U}(\tau)\}_{\tau\in\mathbb{R}_0^+}$ on $\mathfrak{X}_T^C$ and the set of all solution operators $\{U(t,s)\}_{(t,s)\in\Delta}$ on $\mathfrak{X}$, there exists a one-to-one correspondence such that by* (2.1.24) *the Howland-Evans-Neidhardt equation*

$$(\mathcal{U}(\tau)f)(t) = U(t,t-\tau)\,\chi_{[0,T]}(t-\tau)\,f(t-\tau), \quad 0 \leq \tau \leq t \leq T\,, \tag{F.1.30}$$

holds for any $f \in \mathfrak{X}_T^C$ and $t \in [0,T]$. Seeing that solution operators are exponentially bounded (2.1.21) *and because of the nilpotence of evolution semigroup* (2.1.27) *one gets estimate*

$$\|\mathrm{e}^{-\tau\mathcal{K}} f\|_{\mathfrak{X}_T^C} \leq M\,\mathrm{e}^{\beta T}\,\|f\|_{\mathfrak{X}_T^C}\,, \quad f \in \mathfrak{X}_T^C\,, \quad \tau \geq 0\,. \tag{F.1.31}$$

Therefore, generator $\mathcal{K}$ of the evolution semigroup belongs to the class $\mathscr{G}(M\,\mathrm{e}^{\beta T},0)$.

Thus, the statement of Theorem F.1.6 proves assertions of the Theorem 2.1.18; cf. [Nei81, Theorem 3.8 and Theorem 3.10].

Section F.1.4

Proof of Theorem 2.1.17.

In the proof below we follow essentially [Nei81, Theorem 2.8].

The property of a C_0-semigroup to be an evolution semigroup has some consequences for the generator. By Definition F.1.4 an evolution semigroup is selected from C_0-semigroups by condition (F.1.24). Below we elucidate the necessary and sufficient condition for that on the level of *generators*; see Definition 2.1.16.

Theorem F.1.7 *Let operator $\mathcal{K}$ be evolution pre-generator* (2.1.26) *in Banach space $\mathfrak{X}_T^C$. Then $\mathcal{K}$ is evolution generator of semigroup $\{\mathcal{U}(\tau)\}_{\tau\geq 0}$ (Definition* (F.1.4)) *on $\mathfrak{X}_T^C$ if and only if operator $\mathcal{K}$ is generator of a strongly continuous on $\mathfrak{X}_T^C$ semigroup, that is, $\{\mathcal{U}(\tau) = \mathrm{e}^{-\tau\mathcal{K}}\}_{\tau\geq 0}$.*

Proof Sufficiency. Let evolution pre-generator $\mathcal{K}$ (2.1.26) be generator of a C_0-semigroup $\{\mathrm{e}^{-\tau\mathcal{K}}\}_{\tau\geq 0}$ on $\mathfrak{X}_T^C$. Then the bounded operator-valued strongly continuous function:

$$\tau \mapsto G(\tau) := \mathrm{e}^{-\tau\mathcal{K}}\, \mathcal{M}(\mathcal{S}_L(\tau)\phi) - \mathcal{M}(\phi)\, \mathrm{e}^{-\tau\mathcal{K}}, \quad \tau \geq 0\,, \tag{F.1.32}$$

is differentiable for any $f \in \mathrm{dom}\,(\mathcal{K})$ and $\phi \in C_T^1([0, T])$. Thanks to conditions (2.1.26) we obtain

$$\partial_\tau\, G(\tau)\, f = -\, G(\tau)\, \mathcal{K}\, f\,.$$

Note that in turn the bounded strongly continuous function

$$[0, \eta] \ni \tau \mapsto H(\tau) := G(\eta - \tau)\, \mathrm{e}^{-\tau\mathcal{K}}$$

is differentiable for any $f \in \mathrm{dom}\,(\mathcal{K})$ and $\partial_\tau\, H(\tau)\, f = 0$. As a consequence, $H(\tau) = H(\eta) = 0$ and thus $G(\eta) = H(0) = 0$. Seeing that $\eta \geq 0$ is arbitrary, this implies $G(\tau) = 0$. On that account by definition (F.1.32)

$$\mathrm{e}^{-\tau\mathcal{K}}\, \mathcal{M}(\mathcal{S}_L(\tau)\phi) = \mathcal{M}(\phi)\, \mathrm{e}^{-\tau\mathcal{K}}, \quad \tau \geq 0\,. \tag{F.1.33}$$

By Definition F.1.4 and by (F.1.33) semigroup $\{\mathrm{e}^{-\tau\mathcal{K}}\}_{\tau\geq 0}$ satisfies equation (F.1.24). Hence, $\{\mathcal{U}(\tau) = \mathrm{e}^{-\tau\mathcal{K}}\}_{\tau\geq 0}$ is evolution semigroup.

Necessity. Let operator $\mathcal{K}$ generate on $\mathfrak{X}_T^C$ evolution semigroup $\{\mathcal{U}(\tau)\}_{\tau\geq 0}$. By Definition F.1.4 it is strongly continuous, that is, $\{\mathcal{U}(\tau) = \mathrm{e}^{-\tau\mathcal{K}}\}_{\tau\geq 0}$, and satisfies equation (F.1.24). Seeing that, we obtain representation:

$$\begin{aligned} &\frac{1}{\tau}\, \mathrm{e}^{-\tau\mathcal{K}}\, (\mathcal{M}(\mathcal{S}_L(\tau)\phi) - \mathcal{M}(\phi))\, f + \frac{1}{\tau}\, (\mathrm{e}^{-\tau\mathcal{K}} - \mathcal{I})\mathcal{M}(\phi)\, f = \\ &\frac{1}{\tau}\, \mathcal{M}(\phi)\, (\mathrm{e}^{-\tau\mathcal{K}} - \mathcal{I})\, f\,, \quad \tau > 0\,, \quad f \in \mathrm{dom}\,(\mathcal{K})\,. \end{aligned} \tag{F.1.34}$$

Then strong continuity of the evolution semigroup yields

$$\lim_{\tau\to+0} \frac{1}{\tau}\, (\mathrm{e}^{-\tau\mathcal{K}} - \mathcal{I})\, f = -\,\mathcal{K}\, f\,, \quad f \in \mathrm{dom}\,(\mathcal{K})\,. \tag{F.1.35}$$

Taking into account (F.1.11) one gets

$$e^{-\tau\mathcal{K}}(\mathcal{M}(\mathcal{S}_L(\tau)\phi)-\mathcal{M}(\phi)) = e^{-\tau\mathcal{K}} P_{LR}(\tau)(\mathcal{M}(\mathcal{S}_L(\tau)\phi)-\mathcal{M}(\phi)), \quad \tau \geq 0,$$

which provides for $f \in \mathfrak{X}_T^C$:

$$\begin{aligned}&(P_{LR}(\tau)(\mathcal{M}(\mathcal{S}_L(\tau)\phi)-\mathcal{M}(\phi))f)(t)\\&\quad = \chi_{[0,T]}(t+\tau)\left(\int_t^{t+\tau} ds\,\phi'(s)\right) f(t), \quad \tau \geq 0,\end{aligned}$$

and as a result

$$\lim_{\tau\to+0}\frac{1}{\tau} e^{-\tau\mathcal{K}}(\mathcal{M}(\mathcal{S}_L(\tau)\phi)-\mathcal{M}(\phi)) f = \mathcal{M}(\phi') f, \quad f \in \mathfrak{X}_T^C. \tag{F.1.36}$$

By virtue of (F.1.34)–(F.1.36) we obtain existence of the limit

$$\lim_{\tau\to+0}\frac{1}{\tau}(e^{-\tau\mathcal{K}}-\mathcal{I})\mathcal{M}(\phi) f, \quad f \in \mathrm{dom}(\mathcal{K}), \tag{F.1.37}$$

that yields $\mathcal{M}(\phi) f \in \mathrm{dom}(\mathcal{K})$ and, therefore, proves the *first* of the evolution pre-generator properties (2.1.26).

To prove the second of the evolution pre-generator properties (2.1.26), we note that in fact (F.1.37) yields:

$$\lim_{\tau\to+0}\frac{1}{\tau}(e^{-\tau\mathcal{K}}-\mathcal{I})\mathcal{M}(\phi) f = -\mathcal{K}\mathcal{M}(\phi) f, \quad f \in \mathrm{dom}(\mathcal{K}). \tag{F.1.38}$$

Then (F.1.34)–(F.1.36) and (F.1.38) give the *second* of the evolution pre-generator properties (2.1.26). Hence, operator $\mathcal{K}$ is the evolution pre-generator. □

Remark F.1.8 Similarly to Corollary F.1.2 we note that preceding arguments for pre-generator $\mathcal{K}$ in Banach space $\mathfrak{X}_T^C$ are valid for a corresponding assertion about evolution pre-generator $\mathcal{K}_p$, which is defined in Banach space $\mathfrak{X}_T^p$ (see Definition 2.2.8), as well as for a statement concerning evolution semigroup $\{\mathcal{U}_p(\tau)\}_{\tau\geq 0}$. We return to this point in Sect. F.2.

F.2 The Howland-Evans-Neidhardt Formula: L^p-Setting

The *Howland-Evans-Neidhardt* formula (or representation) (2.2.3) for evolution *semigroups* for a general (not necessarily *separable*) Banach space $\mathfrak{X}$ in L^p-setting is due to [Nei79] and [Nei81, Theorem 4.9, Theorem 4.12].

Section F.2.1
We start this subsection with the proof of the *extension* Theorem 2.2.4.

Theorem F.2.1 *For every evolution semigroup* $\{\mathcal{U}(\tau)\}_{\tau\in\mathbb{R}_0^+}$ (2.1.24) *on the Banach space* $\mathfrak{X}_T^C = C_0([0,T],\mathfrak{X})$ *there exists a unique evolution semigroup* $\{\mathcal{U}_p(\tau)\}_{\tau\in\mathbb{R}_0^+}$ (2.2.30) *on the Banach space* $\mathfrak{X}_T^p = L^p([0,T],\mathfrak{X})$ *with the property*

$$(\mathcal{U}_p(\tau)\, f)(t) = (\mathcal{U}(\tau)\, f)(t)\,, \quad f \in C_0([0,T],\mathfrak{X})\,, \quad \textit{for a.a. } t \in [0,T]\,.$$

Proof Since evolution semigroup $\{\mathcal{U}(\tau)\}_{\tau\in\mathbb{R}_0^+}$ (2.1.24) on the Banach space $\mathfrak{X}_T^C$ is nilpotent (2.1.28), the operator norm

$$\|\mathcal{U}(\tau)\|_{\mathcal{L}(\mathfrak{X}_T^C)} \le M\,\mathrm{e}^{\beta T}\,, \quad \tau \ge 0\,,$$

which implies for $f \in \mathfrak{X}_T^C$:

$$\|(\mathcal{U}(\tau)\,\mathcal{M}(\mathcal{S}_L(\tau)\phi)\, f)(t)\|_{\mathfrak{X}} \le M\,\mathrm{e}^{\beta T} \sup_{t\in[0,T-\tau]} \|\phi(t+\tau)\, f(t)\|_{\mathfrak{X}}\,, \quad \tau \ge 0\,. \tag{F.2.1}$$

Given that evolution semigroup $\{\mathcal{U}(\tau)\}_{\tau\in\mathbb{R}_0^+}$ satisfies equation (F.1.24), from (F.2.1) we infer for any $f \in \mathfrak{X}_T^C$ and $\phi \in C_T([0,T])$:

$$\|\phi(t)\,(\mathcal{U}(\tau)\, f)(t)\|_{\mathfrak{X}} \le M\,\mathrm{e}^{\beta T} \sup_{t\in[\tau,T]} \|\phi(t)\, f(t-\tau)\|_{\mathfrak{X}}\,, \quad \tau \ge 0\,,$$

which readily implies

$$\|(\mathcal{U}(\tau)\, f)(t)\|_{\mathfrak{X}} \le M\,\mathrm{e}^{\beta T}\, \chi_{[0,T]}(t-\tau)\, \|f(t-\tau)\|_{\mathfrak{X}}\,, \quad \tau \ge 0\,,$$

Recall that $C_T([0,T],\mathfrak{X}) := \{f \in C([0,T],\mathfrak{X}) : f(T) = 0\}$. Then, as a consequence of definition (2.2.1), we obtain estimate in the norm of the Banach space $\mathfrak{X}_T^p = L^p([0,T],\mathfrak{X})$:

$$\|(\mathcal{U}(\tau)\, f\|_{\mathfrak{X}_T^p} \le M\,\mathrm{e}^{\beta T}\, \|f\|_{\mathfrak{X}_T^p}\,, \quad f \in \mathfrak{X}_T^C = C_0([0,T],\mathfrak{X})\,. \tag{F.2.2}$$

Since the Banach space $\mathfrak{X}_T^C$ is continuously and densely embedded in Banach space $\mathfrak{X}_T^p$, that is, $\mathfrak{X}_T^C \overset{d}{\hookrightarrow} \mathfrak{X}_T^p$ (2.2.8), the estimate (F.2.2) ensures existence of unique continuous *extension* of operator $\mathcal{U}(\tau)$ to bounded operator $\mathcal{U}_p(\tau)$ on $\mathfrak{X}_T^p$ such that restriction

$$\mathcal{U}_p(\tau) \restriction \mathfrak{X}_T^C = \mathcal{U}(\tau)\,, \quad \tau \ge 0\,, \tag{F.2.3}$$

that is, the *part* of operator $\mathcal{U}_p$ on $\mathfrak{X}_T^C$ (see Definition F.2.3), coincides with $\mathcal{U}$.

By construction (F.2.3) the family of operators $\{\mathcal{U}_p(\tau)\}_{\tau\in\mathbb{R}_0^+}$ satisfies equation (F.1.24) and by continuity of embedding (2.2.8) fulfills the estimate

$$\|\mathcal{U}_p(\tau)\, f\|_{\mathfrak{X}_T^p} \leq T^{1/p}\, \|\mathcal{U}_p(\tau)\, f\|_{\mathfrak{X}_T^C}\,, \quad f \in \mathfrak{X}_T^C\,, \quad \tau \geq 0\,. \tag{F.2.4}$$

Then (F.2.4), together with the strong continuity of evolution semigroup $\{\mathcal{U}(\tau)\}_{\tau\in\mathbb{R}_0^+}$ on $\mathfrak{X}_T^C$ and density of embedding (2.2.8), provides that $\{\mathcal{U}_p(\tau)\}_{\tau\in\mathbb{R}_0^+}$ is a C_0-semigroup on $\mathfrak{X}_T^p$ satisfying equation (F.1.15). Consequently, $\{\mathcal{U}_p(\tau)\}_{\tau\in\mathbb{R}_0^+}$ is a unique extension of evolution semigroup $\{\mathcal{U}(\tau)\}_{\tau\in\mathbb{R}_0^+}$, defined on $\mathfrak{X}_T^C$, to evolution semigroup defined on $\mathfrak{X}_T^p$, which satisfies the property (F.2.3). □

Corollary F.2.2 *On account of* (F.1.30) *and Theorem* F.2.1 *we obtain for evolution semigroup* $\{\mathcal{U}_p(\tau)\}_{\tau\in\mathbb{R}_0^+}$ *on* $\mathfrak{X}_T^p$*, induced by evolution semigroup* $\{\mathcal{U}(\tau)\}_{\tau\in\mathbb{R}_0^+}$ *defined on* $\mathfrak{X}_T^p$*, the HEN representation:*

$$(\mathcal{U}_p(\tau) f)(t) = U(t, t-\tau)\, \chi_{[0,T]}(t-\tau)\, f(t-\tau), \quad 0 \leq \tau \leq t \leq T\,, \tag{F.2.5}$$

for any $f \in \mathfrak{X}_T^p$ *and for a.a.* $t \in [0, T]$.

We comment that representation (F.2.5) is proven *only* for evolution semigroup $\{\mathcal{U}_p(\tau)\}_{\tau\in\mathbb{R}_0^+}$ on $\mathfrak{X}_T^p$ *induced* by evolution semigroup $\{\mathcal{U}(\tau)\}_{\tau\in\mathbb{R}_0^+}$ on $\mathfrak{X}_T^C$, which possessed the Howland-Evans-Neidhardt representation (F.1.30). This inheritance does not involve into consideration generators of the both of these C_0-semigroups; see comments in Remark 2.2.9. It also leaves open the question: whether representation (F.2.5) holds for *any* evolution semigroup on $\mathfrak{X}_T^p$ defined by (F.1.15).

In Sect. F.2.3 we shall prove that for *every* evolution semigroup on $\mathfrak{X}_T^p$ with generator defined by Theorem 2.2.10, cf. Theorem F.2.9 (which is also known as the C_{00}-semigroup; see [Nei81], Definition 4.8), there exists a *unique* solution operator such that representation (F.2.5) holds. Note that C_{00}-semigroups are selected from evolution semigroups by condition (iii) in Definition 2.2.8 of pre-generator; see Remark 2.2.9 (e).

Section F.2.2

In this subsection we collect some preliminaries. They are indispensable to answer the question posed in Sect. F.2.1 about the validity of the HEN representation for *any* evolution semigroup $\{\mathcal{U}_p(\tau)\}_{\tau\in\mathbb{R}_0^+}$ on $\mathfrak{X}_T^p$.

Seeing that (F.2.3) relates evolution semigroup $\{\mathcal{U}(\tau)\}_{\tau\in\mathbb{R}_0^+}$ on $\mathfrak{X}_T^C$ with its unique extension $\{\mathcal{U}_p(\tau)\}_{\tau\in\mathbb{R}_0^+}$ on $\mathfrak{X}_T^p$, we are going to recall here some general remarks about C_0-semigroups and their restrictions by embedding $\mathfrak{X}_T^C \overset{d}{\hookrightarrow} \mathfrak{X}_T^p$ (2.2.8). In the abstract part we follow [Kato70, Sections 1 and 2].

Definition F.2.3 Let A be linear operator in a Banach space X with $\mathrm{dom}\,(A) \subset X$. If Y is a subspace of X, then A induces a linear operator $\hat{A}$ such that

$$\mathrm{dom}\,(\hat{A}) := \{x \in \mathrm{dom}\,(A) \cap Y : \; A\,x \in Y \text{ and } \hat{A}\,x = A\,x\}\,. \tag{F.2.6}$$

Operator $\hat{A}$ is called the *part* of operator A in Y.

Recall that Y is a Banach space *continuously* and *densely* embedded into Banach space X if $Y \subset X$, there is $C > 0$ such that $\|y\|_X \leq C\|y\|_Y$ for any $y \in Y$, and Y is dense in X in the $\|\cdot\|_X$-topology. We denote this by $Y \overset{d}{\hookrightarrow} X$ for the *embedding* operator $\iota : Y \to X$.

Definition F.2.4 Let A be generator of a C_0-semigroup $\{\mathrm{e}^{-t\,A}\}_{t\geq 0}$ on Banach space X. Let one has $Y \overset{d}{\hookrightarrow} X$ for a Banach space Y. Then Y is said to be *admissible* with respect to A (or simply A-admissible) if $\mathrm{e}^{-t\,A} Y \subseteq Y$ and restriction $\{\mathrm{e}^{-t\,A} \restriction Y\}_{t\geq 0}$ is a C_0-semigroup on Y.

Remark F.2.5 Seeing that the set $Y \subset X$ is *invariant* under the semigroup $\{\mathrm{e}^{-t\,A}\}_{t\geq 0}$ and that (because of $Y \overset{d}{\hookrightarrow} X$) this set is also *dense* in X, by the *Nelson* theorem (Proposition 1.2.9) we obtain $Y = \mathrm{core}\,(A)$ if $Y \subset \mathrm{dom}\,(A)$.

Proposition F.2.6 *Let $A \in \mathscr{G}(X, M, \beta)$ be generator of a quasi-bounded semigroup on a Banach space X. Let $Y \overset{d}{\hookrightarrow} X$. Then subspace Y is A-admissible if and only if for (sufficiently large) $\lambda > \beta$:*

(i) *Operator $(A + \lambda\mathbb{1})^{-1}$ leaves Y invariant and there are $\hat{M} > 0$, $\hat{\beta} > 0$ such that*

$$\|(A + \lambda\mathbb{1})^{-n}\|_{\mathcal{L}(Y)} \leq \hat{M}(\lambda - \hat{\beta})^{-n}\,, \quad \lambda > \hat{\beta}\,, \quad n \in \mathbb{N}\,.$$

(ii) *The set $(A + \lambda\mathbb{1})^{-1}\,Y$ is dense in Y.*
In this case the part $\hat{A}$ of A in Y with domain (F.2.6) *is generator $\hat{A} \in \mathscr{G}(Y, \hat{M}, \hat{\beta})$ of semigroup $\{\mathrm{e}^{-t\,\hat{A}}\}_{t\geq 0}$, which is the part of the semigroup $\{\mathrm{e}^{-t\,A}\}_{t\geq 0}$ in Y, and $\|\mathrm{e}^{-t\,\hat{A}}\|_Y \leq \hat{M}\,\mathrm{e}^{\hat{\beta}\,t}$.*

Note that if Banach space Y is *reflexive*, then condition (ii) is redundant.

Now, after these general observations we remind (cf. Sect. 2.2.1) that Banach space $\mathfrak{X}_T^C = C_0([0, T], \mathfrak{X})$ can be regarded as *continuously* and *densely* embedded in Banach space $\mathfrak{X}_T^p = L^p([0, T], \mathfrak{X})$:

$$\begin{aligned} &\iota_p : \mathfrak{X}_T^C \to \mathfrak{X}_T^p\,, \quad p \in [1, \infty)\,, \\ &\|f\|_{\mathfrak{X}_T^p} \leq T^{1/p}\,\|f\|_{\mathfrak{X}_T^C}\,, \quad f \in \mathfrak{X}_T^C\,, \end{aligned} \tag{F.2.7}$$

Here ι_p is the *embedding* operator for $C_0([0,T],\mathfrak{X}) \overset{d}{\hookrightarrow} L^P([0,T],\mathfrak{X})$. Then $\ker(\iota_p) = \{0\}$ and also

$$\mathcal{M}_p(\phi)\,\iota_p = \iota_p\,\mathcal{M}(\phi), \quad \text{for} \quad \phi \in C([0,T])\,, \quad p \in [1,\infty)\,; \tag{F.2.8}$$

cf. definition (2.1.25) of $\mathcal{M}(\phi)$.

Corollary F.2.7 *Let* $K \in \mathscr{G}(\mathfrak{X}_T^p, M_p, \beta_p)$ *with domain* $\mathrm{dom}\,(K)$ *denote generator of evolution semigroup* $\{\mathcal{U}_p(\tau)\}_{\tau\in\mathbb{R}_0^+}$, *which is defined by Theorem* F.2.1 *on* $\mathfrak{X}_T^p$ *as an extension induced by evolution semigroup* $\{\mathcal{U}(\tau)\}_{\tau\in\mathbb{R}_0^+}$. *Then Banach* $\mathfrak{X}_T^C$ *is* admissible *with respect to* K.

Proof Since $\mathfrak{X}_T^C \overset{d}{\hookrightarrow} \mathfrak{X}_T^p$ and because of (F.2.3), one gets that $\{e^{-tK} \restriction \mathfrak{X}_T^C\}_{t\geq 0}$ is the *part* of semigroup $\{e^{-tK}\}_{t\geq 0}$, which is a C_0-semigroup on $\mathfrak{X}_T^C$, and that also $e^{-tK}\mathfrak{X}_T^C \subseteq \mathfrak{X}_T^C$. Therefore, assertion follows from Definition F.2.4. □

On account of Proposition F.2.6 and Corollary F.2.7, one also infers the following assertion.

Corollary F.2.8 *Under assumptions of Theorem* F.2.1 *the operator* $\widehat{K}$ *defined by*

$$\begin{aligned} &\mathrm{dom}\,(\widehat{K}) := \{f \in \mathfrak{X}_T^p : f \in \mathfrak{X}_T^C \cap \mathrm{dom}\,K \ \wedge \ Kf \in \mathfrak{X}_T^C\}\,, \\ &\widehat{K} := K \restriction \mathrm{dom}\,(\widehat{K})\,, \end{aligned} \tag{F.2.9}$$

is the generator of the evolution semigroup $\{\mathcal{U}(\tau)\}_{\tau\in\mathbb{R}_0^+}$ *on the Banach space* $\mathfrak{X}_T^C$. *That is,* $\widehat{K} = \mathcal{K}$, *where the definition of operator* $\mathcal{K}$ *is elucidated by Theorem* F.1.7.

In Sect. F.2.3. we use Definitions F.2.3 and F.2.4 and Proposition F.2.6 for Banach spaces $X := \mathfrak{X}_T^p$ and $Y := \mathfrak{X}_T^C$.

Section F.2.3

In this subsection we prove Theorems 2.2.10 and 2.2.11. We point out that in contrast to C-setting in the L^p-setting we have to impose more conditions on evolution generator $\mathcal{K}_p$ to ensure the one-to-one correspondence between solution operators and evolution semigroups by the HEN formula. This supplementary condition is the density of the generator domain cross-section $[\mathrm{dom}\,(\mathcal{K}_p)]_{\mathfrak{X}_T^p}$; see Definition 2.2.5. For L^p-setting we include this condition in Definition 2.2.8 (iii) of the *evolution pre-generator*. The corresponding strongly continuous semigroups are known as *evolution* C_{00}-semigroups; cf. [Nei79, Nei81].

We commence by the proof of Theorem 2.2.10, in which density cross-section condition plays no rôle. It only provides evolution generator $\mathcal{K}_p$ of C_{00}-semigroup with this property.

Theorem F.2.9 *Let operator* $\mathcal{K}_p$ *be evolution pre-generator in Banach space* $\mathfrak{X}_T^p$; *see Definition* 2.2.8. *Then* $\mathcal{K}_p$ *is evolution generator of semigroup* $\{\mathcal{U}_p(\tau)\}_{\tau\geq 0}$

defined on $\mathfrak{X}_T^p$ *by equation* (F.1.15) *if and only if operator* $\mathcal{K}_p$ *is generator of a strongly continuous on* $\mathfrak{X}_T^p$ *semigroup, that is,* $\{\mathcal{U}_p(\tau) = \mathrm{e}^{-\tau\mathcal{K}_p}\}_{\tau\geq 0}$.

Proof We note that dom $(\mathcal{K}_p)$ of evolution pre-generator is *dense* in $\mathfrak{X}_T^p$. This follows from Definition 2.2.8 (i)–(iii) and Lemma 2.2.6. As a consequence and owing to Remark F.1.8, the rest of the proof follows mutatis mutandis the arguments of the proof of Theorem F.1.7. □

For proving Theorem 2.2.11 we begin by preliminaries concerning the evolution C_0-semigroup. That is, in the next statement the generator $\mathcal{K}_p \in \mathcal{G}(M_p, \beta_p)$ is not defined by Theorem F.2.9. In particular, we do not need hypothesis (iii) (see Definition 2.2.8) about density of cross-section $[\mathrm{dom}\,(\mathcal{K}_p)]_{\mathfrak{X}_T^p}$.

Lemma F.2.10 *Let* C_0*-semigroup* $\{\mathcal{U}_p(\tau)\}_{\tau\in\mathbb{R}_0^+}$ *be quasi-bounded evolution semigroup on the Banach space* $\mathfrak{X}_T^p$*, with generator* $\mathcal{K}_p \in \mathcal{G}(M_p, \beta_p)$*, that is,* $\{\mathcal{U}_p(\tau) = \mathrm{e}^{-\tau\mathcal{K}_p}\}_{\tau\in\mathbb{R}_0^+}$. *Then*

$$\|(\mathcal{K}_p + \xi\,\mathcal{I}_p)^{-1} f(t)\|_{\mathfrak{X}} \leq \frac{M_p}{((\xi - \beta_p)\,q)^{1/q}}\,\|f\|_{\mathfrak{X}_T^p}\,, \quad \textit{for a.a. } t \in [0, T]\,, \tag{F.2.10}$$

$\{q : 1/p + 1/q = 1, p \in \mathbb{N}\}$ *and any* $f \in \mathfrak{X}_T^p$.

Proof Let L be a bounded *multiplication* operator (Definition 2.1.23) and $\mathcal{M}_p(\phi)$, for $\phi \in L^\infty([0, T])$, be a *scalar* multiplication operator (2.2.7) on the Banach space $\mathfrak{X}_T^p$. Then

$$\|L\,\mathcal{M}_p(\phi)\,f\|_{\mathfrak{X}_T^p} \leq \|L\|_{\mathcal{L}(\mathfrak{X}_T^p)}\|\mathcal{M}_p(\phi)\,f\|_{\mathfrak{X}_T^p}\,, \quad f \in \mathfrak{X}_T^p\,. \tag{F.2.11}$$

Since operators L and $\mathcal{M}_p(\phi)$ commute, (F.2.11) yields

$$\int_0^\infty dt\,|\phi(t)|^p\,(\|L\|^p_{\mathcal{L}(\mathfrak{X}_T^p)}\,\|f(t)\|^p_{\mathfrak{X}} - \|(L\,f)(t)\|^p_{\mathfrak{X}}) \geq 0\,, \quad p \in \mathbb{N}\,.$$

Seeing that $\phi \in L^\infty([0, T])$ is arbitrary, we obtain for $f \in \mathfrak{X}_T^p$ the estimate

$$\|(L\,f)(t)\|_{\mathfrak{X}} \leq \|L\|_{\mathcal{L}(\mathfrak{X}_T^p)}\,\|f(t)\|_{\mathfrak{X}}\,, \quad \text{for a.a. } t \in [0, T]\,. \tag{F.2.12}$$

Given that evolution semigroup $\{\mathcal{U}_p(\tau)\}_{\tau\in\mathbb{R}_0^+}$ is quasi-bounded with generator $\mathcal{K}_p \in \mathcal{G}(\mathfrak{X}_T^p, M_p, \beta_p)$, one gets

$$\|\mathcal{U}_p(\tau)\|_{\mathcal{L}(\mathfrak{X}_T^p)} \leq M_p\,\mathrm{e}^{\beta_p\,\tau}\,, \quad \tau \geq 0\,. \tag{F.2.13}$$

Recall that operators $\{\mathcal{U}_p(\tau)\,\mathcal{S}_L(\tau)\}_{\tau\geq 0}$ commute with scalar multiplication operators $\{\mathcal{M}_p(\phi)\}_{\phi\in L^\infty([0,T])}$; see (F.1.12). Then by Proposition 2.1.24 we infer that $\{\mathcal{U}_p(\tau)\,\mathcal{S}_L(\tau)\}_{\tau\geq 0}$ are bounded multiplication operators. As a consequence, estimates (F.2.12) and (F.2.13) yield

$$\begin{aligned} \|(\mathcal{U}_p(\tau)\,\mathcal{S}_L(\tau)\,h)(t)\|_{\mathfrak{X}} &\leq \|(\mathcal{U}_p(\tau)\|_{\mathcal{L}(\mathfrak{X}_T^p)}\,\|\mathcal{S}_L(\tau)\|_{\mathcal{L}(\mathfrak{X}_T^p)}\,\|h(t)\|_{\mathfrak{X}} \leq \\ &\leq M_p\,\mathrm{e}^{\beta_p\,\tau}\,\|h(t)\|_{\mathfrak{X}}\,, \quad \tau\geq 0\,, \end{aligned} \tag{F.2.14}$$

for any $h\in\mathfrak{X}_T^p$ and a.a. $t\in[0,T]$. Let $h:=\mathcal{S}_R(\tau)\,f$. Then on account of (F.1.10), (F.1.11) and (F.2.14) we deduce

$$\|(\mathcal{U}_p(\tau)\,f)(t)\|_{\mathfrak{X}} \leq M_p\,\mathrm{e}^{\beta_p\,\tau}\,\chi_{[0,T]}(t-\tau)\,\|f(t-\tau)\|_{\mathfrak{X}}\,, \quad \tau\geq 0\,, \tag{F.2.15}$$

for $f\in\mathfrak{X}_T^p$ and a.a. $t\in[0,T]$.

Owing to $\mathcal{K}_p\in\mathscr{G}(\mathfrak{X}_T^p, M_p, \beta_p)$ one gets representation

$$(\mathcal{K}_p+\xi\,\mathcal{I}_p)^{-1}\,f=\int_0^\infty d\tau\,\mathrm{e}^{-\tau\,\xi}\,\mathrm{e}^{-\tau\mathcal{K}_p}\,f\,, \quad \xi>\beta_p\,, \quad f\in\mathfrak{X}_T^p\,.$$

Then by (F.2.15) and the *Hölder* inequality we obtain for $1/p+1/q=1$, $p\in\mathbb{N}$, and a.a. $t\in[0,T]$ the estimate

$$\begin{aligned} \|(\mathcal{K}_p+\xi\,\mathcal{I}_p)^{-1}\,f(t)\|_{\mathfrak{X}} &\leq \int_0^\infty d\tau\,\mathrm{e}^{-(\xi-\beta_p)\,\tau}\,M_p\,\chi_{[0,T]}(t-\tau)\,\|f(t-\tau)\|_{\mathfrak{X}} \leq \\ \leq \left(\int_0^\infty d\tau\,\mathrm{e}^{-q\,(\xi-\beta_p)\,\tau}\,M_p^q\right)^{1/q} &\left(\int_0^\infty d\tau\,\chi_{[0,T]}(t-\tau)\,\|f(t-\tau)\|_{\mathfrak{X}}^p\right)^{1/p} \leq \\ \leq \frac{M_p}{((\xi-\beta_p)\,q)^{1/q}}\left(\int_0^t ds\,\|f(s)\|_{\mathfrak{X}}^p\right)^{1/p} &\leq \frac{M_p}{((\xi-\beta_p)\,q)^{1/q}}\,\|f\|_{\mathfrak{X}_T^p}\,, \end{aligned} \tag{F.2.16}$$

which proves (F.2.10). □

The next assertion establishes a fundamental one-to-one correspondence connecting quasi-bounded evolution semigroups and strongly continuous quasi-bounded solution operators, which involves the Howland-Evans-Neidhardt formula.

Theorem F.2.11 *Between the set of all quasi-bounded evolution C_{00}-semigroups $\{\mathcal{U}_p(\tau)=\mathrm{e}^{-\tau\mathcal{K}_p}\}_{\tau\in\mathbb{R}_0^+}$ on $\mathfrak{X}_T^p$ (equivalently, all evolution generators $\mathcal{K}_p\in\mathscr{G}(\mathfrak{X}_T^p, M_p, \beta_p)$ defined by Theorem* F.2.9*) and the set of all strongly continuous quasi-bounded solution operators $\{U(t,s)\}_{(t,s)\in\Delta}$ on $\mathfrak{X}$: $\|U(t,s)\|\leq M_p\,\mathrm{e}^{\beta_p(t-s)}$, $M_p\geq 1$, $\beta_p\geq 0$, (Definition* 2.1.4 *and Remark* 2.1.14*), there exists a one-to-one* *correspondence such that the Howland-Evans-Neidhardt representation*

$$(\mathrm{e}^{-\tau\mathcal{K}_p} f)(t) = U(t, t-\tau)\chi_{[0,T]}(t-\tau) f(t-\tau)\,, \quad \tau \geq 0\,, \tag{F.2.17}$$

holds for all $f \in \mathfrak{X}^p_T$ *and for a.a.* $t \in [0, T]$.

Proof Let quasi-bounded evolution C_{00}-semigroup $\{\mathcal{U}_p(\tau) = \mathrm{e}^{-\tau\mathcal{K}_p}\}_{\tau\in\mathbb{R}^+_0}$ on $\mathfrak{X}^p_T$ be defined by evolution generator $\mathcal{K}_p \in \mathscr{G}(\mathfrak{X}^p_T, M_p, \beta_p)$, which is determined in Theorem F.2.9.

Since $(\mathcal{K}_p + \xi\mathcal{I}_p)^{-1}\,\mathfrak{X}^p_T \subseteq \mathrm{dom}\,(\mathcal{K}_p)$, for $\beta_p < \xi$, and $\mathrm{dom}\,(\mathcal{K}_p)$ has dense cross-section $[\mathrm{dom}\,(\mathcal{K}_p)]_{\mathfrak{X}^p_T}$ (Definition 2.2.5 and Definition 2.2.8 (iii)), the operator $(\mathcal{K}_p + \xi\mathcal{I}_p)^{-1}$ leaves the space $\mathfrak{X}^C_T$ embedded into $\mathfrak{X}^p_T$ (F.2.7) invariant:

$$(\mathcal{K}_p + \xi\mathcal{I}_p)^{-1}\,\mathfrak{X}^C_T \subseteq \mathfrak{X}^C_T\,, \quad \beta_p < \xi\,. \tag{F.2.18}$$

Given that $\mathcal{K}_p \in \mathscr{G}(\mathfrak{X}^p_T, M_p, \beta_p)$ we obtain representation

$$(\mathcal{K}_p + \xi\mathcal{I}_p)^{-n} f = \frac{1}{(n-1)!}\int_0^\infty d\tau\, \tau^{n-1}\,\mathrm{e}^{-\tau\xi}\,\mathrm{e}^{-\tau\mathcal{K}_p} f\,, \quad f \in \mathfrak{X}^p_T\,, \quad \beta < \xi\,, \tag{F.2.19}$$

which by (F.2.18) and $\|\mathrm{e}^{-\tau\mathcal{K}_p} f\|_{\mathfrak{X}^C_T} \leq \widehat{M}_p\,\mathrm{e}^{\widehat{\beta}_p\tau}$, for $f \in \mathfrak{X}^C_T$, imply the estimates

$$\|(\mathcal{K}_p + \xi\mathcal{I}_p)^{-n} f\|_{\mathfrak{X}^C_T} \leq \frac{\widehat{M}_p}{(\xi - \widehat{\beta}_p)^n}\|f\|_{\mathfrak{X}^C_T}\,, \quad f \in \mathfrak{X}^C_T\,, \quad \widehat{\beta}_p < \xi\,, \tag{F.2.20}$$

for $n \in \mathbb{N}$. The definition of a quasi-bounded semigroup $\mathrm{e}^{-\tau\mathcal{K}_p}\}_{\tau\in\mathbb{R}^+_0}$ and its restriction to $\mathfrak{X}^C_T$ yields: $\widehat{M}_p \leq M_p$ and $\widehat{\beta}_p \leq \beta_p$. Thus, (F.2.18) and (F.2.20) satisfy condition (i) of Proposition F.2.6.

Taking into account (F.2.18) we introduce in the Banach space $\mathfrak{X}^C_T$ operator $\widehat{\mathcal{K}}_p$ with domain

$$\mathrm{dom}\,(\widehat{\mathcal{K}}_p) := (\mathcal{K}_p + \xi\mathcal{I}_p)^{-1}\,\mathfrak{X}^C_T\,, \quad \beta_p < \xi\,, \tag{F.2.21}$$

such that $\widehat{\mathcal{K}}_p f = \mathcal{K}_p f$ for $f \in \mathrm{dom}\,(\widehat{\mathcal{K}}_p)$. On account of Lemma F.2.10 we obtain

$$\begin{aligned}\|(\mathcal{K}_p + \xi\,\mathcal{I}_p)^{-1} f\|_{\mathfrak{X}^C_T} &= \sup_{t\in[0,T]} \|(\mathcal{K}_p + \xi\,\mathcal{I}_p)^{-1} f(t)\|_{\mathfrak{X}} \leq \\ &\leq \frac{M_p}{((\xi - \beta_p)\,q)^{1/q}}\,\|f\|_{\mathfrak{X}^p_T}\,, \quad f \in \mathfrak{X}^p_T\,, \quad \beta_p < \xi\,, \quad q = p/(p-1)\,.\end{aligned} \tag{F.2.22}$$

Then definition (F.2.21) and estimate (F.2.22) prove that operator $(\widehat{\mathcal{K}}_p + \xi\,\mathcal{I}_p)^{-1}$, on the Banach space $\mathfrak{X}_T^C$, is bounded:

$$\|(\widehat{\mathcal{K}}_p + \xi\,\mathcal{I}_p)^{-1}\|_{\mathcal{L}(\mathfrak{X}_T^C)} \leq \frac{M_p}{((\xi - \beta_p)\,q)^{1/q}}\,T^{1/p}\,, \quad \beta < \xi\,.$$

and consequently closed. Since it is invertible, the operator $\widehat{\mathcal{K}}_p$ is also closed on domain (F.2.21), which is dense in $\mathfrak{X}_T^C$.

Hence, condition (ii) of Proposition F.2.6 is also satisfied and for that reason the Banach space $\mathfrak{X}_T^C$ is *admissible* with respect to evolution generator $\mathcal{K}_p$. As a result the *part* $\widehat{\mathcal{K}}_p$ of $\mathcal{K}_p$ in $\mathfrak{X}_T^C$ generates on the space $\mathfrak{X}_T^C$ the *part* $\{\mathrm{e}^{-\tau\,\widehat{\mathcal{K}}_p}\}_{\tau\in\mathbb{R}_0^+}$ of evolution semigroup $\{\mathrm{e}^{-\tau\,\mathcal{K}_p}\}_{\tau\in\mathbb{R}_0^+}$.

On that account, $\{\widehat{\mathcal{U}}_p(\tau) := \mathrm{e}^{-\tau\,\widehat{\mathcal{K}}_p}\}_{\tau\in\mathbb{R}_0^+}$ is a quasi-bounded C_0-semigroup on $\mathfrak{X}_T^C$ with generator $\widehat{\mathcal{K}}_p \in \mathscr{G}(\mathfrak{X}_T^C, \widehat{M}_p, \widehat{\beta}_p)$. Owing to (F.1.15) and (F.1.24) we infer that the *part* $\{\widehat{\mathcal{U}}_p(\tau)\}_{\tau\in\mathbb{R}_0^+}$ of the evolution C_{00}-semigroup $\{\mathcal{U}_p(\tau)\}_{\tau\in\mathbb{R}_0^+}$ satisfies equation (F.1.24) for evolution C_0-semigroup $\{\mathcal{U}(\tau)\}_{\tau\in\mathbb{R}_0^+}$ on $\mathfrak{X}_T^C$. Therefore, $\{\widehat{\mathcal{U}}_p(\tau) = \mathcal{U}(\tau)\}_{\tau\in\mathbb{R}_0^+}$. By Theorem F.1.5 there exists a *unique* strongly continuous solution operator $\{U(t,s)\}_{(t,s)\in\Delta}$ such that the HEN formula (F.1.25) holds on $\mathfrak{X}_T^C$.

On the other hand, by Theorem F.2.1 there is a *unique induced* by evolution semigroup $\{\mathcal{U}(\tau)\}_{\tau\in\mathbb{R}_0^+}$ evolution semigroup on $\mathfrak{X}_T^p$. By virtue of preceding arguments, this induced evolution semigroup coincides with $\{\mathcal{U}_p(\tau)\}_{\tau\in\mathbb{R}_0^+}$. Then by Corollary F.2.2 it inherits the HEN representation (F.2.17) on $\mathfrak{X}_T^p$ for unique solution operator $\{U(t,s)\}_{(t,s)\in\Delta}$.

For proving the converse statement we suppose that on a Banach space $\mathfrak{X}$ there exists a strongly continuous solution operator $\{U(t,s)\}_{(t,s)\in\Delta}$ and that $\{\mathcal{U}_p(\tau) = \mathrm{e}^{-\tau\,\mathcal{K}_p}\}_{\tau\in\mathbb{R}_0^+}$ is quasi-bounded C_0-semigroup on Banach space $\mathfrak{X}_T^p$ such that formula (F.2.17) holds. Then we prove in few steps that $\{\mathcal{U}_p(\tau)\}_{\tau\in\mathbb{R}_0^+}$ is an evolution C_{00}-semigroup, that is, generator $\mathcal{K}_p$ has dense cross-section $[\mathrm{dom}\,(\mathcal{K}_p)]_{\mathfrak{X}_T^p}$ in $\mathfrak{X}$.

First, we note that by Corollary F.1.2 a one-parameter family $\{\mathcal{U}_p(\tau)\}_{\tau\in\mathbb{R}_0^+}$ that satisfies on $\mathfrak{X}_T^p$ the HEN equation (F.2.17) for strongly continuous solution operator $\{U(t,s)\}_{(t,s)\in\Delta}$ is a (nilpotent) C_0-semigroup. In addition, by Lemma F.1.3 it is also an evolution semigroup.

Second, thanks to (F.2.19), $n = 1$, and formula (F.2.17) one gets for C_0-semigroup $\{\mathcal{U}_p(\tau) = \mathrm{e}^{-\tau\,\mathcal{K}_p}\}_{\tau\in\mathbb{R}_0^+}$, for $f \in \mathfrak{X}_T^p$ and a.a. $t \in [0,T]$:

$$((\mathcal{K}_p + \xi\,\mathcal{I}_p)^{-1} f)(t) = \int_0^\infty d\tau\,\mathrm{e}^{-\tau\,\xi}\,(\mathrm{e}^{-\tau\,\mathcal{K}_p} f)(t) = \tag{F.2.23}$$

$$\int_0^T d\tau\,\mathrm{e}^{-\tau\,\xi}\,U(t,t-\tau)\chi_{[0,T]}(t-\tau)f(t-\tau) = \int_0^t ds\,\mathrm{e}^{-\xi\,(t-s)}\,U(t,s)\,f(s)\,.$$

Since solution operator is strongly continuous, representation (F.2.23) ensures that

$$\operatorname{dom}(\mathcal{K}_p) \subseteq \mathfrak{X}_T^C . \tag{F.2.24}$$

On account of representation (F.2.17), inclusion (F.2.24) and Definition F.2.4, we infer that $\mathfrak{X}_T^C \overset{d}{\hookrightarrow} \mathfrak{X}_T^p$ is admissible with respect to generator $\mathcal{K}_p$.

Third, let us denote by $\mathcal{K}$ the *part* of generator $\mathcal{K}_p$ in $\mathfrak{X}_T^C$. By admissibility and Definition F.2.4 it generates on $\mathfrak{X}_T^C$ a C_0-semigroup $\{e^{-\tau \mathcal{K}}\}_{\tau\in\mathbb{R}_0^+}$, which is a *part* of $\{\mathcal{U}_p(\tau)\}_{\tau\in\mathbb{R}_0^+}$. So, $\operatorname{dom}(\mathcal{K})$ is dense $\mathfrak{X}_T^C$. Hence, by Lemma 2.1.31 domain $\operatorname{dom}(\mathcal{K})$ has dense cross-section $[\operatorname{dom}(\mathcal{K})]_{\mathfrak{X}_T^C}$. By Definition 2.2.5 this implies density of cross-section $[\operatorname{dom}(\mathcal{K}_p)]_{\mathfrak{X}_T^p}$ in $\mathfrak{X}$ (cf. Definition 2.2.5), i.e. $\{e^{-\tau \mathcal{K}_p}\}_{\tau\in\mathbb{R}_0^+}$ is the evolution C_{00}-semigroup. □

Corollary F.2.12 *The semigroup* $\{\mathcal{U}_p(\tau) = e^{-\tau\mathcal{K}_p}\}_{\tau\in\mathbb{R}_0^+}$ *is nilpotent on the space* $\mathfrak{X}_T^p$*, that is,* $\mathcal{U}_p(\tau)|_{\tau>T} = 0$*. Then by definition* (2.2.1) *and representation* (2.2.12) *for quasi-bounded solution operator one gets estimate*

$$\|e^{-\tau\mathcal{K}_p} f\|_{\mathfrak{X}_T^p} \leq M\, e^{\beta T}\, \|f\|_{\mathfrak{X}_T^p}, \quad \tau \geq 0 .$$

As a consequence the nilpotent on $\mathfrak{X}_T^p$ *evolution semigroup is bounded* :

$$\|e^{-\tau\mathcal{K}_p}\|_{\mathcal{L}(\mathfrak{X}_T^p)} \leq M_T , \quad \tau \geq 0 , \tag{F.2.25}$$

where $M_T := M\, e^{\beta T}$.

Appendix G
Integral Equations and the Gronwall-Type Inequalities

G.1 The Gronwall Inequality

Lemma G.1.1 *Let $t \mapsto g(t)$, $t \mapsto \phi(t)$ and $t \mapsto k(t) \geq 0$ be continuous functions for $t \in [a,b] \subset \mathbb{R}$. If $t \mapsto g(t)$ satisfies the inequality*

$$g(t) \leq \phi(t) + \int_a^t ds\, k(s)\, g(s)\,, \quad t \in [a,b]\,, \tag{G.1.1}$$

then this function also satisfies the inequality

$$g(t) \leq \phi(t) + \int_a^t ds\, k(s)\, \phi(s)\, \exp\left(\int_s^t d\tau\, k(\tau)\right), \quad t \in [a,b]\,. \tag{G.1.2}$$

Proof Let $y(t) := \int_a^t ds\, k(s)\, g(s)$ for $t \in [a,b]$. Then $y(a) = 0$ and by (G.1.1) together with $k(t) \geq 0$ one gets

$$\begin{aligned} \partial_t y(t) = k(t)\, g(t) &\leq k(t)\, \phi(t) + k(t) \int_a^t ds\, k(s)\, g(s) \\ &= k(t)\, \phi(t) + k(t)\, y(t)\,, \quad t \in (a,b)\,. \end{aligned}$$

Seeing that $\exp\left(-\int_a^t d\tau\, k(\tau)\right) > 0$, we deduce estimate

$$\partial_t \left(y(t)\, \exp\left(-\int_a^t d\tau\, k(\tau)\right)\right) \leq k(t)\, \phi(t)\, \exp\left(-\int_a^t d\tau\, k(\tau)\right).$$

V. A. Zagrebnov et al., *Trotter-Kato Product Formulæ*, Operator Theory: Advances and Applications 296, https://doi.org/10.1007/978-3-031-56720-9

As a consequence, integration of this estimate over interval $[a, t]$ yields

$$y(t) \exp\left(-\int_a^t d\tau\, k(\tau)\right) \leq \int_a^t ds\, k(s)\, \phi(s) \exp\left(-\int_a^s d\tau\, k(\tau)\right),$$

or equivalently, the inequality

$$y(t) \leq \int_a^t ds\, k(s)\, \phi(s) \exp\left(\int_s^t d\tau\, k(\tau)\right).$$

On account of (G.1.1) and definition of $y(t)$ this inequality provides (G.1.2). □

Corollary G.1.2

(a) *If* $\phi(t) \in C^1([a, b])$, *then*

$$\int_a^t ds\, \partial_s \phi(s) \exp\left(\int_s^t d\tau\, k(\tau)\right) = \phi(t) - \phi(a) \exp\left(\int_a^t d\tau\, k(\tau)\right) + \int_a^t ds\, k(s)\, \phi(s) \exp\left(\int_s^t d\tau\, k(\tau)\right).$$

This identity and (G.1.2) *yield another version of the Gronwall inequality:*

$$g(t) \leq \phi(a) \exp\left(\int_a^t d\tau\, k(\tau)\right) + \int_a^t ds\, \partial_s \phi(s) \exp\left(\int_s^t d\tau\, k(\tau)\right), \tag{G.1.3}$$

for $t \in [a, b]$.

(b) *If* $\phi : [a, b] \to \phi_0$ *in* (G.1.1), *then* (G.1.3) *provides a well-known (simplest) version of the Gronwall inequality:*

$$g(t) \leq \phi_0 \exp\left(\int_a^t d\tau\, k(\tau)\right), \quad t \in (a, b), \tag{G.1.4}$$

for $\phi_0 = \phi(a)$.

(c) *For homogeneous inequality* (G.1.1) *it is instructive to allude to the Volterra integral equation approach; see also Sects.* G.2 *and* G.3. *Namely, iterating the integral* (G.1.1) *for* $\phi = 0$, *we obtain for any* $n \in \mathbb{N}$ *and a measurable on* $[a, b] \subset \mathbb{R}$ *function* : $[a, b] \ni t \mapsto g(t)$, *the estimate*

$$g(t) \leq \int_a^t ds\, k(s) \int_a^s d\tau_1\, k(\tau_1) \ldots \int_a^{\tau_{n-1}} d\tau_n\, k(\tau_n)\, g(\tau_n), \quad \tau_0 = s. \tag{G.1.5}$$

Let $C_k := \sup_{t\in[a,b]} k(t)$ *and* $C_g := \operatorname{ess\,sup}_{t\in[a,b]} g(t)$. *Then* (G.1.5) *yields for any* $n \geq 1$

$$g(t) \leq C_g \,\frac{(C_k\,(t-a))^{n+1}}{(n+1)!}\,, \qquad t \in [a,b]\,,$$

that is, the homogeneous inequality (G.1.1) *has no positive solutions.*

Note that one also infers the conclusion of Corollary G.1.2 (c) from Lemma G.1.1.

G.2 Method of the Volterra Integral Equation

Here we extend the *Gronwall* inequality (G.1.2) to the case when functions $t \mapsto \phi(t) \geq 0$ and $t \mapsto k(t) \geq 0$ are *weakly* singular.

Lemma G.2.1 *Let* $t \mapsto F(t) \geq 0$ *be continuous function in* $\mathbb{R}_0^+$ *satisfying inequality*

$$F(t) \leq c_1\, t^{-\alpha_1} + c_2 \int_0^t ds\,(t-s)^{-\alpha_2} F(s)\,, \quad t \in \mathbb{R}_0^+\,, \tag{G.2.1}$$

for positive constants c_1, c_2 *and for* $\alpha_1, \alpha_2 \in (0,1)$. *Then there exist* $C > 0$ *and* $t_0 > 0$ *such that*

$$F(t) \leq \frac{C}{t^{\alpha_1}}\,, \qquad \alpha_1 \in (0,1)\,, \tag{G.2.2}$$

for $t \in (0, t_0)$.

Proof Consider the auxiliary (*second kind*) Volterra integral equation:

$$f(t) = h_0(t) + c_2 \int_0^t ds\,(t-s)^{-\alpha_2} f(s)\,, \quad t \in \mathbb{R}_0^+\,, \tag{G.2.3}$$

where $h_0(t) := c_1\, t^{-\alpha_1}$. We shall solve (G.2.3) by iterations using the following recurrence relation:

$$\begin{aligned} &f_0(t) = h_0(t)\,, \qquad t \in \mathbb{R}_0^+\,, \\ &f_n(t) = f_0(t) + c_2 \int_0^t d\tau\,(t-\tau)^{-\alpha_2} f_{n-1}(\tau)\,, \quad n \geq 1\,. \end{aligned} \tag{G.2.4}$$

To this aim we introduce $f_n(t) := \sum_{k=0}^{n} h_k(t)$, where

$$
\begin{aligned}
h_0(t) &= c_1\, t^{-\alpha_1}\,,\\
h_1(t) &= c_1\, c_2 \int_0^t \mathrm{d}\tau\ (t-\tau)^{-\alpha_2}\tau^{-\alpha_1} = \frac{c_1\, \Gamma(1-\alpha_1)}{t^{\alpha_1}}\cdot \frac{c_2\, \Gamma(1-\alpha_2)\, t^{1-\alpha_2}}{\Gamma(2-\alpha_1-\alpha_2)}\,,\\
h_2(t) &= c_1\, c_2^2 \int_0^t \mathrm{d}\tau\ (t-\tau)^{-\alpha_2} \int_0^\tau \mathrm{d}\tau'\ (\tau-\tau')^{-\alpha_2}\,(\tau')^{-\alpha_1} =\\
&= \frac{c_1\, \Gamma(1-\alpha_1)}{t^{\alpha_1}}\cdot \frac{[\,c_2\, \Gamma(1-\alpha_2)\, t^{1-\alpha_2}\,]^2}{\Gamma(1-\alpha_1+2(1-\alpha_2))}\,, \qquad \text{etc.}\,,
\end{aligned}
\tag{G.2.5}
$$

and $\Gamma(\cdot)$ denotes the Euler gamma-function; cf. Appendix I. From (G.2.4) and (G.2.5) we infer by induction that

$$
h_k(t) = \frac{c_1\, \Gamma(1-\alpha_1)}{t^{\alpha_1}}\cdot \frac{[\,c_2\, \Gamma(1-\alpha_2)\, t^{1-\alpha_2}\,]^k}{\Gamma(1-\alpha_1+k(1-\alpha_2))}\,, \qquad k \in \mathbb{N}_0\,. \tag{G.2.6}
$$

Recall that for $k \to \infty$ one gets for asymptotic of the Euler gamma-function: $\Gamma(k((1-\alpha)) = O(k^{k\,(1-\alpha)})$; cf. (I.0.7) in Appendix I). Then formulæ (G.2.6) yield that the series:

$$
f(t) := \sum_{k=0}^{\infty} h_k(t)\,, \quad t \in \mathbb{R}_0^+\,, \tag{G.2.7}
$$

converges locally uniformly in $\mathbb{R}_0^+$ to a continuous function $f \in C(\mathbb{R}_0^+)$ such that

$$
f(t) \le \frac{c_1\, \Gamma(1-\alpha_1)}{t^{\alpha_1}} \sum_{k=0}^{\infty} \frac{[\,c_2\, \Gamma(1-\alpha_2)\, T^{1-\alpha_2}\,]^k}{\Gamma(1-\alpha_1+k(1-\alpha_2))} := \frac{C(T)}{t^{\alpha_1}}\,, \qquad t \in (0,T]\,. \tag{G.2.8}
$$

Suppose that $F(t) \ge 0$ is a solution of integral inequality (G.2.1). Let $D(t) := F(t)\, t^{\alpha_1} - f(t)\, t^{\alpha_1}$, which is continuous on $\mathbb{R}_0^+$. Then due to (G.2.1) and (G.2.3) we obtain integral inequality:

$$
D(t) \le c_2\, t^{\alpha_1} \int_0^t ds\, s^{-\alpha_1}(t-s)^{-\alpha_2}\, D(s)\,, \quad t \in \mathbb{R}_0^+\,. \tag{G.2.9}
$$

Note that because of (G.2.5) one gets for integral in (G.2.9) the representation:

$$
\mathfrak{I}(t) := c_2 \int_0^t ds\ \frac{t^{\alpha_1}}{s^{\alpha_1}\,(t-s)^{\alpha_2}} = c_2\, t^{1-\alpha_2}\, \frac{\Gamma(1-\alpha_1)\, \Gamma(1-\alpha_2)}{\Gamma(2-\alpha_1-\alpha_2)}\,. \tag{G.2.10}
$$

As a consequence, owing to monotonicity of (G.2.10) there exists a positive $t_0 := t_0(c_2, \alpha_1, \alpha_2) \leq T$ such that $\Im(t < t_0) < 1/2$. Then by inequality (G.2.9) and by definition (G.2.10) we obtain

$$D(t) \leq \frac{1}{2} \sup_{0<s<t_0} D(s) \, , \quad t \in (0, t_0) \, . \tag{G.2.11}$$

For that reason any solution of inequality (G.2.11) is *non-positive*: $D(t) \leq 0$, and thus $F(t)\, t^{\alpha} \leq f(t)\, t^{\alpha} \leq C(T)$ by virtue of (G.2.8). So, the claim (G.2.2) follows for $t \in (0, t_0)$ and $C = C(T)$. □

G.3 The Operator-Valued Volterra Integral Equations

Here we continue extension and application of the *Gronwall-type* inequalities to analysis of the *operator-valued* Volterra integral equations with weakly singular kernels in the *normed* spaces.

Definition G.3.1 Let $\Delta := \{(t, s) \in [0, T] \times [0, T] \, : \, 0 \leq s \leq t \leq T\}$, and $\Delta_> := \{(t, s) \in \Delta : t > s\}$, for $[0, T] \subset \mathbb{R}_0^+$. We call the operator-valued *strongly continuous* function $\Delta_> \ni (t, s) \to K(t, s)$ in Banach space $\mathfrak{X}$ a *weakly singular* kernel of *order* $\alpha \in [0, 1)$ if its operator norm:

$$\|K(t, s)\| \leq \frac{K_{\alpha}}{(t - s)^{\alpha}} \, , \qquad (t, s) \in \Delta_> \, , \tag{G.3.1}$$

for $K_{\alpha} > 0$.

Theorem G.3.2 *In a Banach space $\mathfrak{X}$ we consider two inhomogeneous operator-valued Volterra equations* :

$$V(t, s) = \Phi_0(t, s) + \int_s^t d\tau \; K(t, \tau)\, V(\tau, s) \, , \quad (t, s) \in \Delta_> \, , \tag{G.3.2}$$

$$W(t, s) = \Phi_0(t, s) + \int_s^t d\tau \; W(t, \tau)\, K(\tau, s) \, , \quad (t, s) \in \Delta_> \, , \tag{G.3.3}$$

for the weakly singular kernel K of the order $\alpha \in [0, 1)$ (G.3.1) *and the operator-valued inhomogeneous term Φ_0, which is in turn a strongly continuous weakly singular kernel of the order $\beta \in [0, 1)$; see Definition* G.3.1. *Then the following statements hold*:

(a) *Solutions $\{V(t, s)\}_{(t,s)\in\Delta_>}$, $\{W(t, s)\}_{(t,s)\in\Delta_>}$ of equations* (G.3.2), (G.3.3) *are operator-valued strongly continuous weakly singular kernels of order β.*
(b) *Solution $\{V(t, s)\}_{(t,s)\in\Delta_>}$ and solution $\{W(t, s)\}_{(t,s)\in\Delta_>}$ are unique.*

Proof

(a) Since the singularity of kernel in (G.3.2), (G.3.3) is integrable, iterating these integral formulæ we can construct solutions $\{V(t,s)\}_{(t,s)\in\Delta_>}$, $\{W(t,s)\}_{(t,s)\in\Delta_>}$ as convergent Dyson-Phillips series. Below we consider only series for $\{V(t,s)\}_{(t,s)\in\Delta_>}$ seeing that the proof for $\{W(t,s)\}_{(t,s)\in\Delta_>}$ is identical.

To this aim we define for iterations of (G.3.2) the recurrence relation:

$$\begin{aligned} &V_0(t,s) = \Phi_0(t,s)\,, \qquad (t,s)\in\Delta_>\,, \\ &V_n(t,s) = V_0(t,s) + \int_s^t \mathrm{d}\tau\ K(t,\tau)\, V_{n-1}(\tau,s)\,, \quad n\geq 1\,, \end{aligned} \tag{G.3.4}$$

and the Dyson-Phillips series $\sum_{k=0}^{\infty}\Phi_k(t,s)$, where $V_n(t,s) := \sum_{k=0}^{n}\Phi_k(t,s)$. Note that the strong continuity of $(t,s)\mapsto\Phi_0(t,s)$ and of the kernel $(t,s)\mapsto K(t,s)$ ensures integrability of the product, which by Definition G.3.1 yields for $(t,s)\in\Delta_>$ the operator-norm estimate

$$\begin{aligned} &\|\Phi_1(t,s)\| = \|\int_s^t d\tau\ K(t,\tau)\, V_0(\tau,s)\| \leq K_\alpha\,\Phi_\beta \int_s^t d\tau\ \frac{1}{(t-\tau)^\alpha\,(\tau-s)^\beta} \leq \\ &\frac{K_\alpha\,\Phi_\beta}{(t-s)^{\alpha+\beta-1}}\int_0^1 dx\ \frac{1}{x^\alpha\,(1-x)^\beta} = \frac{\Phi_\beta\,\Gamma(1-\beta)}{(t-s)^\beta}\cdot\frac{K_\alpha\Gamma(1-\alpha)(t-s)^{1-\alpha}}{\Gamma(2-\alpha-\beta)}\,; \end{aligned} \tag{G.3.5}$$

cf. Appendix G.2 (G.2.5). Then on account of (G.3.4) and (G.3.5), we deduce by iteration of (G.3.4) estimates:

$$\|\Phi_k(t,s)\| \leq \frac{\Phi_\beta\,\Gamma(1-\beta)}{(t-s)^\beta}\cdot\frac{\{K_\alpha\Gamma(1-\alpha)(t-s)^{1-\alpha}\}^k}{\Gamma(1-\beta+k((1-\alpha))}\,, \qquad (t,s)\in\Delta_>\,, \tag{G.3.6}$$

for any $k\in\mathbb{N}_0$; cf. (G.2.6). Seeing that for $k\to\infty$ one gets for asymptotic of the Euler gamma-function $\Gamma(k(1-\alpha)) = O(k^{k\,(1-\alpha)})$ (cf. (I.0.7) in Appendix I), the estimates (G.3.6) yield that the series:

$$V(t,s) := \sum_{k=0}^{\infty}\Phi_k(t,s)\,, \quad (t,s)\in\Delta_>\,, \tag{G.3.7}$$

converges in the *operator-norm* topology to a bounded operator-valued function such that for $(t,s)\in\Delta_>$

$$\|V(t,s)\| \leq \frac{\Phi_\beta}{(t-s)^\beta}\left\{1+\sum_{k=1}^{\infty}\frac{\Gamma(1-\beta)\,\{K_\alpha\Gamma(1-\alpha)\,(t-s)^{1-\alpha}\}^k}{\Gamma(1-\beta+k(1-\alpha))}\right\}. \tag{G.3.8}$$

Note that for any $0 < \delta < T$ the series (G.3.7) converges in operator norm *uniformly* on $\Delta_{>\delta} := \{(t, s) \in \Delta_{>} : t > s + \delta\}$. Then the strong continuity of functions $\Delta_{>} \ni (t, s) \mapsto \Phi_k(t, s)$, for $k = 0, 1, 2, \ldots$, and uniform operator-norm convergence on $\Delta_{>\delta}$ for arbitrary (small) $\delta > 0$ entail the strong continuity of function $(t, s) \mapsto V(t, s)$ in $\Delta_{>}$. Then owing to explicit estimate (G.3.8) the function (G.3.7) is a strongly continuous weakly singular kernel of the order β; cf. Definition G.3.1.

By construction (G.3.4), (G.3.7), the bounded operator-valued function $(t, s) \mapsto V(t, s)$ in $\Delta_{>}$ is a solution of the integral equation (G.3.2). As a consequence,

$$\|V(t, s)\| \leq \|\Phi_0(t, s)\| + \int_s^t d\tau \; \|K(t, \tau)\| \; \|V(\tau, s)\| \,, \quad (t, s) \in \Delta_{>} \,. \tag{G.3.9}$$

By inspection of Lemma G.1.1 the integrability of operator norms involved into (G.3.9) is sufficient to infer the *Gronwall-type* inequality:

$$\|V(t, s)\| \leq \|\Phi_0(t, s)\| + \int_s^t d\tau \; \|K(t, \tau)\| \; \|\Phi_0(\tau, s)\| \exp\left(\int_\tau^t d\eta \; \|K(t, \eta)\|\right) , \tag{G.3.10}$$

for $(t, s) \in \Delta_{>}$. Then on account of property of kernels K, Φ_0 and (G.3.5), the inequality (G.3.10) yields for some $\Phi^*_\beta > 0$

$$\|V(t, s)\| \leq \frac{\Phi^*_\beta}{(t - s)^\beta} \,, \quad (t, s) \in \Delta_{>} \,. \tag{G.3.11}$$

This is a *round-about* way to prove the same result as due to explicit estimate (G.3.8).

(b) Again we consider equation (G.3.2). If it possesses two solutions V_a and V_b, then $V_{ab} := V_a - V_b$ is a solution of homogeneous equation (G.3.2):

$$V_{ab}(t, s) = \int_s^t d\tau \; K(t, \tau) \, V_{ab}(\tau, s) \,, \quad (t, s) \in \Delta_{>} \,.$$

Given that in this case $\Phi_0(t, s) = 0$, the Gronwall-type inequality (G.3.10) ensures the uniqueness: $V_a = V_b$.

The proof for solution W of equation (G.3.3) follows along the same line reasoning as above for V.

□

Corollary G.3.3 *If in condition of Theorem* G.3.2 *the kernels K and Φ_0 are continuous in the operator-norm topology in $\Delta_{>}$, then solutions V and W are also operator-norm continuous in domain $\Delta_{>}$.*

Appendix H
Fractional Power Conditions and Tamura's Example

In [Tam00] *Hiroshi Tamura* proposed a crucial example, which significantly elucidates the question related to the rate of the *operator-norm* convergence of the Trotter product formula. This example shows that the rate of the operator-norm convergence of the Trotter product formula obtained in [NZ99a] is the best possible (*optimal*). It also essentially clarifies (via a *counterexample*) the accuracy of conditions and conclusions in [NZ99a, Theorem 5.5, Corollary 5.6] that allowed later to use the construction [Tam00, Theorem B] in proving optimality of the operator-norm convergence in ITTZ-Theorem 7.1.9, Sect. 7.2 [ITTZ01, Theorem 1, Corollary 1].

To proceed further we remind the corresponding assertion of our paper [NZ99a].

Theorem H.0.1 ([NZ99a, Theorem 5.5, Corollary 5.6]) *If the self-adjoint operators* A *and* B *satisfy, for some* $\alpha \in (1/2, 1)$ *and* $a \in (0, 1)$, *conditions*:

(i) $A \geq \mathbb{1}, \quad B \geq 0$.
(ii) $\mathrm{dom}\,(A^\alpha) \subset \mathrm{dom}\,(B^\alpha)\,, \quad ||B^\alpha x|| \leq a\, ||A^\alpha x||$ *for all* $x \in \mathrm{dom}\,(A^\alpha)$;
(iii) $\mathrm{dom}\,((A \dot{+} B)^\alpha) \subset \mathrm{dom}\,(A^\alpha)$.

Then there is a constant $C > 0$ *such that*

$$\left\| \mathrm{e}^{-t(A \dot{+} B)} - \left(\mathrm{e}^{-tA/n}\mathrm{e}^{-tB/n}\right)^n \right\| \leq \frac{C}{n^{2\alpha-1}} \tag{H.0.1}$$

holds uniformly in $t \geq 0$. *Here self-adjoint operator* $A \dot{+} B$ *is the form-sum of operators* A *and* B.

Proof For the proof of (H.0.1) we refer to [NZ99a, Theorem 5.5, Corollary 5.6], since in Sect. 7.4 (INZ-Theorem 7.4.7) a stronger result with the *same* optimal rate of convergence $O(1/n^{2\alpha-1})$ has been established. □

V. A. Zagrebnov et al., *Trotter-Kato Product Formulæ*, Operator Theory: Advances and Applications 296, https://doi.org/10.1007/978-3-031-56720-9

Here we only mention differences between conditions in [NZ99a, Theorem 5.5] and in INZ-Theorem 7.4.7:

1. In comparison to [NZ99a, Theorem 5.5] the conditions on the operators A and B in INZ-Theorem 7.4.7 are *relaxed*. In [NZ99a, Theorem 5.5] it was assumed that B^α is *small* with respect to A^α, that is, $\operatorname{dom}(A^\alpha) \subseteq \operatorname{dom}(B^\alpha)$ and
$$\|B^\alpha u\| \leq a\, \|A^\alpha u\|\,, \quad u \in \operatorname{dom}(A^\alpha), \tag{H.0.2}$$
for some $\alpha \in (1/2, 1)$ and $a \in (0, 1)$. This condition is reduced in INZ-Theorem 7.4.7 to the *mild* subordination condition: $\operatorname{dom}(A^{1/2}) \subseteq \operatorname{dom}(B^{1/2})$, which obviously implies (H.0.2). The yet *open problem* is to eliminate completely this condition.
2. The conditions on the Kato functions f and g in INZ-Theorem 7.4.7 are essentially relaxed compared with [NZ99a, Theorem 5.5]. Indeed, there the Kato functions f and g have to satisfy at zero a *smoothness* conditions of the type $|f|_2, |g|_2 < \infty$ and $f(x) \sim x^{-2\alpha}$ at *infinity*. They are replaced in INZ-Theorem 7.4.7 by conditions: $|f|_{2\alpha},\ |g|_{2\alpha} < \infty$, $\alpha \in (1/2, 1)$.

Now we clarify the *accuracy* of the result in Theorem H.0.1 by means of Tamura's example. Namely, [Tam00, Theorem B] provides for the Trotter product formula the following assertion (*counterexample*).

Theorem H.0.2 ([Tam00, Theorem B]) *For each $\alpha \in (0, 1)$ and $a \in (0, 1)$, there exist operators A and B on a Hilbert space $\mathfrak{H}$ such that they satisfy conditions* (i), (ii) *and* (iii) *of Theorem* H.0.1, *and the bound*
$$\left\|e^{-t(A\dot{+}B)} - \left(e^{-tA/n}e^{-tB/n}\right)^n\right\| \geq \frac{C_t}{n^{2\alpha-1}} \quad \text{for } \alpha \in (1/2, 1) \text{ and } n > n_0 \text{ (large)},$$
$$\liminf_{n\to\infty}\left\|e^{-t(A\dot{+}B)} - \left(e^{-tA/n}e^{-tB/n}\right)^n\right\| \geq D_t \qquad \text{for} \quad \alpha \in (0, 1/2],$$
where C_t and D_t are positive and continuous in $t \in \mathbb{R}^+$.

Remark H.0.3 Theorem H.0.2 implies that the bound (H.0.1) of the Theorem H.0.1 is the best possible (*optimal*) in its power of $n \in \mathbb{N}$.

Remark H.0.4 The case $\alpha = 1/2$ is an example for which Trotter product formula does not hold in the *operator-norm* convergence sense, while it holds in the *strong* operator topology by Kato's Theorem 5.2.5; see (5.2.56) for exponential Kato functions [Kato78, Theorem, p.186]. In other words, the operator-norm convergence of the product formula *cannot* be extended to the cases $\alpha \leq 1/2$ in this abstract setting, as it has been already conjectured in [NZ99a].

Example H.0.5 (Hiroshi Tamura [Tam00, Section 2, Example])

Before dealing with a concrete example, we recall some basic facts on operators acting in a Hilbert space $\mathfrak{h}^{\oplus}$, which is a *direct sum* of countable family of separable Hilbert spaces.

To this aim let $\{\mathfrak{h}_n\}_{n\in\mathbb{N}}$ be a countable family of Hilbert spaces over $\mathbb{R}$. The inner product in $\mathfrak{h}_n$ is denoted by $(\ ,\ )_n$ for each $n \in \mathbb{N}$. The Hilbert space direct sum $\mathfrak{h}^{\oplus}$ of the Hilbert spaces $\{\mathfrak{h}_n\}_{n\in\mathbb{N}}$ is defined by

$$\mathfrak{h}^{\oplus} = \left\{ x = (x_n)_{n\in\mathbb{N}} \,:\, x_n \in \mathfrak{h}_n \ (\forall n \in \mathbb{N}),\ \sum_{n=1}^{\infty}(x_n, x_n)_n < \infty \right\},$$

with the inner product on $\mathfrak{h}^{\oplus}$ defined by

$$(x, y) := \sum_{n=1}^{\infty}(x_n, y_n)_n \,.$$

Let A_n be a bounded non-negative self-adjoint operator on $\mathfrak{h}_n$ for each $n \in \mathbb{N}$. From the operators $(A_n)_{n\in\mathbb{N}}$, we construct a non-negative self-adjoint operator A in $\mathfrak{h}^{\oplus}$ with domain

$$\operatorname{dom}(A) = \left\{ x = (x_n)_{n\in\mathbb{N}} \in \mathfrak{h}^{\oplus} \,:\, \sum_{n=1}^{\infty}(A_n x_n, A_n x_n)_n < \infty \right\}, \tag{H.0.3}$$

$$(Ax)_n = A_n x_n \qquad \text{for} \qquad x = (x_n)_{n\in\mathbb{N}} \in \operatorname{dom}(A). \tag{H.0.4}$$

Let B_n be another bounded non-negative self-adjoint operator on $\mathfrak{h}_n$ for each $n \in \mathbb{N}$, and B the operator constructed from $(B_n)_{n\in\mathbb{N}}$ as above. Then $\operatorname{dom}(A) \cap \operatorname{dom}(B)$ is dense in $\mathfrak{h}^{\oplus}$ and the sum $A + B$ is *essentially* self-adjoint operator on its domain $\operatorname{dom}(A) \cap \operatorname{dom}(B)$.

By construction (H.0.3), (H.0.4) the *closure* $\overline{A+B}$ of operator $A+B$ on domain $\operatorname{dom}(A) \cap \operatorname{dom}(B)$ is equal to the operator produced from $(A_n + B_n)_{n\in\mathbb{N}}$. As a consequence, the self-adjoint extension $\overline{A+B}$ is *equal* to the self-adjoint *form-sum*: $A \dot{+} B$, of operators A and B. Note that a priori, it is different from the self-adjoint *Friedrichs* extension: $(A+B)_F$; cf. [Kato80, Chapter VI, §2.5].

In that next step, we proceed by considering the strongly continuous contraction self-adjoint semigroup $\{e^{-tA_n}\}_{t\geq 0}$ on $\mathfrak{h}_n$ for each n. Let $T(t)$ be the operator on $\mathfrak{h}^{\oplus}$ defined by

$$(T(t)\,x)_n = e^{-tA_n} x_n \,, \qquad \text{for } \ t \geq 0\,,\ x = (x_n)_{n\in\mathbb{N}} \in \mathfrak{h}^{\oplus}\,.$$

Then $\{T(t)\}_{t\geq 0}$ is equal to the strongly continuous contraction semigroup generated by A, that is, $T(t) = e^{-tA}$. Similarly, A^{α} is equal to the operator constructed from $(A_n^{\alpha})_{n\in\mathbb{N}}$ for $\alpha > 0$.

Now we present a concrete realisation of the construction, which was outlined above. Under the previously mentioned notations, we set $\{\mathfrak{h}_n := \mathbb{R}^2\}_{n\in\mathbb{N}}$ and consider two bounded non-negative self-adjoint operators

$$A_n := \lambda\,(E + n\,(S + E)), \qquad B_n := n\,(S\cos\theta_n + T\sin\theta_n + E)\,,$$

on $\mathfrak{h}_n$ for each $n \in \mathbb{N}$. Here the operators S, T and E are defined by the matrices

$$S = \begin{pmatrix} 1 & 0 \\ 0 & -1 \end{pmatrix}, \qquad T = \begin{pmatrix} 0 & 1 \\ 1 & 0 \end{pmatrix} \qquad E := \mathbb{1}_{\mathbb{C}^2} = \begin{pmatrix} 1 & 0 \\ 0 & 1 \end{pmatrix}. \tag{H.0.5}$$

From $(A_n)_{n\in\mathbb{N}}$ and $(B_n)_{n\in\mathbb{N}}$, we construct, respectively, operators A and B, as above. We also obtain in this manner the semigroup $\{\mathrm{e}^{-t(A+B)}\}_{t\geq 0}$, where we choose parameters λ, $\theta_n \in (0, \pi/2]$ by setting

$$\cos\theta_n := 1 - \epsilon_n, \quad \epsilon_n := 2\,(2\,n)^{-\max\{1,2\alpha\}}, \quad \lambda := (2/a^2)^{1/2\alpha} > 1\,, \tag{H.0.6}$$

for α, $a \in (0, 1)$. Note that operators A and B depend on parameters α and a. These operators have the following properties.

Lemma H.0.6 *The operators A and B, defined above for each α, $a \in (0, 1)$, satisfy*

(i) $A \geq \mathbb{1}$, $\quad B \geq 0$.
(ii) $\mathrm{dom}\,(A^\alpha) \subset \mathrm{dom}\,(B^\alpha)$, $\quad ||B^\alpha x|| \leq a\,||A^\alpha x||$ *for all* $x \in \mathrm{dom}\,(A^\alpha)$.
(iii) $\mathrm{dom}\,((A \dotplus B)^{\alpha}) \subset \mathrm{dom}\,(A^\alpha)$.
Here the self-adjoint form-sum: $A \dotplus B = \overline{A + B}$.

Proof Using definitions (H.0.6) and matrix

$$P_n := \begin{pmatrix} \cos(\theta_n/2) & \sin(\theta_n/2) \\ -\sin(\theta_n/2) & \cos(\theta_n/2) \end{pmatrix},$$

we obtain

$$B_n = P_n^T \begin{pmatrix} 2n & 0 \\ 0 & 0 \end{pmatrix} P_n\,, \qquad A_n = \lambda \begin{pmatrix} 1 + 2n & 0 \\ 0 & 1 \end{pmatrix}.$$

Thus, (i) is obvious and we get (ii) because the bound

$$||B_n^\alpha x_n||_n^2 = (x_n, B_n^{2\alpha} x_n)_n = (2n)^{2\alpha}\left(x_n^1\cos(\theta_n/2) + x_n^2\sin(\theta_n/2)\right)^2$$
$$\leq 2\,(2n)^{2\alpha}(x_n^1)^2 + \epsilon_n\,(2n)^{2\alpha}(x_n^2)^2 \leq a^2\,||A_n^\alpha x_n||_n^2\,,$$

holds for $x_n = (x_n^1, x_n^2)^T \in \mathbb{R}^2$. Here $(\,\cdot\,,\,\cdot\,)^T$ stands for the *vector transposition*.

For (iii), we use the matrix

$$Q_n := \begin{pmatrix} \cos(\varphi_n/2) & \sin(\varphi_n/2) \\ -\sin(\varphi_n/2) & \cos(\varphi_n/2) \end{pmatrix},$$

where $\varphi_n \in (0, \pi/2)$ is defined by

$$\mathrm{tg}\varphi_n = \frac{\sin\theta_n}{\lambda + \cos\theta_n}. \tag{H.0.7}$$

Then we have

$$A_n + B_n = (\lambda + n + n\lambda)\, E + n\sqrt{\lambda^2 + 2\lambda\cos\theta_n + 1}\,(S\cos\varphi_n + T\sin\varphi_n), \tag{H.0.8}$$

and

$$Q_n\,(A_n + B_n)^{2\alpha}\, Q_n^T = \begin{pmatrix} (\lambda + n + \lambda n + n\sqrt{\lambda^2 + 2\lambda\cos\theta_n + 1})^{2\alpha} & 0 \\ 0 & (\lambda + n + \lambda n - n\sqrt{\lambda^2 + 2\lambda\cos\theta_n + 1})^{2\alpha} \end{pmatrix}. \tag{H.0.9}$$

Consequently, we obtain (iii) due to the following bound:

$$||A_n^\alpha x_n||_n^2 = (y_n, Q_n A_n^{2\alpha} Q_n^T y_n)_n = \lambda^{2\alpha}(y_n^1\ y_n^2) \times$$

$$\times \begin{pmatrix} (1+2n)^{2\alpha}\cos^2(\varphi_n/2) + \sin^2(\varphi_n/2) & -\big((1+2n)^{2\alpha} - 1\big)\sin(\varphi_n/2)\cos(\varphi_n/2) \\ -\big((1+2n)^{2\alpha} - 1\big)\sin(\varphi_n/2)\cos(\varphi_n/2) & \cos^2(\varphi_n/2) + (1+2n)^{2\alpha}\sin^2(\varphi_n/2) \end{pmatrix} \begin{pmatrix} y_n^1 \\ y_n^2 \end{pmatrix}$$

$$= \lambda^{2\alpha}\big((1+2n)^{2\alpha}\cos^2(\varphi_n/2) + \sin^2(\varphi_n/2)\big)(y_n^1)^2$$

$$+ \lambda^{\alpha}\Big(\cos^2(\varphi_n/2) + (1+2n)^{2\alpha}\sin^2(\varphi_n/2)\Big)(y_n^2)^2$$

$$+ 2\lambda^{2\alpha} y_n^1 y_n^2 \cos(\varphi_n/2)\sin(\varphi_n/2) - 2\lambda^{2\alpha} y_n^1 y_n^2 (1+2n)^{2\alpha}\cos(\varphi_n/2)\sin(\varphi_n/2) \tag{H.0.10}$$

$$\le 2\lambda^{2\alpha}\big((1+2n)^{2\alpha}\cos^2(\varphi_n/2) + \sin^2(\varphi_n/2)\big)(y_n^1)^2 \tag{H.0.11}$$

$$+ 2\lambda^{2\alpha}\Big(\cos^2(\varphi_n/2) + (1+2n)^{2\alpha}\sin^2(\varphi_n/2)\Big)(y_n^2)^2 \tag{H.0.12}$$

$$\le C\,(y_n, Q_n(A_n + B_n)^{2\alpha} Q_n^T y_n)_n = C\,||(A_n + B_n)^\alpha x_n||_n^2, \tag{H.0.13}$$

where $y_n = (y_n^1, y_n^2)^T = Q_n x_n$ and C is a positive constant independent of n.

We note that above the *first* inequality used for (H.0.10) the elementary formula $2ab \le a^2 + b^2$ to estimate two cross terms $2a_1b_1$ and $-2a_2b_2$. This yields the upper bound (H.0.11), (H.0.12), which is the scalar product that includes a *positive diagonal* matrix.

The *second* inequality aims to estimate (H.0.11), (H.0.12) via scalar product that includes the positive diagonal matrix (H.0.9). This requires some explanations.
– First, we use definitions (H.0.6) and (H.0.7) to obtain asymptotics of (H.0.11) and a reference term:

$$\begin{aligned}
&\lambda^{2\alpha}\big((1+2n)^{2\alpha}\cos^2(\varphi_n/2)+\sin^2(\varphi_n/2)\big)\\
&=\lambda^{2\alpha}(1+2n)^{2\alpha}\left(1-\frac{\epsilon_n}{2(\lambda+1)^2}\right)+O(1/n),\\
&(\lambda+n+\lambda n+n\sqrt{\lambda^2+2\lambda\cos\theta_n+1})^{2\alpha}\\
&=\left(\lambda+2n(\lambda+1)\left(1-\frac{\lambda\,\epsilon_n}{2(\lambda+1)^2}\right)+O(1/n)\right)^{2\alpha},
\end{aligned}$$

for $\alpha \in (0, 1)$ and large $n \in \mathbb{N}$. Then because of $\lambda > 1$ we deduce that there exists $n_0 \in \mathbb{N}$ such that for $n \ge n_0$ one infers inequality

$$\lambda^{2\alpha}\big((1+2n)^{2\alpha}\cos^2(\varphi_n/2)+\sin^2(\varphi_n/2)\big) \le (\lambda+n+\lambda n+n\sqrt{\lambda^2+2\lambda\cos\theta_n+1})^{2\alpha}.$$

Let us put

$$C_1 = \max_{n=1,\dots,n_0-1} \frac{\lambda^{2\alpha}\big((1+2n)^{2\alpha}\cos^2(\varphi_n/2)+\sin^2(\varphi_n/2)\big)}{(\lambda+n+\lambda n+n\sqrt{\lambda^2+2\lambda\cos\theta_n+1})^{2\alpha}},$$

which is bounded. Then by setting $\widetilde{C}_1 := 2\max\{1, C_1\}$ we obtain estimate

$$\begin{aligned}
&\lambda^{2\alpha}\big((1+2n)^{2\alpha}\cos^2(\varphi_n/2)+\sin^2(\varphi_n/2)\big)\\
&\quad\le \widetilde{C}_1\,(\lambda+n+\lambda n+n\sqrt{\lambda^2+2\lambda\cos\theta_n+1})^{2\alpha}.
\end{aligned}\tag{H.0.14}$$

– Second, by virtue of (H.0.6) and (H.0.7) the sequence $\{\lambda^{2\alpha}\big(\cos^2(\varphi_n/2)+(1+2n)^{2\alpha}\sin^2(\varphi_n/2)\big)\}_n$ in (H.0.12) is *bounded*. On the other hand, by inspection one finds that for $\lambda > 1$ (H.0.6) the sequence $\{(\lambda+n+\lambda n-n\sqrt{\lambda^2+2\lambda\cos\theta_n+1})^{2\alpha}\}_n$ is bounded from below by a positive constant. Then we can find an n-independent $\widetilde{C}_2$ such that

$$\begin{aligned}
&\lambda^{2\alpha}\big(\cos^2(\varphi_n/2)+(1+2n)^{2\alpha}\sin^2(\varphi_n/2)\big)\\
&\quad\le \widetilde{C}_2\,(\lambda+n+\lambda n-n\sqrt{\lambda^2+2\lambda\cos\theta_n+1})^{2\alpha},
\end{aligned}\tag{H.0.15}$$

holds for any $n \in \mathbb{N}$. Thus, by virtue of (H.0.9) and owing to (H.0.14), (H.0.15) the *second* inequality is also true.

As a consequence, the estimate (H.0.13) proves claim (iii). □

Lemma H.0.7 *For each $\alpha \in (1/2, 1]$ and $a \in (0, 1)$, there exists $n_0 \in \mathbb{N}$, which depends only on λ such that the bound*

$$\left\| e^{-t(A\dot{+}B)} - \left(e^{-tA/n} e^{-tB/n}\right)^n \right\| \geq C_t\, n\, \epsilon_n\,, \quad \epsilon_n = 2\,(2\,n)^{-\max\{1,2\alpha\}}\,, \tag{H.0.16}$$

holds for all $t > 0$ and $n > n_0$, whereas for each $\alpha \in (0, 1/2]$ and $a \in (0, 1)$, one gets the estimate from below:

$$\liminf_{n\to\infty} \left\| e^{-t(A\dot{+}B)} - \left(e^{-tA/n} e^{-tB/n}\right)^n \right\| \geq D_t\,, \quad t \in \mathbb{R}^+\,. \tag{H.0.17}$$

Here $\{C_t\}_{t>0}$ and $\{D_t\}_{t>0}$ are positive continuous functions for $t \in \mathbb{R}^+$.

Proof Note that the inequalities

$$\begin{aligned} \left\| e^{-t(A+B)} - \left(e^{-tA/n} e^{-tB/n}\right)^n \right\| &\geq \left\| e^{-t(A_n+B_n)} - \left(e^{-tA_n/n} e^{-tB_n/n}\right)^n \right\|_{\mathfrak{h}_n} \\ &\geq \frac{1}{2}\left| \mathrm{Tr} e^{-t(A_n+B_n)} - \mathrm{Tr}\left(e^{-tA_n/n} e^{-tB_n/n}\right)^n \right|, \end{aligned} \tag{H.0.18}$$

hold, where the norm in the left-hand side means the operator norm of bounded operators from $\mathcal{L}(\mathfrak{h}^{\oplus})$, whereas the first member in the right-hand side is the operator norm on $\mathfrak{h}_n = \mathbb{R}^2$ and Tr in the second member is the trace of 2×2 matrices.

For later use, let us prepare the following formulæ:

$$e^{-s\Lambda} = E\,\mathrm{ch}\, s - \Lambda\,\mathrm{sh}\, s, \qquad \mathrm{Tr}\, e^{-s\Lambda} = 2\,\mathrm{ch}\, s\,, \tag{H.0.19}$$

where $\Lambda = S\cos\theta + T\sin\theta$.

To get the first formula in (H.0.19), we use (H.0.5) and expand the exponential using that $\Lambda^2 = E$, which is an immediate consequence of the identities

$$ST + TS = O, \qquad S^2 = T^2 = E\,,$$

where $O := 0 \cdot E$. The second formula in (H.0.19) follows from the first one and from observation: $\mathrm{Tr}\,\Lambda = 0$. Thanks to (H.0.8) and (H.0.19) we infer

$$\begin{aligned} \mathrm{Tr}\, e^{-t(A_n+B_n)} &= 2\, e^{-t(\lambda+n+n\lambda)}\mathrm{ch}\left(t\, n\sqrt{\lambda^2+1+2\lambda\cos\theta_n}\right) = \\ &= 2\, e^{-t(\lambda+n+n\lambda)}\mathrm{ch}\left(tn(\lambda+1) - \frac{t\lambda}{\lambda+1}\, n\,\epsilon_n + O(t\, n\,\epsilon_n^2)\right), \end{aligned} \tag{H.0.20}$$

for $n > n_0$ (large). On the other hand, calculations using the *cyclic* property of the trace and (H.0.19) lead to

$$\mathrm{Tr}\left(\mathrm{e}^{-tA_n/n}\mathrm{e}^{-tB_n/n}\right)^n = \mathrm{Tr}\left(\mathrm{e}^{-tA_n/2n}\mathrm{e}^{-tB_n/n}\mathrm{e}^{-tA_n/2n}\right)^n = \mathrm{e}^{-t(\lambda+n+n\lambda)} \times$$

$$\times\ \mathrm{Tr}\left(e^{-t\lambda S/2}\mathrm{e}^{-t(S\cos\theta_n+T\sin\theta_n)}\mathrm{e}^{-t\lambda S/2}\right)^n$$

$$= \mathrm{e}^{-t(\lambda+n+n\lambda)}\,\mathrm{Tr}\left(a_n E - b_n S - c_n T\right)^n,$$

where a_n, b_n and c_n are real positive numbers defined by

$$a_n := \mathrm{ch}\,t\lambda\ \mathrm{ch}\,t + \mathrm{sh}\,t\lambda\ \mathrm{sh}\,t\ \cos\theta_n = \mathrm{ch}\,t(\lambda+1) - \epsilon_n \mathrm{sh}\,t\lambda\ \mathrm{sh}\,t\ , \quad \text{(H.0.21)}$$

$$b_n := \mathrm{sh}\,t\lambda\ \mathrm{ch}\,t + \mathrm{ch}\,t\lambda\ \mathrm{sh}\ t\cos\theta_n\ ,$$

$$c_n := \mathrm{sh}\,t\ \sin\theta_n\ .$$

Since the identity

$$a_n^2 - b_n^2 - c_n^2 = 1$$

holds, there exist real positive numbers K_n and Θ_n which satisfy

$$a_n = \mathrm{ch}\,K_n, \quad b_n = \mathrm{sh}\,K_n\ \cos\Theta_n, \quad c_n = \mathrm{sh}\,K_n\ \sin\Theta_n\ . \tag{H.0.22}$$

From (H.0.21) and (H.0.22), we get

$$K_n = t(\lambda+1) - \frac{\mathrm{sh}\,t\lambda\ \mathrm{sh}\,t}{\mathrm{sh}\,t(\lambda+1)}\,\epsilon_n + O(t\epsilon_n^2)\,. \tag{H.0.23}$$

Setting in (H.0.19) $s = K_n$ and $\theta = \Theta_n$, we obtain

$$\mathrm{Tr}\left(\mathrm{e}^{-tA_n/n}\,\mathrm{e}^{-tB_n/n}\right)^n = \mathrm{e}^{-t(\lambda+n+n\lambda)}\,\mathrm{Tr}\left(\mathrm{e}^{-K_n(S\cos\Theta_n+T\sin\Theta_n)}\right)^n$$

$$= 2\,\mathrm{e}^{-t(\lambda+n+n\lambda)}\,\mathrm{ch}\,nK_n\,. \tag{H.0.24}$$

First we consider the case $\alpha \in (0, 1/2]$ and $a \in (0, 1)$. Recall that if $\alpha \leq 1/2$, then by virtue of (H.0.6) one gets $n\,\epsilon_n = 1$. To prove the assertion we use estimate (H.0.18) and formulæ (H.0.20), (H.0.24) for (H.0.23). As a consequence, this part of assertion (H.0.17) of the lemma follows for the uniform in $n \in \mathbb{N}$ lower bound:

$$D_t = \frac{e^{-t\lambda}}{2}\left(\exp\left(-\frac{\mathrm{sh}\,t\lambda\ \mathrm{sh}\,t}{\mathrm{sh}\,t(\lambda+1)}\right) - \exp\left(-\frac{t\lambda}{\lambda+1}\right)\right), \quad t > 0\,.$$

It is *positive* because of the following inequality for $a > 0$ and $b > 0$:

$$\frac{\operatorname{sh}(a+b)}{a+b} - \frac{\operatorname{sh} a}{a}\,\frac{\operatorname{sh} b}{b} = \frac{1}{8}\int_{-1}^{1} dx \int_{-1}^{1} dy\, \left(e^{ax} - e^{ay}\right)\left(e^{bx} - e^{by}\right) > 0\,.$$

In order to prove the assertion of the lemma for the case: $1/2 < \alpha \leq 1$ and $a \in (0, 1)$, we again use estimate (H.0.18) and formulæ (H.0.20), (H.0.24) for (H.0.23). Note that now by virtue of (H.0.6) one has $\lim_{n\to\infty} n\,\epsilon_n = 0$.

To proceed with estimate (H.0.18) for formulæ (H.0.20), (H.0.24) and (H.0.23) let us introduce

$$\begin{aligned} \eta_n &:= \lambda + 1 - \sqrt{\lambda^2 + 2\lambda\cos\theta_n + 1} = \lambda + 1 - \sqrt{(\lambda+1)^2 - 2\lambda\epsilon_n}\,, \\ \epsilon_n &= \frac{\lambda+1}{\lambda}\,\eta_n - \frac{1}{2\lambda}\,\eta_n^2\,. \end{aligned} \tag{H.0.25}$$

Then estimate (H.0.18) for formulæ (H.0.20), (H.0.24) and (H.0.23) reduces (up to exponential factors) to estimate of the difference:

$$\Delta_n(t) := \operatorname{ch} nK_n - \operatorname{ch} nt(\lambda + 1 - \eta_n) \quad t > 0\,, \quad n \in \mathbb{N}\,. \tag{H.0.26}$$

To this aim we use for function $\mathbb{R}^+ \ni x \mapsto \operatorname{ch} nx$ and any $n \in \mathbb{N}$ the *Cauchy mean value* theorem on the interval $[x, y]$ in the following form:

$$\operatorname{ch} ny - \operatorname{ch} nx = n\,\frac{\operatorname{sh} n\xi}{\operatorname{sh}\xi}\,(\operatorname{ch} y - \operatorname{ch} x)\,, \quad \xi \in (x, y) \subset \mathbb{R}^+\,. \tag{H.0.27}$$

Applying (H.0.27) to (H.0.26), on account of definitions (H.0.21), (H.0.22), (H.0.23) and inequalities: $\operatorname{sh} nz/\operatorname{sh} z \geq \mathrm{e}^{(n-1)z}$, $(\operatorname{sh} t\lambda\, \operatorname{sh} t)/\operatorname{sh} t(\lambda+1) < 1/2$, for $z, t > 0$, we obtain, on account of $\epsilon_n \operatorname{sh} t\lambda\, \operatorname{sh} t = (\epsilon_n/2)\,(\operatorname{ch} t(\lambda+1) - \operatorname{ch} t(\lambda-1))$ in (H.0.21), that

$$\begin{aligned} \Delta_n(t) &= n\,\frac{\operatorname{sh} nL_n}{\operatorname{sh} L_n}\,\big(\operatorname{ch} K_n - \operatorname{ch} t(\lambda + 1 - \eta_n)\big) \geq \\ &\geq \mathrm{e}^{(n-1)\,L_n}\, n\left(\left(1 - \frac{\epsilon_n}{2}\right)\operatorname{ch} t(\lambda+1) + \frac{\epsilon_n}{2}\operatorname{ch} t(\lambda-1) - \operatorname{ch} t(\lambda+1-\eta_n)\right) \\ &= \mathrm{e}^{t(n-1)(\lambda+1+O(\epsilon_n))} \int_0^t ds\, n\,(t-s)\left(\left(1 - \frac{\epsilon_n}{2}\right)(\lambda+1)^2 \operatorname{ch} s(\lambda+1) + \right. \\ &\left. + \frac{\epsilon_n}{2}\,(\lambda-1)^2 \operatorname{ch} s(\lambda-1) - (\lambda+1-\eta_n)^2 \operatorname{ch} s(\lambda+1-\eta_n)\right)\,. \end{aligned} \tag{H.0.28}$$

Here $\xi := L_n = t\,(\lambda + 1 + O(\epsilon_n))$ is the *intermediate* point in the *Cauchy* mean value theorem (H.0.27) on the interval $(x, y) = (t\,(\lambda + 1 - \eta_n)\,,\ K_n)$. Now, by convexity of $z \mapsto \operatorname{ch} z$, we obtain inequality

$$\begin{aligned} \operatorname{ch} s(\lambda+1-\eta_n) &= \operatorname{ch}\left((1-\frac{\eta_n}{2})\, s(\lambda+1)+\frac{\eta_n}{2}\, s(\lambda-1)\right) \le \\ &\le \left(1-\frac{\eta_n}{2}\right)\operatorname{ch} s(\lambda+1)+\frac{\eta_n}{2}\operatorname{ch} s(\lambda-1)\,. \end{aligned} \tag{H.0.29}$$

Inserting (H.0.29) into (H.0.28) one gets estimate

$$\begin{aligned} \Delta_n(t) \ge\ & e^{t(n-1)(\lambda+1+O(\epsilon_n))} \int_0^t ds\,(t-s)\left(\operatorname{ch} s(\lambda+1)-\operatorname{ch} s(\lambda-1)\right) \times \\ & \times \left(\frac{(\lambda+1)\,(3\lambda-1)}{2\lambda} n\,\eta_n - \frac{3\lambda^2+6\lambda-1}{4\lambda} n\,\eta_n^2 + \frac{n\,\eta_n^3}{2}\right). \end{aligned} \tag{H.0.30}$$

Note that by (H.0.6) and (H.0.25) we infer for $\alpha > 1/2$:

$$\lim_{n\to\infty} \frac{1}{n\,\epsilon_n}\left(\frac{(\lambda+1)\,(3\lambda-1)}{2\lambda} n\,\eta_n - \frac{3\lambda^2+6\lambda-1}{4\lambda} n\,\eta_n^2 + \frac{n\,\eta_n^3}{2}\right) = \frac{3\lambda-1}{2} > 1\,. \tag{H.0.31}$$

As a consequence, on account of (H.0.18), (H.0.20), (H.0.24) and (H.0.26) together with (H.0.30) and (H.0.31), there exists $n_0 > 0$ such that for $n > n_0$ we obtain the estimate (H.0.16) for $C_t = e^{-t\,(2\lambda+1)}$. □

Proof of Theorem H.0.2 Owing to Lemmata H.0.6 and H.0.7 we obtain the *Proof* of Theorem H.0.2 about the existence of operators A and B with desirable properties. □

In conclusion the following *basic* properties of operators A and B constructed in Example H.0.5 are deserving to be accentuated.

Theorem H.0.8 ([Tam00, Proposition 3]) *Operators A and B, constructed above, satisfy* $\operatorname{dom}(A) \not\subset \operatorname{dom}(B)$, $\operatorname{dom}(B) \not\subset \operatorname{dom}(A)$ *and the operator $A+B$, for each $\alpha, a \in (0,1)$, is essentially self-adjoint on domain:* $\operatorname{dom}(A) \cap \operatorname{dom}(B)$, *but not self-adjoint.*

Proof Put $x_n = (0, n^{-1+\alpha/2})^T \in \mathbb{R}^2$, then we have

$$\sum_{n=1}^{\infty}(x_n, x_n)_n < \infty\,, \quad \sum_{n=1}^{\infty}(A_n x_n, A_n x_n)_n < \infty\,, \quad \sum_{n=1}^{\infty}(B_n x_n, B_n x_n)_n = \infty\,,$$

that is,

$$x = (x_n)_{n\in\mathbb{N}} \in \operatorname{dom}(A) \setminus \operatorname{dom}(B)\,.$$

Similarly,

$$y = (y_n)_{n\in\mathbb{N}} \in \mathrm{dom}\,(B) \setminus \mathrm{dom}\,(A)$$

and

$$z = (z_n)_{n\in\mathbb{N}} \in (\mathrm{dom}\,(A \dot{+} B) \setminus \mathrm{dom}\,(A)) \subset \mathrm{dom}\,(A \dot{+} B) \setminus \big(\mathrm{dom}\,(A) \cap \mathrm{dom}\,(B)\big)$$

hold, for $y_n = n^{-1+\alpha/2}(-\sin(\theta_n/2), \cos(\theta_n/2))^T$, $z_n = n^{-1+\alpha/2}(-\sin(\varphi_n/2), \cos(\varphi_n/2))^T$ where θ_n and φ_n are defined by (H.0.7) and $\mathrm{dom}\,(A \dot{+} B)$ is the domain of the *form-sum* of operators A and B. □

Now few remarks are in order to sum up assertions around the Example H.0.5.

Remark H.0.9

1. The basic properties (Theorem H.0.8) of operators A and B ensure that operator $A + B$, for each $\alpha, a \in (0, 1)$, is *only* essentially self-adjoint on domain: $\mathrm{dom}\,(A) \cap \mathrm{dom}\,(B)$ in Hilbert space $\mathfrak{h}^{\oplus}$.
   Now, together with the property that A and B are *generators* of contraction C_0-semigroups (Lemma H.0.6(i)), they satisfy conditions of the *Trotter product theorem* (Theorem 5.1.1). Hence, Trotter product formulæ (5.1.1), (5.1.2) and (5.1.6) hold *only* in the *strong* operator topology.
2. Note that all properties of operators A and B collected in Theorem H.0.8 and in Lemma H.0.6 are not sufficient neither for Rogava's theorem (Proposition 7.5.1) nor for a more general ITTZ-Theorem 7.1.9 since the operator $A + B$, for each $\alpha, a \in (0, 1)$, is *not* self-adjoint, but *only* essentially self-adjoint.
   To prove that *essential* self-adjointness is *not sufficient* for the operator-norm convergence of the Trotter product formula, we used the properties of operators A and B collected in Lemma H.0.6. This allows to prove owing to Lemma H.0.7 and Theorem H.0.8 that for $0 < \alpha \leq 1/2$ and $a \in (0, 1)$ these operators *break* the operator-norm convergence of the Trotter product formula (H.0.17) for *essentially* self-adjoint operator $A + B$.
3. We know that ITTZ-Theorem 7.1.9 guarantees that self-adjointness of $A + B$ on domain $\mathrm{dom}\,(A) \cap \mathrm{dom}\,(B)$ yields the operator-norm convergence of the Trotter product formula, even with the optimal rate of convergence. This does not exclude that it also might happen for *essentially* self-adjoint $A + B$ under some *supplementary conditions*. This is a subject of the next comment.
4. Note that by Lemma H.0.6 operators A, B satisfy for $\alpha \in (1/2, 1)$ and $a \in (0, 1)$ conditions of Theorem H.0.1. Consequently, under these *supplementary conditions* we obtain by Theorem H.0.1 the operator-norm convergence of the Trotter product formula (H.0.1) to the C_0-semigroup for *self-adjoint* generator $A \dot{+} B$, which is the *form-sum* of A and B. Moreover, a posteriori (see point 5.) becomes clear that this converges with the optimal rate.

5. On account of Tamura's example, Theorem H.0.2 implies that the bound (H.0.1) of Theorem H.0.1 is the best possible (*optimal*) in its power of $n \in \mathbb{N}$ (Remark H.0.3)
   Since by construction of A and B in Example H.0.5 the operator $A + B$ is *essentially* self-adjoints, in Theorem H.0.1 and Theorem H.0.2 the self-adjoint form-sum $A \dot{+} B = \overline{A + B} = (A+B)_F$. This is not true in general, for example, in Theorem H.0.1.

Appendix I
Estimate of Integrals $\mathcal{I}_n(t, s)$

To estimate the multiple integrals (cf.(12.2.25)):

$$\mathcal{I}_n(t,s) = \int_s^t d\tau_n \int_s^{\tau_n} d\tau_{n-1} \dots \int_s^{\tau_2} d\tau_1 \cdot$$
$$\cdot (\tau_n - \tau_{n-1})^{-\alpha} (\tau_{n-1} - \tau_{n-2})^{-\alpha} \cdots (\tau_2 - \tau_1)^{-\alpha} (\tau_1 - s)^{-\alpha}, \tag{I.0.1}$$

for $0 \le s \le t \le T$, $n \ge 1$, and for $\alpha \in (0, 1)$, we let $x_k = \tau_k - s$, where $1 \le k \le n$, and define

$$\mathcal{F}_n(x_n) = \int_0^{x_n} dx_{n-1} \int_0^{x_{n-1}} dx_{n-2} \dots \int_0^{x_2} dx_1 \cdot \tag{I.0.2}$$
$$\cdot (x_n - x_{n-1})^{-\alpha} (x_{n-1} - x_{n-2})^{-\alpha} \cdots (x_2 - x_1)^{-\alpha}\, x_1^{-\alpha}.$$

Then $\mathcal{F}_1(x_1) := x_1^{-\alpha}$, for $n = 1$, whereas in general, for $2 \le k \le n$ and $\beta_k := k(1-\alpha) - 1$, we obtain:

$$\mathcal{F}_2(x_2) = \int_0^{x_2} dx_1\, (x_2 - x_1)^{-\alpha}\, \mathcal{F}_1(x_1) = B(\beta_1 + 1, 1 - \alpha)$$
$$x_2^{\beta_2} = \frac{\Gamma(1-\alpha)^2}{\Gamma(2(1-\alpha))}\, x_2^{\beta_2},$$
$$\mathcal{F}_3(x_3) = \int_0^{x_3} dx_2\, (x_3 - x_2)^{-\alpha}\, \mathcal{F}_2(x_2) = \frac{\Gamma(1-\alpha)^2}{\Gamma(2(1-\alpha))} B(\beta_2 + 1, 1 - \alpha)\, x_3^{\beta_3} =$$
$$= \frac{\Gamma(1-\alpha)^3}{\Gamma(3(1-\alpha))}\, x_3^{\beta_3}, \dots,$$

V. A. Zagrebnov et al., *Trotter-Kato Product Formulæ*, Operator Theory: Advances and Applications 296, https://doi.org/10.1007/978-3-031-56720-9

$$\mathcal{F}_n(x_n) = \int_0^{x_n} \mathrm{d}x_{n-1}(x_n - x_{n-1})^{-\alpha}\, \mathcal{F}_{n-1}(x_{n-1})$$
$$= \frac{\Gamma(1-\alpha)^n}{\Gamma(n(1-\alpha))}\, x_n^{\beta_n}\,. \tag{I.0.3}$$

Here $\Gamma(\cdot)$ and $B(\cdot,\cdot)$ denote the *Euler* gamma- and beta-functions. Note that by definitions (I.0.1) and (I.0.2) and owing to (I.0.3),

$$\mathcal{I}_n(t, s) = \int_0^{t-s} \mathrm{d}x_n\, \mathcal{F}_n(x_n) = \frac{\Gamma(1-\alpha)^n}{n\,(1-\alpha)\,\Gamma(n(1-\alpha))}\,(t-s)^{n\,(1-\alpha)}\,, \tag{I.0.4}$$

for $\alpha \in (0, 1), 0 \le s \le t \le T$ and any $n \ge 1$.

To estimate coefficients

$$\kappa_n(\alpha) := \frac{\Gamma(1-\alpha)^n}{n\,(1-\alpha)\,\Gamma(n(1-\alpha))}\,, \qquad n \ge 1\,, \tag{I.0.5}$$

we recall that by the Gauss *multiplication formula* for the gamma-function one gets representations:

$$\Gamma(n\,z) = \frac{n^{n\,z-1/2}}{(\sqrt{2\pi})^{n-1}}\,\Gamma(z)\,\Gamma(z+\frac{1}{n})\,\cdots\,\Gamma(z+\frac{n-1}{n})\,, \quad \mathrm{Re}(z) > 0\,, \tag{I.0.6}$$

for $n \in \mathbb{N}$. Given that $\Gamma(z_*) := \min_{\{1-\alpha \le z \le 2-\alpha\}} \Gamma(z) < 1$, we obtain the estimate

$$\Gamma(1-\alpha)\,\Gamma(1-\alpha+\frac{1}{n})\,\cdots\,\Gamma(1-\alpha+\frac{n-1}{n}) \ge (\Gamma(z_*))^n\,, \quad n \ge 1\,.$$

Then on account of (I.0.6) one deduces for $z = 1-\alpha$:

$$\Gamma(n\,(1-\alpha)) \ge \frac{n^{n\,(1-\alpha)-1/2}}{(\sqrt{2\pi})^{n-1}}\,(\Gamma(z_*))^n\,, \quad n \ge 1\,. \tag{I.0.7}$$

Therefore, owing to (I.0.5) and (I.0.7) we conclude that

$$\kappa_n(\alpha) \le \frac{1}{(1-\alpha)\sqrt{2\pi\,n}}\left(\frac{\sqrt{2\pi}\,\Gamma(1-\alpha)}{\Gamma(z_*)\,n^{(1-\alpha)}}\right)^n\,, \quad \alpha \in (0,1)\,, \quad n \ge 1\,. \tag{I.0.8}$$

References

[AcqTer87] P. Acquistapace, B. Terreni, A unified approach to abstract linear nonautonomous parabolic equations. Rendiconti del Seminario Matematico della Universita di Padova, tome **78**, 47–107 (1987)

[Amr09] W.O. Amrein, *Hilbert Space Methods in Quantum Mechanics* (EPFL Press, Lausanne, 2009)

[Ara90] H. Araki, On an inequality of Lieb and Thirring. Lett. Math. Phys. **19**, 167–170 (1990)

[Are01] W. Arendt, Approximation of degenerate semigroups. Taiwanese. J. Math. **5**, 279–295 (2001)

[ABHN01] W. Arendt, C. Batty, M. Hieber, F. Neubrander, *Vector-valued Laplace Transforms and Cauchy Problems* (Birkhäuser Verlag, Basel–Berlin, 2001)

[ACFP07] W. Arendt, R. Chill, S. Fornaro, C. Poupaud, L^p-maximal regularity for non-autonomous evolution equations. J. Differ. Equ. **237**, 1–26 (2007)

[ArtE12] W. Arendt, A.F.M. ter Elst, *From Forms to Semigroups*, in *Spectral Theory, Mathematical System Theory, Evolution Equations, Differential and Difference Equations.* Operator Theory: Advances and Applications, vol. 221 (Birkhäuser, Basel 2012), pp. 47–69

[ArGu04] A. Arias, S. Gudder, Almost sharp quantum effects. J. Math. Phys. **45**, 4196–4206 (2004)

[Arl87] Yu.M. Arlinskiĭ, A class of contractions in Hilbert space Ukr. Math. Zh. **39**, 691–696 (1987) (in Russian)

[Arl91] Yu.M. Arlinskiĭ, Characteristic functions of operators of the class $C(\alpha)$. Izv. Vyssh. Uchebn. Zaved. Mat. **No.2**, 13–21 (1991) (in Russian)

[Arl05] Yu.M. Arlinskiĭ, On some semigroups on the complex plane. Semigroup Forum **70**, 329–346 (2005)

[Arl12] Yu.M. Arlinskiĭ, Operator-norm approximations of holomorphic one-parameter semigroups of contractions in Hilbert spaces. Methods Funct. Anal. Topol. **18**, 101–110 (2012)

[Arl22] Yu.M. Arlinskiĭ, On the Kato square root problem. J. Math. Anal. Appl. **505**, 125501 (pp. 1–31) (2022)

[ArTr20] Y. Arlinskiĭ, C. Tretter, Everything is possible for the domain intersection dom $(T) \cap$ dom (T^*). Adv. Math. **374**, 107383 (pp. 1–46) (2020)

[ArTs84] Yu.M. Arlinskiĭ, Eh.R. Tsekanovskiĭ, On sectorial extensions of positive Hermitian operators and their resolvents. Dokl. Akad. Nauk Arm. SSR **LXXIX**(5), 199–202 (1984) (in Russian)

V. A. Zagrebnov et al., *Trotter-Kato Product Formulæ*, Operator Theory: Advances and Applications 296, https://doi.org/10.1007/978-3-031-56720-9

[ArTs88] Yu.M. Arlinskiĭ, Eh.R. Tsekanovskiĭ, Quasi-self-adjoint contracting dilations of Hermitian contraction. Teor. Funkts., Funktsional. Anal. Prilpzhen. **50**, 9–16 (1988) (in Russian)

[ArZag10] Yu.M. Arlinskiĭ, V.A. Zagrebnov, Numerical range and quasi-sectorial contractions. J. Math. Anal. Appl. **366**, 33–43 (2010)

[ArZag15] Yu.M. Arlinskiĭ, V.A. Zagrebnov, Around the Van Daele–Schmüdgen theorem. Integral Equ. Oper. Theory **81**, 53–95 (2015)

[AzIch08] Y. Azuma, T. Ichinose, Note on norm and pointwise convergence of exponential products and their integral kernels for the harmonic oscillator. Integr. Equ. Oper. Theory **60**, 151–176 (2008)

[BCD06] C. Badea, M. Crouzeix, B. Delyon, Convex domains and K-spectral sets. Mathematische Zeitschrift **252**, 345–365 (2006)

[Bal76] A.V. Balakrishnan, *Applied Functional Analysis* (Springer, Berlin, 1976)

[Batkai11] A. Bátkai, P. Csomós, B. Farkas, G. Nickel, Operator splitting for non-autonomous evolution equations. J. Funct. Anal. **260**, 2163–2190 (2011)

[BeCr06] B. Beckermann, M. Crouzeix, A lenticular version of a von Neumann inequality. Arch. Math. **86**, 352–355 (2006)

[BDS21] S. Becker, N. Datta, R. Salzmann, Quantum Zeno effect for open quantum systems. Ann. Henri Poincaré **22**, 3795–3840 (2021)

[BenPau04] V. Bentkus, V. Paulauskas, Optimal error estimates in operator-norm approximation of semigroups. Lett. Math. Phys. **68**, 131–138 (2004)

[BN67] A. Beskow, J. Nilsson, The concept of wave function and the irreducible representations of the Poincaré group, II. Unstable systems and the exponential decay law. Arkiv Fys. **34**, 561–569 (1967)

[BiHr00] P. Binding, R. Hryniv, Relative boundedness and relative compactness for linear operators in Banach spaces. Proc. Am. Math. Soc. **128**, 2287–2290 (2000)

[BS87] M.S. Birman, M.Z. Solomyak, *Spectral Theory of Self-adjoint Operators in Hilbert Space*. Mathematics and its Applications (Soviet Series) (D. Reidel Publishing Co., Dordrecht, 1987)

[BEH94] J. Blank, P. Exner, M. Havlíček, *Hilbert Space Operators in Quantum Physics* (Second and Extended Edition) (Springer, Berlin, 2008)

[Bob16] A. Bobrowski, *Convergence of One-Parameter Operator Semigroups* (Cambridge University Press, Cambridge, 2016)

[Bour60] N. Bourbaki, *Topologie générale* (Hermann, Paris, 1960)

[BR79] O. Bratteli, D.W. Robinson, *Operator Algebras and Quantum Statistical Mechanics 1* (Springer, Berlin, 1979)

[BR96] O. Bratteli, D.W. Robinson, *Operator Algebras and Quantum Statistical Mechanics 2* (Springer, Berlin, 1996)

[But20] Ya.A. Butko, The method of Chernoff approximation, pp. 19–46, in *Semigroups of Operators – Theory and Applications, SOTA 2018*, ed. by J. Banasiak et al. Springer Proceedings in Mathematics and Statistics, vol. 325 (Springer, Berlin, 2020)

[Ca04] V. Cachia, Euler's exponential formula for semigroups. Semigroup Forum **68**, 1–24 (2004)

[Ca05] V. Cachia, On a product formula for unitary groups. Bull. Lond. Math. Soc. **37**, 621–626 (2005)

[Ca10] V. Cachia, *La formule de Trotter-Kato*, Éditions universitaires européennes (Südwestduetscher Verlag für Hochschulschriften GmbH & Co. KG, Saarbrücken, 2010)

[CNZ01] V. Cachia, H. Neidhardt, V.A. Zagrebnov, Accretive perturbations and error estimates for the Trotter product formula. Integr. Equ. Oper. Theory **39**, 396–412 (2001)

[CNZ02] V. Cachia, H. Neidhardt, V.A. Zagrebnov, Comments on the Trotter product formula error-bound estimates for nonself-adjoint semigroups. Integr. Equ. Oper. Theory **42**, 425–448 (2002)

[Cr04] M. Crouzeix, Operators with numerical range in parabola. Arch. Math. **82**, 517–527 (2004)

[CZ99] V. Cachia, V.A. Zagrebnov, Operator-norm convergence of the Trotter product formula for sectorial generators. Lett. Math. Phys. **50**, 203–211 (1999)

[CZ01a] V. Cachia, V.A. Zagrebnov, Operator-norm convergence of the Trotter product formula for holomorphic semigroups. J. Oper. Theory **46**, 199–213 (2001)

[CZ01b] V. Cachia, V.A. Zagrebnov, Operator-norm approximation of semigroups by quasi-sectorial contractions. J. Funct. Anal. **180**, 176–194 (2001)

[CZ01c] V. Cachia, V.A. Zagrebnov, Trotter product formula for nonself-adjoint Gibbs semigroups. J. Lond. Math. Soc. **64**, 436–444 (2001)

[CG12] C. Canzi, G. Guerra, A simple counterexample related to the Lie–Trotter product formula. Semigroup Forum **84**, 499–504 (2012)

[CaSu06] A.L. Carey, F.A. Sukochev, Dixmier traces and some applications in non-commutative geometry. Russ. Math. Surv. **61**(6), 1039–1099 (2006)

[Car09] E.A. Carlen, *Trace Inequalities and Quantum Entropy: An introductory course*, in *Entropy and the Quantum*, ed. by R. Sims, D. Ueltschi. Contemporary Mathematics, vol. 529 (AMS, Providence, 2010), pp. 73–140

[CeMe97] P. Cembranos, J. Mendoza. *Banach Spaces of Vector-Valued Functions* (Springer, Berlin, Heidelberg, 1997)

[Cher68] P.R. Chernoff, Note on product formulas for operator semigroups. J. Funct. Anal. **2**, 238–242 (1968)

[Cher70] P.R. Chernoff, Semigroup product formulas and addition of unbounded operators. Bull. Am. Math. Soc. **76**, 395–398 (1970)

[Cher72] P.R. Chernoff, Perturbations of dissipative operators with relative bound one. Proc. Am. Math. Soc. **33**, 72–74 (1972)

[Cher73] P.R. Chernoff, Essential self-adjointness of powers of generators of hyperbolic equations. J. Funct. Anal. **12**, 401–414 (1973)

[Cher74] P.R. Chernoff, Product formulas, nonlinear semigroups and addition of unbounded operators. Memoirs Am. Math. Soc. **140**, 1–121 (1974)

[Cher77] P.R. Chernoff, Schrödinger and Dirac operators with singular potentials and hyperbolic equations. Pacific J. Math. **72**, 361–382 (1977)

[ChSM77] C.B. Chiu, E.C.G. Sudarshan, B. Misra, Time evolution of unstable quantum states and a resolution of Zeno's paradox. Phys. Rev. D **16**, 520–529 (1977)

[Cho85] Y.H. Choe, C0-semigroups on locally convex space. J. Math. Anal. Appl. **106**, 293–320 (1985)

[Con94] A. Connes, *Noncommutative Geometry* (Academic Press, London, 1994)

[CrDe03] M. Crouzeix, B. Delyon, Some estimates for analytic functions of the strip or a sectorial operators. Arch. Math. **81**, 559–566 (2003)

[Dal60] Yu.L. Daletskiĭ, Representability of solutions of operator equations solution by path integrals. Doklady Akad. Nauk USSR **134**, 1013–1016 (1960), in Russian

[Dav69] E.B. Davies, Quantum stochastic processes. Commun. Math. Phys. **15**, 277–304 (1969)

[Dav76] E.B. Davies, *Quantum Theory of Open Systems* (Academic Press, London, 1976)

[Dav80] E.B. Davies, *One-Parameter Semigroups* (Academic Press, London, 1980)

[Dav07] E.B. Davies, *Linear Operators and Their Spectra* (Cambridge University Press, Cambridge, 2007)

[dellA67] G.F. Dell'Antonio, On the limits of sequences of normal states. Commun. Pure Appl. Math. **20**, 413–429 (1967)

[Dem74] B. Dembart, On the theory of semigroups of operators on locally convex spaces. J. Funct. Anal. **16**, 123–160 (1974)

[DS97] B.O. Dia, M. Schatzmann, An estimate on the transfer operator. J. Funct. Anal. **145**, 108–135 (1997)

[Dix66] J. Dixmier, Existence des traces non normales. C. R. Acad. Sci. Paris, Sér. A **262**, 1107–1108 (1966)

[DIT98] A. Doumeki, T. Ichinose, H. Tamura, Error bounds on exponential product formulas for Schrödinger operators. J. Math. Soc. Jpn. **50**, 359–377 (1998)

[DS46] N. Dunford, I.E. Segal, Semigroups of operators and the Weierstrass theorem. Bull. Am. Math. Soc. **52**, 911–914 (1946)

[DLP79] W. Dressler, L. Landau, J.F. Perez, Estimates of critical length and critical temperatures for classical and quantum lattice systems. J. Stat. Phys. **20**, 123–162 (1979)

[DLS78] F. Dyson, E.H. Lieb, B. Simon, Phase transitions in quantum spin systems with isotropic and nonisotropic interactions. J. Stat. Phys. **18**, 335–383 (1978)

[EN00] K.-J. Engel, R. Nagel, *One-Parameter Semigroups for Linear Evolution Equations* (Springer, Berlin, 2000)

[EN06] K.-J. Engel, R. Nagel, *A Short Course on Operator Semigroups* (Springer, Berlin, 2006)

[Ev76] D.E. Evans. Time dependent perturbations and scattering of strongly continuous groups on Banach spaces. Math. Ann. **221**(3), 275–290 (1976)

[Ex85] P. Exner, *Open Quantum Systems and Feynman Integrals*. Fundamental Theories of Physics (D. Reidel Publishing Co., Dordrecht, 1985)

[Ex05] P. Exner, Sufficient conditions for the anti-Zeno effect. J. Phys. A Math. Gen. **38**, L449–L454 (2005)

[Ex13] P. Exner, *Solvable Models of Resonances and Decays*, in *Proceedings of the Conference "Mathematical Physics, Spectral Theory and Stochastic Analysis" Goslar 2011*, ed. by M. Demuth, W. Kirsch (Birkhäuser, Basel, 2013), pp. 165–227

[EI05] P. Exner, T. Ichinose, A product formula related to quantum Zeno dynamics. Ann. Henri Poincaré **6**, 195–215 (2005)

[EI21] P. Exner, T. Ichinose, Note on a product formula related to quantum Zeno dynamics. Ann. Henri Poincaré **22**, 1669–1697 (2021). Correction to: Note on a product formula related to quantum Zeno dynamics. Ann. Henri Poincaré **22**, 1699–1700 (2021)

[EINZ07] P. Exner, T. Ichinose, H. Neidhardt, V.A. Zagrebnov, Zeno product formula revisited. Integr. Equ. Oper. Theory **57**, 67–81 (2007)

[ENZ11] P. Exner, H. Neidhardt, V.A. Zagrebnov, Remarks on the Trotter-Kato product for unitary groups. Integr. Equ. Oper. Theory **69**, 451–478 (2011)

[FP08] P. Facchi, S. Pascazio, Quantum Zeno dynamics: mathematical and physical aspects. J. Phys. A**41**, 493001 (45 p.) (2008)

[FPSS01] P. Facchi, S. Pascazio, A. Scardicchio, L.S. Schulman, Zeno dynamics yields ordinary constraints. Phys. Rev. **A65**, 012108 (2001)

[Fr72] C.N. Friedman, Semigroup product formulas, compressions and continuous observation in quantum mechanics. Indiana Univ. Math. J. **21**, 1001–1011 (1972)

[FH07] I. Faragó, Á. Havasi, Consistency analysis of operator splitting methods for C_0-semigroups. Semigroup Forum **74**, 125–139 (2007)

[Far67a] W.G. Faris, The product formulas for semigroups defined by Friedrichs extensions. Pacific J. Math. **22**, 47–70 (1967)

[Far67b] W.G. Faris, Product formulas for perturbations of linear propagators. J. Funct. Anal. **1**, 93–108 (1967)

[Fat83] H.O. Fattorini, *The Cauchy Problem* (Cambridge University Press, Cambridge, 1983)

[GaRe22] O.E. Galkin, I.D. Remizov, Rate of convergence of chernoff approximations of operator C_0-semigroups. Math. Notes, **111**, 305–307 (2022)

[GK69] I.C. Gohberg, M.G. Kreĭn, *Introduction to the Theory of Linear Nonselfadjoint Operators*. Transl. Math. Monographs, vol. 18 (Amer. Math. Soc., Providence, RI, 1969)

[Gol85] J.A. Goldstein, *Semigroups of Operators and Applications* (Oxford University Press, Oxford, 1985)

[GoTo14] A. Gomilko, Yu. Tomilov, On convergence rates in approximation theory for operator semigroups. J. Funct. Anal. **266**, 3040–3082 (2014)

[GoKoTo19] A. Gomilko, S. Kosowicz, Yu. Tomilov, A general approach to approximation theory of operator semigroups. J. Math. Pures Appl. **127**, 216–267 (2019)

[Gram10] M. Gramegna, Serie di equazioni differenziali lineari ed equazioni integro-differenziali. Atti Reale Accad. Sci. Torino **45**, 291–313 (1910)

[Haa06] M. Haase, *The Functional Calculus for Sectorial Operators*. Operator Theory: Advances and Applications, vol. 169 (Bikhäuser - Springer, Basel, 2006)

[Hel95] B. Helffer, Around the transfer operator and the Trotter-Kato formula. Oper. Theory Adv. Appl. **78**, 161–175 (1995)

[Hille48] E. Hille, *Functional Analysis and Semi-Groups*, vol. XXXI (Amer. Math. Soc. Coll. Publ., Providence, RI, 1948)

[HiPh57] E. Hille, R.S. Phillips, *Functional Analysis and Semi-Groups*, (revised edition) vol. XXXI (Amer. Math. Soc. Coll. Publ., Providence, RI, 1957)

[Ho04] A. Hodges, *What Would Alan Turing Have Done After 1954?*, in *Alan Turing: Life and Legacy of a Great Thinker*, ed. by Ch. Teuscher (Springer, Heidelberg, 2004), pp. 43–58

[How74] J.S. Howland, Stationary scattering theory for time-dependent Hamiltonians. Math. Ann. **207**, 315–335 (1974)

[HuNa01] J.K. Hunter, B. Nachtergaele, *Applied Analysis* (World Scientific, Singapore, 2001)

[Ich80] T. Ichinose, A product formula and its application to the Schrödinger equation. Publ. RIMS, Kyoto Univ. **16**, 585–600 (1980)

[Ich03] T. Ichinose, *Time-Sliced Approximation to Path Integral and Lie-Trotter-Kato Product Formula*, in *A Garden of Quanta: Essays in Honor of Hiroshi Ezawa*, ed. by J. Arafune et al. (World Scientific Publishing, Singapore, 2003), pp. 77–93

[Ich17] T. Ichinose, *The Modified Unitary Trotter-Kato and Zeno Product Formulas Revisited*, in *Functional Analysis and Operator Theory for Quantum Physics*, ed. by J. Dittrich, H. Kovařík, A.Laptev (EMS Publ., Zürich, 2017), pp. 401–417

[INZ03] T. Ichinose, H. Neidhardt, V.A. Zagrebnov, *Operator Norm Convergence of the Trotter-Kato Product Formula*, in *Proceedings of the International Conference on Functional Analysis*, Kiev (August 22–26, 2001) (2003), pp. 100–106

[INZ04] T. Ichinose, H. Neidhardt, V.A. Zagrebnov, Trotter-Kato product formula and fractional powers of self-adjoint generators. J. Funct. Anal. **207**, 33–57 (2004)

[ITak97] T. Ichinose, S. Takanobu, Estimate of the difference between the Kac operator and the Schrödinger semigroup. Commun. Math. Phys. **27**, 167–197 (1997)

[ITak98] T. Ichinose, S. Takanobu, The norm estimate of the difference between the Kac operators and the Schrödinger semigroup: A unified approach to the nonrelativistic and relativistic cases. Nagoya Math. J. **149**, 53–81 (1998)

[ITak00] T. Ichinose, S. Takanobu, The norm estimate of the difference between the Kac operators and the Schrödinger semigroup II: The general case including the relativistic case. Electron. J. Probab. **5**, 1–47 (2000)

[IT97] T. Ichinose, H. Tamura, Error estimate in operator norm for Trotter-Kato product formula. Integr. Equ. Oper. Theory **27**, 195–207 (1997)

[IT98a] T. Ichinose, H. Tamura, Error estimate in operator norm of exponential product formulas for propagators of parabolic evolution equations. Osaka J. Math. **35**, 751–770 (1998)

[IT98b] T. Ichinose, H. Tamura, Error bound in trace norm for Trotter-Kato product formula of Gibbs semigroups. Asymptotic Anal. **17**, 239–266 (1998)

[IT01] T. Ichinose, H. Tamura, The norm convergence of the Trotter-Kato product formula with error bound. Commun. Math. Phys. **217**, 489–502 (2001)

[IT02] T. Ichinose, H. Tamura, On the norm convergence of the self-adjoint Trotter-Kato product formula with error bound. Proc. Indian Acad. Sci. (Math. Sci.) **112**, 99–106 (2002)

[IT04] T. Ichinose, H. Tamura, Sharp error bound on norm convergence of exponential product formula and approximation to kernels of Schrödinger semigroups. Commun. Partial Differ. Equ. **29**, 1905–1918 (2004)

[IT04a] T. Ichinose, H. Tamura, Note on the norm convergence of the unitary Trotter product formula. Lett. Math. Phys. **70**, 65–81 (2004)

[IT05] T. Ichinose, H. Tamura, Erratum: The norm convergence of the Trotter-Kato product formula with error bound. Commun. Math. Phys. **254**, 255 (2005)

[IT09] T. Ichinose, H. Tamura, Results on convergence in norm of exponential product formulas and pointwise of the corresponding integral kernels. Oper. Theory Adv. Appl. **190**, 315–327 (2009)

[ITTZ01] T. Ichinose, H. Tamura, H. Tamura, V.A. Zagrebnov, Note on the paper "The norm convergence of the Trotter-Kato product formula with error bound" by Ichinose and Tamura. Commun. Math. Phys. **221**, 499–510 (2001)

[IZ23] B. Iochum, V.A. Zagrebnov, *Asymptotic Behaviour of Semigroup Traces and Schatten Classes of Resolvents*. arXiv: 2309.05394v1 [math-ph] 11 Sep 2023

[IHBW90] W.M. Itano, D.J. Heinzen, J.J. Bollinger, D.J. Wineland, Quantum Zeno effect. Phys. Rev. **A41**, 2295–2300 (1990)

[JL00] G.W. Johnson, M.L. Lapidus, *The Feynman Integral and Feynman's Operational Calculus*. Oxford Mathematical Monographs (The Clarendon Press Oxford University Press, Oxford, 2000)

[Kato53] T. Kato, Integration of the equation of evolution in a Banach space. J. Math. Soc. Jpn. **5**, 208–234 (1953)

[Kato56] T. Kato, On linear differential equations in Banach spaces. Comm. Pure Appl. Math. **9**, 479–486 (1956)

[Kato59] T. Kato, Remarks on pseudo-resolvents and infinitesimal generators of semi-groups. Proc. Jpn. Acad. **35**, 467–468 (1959)

[Kato61] T. Kato, Abstract evolution equations of parabolic type in Banach and Hilbert spaces. Nagoya Math. J. **19**, 93–125 (1961)

[Kato61a] T. Kato, A generalization of the Heinz inequality. Proc. Jpn. Acad. **37**, 305–308 (1961)

[Kato61b] T. Kato, Fractional powers of dissipative operators. J. Math. Soc. Jpn. **13**, 246–274 (1961)

[Kato62] T. Kato, Fractional powers of dissipative operators II. J. Math. Soc. Jpn. **14**, 242–248 (1962)

[Kato65] T. Kato, Some mapping theorems for the numerical range. Proc. Jpn. Acad. **41**, 652–655 (1965)

[Kato70] T. Kato. Linear evolution equations of "hyperbolic" type. J. Fac. Sci. Univ. Tokyo Sect. I **17**, 241–258 (1970)

[Kato72] T. Kato. Schrödinger operators with singular potentials. Isreal J. Math. **13**, 135–148 (1972)

[Kato73] T. Kato. Linear evolution equations of "hyperbolic" type. II. J. Math. Soc. Jpn. **25**, 648–666 (1973)

[Kato74] T. Kato, On the Trotter-Lie product formula. Proc. Jpn. Acad. **50**, 694–698 (1974)

[Kato78] T. Kato, *Trotter's Product Formula for an Arbitrary Pair of Self-Adjoint Contraction Semigroups*. Topics in Funct. Anal., Adv. Math. Suppl. Studies ed. by I. Gohberg, M. Kac, vol. 3 (Academic Press, New York, 1978)

[Kato80] T. Kato, *Perturbation Theory for Linear Operators* (Corrected Printing of the Second Edition.) Classics in Mathematics (Springer, Berlin, 1980)

[KaTa62] T. Kato, H. Tanabe, On the abstract evolution equation. Osaka Math. J. **14**, 107–133 (1962)

[Kis64] J. Kisyński, Sur les opérateurs de Green des problèmes de Cauchy abstraits. Studia Math. **XXIII**, 285–328 (1964)

[KF89] A.N. Kolmogorov, S.V. Fomin, *Elements of the Theory of Functions and Functional Analysis*. Corrected Printing of the Sixth Edition, Nauka, Moskva, 1989 (in Russian). English translation: (Dover Publications, Mineola, New York 1999)

[Koma64] H. Komatsu, Semigroups of operators in locally convex spaces. J. Math. Soc. Jpn. **16**, 230–262 (1964)

[Komu68] T. Kõmura, Semigroups of operators in locally convex spaces. J. Funct. Anal. **2**, 258–296 (1968)

[KKS56] M.A. Krasnosel'skiĭ, S.G. Kreĭn, P.E. Sobolevskiĭ, On differential equations with unbounded operators in Banach space. Dokl. Akad. Nauk USSR **111**, 19–22 (1956) (in Russian)

[SKre67] S. G. Kreĭn, *Linear Differential Equations in Banach Space* (Nauka, Moskva, 1967) (in Russian). English translation: Transl. Math. Monogr., vol. 29 (Amer. Math. Soc., Providence, RI, 1971)

[KW00] F. Kühnemund, M. Wacker, The Lie-Trotter product formula does not hold for arbitrary sums of generators. Semigroup Forum **60**(3), 478–485 (2000)

[Lap80] M.L. Lapidus, Généralisation de la formule de Trotter-Lie. Comptes Rendus Acad. Sci. Paris **A 291**, 497–500 (1980)

[Lap81] M.L. Lapidus, Generalization of the Trotter-Lie formula. Integr. Equ. Oper. Theory **4**, 366–415 (1981)

[Lap82] M.L. Lapidus, The problem of the Trotter-Lie formula for unitary groups of operators. (Séminaire Choquet: Initiation à l'Analyse), Publ. mathématique université Pierre et Marie Curie (Paris IV) **46**, 1701–1746 (1982)

[Lap85] M.L. Lapidus, Product formula for imaginary resolvents with application to a modified Feynman integral. J. Funct. Anal. **63**, 261–275 (1985)

[Lax02] P. Lax, *Functional Analysis* (Wiley, New York, 2002)

[LeWu11] L.D. Lemle, L.M. Wu, Uniqueness of C_0-semigroups on a general locally convex vector space and an application. Semigroup Forum **82**, 485–496 (2011)

[LM] J.L. Lions, E. Magenes, *Non-Homogeneous Boundary Value Problems and Applications I* (Springer, Berlin–Heidelberg, 1972)

[LHB20] J. Lőrinczi, F. Hiroshima, V. Betz, *Feynman-Kac-Type Theorems and Gibbs Measures on Path Space*. Volume 1: Feynman-Kac-Type Formulæ and Gibbs Measures (2nd edn.) (Walter de Gruyter GmbH, Berlin, 2020)

[LuPh61] G. Lumer, R.S. Phillips, Dissipative operators in a Banach space. Pacific J. Math. **11**, 679–698 (1961)

[LyuSob82] L.A. Lyusternik, V.I. Sobolev, *A Short Course on Functional Analysis*, Textbook (Vysshaya Shkola, Moskva, 1982) (in Russian)

[Mais71] H.D. Maison, Analyticity of partition function for finite quantum systems. Commun. Math. Phys. **22**, 166–172 (1971)

[Mar83] A.I. Markushevich, *The Theory of Analytic Functions, a Brief Course* (Mir Publishers, Moscow, 1983)

[MSh03] M. Matolcsi, R. Shvidkoy, Trotter's product formula for projections. Arch. der Math. **81**, 309–317 (2003)

[MMRS23] S. Mazzucchi, V. Moretti, I. Remizov, O. Smolyanov, Chernoff approximations of Feller semigroups in Riemannian manifolds. Math. Nachr. **296**, 1244–1284 (2023)

[MVZ00] R.A. Minlos, A. Verbeure, V.A. Zagrebnov, A quantum crystal model in the light-mass limit: Gibbs states. Rev. Math. Phys. **12**, 981–1032 (2000)

[MiShe04] Yu.S. Mishura, H.M. Shevchenko, Euler approximations of solutions of abstract equations and their applications in the theory of semigroups. Ukr. Math. J. (Kiev) **56**, 489–503 (2004)

[Mi91] Y. Miyazaki, Domains of square roots of regularly accretive operators. Proc. Jpn. Acad. **67(A)**, 38–42 (1991)

[MS77] B. Misra, E.C.G. Sudarshan, The Zeno's paradox in quantum theory. J. Math. Phys. **18**, 756–763 (1977)

[MoRo23] T. Möbus, C. Rouzé, Optimal convergence rate in the quantum Zeno effect for open quantum systems in infinite dimensions. Ann. Henri Poincaré **24**, 1617–1659 (2023)

[MR2000] S. Monniaux, A. Rhandi, Semigroup method to solve non-autonomous evolution equations. Semigroup Forum **60**, 122–134 (2000)

[Mor22] M.H. Mortad, *Conterexamples in Operator Theory* (Bikhäuser–Springer Nature Switzerland AG, Cham, 2022)

[Nag86] R. Nagel (ed.), *One-Parameter Semigroups of Positive Operators*. Lect. Notes in Math., vol.1184 (Springer, Berlin, 1986)

[NaNi02] R. Nagel, G. Nickel, *Well-Posedness of Nonautonomous Abstract Cauchy Problems*, in: *Progr. Nonlinear Diff.Eqn. and Their Appl.*, ed. by H. Brezis, vol. 50 (Springer Basel AG, Basel, 2002), pp. 279–293

[Nei79] H. Neidhardt. *Integration of Evolutionsgleichungen mit Hilfe von Evolutionshalbgruppen*. Dissertation, AdW der DDR, Berlin, 1979

[Nei81] H. Neidhardt, On abstract linear evolution equations. I. Math. Nachr. **103**, 283–298 (1981)

[Nei81a] H. Neidhardt, On abstract linear evolution equations. II. Preprint Institut für Mathematik, AdW der DDR, P-MATH-07/81, 1–56, Berlin, 1981

[Nei82] H. Neidhardt, On linear evolution equations. III: Hyperbolic case. Preprint Institut für Mathematik, AdW der DDR, P-MATH-05/82, 1–74, Berlin, 1982

[Nel59] E. Nelson, Analytic vectors. Ann. Math. Phys. **70**, 572–614 (1959)

[Nel64] E. Nelson, Feynman integral and the Schrödinger equation. J. Math. Phys. **5**, 332–343 (1964)

[NSZ17] H. Neidhardt, A. Stephan, V.A. Zagrebnov, On convergence rate estimates for approximations of solution operators for linear non-autonomous evolution equations. Nanosyst. Phys. Chem. Math. **8**(2), 201–215 (2017)

[NSZ18a] H. Neidhardt, A. Stephan, V.A. Zagrebnov, Remarks on the operator-norm convergence of the Trotter product formula. Integr. Equ. Oper. Theory **90:15**(2), 1–14 (2018)

[NSZ18b] H. Neidhardt, A. Stephan, V.A. Zagrebnov, *Operator-Norm Convergence of the Trotter Product Formula on Hilbert and Banach Spaces: A Short Survey*, in *Current Research in Nonlinear Analysis*. Springer Optimization and Its Applications, vol. 135 (Springer, Berlin, 2018), pp. 229–247

[NSZ19] H. Neidhardt, A. Stephan, V.A. Zagrebnov, *Trotter Product Formula and Linear Evolution Equations on Hilbert Spaces*, in *Analysis and Operator Theory. Dedicated in Memory of Tosio Kato's 100th Birthday*, vol. 146 (Springer, Berlin, 2019), pp. 271–299

[NSZ20] H. Neidhardt, A. Stephan, V.A. Zagrebnov, Convergence rate estimates for Trotter product approximations of solution operators for non-autonomous Cauchy problems. Publ. RIMS Kyoto Univ. **56**, 83–135 (2020)

[NZ90a] H. Neidhardt, V.A. Zagrebnov, The Trotter product formula for Gibbs semigroups. Commun. Math. Phys. **131**, 333–346 (1990)

[NZ90b] H. Neidhardt, V.A. Zagrebnov, On the Trotter product formula for Gibbs semigroups. Ann. der Physik **47**, 183–191 (1990)

[NZ98] H. Neidhardt, V.A. Zagrebnov, On error estimates for the Trotter-Kato product formula. Lett. Math. Phys. **44**, 169–186 (1998)

[NZ99a] H. Neidhardt, V.A. Zagrebnov, Fractional powers of self-adjoint operators and Trotter-Kato product formula. Integr. Equ. Oper. Theory **35**, 209–231 (1999)

[NZ99b] H. Neidhardt, V.A. Zagrebnov, Trotter-Kato product formula and operator-norm convergence. Commun. Math. Phys. **205**, 129–159 (1999)

[NZ99c] H. Neidhardt, V.A. Zagrebnov, On the operator-norm convergence of the Trotter-Kato product formula. Operator Theory Adv. Appl. **108**, 323–334 (1999)

[NZ99d] H. Neidhardt, V.A. Zagrebnov, Trotter-Kato product formula and symmetrically normed ideals. J. Funct. Anal. **167**, 113–167 (1999)

[NZ09] H. Neidhardt, V.A. Zagrebnov, Linear non-autonomous Cauchy problems and evolution semigroups. Adv. Differ. Equ. **14**, 289–340 (2009)

[Nev58] J. Neveu, Théorie de semi-groupes de Markov. Univ. Calif. Publ. Stat. **2**(14), 319–394 (1958).

[Nick96] G. Nickel, On evolution semigroups and wellposedness of nonautonomous Cauchy problems. Dissertation, Tübingen, 1996, pp.1–114

[Nick97] G. Nickel, Evolution semigroups for nonautonomous Cauchy problems. Abstract Appl. Anal. **2**, 73–95 (1997)

[Nick00] G. Nickel, Evolution semigroups and product formulas for nonautonomous Cauchy problems. Math. Nachr. **212**, 101–116 (2000)

[Ouc73] S. Ōuchi, Semi-groups of operators in locally convex spaces. J. Math. Soc. Jpn. **25**, 265–276 (1973)

[PPW20] P. Pagacz, P.L. Pietrzycki, M.L. Wojtylak, Between the von Neumann inequality and the Crouzeix conjecture. Linear Alg. Appl. **605**, 130–157 (2020)

[Paq79] L. Paquet, *Semi-groupes généralisés et équations d'évolutions*. Seminaire de Théorie du Potententiel. Lecture Notes Math., vol. 713 (Springer, Berlin, 1979), pp. 243–263

[Pau04] V. Paulauskas, On operator-norm approximation of some semigroups by quasi-sectorial operators. J. Funct. Anal. **207**, 58–67 (2004)

[Pau12] V. Paulauskas, A generalization of sectorial and quasi-sectorial operators. J. Funct. Anal. **262**, 2074–2099 (2012)

[Paz83] A. Pazy, *Semigroups of Linear Operators and Applications to Partial Differential Equations* (Springer, Berlin, 1983)

[Pf93] D. Pfeifer, A probabilistic variant of Chernoff's product formula. Semigroup Forum **46**, 279–285 (1993)

[Phi53] R.S. Phillips, Perturbation theory for semi-groups of linear operators. Trans. Am. Math. Soc. **74**, 199–221 (1953)

[Prato1984] G. da Prato, P. Grisvard, Maximal regularity for evolution equations by interpolation and extrapolation. J. Funct. Anal. **58**, 107–124 (1984)

[Pri56] I.I. Priwalow [I.I. Privalov], *Randeigenschaften Analytischer Funktionen* (Deutsch Verlag der Wissenschaften, Berlin, 1956) (Translated from Russian)

[PrSchn01] J. Prüss, R. Schnaubelt, Solvability and maximal regularity of parabolic evolution equations with coefficients continuous in time. J. Math. Anal. Appl. **256**, 405–430 (2001)

[RRSV00] F. Räbiger, A. Rhandi, R. Schnaubelt, J. Voigt, Non-autonomous Miyadera perturbations. Differ. Integr. Equ. **13**, 341–368 (2000)

[Rao04] M.M. Rao, *Measure Theory and Integration*, 2nd edn. (Marcel Dekker Inc., Basel, 2004)

[RS80] M. Reed, B. Simon, *Methods of Modern Mathematical Physics*, I, *Functional Analysis* (Revised and Enlarged Edition) (Academic Press, New York, 1980)

[RS75] M. Reed, B. Simon, *Methods of Modern Mathematical Physics* II, *Fourier-Analysis, Self-Adjointness* (Academic Press, New York, 1975)

[RS78] M. Reed, B. Simon, *Methods of Modern Mathematical Physics* IV, *Analysis of Operators* (Academic Press, New York, 1978)

[RR93] M. Renardy, R.C. Rogers, *An Introduction to Partial Differential Equations* (Springer, Berlin, 1993)

[RBW22] C. Reyes-Bustos, M. Wakayama, Heat kernel for the quantum Rabi model. Adv. Theor. Math. Phys. **26**, 1347–1447 (2022)

[Ri53] R.K. Ritt, A condition that $\lim_{n\to\infty} T^n = 0$. Proc. Am. Math. Soc. **4**, 898–899 (1953)

[Rog93] Dzh. L. Rogava, Error bounds for Trotter-type formulas for self-adjoint operators. Funktsional. Anal.i Prilozhen. **27**(3), 84–86 (1993) (in Russian). Funct. Anal. Appl. **27**, 217–219 (1993)

[Rog22] Dzh. L. Rogava, Approximation of operator semigroups using linear-fractional operator functions and weighted averages. Funktsional. Anal.i Prilozhen. **56**(2), 47–63 (2022) (in Russian). Funct. Anal. Appl. **56**, 116–129 (2022)

[Sad91] V.A. Sadovnichiĭ, *Theory of Operators* (Consultants Bureau, New York, 1991)

[Sch80] H.H. Schaefer, *Topological Vector Spaces* (Springer, Berlin, 1980)

[Schn00] R. Schnaubelt, Semigroups for nonautonomous Cauchy problems, in *One-Parameter Semigroups for Linear Evolution Equations*, ed. by K.J. Engel, R. Nagel (Springer, Berlin, 2000), pp. 477–496

[Schm12] K. Schmüdgen, *Unbounded Self-Adjoint Operators on Hilbert Space* (Springer, Dordrecht, 2012)

[Shim68] N. Shimakura, Sur les domaines des puissances fractionnaires d'opérateurs. Bull. Soc. Math. France **96**, 265–288 (1968)

[Si78] B. Simon, A canonical decomposition for quadratic forms with applications to applications to monotoe convergence theorems. J. Funct. Anal. **28**, 377–385 (1978)

[Si79] B. Simon, *Trace Ideals and Their Applications* (Cambridge University Press, Cambridge, 1979)

[Si05] B. Simon, *Trace Ideals and Their Applications*, 2nd edn. (AMS, Providence, RI, 2005)

[Si09] B. Simon, Schrödinger operators with purely discrete spectrum. Methods Funct. Anal. Topol. **15**, 61–66 (2009)

[Si15] B. Simon, *Real Analysis. A Comprehensive Course of Analysis, Part 1* (AMS, Providence, RI, 2015)

[Si15a] B. Simon, *Basic Complex Analysis. A Comprehensive Course of Analysis, Part 2A* (AMS, Providence, RI, 2015)

[Si15b] B. Simon, *Operator Theory. A Comprehensive Course of Analysis, Part 4* (AMS, Providence, RI, 2015)

[Sin92] Ya.G. Sinai, *Probability Theory. An Introductory Course* (Springer, Berlin–Heidelberg, 1992)

[Sob57] P.E. Sobolevskiĭ, Approximate methods of solving differential equations in Banach space. Dokl. Akad. Nauk USSR **115**, 240–243 (1957) (in Russian)

[Sob58] P.E. Sobolevskiĭ, Generalized solutions of the first order differential equations in Hilbert space. Dokl. Akad. Nauk USSR **122**, 994–996 (1958) (in Russian)

[Sob58a] P.E. Sobolevskiĭ, First order differential equations in Hilbert space with a variable positively defined self-adjoined operator whose fractional power has a constant definition domain. Dokl. Akad. Nauk USSR **123**, 984–987 (1958) (in Russian)

[Sob61] P.E. Sobolevskiĭ, Parabolic equations in a Banach space with an unbounded variable operator, a fractional power of which has a constant domain of definition. Dokl. Akad. Nauk USSR **138**, 59–62 (1961) (in Russian). English translation: Sov. Math. Dokl. **2**, 545–548 (1961)

[Sob61a] P.E. Sobolevskiĭ, Equations of parabolic type in a Banach space. Trudy Moscow Math.Obsch. **10**, 297–350 (1961) (in Russian). English translation: Am. Math. Soc. Transl. **49**, 1–62 (1965)

[Sob62] P.E. Sobolevskiĭ, Second-order differential equations in Banach space. Dokl. Akad. Nauk USSR **146**, 774–777 (1962) (in Russian). English translation: Sov. Math. Dokl. **3**, 1394–1398 (1962)

[Sob61b] P.E. Sobolevskii, Parabolic equations in a Banach space with an unbounded variable operator, a fractional power of which has a constant domain of definition. Dokl. Akad. Nauk USSR **138**(1), 59–62 (1961)

[St70] E. Stein, *Singular Integrals and Differentiability Properties of Functions* (Princeton University Press, Princeton, 1970)

[Sto32] M.H. Stone, On one-parameter unitary groups in Hilbert space. Ann. Math. **33**, 643–648 (1932)

[Str69] G. Strang, Approximating semigroups and the consistency of difference schemes. Proc. Am. Math. Soc. **20**, 1–7 (1969)

[Suz96] M. Suzuki, *Exponential Product Formulas and Quantum Analysis* (collection of papers) (University of Tokyo, January 1996), pp. 1–316

[SzNF70] B.Sz.-Nagy, C. Foiaş, *Harmonic Analysis of Operators on Hilbert Space* (North-Holland Publ. Company - Akadémia Kiadó, Budapest, 1970)

[Tak97] S. Takanobu, On the error estimate of the integral kernel for Trotter product formula operators for Schrödinger operators. Ann. Probab. **25**, 1895–1952 (1997)

[Tam00] H. Tamura, A remark on operator-norm convergence of Trotter-Kato product formula. Integr. Equ. Oper. Theory **37**, 350–356 (2000)

[Tan79] H. Tanabe, *Equations of Evolution* (Pitman (Advanced Publishing Program), Boston, MA, London, 1979)

[Tan60] H. Tanabe, On the equations of evolution in a Banach space. Osaka Math. J. **12**, 365–376 (1960)

[Tan61] H. Tanabe, Evolution equations of parabolic type. Proc. Jpn. Acad. **87**, 610–613 (1961)

[Tan67] H. Tanabe, On regularity of solutions of abstract differential equations of parabolic type in Banach space. J. Math. Soc. Jpn. **19**, 521–542 (1967)

[Tan75] H. Tanabe, *Equations of Evolution* (Iwanami, Tokyo, 1975) (in Japanese). English translation: Pitman Advanced Publishing Program, London (1979)

[Trot58] H.F. Trotter, Approximations of semigroups of operators. Pacific J. Math. **8**, 887–919 (1958)

[Trot59] H.F. Trotter, On the products of semigroups of operators. Proc. Am. Math. Soc. **10**, 545–551 (1959)

[vN29] J. von Neumann, Zur Theorie der unbeschränkten Matrizen. J. Reine Angew. Math. **161**, 208–236 (1929)

[vN32] J. von Neumann, *Mathematische grundlagen der quantenmechanik*, Berlin, 1932. Mit einem Geleitwort von Rudolf Haag (Zweite Auflage) (Springer, Berlin, 1996)

[vN51] J. von Neumann, Eine Spektraltheorie für allgemeine Operatoren eines unitären Raumes. Math. Nachrichten **4**, 258–281 (1951)

[Voi77] J. Voigt, On the perturbation theory for strongly continuous semigroups. Math. Ann. **229**, 161–171 (1977)

[VWZ08] P.-A. Vuillermot, W.F. Wreszinski, V.A. Zagrebnov, A Trotter-Kato product formula for a class of non-autonomous evolution equations. Nonlinear Anal. **69**, 1067–1072 (2008)

[VWZ09] P.-A. Vuillermot, W.F. Wreszinski, V.A. Zagrebnov, A general Trotter-Kato formula for a class of evolution operators. J. Funct. Anal. **257**, 2246–2290 (2009)

[WaSe37] J.L. Walsh, W.E. Sewell, Note on degree of approximation to an integral by Riemann sums. Am. Math. Month. **44**, 155–160 (1937)

[Yagi89] A. Yagi, Parabolic evolution equation in which the coefficients are the generators of infinitely differentiable semigroups I. Funkcialaj Ekvacioj **32**, 107–124 (1989)

[Yagi90] A. Yagi, Parabolic evolution equation in which the coefficients are the generators of infinitely differentiable semigroups II. Funkcialaj Ekvacioj **33**, 139–150 (1990)

[Yos48] K. Yosida, On the differentiability and the representation of one-parameter semigroups of linear operators. J. Math. Soc. Jpn. **1**, 15–21 (1948)

[Yos65] K. Yosida, A perturbation theorem for semi-groups of linear operators. Proc. Jpn. Acad. **41**, 645–647 (1965)

[Yos80] K. Yosida, *Functional Analysis*, 6th edn. (Springer, Berlin, 1980)

[You92] P. You, Characteristic conditions for a C_0-semigroup with continuity in the uniform operator topology for $t > 0$ in Hilbert space. Proc. Am. Math. Soc. **112**, 991–997 (1992)

[Yo68] W. Yourgrau, *A Budget of Paradoxes in Physics*, in *Problems in the Philosophy of Science, Proceedings Int. Colloq. Phil. Science* (London 1965), ed. by I. Lakatos and A.Musgrave (Noth-Holland Publ. Co., Amsterdam, 1968), pp. 178–199

[Zag80a] V.A. Zagrebnov, Singular potentials of interaction in quantum statistical mechanics. Proc. Moscow Math. Soc. **41**, 101–120 (1980) (in Russian)

[Zag80] V.A. Zagrebnov, On the families of Gibbs semigroups. Commun. Math. Phys. **76**, 269–276 (1980)

[Zag88] V.A. Zagrebnov, The Trotter-Lie product formula for Gibbs semigroups. J. Math. Phys. **29**, 888–891 (1988)

[Zag89] V.A. Zagrebnov, Perturbations of Gibbs semigroups. Commun. Math. Phys. **120**, 653–664 (1989)

[Zag95] V.A. Zagrebnov, On the operator-norm error-bound estimate for the Trotter product formula: Comments on Rogava's paper. Preprint CPT-CNRS-Luminy, 1995. (unpublished)

[Zag03a] V.A. Zagrebnov, *Ultimate optimal error bound for the Trotter product formula: Gibbs semigroups*, in *Proceedings of the International Conference on Functional Analysis*, Kiev (August 22–26, 2001) (Naukova Dumka, Kiev, 2003), pp. 94–99

[Zag03b] V.A. Zagrebnov, *Topics in the Theory of Gibbs semigroups*. Leuven Notes in Mathematical and Theoretical Physics, vol. 10 (Series A: Math. Phys.) (Leuven University Press, Leuven, 2003)

[Zag05] V.A. Zagrebnov, *Trotter-Kato product formula: some recent results*, in *Proceedings of the XIVth International Congress on Mathematical Physics*, Lisbon (July 28–August 02, 2003) (World Scientific, Singapore, 2005), pp. 634–641

[Zag08] V.A. Zagrebnov, Quasi-sectorial contractions. J. Funct. Anal. **254**, 2503–2511 (2008)

[Zag17] V.A. Zagrebnov, Comments on the Chernoff $\sqrt{n}$-lemma, in *Functional Analysis and Operator Theory for Quantum Physics* (The Pavel Exner Anniversary Volume) (European Mathematical Society, Zürich, 2017), pp. 565–573

[Zag19] V.A. Zagrebnov, *Trotter-Kato Product Formulæ in Dixmier Ideal*, in *Analysis and Operator Theory*. Dedicated in Memory of Tosio Kato's 100th Birthday. Springer Optimization and Its Applications, vol. 146 (Springer, Berlin, 2019), pp. 395–416

[Zag19a] V.A. Zagrebnov, *Gibbs Semigroups*, Operator Theory Series: Advances and Applications, vol. 273 (Bikhäuser - Springer Nature Switzerland AG, Basel, 2019)

[Zag19b] V.A. Zagrebnov, Approximations of self-adjoint Co-semigroups in the operator-norm topology. Annales Universitatis Mariae Curie-Skłodowska, Lublin Polonia (Section A) vol. **LXXIII**(2), 175–194 (2019)

[Zag19c] V.A. Zagrebnov, Product approximation of solution operators for non-autonomous Cauchy problems. Dyn. Syst. **9**(37) No.4, 321–333 (2019)

[Zag20] V.A. Zagrebnov, Notes on the Chernoff product formula. J. Funct. Anal. **279**, 108696–24pp. (2020)

[Zag20a] V.A. Zagrebnov, Non-autonomous dynamics and product formula approximation of solution operator. Discontin. Nonlinearity Complex. **9**(4), 579–590 (2020)

[Zag21] V.A. Zagrebnov, *Product Formulæ for Non-Autonomous Gibbs Semigroups*, in *Mathematical Analysis in Interdisciplinary Research*. Springer Optimization and Its Applications, vol. 179 (Springer, Berlin, 2021), pp. 1049–1060

[Zag22a] V.A. Zagrebnov, Comments on the Chernoff estimate. Nanosyst. Phys. Chem. Math. **13**(1), 17–23 (2022)

[Zag22b] V.A. Zagrebnov, Product approximations of solution operators for non-autonomous perturbations of Gibbs semigroups. Mathematische Nachrichten **295**(6), 1233–1245 (2022).

[Zag22c] V.A. Zagrebnov, Notes on the Chernoff estimate. Int. J. Mod. Phys. A **37**, 2243925 (18 pages) (2022)

[Zag23] V.A. Zagrebnov, Operator-norm Trotter product formula on Banach spaces. Izvestiya Rossiĭskoĭ Akademii Nauk, Seriya Matematicheskaya **87**, 99–123 (2023) (in Russian). Izvestiya Akademii Nauk: Mathematics **87**(5), 947–971 (2023)

[ZI24] V.A. Zagrebnov, B. Iochum, A characterisation of the Gibbs semigroups. Teor.i Mat. Fiz. **218**, 88–101 (2024) (in Russian). HAL Id: hal-04177061 (3 Aug 2023)

Index

V. A. Zagrebnov et al., *Trotter-Kato Product Formulæ*, Operator Theory: Advances and Applications 296, https://doi.org/10.1007/978-3-031-56720-9

List of Symbols

$a.a.$	almost all	129
$a.e.$	almost everywhere	141
type (A)	holomorphic family	392
type (a)	holomorphic family	392
type (B)	holomorphic family	380
(u, v)	inner product in $\mathfrak{H} : (\alpha u, \beta v) = \alpha\overline{\beta}(u, v)$	25
$\langle x, \phi \rangle$	linear functional on $\mathfrak{X}$ for $x \in \mathfrak{X}$ and $\phi \in \mathfrak{X}^*$	25
$a[u, v]$	sesquilinear form in $\mathfrak{H}$	223
a_z	type (a) holomorphic family	393
$\|A\|$	absolute value of the operator A	287
$A(z)$	type (B) holomorphic family	392
$A \dotplus B$	form-sum of A and B	viii, 221
$A_R := \mathrm{Re}\, A$	real part of A	391
A_t	$A_t := t^{-1}(\mathbb{1} - U_t) \quad \text{for } t > 0$	12
$\mathfrak{X}$	Banach space	3
$\mathfrak{X}^*$	dual space of $\mathfrak{X}$	24
$\mathbb{C}$	complex numbers	3
$\mathbb{C}_+$	open right half-plane of $\mathbb{C}$	30
$\mathbb{C}_-$	open left half-plane of $\mathbb{C}$	424
$\mathcal{C}(\mathfrak{X})$	closed operators in $\mathfrak{X}$	8
$\mathcal{C}(\mathfrak{H})$	closed operators in $\mathfrak{H}$	48
$\mathcal{C}_\infty(\mathfrak{H})$	compact operators on $\mathfrak{H}$	31
$\mathcal{C}_1(\mathfrak{H})$	trace-class operators on $\mathfrak{H}$	32

V. A. Zagrebnov et al., *Trotter-Kato Product Formulæ*, Operator Theory: Advances and Applications 296, https://doi.org/10.1007/978-3-031-56720-9

$\mathcal{C}_p(\mathfrak{H})$ $(1 \leq p < \infty)$	von Neumann-Schatten classes	97
C_n^k	$n!/(k!(n-k)!)$, $k, n \in \mathbb{N}$	5
core (A)	core for the operator A	14, 16
dom (A)	operator domain	4
dom (A^n)	domain of the operator A^n	6, 16
$\overline{\text{dom}\,(A)}$	closure of dom A	12
dom (a)	form domain	223, 389, 701
$d_T = \text{def}\, T$	deficiency (defect) of the operator T	49
$E_A(d\lambda)$	spectral measure of operator A	30, 230
$\{G(t)\}_{t\geq 0}$	Gibbs semigroup	32
$\mathscr{G}(M, \beta)$	generators of quasi-bounded semigroups	17
$\mathfrak{H}$	Hilbert space	viii, 16, 25
$\mathfrak{H} = \mathcal{M} \oplus \mathcal{M}^{\perp}$	orthogonal direct sum	236, 735
$\mathfrak{H}_n^{\perp}$	orthogonal complement of the subspace $\mathfrak{H}_n$	49
$\mathscr{H}(\theta, \beta)$	generators of quasi-bounded holomorphic semigroups	47, 52
$\mathscr{H}_c(\theta, 0)$	generators of contraction holomorphic semigroups	60, 187, 194
$\mathbb{1}$	unit operator	4, 6
$Im(U)$	image of the mapping $U : \mathfrak{X} \to \mathfrak{Y}$	30, 39
Im a	imaginary part of the sesquilinear form a	380, 387
$\mathcal{K}$	generic Kato function	ix , 229
ker (A)	kernel of the operator A	35, 38
$\mathcal{L}(\mathfrak{X}, \mathfrak{Y})$	bounded linear operators mapping $\mathfrak{X} \to \mathfrak{Y}$	3, 54, 769
$\mathcal{L}(\mathfrak{X}, \mathbb{C})$	bounded linear functionals on $\mathfrak{X}$	24
$\mathcal{L}(\mathfrak{H})$	bounded linear operators on $\mathfrak{H}$	784
l^p	space	102
$L^p(\mathbb{R}^d)$	space for $p = 1, 2, \ldots$	39
$L^2_{loc}(\mathbb{R}, \mathfrak{H})$	Fréchet space	140, 662
$\lim_{t\downarrow 0}$	$\lim_{t\to +0} \Leftrightarrow \{\text{limit } t \to 0 \wedge t > 0\}$	4, 38
$\|\cdot\|$- lim	operator-norm limit	22, 261, 725
s−lim	strong operator limit	viii, 28
w−lim	weak limit	70, 259
$\mathcal{P}_{0^+}$	operators with relative bound $a = 0^+$	57
$\mathcal{P}_a$	operators with relative bound $a > 0$	58
P_n	orthogonal projection on subspace $\mathfrak{H}_n \geq 0$	29
ran (A)	range of the operator A	769
Re z	real part of $z \in \mathbb{C}$	14
$\mathbb{R}$	real numbers	4
$\mathbb{R}^-$	negative real semi-axis $(-\infty, 0)$	4
$\mathbb{R}^+ \cup \{0\}$	non-negative real semi-axis $[0, \infty) = \mathbb{R}_0^+$	4
$R_A(\zeta) := (A - \zeta\,\mathbb{1})^{-1}$	resolvent of operator A at $\zeta \in \mathbb{C}$	4
span $\{\cdot\}$	linear hull	131
$\mathfrak{t}_A[u, v]$	sesquiliniear forms	223, 389
S_θ	open sector of semi-angle θ in $\mathbb{C}_+$	40
$\{s_n(A)\}_{n\geq 1}$	singular values of operator A	96
$t \downarrow 0$	denotes $t \to +0$	4, 38
Tr : $\mathcal{C}_1(\mathfrak{H}) \to \mathbb{C}$	trace on $\mathcal{C}_1(\mathfrak{H})$	32

$U_A(t)$	semigroup generated by A	4–5
$\{U(t)\}_{t\geq 0}$	one-parameter semigroup	9
$\hat{U}(z)$	Laplace transform of semigroup	13
$u_n \overset{w}{\to} u$	weak vector convergence	24, 94, 98
$W(A)$	numerical range of the operator A	48, 166, 184
$W^{k,p} \subset L^p$	Sobolev space of order (k, p)	142, 351
$[x]$	integer part of $x \geq 0$	373, 456
$\mathbb{N}$	natural numbers $\{1, 2, \ldots\}$	5, 186
$\mathbb{N}_0 = \mathbb{N} \cup \{0\}$	non-negative integers	17
$\rho(A)$	resolvent set of A	4
$\sigma(A)$	spectrum of operator A	40
$\sigma_{cont}(A)$	continuous spectrum of A	776
$\sigma_p(A)$	point spectrum of A	776
$\sigma_{res}(A)$	residual spectrum of A	776
∂_t	derivative with respect to t	4
$\ominus$	orthogonal difference	236
$\oplus$	orthogonal direct sum	236, 730
$\dot{+}$	form-sum	224
$\lVert \cdot \rVert_p$	norm on $\mathcal{C}_p(\mathfrak{H})$	32
$\lVert \cdot \rVert$	operator norm on $\mathcal{L}(\mathfrak{X})$, or $\mathcal{L}(\mathfrak{H})$	3
$\lVert \cdot \rVert$- ∂_t	operator-norm derivative	4, 5
$\lVert \cdot \rVert$	vector norm on $\mathfrak{X}$	4, 5
$\lVert \cdot \rVert$	vector norm on $\mathfrak{H}$	26, 48
$\lvert\lvert\lvert \cdot \rvert\rvert\rvert$	equivalent norm	19,87
$\Gamma(q)$	Euler Γ-function	36, 598
$\wedge$	and	210, 777
$\vee$	or	776
$\mu \wedge \lambda$	$= \inf(\mu, \lambda)$	631
$\mu \vee \lambda$	$= \sup(\mu, \lambda)$	631
$\Leftrightarrow$	equivalent, if and only if	
$A \Rightarrow B$	A yields (implies) B	
$\restriction$	restriction	15
$:$	such that	3, 4

Printed by Printforce, the Netherlands